Finite Mathematics for the Managerial, Life, and Social Sciences

Custom Edition

ELEVENTH EDITION

Soo T. Tan

CENGAGE Learning

Australia • Brazil • Japan • Korea • Mexico • Singapore • Spain • United Kingdom • United States

CENGAGE
Learning

Finite Mathematics for the Managerial, Life, and Social Sciences, Custom Edition, 11th

Senior Manager, Student Engagement:

Linda deStefano

Janey Moeller

Manager, Student Engagement:

Julie Dierig

Marketing Manager:

Rachael Kloos

Manager, Production Editorial:

Kim Fry

Manager, Intellectual Property Project Manager:

Brian Methe

Senior Manager, Production and Manufacturing:

Donna M. Brown

Manager, Production:

Terri Daley

Printed in the United States of America

Finite Mathematics for the Managerial, Life, and Social Sciences, 11th Edition
Soo T. Tan

ISBN-13: 978-1-305-28383-1

ISBN-10: 1-305-28383-X

WCN: 01-100-101

Cengage Learning

5191 Natorp Boulevard
Mason, Ohio 45040
USA

Cengage Learning is a leading provider of customized learning solutions with office locations around the globe, including Singapore, the United Kingdom, Australia, Mexico, Brazil, and Japan. Locate your local office at: **international.cengage.com/region.**

Cengage Learning products are represented in Canada by Nelson Education, Ltd. For your lifelong learning solutions, visit **www.cengage.com/custom.**
Visit our corporate website at **www.cengage.com.**

WebAssign

WebAssign Student Quick Start Guide

You can use WebAssign to access your homework, quizzes, and tests — whatever your instructor chooses — at any time of day or night, from any computer with a connection to the Internet and a Web browser. Your instructor creates your assignments, schedules them, and decides how many submissions you get. Your instructor also determines if you can have an extension, if you can save your work without submitting it at the time, and how much feedback you get after you submit an assignment.

The WebAssign support staff cannot change your username or password, give extensions, change your score, give you extra submissions, or help you with the content of your assignments.

Logging In

You can log in to WebAssign using any Web browser connected to the Internet. There are two different ways to log in to WebAssign. Each requires information from your teacher. If you are unsure about how to log in, please check with your teacher or another student in your class.

Go to the login page at http://webassign.net/login.html or the web address provided by your teacher. The way you log in depends on how your instructor set up the class:

- If your teacher created a WebAssign account for you, they will provide you with a **Username**, an **Institution** code and a **Password**. Simply enter this information in the boxes provided and click the **Log In** button.

WebAssign.

- If your teacher wants you to **Self-Enroll** in the WebAssign course they will provide you with a **Class Key**. You will create your own username and password. It is important that you remember this information so you can log in for the remainder of the class. In this case, just click the **I have a Class Key** button. You don't need to enter any other information on this page.

  Then, enter the **Class Key** your instructor provided and click **Submit.** Verify you are enrolling in the correct class on the next page.

Class Key

Enter the Class Key that you received from your instructor. You will only need to complete this once. After you have created your account, you can log in on the main page.

Class Key

Class Keys generally start with an institution code, followed by two sets of four digits.

Submit

- Enter your preferred Login and Student information.
- Click the **Create My Account** button to complete the enrollment process.
- A review screen will display, showing your username, institution code, and password. **Retain a copy of this information.** You will need it to log into WebAssign.

Log In Information

Required fields are marked with an asterisk (*).

Preferred Username * Check Availability

Your username may contain letters, numbers, and the following characters: underscore (_), hyphen (-), period (.)

Institution Code **webassign**

Password *

Re-Enter Password *

Passwords are case-sensitive.

Student Information

Required fields are marked with an asterisk (*).

First Name *

Last Name *

Email Address *

Student ID Number

Create My Account

WebAssign.

Access Codes

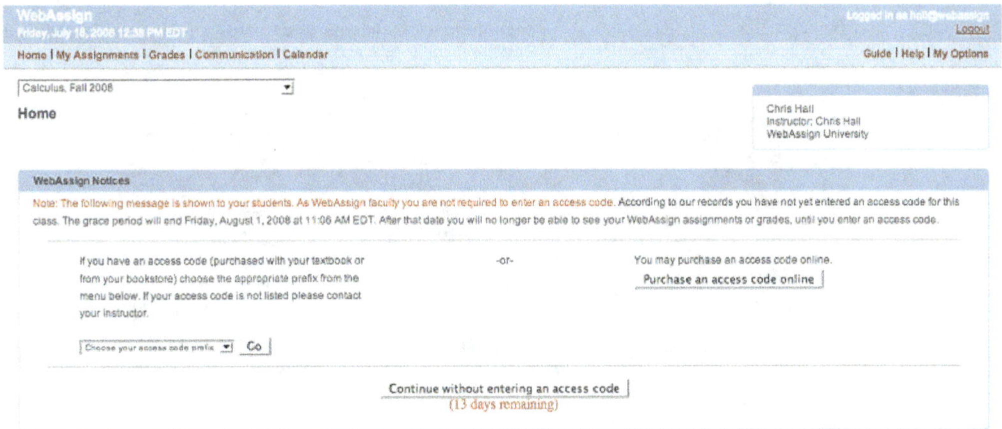

Once you log in, you may see a WebAssign Notice about entering an access code for your class. You can get an Access Code from any of the following places if you need to use one:

- A new textbook you purchased for the class.
- Your bookstore, which may sell Access Code cards.
- Online, where you can purchase an access code with a credit card.

You have a 14 day grace period to use WebAssign, starting with the WebAssign class start date. During this time you can work on and view your WebAssign assignments without registering a code.

After the grace period is over you will only see the code registration message until you submit or purchase a code.

There are two types of WebAssign access code cards. The small card requires you to scratch off the silver surface in order to reveal the complete access code.

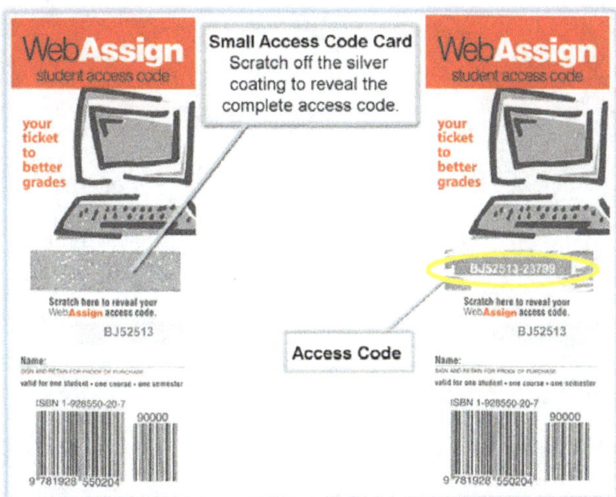

WebAssign

The larger security envelope card requires you to open the card to reveal the access code number.

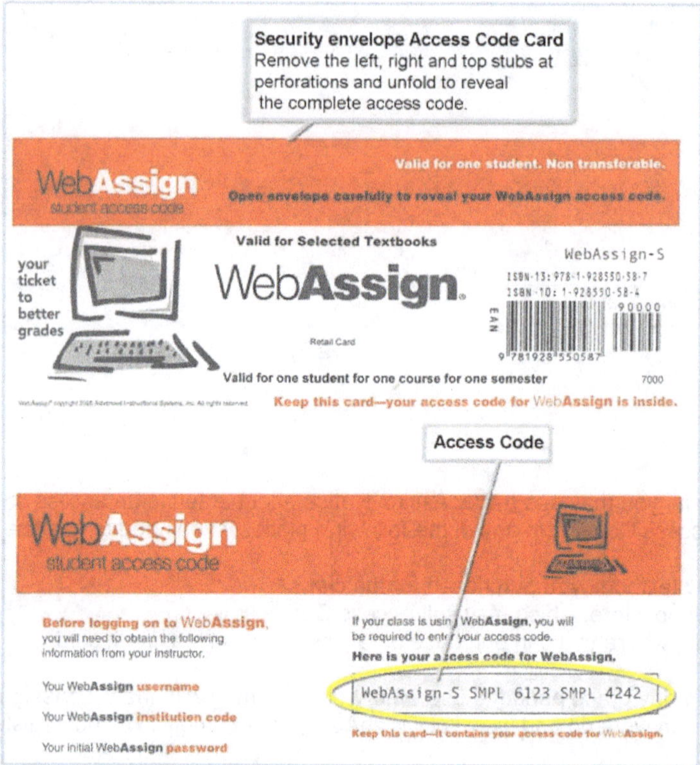

If you would like to purchase an access code directly from WebAssign online, you may do so with a credit card. Your code will be automatically registered to your WebAssign account as soon as the transaction is complete. You will receive an email confirmation. Please keep a copy for your records.

WebAssign

Your WebAssign Home Page

Once you have successfully logged in you will see your WebAssign homepage. If you are taking more than one WebAssign class, you will need to select which class you wish to view first.

The upper right corner features links to a complete Student **Guide**, as well as a link to WebAssign Technical Support under **Help**. If you want to change your password or add or update your email address, simply click **My Options** in the upper right hand corner.

You will see your assignments and due dates listed, as well as any Communications, Grades, and Announcements posted by your teacher.

Answering Questions

WebAssign has a variety of different question types, ranging from multiple choice to fill-in-the-blank to symbolic questions. Here are some things to keep in mind as you work through your assignments:

- Some questions may include numbers or words that appear in red. This signifies that the number or word has been randomized, so that you receive a different version of the same basic question from your classmates.

- Some WebAssign questions check the number of significant figures in your answer. If you enter the correct value with the wrong number of significant figures, you will not receive credit, but you will receive a hint that your number does not have the correct number of significant figures.

- Some questions require entering symbolic notation. Answer symbolic questions by using calculator notation. You must use the exact variables specified in the questions. The order is not important as long as it is mathematically correct. Clicking on the eye button previews the expression you enter in proper mathematical notation. Clicking on the symbolic formatting help button provides tips for using the correct keystrokes.

WebAssign.

- When you click on some WebAssign chemistry or math questions an input palette will open. These palettes, called chemPad and mathPad, will help you enter your answer in proper notation.

- Some questions may require the use of an Active Figure simulation. Active Figures require the free Macromedia Flash Player plug-in, downloadable from www.macromedia.com.

- If your instructor allows it, you can save your work without grading by selecting the Save Work button at the end of the question. After you save your work, it will be available to you the next time you click the assignment.

- Please note that WebAssign will **not** automatically submit your answers for scoring if you only **Save** your work. Your teacher will not be able to see your submissions. Please be sure to **Submit** prior to the due date and time.

- If your instructor allows it, you can submit answers by question part or for the entire assignment. To submit an individual question answer for grading, click the **Submit New Answers to Question __** button at the bottom of each question. To submit the entire assignment for grading, click the **Submit All New Answers** button at the end of the assignment.

Technical Support

If you are having difficulty logging in, please be sure to check with your teacher and verify whether an account has been created for you or whether you need to self-enroll. In either case your teacher needs to provide the appropriate information (username, institution code and password OR Class Key).

To email WebAssign Support go to http://www.webassign.net/info/support/report.html. This page also lists answers to **Common Problems**, and provides links to the **Student Guide**.

August 7, 2008

TO PAT, BILL, AND MICHAEL

CONTENTS

PREFACE

Math plays a vital role in our increasingly complex daily life. *Finite Mathematics for the Managerial, Life, and Social Sciences* attempts to illustrate this point with its applied approach to mathematics. Students have a much greater appreciation of the material if the applications are drawn from their fields of interest and from situations that occur in the real world. This is one reason you will see so many exercises in my texts that are modeled on data gathered from newspapers, magazines, journals, and other media. In addition, many students come into this course with some degree of apprehension. For this reason, I have adopted an intuitive approach in which I try to introduce each abstract mathematical concept through an example drawn from a common life experience. Once the idea has been conveyed, I then proceed to make it precise, thereby ensuring that no mathematical rigor is lost in this intuitive treatment of the subject.

The only prerequisite for understanding this text is one to two years, or the equivalent, of high school algebra. This text offers more than enough material for a one-semester or two-quarter course. The following chapter dependency chart is provided to help the instructor design a course that is most suitable for the intended audience.

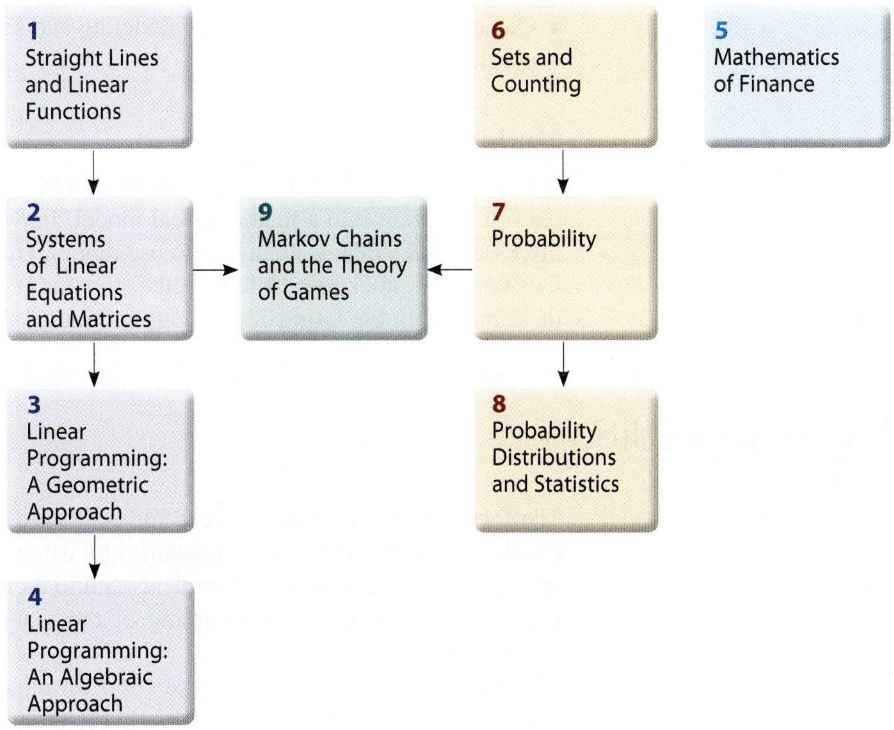

The Approach

Presentation

Consistent with my intuitive approach, I state the results informally. However, I have taken special care to ensure that mathematical precision and accuracy are not compromised.

Motivation

Illustrating the practical value of mathematics in applied areas is an objective of my approach. Concepts are introduced with concrete, real-life examples wherever appropriate. These examples and other applications have been chosen from current topics and issues in the media and serve to answer a question often posed by students: "What will I ever use this for?"

Problem-Solving Emphasis

Special emphasis is placed on helping students formulate, solve, and interpret the results of applied problems. Because students often have difficulty setting up and solving word problems, extra care has been taken to help them master these skills:

- Very early on in the text, students are given practice in solving word problems.
- Guidelines are given to help students formulate and solve word problems.
- One entire section is devoted to modeling and setting up linear programming problems.

Modeling

One important skill that every student should acquire is the ability to translate a real-life problem into a mathematical model. In Section 1.3, the modeling process is discussed, and students are asked to use models (functions) constructed from real-life data to answer questions. Additionally, students get hands-on experience constructing these models in the Using Technology sections.

New to this Edition

The focus of this revision has been the continued emphasis on illustrating the mathematical concepts in *Finite Mathematics* by using more real-life applications that are relevant to the everyday life of students and to their fields of study in the managerial, life, and social sciences. A sampling of these new applications is provided on the inside front cover pages.

Many of the exercise sets have been revamped. In particular, the exercise sets were restructured to follow more closely the order of the presentation of the material in each section and to progress more evenly from easier to more difficult problems in both the rote and applied sections of each exercise set. Additional concept questions, rote exercises, and true-or-false questions were also included.

More Specific Content Changes

Chapter 1 In Section 1.2, parts (b) and (c) of Example 12 illustrate how to determine whether a point lies on a line. A new application, *Smokers in the United States*, has been added to the self-check exercises. In Section 1.3, new data have been used for the *U.S. Health-Care Expenditures* application, and students are shown how the new model for this application is constructed in Section 1.5 using the least-squares method. Also, in Using Technology Section 1.3, Applied Examples 2 and 4, *Drinking and Driving Among High School Students* have been added for the graphing calculator and Excel applications.

Chapters 2–4 A wealth of new application exercises has been added, and many examples and exercises have been updated. Also, in Section 3.1, newly added Example 6 illustrates how to determine whether a point lies in a feasible set of inequalities. This is followed by a new application, Applied Example 7, *A Production Problem,* in which students are shown how they can use a solution set for a given system of inequalities (restrictions) to determine whether certain production goals can be met. Also, in Section 3.1, Exercise 44, we see how the solution of a system of linear equations is obtained by looking at a system of inequalities.

Chapter 5 Interest rate problems throughout the entire chapter were revised to reflect the current interest rate environment. Also, in Section 5.3, two new exercises were added illustrating the new *Ability-to-Repay Rules for Mortgages* adopted by the Consumer Financial Protection Bureau in response to the recent financial crisis.

Chapters 6–8 In the probability and statistics chapters of the text, the emphasis is again placed on providing new real-life application exercises. These chapters deal with the calculations of probabilities and data analysis and the emphasis here was placed on providing data from marketing, economic, consumer, and scientific surveys that was relevant, current, and of interest to students to motivate the mathematical concepts presented. Some of these surveys involve the following questions:

What is the greatest challenge upon starting a new job?

How many years will it take you to fully recover from the Great Recession?

What is the most common cause of on-the-job distraction?

How do workers get ahead on the job?

How many social media accounts do you have?

Have gas prices caused you any financial hardship?

Also, several new examples were added in these chapters: In Section 6.1, Applied Example 13, *Cyber Privacy,* illustrates set operations. In Section 6.2, Example 5 illustrates how the solution of a system of linear equations can sometimes be used to help draw a Venn diagram. In Section 7.5, Example 8 illustrates the difference between mutually exclusive and independent events. Also, Applied Example 12, *Predicting Travel Weather,* illustrates the calculation of the probability of independent events. In Section 8.2, Applied Example 8, *Commuting Times,* illustrates the calculation of expected value for grouped data; and in Section 8.3, Applied Example 4, *Married Men,* illustrates the calculation of standard deviation for grouped data. Using Technology Section 8.1 was expanded to include an example (Applied Example 3, *Time Use of College Students*) and exercises illustrating how Excel can be used to create pie charts.

Features

Real-World Connections

Motivating Applications

Many new applied examples and exercises have been added in the Eleventh Edition. Among the topics of the new applications are Facebook users, satellite TV subscribers, criminal justice, cyber privacy, brand switching among college students, social media accounts, detecting shoplifters, and smartphone ownership.

APPLIED EXAMPLE 13 Cyber Privacy In a poll surveying 1500 registered voters in California, the respondents were asked to rank the following companies on a scale of 0 to 10 in terms of how much they could trust these companies to keep their personal information secure, with zero meaning that they don't trust the company.

Company	Apple	Google	LinkedIn	YouTube	Facebook	Twitter
Rating	4.6	3.8	3.0	2.8	2.7	2.4

Let A denote the set of companies that have a rating higher than 2.5, let B denote the set of companies that have a rating between 2.5 and 4, and let C denote the set of companies that have a rating lower than 3. Find the following sets:

a. A, B, and C **b.** $A \cup B$ **c.** $B \cap C$ **d.** $A^c \cap B$ **e.** $A \cap B^c$

Source: Los Angeles Times.

Portfolios

These interviews share the varied experiences of professionals who use mathematics in the workplace. Among those included are a city manager at a photography company and a technical director at a wireless company who uses his knowledge of game theory to help mobile operators develop and deliver new technologies.

PORTFOLIO Christian Derrick

TITLE Technical Director, Europe
INSTITUTION SpiderCloud Wireless

SpiderCloud Wireless is the start-up company that introduced the first enterprise radio access network (E-RAN) systems into mobile networks. E-RAN technology allows for superior indoor coverage and capacity as well as high-value services for a mobile operator's most important customer, the business user. As the Technical Director, Europe, I am responsible for working with European mobile operators to develop and deliver new wireless technologies.

At SpiderCloud, we draw on our knowledge of mathematics to understand the strategic decisions of our mobile partners. Applying our knowledge of game theory, for example, we can begin to answer a variety of questions such as: How will mobile operators price their voice, data, and SMS plans? And how will suppliers price their cell phone equipment?

Let's consider the first question. If each mobile operator offered identical pricing plans, one would predict that a price reduction by an operator would lead to an increase in that operator's market share. But price, of course, is not the only factor a consumer takes into consideration when selecting a plan. Consumers also consider other variables such as the operator's brand name and reputation, the device that is bundled with the plan, the geographical coverage of the operator's network, and a range of add-on options.

Game theory helps mobile operators to factor the perceived value of these factors into their pricing plans. Operators use game theory to evaluate the likely outcomes of changes in pricing in response to both their competitors' actions and their competitors' responses to their own actions. This helps operators to maximize their profits. Working at SpiderCloud, I have learned that an in-depth knowledge of game theory, including Nash equilibrium and the prisoner's dilemma, is extremely useful for predicting the behavior of players in a complex business environment.

Explorations and Technology

Explore and Discuss

These optional questions can be discussed in class or assigned as homework. They generally require more thought and effort than the usual exercises. They may also be used to add a writing component to the class or as team projects.

Explore and Discuss

Let A, B, and C be nonempty subsets of a set U.

1. Suppose $A \cap B \neq \varnothing$, $A \cap C \neq \varnothing$, and $B \cap C \neq \varnothing$. Can you conclude that $A \cap B \cap C \neq \varnothing$? Explain your answer with an example.
2. Suppose $A \cap B \cap C \neq \varnothing$. Can you conclude that $A \cap B \neq \varnothing$, $A \cap C \neq \varnothing$, and $B \cap C \neq \varnothing$? Explain your answer.

Exploring with Technology

These optional discussions appear throughout the main body of the text and serve to enhance the student's understanding of the concepts and theory presented. Often the solution of an example in the text is augmented with a graphical or numerical solution.

Exploring with TECHNOLOGY

To obtain a visual confirmation of the fact that the expression $\left(1 + \frac{1}{u}\right)^u$ approaches the number $e = 2.71828\ldots$ as u gets larger and larger, plot the graph of $f(x) = \left(1 + \frac{1}{x}\right)^x$ in a suitable viewing window, and observe that $f(x)$ approaches $2.71828\ldots$ as x gets larger and larger. Use **ZOOM** and **TRACE** to find the value of $f(x)$ for large values of x.

Using Technology

Written in the traditional example-exercise format, these optional sections show how to use the graphing calculator and Microsoft Excel 2010 as a tool to solve problems. (Instructions for Microsoft Excel 2007 are given on the companion website.) Illustrations showing graphing calculator screens and spreadsheets are used extensively. In keeping with the theme of motivation through real-life examples, many sourced applications are included.

A *How-To Technology Index* is included at the back of the book for easy reference to Using Technology examples.

APPLIED EXAMPLE 3 Time Use of College Students Use the data given in Table 1 to construct a pie chart.

TABLE T1

Time Used on an Average Weekday for Full-Time University and College Students

Time Use	Time (in hours)
Sleeping	8.5
Leisure and sports	3.7
Working and related activities	2.9
Educational activities	3.3
Eating and drinking	1.0
Grooming	0.7
Traveling	1.5
Other	2.4

Source: Bureau of Labor Statistics.

Solution

We begin by entering the information from Table T1 in Columns A and B on a spreadsheet. Then follow these steps:

Step 1 First, highlight the data in cells A2:A9 and B2:B9 as shown in Figure T4.

	A	B
1	Time Use	Time (in hours)
2	Sleeping	8.5
3	Leisure	
4	Workin	
5	Educati	
6	Eating a	
7	Groomi	
8	Travelir	
9	Other	

FIGURE T4 Completed sp

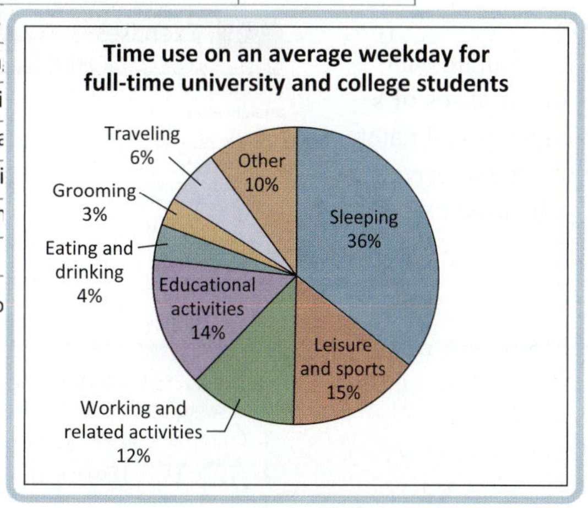

FIGURE T5
The pie chart describing the data in Table T1

Concept Building and Critical Thinking

Self-Check Exercises

Offering students immediate feedback on key concepts, these exercises begin each end-of-section exercise set and contain both rote and word problems (applications). Fully worked-out solutions can be found at the end of each exercise section. If students get stuck while solving these problems, they can get immediate help before attempting to solve the homework exercises. Applications have been included here because students often need extra practice with setting up and solving these problems.

1.2 Self-Check Exercises

1. Determine the number a such that the line passing through the points $(a, 2)$ and $(3, 6)$ is parallel to a line with slope 4.
2. Find an equation of the line that passes through the point $(3, -1)$ and is perpendicular to a line with slope $-\frac{1}{2}$.
3. Does the point $(3, -3)$ lie on the line with equation $2x - 3y - 12 = 0$? Sketch the graph of the line.
4. **Smokers in the United States** The following table gives the percentage of adults in the United States from 2006 through 2010 who smoked in year t. Here, $t = 0$ corresponds to the beginning of 2006.

Year, t	0	1	2	3	4
Percent, y	20.8	20.5	20.1	19.8	19.0

a. Plot the percentage of U.S. adults who smoke (y) versus the year (t) for the given years.
b. Draw the line L through the points $(0, 20.8)$ and $(4, 19.0)$.
c. Find an equation of the line L.
d. Assuming that this trend continues, estimate the percentage of U.S. adults who smoked at the beginning of 2014.

Source: Centers for Disease Control and Prevention.

Solutions to Self-Check Exercises 1.2 can be found on page 24.

Concept Questions

Designed to test students' understanding of the basic concepts discussed in the section, these questions encourage students to explain learned concepts in their own words.

1.4 Concept Questions

1. Explain why you would expect that the intersection of a linear demand curve and a linear supply curve would lie in the first quadrant.
2. In the accompanying figure, $C(x)$ is the cost function and $R(x)$ is the revenue function associated with a certain product.
   a. Plot the break-even point $P(x_0, y_0)$ on the graph.
   b. Identify and mark the break-even quantity, x_0, and the break-even revenue, y_0, on the set of axes.

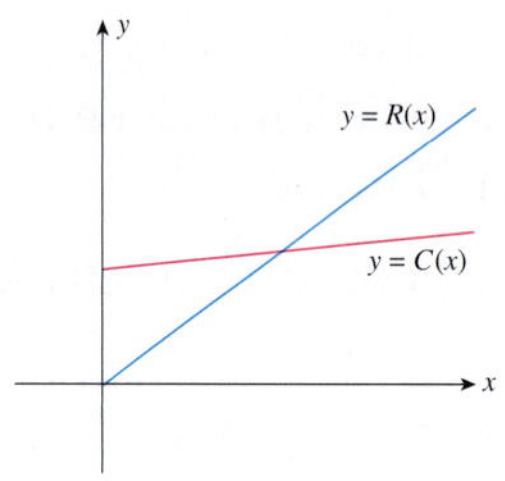

3. The accompanying figure gives the demand curve and the supply curve associated with a certain commodity.

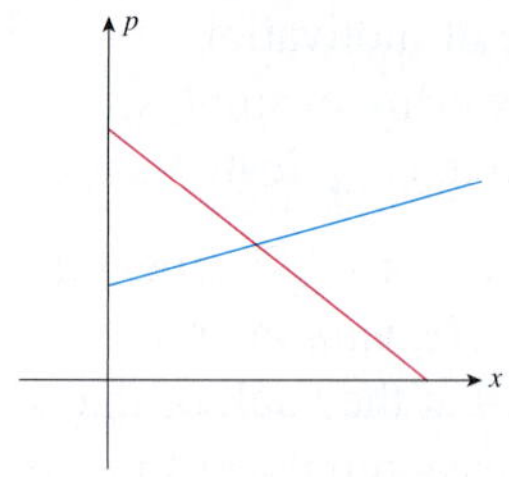

   a. Identify the demand curve and the supply curve.
   b. Plot the point $P(x_0, p_0)$ that corresponds to market equilibrium.
   c. Identify and mark the equilibrium quantity, x_0, and the equilibrium price, p_0, on the set of axes.

Exercises

Each section contains an ample set of exercises of a routine computational nature followed by an extensive set of modern application exercises.

1.4 Exercises

In Exercises 1–6, find the point of intersection of each pair of straight lines.

1. $y = 3x + 4$, $y = -2x + 14$
2. $y = -4x - 7$, $-y = 5x + 10$
3. $2x - 3y = 6$, $3x + 6y = 16$
4. $2x + 4y = 11$, $-5x + 3y = 5$
5. $y = \frac{1}{4}x - 5$, $2x - \frac{3}{2}y = 1$
6. $y = \frac{2}{3}x - 4$, $x + 3y + 3 = 0$

In Exercises 7–10, find the break-even point for the firm whose cost function C and revenue function R are given.

7. $C(x) = 5x + 10{,}000$; $R(x) = 15x$
8. $C(x) = 15x + 12{,}000$; $R(x) = 21x$

a. Find the functions describing the daily cost of leasing from each company.
b. Sketch the graphs of the two functions on the same set of axes.
c. If a customer plans to drive at most 30 mi, from which company should he rent a truck for a single day?
d. If a customer plans to drive at least 60 mi, from which company should he rent a truck for a single day?

15. **Decision Analysis** A product may be made by using Machine I or Machine II. The manufacturer estimates that the monthly fixed costs of using Machine I are \$18,000, whereas the monthly fixed costs of using Machine II are \$15,000. The variable costs of manufacturing 1 unit of the product using Machine I and Machine II are \$15 and \$20, respectively. The product sells for \$50 each.
    a. Find the cost functions associated with using each machine.

Review and Study Tools

Summary of Principal Formulas and Terms

Each review section begins with the Summary, which highlights the important equations and terms, with page numbers given for quick review.

CHAPTER 6 Summary of Principal Formulas and Terms

FORMULAS

1. Commutative laws	$A \cup B = B \cup A$ $A \cap B = B \cap A$
2. Associative laws	$A \cup (B \cup C) = (A \cup B) \cup C$ $A \cap (B \cap C) = (A \cap B) \cap C$
3. Distributive laws	$A \cup (B \cap C) = (A \cup B) \cap (A \cup C)$ $A \cap (B \cup C) = (A \cap B) \cup (A \cap C)$
4. De Morgan's laws	$(A \cup B)^c = A^c \cap B^c$ $(A \cap B)^c = A^c \cup B^c$
5. Number of elements in the union of two finite sets	$n(A \cup B) = n(A) + n(B) - n(A \cap B)$
6. Permutation of n distinct objects, taken r at a time	$P(n, r) = \dfrac{n!}{(n-r)!}$
7. Permutation of n objects, not all distinct, taken n at a time	$\dfrac{n!}{n_1!\, n_2! \cdots n_m!}$
8. Combination of n distinct objects, taken r at a time	$C(n, r) = \dfrac{n!}{r!\,(n-r)!}$

TERMS

set (342)
element of a set (342)
roster notation (342)
set-builder notation (342)
set equality (342)
subset (343)
empty set (343)
universal set (344)
Venn diagram (344)
set union (345)
set intersection (345)
complement of a set (345)
set complementation (346)
multiplication principle (362)
generalized multiplication principle (363)
permutation (368)
n-factorial (370)
combination (374)

Concept Review Questions

These questions give students a chance to check their knowledge of the basic definitions and concepts given in each chapter.

CHAPTER 6 Concept Review Questions

Fill in the blanks.

1. A well-defined collection of objects is called a/an _____. These objects are called _____ of the _____.

2. Two sets having exactly the same elements are said to be _____.

3. If every element of a set A is also an element of a set B, then A is a/an _____ of B.

4. a. The empty set $\varnothing$ is the set containing _____ elements.
b. The universal set is the set containing _____ elements.

Review Exercises

Offering a solid review of the chapter material, the Review Exercises contain routine computational exercises followed by applied problems.

CHAPTER 6 Review Exercises

In Exercises 1–4, list the elements of each set in roster notation.

1. $\{x \mid 3x - 2 = 7 \text{ and } x \text{ is an integer}\}$

2. $\{x \mid x$ is a letter of the word *TALLAHASSEE*$\}$

3. The set whose elements are the even numbers between 3 and 11

4. $\{x \mid (x - 3)(x + 4) = 0 \text{ and } x \text{ is a negative integer}\}$

Let $A = \{a, c, e, r\}$. In Exercises 5–8, determine whether the set is equal to A.

5. $\{r, e, c, a\}$

6. $\{x \mid x$ is a letter of the word *career*$\}$

7. $\{x \mid x$ is a letter of the word *racer*$\}$

8. $\{x \mid x$ is a letter of the word *cares*$\}$

For Exercises 17–20, let

U = {all participants in a consumer-behavior survey conducted by a national polling group}
A = {consumers who avoided buying a product because it is not recyclable}
B = {consumers who used cloth rather than disposable diapers}
C = {consumers who boycotted a company's products because of its record on the environment}
D = {consumers who voluntarily recycled their garbage}

Describe each set in words.

17. $A \cap C$ **18.** $A \cup D$
19. $B^c \cap D$ **20.** $C^c \cup D^c$

Before Moving On . . .
Found at the end of each chapter review, these exercises give students a chance to determine whether they have mastered the basic computational skills developed in the chapter.

CHAPTER 6 Before Moving On . . .

1. Let $U = \{a, b, c, d, e, f, g\}$, $A = \{a, d, f, g\}$, $B = \{d, f, g\}$, and $C = \{b, c, e, f\}$. Find:
   a. $A \cap (B \cup C)$
   b. $(A \cap C) \cup (B \cup C)$
   c. A^c
2. Let A, B, and C be subsets of a universal set U, and suppose that $n(U) = 120$, $n(A) = 20$, $n(A \cap B) = 10$, $n(A \cap C) = 11$, $n(B \cap C) = 9$, and $n(A \cap B \cap C) = 4$. Find $n[A \cap (B \cup C)^c]$.
3. In how many ways can four compact discs be selected from six different compact discs?
4. From a standard 52-card deck, how many 5-card poker hands can be dealt consisting of 3 deuces and 2 face cards?
5. There are six seniors and five juniors in the Chess Club at Madison High School. In how many ways can a team consisting of three seniors and two juniors be selected from the members of the Chess Club?

Action-Oriented Study Tabs
Convenient color-coded study tabs make it easy for students to flag pages that they want to return to later, whether for additional review, exam preparation, online exploration, or identifying a topic to be discussed with the instructor.

Instructor Resources

ENHANCED WEBASSIGN® Enhanced WebAssign
Printed Access Card: 978-1-285-85758-9
Online Access Code: 978-1-285-85761-9
Exclusively from Cengage Learning, Enhanced WebAssign combines the exceptional mathematics content that you know and love with the most powerful online homework solution, WebAssign. Enhanced WebAssign engages students with immediate feedback, rich tutorial content, and interactive, fully customizable e-books (YouBook), helping students to develop a deeper conceptual understanding of their subject matter. Quick Prep and Just In Time exercises provide opportunities for students to review prerequisite skills and content, both at the start of the course and at the beginning of each section. Flexible assignment options give instructors the ability to release assignments conditionally on the basis of students' prerequisite assignment scores. Visit us at **www.cengage.com/ewa** to learn more.

COMPLETE SOLUTIONS MANUAL by Soo T. Tan
Written by the author, the Complete Solutions Manual contains solutions for all exercises in the text, including *Exploring with Technology* and *Explore and Discuss* exercises. The Complete Solutions Manual is available on the Instructor Companion Site.

CENGAGE LEARNING TESTING POWERED BY COGNERO
Cengage Learning Testing Powered by Cognero is a flexible, online system that allows you to author, edit, and manage test bank content from multiple Cengage Learning solutions; create multiple test versions in an instant; and deliver tests from your LMS, your classroom, or wherever you want. Access to Cognero is available on the Instructor Companion Site.

SOLUTION BUILDER (www.cengage.com/solutionbuilder)
This online instructor database offers complete worked-out solutions to all exercises in the text, including *Exploring with Technology* and *Explore and Discuss* questions. Solution Builder allows you to create customized, secure solutions printouts (in PDF format) matched exactly to the problems you assign in class.

INSTRUCTOR COMPANION SITE
Everything you need for your course in one place! This collection of book-specific lecture and class tools is available online at **www.cengage.com/login.** Access and download PowerPoint presentations, images, solutions manual, videos, and more.

Student Resources

STUDENT SOLUTIONS MANUAL by Soo T. Tan (ISBN-13: 978-1-285-84572-2)
Giving you more in-depth explanations, this insightful resource includes fully worked-out solutions for selected exercises in the textbook, as well as problem-solving strategies, additional algebra steps, and review for selected problems.

ENHANCED WEBASSIGN®
Printed Access Card: 978-1-285-85758-9
Online Access Code: 978-1-285-85761-9
Enhanced WebAssign (assigned by the instructor) provides you with instant feedback on homework assignments. This online homework system is easy to use and includes helpful links to textbook sections, video examples, and problem-specific tutorials.

CENGAGEBRAIN.COM
Visit **www.cengagebrain.com** to access additional course materials and companion resources. At the CengageBrain.com home page, search for the ISBN of your title (from the back cover of your book) using the search box at the top of the page. This will take you to the product page where free companion resources can be found.

Acknowledgments

I wish to express my personal appreciation to each of the following reviewers, whose many suggestions have helped make a much improved book.

Zach Abernathy
Winthrop University

Mark Antkowicz
Colorado Technical University

Andrea Brown
Ivy Tech Community College

Ashot Djrbashian
Glendale Community College

Amy Franklin
Jacksonville State University

Jean Johnson
Jacksonville State University

Peter Knopf
Pace University

Thurai Kugan
John Jay College of Criminal Justice, CUNY

Melanie Ledwig
Victoria College

Myra Maxwell
University of Indianapolis

Laurie McManus
St. Louis Community College at Meramec

Markus Pomper
Indiana University East

John Roepke
Doane College

Peter Shenkin
John Jay College of Criminal Justice, CUNY

Beimnet Teclezghi
New Jersey City University

Brenda F. Tiefenbruck
University of St. Thomas

Kenneth V. Turner, Jr.
Anderson University

Francis J. Vasko
Kutztown University

Mary Wright
Southern Illinois University Carbondale

Jill Zarestky
Texas A&M University

I also thank reviewers of previous editions whose comments and suggestions have helped to get the book this far.

I also wish to thank Tao Guo for the excellent job he did as the accuracy checker for this text. I also thank the editorial and production staffs of Cengage Learning—Richard Stratton, Rita Lombard, Laura Wheel, Jennifer Cordoba, Andrew Coppola, Cheryll Linthicum, and Vernon Boes—for all of their help and support during the development and production of this edition. I also thank Martha Emry and Barbara Willette, who both did an excellent job ensuring the accuracy and readability of this edition. Simply stated, the team I have been working with is outstanding, and I truly appreciate all of their hard work and efforts.

S. T. Tan

About the Author

SOO T. TAN received his S.B. degree from Massachusetts Institute of Technology, his M.S. degree from the University of Wisconsin–Madison, and his Ph.D. from the University of California at Los Angeles. He has published numerous papers in optimal control theory, numerical analysis, and mathematics of finance. He is also the author of a series of calculus textbooks.

1 Straight Lines and Linear Functions

THIS CHAPTER INTRODUCES the Cartesian coordinate system, a system that allows us to represent points in the plane in terms of ordered pairs of real numbers. This in turn enables us to compute the distance between two points algebraically. We also study straight lines. *Linear functions,* whose graphs are straight lines, can be used to describe many relationships between two quantities. These relationships can be found in fields of study as diverse as business, economics, the social sciences, physics, and medicine. In addition, we see how some practical problems can be solved by finding the point(s) of intersection of two straight lines. Finally, we learn how to find an algebraic representation of the straight line that "best" fits a set of data points that are scattered about a straight line.

Because the over-65 population will be growing more rapidly in the next few decades, U.S. health-care expenditures are expected to be boosted significantly. What will be the rate of increase of these expenditures over the next few years? How much will health-care expenditures be in 2014? In Example 1, page 31, we use a mathematical model based on figures from the Centers for Medicare & Medicaid Services to answer these questions.

1.1 The Cartesian Coordinate System

The Cartesian Coordinate System

The real number system is made up of the set of real numbers together with the usual operations of addition, subtraction, multiplication, and division. We assume that you are familiar with the rules governing these algebraic operations (see Appendix B).

Real numbers may be represented geometrically by points on a line. This line is called the **real number,** or **coordinate, line.** We can construct the real number line as follows: Arbitrarily select a point on a straight line to represent the number 0. This point is called the **origin.** If the line is horizontal, then choose a point at a convenient distance to the right of the origin to represent the number 1. This determines the scale for the number line. Each positive real number x lies x units to the right of 0, and each negative real number x lies $-x$ units to the left of 0.

In this manner, a one-to-one correspondence is set up between the set of real numbers and the set of points on the number line, with all the positive numbers lying to the right of the origin and all the negative numbers lying to the left of the origin (Figure 1).

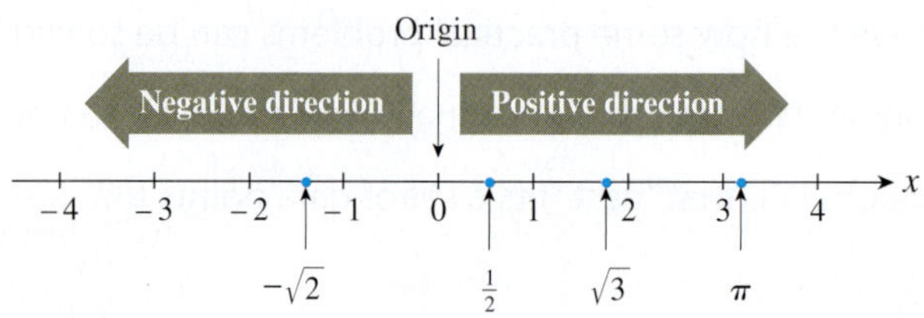

FIGURE 1
The real number line

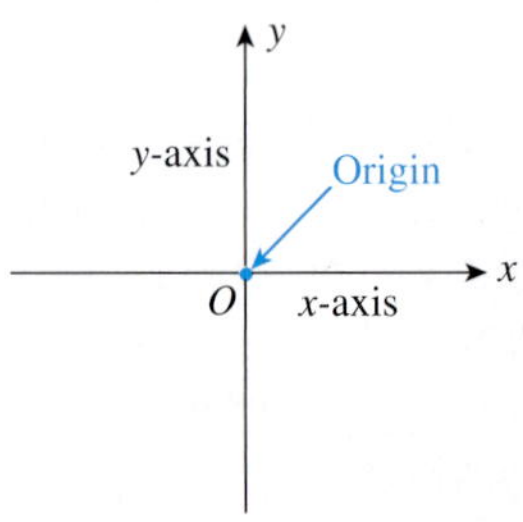

FIGURE 2
The Cartesian coordinate system

In a similar manner, we can represent points in a plane (a two-dimensional space) by using the **Cartesian coordinate system,** which we construct as follows: Take two perpendicular lines, one of which is normally chosen to be horizontal. These lines intersect at a point O, called the **origin** (Figure 2). The horizontal line is called the ***x*-axis,** and the vertical line is called the ***y*-axis.** A number scale is set up along the x-axis, with the positive numbers lying to the right of the origin and the negative numbers lying to the left of it. Similarly, a number scale is set up along the y-axis, with the positive numbers lying above the origin and the negative numbers lying below it.

Note The number scales on the two axes need not be the same. Indeed, in many applications, different quantities are represented by x and y. For example, x may represent the number of smartphones sold, and y may represent the total revenue resulting from the sales. In such cases, it is often desirable to choose different number scales to represent the different quantities. Note, however, that the zeros of both number scales coincide at the origin of the two-dimensional coordinate system.

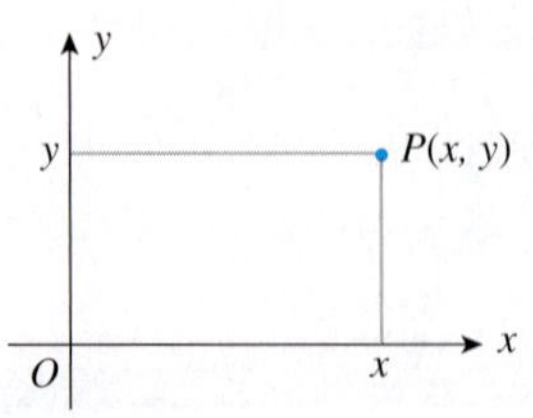

FIGURE 3
An ordered pair in the coordinate plane

We can represent a point in the plane in this coordinate system by an **ordered pair** of numbers—that is, a pair (x, y) in which x is the first number and y is the second. To see this, let P be any point in the plane (Figure 3). Draw perpendicular lines from P to the x-axis and y-axis, respectively. Then the number x is precisely the number that corresponds to the point on the x-axis at which the perpendicular line through P hits the x-axis. Similarly, y is the number that corresponds to the point on the y-axis at which the perpendicular line through P crosses the y-axis.

Conversely, given an ordered pair (x, y) with x as the first number and y as the second, a point P in the plane is uniquely determined as follows: Locate the point on the x-axis represented by the number x, and draw a line through that point perpendicular to the x-axis. Next, locate the point on the y-axis represented by the number y, and draw a line through that point perpendicular to the y-axis. The point of intersection of these two lines is the point P (Figure 3).

In the ordered pair (x, y), x is called the **abscissa,** or ***x*-coordinate;** y is called the **ordinate,** or ***y*-coordinate;** and x and y together are referred to as the **coordinates** of the point P. The point P with x-coordinate equal to a and y-coordinate equal to b is often written $P(a, b)$.

The points $A(2, 3)$, $B(-2, 3)$, $C(-2, -3)$, $D(2, -3)$, $E(3, 2)$, $F(4, 0)$, and $G(0, -5)$ are plotted in Figure 4.

Note In general, $(x, y) \neq (y, x)$. This is illustrated by the points A and E in Figure 4.

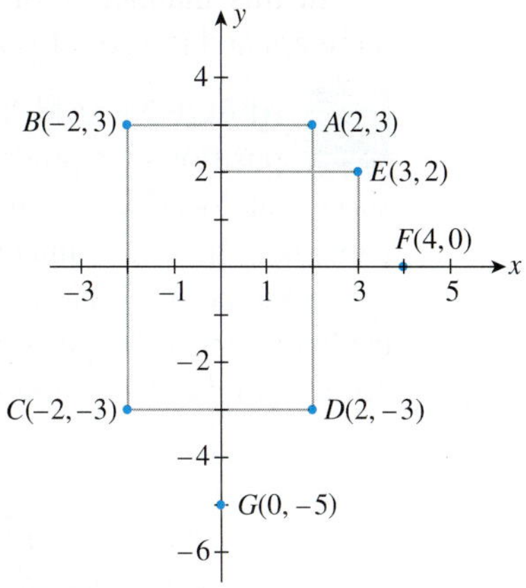

FIGURE 4
Several points in the coordinate plane

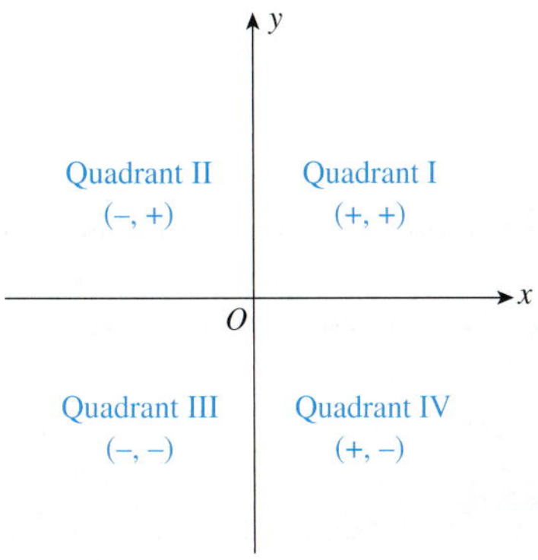

FIGURE 5
The four quadrants in the coordinate plane

The axes divide the plane into four quadrants. Quadrant I consists of the points P with coordinates x and y, denoted by $P(x, y)$, satisfying $x > 0$ and $y > 0$; Quadrant II consists of the points $P(x, y)$ where $x < 0$ and $y > 0$; Quadrant III consists of the points $P(x, y)$ where $x < 0$ and $y < 0$; and Quadrant IV consists of the points $P(x, y)$ where $x > 0$ and $y < 0$ (Figure 5).

The Distance Formula

One immediate benefit that arises from using the Cartesian coordinate system is that the distance between any two points in the plane may be expressed solely in terms of the coordinates of the points. Suppose, for example, (x_1, y_1) and (x_2, y_2) are any two points in the plane (Figure 6). Then we have the following:

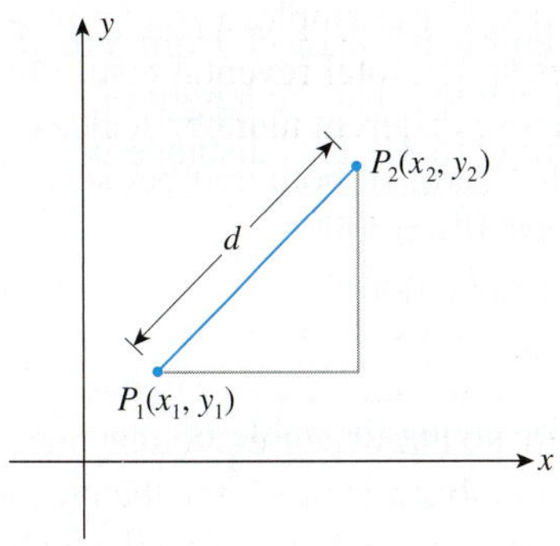

FIGURE 6
The distance between two points in the coordinate plane

Distance Formula

The distance d between two points $P_1(x_1, y_1)$ and $P_2(x_2, y_2)$ in the plane is given by

$$d = \sqrt{(x_2 - x_1)^2 + (y_2 - y_1)^2} \tag{1}$$

For a proof of this result, see Exercise 49, page 9.

Explore and Discuss

Refer to Example 1. Suppose we label the point $(2, 6)$ as P_1 and the point $(-4, 3)$ as P_2.
(1) Show that the distance d between the two points is the same as that obtained in Example 1.
(2) Prove that, in general, the distance d in Formula (1) is independent of the way we label the two points.

In what follows, we give several applications of the distance formula.

EXAMPLE 1 Find the distance between the points $(-4, 3)$ and $(2, 6)$.

Solution Let $P_1(-4, 3)$ and $P_2(2, 6)$ be points in the plane. Then we have

$$x_1 = -4 \quad \text{and} \quad y_1 = 3$$
$$x_2 = 2 \qquad\qquad y_2 = 6$$

Using Formula (1), we have

$$\begin{aligned} d &= \sqrt{[2-(-4)]^2 + (6-3)^2} \\ &= \sqrt{6^2 + 3^2} \\ &= \sqrt{45} \\ &= 3\sqrt{5} \end{aligned}$$

APPLIED EXAMPLE 2 The Cost of Laying Cable In Figure 7, S represents the position of a power relay station located on a straight coastal highway, and M shows the location of a marine biology experimental station on a nearby island. A cable is to be laid connecting the relay station at S with the experimental station at M via the point Q that lies on the x-axis between O and S. If the cost of running the cable on land is \$3 per running foot and the cost of running the cable underwater is \$5 per running foot, find the total cost for laying the cable.

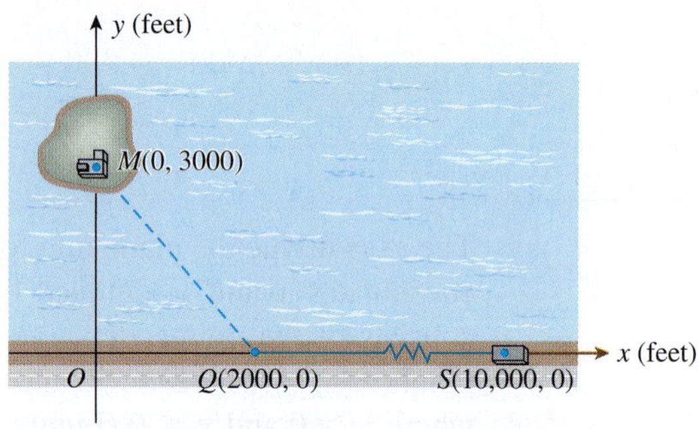

FIGURE 7
The cable will connect the relay station S to the experimental station M.

Solution The length of cable required on land is given by the distance from S to Q. This distance is $(10{,}000 - 2000)$, or 8000 feet. Next, we see that the length of cable required underwater is given by the distance from Q to M. This distance is

$$\begin{aligned} \sqrt{(0-2000)^2 + (3000-0)^2} &= \sqrt{2000^2 + 3000^2} \\ &= \sqrt{13{,}000{,}000} \\ &\approx 3606 \end{aligned}$$

or approximately 3606 feet. Therefore, the total cost for laying the cable is approximately

$$3(8000) + 5(3606) \approx 42{,}030$$

dollars.

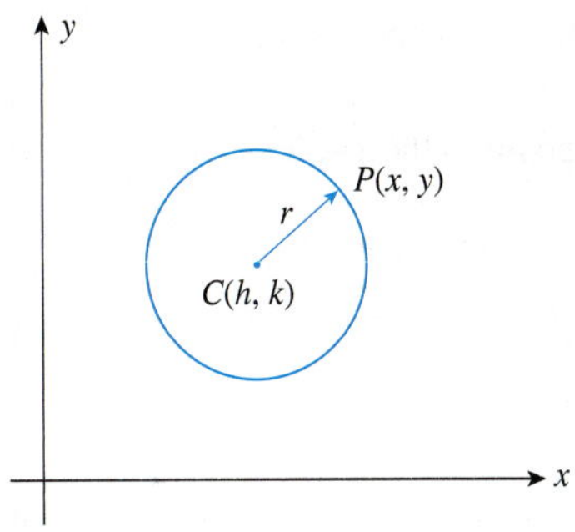

FIGURE 8
A circle with radius r and center $C(h, k)$

EXAMPLE 3 Let $P(x, y)$ denote a point lying on a circle with radius r and center $C(h, k)$ (Figure 8). Find a relationship between x and y.

Solution By the definition of a circle, the distance between $C(h, k)$ and $P(x, y)$ is r. Using Formula (1), we have

$$\sqrt{(x-h)^2 + (y-k)^2} = r$$

which, upon squaring both sides, gives the equation

$$(x-h)^2 + (y-k)^2 = r^2$$

which must be satisfied by the variables x and y.

A summary of the result obtained in Example 3 follows.

Equation of a Circle

An equation of the circle with center $C(h, k)$ and radius r is given by

$$(x-h)^2 + (y-k)^2 = r^2 \qquad \textbf{(2)}$$

EXAMPLE 4 Find an equation of the circle with (a) radius 2 and center $(-1, 3)$ and (b) radius 3 and center located at the origin.

Solution

a. We use Formula (2) with $r = 2$, $h = -1$, and $k = 3$, obtaining

$$[x-(-1)]^2 + (y-3)^2 = 2^2$$
$$(x+1)^2 + (y-3)^2 = 4$$

(Figure 9a).

b. Using Formula (2) with $r = 3$ and $h = k = 0$, we obtain

$$x^2 + y^2 = 3^2$$
$$x^2 + y^2 = 9$$

(Figure 9b).

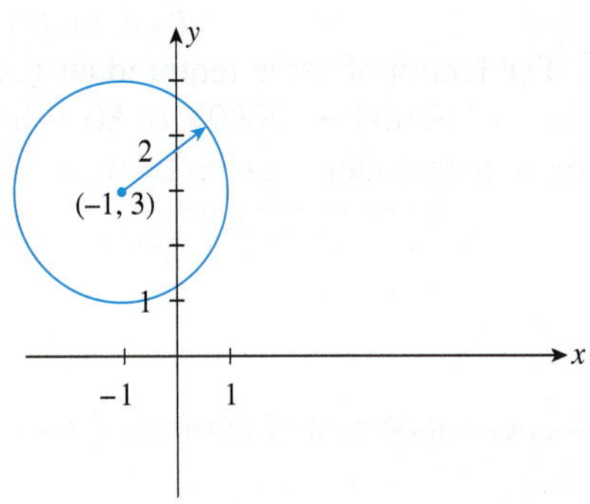

(a) The circle with radius 2 and center $(-1, 3)$

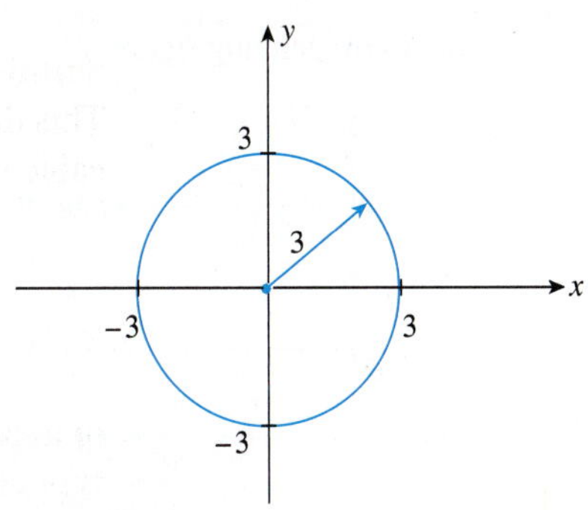

(b) The circle with radius 3 and center $(0, 0)$

FIGURE 9

Explore and Discuss

1. Use the distance formula to help you describe the set of points in the xy-plane satisfying each of the following inequalities, where $r > 0$.
   a. $(x - h)^2 + (y - k)^2 \le r^2$
   b. $(x - h)^2 + (y - k)^2 < r^2$
   c. $(x - h)^2 + (y - k)^2 \ge r^2$
   d. $(x - h)^2 + (y - k)^2 > r^2$
2. Consider the equation $x^2 + y^2 = 4$.
   a. Show that $y = \pm\sqrt{4 - x^2}$.
   b. Describe the set of points (x, y) in the xy-plane satisfying the equation

   (i) $y = \sqrt{4 - x^2}$ (ii) $y = -\sqrt{4 - x^2}$

1.1 Self-Check Exercises

1. **a.** Plot the points $A(4, -2)$, $B(2, 3)$, and $C(-3, 1)$.
   b. Find the distance between the points A and B, between B and C, and between A and C.
   c. Use the Pythagorean Theorem to show that the triangle with vertices A, B, and C is a right triangle.
2. **Fuel Stop Planning** The accompanying figure shows the location of Cities A, B, and C. Suppose a pilot wishes to fly from City A to City C but must make a mandatory stopover in City B. If the single-engine light plane has a range of 650 mi, can the pilot make the trip without refueling in City B?

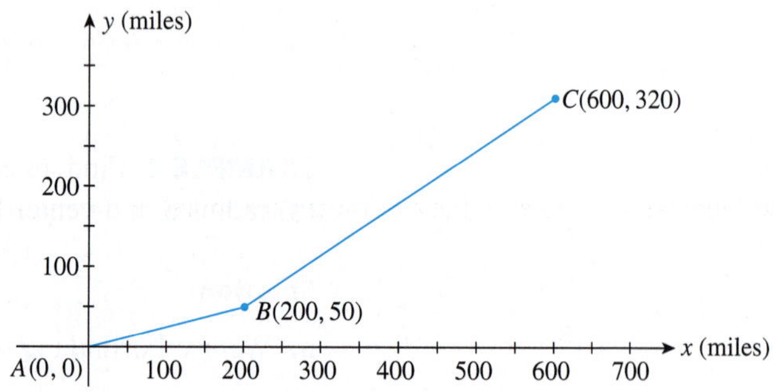

Solutions to Self-Check Exercises 1.1 can be found on page 10.

1.1 Concept Questions

1. What can you say about the signs of a and b if the point $P(a, b)$ lies in (a) the second quadrant? (b) The third quadrant? (c) The fourth quadrant?
2. Refer to the accompanying figure.

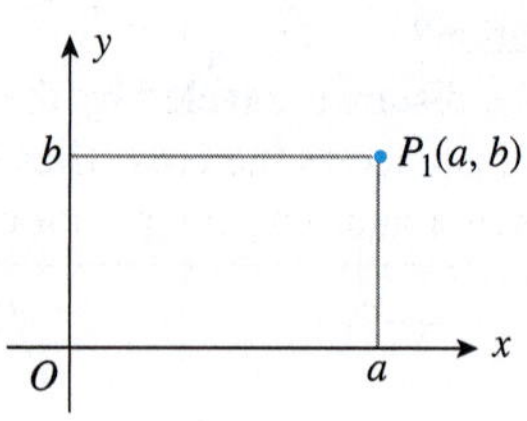

   a. Given the point $P_1(a, b)$, where $a > 0$ and $b > 0$, plot the points $P_2(-a, b)$, $P_3(-a, -b)$, and $P_4(a, -b)$.
   b. What can you say about the distance of the points $P_1(a, b)$, $P_2(-a, b)$, $P_3(-a, -b)$, and $P_4(a, -b)$ from the origin?

1.1 Exercises

In Exercises 1–6, refer to the accompanying figure and determine the coordinates of the point and the quadrant in which it is located.

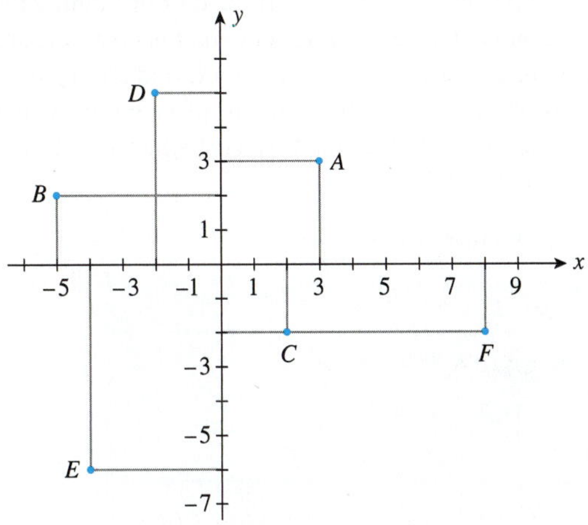

1. *A* **2.** *B* **3.** *C*

4. *D* **5.** *E* **6.** *F*

In Exercises 7–12, refer to the accompanying figure.

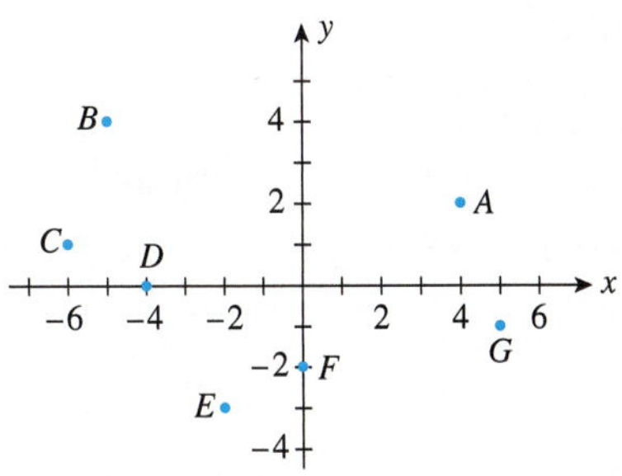

7. Which point is represented by the ordered pair $(4, 2)$?

8. What are the coordinates of point *B*?

9. Which points have negative *y*-coordinates?

10. Which point has a negative *x*-coordinate and a negative *y*-coordinate?

11. Which point has an *x*-coordinate that is equal to zero?

12. Which point has a *y*-coordinate that is equal to zero?

In Exercises 13–20, sketch a set of coordinate axes and then plot the point.

13. $(-2, 5)$ **14.** $(1, 3)$

15. $(3, -1)$ **16.** $(3, -4)$

17. $(8, -\frac{7}{2})$ **18.** $(-\frac{5}{2}, \frac{3}{2})$

19. $(4.5, -4.5)$ **20.** $(1.2, -3.4)$

In Exercises 21–24, find the distance between the points.

21. $(1, 3)$ and $(4, 7)$

22. $(1, 0)$ and $(4, 4)$

23. $(-1, 3)$ and $(4, 9)$

24. $(-2, 1)$ and $(10, 6)$

25. Find the coordinates of the points that are 10 units away from the origin and have a *y*-coordinate equal to -6.

26. Find the coordinates of the points that are 5 units away from the origin and have an *x*-coordinate equal to 3.

27. Show that the points $(3, 4)$, $(-3, 7)$, $(-6, 1)$, and $(0, -2)$ form the vertices of a square.

28. Show that the triangle with vertices $(-5, 2)$, $(-2, 5)$, and $(5, -2)$ is a right triangle.

In Exercises 29–34, find an equation of the circle that satisfies the given conditions.

29. Radius 5 and center $(2, -3)$

30. Radius 3 and center $(-2, -4)$

31. Radius 5 and center at the origin

32. Center at the origin and passes through $(2, 3)$

33. Center $(2, -3)$ and passes through $(5, 2)$

34. Center $(-a, a)$ and radius $2a$

35. **Tracking a Criminal with GPS** After obtaining a warrant, the police attached a GPS tracking device to the car of a murder suspect. Suppose the car was located at the origin of a Cartesian coordinate system when the device was attached. Shortly afterwards, the suspect's car was tracked going 5 mi due east, 4 mi due north, and 1 mi due west before coming to a permanent stop.
   a. What are the coordinates of the suspect's car at its final destination?
   b. What was the distance traveled by the suspect?
   c. What is the distance as the crow flies between the original position and the final position of the suspect's car?

36. Planning a Grand Tour A grand tour of four cities begins at City A and makes successive stops at Cities B, C, and D before returning to City A. If the cities are located as shown in the accompanying figure, find the total distance covered on the tour.

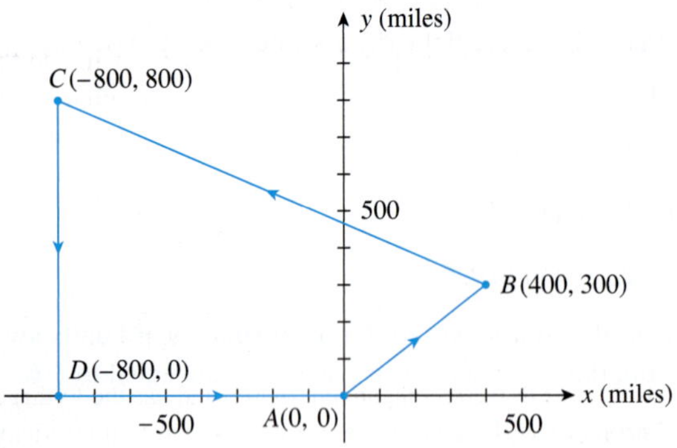

37. Will You Incur a Delivery Charge? A furniture store offers free setup and delivery services to all points within a 25-mi radius of its warehouse distribution center. If you live 20 mi east and 14 mi south of the warehouse, will you incur a delivery charge? Justify your answer.

38. Optimizing Travel Time Towns A, B, C, and D are located as shown in the accompanying figure. Two highways link Town A to Town D. Route 1 runs from Town A to Town D via Town B, and Route 2 runs from Town A to Town D via Town C. If a salesman wishes to drive from Town A to Town D and traffic conditions are such that he could expect to average the same speed on either route, which highway should he take to arrive in the shortest time?

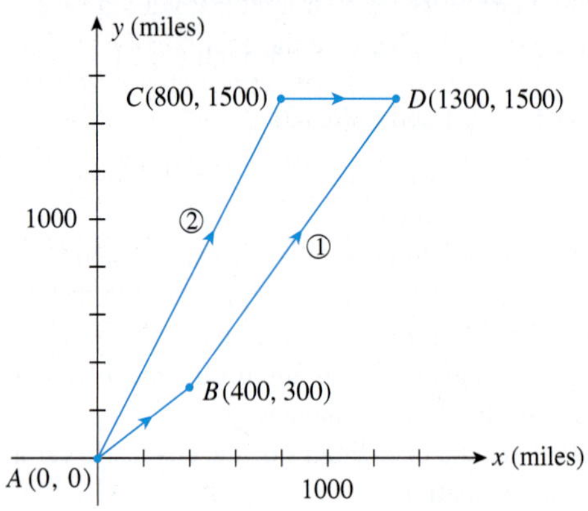

39. Minimizing Shipping Costs for a Fleet of Autos Refer to the figure for Exercise 38. Suppose a fleet of 100 automobiles are to be shipped from an assembly plant in Town A to Town D. They may be shipped either by freight train along Route 1 at a cost of 66¢/mile/automobile or by truck along Route 2 at a cost of 62¢/mile/automobile. Which means of transportation minimizes the shipping cost? What is the net savings?

40. Cost of Laying Cable In the accompanying diagram, S represents the position of a power relay station located on a straight coastal highway, and M shows the location of a marine biology experimental station on a nearby island. A cable is to be laid connecting the relay station at S with the experimental station at M via the point Q that lies on the x-axis between O and S. If the cost of running the cable on land is \$3/running foot and the cost of running cable underwater is \$5/running foot, find an expression in terms of x that gives the total cost of laying the cable. Use this expression to find the total cost when $x = 1500$ and when $x = 2500$.

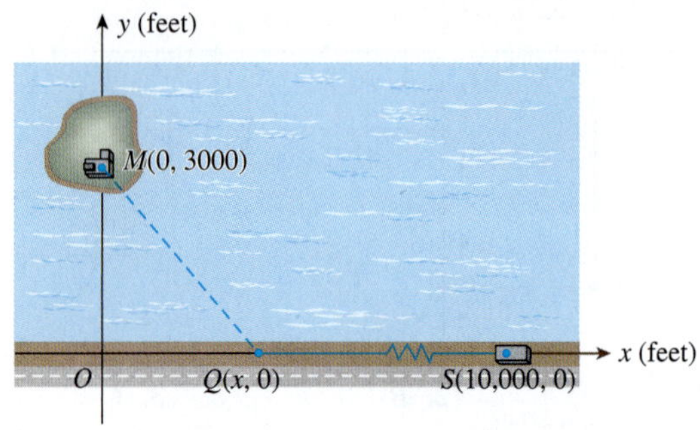

41. Purchasing an HDTV Antenna Will Barclay wishes to determine which HDTV antenna he should purchase for his home. The TV store has supplied him with the following information:

Range in Miles			
VHF	**UHF**	**Model**	**Price**
30	20	A	\$50
45	35	B	60
60	40	C	70
75	55	D	80

Will wishes to receive Channel 17 (VHF), which is located 25 mi east and 35 mi north of his home, and Channel 38 (UHF), which is located 20 mi south and 32 mi west of his home. Which model will allow him to receive both channels at the least cost? (Assume that the terrain between Will's home and both broadcasting stations is flat.)

42. Distance Between Two Cruise Ships Two cruise ships leave port at the same time. Ship A sails north at a speed of 20 mph while Ship B sails east at a speed of 30 mph.

a. Find an expression in terms of the time t (in hours) giving the distance between the two cruise ships.

b. Using the expression obtained in part (a), find the distance between the two cruise ships 2 hr after leaving port.

43. **DISTANCE BETWEEN TWO CARGO SHIPS** Sailing north at a speed of 25 mph, Ship A leaves a port. A half hour later, Ship B leaves the same port, sailing east at a speed of 20 mph. Let t (in hours) denote the time Ship B has been at sea.
  a. Find an expression in terms of t that gives the distance between the two cargo ships.
  b. Use the expression obtained in part (a) to find the distance between the two cargo ships 2 hr after Ship A has left the port.

44. **WATCHING A ROCKET LAUNCH** At a distance of 4000 ft from the launch site, a spectator is observing a rocket being launched. Suppose the rocket lifts off vertically and reaches an altitude of x feet, as shown below:

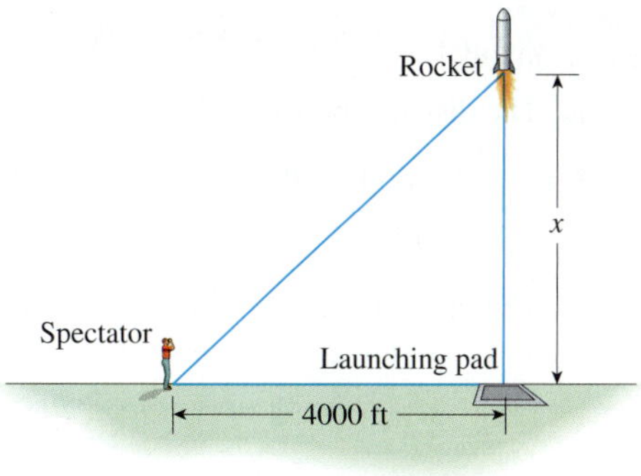

  a. Find an expression giving the distance between the spectator and the rocket.
  b. What is the distance between the spectator and the rocket when the rocket reaches an altitude of 20,000 ft?

45. a. Show that the midpoint of the line segment joining the points $P_1(x_1, y_1)$ and $P_2(x_2, y_2)$ is

$$\left(\frac{x_1 + x_2}{2}, \frac{y_1 + y_2}{2}\right)$$

  b. Use the result of part (a) to find the midpoint of the line segment joining the points $(-3, 2)$ and $(4, -5)$.

46. **A SCAVENGER HUNT** A tree is located 20 yd to the east and 10 yd to the north of a house. A second tree is located 10 yd to the east and 40 yd to the north of the house. The prize of a scavenger hunt is placed exactly midway between the trees.
  a. Place the house at the origin of a Cartesian coordinate system, and draw a diagram depicting the situation.
  b. What are the coordinates of the position of the prize?
  c. How far is the prize from the house?

In Exercises 47 and 48, determine whether the statement is true or false. If it is true, explain why it is true. If it is false, give an example to show why it is false.

47. If the distance between the points $P_1(a, b)$ and $P_2(c, d)$ is D, then the distance between the points $P_1(a, b)$ and $P_3(kc, kd)$ $(k \neq 0)$ is given by $|k|D$.

48. The circle with equation $kx^2 + ky^2 = a^2$ lies inside the circle with equation $x^2 + y^2 = a^2$, provided that $k > 1$ and $a > 0$.

49. Let (x_1, y_1) and (x_2, y_2) be two points lying in the xy-plane. Show that the distance between the two points is given by

$$d = \sqrt{(x_2 - x_1)^2 + (y_2 - y_1)^2}$$

Hint: Refer to the accompanying figure, and use the Pythagorean Theorem.

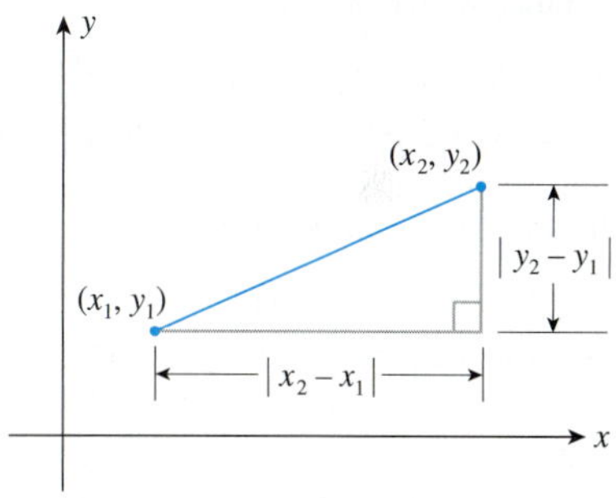

50. In the Cartesian coordinate system, the two axes are perpendicular to each other. Consider a coordinate system in which the x-axis and y-axis are noncollinear (that is, the axes do not lie along a straight line) and are not perpendicular to each other (see the accompanying figure).
  a. Describe how a point is represented in this coordinate system by an ordered pair (x, y) of real numbers. Conversely, show how an ordered pair (x, y) of real numbers uniquely determines a point in the plane.
  b. Suppose you want to find a formula for the distance between two points, $P_1(x_1, y_1)$ and $P_2(x_2, y_2)$, in the plane. What advantage does the Cartesian coordinate system have over the coordinate system under consideration?

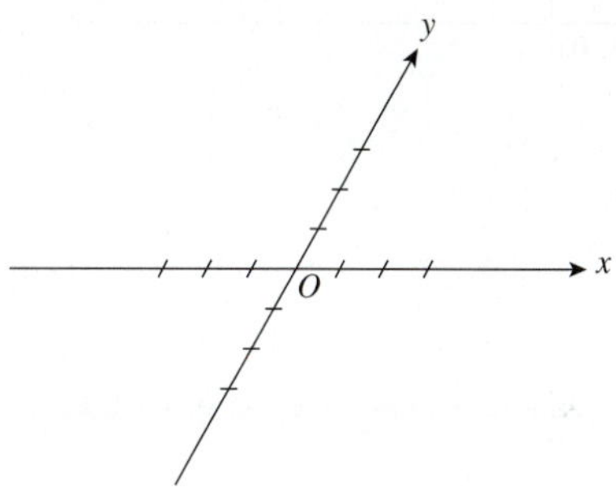

1.1 Solutions to Self-Check Exercises

1. a. The points are plotted in the following figure.

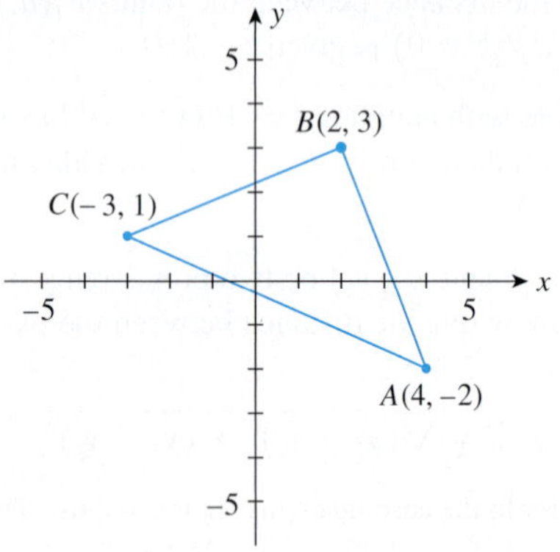

b. The distance between A and B is

$$d(A, B) = \sqrt{(2-4)^2 + [3-(-2)]^2}$$
$$= \sqrt{(-2)^2 + 5^2} = \sqrt{4+25} = \sqrt{29}$$

The distance between B and C is

$$d(B, C) = \sqrt{(-3-2)^2 + (1-3)^2}$$
$$= \sqrt{(-5)^2 + (-2)^2} = \sqrt{25+4} = \sqrt{29}$$

The distance between A and C is

$$d(A, C) = \sqrt{(-3-4)^2 + [1-(-2)]^2}$$
$$= \sqrt{(-7)^2 + 3^2} = \sqrt{49+9} = \sqrt{58}$$

c. We will show that

$$[d(A, C)]^2 = [d(A, B)]^2 + [d(B, C)]^2$$

From part (b), we see that $[d(A, B)]^2 = 29$, $[d(B, C)]^2 = 29$, and $[d(A, C)]^2 = 58$, and the desired result follows.

2. The distance between City A and City B is

$$d(A, B) = \sqrt{200^2 + 50^2} \approx 206$$

or 206 mi. The distance between City B and City C is

$$d(B, C) = \sqrt{(600-200)^2 + (320-50)^2}$$
$$= \sqrt{400^2 + 270^2} \approx 483$$

or 483 mi. Therefore, the total distance the pilot would have to cover is 689 mi, so she must refuel in City B.

1.2 Straight Lines

Businesses may depreciate certain assets such as buildings, machines, furniture, vehicles, and equipment over a period of time for income tax purposes. *Linear depreciation*, or the *straight-line method*, is often used for this purpose. The graph of the straight line shown in Figure 10 describes the book value V of a network server that has an initial value of \$10,000 and that is being depreciated linearly over 5 years with a scrap value of \$3000. Note that only the solid portion of the straight line is of interest here.

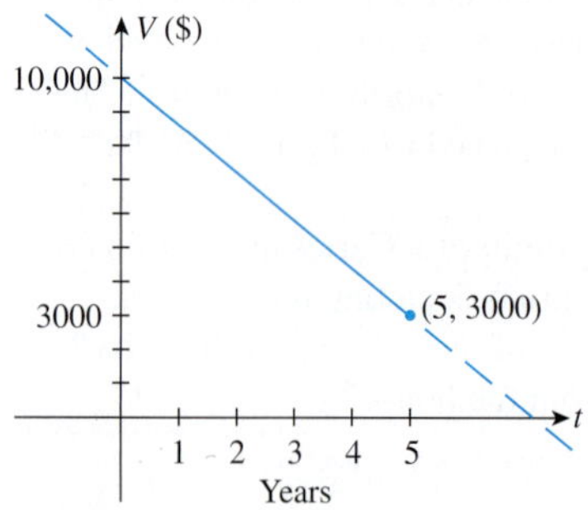

FIGURE **10**
Linear depreciation of a network server

The book value of the server at the end of year t, where t lies between 0 and 5, can be read directly from the graph. But there is one shortcoming in this approach: The result depends on how accurately you draw and read the graph. A better and more accurate method is based on finding an *algebraic* representation of the depreciation line. (We continue our discussion of the linear depreciation problem in Section 1.3.)

To see how a straight line in the xy-plane may be described algebraically, we need first to recall certain properties of straight lines.

Slope of a Line

Let L denote the unique straight line that passes through the two distinct points (x_1, y_1) and (x_2, y_2). If $x_1 \neq x_2$, then we define the slope of L as follows.

Slope of a Nonvertical Line

If (x_1, y_1) and (x_2, y_2) are any two distinct points on a nonvertical line L, then the slope m of L is given by

$$m = \frac{\Delta y}{\Delta x} = \frac{y_2 - y_1}{x_2 - x_1} \tag{3}$$

(Figure 11).

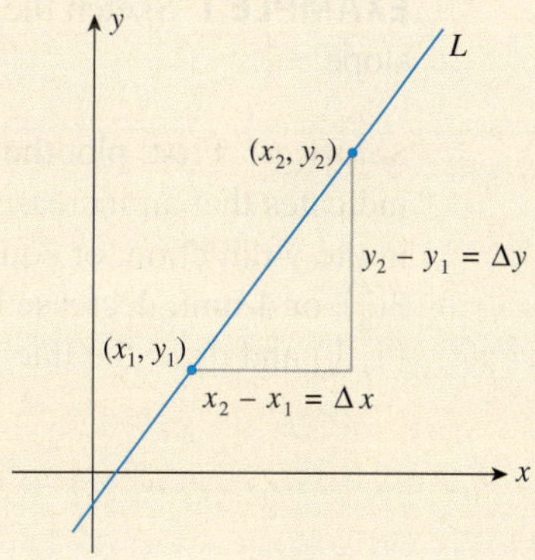

FIGURE 11

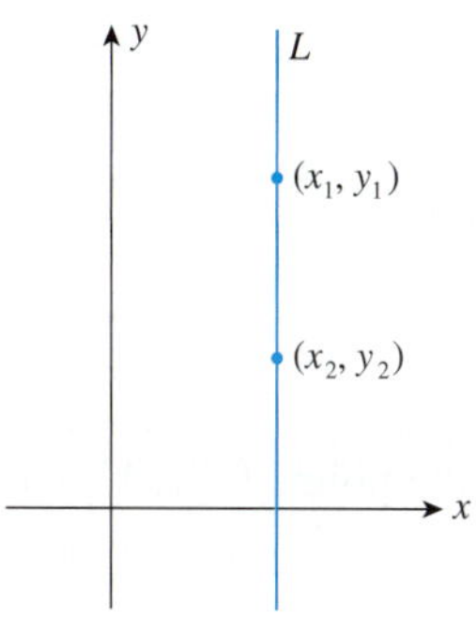

FIGURE 12
The slope of L is undefined if $x_1 = x_2$.

If $x_1 = x_2$, then L is a vertical line (Figure 12). Its slope is undefined, since the denominator in Equation (3) will be zero and division by zero is not allowed.

Observe that the slope of a straight line is a constant whenever it is defined. The number $\Delta y = y_2 - y_1$ (Δy is read "delta y") is a measure of the vertical change in y, and $\Delta x = x_2 - x_1$ is a measure of the horizontal change in x as shown in Figure 11. From this figure, we can see that the slope m of a straight line L is a measure of the *rate of change of y with respect to x*. Furthermore, the slope of a nonvertical straight line is constant, and this tells us that this rate of change is constant.

Figure 13a shows a straight line L_1 with slope 2. Observe that L_1 has the property that a 1-unit increase in x results in a 2-unit increase in y. To see this, let $\Delta x = 1$ in

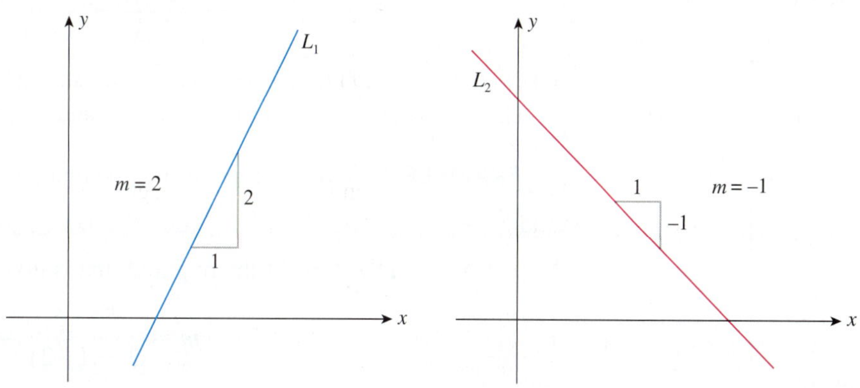

(a) The line rises ($m > 0$).
(b) The line falls ($m < 0$).
FIGURE 13

Equation (3) so that $m = \Delta y$. Since $m = 2$, we conclude that $\Delta y = 2$. Similarly, Figure 13b shows a line L_2 with slope -1. Observe that a straight line with positive slope slants upward from left to right (y increases as x increases), whereas a line with negative slope slants downward from left to right (y decreases as x increases). Finally, Figure 14 shows a family of straight lines passing through the origin with indicated slopes.

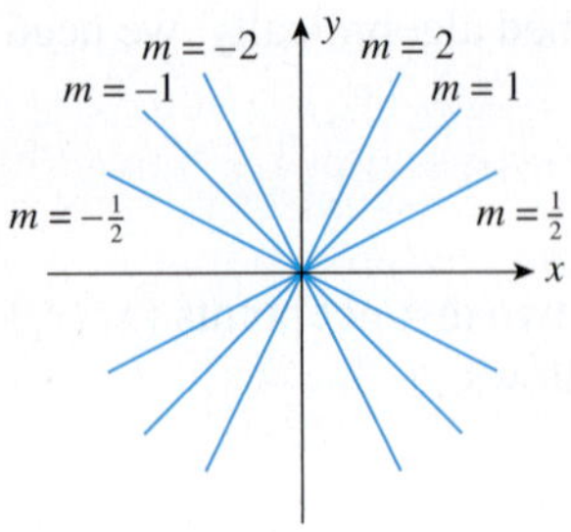

FIGURE 14
A family of straight lines

> **Explore and Discuss**
>
> Show that the slope of a nonvertical line is independent of the two distinct points used to compute it.
> Hint: Pick any two distinct points lying on a line L. Then pick two other distinct points, $P_3(x_3, y_3)$ and $P_4(x_4, y_4)$ lying on L. Draw a picture, and use similar triangles to demonstrate that using P_3 and P_4 gives the same value as that obtained by using P_1 and P_2.

EXAMPLE 1 Sketch the straight line that passes through the point $(-2, 5)$ and has slope $-\frac{4}{3}$.

Solution First, plot the point $(-2, 5)$ (Figure 15). Next, recall that a slope of $-\frac{4}{3}$ indicates that an increase of 1 unit in the x-direction produces a *decrease* of $\frac{4}{3}$ units in the y-direction, or equivalently, a 3-unit increase in the x-direction produces a $3(\frac{4}{3})$, or 4-unit, decrease in the y-direction. Using this information, we plot the point $(1, 1)$ and draw the line through the two points.

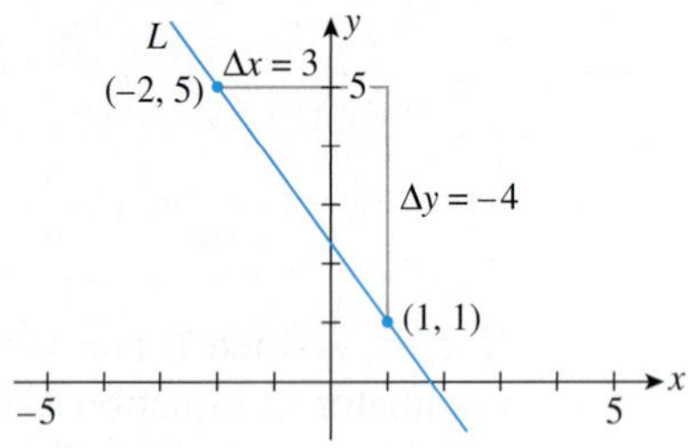

FIGURE 15
L has slope $-\frac{4}{3}$ and passes through $(-2, 5)$.

EXAMPLE 2 Find the slope m of the line that passes through the points $(-1, 1)$ and $(5, 3)$.

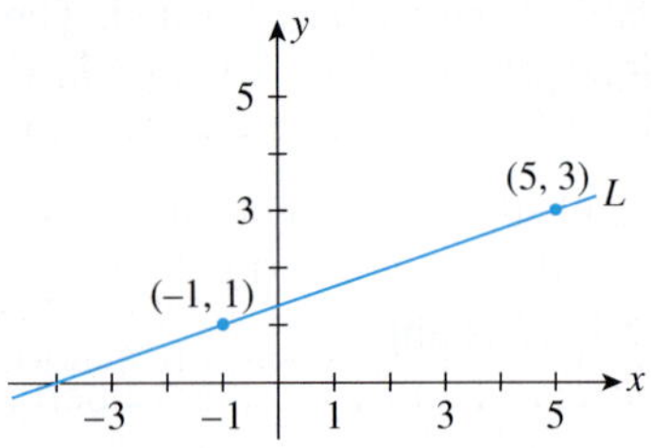

FIGURE 16
L passes through $(5, 3)$ and $(-1, 1)$.

Solution Choose (x_1, y_1) to be the point $(-1, 1)$ and (x_2, y_2) to be the point $(5, 3)$. Then, with $x_1 = -1$, $y_1 = 1$, $x_2 = 5$, and $y_2 = 3$, we find, using Equation (3),

$$m = \frac{y_2 - y_1}{x_2 - x_1} = \frac{3 - 1}{5 - (-1)} = \frac{2}{6} = \frac{1}{3}$$

(Figure 16). You may verify that the result obtained would be the same had we chosen the point $(-1, 1)$ to be (x_2, y_2) and the point $(5, 3)$ to be (x_1, y_1).

EXAMPLE 3 Find the slope of the line that passes through the points $(-2, 5)$ and $(3, 5)$.

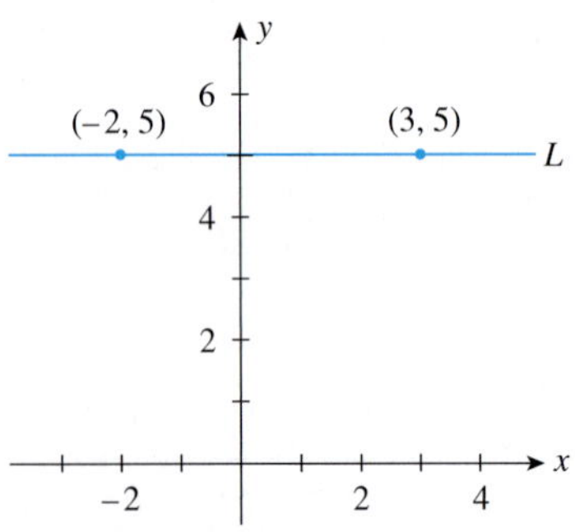

FIGURE 17
The slope of the horizontal line L is zero.

Solution The slope of the required line is given by

$$m = \frac{5 - 5}{3 - (-2)} = \frac{0}{5} = 0$$

(Figure 17).

Note The slope of a horizontal line is zero. ■

We can use the slope of a straight line to determine whether a line is parallel to another line.

Parallel Lines

Two distinct lines are **parallel** if and only if their slopes are equal or their slopes are undefined.

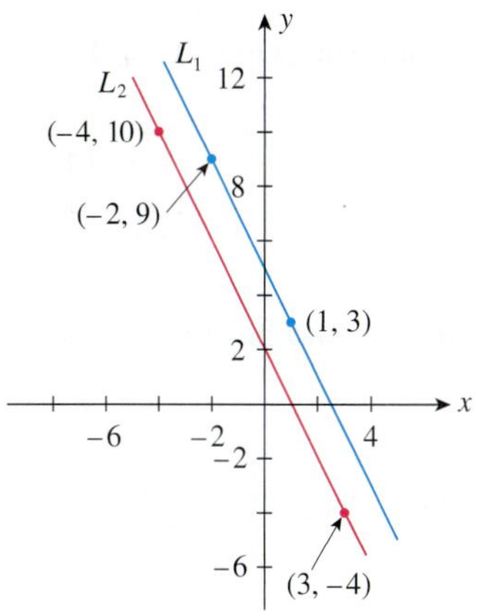

FIGURE 18
L_1 and L_2 have the same slope and hence are parallel.

EXAMPLE 4 Let L_1 be a line that passes through the points $(-2, 9)$ and $(1, 3)$, and let L_2 be the line that passes through the points $(-4, 10)$ and $(3, -4)$. Determine whether L_1 and L_2 are parallel.

Solution The slope m_1 of L_1 is given by

$$m_1 = \frac{3 - 9}{1 - (-2)} = -2$$

The slope m_2 of L_2 is given by

$$m_2 = \frac{-4 - 10}{3 - (-4)} = -2$$

Since $m_1 = m_2$, the lines L_1 and L_2 are in fact parallel (Figure 18). ■

Equations of Lines

We now show that every straight line lying in the xy-plane may be represented by an equation involving the variables x and y. One immediate benefit of this is that problems involving straight lines may be solved algebraically.

Let L be a straight line parallel to the y-axis (perpendicular to the x-axis) (Figure 19). Then L crosses the x-axis at some point $(a, 0)$ with x-coordinate given by $x = a$, where a is some real number. Any other point on L has the form (a, y), where y is an appropriate number. Therefore, the vertical line L is described by the sole condition

$$x = a$$

and this is accordingly an equation of L. For example, the equation $x = -2$ represents a vertical line 2 units to the left of the y-axis, and the equation $x = 3$ represents a vertical line 3 units to the right of the y-axis (Figure 20).

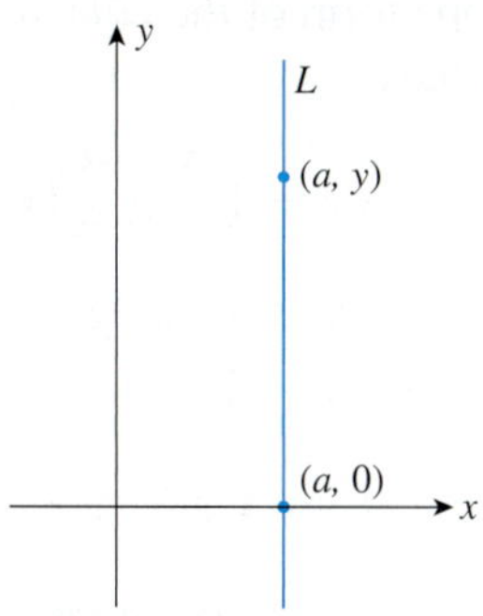

FIGURE 19
The vertical line $x = a$

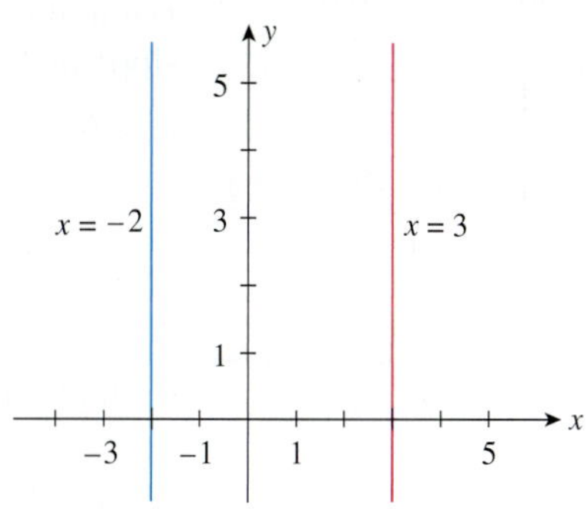

FIGURE 20
The vertical lines $x = -2$ and $x = 3$

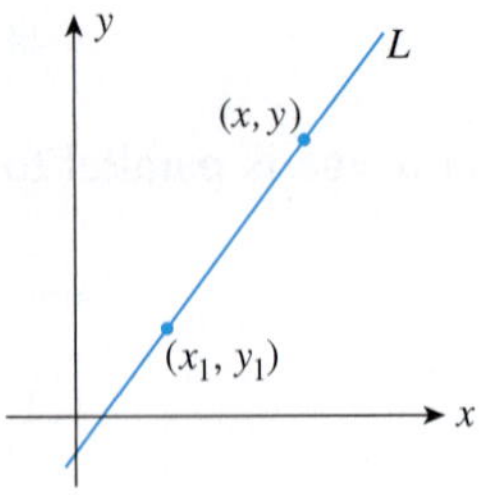

FIGURE 21
L passes through (x_1, y_1) and has slope *m*.

Next, suppose L is a nonvertical line, so it has a well-defined slope m. Suppose (x_1, y_1) is a fixed point lying on L and (x, y) is a variable point on L distinct from (x_1, y_1) (Figure 21). Using Equation (3) with the point $(x_2, y_2) = (x, y)$, we find that the slope of L is given by

$$m = \frac{y - y_1}{x - x_1}$$

Upon multiplying both sides of the equation by $x - x_1$, we obtain Equation (4).

Point-Slope Form of an Equation of a Line

An equation of the line that has slope m and passes through the point (x_1, y_1) is given by

$$y - y_1 = m(x - x_1) \qquad \textbf{(4)}$$

Equation (4) is called the *point-slope form* of an equation of a line because it uses a given point (x_1, y_1) on a line and the slope m of the line.

EXAMPLE 5 Find an equation of the line that passes through the point $(1, 3)$ and has slope 2.

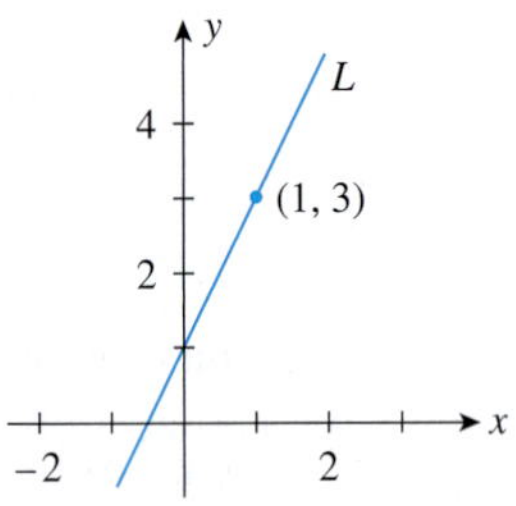

FIGURE 22
L passes through (1, 3) and has slope 2.

Solution Using the point-slope form of the equation of a line with the point $(1, 3)$ and $m = 2$, we obtain

$$y - 3 = 2(x - 1) \qquad y - y_1 = m(x - x_1)$$

which, when simplified, becomes

$$2x - y + 1 = 0$$

(Figure 22).

EXAMPLE 6 Find an equation of the line that passes through the points $(-3, 2)$ and $(4, -1)$.

Solution The slope of the line is given by

$$m = \frac{-1 - 2}{4 - (-3)} = -\frac{3}{7}$$

Using the point-slope form of the equation of a line with the point $(4, -1)$ and the slope $m = -\frac{3}{7}$, we have

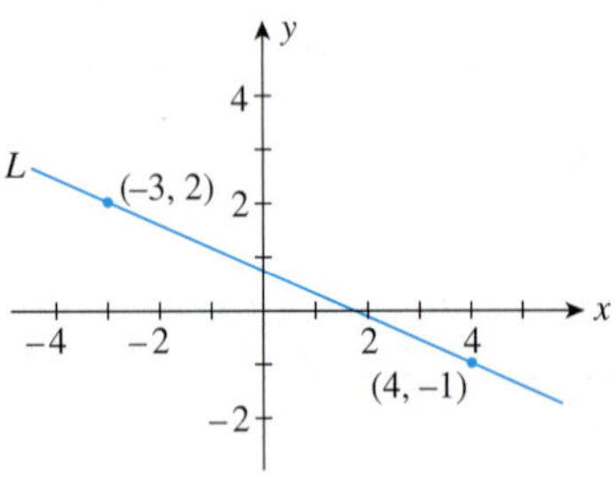

FIGURE 23
L passes through $(-3, 2)$ and $(4, -1)$.

$$y + 1 = -\frac{3}{7}(x - 4) \qquad y - y_1 = m(x - x_1)$$

$$7y + 7 = -3x + 12$$

$$3x + 7y - 5 = 0$$

(Figure 23).

We can use the slope of a straight line to determine whether a line is perpendicular to another line.

Perpendicular Lines

If L_1 and L_2 are two distinct nonvertical lines that have slopes m_1 and m_2, respectively, then L_1 is **perpendicular** to L_2 (written $L_1 \perp L_2$) if and only if

$$m_1 = -\frac{1}{m_2}$$

If the line L_1 is vertical (so that its slope is undefined), then L_1 is perpendicular to another line, L_2, if and only if L_2 is horizontal (so that its slope is zero). For a proof of these results, see Exercise 96, page 24.

EXAMPLE 7 Find an equation of the line that passes through the point $(3, 1)$ and is perpendicular to the line of Example 5.

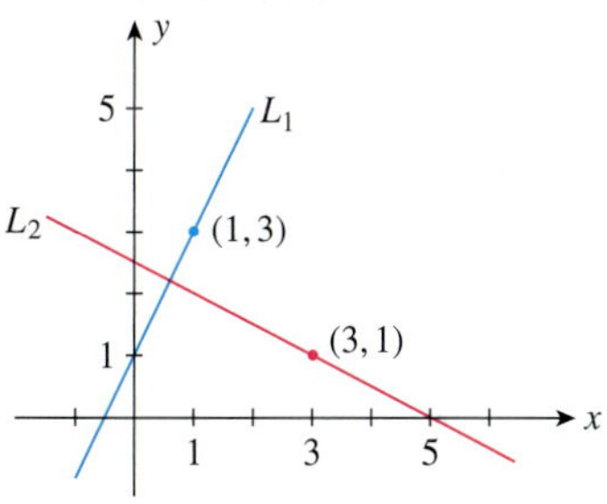

FIGURE 24
L_2 is perpendicular to L_1 and passes through (3, 1).

Solution Since the slope of the line in Example 5 is 2, it follows that the slope of the required line is given by $m = -\frac{1}{2}$, the negative reciprocal of 2. Using the point-slope form of the equation of a line, we obtain

$$\begin{aligned} y - 1 &= -\frac{1}{2}(x - 3) \qquad y - y_1 = m(x - x_1) \\ 2y - 2 &= -x + 3 \\ x + 2y - 5 &= 0 \end{aligned}$$

(Figure 24).

Exploring with TECHNOLOGY

1. Use a graphing utility to plot the straight lines L_1 and L_2 with equations $2x + y - 5 = 0$ and $41x + 20y - 11 = 0$ on the same set of axes, using the standard viewing window.
   a. Can you tell whether the lines L_1 and L_2 are parallel to each other?
   b. Verify your observations by computing the slopes of L_1 and L_2 algebraically.
2. Use a graphing utility to plot the straight lines L_1 and L_2 with equations $x + 2y - 5 = 0$ and $5x - y + 5 = 0$ on the same set of axes, using the standard viewing window.
   a. Can you tell whether the lines L_1 and L_2 are perpendicular to each other?
   b. Verify your observation by computing the slopes of L_1 and L_2 algebraically.

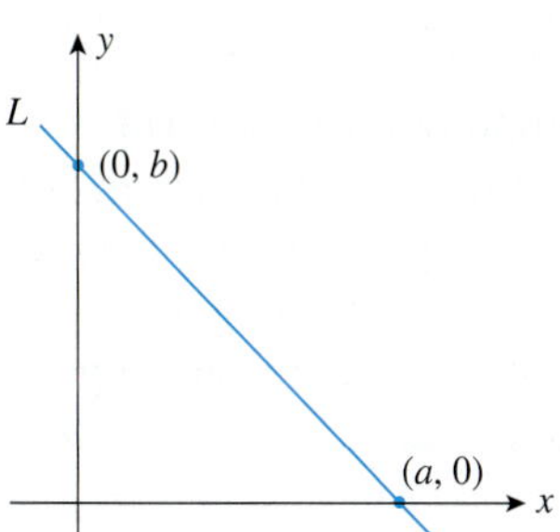

FIGURE 25
The line L has x-intercept a and y-intercept b.

A straight line L that is neither horizontal nor vertical cuts the x-axis and the y-axis at, say, points $(a, 0)$ and $(0, b)$, respectively (Figure 25). The numbers a and b are called the ***x*-intercept** and ***y*-intercept,** respectively, of L.

Now, let L be a line with slope m and y-intercept b. Using Equation (4), the point-slope form of the equation of a line, with the point given by $(0, b)$ and slope m, we have

$$\begin{aligned} y - b &= m(x - 0) \\ y &= mx + b \end{aligned}$$

Slope-Intercept Form of an Equation of a Line

The equation of the line that has slope m and intersects the y-axis at the point $(0, b)$ is given by

$$y = mx + b \tag{5}$$

EXAMPLE 8 Find an equation of the line that has slope 3 and y-intercept -4.

Solution Using Equation (5) with $m = 3$ and $b = -4$, we obtain the required equation:

$$y = 3x - 4$$

EXAMPLE 9 Determine the slope and y-intercept of the line whose equation is $3x - 4y = 8$.

Solution Rewrite the given equation in the slope-intercept form. Thus,

$$\begin{aligned} 3x - 4y &= 8 \\ -4y &= -3x + 8 \\ y &= \frac{3}{4}x - 2 \end{aligned}$$

Comparing this result with Equation (5), we find $m = \frac{3}{4}$ and $b = -2$, and we conclude that the slope and y-intercept of the given line are $\frac{3}{4}$ and -2, respectively.

Exploring with TECHNOLOGY

1. Use a graphing utility to plot the straight lines with equations $y = -2x + 3$, $y = -x + 3$, $y = x + 3$, and $y = 2.5x + 3$ on the same set of axes, using the standard viewing window. What effect does changing the coefficient m of x in the equation $y = mx + b$ have on its graph?
2. Use a graphing utility to plot the straight lines with equations $y = 2x - 2$, $y = 2x - 1$, $y = 2x$, $y = 2x + 1$, and $y = 2x + 4$ on the same set of axes, using the standard viewing window. What effect does changing the constant b in the equation $y = mx + b$ have on its graph?
3. Describe in words the effect of changing both m and b in the equation $y = mx + b$.

APPLIED EXAMPLE 10 Forecasting Sales of a Sporting Goods Store The sales manager of a local sporting goods store plotted sales (in units of ten thousand dollars) versus time for the last 5 years and found the points to lie approximately along a straight line (Figure 26). By using the points corresponding to the first and fifth years, find an equation of the *trend line*. What sales figure can be predicted for the sixth year?

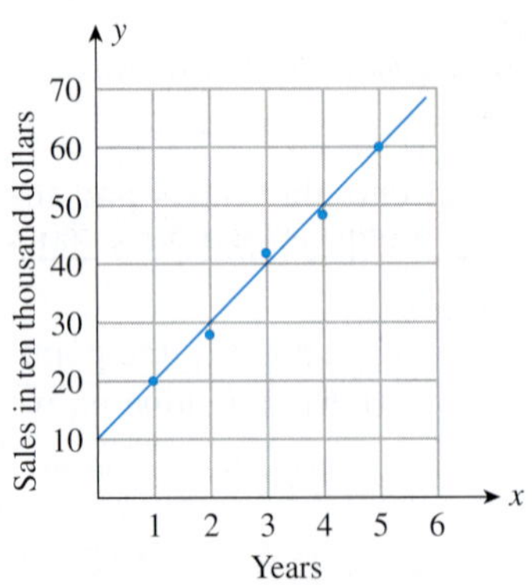

FIGURE 26
Sales of a sporting goods store

Solution Using Equation (3) with the points $(1, 20)$ and $(5, 60)$, we find that the slope of the required line is given by

$$m = \frac{60 - 20}{5 - 1} = 10$$

Next, using the point-slope form of the equation of a line with the point $(1, 20)$ and $m = 10$, we obtain

$$y - 20 = 10(x - 1) \qquad y - y_1 = m(x - x_1)$$
$$y = 10x + 10$$

as the required equation.

The sales figure for the sixth year is obtained by letting $x = 6$ in the last equation, giving

$$y = 10(6) + 10 = 70$$

or \$700,000.

APPLIED EXAMPLE 11 Appreciation in Value of a Painting Suppose a painting purchased for \$50,000 is expected to appreciate in value at a constant rate of \$5000 per year for the next 5 years. Use Equation (5) to write an equation predicting the value of the painting in the next several years. What will be its value 3 years from the purchase date?

Explore and Discuss

Refer to Applied Example 11. Can the equation predicting the value of the painting be used to predict long-term growth?

Solution Let x denote the time (in years) that has elapsed since the purchase date, and let y denote the painting's value (in dollars). Then $y = 50{,}000$ when $x = 0$. Furthermore, the slope of the required equation is given by $m = 5000$, since each unit increase in x (1 year) implies an increase of 5000 units (dollars) in y. Using Equation (5) with $m = 5000$ and $b = 50{,}000$, we obtain

$$y = 5000x + 50{,}000 \qquad y = mx + b$$

Three years from the purchase date, the value of the painting will be given by

$$y = 5000(3) + 50{,}000$$

or \$65,000.

General Form of an Equation of a Line

We have considered several forms of the equation of a straight line in the plane. These different forms of the equation are equivalent to each other. In fact, each is a special case of the following equation.

General Form of a Linear Equation

The equation

$$Ax + By + C = 0 \tag{6}$$

where A, B, and C are constants and A and B are not both zero, is called the general form of a linear equation in the variables x and y.

We now state (without proof) an important result concerning the algebraic representation of straight lines in the plane.

THEOREM 1

An equation of a straight line is a linear equation; conversely, every linear equation represents a straight line.

This result justifies the use of the adjective *linear* in describing Equation (6).

EXAMPLE 12

a. Sketch the straight line represented by the equation

$$3x - 4y - 12 = 0$$

b. Does the point $\left(2, -\frac{3}{2}\right)$ lie on L?
c. Does the point $(1, -2)$ lie on L?

Solution

a. Since every straight line is uniquely determined by two distinct points, we need to find only two points through which the line passes in order to sketch it. For convenience, let's compute the points at which the line crosses the x- and y-axes. Setting $y = 0$, we find $x = 4$, the x-intercept, so the line crosses the x-axis at the point $(4, 0)$. Setting $x = 0$ gives $y = -3$, the y-intercept, so the line crosses the y-axis at the point $(0, -3)$. A sketch of the line appears in Figure 27.

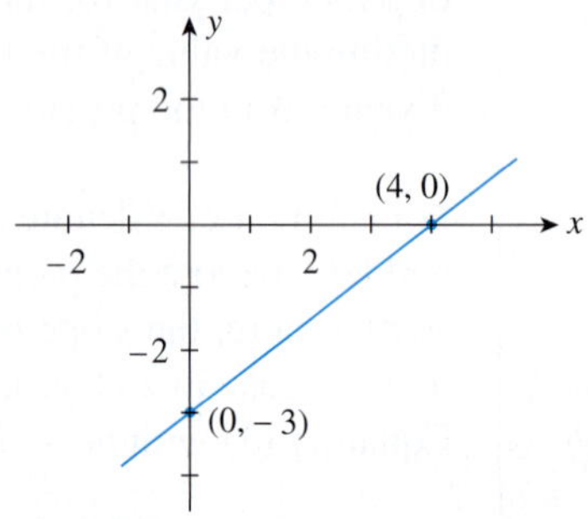

FIGURE 27
To sketch $3x - 4y - 12 = 0$, first find the x-intercept, 4, and the y-intercept, -3.

b. Substituting $x = 2$ and $y = -\frac{3}{2}$ into the left-hand side of the equation $3x - 4y - 12 = 0$ found in part (a), we obtain

$$3(2) - 4\left(-\frac{3}{2}\right) - 12 = 6 + 6 - 12 = 0$$

This shows that the equation is satisfied and we conclude that the point $\left(2, -\frac{3}{2}\right)$ does indeed lie on L.

c. Substituting $x = 1$ and $y = -2$ into the left-hand side of the equation $3x - 4y - 12 = 0$, we obtain

$$3(1) - 4(-2) - 12 = 3 + 8 - 12 = -1$$

which is not equal to zero, the number on the right-hand side of the equation. This shows that the point $(1, -2)$ does not lie on L.

Here is a summary of the common forms of the equations of straight lines discussed in this section.

Equations of Straight Lines

Vertical line:	$x = a$
Horizontal line:	$y = b$
Point-slope form:	$y - y_1 = m(x - x_1)$
Slope-intercept form:	$y = mx + b$
General form:	$Ax + By + C = 0$

1.2 Self-Check Exercises

1. Determine the number a such that the line passing through the points $(a, 2)$ and $(3, 6)$ is parallel to a line with slope 4.

2. Find an equation of the line that passes through the point $(3, -1)$ and is perpendicular to a line with slope $-\frac{1}{2}$.

3. Does the point $(3, -3)$ lie on the line with equation $2x - 3y - 12 = 0$? Sketch the graph of the line.

4. **Smokers in the United States** The following table gives the percentage of adults in the United States from 2006 through 2010 who smoked in year t. Here, $t = 0$ corresponds to the beginning of 2006.

Year, t	0	1	2	3	4
Percent, y	20.8	20.5	20.1	19.8	19.0

a. Plot the percentage of U.S. adults who smoke (y) versus the year (t) for the given years.
b. Draw the line L through the points $(0, 20.8)$ and $(4, 19.0)$.
c. Find an equation of the line L.
d. Assuming that this trend continues, estimate the percentage of U.S. adults who smoked at the beginning of 2014.

Source: Centers for Disease Control and Prevention.

Solutions to Self-Check Exercises 1.2 can be found on page 24.

1.2 Concept Questions

1. What is the slope of a nonvertical line? What can you say about the slope of a vertical line?

2. Give (a) the point-slope form, (b) the slope-intercept form, and (c) the general form of an equation of a line.

3. Let L_1 have slope m_1 and let L_2 have slope m_2. State the conditions on m_1 and m_2 if (a) L_1 is parallel to L_2 and (b) L_1 is perpendicular to L_2.

4. Suppose a line L has equation $Ax + By + C = 0$.
   a. What is the slope of L if $B \neq 0$?
   b. What is the slope of L if $B = 0$ and $A \neq 0$?

1.2 Exercises

In Exercises 1–4, find the slope of the line shown in each figure.

1.

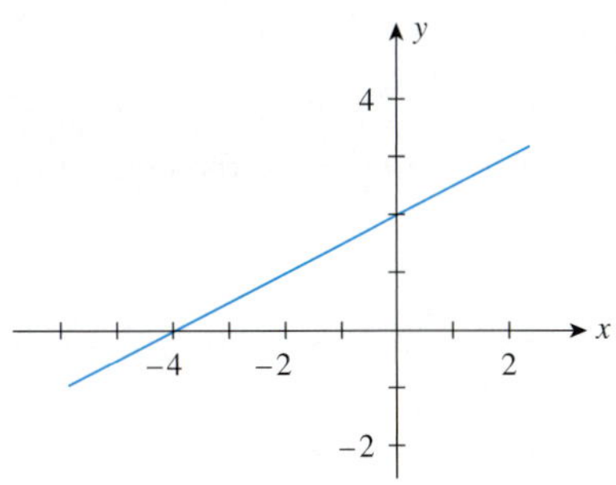

2.

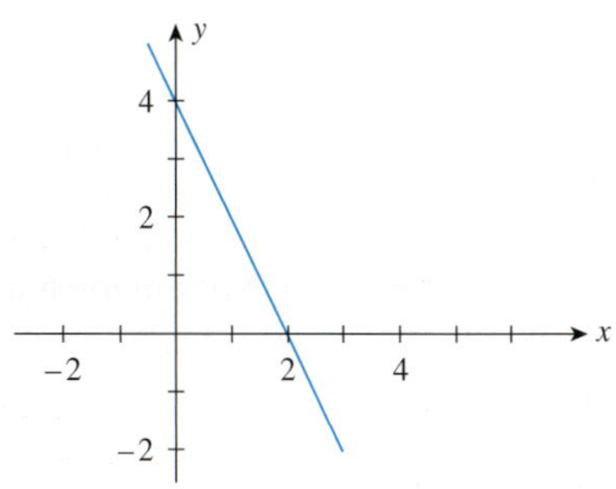

3.

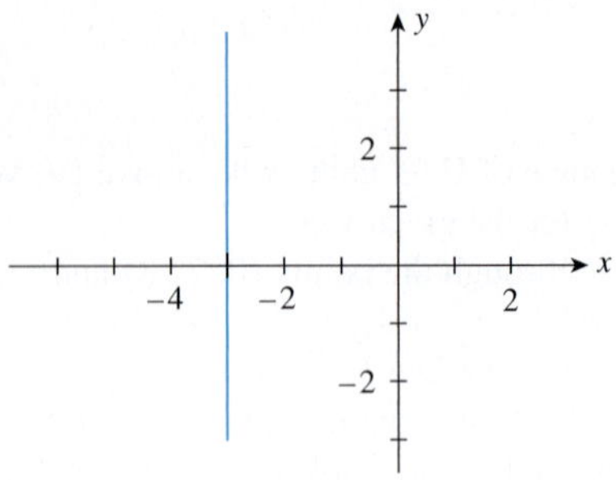

4.

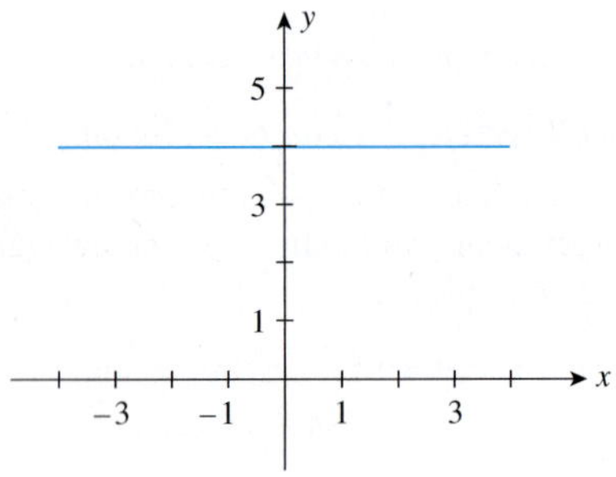

In Exercises 5–10, find the slope of the line that passes through the given pair of points.

5. $(4, 3)$ and $(5, 8)$

6. $(4, 5)$ and $(3, 8)$

7. $(-2, 3)$ and $(4, 8)$

8. $(-2, -2)$ and $(4, -4)$

9. (a, b) and (c, d)

10. $(-a + 1, b - 1)$ and $(a + 1, -b)$

11. Given the equation $y = 4x - 3$, answer the following questions.
   a. If x increases by 1 unit, what is the corresponding change in y?
   b. If x decreases by 2 units, what is the corresponding change in y?

12. Given the equation $2x + 3y = 4$, answer the following questions.
   a. Is the slope of the line described by this equation positive or negative?
   b. As x increases in value, does y increase or decrease?
   c. If x decreases by 2 units, what is the corresponding change in y?

In Exercises 13–18, match the statement with one of the graphs (a)–(f).

13. The slope of the line is zero.

14. The slope of the line is undefined.

15. The slope of the line is positive, and its y-intercept is positive.

16. The slope of the line is positive, and its y-intercept is negative.

17. The slope of the line is negative, and its x-intercept is negative.

18. The slope of the line is negative, and its x-intercept is positive.

(a)

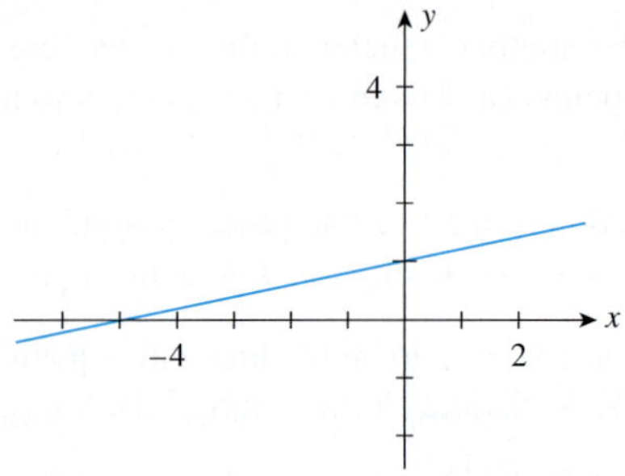

(b)

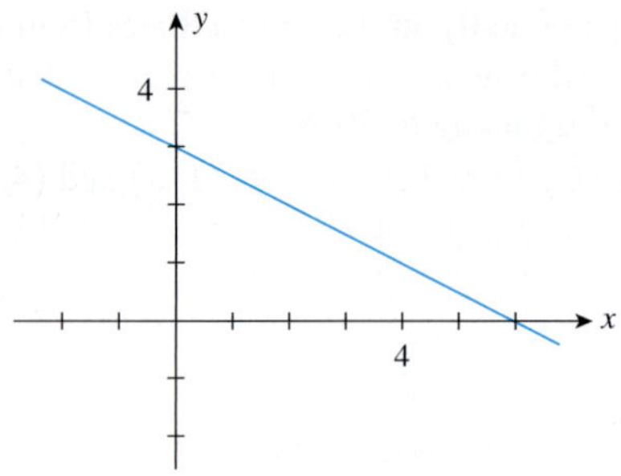

(c)

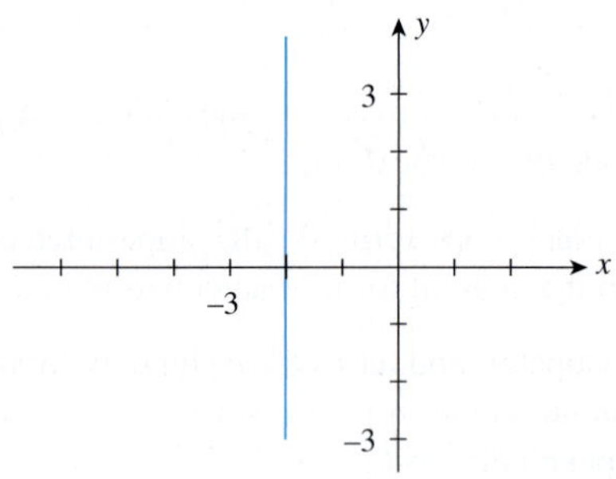

(d)

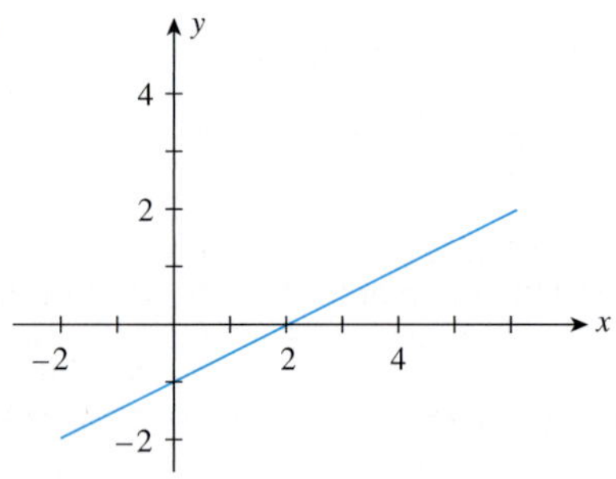

(e)

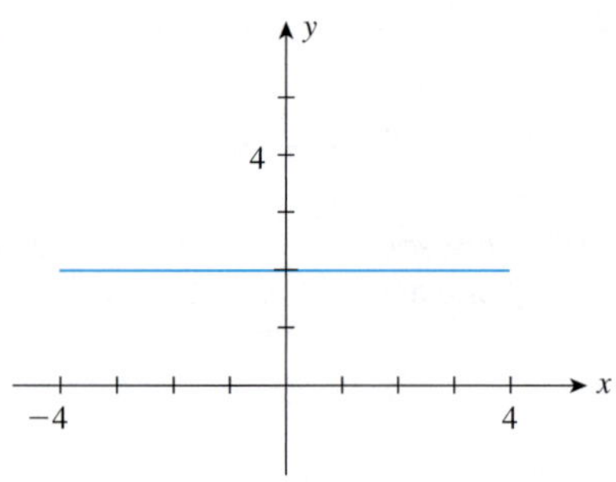

(f)

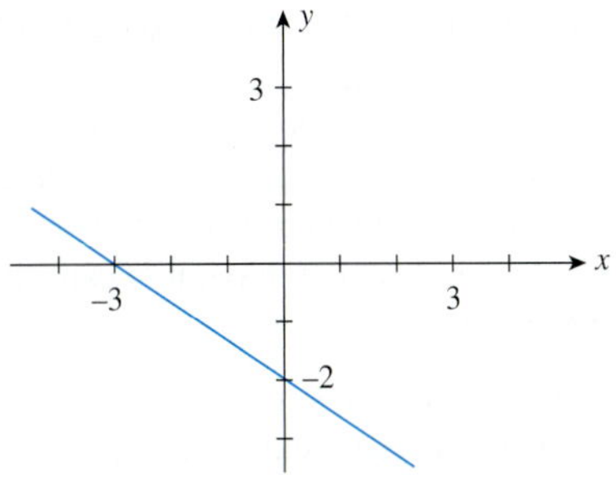

In Exercises 19 and 20, determine whether the lines through the pairs of points are parallel.

19. $A(1, -2)$, $B(-3, -10)$ and $C(1, 5)$, $D(-1, 1)$

20. $A(2, 3)$, $B(2, -2)$ and $C(-2, 4)$, $D(-2, 5)$

21. If the line passing through the points $(1, a)$ and $(4, -2)$ is parallel to the line passing through the points $(2, 8)$ and $(-7, a + 4)$, what is the value of a?

22. If the line passing through the points $(a, 1)$ and $(5, 8)$ is parallel to the line passing through the points $(4, 9)$ and $(a + 2, 1)$, what is the value of a?

In Exercises 23–26, find an equation of the line that passes through the point and has the indicated slope *m*.

23. $(3, -4)$; $m = 2$

24. $(2, 4)$; $m = -1$

25. $(-3, 2)$; $m = 0$

26. $(1, 2)$; $m = -\frac{1}{2}$

In Exercises 27–30, find an equation of the line that passes through the given points.

27. $(2, 4)$ and $(3, 7)$

28. $(2, 1)$ and $(2, 5)$

29. $(1, 2)$ and $(-3, -2)$

30. $(-1, -2)$ and $(3, -4)$

In Exercises 31 and 32, determine whether the lines through the pairs of points are perpendicular.

31. $A(-2, 5)$, $B(4, 2)$ and $C(-1, -2)$, $D(3, 6)$

32. $A(2, 0)$, $B(1, -2)$ and $C(4, 2)$, $D(-8, 4)$

In Exercises 33–36, find an equation of the line that has slope *m* and *y*-intercept *b*.

33. $m = 3$; $b = 4$

34. $m = -2$; $b = -1$

35. $m = 0$; $b = 5$

36. $m = -\frac{1}{2}$; $b = \frac{3}{4}$

In Exercises 37–42, write the equation in the slope-intercept form and then find the slope and *y*-intercept of the corresponding line.

37. $x - 2y = 0$

38. $y - 2 = 0$

39. $2x - 3y - 9 = 0$

40. $3x - 4y + 8 = 0$

41. $2x + 4y = 14$

42. $5x + 8y - 24 = 0$

43. Find an equation of the horizontal line that passes through $(-4, -3)$.

44. Find an equation of the vertical line that passes through $(0, 5)$.

45. Find an equation of the line that passes through the point $(-2, 2)$ and is parallel to the line $2x - 4y - 8 = 0$.

46. Find an equation of the line that passes through the point $(-1, 3)$ and is parallel to the line passing through the points $(-2, -3)$ and $(2, 5)$.

47. Find an equation of the line that passes through the point $(2, 4)$ and is perpendicular to the line $3x + 4y - 22 = 0$.

48. Find an equation of the line that passes through the point $(1, -2)$ and is perpendicular to the line passing through the points $(-2, -1)$ and $(4, 3)$.

49. Find an equation of the line that has slope -2 and passes through the midpoint of the line segment joining the points $P_1(-2, -4)$ and $P_2(3, 6)$.
Hint: See Exercise 45, page 9.

50. Find an equation of the line that passes through the midpoint of the line segment joining the points $P_1(-1, -3)$ and $P_2(3, 3)$ and the midpoint of the line segment joining the points $P_3(-2, 3)$ and $P_4(2, -3)$.
Hint: See Exercise 45, page 9.

In Exercises 51–56, find an equation of the line that satisfies the given condition.

51. The line parallel to the *x*-axis and 6 units below it

52. The line passing through the origin and parallel to the line passing through the points $(2, 4)$ and $(4, 7)$

53. The line passing through the point (a, b) with slope equal to zero

54. The line passing through $(-3, 4)$ and parallel to the *x*-axis

55. The line passing through $(-5, -4)$ and parallel to the line passing through $(-3, 2)$ and $(6, 8)$

56. The line passing through (a, b) with undefined slope

57. Given that the point $P(-3, 5)$ lies on the line $kx + 3y + 9 = 0$, find k.

58. Given that the point $P(2, -3)$ lies on the line $-2x + ky + 10 = 0$, find k.

In Exercises 59–64, sketch the straight line defined by the linear equation by finding the *x*- and *y*-intercepts.
Hint: See Example 12.

59. $3x - 2y + 6 = 0$

60. $2x - 5y + 10 = 0$

61. $x + 2y - 4 = 0$

62. $2x + 3y - 15 = 0$

63. $y + 5 = 0$

64. $-2x - 8y + 24 = 0$

65. Show that an equation of a line through the points $(a, 0)$ and $(0, b)$ with $a \neq 0$ and $b \neq 0$ can be written in the form

$$\frac{x}{a} + \frac{y}{b} = 1$$

(Recall that the numbers a and b are the x- and y-intercepts, respectively, of the line. This form of an equation of a line is called the **intercept form.**)

In Exercises 66–69, use the results of Exercise 65 to find an equation of a line with the x- and y-intercepts.

66. x-intercept 3; y-intercept 4

67. x-intercept -2; y-intercept -4

68. x-intercept $-\frac{1}{2}$; y-intercept $\frac{3}{4}$

69. x-intercept 4; y-intercept $-\frac{1}{2}$

In Exercises 70 and 71, determine whether the points lie on a straight line.

70. $A(-1, 7)$, $B(2, -2)$, and $C(5, -9)$

71. $A(-2, 1)$, $B(1, 7)$, and $C(4, 13)$

72. John claims that the following points lie on a line: $(1.2, -9.04)$, $(2.3, -5.96)$, $(4.8, 1.04)$, and $(7.2, 7.76)$. Prove or disprove his claim.

73. Alison claims that the following points lie on a line: $(1.8, -6.44)$, $(2.4, -5.72)$, $(5.0, -2.72)$, and $(10.4, 3.88)$. Prove or disprove her claim.

74. TEMPERATURE CONVERSION The relationship between the temperature in degrees Fahrenheit (°F) and the temperature in degrees Celsius (°C) is

$$F = \frac{9}{5}C + 32$$

a. Sketch the line with the given equation.
b. What is the slope of the line? What does it represent?
c. What is the F-intercept of the line? What does it represent?

75. NUCLEAR PLANT UTILIZATION The United States is not building many nuclear plants, but the ones it has are running at nearly full capacity. The output (as a percentage of total capacity) of nuclear plants is described by the equation

$$y = 1.9467t + 70.082$$

where t is measured in years, with $t = 0$ corresponding to the beginning of 1990.
a. Sketch the line with the given equation.
b. What are the slope and the y-intercept of the line found in part (a)?
c. Give an interpretation of the slope and the y-intercept of the line found in part (a).
d. If the utilization of nuclear power continued to grow at the same rate and the total capacity of nuclear plants in the United States remained constant, by what year were the plants generating at maximum capacity?
Source: Nuclear Energy Institute.

76. SOCIAL SECURITY CONTRIBUTIONS For wages less than the maximum taxable wage base, Social Security contributions (including those for Medicare) by employees are 7.65% of the employee's wages.
a. Find an equation that expresses the relationship between the wages earned (x) and the Social Security taxes paid (y) by an employee who earns less than the maximum taxable wage base.
b. For each additional dollar that an employee earns, by how much is his or her Social Security contribution increased? (Assume that the employee's wages are less than the maximum taxable wage base.)
c. What Social Security contributions will an employee who earns $65,000 (which is less than the maximum taxable wage base) be required to make?
Source: Social Security Administration.

77. COLLEGE ADMISSIONS Using data compiled by the Admissions Office at Faber University, college admissions officers estimate that 55% of the students who are offered admission to the freshman class at the university will actually enroll.
a. Find an equation that expresses the relationship between the number of students who actually enroll (y) and the number of students who are offered admission to the university (x).
b. If the desired freshman class size for the upcoming academic year is 1100 students, how many students should be admitted?

78. WEIGHT OF WHALES The equation $W = 3.51L - 192$, expressing the relationship between the length L (in feet) and the expected weight W (in British tons) of adult blue whales, was adopted in the late 1960s by the International Whaling Commission.
a. What is the expected weight of an 80-ft blue whale?
b. Sketch the straight line that represents the equation.

79. THE NARROWING GENDER GAP Since the founding of the Equal Employment Opportunity Commission and the passage of equal-pay laws, the gulf between men's and women's earnings has continued to close gradually. At the beginning of 1990 $(t = 0)$, women's wages were 68% of men's wages, and by the beginning of 2000 $(t = 10)$, women's wages were 80% of men's wages. If this gap between women's and men's wages continued to narrow *linearly,* then women's wages were what percentage of men's wages at the beginning of 2004?
Source: Journal of Economic Perspectives.

80. **DECLINING NUMBER OF PAY PHONES** As cell phones proliferate, the number of pay phones continues to drop. The number of pay phones from 2004 through 2009 (in millions) are shown in the following table ($x = 0$ corresponds to 2004):

Year, x	0	1	2	3	4	5
Number of Pay Phones, y	1.30	1.15	1.00	0.84	0.69	0.56

a. Plot the number of pay phones (y) versus the year (x).
b. Draw the straight line L through the points $(0, 1.30)$ and $(5, 0.56)$.
c. Derive an equation for the line L.
d. Assuming that the trend continues, estimate the number of pay phones in 2012.

Source: FCC.

81. **SPENDING ON EQUIPMENT AND SOFTWARE** As the United States continues to slowly recover from the Great Recession, spending on equipment and software is projected to rise. The following table gives the percentage change in equipment and software spending, seasonally adjusted, in 2013 ($x = 0$ corresponds to the first quarter of 2013):

Quarter, x	0	1	2	3
Percent Change, y	1.3	3.8	6.0	8.2

a. Plot the percentage change (y) versus the quarter (x).
b. Draw a straight line L through the points corresponding to the first quarter and the fourth quarter.
c. Derive an equation of the line L.
d. If the trend continues, estimate the percentage change in spending on equipment and software in the first quarter of 2014.

Source: Commerce Department.

82. **IDEAL HEIGHTS AND WEIGHTS FOR WOMEN** The Venus Health Club for Women provides its members with the following table, which gives the average desirable weight (in pounds) for women of a given height (in inches):

Height, x	60	63	66	69	72
Weight, y	108	118	129	140	152

a. Plot the weight (y) versus the height (x).
b. Draw a straight line L through the points corresponding to heights of 5 ft and 6 ft.
c. Derive an equation of the line L.
d. Using the equation of part (c), estimate the average desirable weight for a woman who is 5 ft, 5 in. tall.

83. **COST OF A COMMODITY** A manufacturer obtained the following data relating the cost y (in dollars) to the number of units (x) of a commodity produced:

Units Produced, x	0	20	40	60	80	100
Cost in Dollars, y	200	208	222	230	242	250

a. Plot the cost (y) versus the quantity produced (x).
b. Draw a straight line through the points $(0, 200)$ and $(100, 250)$.
c. Derive an equation of the straight line of part (b).
d. Taking this equation to be an approximation of the relationship between the cost and the level of production, estimate the cost of producing 54 units of the commodity.

84. **CORPORATE FRAUD** The number of pending corporate fraud cases stood at 545 at the beginning of 2008 ($t = 0$) and was 726 at the beginning of 2012. The growth was approximately linear.
a. Derive an equation of the line passing through the points $A(0, 545)$ and $B(4, 726)$.
b. Plot the line with the equation found in part (a).
c. Use the equation found in part (a) to estimate the number of pending corporate fraud cases at the beginning of 2014.

Source: Federal Bureau of Investigation.

85. **SALES GROWTH** Metro Department Store's annual sales (in millions of dollars) during the past 5 years were

Annual Sales, y	5.8	6.2	7.2	8.4	9.0
Year, x	1	2	3	4	5

a. Plot the annual sales (y) versus the year (x).
b. Draw a straight line L through the points corresponding to the first and fifth years.
c. Derive an equation of the line L.
d. Using the equation found in part (c), estimate Metro's annual sales 4 years from now ($x = 9$).

86. **SOCIAL MEDIA** A Nielsen survey of 3000 American moviegoers aged 12–74 found that 27% of them used social media to chat about movies in 2010. The percentage was 29% in 2011 and 31% in 2012. Let $t = 0$, $t = 1$, and $t = 2$ correspond to the years 2010, 2011, and 2012, respectively.
a. Explain why the three points $P_1(0, 27)$, $P_2(1, 29)$, and $P_3(2, 31)$ lie on a straight line L.
b. If the trend continues, what will the percentage of moviegoers who use social media to chat about movies be in 2014?
c. Find an equation of L. Then use this equation to find and reconcile the result obtained in part (b).

Source: Nielsen survey.

87. Is there a difference between the statements "The slope of a straight line is zero" and "The slope of a straight line does not exist (is not defined)"? Explain your answer.

88. Consider the slope-intercept form of a straight line $y = mx + b$. Describe the family of straight lines obtained by keeping
a. the value of m fixed and allowing the value of b to vary.
b. the value of b fixed and allowing the value of m to vary.

In Exercises 89–94, determine whether the statement is true or false. If it is true, explain why it is true. If it is false, give an example to show why it is false.

89. Suppose the slope of a line L is $-\frac{1}{2}$ and P is a given point on L. If Q is the point on L lying 4 units to the left of P, then Q is situated 2 units above P.

90. The point $(1, k)$ lies on the line with equation $3x + 4y = 12$ if and only if $k = \frac{9}{4}$.

91. The line with equation $Ax + By + C = 0$ $(B \neq 0)$ and the line with equation $ax + by + c = 0$ $(b \neq 0)$ are parallel if $Ab - aB = 0$.

92. If the slope of the line L_1 is positive, then the slope of a line L_2 perpendicular to L_1 may be positive or negative.

93. The lines with equations $ax + by + c_1 = 0$ and $bx - ay + c_2 = 0$, where $a \neq 0$ and $b \neq 0$, are perpendicular to each other.

94. If L is the line with equation $Ax + By + C = 0$, where $A \neq 0$, then L crosses the x-axis at the point $(-C/A, 0)$.

95. Show that two distinct lines with equations $a_1x + b_1y + c_1 = 0$ and $a_2x + b_2y + c_2 = 0$, respectively, are parallel if and only if $a_1b_2 - b_1a_2 = 0$.
Hint: Write each equation in the slope-intercept form and compare.

96. Prove that if a line L_1 with slope m_1 is perpendicular to a line L_2 with slope m_2, then $m_1m_2 = -1$.
Hint: Refer to the accompanying figure. Show that $m_1 = b$ and $m_2 = c$. Next, apply the Pythagorean Theorem and the distance formula to the triangles OAC, OCB, and OBA to show that $1 = -bc$.

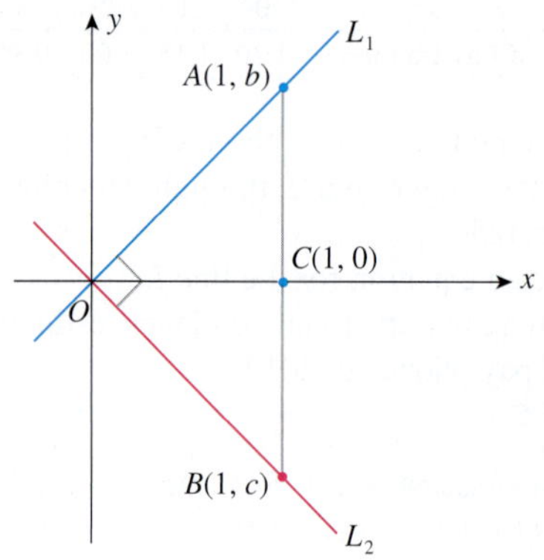

1.2 Solutions to Self-Check Exercises

1. The slope of the line that passes through the points $(a, 2)$ and $(3, 6)$ is

$$m = \frac{6 - 2}{3 - a} = \frac{4}{3 - a}$$

Since this line is parallel to a line with slope 4, m must be equal to 4; that is,

$$\frac{4}{3 - a} = 4$$

or, upon multiplying both sides of the equation by $3 - a$,

$$\begin{aligned} 4 &= 4(3 - a) \\ 4 &= 12 - 4a \\ 4a &= 8 \\ a &= 2 \end{aligned}$$

2. Since the required line L is perpendicular to a line with slope $-\frac{1}{2}$, the slope of L is

$$m = -\frac{1}{-\frac{1}{2}} = 2$$

Next, using the point-slope form of the equation of a line, we have

$$\begin{aligned} y - (-1) &= 2(x - 3) \\ y + 1 &= 2x - 6 \\ y &= 2x - 7 \end{aligned}$$

3. Substituting $x = 3$ and $y = -3$ into the left-hand side of the given equation, we find

$$2(3) - 3(-3) - 12 = 3$$

which is not equal to zero (the right-hand side). Therefore, $(3, -3)$ does not lie on the line with equation $2x - 3y - 12 = 0$. (See the accompanying figure on the next page.)

Setting $x = 0$, we find $y = -4$, the y-intercept. Next, setting $y = 0$ gives $x = 6$, the x-intercept. We now draw the line passing through the points $(0, -4)$ and $(6, 0)$, as shown.

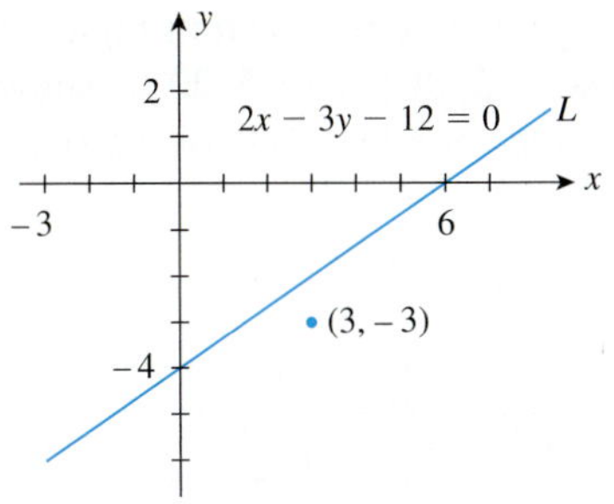

4. a.

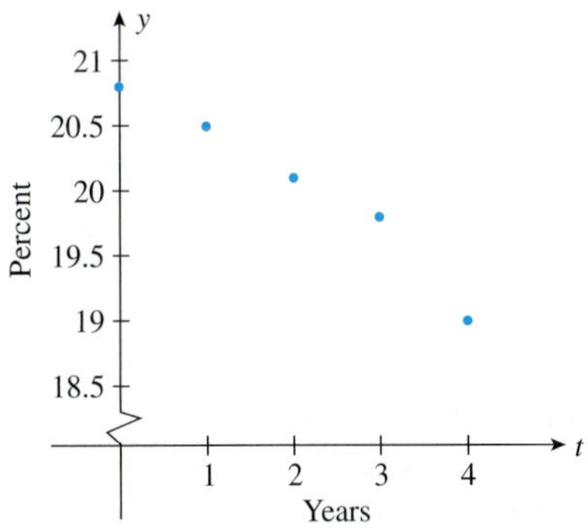

b.

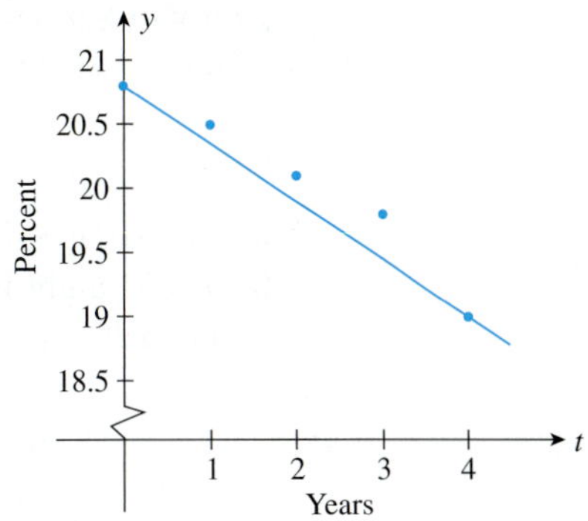

c. The slope of L is

$$m = \frac{19.0 - 20.8}{4 - 0} = -0.45$$

Using the point-slope form of the equation of a line with the point $(0, 20.8)$, we find

$$y - 20.8 = -0.45(t - 0) \quad \text{or} \quad y = -0.45t + 20.8$$

d. The year 2014 corresponds to $t = 8$, so the estimated percentage of U.S. adults who will be smoking is

$$y = -0.45(8) + 20.8 = 17.2$$

or 17.2%.

USING TECHNOLOGY Graphing a Straight Line

Graphing Utility

The first step in plotting a straight line with a graphing utility is to select a suitable viewing window. We usually do this by experimenting. For example, you might first plot the straight line using the **standard viewing window** $[-10, 10] \times [-10, 10]$. If necessary, you then might adjust the viewing window by enlarging it or reducing it to obtain a sufficiently complete view of the line or at least the portion of the line that is of interest.

EXAMPLE 1 Plot the straight line $2x + 3y - 6 = 0$ in the standard viewing window.

Solution The straight line in the standard viewing window is shown in Figure T1.

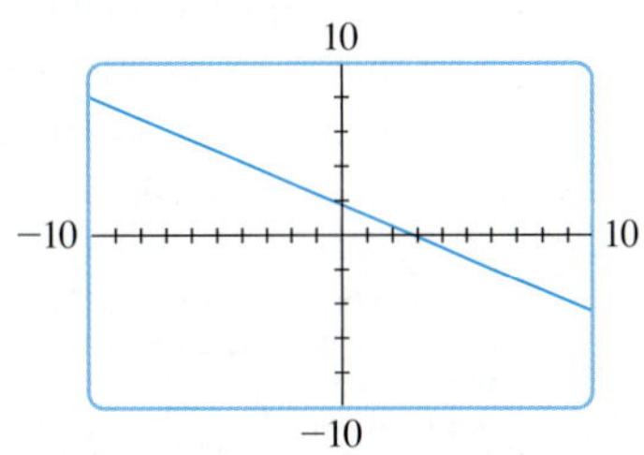

FIGURE **T1**
The straight line $2x + 3y - 6 = 0$ in the standard viewing window

EXAMPLE 2 Plot the straight line $2x + 3y - 30 = 0$ in (a) the standard viewing window and (b) the viewing window $[-5, 20] \times [-5, 20]$.

Solution

a. The straight line in the standard viewing window is shown in Figure T2a.

b. The straight line in the viewing window $[-5, 20] \times [-5, 20]$ is shown in Figure T2b. This figure certainly gives a more complete view of the straight line.

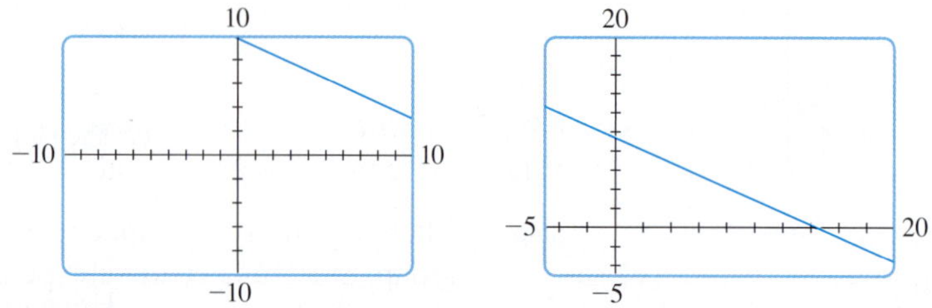

(a) The graph of $2x + 3y - 30 = 0$ in the standard viewing window

(b) The graph of $2x + 3y - 30 = 0$ in the viewing window $[-5, 20] \times [-5, 20]$

FIGURE **T2**

Excel

In the examples and exercises that follow, we assume that you are familiar with the basic features of Microsoft Excel. Please consult your Excel manual or use Excel's Help features to answer questions regarding the standard commands and operating instructions for Excel. Here, we use Microsoft Excel 2010.*

EXAMPLE 3 Plot the graph of the straight line $2x + 3y - 6 = 0$ over the interval $[-10, 10]$.

Solution

1. *Write the equation in the slope-intercept form:*

$$y = -\frac{2}{3}x + 2$$

2. *Create a table of values.* First, enter the input values: Enter the values of the endpoints of the interval over which you are graphing the straight line. (Recall that we need only two distinct data points to draw the graph of a straight line. In general, we select the endpoints of the interval over which the straight line is to be drawn as our data points.) In this case, we enter −10 in cell B1 and 10 in cell C1.

Second, enter the formula for computing the y-values: Here, we enter

```
= −(2/3)*B1+2
```

in cell B2 and then press **Enter**.

Third, evaluate the function at the other input value: To extend the formula to cell C2, move the pointer to the small black box at the lower right corner of cell B2 (the cell containing the formula). Observe that the pointer now appears as a

Note: Boldfaced words/characters enclosed in a box (for example, **Enter**) indicate that an action (click, select, or press) is required. Words/characters printed blue (for example, Chart Type) indicate words/characters that appear on the screen. Words/characters printed in a monospace font (for example, `= (−2/3)*A2+2`) indicate words/characters that need to be typed and entered.

*Instructions for solving these examples and exercises using Microsoft Excel 2007 are given on the companion website.

black + (plus sign). Drag this pointer through cell C2 and then release it. The y-value, -4.66667, corresponding to the x-value in cell C1(10) will appear in cell C2 (Figure T3).

	A	B	C
1	x	-10	10
2	y	8.666667	-4.66667

FIGURE **T3**
Table of values for x and y

3. *Graph the straight line determined by these points.* First, highlight the numerical values in the table. Here, we highlight cells B1:B2 and C1:C2.

Step 1 Click on the **Insert** ribbon tab, and then select **Scatter** from the Charts group. Select the chart subtype in the first row and second column. A chart will then appear on your worksheet.

Step 2 From the Chart Tools group that now appears at the end of the ribbon, click the **Layout** tab and then select **Chart Title** from the Labels group followed by **Above Chart**. Type `y =-(2/3)x + 2` and press **Enter**. Click **Axis Titles** from the Labels group, and select **Primary Horizontal Axis Title** followed by **Title Below Axis**. Type `x` and then press **Enter**. Next, click **Axis Titles** again, and select **Primary Vertical Axis Title** followed by **Vertical Title**. Type `y` and press **Enter**.

Step 3 Click **Series1**, which appears on the right side of the graph and press **Delete**.

The graph shown in Figure T4 will appear.

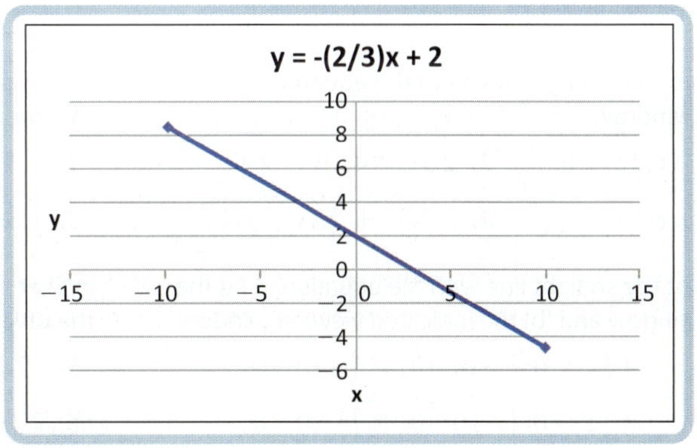

FIGURE **T4**
The graph of $y = -\frac{2}{3}x + 2$ over the interval $[-10, 10]$

If the interval over which the straight line is to be plotted is not specified, then you might have to experiment to find an appropriate interval for the x-values in your graph. For example, you might first plot the straight line over the interval $[-10, 10]$. If necessary, you then might adjust the interval by enlarging it or reducing it to obtain a sufficiently complete view of the line or at least the portion of the line that is of interest.

EXAMPLE 4 Plot the straight line $2x + 3y - 30 = 0$ over the intervals (a) $[-10, 10]$ and (b) $[-5, 20]$.

Solution **a.** and **b.** We first cast the equation in the slope-intercept form, obtaining $y = -\frac{2}{3}x + 10$. Following the procedure given in Example 3, we obtain the graphs shown in Figure T5.

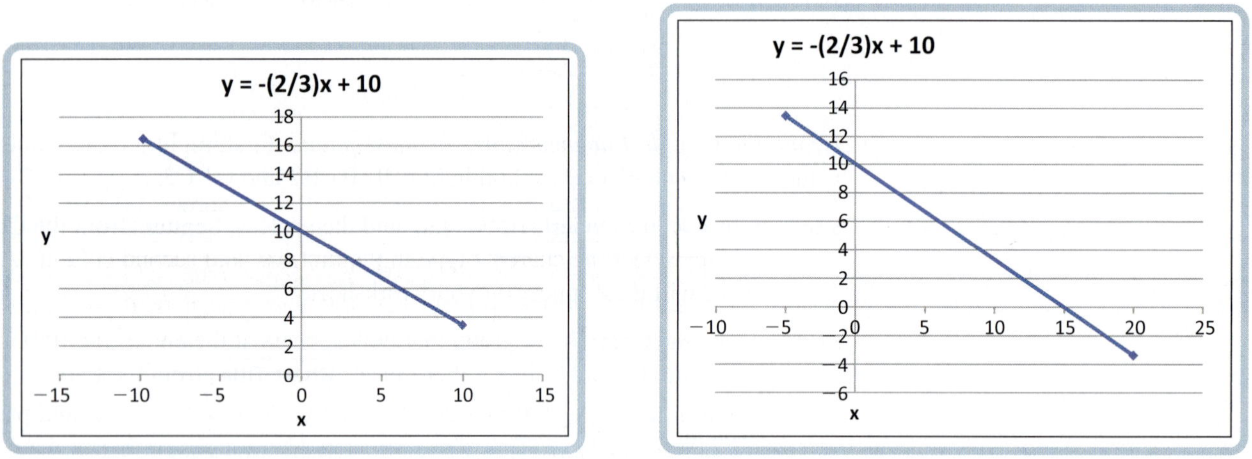

FIGURE **T5**
The graph of $y = -\frac{2}{3}x + 10$ over the intervals (a) $[-10, 10]$ and (b) $[-5, 20]$

Observe that the graph in Figure T5b includes the x- and y-intercepts. This figure certainly gives a more complete view of the straight line.

TECHNOLOGY EXERCISES

Graphing Utility

In Exercises 1–4, plot the straight line with the equation in the standard viewing window.

1. $3.2x + 2.1y - 6.72 = 0$
2. $2.3x - 4.1y - 9.43 = 0$
3. $1.6x + 5.1y = 8.16$
4. $-3.2x + 2.1y = 6.72$

In Exercises 5–8, plot the straight line with the equation in (a) the standard viewing window and (b) the indicated viewing window.

5. $12.1x + 4.1y - 49.61 = 0$; $[-10, 10] \times [-10, 20]$
6. $4.1x - 15.2y - 62.32 = 0$; $[-10, 20] \times [-10, 10]$
7. $20x + 16y = 300$; $[-10, 20] \times [-10, 30]$
8. $32.2x + 21y = 676.2$; $[-10, 30] \times [-10, 40]$

In Exercises 9–12, plot the straight line with the equation in an appropriate viewing window. (*Note*: The answer is *not* unique.)

9. $20x + 30y = 600$
10. $30x - 20y = 600$
11. $22.4x + 16.1y - 352 = 0$
12. $18.2x - 15.1y = 274.8$

Excel

In Exercises 1–4, plot the straight line with the equation over the interval $[-10, 10]$.

1. $3.2x + 2.1y - 6.72 = 0$
2. $2.3x - 4.1y - 9.43 = 0$
3. $1.6x + 5.1y = 8.16$
4. $-3.2x + 2.1y = 6.72$

In Exercises 5–8, plot the straight line with the equation over the given interval.

5. $12.1x + 4.1y - 49.61 = 0$; $[-10, 10]$
6. $4.1x - 15.2y - 62.32 = 0$; $[-10, 20]$
7. $20x + 16y = 300$; $[-10, 20]$
8. $32.2x + 21y = 676.2$; $[-10, 30]$

In Exercises 9–12, plot the straight line with the equation. (*Note*: The answer is *not* unique.)

9. $20x + 30y = 600$
10. $30x - 20y = 600$
11. $22.4x + 16.1y - 352 = 0$
12. $18.2x - 15.1y = 274.8$

1.3 Linear Functions and Mathematical Models

Mathematical Models

Regardless of the field from which a real-world problem is drawn, the problem is solved by analyzing it through a process called **mathematical modeling.** The four steps in this process, as illustrated in Figure 28, follow.

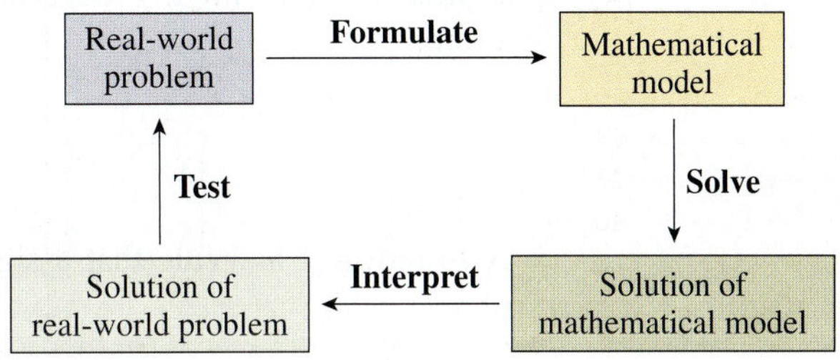

FIGURE 28

Mathematical Modeling

1. **Formulate** Given a real-world problem, we first need to formulate the problem using the language of mathematics. The many techniques that are used in constructing mathematical models range from theoretical consideration of the problem on the one extreme to an interpretation of data associated with the problem on the other. For example, the mathematical model giving the accumulated amount at any time when a certain sum of money is deposited in the bank can be derived theoretically (see Chapter 5). On the other hand, many of the mathematical models in this book are constructed by studying the data associated with the problem. In Section 1.5, we see how linear equations (models) can be constructed from a given set of data points. Also, in the ensuing chapters, we will see how other mathematical models, including statistical and probability models, are used to describe and analyze real-world situations.
2. **Solve** Once a mathematical model has been constructed, we can use the appropriate mathematical techniques, which we will develop throughout the book, to solve the problem.
3. **Interpret** Bearing in mind that the solution obtained in Step 2 is just the solution of the mathematical model, we need to interpret these results in the context of the original real-world problem.
4. **Test** Some mathematical models of real-world applications describe the situations with complete accuracy. For example, the model describing a deposit in a bank account gives the exact accumulated amount in the account at any time. But other mathematical models give, at best, an approximate description of the real-world problem. In this case, we need to test the accuracy of the model by observing how well it describes the original real-world problem and how well it predicts past and/or future behavior. If the results are unsatisfactory, then we may have to reconsider the assumptions made in the construction of the model or, in the worst case, return to Step 1.

We now look at an important way of describing the relationship between two quantities using the notion of a function. As you will see subsequently, many mathematical models are represented by functions.

Functions

A manufacturer would like to know how his company's profit is related to its production level; a biologist would like to know how the population of a certain culture of bacteria will change with time; a psychologist would like to know the relationship between the learning time of an individual and the length of a vocabulary list; and a chemist would like to know how the initial speed of a chemical reaction is related to the amount of substrate used. In each instance, we are concerned with the same question: How does one quantity depend on another? The relationship between two quantities is conveniently described in mathematics by using the concept of a function.

Function

A **function** f is a rule that assigns to each value of x one and only one value of y.

The number y is normally denoted by $f(x)$, read "f of x," emphasizing the dependency of y on x.

An example of a function may be drawn from the familiar relationship between the area of a circle and its radius. Let x and y denote the radius and area of a circle, respectively. From elementary geometry, we have

$$y = \pi x^2$$

This equation defines y as a function of x, since for each admissible value of x (a positive number representing the radius of a certain circle), there corresponds precisely one number $y = \pi x^2$ giving the area of the circle. This *area function* may be written as

$$f(x) = \pi x^2 \tag{7}$$

For example, to compute the area of a circle with a radius of 5 inches, we simply replace x in Equation (7) by the number 5. Thus, the area of the circle is

$$f(5) = \pi 5^2 = 25\pi$$

or 25π square inches.

Suppose we are given the function $y = f(x)$.* The variable x is referred to as the **independent variable**, and the variable y is called the **dependent variable**. The set of all values that may be assumed by x is called the **domain** of the function f, and the set comprising all the values assumed by $y = f(x)$ as x takes on all possible values in its domain is called the **range** of the function f. For the area function (7), the domain of f is the set of all positive numbers x, and the range of f is the set of all positive numbers y.

We now focus our attention on an important class of functions known as linear functions. Recall that a linear equation in x and y has the form $Ax + By + C = 0$, where A, B, and C are constants and A and B are not both zero. If $B \neq 0$, the equation can always be solved for y in terms of x; in fact, as we saw in Section 1.2, the equation may be cast in the slope-intercept form:

$$y = mx + b \qquad (m, b \text{ constants}) \tag{8}$$

*It is customary to refer to a function f as $f(x)$.

Equation (8) defines y as a function of x. The domain and range of this function are the set of all real numbers. Furthermore, the graph of this function, as we saw in Section 1.2, is a straight line in the plane. For this reason, the function $f(x) = mx + b$ is called a linear function.

Linear Function

The function f defined by

$$f(x) = mx + b$$

where m and b are constants, is called a **linear function.**

Linear functions play an important role in the quantitative analysis of business and economic problems. First, many problems that arise in these and other fields are linear in nature or are linear in the intervals of interest and thus can be formulated in terms of linear functions. Second, because linear functions are relatively easy to work with, assumptions involving linearity are often made in the formulation of problems. In many cases, these assumptions are justified, and acceptable mathematical models are obtained that approximate real-life situations.

The following example uses a linear function to model the market for U.S. health-care costs. In Applied Example 3, Section 1.5, we show how this model is constructed using the least-squares technique. (In "Using Technology" on pages 64–67, you will be asked to use a graphing calculator or Excel to construct other mathematical models from raw data.)

APPLIED EXAMPLE 1 U.S. Health-Care Expenditures Because the over-65 population will be growing more rapidly in the next few decades, health-care spending is expected to increase significantly in the coming decades. The following table gives the projected U.S. health-care expenditure (in trillions of dollars) from 2013 through 2018:

Year	2013	2014	2015	2016	2017	2018
Expenditure	2.908	3.227	3.418	3.632	3.850	4.080

A mathematical model giving the approximate U.S. health-care expenditures over the period in question is given by

$$S(t) = 0.226t + 2.954$$

where t is measured in years, with $t = 0$ corresponding to 2013.

a. Sketch the graph of the function S and the given data on the same set of axes.

b. Assuming that the trend continues, how much will U.S. health-care expenditures be in 2019 ($t = 6$)?

c. What is the projected rate of increase of U.S. health-care expenditures over the period in question?

Source: Centers for Medicare & Medicaid Services.

Solution

a. The graph of S is shown in Figure 29.

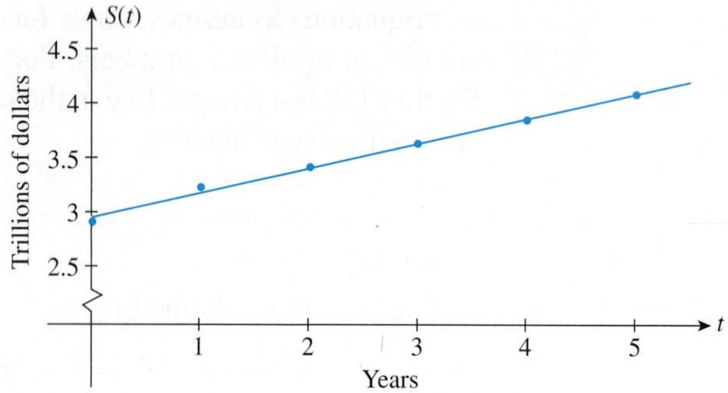

FIGURE 29
Projected U.S. health-care expenditures from 2013 to 2018

b. The projected U.S. health-care expenditure in 2019 is

$$S(6) = 0.226(6) + 2.954 = 4.31$$

or approximately \$4.31 trillion.

c. The function S is linear; hence, we see that the rate of increase of the U.S. health-care expenditures is given by the slope of the straight line represented by S, which is approximately \$0.23 trillion per year.

In the rest of this section, we look at several applications that can be modeled by using linear functions.

Simple Depreciation

We first discussed linear depreciation in the introduction to Section 1.2 as a real-world application of straight lines. The following example illustrates how to derive an equation describing the book value of an asset that is being depreciated linearly.

APPLIED EXAMPLE 2 Linear Depreciation of a Network Server A network server has an original value of \$10,000 and is to be depreciated linearly over 5 years with a \$3000 scrap value. Find an expression giving the book value at the end of year t. What will be the book value of the server at the end of the second year? What is the rate of depreciation of the server?

Solution Let $V(t)$ denote the network server's book value at the end of the tth year. Since the depreciation is linear, V is a linear function of t. Equivalently, the graph of the function is a straight line. To find an equation of the straight line, observe that $V = 10{,}000$ when $t = 0$; this tells us that the line passes through the point $(0, 10{,}000)$. Similarly, the condition that $V = 3000$ when $t = 5$ says that the line also passes through the point $(5, 3000)$. The slope of the line is given by

$$m = \frac{10{,}000 - 3000}{0 - 5} = -\frac{7000}{5} = -1400$$

Using the point-slope form of the equation of a line with the point $(0, 10{,}000)$ and the slope $m = -1400$, we have

$$V - 10{,}000 = -1400(t - 0)$$
$$V = -1400t + 10{,}000$$

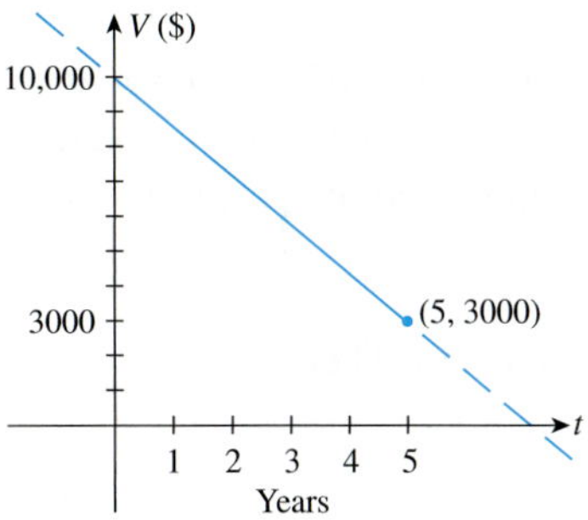

FIGURE 30
Linear depreciation of a network server

the required expression. The book value at the end of the second year is given by

$$V(2) = -1400(2) + 10{,}000 = 7200$$

or $7200. The rate of depreciation of the server is given by the negative of the slope of the depreciation line. Since the slope of the line is $m = -1400$, the rate of depreciation is $1400 per year. The graph of $V = -1400t + 10{,}000$ is sketched in Figure 30.

Linear Cost, Revenue, and Profit Functions

Whether a business is a sole proprietorship or a large corporation, the owner or chief executive must constantly keep track of operating costs, revenue resulting from the sale of products or services, and, perhaps most important, the profits realized. Three functions provide management with a measure of these quantities: the total cost function, the revenue function, and the profit function.

Cost, Revenue, and Profit Functions

Let x denote the number of units of a product manufactured or sold. Then the **total cost function** is

$$C(x) = \text{Total cost of manufacturing } x \text{ units of the product}$$

The **revenue function** is

$$R(x) = \text{Total revenue realized from the sale of } x \text{ units of the product}$$

The **profit function** is

$$P(x) = \text{Total profit realized from manufacturing and selling } x \text{ units of the product}$$

Generally speaking, the total cost, revenue, and profit functions associated with a company will probably be nonlinear (these functions are best studied using the tools of calculus). But *linear* cost, revenue, and profit functions do arise in practice, and we will consider such functions in this section. Before deriving explicit forms of these functions, we need to recall some common terminology.

The costs that are incurred in operating a business are usually classified into two categories. Costs that remain more or less constant regardless of the firm's activity level are called **fixed costs.** Examples of fixed costs are rental fees and executive salaries. Costs that vary with production or sales are called **variable costs.** Examples of variable costs are wages and costs for raw materials.

Suppose a firm has a fixed cost of F dollars, a production cost of c dollars per unit, and a selling price of s dollars per unit. Then the *cost function* $C(x)$, the *revenue function* $R(x)$, and the *profit function* $P(x)$ for the firm are given by

$$\begin{aligned} C(x) &= cx + F \\ R(x) &= sx \\ P(x) &= R(x) - C(x) \qquad \text{Revenue} - \text{cost} \\ &= (s - c)x - F \end{aligned}$$

where x denotes the number of units of the commodity produced and sold. The functions C, R, and P are linear functions of x.

APPLIED EXAMPLE 3 Profit Function for Puritron Water Filters Puritron, a manufacturer of water filters, has a monthly fixed cost of \$20,000, a production cost of \$20 per unit, and a selling price of \$30 per unit. Find the cost function, the revenue function, and the profit function for Puritron.

Solution Let x denote the number of units produced and sold. Then

$$\begin{aligned} C(x) &= 20x + 20{,}000 \\ R(x) &= 30x \\ P(x) &= R(x) - C(x) \\ &= 30x - (20x + 20{,}000) \\ &= 10x - 20{,}000 \end{aligned}$$

Linear Demand and Supply Curves

In a free-market economy, consumer demand for a particular commodity depends on the commodity's unit price. A **demand equation** expresses this relationship between the unit price and the quantity demanded. The corresponding graph of the demand equation is called a **demand curve.** In general, the quantity demanded of a commodity decreases as its unit price increases, and vice versa. Accordingly, a **demand function** defined by $p = f(x)$, where p measures the unit price and x measures the number of units of the commodity, is generally characterized as a decreasing function of x; that is, $p = f(x)$ decreases as x increases.

The simplest demand function is defined by a linear equation in x and p, where both x and p assume only positive values. Its graph is a straight line having a negative slope. Thus, its demand curve is the part of the graph of a straight line that lies in the first quadrant (Figure 31).

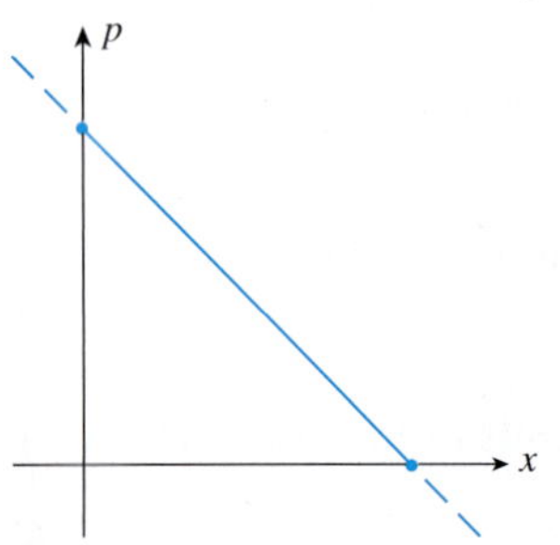

FIGURE **31**
A graph of a linear demand function

APPLIED EXAMPLE 4 Demand Function for Sentinel iPod Alarm Clocks The quantity demanded of the Sentinel iPod™ alarm clock is 48,000 units when the unit price is \$8. At \$12 per unit, the quantity demanded drops to 32,000 units. Find the demand equation, assuming that it is linear. What is the unit price corresponding to a quantity demanded of 40,000 units? What is the quantity demanded if the unit price is \$14?

Solution Let p denote the unit price of an iPod alarm clock (in dollars) and let x (in units of 1000) denote the quantity demanded when the unit price of the clocks is \$$p$. If $p = 8$ then $x = 48$, and the point $(48, 8)$ lies on the demand curve. Similarly, if $p = 12$, then $x = 32$, and the point $(32, 12)$ also lies on the demand curve. Since the demand equation is linear, its graph is a straight line. The slope of the required line is given by

$$m = \frac{12 - 8}{32 - 48} = \frac{4}{-16} = -\frac{1}{4}$$

So, using the point-slope form of an equation of a line with the point $(48, 8)$, we find that

$$p - 8 = -\frac{1}{4}(x - 48)$$

$$p = -\frac{1}{4}x + 20$$

is the required equation. The demand curve is shown in Figure 32.

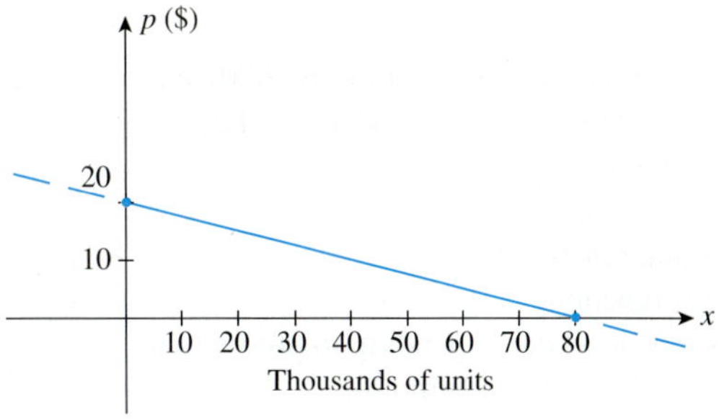

FIGURE 32
The graph of the demand equation $p = -\frac{1}{4}x + 20$

If the quantity demanded is 40,000 units $(x = 40)$, the demand equation yields

$$y = -\frac{1}{4}(40) + 20 = 10$$

and we see that the corresponding unit price is \$10. Next, if the unit price is \$14 $(p = 14)$, the demand equation yields

$$14 = -\frac{1}{4}x + 20$$
$$\frac{1}{4}x = 6$$
$$x = 24$$

so the quantity demanded will be 24,000 iPod alarm clocks. ■

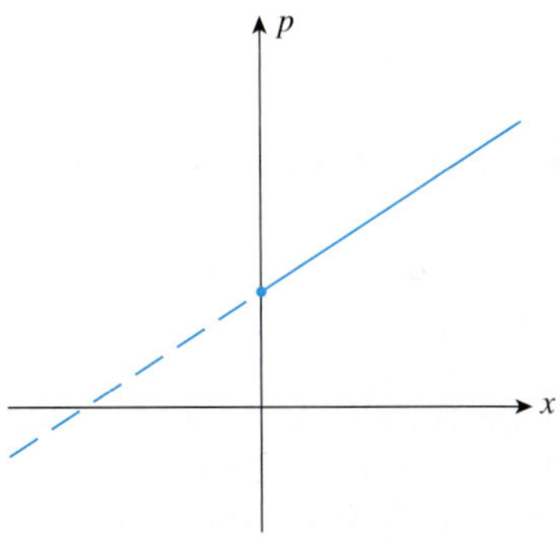

FIGURE 33
A graph of a linear supply function

In a competitive market, a relationship also exists between the unit price of a commodity and its availability in the market. In general, an increase in a commodity's unit price will induce the manufacturer to increase the supply of that commodity. Conversely, a decrease in the unit price generally leads to a drop in the supply. An equation that expresses the relationship between the unit price and the quantity supplied is called a **supply equation,** and the corresponding graph is called a **supply curve.** A **supply function,** defined by $p = f(x)$, is generally characterized by an increasing function of x; that is, $p = f(x)$ increases as x increases.

As in the case of a demand equation, the simplest supply equation is a linear equation in p and x, where p and x have the same meaning as before but its graph is a straight line with positive slope. The supply curve corresponding to a linear supply function is the part of the straight line that lies in the first quadrant (Figure 33).

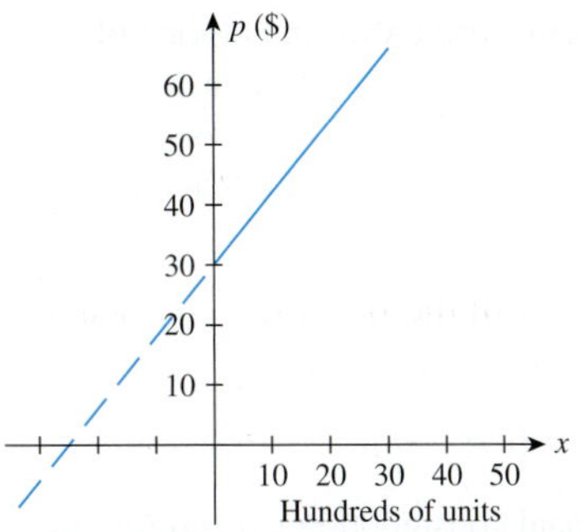

FIGURE 34
The graph of the supply equation $4p - 5x = 120$

APPLIED EXAMPLE 5 Supply Functions The supply equation for a commodity is given by $4p - 5x = 120$, where p is measured in dollars and x is measured in units of 100.

a. Sketch the corresponding supply curve.
b. How many units will be marketed when the unit price is \$55?

Solution

a. Setting $x = 0$, we find the p-intercept to be 30. Next, setting $p = 0$, we find the x-intercept to be -24. The supply curve is sketched in Figure 34.
b. Substituting $p = 55$ in the supply equation, we have $4(55) - 5x = 120$ or $x = 20$, so the amount marketed will be 2000 units. ■

1.3 Self-Check Exercises

1. A manufacturer has a monthly fixed cost of \$60,000 and a production cost of \$10 for each unit produced. The product sells for \$15/unit.
   a. What is the cost function?
   b. What is the revenue function?
   c. What is the profit function?
   d. Compute the profit (loss) corresponding to production levels of 10,000 and 14,000 units/month.

2. **Demand for Designer Dorm Room Rugs** The quantity demanded for a certain 52-in. $\times$ 87-in. designer dorm room rug is 500 when the unit price is \$100. For each \$20 decrease in the unit price, the quantity demanded increases by 500 units. Find the demand equation and sketch its graph.

Solutions to Self-Check Exercises 1.3 can be found on page 39.

1.3 Concept Questions

1. **a.** What is a *function*? Give an example.
   b. What is a *linear function*? Give an example.
   c. What is the domain of a linear function? The range?
   d. What is the graph of a linear function?

2. What is the general form of a linear cost function? A linear revenue function? A linear profit function?

3. Is the slope of a linear demand curve positive or negative? The slope of a linear supply curve?

4. The value of an investment (in dollars) after t years is given by

$$V(t) = 50{,}000 + 4000t \qquad (t \geq 0)$$

   a. What was the initial investment?
   b. What is the rate of growth of the investment?

1.3 Exercises

In Exercises 1–10, determine whether the equation defines y as a linear function of x. If so, write it in the form $y = mx + b$.

1. $2x + 3y = 6$
2. $-2x + 4y = 7$
3. $x = 2y - 4$
4. $2x = 3y + 8$
5. $2x - 4y + 9 = 0$
6. $3x - 6y + 7 = 0$
7. $2x^2 - 8y + 4 = 0$
8. $3\sqrt{x} + 4y = 0$
9. $2x - 3y^2 + 8 = 0$
10. $2x + \sqrt{y} - 4 = 0$

11. A manufacturer has a monthly fixed cost of \$40,000 and a production cost of \$8 for each unit produced. The product sells for \$12/unit.
    a. What is the cost function?
    b. What is the revenue function?
    c. What is the profit function?
    d. Compute the profit (loss) corresponding to production levels of 8000 and 12,000 units.

12. A manufacturer has a monthly fixed cost of \$100,000 and a production cost of \$14 for each unit produced. The product sells for \$20/unit.
    a. What is the cost function?
    b. What is the revenue function?
    c. What is the profit function?
    d. Compute the profit (loss) corresponding to production levels of 12,000 and 20,000 units.

13. Find the constants m and b in the linear function $f(x) = mx + b$ such that $f(0) = 2$ and $f(3) = -1$.

14. Find the constants m and b in the linear function $f(x) = mx + b$ such that $f(2) = 4$ and the straight line represented by f has slope -1.

15. **Linear Depreciation of an Office Building** An office building worth \$1 million when completed in 2008 is being depreciated linearly over 50 years. What was the book value of the building in 2013? What will it be in 2018? (Assume that the scrap value is \$0.)

16. **Linear Depreciation of an Automobile** An automobile purchased for use by the manager of a firm at a price of \$24,000 is to be depreciated using the straight-line method over 5 years. What will be the book value of the automobile at the end of 3 years? (Assume that the scrap value is \$0.)

17. **Consumption Functions** A certain economy's consumption function is given by the equation

$$C(x) = 0.75x + 6$$

    where $C(x)$ is the personal consumption expenditure in billions of dollars and x is the disposable personal income in billions of dollars. Find $C(0)$, $C(50)$, and $C(100)$.

18. **Sales Tax** In a certain state, the sales tax T on the amount of taxable goods is 6% of the value of the goods purchased (x), where both T and x are measured in dollars.
   a. Express T as a function of x.
   b. Find $T(200)$ and $T(5.60)$.

19. **Social Security COLAs** Social Security recipients receive an automatic cost-of-living adjustment (COLA) once each year. Their monthly benefit is increased by the same percentage that consumer prices have increased during the preceding year. Suppose consumer prices have increased by 3.3% during the preceding year.
   a. Express the adjusted monthly benefit of a Social Security recipient as a function of his or her current monthly benefit.
   b. If Carlos Garcia's monthly Social Security benefit is now \$1220, what will be his adjusted monthly benefit?

20. **Profit Function for Digital Timers** AutoTime, a manufacturer of electronic digital timers, has a monthly fixed cost of \$48,000 and a production cost of \$8 for each timer manufactured. The timers sell for \$14 each.
   a. What is the cost function?
   b. What is the revenue function?
   c. What is the profit function?
   d. Compute the profit (loss) corresponding to production levels of 4000, 6000, and 10,000 timers, respectively.

21. **Profit Function for Light Bulbs** The management of TMI finds that the monthly fixed costs attributable to the production of their 100-watt light bulbs is \$12,100.00. If the cost of producing each twin-pack of light bulbs is \$0.60 and each twin-pack sells for \$1.15, find the company's cost function, revenue function, and profit function.

22. **Linear Depreciation of a Textile Machine** In 2010, National Textile installed a new textile machine in one of its factories at a cost of \$250,000. The machine is depreciated linearly over 10 years with a scrap value of \$10,000.
   a. Find an expression for the textile machine's book value in the tth year of use $(0 \le t \le 10)$.
   b. Sketch the graph of the function of part (a).
   c. Find the machine's book value in 2014.
   d. Find the rate at which the machine is being depreciated.

23. **Linear Depreciation of a Workcenter System** A workcenter system purchased at a cost of \$60,000 in 2013 has a scrap value of \$12,000 at the end of 4 years. If the straight-line method of depreciation is used,
   a. Find the rate of depreciation.
   b. Find the linear equation expressing the system's book value at the end of t years.
   c. Sketch the graph of the function of part (b).
   d. Find the system's book value at the end of the third year.

24. **Linear Depreciation** Suppose an asset has an original value of \$$C$ and is depreciated linearly over N years with a scrap value of \$$S$. Show that the asset's book value at the end of the tth year is described by the function

$$V(t) = C - \left(\frac{C - S}{N}\right)t$$

Hint: Find an equation of the straight line passing through the points $(0, C)$ and (N, S). (Why?)

25. **Linear Depreciation of an Office Building** Rework Exercise 15 using the formula derived in Exercise 24.

26. **Linear Depreciation of an Automobile** Rework Exercise 16 using the formula derived in Exercise 24.

27. **Drug Dosages for Children** A method sometimes used by pediatricians to calculate the dosage of medicine for children is based on the child's surface area. If a denotes the adult dosage (in milligrams) and if S is the child's surface area (in square meters), then the child's dosage is given by

$$D(S) = \frac{Sa}{1.7}$$

   a. Show that D is a linear function of S.
   Hint: Think of D as having the form $D(S) = mS + b$. What are the slope m and the y-intercept b?
   b. If the adult dose of a drug is 500 mg, how much should a child whose surface area is 0.4 m^2 receive?

28. **Drug Dosages for Children** Cowling's Rule is a method for calculating pediatric drug dosages. If a denotes the adult dosage (in milligrams) and if t is the child's age (in years), then the child's dosage is given by

$$D(t) = \left(\frac{t + 1}{24}\right)a$$

   a. Show that D is a linear function of t.
   Hint: Think of $D(t)$ as having the form $D(t) = mt + b$. What is the slope m and the y-intercept b?
   b. If the adult dose of a drug is 500 mg, how much should a 4-year-old child receive?

29. **Drinking and Driving Among High School Students** The percentage of high school students who drink and drive was 17.5% at the beginning of 2001 and declined linearly to 10.3% at the beginning of 2011.
   a. Find a linear function $f(t)$ giving the percentage of high school students who drink and drive in year t, where $t = 0$ corresponds to the beginning of 2001.
   b. If the trend continues, what will the percentage of high school students who drink and drive be at the beginning of 2014?

   Source: Centers for Disease Control and Prevention.

30. **California Emissions Caps** The California emissions cap is set at 400 million metric tons of carbon dioxide equivalent in 2015 and is expected to drop by 13.2 million

metric tons of carbon dioxide equivalent per year through 2020.

a. Find a linear function f giving the California emissions cap in year t, where $t = 0$ corresponds to 2015.

b. If the same rate of decline of emissions cap is adopted through 2017, what will the emissions cap be in 2017?

Source: California Air Resource Board.

31. **Erosion of the Middle Class** The idea of a large, stable middle class (defined as those with annual household incomes in 2010 between \$39,000 and \$118,000 for a family of three), is central to America's sense of itself. But the U.S. middle class has been shrinking steadily from 61% of all adults in 1971 ($t = 0$) to 51% in 2011 ($t = 4$), where t is measured in decades. Research has shown that this decline is approximately linear.

a. Find a linear function $f(t)$ giving the percentage of middle-income adults in decade t, where $t = 0$ corresponds to 1971.

b. If this trend continues, what will the percentage of middle-income adults be in 2021?

Source: Pew Research Center.

32. **U.S. Airplane Passenger Projections** In a report issued by the U.S. Department of Transportation in 2012, it was predicted that the number of passengers boarding planes in the United States would grow steadily from the current 0.7 billion boardings/year to 1.2 billion boardings/year in 2032.

a. Find a linear function f giving the projected boardings (in billions) in year t, where $t = 0$ corresponds to 2012.

b. What is the projected annual rate of growth of boardings between 2012 and 2032?

c. How many boardings per year are projected for 2022?

Source: U.S. Department of Transportation.

33. **Celsius and Fahrenheit Temperatures** The relationship between temperature measured on the Celsius scale and on the Fahrenheit scale is linear. The freezing point is 0°C and 32°F, and the boiling point is 100°C and 212°F.

a. Find an equation giving the relationship between the temperature F measured on the Fahrenheit scale and the temperature C measured on the Celsius scale.

b. Find F as a function of C, and use this formula to determine the temperature in Fahrenheit corresponding to a temperature of 20°C.

c. Find C as a function of F, and use this formula to determine the temperature in Celsius corresponding to a temperature of 70°F.

34. **Cricket Chirping and Temperature** Entomologists have discovered that a linear relationship exists between the rate of chirping of crickets of a certain species and the air temperature. When the temperature is 70°F, the crickets chirp at the rate of 120 chirps/min, and when the temperature is 80°F, they chirp at the rate of 160 chirps/min.

a. Find an equation giving the relationship between the air temperature T and the number of chirps per minute N of the crickets.

b. Find N as a function of T, and use this function to determine the rate at which the crickets chirp when the temperature is 102°F.

For each demand equation in Exercises 35–38, where x represents the quantity demanded in units of 1000 and p is the unit price in dollars, (a) sketch the demand curve, and (b) determine the quantity demanded corresponding to the given unit price p.

35. $2x + 3p - 18 = 0;\ p = 4$

36. $5p + 4x - 80 = 0;\ p = 10$

37. $p = -3x + 60;\ p = 30$

38. $p = -0.4x + 120;\ p = 80$

39. **Demand Function for a Commodity** At a unit price of \$55, the quantity demanded of a certain commodity is 1000 units. At a unit price of \$85, the demand drops to 600 units. Given that it is linear, find the demand equation. Above what price will there be no demand? What quantity would be demanded if the commodity were free?

40. **Demand Function for an MP3 Player** The quantity demanded for a certain brand of MP3 players is 200 units when the unit price is set at \$90. The quantity demanded is 1200 units when the unit price is \$40. Find the demand equation, and sketch its graph.

41. **Demand Function for a Commodity** Assume that a certain commodity's demand equation has the form $p = ax + b$, where x is the quantity demanded and p is the unit price in dollars. Suppose the quantity demanded is 1000 units when the unit price is \$9.00 and 6000 when the unit price is \$4.00. What is the quantity demanded when the unit price is \$7.50?

42. **Demand Function for a Sports Watch** The demand equation for the Sicard sports watch is

$$p = -0.025x + 50$$

where x is the quantity demanded per week and p is the unit price in dollars. Sketch the graph of the demand equation. What is the highest price (theoretically) anyone would pay for the watch?

For each supply equation in Exercises 43–46, where x is the quantity supplied in units of 1000 and p is the unit price in dollars, (a) sketch the supply curve, and (b) determine the number of units of the commodity the supplier will make available in the market at the given unit price.

43. $3x - 4p + 24 = 0;\ p = 8$

44. $\frac{1}{2}x - \frac{2}{3}p + 12 = 0;\ p = 24$

45. $p = 2x + 10;\ p = 14$

46. $p = \frac{1}{2}x + 20;\ p = 28$

47. **Supply Function for a Digital Recorder** Suppliers of a certain brand of digital voice recorders will make 10,000 available in the market if the unit price is \$45. At a unit price of \$50, 20,000 units will be made available. Assuming that the relationship between the unit price and the quantity supplied is linear, derive the supply equation. Sketch the supply curve, and determine the quantity suppliers will make available when the unit price is \$70.

48. **Supply Function for a Refrigerator** Producers of a certain brand of refrigerator will make 2000 refrigerators available when the unit price is \$330. At a unit price of \$390, 6000 refrigerators will be marketed. Find the equation relating the unit price of a refrigerator to the quantity supplied if the equation is known to be linear. How many refrigerators will be marketed when the unit price is \$450? What is the lowest price at which a refrigerator will be marketed?

49. **Demand for Wireless Loudspeakers** The demand equation for the BWS Bluetooth wireless loudspeaker is

$$p = -0.05x + 200$$

where x is the quantity demanded per month and p is the unit price in dollars.
   a. Sketch the graph of the demand curve.
   b. What is the highest price (theoretically) anyone would pay for a BWS wireless loudspeaker?
   c. What is the quantity demanded per month when the unit price is \$100?

50. **Demand for Washable Computer Keyboards** The demand equation for the Wilkinson washable computer keyboard is

$$p = -0.02x + 80$$

where x is the quantity demanded per month and p is the unit price in dollars.
   a. Sketch the graph of the demand curve.
   b. What is the highest price (theoretically) anyone would pay for a washable keyboard?
   c. What is the quantity demanded per month when the unit price is \$20?

51. **Supply of Wireless Loudspeakers** The supply equation for the BWS Bluetooth wireless loudspeaker is

$$p = 0.025x + 50$$

where x is the quantity supplied per month and p is the unit price in dollars.
   a. Sketch the graph of the supply curve.
   b. What is the lowest price (theoretically) at which the supplier will make a BWS Bluetooth wireless loudspeaker available in the market?
   c. How many of these loudspeakers will the supplier make available in the market when the unit price is \$100?

52. **Supply of Washable Computer Keyboards** The supply equation for the Wilkinson washable computer keyboard is

$$p = 0.03x + 80$$

where x is the quantity supplied per month and p is the unit price in dollars.
   a. Sketch the graph of the supply curve.
   b. What is the lowest price (theoretically) at which the supplier will make a washable keyboard available in the market?
   c. How many washable keyboards will the supplier make available in the market when the unit price is \$110?

In Exercises 53 and 54, determine whether the statement is true or false. If it is true, explain why it is true. If it is false, give an example to show why it is false.

53. Suppose $C(x) = cx + F$ and $R(x) = sx$ are the cost and revenue functions of a certain firm. Then the firm is making a profit if its level of production is less than $F/(s - c)$.

54. If $p = mx + b$ is a linear demand curve, then it is generally true that $m < 0$.

1.3 Solutions to Self-Check Exercises

1. Let x denote the number of units produced and sold. Then
   a. $C(x) = 10x + 60{,}000$
   b. $R(x) = 15x$
   c. $P(x) = R(x) - C(x) = 15x - (10x + 60{,}000)$
   $= 5x - 60{,}000$
   d. $P(10{,}000) = 5(10{,}000) - 60{,}000$
   $= -10{,}000$
   or a loss of \$10,000 per month.
   $P(14{,}000) = 5(14{,}000) - 60{,}000$
   $= 10{,}000$
   or a profit of \$10,000 per month.

2. Let p denote the price of a rug (in dollars), and let x denote the quantity of rugs demanded when the unit price is \$$p$. The condition that the quantity demanded is 500 when the unit price is \$100 tells us that the demand curve passes through the point $(500, 100)$. Next, the condition that for each \$20 decrease in the unit price, the quantity demanded increases by 500 tells us that the demand curve is linear and that its slope is given by $-\frac{20}{500}$, or $-\frac{1}{25}$. Therefore, letting $m = -\frac{1}{25}$ in the demand equation

$$p = mx + b$$

we find

$$p = -\frac{1}{25}x + b$$

To determine b, use the fact that the straight line passes through (500, 100) to obtain

$$100 = -\frac{1}{25}(500) + b$$

or $b = 120$. Therefore, the required equation is

$$p = -\frac{1}{25}x + 120$$

The graph of the demand curve $p = -\frac{1}{25}x + 120$ is sketched in the following figure.

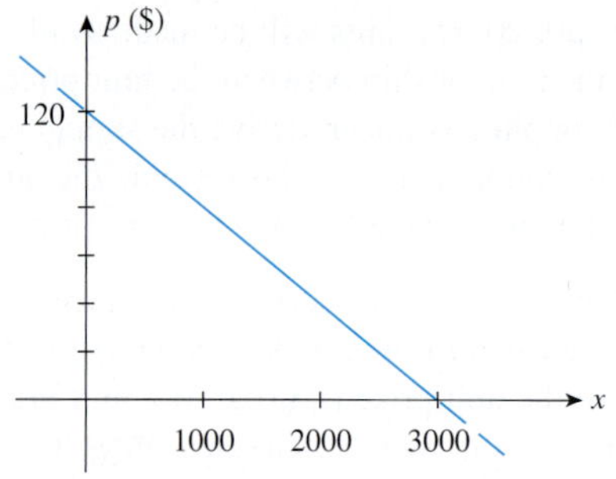

USING TECHNOLOGY Evaluating a Function

Graphing Utility

A graphing utility can be used to find the value of a function f at a given point with minimal effort. However, to find the value of y for a given value of x in a linear equation such as $Ax + By + C = 0$, the equation must first be cast in the slope-intercept form $y = mx + b$, thus revealing the desired rule $f(x) = mx + b$ for y as a function of x.

EXAMPLE 1 Consider the equation $2x + 5y = 7$.

a. Plot the straight line with the given equation in the standard viewing window.
b. Find the value of y when $x = 2$ and verify your result by direct computation.
c. Find the value of y when $x = 1.732$.

Solution

a. The straight line with equation $2x + 5y = 7$ or, equivalently, $y = -\frac{2}{5}x + \frac{7}{5}$ in the standard viewing window is shown in Figure T1.

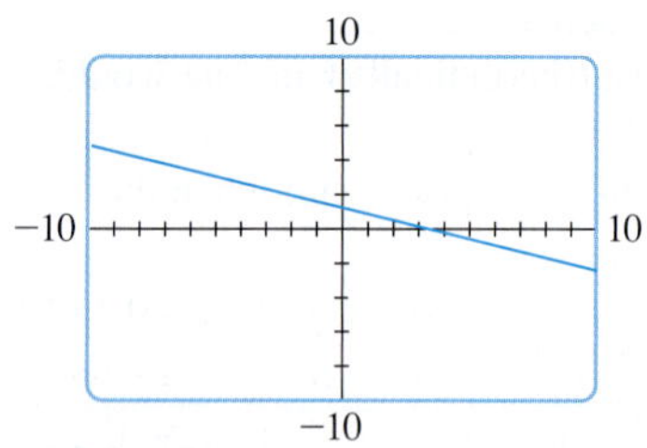

FIGURE **T1**
The straight line $2x + 5y = 7$ in the standard viewing window

b. Using the evaluation function of the graphing utility and the value of 2 for x, we find $y = 0.6$. This result is verified by computing

$$y = -\frac{2}{5}(2) + \frac{7}{5} = -\frac{4}{5} + \frac{7}{5} = \frac{3}{5} = 0.6$$

when $x = 2$.

c. Once again using the evaluation function of the graphing utility, this time with the value 1.732 for x, we find $y = 0.7072$.

When evaluating $f(x)$ at $x = a$, remember that the number a must lie between xMin and xMax.

APPLIED EXAMPLE 2 Drinking and Driving Among High School Students According to the Centers for Disease Control and Prevention, the percentage of high school students who drink and drive stood at 17.5% at the beginning of 2001 ($t = 0$) and dropped steadily in the following years. This percentage is approximated by the function

$$P(t) = -0.73t + 17.5 \qquad (t \geq 0)$$

a. Plot the graph of the function P in the viewing window $[0, 14] \times [0, 25]$.
b. If the trend continues, what will the percentage of high school students who drink and drive be at the beginning of 2014?
c. At what rate was the percentage of students who drink and drive dropping between 2001 and 2011?

Source: Centers for Disease Control and Prevention.

Solution

a. The graph of P is shown in Figure T2.

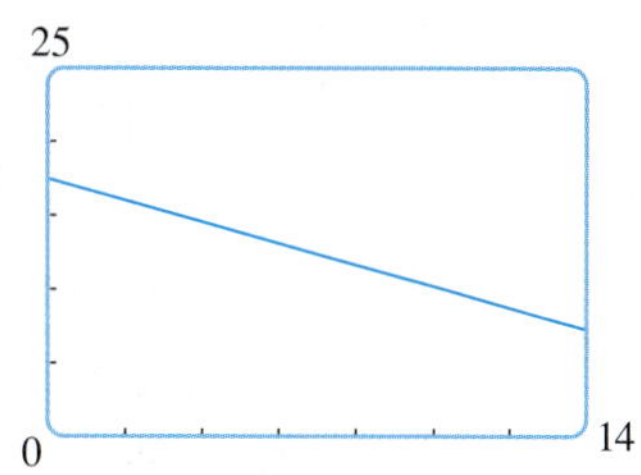

FIGURE **T2**

b. The percentage of students who drink and drive at the beginning of 2014 will be approximately

$$P(13) = 8.01$$

or 8.01%.

c. The rate at which the percentage of students who drink and drive was dropping between 2001 and 2011 is 0.73% per year.

Excel

Excel can be used to find the value of a function at a given value with minimal effort. However, to find the value of y for a given value of x in a linear equation such as $Ax + By + C = 0$, the equation must first be cast in the slope-intercept form $y = mx + b$, thus revealing the desired rule $f(x) = mx + b$ for y as a function of x.

EXAMPLE 3 Consider the equation $2x + 5y = 7$.

a. Find the value of y for $x = 0, 5$, and 10.
b. Plot the straight line with the given equation over the interval $[0, 10]$.

Solution

a. Since this is a linear equation, we first cast the equation in slope-intercept form:

$$y = -\frac{2}{5}x + \frac{7}{5}$$

Next, we create a table of values (Figure T3), following the same procedure outlined in Example 3, pages 26–27. In this case, we use the formula `=(-2/5)*B1+7/5` for the y-values.

	A	B	C	D
1	x	0	5	10
2	y	1.4	-0.6	-2.6

FIGURE **T3**
Table of values for x and y

b. Following the procedure outlined in Example 3, we obtain the graph shown in Figure T4.

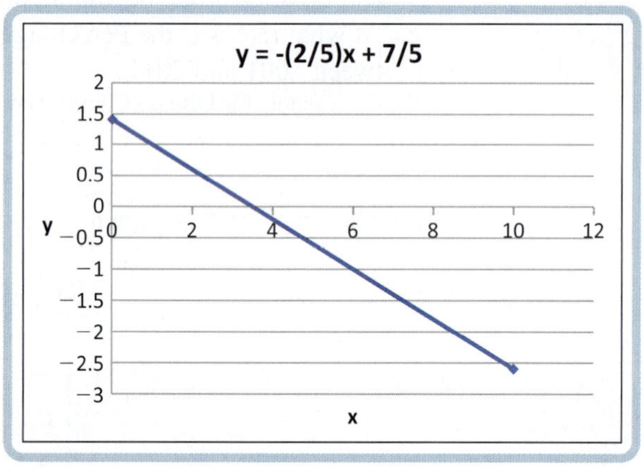

FIGURE **T4**
The graph of $y = -\frac{2}{5}x + \frac{7}{5}$ over the interval [0, 10]

APPLIED EXAMPLE 4 Drinking and Driving Among High School Students According to the Centers for Disease Control and Prevention, the percentage of high school students who drink and drive was 17.5% at the beginning of 2001 ($t = 0$) and dropped steadily in the following years. This percentage is approximated by the function

$$P(t) = -0.73t + 17.5 \qquad (t \geq 0)$$

a. Plot the graph of the function P over the interval $[0, 14]$.
b. If the trend continues, what will the percentage of high school students who drink and drive be at the beginning of 2014?
c. At what rate was the percentage of students who drink and drive dropping between 2001 and 2011?

Source: Centers for Disease Control and Prevention.

Solution

a. Following the instructions given in Example 3, pages 26–27, we obtain the spreadsheet and graph shown in Figure T5. [*Note:* We have made the appropriate entries for the title and x- and y-axis labels. In particular, for Primary Vertical Axis Title, select **Rotated Title** and type `P(t) percent`].

Note: Words/characters printed in a monospace font (for example, `=(-2/3)*A2+2`) indicate words/characters that need to be typed and entered.

	A	B	C
1	t	0	13
2	P(t)	17.5	8.01

(a)

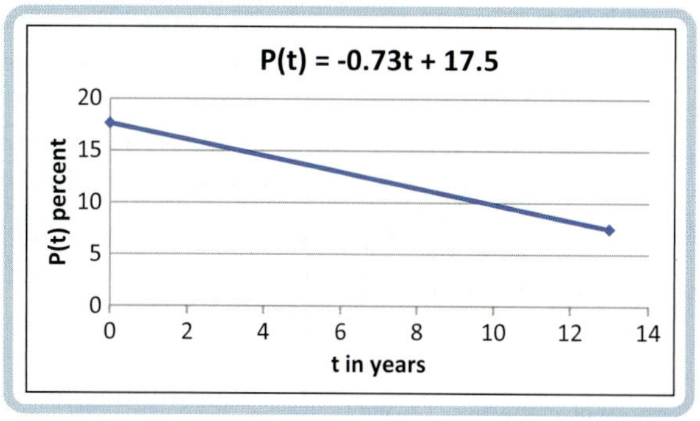

(b)

FIGURE **T5**
(a) The table of values for t and $P(t)$ and (b) the graph showing the percentage of high school students who drink and drive.

b. From the table of values, we see that

$$P(13) = -0.73(13) + 17.5 = 8.01$$

or 8.01%.

c. The rate at which the percentage of students who drink and drive was dropping between 2001 and 2011 is 0.73% per year.

TECHNOLOGY EXERCISES

Find the value of y corresponding to the given value of x.

1. $3.1x + 2.4y - 12 = 0$; $x = 2.1$
2. $1.2x - 3.2y + 8.2 = 0$; $x = 1.2$
3. $2.8x + 4.2y = 16.3$; $x = 1.5$
4. $-1.8x + 3.2y - 6.3 = 0$; $x = -2.1$
5. $22.1x + 18.2y - 400 = 0$; $x = 12.1$
6. $17.1x - 24.31y - 512 = 0$; $x = -8.2$
7. $2.8x = 1.41y - 2.64$; $x = 0.3$
8. $0.8x = 3.2y - 4.3$; $x = -0.4$

1.4 Intersection of Straight Lines

Finding the Point of Intersection

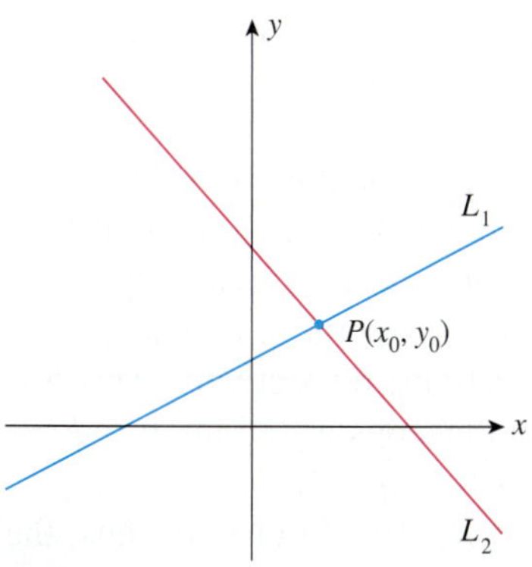

FIGURE **35**
L_1 and L_2 intersect at the point $P(x_0, y_0)$.

The solution of certain practical problems involves finding the point of intersection of two straight lines. To see how such a problem may be solved algebraically, suppose we are given two straight lines L_1 and L_2 with equations

$$y = m_1x + b_1 \quad \text{and} \quad y = m_2x + b_2$$

(where m_1, b_1, m_2, and b_2 are constants) that intersect at the point $P(x_0, y_0)$ (Figure 35).

The point $P(x_0, y_0)$ lies on the line L_1, so it satisfies the equation $y = m_1x + b_1$. It also lies on the line L_2, so it satisfies the equation $y = m_2x + b_2$. Therefore, to find the point of intersection $P(x_0, y_0)$ of the lines L_1 and L_2, we solve the system composed of the two equations

$$y = m_1x + b_1 \quad \text{and} \quad y = m_2x + b_2$$

for x and y.

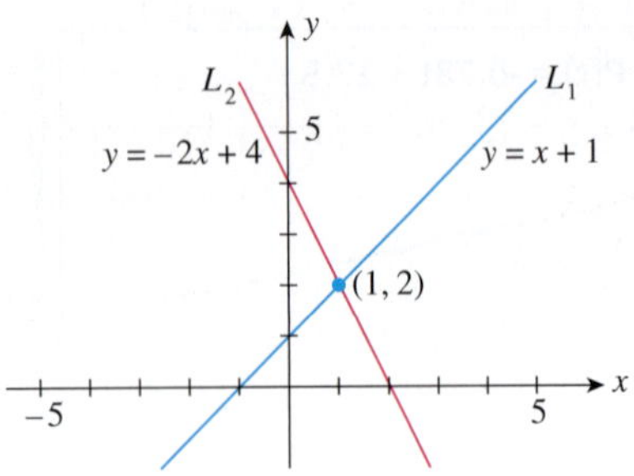

FIGURE 36
The point of intersection of L_1 and L_2 is (1, 2).

EXAMPLE 1 Find the point of intersection of the straight lines that have equations $y = x + 1$ and $y = -2x + 4$.

Solution We solve the given simultaneous equations. Substituting the value of y as given in the first equation into the second, we obtain

$$\begin{aligned} x + 1 &= -2x + 4 \\ 3x &= 3 \\ x &= 1 \end{aligned}$$

Substituting this value of x into either one of the given equations yields $y = 2$. Therefore, the required point of intersection is (1, 2) (Figure 36).

Exploring with TECHNOLOGY

1. Use a graphing utility to plot the straight lines L_1 and L_2 with equations $y = 3x - 2$ and $y = -2x + 3$, respectively, on the same set of axes in the standard viewing window. Then use **TRACE** and **ZOOM** to find the point of intersection of L_1 and L_2. Repeat using the "intersection" function of your graphing utility.
2. Find the point of intersection of L_1 and L_2 algebraically.
3. Comment on the effectiveness of each method.

We now turn to some applications involving the intersections of pairs of straight lines.

Break-Even Analysis

Consider a firm with (linear) cost function $C(x)$, revenue function $R(x)$, and profit function $P(x)$ given by

$$\begin{aligned} C(x) &= cx + F \\ R(x) &= sx \\ P(x) &= R(x) - C(x) = (s - c)x - F \end{aligned}$$

where c denotes the unit cost of production, s the selling price per unit, F the fixed cost incurred by the firm, and x the level of production and sales. The level of production at which the firm neither makes a profit nor sustains a loss is called the **break-even level of operation** and may be determined by solving the equations $y = C(x)$ and $y = R(x)$ simultaneously. At the level of production x_0, the profit is zero, so

$$\begin{aligned} P(x_0) &= R(x_0) - C(x_0) = 0 \\ R(x_0) &= C(x_0) \end{aligned}$$

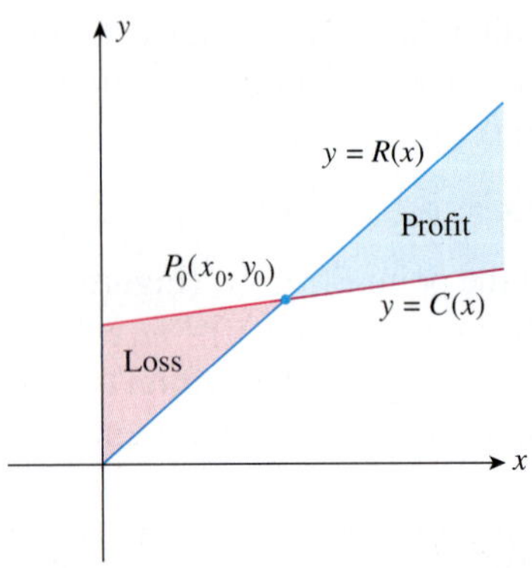

FIGURE 37
P_0 is the break-even point.

The point $P_0(x_0, y_0)$, the solution of the simultaneous equations $y = R(x)$ and $y = C(x)$, is referred to as the **break-even point;** the number x_0 and the number y_0 are called the **break-even quantity** and the **break-even revenue,** respectively.

Geometrically, the break-even point $P_0(x_0, y_0)$ is just the point of intersection of the straight lines representing the cost and revenue functions, respectively. This follows because $P_0(x_0, y_0)$, being the solution of the simultaneous equations $y = R(x)$ and $y = C(x)$, must lie on both these lines simultaneously (Figure 37).

Note that if $x < x_0$, then $R(x) < C(x)$, so $P(x) = R(x) - C(x) < 0$; thus, the firm sustains a loss at this level of production. On the other hand, if $x > x_0$, then $P(x) > 0$, and the firm operates at a profitable level.

APPLIED EXAMPLE 2 Break-Even Level Prescott manufactures its products at a cost of \$4 per unit and sells them for \$10 per unit. If the firm's fixed cost is \$12,000 per month, determine the firm's break-even point.

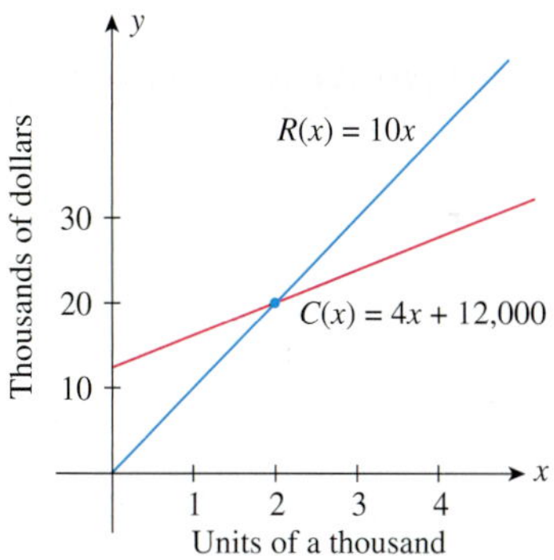

FIGURE **38**
The point at which $R(x) = C(x)$ is the break-even point.

Solution The cost function C and the revenue function R are given by $C(x) = 4x + 12{,}000$ and $R(x) = 10x$, respectively (Figure 38).

Setting $R(x) = C(x)$, we obtain

$$\begin{aligned} 10x &= 4x + 12{,}000 \\ 6x &= 12{,}000 \\ x &= 2000 \end{aligned}$$

Substituting this value of x into $R(x) = 10x$ gives

$$R(2000) = (10)(2000) = 20{,}000$$

So for a break-even operation, the firm should manufacture 2000 units of its product, resulting in a break-even revenue of \$20,000 per month.

APPLIED EXAMPLE 3 Break-Even Analysis Using the data given in Example 2, answer the following questions:

a. What is the loss sustained by the firm if only 1500 units are produced and sold each month?

b. What is the profit if 3000 units are produced and sold each month?

c. How many units should the firm produce to realize a minimum monthly profit of \$9000?

Solution The profit function P is given by the rule

$$\begin{aligned} P(x) &= R(x) - C(x) \\ &= 10x - (4x + 12{,}000) \\ &= 6x - 12{,}000 \end{aligned}$$

a. If 1500 units are produced and sold each month, we have

$$P(1500) = 6(1500) - 12{,}000 = -3000$$

so the firm will sustain a loss of \$3000 per month.

b. If 3000 units are produced and sold each month, we have

$$P(3000) = 6(3000) - 12{,}000 = 6000$$

or a monthly profit of \$6000.

c. Substituting 9000 for $P(x)$ in the equation $P(x) = 6x - 12{,}000$, we obtain

$$\begin{aligned} 9000 &= 6x - 12{,}000 \\ 6x &= 21{,}000 \\ x &= 3500 \end{aligned}$$

Thus, the firm should produce at least 3500 units to realize a \$9000 minimum monthly profit.

APPLIED EXAMPLE 4 Decision Analysis The management of Robertson Controls must decide between two manufacturing processes for its model C electronic thermostat. The monthly cost of the first process is given by

$C_1(x) = 20x + 10{,}000$ dollars, where x is the number of thermostats produced; the monthly cost of the second process is given by $C_2(x) = 10x + 30{,}000$ dollars. If the projected monthly sales are 800 thermostats at a unit price of \$40, which process should management choose in order to maximize the company's profit?

Solution The break-even level of operation using the first process is obtained by solving the equation

$$\begin{aligned} 40x &= 20x + 10{,}000 \\ 20x &= 10{,}000 \\ x &= 500 \end{aligned}$$

giving an output of 500 units. Next, we solve the equation

$$\begin{aligned} 40x &= 10x + 30{,}000 \\ 30x &= 30{,}000 \\ x &= 1000 \end{aligned}$$

giving an output of 1000 units for a break-even operation using the second process. Since the projected sales are 800 units, we conclude that management should choose the first process, which will give the firm a profit.

APPLIED EXAMPLE 5 Decision Analysis Referring to Example 4, decide which process Robertson's management should choose if the projected monthly sales are (a) 1500 units and (b) 3000 units.

Solution In both cases, the production is past the break-even level. Since the revenue is the same regardless of which process is employed, the decision will be based on how much each process costs.

a. If $x = 1500$, then

$$\begin{aligned} C_1(x) &= (20)(1500) + 10{,}000 = 40{,}000 \\ C_2(x) &= (10)(1500) + 30{,}000 = 45{,}000 \end{aligned}$$

Hence, management should choose the first process.

b. If $x = 3000$, then

$$\begin{aligned} C_1(x) &= (20)(3000) + 10{,}000 = 70{,}000 \\ C_2(x) &= (10)(3000) + 30{,}000 = 60{,}000 \end{aligned}$$

In this case, management should choose the second process.

Exploring with TECHNOLOGY

1. Use a graphing utility to plot the straight lines L_1 and L_2 with equations $y = 2x - 1$ and $y = 2.1x + 3$, respectively, on the same set of axes, using the standard viewing window. Do the lines appear to intersect?
2. Plot the straight lines L_1 and L_2, using the viewing window $[-100, 100] \times [-100, 100]$. Do the lines appear to intersect? Can you find the point of intersection using **TRACE** and **ZOOM**? Using the "intersection" function of your graphing utility?
3. Find the point of intersection of L_1 and L_2 algebraically.
4. Comment on the effectiveness of the solution methods in parts 2 and 3.

Market Equilibrium

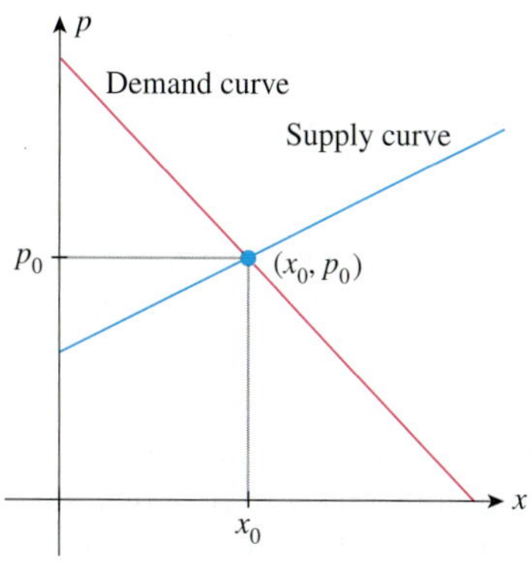

FIGURE 39
Market equilibrium is represented by the point (x_0, p_0).

Under pure competition, the price of a commodity eventually settles at a level dictated by the condition that the supply of the commodity be equal to the demand for it. If the price is too high, consumers will be more reluctant to buy, and if the price is too low, the supplier will be more reluctant to make the product available in the marketplace. **Market equilibrium** is said to prevail when the quantity produced is equal to the quantity demanded. The quantity produced at market equilibrium is called the **equilibrium quantity,** and the corresponding price is called the **equilibrium price.**

From a geometric point of view, market equilibrium corresponds to the point at which the demand curve and the supply curve intersect. In Figure 39, x_0 represents the equilibrium quantity and p_0 the equilibrium price. The point (x_0, p_0) lies on the supply curve and therefore satisfies the supply equation. At the same time, it also lies on the demand curve and therefore satisfies the demand equation. Thus, to find the point (x_0, p_0), and hence the equilibrium quantity and price, we solve the demand and supply equations simultaneously for x and p. For meaningful solutions, x and p must both be positive.

APPLIED EXAMPLE 6 Market Equilibrium The management of ThermoMaster, which manufactures an indoor–outdoor thermometer at its Mexico subsidiary, has determined that the demand equation for its product is

$$5x + 3p - 30 = 0$$

where p is the price of a thermometer in dollars and x is the quantity demanded in units of a thousand. The supply equation for these thermometers is

$$52x - 30p + 45 = 0$$

where x (measured in thousands) is the quantity that ThermoMaster will make available in the market at p dollars each. Find the equilibrium quantity and price.

Solution We need to solve the system of equations

$$\begin{aligned} 5x + 3p - 30 &= 0 \\ 52x - 30p + 45 &= 0 \end{aligned}$$

for x and p. Let us use the *method of substitution* to solve it. As the name suggests, this method calls for choosing one of the equations in the system, solving for one variable in terms of the other, and then substituting the resulting expression into the other equation. This gives an equation in one variable that can then be solved in the usual manner.

Let's solve the first equation for p in terms of x. Thus,

$$\begin{aligned} 3p &= -5x + 30 \\ p &= -\frac{5}{3}x + 10 \end{aligned}$$

Next, we substitute this value of p into the second equation, obtaining

$$\begin{aligned} 52x - 30\left(-\frac{5}{3}x + 10\right) + 45 &= 0 \\ 52x + 50x - 300 + 45 &= 0 \\ 102x - 255 &= 0 \\ x &= \frac{255}{102} = \frac{5}{2} \end{aligned}$$

The corresponding value of p is found by substituting this value of x into the equation for p obtained earlier. Thus,

$$p = -\frac{5}{3}\left(\frac{5}{2}\right) + 10 = -\frac{25}{6} + 10$$
$$= \frac{35}{6} \approx 5.83$$

We conclude that the equilibrium quantity is 2500 units (remember that x is measured in units of a thousand) and the equilibrium price is \$5.83 per thermometer.

APPLIED EXAMPLE 7 Market Equilibrium The quantity demanded of a certain model of DVD player is 8000 units when the unit price is \$260. At a unit price of \$200, the quantity demanded increases to 10,000 units. The manufacturer will not market any players if the price is \$100 or lower. However, for each \$50 increase in the unit price above \$100, the manufacturer will market an additional 1000 units. Both the demand and the supply equations are known to be linear.

a. Find the demand equation.
b. Find the supply equation.
c. Find the equilibrium quantity and price.

Solution Let p denote the unit price in hundreds of dollars, and let x denote the number of units of players in thousands.

a. Since the demand function is linear, the demand curve is a straight line passing through the points $(8, 2.6)$ and $(10, 2)$. Its slope is

$$m = \frac{2 - 2.6}{10 - 8} = -0.3$$

Using the point $(10, 2)$ and the slope $m = -0.3$ in the point-slope form of the equation of a line, we see that the required demand equation is

$$p - 2 = -0.3(x - 10)$$
$$p = -0.3x + 5 \qquad \text{Figure 40}$$

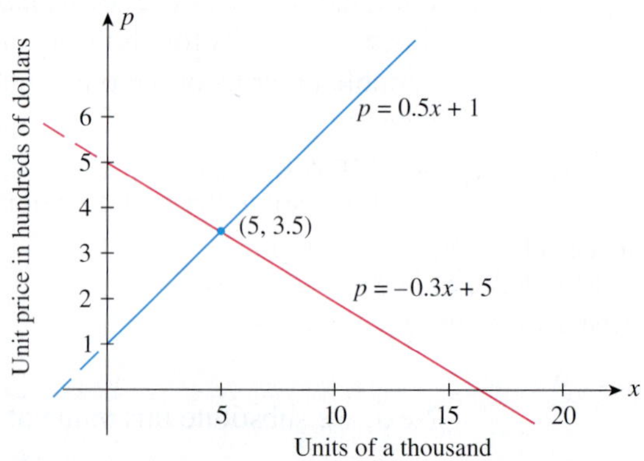

FIGURE 40
Market equilibrium occurs at the point (5, 3.5).

b. The supply curve is the straight line passing through the points $(0, 1)$ and $(1, 1.5)$. Its slope is

$$m = \frac{1.5 - 1}{1 - 0} = 0.5$$

Using the point $(0, 1)$ and the slope $m = 0.5$ in the point-slope form of the equation of a line, we see that the required supply equation is

$$p - 1 = 0.5(x - 0)$$
$$p = 0.5x + 1 \quad \text{Figure 40}$$

c. To find the market equilibrium, we solve simultaneously the system comprising the demand and supply equations obtained in parts (a) and (b)—that is, the system

$$p = -0.3x + 5$$
$$p = 0.5x + 1$$

$$\begin{array}{r} p = 0.5x + 1 \\ -p = 0.3x - 5 \\ \hline 0 = 0.8x - 4 \end{array}$$

Subtracting the first equation from the second gives

$$0.8x - 4 = 0$$

and $x = 5$. Substituting this value of x in the second equation gives $p = 3.5$. Thus, the equilibrium quantity is 5000 units, and the equilibrium price is \$350 (Figure 40).

1.4 Self-Check Exercises

1. Find the point of intersection of the straight lines with equations $2x + 3y = 6$ and $x - 3y = 4$.

2. **MARKET EQUILIBRIUM** There is no demand for a certain model of a disposable camera when the unit price is \$12. However, when the unit price is \$8, the quantity demanded is 8000/week. The suppliers will not market any cameras if the unit price is \$2 or lower. At \$4/camera, however, the manufacturer will market 5000 cameras per week. Both the demand and supply functions are known to be linear.
   a. Find the demand equation.
   b. Find the supply equation.
   c. Find the equilibrium quantity and price.

Solutions to Self-Check Exercises 1.4 can be found on page 52.

1.4 Concept Questions

1. Explain why you would expect that the intersection of a linear demand curve and a linear supply curve would lie in the first quadrant.

2. In the accompanying figure, $C(x)$ is the cost function and $R(x)$ is the revenue function associated with a certain product.
   a. Plot the break-even point $P(x_0, y_0)$ on the graph.
   b. Identify and mark the break-even quantity, x_0, and the break-even revenue, y_0, on the set of axes.

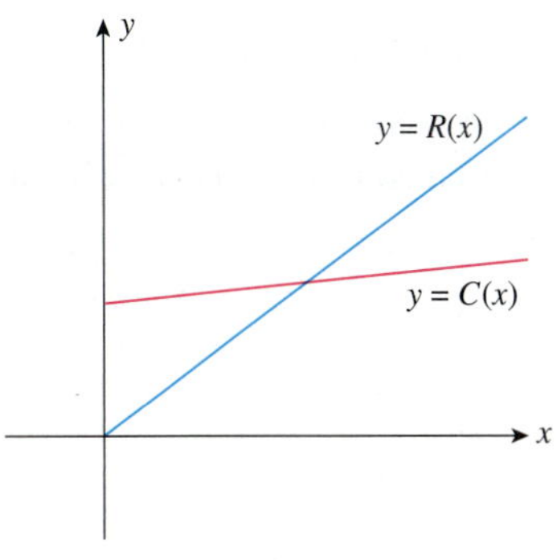

3. The accompanying figure gives the demand curve and the supply curve associated with a certain commodity.

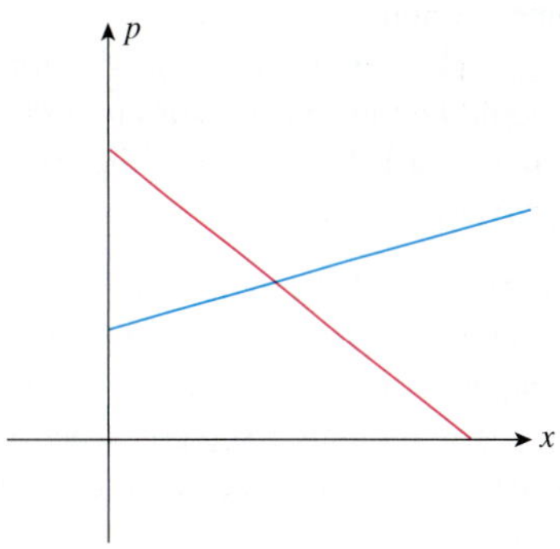

   a. Identify the demand curve and the supply curve.
   b. Plot the point $P(x_0, p_0)$ that corresponds to market equilibrium.
   c. Identify and mark the equilibrium quantity, x_0, and the equilibrium price, p_0, on the set of axes.

1.4 Exercises

In Exercises 1–6, find the point of intersection of each pair of straight lines.

1. $y = 3x + 4$
   $y = -2x + 14$

2. $y = -4x - 7$
   $-y = 5x + 10$

3. $2x - 3y = 6$
   $3x + 6y = 16$

4. $2x + 4y = 11$
   $-5x + 3y = 5$

5. $y = \frac{1}{4}x - 5$
   $2x - \frac{3}{2}y = 1$

6. $y = \frac{2}{3}x - 4$
   $x + 3y + 3 = 0$

In Exercises 7–10, find the break-even point for the firm whose cost function C and revenue function R are given.

7. $C(x) = 5x + 10{,}000;\ R(x) = 15x$

8. $C(x) = 15x + 12{,}000;\ R(x) = 21x$

9. $C(x) = 0.2x + 120;\ R(x) = 0.4x$

10. $C(x) = 150x + 20{,}000;\ R(x) = 270x$

11. **Break-Even Analysis** AutoTime, a manufacturer of 24-hr variable timers, has a monthly fixed cost of \$48,000 and a production cost of \$8 for each timer manufactured. The units sell for \$14 each.
    a. Sketch the graphs of the cost function and the revenue function and thereby find the break-even point graphically.
    b. Find the break-even point algebraically.
    c. Sketch the graph of the profit function.
    d. At what point does the graph of the profit function cross the x-axis? Interpret your result.

12. **Break-Even Analysis** A division of Carter Enterprises produces income tax apps for smartphones. Each income tax app sells for \$8. The monthly fixed costs incurred by the division are \$25,000, and the variable cost of producing each income tax app is \$3.
    a. Find the break-even point for the division.
    b. What should be the level of sales in order for the division to realize a 15% profit over the cost of making the income tax apps?

13. **Break-Even Analysis** A division of the Gibson Corporation manufactures bicycle pumps. Each pump sells for \$9, and the variable cost of producing each unit is 40% of the selling price. The monthly fixed costs incurred by the division are \$50,000. What is the break-even point for the division?

14. **Leasing a Truck** Ace Truck Leasing Company leases a certain size truck for \$25/day and \$.50/mi, whereas Acme Truck Leasing Company leases the same size truck for \$20/day and \$.60/mi.
    a. Find the functions describing the daily cost of leasing from each company.
    b. Sketch the graphs of the two functions on the same set of axes.
    c. If a customer plans to drive at most 30 mi, from which company should he rent a truck for a single day?
    d. If a customer plans to drive at least 60 mi, from which company should he rent a truck for a single day?

15. **Decision Analysis** A product may be made by using Machine I or Machine II. The manufacturer estimates that the monthly fixed costs of using Machine I are \$18,000, whereas the monthly fixed costs of using Machine II are \$15,000. The variable costs of manufacturing 1 unit of the product using Machine I and Machine II are \$15 and \$20, respectively. The product sells for \$50 each.
    a. Find the cost functions associated with using each machine.
    b. Sketch the graphs of the cost functions of part (a) and the revenue functions on the same set of axes.
    c. Which machine should management choose in order to maximize their profit if the projected sales are 450 units? 550 units? 650 units?
    d. What is the profit for each case in part (c)?

16. **Annual Sales of Two Pharmacies** The annual sales of Crimson Pharmacy are expected to be given by $S = 2.3 + 0.4t$ million dollars t years from now, whereas the annual sales of Cambridge Pharmacy are expected to be given by $S = 1.2 + 0.6t$ million dollars t years from now. When will Cambridge's annual sales first surpass Crimson's annual sales?

17. **LCDs Versus CRTs** The global shipments of traditional cathode-ray tube monitors (CRTs) is approximated by the equation

    $$y = -12t + 88 \qquad (0 \le t \le 3)$$

    where y is measured in millions and t in years, with $t = 0$ corresponding to the beginning of 2001. The equation

    $$y = 18t + 13.4 \qquad (0 \le t \le 3)$$

    gives the approximate number (in millions) of liquid crystal displays (LCDs) over the same period. When did the global shipments of LCDs first overtake the global shipments of CRTs?
    Source: International Data Corporation.

18. **Digital Versus Film Cameras** The sales of digital cameras (in millions of units) in year t is given by the function

    $$f(t) = 3.05t + 6.85 \qquad (0 \le t \le 3)$$

where $t = 0$ corresponds to 2001. Over that same period, the sales of film cameras (in millions of units) is given by

$$g(t) = -1.85t + 16.58 \qquad (0 \le t \le 3)$$

a. Show that more film cameras than digital cameras were sold in 2001.

b. When did the sales of digital cameras first exceed those of film cameras?

Source: Popular Science.

19. U.S. Financial Transactions The percentage of U.S. transactions by check between the beginning of 2001 ($t = 0$) and the beginning of 2010 ($t = 9$) is approximated by

$$f(t) = -\frac{11}{9}t + 43 \qquad (0 \le t \le 9)$$

whereas the percentage of transactions done electronically during the same period is approximated by

$$g(t) = \frac{11}{3}t + 23 \qquad (0 \le t \le 9)$$

a. Sketch the graphs of f and g on the same set of axes.

b. Find the time when transactions done electronically first exceeded those done by check.

Source: Foreign Policy.

20. Broadband Versus Dial-Up The number of U.S. broadband Internet households (in millions) between the beginning of 2004 ($t = 0$) and the beginning of 2008 ($t = 4$) is approximated by

$$f(t) = 6.5t + 33 \qquad (0 \le t \le 4)$$

Over the same period, the number of U.S. dial-up Internet households (in millions) is approximated by

$$g(t) = -3.9t + 42.5 \qquad (0 \le t \le 4)$$

a. Sketch the graphs of f and g on the same set of axes.

b. Solve the equation $f(t) = g(t)$, and interpret your result.

Source: Strategic Analytics, Inc.

For each pair of supply-and-demand equations in Exercises 21–24, where *x* represents the quantity demanded in units of 1000 and *p* is the unit price in dollars, find the equilibrium quantity and the equilibrium price.

21. $4x + 3p - 59 = 0$ and $5x - 6p + 14 = 0$

22. $2x + 7p - 56 = 0$ and $3x - 11p + 45 = 0$

23. $p = -2x + 22$ and $p = 3x + 12$

24. $p = -0.3x + 6$ and $p = 0.15x + 1.5$

25. Equilibrium Quantity and Price for DVD Players The quantity demanded of a certain brand of DVD player is 3000/week when the unit price is \$485. For each decrease in unit price of \$20 below \$485, the quantity demanded increases by 250 units. The suppliers will not market any DVD players if the unit price is \$300 or lower. But at a unit price of \$525, they are willing to make available 2500 units in the market. The supply equation is also known to be linear.

a. Find the demand equation.

b. Find the supply equation.

c. Find the equilibrium quantity and price.

26. Equilibrium Quantity and Price for GPS Navigators The demand equation for the Drake GPS Navigator is $x + 4p - 800 = 0$, where x is the quantity demanded per week and p is the wholesale unit price in dollars. The supply equation is $x - 20p + 1000 = 0$, where x is the quantity the supplier will make available in the market each week when the wholesale price is p dollars each. Find the equilibrium quantity and the equilibrium price for the GPS Navigators.

27. Equilibrium Quantity and Price for Tablet Computers The demand equation for the Schmidt-3000 tablet computer is $3x + p - 1500 = 0$, where x is the quantity demanded per week and p is the unit price in dollars. The supply equation is $2x - 3p + 1200 = 0$, where x is the quantity the supplier will make available in the market each week when the unit price is p dollars. Find the equilibrium quantity and the equilibrium price for the tablet computers.

28. Equilibrium Quantity and Price for Espresso Makers The quantity demanded each month of Russo Espresso Makers is 250 when the unit price is \$140; the quantity demanded each month is 1000 when the unit price is \$110. The suppliers will market 750 espresso makers if the unit price is \$60 or higher. At a unit price of \$80, they are willing to market 2250 units. Both the demand and supply equations are known to be linear.

a. Find the demand equation.

b. Find the supply equation.

c. Find the equilibrium quantity and the equilibrium price.

29. Equilibrium Quantity and Price for Wireless Speakers The demand equation for the BWS Bluetooth wireless loudspeaker is

$$p = -0.05x + 200$$

where x is the quantity demanded per month and p is the unit price in dollars. The corresponding supply equation is given by

$$p = 0.025x + 50$$

where x is the quantity demanded per month and p is the unit price in dollars. Find the equilibrium quantity and the equilibrium price for the BWS Bluetooth wireless loudspeakers.

30. **Equilibrium Quantity and Price for Washable Keyboards** The demand equation for the Wilkinson washable keyboards is

$$p = -0.02x + 80$$

where x is the quantity demanded per month and p is the unit price in dollars. The corresponding supply equation is given by

$$p = 0.03x + 20$$

where x is the quantity demanded per month and p is the unit price in dollars. Find the equilibrium quantity and the equilibrium price for the Wilkinson washable keyboards.

31. Suppose the demand and supply equations for a certain commodity are given by $p = ax + b$ and $p = cx + d$, respectively, where $a < 0$, $c > 0$, and $b > d > 0$ (see the following figure).

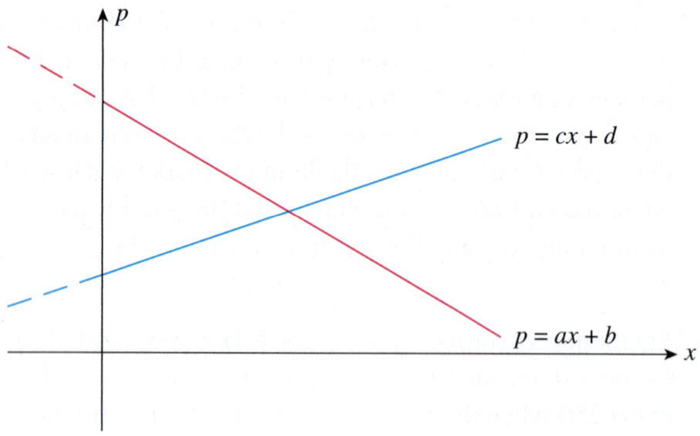

a. Find the equilibrium quantity and equilibrium price in terms of a, b, c, and d.
b. Use part (a) to determine what happens to the market equilibrium if c is increased while a, b, and d remain fixed. Interpret your answer in economic terms.
c. Use part (a) to determine what happens to the market equilibrium if b is decreased while a, c, and d remain fixed. Interpret your answer in economic terms.

32. Suppose the cost function associated with a product is $C(x) = cx + F$ dollars and the revenue function is $R(x) = sx$, where c denotes the unit cost of production, s the unit selling price, F the fixed cost incurred by the firm, and x the level of production and sales. Find the break-even quantity and the break-even revenue in terms of the constants c, s, and F, and interpret your results in economic terms.

In Exercises 33 and 34, determine whether the statement is true or false. If it is true, explain why it is true. If it is false, give an example to show why it is false.

33. Suppose $C(x) = cx + F$ and $R(x) = sx$ are the cost and revenue functions, respectively, of a certain firm. Then the firm is operating at a break-even level of production if its level of production is $F/(s - c)$.

34. If both the demand equation and the supply equation for a certain commodity are linear, then there must be at least one equilibrium point.

35. Let L_1 and L_2 be two nonvertical straight lines in the plane with equations $y = m_1x + b_1$ and $y = m_2x + b_2$, respectively. Find conditions on m_1, m_2, b_1, and b_2 such that (a) L_1 and L_2 do not intersect, (b) L_1 and L_2 intersect at one and only one point, and (c) L_1 and L_2 intersect at infinitely many points.

36. Find conditions on a_1, a_2, b_1, b_2, c_1, and c_2 such that the system of linear equations

$$a_1x + b_1y = c_1$$
$$a_2x + b_2y = c_2$$

has (a) no solution, (b) a unique solution, and (c) infinitely many solutions.
Hint: Use the results of Exercise 35.

1.4 Solutions to Self-Check Exercises

1. The point of intersection of the two straight lines is found by solving the system of linear equations

$$2x + 3y = 6$$
$$x - 3y = 4$$

Solving the first equation for y in terms of x, we obtain

$$y = -\frac{2}{3}x + 2$$

Substituting this expression for y into the second equation, we obtain

$$x - 3\left(-\frac{2}{3}x + 2\right) = 4$$
$$x + 2x - 6 = 4$$
$$3x = 10$$

or $x = \frac{10}{3}$. Substituting this value of x into the expression for y obtained earlier, we find

$$y = -\frac{2}{3}\left(\frac{10}{3}\right) + 2 = -\frac{2}{9}$$

Therefore, the point of intersection is $\left(\frac{10}{3}, -\frac{2}{9}\right)$.

2. a. Let p denote the price per camera, and let x denote the quantity demanded per week. The given conditions imply that $x = 0$ when $p = 12$ and $x = 8000$ when $p = 8$. Since the demand equation is linear, it has the form

$$p = mx + b$$

Now, the first condition implies that

$$12 = m(0) + b \quad \text{or} \quad b = 12$$

Therefore,

$$p = mx + 12$$

Using the second condition, we find

$$8 = 8000m + 12$$

$$m = -\frac{4}{8000} = -0.0005$$

Hence the required demand equation is

$$p = -0.0005x + 12$$

b. Let p denote the price per camera, and let x denote the quantity made available at that price per week. Then, since the supply equation is linear, it also has the form

$$p = mx + b$$

The first condition implies that $x = 0$ when $p = 2$, so we have

$$2 = m(0) + b \quad \text{or} \quad b = 2$$

Therefore,

$$p = mx + 2$$

Next, using the second condition, $x = 5000$ when $p = 4$, we find

$$4 = 5000m + 2$$

giving $m = 0.0004$. So the required supply equation is

$$p = 0.0004x + 2$$

c. The equilibrium quantity and price are found by solving the system of linear equations

$$p = -0.0005x + 12$$
$$p = 0.0004x + 2$$

Equating the two expressions yields

$$-0.0005x + 12 = 0.0004x + 2$$
$$0.0009x = 10$$

or $x \approx 11,111$. Substituting this value of x into either equation in the system yields

$$p \approx 6.44$$

Therefore, the equilibrium quantity is 11,111, and the equilibrium price is \$6.44.

USING TECHNOLOGY Finding the Point(s) of Intersection of Two Graphs

Graphing Utility

A graphing utility can be used to find the point(s) of intersection of the graphs of two functions. Once again, it is important to remember that if the graphs are straight lines, the linear equations defining these lines must first be recast in the slope-intercept form.

EXAMPLE 1 Find the points of intersection of the straight lines with equations $2x + 3y = 6$ and $3x - 4y - 5 = 0$.

Solution **Method I** Solving each equation for y in terms of x, we obtain

$$y = -\frac{2}{3}x + 2 \quad \text{and} \quad y = \frac{3}{4}x - \frac{5}{4}$$

as the respective equations in the slope-intercept form. The graphs of the two straight lines in the standard viewing window are shown in Figure T1.

Then, using **TRACE** and **ZOOM** or the function for finding the point of intersection of two graphs, we find that the point of intersection, accurate to four decimal places, is (2.2941, 0.4706).

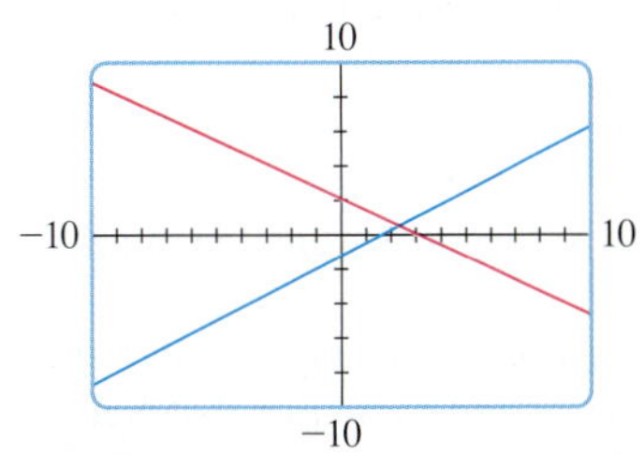

FIGURE **T1**
The straight lines $2x + 3y = 6$ and $3x - 4y - 5 = 0$

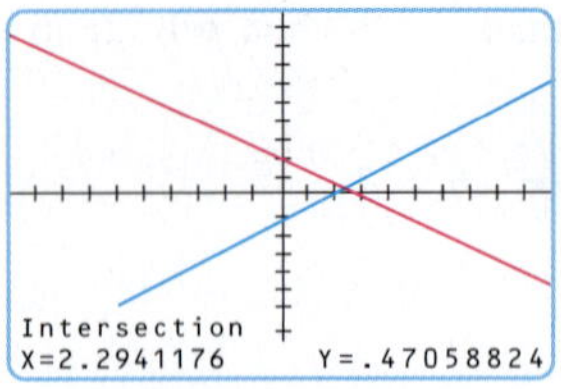

FIGURE **T2**

Method II Proceed as before to obtain the graphs of the two lines. Then use the intersect function of the graphing utility to find the point of intersection, (2.2941176, 0.47058824), of the two straight lines (Figure T2).

Note On the TI-83/84, you can call the **intersect** function by selecting the **CALC** menu and then selecting **5: intersect** (Figure T3a). Press **ENTER** to obtain the graph shown in Figure T3b. Then press **ENTER** to select the first curve (Figure T3c); press **ENTER** again to select the second curve (Figure T3d). Press **ENTER** for the fourth time to find the point of intersection (Figure T3e).

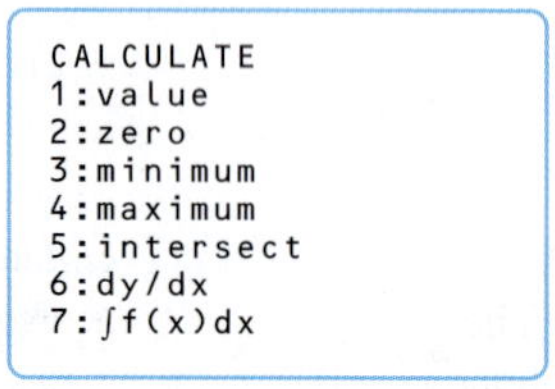

(a) TI-83/84 CALC menu

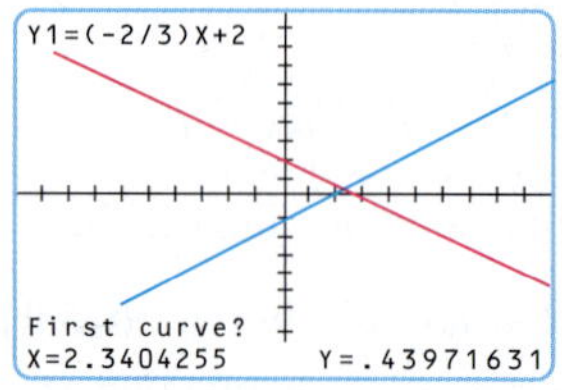

(b) First-graph screen

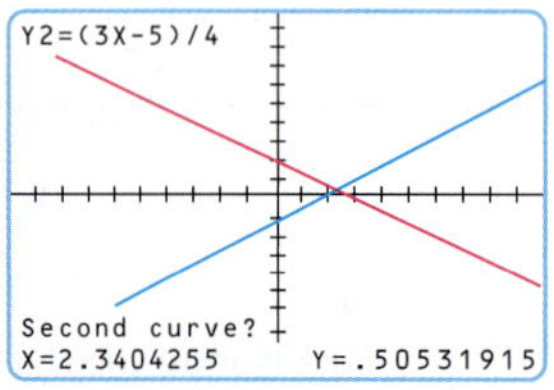

(c) Second-graph screen

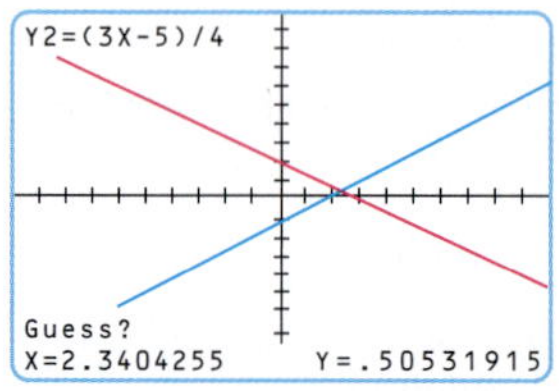

(d) Third-graph screen

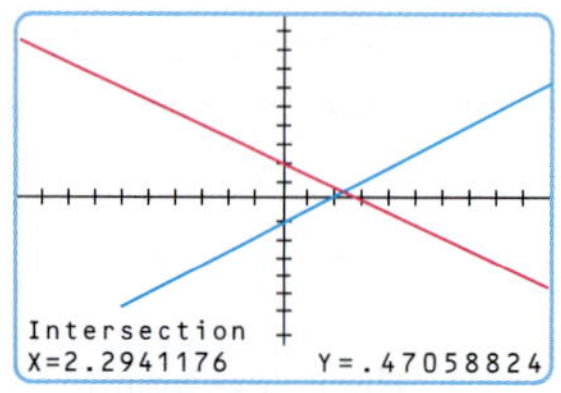

(e) Final-graph screen

FIGURE **T3**

TECHNOLOGY EXERCISES

In Exercises 1–6, find the point of intersection of the pair of straight lines with the given equations. Round your answers to four decimal places.

1. $y = 2x + 5$ and $y = -3x + 8$

2. $y = 1.2x + 6.2$ and $y = -4.3x + 9.1$

3. $2x - 5y = 7$ and $3x + 4y = 12$

4. $1.4x - 6.2y = 8.4$ and $4.1x + 7.3y = 14.4$

5. $2.1x = 5.1y + 71$ and $3.2x = 8.4y + 16.8$

6. $8.3x = 6.2y + 9.3$ and $12.4x = 12.3y + 24.6$

7. Break-Even Analysis PhotoMax makes disposable cameras that sell for \$7.89 each and cost \$3.24 each to produce. The weekly fixed cost for the company is \$16,500.

a. Plot the graphs of the cost function and the revenue function in the viewing window $[0, 6000] \times [0, 60{,}000]$.

b. Find the break-even point by using the viewing window $[0, 6000] \times [-20{,}000, 20{,}000]$.

c. Plot the graph of the profit function and verify the result of part (b) by finding the x-intercept.

8. Break-Even Analysis The Monde Company makes a wine cooler with a capacity of 24 bottles. Each wine cooler sells for \$245. The monthly fixed costs incurred by the company are \$385,000, and the variable cost of producing each wine cooler is \$90.50.

a. Find the break-even point for the company.

b. Find the level of sales needed to ensure that the company will realize a profit of 21% over the cost of producing the wine coolers.

9. Leasing a Truck Ace Truck Leasing Company leases a certain size truck for \$34/day and \$0.18/mi, whereas Acme Truck Leasing Company leases the same size truck for \$28/day and \$0.22/mi.

a. Find the functions describing the daily cost of leasing from each company.

b. Plot the graphs of the two functions using the same viewing window.

c. Find the point of intersection of the graphs of part (b).

d. Use the result of part (c) to find a criterion that a customer can use to help her decide which company to rent the truck from if she knows the maximum distance that she will drive on the day of rental.

10. **Bank Deposits** The total deposits with a branch of Randolph Bank currently stand at \$20.384 million and are projected to grow at the rate of \$1.019 million/year for the next 5 years. The total deposits with a branch of Madison Bank, in the same shopping complex as the Randolph Bank, currently stand at \$18.521 million and are expected to grow at the rate of \$1.482 million/year for the next 5 years.
    a. Find the function describing the total deposits with each bank for the next 5 years.
    b. Plot the graphs of the two functions found in part (a) using the same viewing window.
    c. Do the total deposits of Madison catch up to those of Randolph over the period in question? If so, at what time?

11. **Equilibrium Quantity and Price for Smartphones** The quantity demanded of a certain brand of smartphone is 2000/week when the unit price is \$84. For each decrease in unit price of \$5 below \$84, the quantity demanded increases by 50 units. The supplier will not market any of the smartphones if the unit price is \$60 or less, but the supplier will market 1800/week if the unit price is \$90. The supply equation is also known to be linear.
    a. Find the demand and supply equations.
    b. Plot the graphs of the supply and demand equations and find their point of intersection.
    c. Find the equilibrium quantity and price.

12. **Equilibrium Quantity and Price for Ceramic Heaters** The demand equation for the Miramar ceramic heater is $1.1x + 3.8p - 901 = 0$, where x is the quantity demanded each week and p is the wholesale unit price in dollars. The corresponding supply equation is $0.9x - 20.4p + 1038 = 0$, where x is the quantity the supplier will make available in the market when the wholesale price is p dollars each.
    a. Plot the graphs of the demand and supply equations using the same viewing window.
    b. Find the equilibrium quantity and the equilibrium price for the Miramar heaters.

1.5 The Method of Least Squares

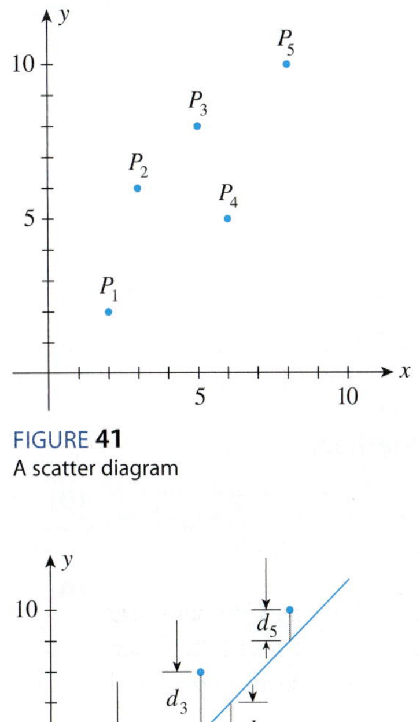

FIGURE 41
A scatter diagram

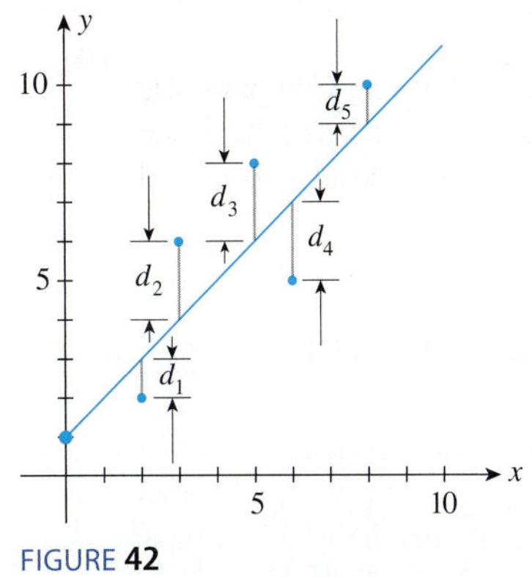

FIGURE 42
d_i is the vertical distance between the straight line and a given data point.

The Method of Least Squares

In Example 10, Section 1.2, we saw how a linear equation may be used to approximate the sales trend for a local sporting goods store. The *trend line*, as we saw, may be used to predict the store's future sales. Recall that we obtained the trend line in Example 10 by requiring that the line pass through two data points, the rationale being that such a line seems to *fit* the data reasonably well.

In this section, we describe a general method known as the **method of least squares** for determining a straight line that, in some sense, best fits a set of data points when the points are scattered about a straight line. To illustrate the principle behind the method of least squares, suppose, for simplicity, that we are given five data points,

$$P_1(x_1, y_1), P_2(x_2, y_2), P_3(x_3, y_3), P_4(x_4, y_4), P_5(x_5, y_5)$$

describing the relationship between the two variables x and y. By plotting these data points, we obtain a graph called a **scatter diagram** (Figure 41).

If we try to *fit* a straight line to these data points, the line will miss the first, second, third, fourth, and fifth data points by the amounts d_1, d_2, d_3, d_4, and d_5, respectively (Figure 42). We can think of the amounts $d_1, d_2, \ldots, d_5$ as the errors made when the values $y_1, y_2, \ldots, y_5$ are approximated by the corresponding values of y lying on the straight line L.

The **principle of least squares** states that the straight line L that fits the data points *best* is the one chosen by requiring that the sum of the squares of $d_1, d_2, \ldots, d_5$—that is,

$$d_1^2 + d_2^2 + d_3^2 + d_4^2 + d_5^2$$

be made as small as possible. In other words, the least-squares criterion calls for minimizing the sum of the squares of the errors. The line L obtained in this manner is called the **least-squares line,** or *regression line.*

PORTFOLIO Melissa Rich

TITLE Owner
INSTITUTION A Rich Experience Massage & Wellness Spa

A Rich Experience Massage & Wellness Spa takes an integrative approach to health-related services in a modern spa environment. The spa provides a variety of services including massage therapy, spa facials and body treatments, Bowenwork, hypnotherapy, and chiropractic care. A Rich Experience also sells retail products and offers continuing education workshops for massage therapists.

As an owner, I closely evaluate the company's fixed and variable revenue streams—generated from our beauty and spa services, retail sales, workshops, and rental space—to determine if I am on track to meet my profitability goals. I plot data points for all the revenue streams and apply the least squares method to see if the best fit line matches my profitability goals. This method allows me to see how increases in specific revenue streams might affect overall profitability.

Bodywork and spa services provide us with our most profitable revenue stream, so I try to maximize the number of these services that we can provide. Based on our available rooms and practitioners, we can provide 60 one-hour services per week. At an average of \$70 per service, this equals \$4200 per week or approximately \$218,440 per year. In order to gross \$109,220 and meet my goals for net profitability, after one year the spa needs to be booking at least 50% of our available time slots. My commissioned practitioners earn 40% on all services they provide, so at 50% productivity, this brings my net total for beauty and spa services to \$65,532.

Applying my college mathematics and accounting courses has helped me create a thriving business. With additional marketing efforts, name recognition, and a positive reputation in the community, I can anticipate that our bookings will continue to increase each year.

Courtesy of A Rich Experience Massage & Wellness Spa; (inset) © Szasz-Fabian Ilka Erica/ShutterStock.com

The method for computing the least-squares lines that best fits a set of data points follows. (We omit the proof.)

The Method of Least Squares

Suppose we are given n data points

$$P_1(x_1, y_1),\ P_2(x_2, y_2),\ P_3(x_3, y_3),\ \ldots,\ P_n(x_n, y_n)$$

Then the least-squares (regression) line for the data is given by the linear equation (function)

$$y = f(x) = mx + b$$

where the constants m and b satisfy the **normal equations**

$$nb + (x_1 + x_2 + \cdots + x_n)m = y_1 + y_2 + \cdots + y_n \tag{9}$$

$$(x_1 + x_2 + \cdots + x_n)b + (x_1^2 + x_2^2 + \cdots + x_n^2)m = x_1y_1 + x_2y_2 + \cdots + x_ny_n \tag{10}$$

simultaneously.

EXAMPLE 1 Find the least-squares line for the data

$$P_1(1, 1),\ P_2(2, 3),\ P_3(3, 4),\ P_4(4, 3),\ P_5(5, 6)$$

Solution Here, we have $n = 5$ and

$$\begin{array}{ccccc} x_1 = 1 & x_2 = 2 & x_3 = 3 & x_4 = 4 & x_5 = 5 \\ y_1 = 1 & y_2 = 3 & y_3 = 4 & y_4 = 3 & y_5 = 6 \end{array}$$

Before using Equations (9) and (10), it is convenient to summarize these data in the form of a table:

	x	y	x^2	xy
	1	1	1	1
	2	3	4	6
	3	4	9	12
	4	3	16	12
	5	6	25	30
Sum	15	17	55	61

Using this table and (9) and (10), we obtain the normal equations

$$5b + 15m = 17 \quad \textbf{(11)}$$
$$15b + 55m = 61 \quad \textbf{(12)}$$

Solving Equation (11) for b gives

$$b = -3m + \frac{17}{5} \quad \textbf{(13)}$$

which, upon substitution into Equation (12), gives

$$\begin{aligned} 15\left(-3m + \frac{17}{5}\right) + 55m &= 61 \\ -45m + 51 + 55m &= 61 \\ 10m &= 10 \\ m &= 1 \end{aligned}$$

Substituting this value of m into Equation (13) gives

$$b = -3 + \frac{17}{5} = \frac{2}{5} = 0.4$$

Therefore, the required least-squares line is

$$y = x + 0.4$$

The scatter diagram and the least-squares line are shown in Figure 43.

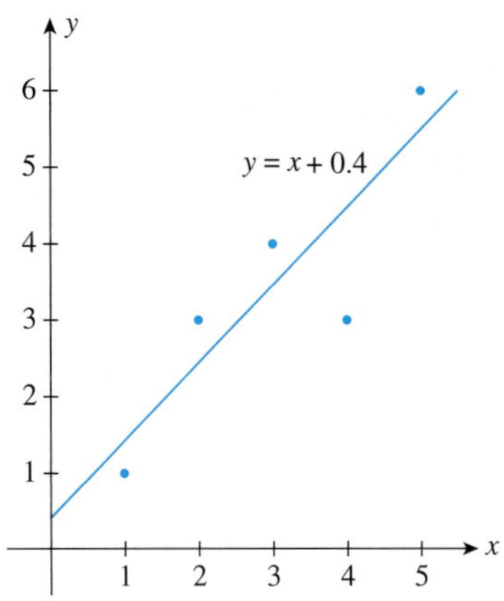

FIGURE 43
The scatter diagram and the least-squares line $y = x + 0.4$

APPLIED EXAMPLE 2 Advertising and Profit The proprietor of Leisure Travel Service compiled the following data relating the firm's annual profit to its annual advertising expenditure (both measured in thousands of dollars):

Annual Advertising Expenditure, x	12	14	17	21	26	30
Annual Profit, y	60	70	90	100	100	120

a. Determine the equation of the least-squares line for these data.
b. Draw a scatter diagram and the least-squares line for these data.
c. Use the result obtained in part (a) to predict Leisure Travel's annual profit if the annual advertising budget is $20,000.

Solution

a. The calculations required for obtaining the normal equations are summarized in the following table:

	x	y	x^2	xy
	12	60	144	720
	14	70	196	980
	17	90	289	1,530
	21	100	441	2,100
	26	100	676	2,600
	30	120	900	3,600
Sum	120	540	2646	11,530

The normal equations are

$$6b + 120m = 540 \tag{14}$$

$$120b + 2646m = 11{,}530 \tag{15}$$

Solving Equation (14) for b gives

$$b = -20m + 90 \tag{16}$$

which, upon substitution into Equation (15), gives

$$\begin{aligned} 120(-20m + 90) + 2646m &= 11{,}530 \\ -2400m + 10{,}800 + 2646m &= 11{,}530 \\ 246m &= 730 \\ m &\approx 2.97 \end{aligned}$$

Substituting this value of m into Equation (16) gives

$$b = -20(2.97) + 90 = 30.6$$

Therefore, the required least-squares line is given by

$$y = f(x) = 2.97x + 30.6$$

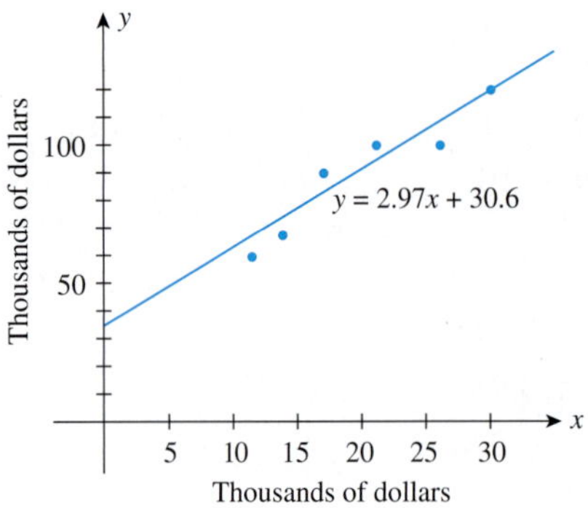

FIGURE **44**
Profit versus advertising expenditure

b. The scatter diagram and the least-squares line are shown in Figure 44.

c. Leisure Travel's predicted annual profit corresponding to an annual budget of \$20,000 is given by

$$f(20) = 2.97(20) + 30.6 = 90$$

or \$90,000.

APPLIED EXAMPLE 3 U.S. Health-Care Expenditures Refer to Example 1 of Section 1.3. Because the over-65 population will be growing more rapidly in the next few decades, health-care spending is expected to increase significantly in the coming decades. The following table gives the projected U.S. health expenditures (in trillions of dollars) from 2013 through 2018, where t is measured in years, with $t = 0$ corresponding to 2013.

Year, t	0	1	2	3	4	5
Expenditure, y	2.91	3.23	3.42	3.63	3.85	4.08

Find a function giving the U.S. health-care spending between 2013 and 2018, using the least-squares technique.
Source: Centers for Medicare & Medicaid Services.

Solution The calculations required for obtaining the normal equations are summarized in the following table:

t	y	t^2	ty
0	2.91	0	0
1	3.23	1	3.23
2	3.42	4	6.84
3	3.63	9	10.89
4	3.85	16	15.40
5	4.08	25	20.40
15	21.12	55	56.76

The normal equations are

$$6b + 15m = 21.12 \tag{17}$$
$$15b + 55m = 56.76 \tag{18}$$

Solving Equation (17) for b gives

$$6b = -15m + 21.12$$
$$b = -2.5m + 3.52 \tag{19}$$

which, upon substitution into Equation (18), gives

$$\begin{aligned} 15(-2.5m + 3.52) + 55m &= 56.76 \\ -37.5m + 52.80 + 55m &= 56.76 \\ 17.5m &= 3.96 \\ m &\approx 0.2263 \end{aligned}$$

Substituting this value of m into Equation (19) gives

$$b \approx -2.5(0.2263) + 3.52 \approx 2.954$$

Therefore, the required function is

$$S(t) = 0.226t + 2.954$$

The scatter diagram and the least-squares lines are shown in Figure 45.

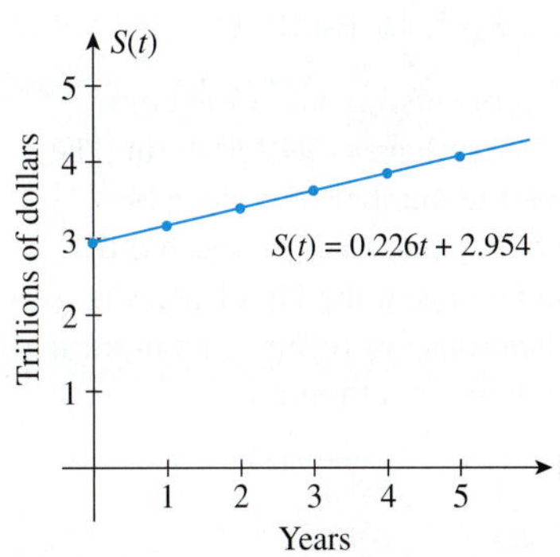

FIGURE 45

1.5 Self-Check Exercises

1. Find an equation of the least-squares line for the data

x	1	3	4	5	7
y	4	10	11	12	16

2. **Box-Office-Hit DVD Sales** In a market research study for Century Communications, the following data were provided based on the projected monthly sales x (in thousands) of a DVD version of a box-office-hit adventure movie with a proposed wholesale unit price of p dollars.

x	2.2	5.4	7.0	11.5	14.6
p	38.0	36.0	34.5	30.0	28.5

Find the demand equation if the demand curve is the least-squares line for these data.

Solutions to Self-Check Exercises 1.5 can be found on page 63.

1.5 Concept Questions

1. Explain the terms (a) *scatter diagram* and (b) *least-squares line.*

2. Explain the principle of least squares in your own words.

1.5 Exercises

In Exercises 1–6, (a) find the equation of the least-squares line for the data, and (b) draw a scatter diagram for the data and graph the least-squares line.

1.

x	1	2	3	4
y	4	6	8	11

2.

x	1	3	5	7	9
y	9	8	6	3	2

3.

x	1	2	3	4	4	6
y	4.5	5	3	2	3.5	1

4.

x	1	1	2	3	4	4	5
y	2	3	3	3.5	3.5	4	5

5. $P_1(1, 3), P_2(2, 5), P_3(3, 5), P_4(4, 7), P_5(5, 8)$

6. $P_1(1, 8), P_2(2, 6), P_3(5, 6), P_4(7, 4), P_5(10, 1)$

7. **College Admissions** The accompanying data were compiled by the admissions office at Faber College during the past 5 years. The data relate the number of college brochures and follow-up letters (x) sent to a preselected list of high school juniors who had taken the PSAT and the number of completed applications (y) received from these students (both measured in units of a thousand).

x	4	4.5	5	5.5	6
y	0.5	0.6	0.8	0.9	1.2

a. Determine the equation of the least-squares line for these data.
b. Draw a scatter diagram and the least-squares line for these data.
c. Use the result obtained in part (a) to predict the number of completed applications expected if 6400 brochures and follow-up letters are sent out during the next year.

8. **Net Sales** The management of Kaldor, a manufacturer of electric motors, submitted the accompanying data in its annual stockholders report. The following table shows the net sales (in millions of dollars) during the 5 years that have elapsed since the new management team took over:

Year, x	1	2	3	4	5
Net Sales, y	426	437	460	473	477

(The first year the firm operated under the new management corresponds to the time period $x = 1$, and the four subsequent years correspond to $x = 2, 3, 4$, and 5.)
a. Determine the equation of the least-squares line for these data.
b. Draw a scatter diagram and the least-squares line for these data.
c. Use the result obtained in part (a) to predict the net sales for the upcoming year.

9. **SAT Verbal Scores** The accompanying data were compiled by the superintendent of schools in a large metropolitan area. The table shows the average SAT verbal scores of high school seniors during the 5 years since the district implemented its "back to basics" program.

Year, x	1	2	3	4	5
Average Score, y	436	438	428	430	426

a. Determine the equation of the least-squares line for these data.
b. Draw a scatter diagram and the least-squares line for these data.
c. Assuming that the trend continued, estimate the average SAT verbal score of high school seniors 2 years from now ($x = 7$).

10. **Cost of Summer Blockbusters** Hollywood is spending more and more to produce its big summer movies each year. The estimated costs of summer big-budget releases (in billions of dollars) for the years 2011 through 2013 are given in the following table:

Year	2011	2012	2013
Spending, y	2.1	2.4	2.7

a. Letting $x = 1$ denote 2011, find an equation of the least-squares line for these data.
b. Use the result of part (a) to estimate the amount of money Hollywood will spend in 2015 to produce its big summer movies for that year, assuming the trend continues.

Source: Los Angeles Times.

11. **Facebook Users** End-of-year data for the number of Facebook users (in millions) from 2008 through 2011 are given in the following table:

Year	2008	2009	2010	2011
Number, y	154.5	381.8	654.5	845.0

a. Letting $x = 0$ denote the end of 2008, find an equation of the least-squares line for these data.
b. Use the result of part (a) to project the number of Facebook users at the end of 2015, assuming that the trend continues.

Source: Company reports.

12. **E-Book Audience** The number of adults (in millions) using e-book devices is expected to climb in the years ahead. The projected number of e-book readers in the United States from 2011 through 2015 is given in the following table:

Year	2011	2012	2013	2014	2015
Number, y	25.3	33.4	39.5	50.0	59.6

a. Letting $x = 0$ denote 2011, find an equation of the least-squares line for these data.
b. Use the result of part (a) to estimate the projected average rate of growth of the number of e-book readers between 2011 and 2015.

Source: Forrester Research, Inc.

13. **Mass-Transit Subsidies** The following table gives the projected state subsidies (in millions of dollars) to the Massachusetts Bay Transit Authority (MBTA) over a 5-year period:

Year, x	1	2	3	4	5
Subsidy, y	20	24	26	28	32

a. Find an equation of the least-squares line for these data.
b. Assuming that the trend continued, estimate the state subsidy to the MBTA for the eighth year ($x = 8$).

Source: Massachusetts Bay Transit Authority.

14. **Percentage of the Population Enrolled in School** The percentage of the population (aged 3 years or older) who were enrolled in school from 2007 through 2011 is given in the following table:

Year	2007	2008	2009	2010	2011
Percent, y	26.2	26.8	27.5	28.3	28.7

a. Letting $x = 0$ denote 2007, find an equation of the least-squares line for these data.
b. Use the result of part (a) to estimate the percentage of the population (aged 3 or older) who were enrolled in school in 2014, assuming that the trend continued.

Source: U.S. Census Bureau.

15. **Global Box-Office Receipts** Global ticket sales have been growing steadily over the years, reflecting the rapid growth in overseas markets, particularly in China. The sales (in billions of dollars) from 2007 through 2011 are summarized in the following table:

Year	2007	2008	2009	2010	2011
Sales, y	26.1	27.2	28.9	31.1	32.6

a. Find an equation of the least-squares line for these data. (Let $x = 1$ represent 2007.)
b. Use the result of part (a) to predict the global ticket sales for 2014, assuming that the trend continued.

Source: Motion Picture Association of America.

16. **Households with Someone Under 18** The percentage of households in which someone was under 18 years old from 2007 through 2011 is given in the following table:

Year	2007	2008	2009	2010	2011
Percent, y	34.4	34.1	33.4	33.1	32.7

a. Letting $x = 0$ denote 2007, find an equation of the least-squares line for these data.

b. Use the result of part (a) to estimate the percentage of households in which someone was under 18 years old in 2013, assuming that the trend continued.

Source: U.S. Census Bureau.

17. **Growth of Credit Unions** Credit union membership is on the rise. The following table gives the number (in millions) of credit union members from 2003 through 2011 in 2-year intervals:

Year	2003	2005	2007	2009	2011
Number, y	82.0	84.7	86.8	89.7	91.8

a. Letting $x = 0$ denote 2003, find an equation of the least-squares line for these data.

b. Assuming that the trend continued, estimate the number of credit union members in 2013 ($x = 5$).

Source: National Credit Union Association.

18. **First-Class Mail Volume** As more and more people turn to using the Internet and phones to pay bills and to communicate, replacing letters, the first-class mail volume is expected to decline until 2020. The following table gives the volume (in billions of pieces) of first-class mail from 2007 through 2011:

Year	2007	2008	2009	2010	2011
Value, y	95.9	91.7	83.8	78.2	73.5

a. Letting $x = 1$ denote 2007, find an equation of the least-squares line for these data.

b. Use the results of part (a) to estimate the volume of first-class mail in 2014, assuming that the trend continued through that year.

Source: U.S. Postal Service.

19. **Satellite TV Subscribers** The number of satellite and telecommunications subscribers continues to grow over the years. The following table gives the number of subscribers (in millions) from 2006 through 2010:

Year	2006	2007	2008	2009	2010
Number, y	29.4	32.2	34.8	37.7	40.4

a. Letting $x = 0$ denote 2006, find an equation of the least-squares line for these data.

b. Use the result of part (a) to estimate the average rate of growth of the number of subscribers between 2006 and 2010.

Source: SNL Ragan.

20. **Online Video Advertising** Although still a small percentage of all online advertising, online video advertising is growing. The following table gives the projected spending on Web video advertising (in billions of dollars) through 2016:

Year	2011	2012	2013	2014	2015	2016
Spending, y	2.0	3.1	4.5	6.3	7.8	9.3

a. Letting $x = 0$ denote 2011, find an equation of the least-squares line for these data.

b. Use the result of part (a) to estimate the projected rate of growth of video advertising from 2011 through 2016.

Source: eMarketer.

21. **U.S. Outdoor Advertising** U.S. outdoor advertising expenditure (in billions of dollars) from 2011 through 2015 is given in the following table ($x = 0$ corresponds to 2011):

Year	2011	2012	2013	2014	2015
Expenditure, y	6.4	6.8	7.1	7.4	7.6

a. Find an equation of the least-squares line for these data.

b. Use the result of part (a) to estimate the rate of change of the advertising expenditures for the period in question.

Source: Outdoor Advertising Association.

22. **Online Sales of Used Autos** The amount (in millions of dollars) of used autos sold online in the United States is expected to grow in accordance with the figures given in the following table ($x = 0$ corresponds to 2011):

Year, x	0	1	2	3	4
Sales, y	12.9	13.9	14.65	15.25	15.85

a. Find an equation of the least-squares line for these data.

b. Use the result of part (a) to estimate the sales of used autos online in 2016, assuming that the predicted trend continued.

Source: comScore Networks, Inc.

23. **Bounced-Check Charges** Overdraft fees have become an important piece of a bank's total fee income. The following table gives the bank revenue from overdraft fees (in billions of dollars) from 2004 through 2009. Here, $x = 4$ corresponds to 2004.

Year, x	4	5	6	7	8	9
Revenue, y	27.5	29	31	34	36	38

a. Find an equation of the least-squares line for these data.

b. Use the result of part (a) to estimate the average rate of increase in overdraft fees over the period under consideration.

c. Assuming that the trend continued, what was the revenue from overdraft fees in 2011?

Source: New York Times.

24. **Male Life Expectancy at 65** The Census Bureau projections of male life expectancy at age 65 in the United States are summarized in the following table ($x = 0$ corresponds to 2000):

Year, x	0	10	20	30	40	50
Years Beyond 65, y	15.9	16.8	17.6	18.5	19.3	20.3

a. Find an equation of the least-squares line for these data.
b. Use the result of part (a) to estimate the life expectancy at 65 of a male in 2040. How does this result compare with the given data for that year?
c. Use the result of part (a) to estimate the life expectancy at 65 of a male in 2030.
Source: U.S. Census Bureau.

25. Home Health-Care and Equipment Spending The following table gives the projected spending on home care and durable medical equipment (in billions of dollars) from 2004 through 2016 ($x = 0$ corresponds to 2004):

Year, x	0	2	4	6	8	10	12
Spending, y	60	74	90	106	118	128	150

a. Find an equation of the least-squares line for these data.
b. Use the result of part (a) to give the approximate projected spending on home care and durable medical equipment in 2015.
c. Use the result of part (a) to estimate the projected rate of change of the spending on home care and durable medical equipment for the period from 2004 through 2016.
Source: National Association of Home Care and Hospice.

26. Global Defense Spending The following table gives the projected global defense spending (in trillions of dollars) from the beginning of 2008 ($t = 0$) through 2015 ($t = 7$):

Year, t	0	1	2	3	4	5	6	7
Spending, y	1.38	1.44	1.49	1.56	1.61	1.67	1.74	1.78

a. Find an equation of the least-squares line for these data.
b. Use the result of part (a) to estimate the rate of change in the projected global defense spending from 2008 through 2015.
c. Assuming that the trend continues, what will the global spending on defense be in 2018?
Source: Homeland Security Research.

In Exercises 27–30, determine whether the statement is true or false. If it is true, explain why it is true. If it is false, give an example to show why it is false.

27. The least-squares line must pass through at least one data point.

28. The error incurred in approximating n data points using the least-squares linear function is zero if and only if the n data points lie on a nonvertical straight line.

29. If the data consist of two distinct points, then the least-squares line is just the line that passes through the two points.

30. A data point lies on the least-squares line if and only if the vertical distance between the point and the line is equal to zero.

1.5 Solutions to Self-Check Exercises

1. The calculations required for obtaining the normal equations may be summarized as follows:

	x	y	x^2	xy
	1	4	1	4
	3	10	9	30
	4	11	16	44
	5	12	25	60
	7	16	49	112
Sum	20	53	100	250

The normal equations are

$$5b + 20m = 53$$
$$20b + 100m = 250$$

Solving the first equation for b gives

$$b = -4m + \frac{53}{5}$$

which, upon substitution into the second equation, yields

$$\begin{aligned} 20\left(-4m + \frac{53}{5}\right) + 100m &= 250 \\ -80m + 212 + 100m &= 250 \\ 20m &= 38 \\ m &= 1.9 \end{aligned}$$

Substituting this value of m into the expression for b found earlier, we find

$$b = -4(1.9) + \frac{53}{5} = 3$$

Therefore, an equation of the least-squares line is

$$y = 1.9x + 3$$

2. The calculations required for obtaining the normal equations may be summarized as follows:

	x	p	x^2	xp
	2.2	38.0	4.84	83.6
	5.4	36.0	29.16	194.4
	7.0	34.5	49.00	241.5
	11.5	30.0	132.25	345.0
	14.6	28.5	213.16	416.1
Sum	40.7	167.0	428.41	1280.6

The normal equations are

$$5b + 40.7m = 167$$
$$40.7b + 428.41m = 1280.6$$

Solving this system of linear equations simultaneously, we find that

$$m \approx -0.81 \quad \text{and} \quad b \approx 40.00$$

Therefore, an equation of the least-squares line is given by

$$p = f(x) = -0.81x + 40$$

which is the required demand equation, provided that

$$0 \leq x \leq 49.38$$

USING TECHNOLOGY Finding an Equation of a Least-Squares Line

Graphing Utility

A graphing utility is especially useful in calculating an equation of the least-squares line for a set of data. We simply enter the given data in the form of lists into the calculator and then use the linear regression function to obtain the coefficients of the required equation.

EXAMPLE 1 Find an equation of the least-squares line for the data

x	1.1	2.3	3.2	4.6	5.8	6.7	8.0
y	−5.8	−5.1	−4.8	−4.4	−3.7	−3.2	−2.5

Plot the scatter diagram and the least-squares line for this data.

Solution First, we enter the data as follows:

$$x_1 = 1.1 \quad y_1 = -5.8 \quad x_2 = 2.3 \quad y_2 = -5.1 \quad x_3 = 3.2$$
$$y_3 = -4.8 \quad x_4 = 4.6 \quad y_4 = -4.4 \quad x_5 = 5.8 \quad y_5 = -3.7$$
$$x_6 = 6.7 \quad y_6 = -3.2 \quad x_7 = 8.0 \quad y_7 = -2.5$$

Then, using the linear regression function from the statistics menu, we obtain the output shown in Figure T1a. Therefore, an equation of the least-squares line ($y = ax + b$) is

$$y = 0.46x - 6.3$$

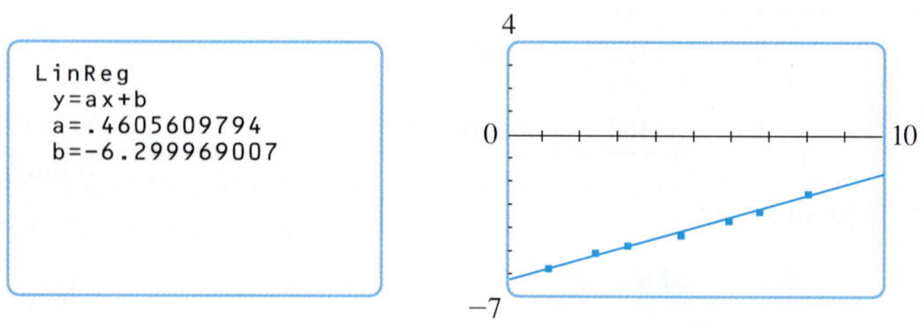

(a) The TI-83/84 linear regression screen

(b) The scatter diagram and least-squares line for the data

FIGURE T1

The graph of the least-squares equation and the scatter diagram for the data are shown in Figure T1b.

APPLIED EXAMPLE 2 Erosion of the Middle Class The idea of a large, stable, middle class (defined as those with annual household incomes in 2010 between \$39,000 and \$118,000 for a family of three), is central to America's sense of itself. The following table gives the percentage of middle-income adults (y) in the United States from 1971 through 2011.

Year	1971	1981	1991	2001	2011
Percent, y	61	59	56	54	51

Let, t be measured in decades with $t = 0$ corresponding to 1971.

a. Find an equation of the least-squares line for these data.
b. If this trend continues, what will the percentage of middle-income adults be in 2021?

Source: Pew Research Center.

Solution

a. First we enter the data as follows:

$$x_1 = 0 \quad y_1 = 61 \quad x_2 = 1 \quad y_2 = 59 \quad x_3 = 2$$
$$y_3 = 56 \quad x_4 = 3 \quad y_4 = 54 \quad x_5 = 4 \quad y_5 = 51$$

Then, using the linear regression function from the statistics menu, we obtain the output shown in Figure T2. Therefore, an equation of this least-squares line is

$$y = -2.5t + 61.2$$

```
LinReg
 y=ax+b
 a=-2.5
 b=61.2
```

FIGURE **T2**
The TI-83/84 linear regression screen

b. The percentage of middle-income adults in 2021 will be

$$y = -(2.5)(5) + 61.2 = 48.7$$

or approximately 48.7%.

Excel

Excel can be used to find an equation of the least-squares line for a set of data and to plot a scatter diagram and the least-squares line for the data.

EXAMPLE 3 Find an equation of the least-squares line for the data given in the following table:

x	1.1	2.3	3.2	4.6	5.8	6.7	8.0
y	−5.8	−5.1	−4.8	−4.4	−3.7	−3.2	−2.5

Plot the scatter diagram and the least-squares line for these data.

Solution

	A	B
1	x	y
2	1.1	-5.8
3	2.3	-5.1
4	3.2	-4.8
5	4.6	-4.4
6	5.8	-3.7
7	6.7	-3.2
8	8	-2.5

FIGURE **T3**
Table of values for x and y

1. *Set up a table of values in two columns on a spreadsheet* (Figure T3).
2. *Plot the scatter diagram.* Highlight the numerical values in the table of values. Follow the procedure given in Example 3, page 26, selecting the first chart subtype instead of the second from the Scatter chart type. The scatter diagram will appear.

Note: Boldfaced words/characters enclosed in a box (for example, **Enter**) indicate that an action (click, select, or press) is required. Words/characters printed blue (for example, Chart sub-type:) indicate words/characters that appear on the screen. Words/characters printed in a monospace font (for example, `= (−2/3)*A2+2`) indicate words/characters that need to be typed and entered.

3. *Insert the least-squares line.* Select the **Layout** tab, and click on **Trendline** in the Analysis group. Next, click on **More Trendline Options...** in the same subgroup. In the Format Trendline dialog box that appears, click on **Display Equation on chart**.

$$y = 0.4606x - 6.3$$

and the least-squares line will appear on the chart (Figure T4).

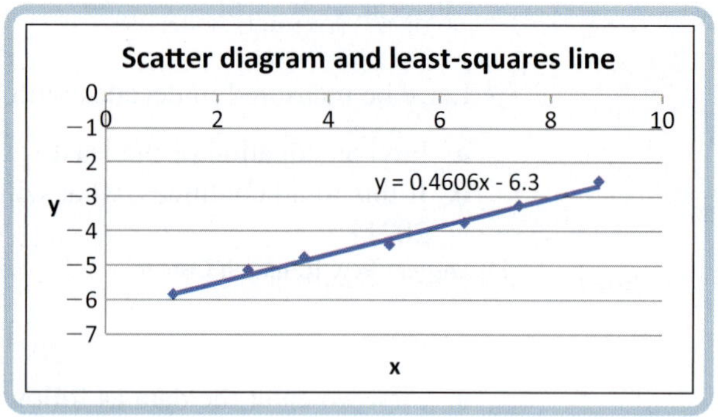

FIGURE T4
Scatter diagram and least-squares line for the given data

The following example requires the Analysis ToolPak. Use Excel's Help function to learn how to install this add-in.

APPLIED EXAMPLE 4 Erosion of the Middle Class The idea of a large, stable, middle class (defined as those with annual household incomes in 2010 between \$39,000 and \$118,000 for a family of three), is central to America's sense of itself. The following table gives the percentage of middle-income adults (y) in the United States from 1971 through 2011.

Year	1971	1981	1991	2001	2011
Percent, y	61	59	56	54	51

Let t be measured in decades with $t = 0$ corresponding to 1971.

a. Find an equation of the least-squares line for these data.
b. If this trend continues, what will the percentage of middle-income adults be in 2021?

Source: Pew Research Center.

Solution

	A	B
1	t	y
2	0	61
3	1	59
4	2	56
5	3	54
6	4	51

FIGURE T5
Table of values for t and y

a. *Set up a table of values on a spreadsheet* (Figure T5).
Find the equation of the least-squares line for the data. Click on **Data Analysis** in the Analysis group of the Data tab. In the Data Analysis dialog box that appears, select **Regression**, and then click **OK**. In the Regression dialog box that appears, select the **Input Y Range:** box, and then enter the y-values by highlighting cells B2:B6. Next select the **Input X Range:** box, and enter the x-values by highlighting cells A2:A6. Click **OK**, and a SUMMARY OUTPUT worksheet

	Coefficients
Intercept	61.2
X Variable 1	−2.5

FIGURE **T6**
Entries in the SUMMARY OUTPUT box

will appear. In the third table, you will see the entries shown in Figure T6. These entries give the value of the y-intercept and the coefficient of x in the equation $y = mx + b$. In our example, we are using the variable t instead of x, so the required equation is

$$y = -2.5t + 61.2$$

b. The percentage of middle-income adults in 2021 will be

$$y = -(2.5)(5) + 61.2 = 48.7$$

or approximately 48.7%.

TECHNOLOGY EXERCISES

In Exercises 1–4, find an equation of the least-squares line for the given data.

1.

x	2.1	3.4	4.7	5.6	6.8	7.2
y	8.8	12.1	14.8	16.9	19.8	21.1

2.

x	1.1	2.4	3.2	4.7	5.6	7.2
y	−0.5	1.2	2.4	4.4	5.7	8.1

3.

x	−2.1	−1.1	0.1	1.4	2.5	4.2	5.1
y	6.2	4.7	3.5	1.9	0.4	−1.4	−2.5

4.

x	−1.12	0.1	1.24	2.76	4.21	6.82
y	7.61	4.9	2.74	−0.47	−3.51	−8.94

5. Modeling with Data The following table gives the projected worldwide consulting spending (in billions of dollars) from 2005 through 2009 ($x = 5$ corresponds to 2005):

Year, x	5	6	7	8	9
Spending, y	254	279	300	320	345

a. Find an equation of the least-squares line for these data.
b. Use the results of part (a) to estimate the average rate of increase of worldwide consulting spending over the period under consideration.
c. Use the results of part (a) to estimate the amount of spending in 2010, assuming that the trend continued.

Source: Kennedy Information.

6. Modeling with Data Moody's Corporation is the holding company for Moody's Investors Service, which has a 40% share in the world credit-rating market. According to *Company Reports,* the total revenue (in billions of dollars) of the company is projected to be as follows ($x = 4$ corresponds to 2004):

Year	2004	2005	2006	2007	2008
Revenue, y	1.42	1.73	1.98	2.32	2.65

a. Find an equation of the least-squares line for these data.
b. Use the results of part (a) to estimate the rate of change of the revenue of the company for the period in question.
c. Use the result of part (a) to estimate the total revenue of the company in 2010, assuming that the trend continued.

Source: Company Reports.

7. Modeling with Data As online attacks persist, spending on information security software continues to rise. The following table gives the forecast for the worldwide sales (in billions of dollars) of information security software through 2007 ($t = 0$ corresponds to 2002):

Year, t	0	1	2	3	4	5
Spending, y	6.8	8.3	9.8	11.3	12.8	14.9

a. Find an equation of the least-squares line for these data.
b. Use the result of part (a) to estimate the spending on information security software in 2008, assuming that the trend continued.

Source: International Data Corporation.

8. Modeling with Data The following table gives the projected U.S. online banking households as a percentage of all U.S. banking households from 2001 ($x = 1$) through 2007 ($x = 7$):

Year, x	1	2	3	4	5	6	7
Percent, y	21.2	26.7	32.2	37.7	43.2	48.7	54.2

a. Find an equation of the least-squares line for these data.
b. Use the result of part (a) to estimate the projected percentage of U.S. online banking households in 2010.

Source: Jupiter Research.

9. **Modeling with Data** The annual sales (in billions of dollars) of global positioning system (GPS) equipment from the year 2000 ($x = 0$) through 2006 are given in the following table:

Year, x	0	1	2	3	4	5	6
Annual Sales, y	7.9	9.6	11.5	13.3	15.2	16.0	18.8

a. Find an equation of the least-squares line for these data.
b. Use the equation found in part (a) to estimate the annual sales of GPS equipment for 2008, assuming that the trend continued.

Source: ABI Research.

10. **Modeling with Data** The convenience of shopping on the Web combined with high-speed broadband access services is spurring online spending. The projected online spending per buyer (in dollars) from 2002 ($x = 0$) through 2008 ($x = 6$) is given in the following table:

Year, x	0	1	2	3	4	5	6
Spending, y	501	540	585	631	680	728	779

a. Find an equation of the least-squares line for these data.
b. Use the result of part (a) to estimate the rate of change of spending per buyer between 2002 and 2008.

Source: U.S. Department of Commerce.

CHAPTER 1 Summary of Principal Formulas and Terms

FORMULAS

1. Distance between two points	$d = \sqrt{(x_2 - x_1)^2 + (y_2 - y_1)^2}$
2. Equation of a circle	$(x - h)^2 + (y - k)^2 = r^2$
3. Slope of a nonvertical line	$m = \dfrac{y_2 - y_1}{x_2 - x_1}$
4. Equation of a vertical line	$x = a$
5. Equation of a horizontal line	$y = b$
6. Point-slope form of the equation of a line	$y - y_1 = m(x - x_1)$
7. Slope-intercept form of the equation of a line	$y = mx + b$
8. General equation of a line	$Ax + By + C = 0$

TERMS

Cartesian coordinate system (2)
ordered pair (2)
coordinates (3)
parallel lines (13)
perpendicular lines (15)
function (30)
independent variable (30)
dependent variable (30)
domain (30)
range (30)
linear function (31)
total cost function (33)
revenue function (33)
profit function (33)
demand function (34)
supply function (35)
break-even point (44)
market equilibrium (47)
equilibrium quantity (47)
equilibrium price (47)

CHAPTER 1 Concept Review Questions

Fill in the blanks.

1. A point in the plane can be represented uniquely by a/an ______ pair of numbers. The first number of the pair is called the ______, and the second number of the pair is called the ______.

2. a. The point $P(a, 0)$ lies on the ______ axis, and the point $P(0, b)$ lies on the ______ axis.
   b. If the point $P(a, b)$ lies in the fourth quadrant, then the point $P(-a, b)$ lies in the ______ quadrant.

3. The distance between two points $P(a, b)$ and $P(c, d)$ is ______.

4. An equation of a circle with center $C(a, b)$ and radius r is given by ______.

5. **a.** If $P_1(x_1, y_1)$ and $P_2(x_2, y_2)$ are any two distinct points on a nonvertical line L, then the slope of L is $m =$ ______.
   b. The slope of a vertical line is ______.
   c. The slope of a horizontal line is ______.
   d. The slope of a line that slants upward is ______.

6. If L_1 and L_2 are distinct nonvertical lines with slopes m_1 and m_2, respectively, then: L_1 is parallel to L_2 if and only if ______; and L_1 is perpendicular to L_2 if and only if ______.

7. **a.** An equation of the line passing through the point $P(x_1, y_1)$ and having slope m is ______. It is called the ______ form of an equation of a line.
   b. An equation of the line that has slope m and y-intercept b is ______. It is called the ______ form of an equation of a line.

8. **a.** The general form of an equation of a line is ______.
   b. If a line has equation $ax + by + c = 0$ $(b \neq 0)$, then its slope is ______.

9. A linear function is a function of the form $f(x) =$ ______.

10. **a.** A demand function expresses the relationship between the unit ______ and the quantity ______ of a commodity. The graph of the demand function is called the ______ curve.
    b. A supply function expresses the relationship between the unit ______ and the quantity ______ of a commodity. The graph of the supply function is called the ______ curve.

11. If $R(x)$ and $C(x)$ denote the total revenue and the total cost incurred in manufacturing x units of a commodity, then the solution of the simultaneous equations $y = C(x)$ and $y = R(x)$ gives the ______ point.

12. The equilibrium quantity and the equilibrium price are found by solving the system composed of the ______ equation and the ______ equation.

CHAPTER 1 Review Exercises

In Exercises 1–4, find the distance between the two points.

1. $(2, 1)$ and $(6, 4)$
2. $(9, 6)$ and $(6, 2)$
3. $(-2, -3)$ and $(1, -7)$
4. $\left(\frac{1}{2}, \sqrt{3}\right)$ and $\left(-\frac{1}{2}, 2\sqrt{3}\right)$

5. Does the point $P(-1, -\frac{5}{4})$ lie on the line $6x - 8y - 16 = 0$? Justify your answer.

In Exercises 6–11, find an equation of the line L that passes through the point $(-2, 4)$ and satisfies the given condition.

6. L is a vertical line.
7. L is a horizontal line.
8. L passes through the point $(3, \frac{7}{2})$.
9. The x-intercept of L is 3.
10. L is parallel to the line $5x - 2y = 6$.
11. L is perpendicular to the line $4x + 3y = 6$.

12. Find an equation of the line with slope $-\frac{1}{2}$ and y-intercept -3.

13. Find the slope and y-intercept of the line with equation $3x - 5y = 6$.

14. Find an equation of the line passing through the point $(2, 3)$ and parallel to the line with equation $3x + 4y - 8 = 0$.

15. Find an equation of the line passing through the point $(-1, 3)$ and parallel to the line joining the points $(-3, 4)$ and $(2, 1)$.

16. Find an equation of the line passing through the point $(-2, -4)$ that is perpendicular to the line with equation $2x - 3y - 24 = 0$.

17. Given that the point $P(2, -4)$ lies on the line $2x + ky = -8$, find k.

18. Find the constants m and b in the linear function $f(x) = mx + b$ such that $f(1) = 3$ and $f(3) = -2$.

In Exercises 19 and 20, sketch the graph of the equation.

19. $3x - 4y = 24$
20. $-2x + 5y = 15$

21. **Sales of MP3 Players** Sales of a certain brand of MP3 players are approximated by the relationship

$$S(x) = 6000x + 30{,}000 \qquad (0 \leq x \leq 5)$$

where $S(x)$ denotes the number of MP3 players sold in year x ($x = 0$ corresponds to the year 2010). Find the expected number of MP3 players to be sold in 2015.

22. **Company Sales** A company's total sales (in millions of dollars) are approximately linear as a function of time (in years). Sales in 2010 were \$2.4 million, whereas sales in 2015 amounted to \$7.4 million.
    a. Letting $x = 0$ correspond to 2010, find a function giving the company's sales in terms of x.
    b. What were the sales in 2013?

23. Show that the triangle with vertices $A(1, 1)$, $B(5, 3)$, and $C(4, 5)$ is a right triangle.

24. Clark's Rule Clark's Rule is a method for calculating pediatric drug dosages based on a child's weight. If a denotes the adult dosage (in milligrams) and if w is the child's weight (in pounds), then the child's dosage is given by

$$D(w) = \frac{aw}{150}$$

a. Show that D is a linear function of w.
b. If the adult dose of a substance is 500 mg, how much should a 35-lb child receive?

25. Linear Depreciation of an Office Building An office building worth \$6 million when it was completed in 2013 is being depreciated linearly over 30 years with a scrap value of \$0.
a. What is the rate of depreciation?
b. What will be the book value of the building in 2023?

26. Linear Depreciation of Construction Machinery In 2010 a home builder purchased construction machinery at a cost of \$300,000. The machinery is depreciated linearly over 12 years with a scrap value of \$30,000.
a. What is the rate of depreciation of the machinery per year?
b. Find an expression for the book value of the machinery in year t $(0 \leq t \leq 12)$.

27. Profit Function for Disposable Cameras A company has a fixed cost of \$30,000 and a production cost of \$6 for each disposable camera it manufactures. Each camera sells for \$10.
a. What is the cost function?
b. What is the revenue function?
c. What is the profit function?
d. Compute the profit (loss) corresponding to production levels of 6000, 8000, and 12,000 units, respectively.

28. Senior Population The percentage of households in which someone is 65 years of age or older was approximately 23.4% in 2007 and grew to approximately 25.2% in 2011.
a. Assuming that the growth was linear, find a function $f(t)$ giving the percentage of households where someone is 65 years of age or older in year t, where $t = 0$ corresponds to 2007.
b. If this trend continues, estimate the percentage of households in which someone is 65 years of age or older in 2013.

Source: U.S. Census Bureau.

29. Cyber Monday Sales The amount (in millions of dollars) spent on Cyber Monday for the years 2009 through 2011 is given in the following table:

Year	2009	2010	2011
y	887	1028	1251

a. Plot the Cyber Monday sales (y) versus the year (t), where $t = 0$ corresponds to 2009.
b. Draw a straight line L through the points $(0, 887)$ and $(2, 1251)$.
c. Derive an equation of the line L.
d. Assuming that the trend continues, use the equation found in part (c) to estimate the amount consumers will spend on Cyber Monday in 2014.

Source: Comscore.

30. Demand Equation for a Commodity There is no demand for a certain commodity when the unit price is \$200 or more, but the demand increases by 200 units for each \$10 decrease in price below \$200. Find the demand equation and sketch its graph.

31. Supply Equation for Bicycles Bicycle suppliers will make 200 bicycles available in the market per month when the unit price is \$50 and 2000 bicycles available per month when the unit price is \$100. Find the supply equation if it is known to be linear.

32. Demand for Earbud Headphones The demand equation for the Primo earbud headphone is

$$p = -0.02x + 40$$

where x is the quantity demanded per week and p is the unit price in dollars.
a. Sketch the graph of the demand curve.
b. What is the highest price (theoretically) anyone would pay for a Primo earbud headphone?
c. What is the quantity demanded per week when the unit price is \$20?

33. Supply of Earbud Headphones The supply equation for the Primo earbud headphone is

$$p = 0.04x + 10$$

where x is the quantity supplied per week and p is the unit price in dollars.
a. Sketch the graph of the supply curve.
b. What is the lowest price (theoretically) at which the supplier will make any headphones available in the market?
c. How many headphones will the supplier make available in the market when the unit price is \$20?

In Exercises 34 and 35, find the point of intersection of the lines with the given equations.

34. $3x + 4y = -6$ and $2x + 5y = -11$

35. $y = \frac{3}{4}x + 6$ and $3x - 2y + 3 = 0$

36. **Break-Even Analysis** The cost function and the revenue function for a certain firm are given by $C(x) = 12x + 20{,}000$ and $R(x) = 20x$, respectively. Find the break-even point for the company.

37. **Market Equilibrium** Given the demand equation $3x + p - 40 = 0$ and the supply equation $2x - p + 10 = 0$, where p is the unit price in dollars and x represents the quantity demanded in units of a thousand, determine the equilibrium quantity and the equilibrium price.

38. **College Admissions** The accompanying data were compiled by the Admissions Office of Carter College during the past 5 years. The data relate the number of college brochures and follow-up letters (x) sent to a preselected list of high school juniors who took the PSAT and the number of completed applications (y) received from these students (both measured in thousands).

Brochures Sent, x	1.8	2	3.2	4	4.8
Applications Completed, y	0.4	0.5	0.7	1	1.3

a. Derive an equation of the straight line L that passes through the points $(2, 0.5)$ and $(4, 1)$.
b. Use this equation to predict the number of completed applications that might be expected if 6400 brochures and follow-up letters are sent out during the next year.

39. **Market Equilibrium for Refrigerators** The demand equation for the Cold Spot compact refrigerator is $2x + 7p - 1760 = 0$, where x is the quantity demanded each week and p is the unit price in dollars. The supply equation for these refrigerators is $3x - 56p + 2680 = 0$, where x is the quantity the supplier will make available in the market when the wholesale price is p dollars each. Find the equilibrium quantity and the equilibrium price for the Cold Spot compact refrigerators.

40. **Social Security Wage Base** The Social Security (FICA) wage base (in thousands of dollars) from 2004 to 2009 is given in the accompanying table ($x = 1$ corresponds to 2004):

Year	2004	2005	2006	2007	2008	2009
Expenditure, y	87.9	90.0	94.2	97.5	102.6	106.8

a. Find an equation of the least-squares line for these data.
b. Use the result of part (a) to estimate the FICA wage base in 2012.

Source: The World Almanac.

41. **Equilibrium Quantity and Price for Earbud Headphones** The demand equation for the Primo earbud headphone is

$$p = -0.02x + 40$$

where x is the quantity demanded per week and p is the unit price in dollars. The corresponding supply equation is given by

$$p = 0.04x + 10$$

where x is the quantity supplied per month and p is the unit price in dollars. Find the equilibrium quantity and the equilibrium price for the Primo earbud headphones.

42. **Female Life Expectancy** The Census Bureau projections of female life expectancy at age 65 in the United States are summarized in the following table ($x = 0$ corresponds to 2000):

Year, x	0	10	20	30	40	50
Years Beyond 65, y	19.5	20.0	20.6	21.2	21.8	22.4

a. Find an equation of the least-squares line for these data.
b. Use the result of part (a) to estimate the life expectancy at 65 of a female in 2040. How does this result compare with the given data for that year?
c. Use the result of part (a) to estimate the life expectancy at 65 of a female in 2030.

Source: U.S. Census Bureau.

The problem-solving skills that you learn in each chapter are building blocks for the rest of the course. Therefore, it is a good idea to make sure that you have mastered these skills before moving on to the next chapter. The Before Moving On exercises that follow are designed for that purpose. After completing these exercises, you can identify the skills that you should review before starting the next chapter.

CHAPTER 1 Before Moving On . . .

1. Plot the points $A(-2, 1)$ and $B(3, 4)$ on the same set of axes, and find the distance between A and B.

2. Find an equation of the line passing through the point $(3, 1)$ and parallel to the line $3x - y - 4 = 0$.

3. Let L be the line passing through the points $(1, 2)$ and $(3, 5)$. Is L perpendicular to the line $2x + 3y = 10$?

4. The monthly total revenue function and total cost function for a company are $R(x) = 18x$ and $C(x) = 15x + 22{,}000$, respectively, where x is the

number of units produced and both $R(x)$ and $C(x)$ are measured in dollars.

a. What is the unit cost for producing the product?
b. What is the monthly fixed cost for the company?
c. What is the selling price for each unit of the product?

5. Find the point of intersection of the lines $2x - 3y = -2$ and $9x + 12y = 25$.

6. **Sales Comparisons** The annual sales of Best Furniture Store are expected to be given by $S_1 = 4.2 + 0.4t$ million dollars t years from now, whereas the annual sales of Lowe's Furniture Store are expected to be given by $S_2 = 2.2 + 0.8t$ million dollars t years from now. When will Lowe's annual sales first surpass Best's annual sales?

2 Systems of Linear Equations and Matrices

THE LINEAR EQUATIONS in two variables that we studied in Chapter 1 are readily extended to cases involving more than two variables. For example, a linear equation in three variables represents a plane in three-dimensional space. In this chapter, we see how some real-world problems can be formulated in terms of systems of linear equations, and we develop two methods for solving these equations.

In addition, we see how *matrices* (rectangular arrays of numbers) can be used to write systems of linear equations in compact form. We then go on to consider some real-life applications of matrices. Finally, we show how matrices can be used to describe the Leontief input–output model, an important tool used by economists. For his work in formulating this model, Wassily Leontief was awarded the Nobel Prize in 1973.

Checkers Rent-A-Car is planning to expand its fleet of cars next quarter. How should the company use its budget of $18 million to meet the expected additional demand for compact and full-size cars? In Example 5, page 147, we will see how we can find the solution to this problem by solving a system of equations.

2.1 Systems of Linear Equations: An Introduction

Systems of Equations

Recall that in Section 1.4, we had to solve two simultaneous linear equations to find the *break-even point* and the *equilibrium point*. These are two examples of real-world problems that call for the solution of a **system of linear equations** in two or more variables. In this chapter, we take up a more systematic study of such systems.

We begin by considering a system of two linear equations in two variables. Recall that such a system may be written in the general form

$$\begin{aligned} ax + by &= h \\ cx + dy &= k \end{aligned} \tag{1}$$

where a, b, c, d, h, and k are real constants and neither a and b nor c and d are both zero.

Now let's study the nature of the **solution of a system of linear equations** in more detail. Recall that the graph of each equation in System (1) is a straight line in the plane, so geometrically, the solution to the system is the point(s) of intersection of the two straight lines L_1 and L_2, represented by the first and second equations of the system.

Given two lines L_1 and L_2, *one and only one* of the following may occur:

a. L_1 and L_2 intersect at exactly one point.
b. L_1 and L_2 are parallel and coincident.
c. L_1 and L_2 are parallel and distinct.

(See Figure 1.) In the first case, the system has a unique solution corresponding to the single point of intersection of the two lines. In the second case, the system has infinitely many solutions corresponding to the points lying on the same line. Finally, in the third case, the system has no solution because the two lines do not intersect.

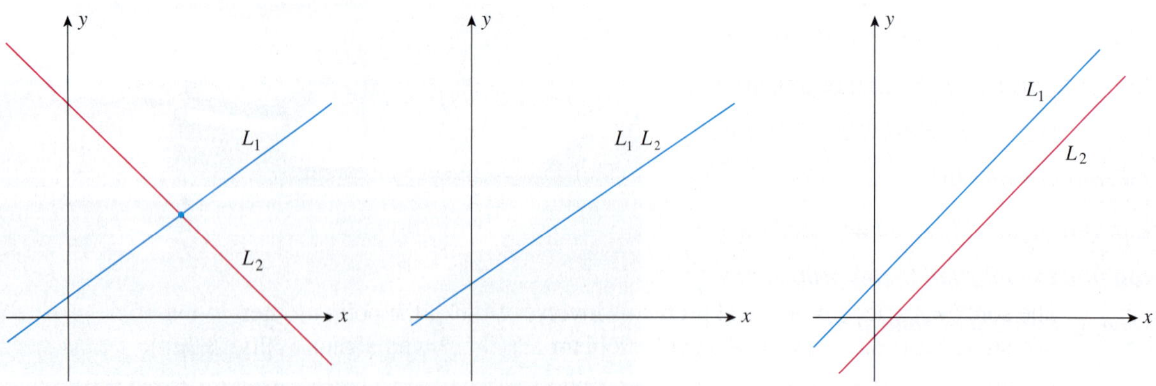

(a) Unique solution **(b)** Infinitely many solutions **(c)** No solution

FIGURE **1**

> *Explore and Discuss*
>
> Generalize the discussion on this page to the case in which there are three straight lines in the plane defined by three linear equations. What if there are n lines defined by n equations?

Let's illustrate each of these possibilities by considering some specific examples.

1. A system of equations with exactly one solution Consider the system

$$\begin{aligned} 2x - y &= 1 \\ 3x + 2y &= 12 \end{aligned}$$

Solving the first equation for y in terms of x, we obtain the equation

$$y = 2x - 1$$

Substituting this expression for y into the second equation yields

$$\begin{aligned} 3x + 2(2x - 1) &= 12 \\ 3x + 4x - 2 &= 12 \\ 7x &= 14 \\ x &= 2 \end{aligned}$$

Finally, substituting this value of x into the expression for y obtained earlier gives

$$y = 2(2) - 1 = 3$$

Therefore, the unique solution of the system is given by $x = 2$ and $y = 3$. Geometrically, the two lines represented by the two linear equations that make up the system intersect at the point $(2, 3)$ (Figure 2).

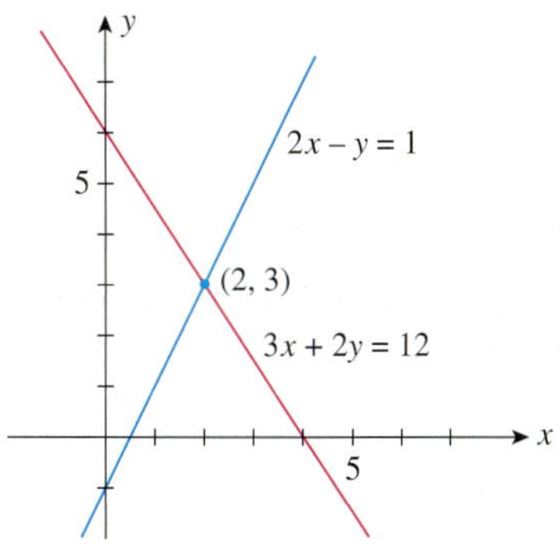

FIGURE 2
A system of equations with one solution

Note We can check our result by substituting the values $x = 2$ and $y = 3$ into the equations. Thus,

$$\begin{aligned} 2(2) - (3) &= 1 \ \checkmark \\ 3(2) + 2(3) &= 12 \ \checkmark \end{aligned}$$

From the geometric point of view, we have just verified that the point $(2, 3)$ lies on both lines.

2. A system of equations with infinitely many solutions Consider the system

$$\begin{aligned} 2x - y &= 1 \\ 6x - 3y &= 3 \end{aligned}$$

Solving the first equation for y in terms of x, we obtain the equation

$$y = 2x - 1$$

Substituting this expression for y into the second equation gives

$$\begin{aligned} 6x - 3(2x - 1) &= 3 \\ 6x - 6x + 3 &= 3 \\ 0 &= 0 \end{aligned}$$

which is a true statement. This result follows from the fact that the second equation is equivalent to the first. (To see this, just multiply both sides of the first equation by 3.) Our computations have revealed that the system of two equations is equivalent to the single equation $2x - y = 1$. Thus, any ordered pair of numbers (x, y) satisfying the equation $2x - y = 1$ (or $y = 2x - 1$) constitutes a solution to the system.

In particular, by assigning the value t to x, where t is any real number, we find that $y = 2t - 1$, so the ordered pair $(t, 2t - 1)$ is a solution of the system. The variable t is called a **parameter.** For example, setting $t = 0$ gives the point $(0, -1)$ as a solution of the system, and setting $t = 1$ gives the point $(1, 1)$ as another solution. Since t represents any real number, there are infinitely many solutions of the

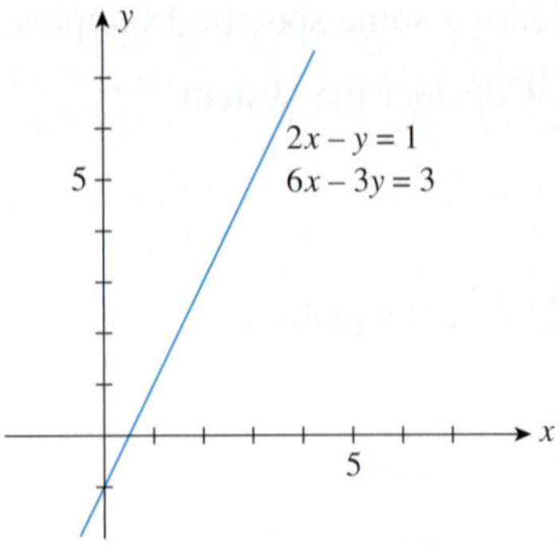

FIGURE 3
A system of equations with infinitely many solutions; each point on the line is a solution.

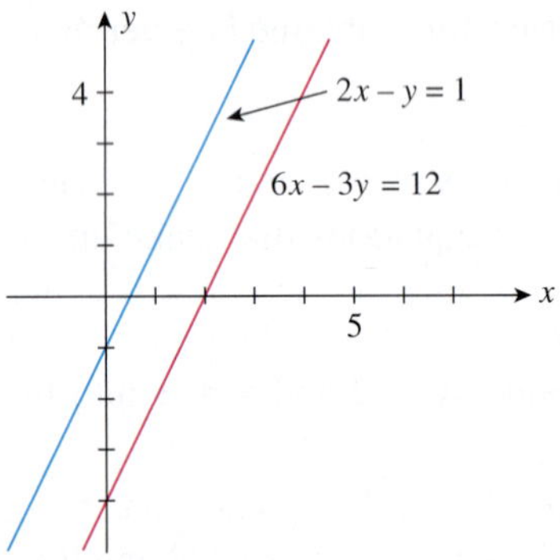

FIGURE 4
A system of equations with no solution

system. Geometrically, the two equations in the system represent the same line, and all solutions of the system are points lying on the line (Figure 3). Such a system is said to be **dependent.**

3. **A system of equations that has no solution** Consider the system

$$\begin{aligned} 2x - y &= 1 \\ 6x - 3y &= 12 \end{aligned}$$

The first equation is equivalent to $y = 2x - 1$. Substituting this expression for y into the second equation gives

$$\begin{aligned} 6x - 3(2x - 1) &= 12 \\ 6x - 6x + 3 &= 12 \\ 0 &= 9 \end{aligned}$$

which is clearly impossible. Thus, there is no solution to the system of equations. To interpret this situation geometrically, cast both equations in the slope-intercept form, obtaining

$$\begin{aligned} y &= 2x - 1 \\ y &= 2x - 4 \end{aligned}$$

We see at once that the lines represented by these equations are parallel (each has slope 2) and distinct, since the first has y-intercept -1 and the second has y-intercept -4 (Figure 4). Systems with no solutions, such as this one, are said to be **inconsistent.**

Explore and Discuss

1. Consider a system composed of two linear equations in two variables. Can the system have exactly two solutions? Exactly three solutions? Exactly a finite number of solutions?
2. Suppose at least one of the equations in a system composed of two equations in two variables is nonlinear. Can the system have no solution? Exactly one solution? Exactly two solutions? Exactly a finite number of solutions? Infinitely many solutions? Illustrate each answer with a sketch.

Note We have used the method of substitution in solving each of these systems. If you are familiar with the method of elimination, you might want to re-solve each of these systems using this method. We will study the method of elimination in detail in Section 2.2.

In Section 1.4, we presented some real-world applications of systems involving two linear equations in two variables. Here is an example involving a system of three linear equations in three variables.

APPLIED EXAMPLE 1 Production Scheduling Ace Novelty wishes to produce three types of souvenirs: Types A, B, and C. To manufacture a Type A souvenir requires 2 minutes on Machine I, 1 minute on Machine II, and 2 minutes on Machine III. A Type B souvenir requires 1 minute on Machine I, 3 minutes on Machine II, and 1 minute on Machine III. A Type C souvenir requires 1 minute on Machine I and 2 minutes each on Machines II and III. There are 3 hours available on Machine I, 5 hours available on Machine II, and 4 hours available on Machine III for processing the order. How many souvenirs of each type should Ace Novelty make in order to use all of the available time? Formulate but do not solve the problem. (We will solve this problem in Example 7, Section 2.2.)

Solution The given information may be tabulated as follows:

	Type *A*	Type *B*	Type *C*	Time Available (min)
Machine I	2	1	1	180
Machine II	1	3	2	300
Machine III	2	1	2	240

We have to determine the number of each of *three* types of souvenirs to be made. So let x, y, and z denote the respective numbers of Type *A*, Type *B*, and Type *C* souvenirs to be made. The total amount of time that Machine I is used is given by $2x + y + z$ minutes and must equal 180 minutes. This leads to the equation

$$2x + y + z = 180 \quad \text{Time spent on Machine I}$$

Similar considerations on the use of Machines II and III lead to the following equations:

$$\begin{aligned} x + 3y + 2z &= 300 &\quad \text{Time spent on Machine II} \\ 2x + y + 2z &= 240 &\quad \text{Time spent on Machine III} \end{aligned}$$

Since the variables x, y, and z must satisfy simultaneously the three conditions represented by the three equations, the solution to the problem is found by solving the following system of linear equations:

$$\begin{aligned} 2x + y + z &= 180 \\ x + 3y + 2z &= 300 \\ 2x + y + 2z &= 240 \end{aligned}$$

Solutions of Systems of Equations

We will complete the solution of the problem posed in Example 1 later on (page 92). For the moment, let's look at the geometric interpretation of a system of linear equations, such as the system in Example 1, to gain some insight into the nature of the solution.

A linear system composed of three linear equations in three variables x, y, and z has the general form

$$\begin{aligned} a_1x + b_1y + c_1z &= d_1 \\ a_2x + b_2y + c_2z &= d_2 \\ a_3x + b_3y + c_3z &= d_3 \end{aligned} \tag{2}$$

Just as a linear equation in two variables represents a straight line in the plane, it can be shown that a linear equation $ax + by + cz = d$ (a, b, and c not all equal to zero) in three variables represents a plane in three-dimensional space. Thus, each equation in System (2) represents a *plane* in three-dimensional space, and the *solution(s) of the system* is precisely the point(s) of intersection of the three planes defined by the three linear equations that make up the system. As before, the system has one and only one solution, infinitely many solutions, or no solution, depending on whether and how the planes intersect one another. Figure 5 illustrates each of these possibilities.

In Figure 5a, the three planes intersect at a point corresponding to the situation in which System (2) has a unique solution. Figure 5b depicts a situation in which there are infinitely many solutions to the system. Here, the three planes intersect along a line, and the solutions are represented by the infinitely many points lying on this line. In Figure 5c, the three planes are parallel and distinct, so there is no point common to all three planes; System (2) has no solution in this case.

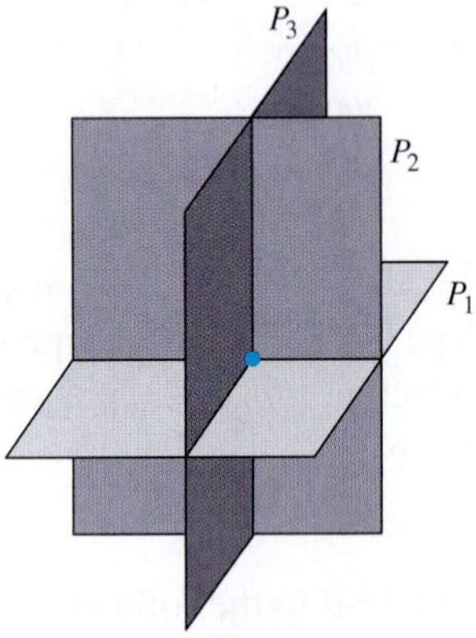

(a) A unique solution

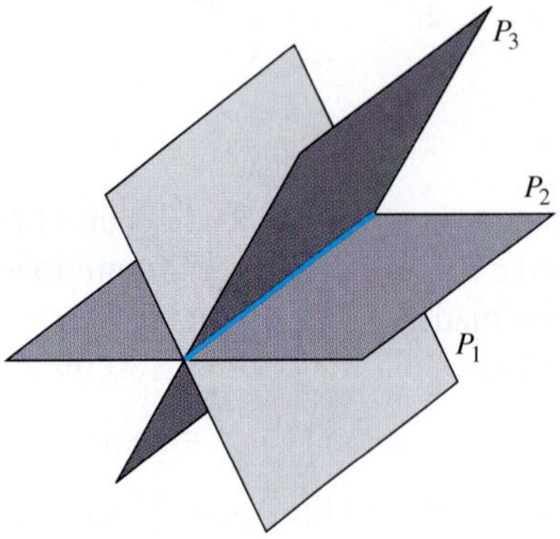

(b) Infinitely many solutions

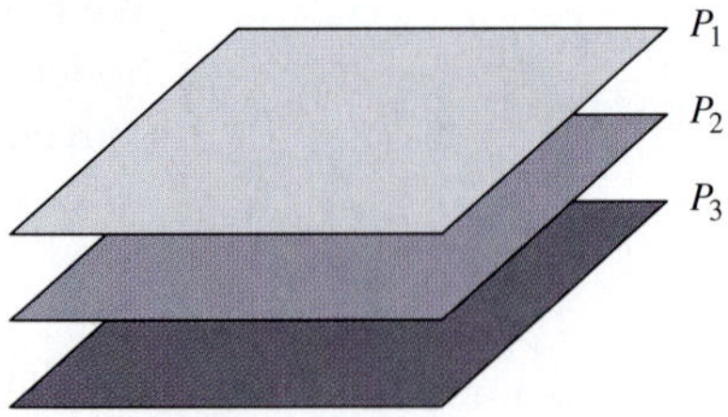

(c) No solution

FIGURE **5**

Note The depictions in Figure 5 are by no means exhaustive. You may consider various other orientations of the three planes that would illustrate the three possible outcomes in solving a system of linear equations involving three variables.

Linear Equations in *n* Variables

A linear equation in n variables, $x_1, x_2, \ldots, x_n$ is an equation of the form

$$a_1x_1 + a_2x_2 + \cdots + a_nx_n = c$$

where $a_1, a_2, \ldots, a_n$ (not all zero) and c are constants.

For example, the equation

$$3x_1 + 2x_2 - 4x_3 + 6x_4 = 8$$

is a linear equation in the four variables, x_1, x_2, x_3, and x_4.

When the number of variables involved in a linear equation exceeds three, we no longer have the geometric interpretation we had for the lower-dimensional spaces. Nevertheless, the algebraic concepts of the lower-dimensional spaces generalize to higher dimensions. For this reason, a linear equation in n variables, $a_1x_1 + a_2x_2 + \cdots + a_nx_n = c$, where $a_1, a_2, \ldots, a_n$ are not all zero, is referred to as an *n-dimensional hyperplane*. We may interpret the solution(s) to a system comprising a finite number of such linear equations to be the *point(s) of intersection* of the hyperplanes defined by the equations that make up the system. As in the case of systems involving two or three variables, it can be shown that only three possibilities exist regarding the nature of the solution of such a system: (1) a unique solution, (2) infinitely many solutions, or (3) no solution.

Explore and Discuss

Refer to the Note above.

Using the orientation of three planes, illustrate the outcomes in solving a system of three linear equations in three variables that result in no solution or infinitely many solutions.

2.1 Self-Check Exercises

1. Determine whether the system of linear equations

$$2x - 3y = 12$$
$$x + 2y = 6$$

has (a) a unique solution, (b) infinitely many solutions, or (c) no solution. Find all solutions whenever they exist. Make a sketch of the set of lines described by the system.

2. **Crop Planning** A farmer has 200 acres of land suitable for cultivating Crops *A*, *B*, and *C*. The cost per acre of cultivating Crops *A*, *B*, and *C* is \$40, \$60, and \$80, respectively. The farmer has \$12,600 available for cultivation. Each acre of Crop *A* requires 20 labor-hours, each acre of Crop *B* requires 25 labor-hours, and each acre of Crop *C* requires 40 labor-hours. The farmer has a maximum of 5950 labor-hours available. If she wishes to use all of her cultivatable land, the entire budget, and all the labor available, how many acres of each crop should she plant? Formulate but do not solve the problem.

Solutions to Self-Check Exercises 2.1 can be found on page 82.

2.1 Concept Questions

1. Suppose you are given a system of two linear equations in two variables.
   a. What can you say about the solution(s) of the system of equations?
   b. Give a geometric interpretation of your answers to the question in part (a). Illustrate each answer with a sketch.

2. Suppose you are given a system of two linear equations in two variables.
   a. Explain what it means for the system to be (i) dependent and (ii) inconsistent.
   b. Illustrate each answer with a sketch.

2.1 Exercises

In Exercises 1–16, determine whether each system of linear equations has (a) one and only one solution, (b) infinitely many solutions, or (c) no solution. Find all solutions whenever they exist.

1. $x - 3y = -1$
   $4x + 3y = 11$

2. $2x - 4y = -10$
   $3x + 2y = 1$

3. $x + 4y = 7$
   $\frac{1}{2}x + 2y = 5$

4. $3x - 4y = 7$
   $9x - 12y = 14$

5. $x + 2y = 7$
   $2x - y = 4$

6. $\frac{3}{2}x - 2y = 4$
   $x + \frac{1}{3}y = 2$

7. $2x - 5y = 10$
   $6x - 15y = 30$

8. $5x - 6y = 8$
   $10x - 12y = 16$

9. $4x - 5y = 14$
   $2x + 3y = -4$

10. $\frac{5}{4}x - \frac{2}{3}y = 3$
    $\frac{1}{4}x + \frac{5}{3}y = 6$

11. $2x - 3y = 6$
    $6x - 9y = 12$

12. $\frac{2}{3}x + y = 5$
    $\frac{1}{2}x + \frac{3}{4}y = \frac{15}{4}$

13. $-3x + 5y = 1$
    $2x - 4y = -1$

14. $-10x + 15y = -3$
    $4x - 6y = -3$

15. $3x - 6y = 2$
    $-\frac{3}{2}x + 3y = -1$

16. $\frac{3}{2}x - \frac{1}{2}y = 1$
    $-x + \frac{1}{3}y = -\frac{2}{3}$

17. $0.2x + y = 1.8$
    $0.4x + 0.3y = 0.2$

18. $0.3x - 0.4y = 0.2$
    $-0.2x + 0.5y = 0.1$

19. Determine the value of *k* for which the system of linear equations

$$2x - y = 3$$
$$4x + ky = 4$$

has no solution.

20. Determine the value of k for which the system of linear equations

$$\begin{aligned} 3x + 4y &= 12 \\ x + ky &= 4 \end{aligned}$$

has infinitely many solutions. Then find all solutions corresponding to this value of k.

21. Determine the conditions on a and b for which the system of linear equations

$$\begin{aligned} ax - by &= c \\ ax + by &= d \end{aligned}$$

has a unique solution. What is the solution?

22. Determine the conditions on a, b, c, and d for which the system of linear equations

$$\begin{aligned} ax + by &= e \\ cx + dy &= f \end{aligned}$$

has a unique solution. What is the solution?

In Exercises 23–44, formulate but do not solve the problem. You will be asked to solve these problems in Section 2.2.

23. **Crop Planning** The Johnson Farm has 500 acres of land allotted for cultivating corn and wheat. The cost of cultivating corn and wheat (including seeds and labor) is \$42 and \$30 per acre, respectively. Jacob Johnson has \$18,600 available for cultivating these crops. If he wishes to use all the allotted land and his entire budget for cultivating these two crops, how many acres of each crop should he plant?

24. **Investments** Michael Perez has a total of \$2000 on deposit with two savings institutions. One pays interest at the rate of 3%/year; the other pays interest at the rate of 4%/year. If Michael earned a total of \$72 in interest during a single year, how much does he have on deposit in each institution?

25. **Blended Coffee Mixtures** The Coffee Shoppe sells a gourmet coffee blend made from two coffees, one costing \$8/lb and the other costing \$9/lb. If the blended coffee sells for \$8.60/lb, find how much of each coffee is used to obtain the desired blend. Assume that the weight of the blended coffee is 100 lb.

26. **Municipal Bonds** Kelly Fisher has a total of \$30,000 invested in two municipal bonds that have yields of 4% and 5% interest per year, respectively. If the interest Kelly receives from the bonds in a year is \$1320, how much does she have invested in each bond?

27. **Metro Bus Ridership** The total number of passengers riding a certain city bus during the morning shift is 1000. If the child's fare is \$0.50, the adult fare is \$1.50, and the total revenue from the fares in the morning shift is \$1300, how many children and how many adults rode the bus during the morning shift?

28. **Apartment Complex Development** Cantwell Associates, a real estate developer, is planning to build a new apartment complex consisting of one-bedroom units and two- and three-bedroom townhouses. A total of 192 units is planned, and the number of family units (two- and three-bedroom townhouses) will equal the number of one-bedroom units. If the number of one-bedroom units will be 3 times the number of three-bedroom units, find how many units of each type will be in the complex.

29. A ball and a bat cost a total of \$110. The bat costs \$100 more than the ball. How much does the ball cost?

30. **Investments** Josh has invested \$70,000 in two projects. The amount invested in project A exceeds that invested in project B by \$20,000. How much has Josh invested in each project?

31. **Investment Planning** The annual returns on Sid Carrington's three investments amounted to \$21,600: 6% on a savings account, 8% on mutual funds, and 12% on bonds. The amount of Sid's investment in bonds was twice the amount of his investment in the savings account, and the interest earned from his investment in bonds was equal to the dividends he received from his investment in mutual funds. Find how much money he placed in each type of investment.

32. **Investment Risk and Return** A private investment club has \$200,000 earmarked for investment in stocks. To arrive at an acceptable overall level of risk, the stocks that management is considering have been classified into three categories: high-risk, medium-risk, and low-risk. Management estimates that high-risk stocks will have a rate of return of 15%/year; medium-risk stocks, 10%/year; and low-risk stocks, 6%/year. The members have decided that the investment in low-risk stocks should be equal to the sum of the investments in the stocks of the other two categories. Determine how much the club should invest in each type of stock if the investment goal is to have a return of \$20,000/year on the total investment. (Assume that all the money available for investment is invested.)

33. **Using Digital Technology** A survey of 500 college students found that the percentage of students who went without using digital technology for up to 1 hr was 67%. The survey also determined that the percentage of students who went without using digital technology for up to 30 min exceeded the percentage of students who went without using digital technology for over 1 hr by 17%. Let x, y, and z represent the percentage of the students in the survey who went without using digital technology (a) for up to 30 min, (b) for more than 30 min but not more than 60 min, and (c) for more than 60 min, respectively. Find the values of x, y, and z.
Source: CourseSmart.

34. **Trustworthiness of Online Reviews** In a survey of 1000 adults aged 18 and older, the following question was

posed: "Are other travelers' online reviews trustworthy?" The participants were asked to answer "yes," "no," or "not sure." The survey revealed that 370 answered "no" or "not sure." It also showed that the number of those who answered "yes" exceeded the number of those who answered "no" by 340. What percentage of respondents answered (a) "yes," (b) "no," and (c) "not sure"?
Source: Alliance Global Assistance.

35. **Lawn Fertilizers** Lawnco produces three grades of commercial fertilizers. A 100-lb bag of grade A fertilizer contains 18 lb of nitrogen, 4 lb of phosphate, and 5 lb of potassium. A 100-lb bag of grade B fertilizer contains 20 lb of nitrogen and 4 lb each of phosphate and potassium. A 100-lb bag of grade C fertilizer contains 24 lb of nitrogen, 3 lb of phosphate, and 6 lb of potassium. How many 100-lb bags of each of the three grades of fertilizers should Lawnco produce if 26,400 lb of nitrogen, 4900 lb of phosphate, and 6200 lb of potassium are available and all the nutrients are used?

36. **Box-Office Receipts** A theater has a seating capacity of 900 and charges $4 for children, $6 for students, and $8 for adults. At a certain screening with full attendance, there were half as many adults as children and students combined. The receipts totaled $5600. How many children attended the show?

37. **Budget Allocation for Auto Fleet** The management of Hartman Rent-A-Car has allocated $2.25 million to buy a fleet of new automobiles consisting of compact, intermediate-size, and full-size cars. Compacts cost $18,000 each, intermediate-size cars cost $27,000 each, and full-size cars cost $36,000 each. If Hartman purchases twice as many compacts as intermediate-size cars and the total number of cars to be purchased is 100, determine how many cars of each type will be purchased. (Assume that the entire budget will be used.)

38. **Investment Risk and Return** The management of a private investment club has a fund of $200,000 earmarked for investment in stocks. To arrive at an acceptable overall level of risk, the stocks that management is considering have been classified into three categories: high-risk, medium-risk, and low-risk. Management estimates that high-risk stocks will have a rate of return of 15%/year; medium-risk stocks, 10%/year; and low-risk stocks, 6%/year. The investment in low-risk stocks is to be twice the sum of the investments in stocks of the other two categories. If the investment goal is to have an average rate of return of 9%/year on the total investment, determine how much the club should invest in each type of stock. (Assume that all the money available for investment is invested.)

39. **Diet Planning** A dietitian wishes to plan a meal around three foods. The percentages of the daily requirements of proteins, carbohydrates, and iron contained in each ounce of the three foods are summarized in the following table:

	Food I	Food II	Food III
Proteins (%)	10	6	8
Carbohydrates (%)	10	12	6
Iron (%)	5	4	12

Determine how many ounces of each food the dietitian should include in the meal to meet exactly the daily requirement of proteins, carbohydrates, and iron (100% of each).

40. **Asset Allocation** Mr. and Mrs. Garcia have a total of $100,000 to be invested in stocks, bonds, and a money market account. The stocks have a rate of return of 12%/year, while the bonds and the money market account pay 8%/year and 4%/year, respectively. The Garcias have stipulated that the amount invested in the money market account should be equal to the sum of 20% of the amount invested in stocks and 10% of the amount invested in bonds. How should the Garcias allocate their resources if they require an annual income of $10,000 from their investments?

41. **Box-Office Receipts** For the opening night at the Opera House, a total of 1000 tickets were sold. Front orchestra seats cost $80 apiece, rear orchestra seats cost $60 apiece, and front balcony seats cost $50 apiece. The combined number of tickets sold for the front orchestra and rear orchestra exceeded twice the number of front balcony tickets sold by 400. The total receipts for the performance were $62,800. Determine how many tickets of each type were sold.

42. **Production Scheduling** A manufacturer of women's blouses makes three types of blouses: sleeveless, short-sleeve, and long-sleeve. The time (in minutes) required by each department to produce a dozen blouses of each type is shown in the following table:

	Sleeveless	Short-sleeve	Long-sleeve
Cutting	9	12	15
Sewing	22	24	28
Packaging	6	8	8

The cutting, sewing, and packaging departments have available a maximum of 80, 160, and 48 labor-hours, respectively, per day. How many dozens of each type of blouse can be produced each day if the plant is operated at full capacity?

43. **Business Travel Expenses** An executive of Trident Communications recently traveled to London, Paris, and Rome. He paid $280, $330, and $260 per night for lodging in London, Paris, and Rome, respectively, and his hotel bills totaled $4060. He spent $130, $140, and $110 per day for his meals in London, Paris, and Rome, respectively, and his expenses for meals totaled $1800. If he spent as many days in London as he did in Paris and Rome combined, how many days did he stay in each city?

44. **Vacation Costs** Joan and Dick spent 2 weeks (14 nights) touring four cities on the East Coast—Boston, New York, Philadelphia, and Washington. They paid \$240, \$400, \$160, and \$200 per night for lodging in each city, respectively, and their total hotel bill came to \$4040. The number of days they spent in New York was the same as the total number of days they spent in Boston and Washington, and the couple spent 3 times as many days in New York as they did in Philadelphia. How many days did Joan and Dick stay in each city?

In Exercises 45–48, determine whether the statement is true or false. If it is true, explain why it is true. If it is false, give an example to show why it is false.

45. A system composed of two linear equations must have at least one solution if the straight lines represented by these equations are nonparallel.

46. Suppose the straight lines represented by a system of three linear equations in two variables are parallel to each other. Then the system has no solution, or it has infinitely many solutions.

47. If at least two of the three lines represented by a system of three linear equations in two variables are parallel, then the system has no solution.

48. If at least two of the four lines represented by a system of four linear equations in two variables are parallel and distinct, then the system has no solution.

2.1 Solutions to Self-Check Exercises

1. Solving the first equation for y in terms of x, we obtain

$$\begin{aligned} -3y &= -2x + 12 \\ y &= \frac{2}{3}x - 4 \end{aligned}$$

Next, substituting this result into the second equation of the system, we find

$$\begin{aligned} x + 2\left(\frac{2}{3}x - 4\right) &= 6 \\ x + \frac{4}{3}x - 8 &= 6 \\ \frac{7}{3}x &= 14 \\ x &= 6 \end{aligned}$$

Substituting this value of x into the expression for y obtained earlier, we have

$$y = \frac{2}{3}(6) - 4 = 0$$

Therefore, the system has the unique solution $x = 6$ and $y = 0$. Both lines are shown in the accompanying figure.

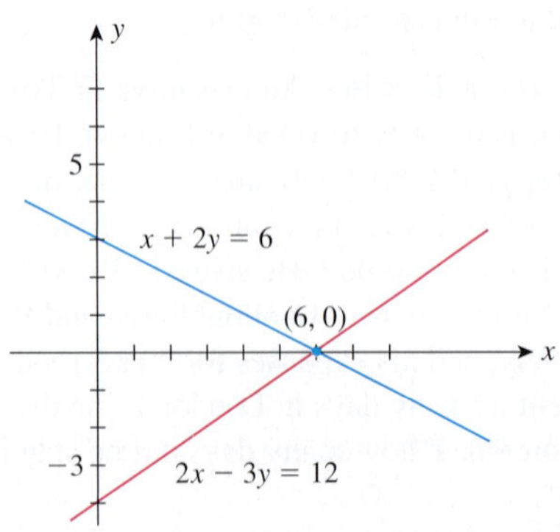

2. Let x, y, and z denote the number of acres of Crop A, Crop B, and Crop C, respectively, to be cultivated. Then the condition that all the cultivatable land be used translates into the equation

$$x + y + z = 200$$

Next, the total cost incurred in cultivating all three crops is $40x + 60y + 80z$ dollars, and since the entire budget is to be expended, we have

$$40x + 60y + 80z = 12{,}600$$

Finally, the amount of labor required to cultivate all three crops is $20x + 25y + 40z$ hours, and since all the available labor is to be used, we have

$$20x + 25y + 40z = 5950$$

Thus, the solution is found by solving the following system of linear equations:

$$\begin{aligned} x + y + z &= 200 \\ 40x + 60y + 80z &= 12{,}600 \\ 20x + 25y + 40z &= 5{,}950 \end{aligned}$$

2.2 Systems of Linear Equations: Unique Solutions

The Method of Elimination

The method of substitution used in Section 2.1 is well suited to solving a system of linear equations when the number of linear equations and variables is small. But for large systems, the steps involved in the procedure become difficult to manage.

The method of elimination is a suitable technique for solving systems of linear equations of any size. One advantage of this technique is its adaptability to the computer. This method involves a sequence of operations on a system of linear equations to obtain at each stage an **equivalent system**—that is, a system having the same solution as the original system. The reduction is complete when the original system has been transformed so that it is in a certain standard form from which the solution can be easily read.

The operations of the method of elimination are as follows:

1. Interchange any two equations.
2. Replace an equation by a nonzero constant multiple of itself.
3. Replace an equation by the sum of that equation and a constant multiple of any other equation.

To illustrate the method of elimination for solving systems of linear equations, let's apply it to the solution of the following system:

$$\begin{aligned} 2x + 4y &= 8 \\ 3x - 2y &= 4 \end{aligned}$$

We begin by working with the first, or x, column. First, we transform the system into an equivalent system in which the coefficient of x in the first equation is 1:

$$\begin{aligned} 2x + 4y &= 8 \\ 3x - 2y &= 4 \end{aligned} \qquad \textbf{(3a)}$$

$$\begin{aligned} x + 2y &= 4 \\ 3x - 2y &= 4 \end{aligned} \qquad \textbf{(3b)}$$

Multiply the first equation in System (3a) by $\frac{1}{2}$ (operation 2).

Next, we eliminate x from the second equation:

$$\begin{aligned} x + 2y &= 4 \\ -8y &= -8 \end{aligned} \qquad \textbf{(3c)}$$

Replace the second equation in System (3b) by the sum of $-3 \times$ the first equation and the second equation (operation 3):

$$\begin{array}{r} -3x - 6y = -12 \\ 3x - 2y = 4 \\ \hline -8y = -8 \end{array}$$

Then we obtain the following equivalent system, in which the coefficient of y in the second equation is 1:

$$\begin{aligned} x + 2y &= 4 \\ y &= 1 \end{aligned} \qquad \textbf{(3d)}$$

Multiply the second equation in System (3c) by $-\frac{1}{8}$ (operation 2).

Next, we eliminate y in the first equation:

$$\begin{aligned} x \qquad &= 2 \\ y &= 1 \end{aligned}$$

Replace the first equation in System (3d) by the sum of $-2 \times$ the second equation and the first equation (operation 3):

$$\begin{array}{r} x + 2y = 4 \\ -2y = -2 \\ \hline x \qquad = 2 \end{array}$$

This system is now in standard form, and we can read off the solution to System (3a) as $x = 2$ and $y = 1$. We can also express this solution as $(2, 1)$ and interpret it geometrically as the point of intersection of the two lines represented by the two linear equations that make up the given system of equations.

The next example involves a system of three linear equations and three variables.

EXAMPLE 1 Solve the following system of equations:

$$\begin{aligned} 2x + 4y + 6z &= 22 \\ 3x + 8y + 5z &= 27 \\ -x + y + 2z &= 2 \end{aligned}$$

Solution First, we transform this system into an equivalent system in which the coefficient of x in the first equation is 1:

$$\begin{aligned} 2x + 4y + 6z &= 22 \\ 3x + 8y + 5z &= 27 \\ -x + y + 2z &= 2 \end{aligned} \qquad \textbf{(4a)}$$

$$\begin{aligned} x + 2y + 3z &= 11 \\ 3x + 8y + 5z &= 27 \\ -x + y + 2z &= 2 \end{aligned} \qquad \textbf{(4b)}$$

Multiply the first equation in System (4a) by $\frac{1}{2}$.

Next, we eliminate the variable x from all equations except the first:

$$\begin{aligned} x + 2y + 3z &= 11 \\ 2y - 4z &= -6 \\ -x + y + 2z &= 2 \end{aligned} \qquad \textbf{(4c)}$$

Replace the second equation in System (4b) by the sum of $(-3) \times$ the first equation and the second equation:

$$\begin{array}{r} -3x - 6y - 9z = -33 \\ 3x + 8y + 5z = 27 \\ \hline 2y - 4z = -6 \end{array}$$

$$\begin{aligned} x + 2y + 3z &= 11 \\ 2y - 4z &= -6 \\ 3y + 5z &= 13 \end{aligned} \qquad \textbf{(4d)}$$

Replace the third equation in System (4c) by the sum of the first equation and the third equation:

$$\begin{array}{r} x + 2y + 3z = 11 \\ -x + y + 2z = 2 \\ \hline 3y + 5z = 13 \end{array}$$

Then we transform System (4d) into yet another equivalent system, in which the coefficient of y in the second equation is 1:

$$\begin{aligned} x + 2y + 3z &= 11 \\ y - 2z &= -3 \\ 3y + 5z &= 13 \end{aligned} \qquad \textbf{(4e)}$$

Multiply the second equation in System (4d) by $\frac{1}{2}$.

We now eliminate y from all equations except the second, using operation 3 of the elimination method:

$$\begin{aligned} x \qquad + 7z &= 17 \\ y - 2z &= -3 \\ 3y + 5z &= 13 \end{aligned} \qquad \textbf{(4f)}$$

Replace the first equation in System (4e) by the sum of the first equation and $(-2) \times$ the second equation:

$$\begin{array}{r} x + 2y + 3z = 11 \\ -2y + 4z = 6 \\ \hline x \qquad + 7z = 17 \end{array}$$

$$\begin{aligned} x \qquad + 7z &= 17 \\ y - 2z &= -3 \\ 11z &= 22 \end{aligned} \qquad \textbf{(4g)}$$

Replace the third equation in System (4f) by the sum of $(-3) \times$ the second equation and the third equation:

$$\begin{array}{r} -3y + 6z = 9 \\ 3y + 5z = 13 \\ \hline 11z = 22 \end{array}$$

Multiplying the third equation by $\frac{1}{11}$ in (4g) leads to the system

$$\begin{aligned} x \qquad + 7z &= 17 \\ y - 2z &= -3 \\ z &= 2 \end{aligned}$$

Eliminating z from all equations except the third (try it!) then leads to the system

$$\begin{aligned} x \qquad &= 3 \\ y \quad &= 1 \\ z &= 2 \end{aligned} \tag{4h}$$

In its final form, the solution to the given system of equations can be easily read off! We have $x = 3$, $y = 1$, and $z = 2$. Geometrically, the point $(3, 1, 2)$ is the intersection of the three planes described by the three equations comprising the given system.

Augmented Matrices

Observe from the preceding example that the variables x, y, and z play no significant role in each step of the reduction process, except as a reminder of the position of each coefficient in the system. With the aid of **matrices,** which are rectangular arrays of numbers, we can eliminate writing the variables at each step of the reduction and thus save ourselves a great deal of work. For example, the system

$$\begin{aligned} 2x + 4y + 6z &= 22 \\ 3x + 8y + 5z &= 27 \\ -x + y + 2z &= 2 \end{aligned} \tag{5}$$

may be represented by the matrix

$$\left[\begin{array}{rrr|r} 2 & 4 & 6 & 22 \\ 3 & 8 & 5 & 27 \\ -1 & 1 & 2 & 2 \end{array}\right] \tag{6}$$

The augmented matrix representing System (5)

The submatrix consisting of the first three columns of Matrix (6) is called the **coefficient matrix** of System (5). The matrix itself, Matrix (6), is referred to as the **augmented matrix** of System (5), since it is obtained by joining the matrix of coefficients to the column (matrix) of constants. The vertical line separates the column of constants from the matrix of coefficients.

The next example shows how much work you can save by using matrices instead of the standard representation of the systems of linear equations.

EXAMPLE 2 Write the augmented matrix corresponding to each equivalent system given in Systems (4a) through (4h).

Solution The required sequence of augmented matrices follows.

Equivalent System	Augmented Matrix	
a. $\begin{aligned} 2x + 4y + 6z &= 22 \\ 3x + 8y + 5z &= 27 \\ -x + y + 2z &= 2 \end{aligned}$	$\left[\begin{array}{rrr\|r} 2 & 4 & 6 & 22 \\ 3 & 8 & 5 & 27 \\ -1 & 1 & 2 & 2 \end{array}\right]$	**(7a)**
b. $\begin{aligned} x + 2y + 3z &= 11 \\ 3x + 8y + 5z &= 27 \\ -x + y + 2z &= 2 \end{aligned}$	$\left[\begin{array}{rrr\|r} 1 & 2 & 3 & 11 \\ 3 & 8 & 5 & 27 \\ -1 & 1 & 2 & 2 \end{array}\right]$	**(7b)**

c. $\begin{aligned} x + 2y + 3z &= 11 \\ 2y - 4z &= -6 \\ -x + y + 2z &= 2 \end{aligned}$ $\quad \left[\begin{array}{rrr|r} 1 & 2 & 3 & 11 \\ 0 & 2 & -4 & -6 \\ -1 & 1 & 2 & 2 \end{array}\right]$ **(7c)**

d. $\begin{aligned} x + 2y + 3z &= 11 \\ 2y - 4z &= -6 \\ 3y + 5z &= 13 \end{aligned}$ $\quad \left[\begin{array}{rrr|r} 1 & 2 & 3 & 11 \\ 0 & 2 & -4 & -6 \\ 0 & 3 & 5 & 13 \end{array}\right]$ **(7d)**

e. $\begin{aligned} x + 2y + 3z &= 11 \\ y - 2z &= -3 \\ 3y + 5z &= 13 \end{aligned}$ $\quad \left[\begin{array}{rrr|r} 1 & 2 & 3 & 11 \\ 0 & 1 & -2 & -3 \\ 0 & 3 & 5 & 13 \end{array}\right]$ **(7e)**

f. $\begin{aligned} x \quad + 7z &= 17 \\ y - 2z &= -3 \\ 3y + 5z &= 13 \end{aligned}$ $\quad \left[\begin{array}{rrr|r} 1 & 0 & 7 & 17 \\ 0 & 1 & -2 & -3 \\ 0 & 3 & 5 & 13 \end{array}\right]$ **(7f)**

g. $\begin{aligned} x \quad + 7z &= 17 \\ y - 2z &= -3 \\ 11z &= 22 \end{aligned}$ $\quad \left[\begin{array}{rrr|r} 1 & 0 & 7 & 17 \\ 0 & 1 & -2 & -3 \\ 0 & 0 & 11 & 22 \end{array}\right]$ **(7g)**

h. $\begin{aligned} x &= 3 \\ y &= 1 \\ z &= 2 \end{aligned}$ $\quad \left[\begin{array}{rrr|r} 1 & 0 & 0 & 3 \\ 0 & 1 & 0 & 1 \\ 0 & 0 & 1 & 2 \end{array}\right]$ **(7h)**

The augmented matrix in (7h) is an example of a matrix in row-reduced form. In general, an augmented matrix with m rows and n columns (called an $m \times n$ matrix) is in **row-reduced form** if it satisfies the following conditions.

Row-Reduced Form of a Matrix

1. Each row consisting entirely of zeros lies below all rows having nonzero entries.
2. The first nonzero entry in each (nonzero) row is 1 (called a **leading 1**).
3. In any two successive (nonzero) rows, the leading 1 in the lower row lies to the right of the leading 1 in the upper row.
4. If a column in the coefficient matrix contains a leading 1, then the other entries in that column are zeros.

EXAMPLE 3 Determine which of the following matrices are in row-reduced form. If a matrix is not in row-reduced form, state the condition that is violated.

a. $\left[\begin{array}{rrr|r} 1 & 0 & 0 & 0 \\ 0 & 1 & 0 & 0 \\ 0 & 0 & 1 & 3 \end{array}\right]$ **b.** $\left[\begin{array}{rrr|r} 1 & 0 & 0 & 4 \\ 0 & 1 & 0 & 3 \\ 0 & 0 & 0 & 0 \end{array}\right]$ **c.** $\left[\begin{array}{rrr|r} 1 & 2 & 0 & 0 \\ 0 & 0 & 1 & 0 \\ 0 & 0 & 0 & 1 \end{array}\right]$

d. $\left[\begin{array}{rrr|r} 0 & 1 & 2 & -2 \\ 1 & 0 & 0 & 3 \\ 0 & 0 & 1 & 2 \end{array}\right]$ **e.** $\left[\begin{array}{rrr|r} 1 & 2 & 0 & 0 \\ 0 & 0 & 1 & 3 \\ 0 & 0 & 2 & 1 \end{array}\right]$ **f.** $\left[\begin{array}{rr|r} 1 & 0 & 4 \\ 0 & 3 & 0 \\ 0 & 0 & 0 \end{array}\right]$

g. $\left[\begin{array}{rrr|r} 0 & 0 & 0 & 0 \\ 1 & 0 & 0 & 3 \\ 0 & 1 & 0 & 2 \end{array}\right]$

Solution The matrices in parts (a)–(c) are in row-reduced form.

d. This matrix is not in row-reduced form. Conditions 3 and 4 are violated: The leading 1 in row 2 lies to the left of the leading 1 in row 1. Also, column 3 contains a leading 1 in row 3 and a nonzero element above it.

e. This matrix is not in row-reduced form. Conditions 2 and 4 are violated: The first nonzero entry in row 3 is a 2, not a 1. Also, column 3 contains a leading 1 and has a nonzero entry below it.

f. This matrix is not in row-reduced form. Condition 2 is violated: The first nonzero entry in row 2 is not a leading 1.

g. This matrix is not in row-reduced form. Condition 1 is violated: Row 1 consists of all zeros and does not lie below the nonzero rows.

The Gauss–Jordan Method

The foregoing discussion suggests the following method, called the **Gauss–Jordan elimination method,** for solving systems of linear equations using matrices. The three operations on the equations of a system (see page 83) translate into the following **row operations** on the corresponding augmented matrices.

Row Operations

1. Interchange any two rows.
2. Replace any row by a nonzero constant multiple of itself.
3. Replace any row by the sum of that row and a constant multiple of any other row.

We obtained the augmented matrices in Example 2 by using the same operations that we used on the equivalent system of equations in Example 1.

To help us describe the Gauss–Jordan elimination method using matrices, let's introduce some terminology. We begin by defining what is meant by a **unit column.**

Unit Column

A column in a coefficient matrix is called a **unit column** if one of the entries in the column is a 1 and the other entries are zeros.

For example, in the coefficient matrix of (7d) in Example 2, page 86, only the first column is in unit form; in the coefficient matrix of (7h), all three columns are in unit form. Now, the sequence of row operations that transforms the augmented matrix (7a) into the equivalent matrix (7d) in which the first column

$$\begin{array}{c} 2 \\ 3 \\ -1 \end{array}$$

of (7a) is transformed into the unit column

$$\begin{array}{c} 1 \\ 0 \\ 0 \end{array}$$

is called **pivoting** the matrix about the element (number) 2. Similarly, we have pivoted about the element 2 in the second column of (7d), shown circled,

$$\begin{matrix} 2 \\ \textcircled{2} \\ 3 \end{matrix}$$

to obtain the augmented matrix (7g), in which the second column

$$\begin{matrix} 0 \\ 1 \\ 0 \end{matrix}$$

is a unit column. Finally, pivoting about the element 11 in column 3 of (7g)

$$\begin{matrix} 7 \\ -2 \\ \textcircled{11} \end{matrix}$$

leads to the augmented matrix (7h), in which the third column

$$\begin{matrix} 0 \\ 0 \\ 1 \end{matrix}$$

is a unit column. Observe that in the final augmented matrix, all three columns to the left of the vertical line are in unit form. The element about which a matrix is pivoted is called the **pivot element.**

Before looking at the next example, let's introduce the following notation for the three types of row operations.

> **Notation for Row Operations**
>
> Letting R_i denote the ith row of a matrix, we write:
>
> **Operation 1** $R_i \leftrightarrow R_j$ to mean: Interchange row i with row j.
>
> **Operation 2** cR_i to mean: Replace row i with c times row i.
>
> **Operation 3** $R_i + aR_j$ to mean: Replace row i with the sum of row i and a times row j.

EXAMPLE 4 Pivot the matrix about the circled element.

$$\left[\begin{array}{cc|c} \textcircled{3} & 5 & 9 \\ 2 & 3 & 5 \end{array}\right]$$

Solution We need a **1** in row 1 where the pivot element (the circled **3**) is. One way of doing this is to replace row 1 by $\frac{1}{3}$ times R_1. In other words, we use operation 2. Thus,

$$\left[\begin{array}{cc|c} 3 & 5 & 9 \\ 2 & 3 & 5 \end{array}\right] \xrightarrow{\frac{1}{3}R_1} \left[\begin{array}{cc|c} 1 & \frac{5}{3} & 3 \\ 2 & 3 & 5 \end{array}\right]$$

Next, we need to replace row 2 by a row with a **0** in the position that is currently occupied by the number **2**. This can be accomplished by replacing row 2 by the sum of row 2 and -2 times row 1. In other words, we use operation 3. Thus,

$$\left[\begin{array}{cc|c} 1 & \frac{5}{3} & 3 \\ 2 & 3 & 5 \end{array}\right] \xrightarrow{R_2 - 2R_1} \left[\begin{array}{cc|c} 1 & \frac{5}{3} & 3 \\ 0 & -\frac{1}{3} & -1 \end{array}\right]$$

Putting these two steps together, we can write the required operations as follows:

$$\left[\begin{array}{cc|c} 3 & 5 & 9 \\ 2 & 3 & 5 \end{array}\right] \xrightarrow{\frac{1}{3}R_1} \left[\begin{array}{cc|c} 1 & \frac{5}{3} & 3 \\ 2 & 3 & 5 \end{array}\right] \xrightarrow{R_2 - 2R_1} \left[\begin{array}{cc|c} 1 & \frac{5}{3} & 3 \\ 0 & -\frac{1}{3} & -1 \end{array}\right]$$

The first column, which originally contained the entry 3, is now in unit form, with a 1 where the pivot element used to be, and we are done.

Alternative Solution In the first solution, we used operation 2 to obtain a 1 where the pivot element was originally. Alternatively, we can use operation 3 as follows:

$$\left[\begin{array}{cc|c} 3 & 5 & 9 \\ 2 & 3 & 5 \end{array}\right] \xrightarrow{R_1 - R_2} \left[\begin{array}{cc|c} 1 & 2 & 4 \\ 2 & 3 & 5 \end{array}\right] \xrightarrow{R_2 - 2R_1} \left[\begin{array}{cc|c} 1 & 2 & 4 \\ 0 & -1 & -3 \end{array}\right]$$

Note In Example 4, the two matrices

$$\left[\begin{array}{cc|c} 1 & \frac{5}{3} & 3 \\ 0 & -\frac{1}{3} & -1 \end{array}\right] \quad \text{and} \quad \left[\begin{array}{cc|c} 1 & 2 & 4 \\ 0 & -1 & -3 \end{array}\right]$$

look quite different, but they are in fact equivalent. You can verify this by observing that they represent the systems of equations

$$\begin{aligned} x + \frac{5}{3}y &= 3 \\ -\frac{1}{3}y &= -1 \end{aligned} \qquad \text{and} \qquad \begin{aligned} x + 2y &= 4 \\ -y &= -3 \end{aligned}$$

respectively, and both have the same solution: $x = -2$ and $y = 3$. Example 4 also shows that we can sometimes avoid working with fractions by using an appropriate row operation.

A summary of the Gauss–Jordan method follows.

The Gauss–Jordan Elimination Method

1. Write the augmented matrix corresponding to the linear system.
2. Interchange rows (operation 1), if necessary, to obtain an augmented matrix in which the first entry in the first row is nonzero. Then pivot the matrix about this entry.
3. Interchange the second row with any row below it, if necessary, to obtain an augmented matrix in which the second entry in the second row is nonzero. Pivot the matrix about this entry.
4. Continue until the final matrix is in row-reduced form.

Before writing the augmented matrix, be sure to write all equations with the variables on the left and constant terms on the right of the equal sign. Also, make sure that the variables are in the same order in all equations.

EXAMPLE 5 Solve the system of linear equations given by

$$\begin{aligned} 3x - 2y + 8z &= 9 \\ -2x + 2y + z &= 3 \\ x + 2y - 3z &= 8 \end{aligned} \tag{8}$$

Solution Using the Gauss–Jordan elimination method, we obtain the following sequence of equivalent augmented matrices:

$$\left[\begin{array}{rrr|r} \textcircled{3} & -2 & 8 & 9 \\ -2 & 2 & 1 & 3 \\ 1 & 2 & -3 & 8 \end{array}\right] \xrightarrow{R_1 + R_2} \left[\begin{array}{rrr|r} 1 & 0 & 9 & 12 \\ -2 & 2 & 1 & 3 \\ 1 & 2 & -3 & 8 \end{array}\right]$$

$$\xrightarrow[R_3 - R_1]{R_2 + 2R_1} \left[\begin{array}{rrr|r} 1 & 0 & 9 & 12 \\ 0 & 2 & 19 & 27 \\ 0 & 2 & -12 & -4 \end{array}\right]$$

$$\xrightarrow{R_2 \leftrightarrow R_3} \left[\begin{array}{rrr|r} 1 & 0 & 9 & 12 \\ 0 & \textcircled{2} & -12 & -4 \\ 0 & 2 & 19 & 27 \end{array}\right]$$

$$\xrightarrow{\frac{1}{2}R_2} \left[\begin{array}{rrr|r} 1 & 0 & 9 & 12 \\ 0 & 1 & -6 & -2 \\ 0 & 2 & 19 & 27 \end{array}\right]$$

$$\xrightarrow{R_3 - 2R_2} \left[\begin{array}{rrr|r} 1 & 0 & 9 & 12 \\ 0 & 1 & -6 & -2 \\ 0 & 0 & \textcircled{31} & 31 \end{array}\right]$$

$$\xrightarrow{\frac{1}{31}R_3} \left[\begin{array}{rrr|r} 1 & 0 & 9 & 12 \\ 0 & 1 & -6 & -2 \\ 0 & 0 & 1 & 1 \end{array}\right]$$

$$\xrightarrow[R_2 + 6R_3]{R_1 - 9R_3} \left[\begin{array}{rrr|r} 1 & 0 & 0 & 3 \\ 0 & 1 & 0 & 4 \\ 0 & 0 & 1 & 1 \end{array}\right]$$

The solution to System (8) is given by $x = 3$, $y = 4$, and $z = 1$. This may be verified by substitution into System (8) as follows:

$$\begin{aligned} 3(3) - 2(4) + 8(1) &= 9 \quad ✓ \\ -2(3) + 2(4) + 1 &= 3 \quad ✓ \\ 3 + 2(4) - 3(1) &= 8 \quad ✓ \end{aligned}$$

⚠ When you are searching for an element to serve as a pivot, it is important to keep in mind that you may work only with the row containing the potential pivot or any row *below* it. To see what can go wrong if this caution is not heeded, consider the following augmented matrix for some linear system:

$$\left[\begin{array}{rrr|r} 1 & 1 & 2 & 3 \\ 0 & 0 & 3 & 1 \\ 0 & 2 & 1 & -2 \end{array}\right]$$

Observe that column 1 is in unit form. The next step in the Gauss–Jordan elimination procedure calls for obtaining a nonzero element in the second position of row 2. If you use row 1 (which is *above* the row under consideration) to help you obtain the pivot, you might proceed as follows:

$$\left[\begin{array}{rrr|r} 1 & 1 & 2 & 3 \\ 0 & 0 & 3 & 1 \\ 0 & 2 & 1 & -2 \end{array}\right] \xrightarrow{R_2 \leftrightarrow R_1} \left[\begin{array}{rrr|r} 0 & 0 & 3 & 1 \\ 1 & 1 & 2 & 3 \\ 0 & 2 & 1 & -2 \end{array}\right]$$

As you can see, not only have we obtained a nonzero element to serve as the next pivot, but it is already a 1, thus obviating the next step. This seems like a good move. But beware—we have undone some of our earlier work: Column 1 is no longer a unit column in which a 1 appears first. The correct move in this case is to interchange row 2 with row 3 in the first augmented matrix.

Explore and Discuss

1. Can the phrase "a nonzero constant multiple of itself" in a type 2 row operation be replaced by "a constant multiple of itself"? Explain.
2. Can a row of an augmented matrix be replaced by a row obtained by adding a constant to every element in that row without changing the solution of the system of linear equations? Explain.

The next example illustrates how to handle a situation in which the first entry in row 1 of the augmented matrix is zero.

EXAMPLE 6 Solve the system of linear equations given by

$$\begin{aligned} 2y + 3z &= 7 \\ 3x + 6y - 12z &= -3 \\ 5x - 2y + 2z &= -7 \end{aligned}$$

Solution Using the Gauss–Jordan elimination method, we obtain the following sequence of equivalent augmented matrices:

$$\left[\begin{array}{rrr|r} 0 & 2 & 3 & 7 \\ 3 & 6 & -12 & -3 \\ 5 & -2 & 2 & -7 \end{array}\right] \xrightarrow{R_1 \leftrightarrow R_2} \left[\begin{array}{rrr|r} \textcircled{3} & 6 & -12 & -3 \\ 0 & 2 & 3 & 7 \\ 5 & -2 & 2 & -7 \end{array}\right]$$

$$\xrightarrow{\frac{1}{3}R_1} \left[\begin{array}{rrr|r} 1 & 2 & -4 & -1 \\ 0 & 2 & 3 & 7 \\ 5 & -2 & 2 & -7 \end{array}\right]$$

$$\xrightarrow{R_3 - 5R_1} \left[\begin{array}{rrr|r} 1 & 2 & -4 & -1 \\ 0 & \textcircled{2} & 3 & 7 \\ 0 & -12 & 22 & -2 \end{array}\right]$$

$$\xrightarrow{\frac{1}{2}R_2} \left[\begin{array}{rrr|r} 1 & 2 & -4 & -1 \\ 0 & 1 & \frac{3}{2} & \frac{7}{2} \\ 0 & -12 & 22 & -2 \end{array}\right]$$

$$\xrightarrow[R_3 + 12R_2]{R_1 - 2R_2} \left[\begin{array}{rrr|r} 1 & 0 & -7 & -8 \\ 0 & 1 & \frac{3}{2} & \frac{7}{2} \\ 0 & 0 & \textcircled{40} & 40 \end{array}\right]$$

$$\xrightarrow{\frac{1}{40}R_3} \left[\begin{array}{rrr|r} 1 & 0 & -7 & -8 \\ 0 & 1 & \frac{3}{2} & \frac{7}{2} \\ 0 & 0 & 1 & 1 \end{array}\right]$$

$$\xrightarrow[R_2 - \frac{3}{2}R_3]{R_1 + 7R_3} \left[\begin{array}{rrr|r} 1 & 0 & 0 & -1 \\ 0 & 1 & 0 & 2 \\ 0 & 0 & 1 & 1 \end{array}\right]$$

The solution to the system is given by $x = -1$, $y = 2$, and $z = 1$; this may be verified by substituting these values into each equation of the system.

APPLIED EXAMPLE 7 Production Scheduling Complete the solution to Example 1 in Section 2.1, page 77.

Solution To complete the solution of the problem posed in Example 1, recall that the mathematical formulation of the problem led to the following system of linear equations:

$$\begin{aligned} 2x + y + z &= 180 \\ x + 3y + 2z &= 300 \\ 2x + y + 2z &= 240 \end{aligned}$$

where x, y, and z denote the respective numbers of Type A, Type B, and Type C souvenirs to be made.

Solving the foregoing system of linear equations by the Gauss–Jordan elimination method, we obtain the following sequence of equivalent augmented matrices:

$$\left[\begin{array}{ccc|c} 2 & 1 & 1 & 180 \\ 1 & 3 & 2 & 300 \\ 2 & 1 & 2 & 240 \end{array}\right] \xrightarrow{R_1 \leftrightarrow R_2} \left[\begin{array}{ccc|c} 1 & 3 & 2 & 300 \\ 2 & 1 & 1 & 180 \\ 2 & 1 & 2 & 240 \end{array}\right]$$

$$\xrightarrow[R_3 - 2R_1]{R_2 - 2R_1} \left[\begin{array}{ccc|c} 1 & 3 & 2 & 300 \\ 0 & -5 & -3 & -420 \\ 0 & -5 & -2 & -360 \end{array}\right]$$

$$\xrightarrow{-\frac{1}{5}R_2} \left[\begin{array}{ccc|c} 1 & 3 & 2 & 300 \\ 0 & 1 & \frac{3}{5} & 84 \\ 0 & -5 & -2 & -360 \end{array}\right]$$

$$\xrightarrow[R_3 + 5R_2]{R_1 - 3R_2} \left[\begin{array}{ccc|c} 1 & 0 & \frac{1}{5} & 48 \\ 0 & 1 & \frac{3}{5} & 84 \\ 0 & 0 & 1 & 60 \end{array}\right]$$

$$\xrightarrow[R_2 - \frac{3}{5}R_3]{R_1 - \frac{1}{5}R_3} \left[\begin{array}{ccc|c} 1 & 0 & 0 & 36 \\ 0 & 1 & 0 & 48 \\ 0 & 0 & 1 & 60 \end{array}\right]$$

Thus, $x = 36$, $y = 48$, and $z = 60$; that is, Ace Novelty should make 36 Type A souvenirs, 48 Type B souvenirs, and 60 Type C souvenirs in order to use all available machine time.

2.2 Self-Check Exercises

1. Solve the system of linear equations

$$\begin{aligned} 2x + 3y + z &= 6 \\ x - 2y + 3z &= -3 \\ 3x + 2y - 4z &= 12 \end{aligned}$$

using the Gauss–Jordan elimination method.

2. **Crop Planning** A farmer has 200 acres of land suitable for cultivating Crops A, B, and C. The cost per acre of cultivating Crop A, Crop B, and Crop C is \$40, \$60, and \$80, respectively. The farmer has \$12,600 available for land cultivation. Each acre of Crop A requires 20 labor-hours, each acre of Crop B requires 25 labor-hours, and each acre of Crop C requires 40 labor-hours. The farmer has a maximum of 5950 labor-hours available. If she wishes to use all of her cultivatable land, the entire budget, and all the labor available, how many acres of each crop should she plant?

Solutions to Self-Check Exercises 2.2 can be found on page 97.

2.2 Concept Questions

1. **a.** Explain what it means for two systems of linear equations to be equivalent to each other.
   b. Give the meaning of the following notation used for row operations in the Gauss–Jordan elimination method:

   i. $R_i \leftrightarrow R_j$ **ii.** cR_i **iii.** $R_i + aR_j$

2. **a.** What is an augmented matrix? A coefficient matrix? A unit column?
   b. Explain what is meant by a pivot operation.

3. Suppose that a matrix is in row-reduced form.
   a. What is the position of a row consisting entirely of zeros relative to the nonzero rows?
   b. What is the first nonzero entry in each row?
   c. What is the position of the leading 1s in successive nonzero rows?
   d. If a column contains a leading 1, then what is the value of the other entries in that column?

2.2 Exercises

In Exercises 1–4, write the augmented matrix corresponding to each system of equations.

1. $\begin{aligned} 2x - 3y &= 7 \\ 3x + y &= 4 \end{aligned}$

2. $\begin{aligned} 3x + 7y - 8z &= 5 \\ x \qquad + 3z &= -2 \\ 4x - 3y \qquad &= 7 \end{aligned}$

3. $\begin{aligned} -y + 2z &= 5 \\ 2x + 2y - 8z &= 4 \\ 3y + 4z &= 0 \end{aligned}$

4. $\begin{aligned} 3x_1 + 2x_2 \qquad &= 0 \\ x_1 - x_2 + 2x_3 &= 4 \\ 2x_2 - 3x_3 &= 5 \end{aligned}$

In Exercises 5–8, write the system of equations corresponding to each augmented matrix.

5. $\left[\begin{array}{cc|c} 3 & 2 & -4 \\ 1 & -1 & 5 \end{array}\right]$

6. $\left[\begin{array}{ccc|c} 0 & 3 & 2 & 4 \\ 1 & -1 & -2 & -3 \\ 4 & 0 & 3 & 2 \end{array}\right]$

7. $\left[\begin{array}{ccc|c} 1 & 3 & 2 & 4 \\ 2 & 0 & 0 & 5 \\ 3 & -3 & 2 & 6 \end{array}\right]$

8. $\left[\begin{array}{ccc|c} 2 & 3 & 1 & 6 \\ 4 & 3 & 2 & 5 \\ 0 & 0 & 0 & 0 \end{array}\right]$

In Exercises 9–18, indicate whether the matrix is in row-reduced form.

9. $\left[\begin{array}{cc|c} 1 & 0 & 3 \\ 0 & 1 & -2 \end{array}\right]$

10. $\left[\begin{array}{cc|c} 1 & 1 & 3 \\ 0 & 0 & 0 \end{array}\right]$

11. $\left[\begin{array}{cc|c} 0 & 1 & 3 \\ 1 & 0 & 5 \end{array}\right]$

12. $\left[\begin{array}{cc|c} 0 & 1 & 3 \\ 0 & 0 & 5 \end{array}\right]$

13. $\left[\begin{array}{ccc|c} 1 & 0 & 0 & 3 \\ 0 & 1 & 0 & 4 \\ 0 & 0 & 1 & 5 \end{array}\right]$

14. $\left[\begin{array}{ccc|c} 1 & 0 & 0 & -1 \\ 0 & 1 & 0 & -2 \\ 0 & 0 & 2 & -3 \end{array}\right]$

15. $\left[\begin{array}{ccc|c} 1 & 0 & 1 & 3 \\ 0 & 1 & 0 & 4 \\ 0 & 0 & -1 & 6 \end{array}\right]$

16. $\left[\begin{array}{cc|c} 1 & 0 & -10 \\ 0 & 1 & 2 \\ 0 & 0 & 0 \end{array}\right]$

17. $\left[\begin{array}{ccc|c} 0 & 0 & 0 & 0 \\ 0 & 1 & 2 & 4 \\ 0 & 0 & 0 & 0 \end{array}\right]$

18. $\left[\begin{array}{ccc|c} 1 & 0 & 0 & 3 \\ 0 & 1 & 0 & 6 \\ 0 & 0 & 0 & 4 \\ 0 & 0 & 1 & 5 \end{array}\right]$

In Exercises 19–26, pivot the system about the circled element.

19. $\left[\begin{array}{cc|c} \textcircled{1} & 3 & 4 \\ 2 & 4 & 6 \end{array}\right]$

20. $\left[\begin{array}{cc|c} \textcircled{2} & 4 & 8 \\ 3 & 1 & 2 \end{array}\right]$

21. $\left[\begin{array}{cc|c} \textcircled{-1} & 2 & 3 \\ 6 & 8 & 2 \end{array}\right]$

22. $\left[\begin{array}{cc|c} 3 & 2 & 6 \\ \textcircled{4} & 2 & 5 \end{array}\right]$

23. $\left[\begin{array}{ccc|c} \textcircled{2} & 4 & 6 & 12 \\ 2 & 3 & 1 & 5 \\ 3 & -1 & 2 & 4 \end{array}\right]$

24. $\left[\begin{array}{ccc|c} 1 & 3 & 2 & 4 \\ \textcircled{2} & 4 & 8 & 6 \\ -1 & 2 & 3 & 4 \end{array}\right]$

25. $\left[\begin{array}{ccc|c} 0 & 1 & 3 & 4 \\ 2 & 4 & \textcircled{1} & 3 \\ 5 & 6 & 2 & -4 \end{array}\right]$

26. $\left[\begin{array}{ccc|c} 1 & 2 & 3 & 5 \\ 0 & \textcircled{-3} & 3 & 2 \\ 0 & 4 & -1 & 3 \end{array}\right]$

In Exercises 27–30, fill in the missing entries by performing the indicated row operations to obtain the row-reduced matrices.

27. $\left[\begin{array}{cc|c} 3 & 9 & 6 \\ 2 & 1 & 4 \end{array}\right] \xrightarrow{\frac{1}{3}R_1} \left[\begin{array}{cc|c} \cdot & \cdot & \cdot \\ 2 & 1 & 4 \end{array}\right] \xrightarrow{R_2 - 2R_1}$

$\left[\begin{array}{cc|c} 1 & 3 & 2 \\ \cdot & \cdot & \cdot \end{array}\right] \xrightarrow{-\frac{1}{5}R_2} \left[\begin{array}{cc|c} 1 & 3 & 2 \\ \cdot & \cdot & \cdot \end{array}\right] \xrightarrow{R_1 - 3R_2} \left[\begin{array}{cc|c} 1 & 0 & 2 \\ 0 & 1 & 0 \end{array}\right]$

28. $$\left[\begin{array}{cc|c} 1 & 2 & 1 \\ 2 & 3 & -1 \end{array}\right] \xrightarrow{R_2 - 2R_1} \left[\begin{array}{cc|c} 1 & 2 & 1 \\ \cdot & \cdot & \cdot \end{array}\right] \xrightarrow{-R_2}$$

$$\left[\begin{array}{cc|c} 1 & 2 & 1 \\ \cdot & \cdot & \cdot \end{array}\right] \xrightarrow{R_1 - 2R_2} \left[\begin{array}{cc|c} 1 & 0 & -5 \\ 0 & 1 & 3 \end{array}\right]$$

29. $$\left[\begin{array}{ccc|c} 1 & 3 & 1 & 3 \\ 3 & 8 & 3 & 7 \\ 2 & -3 & 1 & -10 \end{array}\right] \xrightarrow[R_3 - 2R_1]{R_2 - 3R_1} \left[\begin{array}{ccc|c} 1 & 3 & 1 & 3 \\ \cdot & \cdot & \cdot & \cdot \\ \cdot & \cdot & \cdot & \cdot \end{array}\right] \xrightarrow{-R_2}$$

$$\left[\begin{array}{ccc|c} 1 & 3 & 1 & 3 \\ \cdot & \cdot & \cdot & \cdot \\ 0 & -9 & -1 & -16 \end{array}\right] \xrightarrow[R_3 + 9R_2]{R_1 - 3R_2}$$

$$\left[\begin{array}{ccc|c} \cdot & \cdot & \cdot & \cdot \\ 0 & 1 & 0 & 2 \\ \cdot & \cdot & \cdot & \cdot \end{array}\right] \xrightarrow[-R_3]{R_1 + R_3} \left[\begin{array}{ccc|c} 1 & 0 & 0 & -1 \\ 0 & 1 & 0 & 2 \\ 0 & 0 & 1 & -2 \end{array}\right]$$

30. $$\left[\begin{array}{ccc|c} 0 & 1 & 3 & -4 \\ 1 & 2 & 1 & 7 \\ 1 & -2 & 0 & 1 \end{array}\right] \xrightarrow{R_1 \leftrightarrow R_2} \left[\begin{array}{ccc|c} \cdot & \cdot & \cdot & \cdot \\ \cdot & \cdot & \cdot & \cdot \\ 1 & -2 & 0 & 1 \end{array}\right]$$

$$\xrightarrow{R_3 - R_1} \left[\begin{array}{ccc|c} 1 & 2 & 1 & 7 \\ 0 & 1 & 3 & -4 \\ \cdot & \cdot & \cdot & \cdot \end{array}\right] \xrightarrow[R_3 + 4R_2]{R_1 - 2R_2} \left[\begin{array}{ccc|c} \cdot & \cdot & \cdot & \cdot \\ 0 & 1 & 3 & -4 \\ \cdot & \cdot & \cdot & \cdot \end{array}\right]$$

$$\xrightarrow{\frac{1}{11}R_3} \left[\begin{array}{ccc|c} 1 & 0 & -5 & 15 \\ 0 & 1 & 3 & -4 \\ \cdot & \cdot & \cdot & \cdot \end{array}\right] \xrightarrow[R_2 - 3R_3]{R_1 + 5R_3} \left[\begin{array}{ccc|c} 1 & 0 & 0 & 5 \\ 0 & 1 & 0 & 2 \\ 0 & 0 & 1 & -2 \end{array}\right]$$

31. Write a system of linear equations for the augmented matrix of Exercise 27. Using the results of Exercise 27, determine the solution of the system.

32. Repeat Exercise 31 for the augmented matrix of Exercise 28.

33. Repeat Exercise 31 for the augmented matrix of Exercise 29.

34. Repeat Exercise 31 for the augmented matrix of Exercise 30.

In Exercises 35–56, solve the system of linear equations using the Gauss–Jordan elimination method.

35. $$\begin{aligned} x + y &= 3 \\ 2x - y &= 3 \end{aligned}$$

36. $$\begin{aligned} x - 2y &= -3 \\ 2x + 3y &= 8 \end{aligned}$$

37. $$\begin{aligned} x - 2y &= 8 \\ 3x + 4y &= 4 \end{aligned}$$

38. $$\begin{aligned} 3x + y &= 1 \\ -7x - 2y &= -1 \end{aligned}$$

39. $$\begin{aligned} 2x - 3y &= -8 \\ 4x + y &= -2 \end{aligned}$$

40. $$\begin{aligned} 5x + 3y &= 9 \\ -2x + y &= -8 \end{aligned}$$

41. $$\begin{aligned} 6x + 8y &= 15 \\ 2x - 4y &= -5 \end{aligned}$$

42. $$\begin{aligned} 2x + 10y &= 1 \\ -4x + 6y &= 11 \end{aligned}$$

43. $$\begin{aligned} 3x - 2y &= 1 \\ 2x + 4y &= 2 \end{aligned}$$

44. $$\begin{aligned} x - \frac{1}{2}y &= \frac{7}{6} \\ -\frac{1}{2}x + 4y &= \frac{2}{3} \end{aligned}$$

45. $$\begin{aligned} 2x + y - 2z &= 4 \\ x + 3y - z &= -3 \\ 3x + 4y - z &= 7 \end{aligned}$$

46. $$\begin{aligned} x + y + z &= 0 \\ 2x - y + z &= 1 \\ x + y - 2z &= 2 \end{aligned}$$

47. $$\begin{aligned} 2x + 2y + z &= 9 \\ x + z &= 4 \\ 4y - 3z &= 17 \end{aligned}$$

48. $$\begin{aligned} 2x + 3y - 2z &= 10 \\ 3x - 2y + 2z &= 0 \\ 4x - y + 3z &= -1 \end{aligned}$$

49. $$\begin{aligned} -x_2 + x_3 &= 2 \\ 4x_1 - 3x_2 + 2x_3 &= 16 \\ 3x_1 + 2x_2 + x_3 &= 11 \end{aligned}$$

50. $$\begin{aligned} 2x + 4y - 6z &= 38 \\ x + 2y + 3z &= 7 \\ 3x - 4y + 4z &= -19 \end{aligned}$$

51. $$\begin{aligned} x_1 - 2x_2 + x_3 &= 6 \\ 2x_1 + x_2 - 3x_3 &= -3 \\ x_1 - 3x_2 + 3x_3 &= 10 \end{aligned}$$

52. $$\begin{aligned} 2x + 3y - 6z &= -11 \\ x - 2y + 3z &= 9 \\ 3x + y &= 7 \end{aligned}$$

53. $$\begin{aligned} 2x + 3z &= -1 \\ 3x - 2y + z &= 9 \\ x + y + 4z &= 4 \end{aligned}$$

54. $$\begin{aligned} 2x_1 - x_2 + 3x_3 &= -4 \\ x_1 - 2x_2 + x_3 &= -1 \\ x_1 - 5x_2 + 2x_3 &= -3 \end{aligned}$$

55. $$\begin{aligned} x_1 - x_2 + 3x_3 &= 14 \\ x_1 + x_2 + x_3 &= 6 \\ -2x_1 - x_2 + x_3 &= -4 \end{aligned}$$

56. $$\begin{aligned} 2x_1 - x_2 - x_3 &= 0 \\ 3x_1 + 2x_2 + x_3 &= 7 \\ x_1 + 2x_2 + 2x_3 &= 5 \end{aligned}$$

57. Determine the value(s) of k such that the following system of linear equations has a unique solution, and then find the solution in terms of k:

$$\begin{aligned} 4x + 5y &= 3 \\ 3x + ky &= 10 \end{aligned}$$

58. Determine the value(s) of k such that the following system of linear equations has a unique solution:

$$\begin{aligned} x + 3y + z &= 8 \\ 3x + 2y - 2z &= 5 \\ 4x - 3y + kz &= 0 \end{aligned}$$

The problems in Exercises 59–80 correspond to those in Exercises 23–44, Section 2.1. Use the results of your previous work to help you solve these problems.

59. **Crop Planning** The Johnson Farm has 500 acres of land allotted for cultivating corn and wheat. The cost of cultivating corn and wheat (including seeds and labor) is \$42 and \$30 per acre, respectively. Jacob Johnson has \$18,600 available for cultivating these crops. If he wishes to use all the allotted land and his entire budget for cultivating these two crops, how many acres of each crop should he plant?

60. **Investments** Michael Perez has a total of \$2000 on deposit with two savings institutions. One pays interest at the rate of 3%/year, whereas the other pays interest at the rate of 4%/year. If Michael earned a total of \$72 in interest during a single year, how much does he have on deposit in each institution?

61. **Blended Coffee Mixtures** The Coffee Shoppe sells a gourmet coffee blend made from two coffees, one costing \$8/lb and the other costing \$9/lb. If the blended coffee sells for \$8.60/lb, find how much of each coffee is used to obtain the desired blend. Assume that the weight of the blended coffee is 100 lb.

62. **Municipal Bonds** Kelly Fisher has a total of \$30,000 invested in two municipal bonds that have yields of 4% and 5% interest per year, respectively. If the interest Kelly receives from the bonds in a year is \$1320, how much does she have invested in each bond?

63. **Metro Bus Ridership** The total number of passengers riding a certain city bus during the morning shift is 1000. If the child's fare is \$0.50, the adult fare is \$1.50, and the total revenue from the fares in the morning shift is \$1300, how many children and how many adults rode the bus during the morning shift?

64. **Apartment Complex Development** Cantwell Associates, a real estate developer, is planning to build a new apartment complex consisting of one-bedroom units and two- and three-bedroom townhouses. A total of 192 units is planned, and the number of family units (two- and three-bedroom townhouses) will equal the number of one-bedroom units. If the number of one-bedroom units will be 3 times the number of three-bedroom units, find how many units of each type will be in the complex.

65. A ball and a bat cost a total of \$110. The bat costs \$100 more than the ball. How much does the ball cost?

66. **Investments** Josh has invested \$70,000 in two projects. The amount invested in project A exceeds that invested in project B by \$20,000. How much has Josh invested in each project?

67. **Investment Planning** The annual returns on Sid Carrington's three investments amounted to \$21,600: 6% on a savings account, 8% on mutual funds, and 12% on bonds. The amount of Sid's investment in bonds was twice the amount of his investment in the savings account, and the interest earned from his investment in bonds was equal to the dividends he received from his investment in mutual funds. Find how much money he placed in each type of investment.

68. **Investment Risk and Return** A private investment club has \$200,000 earmarked for investment in stocks. To arrive at an acceptable overall level of risk, the stocks that management is considering have been classified into three categories: high-risk, medium-risk, and low-risk. Management estimates that high-risk stocks will have a rate of return of 15%/year; medium-risk stocks, 10%/year; and low-risk stocks, 6%/year. The members have decided that the investment in low-risk stocks should be equal to the sum of the investments in the stocks of the other two categories. Determine how much the club should invest in each type of stock if the investment goal is to have a return of \$20,000/year on the total investment. (Assume that all the money available for investment is invested.)

69. **Using Digital Technology** A survey of 500 college students found that the percentage of students who went without using digital technology for up to 1 hr was 67%. The survey also determined that the percentage of students who went without using digital technology for up to 30 min exceeded the percentage of students who went without using digital technology for over 1 hr by 17%. Let x, y, and z represent the percentage of the students in the survey who went without using digital technology (a) for up to 30 min, (b) for more than 30 min but not more than 60 min, and (c) for more than 60 min, respectively. Find the values of x, y, and z.
Source: CourseSmart.

70. **Trustworthiness of Online Reviews** In a survey of 1000 adults aged 18 and older, the following question was posed: "Are other travelers' online reviews trustworthy?" The participants were asked to answer "yes," "no," or "not sure." The survey revealed that 370 answered "no" or "not sure." It also showed that the number of those who answered "yes" exceeded the number of those who answered "no" by 340. What percentage of respondents answered (a) "yes," (b) "no," and (c) "not sure"?
Source: Alliance Global Assistance.

71. **Lawn Fertilizers** Lawnco produces three grades of commercial fertilizers. A 100-lb bag of grade A fertilizer contains 18 lb of nitrogen, 4 lb of phosphate, and 5 lb of potassium. A 100-lb bag of grade B fertilizer contains 20 lb of nitrogen and 4 lb each of phosphate and potassium. A 100-lb bag of grade C fertilizer contains 24 lb of nitrogen, 3 lb of phosphate, and 6 lb of potassium. How many 100-lb bags of each of the three grades of fertilizers should Lawnco produce if 26,400 lb of nitrogen, 4900 lb of phosphate, and 6200 lb of potassium are available and all the nutrients are used?

72. **Box-Office Receipts** A theater has a seating capacity of 900 and charges \$4 for children, \$6 for students, and \$8 for adults. At a certain screening with full attendance, there were half as many adults as children and students combined. The receipts totaled \$5600. How many children attended the show?

73. **Budget Allocation for Auto Fleet** The management of Hartman Rent-A-Car has allocated \$2.25 million to buy a fleet of new automobiles consisting of compact, intermediate-size, and full-size cars. Compacts cost \$18,000 each, intermediate-size cars cost \$27,000 each, and full-size cars cost \$36,000 each. If Hartman purchases twice as many compacts as intermediate-size cars and the total number of cars to be purchased is 100, determine how many cars of each type will be purchased. (Assume that the entire budget will be used.)

74. **INVESTMENT RISK AND RETURN** The management of a private investment club has a fund of \$200,000 earmarked for investment in stocks. To arrive at an acceptable overall level of risk, the stocks that management is considering have been classified into three categories: high-risk, medium-risk, and low-risk. Management estimates that high-risk stocks will have a rate of return of 15%/year; medium-risk stocks, 10%/year; and low-risk stocks, 6%/year. The investment in low-risk stocks is to be twice the sum of the investments in stocks of the other two categories. If the investment goal is to have an average rate of return of 9%/year on the total investment, determine how much the club should invest in each type of stock. (Assume that all of the money available for investment is invested.)

75. **DIET PLANNING** A dietitian wishes to plan a meal around three foods. The percentages of the daily requirements of proteins, carbohydrates, and iron contained in each ounce of the three foods are summarized in the following table:

	Food I	Food II	Food III
Proteins (%)	10	6	8
Carbohydrates (%)	10	12	6
Iron (%)	5	4	12

Determine how many ounces of each food the dietitian should include in the meal to meet exactly the daily requirement of proteins, carbohydrates, and iron (100% of each).

76. **ASSET ALLOCATION** Mr. and Mrs. Garcia have a total of \$100,000 to be invested in stocks, bonds, and a money market account. The stocks have a rate of return of 12%/year, while the bonds and the money market account pay 8%/year and 4%/year, respectively. The Garcias have stipulated that the amount invested in the money market account should be equal to the sum of 20% of the amount invested in stocks and 10% of the amount invested in bonds. How should the Garcias allocate their resources if they require an annual income of \$10,000 from their investments?

77. **BOX-OFFICE RECEIPTS** For the opening night at the Opera House, a total of 1000 tickets were sold. Front orchestra seats cost \$80 apiece, rear orchestra seats cost \$60 apiece, and front balcony seats cost \$50 apiece. The combined number of tickets sold for the front orchestra and rear orchestra exceeded twice the number of front balcony tickets sold by 400. The total receipts for the performance were \$62,800. Determine how many tickets of each type were sold.

78. **PRODUCTION SCHEDULING** A manufacturer of women's blouses makes three types of blouses: sleeveless, short-sleeve, and long-sleeve. The time (in minutes) required by each department to produce a dozen blouses of each type is shown in the following table:

	Sleeveless	Short-sleeve	Long-sleeve
Cutting	9	12	15
Sewing	22	24	28
Packaging	6	8	8

The cutting, sewing, and packaging departments have available a maximum of 80, 160, and 48 labor-hours, respectively, per day. How many dozens of each type of blouse can be produced each day if the plant is operated at full capacity?

79. **BUSINESS TRAVEL EXPENSES** An executive of Trident Communications recently traveled to London, Paris, and Rome. He paid \$280, \$330, and \$260 per night for lodging in London, Paris, and Rome, respectively, and his hotel bills totaled \$4060. He spent \$130, \$140, and \$110 per day for his meals in London, Paris, and Rome, respectively, and his expenses for meals totaled \$1800. If he spent as many days in London as he did in Paris and Rome combined, how many days did he stay in each city?

80. **VACATION COSTS** Joan and Dick spent 2 weeks (14 nights) touring four cities on the East Coast—Boston, New York, Philadelphia, and Washington. They paid \$240, \$400, \$160, and \$200 per night for lodging in each city, respectively, and their total hotel bill came to \$4040. The number of days they spent in New York was the same as the total number of days they spent in Boston and Washington, and the couple spent 3 times as many days in New York as they did in Philadelphia. How many days did Joan and Dick stay in each city?

In Exercises 81 and 82, determine whether the statement is true or false. If it is true, explain why it is true. If it is false, give an example to show why it is false.

81. An equivalent system of linear equations can be obtained from a system of equations by replacing one of its equations by any constant multiple of itself.

82. If the augmented matrix corresponding to a system of three linear equations in three variables has a row of the form $[0 \quad 0 \quad 0 \mid a]$, where a is a nonzero number, then the system has no solution.

2.2 Solutions to Self-Check Exercises

1. We obtain the following sequence of equivalent augmented matrices:

$$\left[\begin{array}{rrr|r} 2 & 3 & 1 & 6 \\ 1 & -2 & 3 & -3 \\ 3 & 2 & -4 & 12 \end{array}\right] \xrightarrow{R_1 \leftrightarrow R_2} \left[\begin{array}{rrr|r} \textcircled{1} & -2 & 3 & -3 \\ 2 & 3 & 1 & 6 \\ 3 & 2 & -4 & 12 \end{array}\right]$$

$$\xrightarrow[R_3 - 3R_1]{R_2 - 2R_1} \left[\begin{array}{rrr|r} 1 & -2 & 3 & -3 \\ 0 & 7 & -5 & 12 \\ 0 & 8 & -13 & 21 \end{array}\right] \xrightarrow{R_2 \leftrightarrow R_3}$$

$$\left[\begin{array}{rrr|r} 1 & -2 & 3 & -3 \\ 0 & \textcircled{8} & -13 & 21 \\ 0 & 7 & -5 & 12 \end{array}\right] \xrightarrow{R_2 - R_3} \left[\begin{array}{rrr|r} 1 & -2 & 3 & -3 \\ 0 & 1 & -8 & 9 \\ 0 & 7 & -5 & 12 \end{array}\right]$$

$$\xrightarrow[R_3 - 7R_2]{R_1 + 2R_2} \left[\begin{array}{rrr|r} 1 & 0 & -13 & 15 \\ 0 & 1 & -8 & 9 \\ 0 & 0 & 51 & -51 \end{array}\right] \xrightarrow{\frac{1}{51}R_3} \left[\begin{array}{rrr|r} 1 & 0 & -13 & 15 \\ 0 & 1 & -8 & 9 \\ 0 & 0 & \textcircled{1} & -1 \end{array}\right]$$

$$\xrightarrow[R_2 + 8R_3]{R_1 + 13R_3} \left[\begin{array}{rrr|r} 1 & 0 & 0 & 2 \\ 0 & 1 & 0 & 1 \\ 0 & 0 & 1 & -1 \end{array}\right]$$

The solution to the system is $x = 2$, $y = 1$, and $z = -1$.

2. Referring to the solution of Exercise 2, Self-Check Exercises 2.1, we see that the problem reduces to solving the following system of linear equations:

$$\begin{aligned} x + y + z &= 200 \\ 40x + 60y + 80z &= 12{,}600 \\ 20x + 25y + 40z &= 5{,}950 \end{aligned}$$

Using the Gauss–Jordan elimination method, we have

$$\left[\begin{array}{rrr|r} \textcircled{1} & 1 & 1 & 200 \\ 40 & 60 & 80 & 12{,}600 \\ 20 & 25 & 40 & 5{,}950 \end{array}\right] \xrightarrow[R_3 - 20R_1]{R_2 - 40R_1} \left[\begin{array}{rrr|r} 1 & 1 & 1 & 200 \\ 0 & \textcircled{20} & 40 & 4600 \\ 0 & 5 & 20 & 1950 \end{array}\right]$$

$$\xrightarrow{\frac{1}{20}R_2} \left[\begin{array}{rrr|r} 1 & 1 & 1 & 200 \\ 0 & 1 & 2 & 230 \\ 0 & 5 & 20 & 1950 \end{array}\right] \xrightarrow[R_3 - 5R_2]{R_1 - R_2} \left[\begin{array}{rrr|r} 1 & 0 & -1 & -30 \\ 0 & 1 & 2 & 230 \\ 0 & 0 & \textcircled{10} & 800 \end{array}\right]$$

$$\xrightarrow{\frac{1}{10}R_3} \left[\begin{array}{rrr|r} 1 & 0 & -1 & -30 \\ 0 & 1 & 2 & 230 \\ 0 & 0 & 1 & 80 \end{array}\right] \xrightarrow[R_2 - 2R_3]{R_1 + R_3} \left[\begin{array}{rrr|r} 1 & 0 & 0 & 50 \\ 0 & 1 & 0 & 70 \\ 0 & 0 & 1 & 80 \end{array}\right]$$

From the last augmented matrix in reduced form, we see that $x = 50$, $y = 70$, and $z = 80$. Therefore, the farmer should plant 50 acres of Crop A, 70 acres of Crop B, and 80 acres of Crop C.

USING TECHNOLOGY Systems of Linear Equations: Unique Solutions

Solving a System of Linear Equations Using the Gauss–Jordan Method

The three matrix operations can be performed on a matrix by using a graphing utility. The commands are summarized in the following table.

Operation	Calculator Function: TI-83/84	Calculator Function: TI-86	
$R_i \leftrightarrow R_j$	**rowSwap**([A], i, j)	**rSwap**(A, i, j)	or equivalent
cR_i	***row**(c, [A], i)	**multR**(c, A, i)	or equivalent
$R_i + aR_j$	***row+**(a, [A], j, i)	**mRAdd**(a, A, j, i)	or equivalent

When a row operation is performed on a matrix, the result is stored as an answer in the calculator. If another operation is performed on this matrix, then the matrix is erased. Should a mistake be made in the operation, the previous matrix may be lost. For this reason, you should store the results of each operation. We do this by pressing **STO**, followed by the name of a matrix, and then **ENTER**. We use this process in the following example.

EXAMPLE 1 Use a graphing utility to solve the following system of linear equations by the Gauss–Jordan method (see Example 5 in Section 2.2):

$$\begin{aligned} 3x - 2y + 8z &= 9 \\ -2x + 2y + z &= 3 \\ x + 2y - 3z &= 8 \end{aligned}$$

Solution Using the Gauss–Jordan method, we obtain the following sequence of equivalent matrices.

$$\left[\begin{array}{rrr|r} 3 & -2 & 8 & 9 \\ -2 & 2 & 1 & 3 \\ 1 & 2 & -3 & 8 \end{array}\right] \xrightarrow{\text{*row+ (1, [A], 2, 1) ▸ } B}$$

$$\left[\begin{array}{rrr|r} 1 & 0 & 9 & 12 \\ -2 & 2 & 1 & 3 \\ 1 & 2 & -3 & 8 \end{array}\right] \xrightarrow{\text{*row+ (2, [B], 1, 2) ▸ } C}$$

$$\left[\begin{array}{rrr|r} 1 & 0 & 9 & 12 \\ 0 & 2 & 19 & 27 \\ 1 & 2 & -3 & 8 \end{array}\right] \xrightarrow{\text{*row+ }(-1,\text{ [C], 1, 3) ▸ } B}$$

$$\left[\begin{array}{rrr|r} 1 & 0 & 9 & 12 \\ 0 & 2 & 19 & 27 \\ 0 & 2 & -12 & -4 \end{array}\right] \xrightarrow{\text{*row}(\frac{1}{2},\text{ [B], 2) ▸ } C}$$

$$\left[\begin{array}{rrr|r} 1 & 0 & 9 & 12 \\ 0 & 1 & 9.5 & 13.5 \\ 0 & 2 & -12 & -4 \end{array}\right] \xrightarrow{\text{*row+ }(-2,\text{ [C], 2, 3) ▸ } B}$$

$$\left[\begin{array}{rrr|r} 1 & 0 & 9 & 12 \\ 0 & 1 & 9.5 & 13.5 \\ 0 & 0 & -31 & -31 \end{array}\right] \xrightarrow{\text{*row}(-\frac{1}{31},\text{ [B], 3) ▸ } C}$$

$$\left[\begin{array}{rrr|r} 1 & 0 & 9 & 12 \\ 0 & 1 & 9.5 & 13.5 \\ 0 & 0 & 1 & 1 \end{array}\right] \xrightarrow{\text{*row+ }(-9,\text{ [C], 3, 1) ▸ } B}$$

$$\left[\begin{array}{rrr|r} 1 & 0 & 0 & 3 \\ 0 & 1 & 9.5 & 13.5 \\ 0 & 0 & 1 & 1 \end{array}\right] \xrightarrow{\text{*row+ }(-9.5,\text{ [B], 3, 2) ▸ } C} \left[\begin{array}{rrr|r} 1 & 0 & 0 & 3 \\ 0 & 1 & 0 & 4 \\ 0 & 0 & 1 & 1 \end{array}\right]$$

The last matrix is in row-reduced form, and we see that the solution of the system is $x = 3$, $y = 4$, and $z = 1$.

Using rref (TI-83/84 and TI-86) to Solve a System of Linear Equations

The operation **rref** (or equivalent function in your utility, if there is one) will transform an augmented matrix into one that is in row-reduced form. For example, using **rref**, we find

$$\left[\begin{array}{rrr|r} 3 & -2 & 8 & 9 \\ -2 & 2 & 1 & 3 \\ 1 & 2 & -3 & 8 \end{array}\right] \xrightarrow{\textbf{rref}} \left[\begin{array}{rrr|r} 1 & 0 & 0 & 3 \\ 0 & 1 & 0 & 4 \\ 0 & 0 & 1 & 1 \end{array}\right]$$

as obtained earlier!

Using SIMULT (TI-86) to Solve a System of Equations

The operation **SIMULT** (or equivalent operation on your utility, if there is one) of a graphing utility can be used to solve a system of n linear equations in n variables, where n is an integer between 2 and 30, inclusive.

EXAMPLE 2 Use the **SIMULT** operation to solve the system of Example 1.

Solution Call for the **SIMULT** operation. Since the system under consideration has three equations in three variables, enter $n = 3$. Next, enter a1, 1 = 3, a1, 2 = −2, a1, 3 = 8, b1 = 9, a2, 1 = −2, . . . , b3 = 8. Select <**SOLVE**>, and the display

```
x1 = 3
x2 = 4
x3 = 1
```

appears on the screen, giving $x = 3$, $y = 4$, and $z = 1$ as the required solution. ■

TECHNOLOGY EXERCISES

Use a graphing utility to solve the system of equations (a) by the Gauss–Jordan method, (b) using the rref operation, and (c) using SIMULT.

1.
$$\begin{aligned} x_1 - 2x_2 + 2x_3 - 3x_4 &= -7 \\ 3x_1 + 2x_2 - x_3 + 5x_4 &= 22 \\ 2x_1 - 3x_2 + 4x_3 - x_4 &= -3 \\ 3x_1 - 2x_2 - x_3 + 2x_4 &= 12 \end{aligned}$$

2.
$$\begin{aligned} 2x_1 - x_2 + 3x_3 - 2x_4 &= -2 \\ x_1 - 2x_2 + x_3 - 3x_4 &= 2 \\ x_1 - 5x_2 + 2x_3 + 3x_4 &= -6 \\ -3x_1 + 3x_2 - 4x_3 - 4x_4 &= 9 \end{aligned}$$

3.
$$\begin{aligned} 2x_1 + x_2 + 3x_3 - x_4 &= 9 \\ -x_1 - 2x_2 \qquad\quad - 3x_4 &= -1 \\ x_1 \qquad\quad - 3x_3 + x_4 &= 10 \\ x_1 - x_2 - x_3 - x_4 &= 8 \end{aligned}$$

4.
$$\begin{aligned} x_1 - 2x_2 - 2x_3 + x_4 &= 1 \\ 2x_1 - x_2 + 2x_3 + 3x_4 &= -2 \\ -x_1 - 5x_2 + 7x_3 - 2x_4 &= 3 \\ 3x_1 - 4x_2 + 3x_3 + 4x_4 &= -4 \end{aligned}$$

5.
$$\begin{aligned} 2x_1 - 2x_2 + 3x_3 - x_4 + 2x_5 &= 16 \\ 3x_1 + x_2 - 2x_3 + x_4 - 3x_5 &= -11 \\ x_1 + 3x_2 - 4x_3 + 3x_4 - x_5 &= -13 \\ 2x_1 - x_2 + 3x_3 - 2x_4 + 2x_5 &= 15 \\ 3x_1 + 4x_2 - 3x_3 + 5x_4 - x_5 &= -10 \end{aligned}$$

6.
$$\begin{aligned} 2.1x_1 - 3.2x_2 + 6.4x_3 + 7x_4 - 3.2x_5 &= 54.3 \\ 4.1x_1 + 2.2x_2 - 3.1x_3 - 4.2x_4 + 3.3x_5 &= -20.81 \\ 3.4x_1 - 6.2x_2 + 4.7x_3 + 2.1x_4 - 5.3x_5 &= 24.7 \\ 4.1x_1 + 7.3x_2 + 5.2x_3 + 6.1x_4 - 8.2x_5 &= 29.25 \\ 2.8x_1 + 5.2x_2 + 3.1x_3 + 5.4x_4 + 3.8x_5 &= 43.72 \end{aligned}$$

2.3 Systems of Linear Equations: Underdetermined and Overdetermined Systems

In this section, we continue our study of systems of linear equations. More specifically, we look at systems that have infinitely many solutions and those that have no solution. We also study systems of linear equations in which the number of variables is not equal to the number of equations in the system.

Solution(s) of Linear Equations

Our first two examples illustrate the situation in which a system of linear equations has infinitely many solutions.

EXAMPLE 1 A System of Equations with an Infinite Number of Solutions Solve the system of linear equations given by

$$\begin{aligned} x + 2y &= 4 \\ 3x + 6y &= 12 \end{aligned} \tag{9}$$

Solution Using the Gauss–Jordan elimination method, we obtain the following system of equivalent matrices:

$$\left[\begin{array}{cc|c} \textcircled{1} & 2 & 4 \\ 3 & 6 & 12 \end{array}\right] \xrightarrow{R_2 - 3R_1} \left[\begin{array}{cc|c} 1 & 2 & 4 \\ 0 & 0 & 0 \end{array}\right]$$

The last augmented matrix is in row-reduced form. Interpreting it as a system of linear equations, we see that the given System (9) is equivalent to the single equation

$$x + 2y = 4 \quad \text{or} \quad x = 4 - 2y$$

If we assign a particular value to y—say, $y = 0$—we obtain $x = 4$, giving the solution $(4, 0)$ to System (9). By setting $y = 1$, we obtain the solution $(2, 1)$. In general, if we set $y = t$, where t represents some real number (called a parameter), we obtain the solution given by $(4 - 2t, t)$. Since the parameter t may be any real number, we see that System (9) has infinitely many solutions. Geometrically, the solutions of System (9) lie on the line on the plane with equation $x + 2y = 4$. The two equations in the system have the same graph (straight line), which you can verify graphically. ■

EXAMPLE 2 A System of Equations with an Infinite Number of Solutions Solve the system of linear equations given by

$$\begin{aligned} x + 2y - 3z &= -2 \\ 3x - y - 2z &= 1 \\ 2x + 3y - 5z &= -3 \end{aligned} \tag{10}$$

Solution Using the Gauss–Jordan elimination method, we obtain the following sequence of equivalent augmented matrices:

$$\left[\begin{array}{ccc|c} \textcircled{1} & 2 & -3 & -2 \\ 3 & -1 & -2 & 1 \\ 2 & 3 & -5 & -3 \end{array}\right] \xrightarrow[R_3 - 2R_1]{R_2 - 3R_1} \left[\begin{array}{ccc|c} 1 & 2 & -3 & -2 \\ 0 & \textcircled{-7} & 7 & 7 \\ 0 & -1 & 1 & 1 \end{array}\right] \xrightarrow{-\frac{1}{7}R_2}$$

$$\left[\begin{array}{ccc|c} 1 & 2 & -3 & -2 \\ 0 & 1 & -1 & -1 \\ 0 & -1 & 1 & 1 \end{array}\right] \xrightarrow[R_3 + R_2]{R_1 - 2R_2} \left[\begin{array}{ccc|c} 1 & 0 & -1 & 0 \\ 0 & 1 & -1 & -1 \\ 0 & 0 & 0 & 0 \end{array}\right]$$

The last augmented matrix is in row-reduced form. Interpreting it as a system of linear equations gives

$$\begin{aligned} x - z &= 0 \\ y - z &= -1 \end{aligned}$$

a system of two equations in the three variables x, y, and z.

Let's now single out one variable—say, z—and solve for x and y in terms of it. We obtain

$$\begin{aligned} x &= z \\ y &= z - 1 \end{aligned}$$

If we set $z = t$, where t is a parameter, then System (10) has infinitely many solutions given by $(t, t - 1, t)$. For example, letting $t = 0$ gives the solution $(0, -1, 0)$, and letting $t = 1$ gives the solution $(1, 0, 1)$. Geometrically, the solutions of System (10) lie on the straight line in three-dimensional space given by the intersection of the three planes determined by the three equations in the system.

Note In Example 2, we chose the parameter to be z because it is more convenient to solve for x and y (both the x- and y-columns are in unit form) in terms of z.

The next example shows what happens in the elimination procedure when the system does not have a solution.

EXAMPLE 3 A System of Equations That Has No Solution Solve the system of linear equations given by

$$\begin{aligned} x + y + z &= 1 \\ 3x - y - z &= 4 \\ x + 5y + 5z &= -1 \end{aligned} \tag{11}$$

Solution Using the Gauss–Jordan elimination method, we obtain the following sequence of equivalent augmented matrices:

$$\left[\begin{array}{ccc|c} \textcircled{1} & 1 & 1 & 1 \\ 3 & -1 & -1 & 4 \\ 1 & 5 & 5 & -1 \end{array}\right] \xrightarrow[R_3 - R_1]{R_2 - 3R_1} \left[\begin{array}{ccc|c} 1 & 1 & 1 & 1 \\ 0 & -4 & -4 & 1 \\ 0 & 4 & 4 & -2 \end{array}\right]$$

$$\xrightarrow{R_3 + R_2} \left[\begin{array}{ccc|c} 1 & 1 & 1 & 1 \\ 0 & -4 & -4 & 1 \\ 0 & 0 & 0 & -1 \end{array}\right]$$

Observe that row 3 in the last matrix reads $0x + 0y + 0z = -1$—that is, $0 = -1$! We therefore conclude that System (11) is inconsistent and has no solution. Geometrically, we have a situation in which two of the planes intersect in a straight line but the third plane is parallel to this line of intersection of the two planes and does not intersect it. Consequently, there is no point of intersection of the three planes.

Example 3 illustrates the following more general result of using the Gauss–Jordan elimination procedure.

> **Systems with No Solution**
>
> If there is a row in an augmented matrix containing all zeros to the left of the vertical line and a nonzero entry to the right of the line, then the corresponding system of equations has no solution.

It may have dawned on you that in all the previous examples, we have dealt only with systems involving exactly the same number of linear equations as there are variables. However, systems in which the number of equations differs from the number of variables also occur in practice. Indeed, we will consider such systems in Examples 4 and 5.

The following theorem provides us with some preliminary information on a system of linear equations.

THEOREM 1

a. If the number of equations is greater than or equal to the number of variables in a linear system, then one of the following is true:

i. The system has no solution.
ii. The system has exactly one solution.
iii. The system has infinitely many solutions.

b. If there are fewer equations than variables in a linear system, then the system either has no solution or has infinitely many solutions.

Note Theorem 1 may be used to tell us, before we even begin to solve a problem, what the nature of the solution may be.

Although we will not prove this theorem, you should recall that we have illustrated geometrically part (a) for the case in which there are exactly as many equations (three) as there are variables. To show the validity of part (b), let us once again consider the case in which a system has three variables. Now, if there is only one equation in the system, then it is clear that there are infinitely many solutions corresponding geometrically to all the points lying on the plane represented by the equation.

Next, if there are two equations in the system, then *only* the following possibilities exist:

1. The two planes are parallel and distinct (Figure 6a).
2. The two planes intersect in a straight line (Figure 6b).
3. The two planes are coincident (the two equations define the same plane) (Figure 6c).

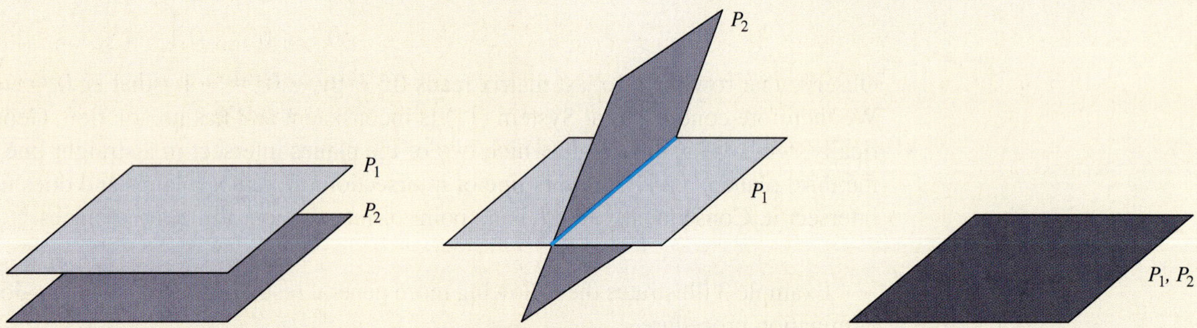

(a) No solution **(b)** Infinitely many solutions **(c)** Infinitely many solutions

FIGURE **6**

Thus, either there is no solution or there are infinitely many solutions corresponding to the points lying on a line of intersection of the two planes or on a single plane determined by the two equations. In the case in which two planes intersect in a straight line, the solutions will involve one parameter, and in the case in which the two planes are coincident, the solutions will involve two parameters.

Explore and Discuss

Give a geometric interpretation of Theorem 1 for a linear system composed of equations involving two variables. Specifically, illustrate what can happen if there are three linear equations in the system (the case involving two linear equations was discussed in Section 2.1). What if there are four linear equations? What if there is only one linear equation in the system?

EXAMPLE 4 A System with More Equations Than Variables Solve the following system of linear equations:

$$\begin{aligned} x + 2y &= 4 \\ x - 2y &= 0 \\ 4x + 3y &= 12 \end{aligned}$$

Solution We obtain the following sequence of equivalent augmented matrices:

$$\left[\begin{array}{cc|c} \textcircled{1} & 2 & 4 \\ 1 & -2 & 0 \\ 4 & 3 & 12 \end{array}\right] \xrightarrow[R_3 - 4R_1]{R_2 - R_1} \left[\begin{array}{cc|c} 1 & 2 & 4 \\ 0 & \textcircled{-4} & -4 \\ 0 & -5 & -4 \end{array}\right] \xrightarrow{-\frac{1}{4}R_2}$$

$$\left[\begin{array}{cc|c} 1 & 2 & 4 \\ 0 & 1 & 1 \\ 0 & -5 & -4 \end{array}\right] \xrightarrow[R_3 + 5R_2]{R_1 - 2R_2} \left[\begin{array}{cc|c} 1 & 0 & 2 \\ 0 & 1 & 1 \\ 0 & 0 & 1 \end{array}\right]$$

The last row of the row-reduced augmented matrix implies that $0 = 1$, which is impossible, so we conclude that the given system has no solution. Geometrically, the three lines defined by the three equations in the system do not intersect at a point. (To see this for yourself, draw the graphs of these equations.) ■

EXAMPLE 5 A System with More Variables Than Equations Solve the following system of linear equations:

$$\begin{aligned} x + 2y - 3z + w &= -2 \\ 3x - y - 2z - 4w &= 1 \\ 2x + 3y - 5z + w &= -3 \end{aligned}$$

Solution First, observe that the given system consists of three equations in four variables, so by Theorem 1b, either the system has no solution or it has infinitely many solutions. To solve it, we use the Gauss–Jordan method and obtain the following sequence of equivalent augmented matrices:

$$\left[\begin{array}{cccc|c} \textcircled{1} & 2 & -3 & 1 & -2 \\ 3 & -1 & -2 & -4 & 1 \\ 2 & 3 & -5 & 1 & -3 \end{array}\right] \xrightarrow[R_3 - 2R_1]{R_2 - 3R_1} \left[\begin{array}{cccc|c} 1 & 2 & -3 & 1 & -2 \\ 0 & \textcircled{-7} & 7 & -7 & 7 \\ 0 & -1 & 1 & -1 & 1 \end{array}\right] \xrightarrow{-\frac{1}{7}R_2}$$

$$\left[\begin{array}{cccc|c} 1 & 2 & -3 & 1 & -2 \\ 0 & 1 & -1 & 1 & -1 \\ 0 & -1 & 1 & -1 & 1 \end{array}\right] \xrightarrow[R_3 + R_2]{R_1 - 2R_2} \left[\begin{array}{cccc|c} 1 & 0 & -1 & -1 & 0 \\ 0 & 1 & -1 & 1 & -1 \\ 0 & 0 & 0 & 0 & 0 \end{array}\right]$$

The last augmented matrix is in row-reduced form. Observe that the given system is equivalent to the system

$$\begin{aligned} x - z - w &= 0 \\ y - z + w &= -1 \end{aligned}$$

of two equations in four variables. Thus, we may solve for two of the variables in terms of the other two. Letting $z = s$ and $w = t$ (where s and t are any real numbers), we find that

$$\begin{aligned} x &= s + t \\ y &= s - t - 1 \\ z &= s \\ w &= t \end{aligned}$$

The solutions may be written in the form $(s + t, s - t - 1, s, t)$. Geometrically, the three equations in the system represent three hyperplanes in four-dimensional space (since there are four variables), and their "points" of intersection lie in a two-dimensional subspace of four-space (since there are two parameters).

Note In Example 5, we assigned parameters to z and w rather than to x and y because x and y are readily solved in terms of z and w.

The following example illustrates a situation in which a system of linear equations has infinitely many solutions.

APPLIED EXAMPLE 6 Traffic Control Figure 7 shows the flow of downtown traffic in a certain city during the rush hours on a typical weekday. The arrows indicate the direction of traffic flow on each one-way road, and the average number of vehicles per hour entering and leaving each intersection appears beside each road. 5th Avenue and 6th Avenue can each handle up to 2000 vehicles per hour without causing congestion, whereas the maximum capacity of both 4th Street and 5th Street is 1000 vehicles per hour. The flow of traffic is controlled by traffic lights installed at each of the four intersections.

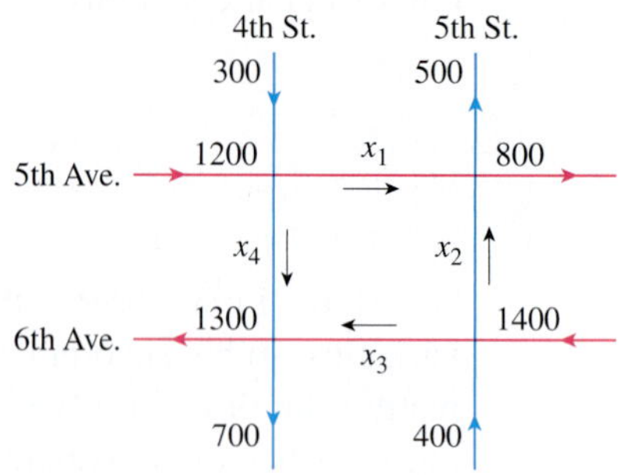

FIGURE 7

a. Write a general expression involving the rates of flow—x_1, x_2, x_3, x_4—and suggest two possible traffic-flow patterns that will ensure no traffic congestion.

b. Suppose that the part of 4th Street between 5th Avenue and 6th Avenue is to be resurfaced and that traffic flow between the two junctions must therefore be reduced to at most 300 vehicles per hour. Find two possible traffic-flow patterns that will result in a smooth flow of traffic.

Solution

a. To avoid congestion, all traffic entering an intersection must also leave that intersection. Applying this condition to each of the four intersections in a clockwise direction beginning with the 5th Avenue and 4th Street intersection, we obtain the following equations:

$$\begin{aligned} 1500 &= x_1 + x_4 \\ 1300 &= x_1 + x_2 \\ 1800 &= x_2 + x_3 \\ 2000 &= x_3 + x_4 \end{aligned}$$

This system of four linear equations in the four variables x_1, x_2, x_3, x_4 may be rewritten in the more standard form

$$\begin{aligned} x_1 \qquad\qquad\quad + x_4 &= 1500 \\ x_1 + x_2 \qquad\qquad\quad &= 1300 \\ x_2 + x_3 \qquad &= 1800 \\ x_3 + x_4 &= 2000 \end{aligned}$$

Using the Gauss–Jordan elimination method to solve the system, we obtain

$$\left[\begin{array}{cccc|c} 1 & 0 & 0 & 1 & 1500 \\ 1 & 1 & 0 & 0 & 1300 \\ 0 & 1 & 1 & 0 & 1800 \\ 0 & 0 & 1 & 1 & 2000 \end{array}\right] \xrightarrow{R_2 - R_1} \left[\begin{array}{cccc|c} 1 & 0 & 0 & 1 & 1500 \\ 0 & 1 & 0 & -1 & -200 \\ 0 & 1 & 1 & 0 & 1800 \\ 0 & 0 & 1 & 1 & 2000 \end{array}\right]$$

$$\xrightarrow{R_3 - R_2} \left[\begin{array}{cccc|c} 1 & 0 & 0 & 1 & 1500 \\ 0 & 1 & 0 & -1 & -200 \\ 0 & 0 & 1 & 1 & 2000 \\ 0 & 0 & 1 & 1 & 2000 \end{array}\right]$$

$$\xrightarrow{R_4 - R_3} \left[\begin{array}{cccc|c} 1 & 0 & 0 & 1 & 1500 \\ 0 & 1 & 0 & -1 & -200 \\ 0 & 0 & 1 & 1 & 2000 \\ 0 & 0 & 0 & 0 & 0 \end{array}\right]$$

The last augmented matrix is in row-reduced form and is equivalent to a system of three linear equations in the four variables x_1, x_2, x_3, x_4. Thus, we may express three of the variables—say, x_1, x_2, x_3—in terms of the fourth, x_4. Setting $x_4 = t$ (t a parameter), we may write the infinitely many solutions of the system as

$$\begin{aligned} x_1 &= 1500 - t \\ x_2 &= -200 + t \\ x_3 &= 2000 - t \\ x_4 &= t \end{aligned}$$

Observe that for a meaningful solution, we must have $200 \leq t \leq 1000$, since x_1, x_2, x_3, and x_4 must all be nonnegative and the maximum capacity of a street is 1000. For example, picking $t = 300$ gives the flow pattern

$$x_1 = 1200 \qquad x_2 = 100 \qquad x_3 = 1700 \qquad x_4 = 300$$

Selecting $t = 500$ gives the flow pattern

$$x_1 = 1000 \qquad x_2 = 300 \qquad x_3 = 1500 \qquad x_4 = 500$$

b. In this case, x_4 must not exceed 300. Again, using the results of part (a), we find, upon setting $x_4 = t = 300$, the flow pattern

$$x_1 = 1200 \qquad x_2 = 100 \qquad x_3 = 1700 \qquad x_4 = 300$$

obtained earlier. Picking $t = 250$ gives the flow pattern

$$x_1 = 1250 \qquad x_2 = 50 \qquad x_3 = 1750 \qquad x_4 = 250$$ ■

2.3 Self-Check Exercises

1. The following augmented matrix in row-reduced form is equivalent to the augmented matrix of a certain system of linear equations. Use this result to solve the system of equations.

$$\left[\begin{array}{ccc|c} 1 & 0 & -1 & 3 \\ 0 & 1 & 5 & -2 \\ 0 & 0 & 0 & 0 \end{array}\right]$$

2. Solve the system of linear equations

$$\begin{aligned} 2x - 3y + z &= 6 \\ x + 2y + 4z &= -4 \\ x - 5y - 3z &= 10 \end{aligned}$$

using the Gauss–Jordan elimination method.

3. Solve the system of linear equations

$$\begin{aligned} x - 2y + 3z &= 9 \\ 2x + 3y - z &= 4 \\ x + 5y - 4z &= 2 \end{aligned}$$

using the Gauss–Jordan elimination method.

Solutions to Self-Check Exercises 2.3 can be found on page 109.

2.3 Concept Questions

1. If a system of linear equations has the same number of equations or more equations than variables, what can you say about the nature of its solution(s)?

2. If a system of linear equations has fewer equations than variables, what can you say about the nature of its solution(s)?

3. A system consists of three linear equations in four variables. Can the system have a unique solution?

2.3 Exercises

In Exercises 1–14, given that the augmented matrix in row-reduced form is equivalent to the augmented matrix of a system of linear equations, (a) determine whether the system has a solution and (b) find the solution or solutions to the system, if they exist.

1. $\left[\begin{array}{ccc|c} 1 & 0 & 0 & 3 \\ 0 & 1 & 0 & -1 \\ 0 & 0 & 1 & 2 \end{array}\right]$

2. $\left[\begin{array}{ccc|c} 1 & 0 & 0 & 3 \\ 0 & 1 & 0 & -2 \\ 0 & 0 & 1 & 1 \end{array}\right]$

3. $\left[\begin{array}{cc|c} 1 & 0 & 2 \\ 0 & 1 & 5 \\ 0 & 0 & 0 \end{array}\right]$

4. $\left[\begin{array}{ccc|c} 1 & 0 & 0 & 3 \\ 0 & 1 & 0 & 1 \\ 0 & 0 & 0 & 0 \end{array}\right]$

5. $\left[\begin{array}{cc|c} 1 & 0 & 2 \\ 0 & 1 & 3 \\ 0 & 0 & -1 \end{array}\right]$

6. $\left[\begin{array}{ccc|c} 1 & 0 & 0 & 2 \\ 0 & 0 & 0 & 1 \end{array}\right]$

7. $\left[\begin{array}{ccc|c} 1 & 0 & 1 & 4 \\ 0 & 1 & 0 & -2 \end{array}\right]$

8. $\left[\begin{array}{cccc|c} 1 & 0 & 0 & 0 & 3 \\ 0 & 1 & 1 & 0 & -1 \\ 0 & 0 & 0 & 1 & 2 \end{array}\right]$

9. $\left[\begin{array}{cccc|c} 1 & 0 & 0 & 0 & 2 \\ 0 & 1 & 0 & 0 & 1 \\ 0 & 0 & 1 & 0 & 3 \\ 0 & 0 & 0 & 0 & 1 \end{array}\right]$

10. $\left[\begin{array}{ccc|c} 1 & 0 & 0 & 4 \\ 0 & 1 & 0 & -1 \\ 0 & 0 & 1 & 3 \\ 0 & 0 & 0 & 1 \end{array}\right]$

11. $\left[\begin{array}{cccc|c} 1 & 0 & 0 & 0 & 4 \\ 0 & 1 & 0 & 0 & -1 \\ 0 & 0 & 1 & 1 & 3 \\ 0 & 0 & 0 & 0 & 0 \end{array}\right]$

12. $\left[\begin{array}{cccc|c} 0 & 1 & 0 & 1 & 3 \\ 0 & 0 & 1 & -2 & 4 \\ 0 & 0 & 0 & 0 & 0 \\ 0 & 0 & 0 & 0 & 0 \end{array}\right]$

13. $\left[\begin{array}{cccc|c} 1 & 0 & 3 & 0 & 2 \\ 0 & 1 & -1 & 0 & 1 \\ 0 & 0 & 0 & 0 & 0 \\ 0 & 0 & 0 & 0 & 0 \end{array}\right]$

14. $\left[\begin{array}{cccc|c} 1 & 0 & 3 & -1 & 4 \\ 0 & 1 & -2 & 3 & 2 \\ 0 & 0 & 0 & 0 & 0 \\ 0 & 0 & 0 & 0 & 0 \end{array}\right]$

In Exercises 15–36, solve the system of linear equations, using the Gauss–Jordan elimination method.

15. $\begin{aligned} 2x - y &= 3 \\ x + 2y &= 4 \\ 2x + 3y &= 7 \end{aligned}$

16. $\begin{aligned} x + 2y &= 3 \\ 2x - 3y &= -8 \\ x - 4y &= -9 \end{aligned}$

17. $$\begin{aligned} 3x - 2y &= -3 \\ 2x + y &= 3 \\ x - 2y &= -5 \end{aligned}$$

18. $$\begin{aligned} 2x + 3y &= 2 \\ x + 3y &= -2 \\ x - y &= 3 \end{aligned}$$

19. $$\begin{aligned} 3x - 2y &= 5 \\ -x + 3y &= -4 \\ 2x - 4y &= 6 \end{aligned}$$

20. $$\begin{aligned} 4x + 6y &= 8 \\ 3x - 2y &= -7 \\ x + 3y &= 5 \end{aligned}$$

21. $$\begin{aligned} x - 2y &= 2 \\ 7x - 14y &= 14 \\ 3x - 6y &= 6 \end{aligned}$$

22. $$\begin{aligned} 3x - y + 2z &= 5 \\ x - y + 2z &= 1 \\ 5x - 2y + 4z &= 12 \end{aligned}$$

23. $$\begin{aligned} x + 2y + z &= -2 \\ -2x - 3y - z &= 1 \\ 2x + 4y + 2z &= -4 \end{aligned}$$

24. $$\begin{aligned} 3y + 2z &= 4 \\ 2x - y - 3z &= 3 \\ 2x + 2y - z &= 7 \end{aligned}$$

25. $$\begin{aligned} 3x + 2y &= 4 \\ -\tfrac{3}{2}x - y &= -2 \\ 6x + 4y &= 8 \end{aligned}$$

26. $$\begin{aligned} 2x_1 - x_2 + x_3 &= -4 \\ 3x_1 - \tfrac{3}{2}x_2 + \tfrac{3}{2}x_3 &= -6 \\ -6x_1 + 3x_2 - 3x_3 &= 12 \end{aligned}$$

27. $$\begin{aligned} x + y - 2z &= -3 \\ 2x - y + 3z &= 7 \\ x - 2y + 5z &= 0 \end{aligned}$$

28. $$\begin{aligned} 2x_1 + 6x_2 - 5x_3 &= 5 \\ x_1 + 3x_2 + x_3 + 7x_4 &= -1 \\ 3x_1 + 9x_2 - x_3 + 13x_4 &= 1 \end{aligned}$$

29. $$\begin{aligned} x - 2y + 3z &= 4 \\ 2x + 3y - z &= 2 \\ x + 2y - 3z &= -6 \end{aligned}$$

30. $$\begin{aligned} x_1 - 2x_2 + x_3 &= -3 \\ 2x_1 + x_2 - 2x_3 &= 2 \\ x_1 + 3x_2 - 3x_3 &= 5 \end{aligned}$$

31. $$\begin{aligned} 4x + y - z &= 4 \\ 8x + 2y - 2z &= 8 \end{aligned}$$

32. $$\begin{aligned} x_1 + 2x_2 + 4x_3 &= 2 \\ x_1 + x_2 + 2x_3 &= 1 \end{aligned}$$

33. $$\begin{aligned} 2x + y - 3z &= 1 \\ x - y + 2z &= 1 \\ 5x - 2y + 3z &= 6 \end{aligned}$$

34. $$\begin{aligned} 3x - 9y + 6z &= -12 \\ x - 3y + 2z &= -4 \\ 2x - 6y + 4z &= 8 \end{aligned}$$

35. $$\begin{aligned} x + 2y - z &= -4 \\ 2x + y + z &= 7 \\ x + 3y + 2z &= 7 \\ x - 3y + z &= 9 \end{aligned}$$

36. $$\begin{aligned} 3x - 2y + z &= 4 \\ x + 3y - 4z &= -3 \\ 2x - 3y + 5z &= 7 \\ x - 8y + 9z &= 10 \end{aligned}$$

37. Management Decisions The management of Hartman Rent-A-Car has allocated $1,512,000 to purchase 60 new automobiles to add to the existing fleet of rental cars. The company will choose from compact, mid-sized, and full-sized cars costing $18,000, $28,800, and $39,600 each, respectively. Find formulas giving the options available to the company. Give two specific options. (*Note*: Your answers will *not* be unique.)

38. Diet Planning A dietitian wishes to plan a meal around three foods. The meal is to include 8800 units of vitamin A, 3380 units of vitamin C, and 1020 units of calcium. The number of units of the vitamins and calcium in each ounce of the foods is summarized in the following table:

	Food I	Food II	Food III
Vitamin A	400	1200	800
Vitamin C	110	570	340
Calcium	90	30	60

Determine the amount of each food the dietitian should include in the meal in order to meet the vitamin and calcium requirements.

39. Diet Planning Refer to Exercise 38. In planning for another meal, the dietitian changes the requirement of vitamin C from 3380 units to 2160 units. All other requirements remain the same. Show that such a meal cannot be planned around the same foods.

40. Production Scheduling Ace Novelty manufactures Giant Pandas, Saint Bernards, and Big Birds. Each Giant Panda requires 1.5 yd^2 of plush, 30 ft^3 of stuffing, and 5 pieces of trim; each Saint Bernard requires 2 yd^2 of plush, 35 ft^3 of stuffing, and 8 pieces of trim; and each Big Bird requires 2.5 yd^2 of plush, 25 ft^3 of stuffing, and 15 pieces of trim. If 4700 yd^2 of plush, 65,000 ft^3 of stuffing, and 23,400 pieces of trim are available, how many of each of the stuffed animals should the company manufacture if all the material is to be used? Give two specific options.

41. Asset Allocation Mr. and Mrs. Garcia have a total of $100,000 to be invested in stocks, bonds, and a money market account. The stocks have a rate of return of 6%/year, while the bonds and the money market account pay 4%/year and 2%/year, respectively. The Garcias have stipulated that the amount invested in stocks should be equal to the sum of the amount invested in bonds and 3 times the amount invested in the money market account. How should the Garcias allocate their resources if they require an annual income of $5,000 from their investments? Give two specific options.

42. Traffic Control The accompanying figure shows the flow of traffic near a city's Civic Center during the rush hours on a typical weekday. Each road can handle a maximum of 1000 cars/hour without causing congestion. The flow of traffic is controlled by traffic lights at each of the five intersections.

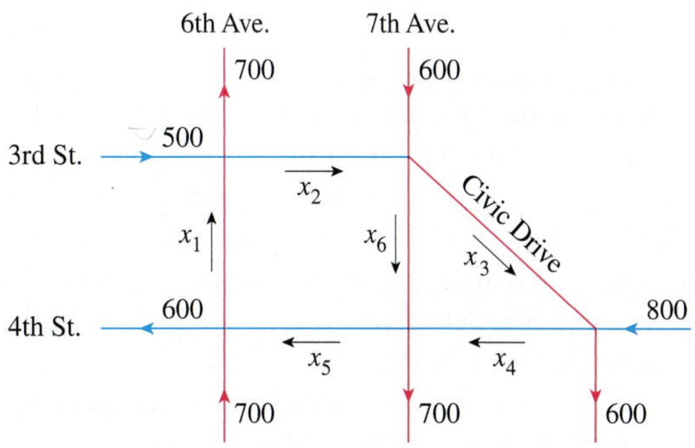

a. Set up a system of linear equations describing the traffic flow.

b. Solve the system devised in part (a), and suggest two possible traffic-flow patterns that will ensure no traffic congestion.

c. Suppose 7th Avenue between 3rd and 4th Streets is soon to be closed for road repairs. Find one possible traffic-flow pattern that will result in a smooth flow of traffic.

43. **Traffic Control** The accompanying figure shows the flow of downtown traffic during the rush hours on a typical weekday. Each avenue can handle up to 1500 vehicles/hour without causing congestion, whereas the maximum capacity of each street is 1000 vehicles/hour. The flow of traffic is controlled by traffic lights at each of the six intersections.

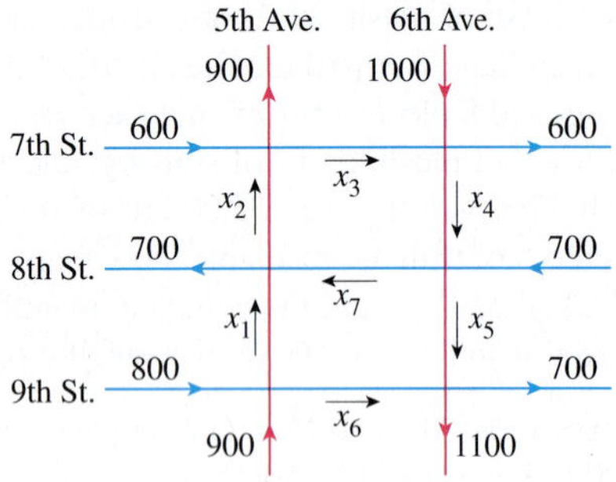

a. Set up a system of linear equations describing the traffic flow.

b. Solve the system devised in part (a), and suggest two possible traffic-flow patterns that will ensure no traffic congestion.

c. Suppose the traffic flow along 9th Street between 5th and 6th Avenues, x_6, is restricted because of sewer construction. What is the minimum permissible traffic flow along this road that will not result in traffic congestion?

44. **Criminal Justice** The body of Alisha was discovered in the basement of the company where she worked. The medical examiner determined that she was killed between 9 P.M. and 10 P.M. After a preliminary investigation, homicide detectives decided to question three of her coworkers: Alex, Bob, and Charlie. The detectives were told by the suspects that Alex left work at 9:18 P.M. on the day of the homicide, walked 1000 ft to his car, and drove 16 mi to his house, arriving home at 9:40 P.M. Bob left work at 9:12 P.M., walked 400 ft to his car, and drove 12 mi to his house, arriving home at 9:30 P.M. Charlie punched out at 9:35:30 P.M., walked 600 ft to his car, and drove 20 mi to his house, arriving home at 10 P.M. After analyzing this information, the detectives singled out Bob for further questioning. Explain why.

Hint: Suppose that each man walks at an average speed of v ft/sec and drives his car at an average speed of w ft/sec.

(i) Show that this leads to the equations

$$\frac{1000}{v} + \frac{84{,}480}{w} = 1320 \qquad \frac{400}{v} + \frac{63{,}360}{w} = 1080$$

$$\frac{600}{v} + \frac{105{,}600}{w} = 1470$$

(ii) Show that these equations can be expressed in the form

$$1000x + 84{,}480y = 1320 \qquad 400x + 63{,}360y = 1080$$

$$600x + 105{,}600y = 1470$$

Then solve each possible pair of equations in System (ii).

45. Determine the value of k such that the following system of linear equations has a solution, and then find the solution:

$$\begin{aligned} 2x + 3y &= 2 \\ x + 4y &= 6 \\ 5x + ky &= 2 \end{aligned}$$

46. Determine the value of k such that the following system of linear equations has infinitely many solutions, and then find the solutions:

$$\begin{aligned} 3x + 4y &= 2 \\ kx + 8y &= 4 \\ 12x + 16y &= 8 \end{aligned}$$

47. Determine the value of k such that the following system of linear equations has infinitely many solutions, and then find the solutions:

$$\begin{aligned} 3x + 2y - z &= 8 \\ 2x \qquad - 2z &= 4 \\ y + kz &= 1 \end{aligned}$$

48. Determine the value of k such that the following system of linear equations has infinitely many solutions, and then find the solutions:

$$\begin{aligned} 3x - 2y + 4z &= 12 \\ -9x + 6y - 12z &= k \end{aligned}$$

49. Determine the value(s) of k such that the following system of linear equations has no solution:

$$\begin{aligned} 2x - 3y + 4z &= 12 \\ 6x - 9y + kz &= 36 \end{aligned}$$

50. Solve the system

$$\begin{aligned} x^2 + 2y^2 &= 9 \\ 2x^2 - 3y^2 &= -10 \end{aligned}$$

51. Solve the equation

$$\begin{aligned} \frac{3}{x} - \frac{4}{y} &= -15 \\ \frac{5}{x} + \frac{6}{y} &= 13 \end{aligned}$$

52. Solve the system

$$\begin{aligned} \frac{1}{x} + \frac{1}{y} + \frac{1}{z} &= -1 \\ \frac{2}{x} + \frac{3}{y} + \frac{2}{z} &= 3 \\ \frac{2}{x} + \frac{1}{y} + \frac{2}{z} &= -7 \end{aligned}$$

In Exercises 53 and 54, determine whether the statement is true or false. If it is true, explain why it is true. If it is false, give an example to show why it is false.

53. A system of linear equations having fewer equations than variables has no solution, a unique solution, or infinitely many solutions.

54. A system of linear equations having more equations than variables has no solution, a unique solution, or infinitely many solutions.

2.3 Solutions to Self-Check Exercises

1. Let x, y, and z denote the variables. Then the given row-reduced augmented matrix tells us that the system of linear equations is equivalent to the two equations

$$\begin{aligned} x \quad - z &= 3 \\ y + 5z &= -2 \end{aligned}$$

Letting $z = t$, where t is a parameter, we find the infinitely many solutions given by

$$\begin{aligned} x &= t + 3 \\ y &= -5t - 2 \\ z &= t \end{aligned}$$

2. We obtain the following sequence of equivalent augmented matrices:

$$\left[\begin{array}{rrr|r} 2 & -3 & 1 & 6 \\ 1 & 2 & 4 & -4 \\ 1 & -5 & -3 & 10 \end{array}\right] \xrightarrow{R_1 \leftrightarrow R_2}$$

$$\left[\begin{array}{rrr|r} \textcircled{1} & 2 & 4 & -4 \\ 2 & -3 & 1 & 6 \\ 1 & -5 & -3 & 10 \end{array}\right] \xrightarrow[R_3 - R_1]{R_2 - 2R_1}$$

$$\left[\begin{array}{rrr|r} 1 & 2 & 4 & -4 \\ 0 & \textcircled{-7} & -7 & 14 \\ 0 & -7 & -7 & 14 \end{array}\right] \xrightarrow{-\frac{1}{7}R_2}$$

$$\left[\begin{array}{rrr|r} 1 & 2 & 4 & -4 \\ 0 & 1 & 1 & -2 \\ 0 & -7 & -7 & 14 \end{array}\right] \xrightarrow[R_3 + 7R_2]{R_1 - 2R_2} \left[\begin{array}{rrr|r} 1 & 0 & 2 & 0 \\ 0 & 1 & 1 & -2 \\ 0 & 0 & 0 & 0 \end{array}\right]$$

The last augmented matrix, which is in row-reduced form, tells us that the given system of linear equations is equivalent to the following system of two equations:

$$\begin{aligned} x \quad + 2z &= 0 \\ y + z &= -2 \end{aligned}$$

Letting $z = t$, where t is a parameter, we see that the infinitely many solutions are given by

$$\begin{aligned} x &= -2t \\ y &= -t - 2 \\ z &= t \end{aligned}$$

3. We obtain the following sequence of equivalent augmented matrices:

$$\left[\begin{array}{rrr|r} \textcircled{1} & -2 & 3 & 9 \\ 2 & 3 & -1 & 4 \\ 1 & 5 & -4 & 2 \end{array}\right] \xrightarrow[R_3 - R_1]{R_2 - 2R_1}$$

$$\left[\begin{array}{rrr|r} 1 & -2 & 3 & 9 \\ 0 & 7 & -7 & -14 \\ 0 & 7 & -7 & -7 \end{array}\right] \xrightarrow{R_3 - R_2} \left[\begin{array}{rrr|r} 1 & -2 & 3 & 9 \\ 0 & 7 & -7 & -14 \\ 0 & 0 & 0 & 7 \end{array}\right]$$

Since the last row of the final augmented matrix is equivalent to the equation $0 = 7$, a contradiction, we conclude that the given system has no solution.

USING TECHNOLOGY Systems of Linear Equations: Underdetermined and Overdetermined Systems

We can use the row operations of a graphing utility to solve a system of m linear equations in n unknowns by the Gauss–Jordan method, as we did in the previous technology section. We can also use the **rref** or equivalent operation to obtain the row-reduced form without going through all the steps of the Gauss–Jordan method. The **SIMULT** function, however, cannot be used to solve a system in which the number of equations and the number of variables are not the same.

EXAMPLE 1 Solve the system

$$\begin{aligned} x_1 - 2x_2 + 4x_3 &= 2 \\ 2x_1 + x_2 - 2x_3 &= -1 \\ 3x_1 - x_2 + 2x_3 &= 1 \\ 2x_1 + 6x_2 - 12x_3 &= -6 \end{aligned}$$

Solution First, we enter the augmented matrix A into the calculator as

$$A = \left[\begin{array}{rrr|r} 1 & -2 & 4 & 2 \\ 2 & 1 & -2 & -1 \\ 3 & -1 & 2 & 1 \\ 2 & 6 & -12 & -6 \end{array}\right]$$

Then using the **rref** or equivalent operation, we obtain the equivalent matrix

$$\left[\begin{array}{rrr|r} 1 & 0 & 0 & 0 \\ 0 & 1 & -2 & -1 \\ 0 & 0 & 0 & 0 \\ 0 & 0 & 0 & 0 \end{array}\right]$$

in reduced form. Thus, the given system is equivalent to

$$\begin{aligned} x_1 \qquad\qquad &= 0 \\ x_2 - 2x_3 &= -1 \end{aligned}$$

If we let $x_3 = t$, where t is a parameter, then we find that the solutions are $(0, 2t - 1, t)$.

TECHNOLOGY EXERCISES

Use a graphing utility to solve the system of equations using the rref or equivalent operation.

1. $$\begin{aligned} 2x_1 - x_2 - x_3 &= 0 \\ 3x_1 - 2x_2 - x_3 &= -1 \\ -x_1 + 2x_2 - x_3 &= 3 \\ 2x_2 - 2x_3 &= 4 \end{aligned}$$

2. $$\begin{aligned} 3x_1 + x_2 - 4x_3 &= 5 \\ 2x_1 - 3x_2 + 2x_3 &= -4 \\ -x_1 - 2x_2 + 4x_3 &= 6 \\ 4x_1 + 3x_2 - 5x_3 &= 9 \end{aligned}$$

3. $$\begin{aligned} 2x_1 + 3x_2 + 2x_3 + x_4 &= -1 \\ x_1 - x_2 + x_3 - 2x_4 &= -8 \\ 5x_1 + 6x_2 - 2x_3 + 2x_4 &= 11 \\ x_1 + 3x_2 + 8x_3 + x_4 &= -14 \end{aligned}$$

4. $$\begin{aligned} x_1 - x_2 + 3x_3 - 6x_4 &= 2 \\ x_1 + x_2 + x_3 - 2x_4 &= 2 \\ -2x_1 - x_2 + x_3 + 2x_4 &= 0 \end{aligned}$$

5. $$\begin{aligned} x_1 + x_2 - x_3 - x_4 &= -1 \\ x_1 - x_2 + x_3 + 4x_4 &= -6 \\ 3x_1 + x_2 - x_3 + 2x_4 &= -4 \\ 5x_1 + x_2 - 3x_3 + x_4 &= -9 \end{aligned}$$

6. $$\begin{aligned} 1.2x_1 - 2.3x_2 + 4.2x_3 + 5.4x_4 - 1.6x_5 &= 4.2 \\ 2.3x_1 + 1.4x_2 - 3.1x_3 + 3.3x_4 - 2.4x_5 &= 6.3 \\ 1.7x_1 + 2.6x_2 - 4.3x_3 + 7.2x_4 - 1.8x_5 &= 7.8 \\ 2.6x_1 - 4.2x_2 + 8.3x_3 - 1.6x_4 + 2.5x_5 &= 6.4 \end{aligned}$$

2.4 Matrices

Using Matrices to Represent Data

Many practical problems are solved by using arithmetic operations on the data associated with the problems. By properly organizing the data into *blocks* of numbers, we can then carry out these arithmetic operations in an orderly and efficient manner. In particular, this systematic approach enables us to use the computer to full advantage.

Let's begin by considering how the monthly output data of a manufacturer may be organized. The Acrosonic Company manufactures four different loudspeaker systems at three separate locations. The company's May output is described in Table 1.

TABLE 1

	Model A	Model B	Model C	Model D
Location I	320	280	460	280
Location II	480	360	580	0
Location III	540	420	200	880

Now, if we agree to preserve the relative location of each entry in Table 1, we can summarize the set of data as follows:

$$\begin{bmatrix} 320 & 280 & 460 & 280 \\ 480 & 360 & 580 & 0 \\ 540 & 420 & 200 & 880 \end{bmatrix}$$

A matrix summarizing the data in Table 1

The array of numbers displayed here is an example of a matrix. Observe that the numbers in row 1 give the output of models A, B, C, and D of Acrosonic loudspeaker systems manufactured at Location I; similarly, the numbers in rows 2 and 3 give the respective outputs of these loudspeaker systems at Locations II and III. The numbers in each column of the matrix give the outputs of a particular model of loudspeaker system manufactured at each of the company's three manufacturing locations.

More generally, a matrix is a rectangular array of real numbers. For example, each of the following arrays is a matrix:

$$A = \begin{bmatrix} 3 & 0 & -1 \\ 2 & 1 & 4 \end{bmatrix} \qquad B = \begin{bmatrix} 3 & 2 \\ 0 & 1 \\ -1 & 4 \end{bmatrix} \qquad C = \begin{bmatrix} 1 \\ 2 \\ 4 \\ 0 \end{bmatrix} \qquad D = \begin{bmatrix} 1 & 3 & 0 & 1 \end{bmatrix}$$

The real numbers that make up the array are called the **entries,** or *elements*, of the matrix. The entries in a row in the array are referred to as a **row** of the matrix, whereas the entries in a column in the array are referred to as a **column** of the matrix. Matrix A, for example, has two rows and three columns, which may be identified as follows:

$$\begin{array}{cc} & \begin{array}{ccc} \text{Column 1} & \text{Column 2} & \text{Column 3} \end{array} \\ \begin{array}{c} \text{Row 1} \\ \text{Row 2} \end{array} & \begin{bmatrix} 3 & \quad 0 \quad & -1 \\ 2 & \quad 1 \quad & 4 \end{bmatrix} \end{array}$$

A 2×3 matrix

The **size,** or *dimension*, **of a matrix** is described in terms of the number of rows and columns of the matrix. For example, matrix A has two rows and three columns and is said to have size 2 by 3, denoted 2×3. In general, a matrix having m rows and n columns is said to have size $m \times n$.

Matrix

A **matrix** is an ordered rectangular array of numbers. A matrix with m rows and n columns has size $m \times n$. The entry in the ith row and jth column of a matrix A is denoted by a_{ij}.

A matrix of size $1 \times n$—a matrix having one row and n columns—is referred to as a **row matrix,** or *row vector*, of dimension n. For example, the matrix D is a row vector of dimension 4. Similarly, a matrix having m rows and one column is referred to as a **column matrix,** or *column vector*, of dimension m. The matrix C is a column vector of dimension 4. Finally, an $n \times n$ matrix—that is, a matrix having the same number of rows as columns—is called a **square matrix.** For example, the matrix

$$\begin{bmatrix} -3 & 8 & 6 \\ 2 & \frac{1}{4} & 4 \\ 1 & 3 & 2 \end{bmatrix}$$

A 3×3 square matrix

is a square matrix of size 3×3, or simply of size 3. Here, $a_{23} = 4$.

APPLIED EXAMPLE 1 Organizing Production Data Consider the matrix

$$P = \begin{bmatrix} 320 & 280 & 460 & 280 \\ 480 & 360 & 580 & 0 \\ 540 & 420 & 200 & 880 \end{bmatrix}$$

representing the output of loudspeaker systems of the Acrosonic Company discussed earlier (see Table 1).

a. What is the size of the matrix P?
b. Find p_{24} (the entry in row 2 and column 4 of the matrix P), and give an interpretation of this number.
c. Find the sum of the entries that make up row 1 of P, and interpret the result.
d. Find the sum of the entries that make up column 4 of P, and interpret the result.

Solution

a. The matrix P has three rows and four columns and hence has size 3×4.
b. The required entry lies in row 2 and column 4 and is the number 0. This means that no model D loudspeaker system was manufactured at Location II in May.
c. The required sum is given by

$$320 + 280 + 460 + 280 = 1340$$

which gives the total number of loudspeaker systems manufactured at Location I in May as 1340 units.
d. The required sum is given by

$$280 + 0 + 880 = 1160$$

giving the output of model D loudspeaker systems at all locations of the company in May as 1160 units.

Equality of Matrices

Two matrices are said to be *equal* if they have the same size and their corresponding entries are equal. For example,

$$\begin{bmatrix} 2 & 3 & 1 \\ 4 & 6 & 2 \end{bmatrix} = \begin{bmatrix} (3-1) & 3 & 1 \\ 4 & (4+2) & 2 \end{bmatrix}$$

Also,

$$\begin{bmatrix} 1 & 3 & 5 \\ 2 & 4 & 3 \end{bmatrix} \neq \begin{bmatrix} 1 & 2 \\ 3 & 4 \\ 5 & 3 \end{bmatrix}$$

since the matrix on the left has size 2×3, whereas the matrix on the right has size 3×2, and

$$\begin{bmatrix} 2 & 3 \\ 4 & 6 \end{bmatrix} \neq \begin{bmatrix} 2 & 3 \\ 4 & 7 \end{bmatrix}$$

since the corresponding elements in row 2 and column 2 of the two matrices are not equal.

Equality of Matrices

Two matrices are equal if they have the same size and their corresponding entries are equal.

EXAMPLE 2 Solve the following matrix equation for x, y, and z:

$$\begin{bmatrix} 1 & x & 3 \\ 2 & y-1 & 2 \end{bmatrix} = \begin{bmatrix} 1 & 4 & z \\ 2 & 1 & 2 \end{bmatrix}$$

Solution Since the corresponding elements of the two matrices must be equal, we find that $x = 4$, $z = 3$, and $y - 1 = 1$, or $y = 2$.

Addition and Subtraction

Two matrices A and B of the *same size* can be added or subtracted to produce a matrix of the same size. This is done by adding or subtracting the corresponding entries in the two matrices. For example,

$$\begin{bmatrix} 1 & 3 & 4 \\ -1 & 2 & 0 \end{bmatrix} + \begin{bmatrix} 1 & 4 & 3 \\ 6 & 1 & -2 \end{bmatrix} = \begin{bmatrix} 1+1 & 3+4 & 4+3 \\ -1+6 & 2+1 & 0+(-2) \end{bmatrix} = \begin{bmatrix} 2 & 7 & 7 \\ 5 & 3 & -2 \end{bmatrix}$$

Adding two matrices of the same size

and

$$\begin{bmatrix} 1 & 2 \\ -1 & 3 \\ 4 & 0 \end{bmatrix} - \begin{bmatrix} 2 & -1 \\ 3 & 2 \\ -1 & 0 \end{bmatrix} = \begin{bmatrix} 1-2 & 2-(-1) \\ -1-3 & 3-2 \\ 4-(-1) & 0-0 \end{bmatrix} = \begin{bmatrix} -1 & 3 \\ -4 & 1 \\ 5 & 0 \end{bmatrix}$$

Subtracting two matrices of the same size

Addition and Subtraction of Matrices

If A and B are two matrices of the same size, then:

1. The *sum* $A + B$ is the matrix obtained by adding the corresponding entries in the two matrices.
2. The *difference* $A - B$ is the matrix obtained by subtracting the corresponding entries in B from those in A.

APPLIED EXAMPLE 3 Organizing Production Data The total output of Acrosonic for June is shown in Table 2.

TABLE 2

	Model A	Model B	Model C	Model D
Location I	210	180	330	180
Location II	400	300	450	40
Location III	420	280	180	740

The output for May was given earlier, in Table 1. Find the total output of the company for May and June.

Solution As we saw earlier, the production matrix for Acrosonic in May is given by

$$A = \begin{bmatrix} 320 & 280 & 460 & 280 \\ 480 & 360 & 580 & 0 \\ 540 & 420 & 200 & 880 \end{bmatrix}$$

Next, from Table 2, we see that the production matrix for June is given by

$$B = \begin{bmatrix} 210 & 180 & 330 & 180 \\ 400 & 300 & 450 & 40 \\ 420 & 280 & 180 & 740 \end{bmatrix}$$

Finally, the total output of Acrosonic for May and June is given by the matrix

$$A + B = \begin{bmatrix} 320 & 280 & 460 & 280 \\ 480 & 360 & 580 & 0 \\ 540 & 420 & 200 & 880 \end{bmatrix} + \begin{bmatrix} 210 & 180 & 330 & 180 \\ 400 & 300 & 450 & 40 \\ 420 & 280 & 180 & 740 \end{bmatrix}$$

$$= \begin{bmatrix} 530 & 460 & 790 & 460 \\ 880 & 660 & 1030 & 40 \\ 960 & 700 & 380 & 1620 \end{bmatrix}$$

The following laws hold for matrix addition.

Laws for Matrix Addition

If A, B, and C are matrices of the same size, then

1. $A + B = B + A$ — Commutative law
2. $(A + B) + C = A + (B + C)$ — Associative law

The *commutative law* for matrix addition states that the order in which matrix addition is performed is immaterial. The *associative law* states that, when adding three matrices together, we may first add A and B and then add the resulting sum to C. Equivalently, we can add A to the sum of B and C.

EXAMPLE 4 Let

$$A = \begin{bmatrix} 2 & 1 \\ 3 & -2 \\ 1 & 0 \end{bmatrix} \qquad B = \begin{bmatrix} -1 & 2 \\ 3 & 0 \\ 2 & 4 \end{bmatrix} \qquad C = \begin{bmatrix} 1 & 1 \\ 2 & 3 \\ 0 & -1 \end{bmatrix}$$

a. Show that $A + B = B + A$.
b. Show that $(A + B) + C = A + (B + C)$.

Solution

a. $A + B = \begin{bmatrix} 2 & 1 \\ 3 & -2 \\ 1 & 0 \end{bmatrix} + \begin{bmatrix} -1 & 2 \\ 3 & 0 \\ 2 & 4 \end{bmatrix} = \begin{bmatrix} 2 + (-1) & 1 + 2 \\ 3 + 3 & -2 + 0 \\ 1 + 2 & 0 + 4 \end{bmatrix} = \begin{bmatrix} 1 & 3 \\ 6 & -2 \\ 3 & 4 \end{bmatrix}$

On the other hand,

$$B + A = \begin{bmatrix} -1 & 2 \\ 3 & 0 \\ 2 & 4 \end{bmatrix} + \begin{bmatrix} 2 & 1 \\ 3 & -2 \\ 1 & 0 \end{bmatrix} = \begin{bmatrix} -1 + 2 & 2 + 1 \\ 3 + 3 & 0 + -2 \\ 2 + 1 & 4 + 0 \end{bmatrix} = \begin{bmatrix} 1 & 3 \\ 6 & -2 \\ 3 & 4 \end{bmatrix}$$

so $A + B = B + A$, as was to be shown.

b. Using the results of part (a), we have

$$(A + B) + C = \begin{bmatrix} 1 & 3 \\ 6 & -2 \\ 3 & 4 \end{bmatrix} + \begin{bmatrix} 1 & 1 \\ 2 & 3 \\ 0 & -1 \end{bmatrix} = \begin{bmatrix} 2 & 4 \\ 8 & 1 \\ 3 & 3 \end{bmatrix}$$

Next,

$$B + C = \begin{bmatrix} -1 & 2 \\ 3 & 0 \\ 2 & 4 \end{bmatrix} + \begin{bmatrix} 1 & 1 \\ 2 & 3 \\ 0 & -1 \end{bmatrix} = \begin{bmatrix} 0 & 3 \\ 5 & 3 \\ 2 & 3 \end{bmatrix}$$

so

$$A + (B + C) = \begin{bmatrix} 2 & 1 \\ 3 & -2 \\ 1 & 0 \end{bmatrix} + \begin{bmatrix} 0 & 3 \\ 5 & 3 \\ 2 & 3 \end{bmatrix} = \begin{bmatrix} 2 & 4 \\ 8 & 1 \\ 3 & 3 \end{bmatrix}$$

This shows that $(A + B) + C = A + (B + C)$.

A *zero matrix* is one in which all entries are zero. A zero matrix O has the property that

$$A + O = O + A = A$$

for any matrix A having the same size as that of O. For example, the zero matrix of size 3×2 is

$$O = \begin{bmatrix} 0 & 0 \\ 0 & 0 \\ 0 & 0 \end{bmatrix}$$

If A is any 3×2 matrix, then

$$A + O = \begin{bmatrix} a_{11} & a_{12} \\ a_{21} & a_{22} \\ a_{31} & a_{32} \end{bmatrix} + \begin{bmatrix} 0 & 0 \\ 0 & 0 \\ 0 & 0 \end{bmatrix} = \begin{bmatrix} a_{11} & a_{12} \\ a_{21} & a_{22} \\ a_{31} & a_{32} \end{bmatrix} = A$$

where a_{ij} denotes the entry in the ith row and jth column of the matrix A.

The matrix that is obtained by interchanging the rows and columns of a given matrix A is called the *transpose* of A and is denoted A^T. For example, if

$$A = \begin{bmatrix} 1 & 2 & 3 \\ 4 & 5 & 6 \\ 7 & 8 & 9 \end{bmatrix}$$

then

$$A^T = \begin{bmatrix} 1 & 4 & 7 \\ 2 & 5 & 8 \\ 3 & 6 & 9 \end{bmatrix}$$

Transpose of a Matrix

If A is an $m \times n$ matrix with elements a_{ij}, then the **transpose** of A is the $n \times m$ matrix A^T with elements a_{ji}.

Scalar Multiplication

A matrix A may be multiplied by a real number c, called a **scalar.** The scalar product, denoted by cA, is a matrix obtained by multiplying each entry of A by c. For example, the scalar product of the matrix

$$A = \begin{bmatrix} 3 & -1 & 2 \\ 0 & 1 & 4 \end{bmatrix}$$

and the scalar 3 is the matrix

$$3A = 3\begin{bmatrix} 3 & -1 & 2 \\ 0 & 1 & 4 \end{bmatrix} = \begin{bmatrix} 9 & -3 & 6 \\ 0 & 3 & 12 \end{bmatrix}$$

Scalar Product

If A is a matrix and c is a real number, then the **scalar product** cA is the matrix obtained by multiplying each entry of A by c.

EXAMPLE 5 Given

$$A = \begin{bmatrix} 3 & 4 \\ -1 & 2 \end{bmatrix} \quad \text{and} \quad B = \begin{bmatrix} 3 & 2 \\ -1 & 2 \end{bmatrix}$$

find a matrix X satisfying the *matrix equation* $2X + B = 3A$.

Solution From the given equation $2X + B = 3A$, we find that

$$\begin{aligned} 2X &= 3A - B \\ &= 3\begin{bmatrix} 3 & 4 \\ -1 & 2 \end{bmatrix} - \begin{bmatrix} 3 & 2 \\ -1 & 2 \end{bmatrix} \\ &= \begin{bmatrix} 9 & 12 \\ -3 & 6 \end{bmatrix} - \begin{bmatrix} 3 & 2 \\ -1 & 2 \end{bmatrix} = \begin{bmatrix} 6 & 10 \\ -2 & 4 \end{bmatrix} \\ X &= \frac{1}{2}\begin{bmatrix} 6 & 10 \\ -2 & 4 \end{bmatrix} = \begin{bmatrix} 3 & 5 \\ -1 & 2 \end{bmatrix} \end{aligned}$$

APPLIED EXAMPLE 6 Production Planning The management of Acrosonic has decided to increase its July production of loudspeaker systems by 10% (over its June output). Find a matrix giving the targeted production for July.

Solution From the results of Example 3, we see that Acrosonic's total output for June may be represented by the matrix

$$B = \begin{bmatrix} 210 & 180 & 330 & 180 \\ 400 & 300 & 450 & 40 \\ 420 & 280 & 180 & 740 \end{bmatrix}$$

The required matrix is given by

$$(1.1)B = 1.1\begin{bmatrix} 210 & 180 & 330 & 180 \\ 400 & 300 & 450 & 40 \\ 420 & 280 & 180 & 740 \end{bmatrix}$$

$$= \begin{bmatrix} 231 & 198 & 363 & 198 \\ 440 & 330 & 495 & 44 \\ 462 & 308 & 198 & 814 \end{bmatrix}$$

and is interpreted in the usual manner.

2.4 Self-Check Exercises

1. Perform the indicated operations:

$$\begin{bmatrix} 1 & 3 & 2 \\ -1 & 4 & 7 \end{bmatrix} - 3\begin{bmatrix} 2 & 1 & 0 \\ 1 & 3 & 4 \end{bmatrix}$$

2. Solve the following matrix equation for x, y, and z:

$$\begin{bmatrix} x & 3 \\ z & 2 \end{bmatrix} + \begin{bmatrix} 2-y & z \\ 2-z & -x \end{bmatrix} = \begin{bmatrix} 3 & 7 \\ 2 & 0 \end{bmatrix}$$

3. **Gasoline Sales** Jack owns two gas stations, one downtown and the other in the Wilshire district. Over two consecutive days, his gas stations recorded gasoline sales represented by the following matrices:

$A =$

	Regular	Regular plus	Premium
Downtown	1200	750	650
Wilshire	1100	850	600

and

$B =$

	Regular	Regular plus	Premium
Downtown	1250	825	550
Wilshire	1150	750	750

Find a matrix representing the total sales of the two gas stations over the 2-day period.

Solutions to Self-Check Exercises 2.4 can be found on page 121.

2.4 Concept Questions

1. Define (a) a matrix, (b) the size of a matrix, (c) a row matrix, (d) a column matrix, and (e) a square matrix.

2. When are two matrices equal? Give an example of two matrices that are equal.

3. **a.** What condition on the size of two matrices A and B ensures that the sum of A and B exist?

   b. What condition on the size of a matrix A ensures that the scalar product of cA, where c is a real number, exists?

4. Construct a 3×3 matrix A having the property that $A = A^T$. What special characteristic does A have?

2.4 Exercises

In Exercises 1–6, refer to the following matrices:

$$A = \begin{bmatrix} 2 & -3 & 9 & -4 \\ -11 & 2 & 6 & 7 \\ 6 & 0 & 2 & 9 \\ 5 & 1 & 5 & -8 \end{bmatrix} \qquad B = \begin{bmatrix} 3 & -1 & 2 \\ 0 & 1 & 4 \\ 3 & 2 & 1 \\ -1 & 0 & 8 \end{bmatrix}$$

$$C = \begin{bmatrix} 1 & 0 & 3 & 4 & 5 \end{bmatrix} \qquad D = \begin{bmatrix} 1 \\ 3 \\ -2 \\ 0 \end{bmatrix}$$

1. What is the size of A? Of B? Of C? Of D?
2. Find a_{14}, a_{21}, a_{31}, and a_{43}.
3. Find b_{13}, b_{31}, and b_{43}.
4. Identify the row matrix. What is its transpose?
5. Identify the column matrix. What is its transpose?
6. Identify the square matrix. What is its transpose?

In Exercises 7–12, refer to the following matrices:

$$A = \begin{bmatrix} -1 & 2 \\ 3 & -2 \\ 4 & 0 \end{bmatrix} \qquad B = \begin{bmatrix} 2 & 4 \\ 3 & 1 \\ -2 & 2 \end{bmatrix}$$

$$C = \begin{bmatrix} 3 & -1 & 0 \\ 2 & -2 & 3 \\ 4 & 6 & 2 \end{bmatrix} \qquad D = \begin{bmatrix} 2 & -2 & 4 \\ 3 & 6 & 2 \\ -2 & 3 & 1 \end{bmatrix}$$

7. What is the size of A? Of B? Of C? Of D?
8. Explain why the matrix $A + C$ does *not* exist.
9. Compute $A + B$.
10. Compute $2A - 3B$.
11. Compute $C - D$.
12. Compute $4D - 2C$.

In Exercises 13–22, perform the indicated operations.

13. $\begin{bmatrix} 2 & -1 & 3 \\ 9 & 2 & 1 \end{bmatrix} + \begin{bmatrix} -1 & 2 & -1 \\ -6 & 4 & 2 \end{bmatrix}$

14. $\begin{bmatrix} 6 & 3 & 8 \\ 4 & 5 & 6 \end{bmatrix} - \begin{bmatrix} 3 & -2 & -1 \\ 0 & -5 & -7 \end{bmatrix}$

15. $\begin{bmatrix} 2 & -3 & 4 & -1 \\ 3 & 1 & 0 & 0 \end{bmatrix} + \begin{bmatrix} 4 & 3 & -2 & -4 \\ 6 & 2 & 0 & -3 \end{bmatrix}$

16. $\begin{bmatrix} 1 & 4 & -5 \\ 3 & -8 & 6 \end{bmatrix} + \begin{bmatrix} 4 & 0 & -2 \\ 3 & 6 & 5 \end{bmatrix} - \begin{bmatrix} 2 & 8 & 9 \\ -11 & 2 & -5 \end{bmatrix}$

17. $\begin{bmatrix} 1.2 & 4.5 & -4.2 \\ 8.2 & 6.3 & -3.2 \end{bmatrix} - \begin{bmatrix} 3.1 & 1.5 & -3.6 \\ 2.2 & -3.3 & -4.4 \end{bmatrix}$

18. $\begin{bmatrix} 0.06 & 0.12 \\ 0.43 & 1.11 \\ 1.55 & -0.43 \end{bmatrix} - \begin{bmatrix} 0.77 & -0.75 \\ 0.22 & -0.65 \\ 1.09 & -0.57 \end{bmatrix}$

19. $3\begin{bmatrix} 1 & 1 & -3 \\ 3 & 2 & 3 \\ 7 & -1 & 6 \end{bmatrix} + 4\begin{bmatrix} -2 & -1 & 8 \\ 4 & 2 & 2 \\ 3 & 6 & 3 \end{bmatrix}$

20. $2\begin{bmatrix} 1 & -2 \\ 2 & -1 \\ 3 & 0 \end{bmatrix} - 3\begin{bmatrix} 2 & 3 \\ 1 & -2 \\ 2 & 4 \end{bmatrix}$

21. $\dfrac{1}{2}\begin{bmatrix} 1 & 0 & 0 & -4 \\ 3 & 0 & -1 & 6 \\ -2 & 1 & -4 & 2 \end{bmatrix} + \dfrac{4}{3}\begin{bmatrix} 3 & 0 & -1 & 4 \\ -2 & 1 & -6 & 2 \\ 8 & 2 & 0 & -2 \end{bmatrix} - \dfrac{1}{3}\begin{bmatrix} 3 & -9 & -1 & 0 \\ 6 & 2 & 0 & -6 \\ 0 & 1 & -3 & 1 \end{bmatrix}$

22. $0.5\begin{bmatrix} 1 & 3 & 5 \\ 5 & 2 & -1 \\ -2 & 0 & 1 \end{bmatrix} - 0.2\begin{bmatrix} 2 & 3 & 4 \\ -1 & 1 & -4 \\ 3 & 5 & -5 \end{bmatrix} + 0.6\begin{bmatrix} 3 & 4 & -1 \\ 4 & 5 & 1 \\ 1 & 0 & 0 \end{bmatrix}$

In Exercises 23–26, solve for u, x, y, and z in the given matrix equation.

23. $\begin{bmatrix} 2x-2 & 3 & 2 \\ 2 & 4 & y-2 \\ 2z & -3 & 2 \end{bmatrix} = \begin{bmatrix} 3 & u & 2 \\ 2 & 4 & 5 \\ 4 & -3 & 2 \end{bmatrix}$

24. $\begin{bmatrix} x & -2 \\ 3 & y \end{bmatrix} + \begin{bmatrix} -2 & z \\ -1 & 2 \end{bmatrix} = \begin{bmatrix} 4 & -2 \\ 2u & 4 \end{bmatrix}$

25. $\begin{bmatrix} 1 & x \\ 2y & -3 \end{bmatrix} - 4\begin{bmatrix} 2 & -2 \\ 0 & 3 \end{bmatrix} = \begin{bmatrix} 3z & 10 \\ 4 & -u \end{bmatrix}$

26. $\begin{bmatrix} 1 & 2 \\ 3 & 4 \\ x & -1 \end{bmatrix} - 3\begin{bmatrix} y-1 & 2 \\ 1 & 2 \\ 4 & 2z+1 \end{bmatrix} = 2\begin{bmatrix} -4 & -u \\ 0 & -1 \\ 4 & 4 \end{bmatrix}$

In Exercises 27 and 28, let

$$A = \begin{bmatrix} -2 & 1 \\ 0 & 3 \end{bmatrix} \quad \text{and} \quad B = \begin{bmatrix} 2 & -3 \\ 1 & -2 \end{bmatrix}$$

27. Find a matrix X satisfying the matrix equation $2A + X = 3B$.
28. Find a matrix X satisfying the matrix equation $3X - A + 2B = 0$.

In Exercises 29 and 30, let

$$A = \begin{bmatrix} 2 & -4 & 3 \\ 4 & 2 & 1 \end{bmatrix} \quad B = \begin{bmatrix} 4 & 3 & 2 \\ 1 & 0 & 4 \end{bmatrix} \quad C = \begin{bmatrix} 1 & 0 & 2 \\ 3 & -2 & 1 \end{bmatrix}$$

29. Verify by direct computation the validity of the commutative law for matrix addition.

30. Verify by direct computation the validity of the associative law for matrix addition.

In Exercises 31–34, let

$$A = \begin{bmatrix} 3 & 1 \\ 2 & 4 \\ -4 & 0 \end{bmatrix} \quad \text{and} \quad B = \begin{bmatrix} 1 & 2 \\ -1 & 0 \\ 3 & 2 \end{bmatrix}$$

Verify each equation by direct computation.

31. $(3 + 5)A = 3A + 5A$ **32.** $2(4A) = (2 \cdot 4)A = 8A$

33. $4(A + B) = 4A + 4B$ **34.** $2(A - 3B) = 2A - 6B$

In Exercises 35–38, find the transpose of each matrix.

35. $\begin{bmatrix} 3 & 2 & -1 & 5 \end{bmatrix}$ **36.** $\begin{bmatrix} 4 & 2 & 0 & -1 \\ 3 & 4 & -1 & 5 \end{bmatrix}$

37. $\begin{bmatrix} 1 & -1 & 2 \\ 3 & 4 & 2 \\ 0 & 1 & 0 \end{bmatrix}$ **38.** $\begin{bmatrix} 1 & 2 & 6 & 4 \\ 2 & 3 & 2 & 5 \\ 6 & 2 & 3 & 0 \\ 4 & 5 & 0 & 2 \end{bmatrix}$

39. **Cholesterol Levels** Mr. Cross, Mr. Jones, and Mr. Smith all suffer from coronary heart disease. As part of their treatment, they were put on special low-cholesterol diets: Cross on Diet I, Jones on Diet II, and Smith on Diet III. Progressive records of each patient's cholesterol level were kept. At the beginning of the first, second, third, and fourth months, the cholesterol levels of the three patients were:

Cross: 220, 215, 210, and 205
Jones: 220, 210, 200, and 195
Smith: 215, 205, 195, and 190

Represent this information in a 3×4 matrix.

40. **Mortgage Interest Rates** With interest rates low, refinancings account for a large chunk of all mortgage applications. The types of mortgages used for refinancings and home purchases in March 2012 applications are as follows:

	30-year fixed	15-year fixed	Fixed (other)	Adjustable
Refinancing (%)	56.6	24.7	14.6	4.1
Home Purchase (%)	85.4	7.0	1.7	5.9

a. Write a 2×4 matrix A to represent the information.
b. Compare a_{11} with a_{12}, and interpret your result.
c. Compare a_{11} with a_{21}, and interpret your result.
Source: Mortgage Bankers Association.

41. **Life Expectancy** Figures for life expectancy at birth of Massachusetts residents in 2008 are 82.6, 80.5, and 91.2 years for white, black, and Hispanic women, respectively, and 78.0, 73.9, and 84.8 years for white, black, and Hispanic men, respectively. Express this information using a 2×3 matrix and a 3×2 matrix.
Source: Massachusetts Department of Public Health.

42. **Terrorism Poll** In a poll surveying 1508 registered California voters in August 2011, the following questions were asked: (a) Has the federal government gone too far in restricting American citizens' civil liberties to keep the country safe from terrorism? (b) Are the people responsible for airline safety in the U.S. going too far to protect airline passengers? The results of the poll follow.

Question (a): (I) about right (46%), (II) gone too far (33%), (III) not far enough (15%), (IV) don't know (6%).

Question (b): (I) about right (53%), (II) gone too far (24%), (III) not far enough (18%), (IV) don't know (5%).

Express this information using a 2×4 matrix.
Source: Los Angeles Times.

43. **Average CD Yields** The average certificate of deposit (CD) yields as of January 14, 2013, according to Bankrates' National Survey, were as follows. For the current week: 6-month CD, 0.17%; 1-year CD, 0.27%; $2\frac{1}{2}$-year CD, 0.41%; 5-year CD, 0.87%. For the previous week: 6-month CD, 0.17%; 1-year CD, 0.27%; $2\frac{1}{2}$-year CD, 0.42%; 5-year CD, 0.88%. For a year ago: 6-month CD, 0.22%; 1-year CD, 0.34%; $2\frac{1}{2}$-year CD, 0.52%; 5-year CD, 1.15%.
a. Express this information using a 3×4 matrix A.
b. What are a_{12} and a_{22}? Interpret your result.
c. What are a_{13} and a_{23}? Interpret your result.
d. What are a_{33} and a_{34}? Interpret your result.
Source: Bankrate.com.

44. **Investment Portfolios** The following table gives the number of shares of certain corporations held by Leslie and Tom in their respective IRA accounts at the beginning of the year:

	IBM	Facebook	Ford	Wal-Mart
Leslie	500	350	200	400
Tom	400	450	300	200

Over the year, they added more shares to their accounts, as shown in the following table:

	IBM	Facebook	Ford	Wal-Mart
Leslie	50	50	0	100
Tom	0	80	100	50

a. Write a matrix A giving the holdings of Leslie and Tom at the beginning of the year and a matrix B giving the shares they have added to their portfolios.
b. Find a matrix C giving their total holdings at the end of the year.

45. Bookstore Inventories The Campus Bookstore's inventory of books is:

Hardcover: textbooks, 5280; fiction, 1680; nonfiction, 2320; reference, 1890

Paperback: fiction, 2810; nonfiction, 1490; reference, 2070; textbooks, 1940

The College Bookstore's inventory of books is

Hardcover: textbooks, 6340; fiction, 2220; nonfiction, 1790; reference, 1980

Paperback: fiction, 3100; nonfiction, 1720; reference, 2710; textbooks, 2050

a. Represent Campus's inventory as a matrix A.
b. Represent College's inventory as a matrix B.
c. Suppose that the two companies decide to merge. Write a matrix C that represents the total inventory of the newly amalgamated company.

46. Market Share of Motorcycles The market share of motorcycles in the United States in 2011 follows: Hero MotoCorp, 44.8%; Bajaj Auto, 20.5%; TVS Motor Company, 15.0%; Honda, 13.2%; Suzuki, 2.4%; Yamaha, 2.4%; and others, 1.7%. The corresponding figures for 2012 are 45.2%, 19.1%, 14.1%, 14.9%, 2.5%, 2.6%, and 1.6%, respectively. Express this information in a 2×7 matrix. What is the sum of all the elements in the first row? In the second row? Is this expected? Which company gained the most market share between 2011 and 2012?
Source: Motorcycle Industry Council.

47. Model Investment Portfolios The following table gives Schwab's five model portfolios.

Portfolio Investment (%)	Large-cap	Small-cap	International	Bonds	Cash
Conservative	15	0	5	50	30
Moderately conservative	25	5	10	50	10
Moderate	35	10	15	35	5
Moderately aggressive	45	15	20	15	5
Aggressive	50	20	25	0	5

a. Write a matrix A representing the above data. (Use the order given in the table.)
b. What is a_{12}? Interpret this number.
c. What are a_{13}, a_{23}, a_{33}, a_{43}, and a_{53}? What can you conclude from this?
d. Find the sum of each row of matrix A. Are the results expected? Explain.
Source: Schwab Center for Financial Research.

48. Student Test Scores For simplicity, suppose there are five students in a class. Suppose the scores of these students on three tests are as follows:

Student	Test 1	Test 2	Test 3
A	80	84	92
B	84	86	88
C	78	76	82
D	86	82	78
E	92	94	88

a. Write a 5×3 matrix A to represent the information.
b. What are a_{11}, a_{12}, and a_{13}? What is $\frac{1}{3}(a_{11} + a_{12} + a_{13})$, and what does this quantity represent?
c. What are a_{12}, a_{22}, a_{32}, a_{42}, and a_{52}? What is $\frac{1}{5}(a_{12} + a_{22} + a_{32} + a_{42} + a_{52})$, and what does this quantity represent?
d. What is the average score of student E based on the three tests?
e. What is the average score of the five students for the third test?

49. Mortgage Rates The mortgage rates for the week ended February 15, 2013, in the New York region follow:

For the 30-year fixed: 3.83% in New York, 3.67% in the New York Co-ops, 3.78% in New Jersey, and 3.79% in Connecticut

For the 15-year fixed: 3.16% in New York, 2.98% in the New York Co-ops, 3.03% in New Jersey, and 3.03% in Connecticut

For the adjustable: 2.99% in New York, 2.96% in the New York Co-ops, 2.97% in New Jersey, and 2.45% in Connecticut

The mortgage rates for the week ended February 22, 2013, in the same New York region follow:

For the 30-year fixed: 3.84% in New York, 3.75% in the New York Co-ops, 3.81% in New Jersey, and 3.80% in Connecticut

For the 15-year fixed: 3.15% in New York, 2.97% in the New York Co-ops, 3.04% in New Jersey, and 3.02% in Connecticut

For the adjustable: 2.99% in New York, 2.95% in the New York Co-ops, 2.96% in New Jersey, and 2.44% in Connecticut

a. Write two 3×4 matrices, A and B, giving the mortgage rates for the three types of loans in the New York region for the week ended February 15, 2013, and for the week ended February 22, 2013, respectively.
b. What are a_{12} and b_{12}? Interpret your result.
c. What are a_{33} and b_{33}? Interpret your result.
d. What is $\frac{1}{4}(a_{11} + a_{12} + a_{13} + a_{14})$? Interpret your result.
e. What is $\frac{1}{2}(a_{34} + b_{34})$? Interpret your result.
Source: HSH.com.

50. Banking The numbers of three types of bank accounts on January 1 at the Central Bank and its branches are represented by matrix A:

$$A = \begin{array}{l} \text{Main office} \\ \text{Westside branch} \\ \text{Eastside branch} \end{array} \overset{\begin{array}{ccc} \text{Checking accounts} & \text{Savings accounts} & \text{Fixed-deposit accounts} \end{array}}{\begin{bmatrix} 2820 & 1470 & 1120 \\ 1030 & 520 & 480 \\ 1170 & 540 & 460 \end{bmatrix}}$$

The number and types of accounts opened during the first quarter are represented by matrix B, and the number and types of accounts closed during the same period are represented by matrix C. Thus,

$$B = \begin{bmatrix} 260 & 120 & 110 \\ 140 & 60 & 50 \\ 120 & 70 & 50 \end{bmatrix} \quad \text{and} \quad C = \begin{bmatrix} 120 & 80 & 80 \\ 70 & 30 & 40 \\ 60 & 20 & 40 \end{bmatrix}$$

a. Find matrix D, which represents the number of each type of account at the end of the first quarter at each location.

b. Because a new manufacturing plant is opening in the immediate area, it is anticipated that there will be a 10% increase in the number of accounts at each location during the second quarter. Write a matrix $E = 1.1D$ to reflect this anticipated increase.

51. Home Sales K & R Builders build three models of houses, M_1, M_2, and M_3, in three subdivisions I, II, and III located in three different areas of a city. The prices of the houses (in thousands of dollars) are given in matrix A:

$$A = \begin{array}{c} \text{I} \\ \text{II} \\ \text{III} \end{array} \overset{\begin{array}{ccc} M_1 & M_2 & M_3 \end{array}}{\begin{bmatrix} 340 & 360 & 380 \\ 410 & 430 & 440 \\ 620 & 660 & 700 \end{bmatrix}}$$

K & R Builders has decided to raise the price of each house by 3% next year. Write a matrix B giving the new prices of the houses.

52. Home Sales K & R Builders build three models of houses, M_1, M_2, and M_3, in three subdivisions I, II, and III located in three different areas of a city. The prices of the homes (in thousands of dollars) are given in matrix A:

$$A = \begin{array}{c} \text{I} \\ \text{II} \\ \text{III} \end{array} \overset{\begin{array}{ccc} M_1 & M_2 & M_3 \end{array}}{\begin{bmatrix} 340 & 360 & 380 \\ 410 & 430 & 440 \\ 620 & 660 & 700 \end{bmatrix}}$$

The new price schedule for next year, reflecting a uniform percentage increase in each house, is given by matrix B:

$$B = \begin{array}{c} \text{I} \\ \text{II} \\ \text{III} \end{array} \overset{\begin{array}{ccc} M_1 & M_2 & M_3 \end{array}}{\begin{bmatrix} 357 & 378 & 399 \\ 430.5 & 451.5 & 462 \\ 651 & 693 & 735 \end{bmatrix}}$$

What was the percentage increase in the prices of the houses?
Hint: Find r such that $(1 + 0.01r)A = B$.

In Exercises 53–56, determine whether the statement is true or false. If it is true, explain why it is true. If it is false, give an example to show why it is false.

53. If A and B are matrices of the same size and c is a scalar, then $c(A + B) = cA + cB$.

54. If A and B are matrices of the same size, then $A - B = A + (-1)B$.

55. If A is a matrix and c is a nonzero scalar, then $(cA)^T = (1/c)A^T$.

56. If A is a matrix, then $(A^T)^T = A$.

2.4 Solutions to Self-Check Exercises

1. $$\begin{bmatrix} 1 & 3 & 2 \\ -1 & 4 & 7 \end{bmatrix} - 3\begin{bmatrix} 2 & 1 & 0 \\ 1 & 3 & 4 \end{bmatrix} = \begin{bmatrix} 1 & 3 & 2 \\ -1 & 4 & 7 \end{bmatrix} - \begin{bmatrix} 6 & 3 & 0 \\ 3 & 9 & 12 \end{bmatrix}$$
$$= \begin{bmatrix} -5 & 0 & 2 \\ -4 & -5 & -5 \end{bmatrix}$$

2. We are given

$$\begin{bmatrix} x & 3 \\ z & 2 \end{bmatrix} + \begin{bmatrix} 2 - y & z \\ 2 - z & -x \end{bmatrix} = \begin{bmatrix} 3 & 7 \\ 2 & 0 \end{bmatrix}$$

Performing the indicated operation on the left-hand side, we obtain

$$\begin{bmatrix} 2 + x - y & 3 + z \\ 2 & 2 - x \end{bmatrix} = \begin{bmatrix} 3 & 7 \\ 2 & 0 \end{bmatrix}$$

By the equality of matrices, we have

$$\begin{aligned} 2 + x - y &= 3 \\ 3 + z &= 7 \\ 2 - x &= 0 \end{aligned}$$

from which we deduce that $x = 2$, $y = 1$, and $z = 4$.

3. The required matrix is

$$A + B = \begin{bmatrix} 1200 & 750 & 650 \\ 1100 & 850 & 600 \end{bmatrix} + \begin{bmatrix} 1250 & 825 & 550 \\ 1150 & 750 & 750 \end{bmatrix}$$
$$= \begin{bmatrix} 2450 & 1575 & 1200 \\ 2250 & 1600 & 1350 \end{bmatrix}$$

USING TECHNOLOGY Matrix Operations

Graphing Utility

A graphing utility can be used to perform matrix addition, matrix subtraction, and scalar multiplication. It can also be used to find the transpose of a matrix.

EXAMPLE 1 Let

$$A = \begin{bmatrix} 1.2 & 3.1 \\ -2.1 & 4.2 \\ 3.1 & 4.8 \end{bmatrix} \quad \text{and} \quad B = \begin{bmatrix} 4.1 & 3.2 \\ 1.3 & 6.4 \\ 1.7 & 0.8 \end{bmatrix}$$

Find (a) $A + B$, (b) $2.1A - 3.2B$, and (c) $(2.1A + 3.2B)^T$.

Solution We first enter the matrices A and B into the calculator.

a. Using matrix operations, we enter the expression $A + B$ and obtain

$$A + B = \begin{bmatrix} 5.3 & 6.3 \\ -0.8 & 10.6 \\ 4.8 & 5.6 \end{bmatrix}$$

b. Using matrix operations, we enter the expression $2.1A - 3.2B$ and obtain

$$2.1A - 3.2B = \begin{bmatrix} -10.6 & -3.73 \\ -8.57 & -11.66 \\ 1.07 & 7.52 \end{bmatrix}$$

c. Using matrix operations, we enter the expression $(2.1A + 3.2B)^T$ and obtain

$$(2.1A + 3.2B)^T = \begin{bmatrix} 15.64 & -0.25 & 11.95 \\ 16.75 & 29.3 & 12.64 \end{bmatrix}$$

APPLIED EXAMPLE 2 Gas Station Sales John operates three gas stations at three locations, I, II, and III. Over two consecutive days, his gas stations recorded the following fuel sales (in gallons):

	Day 1			
	Regular	**Regular Plus**	**Premium**	**Diesel**
Location I	1400	1200	1100	200
Location II	1600	900	1200	300
Location III	1200	1500	800	500

	Day 2			
	Regular	**Regular Plus**	**Premium**	**Diesel**
Location I	1000	900	800	150
Location II	1800	1200	1100	250
Location III	800	1000	700	400

Find a matrix representing the total fuel sales at John's gas stations.

Solution The fuel sales can be represented by the matrix A (day 1) and matrix B (day 2):

$$A = \begin{bmatrix} 1400 & 1200 & 1100 & 200 \\ 1600 & 900 & 1200 & 300 \\ 1200 & 1500 & 800 & 500 \end{bmatrix} \quad \text{and} \quad B = \begin{bmatrix} 1000 & 900 & 800 & 150 \\ 1800 & 1200 & 1100 & 250 \\ 800 & 1000 & 700 & 400 \end{bmatrix}$$

We enter the matrices A and B into the calculator. Using matrix operations, we enter the expression $A + B$ and obtain

$$A + B = \begin{bmatrix} 2400 & 2100 & 1900 & 350 \\ 3400 & 2100 & 2300 & 550 \\ 2000 & 2500 & 1500 & 900 \end{bmatrix}$$

Excel

First, we show how basic operations on matrices can be carried out by using Excel.

EXAMPLE 3 Given the following matrices,

$$A = \begin{bmatrix} 1.2 & 3.1 \\ -2.1 & 4.2 \\ 3.1 & 4.8 \end{bmatrix} \quad \text{and} \quad B = \begin{bmatrix} 4.1 & 3.2 \\ 1.3 & 6.4 \\ 1.7 & 0.8 \end{bmatrix}$$

a. Compute $A + B$. **b.** Compute $2.1A - 3.2B$.

Solution

a. First, represent the matrices A and B in a spreadsheet. Enter the elements of each matrix in a block of cells as shown in Figure T1.

	A	B	C	D	E
1		A			B
2	1.2	3.1		4.1	3.2
3	-2.1	4.2		1.3	6.4
4	3.1	4.8		1.7	0.8

FIGURE **T1**
The elements of matrix *A* and matrix *B* in a spreadsheet

Second, compute the sum of matrix A and matrix B. Highlight the cells that will contain matrix $A + B$, type `=`, highlight the cells in matrix A, type `+`, highlight the cells in matrix B, and press **Ctrl-Shift-Enter**. The resulting matrix $A + B$ is shown in Figure T2.

	A	B
8		A + B
9	5.3	6.3
10	-0.8	10.6
11	4.8	5.6

FIGURE **T2**
The matrix $A + B$

Note: Boldfaced words/characters enclosed in a box (for example, **Enter**) indicate that an action (click, select, or press) is required. Words/characters printed blue (for example, Chart sub-type:) indicate words/characters that appear on the screen. Words/characters printed in a monospace font (for example, `= (-2/3) *A2+2`) indicate words/characters that need to be typed and entered.

b. Highlight the cells that will contain matrix $2.1A - 3.2B$. Type `= 2.1*`, highlight matrix A, type `−3.2*`, highlight the cells in matrix B, and press **Ctrl-Shift-Enter**. The resulting matrix $2.1A - 3.2B$ is shown in Figure T3.

	A	B
13		2.1A - 3.2B
14	-10.6	-3.73
15	-8.57	-11.66
16	1.07	7.52

FIGURE **T3**
The matrix $2.1A - 3.2B$

APPLIED EXAMPLE 4 Gas Station Sales John operates three gas stations at three locations I, II, and III. Over two consecutive days, his gas stations recorded the following fuel sales (in gallons):

	Day 1			
	Regular	**Regular Plus**	**Premium**	**Diesel**
Location I	1400	1200	1100	200
Location II	1600	900	1200	300
Location III	1200	1500	800	500

	Day 2			
	Regular	**Regular Plus**	**Premium**	**Diesel**
Location I	1000	900	800	150
Location II	1800	1200	1100	250
Location III	800	1000	700	400

Find a matrix representing the total fuel sales at John's gas stations.

Solution The fuel sales can be represented by the matrices A (day 1) and B (day 2):

$$A = \begin{bmatrix} 1400 & 1200 & 1100 & 200 \\ 1600 & 900 & 1200 & 300 \\ 1200 & 1500 & 800 & 500 \end{bmatrix} \quad \text{and} \quad B = \begin{bmatrix} 1000 & 900 & 800 & 150 \\ 1800 & 1200 & 1100 & 250 \\ 800 & 1000 & 700 & 400 \end{bmatrix}$$

We first enter the elements of the matrices A and B onto a spreadsheet. Next, we highlight the cells that will contain the matrix $A + B$, type =, highlight A, type +, highlight B, and then press **Ctrl-Shift-Enter**. The resulting matrix $A + B$ is shown in Figure T4.

	A	B	C	D
23		A + B		
24	2400	2100	1900	350
25	3400	2100	2300	550
26	2000	2500	1500	900

FIGURE **T4**
The matrix $A + B$

TECHNOLOGY EXERCISES

Refer to the following matrices and perform the indicated operations.

$$A = \begin{bmatrix} 1.2 & 3.1 & -5.4 & 2.7 \\ 4.1 & 3.2 & 4.2 & -3.1 \\ 1.7 & 2.8 & -5.2 & 8.4 \end{bmatrix}$$

$$B = \begin{bmatrix} 6.2 & -3.2 & 1.4 & -1.2 \\ 3.1 & 2.7 & -1.2 & 1.7 \\ 1.2 & -1.4 & -1.7 & 2.8 \end{bmatrix}$$

1. $12.5A$

2. $-8.4B$

3. $A - B$

4. $B - A$

5. $1.3A + 2.4B$

6. $2.1A - 1.7B$

7. $3(A + B)$

8. $1.3(4.1A - 2.3B)$

2.5 Multiplication of Matrices

Matrix Product

In Section 2.4, we saw how matrices of the same size may be added or subtracted and how a matrix may be multiplied by a scalar (real number), an operation referred to as scalar multiplication. In this section we see how, with certain restrictions, one matrix may be multiplied by another matrix.

To define matrix multiplication, let's consider the following problem. On a certain day, Al's Service Station sold 1600 gallons of regular, 1000 gallons of regular plus, and 800 gallons of premium gasoline. If the price of gasoline on this day was \$3.79 for regular, \$3.89 for regular plus, and \$3.99 for premium gasoline, find the total revenue realized by Al's for that day.

The day's sale of gasoline may be represented by the matrix

$$A = \begin{bmatrix} 1600 & 1000 & 800 \end{bmatrix} \quad \text{Row matrix } (1 \times 3)$$

Next, we let the unit selling price of regular, regular plus, and premium gasoline be the entries in the matrix

$$B = \begin{bmatrix} 3.79 \\ 3.89 \\ 3.99 \end{bmatrix} \quad \text{Column matrix } (3 \times 1)$$

The first entry in matrix A gives the number of gallons of regular gasoline sold, and the first entry in matrix B gives the selling price for each gallon of regular gasoline, so their product $(1600)(3.79)$ gives the revenue realized from the sale of regular gasoline for the day. A similar interpretation of the second and third entries in the two matrices suggests that we multiply the corresponding entries to obtain the respective revenues realized from the sale of regular, regular plus, and premium gasoline. Finally, the total revenue realized by Al's from the sale of gasoline is given by adding these products to obtain

$$(1600)(3.79) + (1000)(3.89) + (800)(3.99) = 13{,}146$$

or \$13,146.

This example suggests that if we have a row matrix of size $1 \times n$,

$$A = \begin{bmatrix} a_1 & a_2 & a_3 & \cdots & a_n \end{bmatrix}$$

and a column matrix of size $n \times 1$,

$$B = \begin{bmatrix} b_1 \\ b_2 \\ b_3 \\ \vdots \\ b_n \end{bmatrix}$$

then we may define the **matrix product** of A and B, written AB, by

$$AB = [a_1 \;\; a_2 \;\; a_3 \cdots a_n] \begin{bmatrix} b_1 \\ b_2 \\ b_3 \\ \vdots \\ b_n \end{bmatrix} = a_1b_1 + a_2b_2 + a_3b_3 + \cdots + a_nb_n \qquad \textbf{(12)}$$

EXAMPLE 1 Let

$$A = [1 \quad -2 \quad 3 \quad 5] \quad \text{and} \quad B = \begin{bmatrix} 2 \\ 3 \\ 0 \\ -1 \end{bmatrix}$$

Then

$$AB = [1 \quad -2 \quad 3 \quad 5] \begin{bmatrix} 2 \\ 3 \\ 0 \\ -1 \end{bmatrix} = (1)(2) + (-2)(3) + (3)(0) + (5)(-1) = -9$$

APPLIED EXAMPLE 2 Stock Transactions Judy's stock holdings are given by the matrix

$$A = \begin{array}{ccc} \text{GM} & \text{IBM} & \text{AAPL} \\ [700 & 400 & 200] \end{array}$$

At the close of trading on a certain day, the prices (in dollars per share) of these stocks are

$$B = \begin{array}{c} \text{GM} \\ \text{IBM} \\ \text{AAPL} \end{array} \begin{bmatrix} 27 \\ 205 \\ 420 \end{bmatrix}$$

What is the total value of Judy's holdings as of that day?

Solution Judy's holdings are worth

$$AB = [700 \quad 400 \quad 200] \begin{bmatrix} 27 \\ 205 \\ 420 \end{bmatrix} = (700)(27) + (400)(205) + (200)(420)$$

$$= 184{,}900$$

or \$184,900.

Returning once again to the matrix product AB in Equation (12), observe that the number of columns of the row matrix A is *equal* to the number of rows of the column matrix B. Observe further that the product matrix AB has size 1×1 (a real number may be thought of as a 1×1 matrix). Schematically,

Size of A — Size of B

$1 \times n$ — $n \times 1$

(1×1)

Size of AB

More generally, if A is a matrix of size $m \times n$ and B is a matrix of size $n \times p$ (the number of columns of A equals the numbers of rows of B), then the *matrix product* of A and B, AB, is defined and is a matrix of size $m \times p$. Schematically,

Size of A — Size of B

$m \times n$ — $n \times p$

$(m \times p)$

Size of AB

Next, let's illustrate the mechanics of matrix multiplication by computing the product of a 2×3 matrix A and a 3×4 matrix B. Suppose

$$A = \begin{bmatrix} a_{11} & a_{12} & a_{13} \\ a_{21} & a_{22} & a_{23} \end{bmatrix}$$

$$B = \begin{bmatrix} b_{11} & b_{12} & b_{13} & b_{14} \\ b_{21} & b_{22} & b_{23} & b_{24} \\ b_{31} & b_{32} & b_{33} & b_{34} \end{bmatrix}$$

From the schematic

Same

Size of A 2×3 — 3×4 Size of B

(2×4)

Size of AB

we see that the matrix product $C = AB$ is defined (since the number of columns of A equals the number of rows of B) and has size 2×4. Thus,

$$C = \begin{bmatrix} c_{11} & c_{12} & c_{13} & c_{14} \\ c_{21} & c_{22} & c_{23} & c_{24} \end{bmatrix}$$

The entries of C are computed as follows: The entry c_{11} (the entry in the *first* row, *first* column of C) is the product of the row matrix composed of the entries from the *first* row of A and the column matrix composed of the *first* column of B. Thus,

$$c_{11} = \begin{bmatrix} a_{11} & a_{12} & a_{13} \end{bmatrix} \begin{bmatrix} b_{11} \\ b_{21} \\ b_{31} \end{bmatrix} = a_{11}b_{11} + a_{12}b_{21} + a_{13}b_{31}$$

The entry c_{12} (the entry in the *first* row, *second* column of C) is the product of the row matrix composed of the *first* row of A and the column matrix composed of the *second* column of B. Thus,

$$c_{12} = \begin{bmatrix} a_{11} & a_{12} & a_{13} \end{bmatrix} \begin{bmatrix} b_{12} \\ b_{22} \\ b_{32} \end{bmatrix} = a_{11}b_{12} + a_{12}b_{22} + a_{13}b_{32}$$

The other entries in C are computed in a similar manner.

EXAMPLE 3 Let

$$A = \begin{bmatrix} 3 & 1 & 4 \\ -1 & 2 & 3 \end{bmatrix} \quad \text{and} \quad B = \begin{bmatrix} 1 & 3 & -3 \\ 4 & -1 & 2 \\ 2 & 4 & 1 \end{bmatrix}$$

Compute AB.

Solution The size of matrix A is 2×3, and the size of matrix B is 3×3. Since the number of columns of matrix A is equal to the number of rows of matrix B, the matrix product $C = AB$ is defined. Furthermore, the size of matrix C is 2×3. Thus,

$$\begin{bmatrix} 3 & 1 & 4 \\ -1 & 2 & 3 \end{bmatrix}\begin{bmatrix} 1 & 3 & -3 \\ 4 & -1 & 2 \\ 2 & 4 & 1 \end{bmatrix} = \begin{bmatrix} c_{11} & c_{12} & c_{13} \\ c_{21} & c_{22} & c_{23} \end{bmatrix}$$

It remains now to determine the entries c_{11}, c_{12}, c_{13}, c_{21}, c_{22}, and c_{23}. We have

$$c_{11} = [3 \quad 1 \quad 4]\begin{bmatrix} 1 \\ 4 \\ 2 \end{bmatrix} = (3)(1) + (1)(4) + (4)(2) = 15$$

$$c_{12} = [3 \quad 1 \quad 4]\begin{bmatrix} 3 \\ -1 \\ 4 \end{bmatrix} = (3)(3) + (1)(-1) + (4)(4) = 24$$

$$c_{13} = [3 \quad 1 \quad 4]\begin{bmatrix} -3 \\ 2 \\ 1 \end{bmatrix} = (3)(-3) + (1)(2) + (4)(1) = -3$$

$$c_{21} = [-1 \quad 2 \quad 3]\begin{bmatrix} 1 \\ 4 \\ 2 \end{bmatrix} = (-1)(1) + (2)(4) + (3)(2) = 13$$

$$c_{22} = [-1 \quad 2 \quad 3]\begin{bmatrix} 3 \\ -1 \\ 4 \end{bmatrix} = (-1)(3) + (2)(-1) + (3)(4) = 7$$

$$c_{23} = [-1 \quad 2 \quad 3]\begin{bmatrix} -3 \\ 2 \\ 1 \end{bmatrix} = (-1)(-3) + (2)(2) + (3)(1) = 10$$

so the required product AB is given by

$$AB = \begin{bmatrix} 15 & 24 & -3 \\ 13 & 7 & 10 \end{bmatrix}$$

EXAMPLE 4 Let

$$A = \begin{bmatrix} 3 & 2 & 1 \\ -1 & 2 & 3 \\ 3 & 1 & 4 \end{bmatrix} \quad \text{and} \quad B = \begin{bmatrix} 1 & 3 & 4 \\ 2 & 4 & 1 \\ -1 & 2 & 3 \end{bmatrix}$$

Then

$$AB = \begin{bmatrix} 3\cdot 1 + 2\cdot 2 + 1\cdot(-1) & 3\cdot 3 + 2\cdot 4 + 1\cdot 2 & 3\cdot 4 + 2\cdot 1 + 1\cdot 3 \\ (-1)\cdot 1 + 2\cdot 2 + 3\cdot(-1) & (-1)\cdot 3 + 2\cdot 4 + 3\cdot 2 & (-1)\cdot 4 + 2\cdot 1 + 3\cdot 3 \\ 3\cdot 1 + 1\cdot 2 + 4\cdot(-1) & 3\cdot 3 + 1\cdot 4 + 4\cdot 2 & 3\cdot 4 + 1\cdot 1 + 4\cdot 3 \end{bmatrix} = \begin{bmatrix} 6 & 19 & 17 \\ 0 & 11 & 7 \\ 1 & 21 & 25 \end{bmatrix}$$

$$BA = \begin{bmatrix} 1\cdot 3 + 3\cdot(-1) + 4\cdot 3 & 1\cdot 2 + 3\cdot 2 + 4\cdot 1 & 1\cdot 1 + 3\cdot 3 + 4\cdot 4 \\ 2\cdot 3 + 4\cdot(-1) + 1\cdot 3 & 2\cdot 2 + 4\cdot 2 + 1\cdot 1 & 2\cdot 1 + 4\cdot 3 + 1\cdot 4 \\ (-1)\cdot 3 + 2\cdot(-1) + 3\cdot 3 & (-1)\cdot 2 + 2\cdot 2 + 3\cdot 1 & (-1)\cdot 1 + 2\cdot 3 + 3\cdot 4 \end{bmatrix} = \begin{bmatrix} 12 & 12 & 26 \\ 5 & 13 & 18 \\ 4 & 5 & 17 \end{bmatrix}$$

The preceding example shows that, in general, $AB \neq BA$ for two square matrices A and B. However, the following laws are valid for matrix multiplication.

Laws for Matrix Multiplication

If the products and sums are defined for the matrices A, B, and C, then

1. $(AB)C = A(BC)$ Associative law

2. $A(B + C) = AB + AC$ Distributive law

The square matrix of size n having 1s along the main diagonal and 0s elsewhere is called the identity matrix of size n.

Identity Matrix

The **identity matrix** of size n is given by

$$I_n = \begin{bmatrix} 1 & 0 & \cdot & \cdot & \cdot & 0 \\ 0 & 1 & \cdot & \cdot & \cdot & 0 \\ \cdot & \cdot & \cdot & \cdot & \cdot & \cdot \\ \cdot & \cdot & \cdot & \cdot & \cdot & \cdot \\ \cdot & \cdot & \cdot & \cdot & \cdot & \cdot \\ 0 & 0 & \cdot & \cdot & \cdot & 1 \end{bmatrix} \quad n \text{ rows}$$

n columns

The identity matrix has the properties that $I_nA = A$ for every $n \times r$ matrix A and $BI_n = B$ for every $s \times n$ matrix B. In particular, if A is a square matrix of size n, then

$$I_nA = AI_n = A$$

EXAMPLE 5 Let

$$A = \begin{bmatrix} 1 & 3 & 1 \\ -4 & 3 & 2 \\ 1 & 0 & 1 \end{bmatrix}$$

Then

$$I_3A = \begin{bmatrix} 1 & 0 & 0 \\ 0 & 1 & 0 \\ 0 & 0 & 1 \end{bmatrix} \begin{bmatrix} 1 & 3 & 1 \\ -4 & 3 & 2 \\ 1 & 0 & 1 \end{bmatrix} = \begin{bmatrix} 1 & 3 & 1 \\ -4 & 3 & 2 \\ 1 & 0 & 1 \end{bmatrix} = A$$

$$AI_3 = \begin{bmatrix} 1 & 3 & 1 \\ -4 & 3 & 2 \\ 1 & 0 & 1 \end{bmatrix} \begin{bmatrix} 1 & 0 & 0 \\ 0 & 1 & 0 \\ 0 & 0 & 1 \end{bmatrix} = \begin{bmatrix} 1 & 3 & 1 \\ -4 & 3 & 2 \\ 1 & 0 & 1 \end{bmatrix} = A$$

so $I_3A = AI_3 = A$, confirming our result for this special case.

APPLIED EXAMPLE 6 **Production Planning** Ace Novelty received an order from Magic World Amusement Park for 900 Giant Pandas, 1200 Saint Bernards, and 2000 Big Birds. Ace's management decided that 500 Giant Pandas, 800 Saint Bernards, and 1300 Big Birds could be manufactured in their Los Angeles plant, and the balance of the order could be filled by their Seattle plant. Each Panda requires 1.5 square yards of plush, 30 cubic feet of stuffing, and 5 pieces of trim; each Saint Bernard requires 2 square yards of plush, 35 cubic feet of stuffing, and 8 pieces of trim; and each Big Bird requires 2.5 square yards of plush, 25 cubic feet of stuffing, and 15 pieces of trim. The plush costs \$4.50 per square yard, the stuffing costs 10 cents per cubic foot, and the trim costs 25 cents per unit.

a. How much of each type of material must be purchased for each plant?

b. What is the total cost of materials incurred by each plant and the total cost of materials incurred by Ace Novelty in filling the order?

Solution The quantities of each type of stuffed animal to be produced at each plant location may be expressed as a 2×3 *production matrix P*. Thus,

$$\begin{array}{ll} & \begin{array}{ccc} \text{Pandas} & \text{St. Bernards} & \text{Birds} \end{array} \\ P = \begin{array}{l} \text{L.A.} \\ \text{Seattle} \end{array} & \begin{bmatrix} 500 & 800 & 1300 \\ 400 & 400 & 700 \end{bmatrix} \end{array}$$

Similarly, we may represent the amount and type of material required to manufacture each type of animal by a 3×3 *activity matrix A*. Thus,

$$\begin{array}{ll} & \begin{array}{ccc} \text{Plush} & \text{Stuffing} & \text{Trim} \end{array} \\ A = \begin{array}{l} \text{Pandas} \\ \text{St. Bernards} \\ \text{Birds} \end{array} & \begin{bmatrix} 1.5 & 30 & 5 \\ 2 & 35 & 8 \\ 2.5 & 25 & 15 \end{bmatrix} \end{array}$$

Finally, the unit cost for each type of material may be represented by the 3×1 *cost matrix C*.

$$C = \begin{array}{l} \text{Plush} \\ \text{Stuffing} \\ \text{Trim} \end{array} \begin{bmatrix} 4.50 \\ 0.10 \\ 0.25 \end{bmatrix}$$

a. The amount of each type of material required for each plant is given by the matrix PA. Thus,

$$PA = \begin{bmatrix} 500 & 800 & 1300 \\ 400 & 400 & 700 \end{bmatrix} \begin{bmatrix} 1.5 & 30 & 5 \\ 2 & 35 & 8 \\ 2.5 & 25 & 15 \end{bmatrix}$$

$$\begin{array}{ll} & \begin{array}{ccc} \text{Plush} & \text{Stuffing} & \text{Trim} \end{array} \\ = \begin{array}{l} \text{L.A.} \\ \text{Seattle} \end{array} & \begin{bmatrix} 5600 & 75{,}500 & 28{,}400 \\ 3150 & 43{,}500 & 15{,}700 \end{bmatrix} \end{array}$$

b. The total cost of materials for each plant is given by the matrix PAC:

$$PAC = \begin{bmatrix} 5600 & 75{,}500 & 28{,}400 \\ 3150 & 43{,}500 & 15{,}700 \end{bmatrix} \begin{bmatrix} 4.50 \\ 0.10 \\ 0.25 \end{bmatrix}$$

$$= \begin{array}{l} \text{L.A.} \\ \text{Seattle} \end{array} \begin{bmatrix} 39{,}850 \\ 22{,}450 \end{bmatrix}$$

or \$39,850 for the L.A. plant and \$22,450 for the Seattle plant. Thus, the total cost of materials incurred by Ace Novelty is \$62,300.

Matrix Representation

Example 7 shows how a system of linear equations may be written in a compact form with the help of matrices. (We will use this matrix equation representation in Section 2.6.)

EXAMPLE 7 Write the following system of linear equations in matrix form.

$$\begin{aligned} 2x - 4y + z &= 6 \\ -3x + 6y - 5z &= -1 \\ x - 3y + 7z &= 0 \end{aligned}$$

Solution Let's write

$$A = \begin{bmatrix} 2 & -4 & 1 \\ -3 & 6 & -5 \\ 1 & -3 & 7 \end{bmatrix} \qquad X = \begin{bmatrix} x \\ y \\ z \end{bmatrix} \qquad B = \begin{bmatrix} 6 \\ -1 \\ 0 \end{bmatrix}$$

Note that A is just the 3×3 matrix of coefficients of the system, X is the 3×1 column matrix of unknowns (variables), and B is the 3×1 column matrix of constants. We now show that the required matrix representation of the system of linear equations is

$$AX = B$$

To see this, observe that

$$AX = \begin{bmatrix} 2 & -4 & 1 \\ -3 & 6 & -5 \\ 1 & -3 & 7 \end{bmatrix}\begin{bmatrix} x \\ y \\ z \end{bmatrix} = \begin{bmatrix} 2x - 4y + z \\ -3x + 6y - 5z \\ x - 3y + 7z \end{bmatrix}$$

Equating this 3×1 matrix with matrix B now gives

$$\begin{bmatrix} 2x - 4y + z \\ -3x + 6y - 5z \\ x - 3y + 7z \end{bmatrix} = \begin{bmatrix} 6 \\ -1 \\ 0 \end{bmatrix}$$

which, by matrix equality, is easily seen to be equivalent to the given system of linear equations.

2.5 Self-Check Exercises

1. Compute

$$\begin{bmatrix} 1 & 3 & 0 \\ 2 & 4 & -1 \end{bmatrix}\begin{bmatrix} 3 & 1 & 4 \\ 2 & 0 & 3 \\ 1 & 2 & -1 \end{bmatrix}$$

2. Write the following system of linear equations in matrix form:

$$\begin{aligned} y - 2z &= 1 \\ 2x - y + 3z &= 0 \\ x \qquad + 4z &= 7 \end{aligned}$$

3. **Stock Transactions** On June 1, the stock holdings of Ash and Joan Robinson were given by the matrix

$$A = \begin{array}{c} \\ \text{Ash} \\ \text{Joan} \end{array}\begin{array}{c} \begin{array}{cccc} \text{T} & \text{TWX} & \text{IBM} & \text{GM} \end{array} \\ \begin{bmatrix} 2000 & 1000 & 500 & 5000 \\ 1000 & 2500 & 1000 & 0 \end{bmatrix} \end{array}$$

and the closing prices of T, TWX, IBM, and GM were \$36, \$54, \$205, and \$27 per share, respectively. Use matrix multiplication to determine the separate values of Ash's and Joan's stock holdings as of that date.

Solutions to Self-Check Exercises 2.5 can be found on page 138.

2.5 Concept Questions

1. What is the difference between scalar multiplication and matrix multiplication? Give examples of each operation.

2. a. Suppose A and B are matrices whose products AB and BA are both defined. What can you say about the sizes of A and B?

   b. If A, B, and C are matrices such that $A(B + C)$ is defined, what can you say about the relationship between the number of columns of A and the number of rows of C? Explain.

2.5 Exercises

In Exercises 1–4, the sizes of matrices A and B are given. Find the size of AB and BA whenever they are defined.

1. A is of size 2×3, and B is of size 3×5.

2. A is of size 3×4, and B is of size 4×3.

3. A is of size 1×7, and B is of size 7×1.

4. A is of size 4×4, and B is of size 4×4.

5. Let A be a matrix of size $m \times n$ and B be a matrix of size $s \times t$. Find conditions on m, n, s, and t such that both matrix products AB and BA are defined.

6. Find condition(s) on the size of a matrix A such that A^2 (that is, AA) is defined.

In Exercises 7–24, compute the indicated products.

7. $\begin{bmatrix} 1 & 2 \\ 3 & 0 \end{bmatrix}\begin{bmatrix} 1 \\ -1 \end{bmatrix}$

8. $\begin{bmatrix} -1 & 3 \\ 5 & 0 \end{bmatrix}\begin{bmatrix} 7 \\ 2 \end{bmatrix}$

9. $\begin{bmatrix} 4 & 1 & 2 \\ -1 & 2 & 4 \end{bmatrix}\begin{bmatrix} 4 \\ 1 \\ -2 \end{bmatrix}$

10. $\begin{bmatrix} 3 & 2 & -1 \\ 4 & -1 & 0 \\ -5 & 2 & 1 \end{bmatrix}\begin{bmatrix} 3 \\ -2 \\ 0 \end{bmatrix}$

11. $\begin{bmatrix} -1 & 2 \\ 3 & 1 \end{bmatrix}\begin{bmatrix} 2 & 4 \\ 3 & 1 \end{bmatrix}$

12. $\begin{bmatrix} 1 & 3 \\ -1 & 2 \end{bmatrix}\begin{bmatrix} 1 & 3 & 0 \\ 3 & 0 & 2 \end{bmatrix}$

13. $\begin{bmatrix} 2 & 1 & 2 \\ 3 & 2 & 4 \end{bmatrix}\begin{bmatrix} -1 & 2 \\ 4 & 3 \\ 0 & 1 \end{bmatrix}$

14. $\begin{bmatrix} -1 & 2 \\ 4 & 3 \\ 0 & 1 \end{bmatrix}\begin{bmatrix} 2 & 1 & 2 \\ 3 & 2 & 4 \end{bmatrix}$

15. $\begin{bmatrix} 0.1 & 0.9 \\ 0.2 & 0.8 \end{bmatrix}\begin{bmatrix} 1.2 & 0.4 \\ 0.5 & 2.1 \end{bmatrix}$

16. $\begin{bmatrix} 1.2 & 0.3 \\ 0.4 & 0.5 \end{bmatrix}\begin{bmatrix} 0.2 & 0.6 \\ 0.4 & -0.5 \end{bmatrix}$

17. $\begin{bmatrix} 6 & -3 & 0 \\ -2 & 1 & -8 \\ 4 & -4 & 9 \end{bmatrix}\begin{bmatrix} 1 & 0 & 0 \\ 0 & 1 & 0 \\ 0 & 0 & 1 \end{bmatrix}$

18. $\begin{bmatrix} 2 & 4 \\ -1 & -5 \\ 3 & -1 \end{bmatrix}\begin{bmatrix} 2 & -2 & 4 \\ 1 & 3 & -1 \end{bmatrix}$

19. $\begin{bmatrix} 3 & 0 & -2 & 1 \\ 1 & 2 & 0 & -1 \end{bmatrix}\begin{bmatrix} 2 & 1 & -2 \\ -1 & 2 & 0 \\ 0 & 0 & 1 \\ -1 & -2 & 2 \end{bmatrix}$

20. $\begin{bmatrix} 2 & 1 & -3 & 0 \\ 4 & -2 & -1 & 1 \\ -1 & 2 & 0 & 1 \end{bmatrix}\begin{bmatrix} 2 & -1 \\ 1 & 4 \\ 3 & -3 \\ 0 & -5 \end{bmatrix}$

21. $4\begin{bmatrix} 1 & -2 & 0 \\ 2 & -1 & 1 \\ 3 & 0 & -1 \end{bmatrix}\begin{bmatrix} 1 & 3 & 1 \\ 1 & 4 & 0 \\ 0 & 1 & -2 \end{bmatrix}$

22. $3\begin{bmatrix} 2 & -1 & 0 \\ 2 & 1 & 2 \\ 1 & 0 & -1 \end{bmatrix}\begin{bmatrix} 2 & 3 & 1 \\ 3 & -3 & 0 \\ 0 & 1 & -1 \end{bmatrix}$

23. $\begin{bmatrix} 1 & 0 \\ 0 & 1 \end{bmatrix}\begin{bmatrix} 4 & -3 & 2 \\ 7 & 1 & -5 \end{bmatrix}\begin{bmatrix} 1 & 0 & 0 \\ 0 & 1 & 0 \\ 0 & 0 & 1 \end{bmatrix}$

24. $2\begin{bmatrix} 3 & 2 & -1 \\ 0 & 1 & 3 \\ 2 & 0 & 3 \end{bmatrix}\begin{bmatrix} 1 & 0 & 0 \\ 0 & 1 & 0 \\ 0 & 0 & 1 \end{bmatrix}\begin{bmatrix} 1 & 2 & 0 \\ 0 & -1 & -2 \\ 1 & 3 & 1 \end{bmatrix}$

In Exercises 25 and 26, let

$$A = \begin{bmatrix} 1 & 0 & -2 \\ 1 & -3 & 2 \\ -2 & 1 & 1 \end{bmatrix} \quad B = \begin{bmatrix} 3 & 1 & 0 \\ 2 & 2 & 0 \\ 1 & -3 & -1 \end{bmatrix}$$

$$C = \begin{bmatrix} 2 & -1 & 0 \\ 1 & -1 & 2 \\ 3 & -2 & 1 \end{bmatrix}$$

25. Verify the validity of the associative law for matrix multiplication.

26. Verify the validity of the distributive law for matrix multiplication.

27. Let

$$A = \begin{bmatrix} 1 & 2 \\ 3 & 4 \end{bmatrix} \quad \text{and} \quad B = \begin{bmatrix} 2 & 1 \\ 4 & 3 \end{bmatrix}$$

Compute AB and BA, and hence deduce that matrix multiplication is, in general, not commutative.

28. Let

$$A = \begin{bmatrix} 0 & 3 & 0 \\ 1 & 0 & 1 \\ 0 & 2 & 0 \end{bmatrix} \quad B = \begin{bmatrix} 2 & 4 & 5 \\ 3 & -1 & -6 \\ 4 & 3 & 4 \end{bmatrix}$$

$$C = \begin{bmatrix} 4 & 5 & 6 \\ 3 & -1 & -6 \\ 2 & 2 & 3 \end{bmatrix}$$

a. Compute AB.
b. Compute AC.
c. Using the results of parts (a) and (b), conclude that $AB = AC$ does *not* imply that $B = C$.

29. Let

$$A = \begin{bmatrix} 3 & 0 \\ 8 & 0 \end{bmatrix} \quad \text{and} \quad B = \begin{bmatrix} 0 & 0 \\ 4 & 5 \end{bmatrix}$$

Show that $AB = 0$, thereby demonstrating that for matrix multiplication, the equation $AB = 0$ does not imply that one or both of the matrices A and B must be the zero matrix.

30. Let

$$A = \begin{bmatrix} 2 & 2 \\ -2 & -2 \end{bmatrix}$$

Show that $A^2 = 0$. Compare this with the equation $a^2 = 0$, where a is a real number.

31. Find the matrix A such that

$$A\begin{bmatrix} 1 & 0 \\ -1 & 3 \end{bmatrix} = \begin{bmatrix} -1 & -3 \\ 3 & 6 \end{bmatrix}$$

Hint: Let $A = \begin{bmatrix} a & b \\ c & d \end{bmatrix}$.

32. Find the matrix A such that

$$\begin{bmatrix} 1 & 0 \\ -1 & 3 \end{bmatrix} A = \begin{bmatrix} -1 & -3 \\ 3 & 6 \end{bmatrix}$$

Hint: Let $A = \begin{bmatrix} a & b \\ c & d \end{bmatrix}$.

33. Find a matrix B such that $AB = I$, where

$$A = \begin{bmatrix} 2 & 1 \\ -2 & 2 \end{bmatrix} \quad \text{and} \quad I = \begin{bmatrix} 1 & 0 \\ 0 & 1 \end{bmatrix}$$

Hint: See Exercises 31 and 32.

34. Find a matrix B such that $AB = I$, where

$$A = \begin{bmatrix} 1 & 2 \\ 4 & 3 \end{bmatrix} \quad \text{and} \quad I = \begin{bmatrix} 1 & 0 \\ 0 & 1 \end{bmatrix}$$

Hint: See Exercises 31 and 32.

35. A square matrix is called an *upper triangular matrix* if all its entries below the main diagonal are zero. For example, the matrix

$$A = \begin{bmatrix} a & b \\ 0 & d \end{bmatrix}$$

is a 2×2 *upper triangular matrix.*
a. Show that the sum and the product of two upper triangular matrices of size two are upper triangular matrices.
b. If A and B are two upper triangular matrices of size two, then is it true that $AB = BA$, in general?

36. Let

$$A = \begin{bmatrix} 3 & 1 \\ 0 & 2 \end{bmatrix} \quad \text{and} \quad B = \begin{bmatrix} 4 & -2 \\ 2 & 1 \end{bmatrix}$$

a. Compute $(A + B)^2$.
b. Compute $A^2 + 2AB + B^2$.
c. From the results of parts (a) and (b), show that in general, $(A + B)^2 \neq A^2 + 2AB + B^2$.

37. Let

$$A = \begin{bmatrix} 2 & 4 \\ 5 & -6 \end{bmatrix} \quad \text{and} \quad B = \begin{bmatrix} 4 & 8 \\ -7 & 3 \end{bmatrix}$$

a. Find A^T and show that $(A^T)^T = A$.
b. Show that $(A + B)^T = A^T + B^T$.
c. Show that $(AB)^T = B^T A^T$.

38. Let

$$A = \begin{bmatrix} 1 & 3 \\ -2 & -1 \end{bmatrix} \quad \text{and} \quad B = \begin{bmatrix} 3 & -4 \\ 2 & -2 \end{bmatrix}$$

a. Find A^T and show that $(A^T)^T = A$.
b. Show that $(A + B)^T = A^T + B^T$.
c. Show that $(AB)^T = B^T A^T$.

In Exercises 39–44, write the given system of linear equations in matrix form.

39. $\begin{aligned} 2x - 3y &= 7 \\ 3x - 4y &= 8 \end{aligned}$

40. $\begin{aligned} 2x \phantom{{}-2y} &= 7 \\ 3x - 2y &= 12 \end{aligned}$

41. $\begin{aligned} 2x - 3y + 4z &= 6 \\ 2y - 3z &= 7 \\ x - y + 2z &= 4 \end{aligned}$

42. $\begin{aligned} x - 2y + 3z &= -1 \\ 3x + 4y - 2z &= 1 \\ 2x - 3y + 7z &= 6 \end{aligned}$

43. $\begin{aligned} -x_1 + x_2 + x_3 &= 0 \\ 2x_1 - x_2 - x_3 &= 2 \\ -3x_1 + 2x_2 + 4x_3 &= 4 \end{aligned}$

44. $\begin{aligned} 3x_1 - 5x_2 + 4x_3 &= 10 \\ 4x_1 + 2x_2 - 3x_3 &= -12 \\ -x_1 \phantom{{}+2x_2} + x_3 &= -2 \end{aligned}$

45. **Stock Transactions** Olivia's and Isabella's stock holdings are given by the matrix

$$A = \begin{array}{c} \\ \text{Olivia} \\ \text{Isabella} \end{array} \begin{array}{c} \begin{array}{cccc} \text{FB} & \text{HD} & \text{PG} & \text{SBUX} \end{array} \\ \begin{bmatrix} 200 & 300 & 100 & 200 \\ 100 & 200 & 400 & 0 \end{bmatrix} \end{array}$$

At the close of trading on a certain day, the prices (in dollars per share) of the stocks are given by the matrix

$$B = \begin{array}{c} \text{FB} \\ \text{HD} \\ \text{PG} \\ \text{SBUX} \end{array} \begin{bmatrix} 27 \\ 24 \\ 63 \\ 56 \end{bmatrix}$$

a. Find AB.
b. Explain the meaning of the entries in the matrix AB.

46. **Airline Flight Scheduling** Pacific Airlines operates three flights between Los Angeles and Hong Kong. Matrix A gives the number of seats in each of three cabin classes on each flight.

$$A = \begin{array}{c} \\ \text{Flight I} \\ \text{Flight II} \\ \text{Flight III} \end{array} \begin{array}{c} \begin{array}{ccc} \text{First class} & \text{Premium economy} & \text{Economy} \end{array} \\ \begin{bmatrix} 60 & 80 & 160 \\ 50 & 60 & 190 \\ 40 & 50 & 210 \end{bmatrix} \end{array}$$

The fares per passenger (in dollars) for each class of seats are given by matrix B:

$$B = \begin{array}{c} \text{First class} \\ \text{Premium economy} \\ \text{Economy} \end{array} \begin{bmatrix} 6000 \\ 3500 \\ 1500 \end{bmatrix}$$

The number of each type of flight operated by Pacific airlines in June is given by matrix C:

$$C = \begin{array}{c} \text{Flight} \\ \begin{array}{ccc} \text{I} & \text{II} & \text{III} \end{array} \\ \begin{bmatrix} 8 & 6 & 6 \end{bmatrix} \end{array}$$

a. Compute AB, and explain what it represents.
b. Compute CAB, and explain its meaning.

47. **River Cruise Schedule Planning** Nordic River Cruises operates four cruises between several cities in Europe. The number of each type of cruise planned for 2015 is given by matrix A:

$$A = \begin{array}{c} \text{Cruise} \\ \begin{array}{cccc} \text{I} & \text{II} & \text{III} & \text{IV} \end{array} \\ \begin{bmatrix} 12 & 14 & 20 & 10 \end{bmatrix} \end{array}$$

For each cruise, the classes of cabins are classified into three categories. The number of cabins in each category for each type of cruise are given by matrix B:

$$B = \begin{array}{c} \\ \text{Cruise I} \\ \text{Cruise II} \\ \text{Cruise III} \\ \text{Cruise IV} \end{array} \begin{array}{c} \text{Category} \\ \begin{array}{ccc} \text{A} & \text{B} & \text{C} \end{array} \\ \begin{bmatrix} 20 & 30 & 40 \\ 20 & 20 & 55 \\ 15 & 35 & 45 \\ 25 & 30 & 40 \end{bmatrix} \end{array}$$

The fares per passenger (in dollars) for each category of cabins are given by matrix C:

$$C = \begin{array}{c} \text{Category A} \\ \text{Category B} \\ \text{Category C} \end{array} \begin{bmatrix} 8000 \\ 10000 \\ 7000 \end{bmatrix}$$

a. Compute AB, and explain what it represents.
b. Compute ABC, and explain its meaning.

48. **Foreign Exchange** Mason has just returned to the United States from a Southeast Asian trip and wishes to exchange the various foreign currencies that he has accumulated for U.S. dollars. He has 1200 Thai bahts, 80,000 Indonesian rupiahs, 42 Malaysian ringgits, and 36 Singapore dollars. Suppose the foreign exchange rates are U.S. \$0.03 for one baht, U.S. \$0.0001 for one rupiah, U.S. \$0.322 for one Malaysian ringgit, and U.S. \$0.806 for one Singapore dollar.

a. Write a row matrix A giving the value of the various currencies that Mason holds. (*Note:* The answer is *not* unique.)
b. Write a column matrix B giving the exchange rates for the various currencies.
c. If Mason exchanges all of his foreign currencies for U.S. dollars, how many dollars will he have?

49. **Investments** Ashley's stock holdings are given by matrix A:

$$A = \begin{array}{c} \begin{array}{cccc} \text{Google} & \text{Ebay} & \text{Priceline} & \text{Netflix} \end{array} \\ \begin{bmatrix} 200 & 300 & 240 & 120 \end{bmatrix} \end{array}$$

At the close of trading on Monday, Tuesday, and Wednesday of a certain week, the prices (in dollars per share) of the stocks were given by matrix B:

$$B = \begin{array}{c} \\ \text{Google} \\ \text{Ebay} \\ \text{Priceline} \\ \text{Netflix} \end{array} \begin{array}{c} \begin{array}{ccc} \text{Mon.} & \text{Tues.} & \text{Wed.} \end{array} \\ \begin{bmatrix} 821.50 & 838.60 & 831.38 \\ 55.48 & 55.26 & 53.37 \\ 714.01 & 718.41 & 718.90 \\ 181.21 & 181.73 & 182.94 \end{bmatrix} \end{array}$$

Use matrix multiplication to find a matrix C giving the total value of Ashley's stock holdings on each of the three days.

50. **Comparative Shopping** Laura is planning to buy two 5-lb bags of sugar, three 5-lb bags of flour, two 1-gal cartons of milk, and three 1-dozen cartons of large eggs. The prices of these items in three neighborhood supermarkets are as follows:

	Sugar (5-lb bag)	Flour (5-lb bag)	Milk (1-gal carton)	Eggs (1-dozen carton)
Supermarket I	\$3.15	\$3.79	\$2.99	\$3.49
Supermarket II	\$2.99	\$2.89	\$2.79	\$3.29
Supermarket III	\$3.74	\$2.98	\$2.89	\$2.99

a. Write a 3×4 matrix A to represent the prices of the items in the three supermarkets.

b. Write a 4×1 matrix B to represent the quantities of the items that Laura plans to purchase in the three supermarkets.

c. Use matrix multiplication to find a matrix C that represents Laura's total outlay at each supermarket. At which supermarket should she make her purchase if she wants to minimize her cost? (Assume that she will shop at only one supermarket.)

51. Foreign Exchange Ava and her friend Ella have returned to the United States from a tour of four cities: Oslo, Stockholm, Copenhagen, and Saint Petersburg. They now wish to exchange the various foreign currencies that they have accumulated for U.S. dollars. Ava has 82 Norwegian kroner, 68 Swedish kronor, 62 Danish kroner, and 1200 Russian rubles. Ella has 64 Norwegian kroner, 74 Swedish kronor, 44 Danish kroner, and 1600 Russian rubles. Suppose the exchange rates are U.S. \$0.1751 for one Norwegian krone, U.S. \$0.1560 for one Swedish krona, U.S. \$0.1747 for one Danish krone, and U.S. \$0.0325 for one Russian ruble.

a. Write a 2×4 matrix A giving the values of the various foreign currencies held by Ava and Ella. (*Note:* The answer is *not* unique.)

b. Write a column matrix B giving the exchange rate for the various currencies.

c. If both Ava and Ella exchange all their foreign currencies for U.S. dollars, how many dollars will each have?

52. Real Estate Bond Brothers, a real estate developer, builds houses in three states. The projected number of units of each model to be built in each state is given by the matrix

$$A = \begin{array}{c} \\ \\ \text{NY} \\ \text{CT} \\ \text{MA} \end{array} \begin{array}{c} \text{Model} \\ \begin{array}{cccc} \text{I} & \text{II} & \text{III} & \text{IV} \end{array} \\ \begin{bmatrix} 60 & 80 & 120 & 40 \\ 20 & 30 & 60 & 10 \\ 10 & 15 & 30 & 5 \end{bmatrix} \end{array}$$

The profits to be realized are \$60,000, \$66,000, \$75,000, and \$90,000, respectively, for each Model I, II, III, and IV house sold.

a. Write a column matrix B representing the profit for each type of house.

b. Find the total profit Bond Brothers expects to earn in each state if all the houses are sold.

53. Real Estate Refer to Exercise 52. Let $B = [1\ 1\ 1]$ and $C = [1\ 1\ 1\ 1]$.

a. Compute BA, and explain what the entries of the matrix represent.

b. Compute AC^T, and give an interpretation of the matrix.

54. Charities The amount of money raised by Charity I, Charity II, and Charity III (in millions of dollars) in each of the years 2013, 2014, and 2015 is represented by the matrix A:

$$A = \begin{array}{c} \\ \\ 2013 \\ 2014 \\ 2015 \end{array} \begin{array}{c} \text{Charity} \\ \begin{array}{ccc} \text{I} & \text{II} & \text{III} \end{array} \\ \begin{bmatrix} 18.2 & 28.2 & 40.5 \\ 19.6 & 28.6 & 42.6 \\ 20.8 & 30.4 & 46.4 \end{bmatrix} \end{array}$$

On average, Charity I puts 78% toward program cost, Charity II puts 88% toward program cost, and Charity III puts 80% toward program cost. Write a 3×1 matrix B reflecting the percentage put toward program cost by the charities. Then use matrix multiplication to find the total amount of money put toward program cost in each of the 3 years by the charities under consideration.

55. Box-Office Receipts The Cinema Center consists of four theaters: Cinemas I, II, III, and IV. The admission price for one feature at the Center is \$4 for children, \$6 for students, and \$8 for adults. The attendance for the Sunday matinee is given by the matrix

$$A = \begin{array}{c} \\ \text{Cinema I} \\ \text{Cinema II} \\ \text{Cinema III} \\ \text{Cinema IV} \end{array} \begin{array}{c} \begin{array}{ccc} \text{Children} & \text{Students} & \text{Adults} \end{array} \\ \begin{bmatrix} 225 & 110 & 50 \\ 75 & 180 & 225 \\ 280 & 85 & 110 \\ 0 & 250 & 225 \end{bmatrix} \end{array}$$

Write a column vector B representing the admission prices. Then compute AB, the column vector showing the gross receipts for each theater. Finally, find the total revenue collected at the Cinema Center for admission that Sunday afternoon.

56. Box-Office Receipts Refer to Exercise 55.

a. Find a 1×4 matrix B such that the entries in BA give the total number of children, the total number of students, and the total number of adults who attended the Sunday matinee. Compute BA.

b. Find a 1×3 matrix C such that the entries in AC^T give the total number of people (children, students, and adults) who attended Cinema I, Cinema II, Cinema III, and Cinema IV. Compute AC^T.

57. Voter Affiliation by Age Matrix A gives the percentage of eligible voters in the city of Newton, classified according to party affiliation and age group.

$$A = \begin{array}{c} \\ \text{Under 30} \\ \text{30 to 50} \\ \text{Over 50} \end{array} \begin{array}{c} \begin{array}{ccc} \text{Dem.} & \text{Rep.} & \text{Ind.} \end{array} \\ \begin{bmatrix} 0.50 & 0.30 & 0.20 \\ 0.45 & 0.40 & 0.15 \\ 0.40 & 0.50 & 0.10 \end{bmatrix} \end{array}$$

The population of eligible voters in the city by age group is given by the matrix B:

$$B = \begin{array}{c} \begin{array}{ccc} \text{Under 30} & \text{30 to 50} & \text{Over 50} \end{array} \\ \begin{bmatrix} 30{,}000 & 40{,}000 & 20{,}000 \end{bmatrix} \end{array}$$

Find a matrix giving the total number of eligible voters in the city who will vote Democratic, Republican, and Independent.

58. 401(k) Retirement Plans Three network consultants, Alan, Maria, and Steven, each received a year-end bonus of \$10,000, which they decided to invest in a 401(k) retirement plan sponsored by their employer. Under this plan, employees are allowed to place their investments in three funds: an equity index fund (I), a growth fund (II), and a global equity fund (III). The allocations of the investments (in dollars) of the three employees at the beginning of the year are summarized in the matrix

$$A = \begin{array}{l} \text{Alan} \\ \text{Maria} \\ \text{Steven} \end{array} \overset{\begin{array}{ccc} & \text{Fund} & \\ \text{I} & \text{II} & \text{III} \end{array}}{\begin{bmatrix} 4000 & 3000 & 3000 \\ 2000 & 5000 & 3000 \\ 2000 & 3000 & 5000 \end{bmatrix}}$$

The returns of the three funds after 1 year are given in the matrix

$$B = \begin{array}{l} \text{Fund I} \\ \text{Fund II} \\ \text{Fund III} \end{array} \begin{bmatrix} 0.18 \\ 0.24 \\ 0.12 \end{bmatrix}$$

Which employee realized the best return on his or her investment for the year in question? The worst return?

59. College Admissions A university admissions committee anticipates an enrollment of 8000 students in its freshman class next year. To satisfy admission quotas, incoming students have been categorized according to their sex and place of residence. The number of students in each category is given by the matrix

$$A = \begin{array}{l} \text{In-state} \\ \text{Out-of-state} \\ \text{Foreign} \end{array} \overset{\begin{array}{cc} \text{Male} & \text{Female} \end{array}}{\begin{bmatrix} 2700 & 3000 \\ 800 & 700 \\ 500 & 300 \end{bmatrix}}$$

By using data accumulated in previous years, the admissions committee has determined that these students will elect to enter the College of Letters and Science, the College of Fine Arts, the School of Business Administration, and the School of Engineering according to the percentages that appear in the following matrix:

$$B = \begin{array}{l} \text{Male} \\ \text{Female} \end{array} \overset{\begin{array}{cccc} \text{L. \& S.} & \text{Fine Arts} & \text{Bus. Ad.} & \text{Eng.} \end{array}}{\begin{bmatrix} 0.25 & 0.20 & 0.30 & 0.25 \\ 0.30 & 0.35 & 0.25 & 0.10 \end{bmatrix}}$$

Find the matrix AB that shows the number of in-state, out-of-state, and foreign students expected to enter each discipline.

60. Production Planning Refer to Example 6 in this section. Suppose Ace Novelty received an order from another amusement park for 1200 Pink Panthers, 1800 Giant Pandas, and 1400 Big Birds. The quantity of each type of stuffed animal to be produced at each plant is shown in the following production matrix:

$$P = \begin{array}{l} \text{L.A.} \\ \text{Seattle} \end{array} \overset{\begin{array}{ccc} \text{Panthers} & \text{Pandas} & \text{Birds} \end{array}}{\begin{bmatrix} 700 & 1000 & 800 \\ 500 & 800 & 600 \end{bmatrix}}$$

Each Panther requires 1.3 yd^2 of plush, 20 ft^3 of stuffing, and 12 pieces of trim. Assume that the materials required to produce the other two stuffed animals and the unit cost for each type of material are as given in Example 6.

a. How much of each type of material must be purchased for each plant?
b. What is the total cost of materials that will be incurred at each plant?
c. What is the total cost of materials incurred by Ace Novelty in filling the order?

61. Computing Phone Bills Cindy regularly makes long-distance phone calls to three foreign cities: London, Tokyo, and Hong Kong. The matrices A and B give the lengths (in minutes) of her calls during peak and nonpeak hours, respectively, to each of these three cities during the month of June.

$$A = \overset{\begin{array}{ccc} \text{London} & \text{Tokyo} & \text{Hong Kong} \end{array}}{\begin{bmatrix} 80 & 60 & 40 \end{bmatrix}}$$

and

$$B = \overset{\begin{array}{ccc} \text{London} & \text{Tokyo} & \text{Hong Kong} \end{array}}{\begin{bmatrix} 300 & 150 & 250 \end{bmatrix}}$$

The costs for the calls (in dollars per minute) for the peak and nonpeak periods in the month in question are given, respectively, by the matrices

$$C = \begin{array}{l} \text{London} \\ \text{Tokyo} \\ \text{Hong Kong} \end{array} \begin{bmatrix} 0.17 \\ 0.21 \\ 0.24 \end{bmatrix}$$

and

$$D = \begin{array}{l} \text{London} \\ \text{Tokyo} \\ \text{Hong Kong} \end{array} \begin{bmatrix} 0.12 \\ 0.15 \\ 0.17 \end{bmatrix}$$

Compute the matrix $AC + BD$, and explain what it represents.

62. Production Planning The total output of loudspeaker systems of the Acrosonic Company at their three production facilities for May and June is given by the matrices A and B, respectively, where

$$A = \begin{array}{l} \text{Location I} \\ \text{Location II} \\ \text{Location III} \end{array} \overset{\begin{array}{cccc} & \text{Model} & & \\ \text{A} & \text{B} & \text{C} & \text{D} \end{array}}{\begin{bmatrix} 320 & 280 & 460 & 280 \\ 480 & 360 & 580 & 0 \\ 540 & 420 & 200 & 880 \end{bmatrix}}$$

$$B = \begin{array}{c} \text{Location I} \\ \text{Location II} \\ \text{Location III} \end{array} \overset{\begin{array}{cccc} & \text{Model} & & \\ \text{A} & \text{B} & \text{C} & \text{D} \end{array}}{\begin{bmatrix} 210 & 180 & 330 & 180 \\ 400 & 300 & 450 & 40 \\ 420 & 280 & 180 & 740 \end{bmatrix}}$$

The unit production costs and selling prices for these loudspeakers are given by matrices C and D, respectively, where

$$C = \begin{array}{c} \text{Model A} \\ \text{Model B} \\ \text{Model C} \\ \text{Model D} \end{array} \begin{bmatrix} 120 \\ 180 \\ 260 \\ 500 \end{bmatrix}$$

and

$$D = \begin{array}{c} \text{Model A} \\ \text{Model B} \\ \text{Model C} \\ \text{Model D} \end{array} \begin{bmatrix} 160 \\ 250 \\ 350 \\ 700 \end{bmatrix}$$

Compute the following matrices, and explain the meaning of the entries in each matrix.

a. AC **b.** AD **c.** BC **d.** BD **e.** $(A + B)C$
f. $(A + B)D$ **g.** $A(D - C)$
h. $B(D - C)$ **i.** $(A + B)(D - C)$

63. Diet Planning A dietitian plans a meal around three foods. The number of units of vitamin A, vitamin C, and calcium in each ounce of these foods is represented by the matrix M, where

$$M = \begin{array}{c} \text{Vitamin A} \\ \text{Vitamin C} \\ \text{Calcium} \end{array} \overset{\begin{array}{ccc} & \text{Food} & \\ \text{I} & \text{II} & \text{III} \end{array}}{\begin{bmatrix} 400 & 1200 & 800 \\ 110 & 570 & 340 \\ 90 & 30 & 60 \end{bmatrix}}$$

The matrices A and B represent the amount of each food (in ounces) consumed by a girl at two different meals, where

$$A = \overset{\begin{array}{ccc} & \text{Food} & \\ \text{I} & \text{II} & \text{III} \end{array}}{[7 \quad 1 \quad 6]}$$

and

$$B = \overset{\begin{array}{ccc} & \text{Food} & \\ \text{I} & \text{II} & \text{III} \end{array}}{[9 \quad 3 \quad 2]}$$

Calculate the following matrices, and explain the meaning of the entries in each matrix.

a. MA^T **b.** MB^T **c.** $M(A + B)^T$

64. Sales Forecasting Hartman Lumber Company has two branches in the city. The sales of four of its products for the last year (in thousands of dollars) are represented by the matrix

$$A = \begin{array}{c} \text{Branch I} \\ \text{Branch II} \end{array} \overset{\begin{array}{cccc} & \text{Product} & & \\ \text{A} & \text{B} & \text{C} & \text{D} \end{array}}{\begin{bmatrix} 5 & 2 & 8 & 10 \\ 3 & 4 & 6 & 8 \end{bmatrix}}$$

For the present year, management has projected that the sales of the four products in Branch I will be 10% more than the corresponding sales for last year and the sales of the four products in Branch II will be 15% more than the corresponding sales for last year.

a. Show that the sales of the four products in the two branches for the current year are given by the matrix AB, where

$$A = \begin{bmatrix} 1.1 & 0 \\ 0 & 1.15 \end{bmatrix}$$

Compute AB.

b. Hartman has m branches nationwide. The sales of n of its products (in thousands of dollars) last year are represented by the matrix

$$B = \begin{array}{c} \text{Branch 1} \\ \text{Branch 2} \\ \vdots \\ \text{Branch } m \end{array} \overset{\begin{array}{ccccc} & & \text{Product} & & \\ 1 & 2 & 3 & \cdots & n \end{array}}{\begin{bmatrix} a_{11} & a_{12} & a_{13} & \cdots & a_{1n} \\ a_{21} & a_{22} & a_{23} & \cdots & a_{2n} \\ \vdots & \vdots & \vdots & \vdots & \vdots \\ a_{m1} & a_{m2} & a_{m3} & \cdots & a_{mn} \end{bmatrix}}$$

Also, management has projected that the sales of the n products in Branch 1, Branch 2, . . . , Branch m will be $r_1\%$, $r_2\%$, . . . , $r_m\%$, respectively, more than the corresponding sales for last year. Write the matrix A such that AB gives the sales of the n products in the m branches for the current year.

In Exercises 65–68, determine whether the statement is true or false. If it is true, explain why it is true. If it is false, give an example to show why it is false.

65. If A and B are matrices such that AB and BA are both defined, then A and B must be square matrices of the same order.

66. If A and B are matrices such that AB is defined and if c is a scalar, then $(cA)B = A(cB) = cAB$.

67. If A, B, and C are matrices and $A(B + C)$ is defined, then B must have the same size as C, and the number of columns of A must be equal to the number of rows of B.

68. If A is a 2×4 matrix and B is a matrix such that ABA is defined, then the size of B must be 4×2.

2.5 Solutions to Self-Check Exercises

1. We compute

$$\begin{bmatrix} 1 & 3 & 0 \\ 2 & 4 & -1 \end{bmatrix}\begin{bmatrix} 3 & 1 & 4 \\ 2 & 0 & 3 \\ 1 & 2 & -1 \end{bmatrix} = \begin{bmatrix} 1(3)+3(2)+0(1) & 1(1)+3(0)+0(2) & 1(4)+3(3)+0(-1) \\ 2(3)+4(2)-1(1) & 2(1)+4(0)-1(2) & 2(4)+4(3)-1(-1) \end{bmatrix}$$

$$= \begin{bmatrix} 9 & 1 & 13 \\ 13 & 0 & 21 \end{bmatrix}$$

2. Let

$$A = \begin{bmatrix} 0 & 1 & -2 \\ 2 & -1 & 3 \\ 1 & 0 & 4 \end{bmatrix} \quad X = \begin{bmatrix} x \\ y \\ z \end{bmatrix} \quad B = \begin{bmatrix} 1 \\ 0 \\ 7 \end{bmatrix}$$

Then the given system may be written as the matrix equation

$$AX = B$$

3. Write

$$B = \begin{matrix} \text{T} \\ \text{TWX} \\ \text{IBM} \\ \text{GM} \end{matrix}\begin{bmatrix} 36 \\ 54 \\ 205 \\ 27 \end{bmatrix}$$

and compute the following:

$$AB = \begin{matrix} \text{Ash} \\ \text{Joan} \end{matrix}\begin{bmatrix} 2000 & 1000 & 500 & 5000 \\ 1000 & 2500 & 1000 & 0 \end{bmatrix}\begin{bmatrix} 36 \\ 54 \\ 205 \\ 27 \end{bmatrix}$$

$$= \begin{matrix} \text{Ash} \\ \text{Joan} \end{matrix}\begin{bmatrix} 363{,}500 \\ 376{,}000 \end{bmatrix}$$

We conclude that Ash's stock holdings were worth \$363,500 and Joan's stock holdings were worth \$376,000 on June 1.

USING TECHNOLOGY Matrix Multiplication

Graphing Utility

A graphing utility can be used to perform matrix multiplication.

EXAMPLE 1 Let

$$A = \begin{bmatrix} 1.2 & 3.1 & -1.4 \\ 2.7 & 4.2 & 3.4 \end{bmatrix} \quad B = \begin{bmatrix} 0.8 & 1.2 & 3.7 \\ 6.2 & -0.4 & 3.3 \end{bmatrix} \quad C = \begin{bmatrix} 1.2 & 2.1 & 1.3 \\ 4.2 & -1.2 & 0.6 \\ 1.4 & 3.2 & 0.7 \end{bmatrix}$$

Find (a) AC and (b) $(1.1A + 2.3B)C$.

Solution First, we enter the matrices A, B, and C into the calculator.

a. Using matrix operations, we enter the expression $A*C$. We obtain the matrix

$$\begin{bmatrix} 12.5 & -5.68 & 2.44 \\ 25.64 & 11.51 & 8.41 \end{bmatrix}$$

(You might need to scroll the display on the screen to obtain the complete matrix.)

b. Using matrix operations, we enter the expression $(1.1A + 2.3B)C$. We obtain the matrix

$$\begin{bmatrix} 39.464 & 21.536 & 12.689 \\ 52.078 & 67.999 & 32.55 \end{bmatrix}$$

Excel

We use the **MMULT** function in Excel to perform matrix multiplication.

EXAMPLE 2 Let

$$A = \begin{bmatrix} 1.2 & 3.1 & -1.4 \\ 2.7 & 4.2 & 3.4 \end{bmatrix} \quad B = \begin{bmatrix} 0.8 & 1.2 & 3.7 \\ 6.2 & -0.4 & 3.3 \end{bmatrix} \quad C = \begin{bmatrix} 1.2 & 2.1 & 1.3 \\ 4.2 & -1.2 & 0.6 \\ 1.4 & 3.2 & 0.7 \end{bmatrix}$$

Find (a) AC and (b) $(1.1A + 2.3B)C$.

Solution

a. First, enter the matrices A, B, and C onto a spreadsheet (Figure T1).

	A	B	C	D	E	F	G
1		A				B	
2	1.2	3.1	-1.4		0.8	1.2	3.7
3	2.7	4.2	3.4		6.2	-0.4	3.3
4							
5		C					
6	1.2	2.1	1.3				
7	4.2	-1.2	0.6				
8	1.4	3.2	0.7				

FIGURE **T1**
Spreadsheet showing the matrices *A*, *B*, and *C*

Second, compute AC. Highlight the cells that will contain the matrix product AC, which has order 2×3. Type `=MMULT(`, highlight the cells in matrix A, type `,`, highlight the cells in matrix C, type `)`, and press **Ctrl-Shift-Enter**. The matrix product AC shown in Figure T2 will appear on your spreadsheet.

	A	B	C
10		AC	
11	12.5	-5.68	2.44
12	25.64	11.51	8.41

FIGURE **T2**
The matrix product *AC*

b. Compute $(1.1A + 2.3B)C$. Highlight the cells that will contain the matrix product $(1.1A + 2.3B)C$. Next, type `=MMULT(1.1*`, highlight the cells in matrix A, type `+2.3*`, highlight the cells in matrix B, type `,`, highlight the cells in matrix C, type `)`, and then press **Ctrl-Shift-Enter**. The matrix product shown in Figure T3 will appear on your spreadsheet.

	A	B	C
13		(1.1A + 2.3B)C	
14	39.464	21.536	12.689
15	52.078	67.999	32.55

FIGURE **T3**
The matrix product $(1.1A + 2.3B)C$

Note: Boldfaced words/characters enclosed in a box (for example, **Enter**) indicate that an action (click, select, or press) is required. Words/characters printed blue (for example, Chart sub-type:) indicate words/characters that appear on the screen. Words/characters printed in a monospace font (for example, `=(−2/3)*A2+2`) indicate words/characters that need to be typed and entered.

TECHNOLOGY EXERCISES

In Exercises 1–8, refer to the following matrices, and perform the indicated operations. Round your answers to two decimal places.

$$A = \begin{bmatrix} 1.2 & 3.1 & -1.2 & 4.3 \\ 7.2 & 6.3 & 1.8 & -2.1 \\ 0.8 & 3.2 & -1.3 & 2.8 \end{bmatrix}$$

$$B = \begin{bmatrix} 0.7 & 0.3 & 1.2 & -0.8 \\ 1.2 & 1.7 & 3.5 & 4.2 \\ -3.3 & -1.2 & 4.2 & 3.2 \end{bmatrix}$$

$$C = \begin{bmatrix} 0.8 & 7.1 & 6.2 \\ 3.3 & -1.2 & 4.8 \\ 1.3 & 2.8 & -1.5 \\ 2.1 & 3.2 & -8.4 \end{bmatrix}$$

1. AC

2. CB

3. $(A + B)C$

4. $(2A + 3B)C$

5. $(2A - 3.1B)C$

6. $C(2.1A + 3.2B)$

7. $(4.1A + 2.7B)1.6C$

8. $2.5C(1.8A - 4.3B)$

In Exercises 9–12, refer to the following matrices, and perform the indicated operations. Round your answers to two decimal places.

$$A = \begin{bmatrix} 2 & 5 & -4 & 2 & 8 \\ 6 & 7 & 2 & 9 & 6 \\ 4 & 5 & 4 & 4 & 4 \\ 9 & 6 & 8 & 3 & 2 \end{bmatrix}$$

$$B = \begin{bmatrix} 2 & 6 & 7 & 5 \\ 3 & 4 & 6 & 2 \\ -5 & 8 & 4 & 3 \\ 8 & 6 & 9 & 5 \\ 4 & 7 & 8 & 8 \end{bmatrix}$$

$$C = \begin{bmatrix} 6.2 & 7.3 & -4.0 & 7.1 & 9.3 \\ 4.8 & 6.5 & 8.4 & -6.3 & 8.4 \\ 5.4 & 3.2 & 6.3 & 9.1 & -2.8 \\ 8.2 & 7.3 & 6.5 & 4.1 & 9.8 \\ 10.3 & 6.8 & 4.8 & -9.1 & 20.4 \end{bmatrix}$$

$$D = \begin{bmatrix} 4.6 & 3.9 & 8.4 & 6.1 & 9.8 \\ 2.4 & -6.8 & 7.9 & 11.4 & 2.9 \\ 7.1 & 9.4 & 6.3 & 5.7 & 4.2 \\ 3.4 & 6.1 & 5.3 & 8.4 & 6.3 \\ 7.1 & -4.2 & 3.9 & -6.4 & 7.1 \end{bmatrix}$$

9. Find AB and BA.

10. Find CD and DC. Is $CD = DC$?

11. Find $AC + AD$.

12. Find:
a. AC **b.** AD **c.** $A(C + D)$
d. Is $A(C + D) = AC + AD$?

2.6 The Inverse of a Square Matrix

The Inverse of a Square Matrix

In this section, we discuss a procedure for finding the inverse of a matrix, and we show how the inverse can be used to help us solve a system of linear equations. The inverse of a matrix also plays a central role in the Leontief input–output model, which we discuss in Section 2.7.

Recall that if a is a nonzero real number, then there exists a unique real number a^{-1} (that is, $\frac{1}{a}$) such that

$$a^{-1}a = \left(\frac{1}{a}\right)(a) = 1$$

The use of the (multiplicative) inverse of a real number enables us to solve algebraic equations of the form

$$ax = b \tag{13}$$

Multiplying both sides of (13) by a^{-1}, we have

$$\begin{aligned} a^{-1}(ax) &= a^{-1}b \\ \left(\frac{1}{a}\right)(ax) &= \frac{1}{a}(b) \\ x &= \frac{b}{a} \end{aligned}$$

For example, since the inverse of 2 is $2^{-1} = \frac{1}{2}$, we can solve the equation

$$2x = 5$$

by multiplying both sides of the equation by $2^{-1} = \frac{1}{2}$, giving

$$\begin{aligned} 2^{-1}(2x) &= 2^{-1} \cdot 5 \\ x &= \frac{5}{2} \end{aligned}$$

We can use a similar procedure to solve the matrix equation

$$AX = B$$

where A, X, and B are matrices of the proper sizes. To do this we need the matrix equivalent of the inverse of a real number. Such a matrix, whenever it exists, is called the **inverse of a matrix.**

> **Inverse of a Matrix**
>
> Let A be a square matrix of size n. A square matrix A^{-1} of size n such that
>
> $$A^{-1}A = AA^{-1} = I_n$$
>
> is called the inverse of A.

Let's show that the matrix

$$A = \begin{bmatrix} 1 & 2 \\ 3 & 4 \end{bmatrix}$$

has the matrix

$$A^{-1} = \begin{bmatrix} -2 & 1 \\ \frac{3}{2} & -\frac{1}{2} \end{bmatrix}$$

as its inverse. Since

$$AA^{-1} = \begin{bmatrix} 1 & 2 \\ 3 & 4 \end{bmatrix}\begin{bmatrix} -2 & 1 \\ \frac{3}{2} & -\frac{1}{2} \end{bmatrix} = \begin{bmatrix} 1 & 0 \\ 0 & 1 \end{bmatrix} = I$$

$$A^{-1}A = \begin{bmatrix} -2 & 1 \\ \frac{3}{2} & -\frac{1}{2} \end{bmatrix}\begin{bmatrix} 1 & 2 \\ 3 & 4 \end{bmatrix} = \begin{bmatrix} 1 & 0 \\ 0 & 1 \end{bmatrix} = I$$

we see that A^{-1} is the inverse of A, as asserted.

> **Explore and Discuss**
>
> In defining the inverse of a matrix A, why is it necessary to require that A be a square matrix?

Not every square matrix has an inverse. A square matrix that has an inverse is said to be **nonsingular**. A matrix that does not have an inverse is said to be **singular**. An example of a singular matrix is given by

$$B = \begin{bmatrix} 0 & 1 \\ 0 & 0 \end{bmatrix}$$

If B had an inverse given by

$$B^{-1} = \begin{bmatrix} a & b \\ c & d \end{bmatrix}$$

where a, b, c, and d are some appropriate numbers, then by the definition of an inverse, we would have $BB^{-1} = I$; that is,

$$\begin{bmatrix} 0 & 1 \\ 0 & 0 \end{bmatrix}\begin{bmatrix} a & b \\ c & d \end{bmatrix} = \begin{bmatrix} 1 & 0 \\ 0 & 1 \end{bmatrix}$$

$$\begin{bmatrix} c & d \\ 0 & 0 \end{bmatrix} = \begin{bmatrix} 1 & 0 \\ 0 & 1 \end{bmatrix}$$

which implies that $0 = 1$—an impossibility! This contradiction shows that B does not have an inverse.

A Method for Finding the Inverse of a Square Matrix

The methods of Section 2.5 can be used to find the inverse of a nonsingular matrix. To discover such an algorithm, let's find the inverse of the matrix

$$A = \begin{bmatrix} 1 & 2 \\ -1 & 3 \end{bmatrix}$$

Suppose A^{-1} exists and is given by

$$A^{-1} = \begin{bmatrix} a & b \\ c & d \end{bmatrix}$$

where a, b, c, and d are to be determined. By the definition of an inverse, we have $AA^{-1} = I$; that is,

$$\begin{bmatrix} 1 & 2 \\ -1 & 3 \end{bmatrix}\begin{bmatrix} a & b \\ c & d \end{bmatrix} = \begin{bmatrix} 1 & 0 \\ 0 & 1 \end{bmatrix}$$

which simplifies to

$$\begin{bmatrix} a + 2c & b + 2d \\ -a + 3c & -b + 3d \end{bmatrix} = \begin{bmatrix} 1 & 0 \\ 0 & 1 \end{bmatrix}$$

But this matrix equation is equivalent to the two systems of linear equations

$$\begin{cases} a + 2c = 1 \\ -a + 3c = 0 \end{cases} \quad \text{and} \quad \begin{cases} b + 2d = 0 \\ -b + 3d = 1 \end{cases}$$

with augmented matrices given by

$$\left[\begin{array}{rr|r} 1 & 2 & 1 \\ -1 & 3 & 0 \end{array}\right] \quad \text{and} \quad \left[\begin{array}{rr|r} 1 & 2 & 0 \\ -1 & 3 & 1 \end{array}\right]$$

Note that the matrices of coefficients of the two systems are identical. This suggests that we solve the two systems of simultaneous linear equations by writing the following augmented matrix, which we obtain by joining the coefficient matrix and the two columns of constants:

$$\left[\begin{array}{rr|rr} 1 & 2 & 1 & 0 \\ -1 & 3 & 0 & 1 \end{array}\right]$$

Using the Gauss–Jordan elimination method, we obtain the following sequence of equivalent matrices:

$$\left[\begin{array}{rr|rr} 1 & 2 & 1 & 0 \\ -1 & 3 & 0 & 1 \end{array}\right] \xrightarrow{R_2 + R_1} \left[\begin{array}{rr|rr} 1 & 2 & 1 & 0 \\ 0 & 5 & 1 & 1 \end{array}\right] \xrightarrow{-\frac{1}{5}R_2}$$

$$\left[\begin{array}{rr|rr} 1 & 2 & 1 & 0 \\ 0 & 1 & \frac{1}{5} & \frac{1}{5} \end{array}\right] \xrightarrow{R_1 - 2R_2} \left[\begin{array}{rr|rr} 1 & 0 & \frac{3}{5} & -\frac{2}{5} \\ 0 & 1 & \frac{1}{5} & \frac{1}{5} \end{array}\right]$$

Thus, $a = \frac{3}{5}$, $b = -\frac{2}{5}$, $c = \frac{1}{5}$, and $d = \frac{1}{5}$, giving

$$A^{-1} = \begin{bmatrix} \frac{3}{5} & -\frac{2}{5} \\ \frac{1}{5} & \frac{1}{5} \end{bmatrix}$$

The following computations verify that A^{-1} is indeed the inverse of A:

$$\begin{bmatrix} 1 & 2 \\ -1 & 3 \end{bmatrix}\begin{bmatrix} \frac{3}{5} & -\frac{2}{5} \\ \frac{1}{5} & \frac{1}{5} \end{bmatrix} = \begin{bmatrix} 1 & 0 \\ 0 & 1 \end{bmatrix} = \begin{bmatrix} \frac{3}{5} & -\frac{2}{5} \\ \frac{1}{5} & \frac{1}{5} \end{bmatrix}\begin{bmatrix} 1 & 2 \\ -1 & 3 \end{bmatrix}$$

The preceding example suggests a general algorithm for computing the inverse of a square matrix of size n when it exists.

Finding the Inverse of a Matrix

Given the $n \times n$ matrix A:

1. Adjoin the $n \times n$ identity matrix I to obtain the augmented matrix

$$[A \mid I]$$

2. Use a sequence of row operations to reduce $[A \mid I]$ to the form

$$[A \mid B]$$

   if possible.

Then the matrix B is the inverse of A.

Note Although matrix multiplication is not generally commutative, it is possible to prove that if A has an inverse and $AB = I$, then $BA = I$ also. Hence to verify that B is the inverse of A, it suffices to show that $AB = I$.

EXAMPLE 1 Find the inverse of the matrix

$$A = \begin{bmatrix} 2 & 1 & 1 \\ 3 & 2 & 1 \\ 2 & 1 & 2 \end{bmatrix}$$

Solution We form the augmented matrix

$$\left[\begin{array}{rrr|rrr} 2 & 1 & 1 & 1 & 0 & 0 \\ 3 & 2 & 1 & 0 & 1 & 0 \\ 2 & 1 & 2 & 0 & 0 & 1 \end{array}\right]$$

and use the Gauss–Jordan elimination method to reduce it to the form $[I \mid B]$:

$$\left[\begin{array}{ccc|ccc} 2 & 1 & 1 & 1 & 0 & 0 \\ 3 & 2 & 1 & 0 & 1 & 0 \\ 2 & 1 & 2 & 0 & 0 & 1 \end{array}\right] \xrightarrow{R_1 - R_2} \left[\begin{array}{ccc|ccc} -1 & -1 & 0 & 1 & -1 & 0 \\ 3 & 2 & 1 & 0 & 1 & 0 \\ 2 & 1 & 2 & 0 & 0 & 1 \end{array}\right]$$

$$\xrightarrow[\substack{R_2 + 3R_1 \\ R_3 + 2R_1}]{-R_1} \left[\begin{array}{ccc|ccc} 1 & 1 & 0 & -1 & 1 & 0 \\ 0 & -1 & 1 & 3 & -2 & 0 \\ 0 & -1 & 2 & 2 & -2 & 1 \end{array}\right]$$

$$\xrightarrow[\substack{-R_2 \\ R_3 - R_2}]{R_1 + R_2} \left[\begin{array}{ccc|ccc} 1 & 0 & 1 & 2 & -1 & 0 \\ 0 & 1 & -1 & -3 & 2 & 0 \\ 0 & 0 & 1 & -1 & 0 & 1 \end{array}\right]$$

$$\xrightarrow[R_2 + R_3]{R_1 - R_3} \left[\begin{array}{ccc|ccc} 1 & 0 & 0 & 3 & -1 & -1 \\ 0 & 1 & 0 & -4 & 2 & 1 \\ 0 & 0 & 1 & -1 & 0 & 1 \end{array}\right]$$

The inverse of A is the matrix

$$A^{-1} = \begin{bmatrix} 3 & -1 & -1 \\ -4 & 2 & 1 \\ -1 & 0 & 1 \end{bmatrix}$$

We leave it to you to verify these results.

Example 2 illustrates what happens to the reduction process when a matrix A does *not* have an inverse.

EXAMPLE 2 Find the inverse of the matrix

$$A = \begin{bmatrix} 1 & 2 & 3 \\ 2 & 1 & 2 \\ 3 & 3 & 5 \end{bmatrix}$$

Solution We form the augmented matrix

$$\left[\begin{array}{ccc|ccc} 1 & 2 & 3 & 1 & 0 & 0 \\ 2 & 1 & 2 & 0 & 1 & 0 \\ 3 & 3 & 5 & 0 & 0 & 1 \end{array}\right]$$

and use the Gauss–Jordan elimination method:

$$\left[\begin{array}{ccc|ccc} 1 & 2 & 3 & 1 & 0 & 0 \\ 2 & 1 & 2 & 0 & 1 & 0 \\ 3 & 3 & 5 & 0 & 0 & 1 \end{array}\right] \xrightarrow[R_3 - 3R_1]{R_2 - 2R_1} \left[\begin{array}{ccc|ccc} 1 & 2 & 3 & 1 & 0 & 0 \\ 0 & -3 & -4 & -2 & 1 & 0 \\ 0 & -3 & -4 & -3 & 0 & 1 \end{array}\right]$$

$$\xrightarrow[R_3 - R_2]{-R_2} \left[\begin{array}{ccc|ccc} 1 & 2 & 3 & 1 & 0 & 0 \\ 0 & 3 & 4 & 2 & -1 & 0 \\ 0 & 0 & 0 & -1 & -1 & 1 \end{array}\right]$$

> ***Explore and Discuss***
>
> Explain in terms of solutions to systems of linear equations why the final augmented matrix in Example 2 implies that A has no inverse.
>
> Hint: See the discussion on page 142.

Since the entries in the last row of the 3×3 submatrix that comprises the left-hand side of the augmented matrix just obtained are all equal to zero, the latter cannot be reduced to the form $[I \mid B]$. Accordingly, we draw the conclusion that A is singular—that is, does not have an inverse.

More generally, we have the following criterion for determining when the inverse of a matrix does not exist.

Matrices That Have No Inverses

If there is a row to the left of the vertical line in the augmented matrix containing all zeros, then the matrix does not have an inverse.

A Formula for the Inverse of a 2×2 Matrix

Before turning to some applications, we show an alternative method that employs a formula for finding the inverse of a 2×2 matrix. This method will prove useful in many situations; we will see an application in Example 5. The derivation of this formula is left as an exercise (Exercise 52).

Formula for the Inverse of a 2×2 Matrix

Let

$$A = \begin{bmatrix} a & b \\ c & d \end{bmatrix}$$

Suppose $D = ad - bc$ is not equal to zero. Then A^{-1} exists and is given by

$$A^{-1} = \frac{1}{D}\begin{bmatrix} d & -b \\ -c & a \end{bmatrix} \qquad \textbf{(14)}$$

Note As an aid to memorizing the formula, note that D is the product of the elements along the main diagonal minus the product of the elements along the other diagonal:

$$\begin{bmatrix} a & b \\ c & d \end{bmatrix} \qquad D = ad - bc$$

Main diagonal

Next, the matrix

$$\begin{bmatrix} d & -b \\ -c & a \end{bmatrix}$$

is obtained by interchanging a and d and reversing the signs of b and c. Finally, A^{-1} is obtained by dividing this matrix by D.

Explore and Discuss

Suppose A is a square matrix with the property that one of its rows is a nonzero constant multiple of another row. What can you say about the existence or nonexistence of A^{-1}? Explain your answer.

EXAMPLE 3 Find the inverse of

$$A = \begin{bmatrix} 1 & 2 \\ 3 & 4 \end{bmatrix}$$

Solution We first compute $D = (1)(4) - (2)(3) = 4 - 6 = -2$. Next, we rewrite the given matrix, obtaining

$$\begin{bmatrix} 4 & -2 \\ -3 & 1 \end{bmatrix}$$

Finally, dividing this matrix by D, we obtain

$$A^{-1} = \frac{1}{-2}\begin{bmatrix} 4 & -2 \\ -3 & 1 \end{bmatrix} = \begin{bmatrix} -2 & 1 \\ \frac{3}{2} & -\frac{1}{2} \end{bmatrix}$$

Solving Systems of Equations with Inverses

We now show how the inverse of a matrix may be used to solve certain systems of linear equations in which the number of equations in the system is equal to the number of variables. For simplicity, let's illustrate the process for a system of three linear equations in three variables:

$$\begin{aligned} a_{11}x_1 + a_{12}x_2 + a_{13}x_3 &= b_1 \\ a_{21}x_1 + a_{22}x_2 + a_{23}x_3 &= b_2 \\ a_{31}x_1 + a_{32}x_2 + a_{33}x_3 &= b_3 \end{aligned} \qquad \textbf{(15)}$$

Let's write

$$A = \begin{bmatrix} a_{11} & a_{12} & a_{13} \\ a_{21} & a_{22} & a_{23} \\ a_{31} & a_{32} & a_{33} \end{bmatrix} \qquad X = \begin{bmatrix} x_1 \\ x_2 \\ x_3 \end{bmatrix} \qquad B = \begin{bmatrix} b_1 \\ b_2 \\ b_3 \end{bmatrix}$$

You should verify that System (15) of linear equations may be written in the form of the matrix equation

$$AX = B \qquad \textbf{(16)}$$

If A is nonsingular, then the method of this section may be used to compute A^{-1}. Next, multiplying both sides of Equation (16) by A^{-1} (on the left), we obtain

$$A^{-1}AX = A^{-1}B \quad \text{or} \quad IX = A^{-1}B \quad \text{or} \quad X = A^{-1}B$$

the desired solution to the problem.

In the case of a system of n equations with n unknowns, we have the following more general result.

Using Inverses to Solve Systems of Equations

If $AX = B$ is a linear system of n equations in n unknowns and if A^{-1} exists, then

$$X = A^{-1}B$$

is the unique solution of the system.

The use of inverses to solve systems of equations is particularly advantageous when we are required to solve more than one system of equations, $AX = B$, involving the same coefficient matrix, A, and different matrices of constants, B. As you will see in Examples 4 and 5, we need to compute A^{-1} just once in each case.

EXAMPLE 4 Solve the following systems of linear equations:

a. $\begin{aligned} 2x + y + z &= 1 \\ 3x + 2y + z &= 2 \\ 2x + y + 2z &= -1 \end{aligned}$ **b.** $\begin{aligned} 2x + y + z &= 2 \\ 3x + 2y + z &= -3 \\ 2x + y + 2z &= 1 \end{aligned}$

Solution We may write the given systems of equations in the form

$$AX = B \quad \text{and} \quad AX = C$$

respectively, where

$$A = \begin{bmatrix} 2 & 1 & 1 \\ 3 & 2 & 1 \\ 2 & 1 & 2 \end{bmatrix} \qquad X = \begin{bmatrix} x \\ y \\ z \end{bmatrix} \qquad B = \begin{bmatrix} 1 \\ 2 \\ -1 \end{bmatrix} \qquad C = \begin{bmatrix} 2 \\ -3 \\ 1 \end{bmatrix}$$

The inverse of the matrix A,

$$A^{-1} = \begin{bmatrix} 3 & -1 & -1 \\ -4 & 2 & 1 \\ -1 & 0 & 1 \end{bmatrix}$$

was found in Example 1. Using this result, we find that the solution of the first system (a) is

$$X = A^{-1}B = \begin{bmatrix} 3 & -1 & -1 \\ -4 & 2 & 1 \\ -1 & 0 & 1 \end{bmatrix}\begin{bmatrix} 1 \\ 2 \\ -1 \end{bmatrix}$$

$$= \begin{bmatrix} (3)(1) + (-1)(2) + (-1)(-1) \\ (-4)(1) + (2)(2) + (1)(-1) \\ (-1)(1) + (0)(2) + (1)(-1) \end{bmatrix} = \begin{bmatrix} 2 \\ -1 \\ -2 \end{bmatrix}$$

or $x = 2$, $y = -1$, and $z = -2$.

The solution of the second system (b) is

$$X = A^{-1}C = \begin{bmatrix} 3 & -1 & -1 \\ -4 & 2 & 1 \\ -1 & 0 & 1 \end{bmatrix}\begin{bmatrix} 2 \\ -3 \\ 1 \end{bmatrix} = \begin{bmatrix} 8 \\ -13 \\ -1 \end{bmatrix}$$

or $x = 8$, $y = -13$, and $z = -1$.

APPLIED EXAMPLE 5 Capital Expenditures The management of Checkers Rent-A-Car plans to expand its fleet of rental cars for the next quarter by purchasing compact and full-size cars. The average cost of a compact car is \$15,000, and the average cost of a full-size car is \$36,000.

a. If a total of 800 cars is to be purchased with a budget of \$18 million, how many cars of each size will be acquired?

b. If the predicted demand calls for a total purchase of 1000 cars with a budget of \$21 million, how many cars of each type will be acquired?

Solution Let x and y denote the number of compact and full-size cars to be purchased. Furthermore, let n denote the total number of cars to be acquired and b the amount of money budgeted for the purchase of these cars. Then

$$\begin{aligned} x + y &= n \\ 15{,}000x + 36{,}000y &= b \end{aligned}$$

This system of two equations in two variables may be written in the matrix form

$$AX = B$$

where

$$A = \begin{bmatrix} 1 & 1 \\ 15{,}000 & 36{,}000 \end{bmatrix} \quad X = \begin{bmatrix} x \\ y \end{bmatrix} \quad B = \begin{bmatrix} n \\ b \end{bmatrix}$$

Therefore,

$$X = A^{-1}B$$

Since A is a 2×2 matrix, its inverse may be found by using Formula (14). We find $D = (1)(36{,}000) - (1)(15{,}000) = 21{,}000$, so

$$A^{-1} = \frac{1}{21{,}000}\begin{bmatrix} 36{,}000 & -1 \\ -15{,}000 & 1 \end{bmatrix} = \begin{bmatrix} \frac{36{,}000}{21{,}000} & -\frac{1}{21{,}000} \\ -\frac{15{,}000}{21{,}000} & \frac{1}{21{,}000} \end{bmatrix}$$

Thus,

$$X = \begin{bmatrix} \frac{12}{7} & -\frac{1}{21,000} \\ -\frac{5}{7} & \frac{1}{21,000} \end{bmatrix} \begin{bmatrix} n \\ b \end{bmatrix}$$

a. Here, $n = 800$ and $b = 18{,}000{,}000$, so

$$X = A^{-1}B = \begin{bmatrix} \frac{12}{7} & -\frac{1}{21,000} \\ -\frac{5}{7} & \frac{1}{21,000} \end{bmatrix} \begin{bmatrix} 800 \\ 18,000,000 \end{bmatrix} \approx \begin{bmatrix} 514.3 \\ 285.7 \end{bmatrix}$$

Therefore, 514 compact cars and 286 full-size cars will be acquired in this case.

b. Here, $n = 1000$ and $b = 21{,}000{,}000$, so

$$X = A^{-1}B = \begin{bmatrix} \frac{12}{7} & -\frac{1}{21,000} \\ -\frac{5}{7} & \frac{1}{21,000} \end{bmatrix} \begin{bmatrix} 1000 \\ 21,000,000 \end{bmatrix} \approx \begin{bmatrix} 714.3 \\ 285.7 \end{bmatrix}$$

Therefore, 714 compact cars and 286 full-size cars will be purchased in this case.

2.6 Self-Check Exercises

1. Find the inverse of the matrix

$$A = \begin{bmatrix} 2 & 1 & -1 \\ 1 & 1 & -1 \\ -1 & -2 & 3 \end{bmatrix}$$

if it exists.

2. Solve the system of linear equations

$$\begin{aligned} 2x + y - z &= b_1 \\ x + y - z &= b_2 \\ -x - 2y + 3z &= b_3 \end{aligned}$$

where (a) $b_1 = 5, b_2 = 4, b_3 = -8$ and (b) $b_1 = 2, b_2 = 0, b_3 = 5$, by finding the inverse of the coefficient matrix.

3. **Tour Ticketing** Grand Canyon Tours offers air and ground scenic tours of the Grand Canyon. Tickets for the $7\frac{1}{2}$-hour tour cost \$169 for an adult and \$129 for a child, and each tour group is limited to 19 people. On three recent fully booked tours, total receipts were \$2931 for the first tour, \$3011 for the second tour, and \$2771 for the third tour. Determine how many adults and how many children were in each tour.

Solutions to Self-Check Exercises 2.6 can be found on page 152.

2.6 Concept Questions

1. What is the inverse of a matrix A?

2. Explain how you would find the inverse of a nonsingular matrix.

3. Give the formula for the inverse of the 2×2 matrix

$$A = \begin{bmatrix} a & b \\ c & d \end{bmatrix}$$

4. Explain how the inverse of a matrix can be used to solve a system of n linear equations in n unknowns. Does the method work for a system of m linear equations in n unknowns with $m \neq n$? Explain.

2.6 Exercises

In Exercises 1–4, show that the matrices are inverses of each other by showing that their product is the identity matrix I.

1. $\begin{bmatrix} 1 & -3 \\ 1 & -2 \end{bmatrix}$ and $\begin{bmatrix} -2 & 3 \\ -1 & 1 \end{bmatrix}$

2. $\begin{bmatrix} 4 & 5 \\ 2 & 3 \end{bmatrix}$ and $\begin{bmatrix} \frac{3}{2} & -\frac{5}{2} \\ -1 & 2 \end{bmatrix}$

3. $\begin{bmatrix} 3 & 2 & 3 \\ 2 & 2 & 1 \\ 2 & 1 & 1 \end{bmatrix}$ and $\begin{bmatrix} -\frac{1}{3} & -\frac{1}{3} & \frac{4}{3} \\ 0 & 1 & -1 \\ \frac{2}{3} & -\frac{1}{3} & -\frac{2}{3} \end{bmatrix}$

4. $\begin{bmatrix} 2 & 4 & -2 \\ -4 & -6 & 1 \\ 3 & 5 & -1 \end{bmatrix}$ and $\begin{bmatrix} \frac{1}{2} & -3 & -4 \\ -\frac{1}{2} & 2 & 3 \\ -1 & 1 & 2 \end{bmatrix}$

In Exercises 5–16, find the inverse of the matrix, if it exists. Verify your answer.

5. $\begin{bmatrix} 2 & 5 \\ 1 & 3 \end{bmatrix}$

6. $\begin{bmatrix} 2 & 3 \\ 3 & 5 \end{bmatrix}$

7. $\begin{bmatrix} 3 & -3 \\ -2 & 2 \end{bmatrix}$

8. $\begin{bmatrix} 4 & 2 \\ 6 & 3 \end{bmatrix}$

9. $\begin{bmatrix} 2 & -3 & -4 \\ 0 & 0 & -1 \\ 1 & -2 & 1 \end{bmatrix}$

10. $\begin{bmatrix} 1 & -1 & 3 \\ 2 & 1 & 2 \\ -2 & -2 & 1 \end{bmatrix}$

11. $\begin{bmatrix} 4 & 2 & 2 \\ -1 & -3 & 4 \\ 3 & -1 & 6 \end{bmatrix}$

12. $\begin{bmatrix} 1 & 2 & 0 \\ -3 & 4 & -2 \\ -5 & 0 & -2 \end{bmatrix}$

13. $\begin{bmatrix} 1 & 4 & -1 \\ 2 & 3 & -2 \\ -1 & 2 & 3 \end{bmatrix}$

14. $\begin{bmatrix} 3 & -2 & 7 \\ -2 & 1 & 4 \\ 6 & -5 & 8 \end{bmatrix}$

15. $\begin{bmatrix} 1 & 1 & -1 & 1 \\ 2 & 1 & 1 & 0 \\ 2 & 1 & 0 & 1 \\ 2 & -1 & -1 & 3 \end{bmatrix}$

16. $\begin{bmatrix} 1 & 1 & 2 & 3 \\ 2 & 3 & 0 & -1 \\ 0 & 2 & -1 & 1 \\ 1 & 2 & 1 & 1 \end{bmatrix}$

In Exercises 17–24, (a) write a matrix equation that is equivalent to the system of linear equations, and (b) solve the system using the inverses found in Exercises 5–16.

17. $\begin{aligned} 2x + 5y &= 3 \\ x + 3y &= 2 \end{aligned}$
(See Exercise 5.)

18. $\begin{aligned} 2x + 3y &= 5 \\ 3x + 5y &= 8 \end{aligned}$
(See Exercise 6.)

19. $\begin{aligned} 2x - 3y - 4z &= 4 \\ -z &= 3 \\ x - 2y + z &= -8 \end{aligned}$
(See Exercise 9.)

20. $\begin{aligned} x_1 - x_2 + 3x_3 &= 2 \\ 2x_1 + x_2 + 2x_3 &= 2 \\ -2x_1 - 2x_2 + x_3 &= 3 \end{aligned}$
(See Exercise 9.)

21. $\begin{aligned} x + 4y - z &= 3 \\ 2x + 3y - 2z &= 1 \\ -x + 2y + 3z &= 7 \end{aligned}$
(See Exercise 13.)

22. $\begin{aligned} 3x_1 - 2x_2 + 7x_3 &= 6 \\ -2x_1 + x_2 + 4x_3 &= 4 \\ 6x_1 - 5x_2 + 8x_3 &= 4 \end{aligned}$
(See Exercise 14.)

23. $\begin{aligned} x_1 + x_2 - x_3 + x_4 &= 6 \\ 2x_1 + x_2 + x_3 &= 4 \\ 2x_1 + x_2 + x_4 &= 7 \\ 2x_1 - x_2 - x_3 + 3x_4 &= 9 \end{aligned}$
(See Exercise 15.)

24. $\begin{aligned} x_1 + x_2 + 2x_3 + 3x_4 &= 4 \\ 2x_1 + 3x_2 - x_4 &= 11 \\ 2x_2 - x_3 + x_4 &= 7 \\ x_1 + 2x_2 + x_3 + x_4 &= 6 \end{aligned}$
(See Exercise 16.)

In Exercises 25–32, (a) write each system of equations as a matrix equation, and (b) solve the system of equations by using the inverse of the coefficient matrix.

25. $\begin{aligned} x + 2y &= b_1 \\ 2x - y &= b_2 \end{aligned}$

where (i) $b_1 = 14, b_2 = 5$
and (ii) $b_1 = 4, b_2 = -1$

26. $\begin{aligned} 3x - 2y &= b_1 \\ 4x + 3y &= b_2 \end{aligned}$

where (i) $b_1 = -6, b_2 = 10$
and (ii) $b_1 = 3, b_2 = -2$

27. $\begin{aligned} x + 2y + z &= b_1 \\ x + y + z &= b_2 \\ 3x + y + z &= b_3 \end{aligned}$

where (i) $b_1 = 7, b_2 = 4, b_3 = 2$
and (ii) $b_1 = 5, b_2 = -3, b_3 = -1$

28. $\begin{aligned} x_1 + x_2 + x_3 &= b_1 \\ x_1 - x_2 + x_3 &= b_2 \\ x_1 - 2x_2 - x_3 &= b_3 \end{aligned}$

where (i) $b_1 = 5, b_2 = -3, b_3 = -1$
and (ii) $b_1 = 1, b_2 = 4, b_3 = -2$

29. $\begin{aligned} 3x + 2y - z &= b_1 \\ 2x - 3y + z &= b_2 \\ x - y - z &= b_3 \end{aligned}$

where (i) $b_1 = 2, b_2 = -2, b_3 = 4$
and (ii) $b_1 = 8, b_2 = -3, b_3 = 6$

30. $\begin{aligned} 2x_1 + x_2 + x_3 &= b_1 \\ x_1 - 3x_2 + 4x_3 &= b_2 \\ -x_1 + x_3 &= b_3 \end{aligned}$

where (i) $b_1 = 1, b_2 = 4, b_3 = -3$
and (ii) $b_1 = 2, b_2 = -5, b_3 = 0$

31. $\begin{aligned} x_1 + x_2 + x_3 + x_4 &= b_1 \\ x_1 - x_2 - x_3 + x_4 &= b_2 \\ x_2 + 2x_3 + 2x_4 &= b_3 \\ x_1 + 2x_2 + x_3 - 2x_4 &= b_4 \end{aligned}$

where (i) $b_1 = 1, b_2 = -1, b_3 = 4, b_4 = 0$
and (ii) $b_1 = 2, b_2 = 8, b_3 = 4, b_4 = -1$

32.
$$\begin{aligned} x_1 + x_2 + 2x_3 + x_4 &= b_1 \\ 4x_1 + 5x_2 + 9x_3 + x_4 &= b_2 \\ 3x_1 + 4x_2 + 7x_3 + x_4 &= b_3 \\ 2x_1 + 3x_2 + 4x_3 + 2x_4 &= b_4 \end{aligned}$$

where (i) $b_1 = 3, b_2 = 6, b_3 = 5, b_4 = 7$
and (ii) $b_1 = 1, b_2 = -1, b_3 = 0, b_4 = -4$

33. Let

$$A = \begin{bmatrix} 2 & 3 \\ -4 & -5 \end{bmatrix}$$

a. Find A^{-1}.
b. Show that $(A^{-1})^{-1} = A$.

34. Let

$$A = \begin{bmatrix} 6 & -4 \\ -4 & 3 \end{bmatrix}$$

and

$$B = \begin{bmatrix} 3 & -5 \\ 4 & -7 \end{bmatrix}$$

a. Find AB, A^{-1}, and B^{-1}.
b. Show that $(AB)^{-1} = B^{-1}A^{-1}$.

35. Let

$$A = \begin{bmatrix} 2 & -5 \\ 1 & -3 \end{bmatrix} \quad B = \begin{bmatrix} 4 & 3 \\ 1 & 1 \end{bmatrix} \quad C = \begin{bmatrix} 2 & 3 \\ -2 & 1 \end{bmatrix}$$

a. Find ABC, A^{-1}, B^{-1}, and C^{-1}.
b. Show that $(ABC)^{-1} = C^{-1}B^{-1}A^{-1}$.

36. Find the matrix A if

$$\begin{bmatrix} 2 & 1 \\ -1 & 3 \end{bmatrix} A = \begin{bmatrix} 3 & 2 \\ 1 & 4 \end{bmatrix}$$

37. Find the matrix A if

$$A \begin{bmatrix} 1 & 2 \\ 3 & -1 \end{bmatrix} = \begin{bmatrix} 2 & 1 \\ 3 & -2 \end{bmatrix}$$

38. **Ticket Revenues** Rainbow Harbor Cruises charges \$16/adult and \$8/child for a round-trip ticket. The records show that, on a certain weekend, 1000 people took the cruise on Saturday, and 800 people took the cruise on Sunday. The total receipts for Saturday were \$12,800, and the total receipts for Sunday were \$9,600. Determine how many adults and children took the cruise on Saturday and on Sunday.

39. **Pricing Personal Planners** BelAir Publishing publishes a deluxe leather edition and a standard edition of its daily organizer. The company's marketing department estimates that x copies of the deluxe edition and y copies of the standard edition will be demanded per month when the unit prices are p dollars and q dollars, respectively, where x, y, p, and q are related by the following system of linear equations:

$$\begin{aligned} 5x + y &= 1000(70 - p) \\ x + 3y &= 1000(40 - q) \end{aligned}$$

Find the monthly demand for the deluxe edition and the standard edition when the unit prices are set according to the following schedules:
a. $p = 50$ and $q = 25$
b. $p = 45$ and $q = 25$
c. $p = 45$ and $q = 20$

40. **Diet Planning** Bob, a nutritionist who works for the University Medical Center, has been asked to prepare special diets for two patients, Susan and Tom. Bob has decided that Susan's meals should contain at least 400 mg of calcium, 20 mg of iron, and 50 mg of vitamin C, whereas Tom's meals should contain at least 350 mg of calcium, 15 mg of iron, and 40 mg of vitamin C. Bob has also decided that the meals are to be prepared from three basic foods: Food *A*, Food *B*, and Food *C*. The special nutritional contents of these foods are summarized in the accompanying table. Find how many ounces of each type of food should be used in a meal so that the minimum requirements of calcium, iron, and vitamin C are met for each patient's meals.

	Contents (mg/oz)		
	Calcium	**Iron**	**Vitamin C**
Food *A*	30	1	2
Food *B*	25	1	5
Food *C*	20	2	4

41. **Crop Planning** Jackson Farms has allotted a certain amount of land for cultivating soybeans, corn, and wheat. Cultivating 1 acre of soybeans requires 2 labor-hours, and cultivating 1 acre of corn or wheat requires 6 labor-hours. The cost of seeds for 1 acre of soybeans is \$12, the cost for 1 acre of corn is \$20, and the cost for 1 acre of wheat is \$8. If all resources are to be used, how many acres of each crop should be cultivated if the following hold?
a. 1000 acres of land are allotted, 4400 labor-hours are available, and \$13,200 is available for seeds.
b. 1200 acres of land are allotted, 5200 labor-hours are available, and \$16,400 is available for seeds.

42. **Lawn Fertilizers** Lawnco produces three grades of commercial fertilizers. A 100-lb bag of grade A fertilizer contains 18 lb of nitrogen, 4 lb of phosphate, and 5 lb of potassium. A 100-lb bag of grade B fertilizer contains 20 lb of nitrogen and 4 lb each of phosphate and potassium. A 100-lb bag of grade C fertilizer contains 24 lb of nitrogen, 3 lb of phosphate, and 6 lb of potassium. How many 100-lb bags of each of the three grades of fertilizers should Lawnco produce if:

a. 26,400 lb of nitrogen, 4900 lb of phosphate, and 6200 lb of potassium are available and all the nutrients are used?
b. 21,800 lb of nitrogen, 4200 lb of phosphate, and 5300 lb of potassium are available and all the nutrients are used?

43. **Investment Risk and Return** A private investment club has a certain amount of money earmarked for investment in stocks. To arrive at an acceptable overall level of risk, the stocks that management is considering have been classified into three categories: high-risk, medium-risk, and low-risk. Management estimates that high-risk stocks will have a rate of return of 15%/year; medium-risk stocks, 10%/year; and low-risk stocks, 6%/year. The members have decided that the investment in low-risk stocks should be equal to the sum of the investments in the stocks of the other two categories. Determine how much the club should invest in each type of stock in each of the following scenarios. (In all cases, assume that the entire sum available for investment is invested.)
a. The club has $200,000 to invest, and the investment goal is to have a return of $20,000/year on the total investment.
b. The club has $220,000 to invest, and the investment goal is to have a return of $22,000/year on the total investment.
c. The club has $240,000 to invest, and the investment goal is to have a return of $22,000/year on the total investment.

44. **Research Funding** The Carver Foundation funds three nonprofit organizations engaged in alternative-energy research activities. From past data, the proportion of funds spent by each organization in research on solar energy, energy from harnessing the wind, and energy from the motion of ocean tides is given in the accompanying table.

	Proportion of Money Spent		
	Solar	Wind	Tides
Organization I	0.6	0.3	0.1
Organization II	0.4	0.3	0.3
Organization III	0.2	0.6	0.2

Find the amount awarded to each organization if the total amount spent by all three organizations on solar, wind, and tidal research is:
a. $9.2 million, $9.6 million, and $5.2 million, respectively.
b. $8.2 million, $7.2 million, and $3.6 million, respectively.

45. Find the value(s) of k such that

$$A = \begin{bmatrix} 1 & 2 \\ k & 3 \end{bmatrix}$$

has an inverse. What is the inverse of A?
Hint: Use Formula (14).

46. Find the value(s) of k such that

$$A = \begin{bmatrix} 1 & 0 & 1 \\ -2 & 1 & k \\ -1 & 2 & k^2 \end{bmatrix}$$

has an inverse.
Hint: Find the value(s) of k such that the augmented matrix $[A \mid I]$ can be reduced to the form $[I \mid B]$.

47. Find conditions on a and d such that the matrix

$$A = \begin{bmatrix} a & 0 \\ 0 & d \end{bmatrix}$$

has an inverse. A square matrix is said to be a *diagonal matrix* if all the entries not lying on the main diagonal are zero. Discuss the existence of the inverse matrix of a diagonal matrix of size $n \times n$.

48. Find conditions a, b, and d such that the 2×2 upper triangular matrix

$$A = \begin{bmatrix} a & b \\ 0 & d \end{bmatrix}$$

has an inverse. A square matrix is said to be an *upper triangular matrix* if all its entries below the main diagonal are zero. Discuss the existence of the inverse of an upper triangular matrix of size $n \times n$.

In Exercises 49–51, determine whether the statement is true or false. If it is true, explain why it is true. If it is false, give an example to show why it is false.

49. If A is a square matrix with inverse A^{-1} and c is a nonzero real number, then

$$(cA)^{-1} = \left(\frac{1}{c}\right)A^{-1}$$

50. The matrix

$$A = \begin{bmatrix} a & b \\ c & d \end{bmatrix}$$

has an inverse if and only if $ad - bc = 0$.

51. If A^{-1} does not exist, then the system $AX = B$ of n linear equations in n unknowns does not have a unique solution.

52. Let

$$A = \begin{bmatrix} a & b \\ c & d \end{bmatrix}$$

a. Find A^{-1} if it exists.
b. Find a necessary condition for A to be nonsingular.
c. Verify that $AA^{-1} = A^{-1}A = I$.

2.6 Solutions to Self-Check Exercises

1. We form the augmented matrix

$$\left[\begin{array}{rrr|rrr} 2 & 1 & -1 & 1 & 0 & 0 \\ 1 & 1 & -1 & 0 & 1 & 0 \\ -1 & -2 & 3 & 0 & 0 & 1 \end{array}\right]$$

and row-reduce as follows:

$$\left[\begin{array}{rrr|rrr} 2 & 1 & -1 & 1 & 0 & 0 \\ 1 & 1 & -1 & 0 & 1 & 0 \\ -1 & -2 & 3 & 0 & 0 & 1 \end{array}\right] \xrightarrow{R_1 \leftrightarrow R_2}$$

$$\left[\begin{array}{rrr|rrr} 1 & 1 & -1 & 0 & 1 & 0 \\ 2 & 1 & -1 & 1 & 0 & 0 \\ -1 & -2 & 3 & 0 & 0 & 1 \end{array}\right] \xrightarrow[R_3 + R_1]{R_2 - 2R_1}$$

$$\left[\begin{array}{rrr|rrr} 1 & 1 & -1 & 0 & 1 & 0 \\ 0 & -1 & 1 & 1 & -2 & 0 \\ 0 & -1 & 2 & 0 & 1 & 1 \end{array}\right] \xrightarrow[\substack{-R_2 \\ R_3 - R_2}]{R_1 + R_2}$$

$$\left[\begin{array}{rrr|rrr} 1 & 0 & 0 & 1 & -1 & 0 \\ 0 & 1 & -1 & -1 & 2 & 0 \\ 0 & 0 & 1 & -1 & 3 & 1 \end{array}\right] \xrightarrow{R_2 + R_3}$$

$$\left[\begin{array}{rrr|rrr} 1 & 0 & 0 & 1 & -1 & 0 \\ 0 & 1 & 0 & -2 & 5 & 1 \\ 0 & 0 & 1 & -1 & 3 & 1 \end{array}\right]$$

From the preceding results, we see that

$$A^{-1} = \begin{bmatrix} 1 & -1 & 0 \\ -2 & 5 & 1 \\ -1 & 3 & 1 \end{bmatrix}$$

2. a. We write the systems of linear equations in the matrix form

$$AX = B_1$$

where

$$A = \begin{bmatrix} 2 & 1 & -1 \\ 1 & 1 & -1 \\ -1 & -2 & 3 \end{bmatrix} \qquad X = \begin{bmatrix} x \\ y \\ z \end{bmatrix} \qquad B_1 = \begin{bmatrix} 5 \\ 4 \\ -8 \end{bmatrix}$$

Now, using the results of Exercise 1, we have

$$X = \begin{bmatrix} x \\ y \\ z \end{bmatrix} = A^{-1}B_1 = \begin{bmatrix} 1 & -1 & 0 \\ -2 & 5 & 1 \\ -1 & 3 & 1 \end{bmatrix}\begin{bmatrix} 5 \\ 4 \\ -8 \end{bmatrix} = \begin{bmatrix} 1 \\ 2 \\ -1 \end{bmatrix}$$

Therefore, $x = 1$, $y = 2$, and $z = -1$.

b. Here, A and X are as in part (a), but

$$B_2 = \begin{bmatrix} 2 \\ 0 \\ 5 \end{bmatrix}$$

Therefore,

$$X = \begin{bmatrix} x \\ y \\ z \end{bmatrix} = A^{-1}B_2 = \begin{bmatrix} 1 & -1 & 0 \\ -2 & 5 & 1 \\ -1 & 3 & 1 \end{bmatrix}\begin{bmatrix} 2 \\ 0 \\ 5 \end{bmatrix} = \begin{bmatrix} 2 \\ 1 \\ 3 \end{bmatrix}$$

or $x = 2$, $y = 1$, and $z = 3$.

3. Let x denote the number of adults, and let y denote the number of children on a tour. Since the tours are filled to capacity, we have

$$x + y = 19$$

Next, since the total receipts for the first tour were \$2931, we have

$$169x + 129y = 2931$$

Therefore, the number of adults and the number of children in the first tour are found by solving the system of linear equations

$$\begin{aligned} x + \quad y &= 19 \\ 169x + 129y &= 2931 \end{aligned} \qquad \text{(a)}$$

Similarly, we see that the number of adults and the number of children in the second and third tours are found by solving the systems

$$\begin{aligned} x + \quad y &= 19 \\ 169x + 129y &= 3011 \end{aligned} \qquad \text{(b)}$$

$$\begin{aligned} x + \quad y &= 19 \\ 169x + 129y &= 2771 \end{aligned} \qquad \text{(c)}$$

These systems may be written in the form

$$AX = B_1 \qquad AX = B_2 \qquad AX = B_3$$

where

$$A = \begin{bmatrix} 1 & 1 \\ 169 & 129 \end{bmatrix} \qquad X = \begin{bmatrix} x \\ y \end{bmatrix}$$

$$B_1 = \begin{bmatrix} 19 \\ 2931 \end{bmatrix} \qquad B_2 = \begin{bmatrix} 19 \\ 3011 \end{bmatrix} \qquad B_3 = \begin{bmatrix} 19 \\ 2771 \end{bmatrix}$$

To solve these systems, we first find A^{-1}. Using Formula (14) with $D = (1)(129) - (1)(169) = -40$, we obtain

$$A^{-1} = \frac{-1}{40}\begin{bmatrix} 129 & -1 \\ -169 & 1 \end{bmatrix} = \begin{bmatrix} -\frac{129}{40} & \frac{1}{40} \\ \frac{169}{40} & -\frac{1}{40} \end{bmatrix}$$

Then, solving each system, we find

$$X = \begin{bmatrix} x \\ y \end{bmatrix} = A^{-1}B_1$$

$$= \begin{bmatrix} -\frac{129}{40} & \frac{1}{40} \\ \frac{169}{40} & -\frac{1}{40} \end{bmatrix}\begin{bmatrix} 19 \\ 2931 \end{bmatrix} = \begin{bmatrix} 12 \\ 7 \end{bmatrix} \qquad \text{(a)}$$

$$X = \begin{bmatrix} x \\ y \end{bmatrix} = A^{-1}B_2$$
$$= \begin{bmatrix} -\frac{129}{40} & \frac{1}{40} \\ \frac{169}{40} & -\frac{1}{40} \end{bmatrix} \begin{bmatrix} 19 \\ 3011 \end{bmatrix}$$
$$= \begin{bmatrix} 14 \\ 5 \end{bmatrix} \qquad \textbf{(b)}$$

$$X = \begin{bmatrix} x \\ y \end{bmatrix} = A^{-1}B_3$$
$$= \begin{bmatrix} -\frac{129}{40} & \frac{1}{40} \\ \frac{169}{40} & -\frac{1}{40} \end{bmatrix} \begin{bmatrix} 19 \\ 2771 \end{bmatrix} = \begin{bmatrix} 8 \\ 11 \end{bmatrix} \qquad \textbf{(c)}$$

We conclude that there were:

a. 12 adults and 7 children on the first tour.
b. 14 adults and 5 children on the second tour.
c. 8 adults and 11 children on the third tour.

USING TECHNOLOGY Finding the Inverse of a Square Matrix

Graphing Utility

A graphing utility can be used to find the inverse of a square matrix.

EXAMPLE 1 Use a graphing utility to find the inverse of

$$\begin{bmatrix} 1 & 3 & 5 \\ -2 & 2 & 4 \\ 5 & 1 & 3 \end{bmatrix}$$

Solution We first enter the given matrix as

$$A = \begin{bmatrix} 1 & 3 & 5 \\ -2 & 2 & 4 \\ 5 & 1 & 3 \end{bmatrix}$$

Then, recalling the matrix A and using the $\boxed{x^{-1}}$ key, we find

$$A^{-1} = \begin{bmatrix} 0.1 & -0.2 & 0.1 \\ 1.3 & -1.1 & -0.7 \\ -0.6 & 0.7 & 0.4 \end{bmatrix}$$

EXAMPLE 2 Use a graphing utility to solve the system

$$\begin{aligned} x + 3y + 5z &= 4 \\ -2x + 2y + 4z &= 3 \\ 5x + y + 3z &= 2 \end{aligned}$$

by using the inverse of the coefficient matrix.

Solution The given system can be written in the matrix form $AX = B$, where

$$A = \begin{bmatrix} 1 & 3 & 5 \\ -2 & 2 & 4 \\ 5 & 1 & 3 \end{bmatrix} \qquad X = \begin{bmatrix} x \\ y \\ z \end{bmatrix} \qquad B = \begin{bmatrix} 4 \\ 3 \\ 2 \end{bmatrix}$$

The solution is $X = A^{-1}B$. Entering the matrices A and B in the graphing utility and using the matrix multiplication capability of the utility gives the output shown in Figure T1—that is, $x = 0$, $y = 0.5$, and $z = 0.5$.

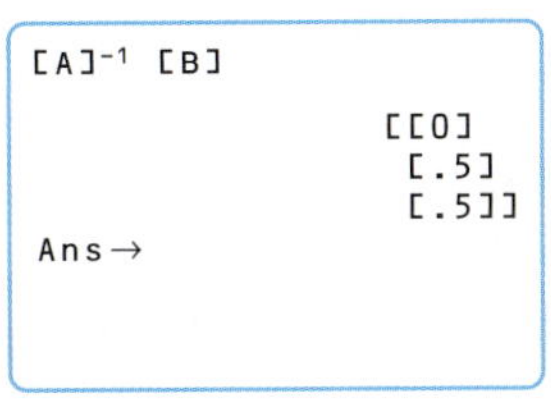

FIGURE **T1**
The TI-83/84 screen showing $A^{-1}B$

Excel

We use the function **MINVERSE** to find the inverse of a square matrix using Excel.

EXAMPLE 3 Find the inverse of

$$A = \begin{bmatrix} 1 & 3 & 5 \\ -2 & 2 & 4 \\ 5 & 1 & 3 \end{bmatrix}$$

Solution

1. Enter the elements of matrix A onto a spreadsheet (Figure T2).
2. Compute the inverse of the matrix A: Highlight the cells that will contain the inverse matrix A^{-1}, type `= MINVERSE(`, highlight the cells containing matrix A, type `)`, and press **Ctrl-Shift-Enter**. The desired matrix will appear in your spreadsheet (Figure T2).

	A	B	C
1		Matrix A	
2	1	3	5
3	-2	2	4
4	5	1	3
5			
6		Matrix A^{-1}	
7	0.1	-0.2	0.1
8	1.3	-1.1	-0.7
9	-0.6	0.7	0.4

FIGURE **T2**
Matrix A and its inverse, matrix A^{-1}

EXAMPLE 4 Solve the system

$$\begin{aligned} x + 3y + 5z &= 4 \\ -2x + 2y + 4z &= 3 \\ 5x + y + 3z &= 2 \end{aligned}$$

by using the inverse of the coefficient matrix.

Solution The given system can be written in the matrix form $AX = B$, where

$$A = \begin{bmatrix} 1 & 3 & 5 \\ -2 & 2 & 4 \\ 5 & 1 & 3 \end{bmatrix} \qquad X = \begin{bmatrix} x \\ y \\ z \end{bmatrix} \qquad B = \begin{bmatrix} 4 \\ 3 \\ 2 \end{bmatrix}$$

The solution is $X = A^{-1}B$.

1. Enter the matrix B on a spreadsheet.
2. Compute $A^{-1}B$. Highlight the cells that will contain the matrix X, and then type `=MMULT(`, highlight the cells in the matrix A^{-1}, type `,`, highlight the cells in the matrix B, type `)`, and press **Ctrl-Shift-Enter**. (*Note*: The matrix A^{-1} was found in Example 3.) The matrix X shown in Figure T3 will appear on your spreadsheet. Thus, $x = 0$, $y = 0.5$, and $z = 0.5$.

	A
12	Matrix X
13	5.55112E-17
14	0.5
15	0.5

FIGURE **T3**
Matrix X gives the solution to the problem.

Note: Boldfaced words/characters enclosed in a box (for example, **Enter**) indicate that an action (click, select, or press) is required. Words/characters printed blue (for example, Chart sub-type:) indicate words/characters that appear on the screen. Words/characters printed in a monospace font (for example, `= (-2/3)*A2+2)`) indicate words/characters that need to be typed and entered.

TECHNOLOGY EXERCISES

In Exercises 1–6, find the inverse of the matrix. Round your answers to two decimal places.

1. $\begin{bmatrix} 1.2 & 3.1 & -2.1 \\ 3.4 & 2.6 & 7.3 \\ -1.2 & 3.4 & -1.3 \end{bmatrix}$

2. $\begin{bmatrix} 4.2 & 3.7 & 4.6 \\ 2.1 & -1.3 & -2.3 \\ 1.8 & 7.6 & -2.3 \end{bmatrix}$

3. $\begin{bmatrix} 1.1 & 2.3 & 3.1 & 4.2 \\ 1.6 & 3.2 & 1.8 & 2.9 \\ 4.2 & 1.6 & 1.4 & 3.2 \\ 1.6 & 2.1 & 2.8 & 7.2 \end{bmatrix}$

4. $\begin{bmatrix} 2.1 & 3.2 & -1.4 & -3.2 \\ 6.2 & 7.3 & 8.4 & 1.6 \\ 2.3 & 7.1 & 2.4 & -1.3 \\ -2.1 & 3.1 & 4.6 & 3.7 \end{bmatrix}$

5. $\begin{bmatrix} 2 & -1 & 3 & 2 & 4 \\ 3 & 2 & -1 & 4 & 1 \\ 3 & 2 & 6 & 4 & -1 \\ 2 & 1 & -1 & 4 & 2 \\ 3 & 4 & 2 & 5 & 6 \end{bmatrix}$

6. $\begin{bmatrix} 1 & 4 & 2 & 3 & 1.4 \\ 6 & 2.4 & 5 & 1.2 & 3 \\ 4 & 1 & 2 & 3 & 1.2 \\ -1 & 2 & -3 & 4 & 2 \\ 1.1 & 2.2 & 3 & 5.1 & 4 \end{bmatrix}$

In Exercises 7–10, solve the system of linear equations by first writing the system in the form $AX = B$ and then solving the resulting system by using A^{-1}. Round your answers to two decimal places.

7. $\begin{aligned} 2x - 3y + 4z &= 2.4 \\ 3x + 2y - 7z &= -8.1 \\ x + 4y - 2z &= 10.2 \end{aligned}$

8. $\begin{aligned} 3.2x - 4.7y + 3.2z &= 7.1 \\ 2.1x + 2.6y + 6.2z &= 8.2 \\ 5.1x - 3.1y - 2.6z &= -6.5 \end{aligned}$

9. $\begin{aligned} 3x_1 - 2x_2 + 4x_3 - 8x_4 &= 8 \\ 2x_1 + 3x_2 - 2x_3 + 6x_4 &= 4 \\ 3x_1 + 2x_2 - 6x_3 - 7x_4 &= -2 \\ 4x_1 - 7x_2 + 4x_3 + 6x_4 &= 22 \end{aligned}$

10. $\begin{aligned} 1.2x_1 + 2.1x_2 - 3.2x_3 + 4.6x_4 &= 6.2 \\ 3.1x_1 - 1.2x_2 + 4.1x_3 - 3.6x_4 &= -2.2 \\ 1.8x_1 + 3.1x_2 - 2.4x_3 + 8.1x_4 &= 6.2 \\ 2.6x_1 - 2.4x_2 + 3.6x_3 - 4.6x_4 &= 3.6 \end{aligned}$

2.7 Leontief Input–Output Model

Input–Output Analysis

One of the many important applications of matrix theory to the field of economics is the study of the relationship between industrial production and consumer demand. At the heart of this analysis is the Leontief input–output model pioneered by Wassily Leontief, who was awarded a Nobel Prize in economics in 1973 for his contributions to the field.

To illustrate this concept, let's consider an oversimplified economy consisting of three sectors: agriculture (A), manufacturing (M), and service (S). In general, part of the output of one sector is absorbed by another sector through interindustry purchases, with the excess available to fulfill consumer demands. The relationship governing both intraindustrial and interindustrial sales and purchases is conveniently represented by means of an **input–output matrix**:

Output (amount produced)

Input (amount used in production)

$$\begin{array}{c} \\ A \\ M \\ S \end{array} \begin{array}{c} \begin{array}{ccc} A & M & S \end{array} \\ \begin{bmatrix} 0.2 & 0.2 & 0.1 \\ 0.2 & 0.4 & 0.1 \\ 0.1 & 0.2 & 0.3 \end{bmatrix} \end{array} \tag{17}$$

The first column (read from top to bottom) tells us that the production of 1 unit of agricultural products requires the consumption of 0.2 unit of agricultural products, 0.2 unit of manufactured goods, and 0.1 unit of services. The second column tells us that the production of 1 unit of manufactured goods requires the consumption of 0.2 unit of agricultural products, 0.4 unit of manufactured goods, and 0.2 unit of services. Finally, the third column tells us that the production of 1 unit of services requires the consumption of 0.1 unit each of agricultural products and manufactured goods and 0.3 unit of services.

APPLIED EXAMPLE 1 Input–Output Analysis Refer to the input–output matrix (17).

a. If the units are measured in millions of dollars, determine the amount of agricultural products consumed in the production of $100 million worth of manufactured goods.

b. Determine the dollar amount of manufactured goods required to produce $200 million worth of all goods and services in the economy.

Solution

a. The production of 1 unit requires the consumption of 0.2 unit of agricultural products. Thus, the amount of agricultural products consumed in the production of $100 million worth of manufactured goods is given by $(100)(0.2)$, or $20 million.

b. The amount of manufactured goods required to produce 1 unit of all goods and services in the economy is given by adding the numbers of the second row of the input–output matrix—that is, $0.2 + 0.4 + 0.1$, or 0.7 unit. Therefore, the production of $200 million worth of all goods and services in the economy requires $200(0.7) million, or $140 million, worth of manufactured goods.

Next, suppose the total output of goods of the agriculture and manufacturing sectors and the total output from the service sector of the economy are given by x, y, and z units, respectively. What is the value of agricultural products consumed in the internal process of producing this total output of various goods and services?

To answer this question, we first note, by examining the input–output matrix

$$\begin{array}{cc} & \text{Output} \\ & \begin{array}{ccc} A & M & S \end{array} \\ \text{Input} \begin{array}{c} A \\ M \\ S \end{array} & \begin{bmatrix} 0.2 & 0.2 & 0.1 \\ 0.2 & 0.4 & 0.1 \\ 0.1 & 0.2 & 0.3 \end{bmatrix} \end{array}$$

that 0.2 unit of agricultural products is required to produce 1 unit of agricultural products, so the amount of agricultural goods required to produce x units of agricultural products is given by $0.2x$ unit. Next, again referring to the input–output matrix, we see that 0.2 unit of agricultural products is required to produce 1 unit of manufactured goods, so the requirement for producing y units of the latter is $0.2y$ unit of agricultural products. Finally, we see that 0.1 unit of agricultural goods is required to produce 1 unit of services, so the amount of agricultural products required to produce z units of services is $0.1z$ unit. Thus, the total amount of agricultural products required to produce the total output of goods and services in the economy is

$$0.2x + 0.2y + 0.1z$$

units. In a similar manner, we see that the total amount of manufactured goods and the total value of services required to produce the total output of goods and services in the economy are given by

$$0.2x + 0.4y + 0.1z$$
$$0.1x + 0.2y + 0.3z$$

respectively.

These results could also be obtained by using matrix multiplication. To see this, write the total output of goods and services x, y, and z as a 3×1 matrix:

$$X = \begin{bmatrix} x \\ y \\ z \end{bmatrix} \quad \text{Total output matrix}$$

The matrix X is called the **total output matrix.** Letting A denote the input–output matrix, we have

$$A = \begin{bmatrix} 0.2 & 0.2 & 0.1 \\ 0.2 & 0.4 & 0.1 \\ 0.1 & 0.2 & 0.3 \end{bmatrix} \quad \text{Input–output matrix}$$

Then the product

$$AX = \begin{bmatrix} 0.2 & 0.2 & 0.1 \\ 0.2 & 0.4 & 0.1 \\ 0.1 & 0.2 & 0.3 \end{bmatrix}\begin{bmatrix} x \\ y \\ z \end{bmatrix}$$
$$= \begin{bmatrix} 0.2x + 0.2y + 0.1z \\ 0.2x + 0.4y + 0.1z \\ 0.1x + 0.2y + 0.3z \end{bmatrix} \quad \text{Internal consumption matrix}$$

is a 3×1 matrix whose entries represent the respective values of the agricultural products, manufactured goods, and services consumed in the internal process of production. The matrix AX is referred to as the **internal consumption matrix.**

Now, since X gives the total production of goods and services in the economy, and AX, as we have just seen, gives the amount of goods and services consumed in the production of these goods and services, it follows that the 3×1 matrix $X - AX$ gives the net output of goods and services that is exactly enough to satisfy consumer demands. Letting matrix D represent these consumer demands, we are led to the following matrix equation:

$$X - AX = D$$
$$(I - A)X = D$$

where I is the 3×3 identity matrix.

Assuming that the inverse of $(I - A)$ exists, multiplying both sides of the last equation by $(I - A)^{-1}$ on the left yields

$$X = (I - A)^{-1}D$$

Leontief Input–Output Model

In a **Leontief input–output model,** the matrix equation giving the net output of goods and services needed to satisfy consumer demand is

$$\underset{\text{Total output}}{X} - \underset{\text{Internal consumption}}{AX} = \underset{\text{Consumer demand}}{D}$$

(continued)

where X is the total output matrix, A is the input–output matrix, and D is the matrix representing consumer demand.

The solution to this equation is

$$X = (I - A)^{-1}D \qquad \text{Assuming that } (I - A)^{-1} \text{ exists} \tag{18}$$

which gives the amount of goods and services that must be produced to satisfy consumer demand.

Equation (18) gives us a means of finding the amount of goods and services to be produced in order to satisfy a given level of consumer demand, as illustrated by the following example.

APPLIED EXAMPLE 2 An Input–Output Model for a Three-Sector Economy For the three-sector economy with input–output matrix given by Matrix (17), which is reproduced here:

$$A = \begin{bmatrix} 0.2 & 0.2 & 0.1 \\ 0.2 & 0.4 & 0.1 \\ 0.1 & 0.2 & 0.3 \end{bmatrix} \qquad \text{Each unit equals \$1 million.}$$

a. Find the total output of goods and services needed to satisfy a consumer demand of $100 million worth of agricultural products, $80 million worth of manufactured goods, and $50 million worth of services.

b. Find the value of the goods and services consumed in the internal process of production to meet this total output.

Solution

a. We are required to determine the total output matrix

$$X = \begin{bmatrix} x \\ y \\ z \end{bmatrix}$$

where x, y, and z denote the value of the agricultural products, the manufactured goods, and services, respectively. The matrix representing the consumer demand is given by

$$D = \begin{bmatrix} 100 \\ 80 \\ 50 \end{bmatrix}$$

Next, we compute

$$I - A = \begin{bmatrix} 1 & 0 & 0 \\ 0 & 1 & 0 \\ 0 & 0 & 1 \end{bmatrix} - \begin{bmatrix} 0.2 & 0.2 & 0.1 \\ 0.2 & 0.4 & 0.1 \\ 0.1 & 0.2 & 0.3 \end{bmatrix} = \begin{bmatrix} 0.8 & -0.2 & -0.1 \\ -0.2 & 0.6 & -0.1 \\ -0.1 & -0.2 & 0.7 \end{bmatrix}$$

Using the method of Section 2.6, we find (to two decimal places)

$$(I - A)^{-1} = \begin{bmatrix} 1.43 & 0.57 & 0.29 \\ 0.54 & 1.96 & 0.36 \\ 0.36 & 0.64 & 1.57 \end{bmatrix}$$

Finally, using Equation (18), we find

$$X = (I - A)^{-1}D = \begin{bmatrix} 1.43 & 0.57 & 0.29 \\ 0.54 & 1.96 & 0.36 \\ 0.36 & 0.64 & 1.57 \end{bmatrix} \begin{bmatrix} 100 \\ 80 \\ 50 \end{bmatrix} = \begin{bmatrix} 203.1 \\ 228.8 \\ 165.7 \end{bmatrix}$$

To fulfill consumer demand, \$203 million worth of agricultural products, \$229 million worth of manufactured goods, and \$166 million worth of services should be produced.

b. The amount of goods and services consumed in the internal process of production is given by AX or, equivalently, by $X - D$. In this case, it is more convenient to use the latter, which gives the required result of

$$\begin{bmatrix} 203.1 \\ 228.8 \\ 165.7 \end{bmatrix} - \begin{bmatrix} 100 \\ 80 \\ 50 \end{bmatrix} = \begin{bmatrix} 103.1 \\ 148.8 \\ 115.7 \end{bmatrix}$$

or \$103 million worth of agricultural products, \$149 million worth of manufactured goods, and \$116 million worth of services.

APPLIED EXAMPLE 3 An Input–Output Model for a Three-Product Company TKK Corporation, a large conglomerate, has three subsidiaries engaged in producing raw rubber, manufacturing tires, and manufacturing other rubber-based goods. The production of 1 unit of raw rubber requires the consumption of 0.08 unit of rubber, 0.04 unit of tires, and 0.02 unit of other rubber-based goods. To produce 1 unit of tires requires 0.6 unit of raw rubber, 0.02 unit of tires, and 0 unit of other rubber-based goods. To produce 1 unit of other rubber-based goods requires 0.3 unit of raw rubber, 0.01 unit of tires, and 0.06 unit of other rubber-based goods. Market research indicates that the demand for the following year will be \$200 million for raw rubber, \$800 million for tires, and \$120 million for other rubber-based products. Find the level of production for each subsidiary in order to satisfy this demand.

Solution View the corporation as an economy having three sectors and with an input–output matrix given by

$$A = \begin{array}{l} \text{Raw rubber} \\ \text{Tires} \\ \text{Goods} \end{array} \overset{\begin{array}{ccc} \text{Raw rubber} & \text{Tires} & \text{Goods} \end{array}}{\begin{bmatrix} 0.08 & 0.60 & 0.30 \\ 0.04 & 0.02 & 0.01 \\ 0.02 & 0 & 0.06 \end{bmatrix}}$$

Using Equation (18), we find that the required level of production is given by

$$X = \begin{bmatrix} x \\ y \\ z \end{bmatrix} = (I - A)^{-1}D$$

where x, y, and z denote the outputs of raw rubber, tires, and other rubber-based goods and where

$$D = \begin{bmatrix} 200 \\ 800 \\ 120 \end{bmatrix}$$

Now,

$$I - A = \begin{bmatrix} 0.92 & -0.60 & -0.30 \\ -0.04 & 0.98 & -0.01 \\ -0.02 & 0 & 0.94 \end{bmatrix}$$

You are asked to verify that

$$(I - A)^{-1} = \begin{bmatrix} 1.12 & 0.69 & 0.37 \\ 0.05 & 1.05 & 0.03 \\ 0.02 & 0.01 & 1.07 \end{bmatrix}$$

See Exercise 7.

Therefore,

$$X = (I - A)^{-1}D = \begin{bmatrix} 1.12 & 0.69 & 0.37 \\ 0.05 & 1.05 & 0.03 \\ 0.02 & 0.01 & 1.07 \end{bmatrix} \begin{bmatrix} 200 \\ 800 \\ 120 \end{bmatrix} = \begin{bmatrix} 820.4 \\ 853.6 \\ 140.4 \end{bmatrix}$$

To fulfill the predicted demand, \$820 million worth of raw rubber, \$854 million worth of tires, and \$140 million worth of other rubber-based goods should be produced.

2.7 Self-Check Exercises

1. Solve the matrix equation $(I - A)X = D$ for x and y, given that

$$A = \begin{bmatrix} 0.4 & 0.1 \\ 0.2 & 0.2 \end{bmatrix} \quad X = \begin{bmatrix} x \\ y \end{bmatrix} \quad D = \begin{bmatrix} 50 \\ 10 \end{bmatrix}$$

2. **Input–Output Analysis** A simple economy consists of two sectors: agriculture (A) and transportation (T). The input–output matrix for this economy is given by

$$A = \begin{array}{c} \\ A \\ T \end{array}\!\!\begin{array}{c} \begin{array}{cc} A & T \end{array} \\ \begin{bmatrix} 0.4 & 0.1 \\ 0.2 & 0.2 \end{bmatrix} \end{array}$$

a. Find the gross output of agricultural products needed to satisfy a consumer demand for \$50 million worth of agricultural products and \$10 million worth of transportation.

b. Find the value of agricultural products and transportation consumed in the internal process of production in order to meet the gross output.

Solutions to Self-Check Exercises 2.7 can be found on page 162.

2.7 Concept Questions

1. What do the quantities X, AX, and D represent in the matrix equation $X - AX = D$ for a Leontief input–output model?

2. What is the solution to the matrix equation $X - AX = D$? Does the solution to this equation always exist? Why or why not?

2.7 Exercises

1. **An Input–Output Model for a Three-Sector Economy** A simple economy consists of three sectors: agriculture (A), manufacturing (M), and transportation (T). The input–output matrix for this economy is given by

$$\begin{array}{c} \\ A \\ M \\ T \end{array}\!\!\begin{array}{c} \begin{array}{ccc} A & M & T \end{array} \\ \begin{bmatrix} 0.4 & 0.1 & 0.1 \\ 0.1 & 0.4 & 0.3 \\ 0.2 & 0.2 & 0.2 \end{bmatrix} \end{array}$$

a. Determine the amount of agricultural products consumed in the production of \$100 million worth of manufactured goods.

b. Determine the dollar amount of manufactured goods required to produce \$200 million worth of all goods in the economy.

c. Which sector consumes the greatest amount of agricultural products in the production of a unit of goods in that sector? The least?

2. **An Input–Output Model for a Four-Sector Economy** The relationship governing the intraindustrial and interindustrial sales and purchases of four basic industries—agriculture (A), manufacturing (M), transportation (T), and energy (E)—of a certain economy is given by the following input–output matrix.

$$\begin{array}{c} \\ A \\ M \\ T \\ E \end{array} \begin{array}{c} \begin{array}{cccc} A & M & T & E \end{array} \\ \begin{bmatrix} 0.3 & 0.2 & 0 & 0.1 \\ 0.2 & 0.3 & 0.2 & 0.1 \\ 0.2 & 0.2 & 0.1 & 0.3 \\ 0.1 & 0.2 & 0.3 & 0.2 \end{bmatrix} \end{array}$$

   a. How many units of energy are required to produce 1 unit of manufactured goods?
   b. How many units of energy are required to produce 3 units of all goods in the economy?
   c. Which sector of the economy is least dependent on the cost of energy?
   d. Which sector of the economy has the smallest intraindustry purchases (sales)?

In Exercises 3–6, use the input–output matrix A and the consumer demand matrix D to solve the matrix equation $(I - A)X = D$ for the total output matrix X.

3. $A = \begin{bmatrix} 0.4 & 0.2 \\ 0.3 & 0.1 \end{bmatrix}$ and $D = \begin{bmatrix} 10 \\ 12 \end{bmatrix}$

4. $A = \begin{bmatrix} 0.2 & 0.3 \\ 0.5 & 0.2 \end{bmatrix}$ and $D = \begin{bmatrix} 4 \\ 8 \end{bmatrix}$

5. $A = \begin{bmatrix} 0.5 & 0.2 \\ 0.2 & 0.5 \end{bmatrix}$ and $D = \begin{bmatrix} 10 \\ 20 \end{bmatrix}$

6. $A = \begin{bmatrix} 0.6 & 0.2 \\ 0.1 & 0.4 \end{bmatrix}$ and $D = \begin{bmatrix} 8 \\ 12 \end{bmatrix}$

7. Let

$$A = \begin{bmatrix} 0.08 & 0.60 & 0.30 \\ 0.04 & 0.02 & 0.01 \\ 0.02 & 0 & 0.06 \end{bmatrix}$$

   Show that

$$(I - A)^{-1} = \begin{bmatrix} 1.12 & 0.69 & 0.37 \\ 0.05 & 1.05 & 0.03 \\ 0.02 & 0.01 & 1.07 \end{bmatrix}$$

8. **An Input–Output Model for a Two-Sector Economy** A simple economy consists of two industries: agriculture and manufacturing. The production of 1 unit of agricultural products requires the consumption of 0.2 unit of agricultural products and 0.3 unit of manufactured goods. The production of 1 unit of manufactured goods requires the consumption of 0.4 unit of agricultural products and 0.3 unit of manufactured goods.
   a. Find the total output of goods needed to satisfy a consumer demand for \$100 million worth of agricultural products and \$150 million worth of manufactured goods.
   b. Find the value of the goods consumed in the internal process of production in order to meet the gross output.

9. **An Input–Output Model for a Two-Sector Economy** Rework Exercise 8 if the consumer demand for the output of agricultural products and the consumer demand for manufactured goods are \$120 million and \$140 million, respectively.

10. **An Input–Output Model for a Two-Sector Economy** Refer to Example 3. Suppose the demand for raw rubber increases by 10%, the demand for tires increases by 20%, and the demand for other rubber-based products decreases by 10%. Find the level of production for each subsidiary in order to meet this demand.

11. **An Input–Output Model for a Three-Sector Economy** Consider the economy of Exercise 1, consisting of three sectors: agriculture (A), manufacturing (M), and transportation (T), with an input–output matrix given by

$$\begin{array}{c} \\ A \\ M \\ T \end{array} \begin{array}{c} \begin{array}{ccc} A & M & T \end{array} \\ \begin{bmatrix} 0.4 & 0.1 & 0.1 \\ 0.1 & 0.4 & 0.3 \\ 0.2 & 0.2 & 0.2 \end{bmatrix} \end{array}$$

   a. Find the total output of goods needed to satisfy a consumer demand for \$200 million worth of agricultural products, \$100 million worth of manufactured goods, and \$60 million worth of transportation.
   b. Find the value of goods and transportation consumed in the internal process of production in order to meet this total output.

12. **An Input–Output Model for a Three-Sector Economy** Consider a simple economy consisting of three sectors: food, clothing, and shelter. The production of 1 unit of food requires the consumption of 0.4 unit of food, 0.2 unit of clothing, and 0.2 unit of shelter. The production of 1 unit of clothing requires the consumption of 0.1 unit of food, 0.2 unit of clothing, and 0.3 unit of shelter. The production of 1 unit of shelter requires the consumption of 0.3 unit of food, 0.1 unit of clothing, and 0.1 unit of shelter. Find the level of production for each sector to satisfy the demand for \$100 million worth of food, \$30 million worth of clothing, and \$250 million worth of shelter.

In Exercises 13–16, matrix A is an input–output matrix associated with an economy, and matrix D (units in millions of dollars) is a demand vector. In each problem, find the final outputs of each industry such that the demands of industry and the consumer sector are met.

13. $A = \begin{bmatrix} 0.4 & 0.2 \\ 0.3 & 0.5 \end{bmatrix}$ and $D = \begin{bmatrix} 12 \\ 24 \end{bmatrix}$

14. $A = \begin{bmatrix} 0.1 & 0.4 \\ 0.3 & 0.2 \end{bmatrix}$ and $D = \begin{bmatrix} 5 \\ 10 \end{bmatrix}$

15. $A = \begin{bmatrix} \frac{1}{5} & \frac{2}{5} & \frac{1}{5} \\ \frac{1}{2} & 0 & \frac{1}{2} \\ 0 & \frac{1}{5} & 0 \end{bmatrix}$ and $D = \begin{bmatrix} 10 \\ 5 \\ 15 \end{bmatrix}$

16. $A = \begin{bmatrix} 0.2 & 0.4 & 0.1 \\ 0.3 & 0.2 & 0.1 \\ 0.1 & 0.2 & 0.2 \end{bmatrix}$ and $D = \begin{bmatrix} 6 \\ 8 \\ 10 \end{bmatrix}$

2.7 Solutions to Self-Check Exercises

1. Multiplying both sides of the given equation on the left by $(I - A)^{-1}$, we see that

$$X = (I - A)^{-1}D$$

Now,

$$I - A = \begin{bmatrix} 1 & 0 \\ 0 & 1 \end{bmatrix} - \begin{bmatrix} 0.4 & 0.1 \\ 0.2 & 0.2 \end{bmatrix} = \begin{bmatrix} 0.6 & -0.1 \\ -0.2 & 0.8 \end{bmatrix}$$

Next, we use the Gauss–Jordan elimination method to compute $(I - A)^{-1}$ (to two decimal places):

$$\left[\begin{array}{rr|rr} 0.6 & -0.1 & 1 & 0 \\ -0.2 & 0.8 & 0 & 1 \end{array}\right] \xrightarrow{\frac{1}{0.6}R_1}$$

$$\left[\begin{array}{rr|rr} 1 & -0.17 & 1.67 & 0 \\ -0.2 & 0.8 & 0 & 1 \end{array}\right] \xrightarrow{R_2 + 0.2R_1}$$

$$\left[\begin{array}{rr|rr} 1 & -0.17 & 1.67 & 0 \\ 0 & 0.77 & 0.33 & 1 \end{array}\right] \xrightarrow{\frac{1}{0.77}R_2}$$

$$\left[\begin{array}{rr|rr} 1 & -0.17 & 1.67 & 0 \\ 0 & 1 & 0.43 & 1.30 \end{array}\right] \xrightarrow{R_1 + 0.17R_2}$$

$$\left[\begin{array}{rr|rr} 1 & 0 & 1.74 & 0.22 \\ 0 & 1 & 0.43 & 1.30 \end{array}\right]$$

giving

$$(I - A)^{-1} = \begin{bmatrix} 1.74 & 0.22 \\ 0.43 & 1.30 \end{bmatrix}$$

Therefore,

$$X = \begin{bmatrix} x \\ y \end{bmatrix} = (I - A)^{-1}D = \begin{bmatrix} 1.74 & 0.22 \\ 0.43 & 1.30 \end{bmatrix}\begin{bmatrix} 50 \\ 10 \end{bmatrix} = \begin{bmatrix} 89.2 \\ 34.5 \end{bmatrix}$$

or $x = 89.2$ and $y = 34.5$.

2. a. Let

$$X = \begin{bmatrix} x \\ y \end{bmatrix}$$

denote the total output matrix, where x denotes the value of the agricultural products and y denotes the value of transportation. Also, let

$$D = \begin{bmatrix} 50 \\ 10 \end{bmatrix}$$

denote the consumer demand. Then

$$(I - A)X = D$$

or, equivalently,

$$X = (I - A)^{-1}D$$

Using the results of Exercise 1, we find that $x = 89.2$ and $y = 34.5$. That is, to fulfill consumer demands, \$89.2 million worth of agricultural products must be produced, and \$34.5 million worth of transportation services must be used.

b. The amounts of agricultural products consumed and transportation services used are given by

$$X - D = \begin{bmatrix} 89.2 \\ 34.5 \end{bmatrix} - \begin{bmatrix} 50 \\ 10 \end{bmatrix} = \begin{bmatrix} 39.2 \\ 24.5 \end{bmatrix}$$

or \$39.2 million worth of agricultural products and \$24.5 million worth of transportation services.

USING TECHNOLOGY The Leontief Input–Output Model

Graphing Utility

Since the solution to a problem involving a Leontief input–output model often involves several matrix operations, a graphing utility can be used to facilitate the necessary computations.

APPLIED EXAMPLE 1 Input–Output Analysis Suppose that the input–output matrix associated with an economy is given by matrix A and that the matrix D is a demand vector, where

$$A = \begin{bmatrix} 0.2 & 0.4 & 0.15 \\ 0.3 & 0.1 & 0.4 \\ 0.25 & 0.4 & 0.2 \end{bmatrix} \quad \text{and} \quad D = \begin{bmatrix} 20 \\ 15 \\ 40 \end{bmatrix}$$

Find the final outputs of each industry such that the demands of industry and the consumer sector are met.

Solution First, we enter the matrices I (the identity matrix), A, and D. We are required to compute the output matrix $X = (I - A)^{-1}D$. Using the matrix operations of the graphing utility, we find (to two decimal places)

$$X = (I - A)^{-1} {*} D = \begin{bmatrix} 110.28 \\ 116.95 \\ 142.94 \end{bmatrix}$$

Hence, the final outputs of the first, second, and third industries are 110.28, 116.95, and 142.94 units, respectively.

Excel

Here, we show how to solve a problem involving a Leontief input–output model using matrix operations on a spreadsheet.

APPLIED EXAMPLE 2 Input–Output Analysis Suppose that the input–output matrix associated with an economy is given by matrix A and that matrix D is a demand vector, where

$$A = \begin{bmatrix} 0.2 & 0.4 & 0.15 \\ 0.3 & 0.1 & 0.4 \\ 0.25 & 0.4 & 0.2 \end{bmatrix} \quad \text{and} \quad D = \begin{bmatrix} 20 \\ 15 \\ 40 \end{bmatrix}$$

Find the final outputs of each industry such that the demands of industry and the consumer sector are met.

Solution

1. Enter the elements of the matrix A and D onto a spreadsheet (Figure T1).

	A	B	C	D	E
1		Matrix A			Matrix D
2	0.2	0.4	0.15		20
3	0.3	0.1	0.4		15
4	0.25	0.4	0.2		40

FIGURE **T1**
Spreadsheet showing matrix A and matrix D

Note: Boldfaced words/characters enclosed in a box (for example, **Enter**) indicate that an action (click, select, or press) is required. Words/characters printed blue (for example, Chart sub-type:) indicate words/characters that appear on the screen. Words/characters printed in a monospace font (for example, `= (−2/3)*A2+2`) indicate words/characters that need to be typed and entered.

2. Find $(I - A)^{-1}$. Enter the elements of the 3×3 identity matrix I onto a spreadsheet. Highlight the cells that will contain the matrix $(I - A)^{-1}$. Type `=MINVERSE(`, highlight the cells containing the matrix I; type `-`, highlight the cells containing the matrix A; type `)`, and press **Ctrl-Shift-Enter**. These results are shown in Figure T2.

	A	B	C
6		Matrix I	
7	1	0	0
8	0	1	0
9	0	0	1
10			
11		Matrix (I - A)$^{-1}$	
12	2.151777137	1.460134486	1.133525456
13	1.306436119	2.315081652	1.402497598
14	1.325648415	1.613832853	2.305475504

FIGURE **T2**
Matrix I and matrix $(I - A)^{-1}$

3. Compute $(I - A)^{-1}*D$. Highlight the cells that will contain the matrix $(I - A)^{-1}*D$. Type `=MMULT(`, highlight the cells containing the matrix $(I - A)^{-1}$, type `,`, highlight the cells containing matrix D, type `)`, and press **Ctrl-Shift-Enter**. The resulting matrix is shown in Figure T3. So, the final outputs of the first, second, and third industries are 110.28, 116.95, and 142.94, respectively.

	A
16	Matrix (I - A)$^{-1}$*D
17	110.2785783
18	116.9548511
19	142.9394813

FIGURE **T3**
Matrix $(I - A)^{-1}*D$

TECHNOLOGY EXERCISES

In Exercises 1–4, A is an input–output matrix associated with an economy, and D (in units of million dollars) is a demand vector. Find the final outputs of each industry such that the demands of industry and the consumer sector are met.

1. $A = \begin{bmatrix} 0.3 & 0.2 & 0.4 & 0.1 \\ 0.2 & 0.1 & 0.2 & 0.3 \\ 0.3 & 0.1 & 0.2 & 0.3 \\ 0.4 & 0.2 & 0.1 & 0.2 \end{bmatrix}$ and $D = \begin{bmatrix} 40 \\ 60 \\ 70 \\ 20 \end{bmatrix}$

2. $A = \begin{bmatrix} 0.12 & 0.31 & 0.40 & 0.05 \\ 0.31 & 0.22 & 0.12 & 0.20 \\ 0.18 & 0.32 & 0.05 & 0.15 \\ 0.32 & 0.14 & 0.22 & 0.05 \end{bmatrix}$ and $D = \begin{bmatrix} 50 \\ 20 \\ 40 \\ 60 \end{bmatrix}$

3. $A = \begin{bmatrix} 0.2 & 0.2 & 0.3 & 0.05 \\ 0.1 & 0.1 & 0.2 & 0.3 \\ 0.3 & 0.2 & 0.1 & 0.4 \\ 0.2 & 0.05 & 0.2 & 0.1 \end{bmatrix}$ and $D = \begin{bmatrix} 25 \\ 30 \\ 50 \\ 40 \end{bmatrix}$

4. $A = \begin{bmatrix} 0.2 & 0.4 & 0.3 & 0.1 \\ 0.1 & 0.2 & 0.1 & 0.3 \\ 0.2 & 0.1 & 0.4 & 0.05 \\ 0.3 & 0.1 & 0.2 & 0.05 \end{bmatrix}$ and $D = \begin{bmatrix} 40 \\ 20 \\ 30 \\ 60 \end{bmatrix}$

CHAPTER 2 Summary of Principal Formulas and Terms

FORMULAS

1. Laws for matrix addition	
a. Commutative law	$A + B = B + A$
b. Associative law	$(A + B) + C = A + (B + C)$
2. Laws for matrix multiplication	
a. Associative law	$(AB)C = A(BC)$
b. Distributive law	$A(B + C) = AB + AC$
3. Inverse of a 2×2 matrix	If $A = \begin{bmatrix} a & b \\ c & d \end{bmatrix}$ and $D = ad - bc \neq 0$ then $A^{-1} = \frac{1}{D}\begin{bmatrix} d & -b \\ -c & a \end{bmatrix}$
4. Solution of system $AX = B$ (A nonsingular)	$X = A^{-1}B$

TERMS

system of linear equations (74)
solution of a system of linear equations (74)
parameter (75)
dependent system (76)
inconsistent system (76)
equivalent system (83)
coefficient matrix (85)
augmented matrix (85)
row-reduced form of a matrix (86)
Gauss–Jordan elimination method (87)
row operations (87)
unit column (87)
pivoting (88)
pivot element (88)
size of a matrix (111)
matrix (112)
row matrix (112)
column matrix (112)
square matrix (112)
transpose of a matrix (116)
scalar (116)
scalar product (116)
matrix product (126)
identity matrix (129)
inverse of a matrix (141)
nonsingular matrix (142)
singular matrix (142)
input–output matrix (155)
total output matrix (157)
internal consumption matrix (157)
Leontief input–output model (157)

CHAPTER 2 Concept Review Questions

Fill in the blanks.

1. a. Two lines in the plane can intersect at (a) exactly ______ point, (b) infinitely ______ points, or (c) ______ point.

b. A system of two linear equations in two variables can have (a) exactly ______ solution, (b) infinitely ______ solutions, or (c) ______ solution.

2. To find the point(s) of intersection of two lines, we solve the system of ______ describing the two lines.

3. The row operations used in the Gauss–Jordan elimination method are denoted by ______, ______, and ______. The use of each of these operations does not alter the ______ of the system of linear equations.

4. a. A system of linear equations with fewer equations than variables cannot have a/an ______ solution.

b. A system of linear equations with at least as many equations as variables may have ______ solution, ______ ______ solutions, or a/an ______ solution.

5. Two matrices are equal provided that they have the same ______ and their corresponding ______ are equal.

6. Two matrices may be added (subtracted) if they both have the same ______. To add or subtract two matrices, we add or subtract their ______ entries.

7. The transpose of a/an ______ matrix with elements a_{ij} is the matrix of size ______ with entries ______.

8. The scalar product of a matrix A by the scalar c is the matrix ______ obtained by multiplying each entry of A by ______.

9. **a.** For the product AB of two matrices A and B to be defined, the number of ______ of A must be equal to the number of ______ of B.
   b. If A is an $m \times n$ matrix and B is an $n \times p$ matrix, then the size of AB is ______.

10. **a.** If the products and sums are defined for the matrices A, B, and C, then the associative law states that $(AB)C =$ ______; the distributive law states that $A(B + C) =$ ______.
    b. If I is an identity matrix of size n, then $IA = A$ if A is any matrix of size ______.

11. A matrix A is nonsingular if there exists a matrix A^{-1} such that ______ $=$ ______ $= I$. If A^{-1} does not exist, then A is said to be ______.

12. A system of n linear equations in n variables written in the form $AX = B$ has a unique solution given by $X =$ ______ if A has an inverse.

CHAPTER 2 Review Exercises

In Exercises 1–4, perform the operations if possible.

1. $\begin{bmatrix} 1 & 2 \\ -1 & 3 \\ 2 & 1 \end{bmatrix} + \begin{bmatrix} 1 & 0 \\ 0 & 1 \\ 1 & 2 \end{bmatrix}$

2. $\begin{bmatrix} -1 & 2 \\ 3 & 4 \end{bmatrix} - \begin{bmatrix} 1 & 2 \\ 5 & -2 \end{bmatrix}$

3. $\begin{bmatrix} -3 & 2 & 1 \end{bmatrix} \begin{bmatrix} 2 & 1 \\ -1 & 0 \\ 2 & 1 \end{bmatrix}$

4. $\begin{bmatrix} 1 & 3 & 2 \\ -1 & 2 & 3 \end{bmatrix} \begin{bmatrix} 1 \\ 4 \\ 2 \end{bmatrix}$

In Exercises 5–8, find the values of the variables.

5. $\begin{bmatrix} 1 & x \\ y & 3 \end{bmatrix} = \begin{bmatrix} z & 2 \\ 3 & w \end{bmatrix}$

6. $\begin{bmatrix} 3 & x \\ y & 3 \end{bmatrix} \begin{bmatrix} 1 \\ 2 \end{bmatrix} = \begin{bmatrix} 7 \\ 4 \end{bmatrix}$

7. $\begin{bmatrix} 3 & a+3 \\ -1 & b \\ c+1 & d \end{bmatrix} = \begin{bmatrix} 3 & 6 \\ e+2 & 4 \\ -1 & 2 \end{bmatrix}$

8. $\begin{bmatrix} x & 3 & 1 \\ 0 & y & 2 \end{bmatrix} \begin{bmatrix} 1 & 1 \\ 3 & z \\ 4 & 2 \end{bmatrix} = \begin{bmatrix} 12 & 4 \\ 2 & 2 \end{bmatrix}$

In Exercises 9–16, compute the expressions if possible, given that

$$A = \begin{bmatrix} 1 & 3 & 1 \\ -2 & 1 & 3 \\ 4 & 0 & 2 \end{bmatrix} \qquad B = \begin{bmatrix} 2 & 1 & 3 \\ -2 & -1 & -1 \\ 1 & 4 & 2 \end{bmatrix}$$

$$C = \begin{bmatrix} 3 & -1 & 2 \\ 1 & 6 & 4 \\ 2 & 1 & 3 \end{bmatrix}$$

9. $2A + 3B$

10. $3A - 2B$

11. $2(3A)$

12. $2(3A - 4B)$

13. $A(B - C)$

14. $AB + AC$

15. $A(BC)$

16. $\frac{1}{2}(CA - CB)$

In Exercises 17–24, solve the system of linear equations using the Gauss–Jordan elimination method.

17. $\begin{aligned} 2x - 3y &= 5 \\ 3x + 4y &= -1 \end{aligned}$

18. $\begin{aligned} 3x + 2y &= 3 \\ 2x - 4y &= -14 \end{aligned}$

19. $\begin{aligned} x - y + 2z &= 5 \\ 3x + 2y + z &= 10 \\ 2x - 3y - 2z &= -10 \end{aligned}$

20. $\begin{aligned} 3x - 2y + 4z &= 16 \\ 2x + y - 2z &= -1 \\ x + 4y - 8z &= -18 \end{aligned}$

21. $\begin{aligned} 3x - 2y + 4z &= 11 \\ 2x - 4y + 5z &= 4 \\ x + 2y - z &= 10 \end{aligned}$

22. $\begin{aligned} x - 2y + 3z + 4w &= 17 \\ 2x + y - 2z - 3w &= -9 \\ 3x - y + 2z - 4w &= 0 \\ 4x + 2y - 3z + w &= -2 \end{aligned}$

23. $\begin{aligned} 3x - 2y + z &= 4 \\ x + 3y - 4z &= -3 \\ 2x - 3y + 5z &= 7 \\ x - 8y + 9z &= 10 \end{aligned}$

24. $\begin{aligned} 2x - 3y + z &= 10 \\ 3x + 2y - 2z &= -2 \\ x - 3y - 4z &= -7 \\ 4x + y - z &= 4 \end{aligned}$

In Exercises 25–32, find the inverse of the matrix (if it exists).

25. $A = \begin{bmatrix} 3 & 1 \\ 1 & 2 \end{bmatrix}$

26. $A = \begin{bmatrix} 2 & 4 \\ 1 & 6 \end{bmatrix}$

27. $A = \begin{bmatrix} 3 & 4 \\ 2 & 2 \end{bmatrix}$

28. $A = \begin{bmatrix} 2 & 4 \\ 1 & -2 \end{bmatrix}$

29. $A = \begin{bmatrix} 2 & 3 & 1 \\ 1 & -1 & 2 \\ 1 & 2 & 1 \end{bmatrix}$

30. $A = \begin{bmatrix} 1 & 2 & 4 \\ 2 & 1 & 3 \\ -1 & 0 & 2 \end{bmatrix}$

31. $A = \begin{bmatrix} 1 & 2 & 4 \\ 3 & 1 & 2 \\ 1 & 0 & -6 \end{bmatrix}$

32. $A = \begin{bmatrix} 2 & 1 & -3 \\ 1 & 2 & -4 \\ 3 & 1 & -2 \end{bmatrix}$

In Exercises 33–36, compute the value of the expressions if possible, given that

$$A = \begin{bmatrix} 1 & 2 \\ -1 & 2 \end{bmatrix} \quad B = \begin{bmatrix} 3 & 1 \\ 4 & 2 \end{bmatrix} \quad C = \begin{bmatrix} 1 & 1 \\ -1 & 2 \end{bmatrix}$$

33. $(A^{-1}B)^{-1}$

34. $(ABC)^{-1}$

35. $(2A - C)^{-1}$

36. $(A + B)^{-1}$

In Exercises 37–40, write each system of linear equations in the form $AX = C$. Find A^{-1} and use the result to solve the system.

37. $\begin{aligned} 2x + 3y &= -8 \\ x - 2y &= 3 \end{aligned}$

38. $\begin{aligned} x - 3y &= -1 \\ 2x + 4y &= 8 \end{aligned}$

39. $\begin{aligned} x - 2y + 4z &= 13 \\ 2x + 3y - 2z &= 0 \\ x + 4y - 6z &= -15 \end{aligned}$

40. $\begin{aligned} 2x - 3y + 4z &= 17 \\ x + 2y - 4z &= -7 \\ 3x - y + 2z &= 14 \end{aligned}$

41. **Gasoline Sales** Gloria Newburg operates three self-service gasoline stations in different parts of town. On a certain day, station A sold 600 gal of premium, 800 gal of super, 1000 gal of regular gasoline, and 700 gal of diesel fuel; station B sold 700 gal of premium, 600 gal of super, 1200 gal of regular gasoline, and 400 gal of diesel fuel; station C sold 900 gal of premium, 700 gal of super, 1400 gal of regular gasoline, and 800 gal of diesel fuel. Assume that the price of gasoline was \$3.80/gal for premium, \$3.60/gal for super, and \$3.40/gal for regular and that diesel fuel sold for \$3.70/gal. Use matrix algebra to find the total revenue at each station.

42. **Stock Transactions** Jack Spaulding bought 10,000 shares of Stock *X*, 20,000 shares of Stock *Y*, and 30,000 shares of Stock *Z* at a unit price of \$20, \$30, and \$50 per share, respectively. Six months later, the closing prices of Stocks *X*, *Y*, and *Z* were \$22, \$35, and \$51 per share, respectively. Jack made no other stock transactions during the period in question. Compare the value of Jack's stock holdings at the time of purchase and 6 months later.

43. **Investments** Josh's and Hannah's stock holdings are given in the following table:

	BAC	GM	IBM	ORCL
Josh	800	1200	250	1500
Hannah	600	1400	300	1200

The prices (in dollars per share) of the stocks of BAC, GM, IBM, and ORCL at the close of the stock market on a certain day are \$12.57, \$28.21, \$214.92, and \$36.34, respectively.

a. Write a 2×4 matrix A giving the stock holdings of Josh and Hannah.

b. Write a 4×1 matrix B giving the closing prices of the stocks of BAC, GM, IBM, and ORCL.

c. Use matrix multiplication to find the total value of the stock holdings of Josh and Hannah at the market close.

44. **Investment Portfolios** The following table gives the number of shares of certain corporations held by Jennifer and Max in their stock portfolios at the beginning of September and at the beginning of October:

September	IBM	Google	Boeing	GM
Jennifer	800	500	1200	1500
Max	500	600	2000	800

October	IBM	Google	Boeing	GM
Jennifer	900	600	1000	1200
Max	700	500	2100	900

a. Write matrices A and B giving the stock portfolios of Jennifer and Max at the beginning of September and at the beginning of October, respectively.

b. Find a matrix C reflecting the change in the stock portfolios of Jennifer and Max between the beginning of September and the beginning of October.

45. **Production Scheduling** Desmond Jewelry wishes to produce three types of pendants: Type *A*, Type *B*, and Type *C*. To manufacture a Type *A* pendant requires 2 min on Machines I and II and 3 min on Machine III. A Type *B* pendant requires 2 min on Machine I, 3 min on Machine II, and 4 min on Machine III. A Type *C* pendant requires 3 min on Machine I, 4 min on Machine II, and 3 min on Machine III. There are $3\frac{1}{2}$ hr available on Machine I, $4\frac{1}{2}$ hr available on Machine II, and 5 hr available on Machine III. How many pendants of each type should Desmond make to use all the available time?

46. **Petroleum Production** Wildcat Oil Company has two refineries, one located in Houston and the other in Tulsa. The Houston refinery ships 60% of its petroleum to a Chicago distributor and 40% of its petroleum to a Los Angeles distributor. The Tulsa refinery ships 30% of its petroleum to the Chicago distributor and 70% of its petroleum to the Los Angeles distributor. Assume that, over the year, the Chicago distributor received 240,000 gal of petroleum and the Los Angeles distributor received

460,000 gal of petroleum. Find the amount of petroleum produced at each of Wildcat's refineries.

47. **Input–Output Matrices** The input–output matrix associated with an economy based on agriculture (A) and manufacturing (M) is given by

$$\begin{array}{cc} & \begin{array}{cc} A & M \end{array} \\ \begin{array}{c} A \\ M \end{array} & \begin{bmatrix} 0.2 & 0.15 \\ 0.1 & 0.15 \end{bmatrix} \end{array}$$

a. Determine the amount of agricultural products consumed in the production of \$200 million worth of manufactured goods.
b. Determine the dollar amount of manufactured goods required to produce \$300 million worth of all goods in the economy.
c. Which sector consumes the greater amount of agricultural products in the production of 1 unit of goods in that sector? The lesser?

48. **Input–Output Matrices** The matrix

$$A = \begin{bmatrix} 0.1 & 0.3 & 0.4 \\ 0.3 & 0.1 & 0.2 \\ 0.2 & 0.1 & 0.3 \end{bmatrix}$$

is an input–output matrix associated with an economy and the matrix

$$D = \begin{bmatrix} 4 \\ 12 \\ 16 \end{bmatrix}$$

(units in millions of dollars) is a demand vector. Find the final outputs of each industry such that the demands of industry and the consumer sector are met.

49. **Input–Output Matrices** The input–output matrix associated with an economy based on agriculture (A) and manufacturing (M) is given by

$$\begin{array}{cc} & \begin{array}{cc} A & M \end{array} \\ \begin{array}{c} A \\ M \end{array} & \begin{bmatrix} 0.2 & 0.15 \\ 0.1 & 0.15 \end{bmatrix} \end{array}$$

a. Find the gross output of goods needed to satisfy a consumer demand for \$100 million worth of agricultural products and \$80 million worth of manufactured goods.
b. Find the value of agricultural products and manufactured goods consumed in the internal process of production to meet this gross output.

The problem-solving skills that you learn in each chapter are building blocks for the rest of the course. Therefore, it is a good idea to make sure that you have mastered these skills before moving on to the next chapter. The Before Moving On exercises that follow are designed for that purpose. After completing these exercises, you can identify the skills that you should review before starting the next chapter.

CHAPTER 2 Before Moving On . . .

1. Solve the following system of linear equations, using the Gauss–Jordan elimination method:

$$\begin{aligned} 2x + y - z &= -1 \\ x + 3y + 2z &= 2 \\ 3x + 3y - 3z &= -5 \end{aligned}$$

2. Find the solution(s), if it exists, of the system of linear equations whose augmented matrix in reduced form follows.

a. $\left[\begin{array}{ccc|c} 1 & 0 & 0 & 2 \\ 0 & 1 & 0 & -3 \\ 0 & 0 & 1 & 1 \end{array}\right]$ b. $\left[\begin{array}{ccc|c} 1 & 0 & 0 & 3 \\ 0 & 1 & 0 & 0 \\ 0 & 0 & 0 & 1 \end{array}\right]$

c. $\left[\begin{array}{ccc|c} 1 & 0 & 0 & 2 \\ 0 & 1 & 3 & 1 \\ 0 & 0 & 0 & 0 \end{array}\right]$ d. $\left[\begin{array}{cccc|c} 1 & 0 & 0 & 0 & 0 \\ 0 & 1 & 0 & 0 & 0 \\ 0 & 0 & 1 & 0 & 0 \\ 0 & 0 & 0 & 1 & 0 \end{array}\right]$

e. $\left[\begin{array}{ccc|c} 1 & 0 & -1 & 2 \\ 0 & 1 & 2 & 3 \end{array}\right]$

3. Solve each system of linear equations using the Gauss–Jordan elimination method.

a. $\begin{aligned} x + 2y &= 3 \\ 3x - y &= -5 \\ 4x + y &= -2 \end{aligned}$ b. $\begin{aligned} x - 2y + 4z &= 2 \\ 3x + y - 2z &= 1 \end{aligned}$

4. Let

$$A = \begin{bmatrix} 1 & -2 & 4 \\ 3 & 0 & 1 \end{bmatrix} \quad B = \begin{bmatrix} 1 & -1 & 2 \\ 3 & 1 & -1 \\ 2 & 1 & 0 \end{bmatrix} \quad C = \begin{bmatrix} 2 & -2 \\ 1 & 1 \\ 3 & 4 \end{bmatrix}$$

Find (a) AB, (b) $(A + C^T)B$, and (c) $C^TB - AB^T$.

5. Find A^{-1} if

$$A = \begin{bmatrix} 2 & 1 & 2 \\ 0 & -1 & 3 \\ 1 & 1 & 0 \end{bmatrix}$$

6. Solve the system

$$\begin{aligned} 2x \phantom{{}+y} + z &= 4 \\ 2x + y - z &= -1 \\ 3x + y - z &= 0 \end{aligned}$$

by first writing it in the matrix form $AX = B$ and then finding A^{-1}.

3 Linear Programming: A Geometric Approach

MANY PRACTICAL PROBLEMS involve maximizing or minimizing a function subject to certain constraints. For example, we might wish to maximize a profit function subject to certain limitations on the amount of material and labor available. Maximization or minimization problems that can be formulated in terms of a *linear* objective function and constraints in the form of linear inequalities are called *linear programming problems*. In this chapter, we look at linear programming problems involving two variables. These problems are amenable to geometric analysis, and the method of solution introduced here will shed much light on the basic nature of a linear programming problem.

How many souvenirs should Ace Novelty make in order to maximize its profit? The company produces two types of souvenirs, each of which requires a certain amount of time on two different machines. Each machine can be operated for only a certain number of hours per day. In Example 1, page 181, we show how this production problem can be formulated as a linear programming problem, and in Example 1, page 192, we solve this linear programming problem.

3.1 Graphing Systems of Linear Inequalities in Two Variables

Graphing Linear Inequalities

In Chapter 1, we saw that a linear equation in two variables x and y

$$ax + by + c = 0 \qquad a, b \text{ not both equal to zero}$$

has a *solution set* that may be exhibited graphically as points on a straight line in the xy-plane. We now show that there is also a simple graphical representation for **linear inequalities** in two variables:

$$ax + by + c < 0 \qquad ax + by + c \leq 0$$
$$ax + by + c > 0 \qquad ax + by + c \geq 0$$

Before turning to a general procedure for graphing such inequalities, let's consider a specific example. Suppose we wish to graph

$$2x + 3y < 6 \tag{1}$$

We first graph the equation $2x + 3y = 6$, which is obtained by replacing the given inequality "$<$" with an equality "$=$" (Figure 1).

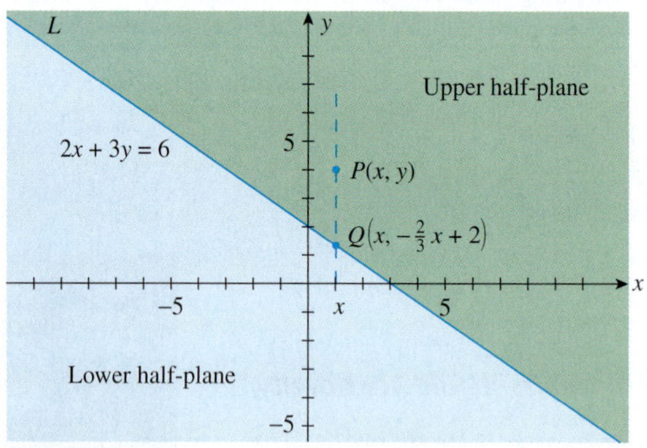

FIGURE 1
A straight line divides the xy-plane into two half-planes.

Observe that this line divides the xy-plane into two half-planes: an upper half-plane and a lower half-plane. Let's show that the upper half-plane is the graph of the linear inequality

$$2x + 3y > 6 \tag{2}$$

whereas the lower half-plane is the graph of the linear inequality

$$2x + 3y < 6 \tag{3}$$

To see this, let's write Inequalities (2) and (3) in the equivalent forms

$$y > -\frac{2}{3}x + 2 \tag{4}$$

and

$$y < -\frac{2}{3}x + 2 \tag{5}$$

The equation of the line itself is

$$y = -\frac{2}{3}x + 2 \tag{6}$$

Now pick any point $P(x, y)$ lying above the line L. Let Q be the point lying on L and directly below P (see Figure 1). Since Q lies on L, its coordinates must satisfy Equation (6). In other words, Q has representation $Q(x, -\frac{2}{3}x + 2)$. Comparing the y-coordinates of P and Q and recalling that P lies above Q, so that its y-coordinate must be larger than that of Q, we have

$$y > -\frac{2}{3}x + 2$$

But this inequality is just Inequality (4) or, equivalently, Inequality (2). Similarly, we can show that every point lying below L must satisfy Inequality (5) and therefore Inequality (3).

This analysis shows that the lower half-plane provides a solution to our problem (Figure 2). (By convention, we draw the line as a dashed line to show that the points on L do not belong to the solution set.) Observe that the two half-planes in question are disjoint; that is, they do not have any points in common.

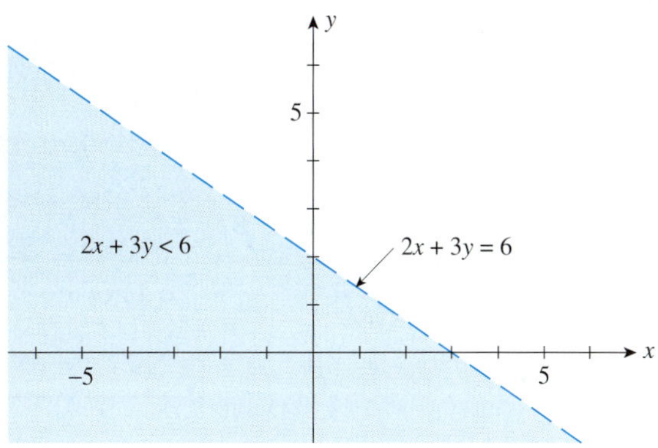

FIGURE **2**
The set of points lying below the dashed line satisfies the given inequality.

Alternatively, there is a simpler method for determining the half-plane that provides the solution to the problem. To determine the required half-plane, let's pick *any* point lying in one of the half-planes. For simplicity, pick the origin $(0, 0)$, which lies in the lower half-plane. Substituting $x = 0$ and $y = 0$ (the coordinates of this point) into the given Inequality (1), we find

$$2(0) + 3(0) < 6$$

or $0 < 6$, which is certainly true. This tells us that the required half-plane is the one containing the test point—namely, the lower half-plane.

Next, let's see what happens if we choose the point $(2, 3)$, which lies in the upper half-plane. Substituting $x = 2$ and $y = 3$ into the given inequality, we find

$$2(2) + 3(3) < 6$$

or $13 < 6$, which is false. This tells us that the upper half-plane is *not* the required half-plane, as expected. Note, too, that no point $P(x, y)$ lying on the line constitutes a solution to our problem, given the *strict* inequality $<$.

This discussion suggests the following procedure for graphing a linear inequality in two variables.

Procedure for Graphing Linear Inequalities

1. Draw the graph of the equation obtained for the given inequality by replacing the inequality sign with an equal sign. Use a dashed or dotted line if the problem involves a strict inequality, $<$ or $>$. Otherwise, use a solid line to indicate that the line itself constitutes part of the solution.
2. Pick a test point (a, b) lying in one of the half-planes determined by the line sketched in Step 1 and substitute the numbers a and b for the values of x and y in the given inequality. For simplicity, use the origin whenever possible.
3. If the inequality is satisfied, the graph of the solution to the inequality is the half-plane containing the test point. Otherwise, the solution is the half-plane not containing the test point.

EXAMPLE 1 Determine the solution set for the inequality $2x + 3y \geq 6$.

Solution Replacing the inequality $\geq$ with an equality $=$, we obtain the equation $2x + 3y = 6$, whose graph is the straight line shown in Figure 3.

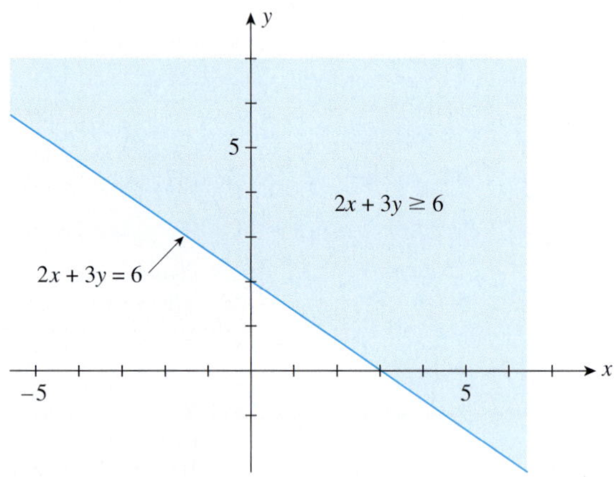

FIGURE 3
The set of points lying on the line and in the upper half-plane satisfies the given inequality.

Instead of a dashed line as before, we use a solid line to show that all points on the line are also solutions to the inequality. Picking the origin as our test point, we find $2(0) + 3(0) \geq 6$, or $0 \geq 6$, which is false. So we conclude that the solution set is made up of the half-plane that does not contain the origin, including (in this case) the line given by $2x + 3y = 6$.

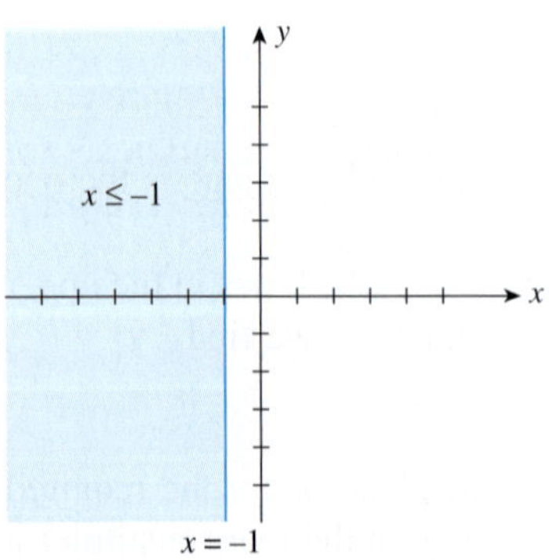

FIGURE 4
The set of points lying on the line $x = -1$ and in the left half-plane satisfies the given inequality.

EXAMPLE 2 Graph $x \leq -1$.

Solution The graph of $x = -1$ is the vertical line shown in Figure 4. Picking the origin $(0, 0)$ as a test point, we find $0 \leq -1$, which is false. Therefore, the required solution is the *left* half-plane, which does not contain the origin.

EXAMPLE 3 Graph $x - 2y > 0$.

Solution We first graph the equation $x - 2y = 0$, or $y = \frac{1}{2}x$ (Figure 5). Since the origin lies on the line, we may not use it as a test point. (Why?) Let's pick $(1, 2)$

as a test point. Substituting $x = 1$ and $y = 2$ into the given inequality, we find $1 - 2(2) > 0$, or $-3 > 0$, which is false. Therefore, the required solution is the half-plane that does not contain the test point—namely, the lower half-plane.

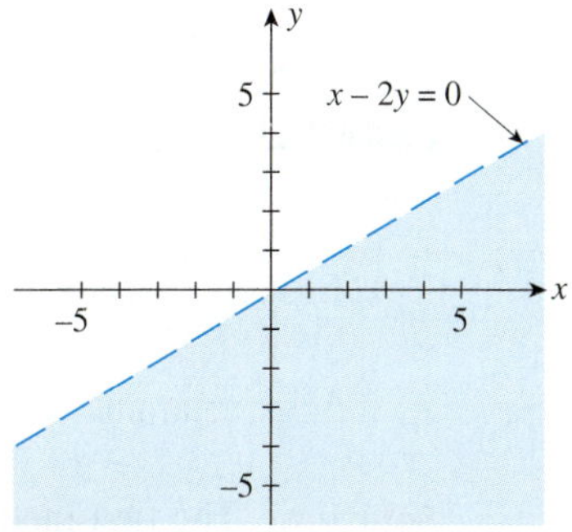

FIGURE **5**
The set of points in the lower half-plane satisfies $x - 2y > 0$.

Exploring with TECHNOLOGY

A graphing utility can be used to plot the graph of a linear inequality. For example, to plot the solution set for Example 1, first rewrite the equation $2x + 3y = 6$ in the form $y = 2 - \frac{2}{3}x$. Next, enter this expression for Y_1 in the calculator, and move the cursor to the left of Y_1 (see Figure a). Then press **ENTER** repeatedly, and select the icon that indicates the shading option desired. The required graph follows (see Figure b).

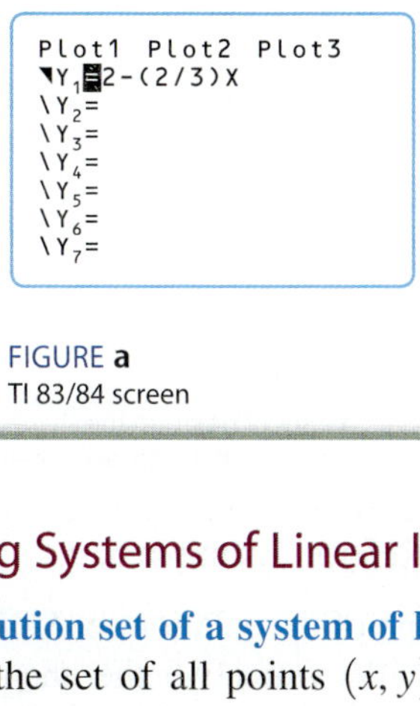

FIGURE **a**
TI 83/84 screen

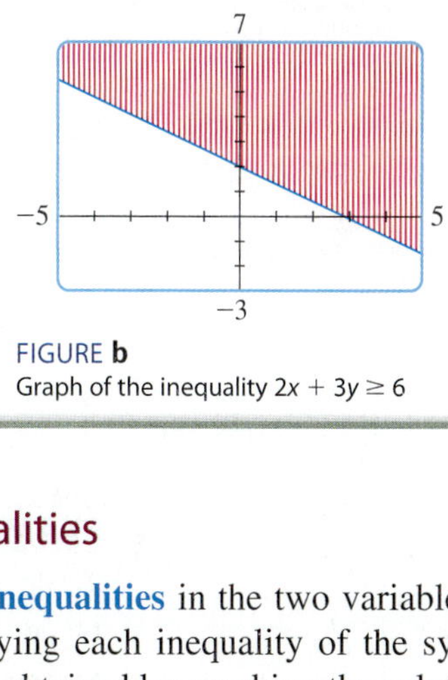

FIGURE **b**
Graph of the inequality $2x + 3y \geq 6$

Graphing Systems of Linear Inequalities

By the **solution set of a system of linear inequalities** in the two variables x and y, we mean the set of all points (x, y) satisfying each inequality of the system. The graphical solution of such a system may be obtained by graphing the solution set for each inequality independently and then determining the region in common with each solution set.

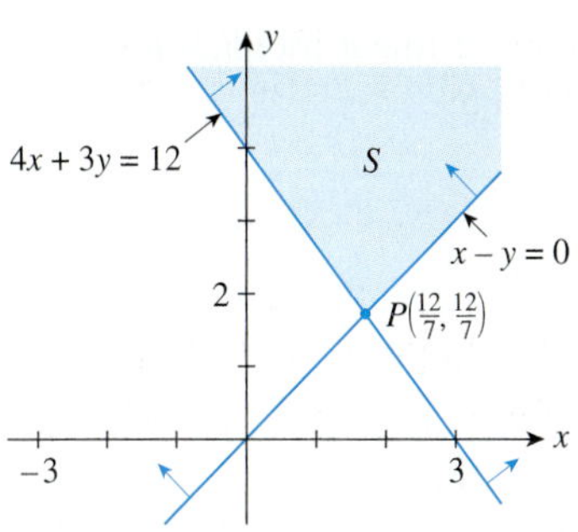

FIGURE **6**
The set of points in the shaded area satisfies the system

$$\begin{aligned} 4x + 3y &\geq 12 \\ x - y &\leq 0 \end{aligned}$$

EXAMPLE 4 Determine the solution set for the system

$$\begin{aligned} 4x + 3y &\geq 12 \\ x - y &\leq 0 \end{aligned}$$

Solution Proceeding as in the previous examples, you should have no difficulty locating the half-planes determined by each of the linear inequalities that make up the system. These half-planes are shown in Figure 6. The intersection of the two half-planes is the shaded region. A point in this region is an element of the solution set for the given system. The point $P(\frac{12}{7}, \frac{12}{7})$, the intersection of the two straight lines determined by the equations, is found by solving the simultaneous equations

$$\begin{aligned} 4x + 3y &= 12 \\ x - y &= 0 \end{aligned}$$

EXAMPLE 5 Sketch the solution set for the system

$$\begin{aligned} x &\geq 0 \\ y &\geq 0 \\ x + y - 6 &\leq 0 \\ 2x + y - 8 &\leq 0 \end{aligned}$$

Solution The first inequality in the system defines the right half-plane—all points to the right of the y-axis plus all points lying on the y-axis itself. The second inequality in the system defines the upper half-plane, including the x-axis. The half-planes defined by the third and fourth inequalities are indicated by arrows in Figure 7. Thus, the required region—the intersection of the four half-planes defined by the four inequalities in the given system of linear inequalities—is the shaded region. The point P is found by solving the simultaneous equations $x + y - 6 = 0$ and $2x + y - 8 = 0$.

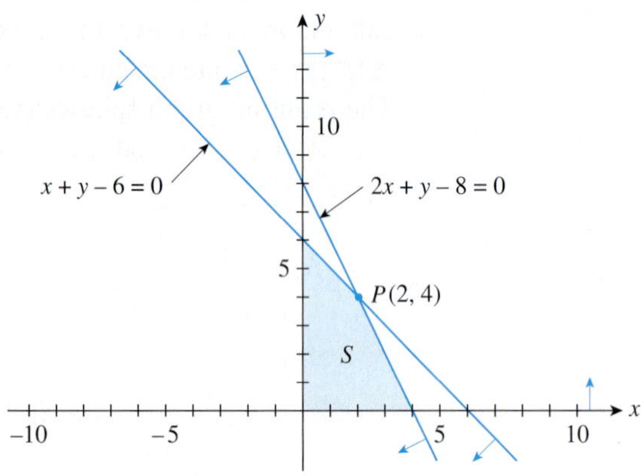

FIGURE 7
The set of points in the shaded region S, including the x- and y-axes, satisfies the given inequalities.

EXAMPLE 6 Refer to Example 5.

a. Use the graph of the solution set S of the given system of linear inequalities (Figure 7) to determine whether the point $A(1, 3)$lies in S.

b. Repeat part (a) for the point $B(5, 3)$.

Solution

a. Referring to Figure 7, we see that $A(1, 3)$ lies in the solution set S of the given system. To prove the result algebraically, we substitute $x = 1$ and $y = 3$ into each inequality in the system. Thus,

$$\begin{aligned} 1 &\geq 0 \\ 3 &\geq 0 \\ 1 + 3 - 6 = -2 &\leq 0 \\ 2(1) + 3 - 8 = -3 &\leq 0 \end{aligned}$$

Since all of these statements are true, it follows that $A(1, 3)$ does lie in S.

b. Referring to Figure 7 once again, we see that $B(5, 3)$ does not lie in S. To prove this assertion algebraically, we substitute $x = 5$ and $y = 3$ into the system of inequalities, obtaining

$$\begin{aligned} 5 &\geq 0 \\ 3 &\geq 0 \\ 5 + 3 - 6 = 2 &\leq 0 \\ 2(5) + 3 - 8 = 5 &\leq 0 \end{aligned}$$

Both the third and fourth inequalities are not true. Therefore, $B(5, 3)$ does not lie in S.

APPLIED EXAMPLE 7 A Production Problem Sonoma Company manufactures a 24-bottle wooden wine rack in two versions: a standard rack and a deluxe rack. Each standard rack requires 12 minutes of fabrication time and 4 minutes of finishing time. Each deluxe rack requires 8 minutes of fabrication time and 16 minutes of finishing time. There are 6 hours of time available for fabrication and 8 hours available for finishing each day.

a. Write a system of linear inequalities that gives the restrictions placed on the number of each type of wine rack manufactured by Sonoma.

b. Graph the solution set.

c. Can Sonoma manufacture 20 each of the standard and deluxe wine racks per day? Prove your assertion.

d. Can Sonoma manufacture 15 each of the standard and 20 deluxe wine racks per day? Prove your assertion.

Solution

a. As a first step toward setting up the system of inequalities, we tabulate the given information (see Table 1).

TABLE 1

	Standard	Deluxe	Time Available
Fabrication	12 min	8 min	360 min (6 hr)
Finishing	4 min	16 min	480 min (8 hr)

Let x denote the number of standard wine racks, and let y denote the number of deluxe wine racks to be manufactured per day. The amount of time required for fabricating these wine racks is $12x + 8y$ minutes and must not exceed 360 minutes. Thus, we have the inequality

$$12x + 8y \leq 360 \quad \text{or} \quad 3x + 2y \leq 90$$

Similarly, the total amount of time required for finishing these wine racks is $4x + 16y$ minutes and must not exceed 480 minutes. This condition leads to the inequality

$$4x + 16y \leq 480 \quad \text{or} \quad x + 4y \leq 120$$

Finally, neither x nor y can be negative, so

$$\begin{aligned} x &\geq 0 \\ y &\geq 0 \end{aligned}$$

Therefore, the desired system of linear inequalities is

$$\begin{aligned} 3x + 2y &\le 90 \\ x + 4y &\le 120 \\ x &\ge 0 \\ y &\ge 0 \end{aligned}$$

b. The solution set is graphed in Figure 8.

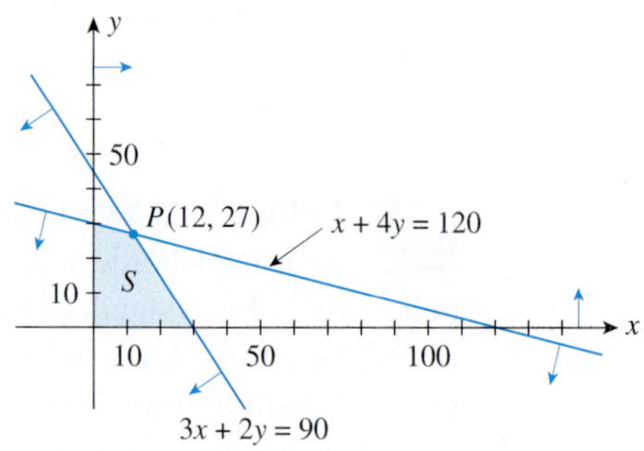

FIGURE **8**

The point P is found by solving the simultaneous equations $3x + 2y = 90$ and $x + 4y = 120$.

c. A visual inspection of the solution set S suggests that this is not possible. In fact, putting $x = y = 20$ into the first inequality in the system of linear inequalities, we find

$$3(20) + 2(20) = 100 \not\le 90$$

This verifies the result.

d. By visual inspection of S, we see that Sonoma can manufacture 15 standard and 20 deluxe wine racks per day. To prove this, we substitute $x = 15$ and $y = 20$ into each inequality in the system, obtaining

$$\begin{aligned} 3(15) + 2(20) = 85 &\le 90 \\ 15 + 4(20) = 95 &\le 120 \\ 15 &\ge 0 \\ 20 &\ge 0 \end{aligned}$$

We see that all of the inequalities in the system are satisfied.

The solution set S, shown in Figure 8, is an example of a bounded set. Observe that the set can be enclosed by a circle. For example, if you draw a circle of radius 30 with center at the origin, you will see that the set lies entirely inside the circle. On the other hand, the solution set S, shown in Figure 6, page 173, that was found in Example 4 cannot be enclosed by a circle and is said to be unbounded.

Bounded and Unbounded Solution Sets

The solution set of a system of linear inequalities is **bounded** if it can be enclosed by a circle. Otherwise, it is **unbounded.**

EXAMPLE 8 Determine the graphical solution set for the following system of linear inequalities:

$$\begin{aligned} 2x + y &\geq 50 \\ x + 2y &\geq 40 \\ x &\geq 0 \\ y &\geq 0 \end{aligned}$$

Solution The required solution set is the unbounded region shown in Figure 9.

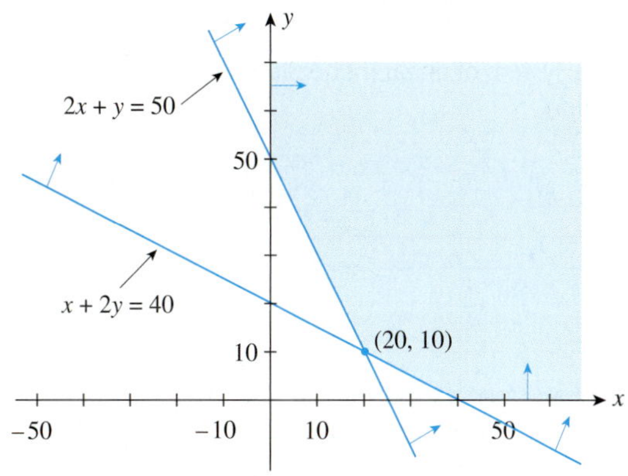

FIGURE **9**
The solution set is an unbounded region.

3.1 Self-Check Exercises

1. Determine graphically the solution set for the following system of inequalities:

$$\begin{aligned} x + 2y &\leq 10 \\ 5x + 3y &\leq 30 \\ x \geq 0&, y \geq 0 \end{aligned}$$

2. Determine graphically the solution set for the following system of inequalities:

$$\begin{aligned} 5x + 3y &\geq 30 \\ x - 3y &\leq 0 \\ x &\geq 2 \end{aligned}$$

Solutions to Self-Check Exercises 3.1 can be found on page 180.

3.1 Concept Questions

1. **a.** What is the difference, geometrically, between the solution set of $ax + by < c$ and the solution set of $ax + by \leq c$?
   b. Describe the set that is obtained by intersecting the solution set of $ax + by \leq c$ with the solution set of $ax + by \geq c$.

2. **a.** What is the solution set of a system of linear inequalities?
   b. How do you find the solution of a system of linear inequalities graphically?

3.1 Exercises

In Exercises 1–10, find the graphical solution of each inequality.

1. $4x - 8 < 0$

2. $3y + 2 > 0$

3. $x - y \leq 0$

4. $3x + 4y \leq -2$

5. $x \leq -3$

6. $y \geq -1$

7. $2x + y \leq 4$

8. $-3x + 6y \geq 12$

9. $4x - 3y \leq -24$

10. $5x - 3y \geq 15$

In Exercises 11–18, write a system of linear inequalities that describes the shaded region.

11.

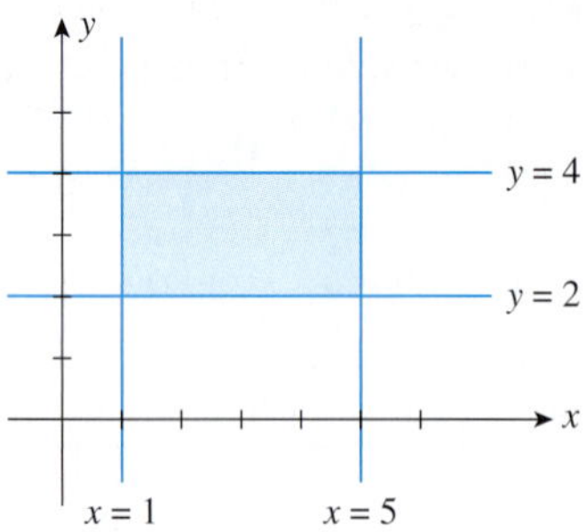

12.

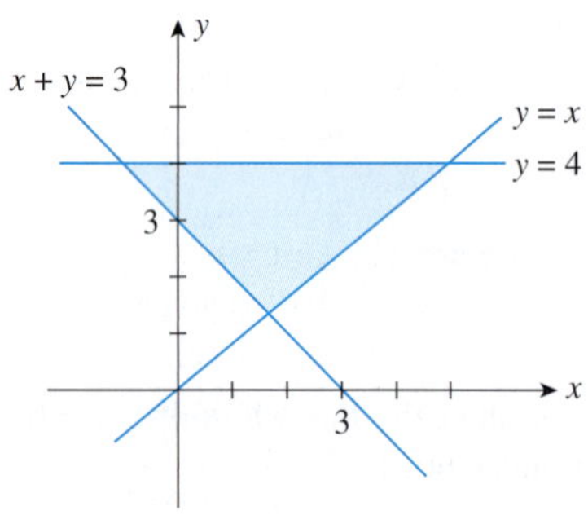

13.

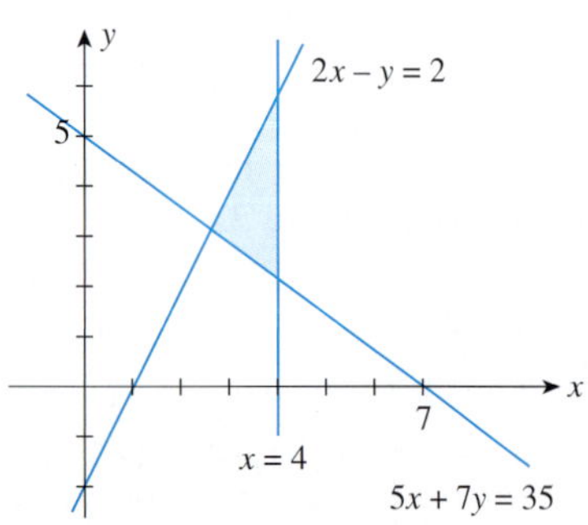

14.

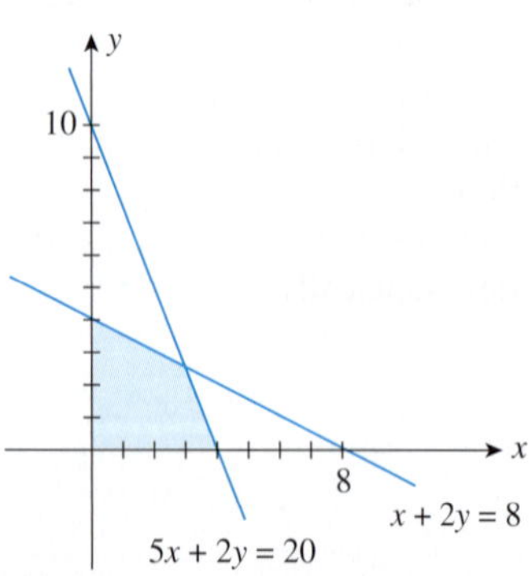

15.

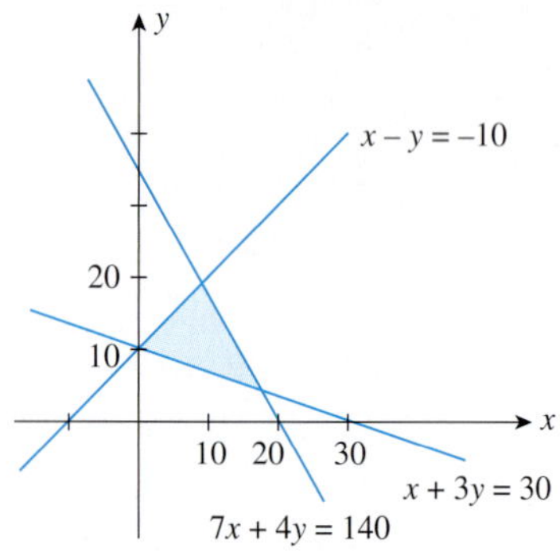

16.

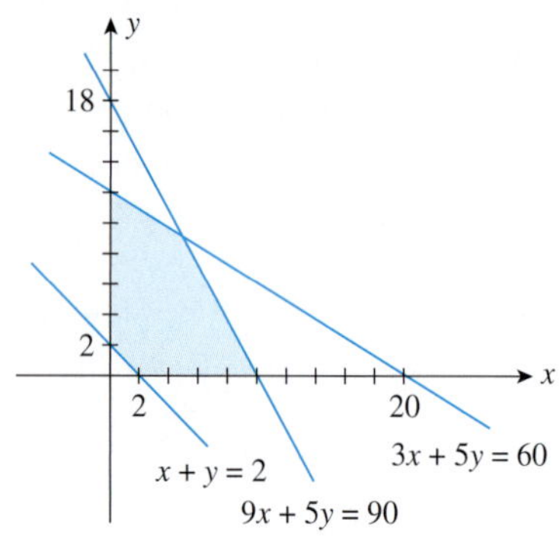

17.

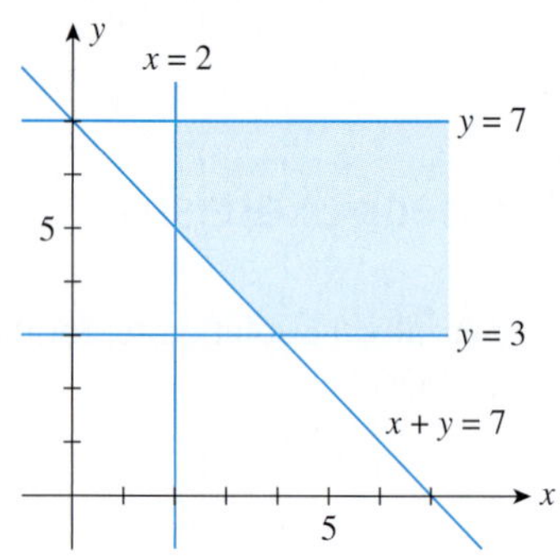

18.

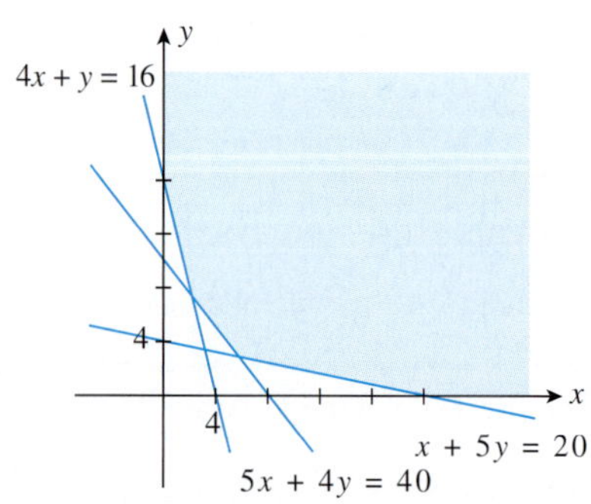

Exercises 19–22 make use of the results of Exercises 13–16.

19. Refer to the figure accompanying Exercise 13. By inspecting the figure, determine whether the point $(3, 3)$ lies in the set depicted there. Use the results of Exercise 13 to prove your assertion.

20. Refer to the figure accompanying Exercise 14. By inspecting the figure, determine whether the point $(5, 1)$ lies in the set depicted there. Use the results of Exercise 14 to prove your assertion.

21. Refer to the figure accompanying Exercise 15. By inspecting the figure, determine whether the point $(10, 10)$ lies in the set depicted there. Use the results of Exercise 15 to prove your assertion.

22. Refer to the figure accompanying Exercise 16. By inspecting the figure, determine whether the point $(8, \frac{18}{5})$ lies in the set depicted there. Use the results of Exercise 16 to prove your assertion.

In Exercises 23–40, determine graphically the solution set for each system of inequalities and indicate whether the solution set is bounded or unbounded.

23. $2x + 4y > 16$, $-x + 3y \geq 7$

24. $3x - 2y > -13$, $-x + 2y > 5$

25. $x - y \leq 0$, $2x + 3y \geq 10$

26. $x + y \geq -2$, $3x - y \leq 6$

27. $x + 2y \geq 3$, $2x + 4y \leq -2$

28. $2x - y \geq 4$, $4x - 2y < -2$

29. $x + y \leq 6$, $0 \leq x \leq 3$, $y \geq 0$

30. $4x - 3y \leq 12$, $5x + 2y \leq 10$, $x \geq 0, y \geq 0$

31. $3x - 6y \leq 12$, $-x + 2y \leq 4$, $x \geq 0, y \geq 0$

32. $x + y \geq 20$, $x + 2y \geq 40$, $x \geq 0, y \geq 0$

33. $3x - 7y \geq -24$, $x + 3y \geq 8$, $x \geq 0, y \geq 0$

34. $3x + 4y \geq 12$, $2x - y \geq -2$, $0 \leq y \leq 3$, $x \geq 0$

35. $x + 2y \geq 3$, $5x - 4y \leq 16$, $0 \leq y \leq 2$, $x \geq 0$

36. $x + y \leq 4$, $2x + y \leq 6$, $2x - y \geq -1$, $x \geq 0, y \geq 0$

37. $6x + 5y \leq 30$, $3x + y \geq 6$, $x + y \geq 4$, $x \geq 0, y \geq 0$

38. $6x + 7y \leq 84$, $12x - 11y \leq 18$, $6x - 7y \leq 28$, $x \geq 0, y \geq 0$

39. $x - y \geq -6$, $x - 2y \leq -2$, $x + 2y \geq 6$, $x - 2y \geq -14$, $x \geq 0, y \geq 0$

40. $x - 3y \geq -18$, $3x - 2y \geq 2$, $x - 3y \leq -4$, $3x - 2y \leq 16$, $x \geq 0, y \geq 0$

41. Concert Attendance The Peninsula Brass Band will hold its semiannual concert in a community center that has a seating capacity of 500. The band expects that at least 200 season ticket holders will attend the concert and that at least 100 nonseason ticket holders will also attend it.

a. Write a system of linear inequalities that gives the restrictions on the number of each type of ticket holder at the concert.
b. Graph the solution set S for the system of linear inequalities found in part (a).
c. Assuming that the attendance will be as expected, is it possible for 300 season ticket holders and 150 non-season ticket holders to attend the concert? Prove your assertion.

42. Manufacturing Fertilizers Agro Products makes two types of fertilizers that are sold in 50-lb bags. A 50-lb bag of Fertilizer A contains 5 lb of nitrogen, 10 lb of phosphorus, and 20 lb of potassium. A 50-lb bag of Fertilizer B contains 6 lb of nitrogen, 4 lb of phosphorus, and 4 lb of potassium. Agro Products has 1800 lb of nitrogen, 1600 lb of phosphorus, and 2600 lb of potassium on hand.

a. Write a system of linear inequalities that gives the restrictions to be placed on the number of bags of each type of fertilizer that Agro Products can manufacture.
b. Graph the solution set S for the system of linear inequalities found in part (a).
c. Is it possible for Agro Products to make 100 50-lb bags of Fertilizer A and 200 50-lb bags of Fertilizer B? Prove your assertion.

43. Investments Louisa has earmarked at most \$250,000 for investing in two companies involved in the production of renewable energy: Solaron Corporation and Windmill Corporation. She specifies that at least \$50,000 must be invested in each company and that the amount invested in Solaron Corporation must not exceed 120% of that invested in Windmill Corporation.

a. Write a system of linear inequalities that gives the restrictions placed upon Louisa's investments.
b. Graph the solution set S for the system of linear inequalities found in part (a).
c. Is it possible for Louisa to invest \$150,000 in Solaron corporation and \$100,000 in Windmill Corporation? Prove your assertion.

44. Diet Planning A dietitian wishes to plan a meal around four foods. The meal is to include 600 mg of phosphorus and 400 mg of magnesium. The number of units of the nutrients in each ounce of the foods (in milligrams) is summarized in the following table:

	Food A	Food B	Food C	Food D
Phosphorus	30	90	30	45
Magnesium	40	20	30	20

a. Let x_1, x_2, x_3, and x_4 denote the amount of Food *A*, Food *B*, Food *C*, and Food *D*, respectively, in a meal. Write a system of linear equations in the four given variables to describe the requirements of the dietitian.

b. Use the Gauss–Jordan elimination method to solve the system of part (a) for x_1 and x_2 in terms of x_3 and x_4 (see Section 2.3).

c. Use the fact that x_1 and x_2 must be nonnegative to write a system of two linear inequalities in the two variables x_3 and x_4, and then determine graphically the solution set for the system.

d. Combine the results of parts (b) and (c) to write the solutions to the problem.

e. Suggest three possible meals that meet the nutritional requirements. Include at least one in which all four foods are used. (*Note:* Your answer is not unique.)

In Exercises 45–48, determine whether the statement is true or false. If it is true, explain why it is true. If it is false, give an example to show why it is false.

45. The solution set of a linear inequality involving two variables is either a half-plane or a straight line.

46. The solution set of the inequality $ax + by + c \leq 0$ is either a left half-plane or a lower half-plane.

47. The solution set of a system of linear inequalities in two variables is bounded if it can be enclosed by a rectangle.

48. The solution set of the system

$$ax + by \leq e$$
$$cx + dy \leq f$$
$$x \geq 0, y \geq 0$$

where a, b, c, d, e, and f are positive real numbers, is a bounded set.

3.1 Solutions to Self-Check Exercises

1. The required solution set is shown in the following figure:

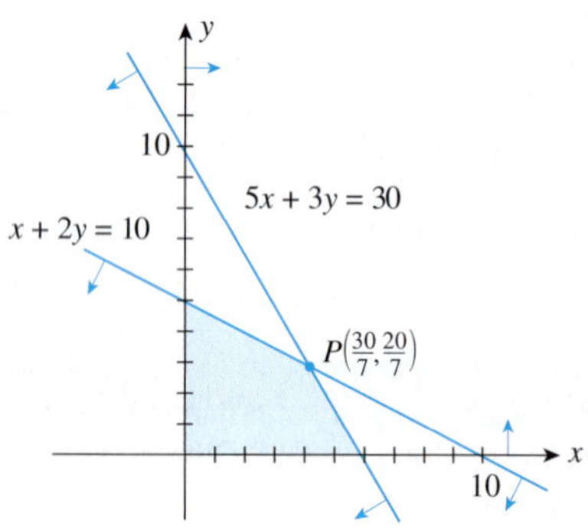

The point *P* is found by solving the system of equations

$$x + 2y = 10$$
$$5x + 3y = 30$$

Solving the first equation for x in terms of y gives

$$x = 10 - 2y$$

Substituting this value of x into the second equation of the system gives

$$\begin{aligned} 5(10 - 2y) + 3y &= 30 \\ 50 - 10y + 3y &= 30 \\ -7y &= -20 \end{aligned}$$

so $y = \frac{20}{7}$. Substituting this value of y into the expression for x found earlier, we obtain

$$x = 10 - 2\left(\frac{20}{7}\right) = \frac{30}{7}$$

giving the point of intersection as $\left(\frac{30}{7}, \frac{20}{7}\right)$.

2. The required solution set is shown in the following figure:

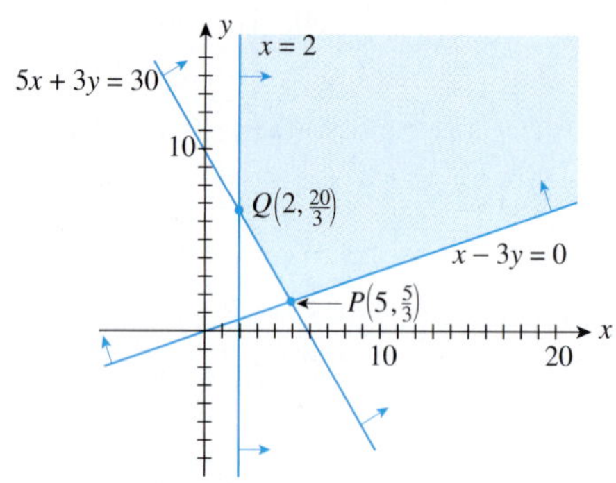

To find the coordinates of *P*, we solve the system

$$5x + 3y = 30$$
$$x - 3y = 0$$

Solving the second equation for x in terms of y and substituting this value of x in the first equation gives

$$5(3y) + 3y = 30$$

or $y = \frac{5}{3}$. Substituting this value of y into the second equation gives $x = 5$. Next, the coordinates of *Q* are found by solving the system

$$5x + 3y = 30$$
$$x = 2$$

yielding $x = 2$ and $y = \frac{20}{3}$.

3.2 Linear Programming Problems

In many business and economic problems, we are asked to optimize (maximize or minimize) a function subject to a system of equalities or inequalities. The function to be optimized is called the **objective function.** Profit functions and cost functions are examples of objective functions. The system of equalities or inequalities to which the objective function is subjected reflects the constraints (for example, limitations on resources such as materials and labor) imposed on the solution(s) to the problem. Problems of this nature are called **mathematical programming problems.** In particular, problems in which both the objective function and the constraints are expressed as linear equations or inequalities are called linear programming problems.

Linear Programming Problem

A **linear programming problem** consists of a linear objective function to be maximized or minimized subject to certain constraints in the form of linear equations or inequalities.

A Maximization Problem

As an example of a linear programming problem in which the objective function is to be maximized, let's consider the following simplified version of a production problem involving two variables.

APPLIED EXAMPLE 1 A Production Problem Ace Novelty wishes to produce two types of souvenirs: Type A and Type B. Each Type A souvenir will result in a profit of \$1, and each Type B souvenir will result in a profit of \$1.20. To manufacture a Type A souvenir requires 2 minutes on Machine I and 1 minute on Machine II. A Type B souvenir requires 1 minute on Machine I and 3 minutes on Machine II. There are 3 hours available on Machine I and 5 hours available on Machine II. How many souvenirs of each type should Ace make to maximize its profit?

Solution As a first step toward the mathematical formulation of this problem, we tabulate the given information (see Table 2).

TABLE 2

	Type A	Type B	Time Available
Machine I	2 min	1 min	180 min
Machine II	1 min	3 min	300 min
Profit/Unit	\$1	\$1.20	

Let x be the number of Type A souvenirs, and let y be the number of Type B souvenirs to be made. Then, the total profit P (in dollars) is given by

$$P = x + 1.2y$$

which is the objective function to be maximized.

The total amount of time that Machine I is used is given by $2x + y$ minutes and must not exceed 180 minutes. Thus, we have the inequality

$$2x + y \leq 180$$

Similarly, the total amount of time that Machine II is used is $x + 3y$ minutes and cannot exceed 300 minutes, so we are led to the inequality

$$x + 3y \leq 300$$

Finally, neither x nor y can be negative, so

$$x \geq 0$$
$$y \geq 0$$

To summarize, the problem here is to maximize the objective function $P = x + 1.2y$ subject to the system of inequalities

$$\begin{aligned} 2x + y &\leq 180 \\ x + 3y &\leq 300 \\ x \geq 0, y &\geq 0 \end{aligned}$$

The solution to this problem will be completed in Example 1, Section 3.3.

Minimization Problems

In the following linear programming problems, the objective function is to be minimized.

APPLIED EXAMPLE 2 A Nutrition Problem A nutritionist advises an individual who is suffering from iron and vitamin B deficiency to take at least 2400 milligrams (mg) of iron, 2100 mg of vitamin B_1 (thiamine), and 1500 mg of vitamin B_2 (riboflavin) over a period of time. Two vitamin pills are suitable, Brand *A* and Brand *B*. Each Brand *A* pill costs 6 cents and contains 40 mg of iron, 10 mg of vitamin B_1, and 5 mg of vitamin B_2. Each Brand *B* pill costs 8 cents and contains 10 mg of iron and 15 mg each of vitamins B_1 and B_2 (Table 3). What combination of pills should the individual purchase to meet the minimum iron and vitamin requirements at the lowest cost?

TABLE 3

	Brand *A*	Brand *B*	Minimum Requirement
Iron	40 mg	10 mg	2400 mg
Vitamin B_1	10 mg	15 mg	2100 mg
Vitamin B_2	5 mg	15 mg	1500 mg
Cost/Pill	6¢	8¢	

Solution Let x be the number of Brand *A* pills, and let y be the number of Brand *B* pills to be purchased. The cost C (in cents) is given by

$$C = 6x + 8y$$

and is the objective function to be minimized.

The amount of iron contained in x Brand *A* pills and y Brand *B* pills is given by $40x + 10y$ mg, and this must be greater than or equal to 2400 mg. This translates into the inequality

$$40x + 10y \geq 2400$$

Similar considerations involving the minimum requirements of vitamins B_1 and B_2 lead to the inequalities

$$\begin{aligned} 10x + 15y &\geq 2100 \\ 5x + 15y &\geq 1500 \end{aligned}$$

respectively. Thus, the problem here is to minimize $C = 6x + 8y$ subject to

$$\begin{aligned} 40x + 10y &\geq 2400 \\ 10x + 15y &\geq 2100 \\ 5x + 15y &\geq 1500 \\ x \geq 0, y &\geq 0 \end{aligned}$$

The solution to this problem will be completed in Example 2, Section 3.3.

APPLIED EXAMPLE 3 A Transportation Problem Curtis-Roe Aviation Industries has two plants, I and II, that produce the Zephyr jet engines used in their light commercial airplanes. There are 100 units of the engines in Plant I and 110 units in Plant II. The engines are shipped to two of Curtis-Roe's main assembly plants, *A* and *B*. The shipping costs (in dollars) per engine from Plants I and II to the Main Assembly Plants *A* and *B* are as follows:

	To Assembly Plant	
From	***A***	***B***
Plant I	100	60
Plant II	120	70

In a certain month, Assembly Plant *A* needs 80 engines, whereas Assembly Plant *B* needs 70 engines. Find how many engines should be shipped from each plant to each main assembly plant if shipping costs are to be kept to a minimum.

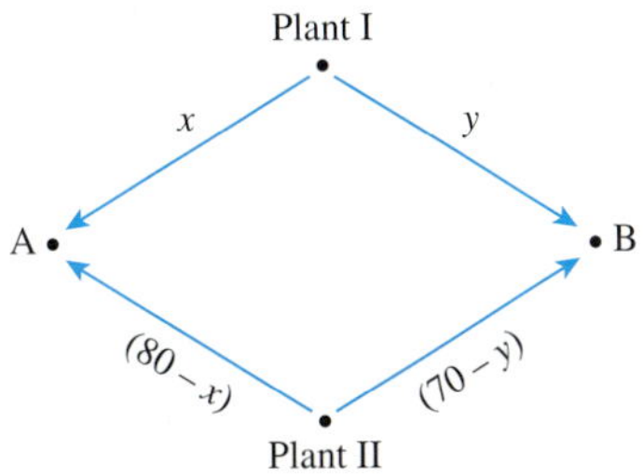

FIGURE 10

Solution Let x denote the number of engines shipped from Plant I to Assembly Plant *A*, and let y denote the number of engines shipped from Plant I to Assembly Plant *B*. Since the requirements of Assembly Plants *A* and *B* are 80 and 70 engines, respectively, the number of engines shipped from Plant II to Assembly Plants *A* and *B* are $(80 - x)$ and $(70 - y)$, respectively. These numbers may be displayed in a schematic. With the aid of the accompanying schematic (Figure 10) and the shipping cost schedule, we find that the total shipping cost incurred by Curtis-Roe is given by

$$\begin{aligned} C &= 100x + 60y + 120(80 - x) + 70(70 - y) \\ &= 14{,}500 - 20x - 10y \end{aligned}$$

Next, the production constraints on Plants I and II lead to the inequalities

$$\begin{aligned} x + y &\leq 100 \\ (80 - x) + (70 - y) &\leq 110 \end{aligned}$$

The last inequality simplifies to

$$x + y \geq 40$$

Also, the requirements of the two main assembly plants lead to the inequalities

$$x \geq 0 \qquad y \geq 0 \qquad 80 - x \geq 0 \qquad 70 - y \geq 0$$

The last two may be written as $x \leq 80$ and $y \leq 70$.

Summarizing, we have the following linear programming problem: Minimize the objective (cost) function $C = 14{,}500 - 20x - 10y$ subject to the constraints

$$\begin{aligned} x + y &\geq 40 \\ x + y &\leq 100 \\ x &\leq 80 \\ y &\leq 70 \end{aligned}$$

where $x \geq 0$ and $y \geq 0$.

You will be asked to complete the solution to this problem in Exercise 49, Section 3.3.

APPLIED EXAMPLE 4 A Warehouse Problem Acrosonic manufactures its Brentwood loudspeaker systems in two separate locations, Plant I and Plant II. The output at Plant I is at most 400 per month, whereas the output at Plant II is at most 600 per month. These loudspeaker systems are shipped to three warehouses that serve as distribution centers for the company. For the warehouses to meet their orders, the minimum monthly requirements of Warehouses *A*, *B*, and *C* are 200, 300, and 400 systems, respectively. Shipping costs from Plant I to Warehouses *A*, *B*, and *C* are \$20, \$8, and \$10 per loudspeaker system, respectively, and shipping costs from Plant II to each of these warehouses are \$12, \$22, and \$18, respectively. What should the shipping schedule be if Acrosonic wishes to meet the requirements of the distribution centers and at the same time keep its shipping costs to a minimum?

TABLE 4

	Warehouse		
Plant	***A***	***B***	***C***
I	20	8	10
II	12	22	18

Solution The respective shipping costs (in dollars) per loudspeaker system may be tabulated as in Table 4. Letting x_1 denote the number of loudspeaker systems shipped from Plant I to Warehouse *A*, x_2 the number shipped from Plant I to Warehouse *B*, and so on leads to Table 5.

TABLE 5

	Warehouse			
Plant	***A***	***B***	***C***	**Max. Prod.**
I	x_1	x_2	x_3	400
II	x_4	x_5	x_6	600
Min. Req.	200	300	400	

From Tables 4 and 5, we see that the cost of shipping x_1 loudspeaker systems from Plant I to Warehouse *A* is $\$20x_1$, the cost of shipping x_2 loudspeaker systems from Plant I to Warehouse *B* is $\$8x_2$, and so on. Thus, the total monthly shipping cost (in dollars) incurred by Acrosonic is given by

$$C = 20x_1 + 8x_2 + 10x_3 + 12x_4 + 22x_5 + 18x_6$$

Next, the production constraints on Plants I and II lead to the inequalities

$$\begin{aligned} x_1 + x_2 + x_3 &\leq 400 \\ x_4 + x_5 + x_6 &\leq 600 \end{aligned}$$

(see Table 5). Also, the minimum requirements of each of the three warehouses lead to the three inequalities

$$\begin{aligned} x_1 + x_4 &\geq 200 \\ x_2 + x_5 &\geq 300 \\ x_3 + x_6 &\geq 400 \end{aligned}$$

Summarizing, we have the following linear programming problem:

$$\begin{aligned} \text{Minimize} \quad & C = 20x_1 + 8x_2 + 10x_3 + 12x_4 + 22x_5 + 18x_6 \\ \text{subject to} \quad & x_1 + x_2 + x_3 \le 400 \\ & x_4 + x_5 + x_6 \le 600 \\ & x_1 + x_4 \ge 200 \\ & x_2 + x_5 \ge 300 \\ & x_3 + x_6 \ge 400 \\ & x_1 \ge 0, x_2 \ge 0, \ldots, x_6 \ge 0 \end{aligned}$$

The solution to this problem will be completed in Example 5, Section 4.2.

3.2 Self-Check Exercise

OPTIMIZING ADVERTISING EXPOSURE Gino Balduzzi, proprietor of Luigi's Pizza Palace, allocates $9000 a month for advertising in two newspapers, the *City Tribune* and the *Daily News*. The *City Tribune* charges $300 for a certain advertisement, whereas the *Daily News* charges $100 for the same ad. Gino has stipulated that the ad is to appear in at least 15 but no more than 30 editions of the *Daily News* per month. The *City Tribune* has a daily circulation of 50,000, and the *Daily News* has a circulation of 20,000. Under these conditions, determine how many ads Gino should place in each newspaper to reach the largest number of readers. Formulate but do not solve the problem. (The solution to this problem can be found in Exercise 3 of Solutions to Self-Check Exercises 3.3.)

The solution to Self-Check Exercise 3.2 can be found on page 189.

3.2 Concept Questions

1. What is a linear programming problem?
2. Suppose you are asked to formulate a linear programming problem in two variables x and y. How would you express the fact that x and y are nonnegative? Why are these conditions often required in practical problems?
3. What is the difference between a maximization linear programming problem and a minimization linear programming problem?

3.2 Exercises

Formulate but do not solve each of the following exercises as a linear programming problem. You will be asked to solve these problems later.

1. **PRODUCTION SCHEDULING** A company manufactures two products, *A* and *B*, on two machines, I and II. It has been determined that the company will realize a profit of $3 on each unit of Product *A* and a profit of $4 on each unit of Product *B*. To manufacture a unit of Product *A* requires 6 min on Machine I and 5 min on Machine II. To manufacture a unit of Product *B* requires 9 min on Machine I and 4 min on Machine II. There are 5 hr of machine time available on Machine I and 3 hr of machine time available on Machine II in each work shift. How many units of each product should be produced in each shift to maximize the company's profit?
2. **PRODUCTION SCHEDULING** National Business Machines manufactures two models of portable printers: A and B. Each model A costs $100 to make, and each model B costs $150. The profits are $30 for each model A and $40 for each model B portable printer. If the total number of portable printers demanded per month does not exceed 2500 and the company has earmarked no more than $600,000/month for manufacturing costs, how many units of each model should National make each month to maximize its monthly profit?

3. **Production Scheduling** Kane Manufacturing has a division that produces two models of fireplace grates, model A and model B. To produce each model A grate requires 3 lb of cast iron and 6 min of labor. To produce each model B grate requires 4 lb of cast iron and 3 min of labor. The profit for each model A grate is \$2.00, and the profit for each model B grate is \$1.50. If 1000 lb of cast iron and 20 hr of labor are available for the production of grates per day, how many grates of each model should the division produce per day to maximize Kane's profits?

4. **Production Scheduling** Refer to Exercise 3. Because of a backlog of orders for model A grates, the manager of Kane Manufacturing has decided to produce at least 150 of these grates a day. Operating under this additional constraint, how many grates of each model should Kane produce to maximize profit?

5. **Production Scheduling** A division of the Winston Furniture Company manufactures dining tables and chairs. Each table requires 40 board feet of wood and 3 labor-hours. Each chair requires 16 board feet of wood and 4 labor-hours. The profit for each table is \$45, and the profit for each chair is \$20. In a certain week, the company has 3200 board feet of wood available and 520 labor-hours. How many tables and chairs should Winston manufacture to maximize its profits?

6. **Production Scheduling** Refer to Exercise 5. If the profit for each table is \$50 and the profit for each chair is \$18, how many tables and chairs should Winston manufacture to maximize its profits?

7. **Allocation of Funds** Madison Finance has a total of \$20 million earmarked for homeowner loans and auto loans. On the average, homeowner loans have a 10% annual rate of return, whereas auto loans yield a 12% annual rate of return. Management has also stipulated that the total amount of homeowner loans should be greater than or equal to 4 times the total amount of automobile loans. Determine the total amount of loans of each type Madison should extend to each category to maximize its returns.

8. **Asset Allocation** A financier plans to invest up to \$500,000 in two projects. Project *A* yields a return of 10% on the investment whereas Project *B* yields a return of 15% on the investment. Because the investment in Project *B* is riskier than the investment in Project *A*, the financier has decided that the investment in Project *B* should not exceed 40% of the total investment. How much should she invest in each project to maximize the return on her investment?

9. **Asset Allocation** Justin has decided to invest at most \$60,000 in medium-risk and high-risk stocks. He has further decided that the medium-risk stocks should make up at least 40% of the total investment, while the high-risk stocks should make up at least 20% of the total investment. He expects that the medium-risk stocks will appreciate by 12% and the high-risk stocks by 20% within a year. How much money should Justin invest in each type of stock to maximize the value of his investment?

10. **Crop Planning** A farmer plans to plant two crops, *A* and *B*. The cost of cultivating Crop *A* is \$40/acre, whereas the cost of cultivating Crop *B* is \$60/acre. The farmer has a maximum of \$7400 available for land cultivation. Each acre of Crop *A* requires 20 labor-hours, and each acre of Crop *B* requires 25 labor-hours. The farmer has a maximum of 3300 labor-hours available. If she expects to make a profit of \$150/acre on Crop *A*, and \$200/acre on Crop *B*, how many acres of each crop should she plant to maximize her profit?

11. **Minimizing Mining Costs** Perth Mining Company operates two mines for the purpose of extracting gold and silver. The Saddle Mine costs \$14,000/day to operate, and it yields 50 oz of gold and 3000 oz of silver each day. The Horseshoe Mine costs \$16,000/day to operate, and it yields 75 oz of gold and 1000 oz of silver each day. Company management has set a target of at least 650 oz of gold and 18,000 oz of silver. How many days should each mine be operated so that the target can be met at a minimum cost?

12. **Minimizing Cruise Line Costs** Deluxe River Cruises operates a fleet of river vessels. The fleet has two types of vessels: A type A vessel has 60 deluxe cabins and 160 standard cabins, whereas a type B vessel has 80 deluxe cabins and 120 standard cabins. Under a charter agreement with Odyssey Travel Agency, Deluxe River Cruises is to provide Odyssey with a minimum of 360 deluxe and 680 standard cabins for their 15-day cruise in May. It costs \$44,000 to operate a type A vessel and \$54,000 to operate a type B vessel for that period. How many of each type vessel should be used to keep the operating costs to a minimum?

13. **Production Scheduling** Acoustical Company manufactures a DVD storage cabinet that can be bought fully assembled or as a kit. Each cabinet is processed in the fabrication department and the assembly department. If the fabrication department manufactures only fully assembled cabinets, it can produce 200 units/day; and if it manufactures only kits, it can produce 200 units/day. If the assembly department produces only fully assembled cabinets, it can produce 100 units/day; but if it produces only kits, then it can produce 300 units/day. Each fully assembled cabinet contributes \$50 to the profits of the company, whereas each kit contributes \$40 to its profits. How many fully assembled units and how many kits should the company produce per day to maximize its profits?

14. **Fertilizers** A farmer uses two types of fertilizers. A 50-lb bag of Fertilizer *A* contains 8 lb of nitrogen, 2 lb of phosphorus, and 4 lb of potassium. A 50-lb bag of Fertilizer *B* contains 5 lb each of nitrogen, phosphorus, and potassium. The minimum requirements for a field are 440 lb of nitrogen, 260 lb of phosphorus, and 360 lb of potassium. If a 50-lb bag of Fertilizer *A* costs \$30 and a 50-lb bag of

Fertilizer B costs \$20, find the amount of each type of fertilizer the farmer should use to minimize his cost while still meeting the minimum requirements.

15. **Minimizing City Water Costs** The water-supply manager for a Midwestern city needs to supply the city with at least 10 million gallons of potable (drinkable) water per day. The supply may be drawn from the local reservoir or from a pipeline to an adjacent town. The local reservoir has a maximum daily yield of 5 million gallons of potable water, and the pipeline has a maximum daily yield of 10 million gallons. By contract, the pipeline is required to supply a minimum of 6 million gallons/day. If the cost for 1 million gallons of reservoir water is \$300 and that for pipeline water is \$500, how much water should the manager get from each source to minimize daily water costs for the city?

16. **Production Scheduling** Ace Novelty manufactures Giant Pandas and Saint Bernards. Each Panda requires 1.5 yd^2 of plush, 30 ft^3 of stuffing, and 5 pieces of trim; each Saint Bernard requires 2 yd^2 of plush, 35 ft^3 of stuffing, and 8 pieces of trim. The profit for each Panda is \$10 and the profit for each Saint Bernard is \$15. If 3600 yd^2 of plush, 66,000 ft^3 of stuffing and 13,600 pieces of trim are available, how many of each of the stuffed animals should the company manufacture to maximize profit?

17. **Diet Planning** A nutritionist at the Medical Center has been asked to prepare a special diet for certain patients. She has decided that the meals should contain a minimum of 400 mg of calcium, 10 mg of iron, and 40 mg of vitamin C. She has further decided that the meals are to be prepared from Foods A and B. Each ounce of Food A contains 30 mg of calcium, 1 mg of iron, 2 mg of vitamin C, and 2 mg of cholesterol. Each ounce of Food B contains 25 mg of calcium, 0.5 mg of iron, 5 mg of vitamin C, and 5 mg of cholesterol. How many ounces of each type of food should be used in a meal so that the cholesterol content is minimized and the minimum requirements of calcium, iron, and vitamin C are met?

18. **Optimizing Advertising Exposure** Everest Deluxe World Travel has decided to advertise in the Sunday editions of two major newspapers in town. These advertisements are directed at three groups of potential customers. Each advertisement in Newspaper I is seen by 70,000 Group A customers, 40,000 Group B customers, and 20,000 Group C customers. Each advertisement in Newspaper II is seen by 10,000 Group A, 20,000 Group B, and 40,000 Group C customers. Each advertisement in Newspaper I costs \$1000, and each advertisement in Newspaper II costs \$800. Everest would like their advertisements to be read by at least 2 million people from Group A, 1.4 million people from Group B, and 1 million people from Group C. How many advertisements should Everest place in each newspaper to achieve its advertising goals at a minimum cost?

19. **Minimizing Shipping Costs** TMA manufactures 37-in. high-definition LCD televisions in two separate locations: Location I and Location II. The output at Location I is at most 6000 televisions/month, whereas the output at Location II is at most 5000 televisions/month. TMA is the main supplier of televisions to Pulsar Corporation, its holding company, which has priority in having all its requirements met. In a certain month, Pulsar placed orders for 3000 and 4000 televisions to be shipped to two of its factories located in City A and City B, respectively. The shipping costs (in dollars) per television from the two TMA plants to the two Pulsar factories are as follows:

	To Pulsar Factories	
From TMA	**City *A***	**City *B***
Location I	\$6	\$4
Location II	\$8	\$10

Find a shipping schedule that meets the requirements of both companies while keeping costs to a minimum.

20. **Social Programs Planning** AntiFam, a hunger-relief organization, has earmarked between \$2 million and \$2.5 million (inclusive) for aid to two African countries, Country A and Country B. Country A is to receive between \$1 million and \$1.5 million (inclusive), and Country B is to receive at least \$0.75 million. It has been estimated that each dollar spent in Country A will yield an effective return of \$0.60, whereas a dollar spent in Country B will yield an effective return of \$0.80. How should the aid be allocated if the money is to be utilized most effectively according to these criteria?
Hint: If x and y denote the amount of money to be given to Country A and Country B, respectively, then the objective function to be maximized is $P = 0.6x + 0.8y$.

21. **Production Scheduling** A company manufactures Products A, B, and C. Each product is processed in three departments: I, II, and III. The total available labor-hours per week for Departments I, II, and III are 900, 1080, and 840, respectively. The time requirements (in hours per unit) and profit per unit for each product are as follows:

	Product *A*	Product *B*	Product *C*
Dept. I	2	1	2
Dept. II	3	1	2
Dept. III	2	2	1
Profit	\$18	\$12	\$15

How many units of each product should the company produce to maximize its profit?

22. **Optimizing Advertising Exposure** As part of a campaign to promote its annual clearance sale, the Excelsior Company decided to buy television advertising time on Station KAOS. Excelsior's advertising budget is \$102,000. Morning time costs \$3000/minute, afternoon time

costs \$1000/minute, and evening (prime) time costs \$12,000/minute. Because of previous commitments, KAOS cannot offer Excelsior more than 6 min of prime time or more than a total of 25 min of advertising time over the 2 weeks in which the commercials are to be run. KAOS estimates that morning commercials are seen by 200,000 people, afternoon commercials are seen by 100,000 people, and evening commercials are seen by 600,000 people. How much morning, afternoon, and evening advertising time should Excelsior buy to maximize exposure of its commercials?

23. **Production Scheduling** Custom Office Furniture Company is introducing a new line of executive desks made from a specially selected grade of walnut. Initially, three different models—A, B, and C—are to be marketed. Each model A desk requires $1\frac{1}{4}$ hr for fabrication, 1 hr for assembly, and 1 hr for finishing; each model B desk requires $1\frac{1}{2}$ hr for fabrication, 1 hr for assembly, and 1 hr for finishing; each model C desk requires $1\frac{1}{2}$ hr, $\frac{3}{4}$ hr, and $\frac{1}{2}$ hr for fabrication, assembly, and finishing, respectively. The profit on each model A desk is \$26, the profit on each model B desk is \$28, and the profit on each model C desk is \$24. The total time available in the fabrication department, the assembly department, and the finishing department in the first month of production is 310 hr, 205 hr, and 190 hr, respectively. To maximize Custom's profit, how many desks of each model should be made in the month?

24. **Asset Allocation** A financier plans to invest up to \$2 million in three projects. She estimates that Project *A* will yield a return of 10% on her investment, Project *B* will yield a return of 15% on her investment, and Project *C* will yield a return of 20% on her investment. Because of the risks associated with the investments, she decided to put not more than 20% of her total investment in Project *C*. She also decided that her investments in Projects *B* and *C* should not exceed 60% of her total investment. Finally, she decided that her investment in Project *A* should be at least 60% of her investments in Projects *B* and *C*. How much should the financier invest in each project if she wishes to maximize the total returns on her investments?

25. **Asset Allocation** Ashley has earmarked at most \$250,000 for investment in three mutual funds: a money market fund, an international equity fund, and a growth-and-income fund. The money market fund has a rate of return of 6%/year, the international equity fund has a rate of return of 10%/year, and the growth-and-income fund has a rate of return of 15%/year. Ashley has stipulated that no more than 25% of her total portfolio should be in the growth-and-income fund and that no more than 50% of her total portfolio should be in the international equity fund. To maximize the return on her investment, how much should Ashley invest in each type of fund?

26. **Optimizing Prefabricated Housing Production** Boise Lumber has decided to enter the lucrative prefabricated housing business. Initially, it plans to offer three models: standard, deluxe, and luxury. Each house is prefabricated and partially assembled in the factory, and the final assembly is completed on site. The dollar amount of building material required, the amount of labor required in the factory for prefabrication and partial assembly, the amount of on-site labor required, and the profit per unit are as follows:

	Standard Model	Deluxe Model	Luxury Model
Material	\$6,000	\$8,000	\$10,000
Factory Labor (hr)	240	220	200
On-site Labor (hr)	180	210	300
Profit	\$3,400	\$4,000	\$5,000

For the first year's production, a sum of \$8.2 million is budgeted for the building material; the number of labor-hours available for work in the factory (for prefabrication and partial assembly) is not to exceed 218,000 hr; and the amount of labor for on-site work is to be less than or equal to 237,000 labor-hours. Determine how many houses of each type Boise should produce (market research has confirmed that there should be no problems with sales) to maximize its profit from this new venture.

27. **Minimizing Shipping Costs** Acrosonic of Example 4 also manufactures a model G loudspeaker system in plants I and II. The output at Plant I is at most 800 systems/month whereas the output at Plant II is at most 600/month. These loudspeaker systems are also shipped to three warehouses—*A*, *B*, and *C*—whose minimum monthly requirements are 500, 400, and 400, respectively. Shipping costs from Plant I to Warehouse *A*, Warehouse *B*, and Warehouse *C* are \$16, \$20, and \$22 per system, respectively, and shipping costs from Plant II to each of these warehouses are \$18, \$16, and \$14 per system, respectively. What shipping schedule will enable Acrosonic to meet the warehouses' requirements and at the same time keep its shipping costs to a minimum?

28. **Optimizing Production of Cold Formulas** Beyer Pharmaceutical produces three kinds of cold formulas: Formula I, Formula II, and Formula III. It takes 2.5 hr to produce 1000 bottles of Formula I, 3 hr to produce 1000 bottles of Formula II, and 4 hr to produce 1000 bottles of Formula III. The profits for each 1000 bottles of Formula I, Formula II, and Formula III are \$180, \$200, and \$300, respectively. For a certain production run, there are enough ingredients on hand to make at most 9000 bottles of Formula I, 12,000 bottles of Formula II, and 6000 bottles of Formula III. Furthermore, the time for the production run is limited to a maximum of 70 hr. How many bottles of each formula should be produced in this production run so that the profit is maximized?

29. **Optimizing Production of Blended Juices** CalJuice Company has decided to introduce three fruit juices made from blending two or more concentrates. These juices

will be packaged in 2-qt (64-oz) cartons. One carton of pineapple–orange juice requires 8 oz each of pineapple and orange juice concentrates. One carton of orange–banana juice requires 12 oz of orange juice concentrate and 4 oz of banana pulp concentrate. Finally, one carton of pineapple–orange–banana juice requires 4 oz of pineapple juice concentrate, 8 oz of orange juice concentrate, and 4 oz of banana pulp. The company has decided to allot 16,000 oz of pineapple juice concentrate, 24,000 oz of orange juice concentrate, and 5000 oz of banana pulp concentrate for the initial production run. The company has also stipulated that the production of pineapple–orange–banana juice should not exceed 800 cartons. Its profit on one carton of pineapple–orange juice is \$1.00, its profit on one carton of orange–banana juice is \$.80, and its profit on one carton of pineapple–orange–banana juice is \$.90. To realize a maximum profit, how many cartons of each blend should the company produce?

30. **Minimizing Shipping Costs** Steinwelt Piano manufactures upright and console pianos in two plants, Plant I and Plant II. The output of Plant I is at most 300/month, whereas the output of Plant II is at most 250/month. These pianos are shipped to three warehouses, which serve as distribution centers for the company. To fill current and projected future orders, Warehouse A requires at least 200 pianos/month, Warehouse B requires at least 150 pianos/month, and Warehouse C requires at least 200 pianos/month. The shipping cost of each piano from Plant I to Warehouse A, Warehouse B, and Warehouse C is \$60, \$60, and \$80, respectively, and the shipping cost of each piano from Plant II to Warehouse A, Warehouse B, and Warehouse C is \$80, \$70, and \$50, respectively. What shipping schedule will enable Steinwelt to meet the warehouses' requirements while keeping shipping costs to a minimum?

In Exercises 31 and 32, determine whether the statement is true or false. If it is true, explain why it is true. If it is false, give an example to show why it is false.

31. The problem

$$\begin{aligned} \text{Maximize} \quad & P = xy \\ \text{subject to} \quad & 2x + 3y \leq 12 \\ & 2x + y \leq 8 \\ & x \geq 0, y \geq 0 \end{aligned}$$

is a linear programming problem.

32. The problem

$$\begin{aligned} \text{Minimize} \quad & C = 2x + 3y \\ \text{subject to} \quad & 2x + 3y \leq 6 \\ & x - y = 0 \\ & x \geq 0, y \geq 0 \end{aligned}$$

is a linear programming problem.

3.2 Solution to Self-Check Exercise

Let x denote the number of ads to be placed in the *City Tribune,* and let y denote the number to be placed in the *Daily News*. The total cost for placing x ads in the *City Tribune* and y ads in the *Daily News* is $300x + 100y$ dollars, and since the monthly budget is \$9000, we must have

$$300x + 100y \leq 9000$$

Next, the condition that the ad must appear in at least 15 but no more than 30 editions of the *Daily News* translates into the inequalities

$$\begin{aligned} y &\geq 15 \\ y &\leq 30 \end{aligned}$$

Finally, the objective function to be maximized is

$$P = 50{,}000x + 20{,}000y$$

To summarize, we have the following linear programming problem:

$$\begin{aligned} \text{Maximize} \quad & P = 50{,}000x + 20{,}000y \\ \text{subject to} \quad & 300x + 100y \leq 9000 \\ & y \geq 15 \\ & y \leq 30 \\ & x \geq 0, y \geq 0 \end{aligned}$$

3.3 Graphical Solution of Linear Programming Problems

The Graphical Method

Linear programming problems in two variables have relatively simple geometric interpretations. For example, the system of linear constraints associated with a two-dimensional linear programming problem, unless it is inconsistent, defines a planar region or a line segment whose boundary is composed of straight-line segments and/or half-lines. Such problems are therefore amenable to graphical analysis.

Consider the following two-dimensional linear programming problem:

$$\begin{aligned} \text{Maximize} \quad & P = 3x + 2y \\ \text{subject to} \quad & 2x + 3y \leq 12 \\ & 2x + y \leq 8 \\ & x \geq 0, y \geq 0 \end{aligned} \tag{7}$$

The system of linear inequalities in (7) defines the planar region S shown in Figure 11. Each point in S is a candidate for the solution of the problem at hand and is referred to as a **feasible solution.** The set S itself is referred to as a **feasible set.** Our goal is to find, from among all the points in the set S, the point(s) that optimizes the objective function P. Such a feasible solution is called an **optimal solution** and constitutes the solution to the linear programming problem under consideration.

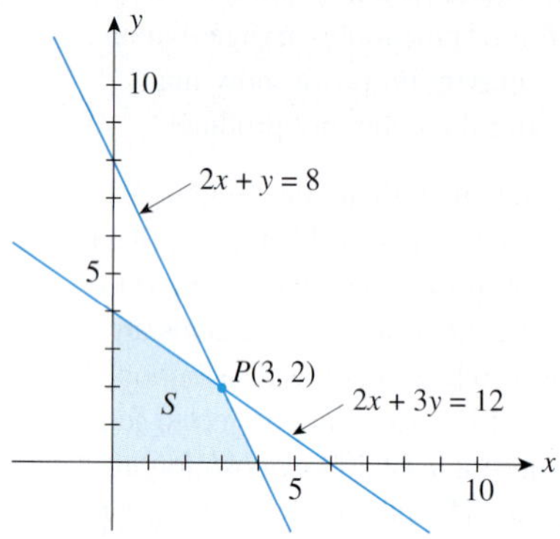

FIGURE **11**
Each point in the feasible set S is a candidate for the optimal solution.

As was noted earlier, each point $P(x, y)$ in S is a candidate for the optimal solution to the problem at hand. For example, the point $(1, 3)$ is easily seen to lie in S and is therefore in the running. The value of the objective function P at the point $(1, 3)$ is given by $P = 3(1) + 2(3) = 9$. Now, if we could compute the value of P corresponding to each point in S, then the point(s) in S that gave the largest value to P would constitute the solution set sought. Unfortunately, in most problems, the number of candidates either is too large or, as in this problem, is infinite. Therefore, this method is at best unwieldy and at worst impractical.

Let's turn the question around. Instead of asking for the value of the objective function P at a feasible point, let's assign a value to the objective function P and ask whether there are feasible points that would correspond to the given value of P. Toward this end, suppose we assign a value of 6 to P. Then the objective function P becomes $3x + 2y = 6$, a linear equation in x and y; thus, it has a graph that is a straight line L_1 in the plane. In Figure 12, we have drawn the graph of this straight line superimposed on the feasible set S.

FIGURE **12**
A family of parallel lines that intersect the feasible set S

It is clear that each point on the straight-line segment given by the intersection of the straight line L_1 and the feasible set S corresponds to the given value, 6, of P. For this reason, the line L_1 is called an **isoprofit line.** Let's repeat the process, this time assigning a value of 10 to P. We obtain the equation $3x + 2y = 10$ and the line L_2 (see Figure 12), which suggests that there are feasible points that correspond to a larger value of P. Observe that the line L_2 is parallel to the line L_1 because both lines have slope equal to $-\frac{3}{2}$, which is easily seen by casting the corresponding equations in the slope-intercept form.

In general, by assigning different values to the objective function, we obtain a family of parallel lines, each with slope equal to $-\frac{3}{2}$. Furthermore, a line corresponding to a larger value of P lies farther away from the origin than one with a smaller value of P. The implication is clear. To obtain the optimal solution(s) to the problem at hand, find the straight line, from this family of straight lines, that is farthest from the origin and still intersects the feasible set S. The required line is the one that passes through the point $P(3, 2)$ (see Figure 12), so the solution to the problem is given by $x = 3$, $y = 2$, resulting in a maximum value of $P = 3(3) + 2(2) = 13$.

That the optimal solution to this problem was found to occur at a vertex of the feasible set S is no accident. In fact, the result is a consequence of the following basic theorem on linear programming, which we state without proof.

THEOREM 1

Solution(s) of Linear Programming Problems

If a linear programming problem has a solution, then it must occur at a vertex, or corner point, of the feasible set S associated with the problem.

Furthermore, if the objective function P is optimized at two adjacent vertices of S, then it is optimized at every point on the line segment joining these vertices, in which case there are infinitely many solutions to the problem.

Theorem 1 tells us that our search for the solution(s) to a linear programming problem may be restricted to the examination of the set of vertices of the feasible set S associated with the problem. Since a feasible set S has finitely many vertices, the theorem suggests that the solution(s) to the linear programming problem may be found by inspecting the values of the objective function P at these vertices.

Although Theorem 1 sheds some light on the nature of the solution of a linear programming problem, it does not tell us when a linear programming problem has a solution. The following theorem states conditions that guarantee when a solution exists.

THEOREM 2

Existence of a Solution

Suppose we are given a linear programming problem with a feasible set S and an objective function $P = ax + by$.

a. If S is bounded, then P has both a maximum and a minimum value on S.

b. If S is unbounded and both a and b are nonnegative, then P has a minimum value on S provided that the constraints defining S include the inequalities $x \geq 0$ and $y \geq 0$.

c. If S is the empty set, then the linear programming problem has no solution; that is, P has neither a maximum nor a minimum value.

The **method of corners,** a simple procedure for solving linear programming problems based on Theorem 1, follows.

The Method of Corners

1. Graph the feasible set.
2. Find the coordinates of all corner points (vertices) of the feasible set.
3. Evaluate the objective function at each corner point.
4. Find the vertex that renders the objective function a maximum (minimum). If there is only one such vertex, then this vertex constitutes a unique solution to the problem. If the objective function is maximized (minimized) at two adjacent corner points of S, there are infinitely many optimal solutions given by the points on the line segment determined by these two vertices.

APPLIED EXAMPLE 1 Maximizing Profit We are now in a position to complete the solution to the production problem posed in Example 1, Section 3.2. Recall that the mathematical formulation led to the following linear programming problem:

$$\begin{aligned} \text{Maximize} \quad & P = x + 1.2y \\ \text{subject to} \quad & 2x + y \le 180 \\ & x + 3y \le 300 \\ & x \ge 0, y \ge 0 \end{aligned}$$

Solution The feasible set S for the problem is shown in Figure 13.

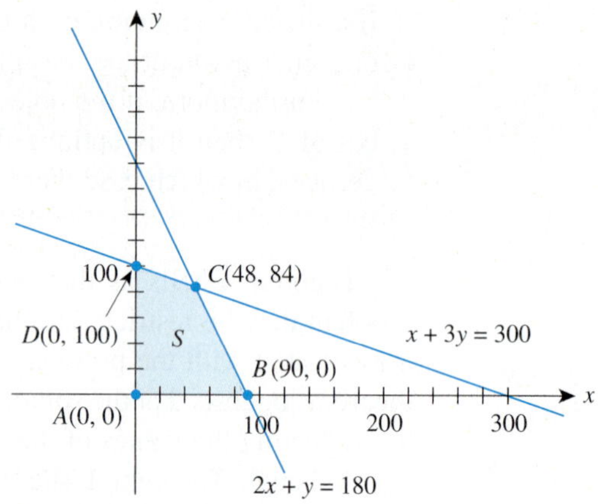

FIGURE 13
The corner point that yields the maximum profit is $C(48, 84)$.

The vertices of the feasible set are $A(0, 0)$, $B(90, 0)$, $C(48, 84)$, and $D(0, 100)$. The values of P at these vertices may be tabulated as follows:

Vertex	$P = x + 1.2y$
$A(0, 0)$	0
$B(90, 0)$	90
$C(48, 84)$	148.8
$D(0, 100)$	120

From the table, we see that the maximum of $P = x + 1.2y$ occurs at the vertex $(48, 84)$ and has a value of 148.8. Recalling what the symbols x, y, and P represent, we conclude that Ace Novelty would maximize its profit ($148.80) by producing 48 Type A souvenirs and 84 Type B souvenirs.

Explore and Discuss

Consider the linear programming problem

$$\begin{aligned} \text{Maximize} \quad & P = 4x + 3y \\ \text{subject to} \quad & 2x + y \le 10 \\ & 2x + 3y \le 18 \\ & x \ge 0, y \ge 0 \end{aligned}$$

1. Sketch the feasible set S for the linear programming problem.
2. Draw the isoprofit lines superimposed on S corresponding to $P = 12$, 16, 20, and 24, and show that these lines are parallel to each other.
3. Show that the solution to the linear programming problem is $x = 3$ and $y = 4$. Is this result the same as that found by using the method of corners?

APPLIED EXAMPLE 2 A Nutrition Problem Complete the solution of the nutrition problem posed in Example 2, Section 3.2.

Solution Recall that the mathematical formulation of the problem led to the following linear programming problem in two variables:

$$\begin{aligned} \text{Minimize} \quad & C = 6x + 8y \\ \text{subject to} \quad & 40x + 10y \geq 2400 \\ & 10x + 15y \geq 2100 \\ & 5x + 15y \geq 1500 \\ & x \geq 0, y \geq 0 \end{aligned}$$

The feasible set S defined by the system of constraints is shown in Figure 14.

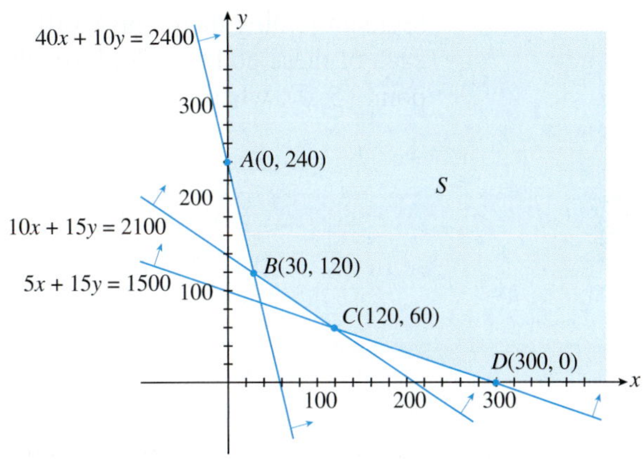

FIGURE **14**
The corner point that yields the minimum cost is $B(30, 120)$.

The vertices of the feasible set S are $A(0, 240)$, $B(30, 120)$, $C(120, 60)$, and $D(300, 0)$. The values of the objective function C at these vertices are given in the following table:

Vertex	$C = 6x + 8y$
$A(0, 240)$	1920
$B(30, 120)$	1140
$C(120, 60)$	1200
$D(300, 0)$	1800

From the table, we can see that the minimum for the objective function $C = 6x + 8y$ occurs at the vertex $B(30, 120)$ and has a value of 1140. Thus, the individual should purchase 30 Brand A pills and 120 Brand B pills at a minimum cost of $11.40.

EXAMPLE 3 A Linear Programming Problem with Multiple Solutions Find the maximum and minimum of $P = 2x + 3y$ subject to the following system of linear inequalities:

$$\begin{aligned} 2x + 3y &\leq 30 \\ -x + y &\leq 5 \\ x + y &\geq 5 \\ x &\leq 10 \\ x \geq 0, y &\geq 0 \end{aligned}$$

Solution The feasible set S is shown in Figure 15. The vertices of the feasible set S are $A(5, 0)$, $B(10, 0)$, $C(10, \frac{10}{3})$, $D(3, 8)$, and $E(0, 5)$. The values of the objective function P at these vertices are given in the following table:

Vertex	$P = 2x + 3y$
$A(5, 0)$	10
$B(10, 0)$	20
$C(10, \frac{10}{3})$	30
$D(3, 8)$	30
$E(0, 5)$	15

From the table, we see that the maximum for the objective function $P = 2x + 3y$ occurs at the vertices $C(10, \frac{10}{3})$ and $D(3, 8)$. This tells us that every point on the line segment joining the points $C(10, \frac{10}{3})$ and $D(3, 8)$ maximizes P. The value of P at each of these points is 30. From the table, it is also clear that P is minimized at the point $(5, 0)$, where it attains a value of 10.

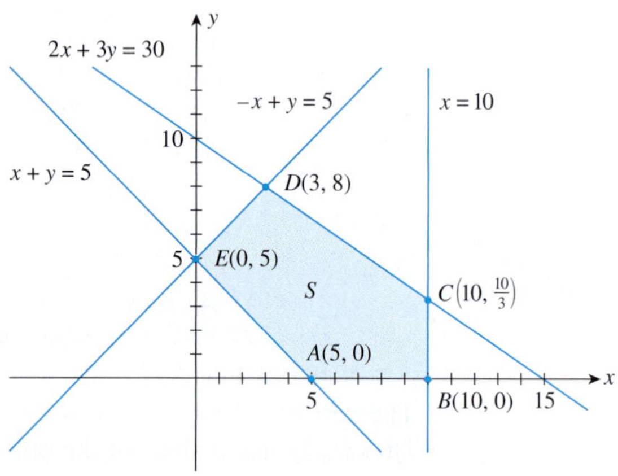

FIGURE **15**
Every point lying on the line segment joining C and D maximizes P.

Explore and Discuss

Consider the linear programming problem

$$\begin{aligned} \text{Maximize} \quad & P = 2x + 3y \\ \text{subject to} \quad & 2x + y \leq 10 \\ & 2x + 3y \leq 18 \\ & x \geq 0, y \geq 0 \end{aligned}$$

1. Sketch the feasible set S for the linear programming problem.
2. Draw the isoprofit lines superimposed on S corresponding to $P = 6, 8, 12$, and 18, and show that these lines are parallel to each other.
3. Show that there are infinitely many solutions to the problem. Is this result as predicted by the method of corners?

We close this section by examining two situations in which a linear programming problem may have no solution.

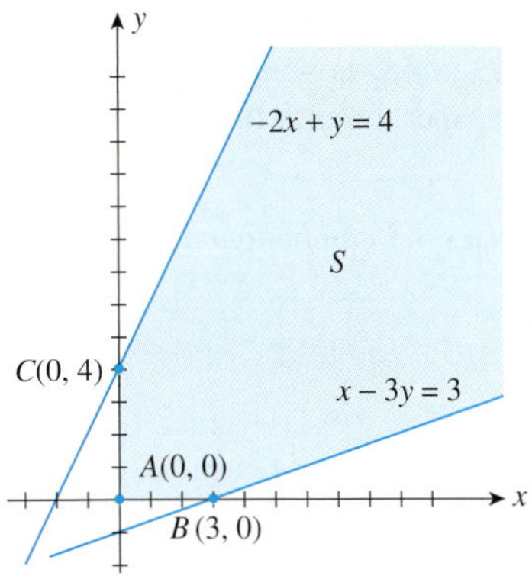

FIGURE 16
This maximization problem has no solution because the feasible set is unbounded.

EXAMPLE 4 An Unbounded Linear Programming Problem with No Solution
Solve the following linear programming problem:

$$\begin{aligned} \text{Maximize} \quad & P = x + 2y \\ \text{subject to} \quad & -2x + y \le 4 \\ & x - 3y \le 3 \\ & x \ge 0, y \ge 0 \end{aligned}$$

Solution The feasible set S for this problem is shown in Figure 16. Since the set S is unbounded (both x and y can take on arbitrarily large positive values), we see that we can make P as large as we please by choosing x and y large enough. This problem has no solution. The problem is said to be unbounded.

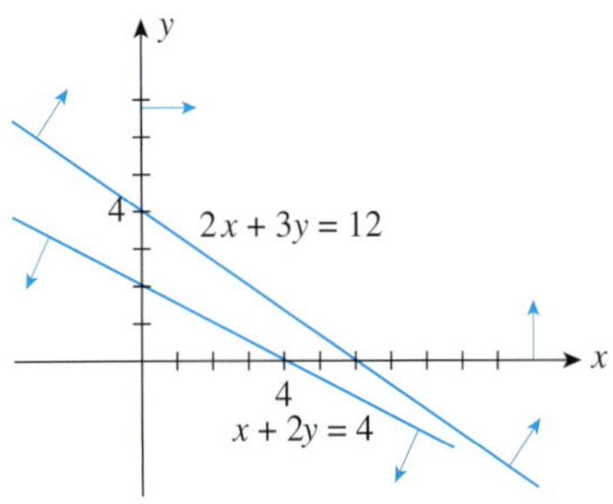

FIGURE 17
This problem is inconsistent because there is no point that satisfies all of the given inequalities.

EXAMPLE 5 An Infeasible Linear Programming Problem Solve the following linear programming problem:

$$\begin{aligned} \text{Maximize} \quad & P = x + 2y \\ \text{subject to} \quad & x + 2y \le 4 \\ & 2x + 3y \ge 12 \\ & x \ge 0, y \ge 0 \end{aligned}$$

Solution The half-planes described by the constraints (inequalities) have no points in common (Figure 17). Hence, there are no feasible points, and the problem has no solution. In this situation, we say that the problem is **infeasible,** or **inconsistent.**

The situations described in Examples 4 and 5 are unlikely to occur in well-posed problems arising from practical applications of linear programming.

The method of corners is particularly effective in solving two-variable linear programming problems with a small number of constraints, as the preceding examples have amply demonstrated. However, its effectiveness decreases rapidly as the number of variables and/or constraints increases. For example, it may be shown that a linear programming problem in three variables and five constraints may have up to ten feasible corner points. The determination of the feasible corner points calls for the solution of ten 3×3 systems of linear equations and then the verification—by the substitution of each of these solutions into the system of constraints—to see whether it is, in fact, a feasible point. When the number of variables and constraints goes up to five and ten, respectively (still a very small system from the standpoint of applications in economics), the number of vertices to be found and checked for feasible corner points increases dramatically to 252, and each of these vertices is found by solving a 5×5 linear system! For this reason, the method of corners is seldom used to solve linear programming problems; its redeeming value lies in the fact that much insight is gained into the nature of the solutions of linear programming problems through its use in solving two-variable problems.

3.3 Self-Check Exercises

1. Use the method of corners to solve the following linear programming problem:

$$\begin{aligned} \text{Maximize} \quad & P = 4x + 5y \\ \text{subject to} \quad & x + 2y \le 10 \\ & 5x + 3y \le 30 \\ & x \ge 0, y \ge 0 \end{aligned}$$

2. Use the method of corners to solve the following linear programming problem:

$$\begin{aligned} \text{Minimize} \quad & C = 5x + 3y \\ \text{subject to} \quad & 5x + 3y \ge 30 \\ & x - 3y \le 0 \\ & x \ge 2 \end{aligned}$$

3. **Optimizing Advertising Exposure** Gino Balduzzi, proprietor of Luigi's Pizza Palace, allocates $9000 a month for advertising in two newspapers, the *City Tribune* and the *Daily News*. The *City Tribune* charges $300 for a certain advertisement, whereas the *Daily News* charges $100 for the same ad. Gino has stipulated that the ad is to appear in at least 15 but no more than 30 editions of the *Daily News* per month. The *City Tribune* has a daily circulation of 50,000, and the *Daily News* has a circulation of 20,000. Under these conditions, determine how many ads Gino should place in each newspaper to reach the largest number of readers.

Solutions to Self-Check Exercises 3.3 can be found on page 201.

3.3 Concept Questions

1. **a.** What is the feasible set associated with a linear programming problem?
   b. What is a feasible solution of a linear programming problem?
   c. What is an optimal solution of a linear programming problem?
2. Describe the method of corners.

3.3 Exercises

In Exercises 1–6, find the maximum and/or minimum value(s) of the objective function on the feasible set S.

1. $Z = 2x + 3y$

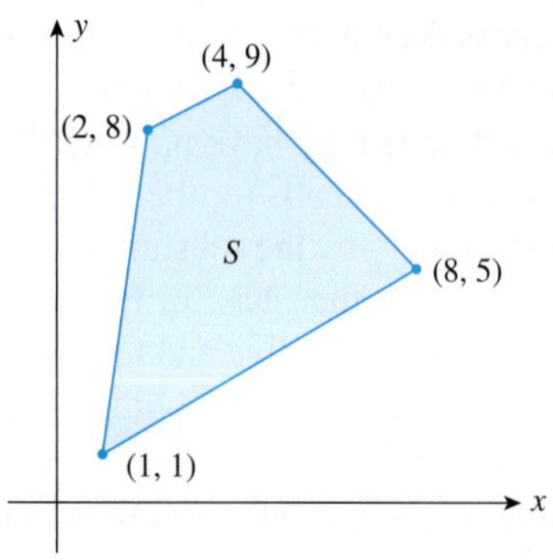

2. $Z = 3x - y$

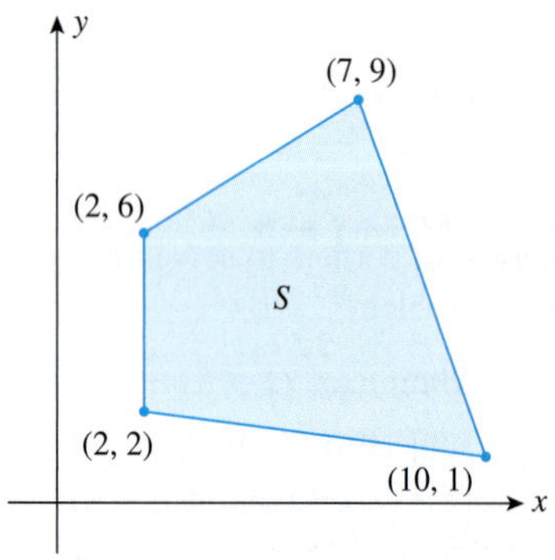

3. $Z = 2x + 3y$

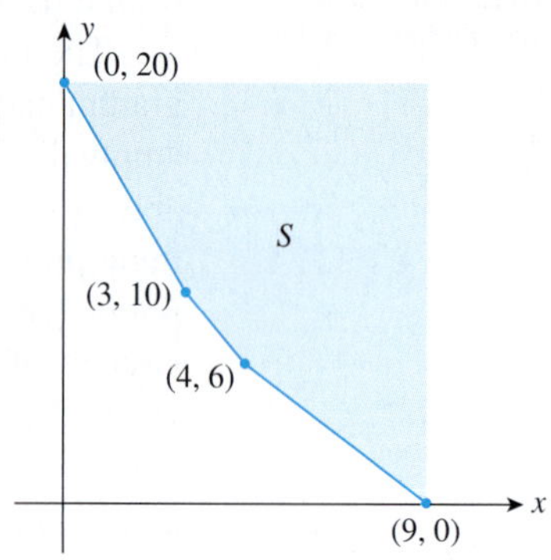

4. $Z = 7x + 9y$

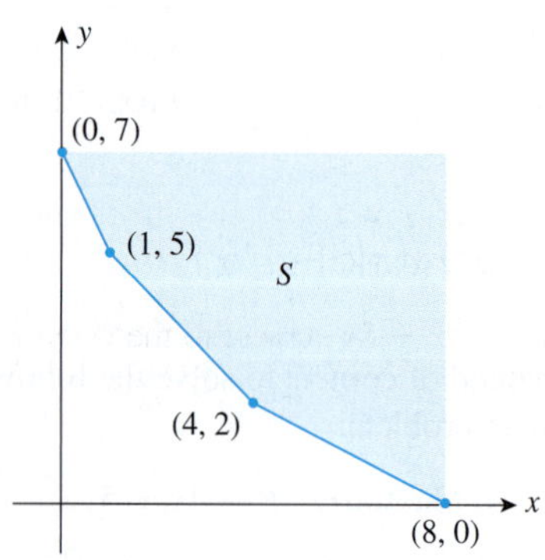

5. $Z = x + 4y$

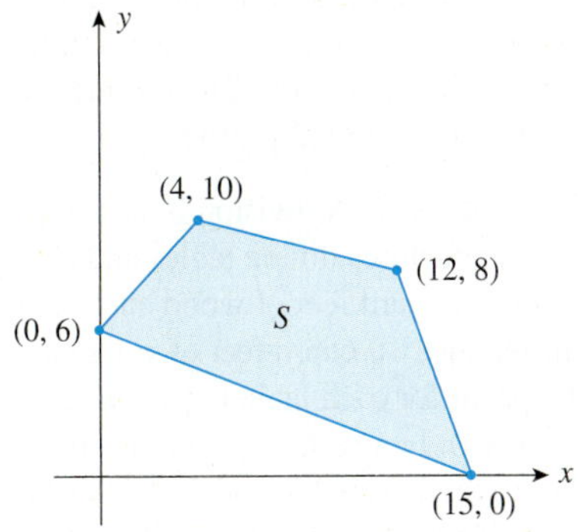

6. $Z = 3x + 2y$

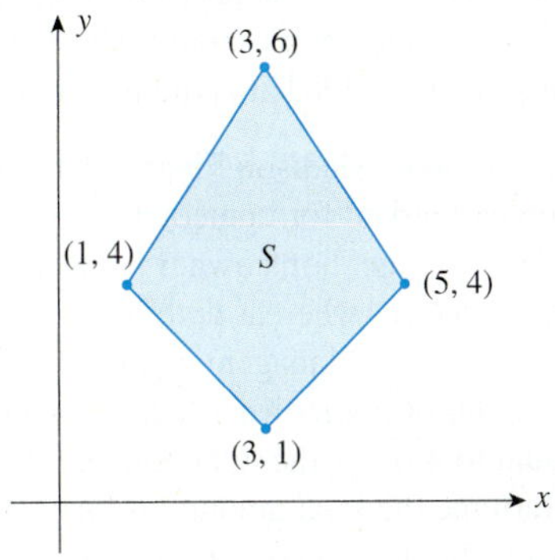

In Exercises 7–28, solve each linear programming problem by the method of corners.

7. Maximize $P = 3x + 2y$
subject to
$$\begin{aligned} x + y &\le 6 \\ x &\le 3 \\ x \ge 0, y &\ge 0 \end{aligned}$$

8. Maximize $P = x + 2y$
subject to
$$\begin{aligned} x + y &\le 4 \\ 2x + y &\le 5 \\ x \ge 0, y &\ge 0 \end{aligned}$$

9. Maximize $P = 2x + y$ subject to the constraints of Exercise 8.

10. Maximize $P = 4x + 2y$
subject to
$$\begin{aligned} x + y &\le 8 \\ 2x + y &\le 10 \\ x \ge 0, y &\ge 0 \end{aligned}$$

11. Maximize $P = x + 8y$ subject to the constraints of Exercise 10.

12. Maximize $P = 3x - 4y$
subject to
$$\begin{aligned} x + 3y &\le 15 \\ 4x + y &\le 16 \\ x \ge 0, y &\ge 0 \end{aligned}$$

13. Maximize $P = x + 3y$
subject to
$$\begin{aligned} 2x + y &\le 6 \\ x + y &\le 4 \\ x &\le 1 \\ x \ge 0, y &\ge 0 \end{aligned}$$

14. Maximize $P = 2x + 5y$
subject to
$$\begin{aligned} 2x + y &\le 16 \\ 2x + 3y &\le 24 \\ y &\le 6 \\ x \ge 0, y &\ge 0 \end{aligned}$$

15. Minimize $C = 2x + 5y$
subject to
$$\begin{aligned} x + y &\ge 3 \\ x + 2y &\ge 4 \\ x \ge 0, y &\ge 0 \end{aligned}$$

16. Minimize $C = 2x + 4y$ subject to the constraints of Exercise 15.

17. Minimize $C = 3x + 6y$
subject to
$$\begin{aligned} x + 2y &\ge 40 \\ x + y &\ge 30 \\ x \ge 0, y &\ge 0 \end{aligned}$$

18. Minimize $C = 3x + y$ subject to the constraints of Exercise 17.

19. Minimize $C = 2x + 10y$
subject to
$$\begin{aligned} 5x + 2y &\ge 40 \\ x + 2y &\ge 20 \\ y \ge 3, x &\ge 0 \end{aligned}$$

20. Minimize $C = 2x + 5y$
subject to
$$\begin{aligned} 4x + y &\ge 40 \\ 2x + y &\ge 30 \\ x + 3y &\ge 30 \\ x \ge 0, y &\ge 0 \end{aligned}$$

21. Minimize $C = 10x + 15y$
subject to
$$\begin{aligned} x + y &\le 10 \\ 3x + y &\ge 12 \\ -2x + 3y &\ge 3 \\ x \ge 0, y &\ge 0 \end{aligned}$$

22. Maximize $P = 2x + 5y$ subject to the constraints of Exercise 21.

23. Maximize $P = 3x + 4y$
subject to
$$\begin{aligned} x + 2y &\le 50 \\ 5x + 4y &\le 145 \\ 2x + y &\ge 25 \\ y \ge 5, x &\ge 0 \end{aligned}$$

24. Maximize $P = 4x - 3y$ subject to the constraints of Exercise 23.

25. Maximize $P = 2x + 3y$
subject to
$$\begin{aligned} x + y &\le 48 \\ x + 3y &\ge 60 \\ 9x + 5y &\le 320 \\ x \ge 10, y &\ge 0 \end{aligned}$$

26. Minimize $C = 5x + 3y$ subject to the constraints of Exercise 25.

27. Find the maximum and minimum of $P = 8x + 5y$ subject to

$$5x + 2y \geq 63$$
$$x + y \geq 18$$
$$3x + 2y \leq 51$$
$$x \geq 0, y \geq 0$$

28. Find the maximum and minimum of $P = 4x + 3y$ subject to

$$3x + 5y \geq 20$$
$$3x + y \leq 16$$
$$-2x + y \leq 1$$
$$x \geq 0, y \geq 0$$

The problems in Exercises 29–48 correspond to those in Exercises 1–20, Section 3.2. Use the results of your previous work to help you solve these problems.

29. **Production Scheduling** A company manufactures two products, *A* and *B*, on two machines, I and II. It has been determined that the company will realize a profit of \$3/unit of Product *A* and a profit of \$4/unit of Product *B*. To manufacture a unit of Product *A* requires 6 min on Machine I and 5 min on Machine II. To manufacture a unit of Product *B* requires 9 min on Machine I and 4 min on Machine II. There are 5 hr of machine time available on Machine I and 3 hr of machine time available on Machine II in each work shift. How many units of each product should be produced in each shift to maximize the company's profit? What is the optimal profit?

30. **Production Scheduling** National Business Machines manufactures two models of portable printers: A and B. Each model A costs \$100 to make, and each model B costs \$150. The profits are \$30 for each model A and \$40 for each model B portable printer. If the total number of portable printers demanded per month does not exceed 2500 and the company has earmarked no more than \$600,000/month for manufacturing costs, how many units of each model should National make each month to maximize its monthly profit? What is the optimal profit?

31. **Production Scheduling** Kane Manufacturing has a division that produces two models of fireplace grates, model A and model B. To produce each model A grate requires 3 lb of cast iron and 6 min of labor. To produce each model B grate requires 4 lb of cast iron and 3 min of labor. The profit for each model A grate is \$2.00, and the profit for each model B grate is \$1.50. If 1000 lb of cast iron and 20 labor-hours are available for the production of fireplace grates per day, how many grates of each model should the division produce to maximize Kane's profit? What is the optimal profit?

32. **Production Scheduling** Refer to Exercise 31. Because of a backlog of orders for model A grates, Kane's manager had decided to produce at least 150 of these grates a day. Operating under this additional constraint, how many grates of each model should Kane produce to maximize profit? What is the optimal profit?

33. **Production Scheduling** A division of the Winston Furniture Company manufactures dining tables and chairs. Each table requires 40 board feet of wood and 3 labor-hours. Each chair requires 16 board feet of wood and 4 labor-hours. The profit for each table is \$45, and the profit for each chair is \$20. In a certain week, the company has 3200 board feet of wood available and 520 labor-hours available. How many tables and chairs should Winston manufacture to maximize its profit? What is the maximum profit?

34. **Production Scheduling** Refer to Exercise 33. If the profit for each table is \$50 and the profit for each chair is \$18, how many tables and chairs should Winston manufacture to maximize its profit? What is the maximum profit?

35. **Allocation of Funds** Madison Finance has a total of \$20 million earmarked for homeowner loans and auto loans. On the average, homeowner loans have a 10% annual rate of return, whereas auto loans yield a 12% annual rate of return. Management has also stipulated that the total amount of homeowner loans should be greater than or equal to 4 times the total amount of automobile loans. Determine the total amount of loans of each type that Madison should extend to each category to maximize its returns. What are the optimal returns?

36. **Asset Allocation** A financier plans to invest up to \$500,000 in two projects. Project *A* yields a return of 10% on the investment, whereas Project *B* yields a return of 15% on the investment. Because the investment in Project *B* is riskier than the investment in Project *A*, the financier has decided that the investment in Project *B* should not exceed 40% of the total investment. How much should she invest in each project to maximize the return on her investment? What is the maximum return?

37. **Asset Allocation** Justin has decided to invest at most \$60,000 in medium-risk and high-risk stocks. He has further decided that the medium-risk stocks should make up at least 40% of the total investment, while the high-risk stocks should make up at least 20% of the total investment. He expects that the medium-risk stocks will appreciate by 12% and the high-risk stocks by 20% within a year. How much money should Justin invest in each type of stock to maximize the value of his investment? What is the maximum return?

38. **Crop Planning** A farmer plans to plant two crops, *A* and *B*. The cost of cultivating Crop *A* is \$40/acre whereas the cost of cultivating Crop *B* is \$60/acre. The farmer has a maximum of \$7400 available for land cultivation. Each acre of Crop *A* requires 20 labor-hours, and each acre of Crop *B* requires 25 labor-hours. The farmer has a

maximum of 3300 labor-hours available. If she expects to make a profit of \$150/acre on Crop *A* and \$200/acre on Crop *B*, how many acres of each crop should she plant to maximize her profit? What is the optimal profit?

39. **Minimizing Mining Costs** Perth Mining Company operates two mines for the purpose of extracting gold and silver. The Saddle Mine costs \$14,000/day to operate, and it yields 50 oz of gold and 3000 oz of silver each day. The Horseshoe Mine costs \$16,000/day to operate, and it yields 75 oz of gold and 1000 oz of silver each day. Company management has set a target of at least 650 oz of gold and 18,000 oz of silver. How many days should each mine be operated so that the target can be met at a minimum cost? What is the minimum cost?

40. **Minimizing Cruise Line Costs** Deluxe River Cruises operates a fleet of river vessels. The fleet has two types of vessels: A type A vessel has 60 deluxe cabins and 160 standard cabins, whereas a type B vessel has 80 deluxe cabins and 120 standard cabins. Under a charter agreement with Odyssey Travel Agency, Deluxe River Cruises is to provide Odyssey with a minimum of 360 deluxe and 680 standard cabins for their 15-day cruise in May. It costs \$44,000 to operate a type A vessel and \$54,000 to operate a type B vessel for that period. How many of each type vessel should be used to keep the operating costs to a minimum? What is the minimum cost?

41. **Production Scheduling** Acoustical manufactures a DVD storage cabinet that can be bought fully assembled or as a kit. Each cabinet is processed in the fabrications department and the assembly department. If the fabrication department manufactures only fully assembled cabinets, then it can produce 200 units/day; and if it manufactures only kits, it can produce 200 units/day. If the assembly department produces only fully assembled cabinets, then it can produce 100 units/day; but if it produces only kits, then it can produce 300 units/day. Each fully assembled cabinet contributes \$50 to the profits of the company whereas each kit contributes \$40 to its profits. How many fully assembled units and how many kits should the company produce per day to maximize its profit? What is the optimal profit?

42. **Fertilizers** A farmer uses two types of fertilizers. A 50-lb bag of Fertilizer *A* contains 8 lb of nitrogen, 2 lb of phosphorus, and 4 lb of potassium. A 50-lb bag of fertilizer B contains 5 lb each of nitrogen, phosphorus, and potassium. The minimum requirements for a field are 440 lb of nitrogen, 260 lb of phosphorus, and 360 lb of potassium. If a 50-lb bag of fertilizer A costs \$30 and a 50-lb bag of fertilizer B costs \$20, find the amount of each type of fertilizer the farmer should use to minimize his cost while still meeting the minimum requirements. What is the minimum cost?

43. **Minimizing City Water Costs** The water-supply manager for a Midwestern city needs to supply the city with at least 10 million gallons of potable (drinkable) water per day. The supply may be drawn from the local reservoir or from a pipeline to an adjacent town. The local reservoir has a maximum daily yield of 5 million gallons of potable water, and the pipeline has a maximum daily yield of 10 million gallons. By contract, the pipeline is required to supply a minimum of 6 million gallons/day. If the cost for 1 million gallons of reservoir water is \$300 and that for pipeline water is \$500, how much water should the manager get from each source to minimize daily water costs for the city? What is the minimum daily cost?

44. **Production Scheduling** Ace Novelty manufactures Giant Pandas and Saint Bernards. Each Panda requires 1.5 yd^2 of plush, 30 ft^3 of stuffing, and 5 pieces of trim; each Saint Bernard requires 2 yd^2 of plush, 35 ft^3 of stuffing, and 8 pieces of trim. The profit for each Panda is \$10, and the profit for each Saint Bernard is \$15. If 3600 yd^2 of plush, 66,000 ft^3 of stuffing, and 13,600 pieces of trim are available, how many of each of the stuffed animals should the company manufacture to maximize profit? What is the maximum profit?

45. **Diet Planning** A nutritionist at the Medical Center has been asked to prepare a special diet for certain patients. She has decided that the meals should be prepared from Foods *A* and *B* and that they should contain a minimum of 400 mg of calcium, 10 mg of iron, and 40 mg of vitamin C. Each ounce of Food *A* contains 30 mg of calcium, 1 mg of iron, 2 mg of vitamin C, and 2 mg of cholesterol. Each ounce of Food *B* contains 25 mg of calcium, 0.5 mg of iron, 5 mg of vitamin C, and 5 mg of cholesterol. How many ounces of each type of food should be used in a meal so that the cholesterol content is minimized and the minimum requirements of calcium, iron, and vitamin C are met? What is the minimum cholesterol content?

46. **Optimizing Advertising Exposure** Everest Deluxe World Travel has decided to advertise in the Sunday editions of two major newspapers in town. These advertisements are directed at three groups of potential customers. Each advertisement in Newspaper I is seen by 70,000 Group A customers, 40,000 Group B customers, and 20,000 Group C customers. Each advertisement in Newspaper II is seen by 10,000 Group A, 20,000 Group B, and 40,000 Group C customers. Each advertisement in Newspaper I costs \$1000, and each advertisement in Newspaper II costs \$800. Everest would like their advertisements to be read by at least 2 million people from Group A, 1.4 million people from Group B, and 1 million people from Group C. How many advertisements should Everest place in each newspaper to achieve its advertising goals at a minimum cost? What is the minimum cost?
Hint: Use different scales for drawing the feasible set.

47. **Minimizing—Shipping Costs** TMA manufactures 37-in. high definition LCD televisions in two separate locations, Locations I and II. The output at Location I is at most 6000 televisions/month, whereas the output at Location II

is at most 5000 televisions/month. TMA is the main supplier of televisions to the Pulsar Corporation, its holding company, which has priority in having all its requirements met. In a certain month, Pulsar placed orders for 3000 and 4000 televisions to be shipped to two of its factories located in City *A* and City *B*, respectively. The shipping costs (in dollars) per television from the two TMA plants to the two Pulsar factories are as follows:

	To Pulsar Factories	
From TMA	**City *A***	**City *B***
Location I	\$6	\$4
Location II	\$8	\$10

Find a shipping schedule that meets the requirements of both companies while keeping costs to a minimum.

48. **Social Programs Planning** AntiFam, a hunger-relief organization, has earmarked between \$2 and \$2.5 million (inclusive) for aid to two African countries, Country *A* and Country *B*. Country *A* is to receive between \$1 million and \$1.5 million (inclusive), and Country *B* is to receive at least \$0.75 million. It has been estimated that each dollar spent in Country *A* will yield an effective return of \$0.60, whereas a dollar spent in Country *B* will yield an effective return of \$0.80. How should the aid be allocated if the money is to be utilized most effectively according to these criteria?
Hint: If x and y denote the amount of money to be given to Country *A* and Country *B*, respectively, then the objective function to be maximized is $P = 0.6x + 0.8y$.

49. Complete the solution to Example 3, Section 3.2.

50. **Veterinary Science** A veterinarian has been asked to prepare a diet for a group of dogs to be used in a nutrition study at the School of Animal Science. It has been stipulated that each serving should be no larger than 8 oz and must contain at least 29 units of Nutrient I and 20 units of Nutrient II. The vet has decided that the diet may be prepared from two brands of dog food: Brand *A* and Brand *B*. Each ounce of Brand *A* contains 3 units of Nutrient I and 4 units of Nutrient II. Each ounce of Brand *B* contains 5 units of Nutrient I and 2 units of Nutrient II. Brand *A* costs 3 cents/oz, and Brand *B* costs 4 cents/oz. Determine how many ounces of each brand of dog food should be used per serving to meet the given requirements at a minimum cost.

51. **Maximizing Investment Returns** Patricia has at most \$30,000 to invest in securities in the form of corporate stocks. She has narrowed her choices to two groups of stocks: growth stocks that she assumes will yield a 15% return (dividends and capital appreciation) within a year and speculative stocks that she assumes will yield a 25% return (mainly in capital appreciation) within a year. Determine how much she should invest in each group of stocks to maximize the return on her investments within a year if she has decided to invest at least 3 times as much in growth stocks as in speculative stocks. What is the maximum return?

52. **Production Scheduling** Bata Aerobics manufactures two models of steppers used for aerobic exercises. Manufacturing each luxury model requires 10 lb of plastic and 10 min of labor. Manufacturing each standard model requires 16 lb of plastic and 8 min of labor. The profit for each luxury model is \$40, and the profit for each standard model is \$30. If 6000 lb of plastic and 60 labor-hours are available for the production of the steppers per day, how many steppers of each model should Bata produce each day to maximize its profit? What is the optimal profit?

53. **Market Research** Trendex, a telephone survey company, has been hired to conduct a television-viewing poll among urban and suburban families in the Los Angeles area. The client has stipulated that a maximum of 1500 families is to be interviewed. At least 500 urban families must be interviewed, and at least half of the total number of families interviewed must be from the suburban area. For this service, Trendex will be paid \$6000 plus \$8 for each completed interview. From previous experience, Trendex has determined that it will incur an expense of \$4.40 for each successful interview with an urban family and \$5 for each successful interview with a suburban family. How many urban and suburban families should Trendex interview to maximize its profit? What is the optimal profit?

In Exercises 54–57, determine whether the statement is true or false. If it is true, explain why it is true. If it is false, give an example to show why it is false.

54. An optimal solution of a linear programming problem is a feasible solution, but a feasible solution of a linear programming problem need not be an optimal solution.

55. An optimal solution of a linear programming problem can occur inside the feasible set of the problem.

56. If a maximization problem has no solution, then the feasible set associated with the linear programming problem must be unbounded.

57. Suppose you are given the following linear programming problem: Maximize $P = ax + by$ on the unbounded feasible set S shown in the accompanying figure.

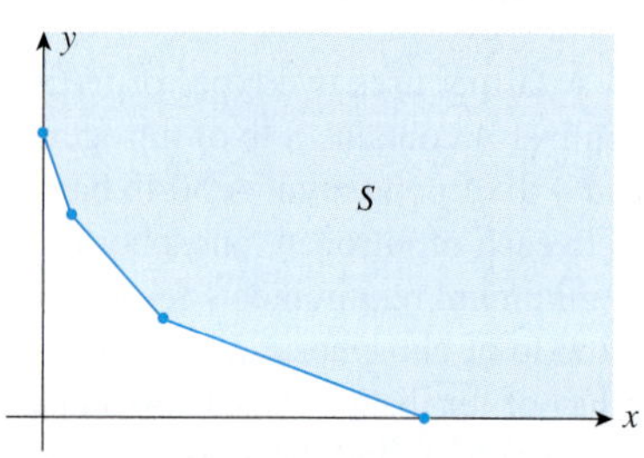

a. If $a > 0$ or $b > 0$, then the linear programming problem has no optimal solution.
b. If $a \le 0$ and $b \le 0$, then the linear programming problem has at least one optimal solution.

58. Suppose you are given the following linear programming problem: Maximize $P = ax + by$, where $a > 0$ and $b > 0$, on the feasible set S shown in the accompanying figure.

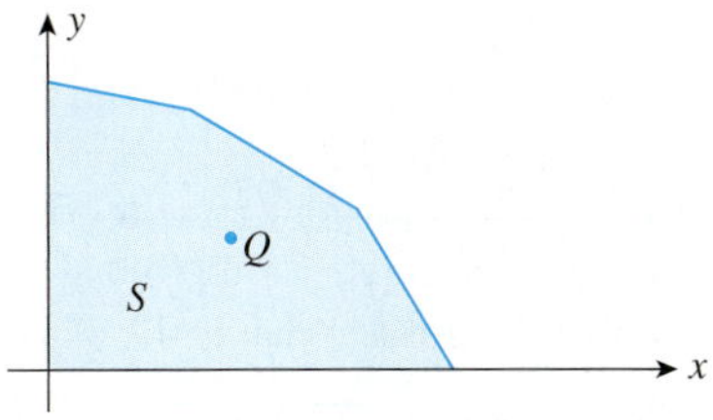

Explain, without using Theorem 1, why the optimal solution of the linear programming problem cannot occur at the point Q.

59. Suppose you are given the following linear programming problem: Maximize $P = ax + by$, where $a > 0$ and $b > 0$, on ... shown in the accompanying figure.

60. Consider the linear programming problem

$$\begin{aligned} \text{Maximize} \quad & P = 2x + 7y \\ \text{subject to} \quad & 2x + y \geq 8 \\ & x + y \geq 6 \\ & x \geq 0, y \geq 0 \end{aligned}$$

a. Sketch the feasible set S.
b. Find the corner points of S.
c. Find the values of P at the corner points of S found in part (b).
d. Show that the linear programming problem has no optimal solution. Does this contradict Theorem 2?

61. Consider the linear programming problem

$$\begin{aligned} \text{Minimize} \quad & C = -2x + 5y \\ \text{subject to} \quad & x + y \leq 3 \\ & 2x + y \leq 4 \\ & 5x + 8y \geq 40 \\ & x \geq 0, y \geq 0 \end{aligned}$$

...tch the feasible set.
...d the solution(s) of the linear programming ...blem, if it exists.

...ider the linear programming problem

$$\begin{aligned} \text{Minimize} \quad & C = x - 4y \\ \text{subject to} \quad & x - 3y \leq -3 \\ & 2x - y \leq 4 \\ & x \geq 0, y \geq 0 \end{aligned}$$

Sketch the feasible set S.
Show that the linear programming problem has an optimal solution. Does this contradict Theorem 2? Explain.

The values of the objective function P at the vertices of S are summarized in the following table:

Vertex	$P = 4x + 5y$
$A(0, 0)$	0
$B(6, 0)$	24
$C(\frac{30}{7}, \frac{20}{7})$	$\frac{220}{7} = 31\frac{3}{7}$
$D(0, 5)$	25

From the table, we see that the maximum for the objective function P is attained at the vertex $C(\frac{30}{7}, \frac{20}{7})$. Therefore, the solution to the problem is $x = \frac{30}{7}$, $y = \frac{20}{7}$, and $P = 31\frac{3}{7}$.

2. The feasible set S for the problem was graphed in the solution to Exercise 2, Self-Check Exercises 3.1. It is reproduced in the following figure.

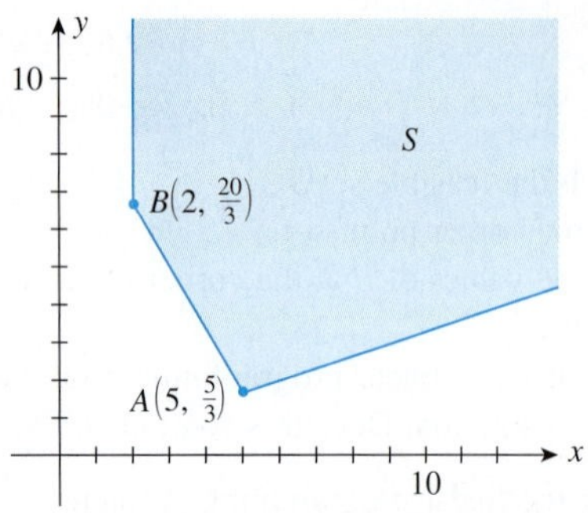

Evaluating the objective function $C = 5x + 3y$ at each corner point, we obtain the following table:

Vertex	$C = 5x + 3y$
$A(5, \frac{5}{3})$	30
$B(2, \frac{20}{3})$	30

We conclude that (i) the objective function is minimized at every point on the line segment joining the points $(5, \frac{5}{3})$ and $(2, \frac{20}{3})$, and (ii) the minimum value of C is 30.

3. Refer to Self-Check Exercise 3.2. The problem is to maximize $P = 50{,}000x + 20{,}000y$ subject to

$$\begin{aligned} 300x + 100y &\leq 9000 \\ y &\geq 15 \\ y &\leq 30 \\ x \geq 0, y &\geq 0 \end{aligned}$$

The feasible set S for the problem is shown in the following figure:

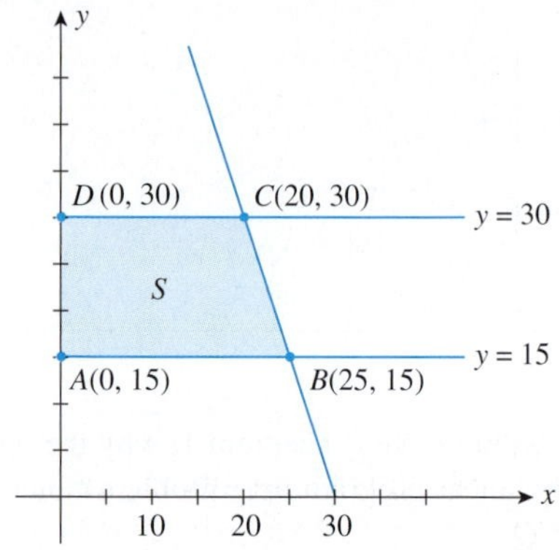

Evaluating the objective function $P = 50{,}000x + 20{,}000y$ at each vertex of S, we obtain the following table:

Vertex	$P = 50{,}000x + 20{,}000y$
$A(0, 15)$	300,000
$B(25, 15)$	1,550,000
$C(20, 30)$	1,600,000
$D(0, 30)$	600,000

From the table, we see that P is maximized when $x = 20$ and $y = 30$. Therefore, Gino should place 20 ads in the *City Tribune* and 30 in the *Daily News*.

3.4 Sensitivity Analysis

In this section, we investigate how changes in the parameters of a linear programming problem affect its optimal solution. This type of analysis is called **sensitivity analysis.** As in the previous sections, we restrict our analysis to the two-variable case, which is amenable to graphical analysis.

Recall the production problem posed in Example 1, Section 3.2, and solved in Example 1, Section 3.3:

Maximize	$P = x + 1.2y$	Objective function
subject to	$2x + y \leq 180$	Constraint 1
	$x + 3y \leq 300$	Constraint 2
	$x \geq 0, y \geq 0$	

where x denotes the number of Type A souvenirs and y denotes the number of Type B souvenirs to be made. The optimal solution of this problem is $x = 48$, $y = 84$ (corresponding to the point C). The optimal value of P is 148.8 (Figure 18).

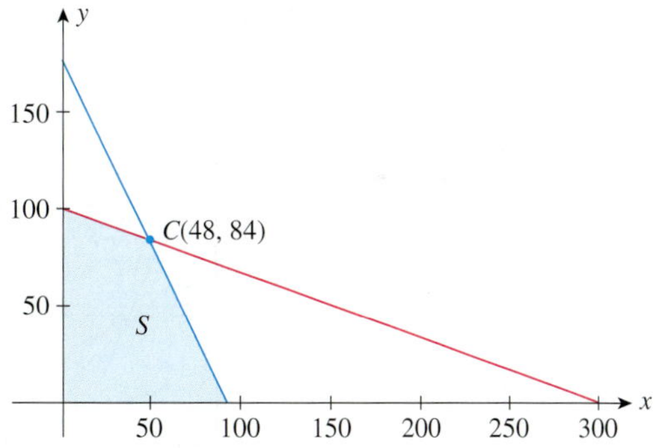

FIGURE 18
The optimal solution occurs at the point $C(48, 84)$.

The following questions arise in connection with this production problem.

1. How do changes made to the coefficients of the objective function affect the optimal solution?
2. How do changes made to the constants on the right-hand side of the constraints affect the optimal solution?

Changes in the Coefficients of the Objective Function

In the production problem under consideration, the objective function is $P = x + 1.2y$. The coefficient of x, which is 1, tells us that the contribution to the profit for each Type A souvenir is \$1.00. The coefficient of y, 1.2, tells us that the contribution to the profit for each Type B souvenir is \$1.20. Now suppose the contribution to the profit for each Type B souvenir remains fixed at \$1.20 per souvenir. By how much can the contribution to the profit for each Type A souvenir vary without affecting the current optimal solution?

To answer this question, suppose the contribution to the profit of each Type A souvenir is \$$c$ so that

$$P = cx + 1.2y \tag{8}$$

We need to determine the range of values of c such that the solution remains optimal.

We begin by rewriting Equation (8) for the isoprofit line in the slope-intercept form. Thus,

$$y = -\frac{c}{1.2}x + \frac{P}{1.2} \tag{9}$$

The slope of the isoprofit line is $-c/1.2$. If the slope of the isoprofit line exceeds that of the line associated with constraint 2, then the optimal solution shifts from point C to point D (see Figure 19 on the next page).

On the other hand, if the slope of the isoprofit line is *less than or equal to* the slope of the line associated with constraint 2, then the optimal solution remains unaffected. (You may verify that $-\frac{1}{3}$ is the slope of the line associated with constraint 2 by writing the equation $x + 3y = 300$ in the slope-intercept form.) In other words, we must have

$$-\frac{c}{1.2} \leq -\frac{1}{3}$$

$$\frac{c}{1.2} \geq \frac{1}{3} \quad \text{Multiplying each side by } -1 \text{ reverses the inequality sign.}$$

$$c \geq \frac{1.2}{3} = 0.4$$

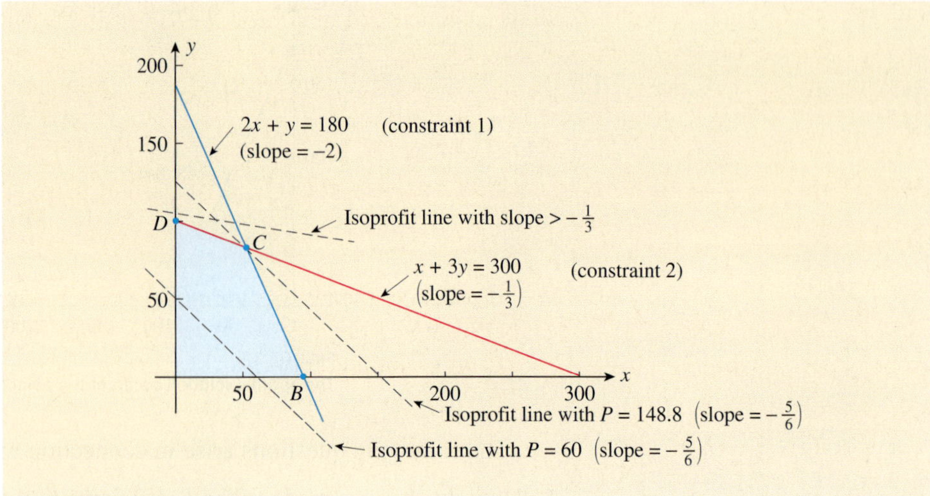

FIGURE 19
Increasing the slope of the isoprofit line $P = cx + 1.2y$ beyond $-\frac{1}{3}$ shifts the optimal solution from point C to point D.

A similar analysis shows that if the slope of the isoprofit line is less than that of the line associated with constraint 1 (see Figure 19), then the optimal solution shifts from point C to point B. Since the slope of the line associated with constraint 1 is -2, we see that point C will remain optimal provided that the slope of the isoprofit line is *greater than or equal to* -2, that is, if

$$-\frac{c}{1.2} \geq -2$$

$$\frac{c}{1.2} \leq 2$$

$$c \leq 2.4$$

Thus, we have shown that if $0.4 \leq c \leq 2.4$, then the optimal solution that we obtained previously remains unaffected.

This result tells us that if the contribution to the profit of each Type *A* souvenir lies between \$0.40 and \$2.40, then Ace Novelty should still make 48 Type *A* souvenirs and 84 Type *B* souvenirs. Of course, the company's profit will change with a change in the value of *c*—it's the product mix that stays the same. For example, if the contribution to the profit of a Type *A* souvenir is \$1.50, then the company's profit will be \$172.80. (See Exercise 1.) Incidentally, our analysis shows that the parameter *c* is not a sensitive parameter.

We leave it as an exercise for you to show that, with the contribution to the profit of Type *A* souvenirs held constant at \$1.00 per souvenir, the contribution to each Type *B* souvenir can vary between \$0.50 and \$3.00 without affecting the product mix for the optimal solution (see Exercise 1).

APPLIED EXAMPLE 1 Profit Function Analysis Kane Manufacturing has a division that produces two models of grates, model A and model B. To produce each model A grate requires 3 pounds of cast iron and 6 minutes of labor. To produce each model B grate requires 4 pounds of cast iron and 3 minutes of labor. The profit for each model A grate is \$2.00, and the profit for each model B grate is \$1.50. Available for grate production each day are 1000 pounds of cast iron and 20 labor-hours. Because of an excess inventory of model A grates, management has decided to limit the production of model A grates to no more than 180 grates per day.

a. Use the method of corners to determine the number of grates of each model Kane should produce to maximize its profit.

b. Find the range of values that the contribution to the profit of a model A grate can assume without changing the optimal solution.

c. Find the range of values that the contribution to the profit of a model B grate can assume without changing the optimal solution.

Solution

a. Let x denote the number of model A grates produced, and let y denote the number of model B grates produced. Then verify that we are led to the following linear programming problem:

$$\begin{aligned} \text{Maximize} \quad & P = 2x + 1.5y \\ \text{subject to} \quad & 3x + 4y \le 1000 \quad \text{Constraint 1} \\ & 6x + 3y \le 1200 \quad \text{Constraint 2} \\ & x \le 180 \quad \text{Constraint 3} \\ & x \ge 0, y \ge 0 \end{aligned}$$

The graph of the feasible set S is shown in Figure 20.

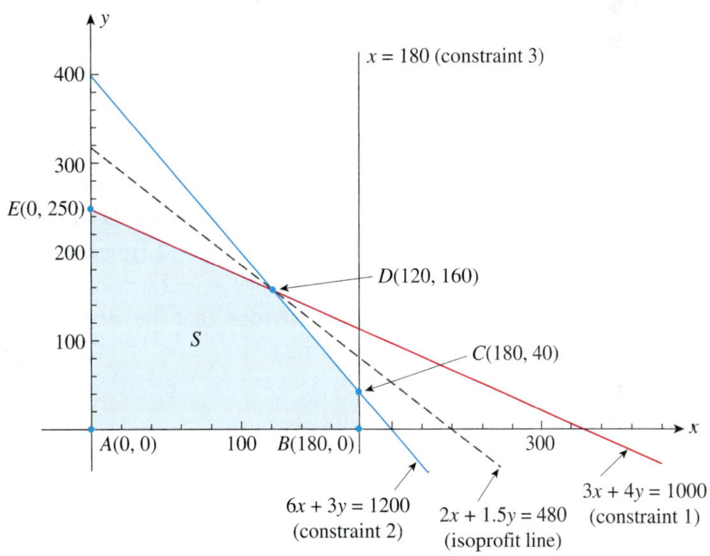

FIGURE 20
The shaded region is the feasible set S. Also shown are the lines of the equations associated with the constraints.

From the following table of values,

Vertex	$P = 2x + 1.5y$
$A(0, 0)$	0
$B(180, 0)$	360
$C(180, 40)$	420
$D(120, 160)$	480
$E(0, 250)$	375

we see that the maximum of $P = 2x + 1.5y$ occurs at the vertex $D(120, 160)$ with a value of 480. Thus, Kane realizes a maximum profit of \$480 per day by producing 120 model A grates and 160 model B grates each day.

b. Let c (in dollars) denote the contribution to the profit of a model A grate. Then $P = cx + 1.5y$ or, upon solving for y,

$$y = -\frac{c}{1.5}x + \frac{P}{1.5}$$
$$= \left(-\frac{2}{3}c\right)x + \frac{2}{3}P$$

Referring to Figure 20, you can see that if the slope of the isoprofit line is greater than the slope of the line associated with constraint 1, then the optimal solution will shift from point D to point E. Thus, for the optimal solution to remain unaffected, the slope of the isoprofit line must be less than or equal to the slope of the line associated with constraint 1. But the slope of the line associated with constraint 1 is $-\frac{3}{4}$, which you can see by rewriting the equation $3x + 4y = 1000$ in the slope-intercept form $y = -\frac{3}{4}x + 250$. Since the slope of the isoprofit line is $-2c/3$, we must have

$$-\frac{2c}{3} \le -\frac{3}{4}$$
$$\frac{2c}{3} \ge \frac{3}{4}$$
$$c \ge \left(\frac{3}{4}\right)\left(\frac{3}{2}\right) = \frac{9}{8} = 1.125$$

Again referring to Figure 20, you can see that if the slope of the isoprofit line is less than that of the line associated with constraint 2, then the optimal solution shifts from point D to point C. Since the slope of the line associated with constraint 2 is -2 (rewrite the equation $6x + 3y = 1200$ in the slope-intercept form $y = -2x + 400$), we see that the optimal solution remains at point D provided that the slope of the isoprofit line is greater than or equal to -2; that is,

$$-\frac{2c}{3} \ge -2$$
$$\frac{2c}{3} \le 2$$
$$c \le (2)\left(\frac{3}{2}\right) = 3$$

We conclude that the contribution to the profit of a model A grate can assume values between \$1.125 and \$3.00 without changing the optimal solution.

c. Let c (in dollars) denote the contribution to the profit of a model B grate. Then

$$P = 2x + cy$$

or, upon solving for y,

$$y = -\frac{2}{c}x + \frac{P}{c}$$

An analysis similar to that performed in part (b) with respect to constraint 1 shows that the optimal solution will remain in effect provided that

$$-\frac{2}{c} \le -\frac{3}{4}$$
$$\frac{2}{c} \ge \frac{3}{4}$$
$$c \le 2\left(\frac{4}{3}\right) = \frac{8}{3} = 2\frac{2}{3}$$

Performing an analysis with respect to constraint 2 shows that the optimal solution will remain in effect, provided that

$$-\frac{2}{c} \ge -2$$
$$\frac{2}{c} \le 2$$
$$c \ge 1$$

Thus, the contribution to the profit of a model B grate can assume values between \$1.00 and \$2.67 without changing the optimal solution.

Changes to the Constants on the Right-Hand Side of the Constraint Inequalities

Let's return to the production problem posed at the beginning of this section:

$$\begin{aligned} \text{Maximize} \quad & P = x + 1.2y \\ \text{subject to} \quad & 2x + y \le 180 \quad \text{Constraint 1} \\ & x + 3y \le 300 \quad \text{Constraint 2} \\ & x \ge 0, y \ge 0 \end{aligned}$$

Now suppose that the time available on Machine I is changed from 180 minutes to $(180 + h)$ minutes, where h is a real number. Then the constraint on Machine I is changed to

$$2x + y \le 180 + h$$

Observe that the line with equation $2x + y = 180 + h$ is parallel to the line $2x + y = 180$ associated with the original constraint 1.

As you can see from Figure 21 (see next page), the result of adding the constant h to the right-hand side of constraint 1 is to shift the current optimal solution from the point C to the new optimal solution occurring at the point C'. To find the coordinates of C', we observe that C' is the point of intersection of the lines with equations

$$2x + y = 180 + h \quad \text{and} \quad x + 3y = 300$$

Thus, the coordinates of the point are found by solving the system of linear equations

$$\begin{aligned} 2x + y &= 180 + h \\ x + 3y &= 300 \end{aligned}$$

The solutions are

$$x = \frac{3}{5}(80 + h) \quad \text{and} \quad y = \frac{1}{5}(420 - h) \tag{10}$$

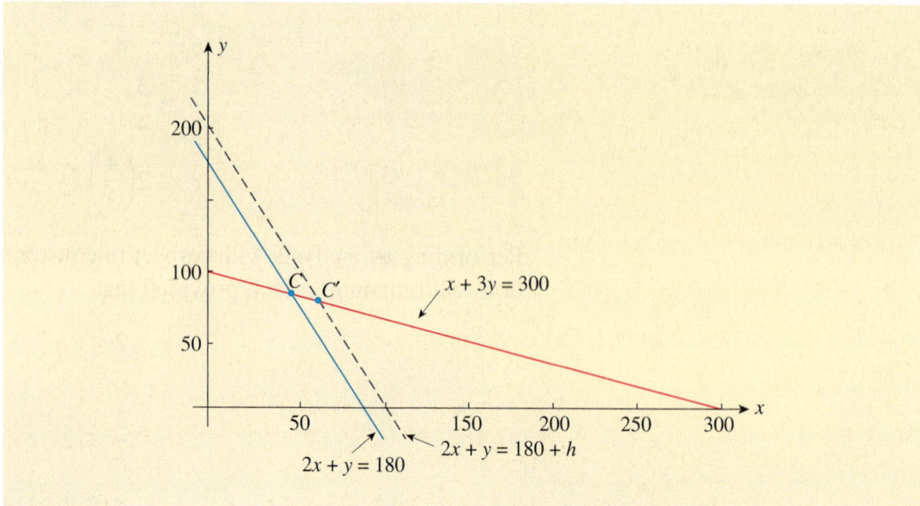

FIGURE 21
The lines with equations $2x + y = 180$ and $2x + y = 180 + h$ are parallel to each other.

The nonnegativity of x implies that

$$\begin{aligned} \frac{3}{5}(80 + h) &\geq 0 \\ 80 + h &\geq 0 \\ h &\geq -80 \end{aligned}$$

Next, the nonnegativity of y implies that

$$\begin{aligned} \frac{1}{5}(420 - h) &\geq 0 \\ 420 - h &\geq 0 \\ h &\leq 420 \end{aligned}$$

Thus, h must satisfy the inequalities $-80 \leq h \leq 420$. Our computations reveal that a meaningful solution will require that the time available for Machine I must range between $(180 - 80)$ and $(180 + 420)$ minutes—that is, between 100 and 600 minutes. Under these conditions, Ace Novelty should produce $\frac{3}{5}(80 + h)$ Type A souvenirs and $\frac{1}{5}(420 - h)$ Type B souvenirs.

For example, if Ace Novelty can manage to increase the time available on Machine I by 10 minutes, then it should produce $\frac{3}{5}(80 + 10)$, or 54, Type A souvenirs and $\frac{1}{5}(420 - 10)$, or 82, Type B souvenirs; the resulting profit is

$$P = x + 1.2y = 54 + (1.2)(82) = 152.4$$

or \$152.40.

We leave it as an exercise for you to show that if the time available on Machine II is changed from 300 minutes to $(300 + k)$ minutes with no change in the maximum capacity for Machine I, then k must satisfy the inequalities $-210 \leq k \leq 240$ (see Exercise 2). Thus, for a meaningful solution to the problem, the time available on Machine II must lie between 90 and 540 min. Furthermore, in this case, Ace Novelty should produce $\frac{1}{5}(240 - k)$ Type A souvenirs and $\frac{1}{5}(420 + 2k)$ Type B souvenirs.

Shadow Prices

We have just seen that if Ace Novelty could increase the maximum available time on Machine I by 10 minutes, then the profit would increase from the original optimal value of \$148.80 to \$152.40. In this case, finding the extra time on Machine I proved beneficial to the company. More generally, to study the economic benefits that can be derived from increasing its resources, a company looks at the shadow prices associated with the respective resources. We define the **shadow price** for the ith resource (associated with the ith constraint of the linear programming problem) to be the amount by which the value of the objective function is improved—increased in a maximization problem and decreased in a minimization problem—if the right-hand side of the ith constraint is changed by 1 unit.

In the Ace Novelty example discussed earlier, we showed that if the right-hand side of constraint 1 is increased by h units, then the optimal solution is given by Equations (10):

$$x = \frac{3}{5}(80 + h) \quad \text{and} \quad y = \frac{1}{5}(420 - h)$$

The resulting profit is calculated as follows:

$$\begin{aligned} P &= x + 1.2y \\ &= x + \frac{6}{5}y \\ &= \frac{3}{5}(80 + h) + \left(\frac{6}{5}\right)\left(\frac{1}{5}\right)(420 - h) \\ &= \frac{3}{25}(1240 + 3h) \end{aligned}$$

Upon setting $h = 1$, we find

$$\begin{aligned} P &= \frac{3}{25}(1240 + 3) \\ &= 149.16 \end{aligned}$$

Since the optimal profit for the original problem is \$148.80, we see that the shadow price for the first resource is 149.16 − 148.80, or \$0.36. To summarize, Ace Novelty's profit increases at the rate of \$0.36 per 1-minute increase in the time available on Machine I.

We leave it as an exercise for you to show that the shadow price for Resource 2 (associated with constraint 2) is \$0.28 (see Exercise 2).

APPLIED EXAMPLE 2 Shadow Prices Consider the problem posed in Example 1:

$$\begin{aligned} \text{Maximize} \quad & P = 2x + 1.5y \\ \text{subject to} \quad & 3x + 4y \le 1000 \quad \text{Constraint 1} \\ & 6x + 3y \le 1200 \quad \text{Constraint 2} \\ & x \le 180 \quad \text{Constraint 3} \\ & x \ge 0, y \ge 0 \end{aligned}$$

a. Find the range of values that Resource 1 (the constant on the right-hand side of constraint 1) can assume.

b. Find the shadow price for Resource 1.

Solution

a. Suppose the right-hand side of constraint 1 is replaced by $1000 + h$, where h is a real number. Then the new optimal solution occurs at the point D' (Figure 22).

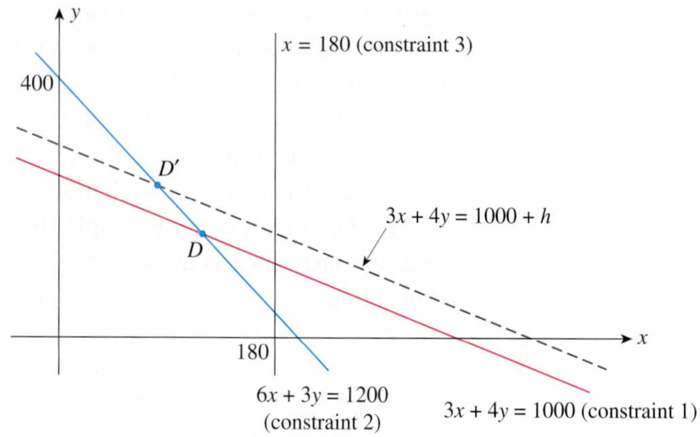

FIGURE 22
As the amount of Resource 1 changes, the point at which the optimal solution occurs shifts from D to D'.

To find the coordinates of D', we solve the system

$$\begin{aligned} 3x + 4y &= 1000 + h \\ 6x + 3y &= 1200 \end{aligned}$$

Multiplying the first equation by -2 and then adding the resulting equation to the second equation, we obtain

$$\begin{aligned} -5y &= -800 - 2h \\ y &= \frac{2}{5}(400 + h) \end{aligned}$$

$$\begin{array}{rl} -6x - 8y = & -2000 - 2h \\ 6x + 3y = & 1200 \\ \hline -5y = & -800 - 2h \end{array}$$

Substituting this value of y into the second equation in the system gives

$$\begin{aligned} 6x + \frac{6}{5}(400 + h) &= 1200 \\ x + \frac{1}{5}(400 + h) &= 200 \\ x &= \frac{1}{5}(600 - h) \end{aligned}$$

The nonnegativity of y implies that $h \geq -400$, and the nonnegativity of x implies that $h \leq 600$. But constraint 3 dictates that x must also satisfy

$$\begin{aligned} x = \frac{1}{5}(600 - h) &\leq 180 \\ 600 - h &\leq 900 \\ -h &\leq 300 \\ h &\geq -300 \end{aligned}$$

Therefore, h must satisfy $-300 \leq h \leq 600$. This tells us that the amount of Resource 1 must lie between $1000 - 300$, or 700, and $1000 + 600$, or 1600—that is, between 700 and 1600 pounds.

b. If we set $h = 1$ in part (a), we obtain

$$x = \frac{1}{5}(600 - 1) = \frac{599}{5}$$

$$y = \frac{2}{5}(400 + 1) = \frac{802}{5}$$

Therefore, the profit realized at this level of production is

$$P = 2x + \frac{3}{2}y = 2\left(\frac{599}{5}\right) + \frac{3}{2}\left(\frac{802}{5}\right)$$
$$= \frac{2401}{5} = 480.2$$

Since the original optimal profit is \$480 (see Example 1), we see that the shadow price for Resource 1 is \$0.20.

If you examine Figure 22, you can see that increasing Resource 3 (the constant on the right-hand side of constraint 3) has no effect on the optimal solution $D(120, 160)$ of the problem at hand. In other words, an increase in the resource associated with constraint 3 has no economic benefit for Kane Manufacturing. The shadow price for this resource is *zero*. There is a *surplus* of this resource. Hence, we say that the constraint $x \leq 180$ is not binding on the optimal solution $D(120, 160)$.

On the other hand, constraints 1 and 2, which *hold with equality* at the optimal solution $D(120, 160)$, are said to be **binding constraints.** The objective function cannot be increased without increasing these resources. They have *positive* shadow prices.

Importance of Sensitivity Analysis

We conclude this section by pointing out the importance of sensitivity analysis in solving real-world problems. The values of the parameters in these problems may change. For example, the management of Ace Novelty might wish to increase the price of a Type A souvenir because of increased demand for the product, or they might want to see how a change in the time available on Machine I affects the (optimal) profit of the company.

When a parameter of a linear programming problem is changed, it is true that one need only re-solve the problem to obtain a new solution to the problem. But since a real-world linear programming problem often involves thousands of parameters, the amount of work involved in finding a new solution is prohibitive. Another disadvantage in using this approach is that it often takes many trials with different values of a parameter to see their effect on the optimal solution of the problem. Thus, a more analytical approach such as that discussed in this section is desirable.

Returning to the discussion of Ace Novelty, our analysis of the changes in the coefficients of the objective (profit) function suggests that if management decides to raise the price of a Type A souvenir, it can do so with the assurance that the optimal solution holds as long as the new price leaves the contribution to the profit of a Type A souvenir between \$0.40 and \$2.40. There is no need to re-solve the linear programming problem for each new price being considered. Also, our analysis of the changes in the parameters on the right-hand side of the constraints suggests, for example, that a meaningful solution to the problem requires that the time available for Machine I lie in the range between 100 and 600 minutes. Furthermore, the analysis tells us how to compute the increase (decrease) in the optimal profit when the resource is adjusted, by using the shadow price associated with that constraint. Again, there is no need to re-solve the linear programming problem each time a change in the resource available is anticipated.

Using Technology examples and exercises that are solved using Excel's Solver can be found on pages 242–245 and 260–262.

3.4 Self-Check Exercises

Consider the linear programming problem:

$$\begin{aligned} \text{Maximize} \quad & P = 2x + 4y \\ \text{subject to} \quad & 2x + 5y \leq 19 \quad \text{Constraint 1} \\ & 3x + 2y \leq 12 \quad \text{Constraint 2} \\ & x \geq 0, y \geq 0 \end{aligned}$$

1. Use the method of corners to solve this problem.
2. Find the range of values that the coefficient of x can assume without changing the optimal solution.
3. Find the range of values that Resource 1 (the constant on the right-hand side of constraint 1) can assume without changing the optimal solution.
4. Find the shadow price for Resource 1.
5. Identify the binding and nonbinding constraints.

Solutions to Self-Check Exercises can be found on page 214.

3.4 Concept Questions

1. Suppose $P = 3x + 4y$ is the objective function in a linear programming (maximization) problem, where x denotes the number of units of Product A and y denotes the number of units of Product B to be made. What does the coefficient of x represent? The coefficient of y?
2. Given the linear programming problem

$$\begin{aligned} \text{Maximize} \quad & P = 3x + 4y \\ \text{subject to} \quad & x + y \leq 4 \quad \text{Resource 1} \\ & 2x + y \leq 5 \quad \text{Resource 2} \end{aligned}$$

   a. Write the inequality that represents an increase of h units in Resource 1.
   b. Write the inequality that represents an increase of k units in Resource 2.
3. Explain the meaning of (a) a shadow price and (b) a binding constraint.

3.4 Exercises

1. **Optimizing Profit** Refer to the production problem discussed on pages 203–204.
   a. Show that the optimal solution holds if the contribution to the profit of a Type B souvenir lies between \$0.50 and \$3.00.
   b. Show that if the contribution to the profit of a Type A souvenir is \$1.50 (with the contribution to the profit of a Type B souvenir held at \$1.20), then the optimal profit of the company will be \$172.80.
   c. What will be the optimal profit of the company if the contribution to the profit of a Type B souvenir is \$2.00 (with the contribution to the profit of a Type A souvenir held at \$1.00)?
2. **Optimizing Profit** Refer to the production problem discussed on pages 207–208.
   a. Show that for a meaningful solution, the time available on Machine II must lie between 90 and 540 min.
   b. Show that if the time available on Machine II is changed from 300 min to $(300 + k)$ min, with no change in the maximum capacity for Machine I, then Ace Novelty's profit is maximized by producing $\frac{1}{5}(240 - k)$ Type A souvenirs and $\frac{1}{5}(420 + 2k)$ Type B souvenirs, where $-210 \leq k \leq 240$.
   c. Show that the shadow price for Resource 2 (associated with constraint 2) is \$0.28.

3. **OPTIMIZING PROFIT** Refer to Example 2.
   a. Find the range of values that Resource 2 can assume.
   b. By how much can the right-hand side of constraint 3 be changed such that the current optimal solution still holds?

4. **SHADOW PRICES** Refer to Example 2.
   a. Find the shadow price for Resource 2.
   b. Identify the binding and nonbinding constraints.

In Exercises 5–10, you are given a linear programming problem.
a. Use the method of corners to solve the problem.
b. Find the range of values that the coefficient of x can assume without changing the optimal solution.
c. Find the range of values that Resource 1 (requirement 1) can assume.
d. Find the shadow price for Resource 1 (requirement 1).
e. Identify the binding and nonbinding constraints.

5. Maximize $P = 3x + 4y$
   subject to
   $2x + 3y \leq 12$ Resource 1
   $2x + y \leq 8$ Resource 2
   $x \geq 0, y \geq 0$

6. Maximize $P = 2x + 5y$
   subject to
   $x + 3y \leq 15$ Resource 1
   $4x + y \leq 16$ Resource 2
   $x \geq 0, y \geq 0$

7. Minimize $C = 2x + 5y$
   subject to
   $x + 2y \geq 4$ Requirement 1
   $x + y \geq 3$ Requirement 2
   $x \geq 0, y \geq 0$

8. Minimize $C = 3x + 4y$
   subject to
   $x + 3y \geq 8$ Requirement 1
   $x + y \geq 4$ Requirement 2
   $x \geq 0, y \geq 0$

9. Maximize $P = 4x + 3y$
   subject to
   $5x + 3y \leq 30$ Resource 1
   $2x + 3y \leq 21$ Resource 2
   $x \leq 4$ Resource 3
   $x \geq 0, y \geq 0$

10. Maximize $P = 4x + 5y$
    subject to
    $x + y \leq 30$ Resource 1
    $x + 2y \leq 40$ Resource 2
    $x \leq 25$ Resource 3
    $x \geq 0, y \geq 0$

11. **PRODUCTION SCHEDULING** A company manufactures two products, *A* and *B*, on Machines I and II. The company will realize a profit of \$3/unit of Product *A* and a profit of \$4/unit of Product *B*. Manufacturing 1 unit of Product *A* requires 6 min on Machine I and 5 min on Machine II. Manufacturing 1 unit of Product *B* requires 9 min on Machine I and 4 min on Machine II. There are 5 hr of time available on Machine I and 3 hr of time available on Machine II in each work shift.
    a. How many units of each product should be produced in each shift to maximize the company's profit?
    b. Find the range of values that the contribution to the profit of 1 unit of Product *A* can assume without changing the optimal solution.
    c. Find the range of values that the resource associated with the time constraint on Machine I can assume.
    d. Find the shadow price for the resource associated with the time constraint on Machine I.

12. **CROP PLANNING** A farmer plans to plant two crops, *A* and *B*. The cost of cultivating Crop *A* is \$40/acre whereas that of Crop *B* is \$60/acre. The farmer has a maximum of \$7400 available for land cultivation. Each acre of Crop *A* requires 20 labor-hours, and each acre of Crop *B* requires 25 labor-hours. The farmer has a maximum of 3300 labor-hours available. If he expects to make a profit of \$150/acre on Crop *A* and \$200/acre on Crop *B*, how many acres of each crop should he plant to maximize his profit?
    a. Find the range of values that the contribution to the profit of an acre of Crop *A* can assume without changing the optimal solution.
    b. Find the range of values that the resource associated with the constraint on the available land can assume.
    c. Find the shadow price for the resource associated with the constraint on the available land.

13. **MINIMIZNG COSTS** Perth Mining Company operates two mines for the purpose of extracting gold and silver. The Saddle Mine costs \$14,000/day to operate, and it yields 50 oz of gold and 3000 oz of silver per day. The Horseshoe Mine costs \$16,000/day to operate, and it yields 75 oz of gold and 1000 ounces of silver per day. Company management has set a target of at least 650 oz of gold and 18,000 oz of silver.
    a. How many days should each mine be operated so that the target can be met at a minimum cost?
    b. Find the range of values that the Saddle Mine's daily operating cost can assume without changing the optimal solution.
    c. Find the range of values that the requirement for gold can assume.
    d. Find the shadow price for the requirement for gold.

14. **MINIMIZING CRUISE LINE COSTS** Deluxe River Cruises operates a fleet of river vessels. The fleet has two types of vessels: a type A vessel has 60 deluxe cabins and 160 standard cabins, whereas a type B vessel has 80 deluxe cabins and 120 standard cabins. Under a charter agreement with the Odyssey Travel Agency, Deluxe River Cruises is to provide Odyssey with a minimum of 360 deluxe and 680 standard cabins for their 15-day cruise in May. It costs \$44,000 to operate a type A vessel and \$54,000 to operate a type B vessel for that period.
    a. How many of each type of vessel should be used to keep the operating costs to a minimum?
    b. Find the range of values that the cost of operating a type A vessel can assume without changing the optimal solution.

c. Find the range of values that the requirement for deluxe cabins can assume.

d. Find the shadow price for the requirement for deluxe cabins.

15. PRODUCTION SCHEDULING Soundex produces two models of satellite radios. Model A requires 15 min of work on Assembly Line I and 10 min of work on Assembly Line II. Model B requires 10 min of work on Assembly Line I and 12 min of work on Assembly Line II. At most 25 hr of assembly time on Line I and 22 hr of assembly time on Line II are available each day. Soundex anticipates a profit of \$12 on model A and \$10 on model B. Because of previous overproduction, management decides to limit the production of model A satellite radios to no more than 80/day.

a. To maximize Soundex's profit, how many satellite radios of each model should be produced each day?

b. Find the range of values that the contribution to the profit of a model A satellite radio can assume without changing the optimal solution.

c. Find the range of values that the resource associated with the time constraint on Assembly Line I can assume.

d. Find the shadow price for the resource associated with the time constraint on Assembly Line I.

e. Identify the binding and nonbinding constraints.

16. PRODUCTION SCHEDULING Refer to Exercise 15.

a. If the contribution to the profit of a model A satellite radio is changed to \$8.50/radio, will the original optimal solution still hold? What will be the optimal profit?

b. If the contribution to the profit of a model A satellite radio is changed to \$14.00/radio, will the original optimal solution still hold? What will be the optimal profit?

17. PRODUCTION SCHEDULING Kane Manufacturing has a division that produces two models of fireplace grates, model A and model B. To produce each model A grate requires 3 lb of cast iron and 6 min of labor. To produce each model B grate requires 4 lb of cast iron and 3 min of labor. The profit for each model A grate is \$2, and the profit for each model B grate is \$1.50. 1000 lb of cast iron and 20 labor-hours are available for the production of grates each day. Because of an excess inventory of model B grates, management has decided to limit the production of model B grates to no more than 200 grates per day. How many grates of each model should the division produce daily to maximize Kane's profit?

a. Use the method of corners to solve the problem.

b. Find the range of values that the contribution to the profit of a model A grate can assume without changing the optimal solution.

c. Find the range of values that the resource for cast iron can assume without changing the optimal solution.

d. Find the shadow price for the resource for cast iron.

e. Identify the binding and nonbinding constraints.

18. PRODUCTION SCHEDULING Refer to Exercise 17.

a. If the contribution to the profit of a model A grate is changed to \$1.75/grate, will the original optimal solution still hold? What will be the new optimal solution?

b. If the contribution to the profit of a model A grate is changed to \$2.50/grate, will the original optimal solution still hold? What will be the new optimal solution?

3.4 Solutions to Self-Check Exercises

1. The feasible set for the problem is shown in the accompanying figure.

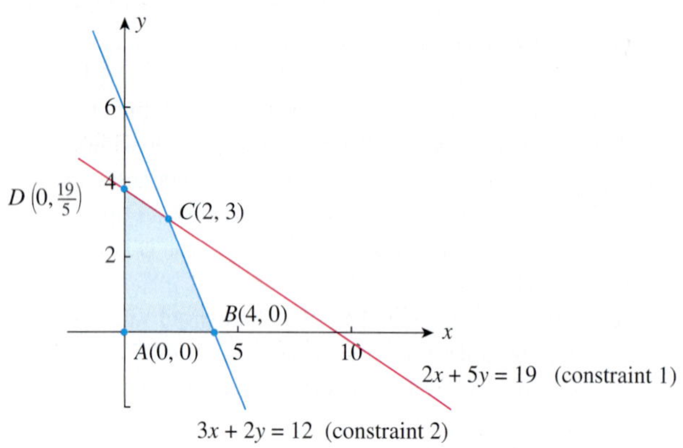

Evaluating the objective function $P = 2x + 4y$ at each feasible corner point, we obtain the following table:

Vertex	$P = 2x + 4y$
$A(0, 0)$	0
$B(4, 0)$	8
$C(2, 3)$	16
$D(0, \frac{19}{5})$	$15\frac{1}{5}$

We conclude that the maximum value of P is 16 attained at the point $(2, 3)$.

2. Assume that $P = cx + 4y$. Then

$$y = -\frac{c}{4}x + \frac{P}{4}$$

The slope of the isoprofit line is $-\frac{c}{4}$ and must be less than or equal to the slope of the line associated with constraint 1; that is,

$$-\frac{c}{4} \le -\frac{2}{5}$$

Solving, we find $c \ge \frac{8}{5}$. A similar analysis shows that the slope of the isoprofit line must be greater than or equal to the slope of the line associated with constraint 2; that is,

$$-\frac{c}{4} \ge -\frac{3}{2}$$

Solving, we find $c \le 6$. Thus, we have shown that if $1.6 \le c \le 6$, then the optimal solution obtained previously remains unaffected.

3. Suppose the right-hand side of constraint 1 is replaced by $19 + h$, where h is a real number. Then the new optimal solution occurs at the point whose coordinates are found by solving the system

$$2x + 5y = 19 + h$$
$$3x + 2y = 12$$

Multiplying the second equation by -5 and adding the resulting equation to 2 times the first equation, we obtain

$$-11x = -60 + 2(19 + h) = -22 + 2h$$
$$x = 2 - \frac{2}{11}h$$

Substituting this value of x into the second equation in the system gives

$$3\left(2 - \frac{2}{11}h\right) + 2y = 12$$
$$2y = 12 - 6 + \frac{6}{11}h$$
$$y = 3 + \frac{3}{11}h$$

The nonnegativity of x implies that $2 - \frac{2}{11}h \ge 0$, or $h \le 11$. The nonnegativity of y implies that $3 + \frac{3}{11}h \ge 0$, or $h \ge -11$. Therefore, h must satisfy $-11 \le h \le 11$. This tells us that the amount used of Resource 1 must lie between $19 - 11$ and $19 + 11$—that is, between 8 and 30.

4. If we set $h = 1$ in Exercise 3, we find that $x = \frac{20}{11}$ and $y = \frac{36}{11}$. Therefore, for these values of x and y,

$$P = 2\left(\frac{20}{11}\right) + 4\left(\frac{36}{11}\right) = \frac{184}{11} = 16\frac{8}{11}$$

Since the original optimal value of P is 16, we see that the shadow price for Resource 1 is $\frac{8}{11}$.

5. Since both constraints hold with equality at the optimal solution $C(2, 3)$, they are binding constraints.

CHAPTER 3 Summary of Principal Terms

TERMS

solution set of a system of linear inequalities (173)
bounded solution set (176)
unbounded solution set (176)
objective function (181)
linear programming problem (181)
feasible solution (190)
feasible set (190)
optimal solution (190)
isoprofit line (190)
method of corners (191)
sensitivity analysis (202)
shadow price (209)
binding constraint (211)

CHAPTER 3 Concept Review Questions

Fill in the blanks.

1. a. The solution set of the inequality $ax + by < c$ (a, b not both zero) is a/an _____ _____ that does not include the _____ with equation $ax + by = c$.

b. If $ax + by < c$ describes the lower half-plane, then the inequality _____ describes the lower half-plane together with the line having equation _____.

2. a. The solution set of a system of linear inequalities in the two variables x and y is the set of all _____ satisfying _____ inequality of the system.

b. The solution set of a system of linear inequalities is _____ if it can be _____ by a circle.

3. A linear programming problem consists of a linear function, called a/an _____ _____ to be _____ or _____ subject to constraints in the form of _____ equations or _____.

4. a. If a linear programming problem has a solution, then it must occur at a/an _____ _____ of the feasible set.

b. If the objective function of a linear programming problem is optimized at two adjacent vertices of the feasible set, then it is optimized at every point on the _____ segment joining these vertices.

5. In sensitivity analysis, we investigate how changes in the _____ of a linear programming problem affect the _____ solution.

6. The shadow price for the ith _____ is the _____ by which the _____ of the objective function is _____ if the right-hand side of the ith constraint is _____ by 1 unit.

CHAPTER 3 Review Exercises

In Exercises 1 and 2, find the optimal value(s) of the objective function on the feasible set *S*.

1. $Z = 2x + 3y$

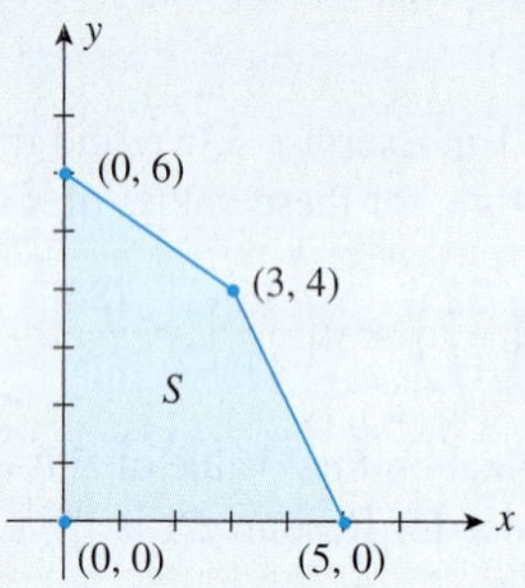

2. $Z = 4x + 3y$

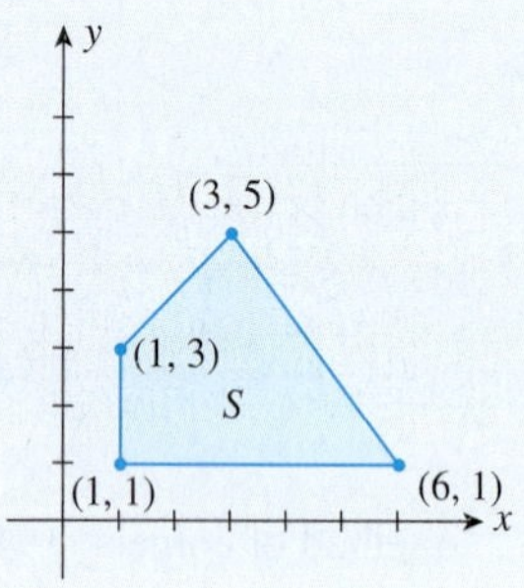

In Exercises 3–14, use the method of corners to solve the linear programming problem.

3. Maximize $P = 3x + 5y$
subject to
$$\begin{aligned} 2x + 3y &\le 12 \\ x + y &\le 5 \\ x \ge 0,\ y &\ge 0 \end{aligned}$$

4. Maximize $P = 2x + 3y$
subject to
$$\begin{aligned} 2x + y &\le 12 \\ x - 2y &\le 1 \\ x \ge 0,\ y &\ge 0 \end{aligned}$$

5. Minimize $C = 2x + 5y$
subject to
$$\begin{aligned} x + 3y &\ge 15 \\ 4x + y &\ge 16 \\ x \ge 0,\ y &\ge 0 \end{aligned}$$

6. Minimize $C = 3x + 4y$
subject to
$$\begin{aligned} 2x + y &\ge 4 \\ 2x + 5y &\ge 10 \\ x \ge 0,\ y &\ge 0 \end{aligned}$$

7. Maximize $P = 3x + 2y$
subject to
$$\begin{aligned} 2x + y &\le 16 \\ 2x + 3y &\le 36 \\ 4x + 5y &\ge 28 \\ x \ge 0,\ y &\ge 0 \end{aligned}$$

8. Maximize $P = 6x + 2y$
subject to
$$\begin{aligned} x + 2y &\le 12 \\ x + y &\le 8 \\ 2x - 3y &\ge 6 \\ x \ge 0,\ y &\ge 0 \end{aligned}$$

9. Minimize $C = 2x + 7y$
subject to
$$\begin{aligned} 3x + 5y &\ge 45 \\ 3x + 10y &\ge 60 \\ x \ge 0,\ y &\ge 0 \end{aligned}$$

10. Minimize $C = 3x + 2y$
subject to
$$\begin{aligned} 2x + y &\ge 8 \\ x + y &\ge 6 \\ 2x + 3y &\ge 14 \\ x \ge 0,\ y &\ge 0 \end{aligned}$$

11. Minimize $C = 4x + y$
subject to
$$\begin{aligned} 6x + y &\ge 18 \\ 2x + y &\ge 10 \\ x + 4y &\ge 12 \\ x \ge 0,\ y &\ge 0 \end{aligned}$$

12. Find the maximum and minimum values of $Q = 3x + 4y$ subject to
$$\begin{aligned} x - y &\ge -10 \\ x + 3y &\ge 30 \\ 7x + 4y &\le 140 \end{aligned}$$

13. Find the maximum and minimum of $Q = x + y$ subject to
$$\begin{aligned} 5x + 2y &\ge 20 \\ x + 2y &\ge 8 \\ x + 4y &\le 22 \\ x \ge 0,\ y &\ge 0 \end{aligned}$$

14. Find the maximum and minimum of $Q = 2x + 5y$ subject to

$$\begin{aligned} x + y &\geq 4 \\ -x + y &\leq 6 \\ x + 3y &\leq 30 \\ x &\leq 12 \\ x \geq 0, y &\geq 0 \end{aligned}$$

15. Financial Analysis An investor has decided to commit no more than \$80,000 to the purchase of the common stocks of two companies, Company *A* and Company *B*. He has also estimated that there is a chance of at most a 1% capital loss on his investment in Company *A* and a chance of at most a 4% loss on his investment in Company *B*, and he has decided that together these losses should not exceed \$2000. On the other hand, he expects to make a 14% profit from his investment in Company *A* and a 20% profit from his investment in Company *B*. Determine how much he should invest in the stock of each company to maximize his investment returns. What is the maximum return?

16. Production Scheduling Soundex produces two models of satellite radios. Model A requires 15 min of work on Assembly Line I and 10 min of work on Assembly Line II. Model B requires 10 min of work on Assembly Line I and 12 min of work on Assembly Line II. At most, 25 labor-hours of assembly time on Line I and 22 labor-hours of assembly time on Line II are available each day. It is anticipated that Soundex will realize a profit of \$12 on model A and \$10 on model B. How many satellite radios of each model should be produced each day to maximize Soundex's profit? What is the maximum profit?

17. Production Scheduling Kane Manufacturing has a division that produces two models of grates, model A and model B. To produce each model A grate requires 3 lb of cast iron and 6 min of labor. To produce each model B grate requires 4 lb of cast iron and 3 min of labor. The profit for each model A grate is \$2.00, and the profit for each model B grate is \$1.50. Available for grate production each day are 1000 lb of cast iron and 20 labor-hours. Because of a backlog of orders for model B grates, Kane's manager has decided to produce at least 180 model B grates per day. How many grates of each model should Kane produce to maximize its profit? What is the maximum profit?

18. Minimizing Shipping Costs A manufacturer of projection TVs must ship a total of at least 1000 TVs to its two central warehouses. Each warehouse can hold a maximum of 750 TVs. The first warehouse already has 150 TVs on hand, whereas the second has 50 TVs on hand. It costs \$8 to ship a TV to the first warehouse, and it costs \$16 to ship a TV to the second warehouse. How many TVs should be shipped to each warehouse to minimize the cost? What is the minimum cost?

The problem-solving skills that you learn in each chapter are building blocks for the rest of the course. Therefore, it is a good idea to make sure that you have mastered these skills before moving on to the next chapter. The Before Moving On exercises that follow are designed for that purpose. After completing these exercises, you can identify the skills that you should review before starting the next chapter.

CHAPTER 3 Before Moving On . . .

1. Determine graphically the solution set for the following systems of inequalities.

a. $$\begin{aligned} 2x + y &\leq 10 \\ x + 3y &\leq 15 \\ x &\leq 4 \\ x \geq 0, y &\geq 0 \end{aligned}$$

b. $$\begin{aligned} 2x + y &\geq 8 \\ 2x + 3y &\geq 15 \\ x &\geq 0 \\ y &\geq 2 \end{aligned}$$

2. Find the maximum and minimum values of $Z = 3x - y$ on the following feasible set.

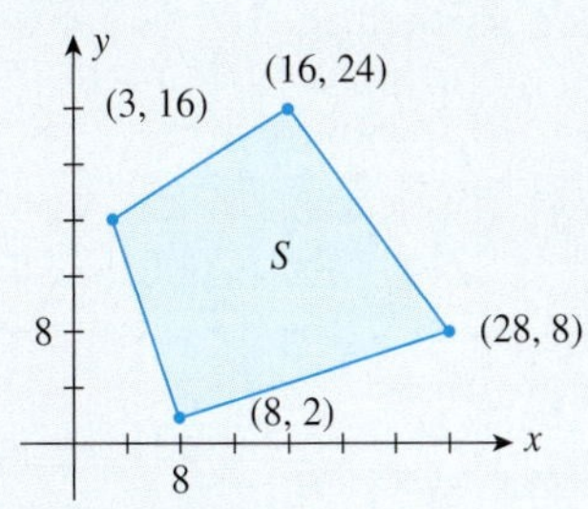

3. Maximize $P = x + 3y$
subject to
$$\begin{aligned} 2x + 3y &\le 11 \\ 3x + 7y &\le 24 \\ x \ge 0, y &\ge 0 \end{aligned}$$

4. Minimize $C = 4x + y$
subject to
$$\begin{aligned} 2x + y &\ge 10 \\ 2x + 3y &\ge 24 \\ x + 3y &\ge 15 \\ x \ge 0, y &\ge 0 \end{aligned}$$

5. **Sensitivity Analysis.** Consider the following linear programming problem:

Maximize $P = 2x + 3y$
subject to
$$\begin{aligned} x + 2y &\le 16 \\ 3x + 2y &\le 24 \\ x \ge 0, y &\ge 0 \end{aligned}$$

a. Solve the problem.
b. Find the range of values that the coefficient of x can assume without changing the optimal solution.
c. Find the range of values that Resource 1 (constraint 1) can assume.
d. Find the shadow price for Resource 1.
e. Identify the binding and nonbinding constraints.

4 Linear Programming: An Algebraic Approach

THE GEOMETRIC APPROACH introduced in Chapter 3 may be used to solve linear programming problems involving two or even three variables. But for linear programming problems involving more than two variables, an algebraic approach is preferred. One such technique, the *simplex method,* was developed by George Dantzig in the late 1940s and remains in wide use to this day.

We begin Chapter 4 by introducing the simplex method for solving *standard maximization problems.* We then see how, thanks to the principle of duality discovered by the great mathematician John von Neumann, this method can be used to solve a restricted class of *standard minimization problems.* Finally, we see how the simplex method can be adapted to solve *nonstandard problems*—problems that do not belong in the aforementioned categories.

How can Acrosonic fill its orders and at the same time keep its shipping costs to a minimum? The company manufactures its loudspeaker systems in two locations and ships them to three warehouses that serve as distribution centers. In Example 5, page 254, we find the shipping schedule that minimizes the shipping costs while meeting the needs of the distribution centers.

4.1 The Simplex Method: Standard Maximization Problems

The Simplex Method

As was mentioned in Chapter 3, the method of corners is not suitable for solving linear programming problems when the number of variables or constraints is large. Its major shortcoming is that a knowledge of all the corner points of the feasible set S associated with the problem is required. What we need is a method of solution that is based on a judicious selection of the corner points of the feasible set S, thereby reducing the number of points to be inspected. One such technique, called the *simplex method*, was developed in the late 1940s by George Dantzig and is based on the Gauss–Jordan elimination method. The simplex method is readily adaptable to the computer, which makes it ideally suitable for solving linear programming problems involving large numbers of variables and constraints.

Basically, the simplex method is an iterative procedure; that is, it is repeated over and over again. Beginning at some initial feasible solution (a corner point of the feasible set S, usually the origin), each iteration brings us to another corner point of S, usually with an improved (but certainly no worse) value of the objective function. The iteration is terminated when the optimal solution is reached (if it exists).

In this section, we describe the simplex method for solving a large class of problems that are referred to as standard maximization problems.

Before stating a formal procedure for solving standard linear programming problems based on the simplex method, let's consider the following analysis of a two-variable problem. The ensuing discussion will clarify the general procedure and at the same time enhance our understanding of the simplex method by examining the motivation that led to the steps of the procedure.

A Standard Linear Programming Problem

A **standard maximization problem** is one in which

1. The objective function is to be maximized.
2. All the variables involved in the problem are nonnegative.
3. All other linear constraints may be written so that the expression involving the variables is less than or equal to a nonnegative constant.

Consider the linear programming problem presented at the beginning of Section 3.3:

$$\text{Maximize } P = 3x + 2y \qquad \textbf{(1)}$$

$$\begin{aligned} \text{subject to } 2x + 3y &\le 12 \\ 2x + y &\le 8 \qquad \textbf{(2)} \\ x \ge 0, y &\ge 0 \end{aligned}$$

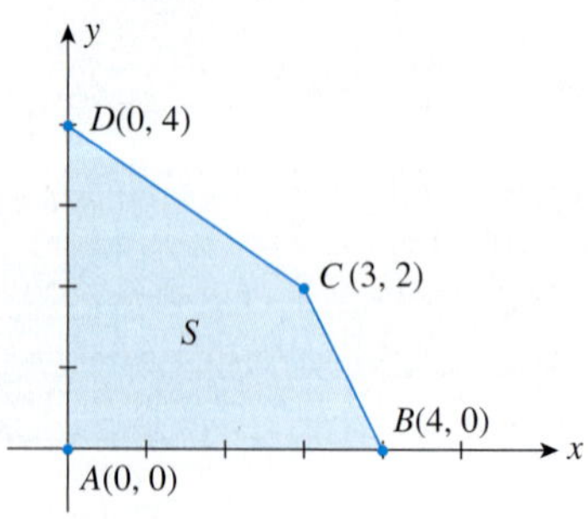

FIGURE 1
The optimal solution occurs at $C(3, 2)$.

You can easily verify that this is a standard maximization problem. The feasible set S associated with this problem is reproduced in Figure 1, where we have labeled the four feasible corner points $A(0, 0)$, $B(4, 0)$, $C(3, 2)$, and $D(0, 4)$. Recall that the optimal solution to the problem occurs at the corner point $C(3, 2)$.

To solve this problem using the simplex method, we first replace the system of inequality constraints (2) with a system of equality constraints. This may be accomplished by using nonnegative variables called **slack variables**. Let's begin by considering the inequality

$$2x + 3y \le 12$$

Observe that the left-hand side of this equation is always less than or equal to the right-hand side. Therefore, by adding a nonnegative variable u to the left-hand side to compensate for this difference, we obtain the equality

$$2x + 3y + u = 12$$

For example, if $x = 1$ and $y = 1$, then $u = 7$, because

$$2(1) + 3(1) + 7 = 12$$

(You can see by referring to Figure 1 that the point $(1, 1)$ is a feasible point of S.) If $x = 2$ and $y = 1$, then $u = 5$, because

$$2(2) + 3(1) + 5 = 12$$

(So the point $(2, 1)$ is also a feasible point of S.) The variable u is a slack variable.

Similarly, the inequality $2x + y \leq 8$ is converted into the equation $2x + y + v = 8$ through the introduction of the slack variable v. System (2) of linear inequalities may now be viewed as the system of linear equations

$$\begin{aligned} 2x + 3y + u &= 12 \\ 2x + y + v &= 8 \end{aligned}$$

where x, y, u, and v are all nonnegative.

Finally, rewriting the objective function (1) in the form $-3x - 2y + P = 0$, where the coefficient of P is $+1$, we are led to the following system of linear equations:

$$\begin{aligned} 2x + 3y + u &= 12 \\ 2x + y + v &= 8 \\ -3x - 2y + P &= 0 \end{aligned} \tag{3}$$

Since System (3) consists of three linear equations in the five variables x, y, u, v, and P, we may solve for three of the variables in terms of the other two. Thus, there are infinitely many solutions to this system expressible in terms of two parameters. Our linear programming problem is now seen to be equivalent to the following: From among all the solutions of System (3) for which x, y, u, and v are nonnegative (such solutions are called **feasible solutions**), determine the solution(s) that maximizes P.

The augmented matrix associated with System (3) is

Nonbasic variables (x, y); Basic variables (u, v, P); Column of constants

$$\begin{array}{c} \\ \\ \end{array}\begin{array}{cc|ccc|c} x & y & u & v & P & \\ \end{array}$$

$$\left[\begin{array}{ccccc|c} 2 & 3 & \boxed{1} & \boxed{0} & \boxed{0} & 12 \\ 2 & 1 & 0 & 1 & 0 & 8 \\ -3 & -2 & 0 & 0 & 1 & 0 \end{array}\right] \tag{4}$$

Observe that each of the u-, v-, and P-columns of the augmented matrix (4) is a unit column (see page 87). The variables associated with unit columns are called **basic variables**; all other variables are called **nonbasic variables.**

Now, the configuration of the augmented matrix (4) suggests that we solve for the basic variables u, v, and P in terms of the nonbasic variables x and y, obtaining

$$\begin{aligned} u &= 12 - 2x - 3y \\ v &= 8 - 2x - y \\ P &= \phantom{12 - {}} 3x + 2y \end{aligned} \tag{5}$$

Of the infinitely many feasible solutions that are obtainable by assigning arbitrary nonnegative values to the parameters x and y, a particular solution is obtained by letting $x = 0$ and $y = 0$. In fact, this solution is given by

$$x = 0 \qquad y = 0 \qquad u = 12 \qquad v = 8 \qquad P = 0$$

Such a solution, obtained by setting the nonbasic variables equal to zero, is called a **basic solution** of the system. This particular solution corresponds to the corner point $A(0, 0)$ of the feasible set associated with the linear programming problem (see Figure 1). Observe that $P = 0$ at this point.

Now, if the value of P cannot be increased, we have found the optimal solution to the problem at hand. To determine whether the value of P can in fact be improved, let's turn our attention to the objective function in Equation (1). Since the coefficients of both x and y are positive, the value of P can be improved by increasing x and/or y—that is, by moving away from the origin. Note that we arrive at the same conclusion by observing that the last row of the augmented matrix (4) contains entries that are *negative*. (Compare the original objective function, $P = 3x + 2y$, with the rewritten objective function, $-3x - 2y + P = 0$.)

Continuing our quest for an optimal solution, our next task is to determine whether it is more profitable to increase the value of x or that of y (increasing x and y simultaneously is more difficult). Since the coefficient of x is greater than that of y, a unit increase in the x-direction will result in a greater increase in the value of the objective function P than will a unit increase in the y-direction. Therefore, we should increase the value of x while holding y constant. How much can x be increased while holding $y = 0$? Upon setting $y = 0$ in the first two equations of System (5), we see that

$$\begin{aligned} u &= 12 - 2x \\ v &= \ \ 8 - 2x \end{aligned} \tag{6}$$

Since u must be nonnegative, the first equation of System (6) implies that x cannot exceed $\frac{12}{2}$, or 6. The second equation of System (6) and the nonnegativity of v imply that x cannot exceed $\frac{8}{2}$, or 4. Thus, we conclude that x can be increased by at most 4.

Now, if we set $y = 0$ and $x = 4$ in System (5), we obtain the solution

$$x = 4 \qquad y = 0 \qquad u = 4 \qquad v = 0 \qquad P = 12$$

which is a basic solution to System (3), this time with y and v as nonbasic variables. (Recall that the nonbasic variables are precisely the variables that are set equal to zero.)

Let's see how this basic solution may be found by working with the augmented matrix of the system. Since x is to replace v as a basic variable, our aim is to find an augmented matrix that is equivalent to the matrix (4) and has a configuration in which the x-column is in the unit form

$$\begin{bmatrix} 0 \\ 1 \\ 0 \end{bmatrix}$$

replacing what is presently the form of the v-column in augmented matrix (4). This may be accomplished by pivoting about the circled number 2.

$$\begin{array}{c} \begin{array}{cccccc} x & y & u & v & P & \text{Const.} \end{array} \\ \left[\begin{array}{ccccc|c} 2 & 3 & 1 & 0 & 0 & 12 \\ \textcircled{2} & 1 & 0 & 1 & 0 & 8 \\ -3 & -2 & 0 & 0 & 1 & 0 \end{array}\right] \end{array} \xrightarrow{\frac{1}{2}R_2} \begin{array}{c} \begin{array}{cccccc} x & y & u & v & P & \text{Const.} \end{array} \\ \left[\begin{array}{ccccc|c} 2 & 3 & 1 & 0 & 0 & 12 \\ \textcircled{1} & \frac{1}{2} & 0 & \frac{1}{2} & 0 & 4 \\ -3 & -2 & 0 & 0 & 1 & 0 \end{array}\right] \end{array} \tag{7}$$

$$\xrightarrow[R_3 + 3R_2]{R_1 - 2R_2}\quad \begin{array}{c} \begin{array}{cccccc} x & y & u & v & P & \text{Const.} \end{array} \\ \left[\begin{array}{ccccc|c} 0 & 2 & 1 & -1 & 0 & 4 \\ 1 & \frac{1}{2} & 0 & \frac{1}{2} & 0 & 4 \\ 0 & -\frac{1}{2} & 0 & \frac{3}{2} & 1 & 12 \end{array}\right] \end{array} \tag{8}$$

Using System (8), we now solve for the basic variables x, u, and P in terms of the nonbasic variables y and v, obtaining

$$\begin{aligned} x &= 4 - \frac{1}{2}y - \frac{1}{2}v \\ u &= 4 - 2y + \phantom{\frac{1}{2}}v \\ P &= 12 + \frac{1}{2}y - \frac{3}{2}v \end{aligned}$$

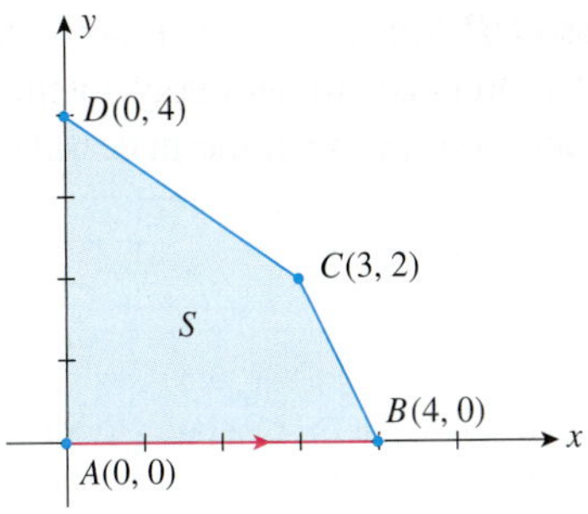

FIGURE **2**
One iteration has taken us from $A(0, 0)$, where $P = 0$, to $B(4, 0)$, where $P = 12$.

Setting the nonbasic variables y and v equal to zero gives

$$x = 4 \qquad y = 0 \qquad u = 4 \qquad v = 0 \qquad P = 12$$

as before.

We have now completed one iteration of the simplex procedure, and our search has brought us from the feasible corner point $A(0, 0)$, where $P = 0$, to the feasible corner point $B(4, 0)$, where P attained a value of 12, which is certainly an improvement! (See Figure 2.)

Before going on, let's introduce the following terminology. In what follows, refer to the augmented matrix (9), which is reproduced from the first augmented matrix in (7):

$$\begin{array}{c} \begin{array}{cccccc} x & y & u & v & P & \text{Const.} \end{array} \\ \text{Pivot row} \rightarrow \left[\begin{array}{ccccc|c} 2 & 3 & 1 & 0 & 0 & 12 \\ \textcircled{2} & 1 & 0 & 1 & 0 & 8 \\ -3 & -2 & 0 & 0 & 1 & 0 \end{array}\right] \end{array} \quad \begin{array}{l} \text{Ratio} \\ \frac{12}{2} = 6 \\ \frac{8}{2} = 4 \quad \text{(smallest ratio)} \\ \text{—} \end{array} \tag{9}$$

↑ Pivot column

(the negative number in the last row to the left of the vertical line with the largest absolute value)

The circled element 2 in the augmented matrix (9), which is to be converted into a 1, is called a *pivot element*. The column containing the pivot element is called the *pivot column*. The pivot column is associated with a nonbasic variable that is to be converted to a basic variable. Note that *the last entry in the pivot column is the negative number to the left of the vertical line in the last row with the largest absolute value*—precisely the criterion for choosing the direction of maximum increase in P.

The row containing the pivot element is called the *pivot row*. The pivot row can also be found by dividing each positive number in the pivot column into the corresponding number in the last column (the column of constants). *The pivot row is the one with the smallest ratio*. In augmented matrix (9), the pivot row is the second row because the ratio $\frac{8}{2}$, or 4, is less than the ratio $\frac{12}{2}$, or 6. (Compare this with the earlier analysis pertaining to the determination of the largest permissible increase in the value of x.) Then pivoting about the pivot element, we obtain the second tableau in (7).

The following is a summary of the procedure for selecting the pivot element.

Selecting the Pivot Element

1. *Select the pivot column.* Locate the most negative entry to the left of the vertical line in the last row. The column containing this entry is the **pivot column.** (If there is more than one such column, choose any one.)

(*continued*)

2. *Select the pivot row.* Divide each positive entry in the pivot column into its corresponding entry in the column of constants. The **pivot row** is the row corresponding to the smallest ratio thus obtained. (If there is more than one such entry, choose any one.)
3. The **pivot element** is the element common to both the pivot column and the pivot row.

Continuing with the solution to our problem, we observe that the last row of the augmented matrix (8) contains a negative number—namely, $-\frac{1}{2}$. This indicates that P is not maximized at the feasible corner point $B(4, 0)$, so another iteration is required. Without once again going into a detailed analysis, we proceed immediately to the selection of a pivot element. In accordance with the rules, we perform the necessary row operations as follows:

$$\begin{array}{c} \text{Pivot row} \rightarrow \\ \\ \\ \end{array} \begin{array}{c} \begin{array}{cccccc} x & y & u & v & P & \end{array} \\ \left[\begin{array}{ccccc|c} 0 & \textcircled{2} & 1 & -1 & 0 & 4 \\ 1 & \frac{1}{2} & 0 & \frac{1}{2} & 0 & 4 \\ 0 & -\frac{1}{2} & 0 & \frac{3}{2} & 1 & 12 \end{array}\right] \\ \uparrow \\ \text{Pivot column} \end{array} \quad \begin{array}{c} \text{Ratio} \\ \frac{4}{2} = 2 \\ \frac{4}{1/2} = 8 \end{array}$$

$$\xrightarrow{\frac{1}{2}R_1} \begin{array}{c} \begin{array}{cccccc} x & y & u & v & P & \end{array} \\ \left[\begin{array}{ccccc|c} 0 & \textcircled{1} & \frac{1}{2} & -\frac{1}{2} & 0 & 2 \\ 1 & \frac{1}{2} & 0 & \frac{1}{2} & 0 & 4 \\ 0 & -\frac{1}{2} & 0 & \frac{3}{2} & 1 & 12 \end{array}\right] \end{array}$$

$$\xrightarrow[R_3 + \frac{1}{2}R_1]{R_2 - \frac{1}{2}R_1} \begin{array}{c} \begin{array}{cccccc} x & y & u & v & P & \end{array} \\ \left[\begin{array}{ccccc|c} 0 & 1 & \frac{1}{2} & -\frac{1}{2} & 0 & 2 \\ 1 & 0 & -\frac{1}{4} & \frac{3}{4} & 0 & 3 \\ 0 & 0 & \frac{1}{4} & \frac{5}{4} & 1 & 13 \end{array}\right] \end{array}$$

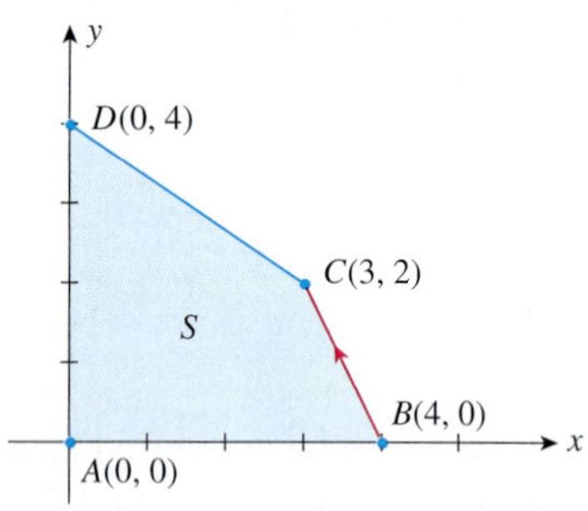

FIGURE 3
The next iteration has taken us from $B(4, 0)$, where $P = 12$, to $C(3, 2)$, where $P = 13$.

Interpreting the last augmented matrix in the usual fashion, we find the basic solution $x = 3$, $y = 2$, and $P = 13$. Since there are no negative entries in the last row, the solution is optimal, and P cannot be increased further. The optimal solution is the feasible corner point $C(3, 2)$ (Figure 3). Observe that this agrees with the solution we found using the method of corners in Section 3.3.

Having seen how the simplex method works, let's list the steps involved in the procedure. The first step is to set up the initial **simplex tableau.**

Setting Up the Initial Simplex Tableau

1. Transform the system of linear inequalities into a system of linear equations by introducing slack variables.
2. Rewrite the objective function

$$P = c_1x_1 + c_2x_2 + \cdots + c_nx_n$$

in the form

$$-c_1x_1 - c_2x_2 - \cdots - c_nx_n + P = 0$$

where all the variables are on the left and the coefficient of P is $+1$. Write this equation below the equations of Step 1.

3. Write the tableau associated with this system of linear equations.

EXAMPLE 1 Set up the initial simplex tableau for the linear programming problem posed in Example 1, Section 3.2.

Solution The problem at hand is to maximize

$$P = x + 1.2y$$

or, equivalently,

$$P = x + \frac{6}{5}y$$

subject to

$$\begin{aligned} 2x + y &\le 180 \\ x + 3y &\le 300 \\ x \ge 0, y &\ge 0 \end{aligned} \tag{10}$$

This is a standard maximization problem and may be solved by the simplex method. Since System (10) has two linear inequalities (other than $x \ge 0$, $y \ge 0$), we introduce the two slack variables u and v to convert it to a system of linear equations:

$$\begin{aligned} 2x + y + u \quad &= 180 \\ x + 3y \quad + v &= 300 \end{aligned}$$

Next, by rewriting the objective function in the form

$$-x - \frac{6}{5}y + P = 0$$

where the coefficient of P is $+1$, and placing it below the system of equations, we obtain the system of linear equations

$$\begin{aligned} 2x + y + u \qquad &= 180 \\ x + 3y \quad + v \quad &= 300 \\ -x - \frac{6}{5}y \qquad + P &= 0 \end{aligned}$$

The initial simplex tableau associated with this system is

x	y	u	v	P	Constant
2	1	1	0	0	180
1	3	0	1	0	300
-1	$-\frac{6}{5}$	0	0	1	0

Before completing the solution to the problem posed in Example 1, let's summarize the main steps of the **simplex method**.

The Simplex Method

1. *Set up the initial simplex tableau.*
2. *Determine whether the optimal solution has been reached by examining all entries in the last row to the left of the vertical line.*
   a. If all the entries are nonnegative, the optimal solution has been reached. Proceed to Step 4.
   b. If there are one or more negative entries, the optimal solution has not been reached. Proceed to Step 3.

(continued)

3. *Perform the pivot operation.* Locate the pivot element and convert it to a 1 by dividing all the elements in the pivot row by the pivot element. Using row operations, convert the pivot column into a unit column by adding suitable multiples of the pivot row to each of the other rows as required. Return to Step 2.
4. *Determine the optimal solution(s).* The value of the variable heading each unit column is given by the entry lying in the column of constants in the row containing the 1. The variables heading columns not in unit form are assigned the value zero.

EXAMPLE 2 Complete the solution to the problem discussed in Example 1.

Solution The first step in our procedure, setting up the initial simplex tableau, was completed in Example 1. We continue with Step 2.

Step 2 *Determine whether the optimal solution has been reached.* First, refer to the initial simplex tableau:

x	y	u	v	P	Constant
2	1	1	0	0	180
1	3	0	1	0	300
-1	$-\frac{6}{5}$	0	0	1	0

(11)

Since there are negative entries in the last row of the initial simplex tableau, the initial solution is not optimal. We proceed to Step 3.

Step 3 *Perform the following iterations.* First, locate the pivot element:

a. Since the entry $-\frac{6}{5}$ is the most negative entry to the left of the vertical line in the last row of the initial simplex tableau, the second column in the tableau is the pivot column.

b. Divide each positive number of the pivot column into the corresponding entry in the column of constants, and compare the ratios thus obtained. We see that the ratio $\frac{300}{3}$ is less than the ratio $\frac{180}{1}$, so row 2 is the pivot row.

c. The entry 3 lying in the pivot column and the pivot row is the pivot element.

	x	y	u	v	P	Constant	Ratio
	2	1	1	0	0	180	$\frac{180}{1} = 180$
Pivot row →	1	(3)	0	1	0	300	$\frac{300}{3} = 100$
	-1	$-\frac{6}{5}$	0	0	1	0	
		↑ Pivot column					

Next, we convert this pivot element into a 1 by multiplying all the entries in the pivot row by $\frac{1}{3}$. Then, using elementary row operations, we complete the conversion of the pivot column into a unit column. The details of the iteration follow:

	x	y	u	v	P	Constant
	2	1	1	0	0	180
$\xrightarrow{\frac{1}{3}R_2}$	$\frac{1}{3}$	(1)	0	$\frac{1}{3}$	0	100
	-1	$-\frac{6}{5}$	0	0	1	0

$\xrightarrow[R_3 + \frac{6}{5}R_2]{R_1 - R_2}$ **(12)**

x	y	u	v	P	Constant
$\frac{5}{3}$	0	1	$-\frac{1}{3}$	0	80
$\frac{1}{3}$	1	0	$\frac{1}{3}$	0	100
$-\frac{3}{5}$	0	0	$\frac{2}{5}$	1	120

This completes one iteration. The last row of the simplex tableau contains a negative number, so an optimal solution has not been reached. Therefore, we repeat the iterative step once again, as follows:

x	y	u	v	P	Constant	Ratio
$\frac{5}{3}$ (pivot)	0	1	$-\frac{1}{3}$	0	80	$\frac{80}{5/3} = 48$
$\frac{1}{3}$	1	0	$\frac{1}{3}$	0	100	$\frac{100}{1/3} = 300$
$-\frac{3}{5}$	0	0	$\frac{2}{5}$	1	120	

Pivot row → first row; ↑ Pivot column → x column

$\xrightarrow{\frac{3}{5}R_1}$

x	y	u	v	P	Constant
(1)	0	$\frac{3}{5}$	$-\frac{1}{5}$	0	48
$\frac{1}{3}$	1	0	$\frac{1}{3}$	0	100
$-\frac{3}{5}$	0	0	$\frac{2}{5}$	1	120

$\xrightarrow[R_3 + \frac{3}{5}R_1]{R_2 - \frac{1}{3}R_1}$ **(13)**

x	y	u	v	P	Constant
1	0	$\frac{3}{5}$	$-\frac{1}{5}$	0	48
0	1	$-\frac{1}{5}$	$\frac{2}{5}$	0	84
0	0	$\frac{9}{25}$	$\frac{7}{25}$	1	$148\frac{4}{5}$

The last row of the simplex tableau (13) contains no negative numbers, so we conclude that the optimal solution has been reached.

Step 4 *Determine the optimal solution.* Locate the basic variables in the final tableau. In this case, the basic variables (those heading unit columns) are x, y, and P. The value assigned to the basic variable x is the number 48, which is the entry lying in the column of constants and in row 1 (the row that contains the 1).

x	y	u	v	P	Constant
(1)	0	$\frac{3}{5}$	$-\frac{1}{5}$	0	48 ←
0	(1)	$-\frac{1}{5}$	$\frac{2}{5}$	0	84 ←
0	0	$\frac{9}{25}$	$\frac{7}{25}$	(1)	$148\frac{4}{5}$ ←

Similarly, we conclude that $y = 84$ and $P = 148.8$. Next, we note that the variables u and v are nonbasic and are accordingly assigned the values $u = 0$ and $v = 0$. These results agree with those obtained in Example 1, Section 3.3.

EXAMPLE 3

Maximize $P = 2x + 2y + z$

subject to

$$\begin{aligned} 2x + y + 2z &\le 14 \\ 2x + 4y + z &\le 26 \\ x + 2y + 3z &\le 28 \\ x \ge 0, y \ge 0, z &\ge 0 \end{aligned}$$

Solution Introducing the slack variables u, v, and w and rewriting the objective function in the standard form gives the system of linear equations

$$\begin{aligned} 2x + y + 2z + u \quad &= 14 \\ 2x + 4y + z \quad + v \quad &= 26 \\ x + 2y + 3z \quad + w \quad &= 28 \\ -2x - 2y - z \quad + P &= 0 \end{aligned}$$

The initial simplex tableau is given by

x	y	z	u	v	w	P	Constant
2	1	2	1	0	0	0	14
2	4	1	0	1	0	0	26
1	2	3	0	0	1	0	28
−2	−2	−1	0	0	0	1	0

Since the most negative entry in the last row (-2) occurs twice, we may choose either the x- or the y-column as the pivot column. Choosing the x-column as the pivot column and proceeding with the first iteration, we obtain the following sequence of tableaus:

	x	y	z	u	v	w	P	Constant	Ratio
Pivot row →	(2)	1	2	1	0	0	0	14	$\frac{14}{2} = 7$
	2	4	1	0	1	0	0	26	$\frac{26}{2} = 13$
	1	2	3	0	0	1	0	28	$\frac{28}{1} = 28$
	−2	−2	−1	0	0	0	1	0	
	↑ Pivot column								

	x	y	z	u	v	w	P	Constant
	(1)	$\frac{1}{2}$	1	$\frac{1}{2}$	0	0	0	7
$\xrightarrow{\frac{1}{2}R_1}$	2	4	1	0	1	0	0	26
	1	2	3	0	0	1	0	28
	−2	−2	−1	0	0	0	1	0

	x	y	z	u	v	w	P	Constant
	1	$\frac{1}{2}$	1	$\frac{1}{2}$	0	0	0	7
$\xrightarrow[R_3 - R_1,\ R_4 + 2R_1]{R_2 - 2R_1}$	0	3	−1	−1	1	0	0	12
	0	$\frac{3}{2}$	2	$-\frac{1}{2}$	0	1	0	21
	0	−1	1	1	0	0	1	14

Since there is a negative number in the last row of the simplex tableau, we perform another iteration, as follows:

	x	y	z	u	v	w	P	Constant	Ratio
	1	$\frac{1}{2}$	1	$\frac{1}{2}$	0	0	0	7	$\frac{7}{1/2} = 14$
Pivot row →	0	(3)	−1	−1	1	0	0	12	$\frac{12}{3} = 4$
	0	$\frac{3}{2}$	2	$-\frac{1}{2}$	0	1	0	21	$\frac{21}{3/2} = 14$
	0	−1	1	1	0	0	1	14	
		↑ Pivot column							

	x	y	z	u	v	w	P	Constant
	1	$\frac{1}{2}$	1	$\frac{1}{2}$	0	0	0	7
$\xrightarrow{\frac{1}{3}R_2}$	0	①	$-\frac{1}{3}$	$-\frac{1}{3}$	$\frac{1}{3}$	0	0	4
	0	$\frac{3}{2}$	2	$-\frac{1}{2}$	0	1	0	21
	0	-1	1	1	0	0	1	14

	x	y	z	u	v	w	P	Constant
	1	0	$\frac{7}{6}$	$\frac{2}{3}$	$-\frac{1}{6}$	0	0	5
$\xrightarrow{R_1 - \frac{1}{2}R_2}$	0	1	$-\frac{1}{3}$	$-\frac{1}{3}$	$\frac{1}{3}$	0	0	4
$R_3 - \frac{3}{2}R_2$ $R_4 + R_2$	0	0	$\frac{5}{2}$	0	$-\frac{1}{2}$	1	0	15
	0	0	$\frac{2}{3}$	$\frac{2}{3}$	$\frac{1}{3}$	0	1	18

All entries in the last row are nonnegative, so we have reached the optimal solution. We conclude that $x = 5$, $y = 4$, $z = 0$, $u = 0$, $v = 0$, $w = 15$, and $P = 18$.

Explore and Discuss

Consider the linear programming problem

$$\begin{aligned} \text{Maximize} \quad & P = x + 2y \\ \text{subject to} \quad & -2x + y \le 4 \\ & x - 3y \le 3 \\ & x \ge 0, y \ge 0 \end{aligned}$$

1. Sketch the feasible set S for the linear programming problem and explain why the problem has an unbounded solution.
2. Use the simplex method to solve the problem as follows:
   a. Perform one iteration on the initial simplex tableau. Interpret your result. Indicate the point on S corresponding to this (nonoptimal) solution.
   b. Show that the simplex procedure breaks down when you attempt to perform another iteration by demonstrating that there is no pivot element.
   c. Describe what happens if you violate the rule for finding the pivot element by allowing the ratios to be negative and proceeding with the iteration.

The following example is constructed to illustrate the geometry associated with the simplex method when used to solve a problem in three-dimensional space. We sketch the feasible set for the problem and show the path dictated by the simplex method in arriving at the optimal solution for the problem. The use of a calculator will help in the arithmetic operations if you wish to verify the steps.

EXAMPLE 4 Geometric Illustration of Simplex Method in 3-Space

$$\begin{aligned} \text{Maximize} \quad & P = 20x + 12y + 18z \\ \text{subject to} \quad & 3x + y + 2z \le 9 \\ & 2x + 3y + z \le 8 \\ & x + 2y + 3z \le 7 \\ & x \ge 0, y \ge 0, z \ge 0 \end{aligned}$$

Solution Introducing the slack variables u, v, and w and rewriting the objective function in standard form, we obtain the following system of linear equations:

$$
\begin{aligned}
3x + y + 2z + u &= 9 \\
2x + 3y + z + v &= 8 \\
x + 2y + 3z + w &= 7 \\
-20x - 12y - 18z + P &= 0
\end{aligned}
$$

The initial simplex tableau is given by

Initial Tableau starts at the point $A(0, 0, 0)$. See Figure 4.

x	y	z	u	v	w	P	Constant
3	1	2	1	0	0	0	9
2	3	1	0	1	0	0	8
1	2	3	0	0	1	0	7
−20	−12	−18	0	0	0	1	0

Since the most negative entry in the last row (-20) occurs in the x-column, we choose the x-column as the pivot column. Proceeding with the first iteration, we obtain the following sequence of tableaus:

	x	y	z	u	v	w	P	Constant	Ratio
Pivot row →	③	1	2	1	0	0	0	9	$\frac{9}{3} = 3$
	2	3	1	0	1	0	0	8	$\frac{8}{2} = 4$
	1	2	3	0	0	1	0	7	$\frac{7}{1} = 7$
	−20 ↑ Pivot column	−12	−18	0	0	0	1	0	

$\xrightarrow{\frac{1}{3}R_1}$

x	y	z	u	v	w	P	Constant
①	$\frac{1}{3}$	$\frac{2}{3}$	$\frac{1}{3}$	0	0	0	3
2	3	1	0	1	0	0	8
1	2	3	0	0	1	0	7
−20	−12	−18	0	0	0	1	0

First Iteration brings us to the point $B(3, 0, 0)$. See Figure 4.

$\xrightarrow[R_4 + 20R_1]{\substack{R_2 - 2R_1 \\ R_3 - R_1}}$

	x	y	z	u	v	w	P	Constant	Ratio
	1	$\frac{1}{3}$	$\frac{2}{3}$	$\frac{1}{3}$	0	0	0	3	9
Pivot row →	0	($\frac{7}{3}$)	$-\frac{1}{3}$	$-\frac{2}{3}$	1	0	0	2	$\frac{6}{7}$
	0	$\frac{5}{3}$	$\frac{7}{3}$	$-\frac{1}{3}$	0	1	0	4	$\frac{12}{5}$
	0	$-\frac{16}{3}$ ↑ Pivot column	$-\frac{14}{3}$	$\frac{20}{3}$	0	0	1	60	

Interpreting this tableau, we see that $x = 3$, $y = 0$, $z = 0$, and $P = 60$. Thus, after one iteration, we are at the point $B(3, 0, 0)$ with $P = 60$. (See Figure 4.)

Since the most negative entry in the last row is $-\frac{16}{3}$, we choose the y-column as the pivot column. Proceeding with this iteration, we obtain

$\xrightarrow{\frac{3}{7}R_2}$

x	y	z	u	v	w	P	Constant
1	$\frac{1}{3}$	$\frac{2}{3}$	$\frac{1}{3}$	0	0	0	3
0	⓵	$-\frac{1}{7}$	$-\frac{2}{7}$	$\frac{3}{7}$	0	0	$\frac{6}{7}$
0	$\frac{5}{3}$	$\frac{7}{3}$	$-\frac{1}{3}$	0	1	0	4
0	$-\frac{16}{3}$	$-\frac{14}{3}$	$\frac{20}{3}$	0	0	1	60

Second Iteration brings us to the point $C\left(\frac{19}{7}, \frac{6}{7}, 0\right)$. See Figure 4.

$\xrightarrow[\substack{R_3 - \frac{5}{3}R_2 \\ R_4 + \frac{16}{3}R_2}]{R_1 - \frac{1}{3}R_2}$

	x	y	z	u	v	w	P	Constant	Ratio
	1	0	$\frac{5}{7}$	$\frac{3}{7}$	$-\frac{1}{7}$	0	0	$\frac{19}{7}$	$\frac{19}{5}$
	0	1	$-\frac{1}{7}$	$-\frac{2}{7}$	$\frac{3}{7}$	0	0	$\frac{6}{7}$	—
Pivot row →	0	0	$\boxed{\frac{18}{7}}$	$\frac{1}{7}$	$-\frac{5}{7}$	1	0	$\frac{18}{7}$	1
	0	0	$-\frac{38}{7}$	$\frac{36}{7}$	$\frac{16}{7}$	0	1	$64\frac{4}{7}$	
			↑ Pivot column						

Interpreting this tableau, we see that $x = \frac{19}{7}$, $y = \frac{6}{7}$, $z = 0$ and $P = 64\frac{4}{7}$. Thus, the second iteration brings us to the point $C\left(\frac{19}{7}, \frac{6}{7}, 0\right)$. (See Figure 4.)

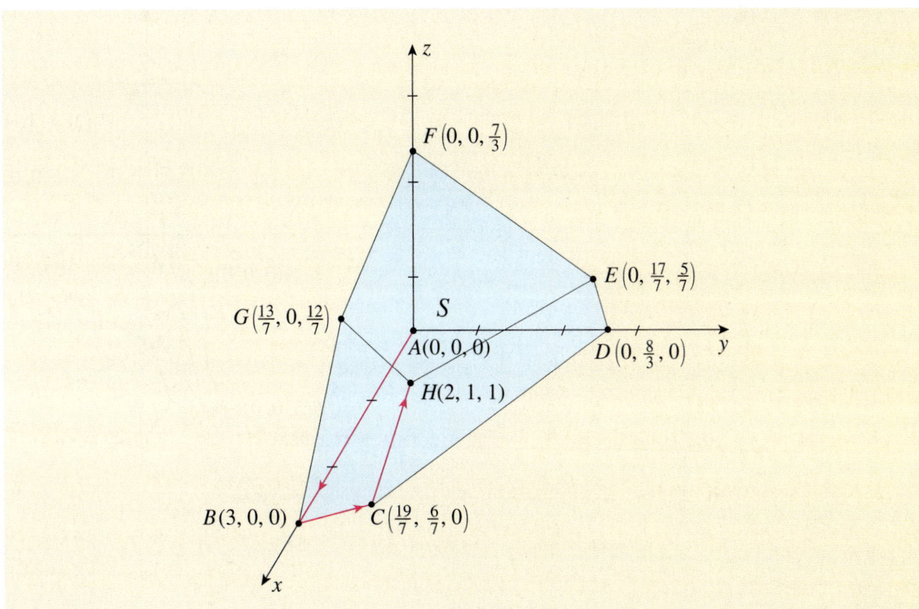

FIGURE **4**
The simplex method brings us from the point A to the point H, at which the objective function is maximized.

Since there is a negative number in the last row of the simplex tableau, we perform yet another iteration, as follows:

$\xrightarrow{\frac{7}{18}R_3}$

x	y	z	u	v	w	P	Constant
1	0	$\frac{5}{7}$	$\frac{3}{7}$	$-\frac{1}{7}$	0	0	$\frac{19}{7}$
0	1	$-\frac{1}{7}$	$-\frac{2}{7}$	$\frac{3}{7}$	0	0	$\frac{6}{7}$
0	0	⓵	$\frac{1}{18}$	$-\frac{5}{18}$	$\frac{7}{18}$	0	1
0	0	$-\frac{38}{7}$	$\frac{36}{7}$	$\frac{16}{7}$	0	1	$64\frac{4}{7}$

Third Iteration brings us to the point $H(2, 1, 1)$. We have reached the optimal solution.

$\xrightarrow[R_4 + \frac{38}{7}R_3]{R_1 - \frac{5}{7}R_3 \quad R_2 + \frac{1}{7}R_3}$

x	y	z	u	v	w	P	Constant
1	0	0	$\frac{7}{18}$	$\frac{1}{18}$	$-\frac{5}{18}$	0	2
0	1	0	$-\frac{5}{18}$	$\frac{7}{18}$	$\frac{1}{18}$	0	1
0	0	1	$\frac{1}{18}$	$-\frac{5}{18}$	$\frac{7}{18}$	0	1
0	0	0	$\frac{49}{9}$	$\frac{7}{9}$	$\frac{19}{9}$	1	70

All entries in the last row are nonnegative, so we have reached the optimal solution (corresponding to the point $H(2, 1, 1)$). We conclude that $x = 2$, $y = 1$, $z = 1$, $u = 0$, $v = 0$, $w = 0$, and $P = 70$.

The feasible set S for the problem is the hexahedron shown in Figure 5. It is the intersection of the half-spaces determined by the planes P_1, P_2, and P_3 with equations $3x + y + 2z = 9$, $2x + 3y + z = 8$, $x + 2y + 3z = 7$, respectively, and the coordinate planes $x = 0$, $y = 0$, and $z = 0$. The portion of the figure showing the feasible set S is shown in Figure 4. Observe that the first iteration of the simplex method brings us from $A(0, 0, 0)$ with $P = 0$ to $B(3, 0, 0)$ with $P = 60$. The second iteration brings us from $B(3, 0, 0)$ to $C\left(\frac{19}{7}, \frac{6}{7}, 0\right)$ with $P = 64\frac{4}{7}$, and the third iteration brings us from $C\left(\frac{19}{7}, \frac{6}{7}, 0\right)$ to the point $H(2, 1, 1)$ with an optimal value of 70 for P.

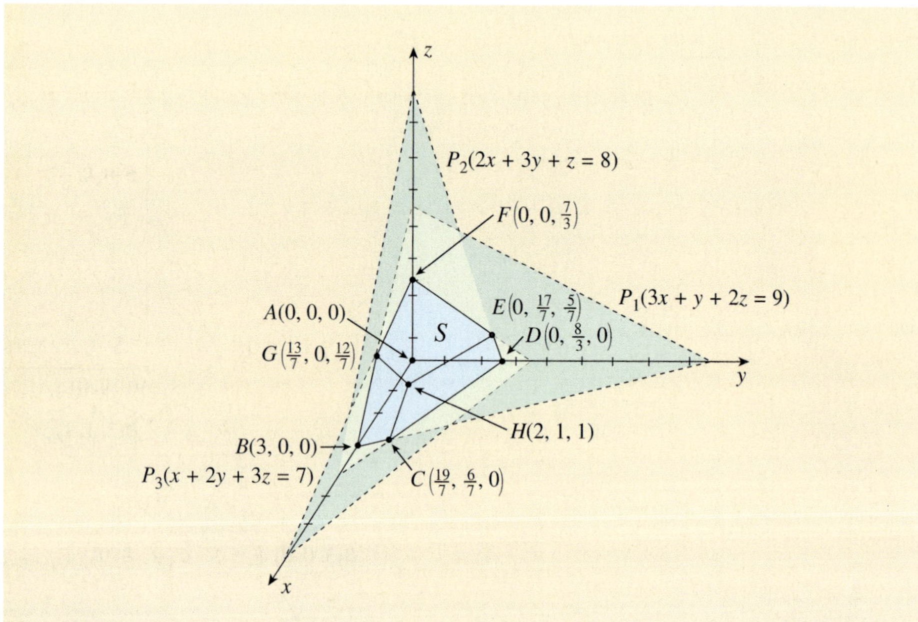

FIGURE 5
The feasible set S is obtained from the intersection of the half-spaces determined by P_1, P_2, and P_3 with the coordinate planes $x = 0$, $y = 0$, and $z = 0$.

APPLIED EXAMPLE 5 Production Planning Ace Novelty Company has determined that the profit for each Type A, Type B, and Type C souvenir that it plans to produce is \$6, \$5, and \$4, respectively. To manufacture a Type A souvenir requires 2 minutes on Machine I, 1 minute on Machine II, and 2 minutes on Machine III. A Type B souvenir requires 1 minute on Machine I, 3 minutes on Machine II, and 1 minute on Machine III. A Type C souvenir requires 1 minute on Machine I and 2 minutes on each of Machines II and III. Each day, there are 3 hours available on Machine I, 5 hours available on Machine II, and 4 hours available on Machine III for manufacturing these souvenirs. How many souvenirs of each type should Ace Novelty make per day to maximize its profit? (Compare with Example 1, Section 2.1.)

Solution The given information is tabulated as follows:

	Type *A*	Type *B*	Type *C*	Time Available (min)
Machine I	2	1	1	180
Machine II	1	3	2	300
Machine III	2	1	2	240
Profit/Unit	\$6	\$5	\$4	

Let x, y, and z denote the respective numbers of Type A, Type B, and Type C souvenirs to be made. The total amount of time that Machine I is used is given by $2x + y + z$ minutes and must not exceed 180 minutes. Thus, we have the inequality

$$2x + y + z \leq 180$$

Similar considerations on the use of Machines II and III lead to the inequalities

$$\begin{aligned} x + 3y + 2z &\leq 300 \\ 2x + y + 2z &\leq 240 \end{aligned}$$

The profit resulting from the sale of the souvenirs produced is given by

$$P = 6x + 5y + 4z$$

The mathematical formulation of this problem leads to the following standard linear programming problem: Maximize the objective (profit) function $P = 6x + 5y + 4z$ subject to

$$\begin{aligned} 2x + y + z &\leq 180 \\ x + 3y + 2z &\leq 300 \\ 2x + y + 2z &\leq 240 \\ x \geq 0, y \geq 0, z &\geq 0 \end{aligned}$$

Introducing the slack variables u, v, and w gives the system of linear equations

$$\begin{aligned} 2x + y + z + u \qquad\qquad &= 180 \\ x + 3y + 2z \quad + v \qquad &= 300 \\ 2x + y + 2z \qquad + w \quad &= 240 \\ -6x - 5y - 4z \qquad\qquad + P &= 0 \end{aligned}$$

The tableaus resulting from the use of the simplex algorithm are as follows:

	x	y	z	u	v	w	P	Constant	Ratio
Pivot row →	(2)	1	1	1	0	0	0	180	$\frac{180}{2} = 90$
	1	3	2	0	1	0	0	300	$\frac{300}{1} = 300$
	2	1	2	0	0	1	0	240	$\frac{240}{2} = 120$
	−6	−5	−4	0	0	0	1	0	
	↑ Pivot column								

	x	y	z	u	v	w	P	Constant
$\xrightarrow{\frac{1}{2}R_1}$	(1)	$\frac{1}{2}$	$\frac{1}{2}$	$\frac{1}{2}$	0	0	0	90
	1	3	2	0	1	0	0	300
	2	1	2	0	0	1	0	240
	−6	−5	−4	0	0	0	1	0

$\xrightarrow[R_3 - 2R_1,\ R_4 + 6R_1]{R_2 - R_1}$

	x	y	z	u	v	w	P	Constant	Ratio
	1	$\frac{1}{2}$	$\frac{1}{2}$	$\frac{1}{2}$	0	0	0	90	$\frac{90}{1/2} = 180$
Pivot row →	0	($\frac{5}{2}$)	$\frac{3}{2}$	$-\frac{1}{2}$	1	0	0	210	$\frac{210}{5/2} = 84$
	0	0	1	−1	0	1	0	60	
	0	−2	−1	3	0	0	1	540	
		↑ Pivot column							

$\xrightarrow{\frac{2}{5}R_2}$

x	y	z	u	v	w	P	Constant
1	$\frac{1}{2}$	$\frac{1}{2}$	$\frac{1}{2}$	0	0	0	90
0	(1)	$\frac{3}{5}$	$-\frac{1}{5}$	$\frac{2}{5}$	0	0	84
0	0	1	−1	0	1	0	60
0	−2	−1	3	0	0	1	540

$\xrightarrow[R_4 + 2R_2]{R_1 - \frac{1}{2}R_2}$

x	y	z	u	v	w	P	Constant
1	0	$\frac{1}{5}$	$\frac{3}{5}$	$-\frac{1}{5}$	0	0	48
0	1	$\frac{3}{5}$	$-\frac{1}{5}$	$\frac{2}{5}$	0	0	84
0	0	1	−1	0	1	0	60
0	0	$\frac{1}{5}$	$\frac{13}{5}$	$\frac{4}{5}$	0	1	708

From the final simplex tableau, we read off the solution

$$x = 48 \qquad y = 84 \qquad z = 0 \qquad u = 0 \qquad v = 0 \qquad w = 60 \qquad P = 708$$

Thus, to maximize its profit, Ace Novelty should produce 48 Type *A* souvenirs, 84 Type *B* souvenirs, and no Type *C* souvenirs. The resulting profit is \$708 per day. The value of the slack variable $w = 60$ tells us that 1 hour of the available time on Machine III is left unused.

Interpreting Our Results Let's compare the results obtained here with those obtained in Example 7, Section 2.2. Recall that to use all available machine time on each of the three machines, Ace Novelty had to produce 36 Type *A*, 48 Type *B*, and 60 Type *C* souvenirs. This would have resulted in a profit of \$696. Example 5 shows how, through the optimal use of equipment, a company can boost its profit while reducing machine wear!

Problems with Multiple Solutions and Problems with No Solutions

As we saw in Section 3.3, a linear programming problem may have infinitely many solutions. We also saw that a linear programming problem may have no solution. How do we spot each of these phenomena when using the simplex method to solve a problem?

A linear programming problem will have infinitely many solutions if, for example, the last row to the left of the vertical line of the final simplex tableau has a zero in a column that is not a unit column, or if the final tableau contains two or more identical unit columns. Also, a linear programming problem will have *no* solution if the simplex method breaks down at some stage. For example, if at some stage there are no nonnegative ratios in our computation, then the linear programming problem has no solution (see Exercise 51).

Explore and Discuss

Consider the linear programming problem

$$\begin{aligned} \text{Maximize} \quad & P = 4x + 6y \\ \text{subject to} \quad & 2x + y \le 10 \\ & 2x + 3y \le 18 \\ & x \ge 0, y \ge 0 \end{aligned}$$

1. Sketch the feasible set for the linear programming problem.
2. Use the method of corners to show that there are infinitely many optimal solutions. What are they?
3. Use the simplex method to solve the problem as follows.
   a. Perform one iteration on the initial simplex tableau and conclude that you have arrived at an optimal solution. What is the value of P, and where is it attained? Compare this result with that obtained in Step 2.
   b. Observe that the tableau obtained in part (a) indicates that there are infinitely many solutions (see the comment above on multiple solutions). Now perform another iteration on the simplex tableau using the x-column as the pivot column. Interpret the final tableau.

4.1 Self-Check Exercises

1. Solve the following linear programming problem by the simplex method:

$$\begin{aligned} \text{Maximize} \quad & P = 2x + 3y + 6z \\ \text{subject to} \quad & 2x + 3y + z \le 10 \\ & x + y + 2z \le 8 \\ & 2y + 3z \le 6 \\ & x \ge 0, y \ge 0, z \ge 0 \end{aligned}$$

2. **Maximizing Profit** The LaCrosse Iron Works makes two models of cast-iron fireplace grates: model A and model B. Producing one model A grate requires 20 lb of cast iron and 20 min of labor, whereas producing one model B grate requires 30 lb of cast iron and 15 min of labor. The profit for a model A grate is \$6, and the profit for a model B grate is \$8. There are 7200 lb of cast iron and 100 labor-hours available each week. Because of a surplus from the previous week, the proprietor has decided to make no more than 150 units of model A grates this week. Determine how many of each model he should make to maximize his profit.

Solutions to Self-Check Exercises 4.1 can be found on page 240.

4.1 Concept Questions

1. Give the three characteristics of a standard maximization linear programming problem.
2. **a.** When the initial simplex tableau is set up, how is the system of linear inequalities transformed into a system of linear equations? How is the objective function $P = c_1x_1 + c_2x_2 + \cdots + c_nx_n$ rewritten?
   b. If you are given a simplex tableau, how do you determine whether the optimal solution has been reached?
3. In the simplex method, how is a pivot column selected? A pivot row? A pivot element?

4.1 Exercises

In Exercises 1–6, (a) write the linear programming problem as a standard maximization problem if it is not already in that form, and (b) write the initial simplex tableau.

1. Maximize $P = 2x + 4y$ subject to the constraints

$$\begin{aligned} x + 4y &\leq 12 \\ x + 3y &\leq 10 \\ x \geq 0, y &\geq 0 \end{aligned}$$

2. Maximize $P = 3x + 5y$ subject to the constraints

$$\begin{aligned} x + 3y &\leq 12 \\ -2x - 3y &\geq -18 \\ x \geq 0, y &\geq 0 \end{aligned}$$

3. Maximize $P = 2x + 3y$ subject to the constraints

$$\begin{aligned} x + y &\leq 10 \\ -x - 2y &\geq -12 \\ 2x + y &\leq 12 \\ x \geq 0, y &\geq 0 \end{aligned}$$

4. Maximize $P = 2x + 5y$ subject to the constraints

$$\begin{aligned} 3x + 8y &\leq 1 \\ 4x - 5y &\leq 4 \\ 2x + 7y &\leq 6 \\ x \geq 0, y &\geq 0 \end{aligned}$$

5. Maximize $P = x + 3y + 4z$ subject to the constraints

$$\begin{aligned} x + 2y + z &\leq 40 \\ -x - y - z &\geq -30 \\ x \geq 0, y \geq 0, z &\geq 0 \end{aligned}$$

6. Maximize $P = 4x + 5y + 6z$ subject to the constraints

$$\begin{aligned} 2x + 3y + z &\leq 900 \\ 3x + y + z &\leq 350 \\ 4x + 2y + z &\leq 400 \\ x \geq 0, y \geq 0, z &\geq 0 \end{aligned}$$

In Exercises 7–16, determine whether the given simplex tableau is in final form. If so, find the solution to the associated regular linear programming problem. If not, find the pivot element to be used in the next iteration of the simplex method.

7.

x	y	u	v	P	Constant
0	1	$\frac{5}{7}$	$-\frac{1}{7}$	0	$\frac{20}{7}$
1	0	$-\frac{3}{7}$	$\frac{2}{7}$	0	$\frac{30}{7}$
0	0	$\frac{13}{7}$	$\frac{3}{7}$	1	$\frac{220}{7}$

8.

x	y	u	v	P	Constant
1	1	1	0	0	6
1	0	−1	1	0	2
3	0	5	0	1	30

9.

x	y	u	v	P	Constant
0	$\frac{1}{2}$	1	$-\frac{1}{2}$	0	2
1	$\frac{1}{2}$	0	$\frac{1}{2}$	0	4
0	$-\frac{1}{2}$	0	$\frac{3}{2}$	1	12

10.

x	y	z	u	v	P	Constant
3	0	5	1	1	0	28
2	1	3	0	1	0	16
2	0	8	0	3	1	48

11.

x	y	z	u	v	w	P	Constant
1	$-\frac{1}{3}$	0	$\frac{1}{3}$	0	$-\frac{2}{3}$	0	$\frac{1}{3}$
0	2	0	0	1	1	0	6
0	$\frac{2}{3}$	1	$\frac{1}{3}$	0	$\frac{1}{3}$	0	$\frac{13}{3}$
0	4	0	1	0	2	1	17

12.

x	y	z	u	v	w	P	Constant
$\frac{1}{2}$	0	$\frac{1}{4}$	1	$-\frac{1}{4}$	0	0	$\frac{19}{2}$
$\frac{1}{2}$	1	$\frac{3}{4}$	0	$\frac{1}{4}$	0	0	$\frac{21}{2}$
2	0	3	0	0	1	0	30
−1	0	$-\frac{1}{2}$	6	$\frac{3}{2}$	0	1	63

13.

x	y	z	s	t	u	v	P	Constant
$\frac{5}{2}$	3	0	1	0	0	−4	0	46
1	0	0	0	1	0	0	0	9
0	1	0	0	0	1	0	0	12
0	0	1	0	0	0	1	0	6
−180	−200	0	0	0	0	300	1	1800

14.

x	y	z	s	t	u	v	P	Constant
1	0	0	$\frac{2}{5}$	0	$-\frac{6}{5}$	$-\frac{8}{5}$	0	4
0	0	0	$-\frac{2}{5}$	1	$\frac{6}{5}$	$\frac{8}{5}$	0	5
0	1	0	0	0	1	0	0	12
0	0	1	0	0	0	1	0	6
0	0	0	72	0	−16	12	1	4920

15.

x	y	z	u	v	P	Constant
1	0	$\frac{3}{5}$	0	$\frac{1}{5}$	0	30
0	1	$-\frac{19}{5}$	1	$-\frac{3}{5}$	0	10
0	0	$\frac{26}{5}$	0	0	1	60

16.

x	y	z	u	v	w	P	Constant
0	$\frac{1}{2}$	0	1	$-\frac{1}{2}$	0	0	2
1	$\frac{1}{2}$	1	0	$\frac{1}{2}$	0	0	13
2	$\frac{1}{2}$	0	0	$-\frac{3}{2}$	1	0	4
-1	3	0	0	1	0	1	26

In Exercises 17–31, solve each linear programming problem by the simplex method.

17. Maximize $P = 3x + 4y$
subject to
$$\begin{aligned} x + y &\le 4 \\ 2x + y &\le 5 \\ x \ge 0, y &\ge 0 \end{aligned}$$

18. Maximize $P = 5x + 3y$
subject to
$$\begin{aligned} x + y &\le 80 \\ 3x \quad &\le 90 \\ x \ge 0, y &\ge 0 \end{aligned}$$

19. Maximize $P = 10x + 12y$
subject to
$$\begin{aligned} x + 2y &\le 12 \\ 3x + 2y &\le 24 \\ x \ge 0, y &\ge 0 \end{aligned}$$

20. Maximize $P = 5x + 4y$
subject to
$$\begin{aligned} 3x + 5y &\le 78 \\ 4x + y &\le 36 \\ x \ge 0, y &\ge 0 \end{aligned}$$

21. Maximize $P = 4x + 6y$
subject to
$$\begin{aligned} 3x + y &\le 24 \\ 2x + y &\le 18 \\ x + 3y &\le 24 \\ x \ge 0, y &\ge 0 \end{aligned}$$

22. Maximize $P = 15x + 12y$
subject to
$$\begin{aligned} x + y &\le 12 \\ 3x + y &\le 30 \\ 10x + 7y &\le 70 \\ x \ge 0, y &\ge 0 \end{aligned}$$

23. Maximize $P = 3x + 4y + z$
subject to
$$\begin{aligned} 3x + 10y + 5z &\le 120 \\ 5x + 2y + 8z &\le 6 \\ 8x + 10y + 3z &\le 105 \\ x \ge 0, y \ge 0, z &\ge 0 \end{aligned}$$

24. Maximize $P = 3x + 3y + 4z$
subject to
$$\begin{aligned} x + y + 3z &\le 15 \\ 4x + 4y + 3z &\le 65 \\ x \ge 0, y \ge 0, z &\ge 0 \end{aligned}$$

25. Maximize $P = 3x + 4y + 5z$
subject to
$$\begin{aligned} x + y + z &\le 8 \\ 3x + 2y + 4z &\le 24 \\ x \ge 0, y \ge 0, z &\ge 0 \end{aligned}$$

26. Maximize $P = x + 4y - 2z$
subject to
$$\begin{aligned} 3x + y - z &\le 80 \\ 2x + y - z &\le 40 \\ -x + y + z &\le 80 \\ x \ge 0, y \ge 0, z &\ge 0 \end{aligned}$$

27. Maximize $P = 4x + 6y + 5z$
subject to
$$\begin{aligned} x + y + z &\le 20 \\ 2x + 4y + 3z &\le 42 \\ 2x \quad + 3z &\le 30 \\ x \ge 0, y \ge 0, z &\ge 0 \end{aligned}$$

28. Maximize $P = x + 2y - z$
subject to
$$\begin{aligned} 2x + y + z &\le 14 \\ 4x + 2y + 3z &\le 28 \\ 2x + 5y + 5z &\le 30 \\ x \ge 0, y \ge 0, z &\ge 0 \end{aligned}$$

29. Maximize $P = 12x + 10y + 5z$
subject to
$$\begin{aligned} 2x + y + z &\le 10 \\ 3x + 5y + z &\le 45 \\ 2x + 5y + z &\le 40 \\ x \ge 0, y \ge 0, z &\ge 0 \end{aligned}$$

30. Maximize $P = 2x + 6y + 6z$
subject to
$$\begin{aligned} 2x + y + 3z &\le 10 \\ 4x + y + 2z &\le 56 \\ 6x + 4y + 3z &\le 126 \\ 2x + y + z &\le 32 \\ x \ge 0, y \ge 0, z &\ge 0 \end{aligned}$$

31. Maximize $P = 24x + 16y + 23z$
subject to
$$\begin{aligned} 2x + y + 2z &\le 7 \\ 2x + 3y + z &\le 8 \\ x + 2y + 3z &\le 7 \\ x \ge 0, y \ge 0, z &\ge 0 \end{aligned}$$

32. Rework Example 3 using the y-column as the pivot column in the first iteration of the simplex method.

33. Show that the linear programming problem

Maximize $P = 2x + 2y - 4z$
subject to
$$\begin{aligned} 3x + 3y - 2z &\le 100 \\ 5x + 5y + 3z &\le 150 \\ x \ge 0, y \ge 0, z &\ge 0 \end{aligned}$$

has infinitely many solutions and give two of them.

34. **Production Scheduling** A company manufactures two products, A and B, on two machines, I and II. It has been determined that the company will realize a profit of \$3/unit on Product A and a profit of \$4/unit on Product B. To manufacture 1 unit of Product A requires 6 min on Machine I and 5 min on Machine II. To manufacture 1 unit of Product B requires 9 min on Machine I and 4 min on Machine II. There are 5 hr of machine time available

on Machine I and 3 hr of machine time available on Machine II in each work shift. How many units of each product should be produced in each shift to maximize the company's profit? What is the largest profit the company can realize? Is there any time left unused on the machines?

35. **Production Scheduling** National Business Machines Corporation manufactures two models of portable printers: A and B. Each model A costs \$100 to make, and each model B costs \$150. The profits are \$30 for each model A and \$40 for each model B portable printer. If the total number of portable printers demanded each month does not exceed 2500 and the company has earmarked no more than \$600,000/month for manufacturing costs, find how many units of each model National should make each month to maximize its monthly profit. What is the largest monthly profit the company can make?

36. **Production Scheduling** Kane Manufacturing has a division that produces two models of hibachis, model A and model B. To produce each model A hibachi requires 3 lb of cast iron and 6 min of labor. To produce each model B hibachi requires 4 lb of cast iron and 3 min of labor. The profit for each model A hibachi is \$2, and the profit for each model B hibachi is \$1.50. If 1000 lb of cast iron and 20 labor-hours are available for the production of hibachis each day, how many hibachis of each model should the division produce in order to maximize Kane's profit? What is the largest profit the company can realize? Is there any raw material left over?

37. **Asset Allocation** Justin has decided to invest at most \$60,000 in medium-risk and high-risk stocks. He has further decided that the medium-risk stocks should make up at least 40% of the total investment, while the high-risk stocks should make up at least 20% of the total investment. He expects that the medium-risk stocks will appreciate by 12% and the high-risk stocks will appreciate by 20% within a year. How much money should Justin invest in each type of stock to maximize the value of his investment?
Hint: Write the linear programming problem in the form of a standard maximization problem.

38. **Asset Allocation** A financier plans to invest up to \$500,000 in two projects. Project *A* yields a return of 10% on the investment, whereas Project *B* yields a return of 15% on the investment. Because the investment in Project *B* is riskier than the investment in Project *A*, the financier has decided that the investment in Project *B* should not exceed 40% of the total investment. How much should she invest in each project to maximize the return on her investment? What is the maximum return?

39. **Production Scheduling** A division of the Winston Furniture Company manufactures dining tables and chairs. Each table requires 40 board feet of wood and 3 labor-hours. Each chair requires 16 board feet of wood and 4 labor-hours. The profit for each table is \$45, and the profit for each chair is \$20. In a certain week, the company has 3200 board feet of wood available and 520 labor-hours available. How many tables and chairs should Winston manufacture to maximize its profit? What is the maximum profit?

40. **Crop Planning** A farmer has 150 acres of land suitable for cultivating Crops *A* and *B*. The cost of cultivating Crop *A* is \$40/acre, whereas the cost of cultivating Crop *B* is \$60/acre. The farmer has a maximum of \$7400 available for land cultivation. Each acre of Crop *A* requires 20 labor-hours, and each acre of Crop *B* requires 25 labor-hours. The farmer has a maximum of 3300 labor-hours available. If he expects to make a profit of \$150/acre on Crop *A* and \$200/acre on Crop *B*, how many acres of each crop should he plant to maximize his profit? What is the largest profit the farmer can realize? Are there any resources left over?

41. **Production Scheduling** A company manufactures Products *A*, *B*, and *C*. Each product is processed in three departments: I, II, and III. The total available labor-hours per week for Departments I, II, and III are 900, 1080, and 840, respectively. The time requirements (in hours per unit) and profit per unit for each product are as follows:

	Product *A*	Product *B*	Product *C*
Dept. I	2	1	2
Dept. II	3	1	2
Dept. III	2	2	1
Profit	\$18	\$12	\$15

How many units of each product should the company produce to maximize its profit? What is the largest profit the company can realize? Are there any resources left over?

42. **Asset Allocation** Ashley has earmarked at most \$250,000 for investment in three mutual funds: a money market fund, an international equity fund, and a growth-and-income fund. The money market fund has a rate of return of 6%/year, the international equity fund has a rate of return of 10%/year, and the growth-and-income fund has a rate of return of 15%/year. Ashley has stipulated that no more than 25% of her total portfolio should be in the growth-and-income fund and that no more than 50% of her total portfolio should be in the international equity fund. To maximize the return on her investment, how much should Ashley invest in each type of fund? What is the maximum return?

43. **Production Scheduling** Ace Novelty manufactures Giant Pandas and Saint Bernards. Each Panda requires 1.5 yd^2 of plush, 30 ft^3 of stuffing, and 5 pieces of trim; each Saint Bernard requires 2 yd^2 of plush, 35 ft^3 of stuffing, and 8 pieces of trim. The profit for each Panda is \$10, and the profit for each Saint Bernard is \$15. If 3600 yd^2 of plush, 66,000 ft^3 of stuffing, and 13,600 pieces of trim are available, how many of each of the stuffed animals should the company manufacture to maximize its profit? What is the maximum profit?

44. **Optimizing Advertising Exposure** As part of a campaign to promote its annual clearance sale, Excelsior Company decided to buy television advertising time on Station KAOS. Excelsior's television advertising budget is \$102,000. Morning time costs \$3000/min, afternoon time costs \$1000/min, and evening (prime) time costs \$12,000/min. Because of previous commitments, KAOS cannot offer Excelsior more than 6 min of prime time or more than a total of 25 min of advertising time over the 2 weeks in which the commercials are to be run. KAOS estimates that morning commercials are seen by 200,000 people, afternoon commercials are seen by 100,000 people, and evening commercials are seen by 600,000 people. How much morning, afternoon, and evening advertising time should Excelsior buy to maximize exposure of its commercials?

45. **Production Scheduling** Custom Office Furniture is introducing a new line of executive desks made from a specially selected grade of walnut. Initially, three models—A, B, and C—are to be marketed. Each model A desk requires $1\frac{1}{4}$ hr for fabrication, 1 hr for assembly, and 1 hr for finishing; each model B desk requires $1\frac{1}{2}$ hr for fabrication, 1 hr for assembly, and 1 hr for finishing; each model C desk requires $1\frac{1}{2}$ hr, $\frac{3}{4}$ hr, and $\frac{1}{2}$ hr for fabrication, assembly, and finishing, respectively. The profit on each model A desk is \$26, the profit on each model B desk is \$28, and the profit on each model C desk is \$24. The total time available in the fabrication department, the assembly department, and the finishing department in the first month of production is 310 hr, 205 hr, and 190 hr, respectively. To maximize Custom's profit, how many desks of each model should be made in the month? What is the largest profit the company can realize? Are there any resources left over?

46. **Optimizing Profit for Prefabricated Housing** Boise Lumber has decided to enter the lucrative prefabricated housing business. Initially, it plans to offer three models: standard, deluxe, and luxury. Each house is prefabricated and partially assembled in the factory, and the final assembly is completed on site. The dollar amount of building material required, the amount of labor required in the factory for prefabrication and partial assembly, the amount of on-site labor required, and the profit per unit are as follows:

	Standard Model	Deluxe Model	Luxury Model
Material	\$6,000	\$8,000	\$10,000
Factory Labor (hr)	240	220	200
On-site Labor (hr)	180	210	300
Profit	\$3,400	\$4,000	\$5,000

For the first year's production, a sum of \$8,200,000 is budgeted for the building material; the number of labor-hours available for work in the factory (for prefabrication and partial assembly) is not to exceed 218,000 hr; and the amount of labor for on-site work is to be less than or equal to 237,000 labor-hours. Determine how many houses of each type Boise should produce to maximize its profit from this new venture. (Market research has confirmed that there should be no problems with sales.)

47. **Asset Allocation** Sharon has a total of \$200,000 to invest in three types of mutual funds: growth, balanced, and income funds. Growth funds have a rate of return of 12%/year, balanced funds have a rate of return of 10%/year, and income funds have a return of 6%/year. The growth, balanced, and income mutual funds are assigned risk factors of 0.1, 0.06, and 0.02, respectively. Sharon has decided that at least 50% of her total portfolio is to be in income funds and at least 25% in balanced funds. She has also decided that the average risk factor for her investment should not exceed 0.05. How much should Sharon invest in each type of fund to realize a maximum return on her investment? What is the maximum return?
Hint: The constraint for the average risk factor for the investment is given by $0.1x + 0.06y + 0.02z \leq 0.05(x + y + z)$.

48. **Optimizing Profit for Blended Juice Drinks** CalJuice Company has decided to introduce three fruit juices made from blending two or more concentrates. These juices will be packaged in 2-qt (64-oz) cartons. One carton of pineapple–orange juice requires 8 oz each of pineapple and orange juice concentrates. One carton of orange–banana juice requires 12 oz of orange juice concentrate and 4 oz of banana pulp concentrate. Finally, one carton of pineapple–orange–banana juice requires 4 oz of pineapple juice concentrate, 8 oz of orange juice concentrate, and 4 oz of banana pulp. The company has decided to allot 16,000 oz of pineapple juice concentrate, 24,000 oz of orange juice concentrate, and 5000 oz of banana pulp concentrate for the initial production run. The company has also stipulated that the production of pineapple–orange–banana juice should not exceed 800 cartons. Its profit on one carton of

pineapple–orange juice is \$1.00, its profit on one carton of orange–banana juice is \$0.80, and its profit on one carton of pineapple–orange–banana juice is \$0.90. To realize a maximum profit, how many cartons of each blend should the company produce? What is the largest profit it can realize? Are there any concentrates left over?

49. **Optimizing Profit for Cold Formulas** Beyer Pharmaceutical produces three kinds of cold formulas: I, II, and III. It takes 2.5 hr to produce 1000 bottles of Formula I, 3 hr to produce 1000 bottles of Formula II, and 4 hr to produce 1000 bottles of Formula III. The profits for each 1000 bottles of Formula I, Formula II, and Formula III are \$180, \$200, and \$300, respectively. Suppose that for a certain production run, there are enough ingredients on hand to make at most 9000 bottles of Formula I, 12,000 bottles of Formula II, and 6000 bottles of Formula III. Furthermore, suppose the time for the production run is limited to a maximum of 70 hr. How many bottles of each formula should be produced in this production run so that the profit is maximized? What is the maximum profit realizable by the company? Are there any resources left over?

50. **Asset Allocation** A financier plans to invest up to \$2 million in three projects. She estimates that Project *A* will yield a return of 10% on her investment, Project *B* will yield a return of 15% on her investment, and Project *C* will yield a return of 20% on her investment. Because of the risks associated with the investments, she decided to put not more than 20% of her total investment in Project *C*. She also decided that her investments in Projects *B* and *C* should not exceed 60% of her total investment. Finally, she decided that her investment in Project *A* should be at least 60% of her investments in Projects *B* and *C*. How much should the financier invest in each project if she wishes to maximize the total returns on her investments? What is the maximum amount she can expect to make from her investments?

51. Consider the linear programming problem

$$\begin{aligned} \text{Maximize} \quad & P = 3x + 2y \\ \text{subject to} \quad & x - y \leq 3 \\ & x \leq 2 \\ & x \geq 0, y \geq 0 \end{aligned}$$

a. Sketch the feasible set for the linear programming problem.
b. Show that the linear programming problem is unbounded.
c. Solve the linear programming problem using the simplex method. How does the method break down?
d. Explain why the result in part (c) implies that no solution exists for the linear programming problem.

In Exercises 52–55, determine whether the statement is true or false. If it is true, explain why it is true. If it is false, give an example to show why it is false.

52. If at least one of the coefficients $a_1, a_2, \ldots, a_n$ of the objective function $P = a_1x_1 + a_2x_2 + \cdots + a_nx_n$ is positive, then $(0, 0, \ldots, 0)$ cannot be the optimal solution of the standard (maximization) linear programming problem.

53. Choosing the pivot row by requiring that the ratio associated with that row be the smallest ensures that the iteration will not take us from a feasible point to a nonfeasible point.

54. Choosing the pivot column by requiring that it be the column associated with the most negative entry to the left of the vertical line in the last row of the simplex tableau ensures that the iteration will result in the greatest increase or, at worse, no decrease in the objective function.

55. If, at any stage of an iteration of the simplex method, it is not possible to compute the ratios (division by zero) or the ratios are negative, then we can conclude that the standard linear programming problem may have no solution.

4.1 Solutions to Self-Check Exercises

1. Introducing the slack variables u, v, and w, we obtain the system of linear equations

$$\begin{aligned} 2x + 3y + z + u \qquad\qquad &= 10 \\ x + y + 2z \qquad + v \qquad &= 8 \\ 2y + 3z \qquad\qquad + w \quad &= 6 \\ -2x - 3y - 6z \qquad\qquad + P &= 0 \end{aligned}$$

The initial simplex tableau and the successive tableaus resulting from the use of the simplex procedure follow:

	x	y	z	u	v	w	P	Constant	Ratio
	2	3	1	1	0	0	0	10	$\frac{10}{1} = 10$
	1	1	2	0	1	0	0	8	$\frac{8}{2} = 4$
Pivot row →	0	2	(3)	0	0	1	0	6	$\frac{6}{3} = 2$
	−2	−3	−6	0	0	0	1	0	
			↑ Pivot column						

$\xrightarrow{\frac{1}{3}R_3}$

x	y	z	u	v	w	P	Constant
2	3	1	1	0	0	0	10
1	1	2	0	1	0	0	8
0	$\frac{2}{3}$	①	0	0	$\frac{1}{3}$	0	2
−2	−3	−6	0	0	0	1	0

$\xrightarrow[\substack{R_2 - 2R_3 \\ R_4 + 6R_3}]{R_1 - R_3}$

	x	y	z	u	v	w	P	Constant	Ratio
	2	$\frac{7}{3}$	0	1	0	$-\frac{1}{3}$	0	8	$\frac{8}{2} = 4$
Pivot row →	①	$-\frac{1}{3}$	0	0	1	$-\frac{2}{3}$	0	4	$\frac{4}{1} = 4$
	0	$\frac{2}{3}$	1	0	0	$\frac{1}{3}$	0	2	—
	−2	1	0	0	0	2	1	12	
	↑ Pivot column								

$\xrightarrow[R_4 + 2R_2]{R_1 - 2R_2}$

x	y	z	u	v	w	P	Constant
0	3	0	1	−2	1	0	0
1	$-\frac{1}{3}$	0	0	1	$-\frac{2}{3}$	0	4
0	$\frac{2}{3}$	1	0	0	$\frac{1}{3}$	0	2
0	$\frac{1}{3}$	0	0	2	$\frac{2}{3}$	1	20

All entries in the last row are nonnegative, and the tableau is final. We conclude that $x = 4$, $y = 0$, $z = 2$, and $P = 20$.

2. Let x denote the number of model A grates, and let y denote the number of model B grates to be made this week. Then the profit function to be maximized is given by

$$P = 6x + 8y$$

The limitations on the availability of material and labor may be expressed by the linear inequalities

$$20x + 30y \leq 7200 \quad \text{or} \quad 2x + 3y \leq 720$$
$$20x + 15y \leq 6000 \quad \text{or} \quad 4x + 3y \leq 1200$$

Finally, the condition that no more than 150 units of model A grates be made this week may be expressed by the linear inequality

$$x \leq 150$$

Thus, we are led to the following linear programming problem:

$$\begin{aligned} \text{Maximize} \quad & P = 6x + 8y \\ \text{subject to} \quad & 2x + 3y \leq 720 \\ & 4x + 3y \leq 1200 \\ & x \leq 150 \\ & x \geq 0, y \geq 0 \end{aligned}$$

To solve this problem, we introduce slack variables u, v, and w and use the simplex method, obtaining the following sequence of simplex tableaus:

	x	y	u	v	w	P	Constant	Ratio
Pivot row →	2	③	1	0	0	0	720	$\frac{720}{3} = 240$
	4	3	0	1	0	0	1200	$\frac{1200}{3} = 400$
	1	0	0	0	1	0	150	—
	−6	−8	0	0	0	1	0	
		↑ Pivot column						

$\xrightarrow{\frac{1}{3}R_1}$

x	y	u	v	w	P	Constant
$\frac{2}{3}$	①	$\frac{1}{3}$	0	0	0	240
4	3	0	1	0	0	1200
1	0	0	0	1	0	150
−6	−8	0	0	0	1	0

$\xrightarrow[R_4 + 8R_1]{R_2 - 3R_1}$

	x	y	u	v	w	P	Constant	Ratio
	$\frac{2}{3}$	1	$\frac{1}{3}$	0	0	0	240	$\frac{240}{2/3} = 360$
	2	0	−1	1	0	0	480	$\frac{480}{2} = 240$
Pivot row →	①	0	0	0	1	0	150	$\frac{150}{1} = 150$
	$-\frac{2}{3}$	0	$\frac{8}{3}$	0	0	1	1920	
	↑ Pivot column							

$\xrightarrow[\substack{R_2 - 2R_3 \\ R_4 + \frac{2}{3}R_3}]{R_1 - \frac{2}{3}R_3}$

x	y	u	v	w	P	Constant
0	1	$\frac{1}{3}$	0	$-\frac{2}{3}$	0	140
0	0	−1	1	−2	0	180
1	0	0	0	1	0	150
0	0	$\frac{8}{3}$	0	$\frac{2}{3}$	1	2020

The last tableau is final, and we see that $x = 150$, $y = 140$, and $P = 2020$. Therefore, LaCrosse should make 150 model A grates and 140 model B grates this week. The profit will be \$2020.

USING TECHNOLOGY The Simplex Method: Solving Maximization Problems

Graphing Utility

A graphing utility can be used to solve a linear programming problem by the simplex method, as illustrated in Example 1.

EXAMPLE 1 (Refer to Example 5, Section 4.1.) The problem reduces to the following linear programming problem:

$$\begin{aligned} \text{Maximize} \quad & P = 6x + 5y + 4z \\ \text{subject to} \quad & 2x + y + z \le 180 \\ & x + 3y + 2z \le 300 \\ & 2x + y + 2z \le 240 \\ & x \ge 0, y \ge 0, z \ge 0 \end{aligned}$$

With u, v, and w as slack variables, we are led to the following sequence of simplex tableaus, where the first tableau is entered as the matrix A:

	x	y	z	u	v	w	P	Constant	Ratio
Pivot row →	(2)	1	1	1	0	0	0	180	$\frac{180}{2} = 90$
	1	3	2	0	1	0	0	300	$\frac{300}{1} = 300$
	2	1	2	0	0	1	0	240	$\frac{240}{2} = 120$
	−6	−5	−4	0	0	0	1	0	
	↑ Pivot column								

$\xrightarrow{\text{*row}(\frac{1}{2}, A, 1) \blacktriangleright B}$

x	y	z	u	v	w	P	Constant
(1)	0.5	0.5	0.5	0	0	0	90
1	3	2	0	1	0	0	300
2	1	2	0	0	1	0	240
−6	−5	−4	0	0	0	1	0

*row+(−1, B, 1, 2) ► C
*row+(−2, C, 1, 3) ► B
*row+(6, B, 1, 4) ► C
→

	x	y	z	u	v	w	P	Constant	Ratio
	1	0.5	0.5	0.5	0	0	0	90	$\frac{90}{0.5} = 180$
Pivot row →	0	(2.5)	1.5	−0.5	1	0	0	210	$\frac{210}{2.5} = 84$
	0	0	1	−1	0	1	0	60	
	0	−2	−1	3	0	0	1	540	
		↑ Pivot column							

$\xrightarrow{\text{*row}(\frac{1}{2.5}, C, 2) \blacktriangleright B}$

x	y	z	u	v	w	P	Constant
1	0.5	0.5	0.5	0	0	0	90
0	(1)	0.6	−0.2	0.4	0	0	84
0	0	1	−1	0	1	0	60
0	−2	−1	3	0	0	1	540

*row+(−0.5, B, 2, 1) ► C
*row+(2, C, 2, 4) ► B
→

x	y	z	u	v	w	P	Constant
1	0	0.2	0.6	−0.2	0	0	48
0	1	0.6	−0.2	0.4	0	0	84
0	0	1	−1	0	1	0	60
0	0	0.2	2.6	0.8	0	1	708

The final simplex tableau is the same as the one obtained earlier. We see that $x = 48$, $y = 84$, $z = 0$, and $P = 708$. Hence, Ace Novelty should produce 48 Type A

souvenirs, 84 Type *B* souvenirs, and no Type *C* souvenirs, resulting in a profit of \$708 per day.

Excel

Solver is an Excel add-in that is used to solve linear programming problems. When you start the Excel program, check the *Tools* menu for the *Solver* command. If it is not there, you will need to install it. (Check your manual for installation instructions.)

EXAMPLE 2 Solve the following linear programming problem:

$$\begin{aligned} \text{Maximize} \quad & P = 6x + 5y + 4z \\ \text{subject to} \quad & 2x + y + z \leq 180 \\ & x + 3y + 2z \leq 300 \\ & 2x + y + 2z \leq 240 \\ & x \geq 0, y \geq 0, z \geq 0 \end{aligned}$$

Solution

1. Enter the data for the linear programming problem onto a spreadsheet. Enter the labels shown in column A and the variables with which we are working under Decision Variables in cells B4:B6, as shown in Figure T1. This optional step will help us to organize our work.

	A	B	C	D	E	F	G	H	I
1	Maximization Problem								
2									
3	Decision Variables								
4		x	0						
5		y	0						
6		z	0						
7									
8	Objective Function		0						
9									
10	Constraints								
11			0	<=	180				
12			0	<=	300				
13			0	<=	240				

Formulas for indicated cells
C8: = 6*C4 + 5*C5 + 4*C6
C11: = 2*C4 + C5 + C6
C12: = C4 + 3*C5 + 2*C6
C13: = 2*C4 + C5 + 2*C6

FIGURE **T1**
Setting up the spreadsheet for Solver

For the moment, the cells that will contain the values of the variables (C4:C6) are left blank. In C8, type the formula for the objective function: `=6*C4+5*C5+4*C6`. In C11, type the formula for the left-hand side of the first constraint: `=2*C4+C5+C6`. In C12, type the formula for the left-hand side of the second constraint: `=C4+3*C5+2*C6`. In C13, type the formula for the left-hand side of the third constraint: `=2*C4+C5+2*C6`. Zeros will then appear in cell C8 and cells C11:C13. In cells D11:D13, type `<=` to indicate that each constraint is of the form $\leq$. Finally, in cells E11:E13, type the right-hand value of each constraint—in this case, `180`, `300`, and `240`, respectively. Note that we need not enter the nonnegativity constraints $x \geq 0$, $y \geq 0$, and $z \geq 0$. The

Note: Boldfaced words/characters enclosed in a box (for example, **Enter**) indicate that an action (click, select, or press) is required. Words/characters printed blue (for example, Chart sub-type:) indicate words/characters that appear on the screen. Words/characters printed in a monospace font (for example, `=(-2/3)*A2+2`) indicate words/characters that need to be typed and entered.

resulting spreadsheet is shown in Figure T1, where the formulas that were entered for the objective function and the constraints are shown in the comment box.

2. Use Solver to solve the problem. Click the **Data** tab, and then click **Solver** in the Analysis group. The Solver Parameters dialog box will appear.
   a. The pointer will be in the Set Objective: box (refer to Figure T2). Highlight the cell on your spreadsheet containing the formula for the objective function—in this case, C8.

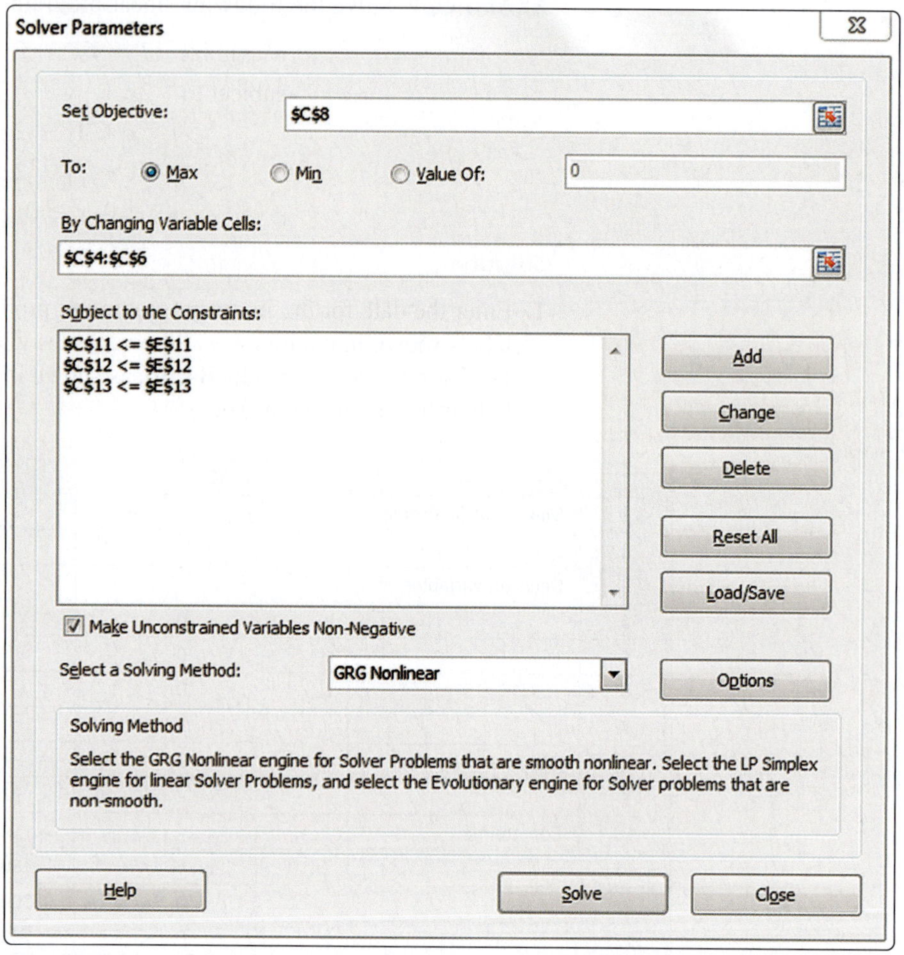

FIGURE **T2**
The completed Solver Parameters dialog box

Then, next to To:, select **Max**. Select the **By Changing Variable Cells:** box, and highlight the cells in your spreadsheet that will contain the values of the variables—in this case, C4:C6. Select the **Subject to the Constraints:** box, and then click **Add**. The Add Constraint dialog box will appear (Figure T3).

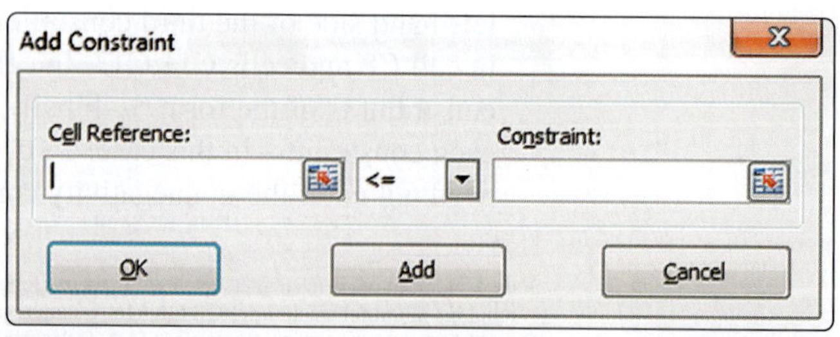

FIGURE **T3**
The Add Constraint dialog box

b. The pointer will appear in the Cell Reference: box. Highlight the cells on your spreadsheet that contain the formula for the left-hand side of the first constraint—in this case, C11. Next, select the symbol for the appropriate constraint—in this case, **<=**. Select the **Constraint:** box, and highlight the value of the right-hand side of the first constraint on your spreadsheet—in this case, 180. Click **Add**, and then follow the same procedure to enter the second and third constraints. Click **OK**. The resulting Solver Parameters dialog box shown in Figure T2 will appear.

c. In the Solver Parameters dialog box, go to the Select a Solving Method: box, and select **Simplex LP**.

d. Next, click **Solve**. A Solver Results dialog box will then appear (see Figure T4), and at the same time, the answers will appear on your spreadsheet (see Figure T5).

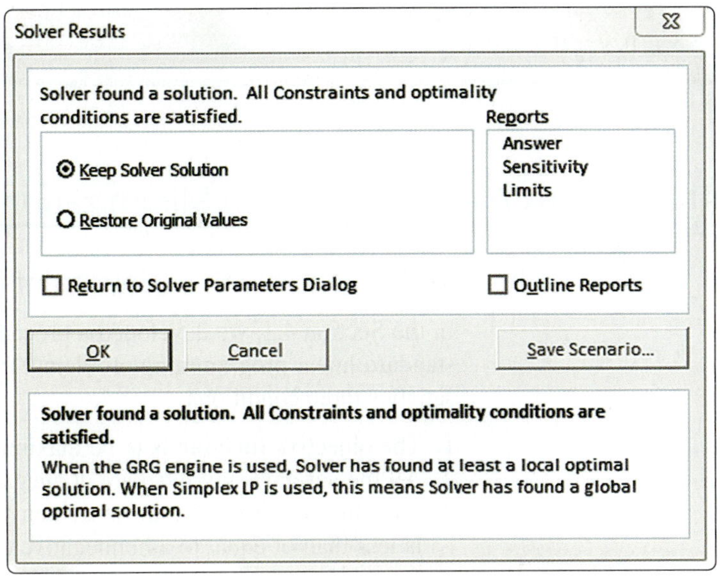

FIGURE **T4**
The Solver Results dialog box

	A	B	C	D	E
1	Maximization Problem				
2					
3	Decision Variables				
4		x	48		
5		y	84		
6		z	0		
7					
8	Objective Function		708		
9					
10	Constraints				
11			180	<=	180
12			300	<=	300
13			240	<=	240

FIGURE **T5**
Completed spreadsheet after using Solver

3. Read off your answers. From the spreadsheet, we see that the objective function attains a maximum value of 708 (cell C8) when $x = 48$, $y = 84$, and $z = 0$ (cells C4:C6).

TECHNOLOGY EXERCISES

Solve the linear programming problems.

1. Maximize $P = 2x + 3y + 4z + 2w$
subject to
$$\begin{aligned} x + 2y + 3z + 2w &\le 6 \\ 2x + 4y + z - w &\le 4 \\ 3x + 2y - 2z + 3w &\le 12 \\ x \ge 0, y \ge 0, z \ge 0, w &\ge 0 \end{aligned}$$

2. Maximize $P = 3x + 2y + 2z + w$
subject to
$$\begin{aligned} 2x + y - z + 2w &\le 8 \\ 2x - y + 2z + 3w &\le 20 \\ x + y + z + 2w &\le 8 \\ 4x - 2y + z + 3w &\le 24 \\ x \ge 0, y \ge 0, z \ge 0, w &\ge 0 \end{aligned}$$

3. Maximize $P = x + y + 2z + 3w$
subject to
$$\begin{aligned} 3x + 6y + 4z + 2w &\le 12 \\ x + 4y + 8z + 4w &\le 16 \\ 2x + y + 4z + w &\le 10 \\ x \ge 0, y \ge 0, z \ge 0, w &\ge 0 \end{aligned}$$

4. Maximize $P = 2x + 4y + 3z + 5w$
subject to
$$\begin{aligned} x - 2y + 3z + 4w &\le 8 \\ 2x + 2y + 4z + 6w &\le 12 \\ 3x + 2y + z + 5w &\le 10 \\ 2x + 8y - 2z + 6w &\le 24 \\ x \ge 0, y \ge 0, z \ge 0, w &\ge 0 \end{aligned}$$

4.2 The Simplex Method: Standard Minimization Problems

Minimization with ≤ Constraints

In the Section 4.1, we developed a procedure, called the simplex method, for solving standard linear programming problems. Recall that a standard maximization problem satisfies three conditions:

1. The objective function is to be maximized.
2. All the variables involved are nonnegative.
3. Each linear constraint may be written so that the expression involving the variables is less than or equal to a nonnegative constant.

In this section, we see how the simplex method may be used to solve certain classes of problems that are not necessarily standard maximization problems. In particular, we see how a modified procedure may be used to solve problems involving the minimization of objective functions.

We begin by considering the class of linear programming problems that calls for the minimization of objective functions but otherwise satisfies conditions 2 and 3 for standard maximization problems. The method that is used to solve these problems is illustrated in the following example.

EXAMPLE 1

$$\begin{aligned} \text{Minimize} \quad C &= -2x - 3y \\ \text{subject to} \quad 5x + 4y &\le 32 \\ x + 2y &\le 10 \\ x \ge 0, y &\ge 0 \end{aligned}$$

Solution This problem involves the minimization of the objective function and is accordingly *not* a standard maximization problem. Note, however, that all other conditions for a standard maximization problem hold true. To solve a problem of this type, we observe that minimizing the objective function C is equivalent to maximizing the objective function $P = -C$. Thus, the solution to this problem may be found by solving the following associated standard maximization problem: Maximize $P = 2x + 3y$ subject to the given constraints. Using the simplex method with u and v as slack variables, we obtain the following sequence of simplex tableaus:

	x	y	u	v	P	Constant	Ratio
	5	4	1	0	0	32	$\frac{32}{4} = 8$
Pivot row →	1	(2)	0	1	0	10	$\frac{10}{2} = 5$
	−2	−3	0	0	1	0	
		↑ Pivot column					

	x	y	u	v	P	Constant
	5	4	1	0	0	32
$\xrightarrow{\frac{1}{2}R_2}$	$\frac{1}{2}$	(1)	0	$\frac{1}{2}$	0	5
	−2	−3	0	0	1	0

	x	y	u	v	P	Constant	Ratio
Pivot row →	(3)	0	1	−2	0	12	$\frac{12}{3} = 4$
$\xrightarrow[R_3 + 3R_2]{R_1 - 4R_2}$	$\frac{1}{2}$	1	0	$\frac{1}{2}$	0	5	$\frac{5}{1/2} = 10$
	$-\frac{1}{2}$	0	0	$\frac{3}{2}$	1	15	
	↑ Pivot column						

	x	y	u	v	P	Constant
$\xrightarrow{\frac{1}{3}R_1}$	(1)	0	$\frac{1}{3}$	$-\frac{2}{3}$	0	4
	$\frac{1}{2}$	1	0	$\frac{1}{2}$	0	5
	$-\frac{1}{2}$	0	0	$\frac{3}{2}$	1	15

	x	y	u	v	P	Constant
$\xrightarrow[R_3 + \frac{1}{2}R_1]{R_2 - \frac{1}{2}R_1}$	1	0	$\frac{1}{3}$	$-\frac{2}{3}$	0	4
	0	1	$-\frac{1}{6}$	$\frac{5}{6}$	0	3
	0	0	$\frac{1}{6}$	$\frac{7}{6}$	1	17

Explore and Discuss

Refer to Example 1.

1. Sketch the feasible set S for the linear programming problem.
2. Solve the problem using the method of corners.
3. Indicate on S the corner points corresponding to each iteration of the simplex procedure and trace the path leading to the optimal solution.

The last tableau is in final form. The solution to the standard maximization problem associated with the given linear programming problem is $x = 4$, $y = 3$, and $P = 17$, so the required solution is given by $x = 4$, $y = 3$, and $C = -17$. You may verify that the solution is correct by using the method of corners.

The Dual Problem

Another special class of linear programming problems we encounter in practical applications is characterized by the following conditions:

1. The objective function is to be *minimized.*
2. All the variables involved are nonnegative.
3. All other linear constraints may be written so that the expression involving the variables is *greater than or equal to* a constant.

Such problems are called **standard minimization problems.**

A convenient method for solving this type of problem is based on the following observation. Each maximization linear programming problem is associated with a

PORTFOLIO Chad Smith

TITLE Manager of Supply Chain Sustainability
INSTITUTION Earthbound Farm

Earthbound Farm, America's largest grower of organic produce, is committed to protecting the environment. As Manager of Supply Chain Sustainability, I run life-cycle analyses to determine the financial and environmental feasibility of decisions relating to sustainability.

I use Microsoft Excel on a daily basis to run these analyses. Sometimes I use Excel to model the project cost or financial return of a project. In other cases, I use the program to calculate energy usage or to measure environmental impacts. For example, before making the decision to convert to 100% PCR PET (post-consumer recycled polyethylene terephthalate PET), I ran a life-cycle analysis of the product. On the basis of Earthbound Farm's use of rigid PET plastic in 2008, I learned that the company's conversion to 100% PCRT PET clamshell packaging will conserve 424,000 million BTUs of energy and 68,307 gallons of water, eliminate 16,191 tons of carbon dioxide, and divert 1.3 million pounds of solid waste from landfills.

When I run these analyses, I use a variety of tools and functions in Excel. The solver tool, for example, is very useful when I am working on these feasibility projects. In essence, the solver tool determines where linear functions intersect. This information allows me to manipulate parameters to maximize or minimize certain outcomes.

Although the solver tool is easy to use, it is important that I understand the mathematics—in this case linear programming—that determine the tool's output. This means that I need to be able to identify the variables, constraints, and parameters of a problem before I run an analysis. When I am helping my company to make important financial and environmental decisions, I have to be confident that I have entered the data correctly and that the results of the analysis make sense.

Courtesy of Earthbound Farm

minimization problem, and vice versa. For the purpose of identification, the given problem is called the **primal problem;** the problem related to it is called the **dual problem.** The following example illustrates the technique for constructing the dual of a given linear programming problem.

EXAMPLE 2 Write the dual problem associated with the following problem:

$$\begin{aligned} &\text{Minimize the objective function} \quad C = 4x + 2y \\ &\text{subject to} \quad 5x + y \geq 5 \\ &\qquad 5x + 3y \geq 10 \\ &\qquad x \geq 0, y \geq 0 \end{aligned} \quad \text{Primal problem}$$

Solution We first write the following tableau for the given primal problem:

x	y	Constant
5	1	5
5	3	10
4	2	

Next, we interchange the columns and rows of the foregoing tableau and head the two columns of the resulting array with the two variables u and v, obtaining the tableau:

u	v	Constant
5	5	4
1	3	2
5	10	

Interpreting the last tableau as if it were part of the initial simplex tableau for a standard maximization problem—with the exception that the signs of the coefficients pertaining to the objective function are not reversed—we construct the required dual problem as follows:

$$\left.\begin{array}{ll}\text{Maximize the objective function} & P = 5u + 10v \\ \text{subject to} & 5u + 5v \le 4 \\ & u + 3v \le 2 \\ & u \ge 0, v \ge 0\end{array}\right\} \text{Dual problem}$$

Since both the primal problem and the dual problem in Example 2 involve two variables, they can be solved graphically. In fact, the feasible sets of the primal problem and the dual problem are shown in Figure 6. From the following tables of values associated with the two problems, we see that the solution to the primal problem is $x = \frac{1}{2}$, $y = \frac{5}{2}$, and $C = 7$, and the solution for the dual problem is $u = \frac{1}{5}$, $v = \frac{3}{5}$, and $P = 7$.

Table for the primal problem

Vertex	$C = 4x + 2y$
$A(2, 0)$	8
$B(\frac{1}{2}, \frac{5}{2})$	7
$C(0, 5)$	10

Table for the dual problem

Vertex	$P = 5u + 10v$
$A(0, 0)$	0
$B(\frac{4}{5}, 0)$	4
$C(\frac{1}{5}, \frac{3}{5})$	7
$D(0, \frac{2}{3})$	$\frac{20}{3}$

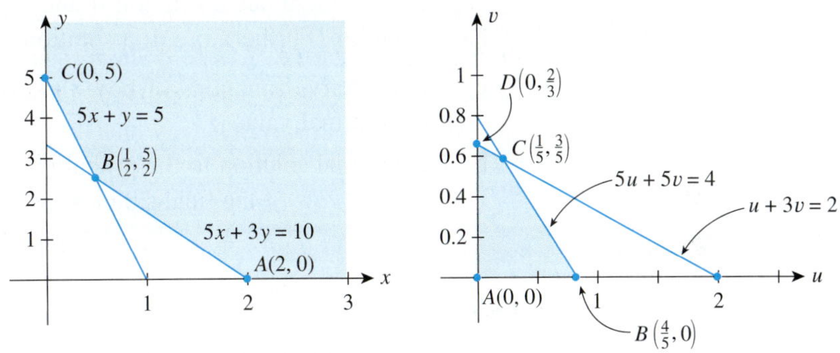

(a) Feasible set for the primal problem

(b) Feasible set for the dual problem

FIGURE **6**

Observe that the *objective functions of both the primal problem and the dual problem attain the same optimal value.*

Next let's see how we can find these solutions using the simplex method. Since the dual problem is a standard maximization problem, we can use the method of Section 4.1. Proceeding, we obtain the following sequence of tableaus. Note that we use the letters x and y (the variables used in the primal problem) to denote the slack variables.

	u	v	x	y	P	Constant	Ratio
	5	5	1	0	0	4	$\frac{4}{5}$
Pivot row →	1	(3)	0	1	0	2	$\frac{2}{3}$
	−5	−10	0	0	1	0	
		↑ Pivot column					

$\xrightarrow{\frac{1}{3}R_2}$

u	v	x	y	P	Constant
5	5	1	0	0	4
$\frac{1}{3}$	1	0	$\frac{1}{3}$	0	$\frac{2}{3}$
-5	-10	0	0	1	0

$\xrightarrow[R_3 + 10R_2]{R_1 - 5R_2}$

u	v	x	y	P	Constant	Ratio
$\left(\frac{10}{3}\right)$	0	1	$-\frac{5}{3}$	0	$\frac{2}{3}$	$\frac{1}{5}$
$\frac{1}{3}$	1	0	$\frac{1}{3}$	0	$\frac{2}{3}$	2
$-\frac{5}{3}$	0	0	$\frac{10}{3}$	1	$\frac{20}{3}$	

$\xrightarrow{\frac{3}{10}R_1}$

u	v	x	y	P	Constant
1	0	$\frac{3}{10}$	$-\frac{1}{2}$	0	$\frac{1}{5}$
$\frac{1}{3}$	1	0	$\frac{1}{3}$	0	$\frac{2}{3}$
$-\frac{5}{3}$	0	0	$\frac{10}{3}$	1	$\frac{20}{3}$

$\xrightarrow[R_3 + \frac{5}{3}R_1]{R_2 - \frac{1}{3}R_1}$

u	v	x	y	P	Constant
1	0	$\frac{3}{10}$	$-\frac{1}{2}$	0	$\frac{1}{5}$
0	1	$-\frac{1}{10}$	$\frac{1}{2}$	0	$\frac{3}{5}$
0	0	$\frac{1}{2}$	$\frac{5}{2}$	1	7

Solution for the primal problem

Interpreting the final tableau in the usual manner, we see that the solution of the dual problem is $u = \frac{1}{5}$, $v = \frac{3}{5}$, and $P = 7$, as obtained earlier by using the graphical method. Next, observe that the solution of the primal problem $x = \frac{1}{2}$ and $y = \frac{5}{2}$ appears under the respective slack variables in the last row of the final tableau, as indicated. That we can always find the solution to both the optimal and dual problems in this manner is the content of the following theorem. The theorem, attributed to John von Neumann (1903–1957), is stated without proof.

THEOREM 1

The Fundamental Theorem of Duality

A primal problem has a solution if and only if the corresponding dual problem has a solution. Furthermore, if a solution exists, then:

a. The objective functions of both the primal and the dual problem attain the same optimal value.

b. The optimal solution to the primal problem appears under the slack variables in the last row of the final simplex tableau associated with the dual problem.

EXAMPLE 3 Write and solve the dual problem associated with the following problem:

$$\left.\begin{aligned} &\text{Minimize the objective function} \quad C = 6x + 8y \\ &\text{subject to} \quad 40x + 10y \geq 2400 \\ &\qquad 10x + 15y \geq 2100 \\ &\qquad 5x + 15y \geq 1500 \\ &\qquad x \geq 0,\ y \geq 0 \end{aligned}\right\} \text{Primal problem}$$

Solution We first write down the following tableau for the given primal problem:

x	y	Constant
40	10	2400
10	15	2100
5	15	1500
6	8	

Next, we interchange the columns and rows of the foregoing tableau and head the three columns of the resulting array with the three variables u, v, and w, obtaining the tableau

u	v	w	Constant
40	10	5	6
10	15	15	8
2400	2100	1500	

Interpreting the last tableau as if it were part of the initial simplex tableau for a standard maximization problem—with the exception that the signs of the coefficients pertaining to the objective function are not reversed—we construct the required dual problem as follows:

$$\left.\begin{array}{l}\text{Maximize the objective function} \quad P = 2400u + 2100v + 1500w \\ \text{subject to} \quad 40u + 10v + 5w \le 6 \\ \qquad\qquad\quad 10u + 15v + 15w \le 8 \\ \qquad\qquad\qquad u \ge 0, v \ge 0, w \ge 0\end{array}\right\} \text{Dual problem}$$

Observe that the dual problem associated with the given (primal) problem is a standard maximization problem. The solution may thus be found by using the simplex algorithm. Introducing the slack variables x and y, we obtain the system of linear equations

$$\begin{aligned} 40u + 10v + 5w + x \quad &= 6 \\ 10u + 15v + 15w \quad + y &= 8 \\ -2400u - 2100v - 1500w \quad + P &= 0 \end{aligned}$$

Continuing with the simplex algorithm, we obtain the following sequence of simplex tableaus:

	u	v	w	x	y	P	Constant	Ratio
Pivot row →	(40)	10	5	1	0	0	6	$\frac{6}{40} = \frac{3}{20}$
	10	15	15	0	1	0	8	$\frac{8}{10} = \frac{4}{5}$
	−2400	−2100	−1500	0	0	1	0	
	↑ Pivot column							

$\xrightarrow{\frac{1}{40}R_1}$	u	v	w	x	y	P	Constant
	(1)	$\frac{1}{4}$	$\frac{1}{8}$	$\frac{1}{40}$	0	0	$\frac{3}{20}$
	10	15	15	0	1	0	8
	−2400	−2100	−1500	0	0	1	0

$\xrightarrow[R_3 + 2400R_1]{R_2 - 10R_1}$	u	v	w	x	y	P	Constant	Ratio
	1	$\frac{1}{4}$	$\frac{1}{8}$	$\frac{1}{40}$	0	0	$\frac{3}{20}$	$\frac{3/20}{1/4} = \frac{3}{5}$
Pivot row →	0	($\frac{25}{2}$)	$\frac{55}{4}$	$-\frac{1}{4}$	1	0	$\frac{13}{2}$	$\frac{13/2}{25/2} = \frac{13}{25}$
	0	−1500	−1200	60	0	1	360	
		↑ Pivot column						

$\xrightarrow{\frac{2}{25}R_2}$	u	v	w	x	y	P	Constant
	1	$\frac{1}{4}$	$\frac{1}{8}$	$\frac{1}{40}$	0	0	$\frac{3}{20}$
	0	(1)	$\frac{11}{10}$	$-\frac{1}{50}$	$\frac{2}{25}$	0	$\frac{13}{25}$
	0	−1500	−1200	60	0	1	360

$$\xrightarrow[R_3 + 1500R_2]{R_1 - \frac{1}{4}R_2}$$

u	v	w	x	y	P	Constant
1	0	$-\frac{3}{20}$	$\frac{3}{100}$	$-\frac{1}{50}$	0	$\frac{1}{50}$
0	1	$\frac{11}{10}$	$-\frac{1}{50}$	$\frac{2}{25}$	0	$\frac{13}{25}$
0	0	450	30	120	1	1140

Solution for the primal problem

The last tableau is final. The fundamental theorem of duality tells us that the solution to the primal problem is $x = 30$ and $y = 120$ with a minimum value for C of 1140. Observe that the solution to the dual (maximization) problem may be read from the simplex tableau in the usual manner: $u = \frac{1}{50}$, $v = \frac{13}{25}$, $w = 0$, and $P = 1140$. Note that the maximum value of P is equal to the minimum value of C, as guaranteed by the fundamental theorem of duality. The solution to the primal problem agrees with the solution of the same problem that we solved in Example 2, Section 3.3, using the method of corners.

Notes

1. The dual of a standard minimization problem is always a standard maximization problem provided that the coefficients of the objective function in the primal problem are all nonnegative. Such problems can always be solved by applying the simplex method to solve the dual problem.
2. Standard minimization problems in which the coefficients of the objective function are not all nonnegative do not necessarily have a dual problem that is a standard maximization problem.

EXAMPLE 4

$$\begin{aligned} \text{Minimize} \quad & C = 3x + 2y \\ \text{subject to} \quad & 8x + y \geq 80 \\ & 8x + 5y \geq 240 \\ & x + 5y \geq 100 \\ & x \geq 0, y \geq 0 \end{aligned}$$

Solution We begin by writing the dual problem associated with the given primal problem. First, we write down the following tableau for the primal problem:

x	y	Constant
8	1	80
8	5	240
1	5	100
3	2	

Next, interchanging the columns and rows of this tableau and heading the three columns of the resulting array with the three variables, u, v, and w, we obtain the tableau

u	v	w	Constant
8	8	1	3
1	5	5	2
80	240	100	

Interpreting the last tableau as if it were part of the initial simplex tableau for a standard maximization problem—with the exception that the signs of the coefficients pertaining to the objective function are not reversed—we construct the dual problem as follows: Maximize the objective function $P = 80u + 240v + 100w$ subject to the constraints

$$\begin{aligned} 8u + 8v + w &\le 3 \\ u + 5v + 5w &\le 2 \end{aligned}$$

where $u \ge 0$, $v \ge 0$, and $w \ge 0$. Having constructed the dual problem, which is a standard maximization problem, we now solve it using the simplex method. Introducing the slack variables x and y, we obtain the system of linear equations

$$\begin{aligned} 8u + 8v + w + x &= 3 \\ u + 5v + 5w + y &= 2 \\ -80u - 240v - 100w + P &= 0 \end{aligned}$$

Continuing with the simplex algorithm, we obtain the following sequence of simplex tableaus:

	u	v	w	x	y	P	Constant	Ratio
Pivot row →	8	(8)	1	1	0	0	3	$\frac{3}{8}$
	1	5	5	0	1	0	2	$\frac{2}{5}$
	−80	−240	−100	0	0	1	0	
		↑ Pivot column						

$\xrightarrow{\frac{1}{8}R_1}$

u	v	w	x	y	P	Constant
1	(1)	$\frac{1}{8}$	$\frac{1}{8}$	0	0	$\frac{3}{8}$
1	5	5	0	1	0	2
−80	−240	−100	0	0	1	0

$\xrightarrow[R_3 + 240R_1]{R_2 - 5R_1}$

	u	v	w	x	y	P	Constant	Ratio
	1	1	$\frac{1}{8}$	$\frac{1}{8}$	0	0	$\frac{3}{8}$	3
Pivot row →	−4	0	($\frac{35}{8}$)	$-\frac{5}{8}$	1	0	$\frac{1}{8}$	$\frac{1}{35}$
	160	0	−70	30	0	1	90	
			↑ Pivot column					

$\xrightarrow{\frac{8}{35}R_2}$

u	v	w	x	y	P	Constant
1	1	$\frac{1}{8}$	$\frac{1}{8}$	0	0	$\frac{3}{8}$
$-\frac{32}{35}$	0	(1)	$-\frac{1}{7}$	$\frac{8}{35}$	0	$\frac{1}{35}$
160	0	−70	30	0	1	90

$\xrightarrow[R_3 + 70R_2]{R_1 - \frac{1}{8}R_2}$

u	v	w	x	y	P	Constant
$\frac{39}{35}$	1	0	$\frac{1}{7}$	$-\frac{1}{35}$	0	$\frac{13}{35}$
$-\frac{32}{35}$	0	1	$-\frac{1}{7}$	$\frac{8}{35}$	0	$\frac{1}{35}$
96	0	0	20	16	1	92

(20 and 16: Solution for the primal problem)

The last tableau is final. The fundamental theorem of duality tells us that the solution to the primal problem is $x = 20$ and $y = 16$ with a minimum value for C of 92. ■

Our last example illustrates how the warehouse problem posed in Section 3.2 may be solved by duality.

APPLIED EXAMPLE 5 A Warehouse Problem Complete the solution to the warehouse problem given in Example 4, Section 3.2 (page 184). Minimize

$$C = 20x_1 + 8x_2 + 10x_3 + 12x_4 + 22x_5 + 18x_6 \tag{14}$$

subject to

$$\begin{array}{l} x_1 + x_2 + x_3 \leq 400 \\ x_4 + x_5 + x_6 \leq 600 \\ x_1 + x_4 \geq 200 \\ x_2 + x_5 \geq 300 \\ x_3 + x_6 \geq 400 \\ x_1 \geq 0, x_2 \geq 0, \ldots, x_6 \geq 0 \end{array} \tag{15}$$

Solution Upon multiplying each of the first two inequalities of (15) by -1, we obtain the following equivalent system of constraints, in which each of the expressions involving the variables is greater than or equal to a constant:

$$\begin{array}{l} -x_1 - x_2 - x_3 \geq -400 \\ -x_4 - x_5 - x_6 \geq -600 \\ x_1 + x_4 \geq 200 \\ x_2 + x_5 \geq 300 \\ x_3 + x_6 \geq 400 \\ x_1 \geq 0, x_2 \geq 0, \ldots, x_6 \geq 0 \end{array}$$

The problem may now be solved by duality. First, we write the tableau

x_1	x_2	x_3	x_4	x_5	x_6	Constant
−1	−1	−1	0	0	0	−400
0	0	0	−1	−1	−1	−600
1	0	0	1	0	0	200
0	1	0	0	1	0	300
0	0	1	0	0	1	400
20	8	10	12	22	18	

Interchanging the rows and columns of this tableau and heading the five columns of the resulting array of numbers by the variables u_1, u_2, u_3, u_4, and u_5, we obtain the tableau

u_1	u_2	u_3	u_4	u_5	Constant
−1	0	1	0	0	20
−1	0	0	1	0	8
−1	0	0	0	1	10
0	−1	1	0	0	12
0	−1	0	1	0	22
0	−1	0	0	1	18
−400	−600	200	300	400	

from which we construct the associated dual problem:

$$\text{Maximize} \quad P = -400u_1 - 600u_2 + 200u_3 + 300u_4 + 400u_5$$

subject to

$$\begin{aligned}
-u_1 \qquad + u_3 \qquad\qquad &\leq 20\\
-u_1 \qquad\qquad + u_4 \qquad &\leq 8\\
-u_1 \qquad\qquad\qquad + u_5 &\leq 10\\
-u_2 + u_3 \qquad\qquad &\leq 12\\
-u_2 \qquad + u_4 \qquad &\leq 22\\
-u_2 \qquad\qquad + u_5 &\leq 18\\
u_1 \geq 0, u_2 \geq 0, \ldots, u_5 &\geq 0
\end{aligned}$$

Solving the standard maximization problem by the simplex algorithm, we obtain the following sequence of tableaus ($x_1, x_2, \ldots, x_6$ are slack variables):

	u_1	u_2	u_3	u_4	u_5	x_1	x_2	x_3	x_4	x_5	x_6	P	Constant	Ratio
	−1	0	1	0	0	1	0	0	0	0	0	0	20	—
	−1	0	0	1	0	0	1	0	0	0	0	0	8	—
Pivot row →	−1	0	0	0	(1)	0	0	1	0	0	0	0	10	10
	0	−1	1	0	0	0	0	0	1	0	0	0	12	—
	0	−1	0	1	0	0	0	0	0	1	0	0	22	—
	0	−1	0	0	1	0	0	0	0	0	1	0	18	18
	400	600	−200	−300	−400	0	0	0	0	0	0	1	0	
					↑ Pivot column									

$\xrightarrow[R_7 + 400R_3]{R_6 - R_3}$

	u_1	u_2	u_3	u_4	u_5	x_1	x_2	x_3	x_4	x_5	x_6	P	Constant	Ratio
	−1	0	1	0	0	1	0	0	0	0	0	0	20	—
Pivot row →	−1	0	0	(1)	0	0	1	0	0	0	0	0	8	8
	−1	0	0	0	1	0	0	1	0	0	0	0	10	—
	0	−1	1	0	0	0	0	0	1	0	0	0	12	—
	0	−1	0	1	0	0	0	0	0	1	0	0	22	22
	1	−1	0	0	0	0	0	−1	0	0	1	0	8	—
	0	600	−200	−300	0	0	0	400	0	0	0	1	4000	
				↑ Pivot column										

$\xrightarrow[R_7 + 300R_2]{R_5 - R_2}$

	u_1	u_2	u_3	u_4	u_5	x_1	x_2	x_3	x_4	x_5	x_6	P	Constant	Ratio
	−1	0	1	0	0	1	0	0	0	0	0	0	20	—
	−1	0	0	1	0	0	1	0	0	0	0	0	8	—
	−1	0	0	0	1	0	0	1	0	0	0	0	10	—
	0	−1	1	0	0	0	0	0	1	0	0	0	12	—
	1	−1	0	0	0	0	−1	0	0	1	0	0	14	14
Pivot row →	(1)	−1	0	0	0	0	0	−1	0	0	1	0	8	8
	−300	600	−200	0	0	0	300	400	0	0	0	1	6400	
	↑ Pivot column													

$\xrightarrow[\substack{R_3 + R_6\\ R_5 - R_6\\ R_7 + 300R_6}]{\substack{R_1 + R_6\\ R_2 + R_6}}$

	u_1	u_2	u_3	u_4	u_5	x_1	x_2	x_3	x_4	x_5	x_6	P	Constant	Ratio
	0	−1	1	0	0	1	0	−1	0	0	1	0	28	28
	0	−1	0	1	0	0	1	−1	0	0	1	0	16	—
	0	−1	0	0	1	0	0	0	0	0	1	0	18	—
Pivot row →	0	−1	(1)	0	0	0	0	0	1	0	0	0	12	12
	0	0	0	0	0	0	−1	1	0	1	−1	0	6	—
	1	−1	0	0	0	0	0	−1	0	0	1	0	8	—
	0	300	−200	0	0	0	300	100	0	0	300	1	8800	
			↑ Pivot column											

$\xrightarrow[R_7 + 200R_4]{R_1 - R_4}$

u_1	u_2	u_3	u_4	u_5	x_1	x_2	x_3	x_4	x_5	x_6	P	Constant
0	0	0	0	0	1	0	−1	−1	0	1	0	16
0	−1	0	1	0	0	1	−1	0	0	1	0	16
0	−1	0	0	1	0	0	0	0	0	1	0	18
0	−1	1	0	0	0	0	0	1	0	0	0	12
0	0	0	0	0	0	−1	1	0	1	−1	0	6
1	−1	0	0	0	0	0	−1	0	0	1	0	8
0	100	0	0	0	0	300	100	200	0	300	1	11,200

The last tableau is final, and we find that

$$x_1 = 0 \qquad x_2 = 300 \qquad x_3 = 100 \qquad x_4 = 200$$
$$x_5 = 0 \qquad x_6 = 300 \qquad P = 11{,}200$$

Thus, to minimize shipping costs, Acrosonic should ship 300 loudspeaker systems from Plant I to Warehouse B, 100 systems from Plant I to Warehouse C, 200 systems from Plant II to Warehouse A, and 300 systems from Plant II to Warehouse C. The company's total shipping cost is \$11,200.

4.2 Self-Check Exercises

1. Write the dual problem associated with the following problem:

$$\begin{aligned} \text{Minimize} \quad & C = 2x + 5y \\ \text{subject to} \quad & 4x + y \geq 40 \\ & 2x + y \geq 30 \\ & x + 3y \geq 30 \\ & x \geq 0, y \geq 0 \end{aligned}$$

2. Solve the primal problem posed in Exercise 1.

Solutions to Self-Check Exercises 4.2 can be found on page 259.

4.2 Concept Questions

1. Suppose you are given the linear programming problem

$$\begin{aligned} \text{Minimize} \quad & C = -3x - 5y \\ \text{subject to} \quad & 5x + 2y \leq 30 \\ & x + 3y \leq 21 \\ & x \geq 0, y \geq 0 \end{aligned}$$

Give the associated standard maximization problem that you would use to solve this linear programming problem via the simplex method.

2. Give three characteristics of a standard minimization linear programming problem.

3. What is the primal problem associated with a standard minimization linear programming problem? The dual problem?

4. **a.** What does the fundamental theorem of duality tell us about the existence of a solution to a primal problem?
b. How are the optimal values of the primal and dual problems related?
c. Given the final simplex tableau associated with a dual problem, how would you determine the optimal solution to the associated primal problem?

4.2 Exercises

In Exercises 1–6, use the technique developed in this section to solve the minimization problem.

1. Minimize $C = -2x + y$
subject to
$$\begin{aligned} x + 2y &\le 6 \\ 3x + 2y &\le 12 \\ x \ge 0, y &\ge 0 \end{aligned}$$

2. Minimize $C = -2x - 3y$
subject to
$$\begin{aligned} 3x + 4y &\le 24 \\ 7x - 4y &\le 16 \\ x \ge 0, y &\ge 0 \end{aligned}$$

3. Minimize $C = -3x - 2y$ subject to the constraints of Exercise 2.

4. Minimize $C = x - 2y + z$
subject to
$$\begin{aligned} x - 2y + 3z &\le 10 \\ 2x + y - 2z &\le 15 \\ 2x + y + 3z &\le 20 \\ x \ge 0, y \ge 0, z &\ge 0 \end{aligned}$$

5. Minimize $C = 2x - 3y - 4z$
subject to
$$\begin{aligned} -x + 2y - z &\le 8 \\ x - 2y + 2z &\le 10 \\ 2x + 4y - 3z &\le 12 \\ x \ge 0, y \ge 0, z &\ge 0 \end{aligned}$$

6. Minimize $C = -3x - 2y - z$ subject to the constraints of Exercise 5.

In Exercises 7–10, you are given the final simplex tableau for the dual problem. Give the solution to the primal problem and the solution to the associated dual problem.

7. Problem: Minimize $C = 8x + 12y$
subject to
$$\begin{aligned} x + 3y &\ge 2 \\ 2x + 2y &\ge 3 \\ x \ge 0, y &\ge 0 \end{aligned}$$

Final tableau:

u	v	x	y	P	Constant
0	1	$\frac{3}{4}$	$-\frac{1}{4}$	0	3
1	0	$-\frac{1}{2}$	$\frac{1}{2}$	0	2
0	0	$\frac{5}{4}$	$\frac{1}{4}$	1	13

8. Problem: Minimize $C = 3x + 2y$
subject to
$$\begin{aligned} 5x + y &\ge 10 \\ 2x + 2y &\ge 12 \\ x + 4y &\ge 12 \\ x \ge 0, y &\ge 0 \end{aligned}$$

Final tableau:

u	v	w	x	y	P	Constant
1	0	$-\frac{3}{4}$	$\frac{1}{4}$	$-\frac{1}{4}$	0	$\frac{1}{4}$
0	1	$\frac{19}{8}$	$-\frac{1}{8}$	$\frac{5}{8}$	0	$\frac{7}{8}$
0	0	9	1	5	1	13

9. Problem: Minimize $C = 10x + 3y + 10z$
subject to
$$\begin{aligned} 2x + y + 5z &\ge 20 \\ 4x + y + z &\ge 30 \\ x \ge 0, y \ge 0, z &\ge 0 \end{aligned}$$

Final tableau:

u	v	x	y	z	P	Constant
0	1	$\frac{1}{2}$	-1	0	0	2
1	0	$-\frac{1}{2}$	2	0	0	1
0	0	2	-9	1	0	3
0	0	5	10	0	1	80

10. Problem: Minimize $C = 2x + 3y$
subject to
$$\begin{aligned} x + 4y &\ge 8 \\ x + y &\ge 5 \\ 2x + y &\ge 7 \\ x \ge 0, y &\ge 0 \end{aligned}$$

Final tableau:

u	v	w	x	y	P	Constant
0	1	$\frac{7}{3}$	$\frac{4}{3}$	$-\frac{1}{3}$	0	$\frac{5}{3}$
1	0	$-\frac{1}{3}$	$-\frac{1}{3}$	$\frac{1}{3}$	0	$\frac{1}{3}$
0	0	2	4	1	1	11

In Exercises 11–20, construct the dual problem associated with the primal problem. Solve the primal problem.

11. Minimize $C = 3x + 2y$
subject to
$$\begin{aligned} 2x + 3y &\ge 90 \\ 3x + 2y &\ge 120 \\ x \ge 0, y &\ge 0 \end{aligned}$$

12. Minimize $C = 2x + 5y$
subject to
$$\begin{aligned} x + 2y &\ge 4 \\ 3x + 2y &\ge 6 \\ x \ge 0, y &\ge 0 \end{aligned}$$

13. Minimize $C = 6x + 4y$
subject to
$$\begin{aligned} 6x + y &\ge 60 \\ 2x + y &\ge 40 \\ x + y &\ge 30 \\ x \ge 0, y &\ge 0 \end{aligned}$$

14. Minimize $C = 10x + y$
subject to
$$\begin{aligned} 4x + y &\ge 16 \\ x + 2y &\ge 12 \\ x &\ge 2 \\ x \ge 0, y &\ge 0 \end{aligned}$$

15. Minimize $C = 200x + 150y + 120z$
subject to
$$\begin{aligned} 20x + 10y + z &\ge 10 \\ x + y + 2z &\ge 20 \\ x \ge 0, y \ge 0, z &\ge 0 \end{aligned}$$

16. Minimize $C = 40x + 30y + 11z$
subject to
$$\begin{aligned} 2x + y + z &\ge 8 \\ x + y - z &\ge 6 \\ x \ge 0, y \ge 0, z &\ge 0 \end{aligned}$$

17. Minimize $C = 6x + 8y + 4z$
subject to
$$\begin{aligned} x + 2y + 2z &\ge 10 \\ 2x + y + z &\ge 24 \\ x + y + z &\ge 16 \\ x \ge 0, y \ge 0, z &\ge 0 \end{aligned}$$

18. Minimize $C = 12x + 4y + 8z$
subject to
$$\begin{aligned} 2x + 4y + z &\geq 6 \\ 3x + 2y + 2z &\geq 2 \\ 4x + y + z &\geq 2 \\ x \geq 0, y \geq 0, z &\geq 0 \end{aligned}$$

19. Minimize $C = 30x + 12y + 20z$
subject to
$$\begin{aligned} 2x + 4y + 3z &\geq 6 \\ 6x + z &\geq 2 \\ 6y + 2z &\geq 4 \\ x \geq 0, y \geq 0, z &\geq 0 \end{aligned}$$

20. Minimize $C = 8x + 6y + 4z$
subject to
$$\begin{aligned} 2x + 3y + z &\geq 6 \\ x + 2y - 2z &\geq 4 \\ x + y + 2z &\geq 2 \\ x \geq 0, y \geq 0, z &\geq 0 \end{aligned}$$

21. **Minimizing Cruise Line Costs** Deluxe River Cruises operates a fleet of river vessels. The fleet has two types of vessels: A type A vessel has 60 deluxe cabins and 160 standard cabins, whereas a type B vessel has 80 deluxe cabins and 120 standard cabins. Under a charter agreement with Odyssey Travel Agency, Deluxe River Cruises is to provide Odyssey with a minimum of 360 deluxe and 680 standard cabins for their 15-day cruise in May. It costs $44,000 to operate a type A vessel and $54,000 to operate a type B vessel for that period. How many of each type vessel should be used to keep the operating costs to a minimum? What is the minimum cost?

22. **Fertilizer Costs** A farmer uses two types of fertilizers. A 50-lb bag of Fertilizer *A* contains 8 lb of nitrogen, 2 lb of phosphorus, and 4 lb of potassium. A 50-lb bag of Fertilizer *B* contains 5 lb each of nitrogen, phosphorus, and potassium. The minimum requirements for a field are 440 lb of nitrogen, 260 lb of phosphorus, and 360 lb of potassium. If a 50-lb bag of Fertilizer *A* costs $30 and a 50-lb bag of Fertilizer *B* costs $20, find the amount of each type of fertilizer the farmer should use to minimize his cost while still meeting the minimum requirements. What is the cost?

23. **Diet Planning** The owner of the Health Juice Bar wishes to prepare a low-calorie fruit juice with a high vitamin A and vitamin C content by blending orange juice and pink grapefruit juice. Each glass of the blended juice is to contain at least 1200 International Units (IU) of vitamin A and 200 IU of vitamin C. One ounce of orange juice contains 60 IU of vitamin A, 16 IU of vitamin C, and 14 calories; each ounce of pink grapefruit juice contains 120 IU of vitamin A, 12 IU of vitamin C, and 11 calories. How many ounces of each juice should a glass of the blend contain if it is to meet the minimum vitamin requirements while containing a minimum number of calories?

24. **Optimizing Advertising Exposure** Everest Deluxe World Travel has decided to advertise in the Sunday editions of two major newspapers in town. These advertisements are directed at three groups of potential customers. Each advertisement in Newspaper I is seen by 70,000 Group *A* customers, 40,000 Group *B* customers, and 20,000 Group *C* customers. Each advertisement in Newspaper II is seen by 10,000 Group *A*, 20,000 Group *B*, and 40,000 Group *C* customers. Each advertisement in Newspaper I costs $1000, and each advertisement in Newspaper II costs $800. Everest would like their advertisements to be read by at least 2 million people from Group *A*, 1.4 million people from Group *B*, and 1 million people from Group *C*. How many advertisements should Everest place in each newspaper to achieve its advertising goals at a minimum cost? What is the minimum cost?

25. **Minimizing Shipping Costs** Acrosonic manufactures a model G loudspeaker system in Plants I and II. The output at Plant I is at most 800/month, and the output at Plant II is at most 600/month. Model G loudspeaker systems are also shipped to the three warehouses—*A*, *B*, and *C*—whose minimum monthly requirements are 500, 400, and 400 systems, respectively. Shipping costs from Plant I to Warehouse *A*, Warehouse *B*, and Warehouse *C* are $16, $20, and $22 per loudspeaker system, respectively, and shipping costs from Plant II to each of these warehouses are $18, $16, and $14, respectively. What shipping schedule will enable Acrosonic to meet the requirements of the warehouses while keeping its shipping costs to a minimum? What is the minimum cost?

26. **Minimizing Shipping Costs** Steinwelt Piano manufactures uprights and consoles in two plants, Plant I and Plant II. The output of Plant I is at most 300/month, and the output of Plant II is at most 250/month. These pianos are shipped to three warehouses that serve as distribution centers for Steinwelt. To fill current and projected future orders, Warehouse *A* requires a minimum of 200 pianos/month, Warehouse *B* requires at least 150 pianos/month, and Warehouse *C* requires at least 200 pianos/month. The shipping cost of each piano from Plant I to Warehouse *A*, Warehouse *B*, and Warehouse *C* is $60, $60, and $80, respectively, and the shipping cost of each piano from Plant II to Warehouse *A*, Warehouse *B*, and Warehouse *C* is $80, $70, and $50, respectively. What shipping schedule will enable Steinwelt to meet the requirements of the warehouses while keeping the shipping costs to a minimum? What is the minimum cost?

27. **Minimizing Oil Refinery Costs** An oil company operates two refineries in a certain city. Refinery I has an output of 200, 100, and 100 barrels of low-, medium-, and high-grade oil per day, respectively. Refinery II has an output of 100, 200, and 600 barrels of low-, medium-, and high-grade oil per day, respectively. The company wishes to produce at least 1000, 1400, and 3000 barrels of low-, medium-, and high-grade oil, respectively, to fill an order. If it costs $200/day to operate Refinery I and $300/day to operate Refinery II, determine how many days each refinery should be operated to meet the production requirements at minimum cost to the company. What is the minimum cost?

In Exercises 28 and 29, determine whether the statement is true or false. If it is true, explain why it is true. If it is false, give an example to show why it is false.

28. If a standard minimization linear programming problem has a unique solution, then so does the corresponding maximization problem with objective function $P = -C$, where $C = a_1x_1 + a_2x_2 + \cdots + a_nx_n$ is the objective function for the minimization problem.

29. The optimal value attained by the objective function of the primal problem may be different from that attained by the objective function of the dual problem.

4.2 Solutions to Self-Check Exercises

1. We first write down the following tableau for the given (primal) problem:

x	y	Constant
4	1	40
2	1	30
1	3	30
2	5	0

Next, we interchange the columns and rows of the tableau and head the three columns of the resulting array with the three variables, u, v, and w, obtaining the tableau

u	v	w	Constant
4	2	1	2
1	1	3	5
40	30	30	0

Interpreting the last tableau as if it were the initial tableau for a standard linear programming problem—with the exception that the signs of the coefficients pertaining to the objective function are not reversed—we construct the required dual problem as follows:

$$\begin{aligned} \text{Maximize} \quad & P = 40u + 30v + 30w \\ \text{subject to} \quad & 4u + 2v + w \le 2 \\ & u + v + 3w \le 5 \\ & u \ge 0, v \ge 0, w \ge 0 \end{aligned}$$

2. We introduce slack variables x and y to obtain the system of linear equations

$$\begin{aligned} 4u + 2v + w + x &= 2 \\ u + v + 3w + y &= 5 \\ -40u - 30v - 30w + P &= 0 \end{aligned}$$

Using the simplex algorithm, we obtain the sequence of simplex tableaus

	u	v	w	x	y	P	Constant	Ratio	
Pivot row →	(4)	2	1	1	0	0	2	$\frac{2}{4} = \frac{1}{2}$	$\xrightarrow{\frac{1}{4}R_1}$
	1	1	3	0	1	0	5	$\frac{5}{1} = 5$	
	−40	−30	−30	0	0	1	0		
	↑ Pivot column								

u	v	w	x	y	P	Constant	
(1)	$\frac{1}{2}$	$\frac{1}{4}$	$\frac{1}{4}$	0	0	$\frac{1}{2}$	$\xrightarrow[R_3 + 40R_1]{R_2 - R_1}$
1	1	3	0	1	0	5	
−40	−30	−30	0	0	1	0	

	u	v	w	x	y	P	Constant	Ratio	
	1	$\frac{1}{2}$	$\frac{1}{4}$	$\frac{1}{4}$	0	0	$\frac{1}{2}$	$\frac{1/2}{1/4} = 2$	$\xrightarrow{\frac{4}{11}R_2}$
Pivot row →	0	$\frac{1}{2}$	($\frac{11}{4}$)	$-\frac{1}{4}$	1	0	$\frac{9}{2}$	$\frac{9/2}{11/4} = \frac{18}{11}$	
	0	−10	−20	10	0	1	20		
			↑ Pivot column						

u	v	w	x	y	P	Constant	
1	$\frac{1}{2}$	$\frac{1}{4}$	$\frac{1}{4}$	0	0	$\frac{1}{2}$	$\xrightarrow[R_3 + 20R_2]{R_1 - \frac{1}{4}R_2}$
0	$\frac{2}{11}$	(1)	$-\frac{1}{11}$	$\frac{4}{11}$	0	$\frac{18}{11}$	
0	−10	−20	10	0	1	20	

	u	v	w	x	y	P	Constant	Ratio	
Pivot row →	1	($\frac{5}{11}$)	0	$\frac{3}{11}$	$-\frac{1}{11}$	0	$\frac{1}{11}$	$\frac{1/11}{5/11} = \frac{1}{5}$	$\xrightarrow{\frac{11}{5}R_1}$
	0	$\frac{2}{11}$	1	$-\frac{1}{11}$	$\frac{4}{11}$	0	$\frac{18}{11}$	$\frac{18/11}{2/11} = 9$	
	0	$-\frac{70}{11}$	0	$\frac{90}{11}$	$\frac{80}{11}$	1	$\frac{580}{11}$		
		↑ Pivot column							

u	v	w	x	y	P	Constant	
$\frac{11}{5}$	(1)	0	$\frac{3}{5}$	$-\frac{1}{5}$	0	$\frac{1}{5}$	$\xrightarrow[R_3 + \frac{70}{11}R_1]{R_2 - \frac{2}{11}R_1}$
0	$\frac{2}{11}$	1	$-\frac{1}{11}$	$\frac{4}{11}$	0	$\frac{18}{11}$	
0	$-\frac{70}{11}$	0	$\frac{90}{11}$	$\frac{80}{11}$	1	$\frac{580}{11}$	

u	v	w	x	y	P	Constant
$\frac{11}{5}$	1	0	$\frac{3}{5}$	$-\frac{1}{5}$	0	$\frac{1}{5}$
$-\frac{2}{5}$	0	1	$-\frac{1}{5}$	$\frac{2}{5}$	0	$\frac{8}{5}$
14	0	0	12	6	1	54
			Solution for the primal problem			

The last tableau is final, and the solution to the primal problem is $x = 12$ and $y = 6$ with a minimum value for C of 54.

USING TECHNOLOGY The Simplex Method: Solving Minimization Problems

Graphing Utility

A graphing utility can be used to solve minimization problems using the simplex method.

EXAMPLE 1

$$\begin{aligned} \text{Minimize} \quad & C = 2x + 3y \\ \text{subject to} \quad & 8x + y \geq 80 \\ & 3x + 2y \geq 100 \\ & x + 4y \geq 80 \\ & x \geq 0, y \geq 0 \end{aligned}$$

Solution We begin by writing the dual problem associated with the given primal problem. From the tableau for the primal problem

x	y	Constant
8	1	80
3	2	100
1	4	80
2	3	

we obtain—upon interchanging the columns and rows of this tableau and heading the three columns of the resulting array with the variables u, v, and w—the tableau

u	v	w	Constant
8	3	1	2
1	2	4	3
80	100	80	

This tells us that the dual problem is

$$\begin{aligned} \text{Maximize} \quad & P = 80u + 100v + 80w \\ \text{subject to} \quad & 8u + 3v + w \leq 2 \\ & u + 2v + 4w \leq 3 \\ & u \geq 0, v \geq 0, w \geq 0 \end{aligned}$$

To solve this standard maximization problem, we proceed as follows:

	u	v	w	x	y	P	Constant	Ratio
Pivot row →	8	③	1	1	0	0	2	$\frac{2}{3}$
	1	2	4	0	1	0	3	$\frac{3}{2}$
	−80	−100	−80	0	0	1	0	
		↑ Pivot column						

$\xrightarrow{\text{*row}(\frac{1}{3}, A, 1) \blacktriangleright B}$

u	v	w	x	y	P	Constant
2.67	①	0.33	0.33	0	0	0.67
1	2	4	0	1	0	3
−80	−100	−80	0	0	1	0

$\xrightarrow[\text{*row+}(100, C, 1, 3) \blacktriangleright B]{\text{*row+}(-2, B, 1, 2) \blacktriangleright C}$

	u	v	w	x	y	P	Constant	Ratio
	2.67	1	0.33	0.33	0	0	0.67	2
Pivot row →	−4.33	0	(3.33)	−0.67	1	0	1.67	0.5
	186.67	0	−46.67	33.33	0	1	66.67	
			↑ Pivot column					

$\xrightarrow{*\textbf{row}(\frac{1}{3.33}, B, 2) \blacktriangleright C}$

u	v	w	x	y	P	Constant
2.67	1	0.33	0.33	0	0	0.67
−1.30	0	1	−0.2	0.3	0	0.5
186.67	0	−46.67	33.33	0	1	66.67

$\xrightarrow[*\textbf{row+}(46.67, B, 2, 3) \blacktriangleright C]{*\textbf{row+}(-0.33, C, 2, 1) \blacktriangleright B}$

u	v	w	x	y	P	Constant
3.1	1	0	0.4	−0.1	0	0.50
−1.3	0	1	−0.2	0.3	0	0.50
125.93	0	0.05	23.99	14.02	1	90.03

(Solution for the primal problem: 23.99, 14.02)

From the last tableau, we see that $x = 23.99$, $y = 14.02$, and $C = 90.03$.

Excel

EXAMPLE 2

$$\begin{aligned} \text{Minimize} \quad & C = 2x + 3y \\ \text{subject to} \quad & 8x + y \geq 80 \\ & 3x + 2y \geq 100 \\ & x + 4y \geq 80 \\ & x \geq 0, y \geq 0 \end{aligned}$$

Solution We use Solver as outlined in Example 2, pages 243–245, to obtain the spreadsheet shown in Figure T1. (In this case, select **Min** next to To: instead of Max because this is a minimization problem. Also select **>=** in the Add Constraint dialog box because the inequalities in the problem are of the form ≥.) From the spreadsheet, we read off the solution: $x = 24$, $y = 14$, and $C = 90$.

	A	B	C	D	E	F	G	H	I	J
1	Minimization Problem									
2										
3	Decision Variables									
4		x	24							
5		y	14							
6										
7										
8	Objective Function		90							
9										
10	Constraints									
11			206	>=	80					
12			100	>=	100					
13			80	>=	80					

Formulas for indicated cells
C8: = 2*C4 + 3*C5
C11: = 8*C4 + C5
C12: = 3*C4 + 2*C5
C13: = C4 + 4*C5

FIGURE **T1**
Completed spreadsheet after using Solver

EXAMPLE 3 Sensitivity Analysis* Solve the following linear programming problem:

$$\begin{aligned} \text{Maximize} \quad & P = 2x + 4y \\ \text{subject to} \quad & 2x + 5y \le 19 \qquad \text{Constraint 1} \\ & 3x + 2y \le 12 \qquad \text{Constraint 2} \\ & x \ge 0,\ y \ge 0 \end{aligned}$$

a. Use Solver to solve the problem.

b. Find the range of values that the coefficient of x can assume without changing the optimal solution.

c. Find the range of values that Resource 1 (the constant on the right-hand side of constraint 1) can assume without changing the optimal solution.

d. Find the shadow price for Resource 1.

e. Identify the binding and nonbinding constraints.

Solution

a. Use Solver as outlined in Example 2, pages 243–245, to obtain the spreadsheet shown in Figure T2.

	A	B	C	D	E	F	G	H	I	J
1	Maximization Problem									
2										
3	Decision Variables									
4		x	2							
5		y	3							
6										
7										
8	Objective Function		16							
9										
10	Constraints									
11			19	<=	19					
12			12	<=	12					
13										

Formulas for indicated cells
C8: = 2*C4 + 4*C5
C11: = 2*C4 + 5*C5
C12: = 3*C4 + 2*C5

FIGURE **T2**
Completed spreadsheet for a maximization problem

b. In the Solver Results dialog box, hold down **Ctrl** while selecting **Answer** and **Sensitivity** under Reports, and then click **OK**. By clicking the **Answer Report 1** and **Sensitivity Report 1** tabs that appear at the bottom of your worksheet, you can obtain the reports shown in Figures T3a and T3b. From the sensitivity report, we see that the value of the Objective Coefficient for x is 2, the Allowable Increase for the coefficient is 4, and the Allowable Decrease for the coefficient is 0.4. Thus, the coefficient can vary from 1.6 to 6 without affecting the optimal solution.

---

*For those who have completed Section 3.4.

Note: Boldfaced words/characters enclosed in a box (for example, **Enter**) indicate that an action (click, select, or press) is required. Words/characters printed blue (for example, Chart sub-type) indicate words/characters that appear on the screen. Words/characters printed in a monospace font (for example, `=(-2/3)*A2+2)`) indicate words/characters that need to be typed and entered.

Objective Cell (Max)

Cell	Name	Original Value	Final Value
$C$8	Objective Function	0	16

Variable Cells

Cell	Name	Original Value	Final Value
$C$4	x	0	2
$C$5	y	0	3

Constraints

Cell	Name	Cell Value	Formula	Status	Slack
$C$11		19	$C$11<=$E$11	Binding	0
$C$12		12	$C$12<=$E$12	Binding	0

(a) The Answer Report

Variable Cells

Cell	Name	Final Value	Reduced Cost	Objective Coefficient	Allowable Increase	Allowable Decrease
$C$4	x	2	0	2	4	0.4
$C$5	y	3	0	4	1	2.666666667

Constraints

Cell	Name	Final Value	Shadow Price	Constraint R.H. Side	Allowable Increase	Allowable Decrease
$C$11		19	0.727272727	19	11	11
$C$12		12	0.181818182	12	16.5	4.4

(b) The Sensitivity Report

FIGURE **T3**
The Solver reports

c. From the sensitivity report, we see that the Final Value of the constraint for Resource 1 is 19 and the Allowable Increase and Allowable Decrease for this value is 11. Thus, the value of the constraint must lie between $19 - 11$ and $19 + 11$—that is, between 8 and 30.

d. From the sensitivity report, we see that the Shadow Price for Resource 1 is 0.727272727.

e. From the answer report, we conclude that both Constraints are binding.

TECHNOLOGY EXERCISES

In Exercises 1–4, solve the linear programming problem by the simplex method.

1. Minimize $C = x + y + 3z$
subject to
$$\begin{aligned} 2x + y + 3z &\geq 6 \\ x + 2y + 4z &\geq 8 \\ 3x + y - 2z &\geq 4 \\ x \geq 0, y \geq 0, z &\geq 0 \end{aligned}$$

2. Minimize $C = 2x + 4y + z$
subject to
$$\begin{aligned} x + 2y + 4z &\geq 7 \\ 3x + y - z &\geq 6 \\ x + 4y + 2z &\geq 24 \\ x \geq 0, y \geq 0, z &\geq 0 \end{aligned}$$

3. Minimize $C = x + 1.2y + 3.5z$
subject to
$$\begin{aligned} 2x + 3y + 5z &\geq 12 \\ 3x + 1.2y - 2.2z &\geq 8 \\ 1.2x + 3y + 1.8z &\geq 14 \\ x \geq 0, y \geq 0, z &\geq 0 \end{aligned}$$

4. Minimize $C = 2.1x + 1.2y + z$
subject to
$$\begin{aligned} x + y - z &\geq 5.2 \\ x - 2.1y + 4.2z &\geq 8.4 \\ x \geq 0, y \geq 0, z &\geq 0 \end{aligned}$$

Exercises 5–8 are for Excel users only. You are given a linear programming problem:

a. Use Solver to solve the problem.
b. Find the range of values that the coefficient of x can assume without changing the optimal solution and the range of values that the coefficient of y can assume without changing the optimal solution.
c. Find the range of values that each resource (requirement) can assume.
d. Find the shadow price for each resource (requirement).
e. Identify the binding and nonbinding constraints.

5. Maximize $P = 3x + 4y$
subject to
$$\begin{aligned} 2x + 3y &\le 12 &\quad &\text{Resource 1}\\ 2x + y &\le 8 & &\text{Resource 2}\\ x \ge 0, y &\ge 0 \end{aligned}$$

6. Maximize $P = 2x + 5y$
subject to
$$\begin{aligned} x + 3y &\le 15 &\quad &\text{Resource 1}\\ 4x + y &\le 16 & &\text{Resource 2}\\ x \ge 0, y &\ge 0 \end{aligned}$$

7. Minimize $C = 2x + 5y$
subject to
$$\begin{aligned} x + 2y &\ge 4 &\quad &\text{Requirement 1}\\ x + y &\ge 3 & &\text{Requirement 2}\\ x \ge 0, y &\ge 0 \end{aligned}$$

8. Minimize $C = 3x + 4y$
subject to
$$\begin{aligned} x + 3y &\ge 8 &\quad &\text{Requirement 1}\\ x + y &\ge 4 & &\text{Requirement 2}\\ x \ge 0, y &\ge 0 \end{aligned}$$

4.3 The Simplex Method: Nonstandard Problems (Optional)

Section 4.1 showed how we can use the simplex method to solve standard maximization problems; Section 4.2 showed how, thanks to duality, we can use it to solve standard minimization problems provided that the coefficients in the objective function are all nonnegative.

In this section, we see how the simplex method can be incorporated into a method for solving **nonstandard problems**—problems that do not fall into either of the two previous categories. We begin by recalling the characteristics of standard problems.

Standard Maximization Problem

1. The objective function is to be maximized.
2. All variables involved in the problem are nonnegative.
3. Each linear constraint may be written so that the expression involving the variables is less than or equal to a nonnegative constant.

Standard Minimization Problem (Restricted Version—See Condition 4)

1. The objective function is to be minimized.
2. All variables involved in the problem are nonnegative.
3. Each linear constraint may be written so that the expression involving the variables is greater than or equal to a constant.
4. *All coefficients in the objective function are nonnegative.*

Note Recall that if all coefficients in the objective function are nonnegative, then a standard minimization problem can be solved by using the simplex method to solve the associated dual problem.

We now give some examples of linear programming problems that do not fit into these two categories of problems.

EXAMPLE 1 Explain why the following linear programming problem is *not* a standard maximization problem.

$$\begin{aligned} \text{Maximize} \quad & P = x + 2y\\ \text{subject to} \quad & 4x + 3y \le 18\\ & -x + 3y \ge 3\\ & x \ge 0, y \ge 0 \end{aligned}$$

Solution This is not a standard maximization problem because the second constraint,

$$-x + 3y \geq 3$$

violates condition 3. Observe that multiplying both sides of this inequality by -1 yields

$$x - 3y \leq -3$$

Recall that multiplying both sides of an inequality by a negative number reverses the inequality sign.

Now the last equation still violates condition 3 because the constant on the right is *negative*.

Observe that the constraints in Example 1 involve both *less than or equal to constraints* ($\leq$) and *greater than or equal to constraints* ($\geq$). Such constraints are called **mixed constraints.** We will solve the problem posed in Example 1 later.

EXAMPLE 2 Explain why the following linear programming problem is *not* a restricted standard minimization problem.

$$\begin{aligned} \text{Minimize} \quad & C = 2x - 3y \\ \text{subject to} \quad & x + y \leq 5 \\ & x + 3y \geq 9 \\ & -2x + y \leq 2 \\ & x \geq 0, y \geq 0 \end{aligned}$$

Solution Observe that the coefficients in the objective function C are not all nonnegative. Also, condition 3 is violated. Therefore, the problem is not a restricted standard minimization problem. By constructing the dual problem, you can convince yourself that the latter is not a standard maximization problem and thus cannot be solved using the methods described in Sections 4.1 and 4.2. We will solve this problem in Example 4.

EXAMPLE 3 Explain why the following linear programming problem is *not* a standard maximization problem. Show that it cannot be rewritten as a restricted standard minimization problem.

$$\begin{aligned} \text{Maximize} \quad & P = x + 2y \\ \text{subject to} \quad & 2x + 3y \leq 12 \\ & -x + 3y = 3 \\ & x \geq 0, y \geq 0 \end{aligned}$$

Solution The constraint equation $-x + 3y = 3$ is equivalent to the two inequalities

$$-x + 3y \leq 3 \quad \text{and} \quad -x + 3y \geq 3$$

If we multiply both sides of the second inequality by -1, then it can be written in the form

$$x - 3y \leq -3$$

Therefore, the two given constraints are equivalent to the three constraints

$$\begin{aligned} 2x + 3y &\leq 12 \\ -x + 3y &\leq 3 \\ x - 3y &\leq -3 \end{aligned}$$

The third inequality violates condition 3 for a standard maximization problem. Next, we see that the given problem is equivalent to the following:

$$\begin{aligned} \text{Minimize} \quad & C = -x - 2y \\ \text{subject to} \quad & -2x - 3y \geq -12 \\ & x - 3y \geq -3 \\ & -x + 3y \geq 3 \\ & x \geq 0, y \geq 0 \end{aligned}$$

Since the coefficients of the objective function are not all nonnegative, we conclude that the given problem cannot be rewritten as a restricted standard minimization problem. You will be asked to solve this problem in Exercise 11.

The Simplex Method for Solving Nonstandard Problems

To describe a technique for solving nonstandard problems, let's consider the problem of Example 1:

$$\begin{aligned} \text{Maximize} \quad & P = x + 2y \\ \text{subject to} \quad & 4x + 3y \leq 18 \\ & -x + 3y \geq 3 \\ & x \geq 0, y \geq 0 \end{aligned}$$

As a first step, we rewrite the constraints so that the second constraint involves a $\leq$ constraint. As in Example 1, we obtain

$$\begin{aligned} 4x + 3y &\leq 18 \\ x - 3y &\leq -3 \\ x \geq 0, y &\geq 0 \end{aligned}$$

Disregarding the fact that the constant on the right of the second inequality is negative, let's attempt to solve the problem using the simplex method for problems in standard form. Introducing the slack variables u and v gives the system of linear equations

$$\begin{aligned} 4x + 3y + u \qquad\qquad &= 18 \\ x - 3y \qquad + v \qquad &= -3 \\ -x - 2y \qquad\qquad + P &= 0 \end{aligned}$$

The initial simplex tableau is

x	y	u	v	P	Constant
4	3	1	0	0	18
1	−3	0	1	0	−3
−1	−2	0	0	1	0

Interpreting the tableau in the usual fashion, we see that

$$x = 0 \qquad y = 0 \qquad u = 18 \qquad v = -3 \qquad P = 0$$

Since the value of the slack variable v is negative, we see that this cannot be a feasible solution (remember, all variables must be nonnegative). In fact, you can see from Figure 7 that the point $(0, 0)$ does not lie in the feasible set associated with the given problem. Since we must start from a feasible point when using the simplex method for problems in standard form, we see that this method is not applicable at this juncture.

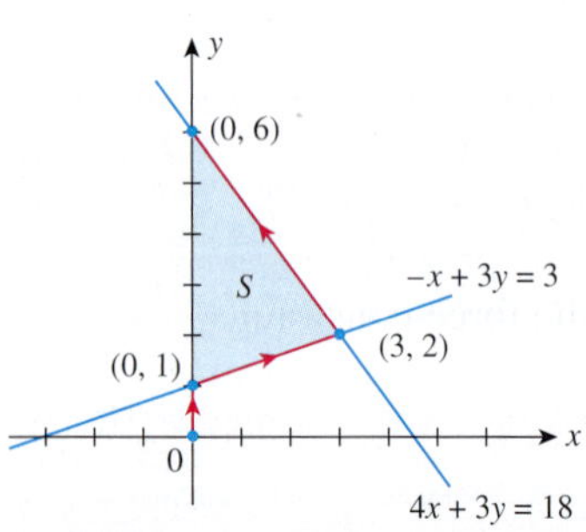

FIGURE 7
S is the feasible set for the problem.

Let's find a way to bring us from the nonfeasible point $(0, 0)$ to *any* feasible point, after which we can switch to the simplex method for problems in standard form. This can be accomplished by pivoting as follows. Referring to the tableau

	x	y	u	v	P	Constant	Ratio
	4	3	1	0	0	18	$\frac{18}{3} = 6$
Pivot row →	1	(-3)	0	1	0	-3	$\frac{-3}{-3} = 1$
	-1	-2	0	0	1	0	
		↑ Pivot column					

notice the negative number -3 lying in the column of constants and above the lower horizontal line. Locate any negative number to the left of this number (there must always be at least one such number if the problem has a solution). For the problem under consideration, there is only one such number: the number -3 in the y-column. This column is designated as the pivot column. To find the pivot element, we form the *positive* ratios of the numbers in the column of constants to the corresponding numbers in the pivot column (above the last row). The pivot row is the row corresponding to the smallest ratio, and the pivot element is the element common to both the pivot row and the pivot column (the number circled in the foregoing tableau). Pivoting about this element, we have

$\xrightarrow{-\frac{1}{3}R_2}$

x	y	u	v	P	Constant
4	3	1	0	0	18
$-\frac{1}{3}$	(1)	0	$-\frac{1}{3}$	0	1
-1	-2	0	0	1	0

$\xrightarrow[R_3 + 2R_2]{R_1 - 3R_2}$

x	y	u	v	P	Constant
5	0	1	1	0	15
$-\frac{1}{3}$	1	0	$-\frac{1}{3}$	0	1
$-\frac{5}{3}$	0	0	$-\frac{2}{3}$	1	2

Interpreting the last tableau in the usual fashion, we see that

$$x = 0 \qquad y = 1 \qquad u = 15 \qquad v = 0 \qquad P = 2$$

Observe that the point $(0, 1)$ is a feasible point (see Figure 7). Our iteration has brought us from a nonfeasible point to a feasible point in one iteration. Observe, too, that all the constants in the column of constants are now nonnegative, reflecting the fact that $(0, 1)$ is a feasible point, as we have just noted.

We can now use the simplex method for problems in standard form to complete the solution to our problem.

	x	y	u	v	P	Constant	Ratio
Pivot row →	(5)	0	1	1	0	15	3
	$-\frac{1}{3}$	1	0	$-\frac{1}{3}$	0	1	—
	$-\frac{5}{3}$	0	0	$-\frac{2}{3}$	1	2	
	↑ Pivot column						

$\xrightarrow{\frac{1}{5}R_1}$

x	y	u	v	P	Constant
(1)	0	$\frac{1}{5}$	$\frac{1}{5}$	0	3
$-\frac{1}{3}$	1	0	$-\frac{1}{3}$	0	1
$-\frac{5}{3}$	0	0	$-\frac{2}{3}$	1	2

$\xrightarrow[R_3 + \frac{5}{3}R_1]{R_2 + \frac{1}{3}R_1}$

	x	y	u	v	P	Constant	Ratio
Pivot row →	1	0	$\frac{1}{5}$	($\frac{1}{5}$)	0	3	15
	0	1	$\frac{1}{15}$	$-\frac{4}{15}$	0	2	—
	0	0	$\frac{1}{3}$	$-\frac{1}{3}$	1	7	
				↑ Pivot column			

$\xrightarrow{5R_1}$

x	y	u	v	P	Constant
5	0	1	(1)	0	15
0	1	$\frac{1}{15}$	$-\frac{4}{15}$	0	2
0	0	$\frac{1}{3}$	$-\frac{1}{3}$	1	7

$\xrightarrow[R_3+\frac{1}{3}R_1]{R_2+\frac{4}{15}R_1}$

x	y	u	v	P	Constant
5	0	1	1	0	15
$\frac{4}{3}$	1	$\frac{1}{3}$	0	0	6
$\frac{5}{3}$	0	$\frac{2}{3}$	0	1	12

All entries in the last row are nonnegative, so the tableau is final. We see that the optimal solution is

$$x = 0 \qquad y = 6 \qquad u = 0 \qquad v = 15 \qquad P = 12$$

Observe that the maximum of P occurs at $(0, 6)$ (see Figure 7). The arrows indicate the path that our search for the maximum of P has taken us on.

Before looking at further examples, let's summarize the method for solving nonstandard problems.

The Simplex Method for Solving Nonstandard Problems

1. If necessary, rewrite the problem as a maximization problem (recall that minimizing C is equivalent to maximizing $-C$).
2. If necessary, rewrite all constraints (except $x \geq 0$, $y \geq 0$, $z \geq 0, \ldots$) using less than or equal to ($\leq$) inequalities.
3. Introduce slack variables and set up the initial simplex tableau.
4. Scan the upper part of the column of constants of the tableau for negative entries.
   a. If there are no negative entries, complete the solution using the simplex method for problems in standard form.
   b. If there are negative entries, proceed to Step 5.
5. a. Pick any negative entry in a row in which a negative entry in the column of constants occurs. The column containing this entry is the pivot column.
   b. Compute the positive ratios of the numbers in the column of constants to the corresponding numbers in the pivot column (above the last row). The pivot row corresponds to the smallest ratio. The intersection of the pivot column and the pivot row determines the pivot element.
   c. Pivot the tableau about the pivot element. Then return to Step 4.

We now apply the method to solve the nonstandard problem posed in Example 2.

EXAMPLE 4 Solve the problem of Example 2:

$$\begin{aligned} \text{Minimize} \quad & C = 2x - 3y \\ \text{subject to} \quad & x + y \leq 5 \\ & x + 3y \geq 9 \\ & -2x + y \leq 2 \\ & x \geq 0, y \geq 0 \end{aligned}$$

Solution We first rewrite the problem as a maximization problem with constraints using $\leq$, which gives the following equivalent problem:

$$\begin{aligned} \text{Maximize} \quad & P = -C = -2x + 3y \\ \text{subject to} \quad & x + y \leq 5 \\ & -x - 3y \leq -9 \\ & -2x + y \leq 2 \\ & x \geq 0, y \geq 0 \end{aligned}$$

Introducing slack variables u, v, and w and following the procedure for solving nonstandard problems outlined previously, we obtain the following sequence of tableaus:

	x	y	u	v	w	P	Constant	Ratio
	1	1	1	0	0	0	5	$\frac{5}{1} = 5$
	−1	−3	0	1	0	0	−9	$\frac{-9}{-3} = 3$
Pivot row →	−2	(1)	0	0	1	0	2	$\frac{2}{1} = 2$
	2	−3	0	0	0	1	0	
		↑ Pivot column						

Column 1 could have been chosen as the pivot column as well.

$\xrightarrow[\substack{R_2 + 3R_3 \\ R_4 + 3R_3}]{R_1 - R_3}$

	x	y	u	v	w	P	Constant	Ratio
	3	0	1	0	−1	0	3	$\frac{3}{3} = 1$
Pivot row →	(−7)	0	0	1	3	0	−3	$\frac{-3}{-7} = \frac{3}{7}$
	−2	1	0	0	1	0	2	
	−4	0	0	0	3	1	6	
	↑ Pivot column							

$\xrightarrow{-\frac{1}{7}R_2}$

x	y	u	v	w	P	Constant
3	0	1	0	−1	0	3
(1)	0	0	$-\frac{1}{7}$	$-\frac{3}{7}$	0	$\frac{3}{7}$
−2	1	0	0	1	0	2
−4	0	0	0	3	1	6

$\xrightarrow[\substack{R_3 + 2R_2 \\ R_4 + 4R_2}]{R_1 - 3R_2}$

	x	y	u	v	w	P	Constant	Ratio
Pivot row →	0	0	1	$\left(\frac{3}{7}\right)$	$\frac{2}{7}$	0	$\frac{12}{7}$	4
	1	0	0	$-\frac{1}{7}$	$-\frac{3}{7}$	0	$\frac{3}{7}$	—
	0	1	0	$-\frac{2}{7}$	$\frac{1}{7}$	0	$\frac{20}{7}$	—
	0	0	0	$-\frac{4}{7}$	$\frac{9}{7}$	1	$\frac{54}{7}$	
				↑ Pivot column				

We now use the simplex method for problems in standard form to complete the problem.

$\xrightarrow{\frac{7}{3}R_1}$

x	y	u	v	w	P	Constant
0	0	$\frac{7}{3}$	(1)	$\frac{2}{3}$	0	4
1	0	0	$-\frac{1}{7}$	$-\frac{3}{7}$	0	$\frac{3}{7}$
0	1	0	$-\frac{2}{7}$	$\frac{1}{7}$	0	$\frac{20}{7}$
0	0	0	$-\frac{4}{7}$	$\frac{9}{7}$	1	$\frac{54}{7}$

$\xrightarrow[\substack{R_3 + \frac{2}{7}R_1 \\ R_4 + \frac{4}{7}R_1}]{R_2 + \frac{1}{7}R_1}$

x	y	u	v	w	P	Constant
0	0	$\frac{7}{3}$	1	$\frac{2}{3}$	0	4
1	0	$\frac{1}{3}$	0	$-\frac{1}{3}$	0	1
0	1	$\frac{2}{3}$	0	$\frac{1}{3}$	0	4
0	0	$\frac{4}{3}$	0	$\frac{5}{3}$	1	10

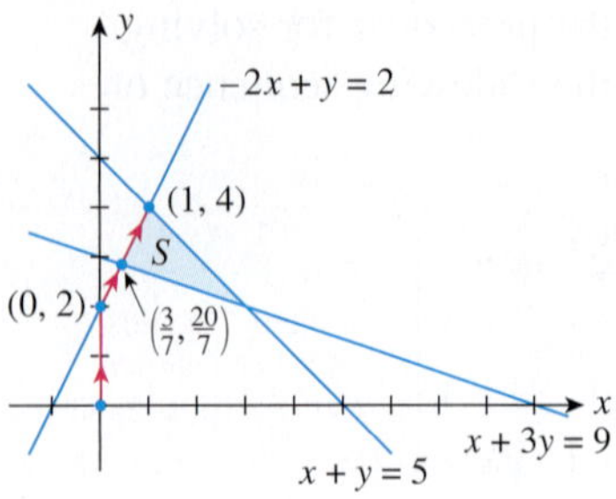

FIGURE **8**
The feasible set S and the path leading from the initial nonfeasible point $(0, 0)$ to the optimal point $(1, 4)$

All the entries in the last row are nonnegative, and hence the tableau is final. We see that the optimal solution is

$$x = 1 \qquad y = 4 \qquad u = 0 \qquad v = 4 \qquad w = 0 \qquad C = -P = -10$$

The feasible set S for this problem is shown in Figure 8. The path leading from the nonfeasible initial point $(0, 0)$ to the optimal point $(1, 4)$ goes through the nonfeasible point $(0, 2)$ and the feasible point $\left(\frac{3}{7}, \frac{20}{7}\right)$, in that order.

APPLIED EXAMPLE 5 Production Planning Rockford manufactures two models of exercise bicycles—a standard model and a deluxe model—in two separate plants, Plant I and Plant II. The maximum output at Plant I is 1200 per month, and the maximum output at Plant II is 1000 per month. The profit per bike for standard and deluxe models manufactured at Plant I is \$40 and \$60, respectively; the profit per bike for standard and deluxe models manufactured at Plant II is \$45 and \$50, respectively.

For the month of May, Rockford received an order for 1000 standard models and 800 deluxe models. If prior commitments dictate that the number of deluxe models manufactured at Plant I not exceed the number of standard models manufactured there by more than 200, find how many of each model should be produced at each plant to satisfy the order and at the same time maximize Rockford's profit.

Solution Let x and y denote the number of standard and deluxe models, respectively, to be manufactured at Plant I. Since the number of standard and deluxe models required are 1000 and 800, respectively, we see that the number of standard and deluxe models to be manufactured at Plant II are $(1000 - x)$ and $(800 - y)$, respectively. Rockford's profit will then be

$$\begin{aligned} P &= 40x + 60y + 45(1000 - x) + 50(800 - y) \\ &= 85{,}000 - 5x + 10y \end{aligned}$$

Since the maximum output of Plant I is 1200, we have the constraint

$$x + y \le 1200$$

Similarly, since the maximum output of Plant II is 1000, we have

$$(1000 - x) + (800 - y) \le 1000$$

or, equivalently,

$$-x - y \le -800$$

Finally, the additional constraints placed on the production schedule at Plants I and II translate into the inequalities

$$\begin{aligned} y - x &\le 200 \\ x &\le 1000 \\ y &\le 800 \end{aligned}$$

To summarize, the problem at hand is the following nonstandard problem:

$$\begin{aligned} \text{Maximize} \quad & P = 85{,}000 - 5x + 10y \\ \text{subject to} \quad & x + y \le 1200 \\ & -x - y \le -800 \\ & -x + y \le 200 \\ & x \le 1000 \\ & y \le 800 \\ & x \ge 0, y \ge 0 \end{aligned}$$

Let's introduce the slack variables, u, v, w, r, and s. Using the simplex method for nonstandard problems, we obtain the following sequence of tableaus:

	x	y	u	v	w	r	s	P	Constant	Ratio
	1	1	1	0	0	0	0	0	1,200	$\frac{1200}{1} = 1200$
Pivot row →	(−1)	−1	0	1	0	0	0	0	−800	$\frac{-800}{-1} = 800$
	−1	1	0	0	1	0	0	0	200	—
	1	0	0	0	0	1	0	0	1,000	$\frac{1000}{1} = 1000$
	0	1	0	0	0	0	1	0	800	—
	5	−10	0	0	0	0	0	1	85,000	
	↑ Pivot column									

$\xrightarrow{-R_2}$

x	y	u	v	w	r	s	P	Constant
1	1	1	0	0	0	0	0	1,200
(1)	1	0	−1	0	0	0	0	800
−1	1	0	0	1	0	0	0	200
1	0	0	0	0	1	0	0	1,000
0	1	0	0	0	0	1	0	800
5	−10	0	0	0	0	0	1	85,000

$\xrightarrow[\substack{R_4 - R_2 \\ R_6 - 5R_2}]{\substack{R_1 - R_2 \\ R_3 + R_2}}$

	x	y	u	v	w	r	s	P	Constant	Ratio
	0	0	1	1	0	0	0	0	400	—
	1	1	0	−1	0	0	0	0	800	$\frac{800}{1} = 800$
Pivot row →	0	(2)	0	−1	1	0	0	0	1,000	$\frac{1000}{2} = 500$
	0	−1	0	1	0	1	0	0	200	—
	0	1	0	0	0	0	1	0	800	$\frac{800}{1} = 800$
	0	−15	0	5	0	0	0	1	81,000	
		↑ Pivot column								

We now use the simplex method for standard problems.

$\xrightarrow{\frac{1}{2}R_3}$

x	y	u	v	w	r	s	P	Constant
0	0	1	1	0	0	0	0	400
1	1	0	−1	0	0	0	0	800
0	(1)	0	$-\frac{1}{2}$	$\frac{1}{2}$	0	0	0	500
0	−1	0	1	0	1	0	0	200
0	1	0	0	0	0	1	0	800
0	−15	0	5	0	0	0	1	81,000

$\xrightarrow[\substack{R_5 - R_3 \\ R_6 + 15R_3}]{\substack{R_2 - R_3 \\ R_4 + R_3}}$

	x	y	u	v	w	r	s	P	Constant	Ratio
Pivot row →	0	0	1	(1)	0	0	0	0	400	$\frac{400}{1} = 400$
	1	0	0	$-\frac{1}{2}$	$-\frac{1}{2}$	0	0	0	300	—
	0	1	0	$-\frac{1}{2}$	$\frac{1}{2}$	0	0	0	500	—
	0	0	0	$\frac{1}{2}$	$\frac{1}{2}$	1	0	0	700	$\frac{700}{1/2} = 1400$
	0	0	0	$\frac{1}{2}$	$-\frac{1}{2}$	0	1	0	300	$\frac{300}{1/2} = 600$
	0	0	0	$-\frac{5}{2}$	$\frac{15}{2}$	0	0	1	88,500	
				↑ Pivot column						

$$\xrightarrow{\substack{R_2 + \frac{1}{2}R_1 \\ R_3 + \frac{1}{2}R_1 \\ R_4 - \frac{1}{2}R_1 \\ R_5 - \frac{1}{2}R_1 \\ R_6 + \frac{5}{2}R_1}}$$

x	y	u	v	w	r	s	P	Constant
0	0	1	1	0	0	0	0	400
1	0	$\frac{1}{2}$	0	$-\frac{1}{2}$	0	0	0	500
0	1	$\frac{1}{2}$	0	$\frac{1}{2}$	0	0	0	700
0	0	$-\frac{1}{2}$	0	$\frac{1}{2}$	1	0	0	500
0	0	$-\frac{1}{2}$	0	$-\frac{1}{2}$	0	1	0	100
0	0	$\frac{5}{2}$	0	$\frac{15}{2}$	0	0	1	89,500

All entries in the last row are nonnegative, and the tableau is final. We see that $x = 500$, $y = 700$, and $P = 89{,}500$. This tells us that Plant I should manufacture 500 standard and 700 deluxe exercise bicycles and that Plant II should manufacture $(1000 - 500)$, or 500, standard and $(800 - 700)$, or 100, deluxe models. Rockford's profit will then be \$89,500.

Explore and Discuss

Refer to Example 5.

1. Sketch the feasible set S for the linear programming problem.
2. Solve the problem using the method of corners.
3. Indicate on S the points (both nonfeasible and feasible) corresponding to each iteration of the simplex method, and trace the path leading to the optimal solution.

4.3 Self-Check Exercises

1. Solve the following nonstandard problem using the method of this section:

$$\begin{aligned} \text{Maximize} \quad & P = 2x + 3y \\ \text{subject to} \quad & x + y \le 40 \\ & -x + 2y \le -10 \\ & x \ge 0, y \ge 0 \end{aligned}$$

2. Crop Planning A farmer has 150 acres of land suitable for cultivating Crops A and B. The cost of cultivating Crop A is \$40/acre, and the cost of cultivating Crop B is \$60/acre. The farmer has a maximum of \$7400 available for land cultivation. Each acre of Crop A requires 20 labor-hours, and each acre of Crop B requires 25 labor-hours. The farmer has a maximum of 3300 labor-hours available. She has also decided that she will cultivate at least 70 acres of Crop A. If she expects to make a profit of \$150/acre on Crop A and \$200/acre on Crop B, how many acres of each crop should she plant to maximize her profit?

Solutions to Self-Check Exercises 4.3 can be found on page 275.

4.3 Concept Questions

1. Explain why the following linear programming problem is *not* a standard maximization problem.

$$\begin{aligned} \text{Maximize} \quad & P = 2x + 3y \\ \text{subject to} \quad & 4x + 2y \le 20 \\ & -2x + 3y \ge 5 \\ & x \ge 0, y \ge 0 \end{aligned}$$

2. Explain why the following linear programming problem is *not* a restricted standard minimization problem.

$$\begin{aligned} \text{Minimize} \quad & C = 3x - y \\ \text{subject to} \quad & x + y \ge 5 \\ & 3x + 5y \ge 16 \\ & x \ge 0, y \ge 0 \end{aligned}$$

3. Explain why the following linear programming problem is *not* a standard maximization problem.

$$\begin{aligned} \text{Maximize} \quad & P = x + 3y \\ \text{subject to} \quad & 3x + 4y \le 16 \\ & 2x - 3y = 5 \\ & x \ge 0, y \ge 0 \end{aligned}$$

Can it be rewritten as a restricted standard minimization problem? Why or why not?

4.3 Exercises

In Exercises 1–4, rewrite each linear programming problem as a maximization problem with constraints involving inequalities of the form $\le$ a constant (with the exception of the inequalities $x \ge 0$, $y \ge 0$, and $z \ge 0$).

1. $$\begin{aligned} \text{Minimize} \quad & C = 2x - 3y \\ \text{subject to} \quad & 3x + 5y \ge 20 \\ & 3x + y \le 16 \\ & -2x + y \le 1 \\ & x \ge 0, y \ge 0 \end{aligned}$$

2. $$\begin{aligned} \text{Minimize} \quad & C = 2x + 3y \\ \text{subject to} \quad & x + y \le 10 \\ & x + 2y \ge 12 \\ & 2x + y \ge 12 \\ & x \ge 0, y \ge 0 \end{aligned}$$

3. $$\begin{aligned} \text{Minimize} \quad & C = 5x + 10y + z \\ \text{subject to} \quad & 2x + y + z \ge 4 \\ & x + 2y + 2z \ge 2 \\ & 2x + 4y + 3z \le 12 \\ & x \ge 0, y \ge 0, z \ge 0 \end{aligned}$$

4. $$\begin{aligned} \text{Maximize} \quad & P = 2x + y - 2z \\ \text{subject to} \quad & x + 2y + z \ge 10 \\ & 3x + 4y + 2z \ge 5 \\ & 2x + 5y + 12z \le 20 \\ & x \ge 0, y \ge 0, z \ge 0 \end{aligned}$$

In Exercises 5–20, use the method of this section to solve each linear programming problem.

5. $$\begin{aligned} \text{Maximize} \quad & P = x + 2y \\ \text{subject to} \quad & 2x + 5y \le 20 \\ & x - 5y \le -5 \\ & x \ge 0, y \ge 0 \end{aligned}$$

6. $$\begin{aligned} \text{Maximize} \quad & P = 2x + 3y \\ \text{subject to} \quad & x + 2y \le 8 \\ & x - y \le -2 \\ & x \ge 0, y \ge 0 \end{aligned}$$

7. $$\begin{aligned} \text{Minimize} \quad & C = -2x + y \\ \text{subject to} \quad & x + 2y \le 6 \\ & 3x + 2y \le 12 \\ & x \ge 0, y \ge 0 \end{aligned}$$

8. $$\begin{aligned} \text{Minimize} \quad & C = -2x + 3y \\ \text{subject to} \quad & x + 3y \le 60 \\ & 2x + y \ge 45 \\ & x \le 40 \\ & x \ge 0, y \ge 0 \end{aligned}$$

9. $$\begin{aligned} \text{Maximize} \quad & P = x + 4y \\ \text{subject to} \quad & x + 3y \le 6 \\ & -2x + 3y \le -6 \\ & x \ge 0, y \ge 0 \end{aligned}$$

10. $$\begin{aligned} \text{Maximize} \quad & P = 5x + y \\ \text{subject to} \quad & 2x + y \le 8 \\ & -x + y \ge 2 \\ & x \ge 0, y \ge 0 \end{aligned}$$

11. $$\begin{aligned} \text{Maximize} \quad & P = x + 2y \\ \text{subject to} \quad & 2x + 3y \le 12 \\ & -x + 3y = 3 \\ & x \ge 0, y \ge 0 \end{aligned}$$

12. $$\begin{aligned} \text{Minimize} \quad & C = x + 2y \\ \text{subject to} \quad & 4x + 7y \le 70 \\ & 2x + y = 20 \\ & x \ge 0, y \ge 0 \end{aligned}$$

13. $$\begin{aligned} \text{Maximize} \quad & P = x - 2y + z \\ \text{subject to} \quad & 2x + 3y + 2z \le 12 \\ & x + 2y - 3z \ge 6 \\ & x \ge 0, y \ge 0, z \ge 0 \end{aligned}$$

14. $$\begin{aligned} \text{Maximize} \quad & P = 5x + 4y + 2z \\ \text{subject to} \quad & x + 2y + 3z \le 24 \\ & x - y + z \ge 6 \\ & x \ge 0, y \ge 0, z \ge 0 \end{aligned}$$

15. $$\begin{aligned} \text{Minimize} \quad & C = 2x - 3y + 4z \\ \text{subject to} \quad & -x + 2y - z \le 8 \\ & x - 2y + 2z \le 10 \\ & 2x + 4y - 3z \le 12 \\ & x \ge 0, y \ge 0, z \ge 0 \end{aligned}$$

16. $$\begin{aligned} \text{Minimize} \quad & C = x - 2y + z \\ \text{subject to} \quad & x - 2y + 3z \le 10 \\ & 2x + y - 2z \le 15 \\ & 2x + y + 3z \le 20 \\ & x \ge 0, y \ge 0, z \ge 0 \end{aligned}$$

17. Maximize $P = 2x + y + z$
subject to
$$\begin{aligned} x + 2y + 3z &\le 28 \\ 2x + 3y - z &\le 6 \\ x - 2y + z &\ge 4 \\ x \ge 0, y \ge 0, &z \ge 0 \end{aligned}$$

18. Minimize $C = 2x - y + 3z$
subject to
$$\begin{aligned} 2x + y + z &\ge 2 \\ x + 3y + z &\ge 6 \\ 2x + y + 2z &\le 12 \\ x \ge 0, y \ge 0, &z \ge 0 \end{aligned}$$

19. Maximize $P = x + 2y + 3z$
subject to
$$\begin{aligned} x + 2y + z &\le 20 \\ 3x + y &\le 30 \\ 2x + y + z &= 10 \\ x \ge 0, y \ge 0, &z \ge 0 \end{aligned}$$

20. Minimize $C = 3x + 2y + z$
subject to
$$\begin{aligned} x + 2y + z &\le 20 \\ 3x + y &\le 30 \\ 2x + y + z &= 10 \\ x \ge 0, y \ge 0, &z \ge 0 \end{aligned}$$

21. **Maximizing Investment Returns** Natsano has at most \$50,000 to invest in the common stocks of two companies. He estimates that an investment in Company *A* will yield a return of 10%, whereas an investment in Company *B*, which he feels is a riskier investment, will yield a return of 20%. If he decides that his investment in the stock of Company *A* is to exceed his investment in the stock of Company *B* by at least \$20,000, determine how much he should invest in the stock of each company to maximize the return on his investment. What is the maximum return?

22. **Veterinary Science** A veterinarian has been asked to prepare a diet for a group of dogs to be used in a nutrition study at the School of Animal Science. It has been stipulated that each serving should be no larger than 8 oz and must contain at least 29 units of Nutrient I and 20 units of Nutrient II. The vet has decided that the diet may be prepared from two brands of dog food: Brand *A* and Brand *B*. Each ounce of Brand *A* contains 3 units of Nutrient I and 4 units of Nutrient II. Each ounce of Brand *B* contains 5 units of Nutrient I and 2 units of Nutrient II. Brand *A* costs 3 cents/oz and Brand *B* costs 4 cents/oz. Determine how many ounces of each brand of dog food should be used per serving to meet the given requirements at minimum cost.

23. **Allocation of Funds** The First Street branch of Capitol Bank has a sum of \$60 million earmarked for home loans and commercial-development loans. The bank expects to realize an 8% annual rate of return on the home loans and a 6% annual rate of return on the commercial-development loans. Management has decided that the total amount of home loans is to be greater than or equal to 3 times the total amount of commercial-development loans. Owing to prior commitments, at least \$10 million of the funds has been designated for commercial-development loans. Determine the amount of each type of loan the bank should extend to maximize its returns. What is the maximum return?

24. **Crop Planning** A farmer has 150 acres of land suitable for cultivating Crops *A* and *B*. The cost of cultivating Crop *A* is \$40/acre, and the cost of cultivating Crop *B* is \$60/acre. The farmer has a maximum of \$7400 available for land cultivation. Each acre of Crop *A* requires 20 labor-hours, and each acre of Crop *B* requires 25 labor-hours. The farmer has a maximum of 3300 labor-hours available. He has also decided that he will cultivate at least 80 acres of Crop *A*. If he expects to make a profit of \$150/acre on Crop *A* and \$200/acre on Crop *B*, how many acres of each crop should he plant to maximize his profit?

25. **Production Scheduling** A company manufactures Products *A*, *B*, and *C*. Each product is processed in three departments: I, II, and III. The total available labor-hours per week for Departments I, II, and III are 900, 1080, and 840, respectively. The time requirements (in hours per unit) and the profit per unit for each product are as follows:

	Product *A*	Product *B*	Product *C*
Dept. I	2	1	2
Dept. II	3	1	2
Dept. III	2	2	1
Profit	\$18	\$12	\$15

If management decides that the number of units of Product *B* manufactured must equal or exceed the number of units of products *A* and *C* manufactured, how many units of each product should the company produce to maximize its profit? What is the maximum profit?

26. **Production Scheduling** Wayland Company manufactures two models of its twin-size futons, standard and deluxe, in two locations, I and II. The maximum output at Location I is 600/week, whereas the maximum output at Location II is 400/week. The profit per futon for standard and deluxe models manufactured at Location I is \$30 and \$20, respectively; the profit per futon for standard and deluxe models manufactured at Location II is \$34 and \$18, respectively. For a certain week, the company has received an order for 600 standard models and 300 deluxe models. If prior commitments dictate that the number of deluxe models manufactured at Location II not exceed the number of standard models manufactured there by more than 50, find how many of each model should be manufactured at each location to satisfy the order and at the same time maximize Wayland's profit. What is the maximum profit?

27. **Diet Planning** A nutritionist at the Medical Center has been asked to prepare a special diet for certain patients. He has decided that the meals should contain a minimum of 400 mg of calcium, 10 mg of iron, and 40 mg of vitamin C. He has further decided that the meals are to be prepared from Foods *A* and *B*. Each ounce of Food *A* contains 30 mg of calcium, 1 mg of iron, 2 mg of vitamin C, and 2 mg of cholesterol. Each ounce of Food *B* contains 25 mg of calcium, 0.5 mg of iron, 5 mg of vitamin C, and 5 mg of cholesterol. Use the method of this section to determine how many ounces of each type of food the nutritionist should use in a meal so that the cholesterol content is minimized and the minimum requirements of calcium, iron, and vitamin C are met.

28. **Minimizing Shipping Costs** Steinwelt Piano manufactures uprights and consoles in two plants, Plant I and Plant II. The output of Plant I is at most 300/month, whereas the output of Plant II is at most 250/month. These pianos are shipped to three warehouses that serve as distribution centers for the company. To fill current and projected orders, Warehouse *A* requires a minimum of 200 pianos/month, Warehouse *B* requires at least 150 pianos/month, and Warehouse *C* requires at least 200 pianos/month. The shipping cost of each piano from Plant I to Warehouse *A*, Warehouse *B*, and Warehouse *C* is \$60, \$60, and \$80, respectively, and the shipping cost of each piano from Plant II to Warehouse *A*, Warehouse *B*, and Warehouse *C* is \$80, \$70, and \$50, respectively. Use the method of this section to determine the shipping schedule that will enable Steinwelt to meet the warehouses' requirements while keeping the shipping costs to a minimum.

4.3 Solutions to Self-Check Exercises

1. We are given the problem

$$\begin{aligned} \text{Maximize} \quad & P = 2x + 3y \\ \text{subject to} \quad & x + y \le 40 \\ & -x + 2y \le -10 \\ & x \ge 0, y \ge 0 \end{aligned}$$

Using the method of this section and introducing the slack variables u and v, we obtain the following tableaus:

	x	y	u	v	P	Constant	Ratio
	1	1	1	0	0	40	$\frac{40}{1} = 40$
Pivot row →	(−1)	2	0	1	0	−10	$\frac{-10}{-1} = 10$
	−2	−3	0	0	1	0	
	↑ Pivot column						

	x	y	u	v	P	Constant
	1	1	1	0	0	40
$\xrightarrow{-R_2}$	(1)	−2	0	−1	0	10
	−2	−3	0	0	1	0

	x	y	u	v	P	Constant	Ratio
$\xrightarrow[R_3 + 2R_2]{R_1 - R_2}$ Pivot row →	0	(3)	1	1	0	30	$\frac{30}{3} = 10$
	1	−2	0	−1	0	10	—
	0	−7	0	−2	1	20	
		↑ Pivot column					

	x	y	u	v	P	Constant
$\xrightarrow{\frac{1}{3}R_1}$	0	(1)	$\frac{1}{3}$	$\frac{1}{3}$	0	10
	1	−2	0	−1	0	10
	0	−7	0	−2	1	20

	x	y	u	v	P	Constant
$\xrightarrow[R_3 + 7R_1]{R_2 + 2R_1}$	0	1	$\frac{1}{3}$	$\frac{1}{3}$	0	10
	1	0	$\frac{2}{3}$	$-\frac{1}{3}$	0	30
	0	0	$\frac{7}{3}$	$\frac{1}{3}$	1	90

The last tableau is in final form, so we have obtained the following solution

$$x = 30 \qquad y = 10 \qquad u = 0 \qquad v = 0 \qquad P = 90$$

2. Let x denote the number of acres of Crop *A* to be cultivated, and let y denote the number of acres of Crop *B* to be cultivated. Since there is a total of 150 acres of land available for cultivation, we have $x + y \le 150$. Next, the restriction on the amount of money available for land cultivation implies that $40x + 60y \le 7400$. Similarly, the restriction on the amount of time available for labor implies that $20x + 25y \le 3300$. Also, since the farmer will cultivate at least 70 acres of Crop *A*, $x \ge 70$. Since the profit on each acre of Crop *A* is \$150 and the profit on each acre of Crop *B* is \$200, we see that the profit realizable by the farmer is $P = 150x + 200y$. Summarizing, we have the following linear programming problem:

$$\begin{aligned} \text{Maximize} \quad & P = 150x + 200y \\ \text{subject to} \quad & x + y \le 150 \\ & 40x + 60y \le 7400 \\ & 20x + 25y \le 3300 \\ & x \ge 70 \\ & x \ge 0, y \ge 0 \end{aligned}$$

To solve this nonstandard problem, we rewrite the fourth inequality in the form

$$-x \le -70$$

Using the method of this section with u, v, w, and z as slack variables, we have the following tableaus:

	x	y	u	v	w	z	P	Constant	Ratio
	1	1	1	0	0	0	0	150	$\frac{150}{1}=150$
	40	60	0	1	0	0	0	7,400	$\frac{7400}{40}=185$
	20	25	0	0	1	0	0	3,300	$\frac{3300}{20}=165$
Pivot row →	(−1)	0	0	0	0	1	0	−70	$\frac{-70}{-1}=70$
	−150	−200	0	0	0	0	1	0	
	↑ Pivot column								

$\xrightarrow{-R_4}$	x	y	u	v	w	z	P	Constant
	1	1	1	0	0	0	0	150
	40	60	0	1	0	0	0	7,400
	20	25	0	0	1	0	0	3,300
	(1)	0	0	0	0	−1	0	70
	−150	−200	0	0	0	0	1	0

$\xrightarrow{R_1 - R_4,\ R_2 - 40R_4,\ R_3 - 20R_4,\ R_5 + 150R_4}$	x	y	u	v	w	z	P	Constant	Ratio
	0	1	1	0	0	1	0	80	$\frac{80}{1}=80$
	0	60	0	1	0	40	0	4,600	$\frac{4600}{60}=76\frac{2}{3}$
Pivot row →	0	(25)	0	0	1	20	0	1,900	$\frac{1900}{25}=76$
	1	0	0	0	0	−1	0	70	—
	0	−200	0	0	0	−150	1	10,500	
		↑ Pivot column							

$\xrightarrow{\frac{1}{25}R_3}$	x	y	u	v	w	z	P	Constant
	0	1	1	0	0	1	0	80
	0	60	0	1	0	40	0	4,600
	0	(1)	0	0	$\frac{1}{25}$	$\frac{4}{5}$	0	76
	1	0	0	0	0	−1	0	70
	0	−200	0	0	0	−150	1	10,500

$\xrightarrow{R_1 - R_3,\ R_2 - 60R_3,\ R_5 + 200R_3}$	x	y	u	v	w	z	P	Constant
	0	0	1	0	$-\frac{1}{25}$	$\frac{1}{5}$	0	4
	0	0	0	1	$-\frac{12}{5}$	−8	0	40
	0	1	0	0	$\frac{1}{25}$	$\frac{4}{5}$	0	76
	1	0	0	0	0	−1	0	70
	0	0	0	0	8	10	1	25,700

This last tableau is in final form, and the solution is

$$x = 70 \qquad y = 76 \qquad u = 4 \qquad v = 40$$
$$w = 0 \qquad z = 0 \qquad P = 25{,}700$$

Thus, by cultivating 70 acres of Crop A and 76 acres of Crop B, the farmer will attain the maximum possible profit of $25,700.

CHAPTER 4 Summary of Principal Terms

TERMS

standard maximization problem (220)
slack variable (220)
basic variable (221)
nonbasic variable (221)
basic solution (222)
pivot column (223)
pivot row (224)
pivot element (224)
simplex tableau (224)
simplex method (225)
standard minimization problem (247)
primal problem (248)
dual problem (248)
nonstandard problem (264)
mixed constraints (265)

CHAPTER 4 Concept Review Questions

Fill in the blanks.

1. In a standard maximization problem: the objective function is to be _____; all the variables involved in the problem are _____; and each linear constraint may be written so that the expression involving the variables is _____ _____ or _____ _____ a nonnegative constant.

2. In setting up the initial simplex tableau, we first transform the system of linear inequalities into a system of linear _____, using_____ _____; the objective function is rewritten so that it has the form _____ and then is placed _____ the system of linear equations obtained earlier. Finally, the initial simplex tableau is the _____ matrix associated with this system of linear equations.

3. In a standard minimization problem: the objective function is to be _____; all the variables involved in the problem are _____; and each linear constraint may be

written so that the expression involving the variables is _____ _____ or _____ _____ a constant.

4. The fundamental theorem of duality states that a primal problem has a solution if and only if the corresponding _____ problem has a solution. If a solution exists, then the _____ functions of both the primal and the dual problem attain the same _____ _____.

CHAPTER 4 Review Exercises

In Exercises 1–8, use the simplex method to solve each linear programming problem.

1. Maximize $P = 3x + 4y$
subject to
$$\begin{aligned} x + 3y &\le 15 \\ 4x + y &\le 16 \\ x \ge 0, y &\ge 0 \end{aligned}$$

2. Maximize $P = 2x + 5y$
subject to
$$\begin{aligned} 2x + y &\le 16 \\ 2x + 3y &\le 24 \\ y &\le 6 \\ x \ge 0, y &\ge 0 \end{aligned}$$

3. Maximize $P = 3x + 2y$
subject to
$$\begin{aligned} x + 3y &\le 18 \\ 3x + 2y &\le 19 \\ 3x + y &\le 15 \\ x \ge 0, y &\ge 0 \end{aligned}$$

4. Maximize $P = 5x + 3y$
subject to
$$\begin{aligned} 3x + 4y &\le 42 \\ x + y &\le 12 \\ 2x + y &\le 20 \\ x \ge 0, y &\ge 0 \end{aligned}$$

5. Maximize $P = 2x + 3y + 5z$
subject to
$$\begin{aligned} x + 2y + 3z &\le 12 \\ x - 3y + 2z &\le 10 \\ x \ge 0, y \ge 0, z &\ge 0 \end{aligned}$$

6. Maximize $P = x + 2y + 3z$
subject to
$$\begin{aligned} 2x + y + z &\le 14 \\ 3x + 2y + 4z &\le 24 \\ 2x + 5y - 2z &\le 10 \\ x \ge 0, y \ge 0, z &\ge 0 \end{aligned}$$

7. Minimize $C = -4x - 7y$
subject to
$$\begin{aligned} 3x + y &\le 8 \\ x + 2y &\le 6 \\ x \ge 0, y &\ge 0 \end{aligned}$$

8. Minimize $C = -x - 3y + 2z$
subject to
$$\begin{aligned} 2x + y + 4z &\le 24 \\ x + 2y + z &\le 8 \\ 2x - y + 2z &\le 16 \\ x \ge 0, y \ge 0, z &\ge 0 \end{aligned}$$

9. Suppose that the primal problem for a linear programming problem is

Minimize $C = 2x + 5y$
subject to
$$\begin{aligned} 3x + 2y &\ge 8 \\ x + 4y &\ge 6 \\ x \ge 0, y &\ge 0 \end{aligned}$$

and the final simplex tableau for the dual problem associated with the primal problem is

u	v	x	y	P	Constant
1	0	$\frac{2}{5}$	$-\frac{1}{10}$	0	$\frac{3}{10}$
0	1	$-\frac{1}{5}$	$\frac{3}{10}$	0	$\frac{11}{10}$
0	0	2	1	1	9

Find the solution to the primal problem and the solution to the dual problem.

10. Construct the dual problem associated with the following primal problem:

Minimize $C = 2x + 4y + 3z$
subject to
$$\begin{aligned} x + 2y + z &\ge 14 \\ 4x + y + 2z &\ge 21 \\ -3x - 2y + 5z &\ge 12 \\ x \ge 0, y \ge 0, z &\ge 0 \end{aligned}$$

Do not solve.

11. Minimize $C = 3x + 2y$
subject to
$$\begin{aligned} 2x + 3y &\ge 6 \\ 2x + y &\ge 4 \\ x \ge 0, y &\ge 0 \end{aligned}$$

12. Minimize $C = x + 2y$
subject to
$$\begin{aligned} 3x + y &\ge 12 \\ x + 4y &\ge 16 \\ x \ge 0, y &\ge 0 \end{aligned}$$

13. Minimize $C = 24x + 18y + 24z$
subject to
$$\begin{aligned} 3x + 2y + z &\ge 4 \\ x + y + 3z &\ge 6 \\ x \ge 0, y \ge 0, z &\ge 0 \end{aligned}$$

14. Minimize $C = 4x + 2y + 6z$
subject to
$$\begin{aligned} x + 2y + z &\ge 4 \\ 2x + y + 2z &\ge 2 \\ 3x + 2y + z &\ge 3 \\ x \ge 0, y \ge 0, z &\ge 0 \end{aligned}$$

15. Maximize $P = 3x - 4y$
subject to
$$\begin{aligned} x + y &\le 45 \\ x - 2y &\ge 10 \\ x \ge 0, y &\ge 0 \end{aligned}$$

16. Minimize $C = 2x + 3y$
subject to
$$\begin{aligned} x + y &\le 10 \\ x + 2y &\ge 12 \\ 2x + y &\ge 12 \\ x \ge 0, y &\ge 0 \end{aligned}$$

17. Maximize $P = 2x + 3y$
subject to
$$\begin{aligned} 2x + 5y &\le 20 \\ -x + 5y &\ge 5 \\ x \ge 0, y &\ge 0 \end{aligned}$$

18. Minimize $C = -3x - 4y$
subject to
$$\begin{aligned} x + y &\le 45 \\ x + y &\ge 15 \\ x &\le 30 \\ y &\le 25 \\ x \ge 0, y &\ge 0 \end{aligned}$$

19. **Minimizing Mining Costs** Perth Mining Company operates two mines for the purpose of extracting gold and silver. The Saddle Mine costs \$14,000/day to operate, and it yields 50 oz of gold and 3000 oz of silver each day. The Horseshoe Mine costs \$16,000/day to operate, and it yields 75 oz of gold and 1000 oz of silver each day. Company management has set a target of at least 650 oz of gold and 18,000 oz of silver. How many days should each mine be operated at so that the target can be met at a minimum cost to the company? What is the minimum cost?

20. **Minimizing River Cruise Costs** Nordic River Cruises operates two ships between two cities along the Rhine. Ship A has 10 suites, 20 deluxe cabins, and 60 regular cabins, whereas Ship B has 15 suites, 40 deluxe cabins, and 70 regular cabins. The cost for running Ship A for one trip is \$240,000, and the cost for running Ship B for one trip is \$350,000. The company estimates that there will be demand for at least 400 suites, 900 deluxe cabins, and 2000 regular cabins during the season. How many trips should each ship make to meet the demand and at the same time minimize the operating cost? What is the cost?

21. **Minimizing City Water Costs** The water-supply manager for a Midwestern city needs to supply the city with at least 10 million gallons of potable (drinkable) water per day. The supply may be drawn from the local reservoir or from a pipeline to an adjacent town. The local reservoir has a maximum daily yield of 5 million gallons of potable water, and the pipeline has a maximum daily yield of 10 million gallons. By contract, the pipeline is required to supply a minimum of 6 million gallons/day. If the cost for 1 million gallons of reservoir water is \$300 and that for pipeline water is \$500, how much water should the manager get from each source to minimize daily water costs for the city?

22. **Investment Analysis** Jorge has decided to invest at most \$100,000 in securities in the form of corporate stocks. He has classified his options into three groups of stocks: blue-chip stocks that he assumes will yield a 10% return (dividends and capital appreciation) within a year, growth stocks that he assumes will yield a 15% return within a year, and speculative stocks that he assumes will yield a 20% return (mainly due to capital appreciation) within a year. Because of the relative risks involved in his investment, Jorge has further decided that no more than 30% of his investment should be in growth and speculative stocks and at least 50% of his investment should be in blue-chip and speculative stocks. Determine how much Jorge should invest in each group of stocks in the hope of maximizing the return on his investments. What is the maximum return?

23. **Maximizing Profit** A company manufactures three products, A, B, and C, on two machines, I and II. It has been determined that the company will realize a profit of \$4/unit of Product A, \$6/unit of Product B, and \$8/unit of Product C. Manufacturing a unit of Product A requires 9 min on Machine I and 6 min on Machine II; manufacturing a unit of Product B requires 12 min on Machine I and 6 min on Machine II; manufacturing a unit of Product C requires 18 min on Machine I and 10 min on Machine II. There are 6 hr of machine time available on Machine I and 4 hr of machine time available on Machine II in each work shift. How many units of each product should be produced in each shift to maximize the company's profit? What is the maximum profit?

24. **Investment Analysis** Sandra has at most \$200,000 to invest in stocks, bonds, and money market funds. She expects annual yields of 15%, 10%, and 8%, respectively, on these investments. If Sandra wants at least \$50,000 to be invested in money market funds and requires that the amount invested in bonds be greater than or equal to the sum of her investments in stocks and money market funds, determine how much she should invest in each vehicle to maximize the return on her investments. What is the maximum return?

The problem-solving skills that you learn in each chapter are building blocks for the rest of the course. Therefore, it is a good idea to make sure that you have mastered these skills before moving on to the next chapter. The Before Moving On exercises that follow are designed for that purpose. After completing these exercises, you can identify the skills that you should review before starting the next chapter.

CHAPTER 4 Before Moving On . . .

1. Consider the following linear programming problem:

$$\begin{aligned} \text{Maximize} \quad & P = x + 2y - 3z \\ \text{subject to} \quad & 2x + y - z \le 3 \\ & x - 2y + 3z \le 1 \\ & 3x + 2y + 4z \le 17 \\ & x \ge 0, y \ge 0, z \ge 0 \end{aligned}$$

Write the initial simplex tableau for the problem and identify the pivot element to be used in the first iteration of the simplex method.

2. The following simplex tableau is in final form. Find the solution to the linear programming problem associated with this tableau.

x	y	z	u	v	w	P	Constant
0	$\frac{1}{2}$	0	1	$-\frac{1}{2}$	0	0	2
0	$\frac{1}{4}$	1	0	$\frac{5}{4}$	$-\frac{1}{2}$	0	11
1	$\frac{1}{4}$	0	0	$-\frac{3}{4}$	$\frac{1}{2}$	0	2
0	$\frac{13}{4}$	0	0	$\frac{1}{4}$	$\frac{1}{2}$	1	28

3. Using the simplex method, solve the following linear programming problem:

$$\begin{aligned} \text{Maximize} \quad & P = 5x + 2y \\ \text{subject to} \quad & 4x + 3y \leq 30 \\ & 2x - 3y \leq 6 \\ & x \geq 0, y \geq 0 \end{aligned}$$

4. Using the simplex method, solve the following linear programming problem:

$$\begin{aligned} \text{Minimize} \quad & C = x + 2y \\ \text{subject to} \quad & x + y \geq 3 \\ & 2x + 3y \geq 6 \\ & x \geq 0, y \geq 0 \end{aligned}$$

5. Using the simplex method, solve the following linear programming problem:

$$\begin{aligned} \text{Maximize} \quad & P = 2x + y \\ \text{subject to} \quad & 2x + 5y \leq 20 \\ & 4x + 3y \geq 16 \\ & x \geq 0, y \geq 0 \end{aligned}$$

5 Mathematics of Finance

INTEREST THAT IS periodically added to the principal and thereafter itself earns interest is called *compound interest.* We begin this chapter by deriving the *compound interest formula,* which gives the amount of money accumulated when an initial amount of money is invested in an account for a fixed term and earns compound interest.

An *annuity* is a sequence of payments made at regular intervals. We derive formulas giving the *future value of an annuity* (what you end up with) and the *present value of an annuity* (the lump sum that, when invested now, will yield the same future value as that of the annuity). Then, using these formulas, we answer questions involving the amortization of certain types of installment loans and questions involving *sinking funds* (funds that are set up to be used for a specific purpose at a future date).

How much can the Jacksons afford to borrow from the bank for the purchase of a home? They have determined that after making a down payment they can afford a monthly payment of $2000. In Example 4, page 315, we learn how to determine the maximum amount they can afford to borrow if they secure a 30-year fixed mortgage at the current rate.

5.1 Compound Interest

Simple Interest

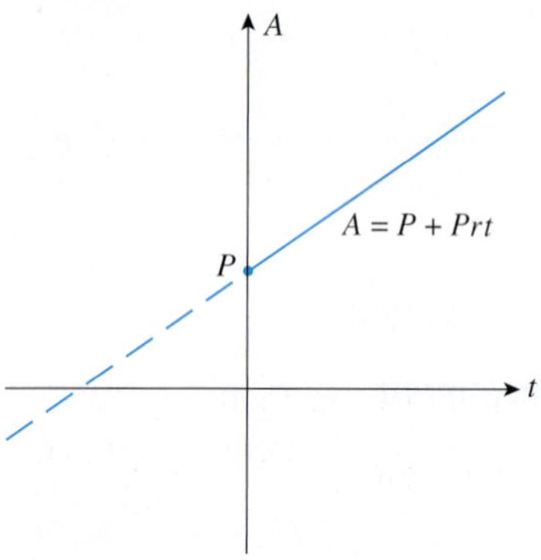

FIGURE 1
The accumulated amount is a linear function of t.

A natural application of linear functions to the business world is found in the computation of **simple interest**—interest that is computed on the original principal only. Thus, if I denotes the interest on a principal P (in dollars) at an interest rate of r per year for t years, we have

$$I = Prt$$

The **accumulated amount** A, the sum of the principal and interest after t years, is given by

$$\begin{aligned} A &= P + I = P + Prt \\ &= P(1 + rt) \end{aligned}$$

and is a linear function of t (see Exercise 42). In business applications, we are normally interested only in the case in which t is positive, so only the part of the line that lies in Quadrant I is of interest to us (Figure 1).

Simple Interest Formulas

Interest: $I = Prt$ **(1a)**

Accumulated amount: $A = P(1 + rt)$ **(1b)**

EXAMPLE 1 A bank pays simple interest at the rate of 8% per year for certain deposits. If a customer deposits $1000 and makes no withdrawals for 3 years, what is the total amount on deposit at the end of 3 years? What is the interest earned in that period of time?

Solution Using Formula (1b) with $P = 1000$, $r = 0.08$, and $t = 3$, we see that the total amount on deposit at the end of 3 years is given by

$$\begin{aligned} A &= P(1 + rt) \\ &= 1000[1 + (0.08)(3)] = 1240 \end{aligned}$$

or $1240.

The interest earned over the 3-year period is given by

$$\begin{aligned} I &= Prt \qquad \text{Use Formula (1a).} \\ &= 1000(0.08)(3) = 240 \end{aligned}$$

or $240.

Exploring with TECHNOLOGY

Refer to Example 1. Use a graphing utility to plot the graph of the function $A = 1000(1 + 0.08t)$, using the viewing window $[0, 10] \times [0, 2000]$.

1. What is the A-intercept of the straight line, and what does it represent?
2. What is the slope of the straight line, and what does it represent? (See Exercise 42.)

APPLIED EXAMPLE 2 Trust Funds An amount of \$2000 is invested in a 10-year trust fund that pays 6% annual simple interest. What is the total amount of the trust fund at the end of 10 years?

Solution The total amount of the trust fund at the end of 10 years is given by

$$\begin{aligned} A &= P(1 + rt) \\ &= 2000[1 + (0.06)(10)] = 3200 \end{aligned}$$

or \$3200.

A Treasury Bill (T-Bill) is a short-term debt obligation (less than or equal to 1 year) backed by the U.S. government. Rather than paying fixed interest payments, T-Bills are sold at a discount from face value. The appreciation of a T-Bill (face value − purchase price) provides the investment return to the holder.

APPLIED EXAMPLE 3 T-Bills Suppose that Jane buys a 26-week T-Bill with a maturity value of \$10,000. If she pays \$9850 for the T-Bill, what will be the rate of return on her investment?

Solution We use Formula (1b) with $A = 10{,}000$, $P = 9850$, and $t = \frac{26}{52} = \frac{1}{2}$. We obtain

$$\begin{aligned} 10{,}000 &= 9850\left(1 + \frac{1}{2}r\right) \\ &= 9850 + 4925r \end{aligned}$$

Solving for r, we find

$$\begin{aligned} 4925r &= 150 \\ r &= \frac{150}{4925} \approx 0.0305 \end{aligned}$$

So Jane's investment will earn simple interest at the rate of approximately 3.05% per year.

Compound Interest

In contrast to simple interest, **compound interest** is earned interest that is periodically added to the principal and thereafter itself earns interest at the same rate. To find a formula for the accumulated amount, let's consider a numerical example. Suppose \$1000 (the principal) is deposited in a bank for a term of 3 years, earning interest at the rate of 8% per year (called the **nominal**, or **stated, rate**) compounded annually. Then, using Formula (1b) with $P = 1000$, $r = 0.08$, and $t = 1$, we see that the accumulated amount at the end of the first year is

$$\begin{aligned} A_1 &= P(1 + rt) \\ &= 1000[1 + (0.08)(1)] = 1000(1.08) = 1080 \end{aligned}$$

or \$1080.

To find the accumulated amount A_2 at the end of the second year, we use Formula (1b) once again, this time with $P = A_1$. (Remember, the principal *and* interest now earn interest over the second year.) We obtain

$$\begin{aligned} A_2 &= P(1 + rt) = A_1(1 + rt) \\ &= 1000[1 + 0.08(1)][1 + 0.08(1)] \\ &= 1000[1 + 0.08]^2 = 1000(1.08)^2 = 1166.40 \end{aligned}$$

or \$1166.40.

Finally, the accumulated amount A_3 at the end of the third year is found using (1b) with $P = A_2$, giving

$$\begin{aligned} A_3 &= P(1 + rt) = A_2(1 + rt) \\ &= 1000[1 + 0.08(1)]^2[1 + 0.08(1)] \\ &= 1000[1 + 0.08]^3 = 1000(1.08)^3 \approx 1259.71 \end{aligned}$$

or approximately \$1259.71.

If you reexamine our calculations, you will see that the accumulated amounts at the end of each year have the following form:

$$\begin{aligned} &\text{First year:} & A_1 &= 1000(1 + 0.08), \text{ or } A_1 = P(1 + r) \\ &\text{Second year:} & A_2 &= 1000(1 + 0.08)^2, \text{ or } A_2 = P(1 + r)^2 \\ &\text{Third year:} & A_3 &= 1000(1 + 0.08)^3, \text{ or } A_3 = P(1 + r)^3 \end{aligned}$$

These observations suggest the following general result: If P dollars is invested over a term of t years, earning interest at the rate of r per year compounded annually, then the accumulated amount is

$$A = P(1 + r)^t \tag{2}$$

Equation (2) was derived under the assumption that interest was compounded *annually*. In practice, however, interest is usually compounded more than once a year. The interval of time between successive interest calculations is called the **conversion period.**

If interest at a nominal rate of r per year is compounded m times a year on a principal of P dollars, then the simple interest rate per conversion period is

$$i = \frac{r}{m} \qquad \frac{\text{Annual interest rate}}{\text{Periods per year}}$$

For example, if the nominal interest rate is 8% per year ($r = 0.08$) and interest is compounded quarterly ($m = 4$), then

$$i = \frac{r}{m} = \frac{0.08}{4} = 0.02$$

or 2% per period.

To find a general formula for the accumulated amount when a principal of P dollars is deposited in a bank for a term of t years and earns interest at the (nominal) rate of r per year compounded m times per year, we proceed as before, using Formula (1b) repeatedly with the interest rate $i = \frac{r}{m}$. We see that the accumulated amount at the end of each period is as follows:

$$\begin{aligned} &\text{First period:} & A_1 &= P(1 + i) \\ &\text{Second period:} & A_2 &= A_1(1 + i) = [P(1 + i)](1 + i) = P(1 + i)^2 \\ &\text{Third period:} & A_3 &= A_2(1 + i) = [P(1 + i)^2](1 + i) = P(1 + i)^3 \\ &\quad\vdots & &\quad\vdots \\ &n\text{th period:} & A_n &= A_{n-1}(1 + i) = [P(1 + i)^{n-1}](1 + i) = P(1 + i)^n \end{aligned}$$

There are $n = mt$ periods in t years (number of conversion periods per year times the term in years). Hence, the accumulated amount at the end of t years is given by

$$A = P(1 + i)^n$$

Compound Interest Formula (Accumulated Amount)

$$A = P(1 + i)^n \tag{3}$$

where $i = \dfrac{r}{m}$, $n = mt$, and

A = Accumulated amount at the end of n conversion periods

P = Principal

r = Nominal interest rate per year

m = Number of conversion periods per year

t = Term (number of years)

Exploring with TECHNOLOGY

Let $A_1(t)$ denote the accumulated amount of \$100 earning simple interest at the rate of 6% per year over t years, and let $A_2(t)$ denote the accumulated amount of \$100 earning interest at the rate of 6% per year compounded monthly over t years.

1. Find expressions for $A_1(t)$ and $A_2(t)$.
2. Use a graphing utility to plot the graphs of A_1 and A_2 on the same set of axes, using the viewing window $[0, 20] \times [0, 400]$.
3. Comment on the growth of $A_1(t)$ and $A_2(t)$ by referring to the graphs of A_1 and A_2.

EXAMPLE 4 Find the accumulated amount after 3 years if \$1000 is invested at 8% per year compounded (a) annually, (b) semiannually, (c) quarterly, (d) monthly, and (e) daily (assume a 365-day year).

Solution

a. Here, $P = 1000$, $r = 0.08$, and $m = 1$. Thus, $i = r = 0.08$ and $n = 3$, so Formula (3) gives

$$\begin{aligned} A &= 1000(1 + 0.08)^3 \\ &\approx 1259.71 \end{aligned}$$

or \$1259.71.

b. Here, $P = 1000$, $r = 0.08$, and $m = 2$. Thus, $i = \frac{0.08}{2}$ and $n = (3)(2) = 6$, so Formula (3) gives

$$\begin{aligned} A &= 1000\left(1 + \frac{0.08}{2}\right)^6 \\ &\approx 1265.32 \end{aligned}$$

or \$1265.32.

c. In this case, $P = 1000$, $r = 0.08$, and $m = 4$. Thus, $i = \frac{0.08}{4}$ and $n = (3)(4) = 12$, so Formula (3) gives

$$A = 1000\left(1 + \frac{0.08}{4}\right)^{12} \approx 1268.24$$

or \$1268.24.

d. Here, $P = 1000$, $r = 0.08$, and $m = 12$. Thus, $i = \frac{0.08}{12}$ and $n = (3)(12) = 36$, so Formula (3) gives

$$A = 1000\left(1 + \frac{0.08}{12}\right)^{36} \approx 1270.24$$

or \$1270.24.

e. Here, $P = 1000$, $r = 0.08$, $m = 365$, and $t = 3$. Thus, $i = \frac{0.08}{365}$ and $n = (3)(365) = 1095$, so Formula (3) gives

$$A = 1000\left(1 + \frac{0.08}{365}\right)^{1095} \approx 1271.22$$

or \$1271.22. These results are summarized in Table 1.

TABLE 1

Nominal Rate, r	Conversion Period	Interest Rate/ Conversion Period	Initial Investment	Accumulated Amount
8%	Annually ($m = 1$)	8%	\$1000	\$1259.71
8	Semiannually ($m = 2$)	4	1000	1265.32
8	Quarterly ($m = 4$)	2	1000	1268.24
8	Monthly ($m = 12$)	2/3	1000	1270.24
8	Daily ($m = 365$)	8/365	1000	1271.22

Exploring with TECHNOLOGY

Investments that are allowed to grow over time can increase in value surprisingly fast. Consider the potential growth of \$10,000 if earnings are reinvested. More specifically, suppose $A_1(t)$, $A_2(t)$, $A_3(t)$, $A_4(t)$, and $A_5(t)$ denote the accumulated values of an investment of \$10,000 over a term of t years and earning interest at the rate of 4%, 6%, 8%, 10%, and 12% per year compounded annually.

1. Find expressions for $A_1(t), A_2(t), \ldots, A_5(t)$.
2. Use a graphing utility to plot the graphs of $A_1, A_2, \ldots, A_5$ on the same set of axes, using the viewing window $[0, 20] \times [0, 100{,}000]$.
3. Use **TRACE** to find $A_1(20), A_2(20), \ldots, A_5(20)$, and then interpret your results.

Continuous Compounding of Interest

One question that arises naturally in the study of compound interest is: What happens to the accumulated amount over a fixed period of time if the interest is computed more and more frequently?

Intuition suggests that the more often interest is compounded, the larger the accumulated amount will be. This is confirmed by the results of Example 4, where we found that the accumulated amounts did in fact increase when we increased the number of conversion periods per year.

This leads us to another question: Does the accumulated amount keep growing without bound, or does it approach a fixed number when the interest is computed more and more frequently over a fixed period of time?

To answer this question, let's look again at the compound interest formula:

$$A = P(1 + i)^n = P\left(1 + \frac{r}{m}\right)^{mt} \tag{4}$$

Recall that m is the number of conversion periods per year. So to find an answer to our question, we should let m get larger and larger in Equation (4). If we let $u = \frac{m}{r}$ so that $m = ru$, then (4) becomes

$$A = P\left(1 + \frac{1}{u}\right)^{urt} \qquad \frac{r}{m} = \frac{1}{u}$$

$$= P\left[\left(1 + \frac{1}{u}\right)^{u}\right]^{rt} \qquad \text{Since } a^{xy} = (a^x)^y$$

Now let's see what happens to the expression

$$\left(1 + \frac{1}{u}\right)^{u}$$

as u gets larger and larger. From Table 2, you can see that as u increases,

$$\left(1 + \frac{1}{u}\right)^{u}$$

seems to approach the number 2.71828 (we have rounded all our calculations to five decimal places).

TABLE 2

u	$\left(1 + \frac{1}{u}\right)^{u}$
10	2.59374
100	2.70481
1000	2.71692
10,000	2.71815
100,000	2.71827
1,000,000	2.71828

It can be shown—although we will not do so here—that as u gets larger and larger, the value of the expression $\left(1 + \frac{1}{u}\right)^u$ approaches the irrational number 2.71828. . . , which we denote by e. (See the Exploring with Technology exercise that follows.)

Exploring with TECHNOLOGY

To obtain a visual confirmation of the fact that the expression $\left(1 + \frac{1}{u}\right)^u$ approaches the number $e = 2.71828\ldots$ as u gets larger and larger, plot the graph of $f(x) = \left(1 + \frac{1}{x}\right)^x$ in a suitable viewing window, and observe that $f(x)$ approaches 2.71828 . . . as x gets larger and larger. Use **ZOOM** and **TRACE** to find the value of $f(x)$ for large values of x.

Using this result, we can see that as m gets larger and larger, A approaches $P(e)^{rt} = Pe^{rt}$. In this situation, we say that interest is *compounded continuously.* Let's summarize this important result.

Continuous Compound Interest Formula

$$A = Pe^{rt} \tag{5}$$

where

P = Principal

r = Nominal interest rate compounded continuously

t = Time in years

A = Accumulated amount at the end of t years

EXAMPLE 5 Find the accumulated amount after 3 years if \$1000 is invested at 8% per year compounded (a) daily (assume a 365-day year) and (b) continuously.

Solution

a. Use Formula (3) with $P = 1000$, $r = 0.08$, $m = 365$, and $t = 3$. Thus, $i = \frac{0.08}{365}$ and $n = (365)(3) = 1095$, so

$$A = 1000\left(1 + \frac{0.08}{365}\right)^{(365)(3)} \approx 1271.22$$

or \$1271.22.

b. Here, we use Formula (5) with $P = 1000$, $r = 0.08$, and $t = 3$, obtaining

$$A = 1000e^{(0.08)(3)}$$
$$\approx 1271.25$$

or \$1271.25.

Observe that the accumulated amounts corresponding to interest compounded daily and interest compounded continuously differ by very little. The continuous compound interest formula is a very important tool in theoretical work in financial analysis.

Effective Rate of Interest

Example 4 showed that the interest actually earned on an investment depends on the frequency with which the interest is compounded. Thus, the stated, or nominal, rate of 8% per year does not reflect the actual rate at which interest is earned. This suggests that we need to find a common basis for comparing interest rates. One such way of comparing interest rates is provided by the use of the *effective rate of interest*. The **effective rate of interest** is the annual rate of interest that, when compounded annually, will yield the same accumulated amount as the nominal rate compounded m times a year (over the same term). Equivalently, the effective rate of interest is the *simple* interest rate that would produce the same accumulated amount in 1 year as the nominal rate compounded m times a year. The effective rate of interest is also called the **annual percentage yield.**

To derive a relationship between the nominal interest rate, r per year compounded m times, and its corresponding effective rate, R per year, let's assume an initial investment of P dollars. Then the accumulated amount after 1 year at a simple interest rate of R per year is

$$A = P(1 + R)$$

Also, the accumulated amount after 1 year at an interest rate of r per year compounded m times a year is

$$A = P(1 + i)^n = P\left(1 + \frac{r}{m}\right)^m \qquad \text{Since } i = \frac{r}{m} \text{ and } t = 1$$

Equating the two expressions gives

$$P(1 + R) = P\left(1 + \frac{r}{m}\right)^m$$

$$1 + R = \left(1 + \frac{r}{m}\right)^m \qquad \text{Divide both sides by } P.$$

If we solve the preceding equation for R, we obtain the following formula for computing the effective rate of interest.

Effective Rate of Interest Formula

$$r_{\text{eff}} = \left(1 + \frac{r}{m}\right)^m - 1 \tag{6}$$

where

r_{eff} = Effective rate of interest

r = Nominal interest rate per year

m = Number of conversion periods per year

EXAMPLE 6 Find the effective rate of interest corresponding to a nominal rate of 8% per year compounded (a) annually, (b) semiannually, (c) quarterly, (d) monthly, and (e) daily.

Solution

a. The effective rate of interest corresponding to a nominal rate of 8% per year compounded annually is, of course, given by 8% per year. This result is also confirmed by using Formula (6) with $r = 0.08$ and $m = 1$. Thus,

$$r_{\text{eff}} = (1 + 0.08) - 1 = 0.08$$

b. Let $r = 0.08$ and $m = 2$. Then Formula (6) yields

$$\begin{aligned} r_{\text{eff}} &= \left(1 + \frac{0.08}{2}\right)^2 - 1 \\ &= (1.04)^2 - 1 \\ &= 0.0816 \end{aligned}$$

so the effective rate is 8.16% per year.

c. Let $r = 0.08$ and $m = 4$. Then Formula (6) yields

$$\begin{aligned} r_{\text{eff}} &= \left(1 + \frac{0.08}{4}\right)^4 - 1 \\ &= (1.02)^4 - 1 \\ &\approx 0.08243 \end{aligned}$$

so the corresponding effective rate in this case is 8.243% per year.

d. Let $r = 0.08$ and $m = 12$. Then Formula (6) yields

$$\begin{aligned} r_{\text{eff}} &= \left(1 + \frac{0.08}{12}\right)^{12} - 1 \\ &\approx 0.08300 \end{aligned}$$

so the corresponding effective rate in this case is 8.3% per year.

e. Let $r = 0.08$ and $m = 365$. Then Formula (6) yields

$$\begin{aligned} r_{\text{eff}} &= \left(1 + \frac{0.08}{365}\right)^{365} - 1 \\ &\approx 0.08328 \end{aligned}$$

so the corresponding effective rate in this case is 8.328% per year. ■

If the effective rate of interest r_{eff} is known, then the accumulated amount after t years on an investment of P dollars may be more readily computed by using the formula

$$A = P(1 + r_{\text{eff}})^t$$

Explore and Discuss

Recall the effective rate of interest formula:

$$r_{\text{eff}} = \left(1 + \frac{r}{m}\right)^m - 1$$

1. Show that

$$r = m[(1 + r_{\text{eff}})^{1/m} - 1]$$

2. A certificate of deposit (CD) is known to have an effective rate of 5.3% per year. If interest is compounded monthly, find the nominal rate of interest by using the result of part 1.

The 1968 Truth in Lending Act passed by Congress requires that the effective rate of interest be disclosed in all contracts involving interest charges. The passage of this act has benefited consumers because they now have a common basis for comparing the various nominal rates quoted by different financial institutions. Furthermore, knowing the effective rate enables consumers to compute the actual charges involved in a transaction. Thus, if the effective rates of interest found in Example 6 were known, then the accumulated values of Example 4 could have been readily found (see Table 3).

TABLE 3

Nominal Rate, r	Frequency of Interest Payment	Effective Rate	Initial Investment	Accumulated Amount After 3 Years
8%	Annually	8%	\$1000	$1000(1 + 0.08)^3 \approx \$1259.71$
8	Semiannually	8.16	1000	$1000(1 + 0.0816)^3 \approx 1265.32$
8	Quarterly	8.243	1000	$1000(1 + 0.08243)^3 \approx 1268.23$
8	Monthly	8.300	1000	$1000(1 + 0.08300)^3 \approx 1270.24$
8	Daily	8.328	1000	$1000(1 + 0.08328)^3 \approx 1271.22$

Present Value

Let's return to the compound interest Formula (3), which expresses the accumulated amount at the end of n periods when interest at the rate of r is compounded m times a year. The principal P in (3) is often referred to as the **present value,** and the accumulated value A is called the **future value,** since it is realized at a future date. In certain instances, an investor might wish to determine how much money he should invest now, at a fixed rate of interest, so that he will realize a certain sum at some future date. This problem may be solved by expressing P in terms of A. Thus, from Formula (3), we find

$$P = A(1 + i)^{-n}$$

Here, as before, $i = \frac{r}{m}$, where m is the number of conversion periods per year.

> **Present Value Formula for Compound Interest**
>
> $$P = A(1 + i)^{-n} \qquad (7)$$

EXAMPLE 7 How much money should be deposited in a bank paying interest at the rate of 6% per year compounded monthly so that at the end of 3 years, the accumulated amount will be \$20,000?

Solution Here, $r = 0.06$ and $m = 12$, so $i = \frac{0.06}{12}$ and $n = (3)(12) = 36$. Thus, the problem is to determine P given that $A = 20{,}000$. Using Formula (7), we obtain

$$P = 20{,}000\left(1 + \frac{0.06}{12}\right)^{-36}$$
$$\approx 16{,}713$$

or \$16,713.

EXAMPLE 8 Find the present value of \$49,158.60 due in 5 years at an interest rate of 10% per year compounded quarterly.

Solution Using Formula (7) with $r = 0.1$ and $m = 4$, so that $i = \frac{0.1}{4}$, $n = (4)(5) = 20$, and $A = 49{,}158.6$, we obtain

$$P = (49{,}158.6)\left(1 + \frac{0.1}{4}\right)^{-20} \approx 30{,}000.07$$

or approximately \$30,000.

If we solve Formula (5) for P, we have

$$P = Ae^{-rt}$$

which gives the present value in terms of the future (accumulated) value for the case of continuous compounding.

Present Value Formula for Continuous Compound Interest

$$P = Ae^{-rt} \tag{8}$$

APPLIED EXAMPLE 9 Real Estate Investment Blakely Investment Company owns an office building located in the commercial district of a city. As a result of the continued success of an urban renewal program, local business is enjoying a miniboom. The market value of Blakely's property is

$$V(t) = 300{,}000e^{\sqrt{t}/2}$$

where $V(t)$ is measured in dollars and t is the time in years from the present. If the expected rate of appreciation is 9% per year compounded continuously for the next 10 years, find an expression for the present value $P(t)$ of the market price of the property that will be valid for the next 10 years. Compute $P(7)$, $P(8)$, and $P(9)$, and then interpret your results.

Solution Using Formula (8) with $A = V(t)$ and $r = 0.09$, we find that the present value of the market price of the property t years from now is

$$\begin{aligned} P(t) &= V(t)e^{-0.09t} \\ &= 300{,}000e^{-0.09t + \sqrt{t}/2} \qquad (0 \le t \le 10) \end{aligned}$$

Letting $t = 7$, 8, and 9, we find

$$\begin{aligned} P(7) &= 300{,}000e^{-0.09(7) + \sqrt{7}/2} \approx 599{,}837, \text{ or } \$599{,}837 \\ P(8) &= 300{,}000e^{-0.09(8) + \sqrt{8}/2} \approx 600{,}640, \text{ or } \$600{,}640 \\ P(9) &= 300{,}000e^{-0.09(9) + \sqrt{9}/2} \approx 598{,}115, \text{ or } \$598{,}115 \end{aligned}$$

respectively. From the results of these computations, we see that the present value of the property's market price seems to decrease after a certain period of growth. This suggests that there is an optimal time for the owners to sell. By plotting the graph of the function P, you can show that the highest present value of the property's market value is approximately \$600,779 and that it occurs at time $t \approx 7.72$ years.

The returns on certain investments such as zero coupon certificates of deposit (CDs) and zero coupon bonds are compared by quoting the time it takes for each investment to triple, or even quadruple. These calculations make use of the compound interest Formula (3).

APPLIED EXAMPLE 10 Investment Options Jane has narrowed her investment options down to two:

1. Purchase a CD that matures in 24 years and pays interest upon maturity at the rate of 5% per year compounded daily (assume 365 days in a year).
2. Purchase a zero coupon CD that will triple her investment in the same period.

Which option will optimize Jane's investment?

Solution Let's compute the accumulated amount under option 1. Here,

$$r = 0.05 \qquad m = 365 \qquad t = 24$$

so $n = 24(365) = 8760$ and $i = \frac{0.05}{365}$. The accumulated amount at the end of 24 years (after 8760 conversion periods) is

$$A = P\left(1 + \frac{0.05}{365}\right)^{8760} \approx 3.32P$$

or $\$3.32P$. If Jane chooses option 2, the accumulated amount of her investment after 24 years will be $\$3P$. Therefore, she should choose option 1.

APPLIED EXAMPLE 11 IRAs Moesha has an Individual Retirement Account (IRA) with a brokerage firm. Her money is invested in a money market mutual fund that pays interest on a daily basis. Over a 2-year period in which no deposits or withdrawals were made, her account grew from \$4500 to \$4792.61. Find the effective rate at which Moesha's account was earning interest over that period (assume 365 days in a year).

Solution Let r_{eff} denote the required effective rate of interest. We have

$$\begin{aligned} 4792.61 &= 4500(1 + r_{\text{eff}})^2 \\ (1 + r_{\text{eff}})^2 &\approx 1.06502 \\ 1 + r_{\text{eff}} &\approx 1.031998 \qquad \text{Take the square root on both sides.} \end{aligned}$$

or $r_{\text{eff}} \approx 0.031998$. Therefore, the effective rate was approximately 3.20% per year.

Using Logarithms to Solve Problems in Finance*

The next two examples show how logarithms can be used to solve problems involving compound interest.

EXAMPLE 12 How long will it take \$10,000 to grow to \$15,000 if the investment earns an interest rate of 8% per year compounded quarterly?

Solution Using Formula (3) with $A = 15{,}000$, $P = 10{,}000$, $r = 0.08$, and $m = 4$, we obtain

$$\begin{aligned} 15{,}000 &= 10{,}000\left(1 + \frac{0.08}{4}\right)^{4t} \\ (1.02)^{4t} &= \frac{15{,}000}{10{,}000} = 1.5 \end{aligned}$$

*Coverage of the next two examples is optional, and the corresponding exercises in Exercise Set 5.1 are marked optional. A review of logarithms and examples in which the properties of logarithms are used to solve equations is given in Appendix C.

Taking the logarithm on each side of the equation gives

$$\begin{aligned} \ln(1.02)^{4t} &= \ln 1.5 \\ 4t \ln 1.02 &= \ln 1.5 \qquad \log_b m^n = n \log_b m \\ 4t &= \frac{\ln 1.5}{\ln 1.02} \\ t &= \frac{\ln 1.5}{4 \ln 1.02} \approx 5.12 \end{aligned}$$

So it will take approximately 5.12 years for the investment to grow from $10,000 to $15,000.

EXAMPLE 13 Find the interest rate needed for an investment of $10,000 to grow to an amount of $18,000 in 5 years if the interest is compounded monthly.

Solution Use Formula (3) with $A = 18{,}000$, $P = 10{,}000$, $m = 12$, and $t = 5$. Thus, $i = \frac{r}{12}$ and $n = (12)(5) = 60$, so

$$18{,}000 = 10{,}000\left(1 + \frac{r}{12}\right)^{12(5)}$$

Dividing both sides of the equation by 10,000 gives

$$\frac{18{,}000}{10{,}000} = \left(1 + \frac{r}{12}\right)^{60}$$

or, upon simplification,

$$\left(1 + \frac{r}{12}\right)^{60} = 1.8$$

Now we take the logarithm on each side of the equation, obtaining

$$\begin{aligned} \ln\left(1 + \frac{r}{12}\right)^{60} &= \ln 1.8 \\ 60 \ln\left(1 + \frac{r}{12}\right) &= \ln 1.8 \\ \ln\left(1 + \frac{r}{12}\right) &= \frac{\ln 1.8}{60} \approx 0.009796 \\ \left(1 + \frac{r}{12}\right) &\approx e^{0.009796} \qquad \ln e^x = x \\ &\approx 1.009844 \end{aligned}$$

and

$$\begin{aligned} \frac{r}{12} &\approx 1.009844 - 1 \\ r &\approx 0.1181 \end{aligned}$$

or 11.81% per year.

5.1 Self-Check Exercises

1. Find the present value of $20,000 due in 3 years at an interest rate of 5.4%/year compounded monthly.

2. **Investment Income** Paul is a retiree living on Social Security and the income from his investment. Currently, his

$100,000 investment in a 1-year CD is yielding 4.6% interest compounded daily. If he reinvests the principal ($100,000) on the due date of the CD in another 1-year CD paying 3.2% interest compounded daily, find the net decrease in his yearly income from his investment.

Solutions to Self-Check Exercises 5.1 can be found on page 297.

5.1 Concept Questions

1. Explain the difference between simple interest and compound interest.
2. What is the difference between the accumulated amount (future value) and the present value of an investment?
3. What is the effective rate of interest?

5.1 Exercises

1. Find the simple interest on a $500 investment made for 2 years at an interest rate of 8%/year. What is the accumulated amount?
2. Find the simple interest on a $1000 investment made for 3 years at an interest rate of 5%/year. What is the accumulated amount?
3. Find the accumulated amount at the end of 9 months on an $800 deposit in a bank paying simple interest at a rate of 6%/year.
4. Find the accumulated amount at the end of 8 months on a $1200 bank deposit paying simple interest at a rate of 7%/year.
5. If the accumulated amount is $1160 at the end of 2 years and the simple rate of interest is 8%/year, what is the principal?
6. A bank deposit paying simple interest at the rate of 5%/year grew to a sum of $3100 in 10 months. Find the principal.
7. How many days will it take for a sum of $1000 to earn $20 interest if it is deposited in a bank paying simple interest at the rate of 2.5%/year? (Use a 365-day year.)
8. How many days will it take for a sum of $1500 to earn $25 interest if it is deposited in a bank paying simple interest at the rate of 5%/year? (Use a 365-day year.)
9. A bank deposit paying simple interest grew from an initial sum of $1000 to a sum of $1075 in 9 months. Find the interest rate.
10. Determine the simple interest rate at which $1200 will grow to $1250 in 8 months.

In Exercises 11–20, find the accumulated amount *A* if the principal *P* is invested at the interest rate of *r*/year for *t* years.

11. $P = \$1000$, $r = 4\%$, $t = 8$, compounded annually
12. $P = \$1000$, $r = 5\frac{1}{2}\%$, $t = 6$, compounded annually
13. $P = \$2500$, $r = 4\%$, $t = 10$, compounded semiannually
14. $P = \$2500$, $r = 6\%$, $t = 10\frac{1}{2}$, compounded semiannually
15. $P = \$12{,}000$, $r = 5\%$, $t = 10\frac{1}{2}$, compounded quarterly
16. $P = \$42{,}000$, $r = 4\frac{3}{4}\%$, $t = 8$, compounded quarterly
17. $P = \$150{,}000$, $r = 4\%$, $t = 4$, compounded monthly
18. $P = \$180{,}000$, $r = 6\%$, $t = 6\frac{1}{4}$, compounded monthly
19. $P = \$150{,}000$, $r = 9\%$, $t = 3$, compounded daily
20. $P = \$200{,}000$, $r = 8\%$, $t = 4$, compounded daily

In Exercises 21–24, find the effective rate corresponding to the given nominal rate.

21. 6%/year compounded semiannually
22. 5%/year compounded quarterly
23. 4%/year compounded monthly
24. 4%/year compounded daily

In Exercises 25–28, find the present value of $40,000 due in 4 years at the given rate of interest.

25. 4%/year compounded semiannually
26. 4%/year compounded quarterly
27. 3%/year compounded monthly
28. 5%/year compounded daily
29. Find the accumulated amount after 4 years if $5000 is invested at 6%/year compounded continuously.
30. Find the accumulated amount after 6 years if $6500 is invested at 5%/year compounded continuously.

31. **CONSUMER DECISIONS** Mitchell has been given the option of either paying his $300 bill now or settling it for $306 after 1 month (30 days). If he chooses to pay after 1 month, find the simple interest rate at which he would be charged.

32. **COURT JUDGMENT** Jennifer was awarded damages of $150,000 in a successful lawsuit she brought against her employer 5 years ago. Simple interest on the judgment accrues at the rate of 12%/year from the date of filing. If the case were settled today, how much would Jennifer receive in the final judgment?

33. **BRIDGE LOANS** To help finance the purchase of a new house, the Abdullahs have decided to apply for a short-term loan (a bridge loan) in the amount of $120,000 for a term of 3 months. If the bank charges simple interest at the rate of 10%/year, how much will the Abdullahs owe the bank at the end of the term?

34. **CORPORATE BONDS** David owns $20,000 worth of 10-year bonds of Ace Corporation. These bonds pay interest every 6 months at the rate of 3%/year (simple interest). How much income will David receive from this investment every 6 months? How much interest will David receive over the life of the bonds?

35. **MUNICIPAL BONDS** Maya paid $10,000 for a 7-year bond issued by a city. She received interest amounting to $3500 over the life of the bonds. What rate of (simple) interest did the bond pay?

36. **TREASURY BILLS** Isabella purchased $20,000 worth of 13-week T-Bills for $19,875. What will be the rate of return on her investment?

37. **TREASURY BILLS** Maxwell purchased $15,000 worth of 52-week T-Bills for $14,650. What will be the rate of return on his investment?

38. **COMPARING INVESTMENT RETURNS** The value of Maria's investments increased by 20% in the first year and by a further 10% in the second year. The value of Laura's investments grew 10% in the first year, followed by a gain of 20% in the second year. Both Maria and Laura started out with a $10,000 investment. Whose investment increased more in the 2-year period? Explain.

39. **INVESTMENTS** The value of Alan's stock portfolio grew by 20% in the first year, followed by a growth of 10% in the second year. It dropped 10% and 20% in the third and fourth years, respectively. Is the value of Alan's stock portfolio after 4 years the same as that when he started out? Explain.

40. **INVESTMENTS** The value of Jack's investment portfolio fell by 20% in the first year but rebounded by 20% in the second year. Did Jack regain all of the money that he lost in the first year at the end of the second year? Explain.

41. **INVESTMENTS** The value of Arabella's stock portfolio dropped 20% in the first year. Find the annual rate of growth, compounded yearly, that she must achieve in the next 2 years to bring the value of her stock portfolio back to its initial value (its value at the beginning of the first year).

42. Write Formula (1b) in the slope-intercept form, and interpret the meaning of the slope and the A-intercept in terms of r and P.
Hint: Refer to Figure 1.

43. **HOSPITAL COSTS** If the cost of a semiprivate room in a hospital was $680/day 5 years ago and hospital costs have risen at the rate of 8%/year since that time, what rate would you expect to pay for a semiprivate room today?

44. **FAMILY FOOD EXPENDITURE** Today, a typical family of four spends $880/month for food. If inflation occurs at the rate of 3%/year over the next 6 years, how much should the typical family of four expect to spend for food 6 years from now?

45. **HOUSING APPRECIATION** The Kwans are planning to buy a house 4 years from now. Housing experts in their area have estimated that the cost of a home will increase at a rate of 5%/year during that period. If this economic prediction holds true, how much can the Kwans expect to pay for a house that currently costs $260,000?

46. **ELECTRICITY CONSUMPTION** A utility company in a western city of the United States expects the consumption of electricity to increase by 8%/year during the next decade, owing mainly to the expected increase in population. If consumption does increase at this rate, find the amount by which the utility company will have to increase its generating capacity in order to meet the needs of the area at the end of the decade.

47. **PENSION FUNDS** The managers of a pension fund have invested $1.5 million in U.S. government certificates of deposit that pay interest at the rate of 2.5%/year compounded semiannually over a period of 10 years. At the end of this period, how much will the investment be worth?

48. **RETIREMENT FUNDS** Five and a half years ago, Chris invested $10,000 in a retirement fund that grew at the rate of 6.82%/year compounded quarterly. What is his account worth today?

49. **MUTUAL FUNDS** Jodie invested $15,000 in a mutual fund 4 years ago. If the fund grew at the rate of 7.8%/year compounded monthly, what would Jodie's account be worth today?

50. **TRUST FUNDS** A young man is the beneficiary of a trust fund established for him 21 years ago at his birth. If the original amount placed in trust was $10,000, how much will he receive if the money has earned interest at the rate of 6%/year compounded annually? Compounded quarterly? Compounded monthly?

51. **INVESTMENT PLANNING** Find how much money should be deposited in a bank paying interest at the rate of 3.5%/year compounded quarterly so that at the end of 5 years, the accumulated amount will be $40,000.

52. **PROMISSORY NOTES** An individual purchased a 4-year, \$10,000 promissory note with an interest rate of 5.5%/year compounded semiannually. How much did the note cost?

53. **FINANCING A COLLEGE EDUCATION** The parents of a child have just come into a large inheritance and wish to establish a trust fund for her college education. If they estimate that they will need \$100,000 in 13 years, how much should they set aside in the trust now if they can invest the money at $4\frac{1}{2}$%/year compounded (a) annually, (b) semiannually, and (c) quarterly?

54. **INVESTMENTS** Anthony invested a sum of money 5 years ago in a savings account that has since paid interest at the rate of 4%/year compounded quarterly. His investment is now worth \$22,289.22. How much did he originally invest?

55. **COMPARING RATES OF RETURN** In the last 5 years, Bendix Mutual Fund grew at the rate of 6.4%/year compounded quarterly. Over the same period, Acme Mutual Fund grew at the rate of 6.5%/year compounded semiannually. Which mutual fund has a better rate of return?

56. **COMPARING RATES OF RETURN** Fleet Street Savings Bank pays interest at the rate of 4.25%/year compounded weekly in a savings account, whereas Washington Bank pays interest at the rate of 4.125%/year compounded daily (assume a 365-day year). Which bank offers a better rate of interest?

57. **LOAN CONSOLIDATION** The proprietors of The Coachmen Inn secured two loans from Union Bank: one for \$8000 due in 3 years and one for \$15,000 due in 6 years, both at an interest rate of 8%/year compounded semiannually. The bank has agreed to allow the two loans to be consolidated into one loan payable in 5 years at the same interest rate. What amount will the proprietors of the inn be required to pay the bank at the end of 5 years?
Hint: Find the present value of the first two loans.

58. **EFFECTIVE RATE OF INTEREST** Find the effective rate of interest corresponding to a nominal rate of 5.5%/year compounded annually, semiannually, quarterly, and monthly.

59. **ZERO COUPON BONDS** Juan is contemplating buying a zero coupon bond that matures in 10 years and has a face value of \$10,000. If the bond yields a return of 5.25%/year, how much should Juan pay for the bond?

60. **REVENUE GROWTH OF A HOME THEATER BUSINESS** Maxwell started a home theater business in 2011. The revenue of his company for that year was \$240,000. The revenue grew by 20% in 2012 and by 30% in 2013. Maxwell projected that the revenue growth for his company in the next 3 years will be at least 25%/year. How much does Maxwell expect his minimum revenue to be for 2016?

61. **ONLINE RETAIL SALES** Online retail sales stood at \$141.4 billion for the year 2004. For the next 2 years, they grew by 24.3% and 14.0% per year, respectively. For the next 3 years, online retail sales were projected to grow at 30.5%, 17.6%, and 10.5% per year, respectively. What were the projected online sales for 2009?
Source: Jupiter Research.

62. **PURCHASING POWER** The inflation rates in the U.S. economy for 2009 through 2012 are 2.7%, 1.5%, 3.0%, and 1.7%, respectively. What was the purchasing power of a dollar at the beginning of 2013 compared to that at the beginning of 2009?
Source: U.S. Bureau of Labor Statistics.

63. **INVESTMENT OPTIONS** Investment A offers an 8%/year return compounded semiannually, and Investment B offers a 7.8%/year return compounded continuously. Which investment has a higher rate of return over a 4-year period?

64. **EFFECT OF INFLATION ON SALARIES** Leonard's current annual salary is \$65,000. Ten years from now, how much will he need to earn to retain his present purchasing power if the rate of inflation over that period is 3%/year compounded continuously?

65. **SAVING FOR COLLEGE** Having received a large inheritance, Jing-mei's parents wish to establish a trust for her college education. Seven years from now they need an estimated \$120,000. How much should they set aside in trust now if they invest the money at 6.6%/year compounded quarterly? Continuously?

66. **PENSIONS** Maria, who is now 50 years old, is employed by a firm that guarantees her a pension of \$40,000/year at age 65. What is the present value of her first year's pension if the inflation rate over the next 15 years is 3%/year compounded continuously? 4%/year compounded continuously? 6%/year compounded continuously?

67. **REAL ESTATE INVESTMENTS** An investor purchased a piece of waterfront property. Because of the development of a marina in the vicinity, the market value of the property is expected to increase according to the rule

$$V(t) = 80{,}000e^{\sqrt{t}/2}$$

where $V(t)$ is measured in dollars and t is the time (in years) from the present. If the rate of appreciation is expected to be 5%/year compounded continuously for the next 8 years, find an expression for the present value $P(t)$ of the property's market price valid for the next 8 years. What is $P(t)$ expected to be in 4 years?

68. The simple interest formula $A = P(1 + rt)$ [Formula (1b)] can be written in the form $A = Prt + P$, which is the slope-intercept form of a straight line with slope Pr and A-intercept P.
    a. Describe the family of straight lines obtained by keeping the value of r fixed and allowing the value of P to vary. Interpret your results.
    b. Describe the family of straight lines obtained by keeping the value of P fixed and allowing the value of r to vary. Interpret your results.

69. **EFFECTIVE RATE OF INTEREST** Suppose an initial investment of \$$P$ grows to an accumulated amount of \$$A$ in t years. Show that the effective rate (annual effective yield) is

$$r_{\text{eff}} = \left(\frac{A}{P}\right)^{1/t} - 1$$

Use the formula given in Exercise 69 to solve Exercises 70–74.

70. **Effective Rate of Interest** Martha invested \$40,000 in a boutique 5 years ago. Her investment is worth \$60,000 today. What is the effective rate (annual effective yield) of her investment?

71. **Housing Appreciation** Georgia purchased a house in September 2008 for \$300,000. In September 2014, she sold the house and made a net profit of \$66,000. Find the effective annual rate of return on her investment over the 6-year period.

72. **Common Stock Transaction** Steven purchased 1000 shares of a certain stock for \$25,250 (including commissions). He sold the shares 2 years later and received \$32,100 after deducting commissions. Find the effective annual rate of return on his investment over the 2-year period.

73. **Zero Coupon Bonds** Nina purchased a zero coupon bond for \$6724.53. The bond matures in 7 years and has a face value of \$10,000. Find the effective annual rate of interest for the bond.
    Hint: Assume that the purchase price of the bond is the initial investment and that the face value of the bond is the accumulated amount.

74. **Money Market Mutual Funds** Carlos invested \$5000 in a money market mutual fund that pays interest on a daily basis. The balance in his account at the end of 8 months (245 days) was \$5070.42. Find the effective rate at which Carlos's account earned interest over this period (assume a 365-day year).

In Exercises 75–78, determine whether the statement is true or false. If it is true, explain why it is true. If it is false, give an example to show why it is false.

75. When simple interest is used, the accumulated amount is a linear function of t.

76. If interest is compounded annually, then the accumulated amount after t years is the same as the accumulated amount under simple interest over t years.

77. If interest is compounded annually, then the effective rate is the same as the nominal rate.

78. Susan's salary increased from \$50,000/year to \$60,000/year over a 4-year period. Therefore, Susan received annual increases of 5% over that period.

(Optional) In Exercises 79–86, use logarithms to solve each problem.

79. How long will it take \$5000 to grow to \$6500 if the investment earns interest at the rate of 6%/year compounded monthly?

80. How long will it take \$12,000 to grow to \$15,000 if the investment earns interest at the rate of 4%/year compounded monthly?

81. How long will it take an investment of \$2000 to double if the investment earns interest at the rate of 5%/year compounded monthly?

82. How long will it take an investment of \$5000 to triple if the investment earns interest at the rate of 5%/year compounded daily?

83. Find the interest rate needed for an investment of \$5000 to grow to an amount of \$6000 in 3 years if interest is compounded continuously.

84. Find the interest rate needed for an investment of \$4000 to double in 8 years if interest is compounded continuously.

85. How long will it take an investment of \$6000 to grow to \$7000 if the investment earns interest at the rate of $5\frac{1}{2}$%/year compounded continuously?

86. How long will it take an investment of \$8000 to double if the investment earns interest at the rate of 8%/year compounded continuously?

5.1 Solutions to Self-Check Exercises

1. Using Formula (7) with $A = 20{,}000$, $r = 0.054$, and $m = 12$ so that $i = \frac{0.054}{12} = 0.0045$ and $n = (3)(12) = 36$, we find the required present value to be

$$P = 20{,}000(1 + 0.0045)^{-36} \approx 17{,}015.01$$

or \$17,015.01.

2. The accumulated amount of Paul's current investment is found by using Formula (3) with $P = 100{,}000$, $r = 0.046$, and $m = 365$. Thus, $i = \frac{0.046}{365}$ and $n = 365$, so the required accumulated amount is given by

$$A_1 = 100{,}000\left(1 + \frac{0.046}{365}\right)^{365} \approx 104{,}707.14$$

or \$104,707.14. Next, we compute the accumulated amount of Paul's reinvestment. Using (3) with $P = 100{,}000$, $r = 0.032$, and $m = 365$ so that $i = \frac{0.032}{365}$ and $n = 365$, we find the required accumulated amount in this case to be

$$A_2 = 100{,}000\left(1 + \frac{0.032}{365}\right)^{365} \approx 103{,}251.61$$

or approximately \$103,251.61. Therefore, Paul can expect to experience a net decrease in yearly income of approximately $104{,}707.14 - 103{,}251.61$, or \$1455.53.

USING TECHNOLOGY

Finding the Accumulated Amount of an Investment, the Effective Rate of Interest, and the Present Value of an Investment

Graphing Utility

Some graphing utilities have built-in routines for solving problems involving the mathematics of finance. For example, the TI-83/84 **TVM SOLVER** function incorporates several functions that can be used to solve the problems that are encountered in Sections 5.1–5.3. To access the **TVM SOLVER** on the TI-83, press **2nd**, press **FINANCE**, and then select **1: TVM Solver**. To access the TVM Solver on the TI-83 plus and the TI-84, press **APPS**, press **1: Finance**, and then select **1: TVM Solver**.

EXAMPLE 1 Finding the Accumulated Amount of an Investment Find the accumulated amount after 10 years if \$5000 is invested at a rate of 10% per year compounded monthly.

Solution Use the TI-83/84 **TVM SOLVER** with the following inputs:

N = 120	(10)(12)
I% = 10	
PV = −5000	Since an investment is an outflow, we enter the negative of the present value.
PMT = 0	
FV = 0	
P/Y = 12	The number of payments each year
C/Y = 12	The number of conversion periods each year
PMT: **END** BEGIN	

```
N=120
I%=10
PV=-5000
PMT=0
■ FV=13535.20745
P/Y=12
C/Y=12
PMT:END BEGIN
```

FIGURE **T1**
The TI-83/84 screen showing the future value (FV) of an investment

Move the cursor up to the FV line and press **ALPHA** **SOLVE**. We obtain the display shown in Figure T1. We conclude that the required accumulated amount is \$13,535.21.

EXAMPLE 2 Finding the Effective Rate of Interest Find the effective rate of interest corresponding to a nominal rate of 10% per year compounded quarterly.

```
▶ Eff(10,4)
      10.38128906
```

FIGURE **T2**
The TI-83/84 screen showing the effective rate of interest (Eff)

Solution Using the TI-83/84 **TVM SOLVER Eff** function, we obtain the display shown in Figure T2. The required effective rate is approximately 10.38% per year.

EXAMPLE 3 Finding the Present Value of an Investment Find the present value of \$20,000 due in 5 years if the interest rate is 7.5% per year compounded daily.

Solution Using the TI-83/84 **TVM SOLVER** with the following inputs:

N = 1825	(5)(365)
I% = 7.5	
PV = 0	
PMT = 0	
FV = 20000	
P/Y = 365	The number of payments each year
C/Y = 365	The number of conversions each year
PMT: **END** BEGIN	

Note: Boldfaced words/characters enclosed in a box (for example, **Enter**) indicate that an action (click, select, or press) is required. Words/characters printed blue (for example, Chart sub-type:) indicate words/characters appearing on the screen.

```
  N=1825
  I%=7.5
■ PV=-13746.3151
  PMT=0
  FV=20000
  P/Y=365
  C/Y=365
  PMT:END BEGIN
```

FIGURE **T3**
The TI-83/84 screen showing the present value (PV) of an investment

By moving the cursor up to the PV line and pressing **ALPHA** **SOLVE**, we obtain the display shown in Figure T3. We see that the required present value is approximately $13,746.32. Note that PV is negative because an investment is an outflow (money is paid out).

Excel

Excel has many built-in functions for solving problems involving the mathematics of finance. Here we illustrate the use of the FV (future value), EFFECT (effective rate), and PV (present value) functions to solve problems of the type that we encountered in Section 5.1.

EXAMPLE 4 Finding the Accumulated Amount of an Investment Find the accumulated amount after 10 years if $5000 is invested at a rate of 10% per year compounded monthly.

Solution Here, we are computing the future value of a lump-sum investment, so we use the FV (future value) function. Click **Financial** from the Function Library on the Formulas tab and select **FV**. The Function Arguments dialog box will appear (see Figure T4). In our example, the mouse cursor is in the edit box headed by Type, so a definition of that term appears near the bottom of the box. Figure T4 shows the entries for each edit box in our example.

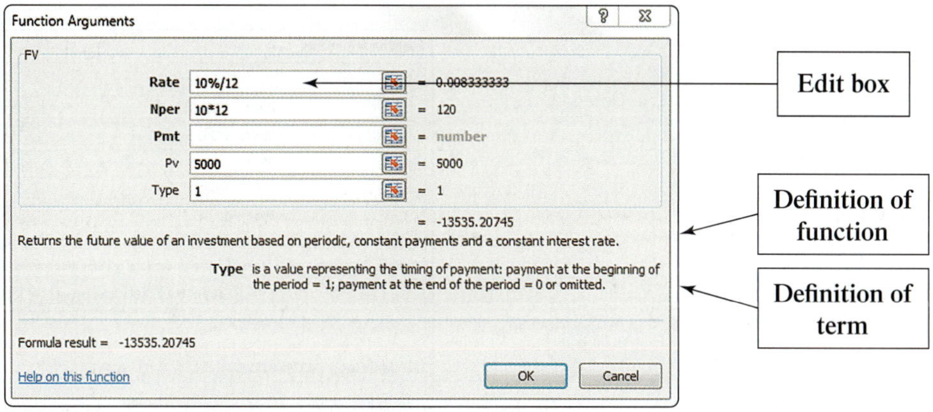

FIGURE **T4**
Excel's dialog box for computing the future value (FV) of an investment

Note that the entry for Nper is given by the total number of periods for which the investment earns interest. The Pmt box is left blank, since no money is added to the original investment. The Pv entry is 5000. The entry for Type is a 1 because the lump-sum payment is made at the beginning of the investment period. The answer, −$13,535.21, is shown at the bottom of the dialog box. It is negative because an investment is considered to be an outflow of money (money is paid out). (Click **OK**, and the answer will also appear on your spreadsheet.)

EXAMPLE 5 Finding the Effective Rate of Interest Find the effective rate of interest corresponding to a nominal rate of 10% per year compounded quarterly.

Solution Here we use the EFFECT function to compute the effective rate of interest. Accessing this function from the Financial function library subgroup and making the required entries, we obtain the Function Arguments dialog box shown in Figure T5. The required effective rate is approximately 10.38% per year.

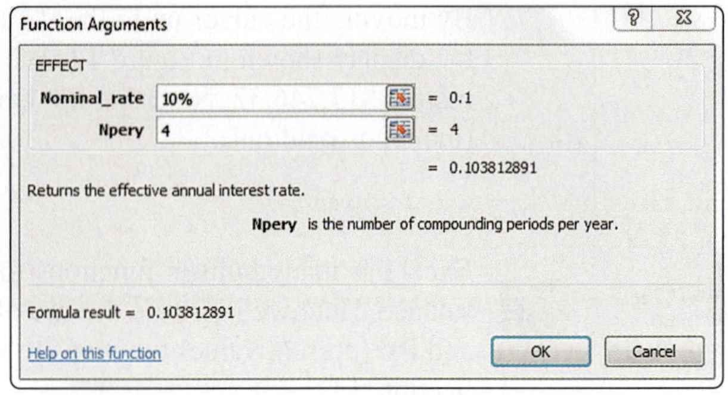

FIGURE **T5**
Excel's dialog box for the effective rate of interest function (EFFECT)

EXAMPLE 6 Finding the Present Value of an Investment Find the present value of \$20,000 due in 5 years if the interest rate is 7.5% per year compounded daily.

Solution We use the PV function to compute the present value of a lump-sum investment. Accessing this function as above and making the required entries, we obtain the PV dialog box shown in Figure T6. Once again, the Pmt edit box is left blank, since no additional money is added to the original investment. The Fv entry is 20000. The answer is negative because an investment is considered to be an outflow of money (money is paid out). We deduce that the required amount is \$13,746.32.

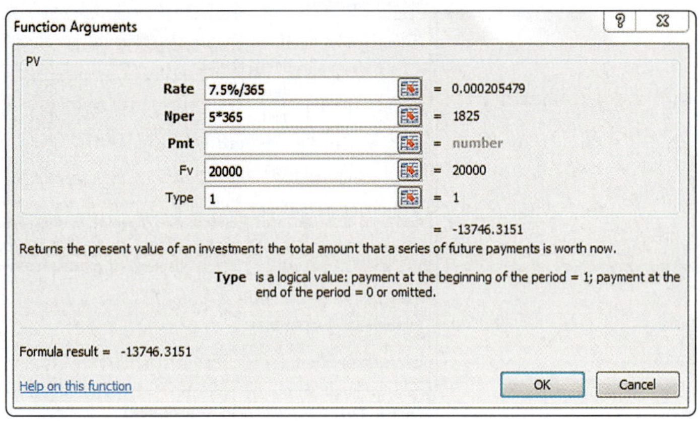

FIGURE **T6**
Excel dialog box for the present value function (PV)

TECHNOLOGY EXERCISES

1. Find the accumulated amount A if \$5000 is invested at the interest rate of $5\frac{3}{8}$%/year compounded monthly for 3 years.

2. Find the accumulated amount A if \$2850 is invested at the interest rate of $6\frac{5}{8}$%/year compounded monthly for 4 years.

3. Find the accumulated amount A if \$327.35 is invested at the interest rate of $5\frac{1}{3}$%/year compounded daily for 7 years.

4. Find the accumulated amount A if \$327.35 is invested at the interest rate of $6\frac{7}{8}$%/year compounded daily for 8 years.

5. Find the effective rate corresponding to $8\frac{2}{3}$%/year compounded quarterly.

6. Find the effective rate corresponding to $10\frac{5}{8}$%/year compounded monthly.

7. Find the effective rate corresponding to $9\frac{3}{4}$%/year compounded monthly.

8. Find the effective rate corresponding to $4\frac{3}{8}$%/year compounded quarterly.

9. Find the present value of \$38,000 due in 3 years at $8\frac{1}{4}$%/year compounded quarterly.

10. Find the present value of \$150,000 due in 5 years at $9\frac{3}{8}$%/year compounded monthly.

11. Find the present value of \$67,456 due in 3 years at $7\frac{7}{8}$%/year compounded monthly.

12. Find the present value of \$111,000 due in 5 years at $11\frac{5}{8}$%/year compounded monthly.

5.2 Annuities

Future Value of an Annuity

An **annuity** is a sequence of payments made at regular time intervals. The time period in which these payments are made is called the **term** of the annuity. Depending on whether the term is given by a *fixed time interval*, a time interval that begins at a definite date but extends indefinitely, or one that is not fixed in advance, an annuity is called an **annuity certain,** a *perpetuity*, or a *contingent annuity*, respectively. In general, the payments in an annuity need not be equal, but in many important applications they are equal. In this section, we assume that annuity payments are equal. Examples of annuities are regular deposits to a savings account, monthly home mortgage payments, and monthly insurance payments.

Annuities are also classified by payment dates. An annuity in which the payments are made at the *end* of each payment period is called an **ordinary annuity,** whereas an annuity in which the payments are made at the beginning of each period is called an *annuity due.* Furthermore, an annuity in which the payment period coincides with the interest conversion period is called a **simple annuity,** whereas an annuity in which the payment period differs from the interest conversion period is called a *complex annuity.*

In this section, we consider ordinary annuities that are certain and simple, with periodic payments that are equal in size. In other words, we study annuities that are subject to the following conditions:

1. The terms are given by fixed time intervals.
2. The periodic payments are equal in size.
3. The payments are made at the *end* of the payment periods.
4. The payment periods coincide with the interest conversion periods.

To find a formula for the accumulated amount S of an annuity, suppose a sum of \$100 is paid into an account at the end of each quarter over a period of 3 years. Furthermore, suppose the account earns interest on the deposit at the rate of 8% per year, compounded quarterly. Then the first payment of \$100 made at the end of the first quarter earns interest at the rate of 8% per year compounded four times a year (or $8/4 = 2\%$ per quarter) over the remaining 11 quarters and therefore, by the compound interest formula, has an accumulated amount of

$$100\left(1 + \frac{0.08}{4}\right)^{11} \quad \text{or} \quad 100(1 + 0.02)^{11}$$

dollars at the end of the term of the annuity (Figure 2).

The second payment of \$100 made at the end of the second quarter earns interest at the same rate over the remaining 10 quarters and therefore has an accumulated amount of

$$100(1 + 0.02)^{10}$$

dollars at the end of the term of the annuity, and so on. The last payment earns no interest because it is due at the end of the term. The amount of the annuity is obtained by adding all the terms in Figure 2. Thus,

$$S = 100 + 100(1 + 0.02) + 100(1 + 0.02)^2 + \cdots + 100(1 + 0.02)^{11}$$

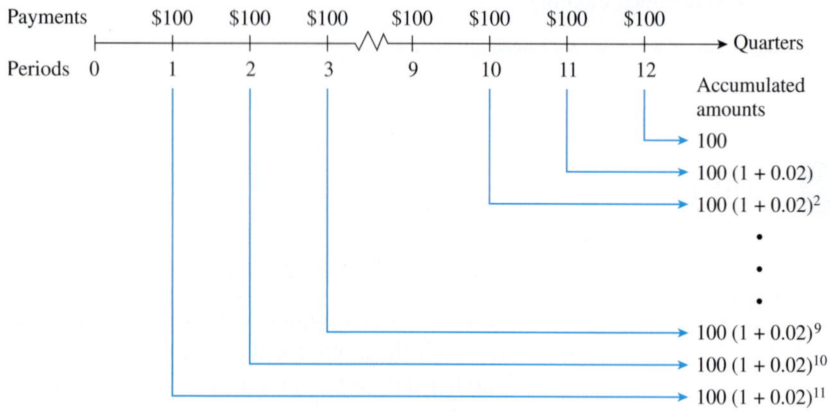

FIGURE 2
The sum of the accumulated amounts is the amount of the annuity.

The sum on the right is the sum of the first n terms of a *geometric progression* with first term 100 and common ratio $(1 + 0.02)$. We show in Section 5.4 that the sum S can be written in the more compact form

$$S = 100\left[\frac{(1 + 0.02)^{12} - 1}{0.02}\right]$$
$$\approx 1341.21$$

or approximately \$1341.21.

To find a general formula for the accumulated amount S of an annuity, suppose that a sum of \$$R$ is paid into an account at the end of each period for n periods and that the account earns interest at the rate of i per period. Then, proceeding as we did with the numerical example, we obtain

$$S = R + R(1 + i) + R(1 + i)^2 + \cdots + R(1 + i)^{n-1}$$
$$= R\left[\frac{(1 + i)^n - 1}{i}\right] \qquad (9)$$

The expression inside the brackets is called the **compound-amount factor**. The quantity S in Formula (9) is realizable at some future date and is accordingly called the future value of an annuity.

Future Value of an Annuity

The **future value S of an annuity** of n payments of R dollars each, paid at the end of each investment period into an account that earns interest at the rate of i per period, is

$$S = R\left[\frac{(1 + i)^n - 1}{i}\right]$$

EXAMPLE 1 Find the amount of an ordinary annuity consisting of 12 monthly payments of \$100 that earn interest at 12% per year compounded monthly.

Solution Since i is the interest rate per *period* and since interest is compounded monthly in this case, we have $i = \frac{0.12}{12} = 0.01$. Using Formula (9) with $R = 100$, $n = 12$, and $i = 0.01$, we have

$$\begin{aligned} S &= 100\left[\frac{(1.01)^{12} - 1}{0.01}\right] \\ &\approx 1268.25 \quad \text{Use a calculator.} \end{aligned}$$

or \$1268.25.

Explore and Discuss

Future Value S of an Annuity Due

1. Consider an annuity satisfying conditions 1, 2, and 4 on page 301 but with condition 3 replaced by the condition that payments are made at the *beginning* of the payment periods. By using an argument similar to that used to establish Formula (9), show that the future value S of an annuity due of n payments of R dollars each, paid at the beginning of each investment into an account that earns interest at the rate of i per period, is

$$S = R(1 + i)\left[\frac{(1 + i)^n - 1}{i}\right]$$

2. Use the result of part 1 to see how large your nest egg will be at age 65 if you start saving \$4000 annually at age 30, assuming a 10% average annual return; if you start saving at 35; if you start saving at 40. [Moral of the story: It is never too early to start saving!]

Exploring with TECHNOLOGY

Refer to the preceding Explore and Discuss problem.

1. Show that if $R = 4000$ and $i = 0.1$, then $S = 44{,}000[(1.1)^n - 1]$. Using a graphing utility, plot the graph of $f(x) = 44{,}000[(1.1)^x - 1]$, using the viewing window $[0, 40] \times [0, 1{,}200{,}000]$.
2. Verify the results of part 1 by evaluating $f(35)$, $f(30)$, and $f(25)$ using the EVAL function.

Present Value of an Annuity

In certain instances, you may want to determine the current value P of a sequence of equal periodic payments that will be made over a certain period of time. After each payment is made, the new balance continues to earn interest at some nominal rate. The amount P is referred to as the present value of an annuity.

To derive a formula for determining the present value P of an annuity, we may argue as follows. The amount P invested now and earning interest at the rate of i per period will have an accumulated value of $P(1 + i)^n$ at the end of n periods. But this must be equal to the future value of the annuity S given by Formula (9). Therefore, equating the two expressions, we have

$$P(1 + i)^n = R\left[\frac{(1 + i)^n - 1}{i}\right]$$

Multiplying both sides of this equation by $(1 + i)^{-n}$ gives

$$
\begin{aligned}
P &= R(1 + i)^{-n}\left[\frac{(1 + i)^n - 1}{i}\right] \\
&= R\left[\frac{(1 + i)^n(1 + i)^{-n} - (1 + i)^{-n}}{i}\right] \\
&= R\left[\frac{1 - (1 + i)^{-n}}{i}\right] \qquad (1 + i)^n(1 + i)^{-n} = 1
\end{aligned}
$$

Present Value of an Annuity

The **present value *P* of an annuity** consisting of *n* payments of *R* dollars each, paid at the end of each investment period into an account that earns interest at the rate of *i* per period, is

$$P = R\left[\frac{1 - (1 + i)^{-n}}{i}\right] \tag{10}$$

EXAMPLE 2 Find the present value of an ordinary annuity consisting of 24 monthly payments of $100 each and earning interest at 3% per year compounded monthly.

Solution Here, $R = 100$, $i = \frac{r}{m} = \frac{0.03}{12} = 0.0025$, and $n = 24$, so by Formula (10), we have

$$
\begin{aligned}
P &= 100\left[\frac{1 - (1.0025)^{-24}}{0.0025}\right] \\
&\approx 2326.60
\end{aligned}
$$

or $2326.60.

APPLIED EXAMPLE 3 Saving for a College Education As a savings program toward Alberto's college education, his parents decide to deposit $100 at the end of every month into a bank account paying interest at the rate of 6% per year compounded monthly. If the savings program began when Alberto was 6 years old, how much money would have accumulated by the time he turns 18?

Solution By the time the child turns 18, the parents would have made 144 deposits into the account. Thus, $n = 144$. Furthermore, we have $R = 100$, $r = 0.06$, and $m = 12$, so $i = \frac{0.06}{12} = 0.005$. Using Formula (9), we find that the amount of money that would have accumulated is given by

$$
\begin{aligned}
S &= 100\left[\frac{(1.005)^{144} - 1}{0.005}\right] \\
&\approx 21{,}015
\end{aligned}
$$

or $21,015.

APPLIED EXAMPLE 4 Financing a Car After making a down payment of $6000 for an automobile, Murphy paid $600 per month for 36 months with interest charged at 6% per year compounded monthly on the unpaid balance. What was the original cost of the car? What portion of Murphy's total car payments went toward interest charges?

Solution The loan taken up by Murphy is given by the present value of the annuity

$$P = 600\left[\frac{1 - (1.005)^{-36}}{0.005}\right]$$
$$\approx 19{,}723$$

or \$19,723. Therefore, the original cost of the automobile is \$25,723 (\$19,723 plus the \$6000 down payment). The interest charges paid by Murphy are given by $(36)(600) - 19{,}723 = 1{,}877$, or \$1,877.

One important application of annuities arises in the area of tax planning. During the 1980s, Congress created many tax-sheltered retirement savings plans, such as Individual Retirement Accounts (IRAs), Keogh plans, and Simplified Employee Pension (SEP) plans. These plans are examples of annuities in which the individual is allowed to make contributions (which are often tax deductible) to an investment account. The amount of the contribution is limited by congressional legislation. The taxes on the contributions and/or the interest accumulated in these accounts are deferred until the money is withdrawn—ideally during retirement, when tax brackets should be lower. In the interim period, the individual has the benefit of tax-free growth on his or her investment.

Suppose, for example, you are eligible to make a fully deductible contribution to an IRA and you are in a marginal tax bracket of 28%. Additionally, suppose you receive a year-end bonus of \$2000 from your employer and have the option of depositing the \$2000 into either an IRA or a regular savings account, where both accounts earn interest at an effective annual rate of 8% per year. If you choose to invest your bonus in a regular savings account, you will first have to pay taxes on the \$2000, leaving \$1440 to invest. At the end of 1 year, you will also have to pay taxes on the interest earned, leaving you with

Accumulated amount	−	Tax on interest	=	Net amount
1555.20	−	32.26	=	1522.94

or \$1522.94.

On the other hand, if you put the money into the IRA, the entire sum will earn interest, and at the end of 1 year, you will have (1.08)(\$2000), or \$2160, in your account. Of course, you will still have to pay taxes on this money when you withdraw it, but you will have gained the advantage of tax-free growth of the larger principal over the years. The disadvantage of this option is that if you withdraw the money before you reach the age of $59\frac{1}{2}$, you will be liable for taxes on both your contributions and the interest earned, *and* you will also have to pay a 10% penalty.

Note In practice, the size of the contributions an individual might make to the various retirement plans might vary from year to year. Also, he or she might make the contributions at different payment periods. To simplify our discussion, we will consider examples in which fixed payments are made at regular intervals.

APPLIED EXAMPLE 5 IRAs Caroline is planning to make a contribution of \$2000 on January 31 of each year into a traditional IRA earning interest at an effective rate of 5% per year.

a. After she makes her 25th payment on January 31 of the year following her retirement at age 66, how much will she have in her IRA?

b. Suppose that Caroline withdraws all of her money from her traditional IRA after she makes her 25th payment in the year following her retirement at age 66 and that her investment is subjected to a tax of 28% at that time. How much money will she end up with after taxes?

Solution

a. The amount of money Caroline will have after her 25th payment into her account is found by using Formula (9) with $R = 2000$, $r = 0.05$, $m = 1$, and $t = 25$, so that $i = \frac{r}{m} = 0.05$ and $n = mt = 25$. The required amount is given by

$$S = 2000\left[\frac{(1.05)^{25} - 1}{0.05}\right] \approx 95{,}454.20$$

or $95,454.20.

b. If she withdraws the entire amount from her account, she will end up with

$$(1 - 0.28)(95{,}454.20) \approx 68{,}727.02$$

that is, she will have approximately $68,727.02 after paying taxes.

After-tax-deferred annuities are another type of investment vehicle that allows an individual to build assets for retirement, college funds, or other future needs. The advantage gained in this type of investment is that the tax on the accumulated interest is deferred to a later date. Note that in this type of investment, the contributions themselves are not tax deductible. At first glance, the advantage thus gained may seem to be relatively inconsequential, but its true effect is illustrated by the next example.

APPLIED EXAMPLE 6 Investment Analysis Both Clark and Colby are salaried individuals, 45 years of age, who are saving for their retirement 20 years from now. Both Clark and Colby are also in the 28% marginal tax bracket. Clark makes a $1000 contribution annually on December 31 into a savings account earning an effective rate of 8% per year. At the same time, Colby makes a $1000 annual payment to an insurance company for an after-tax-deferred annuity. The annuity also earns interest at an effective rate of 8% per year. (Assume that both men remain in the same tax bracket throughout this period, and disregard state income taxes.)

a. Calculate how much each man will have in his investment account at the end of 20 years.

b. Compute the interest earned on each account.

c. Show that even if the interest on Colby's investment were subjected to a tax of 28% upon withdrawal of his investment at the end of 20 years, the net accumulated amount of his investment would still be greater than that of Clark's.

Solution

a. Because Clark is in the 28% marginal tax bracket, the net yield for his investment is (0.72)(8), or 5.76%, per year.

Using Formula (9) with $R = 1000$, $r = 0.0576$, $m = 1$, and $t = 20$, so that $i = 0.0576$ and $n = mt = 20$, we see that Clark's investment will be worth

$$S = 1000\left[\frac{(1 + 0.0576)^{20} - 1}{0.0576}\right] \approx 35{,}850.49$$

or $35,850.49 at his retirement.

Colby has a tax-sheltered investment with an effective yield of 8% per year. Using Formula (9) with $R = 1000$, $r = 0.08$, $m = 1$, and $t = 20$, so that $i = 0.08$ and $n = mt = 20$, we see that Colby's investment will be worth

$$S = 1000\left[\frac{(1 + 0.08)^{20} - 1}{0.08}\right] \approx 45{,}761.96$$

or \$45,761.96 at his retirement.

b. Each man will have paid 20(1000), or \$20,000, into his account. Therefore, the total interest earned in Clark's account will be (35,850.49 − 20,000), or \$15,850.49, whereas the total interest earned in Colby's account will be (45,761.96 − 20,000), or \$25,761.96.

c. From part (b) we see that the total interest earned in Colby's account will be \$25,761.96. If it were taxed at 28%, he would still end up with (0.72)(25,761.96), or \$18,548.61. This amount is larger than the total interest of \$15,850.49 earned by Clark.

In 1997, another type of tax-sheltered retirement savings plan was created by Congress: the Roth IRA. In contrast to traditional IRAs, contributions to Roth IRAs are not tax-deferrable. However, direct contributions to Roth IRAs (but not rollovers) may be withdrawn tax-free at any time. Also, holders of a Roth IRA are not required to take minimum distributions after age $70\frac{1}{2}$.

APPLIED EXAMPLE 7 Roth IRAs Refer to Example 5. Suppose that Caroline decides to invest her money in a Roth IRA instead of a traditional IRA. Also, suppose that she is in the 28% tax bracket and remains in that bracket for the next 25 years until her retirement at age 65. If she pays taxes on \$2000 and then invests the remaining \$1440 into a Roth IRA earning interest at a rate of 5% per year, compounded annually, how much will she have in her Roth IRA after her 25th payment on January 31 of the year following her retirement? (Disregard state and city taxes.) How does this compare with the amount of money she would have if she had stayed with a traditional IRA and withdrawn all of her money from that account at that time? (See Example 5b.)

Solution We use Formula (9) with $R = 1440$, $r = 0.05$, and $n = 25$, obtaining

$$S = 1440\left[\frac{(1.05)^{25} - 1}{0.05}\right] \approx 68{,}727.02$$

that is, she will have approximately \$68,727.02 in her account. This is the same as the amount she ended up with in her traditional IRA after paying taxes (see Example 5b).

5.2 Self-Check Exercises

1. **Traditional IRA Investment** Phyliss opened an IRA on January 31, 2000, with a contribution of \$2000. She plans to make a contribution of \$2000 thereafter on January 31 of each year until her retirement in the year 2019 (20 payments). If the account earns interest at the rate of 8%/year compounded yearly, how much will Phyliss have in her account when she retires?

2. **Securing a Bank Loan** Denver Wildcatting Company has an immediate need for a loan. In an agreement worked out with its banker, Denver assigns its royalty income of \$4800/month for the next 3 years from certain oil properties to the bank, with the first payment due at the end of the first month. If the bank charges interest at the rate of 9%/year compounded monthly, what is the amount of the loan negotiated between the parties?

Solutions to Self-Check Exercises 5.2 can be found on page 310.

5.2 Concept Questions

1. Is the term of an ordinary annuity fixed or variable? Are the periodic payments all of the same size, or do they vary in size? Are the payments made at the beginning or the end of the payment period? Do the payment periods coincide with the interest conversion periods?
2. What is the difference between an ordinary annuity and an annuity due?
3. What is the future value of an annuity? Give an example.
4. What is the present value of an annuity? Give an example.

5.2 Exercises

In Exercises 1–8, find the amount (future value) of each ordinary annuity.

1. \$1000/year for 10 years at 5%/year compounded annually
2. \$1500/semiannual period for 8 years at 4.5%/year compounded semiannually
3. \$500/semiannual period for 12 years at 6%/year compounded semiannually
4. \$1800/quarter for 6 years at 4%/year compounded quarterly
5. \$600/quarter for 9 years at 5%/year compounded quarterly
6. \$150/month for 15 years at 6%/year compounded monthly
7. \$200/month for $20\frac{1}{4}$ years at 6.5%/year compounded monthly
8. \$100/week for $7\frac{1}{2}$ years at 3.5%/year compounded weekly

In Exercises 9–14, find the present value of each ordinary annuity.

9. \$5000/year for 8 years at 6%/year compounded annually
10. \$4000/year for 5 years at 4.5%/year compounded yearly
11. \$1200/semiannual period for 6 years at 5%/year compounded semiannually
12. \$3000/semiannual period for 6 years at 5.5%/year compounded semiannually
13. \$800/quarter for 7 years at 6%/year compounded quarterly
14. \$150/month for 10 years at 4%/year compounded monthly
15. **IRAs** If a merchant deposits \$1500 at the end of each tax year in an IRA paying interest at the rate of 4%/year compounded annually, how much will she have in her account at the end of 25 years?
16. **Savings Accounts** If Jackson deposits \$100 at the end of each month in a savings account earning interest at the rate of 3%/year compounded monthly, how much will he have on deposit in his savings account at the end of 6 years, assuming that he makes no withdrawals during that period?
17. **Savings Accounts** Linda has joined a Christmas Fund Club at her bank. At the end of every month, December through October inclusive, she will make a deposit of \$40 in her fund. If the money earns interest at the rate of 2.5%/year compounded monthly, how much will she have in her account on December 1 of the following year?
18. **Keogh Accounts** Robin, who is self-employed, contributes \$5000/year into a Keogh account. How much will he have in the account after 25 years if the account earns interest at the rate of 4.5%/year compounded yearly?
19. **Investment Analysis** Karen has been depositing \$150 at the end of each month in a tax-free retirement account since she was 25. Matt, who is the same age as Karen, started depositing \$250 at the end of each month in a tax-free retirement account when he was 35. Assuming that both accounts have been and will be earning interest at the rate of 4%/year compounded monthly, who will end up with the larger retirement account at the age of 65?
20. **Retirement Planning** As a fringe benefit for the past 12 years, Colin's employer has contributed \$100 at the end of each month into an employee retirement account for Colin that pays interest at the rate of 5%/year

compounded monthly. Colin has also contributed \$2000 at the end of each of the last 8 years into an IRA that pays interest at the rate of 4.5%/year compounded yearly. How much does Colin have in his retirement fund at this time?

21. **Investment Analysis** Luis has \$150,000 in his retirement account at his present company. Because he is assuming a position with another company, Luis is planning to "roll over" his assets to a new account. Luis also plans to put \$3000/quarter into the new account until his retirement 20 years from now. If the new account earns interest at the rate of 4.5%/year compounded quarterly, how much will Luis have in his account at the time of his retirement?
Hint: Use the compound interest formula and the annuity formula.

22. **Auto Leasing** The Betzes have leased an auto for 2 years at \$450/month. If money is worth 3.5%/year compounded monthly, what is the equivalent cash payment (present value) of this annuity?

23. **Savings Accounts** The Pirerras are planning to go to Europe 3 years from now and have agreed to set aside \$150/month for their trip. If they deposit this money at the end of each month into a savings account paying interest at the rate of 3%/year compounded monthly, how much money will be in their travel fund at the end of the third year?

24. **Installment Plans** Mike's Sporting Goods sells elliptical trainers under two payment plans: cash or installment. Under the installment plan, the customer pays \$22/month over 3 years with interest charged on the balance at a rate of 9%/year compounded monthly. Find the cash price for an elliptical trainer if it is equivalent to the price paid by a customer using the installment plan.

25. **Auto Financing** Lupé made a down payment of \$8000 toward the purchase of a new car. To pay the balance of the purchase price, she has secured a loan from her bank at the rate of 6%/year compounded monthly. Under the terms of her finance agreement, she is required to make payments of \$420/month for 36 months. What is the cash price of the car?

26. **Lottery Payouts** A state lottery commission pays the winner of the Million Dollar lottery 20 installments of \$50,000/year. The commission makes the first payment of \$50,000 immediately and the other $n = 19$ payments at the end of each of the next 19 years. Determine how much money the commission should have in the bank initially to guarantee the payments, assuming that the balance on deposit with the bank earns interest at the rate of 4%/year compounded yearly.
Hint: Find the present value of an annuity.

27. **Purchasing a Home** The Johnsons have accumulated a nest egg of \$40,000 that they intend to use as a down payment toward the purchase of a new house. Because their present gross income has placed them in a relatively high tax bracket, they have decided to invest a minimum of \$2400/month in monthly payments (to take advantage of the tax deduction) toward the purchase of their house. However, because of other financial obligations, their monthly payments should not exceed \$3000. If local mortgage rates are 5.5%/year compounded monthly for a conventional 30-year mortgage, what is the price range of houses that they should consider?

28. **Purchasing a Home** Refer to Exercise 27. If local mortgage rates fell to 5%, how would this affect the price range of houses that the Johnsons should consider?

29. **Purchasing a Home** Refer to Exercise 27. If the Johnsons decide to secure a 15-year mortgage instead of a 30-year mortgage, what is the price range of houses they should consider when the local mortgage rate for this type of loan is 5%?

30. **Savings Plan** Lauren plans to deposit \$5000 into a bank account at the beginning of next month and \$200/month into the same account at the end of that month and at the end of each subsequent month for the next 5 years. If her bank pays interest at the rate of 3%/year compounded monthly, how much will Lauren have in her account at the end of 5 years? (Assume that she makes no withdrawals during the 5-year period.)

31. **Financial Planning** Joe plans to deposit \$200 at the end of each month into a bank account for a period of 2 years, after which he plans to deposit \$300 at the end of each month into the same account for another 3 years. If the bank pays interest at the rate of 3.5%/year compounded monthly, how much will Joe have in his account by the end of 5 years? (Assume that no withdrawals are made during the 5-year period.)

32. **Investment Analysis** From age 25 to age 40, Jessica deposited \$200 at the end of each month into a tax-free retirement account. She made no withdrawals or further contributions until age 65. Alex made deposits of \$300 into his tax-free retirement account from age 40 to age 65. If both accounts earned interest at the rate of 5%/year compounded monthly, who ends up with a bigger nest egg upon reaching the age of 65?
Hint: Use both the annuity formula and the compound interest formula.

33. **Roth IRAs** Suppose that Jacob deposits \$3000/year for 10 years into a Roth IRA earning interest at the rate of 5%/year, compounded annually. During the next 10 years, he makes no withdrawals or no further contributions, but the account continues to earn interest at the same rate. How much will Jacob have in his retirement account at the end of the 20-year period?

34. **Retirement Planning** Suppose that Ramos contributes \$5000/year into a traditional IRA earning interest at the rate of 4%/year compounded annually, every year after age 35 until his retirement at age 65. At the same time, his wife Vanessa deposits \$3600/year into a Roth IRA earning interest at the same rate as that of Ramos and also for a period of 30 years. Suppose that the investments of both Ramos and Vanessa are subjected to a tax of 30% at

the time of their retirement and that they both wish to withdraw all of the money in their IRAs at that time.

a. After all due taxes are paid, who will have the larger amount?

b. How much larger will that amount be?

In Exercises 35 and 36, determine whether the statement is true or false. If it is true, explain why it is true. If it is false, give an example to show why it is false.

35. The future value of an annuity can be found by adding together all the payments that are paid into the account.

36. If the future value of an annuity consisting of n payments of R dollars each—paid at the end of each investment period into an account that earns interest at the rate of i per period—is S dollars, then

$$R = \frac{iS}{(1 + i)^n - 1}$$

5.2 Solutions to Self-Check Exercises

1. The amount Phyliss will have in her account when she retires may be found by using Formula (9) with $R = 2000$, $r = 0.08$, $m = 1$, and $t = 20$, so that $i = r = 0.08$ and $n = mt = 20$. Thus,

$$S = 2000\left[\frac{(1.08)^{20} - 1}{0.08}\right] \approx 91{,}523.93$$

or \$91,523.93.

2. We want to find the present value of an ordinary annuity consisting of 36 monthly payments of \$4800 each and earning interest at 9%/year compounded monthly. Using Formula (10) with $R = 4800$, $m = 12$, and $t = 3$, so that $i = \frac{r}{m} = \frac{0.09}{12} = 0.0075$ and $n = (12)(3) = 36$, we find

$$P = 4800\left[\frac{1 - (1.0075)^{-36}}{0.0075}\right] \approx 150{,}944.67$$

or \$150,944.67, the amount of the loan negotiated.

USING TECHNOLOGY Finding the Amount of an Annuity

Graphing Utility

As was mentioned in Using Technology, Section 5.1, the TI-83/84 can facilitate the solution of problems in finance. We continue to exploit its versatility in this section.

EXAMPLE 1 Finding the Future Value of an Annuity Find the amount of an ordinary annuity of 36 quarterly payments of \$220 each that earn interest at the rate of 10% per year compounded quarterly.

Solution We use the TI-83/84 **TVM SOLVER** with the following inputs:

$$\text{N} = 36$$
$$\text{I\%} = 10$$
$$\text{PV} = 0$$
$$\text{PMT} = -220 \quad \text{Recall that a payment is an outflow.}$$
$$\text{FV} = 0$$
$$\text{P/Y} = 4 \quad \text{The number of payments each year}$$
$$\text{C/Y} = 4 \quad \text{The number of conversion periods each year}$$

PMT: [END] BEGIN

Move the cursor up to the FV line and press [ALPHA] [SOLVE]. The result is displayed in Figure T1. We deduce that the desired amount is \$12,606.31.

```
N=36
I%=10
PV=0
PMT=-220
FV=12606.31078
P/Y=4
C/Y=4
PMT:[END] BEGIN
```

FIGURE **T1**
The TI-83/84 screen showing the future value (FV) of an annuity

EXAMPLE 2 Finding the Present Value of an Annuity Find the present value of an ordinary annuity consisting of 48 monthly payments of \$300 each and earning interest at the rate of 9% per year compounded monthly.

Solution We use the TI-83/84 **TVM SOLVER** with the following inputs:

N = 48
I% = 9
PV = 0
PMT = −300 — A payment is an outflow.
FV = 0
P/Y = 12 — The number of payments each year
C/Y = 12 — The number of conversion periods each year
PMT: **END** BEGIN

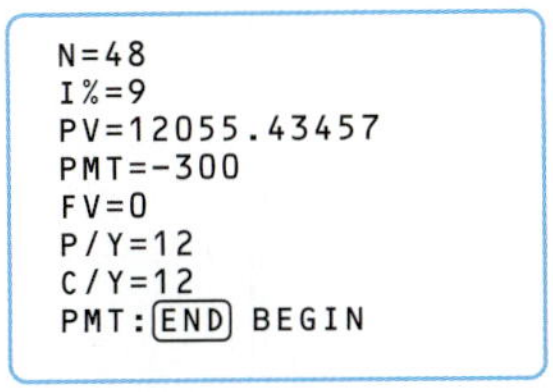

FIGURE **T2**
The TI-83/84 screen showing the present value (PV) of an ordinary annuity

By moving the cursor up to the PV line and pressing **ALPHA** **SOLVE**, we obtain the output displayed in Figure T2. We see that the required present value of the annuity is \$12,055.43.

Excel

Now we show how Excel can be used to solve financial problems involving annuities.

EXAMPLE 3 Finding the Future Value of an Annuity Find the amount of an ordinary annuity of 36 quarterly payments of \$220 each that earn interest at the rate of 10% per year compounded quarterly.

Solution Here we are computing the future value of a series of equal payments, so we use the FV (future value) function. As before, we choose this function from the Financial function library to obtain the Function Arguments dialog box. After making each of the required entries, we obtain the dialog box shown in Figure T3.

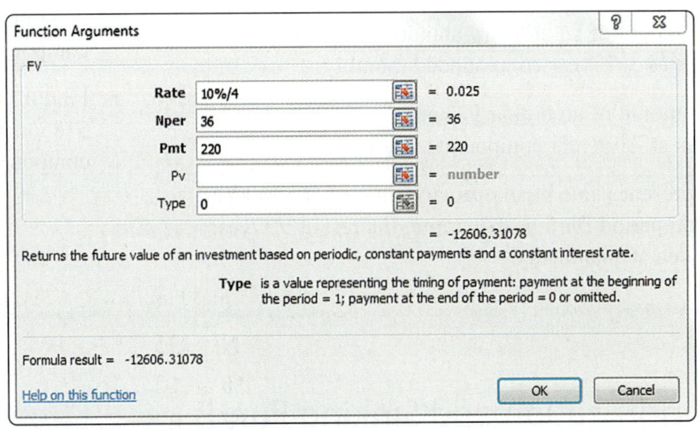

FIGURE **T3**
Excel's dialog box for the future value (FV) of an annuity

Note that a 0 is entered in the Type edit box because payments are made at the end of each payment period. Once again, the answer is negative because cash is paid out. We deduce that the desired amount is \$12,606.31.

Note: Boldfaced words/characters enclosed in a box (for example, **Enter**) indicate that an action (click, select, or press) is required. Words/characters printed blue (for example, Chart sub-type:) indicate words/characters appearing on the screen.

EXAMPLE 4 Finding the Present Value of an Annuity Find the present value of an ordinary annuity consisting of 48 monthly payments of \$300 each and earning interest at the rate of 9% per year compounded monthly.

Solution Here, we use the PV function to compute the present value of an annuity. Accessing the PV (present value) function from the Financial function library and making the required entries, we obtain the PV dialog box shown in Figure T4. We see that the required present value of the annuity is \$12,055.43.

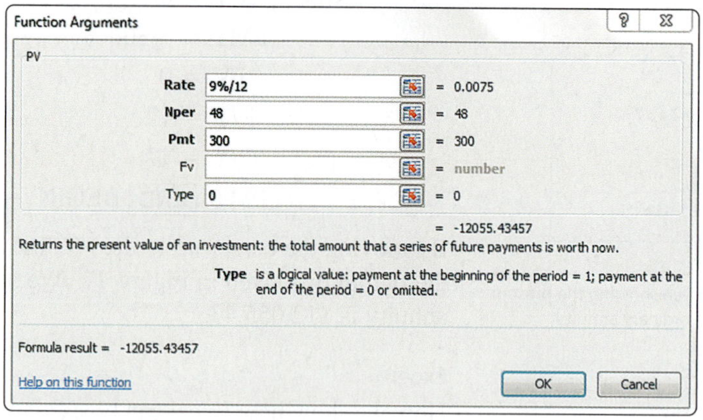

FIGURE **T4**
Excel's dialog box for computing the present value (PV) of an annuity

TECHNOLOGY EXERCISES

1. Find the amount of an ordinary annuity of 20 payments of \$2500/quarter at $7\frac{1}{4}$%/year compounded quarterly.
2. Find the amount of an ordinary annuity of 24 payments of \$1790/quarter at $8\frac{3}{4}$%/year compounded quarterly.
3. Find the amount of an ordinary annuity of \$120/month for 5 years at $6\frac{3}{8}$%/year compounded monthly.
4. Find the amount of an ordinary annuity of \$225/month for 6 years at $7\frac{5}{8}$%/year compounded monthly.
5. Find the present value of an ordinary annuity of \$4500/semiannual period for 5 years earning interest at 9%/year compounded semiannually.
6. Find the present value of an ordinary annuity of \$2100/quarter for 7 years earning interest at $7\frac{1}{8}$%/year compounded quarterly.
7. Find the present value of an ordinary annuity of \$245/month for 6 years earning interest at $8\frac{3}{8}$%/year compounded monthly.
8. Find the present value of an ordinary annuity of \$185/month for 12 years earning interest at $6\frac{5}{8}$%/year compounded monthly.

5.3 Amortization and Sinking Funds

Amortization of Loans

The annuity formulas derived in Section 5.2 may be used to answer questions involving the amortization of certain types of installment loans. For example, in a typical housing loan, the mortgagor makes periodic payments toward reducing his or her indebtedness to the lender, who charges interest at a fixed rate on the unpaid portion of the debt. In practice, the borrower is required to repay the lender in periodic installments, usually of the same size and over a fixed term, so that the loan (principal plus interest charges) is amortized at the end of the term.

By thinking of the monthly loan repayments R as the payments in an annuity, we see that the original amount of the loan is given by P, the present value of the annuity. From Equation (10), Section 5.2, we have

$$P = R\left[\frac{1 - (1 + i)^{-n}}{i}\right] \tag{11}$$

A question a financier might ask is: How much should the monthly installment be so that a loan will be amortized at the end of the term of the loan? To answer this question, we simply solve Equation (11) for R in terms of P, obtaining

$$R = \frac{Pi}{1 - (1 + i)^{-n}}$$

Amortization Formula

The periodic payment R on a loan of P dollars to be amortized over n periods with interest charged at the rate of i per period is

$$R = \frac{Pi}{1 - (1 + i)^{-n}} \tag{12}$$

APPLIED EXAMPLE 1 Amortization Schedule A sum of \$50,000 is to be repaid over a 5-year period through equal installments made at the end of each year. If an interest rate of 8% per year is charged on the unpaid balance and interest calculations are made at the end of each year, determine the size of each installment so that the loan (principal plus interest charges) is amortized at the end of 5 years. Verify the result by displaying the amortization schedule.

Solution Substituting $P = 50{,}000$, $i = r = 0.08$ (here, $m = 1$), and $n = 5$ into Formula (12), we obtain

$$R = \frac{(50{,}000)(0.08)}{1 - (1.08)^{-5}} \approx 12{,}522.82$$

giving the required yearly installment as \$12,522.82.

The amortization schedule is presented in Table 4. The outstanding principal at the end of 5 years is, of course, zero. (The figure of \$0.01 in Table 4 is the result of round-off errors.) Observe that initially the larger portion of the repayment goes toward payment of interest charges, but as time goes by, more and more of the payment goes toward repayment of the principal.

TABLE 4

An Amortization Schedule

End of Period	Interest Charged	Repayment Made	Payment Toward Principal	Outstanding Principal
0	—	—	—	\$50,000.00
1	\$4,000.00	\$12,522.82	\$ 8,522.82	41,477.18
2	3,318.17	12,522.82	9,204.65	32,272.53
3	2,581.80	12,522.82	9,941.02	22,331.51
4	1,786.52	12,522.82	10,736.30	11,595.21
5	927.62	12,522.82	11,595.20	0.01

Financing a Home

APPLIED EXAMPLE 2 Home Mortgage Payments The Blakelys borrowed \$120,000 from a bank to help finance the purchase of a house. The bank charges interest at a rate of 5.4% per year on the unpaid balance, with interest computations made at the end of each month. The Blakelys have agreed to repay the loan in equal monthly installments over 30 years. How much should each payment be if the loan is to be amortized at the end of the term?

Solution Here, $P = 120{,}000$, $i = \frac{r}{m} = \frac{0.054}{12} = 0.0045$, and $n = (30)(12) = 360$. Using Formula (12), we find that the size of each monthly installment required is given by

$$R = \frac{(120{,}000)(0.0045)}{1 - (1.0045)^{-360}} \approx 673.84$$

or \$673.84.

APPLIED EXAMPLE 3 Home Equity Teresa and Raul purchased a house 10 years ago for \$200,000. They made a down payment of 20% of the purchase price and secured a 30-year conventional home mortgage at 6% per year compounded monthly on the unpaid balance. The house is now worth \$380,000. How much equity do Teresa and Raul have in their house now (after making 120 monthly payments)?

Solution Since the down payment was 20%, we know that they secured a loan of 80% of \$200,000, or \$160,000. Furthermore, using Formula (12) with $P = 160{,}000$, $i = \frac{r}{m} = \frac{0.06}{12} = 0.005$ and $n = (30)(12) = 360$, we determine that their monthly installment is

$$R = \frac{(160{,}000)(0.005)}{1 - (1.005)^{-360}} \approx 959.28$$

or \$959.28.

After 120 monthly payments have been made, the outstanding principal is given by the sum of the present values of the remaining installments (that is, $360 - 120 = 240$ installments). But this sum is just the present value of an annuity with $n = 240$, $R = 959.28$, and $i = 0.005$. Using Formula (10), we find

$$P = 959.28\left[\frac{1 - (1 + 0.005)^{-240}}{0.005}\right] \approx 133{,}897.04$$

or approximately \$133,897. Therefore, Teresa and Raul have an equity of $380{,}000 - 133{,}897$, that is, \$246,103.

Explore and Discuss and Exploring with Technology

1. Consider the amortization Formula (12):

$$R = \frac{Pi}{1 - (1 + i)^{-n}}$$

Suppose you know the values of R, P, and n and you wish to determine i. Explain why you can accomplish this task by finding the point of intersection of the graphs of the functions

$$y_1 = R \quad \text{and} \quad y_2 = \frac{Pi}{1 - (1 + i)^{-n}}$$

2. Thalia knows that her monthly repayment on her 30-year conventional home loan of \$150,000 is \$1100.65 per month. Help Thalia determine the interest rate for her loan by verifying or executing the following steps:
   a. Plot the graphs of

$$y_1 = 1100.65 \quad \text{and} \quad y_2 = \frac{150{,}000x}{1 - (1 + x)^{-360}}$$

   using the viewing window $[0, 0.01] \times [0, 1200]$.
   b. Use the ISECT (intersection) function of the graphing utility to find the point of intersection of the graphs of part (a). Explain why this gives the value of i.
   c. Compute r from the relationship $r = 12i$.

Explore and Discuss and Exploring with Technology

1. Suppose you secure a home mortgage loan of \$$P$ with an interest rate of r per year to be amortized over t years through monthly installments of \$$R$. Show that after N installments, your outstanding principal is given by

$$B(N) = P\left[\frac{(1 + i)^n - (1 + i)^N}{(1 + i)^n - 1}\right] \qquad (0 \le N \le n)$$

Hint: $B(N) = R\left[\frac{1 - (1 + i)^{-n+N}}{i}\right]$. To see this, study Example 3, page 314. Replace R using Formula (12).

2. Refer to Example 3, page 314. Using the result of part 1, show that Teresa and Raul's outstanding balance after making N payments is

$$E(N) = \frac{160{,}000(1.005^{360} - 1.005^N)}{1.005^{360} - 1} \qquad (0 \le N \le 360)$$

3. Using a graphing utility, plot the graph of

$$E(x) = \frac{160{,}000(1.005^{360} - 1.005^x)}{1.005^{360} - 1}$$

   using the viewing window $[0, 360] \times [0, 160{,}000]$.
4. Referring to the graph in part 3, observe that the outstanding principal drops off slowly in the early years and accelerates quickly to zero toward the end of the loan. Can you explain why?
5. How long does it take Teresa and Raul to repay half of the loan of \$160,000?
   Hint: See the previous Explore and Discuss and Exploring with Technology box.

APPLIED EXAMPLE 4 Home Affordability The Jacksons have determined that after making a down payment, they could afford at most \$2000 for a monthly house payment. The bank charges interest at the rate of 6% per year on the unpaid balance, with interest computations made at the end of each month. If the loan is to be amortized in equal monthly installments over 30 years, what is the maximum amount that the Jacksons can borrow from the bank?

Solution Here, $i = \frac{r}{m} = \frac{0.06}{12} = 0.005$, $n = (30)(12) = 360$, and $R = 2000$; we are required to find P. From Formula (11), we have

$$P = R\left[\frac{1 - (1 + i)^{-n}}{i}\right]$$

Substituting the numerical values for R, n, and i into this expression for P, we obtain

$$P = 2000\left[\frac{1 - (1.005)^{-360}}{0.005}\right] \approx 333{,}583.23$$

Therefore, the Jacksons can borrow at most \$333,583.

An adjustable-rate mortgage (ARM) is a home loan in which the interest rate is changed periodically based on a financial index. For example, a 5/1 ARM is one that has an initial rate for the first 5 years and thereafter is adjusted every year for the remaining term of the loan. Similarly, a 7/1 ARM is one that has an initial rate for the first 7 years and thereafter is adjusted every year for the remaining term of the loan.

During the housing boom of 2001–2005, lenders aggressively promoted another type of loan—the interest-only mortgage loan—to help prospective buyers qualify for larger mortgages. With an interest-only loan, the homeowner pays only the interest on the mortgage for a fixed term, usually 5 to 7 years. At the end of that period, the borrower usually has an option to convert the loan to one that will be amortized.

APPLIED EXAMPLE 5 Home Affordability Refer to Example 4. Suppose that the bank has also offered the Jacksons (a) a 7/1 ARM with a term of 30 years and an interest rate of 5.70% per year compounded monthly for the first 7 years and (b) an interest-only loan for a term of 30 years and an interest rate of 5.94% per year for the first 7 years. If the Jacksons limit their monthly payment to \$2000 per month, what is the maximum amount they can borrow with each of these mortgages?

Solution

a. Here, $i = \frac{r}{m} = \frac{0.057}{12} = 0.00475$, $n = (30)(12) = 360$, and $R = 2000$, and we want to find P. From Formula (11), we have

$$P = R\left[\frac{1 - (1 + i)^{-n}}{i}\right]$$
$$= 2000\left[\frac{1 - (1.00475)^{-360}}{0.00475}\right]$$
$$\approx 344{,}589.68$$

Therefore, if the Jacksons choose the 7/1 ARM, they can borrow at most \$344,590.

b. If P denotes the maximum amount that the Jacksons can borrow, then the interest per year on their loan is $0.0594P$ dollars. But this is equal to their payments for the year, which are $(12)(2000)$ dollars. So we have

$$0.0594P = (12)(2000)$$

or

$$P = \frac{(12)(2000)}{0.0594} \approx 404{,}040.40$$

So if the Jacksons choose the 7-year interest-only loan, they can borrow at most \$404,040.

APPLIED EXAMPLE 6 Adjustable Rate Mortgages Five years ago, the Campbells secured a 5/1 ARM to help finance the purchase of their home. The amount of the original loan was \$350,000 for a term of 30 years, with interest at the rate of 5.76% per year, compounded monthly for the first 5 years. The Campbells' mortgage is due to reset next month, and the new interest rate will be 6.96% per year, compounded monthly.

a. What was the Campbells' monthly mortgage payment for the first 5 years?
b. What will the Campbells' new monthly mortgage payment be (after the reset)? By how much will the monthly payment increase?

Solution

a. First, we find the Campbells' monthly payment on the original loan amount. Using Formula (12) with $P = 350{,}000$, $i = \frac{r}{m} = \frac{0.0576}{12} = 0.0048$, and $n = mt = (12)(30) = 360$, we find that the monthly payment was

$$R = \frac{350{,}000(0.0048)}{1 - (1 + 0.0048)^{-360}} \approx 2044.729$$

or \$2044.73 for the first 5 years.

b. To find the amount of the Campbells' new mortgage payment, we first need to find their outstanding principal. This is given by the present value of their remaining mortgage payments. Using Formula (10), with $R = 2044.729$, $i = \frac{r}{m} = \frac{0.0576}{12} = 0.0048$, and $n = mt = 360 - 5(12) = 300$, we find that their outstanding principal is

$$P = 2044.729\left[\frac{1 - (1 + 0.0048)^{-300}}{0.0048}\right] \approx 324{,}709.194$$

or \$324,709.19.

Next, we compute the amount of their new mortgage payment for the remaining term (300 months). Using Formula (12) with $P = 324{,}709.194$, $i = \frac{r}{m} = \frac{0.0696}{12} = 0.0058$, and $n = mt = 300$, we find that the monthly payment is

$$R = \frac{324{,}709.194(0.0058)}{1 - (1 + 0.0058)^{-300}} \approx 2286.698$$

or \$2286.70—an increase of \$241.97.

Sinking Funds

Sinking funds are another important application of the annuity formulas. Simply stated, a **sinking fund** is an account that is set up for a specific purpose at some future date. For example, an individual might establish a sinking fund for the purpose of discharging a debt at a future date. A corporation might establish a sinking fund in order to accumulate sufficient capital to replace equipment that is expected to be obsolete at some future date.

By thinking of the amount to be accumulated by a specific date in the future as the future value of an annuity [Formula (9), Section 5.2], we can answer questions about a large class of sinking fund problems.

APPLIED EXAMPLE 7 Sinking Fund The proprietor of Carson Hardware has decided to set up a sinking fund for the purpose of purchasing a truck in 2 years' time. It is expected that the truck will cost \$30,000. If the fund earns 10% interest per year compounded quarterly, determine the size of each (equal) quarterly installment the proprietor should pay into the fund. Verify the result by displaying the schedule.

Solution The problem at hand is to find the size of each quarterly payment R of an annuity, given that its future value is $S = 30{,}000$, the interest earned per conversion period is $i = \frac{r}{m} = \frac{0.1}{4} = 0.025$, and the number of payments is $n = (2)(4) = 8$.

The formula for an annuity,

$$S = R\left[\frac{(1+i)^n - 1}{i}\right]$$

when solved for R yields

$$R = \frac{iS}{(1+i)^n - 1} \tag{13}$$

Substituting the appropriate numerical values for i, S, and n into Equation (13), we obtain the desired quarterly payment

$$R = \frac{(0.025)(30{,}000)}{(1.025)^8 - 1} \approx 3434.02$$

or \$3434.02. Table 5 shows the required schedule.

TABLE 5

A Sinking Fund Schedule

End of Period	Deposit Made	Interest Earned	Addition to Fund	Accumulated Amount in Fund
1	\$3,434.02	0	\$3,434.02	\$ 3,434.02
2	3,434.02	\$ 85.85	3,519.87	6,953.89
3	3,434.02	173.85	3,607.87	10,561.76
4	3,434.02	264.04	3,698.06	14,259.82
5	3,434.02	356.50	3,790.52	18,050.34
6	3,434.02	451.26	3,885.28	21,935.62
7	3,434.02	548.39	3,982.41	25,918.03
8	3,434.02	647.95	4,081.97	30,000.00

The formula derived in Example 7 is restated as follows.

Sinking Fund Payment

The periodic payment R required to accumulate a sum of S dollars over n periods with interest charged at the rate of i per period is

$$R = \frac{iS}{(1+i)^n - 1} \tag{14}$$

APPLIED EXAMPLE 8 Retirement Planning Jason is the owner of a computer consulting firm. He is currently planning to retire in 25 years and wishes to withdraw \$8000 per month from his retirement account for 25 years starting at that time. How much must he contribute each month into a retirement account earning interest at the rate of 6% per year compounded monthly to meet his retirement goal?

Solution In this case, we work backwards. We first calculate the amount of the sinking fund that Jason needs to accumulate to fund his retirement needs. Using Formula (10) with $R = 8000$, $i = \frac{r}{m} = \frac{0.06}{12} = 0.005$, and $n = 300$, we have

$$P = 8000\left[\frac{1 - (1 + 0.005)^{-300}}{0.005}\right]$$
$$\approx 1{,}241{,}654.91$$

Next, we calculate how much Jason should deposit into his retirement account each month to accumulate \$1,241,654.91 in 25 years. Here, we use Formula (14) with $S = 1{,}241{,}654.91$, $i = \frac{r}{m} = \frac{0.06}{12} = 0.005$, and $n = 300$. We have

$$R = 1{,}241{,}654.91\left[\frac{0.005}{(1 + 0.005)^{300} - 1}\right]$$
$$\approx 1791.73$$

or approximately \$1791.73 per month.

Here is a summary of the formulas developed thus far in this chapter:

1. Simple and compound interest; annuities

One deposit

- Simple interest
  $A = P(1 + rt)$
  $I = Prt$
- Compound interest
  $A = P(1 + i)^n$
  $P = A(1 + i)^{-n}$
  $A = Pe^{rt}$ Continuous compounding
  $P = Ae^{-rt}$ Continuous compounding

Periodic deposits

- Future value of an annuity
  $S = R\left[\frac{(1 + i)^n - 1}{i}\right]$
- Present value of an annuity
  $P = R\left[\frac{1 - (1 + i)^{-n}}{i}\right]$

2. Effective rate of interest

$$r_{\text{eff}} = \left(1 + \frac{r}{m}\right)^m - 1$$

3. Amortization

$$R = \frac{Pi}{1 - (1 + i)^{-n}} \quad \text{Periodic payment}$$

$$P = R\left[\frac{1 - (1 + i)^{-n}}{i}\right] \quad \text{Amount amortized}$$

4. Sinking fund

$$R = \frac{iS}{(1 + i)^n - 1} \quad \text{Periodic payment taken out}$$

5.3 Self-Check Exercises

1. **Comparing Loans** The Mendozas wish to borrow \$300,000 from a bank to help finance the purchase of a house. Their banker has offered the following plans for their consideration. In Plan I, the Mendozas have 30 years to repay the loan in monthly installments with interest on the unpaid balance charged at 6.09%/year compounded

monthly. In Plan II, the loan is to be repaid in monthly installments over 15 years with interest on the unpaid balance charged at 5.76%/year compounded monthly.

a. Find the monthly repayment for each plan.

b. What is the difference in total payments made under each plan?

2. **Planning for Retirement** Harris, a self-employed individual who is 46 years old, is setting up a defined-benefit retirement plan. If he wishes to have \$250,000 in this retirement account by age 65, what is the size of each yearly installment he will be required to make into a savings account earning interest at $8\frac{1}{4}$%/year?

Solutions to Self-Check Exercises 5.3 can be found on page 324.

5.3 Concept Questions

1. Write the amortization formula.
   a. If P and i are fixed and n is allowed to increase, what will happen to R?
   b. Interpret the result of part (a).

2. Using the formula for computing a sinking fund payment, show that if the number of payments into a sinking fund increases, then the size of the periodic payment into the sinking fund decreases.

5.3 Exercises

In Exercises 1–8, find the periodic payment R required to amortize a loan of P dollars over t years with interest charged at the rate of r%/year compounded m times a year.

1. $P = 100{,}000$, $r = 6$, $t = 10$, $m = 1$
2. $P = 40{,}000$, $r = 3$, $t = 15$, $m = 2$
3. $P = 5000$, $r = 4$, $t = 3$, $m = 4$
4. $P = 16{,}000$, $r = 5$, $t = 4$, $m = 12$
5. $P = 25{,}000$, $r = 3$, $t = 12$, $m = 4$
6. $P = 80{,}000$, $r = 5.5$, $t = 15$, $m = 12$
7. $P = 80{,}000$, $r = 5.5$, $t = 30$, $m = 12$
8. $P = 100{,}000$, $r = 5.5$, $t = 25$, $m = 12$

In Exercises 9–14, find the periodic payment R required to accumulate a sum of S dollars over t years with interest earned at the rate of r%/year compounded m times a year.

9. $S = 20{,}000$, $r = 4$, $t = 6$, $m = 2$
10. $S = 40{,}000$, $r = 4$, $t = 9$, $m = 4$
11. $S = 100{,}000$, $r = 4.5$, $t = 20$, $m = 6$
12. $S = 120{,}000$, $r = 4.5$, $t = 30$, $m = 6$
13. $S = 250{,}000$, $r = 6.5$, $t = 25$, $m = 12$
14. $S = 350{,}000$, $r = 4.2$, $t = 10$, $m = 12$

15. Suppose payments were made at the end of each quarter into an ordinary annuity earning interest at the rate of 5%/year compounded quarterly. If the future value of the annuity after 5 years is \$50,000, what was the size of each payment?

16. Suppose payments were made at the end of each month into an ordinary annuity earning interest at the rate of 4.5%/year compounded monthly. If the future value of the annuity after 10 years is \$60,000, what was the size of each payment?

17. Suppose payments will be made for $6\frac{1}{2}$ years at the end of each semiannual period into an ordinary annuity earning interest at the rate of 3.5%/year compounded semiannually. If the present value of the annuity is \$35,000, what should be the size of each payment?

18. Suppose payments will be made for $9\frac{1}{4}$ years at the end of each month into an ordinary annuity earning interest at the rate of 3.25%/year compounded monthly. If the present value of the annuity is \$42,000, what should be the size of each payment?

19. **Loan Amortization** A sum of \$100,000 is to be repaid over a 10-year period through equal installments made at the end of each year. If an interest rate of 5%/year is charged on the unpaid balance and interest calculations are made at the end of each year, determine the size of each installment so that the loan (principal plus interest charges) is amortized at the end of 10 years.

20. **Loan Amortization** What monthly payment is required to amortize a loan of \$30,000 over 10 years if interest at the rate of 6%/year is charged on the unpaid balance and interest calculations are made at the end of each month?

21. **Home Mortgages** Complete the following table, which shows the monthly payments on a $100,000, 30-year mortgage at the interest rates shown. Use this information to answer the following questions.

Amount of Mortgage ($)	Interest Rate (%)	Monthly Payment ($)
100,000	3	421.60
100,000	4	. . .
100,000	5	. . .
100,000	6	. . .
100,000	7	. . .
100,000	8	733.76

a. What is the difference in monthly payments between a $100,000, 30-year mortgage secured at 7%/year and one secured at 3%/year?
b. Use the table to calculate the monthly mortgage payments on a $150,000 mortgage at 5%/year over 30 years and a $50,000 mortgage at 5%/year over 30 years.

22. **Financing a Home** The Flemings secured a bank loan of $288,000 to help finance the purchase of a house. The bank charges interest at a rate of 5%/year on the unpaid balance, and interest computations are made at the end of each month. The Flemings have agreed to repay the loan in equal monthly installments over 25 years. What should be the size of each repayment if the loan is to be amortized at the end of the term?

23. **Financing a Car** The price of a new car is $20,000. Assume that an individual makes a down payment of 25% toward the purchase of the car and secures financing for the balance at the rate of 6%/year compounded monthly.
a. What monthly payment will she be required to make if the car is financed over a period of 36 months? Over a period of 48 months?
b. What will the interest charges be if she elects the 36-month plan? The 48-month plan?

24. **Financial Analysis** A group of private investors purchased a condominium complex for $2 million. They made an initial down payment of 10% and obtained financing for the balance. If the loan is to be amortized over 15 years at an interest rate of 6.6%/year compounded quarterly, find the required quarterly payment.

25. **Financing a Home** The Taylors have purchased a $270,000 house. They made an initial down payment of $30,000 and secured a mortgage with interest charged at the rate of 6%/year on the unpaid balance. Interest computations are made at the end of each month. If the loan is to be amortized over 30 years, what monthly payment will the Taylors be required to make? What is their equity (disregarding appreciation) after 5 years? After 10 years? After 20 years?

26. **Financial Planning** Jessica wants to accumulate $10,000 by the end of 5 years in a special bank account, which she had opened for this purpose. To achieve this goal, Jessica plans to deposit a fixed sum of money into the account at the end of each month over the 5-year period. If the bank pays interest at the rate of 5%/year compounded monthly, how much does she have to deposit each month into her account?

27. **Sinking Funds** A city has $2.5 million worth of school bonds that are due in 20 years and has established a sinking fund to retire this debt. If the fund earns interest at the rate of 4%/year compounded annually, what amount must be deposited annually in this fund?

28. **Trust Funds** Carl is the beneficiary of a $20,000 trust fund set up for him by his grandparents. Under the terms of the trust, he is to receive the money over a 5-year period in equal installments at the end of each year. If the fund earns interest at the rate of 5%/year compounded annually, what amount will he receive each year?

29. **Sinking Funds** Lowell Corporation wishes to establish a sinking fund to retire a $200,000 debt that is due in 10 years. If the investment will earn interest at the rate of 9%/year compounded quarterly, find the amount of the quarterly deposit that must be made in order to accumulate the required sum.

30. **Sinking Funds** The management of Gibraltar Brokerage Services anticipates a capital expenditure of $20,000 in 3 years for the purchase of new computers and has decided to set up a sinking fund to finance this purchase. If the fund earns interest at the rate of 5%/year compounded quarterly, determine the size of each (equal) quarterly installment that should be deposited in the fund.

31. **Retirement Accounts** Andrea, a self-employed individual, wishes to accumulate a retirement fund of $250,000. How much should she deposit each month into her retirement account, which pays interest at the rate of 4.5%/year compounded monthly, to reach her goal upon retirement 25 years from now?

32. **Student Loans** Joe secured a loan of $12,000 3 years ago from a bank for use toward his college expenses. The bank charged interest at the rate of 4%/year compounded monthly on his loan. Now that he has graduated from college, Joe wishes to repay the loan by amortizing it through monthly payments over 10 years at the same interest rate. Find the size of the monthly payments he will be required to make.

33. **Retirement Accounts** Robin wishes to accumulate a sum of $450,000 in a retirement account by the time of her retirement 30 years from now. If she wishes to do this through monthly payments into the account that earn interest at the rate of 6%/year compounded monthly, what should be the size of each payment?

34. **Financing College Expenses** Yumi's grandparents presented her with a gift of $20,000 when she was 10 years old to be used for her college education. Over the next 7 years, until she turned 17, Yumi's parents had invested her money in a

tax-free account that had yielded interest at the rate of 3.5%/year compounded monthly. Upon turning 17, Yumi now plans to withdraw her funds in equal annual installments over the next 4 years, starting at age 18. If the college fund is expected to earn interest at the rate of 4%/year, compounded annually, what will be the size of each installment?

35. **IRAs** Martin has deposited \$375 in his IRA at the end of each quarter for the past 20 years. His investment has earned interest at the rate of 4%/year compounded quarterly over this period. Now, at age 60, he is considering retirement. What quarterly payment will he receive over the next 15 years? (Assume that the money is earning interest at the same rate and that payments are made at the end of each quarter.) If he continues working and makes quarterly payments of the same amount in his IRA until age 65, what quarterly payment will he receive from his fund upon retirement over the following 10 years?

36. **Retirement Planning** Jennifer is the owner of a video game and entertainment software retail store. She is currently planning to retire in 30 years and wishes to withdraw \$10,000/month for 20 years from her retirement account starting at that time. How much must she contribute each month for 30 years into a retirement account earning interest at the rate of 4%/year compounded monthly to meet her retirement goal?

37. **Effect of Delaying Retirement on Retirement Funds** Refer to Example 8. Suppose that Jason delays his retirement plans and decides to continue working and contributing to his retirement fund for an additional 5 years. By delaying his retirement, he will need to withdraw only \$8000/month for 20 years. In this case, how much must he contribute each month for 30 years into a retirement account earning interest at the rate of 6%/year compounded monthly to meet his retirement goal?

38. **Financing a Car** Darla purchased a new car during a special sales promotion by the manufacturer. She secured a loan from the manufacturer in the amount of \$16,000 at a rate of 4.9%/year compounded monthly. Her bank is now charging 6.5%/year compounded monthly for new car loans. Assuming that each loan would be amortized by 36 equal monthly installments, determine the amount of interest she would have paid at the end of 3 years for each loan. How much less will she have paid in interest payments over the life of the loan by borrowing from the manufacturer instead of her bank?

39. **Auto Financing** Dan is contemplating trading in his car for a new one. He can afford a monthly payment of at most \$400. If the prevailing interest rate is 4.2%/year compounded monthly for a 48-month loan, what is the most expensive car that Dan can afford, assuming that he will receive \$8000 for his trade-in?

40. **Auto Financing** Paula is considering the purchase of a new car. She has narrowed her search to two cars that are equally appealing to her. Car A costs \$28,000, and Car B costs \$28,200. The manufacturer of Car A is offering 0% financing for 48 months with zero down, while the manufacturer of Car B is offering a rebate of \$2000 at the time of purchase plus financing at the rate of 3%/year compounded monthly over 48 months with zero down. If Paula has decided to buy the car with the lower net cost to her, which car should she purchase?

41. **Financing a Home** Eight years ago, Kim secured a bank loan of \$180,000 to help finance the purchase of a house. The mortgage was for a term of 30 years, with an interest rate of 4.5%/year compounded monthly on the unpaid balance to be amortized through monthly payments. What is the outstanding principal on Kim's house now?

42. **Financing a Home** Sarah secured a bank loan of \$200,000 for the purchase of a house. The mortgage is to be amortized through monthly payments for a term of 15 years, with an interest rate of 3%/year compounded monthly on the unpaid balance. She plans to sell her house in 5 years. How much will Sarah still owe on her house?

43. **Personal Loans** Two years ago, Paul borrowed \$10,000 from his sister Gerri to start a business. Paul agreed to pay Gerri interest for the loan at the rate of 4%/year, compounded continuously. Paul will now begin repaying the amount he owes by amortizing the loan (plus the interest that has accrued over the past 2 years) through monthly payments over the next 5 years at an interest rate of 3%/year compounded monthly. Find the size of the monthly payments Paul will be required to make.

44. **Investment Analysis** Since he was 22 years old, Ben has been depositing \$200 at the end of each month into a tax-free retirement account earning interest at the rate of 3.5%/year compounded monthly. Larry, who is the same age as Ben, decided to open a tax-free retirement account 5 years after Ben opened his. If Larry's account earns interest at the same rate as Ben's, determine how much Larry should deposit each month into his account so that both men will have the same amount of money in their accounts at age 65.

45. **Balloon Payment Mortgages** Emilio is securing a 7-year balloon mortgage for \$280,000 to finance the purchase of his first home. The monthly payments are based on a 30-year amortization. If the prevailing interest rate is 4.5%/year compounded monthly, what will be Emilio's monthly payment? What will be his balloon payment at the end of 7 years?

46. **Balloon Payment Mortgages** Olivia plans to secure a 5-year balloon mortgage of \$200,000 toward the purchase of a condominium. Her monthly payment for the 5 years is calculated on the basis of a 30-year conventional mortgage at the rate of 3%/year compounded monthly. At the end of the 5 years, Olivia is required to pay the balance owed (the "balloon" payment). What will be her monthly payment for the first 5 years, and what will be her balloon payment?

47. **Home Refinancing** Four years ago, Emily secured a bank loan of $200,000 to help finance the purchase of an apartment in Boston. The term of the mortgage is 30 years, and the interest rate is 6.5%/year compounded monthly. Because the interest rate for a conventional 30-year home mortgage has now dropped to 4.75%/year compounded monthly, Emily is thinking of refinancing her property.
   a. What is Emily's current monthly mortgage payment?
   b. What is Emily's current outstanding principal?
   c. If Emily decides to refinance her property by securing a 30-year home mortgage loan in the amount of the current outstanding principal at the prevailing interest rate of 4.75%/year compounded monthly, what will be her monthly mortgage payment?
   d. How much less would Emily's monthly mortgage payment be if she refinances?

48. **Home Refinancing** Five years ago, Diane secured a bank loan of $300,000 to help finance the purchase of a loft in the San Francisco Bay area. The term of the mortgage was 30 years, and the interest rate was 6%/year compounded monthly on the unpaid balance. Because the interest rate for a conventional 30-year home mortgage has now dropped to 4.5%/year compounded monthly, Diane is thinking of refinancing her property.
   a. What is Diane's current monthly mortgage payment?
   b. What is Diane's current outstanding principal?
   c. If Diane decides to refinance her property by securing a 30-year home mortgage loan in the amount of the current outstanding principal at the prevailing interest rate of 4.5%/year compounded monthly, what will be her monthly mortgage payment?
   d. How much less would Diane's monthly mortgage payment be if she refinances?

49. **Refinancing a Home** The Sandersons are planning to refinance their home. The outstanding principal on their original loan is $100,000 and is now to be amortized in 240 equal monthly installments at an interest rate of 5%/year compounded monthly. The new loan they expect to secure is to be amortized over the same period at an interest rate of 4.2%/year compounded monthly. How much less can they expect to pay over the life of the loan in interest payments by refinancing the loan at this time?

50. **Refinancing a Home** Josh purchased a condominium 5 years ago for $180,000. He made a down payment of 20% and financed the balance with a 30-year conventional mortgage to be amortized through monthly payments with an interest rate of 4%/year compounded monthly on the unpaid balance. The condominium is now appraised at $250,000. Josh plans to start his own business and wishes to tap into the equity that he has in the condominium. If Josh can secure a new 30-year conventional mortgage at the same rate to refinance his condominium based on a loan of 80% of the appraised value, how much cash can Josh muster for his business? (Disregard taxes.)

51. **Adjustable-Rate Mortgages** Three years ago, Samantha secured an adjustable-rate mortgage (ARM) loan to help finance the purchase of a house. The amount of the original loan was $150,000 for a term of 30 years, with interest at the rate of 5.5%/year compounded monthly. Currently, the interest rate is 4%/year compounded monthly, and Samantha's monthly payments are due to be recalculated. What will be her new monthly payment?
Hint: Calculate her current outstanding principal. Then, to amortize the loan in the next 27 years, determine the monthly payment based on the current interest rate.

52. **Adjustable-Rate Mortgages** George secured an adjustable-rate mortgage (ARM) loan to help finance the purchase of his home 5 years ago. The amount of the loan was $300,000 for a term of 30 years, with interest at the rate of 6%/year compounded monthly. Currently, the interest rate for his ARM is 4.5%/year compounded monthly, and George's monthly payments are due to be reset. What will be the new monthly payment?

53. **Financing a Home** After making a down payment of $25,000, the Meyers need to secure a loan of $280,000 to purchase a certain house. Their bank's current rate for 25-year home loans is 5.5%/year compounded monthly. The owner has offered to finance the loan at 4.9%/year compounded monthly. Assuming that both loans would be amortized over a 25-year period by 300 equal monthly installments, determine the difference in the amount of interest the Meyers would pay by choosing the seller's financing rather than their bank's.

54. **Refinancing a Home** The Martinezes are planning to refinance their home. The outstanding balance on their original loan is $150,000. Their finance company has offered them two options:

   Option A: A fixed-rate mortgage at an interest rate of 4.5%/year compounded monthly, payable over a 30-year period in 360 equal monthly installments.

   Option B: A fixed-rate mortgage at an interest rate of 4.25%/year compounded monthly, payable over a 15-year period in 180 equal monthly installments.

   a. Find the monthly payment required to amortize each of these loans over the life of the loan.
   b. How much interest would the Martinezes save if they chose the 15-year mortgage instead of the 30-year mortgage?

55. **Ability-to-Repay Rule for Mortgages** The Ability-to-Repay Rule, adopted by the Consumer Financial Protection Bureau in compliance with the Dodd-Frank Wall Street Reform and Consumer Protection Act, requires lenders to determine whether a consumer applying for a Qualified Mortgage can afford to repay the loan. One of the requirements is that the borrower's total monthly debt (including property taxes) cannot exceed 43% of the borrower's monthly pre-tax income. Suppose that the Foleys have applied for a $400,000 Qualified Mortgage with an

interest rate of 4%/year compounded monthly and a term of 30 years. The property tax on the home they wish to purchase is $6000/year. If the Foleys' annual income is $72,000, will they qualify for the mortgage?
Source: Consumer Financial Protection Bureau.

56. **Ability-to-Repay Rule for Mortgages** Refer to Exercise 55. What is the maximum Qualified Mortgage with an interest rate of 3.75%/year compounded monthly and a term of 30 years the Foleys' can qualify for?

57. **Home Affordability** Suppose that the Carlsons have decided that they can afford a maximum of $3000/month for a monthly house payment. The bank has offered them (a) a 5/1 ARM for a term of 30 years with interest at the rate of 4.40%/year compounded monthly for the first 5 years and (b) an interest-only loan for a term of 30 years at the rate of 4.62%/year for the first 5 years. What is the maximum amount that they can borrow with each of these mortgages if they keep to their budget?

58. **Comparing Mortgages** Refer to Example 5. Suppose that the Jacksons choose the 7/1 ARM and borrow the maximum amount of $344,589.68.
   a. By how much will the principal of their loan be reduced at the end of 7 years?
   Hint: See Example 3.
   b. If they choose the 7-year interest-only mortgage, by how much will the principal of their loan be reduced after 7 years?

5.3 Solutions to Self-Check Exercises

1. a. We use Formula (12) in each instance. Under Plan I,

$$P = 300{,}000 \qquad i = \frac{r}{m} = \frac{0.0609}{12} = 0.005075$$
$$n = (30)(12) = 360$$

Therefore, the size of each monthly repayment under Plan I is

$$R = \frac{300{,}000(0.005075)}{1 - (1.005075)^{-360}} \approx 1816.05$$

or $1816.05.

Under Plan II,

$$P = 300{,}000 \qquad i = \frac{r}{m} = \frac{0.0576}{12} = 0.0048$$
$$n = (15)(12) = 180$$

Therefore, the size of each monthly repayment under Plan II is

$$R = \frac{300{,}000(0.0048)}{1 - (1.0048)^{-180}} \approx 2492.84$$

or $2492.84.

b. Under Plan I, the total amount of repayments will be

$$(360)(1816.05) = 653{,}778 \quad \text{Number of payments} \times \text{the size of each installment}$$

or $653,778. Under Plan II, the total amount of repayments will be

$$(180)(2492.84) = 448{,}711.20$$

or $448,711.20. Therefore, the difference in payments is

$$653{,}778 - 448{,}711.20 = 205{,}066.80$$

or $205,066.80.

2. We use Formula (14) with

$$S = 250{,}000$$
$$i = r = 0.0825 \quad \text{Since } m = 1$$
$$n = 20$$

giving the required size of each installment as

$$R = \frac{(0.0825)(250{,}000)}{(1.0825)^{20} - 1} \approx 5313.59$$

or $5313.59.

USING TECHNOLOGY Amortizing a Loan

Graphing Utility

Here, we use the TI-83/84 **TVM SOLVER** function to help us solve problems involving amortization and sinking funds.

APPLIED EXAMPLE 1 **Finding the Payment to Amortize a Loan** The Wongs are considering obtaining a preapproved 30-year loan of \$120,000 to help finance the purchase of a house. The mortgage company charges interest at the rate of 8% per year on the unpaid balance, with interest computations made at the end of each month. What will be the monthly installments if the loan is amortized?

Solution We use the TI-83/84 **TVM SOLVER** with the following inputs:

N = 360 (30)(12)
I% = 8
PV = 120000
PMT = 0
FV = 0
P/Y = 12 The number of payments each year
C/Y = 12 The number of conversion periods each year
PMT: **END** BEGIN

By moving the cursor up to the PMT line and pressing **ALPHA** **SOLVE**, we obtain the output shown in Figure T1. We see that the required payment is \$880.52.

```
  N=360
  I%=8
  PV=120000
■ PMT=-880.51748...
  FV=0
  P/Y=12
  C/Y=12
  PMT:[END] BEGIN
```

FIGURE **T1**
The TI-83/84 screen showing the monthly installment, PMT

APPLIED EXAMPLE 2 **Finding the Payment in a Sinking Fund** Heidi wishes to establish a retirement account that will be worth \$500,000 in 20 years' time. She expects that the account will earn interest at the rate of 11% per year compounded monthly. What should be the monthly contribution into her account each month?

Solution We use the TI-83/84 **TVM SOLVER** with the following inputs:

N = 240 (20)(12)
I% = 11
PV = 0
PMT = 0
FV = 500000
P/Y = 12 The number of payments each year
C/Y = 12 The number of conversion periods each year
PMT: **END** BEGIN

```
  N=240
  I%=11
  PV=0
■ PMT=-577.60862...
  FV=500000
  P/Y=12
  C/Y=12
  PMT:[END] BEGIN
```

FIGURE **T2**
The TI-83/84 screen showing the monthly payment, PMT

By moving the cursor up to the PMT line and pressing **ALPHA** **SOLVE**, we obtain the result displayed in Figure T2. We see that Heidi's monthly contribution should be \$577.61. (*Note:* The display for PMT is negative because it is an outflow.)

Note: Boldfaced words/characters enclosed in a box (for example, **Enter**) indicate that an action (click, select, or press) is required. Words/characters printed blue (for example, Chart sub-type:) indicate words/characters appearing on the screen.

Excel

Here we use Excel to help us solve problems involving amortization and sinking funds.

APPLIED EXAMPLE 3 Finding the Payment to Amortize a Loan The Wongs are considering a preapproved 30-year loan of \$120,000 to help finance the purchase of a house. The mortgage company charges interest at the rate of 8% per year on the unpaid balance, with interest computations made at the end of each month. What will be the monthly installments if the loan is amortized at the end of the term?

Solution We use the PMT function to solve this problem. Accessing this function from the Financial function library subgroup and making the required entries, we obtain the Function Arguments dialog box shown in Figure T3. We see that the desired result is \$880.52. (Recall that cash you pay out is represented by a negative number.)

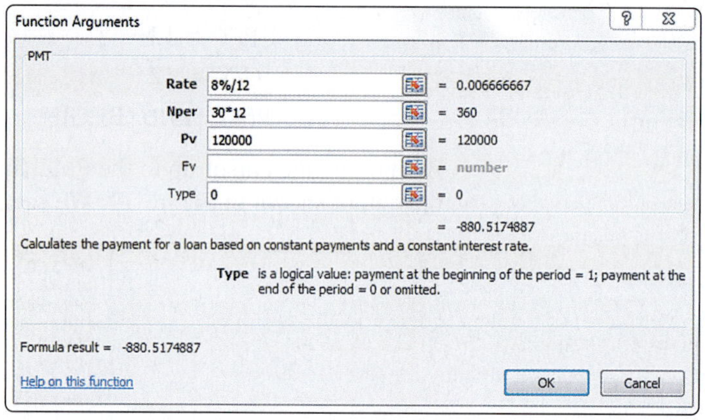

FIGURE **T3**
Excel's dialog box giving the payment function, PMT

APPLIED EXAMPLE 4 Finding the Payment in a Sinking Fund Heidi wishes to establish a retirement account that will be worth \$500,000 in 20 years' time. She expects that the account will earn interest at the rate of 11% per year compounded monthly. What should be the monthly contribution into her account each month?

Solution As in Example 3, we use the PMT function, but this time, we are given the future value of the investment. Accessing the PMT function as before and making the required entries, we obtain the Function Arguments dialog box shown in Figure T4. We see that Heidi's monthly contribution should be \$577.61. (Note that the value for PMT is negative because it is an outflow.)

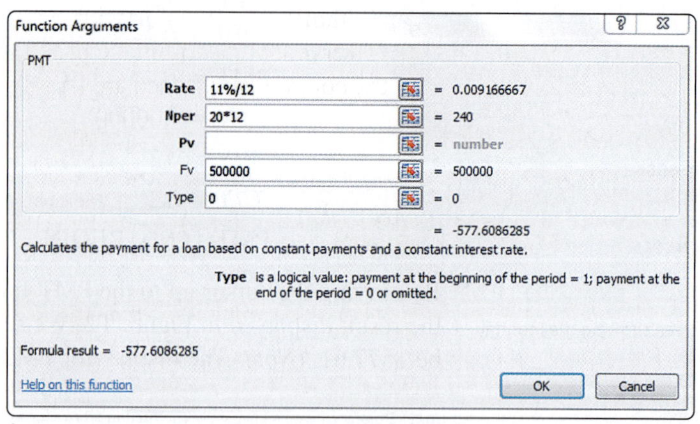

FIGURE **T4**
Excel's dialog box giving the payment function, PMT

TECHNOLOGY EXERCISES

1. Find the periodic payment required to amortize a loan of \$55,000 over 120 months with interest charged at the rate of $6\frac{5}{8}$%/year compounded monthly.

2. Find the periodic payment required to amortize a loan of \$178,000 over 180 months with interest charged at the rate of $5\frac{1}{8}$%/year compounded monthly.

3. Find the periodic payment required to amortize a loan of \$227,000 over 360 months with interest charged at the rate of $6\frac{1}{8}$%/year compounded monthly.

4. Find the periodic payment required to amortize a loan of \$150,000 over 360 months with interest charged at the rate of $4\frac{3}{8}$%/year compounded monthly.

5. Find the periodic payment required to accumulate \$25,000 over 12 quarters with interest earned at the rate of $3\frac{3}{8}$%/year compounded quarterly.

6. Find the periodic payment required to accumulate \$50,000 over 36 quarters with interest earned at the rate of $3\frac{7}{8}$%/year compounded quarterly.

7. Find the periodic payment required to accumulate \$137,000 over 120 months with interest earned at the rate of $4\frac{3}{4}$%/year compounded monthly.

8. Find the periodic payment required to accumulate \$144,000 over 120 months with interest earned at the rate of $4\frac{5}{8}$%/year compounded monthly.

9. A loan of \$120,000 is to be repaid over a 10-year period through equal installments made at the end of each year. If an interest rate of 4.5%/year is charged on the unpaid balance and interest calculations are made at the end of each year, determine the size of each installment such that the loan is amortized at the end of 10 years. Verify the result by displaying the amortization schedule.

10. A loan of \$265,000 is to be repaid over an 8-year period through equal installments made at the end of each year. If an interest rate of 5.4%/year is charged on the unpaid balance and interest calculations are made at the end of each year, determine the size of each installment so that the loan is amortized at the end of 8 years. Verify the result by displaying the amortization schedule.

5.4 Arithmetic and Geometric Progressions

Arithmetic Progressions

An **arithmetic progression** is a sequence of numbers in which each term after the first is obtained by adding a constant d to the preceding term. The constant d is called the **common difference.** For example, the sequence

$$2, 5, 8, 11, \ldots$$

is an arithmetic progression with common difference equal to 3.

Observe that an arithmetic progression is completely determined if the first term and the common difference are known. In fact, if

$$a_1, a_2, a_3, \ldots, a_n, \ldots$$

is an arithmetic progression with the first term given by a and common difference given by d, then by definition we have

$$\begin{aligned}
a_1 &= a \\
a_2 &= a_1 + d = a + d \\
a_3 &= a_2 + d = (a + d) + d = a + 2d \\
a_4 &= a_3 + d = (a + 2d) + d = a + 3d \\
&\vdots \\
a_n &= a_{n-1} + d = a + (n - 2)d + d = a + (n - 1)d
\end{aligned}$$

Thus, we have the following formula:

*n*th Term of an Arithmetic Progression

The nth term of an arithmetic progression with first term a and common difference d is given by

$$a_n = a + (n - 1)d \tag{15}$$

EXAMPLE 1 Find the twelfth term of the arithmetic progression

$$2, 7, 12, 17, 22, \ldots$$

Solution The first term of the arithmetic progression is $a_1 = a = 2$, and the common difference is $d = 5$; so upon setting $n = 12$ in Equation (15), we find

$$a_{12} = 2 + (12 - 1)5 = 57$$

EXAMPLE 2 Write the first five terms of an arithmetic progression whose third and eleventh terms are 21 and 85, respectively.

Solution Using Equation (15), we obtain

$$a_3 = a + 2d = 21$$
$$a_{11} = a + 10d = 85$$

Subtracting the first equation from the second gives $8d = 64$, or $d = 8$. Substituting this value of d into the first equation yields $a + 16 = 21$, or $a = 5$. Thus, the required arithmetic progression is given by the sequence

$$5, 13, 21, 29, 37, \ldots$$

Let S_n denote the sum of the first n terms of an arithmetic progression with first term $a_1 = a$ and common difference d. Then

$$S_n = a + (a + d) + (a + 2d) + \cdots + [a + (n - 1)d] \tag{16}$$

Rewriting the expression for S_n with the terms in reverse order gives

$$S_n = [a + (n - 1)d] + [a + (n - 2)d] + \cdots + (a + d) + a \tag{17}$$

Adding Equations (16) and (17), we obtain

$$\begin{aligned} 2S_n &= [2a + (n - 1)d] + [2a + (n - 1)d] \\ &\quad + \cdots + [2a + (n - 1)d] \\ &= n[2a + (n - 1)d] \\ S_n &= \frac{n}{2}[2a + (n - 1)d] \end{aligned}$$

Sum of Terms in an Arithmetic Progression

The sum of the first n terms of an arithmetic progression with first term a and common difference d is given by

$$S_n = \frac{n}{2}[2a + (n - 1)d] \tag{18}$$

EXAMPLE 3 Find the sum of the first 20 terms of the arithmetic progression of Example 1.

Solution Letting $a = 2$, $d = 5$, and $n = 20$ in Equation (18), we obtain

$$S_{20} = \frac{20}{2}[2 \cdot 2 + 19 \cdot 5] = 990$$

APPLIED EXAMPLE 4 Company Sales Madison Electric Company had sales of \$200,000 in its first year of operation. If the sales increased by \$30,000 per year thereafter, find Madison's sales in the fifth year and its total sales over the first 5 years of operation.

Solution Madison's yearly sales follow an arithmetic progression, with the first term given by $a = 200{,}000$ and the common difference given by $d = 30{,}000$. The sales in the fifth year are found by using Equation (15) with $n = 5$. Thus,

$$a_5 = 200{,}000 + (5 - 1)30{,}000 = 320{,}000$$

or \$320,000.

Madison's total sales over the first 5 years of operation are found by using Equation (18) with $n = 5$. Thus,

$$\begin{aligned} S_5 &= \frac{5}{2}[2(200{,}000) + (5 - 1)30{,}000] \\ &= 1{,}300{,}000 \end{aligned}$$

or \$1,300,000.

Geometric Progressions

A **geometric progression** is a sequence of numbers in which each term after the first is obtained by multiplying the preceding term by a constant r. The constant r is called the **common ratio.**

A geometric progression is completely determined if the first term and the common ratio are known. Thus, if

$$a_1, a_2, a_3, \ldots, a_n, \ldots$$

is a geometric progression with first term a and common ratio r, then by definition, we have

$$\begin{aligned} a_1 &= a \\ a_2 &= a_1 r = ar \\ a_3 &= a_2 r = ar^2 \\ a_4 &= a_3 r = ar^3 \\ &\vdots \\ a_n &= a_{n-1} r = ar^{n-1} \end{aligned}$$

This gives the following:

nth Term of a Geometric Progression

The nth term of a geometric progression with first term a and common ratio r is given by

$$a_n = ar^{n-1} \qquad \textbf{(19)}$$

EXAMPLE 5 Find the eighth term of a geometric progression whose first five terms are 162, 54, 18, 6, and 2.

Solution The common ratio is found by taking the ratio of any term other than the first to the preceding term. Taking the ratio of the fourth term to the third term, for example, gives $r = \frac{6}{18} = \frac{1}{3}$. To find the eighth term of the geometric progression, use Equation (19) with $a = 162$, $r = \frac{1}{3}$, and $n = 8$, obtaining

$$\begin{aligned} a_8 &= 162\left(\frac{1}{3}\right)^7 \\ &= \frac{2}{27} \end{aligned}$$

EXAMPLE 6 Find the tenth term of a geometric progression with positive terms and third term equal to 16 and seventh term equal to 1.

Solution Using Equation (19) with $n = 3$ and $n = 7$, respectively, yields

$$\begin{aligned} a_3 &= ar^2 = 16 \\ a_7 &= ar^6 = 1 \end{aligned}$$

Dividing a_7 by a_3 gives

$$\frac{ar^6}{ar^2} = \frac{1}{16}$$

from which we obtain $r^4 = \frac{1}{16}$, or $r = \frac{1}{2}$. Substituting this value of r into the expression for a_3, we obtain

$$a\left(\frac{1}{2}\right)^2 = 16 \quad \text{or} \quad a = 64$$

Finally, using Equation (19) once again with $a = 64$, $r = \frac{1}{2}$, and $n = 10$ gives

$$a_{10} = 64\left(\frac{1}{2}\right)^9 = \frac{1}{8}$$

To find the sum of the first n terms of a geometric progression with the first term $a_1 = a$ and common ratio r, denote the required sum by S_n. Then

$$S_n = a + ar + ar^2 + \cdots + ar^{n-2} + ar^{n-1} \tag{20}$$

Upon multiplying (20) by r, we obtain

$$rS_n = ar + ar^2 + ar^3 + \cdots + ar^{n-1} + ar^n \tag{21}$$

Subtracting Equation (21) from (20) gives

$$\begin{aligned} S_n - rS_n &= a - ar^n \\ (1 - r)S_n &= a(1 - r^n) \end{aligned}$$

If $r \neq 1$, we may divide both sides of the last equation by $(1 - r)$, obtaining

$$S_n = \frac{a(1 - r^n)}{1 - r}$$

If $r = 1$, then (20) gives

$$\begin{aligned} S_n &= a + a + a + \cdots + a \qquad n \text{ terms} \\ &= na \end{aligned}$$

Thus,

$$S_n = \begin{cases} \dfrac{a(1 - r^n)}{1 - r} & \text{if } r \neq 1 \\ na & \text{if } r = 1 \end{cases}$$

Sum of Terms in a Geometric Progression

The sum of the first n terms of a geometric progression with first term a and common ratio r is given by

$$S_n = \begin{cases} \dfrac{a(1 - r^n)}{1 - r} & \text{if } r \neq 1 \\ na & \text{if } r = 1 \end{cases} \tag{22}$$

EXAMPLE 7 Find the sum of the first six terms of the following geometric progression:

$$3, 6, 12, 24, \ldots$$

Solution Here, $a = 3$, $r = \frac{6}{3} = 2$, and $n = 6$, so Equation (22) gives

$$S_6 = \frac{3(1 - 2^6)}{1 - 2} = 189$$

APPLIED EXAMPLE 8 Company Sales Michaelson Land Development Company had sales of $1 million in its first year of operation. If sales increased by 10% per year thereafter, find Michaelson's sales in the fifth year and its total sales over the first 5 years of operation.

Solution Michaelson's yearly sales follow a geometric progression, with the first term given by $a = 1{,}000{,}000$ and the common ratio given by $r = 1.1$. The sales in the fifth year are found by using Formula (19) with $n = 5$. Thus,

$$a_5 = 1{,}000{,}000(1.1)^4 = 1{,}464{,}100$$

or $1,464,100.

Michaelson's total sales over the first 5 years of operation are found by using Equation (22) with $n = 5$. Thus,

$$S_5 = \frac{1{,}000{,}000[1 - (1.1)^5]}{1 - 1.1} = 6{,}105{,}100$$

or $6,105,100.

Double Declining–Balance Method of Depreciation

In Section 1.3, we discussed the straight-line, or linear, method of depreciating an asset. Linear depreciation assumes that the asset depreciates at a constant rate. For certain assets (such as machines) whose market values drop rapidly in the early years of usage and thereafter less rapidly, another method of depreciation called the **double declining–balance method** is often used. In practice, a business firm normally employs the double declining–balance method for depreciating such assets for a certain number of years and then switches over to the linear method.

To derive an expression for the book value of an asset being depreciated by the double declining–balance method, let C (in dollars) denote the original cost of the asset and let the asset be depreciated over N years. When this method is used, the amount depreciated each year is $\frac{2}{N}$ times the value of the asset at the beginning of that year. Thus, the amount by which the asset is depreciated in its first year of use is given by $\frac{2C}{N}$, so if $V(1)$ denotes the book value of the asset at the end of the first year, then

$$V(1) = C - \frac{2C}{N} = C\left(1 - \frac{2}{N}\right)$$

Next, if $V(2)$ denotes the book value of the asset at the end of the second year, then a similar argument leads to

$$\begin{aligned} V(2) &= C\left(1 - \frac{2}{N}\right) - C\left(1 - \frac{2}{N}\right)\frac{2}{N} \\ &= C\left(1 - \frac{2}{N}\right)\left(1 - \frac{2}{N}\right) \\ &= C\left(1 - \frac{2}{N}\right)^2 \end{aligned}$$

Continuing, we find that if $V(n)$ denotes the book value of the asset at the end of n years, then the terms $C, V(1), V(2), \ldots, V(N)$ form a geometric progression with first term C and common ratio $(1 - \frac{2}{N})$. Consequently, the nth term, $V(n)$, is given by

$$V(n) = C\left(1 - \frac{2}{N}\right)^n \qquad (1 \le n \le N) \tag{23}$$

Also, if $D(n)$ denotes the amount by which the asset has been depreciated by the end of the nth year, then

$$\begin{aligned} D(n) &= C - C\left(1 - \frac{2}{N}\right)^n \\ &= C\left[1 - \left(1 - \frac{2}{N}\right)^n\right] \end{aligned} \tag{24}$$

APPLIED EXAMPLE 9 Depreciation of Equipment A tractor purchased at a cost of \$60,000 is to be depreciated by the double declining–balance method over 10 years. What is the book value of the tractor at the end of 5 years? By what amount has the tractor been depreciated by the end of the fifth year?

Solution We have $C = 60{,}000$ and $N = 10$. Thus, using Equation (23) with $n = 5$ gives the book value of the tractor at the end of 5 years as

$$\begin{aligned} V(5) &= 60{,}000\left(1 - \frac{2}{10}\right)^5 \\ &= 60{,}000\left(\frac{4}{5}\right)^5 = 19{,}660.80 \end{aligned}$$

or \$19,660.80.

The amount by which the tractor has been depreciated by the end of the fifth year is given by

$$60{,}000 - 19{,}660.80 = 40{,}339.20$$

or \$40,339.20. You may verify the last result by using Equation (24) directly.

Exploring with TECHNOLOGY

A tractor purchased at a cost of $60,000 is to be depreciated over 10 years with a residual value of $0. When the double declining–balance method is used, its value at the end of n years is $V_1(n) = 60{,}000(0.8)^n$ dollars. When straight-line depreciation is used, its value at the end of n years is $V_2(n) = 60{,}000 - 6000n$. Use a graphing utility to sketch the graphs of V_1 and V_2 in the viewing window $[0, 10] \times [0, 70{,}000]$. Comment on the relative merits of each method of depreciation.

5.4 Self-Check Exercises

1. Find the sum of the first five terms of the geometric progression with first term -24 and common ratio $-\frac{1}{2}$.

2. **Depreciation of Office Equipment** Office equipment purchased for $75,000 is to be depreciated by the double declining–balance method over 5 years. Find the book value at the end of 3 years.

3. Derive the formula for the future value of an annuity [Formula (9), Section 5.2].

Solutions to Self-Check Exercises 5.4 can be found on page 335.

5.4 Concept Questions

1. Suppose an arithmetic progression has first term a and common difference d.
   a. What is the formula for the nth term of this progression?
   b. What is the formula for the sum of the first n terms of this progression?

2. Suppose a geometric progression has first term a and common ratio r.
   a. What is the formula for the nth term of this progression?
   b. What is the formula for the sum of the first n terms of this progression?

5.4 Exercises

In Exercises 1–4, find the nth term of the arithmetic progression that has the given values of a, d, and n.

1. $a = 6, d = 3, n = 9$
2. $a = -5, d = 3, n = 7$
3. $a = -15, d = \dfrac{3}{2}, n = 8$
4. $a = 1.2, d = 0.4, n = 98$

5. Find the first five terms of the arithmetic progression whose fourth and eleventh terms are 30 and 107, respectively.

6. Find the first five terms of the arithmetic progression whose seventh and twenty-third terms are -5 and -29, respectively.

7. Find the seventh term of the arithmetic progression $x, x + y, x + 2y, \ldots$.

8. Find the eleventh term of the arithmetic progression $a + b, 2a, 3a - b, \ldots$.

9. Find the sum of the first 15 terms of the arithmetic progression $4, 11, 18, \ldots$.

10. Find the sum of the first 20 terms of the arithmetic progression $5, -1, -7, \ldots$.

11. Find the sum of the odd integers between 14 and 58.

12. Find the sum of the even integers between 21 and 99.

13. Find $f(1) + f(2) + f(3) + \cdots + f(22)$, given that $f(x) = 3x - 4$.

14. Find $g(1) + g(2) + g(3) + \cdots + g(50)$, given that $g(x) = 12 - 4x$.

15. Show that Equation (18) can be written as

$$S_n = \frac{n}{2}(a + a_n)$$

where a_n represents the last term of an arithmetic progression. Use this formula to find:

a. The sum of the first 11 terms of the arithmetic progression whose first and eleventh terms are 3 and 47, respectively.

b. The sum of the first 20 terms of the arithmetic progression whose first and twentieth terms are 5 and −33, respectively.

16. **Sales Growth** Moderne Furniture Company had sales of \$1,500,000 during its first year of operation. If the sales increased by \$160,000/year thereafter, find Moderne's sales in the fifth year and its total sales over the first 5 years of operation.

17. **Exercise Program** As part of her fitness program, Karen has taken up jogging. If she jogs 1 mi the first day and increases her daily run by $\frac{1}{4}$ mi every week, when will she reach her goal of 10 mi/day?

18. **Cost of Drilling** A 100-ft oil well is to be drilled. The cost of drilling the first foot is \$10.00, and the cost of drilling each additional foot is \$4.50 more than that of the preceding foot. Find the cost of drilling the entire 100 ft.

19. **Consumer Decisions** Kunwoo wishes to go from the airport to his hotel, which is 25 mi away. The taxi rate is \$2.00 for the first mile and \$1.20 for each additional mile. The airport limousine also goes to his hotel and charges a flat rate of \$15.00. How much money will he save by taking the airport limousine?

20. **Salary Comparisons** Markeeta, a recent college graduate, received two job offers. Company *A* offered her an initial salary of \$48,800 with guaranteed annual increases of \$2000/year for the first 5 years. Company *B* offered an initial salary of \$50,400 with guaranteed annual increases of \$1500/year for the first 5 years.

a. Which company is offering a higher salary for the fifth year of employment?

b. Which company is offering more money for the first 5 years of employment?

21. **Sum-of-the-Years'-Digits Method of Depreciation** One of the methods that the Internal Revenue Service allows for computing depreciation of certain business property is the sum-of-the-years'-digits method. If a property valued at C dollars has an estimated useful life of N years and a salvage value of S dollars, then the amount of depreciation D_n allowed during the nth year is given by

$$D_n = (C - S)\frac{N - (n - 1)}{S_N} \qquad (0 \leq n \leq N)$$

where S_N is the sum of the first N positive integers. Thus,

$$S_N = 1 + 2 + \cdots + N = \frac{N(N + 1)}{2}$$

a. Verify that the sum of the arithmetic progression $S_N = 1 + 2 + \cdots + N$ is given by

$$\frac{N(N + 1)}{2}$$

b. If office furniture worth \$6000 is to be depreciated by this method over $N = 10$ years and the salvage value of the furniture is \$500, find the depreciation for the third year by computing D_3.

22. **Sum-of-the-Years'-Digits Method of Depreciation** The amount of depreciation allowed for a printing machine, which has an estimated useful life of 5 years and an initial value of \$100,000 (with no salvage value), is \$20,000/year using the straight-line method of depreciation. Determine the amount of depreciation that would be allowed for the first year if the printing machine were depreciated using the sum-of-the-years'-digits method described in Exercise 21. Which method would result in a larger depreciation of the asset in its first year of use?

In Exercises 23–28, determine which of the sequences are geometric progressions. For each geometric progression, find the seventh term and the sum of the first seven terms.

23. 4, 8, 16, 32, . . .

24. $1, -\frac{1}{2}, \frac{1}{4}, -\frac{1}{8}, \ldots$

25. $\frac{1}{2}, -\frac{3}{8}, \frac{1}{4}, -\frac{9}{64}, \ldots$

26. 0.004, 0.04, 0.4, 4, . . .

27. 243, 81, 27, 9, . . .

28. −1, 1, 3, 5, . . .

29. Find the twentieth term and sum of the first 20 terms of the geometric progression −3, 3, −3, 3, ...

30. Find the twenty-third term in a geometric progression having the first term $a = 0.1$ and ratio $r = 2$.

31. **Population Growth** It has been projected that the population of a certain city in the southwest will increase by 8% during each of the next 5 years. If the current population is 200,000, what is the expected population after 5 years?

32. **Sales Growth** Metro Cable TV had sales of \$2,500,000 in its first year of operation. If thereafter the sales increased by 12% of the previous year, find the sales of the company in the fifth year and the total sales over the first 5 years of operation.

33. **COLAs** Suppose the cost-of-living index had increased by 3% during each of the past 6 years and that a member of the EUW Union had been guaranteed an annual increase equal to 2% above the increase in the cost-of-living index over that period. What would be the present salary of a union member whose salary 6 years ago was \$42,000?

34. **Savings Plans** The parents of a 9-year-old boy have agreed to deposit \$10 in their son's bank account on his 10th birthday and to double the size of their deposit every year thereafter until his 18th birthday.

a. How much will they have to deposit on his 18th birthday?

b. How much will they have deposited by his 18th birthday?

35. **Salary Comparisons** A Stenton Printing Co. employee whose current annual salary is \$48,000 has the option of taking an annual raise of 8%/year for the next 4 years or a fixed annual raise of \$4000/year. Which option would be more profitable to him considering his total earnings over the 4-year period?

36. **Bacteria Growth** A culture of a certain bacteria is known to double in number every 3 hr. If the culture has an initial count of 20, what will be the population of the culture at the end of 24 hr?

37. **Trust Funds** Sarah is the recipient of a trust fund that she will receive over a period of 6 years. Under the terms of the trust, she is to receive \$10,000 the first year and each succeeding annual payment is to be increased by 15%.

a. How much will she receive during the sixth year?

b. What is the total amount of the six payments she will receive?

In Exercises 38–40, find the book value of office equipment purchased at a cost C at the end of the nth year if it is to be depreciated by the double declining–balance method over 10 years.

38. $C = \$20{,}000$, $n = 4$ **39.** $C = \$150{,}000$, $n = 8$

40. $C = \$80{,}000$, $n = 7$

41. **Double Declining–Balance Method of Depreciation** Restaurant equipment purchased at a cost of \$150,000 is to be depreciated by the double declining–balance method over 10 years. What is the book value of the equipment at the end of 6 years? By what amount has the equipment been depreciated at the end of the sixth year?

42. **Double Declining–Balance Method of Depreciation** Refer to Exercise 22. Recall that a printing machine with an estimated useful life of 5 years and an initial value of \$100,000 (and no salvage value) was to be depreciated. At the end of the first year, the amount of depreciation allowed was \$20,000 using the straight-line method and \$33,333 using the sum-of-the-years'-digits method. Determine the amount of depreciation that would be allowed for the first year if the printing machine were depreciated by the double declining–balance method. Which of these three methods would result in the largest depreciation of the printing machine at the end of its first year of use?

In Exercises 43 and 44, determine whether the statement is true or false. If it is true, explain why it is true. If it is false, give an example to show why it is false.

43. If $a_1, a_2, a_3, \ldots, a_n$ and $b_1, b_2, b_3, \ldots, b_n$ are arithmetic progressions, then $a_1 + b_1, a_2 + b_2, a_3 + b_3, \ldots, a_n + b_n$ is also an arithmetic progression.

44. If $a_1, a_2, a_3, \ldots, a_n$ and $b_1, b_2, b_3, \ldots, b_n$ are geometric progressions, then $a_1b_1, a_2b_2, a_3b_3, \ldots, a_nb_n$ is also a geometric progression.

5.4 Solutions to Self-Check Exercises

1. Use Equation (22) with $a = -24$, $n = 5$, and $r = -\frac{1}{2}$, obtaining

$$S_5 = \frac{-24[1 - (-\frac{1}{2})^5]}{1 - (-\frac{1}{2})} = \frac{-24(1 + \frac{1}{32})}{\frac{3}{2}} = -\frac{33}{2}$$

2. Use Equation (23) with $C = 75{,}000$, $N = 5$, and $n = 3$. The book value of the office equipment at the end of 3 years is

$$V(3) = 75{,}000\left(1 - \frac{2}{5}\right)^3 = 16{,}200$$

or \$16,200.

3. We have

$$S = R + R(1 + i) + R(1 + i)^2 + \cdots + R(1 + i)^{n-1}$$

The sum on the right is easily seen to be the sum of the first n terms of a geometric progression with first term R and common ratio $(1 + i)$, so by virtue of Equation (22), we obtain

$$S = R\left[\frac{1 - (1 + i)^n}{1 - (1 + i)}\right] = R\left[\frac{(1 + i)^n - 1}{i}\right]$$

CHAPTER 5 Summary of Principal Formulas and Terms

FORMULAS

1. Simple interest (accumulated amount)	$A = P(1 + rt)$
2. Compound interest	
a. Accumulated amount	$A = P(1 + i)^n$
b. Present value	$P = A(1 + i)^{-n}$
c. Interest rate per conversion period	$i = \dfrac{r}{m}$
d. Number of conversion periods	$n = mt$
3. Continuous compound interest	
a. Accumulated amount	$A = Pe^{rt}$
b. Present value	$P = Ae^{-rt}$
4. Effective rate of interest	$r_{\text{eff}} = \left(1 + \dfrac{r}{m}\right)^m - 1$
5. Annuities	
a. Future value	$S = R\left[\dfrac{(1 + i)^n - 1}{i}\right]$
b. Present value	$P = R\left[\dfrac{1 - (1 + i)^{-n}}{i}\right]$
6. Amortization payment	$R = \dfrac{Pi}{1 - (1 + i)^{-n}}$
7. Amount amortized	$P = R\left[\dfrac{1 - (1 + i)^{-n}}{i}\right]$
8. Sinking fund payment	$R = \dfrac{iS}{(1 + i)^n - 1}$

TERMS

simple interest (282)
accumulated amount (282)
compound interest (283)
nominal rate (stated rate) (283)
conversion period (284)
effective rate of interest (288)
present value (290)
future value (290)
annuity (301)
annuity certain (301)
ordinary annuity (301)
simple annuity (301)
future value of an annuity (302)
present value of an annuity (304)
sinking fund (317)

CHAPTER 5 Concept Review Questions

Fill in the blanks.

1. a. Simple interest is computed on the _____ principal only. The formula for the accumulated amount using simple interest is $A =$ _____.

b. In calculations using compound interest, earned interest is periodically added to the principal and thereafter itself earns _____. The formula for the accumulated amount using compound interest is $A =$ _____. Solving this equation for P gives the present value formula using compound interest as $P =$ _____.

2. The effective rate of interest is the _____ interest rate that would produce the same accumulated amount in _____ year as the _____ rate compounded _____ times a year. The formula for calculating the effective rate is r_{eff} = _____.

3. A sequence of payments made at regular time intervals is called a/an _____; if the payments are made at the end of each payment period, then it is called a/an _____ _____; if the payment period coincides with the interest conversion period, then it is called a/an _____ _____.

4. The formula for the future value of an annuity is S = _____. The formula for the present value of an annuity is P = _____.

5. The periodic payment R on a loan of P dollars to be amortized over n periods with interest charged at the rate of i per period is R = _____.

6. A sinking fund is an account that is set up for a specific purpose at some _____ date. The periodic payment R required to accumulate a sum of S dollars over n periods with interest charged at the rate of i per period is R = _____.

7. An arithmetic progression is a sequence of numbers in which each term after the first is obtained by adding a/an _____ _____ to the preceding term. The nth term of an arithmetic progression is a_n = _____. The sum of the first n terms of an arithmetic progression is S_n = _____.

8. A geometric progression is a sequence of numbers in which each term after the first is obtained by multiplying the preceding term by a/an _____ _____. The nth term of a geometric progression is a_n = _____. If $r \neq 1$, the sum of the first n terms of a geometric progression is S_n = _____.

CHAPTER 5 Review Exercises

1. Find the accumulated amount after 4 years if \$5000 is invested at 5%/year compounded (a) annually, (b) semiannually, (c) quarterly, and (d) monthly.

2. Find the accumulated amount after 8 years if \$12,000 is invested at 3.5%/year compounded (a) annually, (b) semiannually, (c) quarterly, and (d) monthly.

3. Find the effective rate of interest corresponding to a nominal rate of 6%/year compounded (a) annually, (b) semiannually, (c) quarterly, and (d) monthly.

4. Find the effective rate of interest corresponding to a nominal rate of 5.5%/year compounded (a) annually, (b) semiannually, (c) quarterly, and (d) monthly.

5. Find the present value of \$41,413 due in 5 years at an interest rate of 4.5%/year compounded quarterly.

6. Find the present value of \$64,540 due in 6 years at an interest rate of 4%/year compounded monthly.

7. Find the amount (future value) of an ordinary annuity of \$150/quarter for 7 years at 5%/year compounded quarterly.

8. Find the future value of an ordinary annuity of \$120/month for 10 years at 4.5%/year compounded monthly.

9. Find the present value of an ordinary annuity of 36 payments of \$250 each made monthly and earning interest at 4.5%/year compounded monthly.

10. Find the present value of an ordinary annuity of 60 payments of \$5000 each made quarterly and earning interest at 3.5%/year compounded quarterly.

11. Find the payment R needed to amortize a loan of \$22,000 at 3.5%/year compounded monthly with 36 monthly installments over a period of 3 years.

12. Find the payment R needed to amortize a loan of \$10,000 at 4.6%/year compounded monthly with 36 monthly installments over a period of 3 years.

13. Find the payment R needed to accumulate \$18,000 with 48 monthly installments over a period of 4 years at an interest rate of 3%/year compounded monthly.

14. Find the payment R needed to accumulate \$15,000 with 60 monthly installments over a period of 5 years at an interest rate of 3.6%/year compounded monthly.

15. Find the effective rate of interest corresponding to a nominal rate of 3.6%/year compounded monthly.

16. Find the effective rate of interest corresponding to a nominal rate of 4.8%/year compounded monthly.

17. Find the present value of \$119,346 due in 4 years at an interest rate of 5%/year compounded continuously.

18. **Company Sales** JCN Media had sales of \$1,750,000 in the first year of operation. If the sales increased by 7%/year thereafter, find the company's sales in the fourth year and the total sales over the first 4 years of operation.

19. **CDs** The manager of a money market fund has invested \$4.2 million in certificates of deposit that pay interest at the rate of 5.4%/year compounded quarterly over a period of 5 years. How much will the investment be worth at the end of 5 years?

20. **Savings Accounts** Emily deposited \$2000 into a bank account 5 years ago. The bank paid interest at the rate of 3.2%/year compounded weekly. What is Emily's account worth today?

21. **Savings Accounts** Kim invested a sum of money 4 years ago in a savings account that has since paid interest at the rate of 3.5%/year compounded monthly. Her investment is now worth \$19,440.31. How much did she originally invest?

22. **Savings Accounts** Andrew withdrew \$5470.87 from a savings account, which he closed this morning. The account had earned interest at the rate of 3%/year compounded continuously during the 3-year period that the money was on deposit. How much did Andrew originally deposit into the account?

23. **Mutual Funds** Juan invested \$24,000 in a mutual fund 5 years ago. Today his investment is worth \$34,616. Find the effective annual rate of return on his investment over the 5-year period.

24. **College Savings Program** The Blakes have decided to start a monthly savings program to provide for their son's college education. How much should they deposit at the end of each month in a savings account earning interest at the rate of 3.5%/year compounded monthly so that, at the end of the tenth year, the accumulated amount will be \$40,000?

25. **Retirement Accounts** Mai Lee has contributed \$200 at the end of each month into her company's employee retirement account for the past 10 years. Her employer has matched her contribution each month. If the account has earned interest at the rate of 5%/year compounded monthly over the 10-year period, determine how much Mai Lee now has in her retirement account.

26. **Automobile Leasing** Maria has leased an auto for 4 years at \$300/month. If money is worth 5%/year compounded monthly, what is the equivalent cash payment (present value) of this annuity? (Assume that the payments are made at the end of each month.)

27. **Installment Financing** Peggy made a down payment of \$400 toward the purchase of new furniture. To pay the balance of the purchase price, she has secured a loan from her bank at 6%/year compounded monthly. Under the terms of her finance agreement, she is required to make payments of \$75.32 at the end of each month for 24 months. What was the purchase price of the furniture?

28. **Home Financing** The Turners have purchased a house for \$150,000. They made an initial down payment of \$30,000 and secured a mortgage with interest charged at the rate of 4.5%/year on the unpaid balance. (Interest computations are made at the end of each month.) Assume that the loan is amortized over 30 years.
    a. What monthly payment will the Turners be required to make?
    b. What will be their total interest payment?
    c. What will be their equity (disregard depreciation) after 10 years?

29. **Home Financing** Refer to Exercise 28. If the loan is amortized over 15 years:
    a. What monthly payment will the Turners be required to make?
    b. What will be their total interest payment?
    c. What will be their equity (disregard depreciation) after 10 years?

30. **Sinking Funds** The management of a corporation anticipates a capital expenditure of \$500,000 in 5 years for the purpose of purchasing replacement machinery. To finance this purchase, a sinking fund that earns interest at the rate of 5%/year compounded quarterly will be set up. Determine the amount of each (equal) quarterly installment that should be deposited in the fund. (Assume that the payments are made at the end of each quarter.)

31. **Sinking Funds** The management of a condominium association anticipates a capital expenditure of \$120,000 in 2 years for the purpose of painting the exterior of the condominium. To pay for this maintenance, a sinking fund will be set up that will earn interest at the rate of 5.8%/year compounded monthly. Determine the amount of each (equal) monthly installment the association will be required to deposit into the fund at the end of each month for the next 2 years.

32. **Credit Card Payments** The outstanding balance on Bill's credit card account is \$3200. The bank issuing the credit card is charging 9.3%/year compounded monthly. If Bill decides to pay off this balance in equal monthly installments at the end of each month for the next 18 months, how much will be his monthly payment? What is the effective rate of interest the bank is charging Bill?

33. **Financial Planning** Matt's parents have agreed to contribute \$250/month toward the rent for his apartment in his junior year in college. The plan is for Matt's parents to deposit a lump sum in Matt's bank account on August 1 and then have Matt withdraw \$250 on the first of each month starting on September 1 and ending on May 1 the following year. If the bank pays interest on the balance at the rate of 5%/year compounded monthly, how much should Matt's parents deposit into his account?

CHAPTER 5 Before Moving On . . .

1. Find the accumulated amount at the end of 3 years if \$2000 is deposited in an account paying interest at the rate of 8%/year compounded monthly.

2. Find the effective rate of interest corresponding to a nominal rate of 6%/year compounded daily.

3. Find the future value of an ordinary annuity of \$800/week for 10 years at 6%/year compounded weekly.

4. Find the monthly payment required to amortize a loan of \$100,000 over 10 years with interest charged at the rate of 8%/year compounded monthly.

5. Find the weekly payment required to accumulate a sum of \$15,000 over 6 years with interest earned at the rate of 10%/year compounded weekly.

6. **a.** Find the sum of the first ten terms of the arithmetic progression 3, 7, 11, 15, 19, ...

b. Find the sum of the first eight terms of the geometric progression $\frac{1}{2}$, 1, 2, 4, 8, ...

6 Sets and Counting

WE OFTEN DEAL with well-defined collections of objects called *sets*. In this chapter, we see how sets can be combined algebraically to yield other sets. We also look at some techniques for determining the number of elements in a set and for determining the number of ways in which the elements of a set can be arranged or combined. These techniques enable us to solve many practical problems, as you will see throughout the chapter.

In how many ways can the Futurists (a rock group) plan their concert tour to San Francisco, Los Angeles, San Diego, Denver, and Las Vegas if the three performances in California must be given consecutively? In Example 13, page 376, we will show how to determine the number of possible different itineraries.

6.1 Sets and Set Operations

Set Terminology and Notation

We often deal with collections of different kinds of objects. For example, in conducting a study of the distribution of the weights of newborn infants, we might consider the collection of all infants born in Massachusetts General Hospital during 2014. In a study of the fuel consumption of compact cars, we might be interested in the collection of hybrid cars manufactured by General Motors in the 2014 model year. Such collections are examples of sets. More specifically, a **set** is a well-defined collection of objects. Thus, a set is not just any collection of objects; a set must be well defined in the sense that if we are given an object, then we should be able to determine whether or not it belongs in the collection.

The objects of a set are called the **elements,** or *members,* **of a set** and are usually denoted by lowercase letters $a, b, c, \ldots$; the sets themselves are usually denoted by uppercase letters $A, B, C, \ldots$. The elements of a set can be displayed by listing all the elements between braces. For example, in **roster notation,** the set A consisting of the first three letters of the English alphabet is written

$$A = \{a, b, c\}$$

The set B of all letters of the alphabet can be written

$$B = \{a, b, c, \ldots, z\}$$

Another notation that is commonly used is **set-builder notation.** Here, a rule is given that describes the definite property or properties an object x must satisfy to qualify for membership in the set. In this notation, the set B is written as

$$B = \{x \mid x \text{ is a letter of the English alphabet}\}$$

and is read "B is the set of all elements x such that x is a letter of the English alphabet."

If a is an element of a set A, we write $a \in A$ and read "a belongs to A" or "a is an element of A." If the element a does not belong to the set A, however, then we write $a \notin A$ and read "a does not belong to A." For example, if $A = \{1, 2, 3, 4, 5\}$, then $3 \in A$ but $6 \notin A$.

Explore and Discuss

1. Let A denote the collection of all the days in August 2014 in which the average daily temperature at the San Francisco International Airport was approximately 75°F. Is A a set? Explain your answer.
2. Let B denote the collection of all the days in August 2014 in which the average daily temperature at the San Francisco International Airport was between 73.5°F and 81.2°F, inclusive. Is B a set? Explain your answer.

Set Equality

Two sets A and B are **equal,** written $A = B$, if and only if they have exactly the same elements.

EXAMPLE 1 Let A, B, and C be the sets

$$A = \{a, e, i, o, u\}$$
$$B = \{a, i, o, e, u\}$$
$$C = \{a, e, i, o\}$$

Then $A = B$, since they both contain exactly the same elements. Note that the order in which the elements are displayed is immaterial. Also, $A \neq C$, since $u \in A$ but $u \notin C$. Similarly, we conclude that $B \neq C$.

Subset

If every element of a set A is also an element of a set B, then we say that A is a **subset** of B and write $A \subseteq B$.

By this definition, two sets A and B are equal if and only if (1) $A \subseteq B$, and (2) $B \subseteq A$. You can verify this (see Exercise 70).

EXAMPLE 2 Referring to Example 1, we find that $C \subseteq B$, since every element of C is also an element of B. Also, if D is the set

$$D = \{a, e, i, o, x\}$$

then D is not a subset of A, written $D \not\subseteq A$, since $x \in D$ but $x \notin A$. Observe that $A \not\subseteq D$ as well, since $u \in A$ but $u \notin D$.

If A and B are sets such that $A \subseteq B$ but $A \neq B$, then we say that A is a **proper subset** of B. In other words, a set A is a proper subset of a set B, written $A \subset B$, if (1) $A \subseteq B$ and (2) there exists at least one element in B that is not in A. The second condition states that the set A is properly "smaller" than the set B.

EXAMPLE 3 Let $A = \{1, 2, 3, 4, 5, 6\}$ and $B = \{2, 4, 6\}$. Then B is a proper subset of A because (1) $B \subseteq A$, which is easily verified, and (2) there exists at least one element in A that is not in B—for example, the element 1.

When we refer to sets and subsets, we use the symbols $\subset$, $\subseteq$, $\supset$, and $\supseteq$ to express the idea of "containment." However, when we wish to show that an element is contained in a set, we use the symbol $\in$ to express the idea of "membership." Thus, in Example 3, we would write $1 \in A$ and *not* $\{1\} \in A$.

Empty Set

The set that contains no elements is called the **empty set** and is denoted by $\varnothing$.

The empty set, $\varnothing$, is a subset of every set. To see this, observe that $\varnothing$ has no elements and therefore contains no element that is not also in any set A.

Do not confuse the empty set $\varnothing = \{\ \}$ with the set $\{0\}$, which is a set with one element—the number zero.

EXAMPLE 4 List all subsets of the set $A = \{a, b, c\}$.

Solution There is one subset consisting of no elements, namely, the empty set $\varnothing$. Next, observe that there are three subsets consisting of one element,

$$\{a\}, \{b\}, \{c\}$$

three subsets consisting of two elements,

$$\{a, b\}, \{a, c\}, \{b, c\}$$

and one subset consisting of three elements, the set A itself. Therefore, the subsets of A are

$$\varnothing, \{a\}, \{b\}, \{c\}, \{a, b\}, \{a, c\}, \{b, c\}, \{a, b, c\}$$

In contrast with the empty set, we have, at the other extreme, the notion of a largest, or universal, set. A **universal set** is the set of all elements of interest in a particular discussion. It is the largest in the sense that all sets considered in the discussion of the problem are subsets of the universal set. Of course, different universal sets are associated with different problems, as shown in Example 5.

EXAMPLE 5

a. If the problem is to determine the ratio of female to male students in a college, then a logical choice of a universal set is the set consisting of the whole student body of the college.

b. If the problem is to determine the ratio of female to male students in the business department of the college in part (a), then the set of all students in the business department can be chosen as the universal set.

We can use **Venn diagrams** to obtain a visual representation of sets. Venn diagrams are of considerable help in understanding the concepts introduced earlier as well as in solving problems involving sets. The universal set U is represented by a rectangle, and subsets of U are represented by regions lying inside the rectangle.

EXAMPLE 6 Use Venn diagrams to illustrate the following statements:

a. The sets A and B are equal.

b. The set A is a proper subset of the set B.

c. The sets A and B are not subsets of each other.

Solution The respective Venn diagrams are shown in Figure 1a–c.

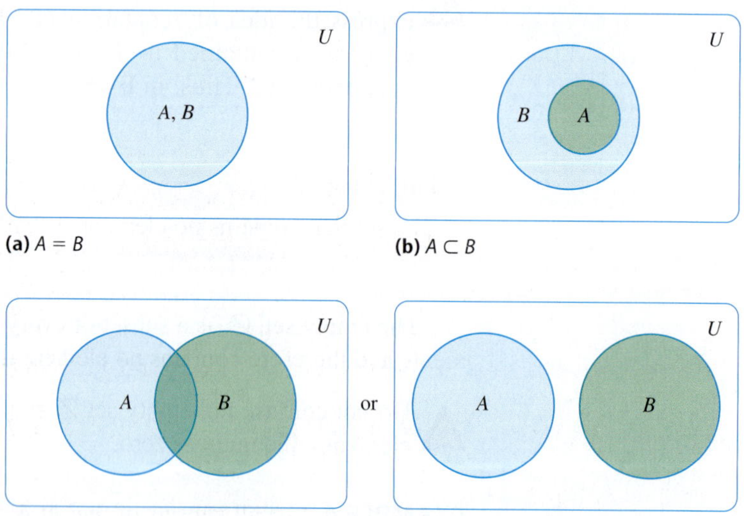

(a) $A = B$ (b) $A \subset B$

(c) $A \not\subseteq B$ and $B \not\subseteq A$

FIGURE 1

Set Operations

Now that we have introduced the concept of a set, our next task is to consider operations on sets—that is, to consider ways in which sets can be combined to yield other sets. These operations enable us to combine sets in much the same way that the operations of addition and multiplication enable us to combine numbers to obtain other numbers. In what follows, all sets are assumed to be subsets of a given universal set U.

Set Union

Let A and B be sets. The **union** of A and B, written $A \cup B$, is the set of all elements that belong to either A or B or both.

$$A \cup B = \{x \mid x \in A \text{ or } x \in B \text{ or both}\}$$

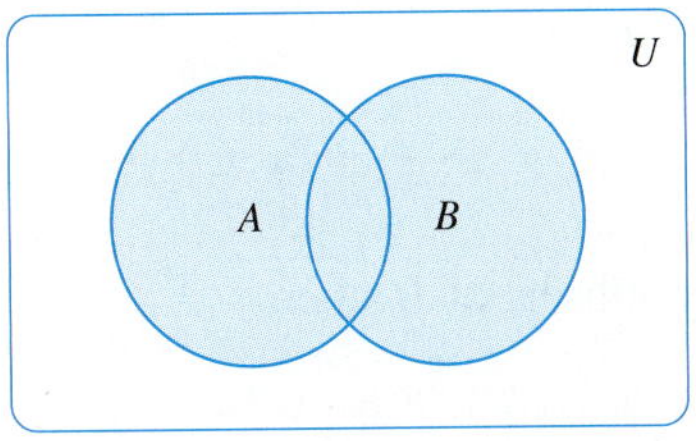

FIGURE 2
Set union $A \cup B$

The shaded portion of the Venn diagram (Figure 2) depicts the set $A \cup B$.

EXAMPLE 7 If $A = \{a, b, c\}$ and $B = \{a, c, d\}$, then $A \cup B = \{a, b, c, d\}$.

Set Intersection

Let A and B be sets. The set of elements common to the sets A and B, written $A \cap B$, is called the **intersection** of A and B.

$$A \cap B = \{x \mid x \in A \text{ and } x \in B\}$$

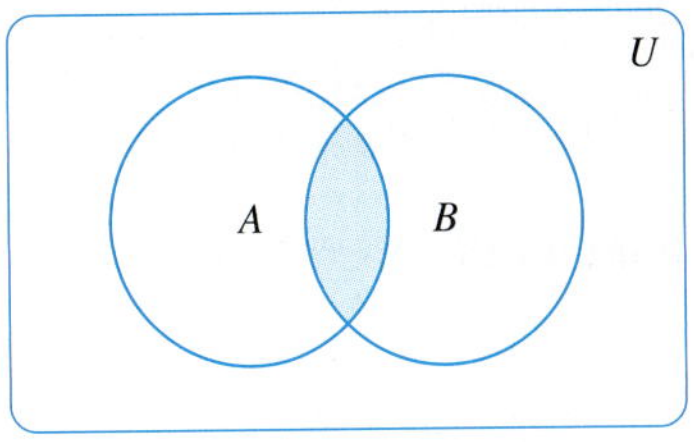

FIGURE 3
Set intersection $A \cap B$

The shaded portion of the Venn diagram (Figure 3) depicts the set $A \cap B$.

EXAMPLE 8 Let $A = \{a, b, c\}$, and let $B = \{a, c, d\}$. Then $A \cap B = \{a, c\}$. (Compare this result with Example 7.)

EXAMPLE 9 Let $A = \{1, 3, 5, 7, 9\}$, and let $B = \{2, 4, 6, 8, 10\}$. Then $A \cap B = \varnothing$.

The two sets of Example 9 have empty, or null, intersection. In general, the sets A and B are said to be **disjoint** if they have no elements in common—that is, if $A \cap B = \varnothing$ (see Figure 4).

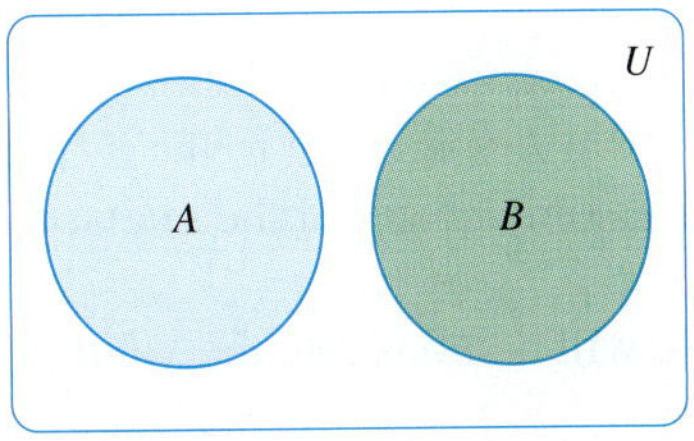

FIGURE 4
A and B are disjoint

EXAMPLE 10 Let U be the set of all students in the classroom. If $M = \{x \in U \mid x \text{ is male}\}$ and $F = \{x \in U \mid x \text{ is female}\}$, then $F \cap M = \varnothing$, so F and M are disjoint.

Complement of a Set

If U is a universal set and A is a subset of U, then the set of all elements in U that are not in A is called the **complement** of A and is denoted A^c.

$$A^c = \{x \mid x \in U \text{ and } x \notin A\}$$

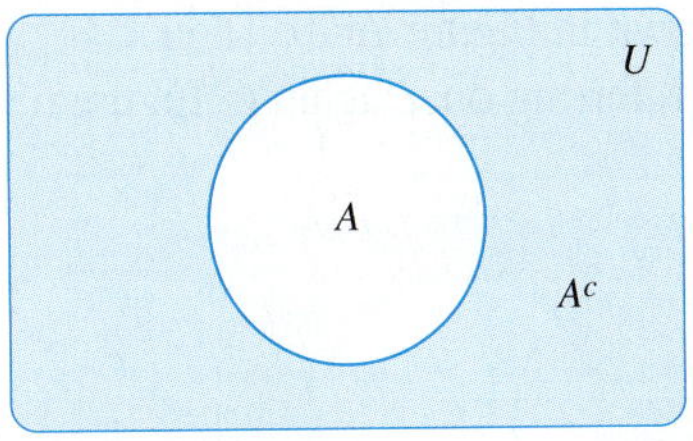

FIGURE 5
Set complementation

The shaded portion of the Venn diagram (Figure 5) shows the set A^c.

EXAMPLE 11 Let $U = \{1, 2, 3, 4, 5, 6, 7, 8, 9, 10\}$, and let $A = \{2, 4, 6, 8, 10\}$. Then $A^c = \{1, 3, 5, 7, 9\}$.

Explore and Discuss

Let A, B, and C be nonempty subsets of a set U.

1. Suppose $A \cap B \neq \varnothing$, $A \cap C \neq \varnothing$, and $B \cap C \neq \varnothing$. Can you conclude that $A \cap B \cap C \neq \varnothing$? Explain your answer with an example.
2. Suppose $A \cap B \cap C \neq \varnothing$. Can you conclude that $A \cap B \neq \varnothing$, $A \cap C \neq \varnothing$, and $B \cap C \neq \varnothing$? Explain your answer.

The following rules hold for the operation of **complementation.** See whether you can verify them.

Set Complementation

If U is a universal set and A is a subset of U, then

a. $U^c = \varnothing$ **b.** $\varnothing^c = U$ **c.** $(A^c)^c = A$
d. $A \cup A^c = U$ **e.** $A \cap A^c = \varnothing$

The operations on sets satisfy the following properties.

Properties of Set Operations

Let U be a universal set. If A, B, and C are arbitrary subsets of U, then

$A \cup B = B \cup A$	Commutative law for union
$A \cap B = B \cap A$	Commutative law for intersection
$A \cup (B \cup C) = (A \cup B) \cup C$	Associative law for union
$A \cap (B \cap C) = (A \cap B) \cap C$	Associative law for intersection
$A \cup (B \cap C) = (A \cup B) \cap (A \cup C)$	Distributive law for union
$A \cap (B \cup C) = (A \cap B) \cup (A \cap C)$	Distributive law for intersection

Two additional properties, called De Morgan's Laws, hold for the operations on sets.

De Morgan's Laws

Let A and B be sets. Then

$$(A \cup B)^c = A^c \cap B^c \quad \textbf{(1)}$$

$$(A \cap B)^c = A^c \cup B^c \quad \textbf{(2)}$$

Equation (1) states that the complement of the union of two sets is equal to the intersection of their complements. Equation (2) states that the complement of the intersection of two sets is equal to the union of their complements.

We will not prove De Morgan's Laws here, but we will demonstrate the validity of Equation (2) in the following example.

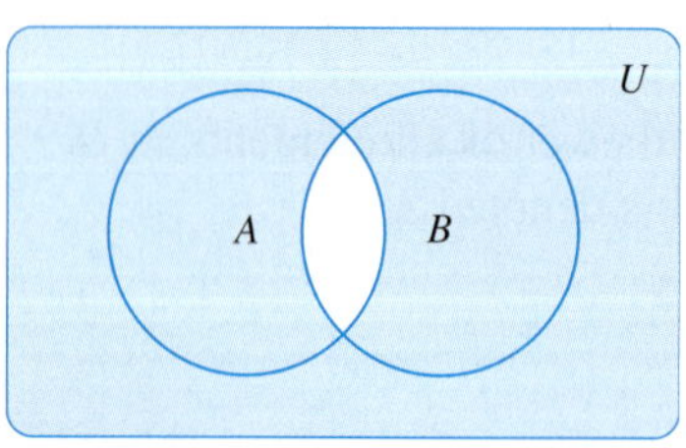

FIGURE 6
$(A \cap B)^c$

EXAMPLE 12 Using Venn diagrams, show that $(A \cap B)^c = A^c \cup B^c$.

Solution $(A \cap B)^c$ is the set of elements in U but not in $A \cap B$ and is therefore the shaded region shown in Figure 6. Next, A^c and B^c are shown in Figure 7a–b. Their union, $A^c \cup B^c$, is easily seen to be equal to $(A \cap B)^c$ by referring once again to Figure 6.

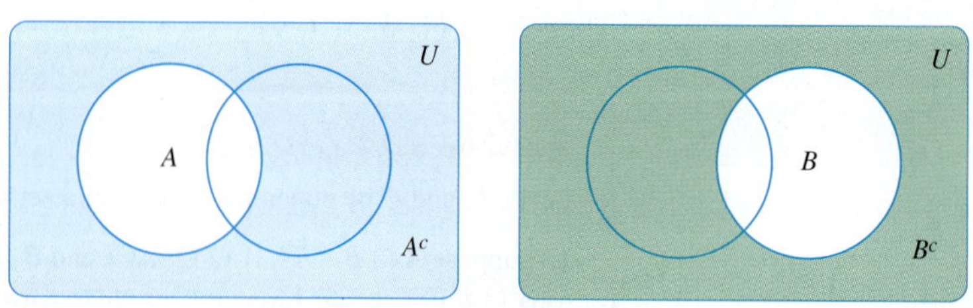

(a) **(b)**

FIGURE 7
$A^c \cup B^c$ is the set obtained by joining (a) and (b).

APPLIED EXAMPLE 13 Cyber Privacy In a poll surveying 1500 registered voters in California, the respondents were asked to rank the following companies on a scale of 0 to 10 in terms of how much they could trust these companies to keep their personal information secure, with zero meaning that they don't trust the company.

Company	Apple	Google	LinkedIn	YouTube	Facebook	Twitter
Rating	4.6	3.8	3.0	2.8	2.7	2.4

Let A denote the set of companies that have a rating higher than 2.5, let B denote the set of companies that have a rating between 2.5 and 4, and let C denote the set of companies that have a rating lower than 3. Find the following sets:

a. A, B, and C **b.** $A \cup B$ **c.** $B \cap C$ **d.** $A^c \cap B$ **e.** $A \cap B^c$

Source: Los Angeles Times.

Solution

a. $A = \{\text{Apple, Google, LinkedIn, YouTube, Facebook}\}$
$B = \{\text{Google, LinkedIn, YouTube, Facebook}\}$
$C = \{\text{YouTube, Facebook, Twitter}\}$

b. $A \cup B = \{\text{Apple, Google, LinkedIn, YouTube, Facebook}\} = A$

c. $B \cap C = \{\text{YouTube, Facebook}\}$

d. $A^c \cap B = \{\text{Twitter}\} \cap \{\text{Google, LinkedIn, YouTube, Facebook}\} = \varnothing$

e. $A \cap B^c = \{\text{Apple, Google, LinkedIn, YouTube, Facebook}\} \cap \{\text{Apple, Twitter}\}$
$= \{\text{Apple}\}$

EXAMPLE 14 Let $U = \{1, 2, 3, 4, 5, 6, 7, 8, 9, 10\}$, $A = \{1, 2, 4, 8, 9\}$, and $B = \{3, 4, 5, 6, 8\}$. Verify by direct computation that $(A \cup B)^c = A^c \cap B^c$.

Solution $A \cup B = \{1, 2, 3, 4, 5, 6, 8, 9\}$, so $(A \cup B)^c = \{7, 10\}$. Moreover, $A^c = \{3, 5, 6, 7, 10\}$ and $B^c = \{1, 2, 7, 9, 10\}$, so $A^c \cap B^c = \{7, 10\}$. The required result follows.

APPLIED EXAMPLE 15 Automobile Options Let U denote the set of all cars in a dealer's lot, and let

$$A = \{x \in U \mid x \text{ is equipped with satellite radio}\}$$
$$B = \{x \in U \mid x \text{ is equipped with a moonroof}\}$$
$$C = \{x \in U \mid x \text{ is equipped with keyless entry}\}$$

Find an expression in terms of A, B, and C for each of the following sets:

a. The set of cars with at least one of the given options
b. The set of cars with exactly one of the given options
c. The set of cars with satellite radio and keyless entry but no moonroof.

Solution

a. The set of cars with at least one of the given options is $A \cup B \cup C$ (Figure 8a).
b. The set of cars with satellite radio only is given by $A \cap B^c \cap C^c$. Similarly, we find that the set of cars with a moonroof only is given by $B \cap C^c \cap A^c$, while the set of cars with keyless entry only is given by $C \cap A^c \cap B^c$. Thus, the set of cars

with exactly one of the given options is $(A \cap B^c \cap C^c) \cup (B \cap C^c \cap A^c) \cup (C \cap A^c \cap B^c)$ (Figure 8b).

c. The set of cars with satellite radio and keyless entry but no moonroof is given by $A \cap C \cap B^c$ (Figure 8c).

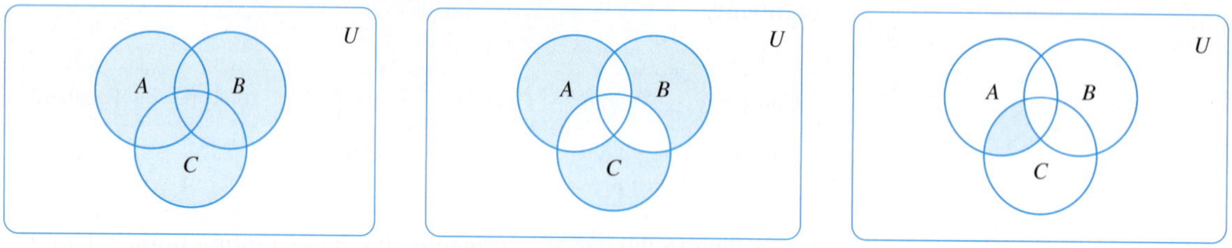

(a) The set of cars with at least one option

(b) The set of cars with exactly one option

(c) The set of cars with satellite radio and keyless entry but no moonroof

FIGURE **8**

6.1 Self-Check Exercises

1. Let $U = \{1, 2, 3, 4, 5, 6, 7\}$, $A = \{1, 2, 3\}$, $B = \{3, 4, 5, 6\}$, and $C = \{2, 3, 4\}$. Find the following sets:

a. A^c **b.** $A \cup B$ **c.** $B \cap C$

d. $(A \cup B) \cap C$ **e.** $(A \cap B) \cup C$ **f.** $A^c \cap (B \cup C)^c$

2. Politics Let U denote the set of all members of the House of Representatives. Let

$$D = \{x \in U \mid x \text{ is a Democrat}\}$$

$$R = \{x \in U \mid x \text{ is a Republican}\}$$

$$F = \{x \in U \mid x \text{ is a female}\}$$

$$L = \{x \in U \mid x \text{ is a lawyer by training}\}$$

Describe each of the following sets in words.

a. $D \cap F$ **b.** $F^c \cap R$ **c.** $D \cap F \cap L^c$

Solutions to Self-Check Exercises 6.1 can be found on page 352.

6.1 Concept Questions

1. a. What is a set? Give an example.

b. When are two sets equal? Give an example of two equal sets.

c. What is the empty set?

2. What can you say about two sets A and B such that

a. $A \cup B \subseteq A$ **b.** $A \cup B = \varnothing$

c. $A \cap B = B$ **d.** $A \cap B = \varnothing$

3. a. If $A \subset B$, what can you say about the relationship between A^c and B^c?

b. If $A^c = \varnothing$, what can you say about A?

6.1 Exercises

In Exercises 1–4, write the set in set-builder notation.

1. The set of gold medalists in the 2014 Winter Olympic Games

2. The set of football teams in the NFL

3. $\{3, 4, 5, 6, 7\}$

4. $\{1, 3, 5, 7, 9, 11, \ldots, 39\}$

In Exercises 5–8, list the elements of the set in roster notation.

5. $\{x \mid x \text{ is a digit in the number } 352{,}646\}$

6. $\{x \mid x$ is a letter in the word *HIPPOPOTAMUS*$\}$

7. $\{x \mid 2 - x = 4 \text{ and } x \text{ is an integer}\}$

8. $\{x \mid 2 - x = 4 \text{ and } x \text{ is a fraction}\}$

In Exercises 9–14, state whether the statements are true or false.

9. **a.** $\{a, b, c\} = \{c, a, b\}$ **b.** $A \in A$

10. **a.** $\varnothing \in A$ **b.** $A \subset A$

11. **a.** $0 \in \varnothing$ **b.** $0 = \varnothing$

12. **a.** $\{\varnothing\} = \varnothing$ **b.** $\{a, b\} \in \{a, b, c\}$

13. {Chevrolet, Cadillac, Buick} $\subset$ $\{x \mid x$ is a division of General Motors$\}$

14. $\{x \mid x$ is a silver medalist in the 2014 Winter Olympic Games$\} = \varnothing$

In Exercises 15 and 16, let $A = \{1, 2, 3, 4, 5\}$. Determine whether the statements are true or false.

15. **a.** $2 \in A$ **b.** $A \subseteq \{2, 4, 6\}$

16. **a.** $0 \in A$ **b.** $\{1, 3, 5\} \in A$

17. Let $A = \{1, 2, 3\}$. Which of the following sets are equal to A?
a. $\{2, 1, 3\}$ **b.** $\{3, 2, 1\}$
c. $\{0, 1, 2, 3\}$

18. Let $A = \{a, e, l, t, r\}$. Which of the following sets are equal to A?
a. $\{x \mid x$ is a letter of the word *later*$\}$
b. $\{x \mid x$ is a letter of the word *latter*$\}$
c. $\{x \mid x$ is a letter of the word *relate*$\}$

19. List all subsets of the following sets:
a. $\{1, 2\}$ **b.** $\{1, 2, 3\}$ **c.** $\{1, 2, 3, 4\}$

20. List all subsets of the set A = {IBM, U.S. Steel, Union Carbide, Boeing}. Which of these are proper subsets of A?

In Exercises 21–24, find the smallest possible set (i.e., the set with the least number of elements) that contains the given sets as subsets.

21. $\{1, 2\}, \{1, 3, 4\}, \{4, 6, 8, 10\}$

22. $\{1, 2, 4\}, \{a, b\}$

23. {Jill, John, Jack}, {Susan, Sharon}

24. {GM, Ford, Chrysler}, {Daimler-Benz, Volkswagen}, {Toyota, Nissan}

25. Use Venn diagrams to represent the following relationships:
a. $A \subset B$ and $B \subset C$
b. $A \subset U$ and $B \subset U$, where A and B have no elements in common
c. The sets A, B, and C are equal.

26. Let U denote the set of all students who applied for admission to the freshman class at Faber College for the upcoming academic year, and let

$A = \{x \in U \mid x$ is a successful applicant$\}$

$B = \{x \in U \mid x$ is a female student who enrolled in the freshman class$\}$

$C = \{x \in U \mid x$ is a male student who enrolled in the freshman class$\}$

a. Use Venn diagrams to represent the sets U, A, B, and C.
b. Determine whether the following statements are true or false.
i. $A \subseteq B$ **ii.** $B \subset A$ **iii.** $C \subset B$

In Exercises 27 and 28, write an expression describing the shaded portion(s) of the Venn diagram.

27. **a.**

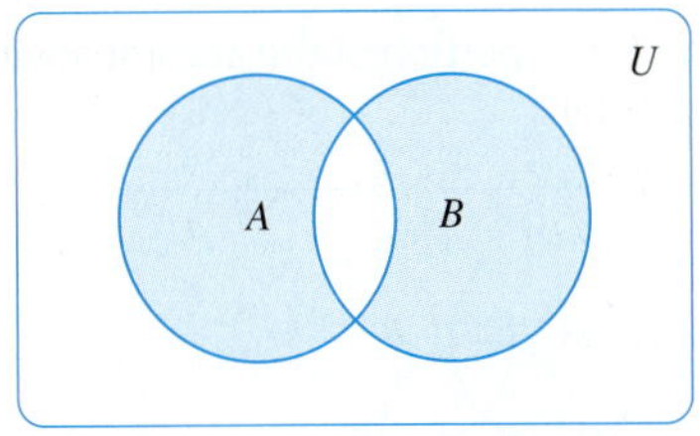

b.

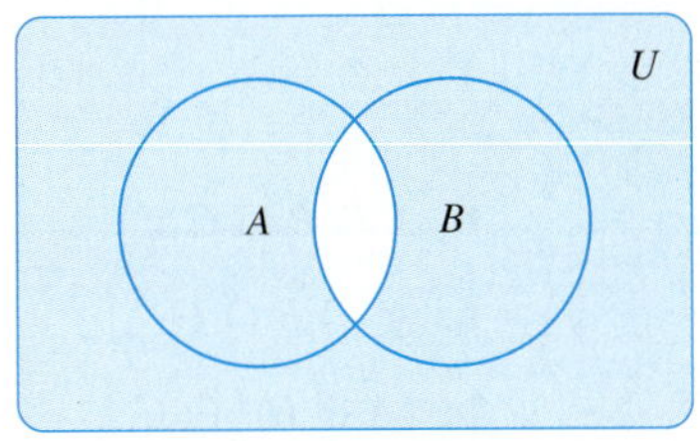

28. **a.**

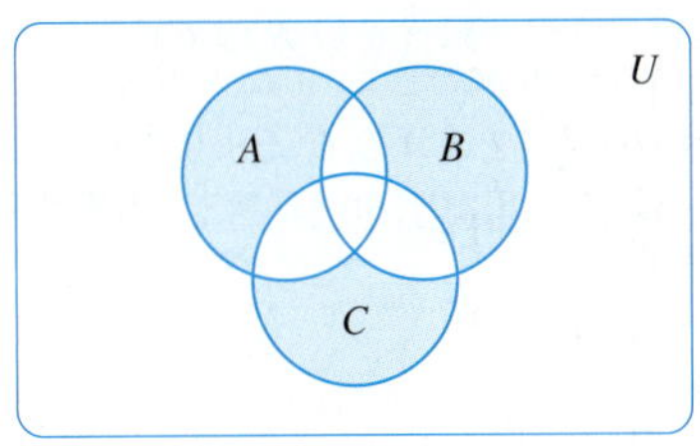

b.

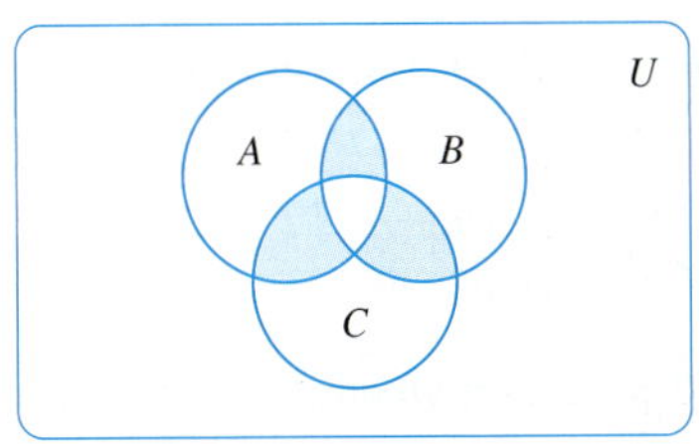

c.

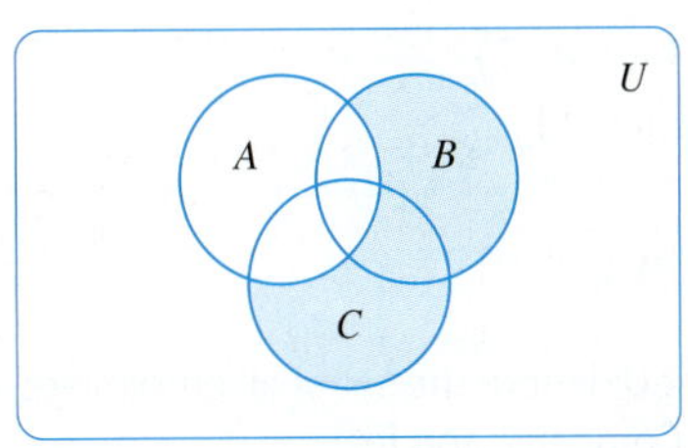

d.

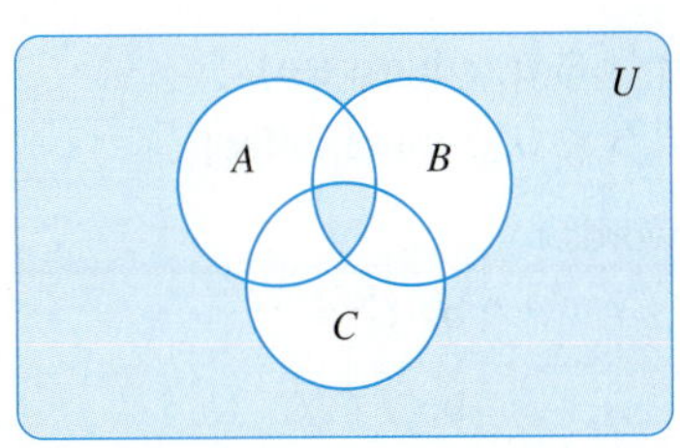

In Exercises 29 and 30, shade the portion of the accompanying figure that represents each set.

29. **a.** $A \cap B^c$
b. $A^c \cap B$

30. **a.** $A^c \cap B^c$
b. $(A \cup B)^c$

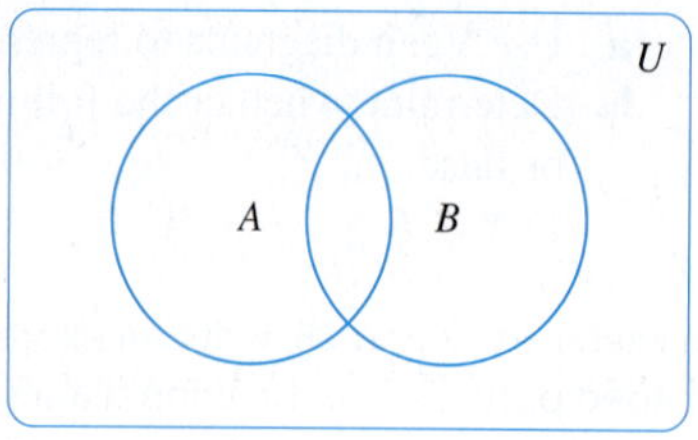

In Exercises 31–34, shade the portion of the accompanying figure that represents each set.

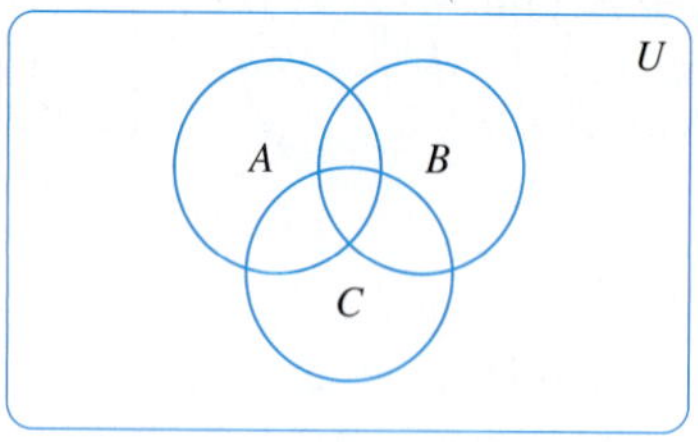

31. **a.** $A \cup B \cup C$ **b.** $A \cap B \cap C$

32. **a.** $A \cap B \cap C^c$ **b.** $A^c \cap B \cap C$

33. **a.** $A^c \cap B^c \cap C^c$ **b.** $(A \cup B)^c \cap C$

34. **a.** $A \cup (B \cap C)^c$ **b.** $(A \cup B \cup C)^c$

In Exercises 35–38, let $U = \{1, 2, 3, 4, 5, 6, 7, 8, 9, 10\}$, $A = \{1, 3, 5, 7, 9\}$, $B = \{2, 4, 6, 8, 10\}$, and $C = \{1, 2, 4, 5, 8, 9\}$. List the elements of each set.

35. **a.** A^c **b.** $B \cup C$ **c.** $C \cup C^c$

36. **a.** $C \cap C^c$ **b.** $(A \cap C)^c$ **c.** $A \cup (B \cap C)$

37. **a.** $(A \cap B) \cup C$ **b.** $(A \cup B \cup C)^c$
c. $(A \cap B \cap C)^c$

38. **a.** $A^c \cap (B \cap C^c)$ **b.** $(A \cup B^c) \cup (B \cap C^c)$
c. $(A \cup B)^c \cap C^c$

In Exercises 39 and 40, determine whether the pairs of sets are disjoint.

39. **a.** $\{1, 2, 3, 4\}, \{4, 5, 6, 7\}$
b. $\{a, c, e, g\}, \{b, d, f\}$

40. **a.** $\varnothing, \{1, 3, 5\}$
b. $\{0, 1, 3, 4\}, \{0, 2, 5, 7\}$

In Exercises 41–44, let U denote the set of all employees at Universal Life Insurance Company, and let

$$T = \{x \in U \mid x \text{ drinks tea}\}$$
$$C = \{x \in U \mid x \text{ drinks coffee}\}$$

Describe each set in words.

41. **a.** T^c **b.** C^c

42. **a.** $T \cup C$ **b.** $T \cap C$

43. **a.** $T \cap C^c$ **b.** $T^c \cap C$

44. **a.** $T^c \cap C^c$ **b.** $(T \cup C)^c$

In Exercises 45–48, let U denote the set of all employees in a hospital, and let

$$N = \{x \in U \mid x \text{ is a nurse}\}$$
$$D = \{x \in U \mid x \text{ is a doctor}\}$$
$$A = \{x \in U \mid x \text{ is an administrator}\}$$
$$M = \{x \in U \mid x \text{ is a male}\}$$
$$F = \{x \in U \mid x \text{ is a female}\}$$

Describe each set in words.

45. **a.** D^c **b.** N^c

46. **a.** $N \cup D$ **b.** $N \cap M$

47. **a.** $D \cap M^c$ **b.** $D \cap A$

48. **a.** $N \cap F$ **b.** $(D \cup N)^c$

In Exercises 49 and 50, let U denote the set of all senators in Congress, and let

$$D = \{x \in U \mid x \text{ is a Democrat}\}$$
$$R = \{x \in U \mid x \text{ is a Republican}\}$$
$$F = \{x \in U \mid x \text{ is a female}\}$$
$$L = \{x \in U \mid x \text{ is a lawyer}\}$$

Use set notation to represent each statement.

49. **a.** The set of all Democrats who are female
b. The set of all Republicans who are male and are not lawyers

50. **a.** The set of all Democrats who are female or are lawyers
b. The set of all senators who are not Democrats or are lawyers

In Exercises 51 and 52, let U denote the set of all students in the business college of a certain university. Let

$$A = \{x \in U \mid x \text{ had taken a course in accounting}\}$$
$$B = \{x \in U \mid x \text{ had taken a course in economics}\}$$
$$C = \{x \in U \mid x \text{ had taken a course in marketing}\}$$

Use set notation to represent each statement.

51. **a.** The set of students who have not had a course in economics
b. The set of students who have had courses in accounting and economics
c. The set of students who have had courses in accounting and economics but not marketing

52. **a.** The set of students who have had courses in economics but not courses in accounting or marketing
b. The set of students who have had at least one of the three courses
c. The set of students who have had all three courses

53. Best U.S. City for Italian Restaurants In a reader survey conducted by *USA Today*, readers were asked to name the best U.S. city for Italian restaurants. The results follow:

City	New York	Chicago	Boston	Las Vegas	San Francisco
Respondents (%)	55	16	15	7	7

Let A denote the set of cities that were voted the best by more than 10% of the respondents, let B be the set of cities that were voted the best by between 10% and 20% of the respondents, and let C denote the set of cities that were voted the best by fewer than 10% of the respondents. Find the following sets:

a. A, B, and C **b.** $A \cup B$ **c.** $A \cap B$
d. $A^c \cap B$ **e.** $A \cap B^c$ **f.** $(A \cup B)^c$

Source: travel.usatoday.com.

54. Inventory Loss The biggest cause of inventory loss, called *shrinkage*, is shoplifting, followed closely by employee theft. In a study conducted by the Center for Retail Research, the nine countries with the highest shrinkage rates, measured in the dollar amount lost for every \$100 in sales, are as follows:

Country	India	Russia	Morocco	South Africa	Brazil	Mexico	Thailand	Turkey
Shrinkage Rate (\$)	2.38	1.74	1.72	1.71	1.69	1.64	1.64	1.63

Let A denote the set of countries that have a shrinkage rate greater than \$1.65, let B be the set of countries that have a shrinkage rate between \$1.65 and \$1.73, and let C be the set of countries that have a shrinkage rate less than \$1.70. Find the following sets:

a. A, B, and C **b.** $A \cap B$ **c.** $A^c \cap B$
d. $A \cap B^c$ **e.** $A^c \cup B^c$

Source: Center for Retail Research Graphics.

In Exercises 55 and 56, refer to the following diagram, where U is the set of all tourists surveyed over a 1-week period in London and where

$$A = \{x \in U \mid x \text{ has taken the underground [subway]}\}$$
$$B = \{x \in U \mid x \text{ has taken a cab}\}$$
$$C = \{x \in U \mid x \text{ has taken a bus}\}$$

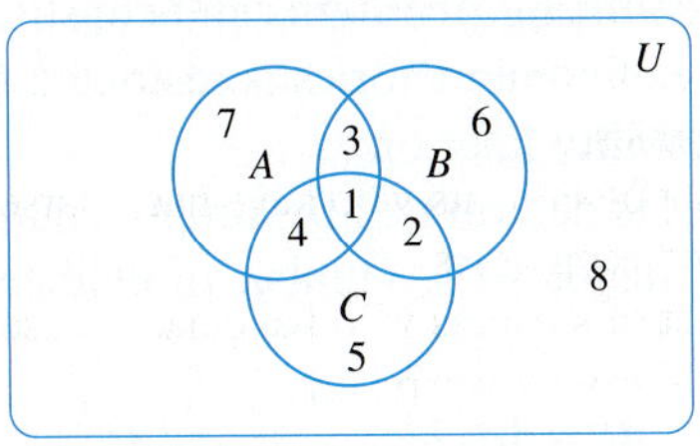

Express the indicated regions in set notation and in words.

55. a. Region 1
b. Regions 1 and 4 together
c. Regions 4, 5, 7, and 8 together

56. a. Region 3
b. Regions 4 and 6 together
c. Regions 5, 6, and 7 together

In Exercises 57–62, use Venn diagrams to illustrate each statement.

57. $A \subseteq A \cup B$; $B \subseteq A \cup B$ **58.** $A \cap B \subseteq A$; $A \cap B \subseteq B$

59. $A \cup (B \cup C) = (A \cup B) \cup C$

60. $A \cap (B \cap C) = (A \cap B) \cap C$

61. $A \cap (B \cup C) = (A \cap B) \cup (A \cap C)$

62. $(A \cup B)^c = A^c \cap B^c$

In Exercises 63 and 64, let

$$U = \{1, 2, 3, 4, 5, 6, 7, 8, 9, 10\}$$
$$A = \{1, 3, 5, 7, 9\}$$
$$B = \{1, 2, 4, 7, 8\}$$
$$C = \{2, 4, 6, 8\}$$

Verify each equation by direct computation.

63. a. $A \cup (B \cup C) = (A \cup B) \cup C$
b. $A \cap (B \cap C) = (A \cap B) \cap C$

64. a. $A \cap (B \cup C) = (A \cap B) \cup (A \cap C)$
b. $(A \cup B)^c = A^c \cap B^c$

In Exercises 65–68, refer to the accompanying figure, and list the points that belong to each set.

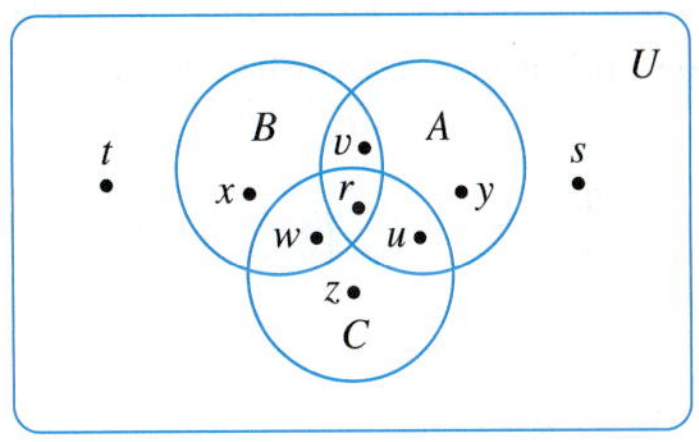

65. a. $A \cup B$ **b.** $A \cap B$

66. a. $A \cap (B \cup C)$ **b.** $(B \cap C)^c$

67. a. $(B \cup C)^c$ **b.** A^c

68. a. $(A \cap B) \cap C^c$ **b.** $(A \cup B \cup C)^c$

69. Suppose $A \subset B$ and $B \subset C$, where A and B are any two sets. What conclusion can be drawn regarding the sets A and C?

70. Verify the assertion that two sets A and B are equal if and only if (1) $A \subseteq B$ and (2) $B \subseteq A$.

In Exercises 71–80, determine whether the statement is true or false. If it is true, explain why it is true. If it is false, give an example to show why it is false.

71. A set is never a subset of itself.

72. A proper subset of a set is itself a subset of the set but not necessarily vice versa.

73. If $A \cup B = \varnothing$, then $A = \varnothing$ and $B = \varnothing$.

74. If $A \cap B = \varnothing$, then either $A = \varnothing$ or $B = \varnothing$.

75. $(A \cup A^c)^c = \varnothing$

76. If $A \subseteq B$, then $A \cap B = A$.

77. If $A \subseteq B$, then $A \cup B = B$.

78. If $A \cup B = A$, then $A \subseteq B$.

79. If $A \subset B$, then $A^c \supset B^c$.

80. $A \cap \varnothing = \varnothing$

6.1 Solutions to Self-Check Exercises

1. a. A^c is the set of all elements in U but not in A. Therefore,

$$A^c = \{4, 5, 6, 7\}$$

b. $A \cup B$ consists of all elements in A and/or B. Hence,

$$A \cup B = \{1, 2, 3, 4, 5, 6\}$$

c. $B \cap C$ is the set of all elements in both B and C. Therefore,

$$B \cap C = \{3, 4\}$$

d. Using the result from part (b), we find

$$\begin{aligned}(A \cup B) \cap C &= \{1, 2, 3, 4, 5, 6\} \cap \{2, 3, 4\} \\ &= \{2, 3, 4\}\end{aligned}$$

e. First, we compute

$$A \cap B = \{3\}$$

Next, since $(A \cap B) \cup C$ is the set of all elements in $(A \cap B)$ and/or C, we conclude that

$$\begin{aligned}(A \cap B) \cup C &= \{3\} \cup \{2, 3, 4\} \\ &= \{2, 3, 4\}\end{aligned}$$

f. From part (a), we have $A^c = \{4, 5, 6, 7\}$. Next, we compute

$$\begin{aligned}B \cup C &= \{3, 4, 5, 6\} \cup \{2, 3, 4\} \\ &= \{2, 3, 4, 5, 6\}\end{aligned}$$

from which we deduce that

$$(B \cup C)^c = \{1, 7\}$$ The set of elements in U but not in $B \cup C$

Finally, using these results, we obtain

$$A^c \cap (B \cup C)^c = \{4, 5, 6, 7\} \cap \{1, 7\} = \{7\}$$

2. a. $D \cap F$ denotes the set of all elements in both D and F. Since an element in D is a Democrat and an element in F is a female representative, we see that $D \cap F$ is the set of all female Democrats in the House of Representatives.

b. Since F^c is the set of male representatives and R is the set of Republicans, it follows that $F^c \cap R$ is the set of male Republicans in the House of Representatives.

c. L^c is the set of representatives who are not lawyers by training. Therefore, $D \cap F \cap L^c$ is the set of female Democratic representatives who are not lawyers by training.

6.2 The Number of Elements in a Finite Set

Counting the Elements in a Set

The solution to some problems in mathematics calls for finding the number of elements in a set. Such problems are called **counting problems** and constitute a field of study known as **combinatorics.** Our study of combinatorics is restricted to the results that will be required for our work in probability later on.

The number of elements in a finite set is determined by simply counting the elements in the set. If A is a set, then $n(A)$ denotes the number of elements in A. For example, if

$$A = \{1, 2, 3, \ldots, 20\} \qquad B = \{a, b\} \qquad C = \{8\}$$

then $n(A) = 20$, $n(B) = 2$, and $n(C) = 1$.

The empty set has no elements in it, so $n(\varnothing) = 0$. Another result that is easily seen to be true is the following: If A and B are disjoint sets, then

$$n(A \cup B) = n(A) + n(B) \tag{3}$$

EXAMPLE 1 If $A = \{a, c, d\}$ and $B = \{b, e, f, g\}$, then $n(A) = 3$ and $n(B) = 4$, so $n(A) + n(B) = 7$. Moreover, $A \cup B = \{a, b, c, d, e, f, g\}$ and $n(A \cup B) = 7$. Thus, Equation (3) holds true in this case. Note that $A \cap B = \varnothing$.

In the general case, A and B need not be disjoint, which leads us to the following rule.

Addition Rule for Sets

If A and B are finite sets, then

$$n(A \cup B) = n(A) + n(B) - n(A \cap B) \tag{4}$$

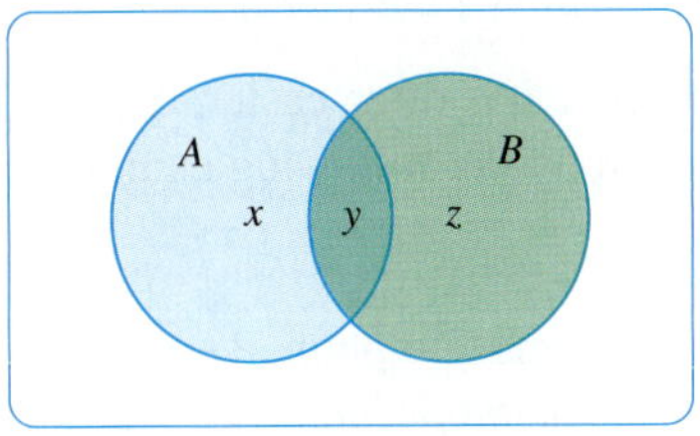

FIGURE 9
$n(A \cup B) = x + y + z$

To see this, we observe that the set $A \cup B$ may be viewed as the union of three mutually disjoint sets with x, y, and z elements, respectively (Figure 9). This figure shows that

$$n(A \cup B) = x + y + z$$

Also,

$$n(A) = x + y \quad \text{and} \quad n(B) = y + z$$

so

$$\begin{aligned} n(A) + n(B) &= (x + y) + (y + z) \\ &= (x + y + z) + y \\ &= n(A \cup B) + n(A \cap B) \qquad n(A \cap B) = y \end{aligned}$$

Solving for $n(A \cup B)$, we obtain

$$n(A \cup B) = n(A) + n(B) - n(A \cap B)$$

which is the desired result.

EXAMPLE 2 Let $A = \{a, b, c, d, e\}$, and let $B = \{b, d, f, h\}$. Verify Equation (4) directly.

Solution

$$A \cup B = \{a, b, c, d, e, f, h\} \quad \text{so} \quad n(A \cup B) = 7$$
$$A \cap B = \{b, d\} \quad \text{so} \quad n(A \cap B) = 2$$

Furthermore,

$$n(A) = 5 \quad \text{and} \quad n(B) = 4$$

so

$$n(A) + n(B) - n(A \cap B) = 5 + 4 - 2 = 7 = n(A \cup B)$$

APPLIED EXAMPLE 3 Consumer Beverage Survey A survey of 100 coffee drinkers found that 70 take sugar, 60 take cream, and 50 take both sugar and cream with their coffee. How many coffee drinkers take sugar or cream with their coffee?

Solution Let U denote the set of 100 coffee drinkers surveyed, and let

$$A = \{x \in U \mid x \text{ takes sugar}\}$$
$$B = \{x \in U \mid x \text{ takes cream}\}$$

Then $n(A) = 70$, $n(B) = 60$, and $n(A \cap B) = 50$. The set of coffee drinkers who take sugar or cream with their coffee is given by $A \cup B$. Using Equation (4), we find

$$\begin{aligned} n(A \cup B) &= n(A) + n(B) - n(A \cap B) \\ &= 70 + 60 - 50 = 80 \end{aligned}$$

Thus, 80 out of the 100 coffee drinkers surveyed take cream or sugar with their coffee.

Explore and Discuss

Prove Equation (5), using an argument similar to that used to prove Equation (4). Another proof is outlined in Exercise 53 on page 361.

An equation similar to Equation (4) can be derived for the case that involves any finite number of finite sets. For example, a relationship involving the number of elements in the sets A, B, and C is given by

$$\begin{aligned} n(A \cup B \cup C) = n(A) + n(B) + n(C) - n(A \cap B) \\ - n(A \cap C) - n(B \cap C) + n(A \cap B \cap C) \end{aligned} \quad (5)$$

As useful as equations such as Equation (5) are, in practice it is often easier to attack a problem directly with the aid of Venn diagrams, as shown by the following example.

APPLIED EXAMPLE 4 Marketing Surveys A leading mobile phone service provider advertises its services in three magazines: *Cosmopolitan*, *People*, and *Time*. A survey of 500 customers by the mobile phone service provider reveals the following information:

180 learned of its services from *Cosmopolitan*.

200 learned of its services from *People*.

192 learned of its services from *Time*.

84 learned of its services from *Cosmopolitan* and *People*.

52 learned of its services from *Cosmopolitan* and *Time*.

64 learned of its services from *People* and *Time*.

38 learned of its services from all three magazines.

How many of the customers saw the manufacturer's advertisement in

a. At least one magazine?
b. Exactly one magazine?

Solution Let U denote the set of all customers surveyed, and let

$$C = \{x \in U \mid x \text{ learned of the services from } Cosmopolitan\}$$
$$P = \{x \in U \mid x \text{ learned of the services from } People\}$$
$$T = \{x \in U \mid x \text{ learned of the services from } Time\}$$

We begin by constructing a Venn diagram for the problem. It is best to start by filling in the number of elements in the region that is the intersection of all three sets (if any), then working with the region(s) that are the intersection of two sets, and so on. In this case, the information that 38 customers learned of the services from all three magazines translates into $n(C \cap P \cap T) = 38$ (Figure 10a). Next, the result that 64 learned of the services from *People* and *Time* translates into $n(P \cap T) = 64$. This leaves

$$64 - 38 = 26$$

who learned of the services only from *People* and *Time* (Figure 10b). Similarly, $n(C \cap T) = 52$, so

$$52 - 38 = 14$$

learned of the services only from *Cosmopolitan* and *Time*, and $n(C \cap P) = 84$, so

$$84 - 38 = 46$$

learned of the services only from *Cosmopolitan* and *People*. These numbers appear in the appropriate regions in Figure 10b.

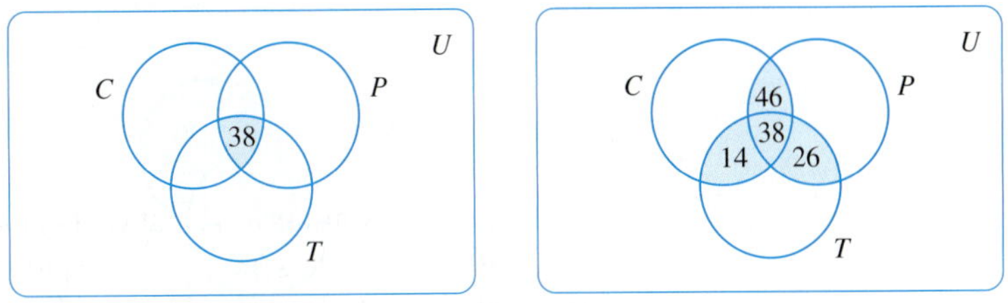

(a) All three magazines **(b)** Two or more magazines

FIGURE **10**

Continuing, we have $n(T) = 192$, so the number who learned of the services only from *Time* is given by

$$192 - 14 - 38 - 26 = 114$$

(Figure 11). Similarly, $n(P) = 200$, so

$$200 - 46 - 38 - 26 = 90$$

learned of the services from only *People*, and $n(C) = 180$, so

$$180 - 14 - 38 - 46 = 82$$

learned of the services from only *Cosmopolitan*. Finally,

$$500 - (90 + 26 + 114 + 14 + 82 + 46 + 38) = 90$$

learned of the services from other sources.

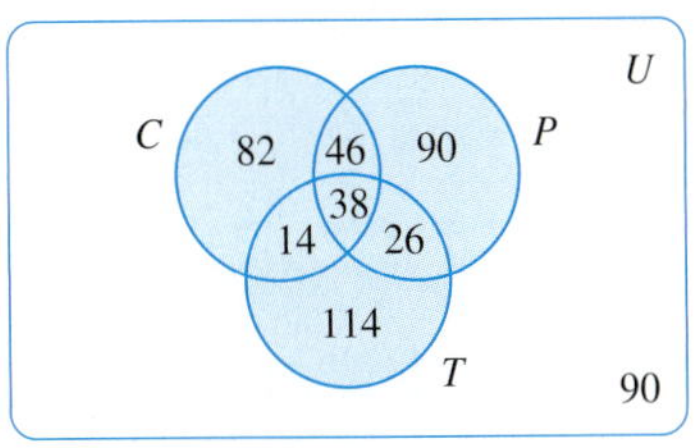

FIGURE **11**
The completed Venn diagram

We are now in a position to answer questions (a) and (b).

a. Referring to Figure 11, we see that the number of customers who learned of the services from at least one magazine is given by

$$n(C \cup P \cup T) = 500 - 90 = 410$$

b. The number of customers who learned of the products from exactly one magazine (Figure 12) is given by

$$n(C \cap P^c \cap T^c) + n(C^c \cap P \cap T^c) + n(C^c \cap P^c \cap T) = 82 + 90 + 114 = 286$$

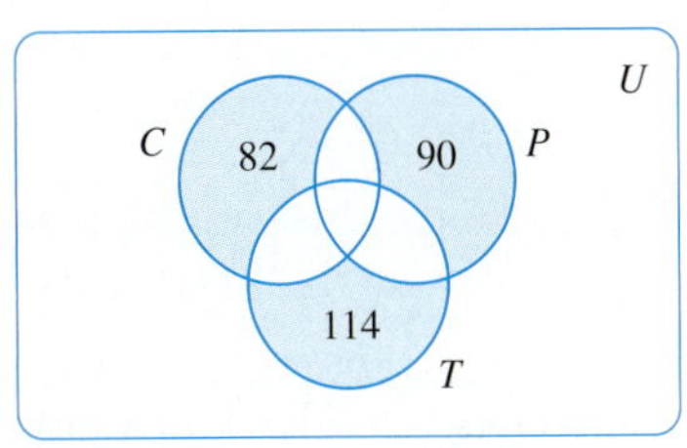

FIGURE **12**
Exactly one magazine

Our last example illustrates how we can sometimes use the solution of a system of linear equations to help us draw a Venn diagram.

EXAMPLE 5 Let A, B, and C be sets in a universal set U, and suppose $n(A \cap B \cap C^c) = 10$, $n(A \cap B^c \cap C) = 5$, $n(A^c \cap B \cap C^c) = 30$, $n(A^c \cap B^c \cap C) = 20$, $n(A^c \cap B^c \cap C^c) = 80$, $n(A) = 22$, $n(B) = 46$, and $n(U) = 156$. Use this information to complete a Venn diagram.

Solution The first five conditions lead to the Venn diagram shown in Figure 13a. Let's denote the number of elements in the three subsets that are yet to be determined by x, y, and z, as shown in Figure 13b.

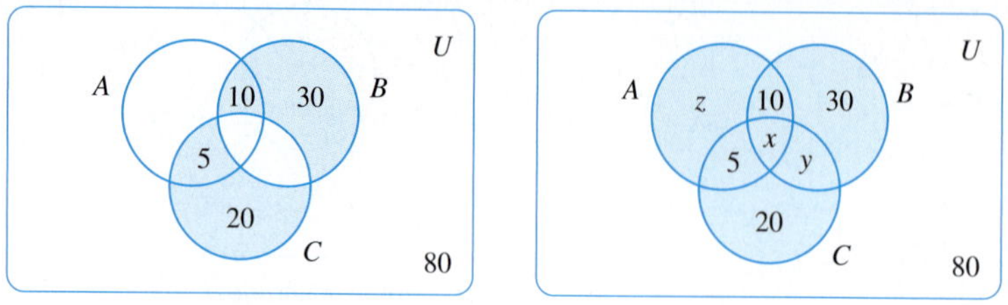

(a) The Venn diagram reflecting the first five conditions

(b) The numbers x, y, and z are to be determined.

FIGURE **13**

Then using the last three of the eight given conditions, we have

$$n(A) = 22 \text{ implies that } x + 5 + z + 10 = 22, \text{ or } x + z = 7$$
$$n(B) = 46 \text{ implies that } x + 10 + 30 + y = 46, \text{ or } x + y = 6$$
$$n(U) = 156 \text{ implies that } x + 5 + z + 10 + y + 30 + 20 + 80 = 156, \text{ or } x + y + z = 11$$

This leads to the system of equations

$$\begin{aligned} x \quad\quad + z &= 7 \\ x + y \quad\quad &= 6 \\ x + y + z &= 11 \end{aligned}$$

Subtracting the second equation from the third gives $z = 5$. Substituting this value of z into the first equation gives $x = 2$. Finally, substituting this value of x into the second equation tells us that $y = 4$. So $x = 2$, $y = 4$, and $z = 5$. This gives the completed Venn diagram shown in Figure 14.

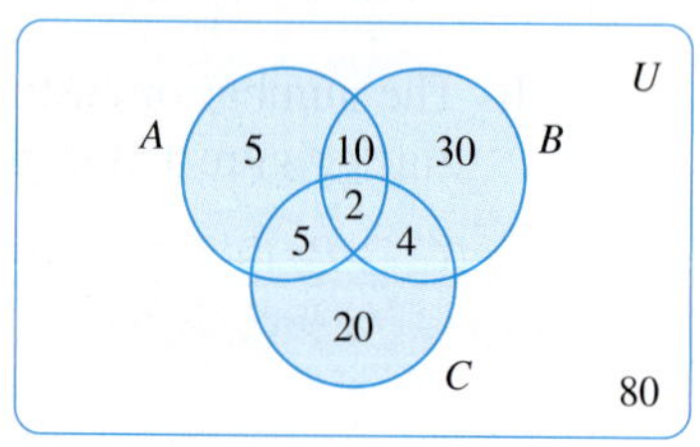

FIGURE **14**
The completed Venn diagram

6.2 Self-Check Exercises

1. Let A and B be subsets of a universal set U, and suppose that $n(U) = 100$, $n(A) = 60$, $n(B) = 40$, and $n(A \cap B) = 20$. Compute:
   a. $n(A \cup B)$ **b.** $n(A \cap B^c)$ **c.** $n(A^c \cap B)$

2. **Readership Survey** A survey of 1000 readers of *Video Magazine* found that 166 own at least one HD player in the HD-DVD format, 161 own at least one HD player in the Blu-ray format, and 22 own HD players in both formats. How many of the readers surveyed own HD players in the HD-DVD format only? How many of the readers surveyed do not own an HD player in either format?

Solutions to Self-Check Exercises 6.2 can be found on page 361.

6.2 Concept Questions

1. **a.** If A and B are sets with $A \cap B = \varnothing$, what can you say about $n(A) + n(B)$? Explain.
   b. If A and B are sets satisfying $n(A \cup B) \neq n(A) + n(B)$, what can you say about $A \cap B$? Explain.

2. Let A and B be subsets of U, the universal set, and suppose that $A \cap B = \varnothing$. Is it true that $n(A) - n(B) = n(B^c) - n(A^c)$? Explain.

6.2 Exercises

In Exercises 1 and 2, verify the equation

$$n(A \cup B) = n(A) + n(B)$$

for the given disjoint sets.

1. $A = \{a, e, i, o, u\}$ and $B = \{g, h, k, l, m\}$

2. $A = \{x \mid x \text{ is a whole number between 0 and 4}\}$
   $B = \{x \mid x \text{ is a negative integer greater than } -4\}$

3. Let $A = \{2, 4, 6, 8\}$ and $B = \{6, 7, 8, 9, 10\}$. Compute:
   a. $n(A)$ **b.** $n(B)$
   c. $n(A \cup B)$ **d.** $n(A \cap B)$

4. Let $U = \{1, 2, 3, 4, 5, 6, 7, a, b, c, d, e\}$. If $A = \{1, 2, a, e\}$ and $B = \{1, 2, 3, 4, a, b, c\}$, find:
   a. $n(A^c)$ **b.** $n(A \cap B^c)$
   c. $n(A \cup B^c)$ **d.** $n(A^c \cap B^c)$

5. Verify directly that $n(A \cup B) = n(A) + n(B) - n(A \cap B)$ for the sets in Exercise 3.

6. Let $A = \{a, e, i, o, u\}$ and $B = \{b, d, e, o, u\}$. Verify by direct computation that $n(A \cup B) = n(A) + n(B) - n(A \cap B)$.

7. If $n(A) = 15$, $n(A \cap B) = 5$, and $n(A \cup B) = 30$, then what is $n(B)$?

8. If $n(A) = 10$, $n(A \cup B) = 15$, and $n(B) = 8$, then what is $n(A \cap B)$?

In Exercises 9 and 10, refer to the following Venn diagram.

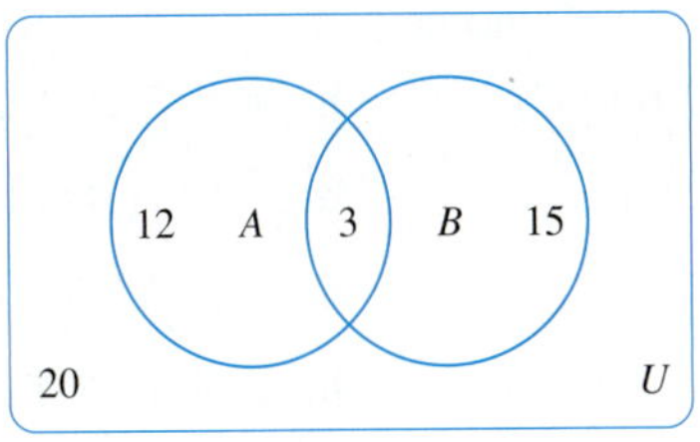

9. Find:
   a. $n(A)$ **b.** $n(A \cup B)$ **c.** $n(A^c \cap B)$
   d. $n(A \cap B^c)$ **e.** $n(U)$ **f.** $n[(A \cup B)^c]$

10. Find:
    a. $n(A \cap B)$ **b.** $n(A^c \cap B^c)$ **c.** $n[(A \cap B)^c]$
    d. $n(A^c \cup B^c)$ **e.** $n[(A \cap B^c) \cup (A^c \cap B)]$ **f.** $n(U^c)$

In Exercises 11 and 12, let A and B be subsets of a universal set U and suppose $n(U) = 200$, $n(A) = 100$, $n(B) = 80$, and $n(A \cap B) = 40$. Compute:

11. **a.** $n(A \cup B)$ **b.** $n(A^c)$ **c.** $n(A \cap B^c)$

12. **a.** $n(A^c \cap B)$ **b.** $n(B^c)$ **c.** $n(A^c \cap B^c)$

13. Find $n(A \cup B)$ given that $n(A) = 6$, $n(B) = 10$, and $n(A \cap B) = 3$.

14. If $n(B) = 6$, $n(A \cup B) = 14$, and $n(A \cap B) = 3$, find $n(A)$.

15. If $n(A) = 4$, $n(B) = 5$, and $n(A \cup B) = 9$, find $n(A \cap B)$.

16. If $n(A) = 16$, $n(B) = 16$, $n(C) = 14$, $n(A \cap B) = 6$, $n(A \cap C) = 5$, $n(B \cap C) = 6$, and $n(A \cup B \cup C) = 31$, find $n(A \cap B \cap C)$.

17. If $n(A) = 12$, $n(B) = 12$, $n(A \cap B) = 5$, $n(A \cap C) = 5$, $n(B \cap C) = 4$, $n(A \cap B \cap C) = 2$, and $n(A \cup B \cup C) = 25$, find $n(C)$.

18. **Newspaper Subscribers** A survey of 1000 subscribers to the *Los Angeles Times* revealed that 900 people subscribe to the daily morning edition and 500 subscribe to both the daily morning and the Sunday editions. How many subscribe to the Sunday edition? How many subscribe to the Sunday edition only?

19. **Jail Inmates** On a certain day, the Wilton County Jail held 190 prisoners accused of a crime (felony and/or misdemeanor). Of these, 130 were accused of felonies and 121 were accused of misdemeanors. How many prisoners were accused of both a felony and a misdemeanor?

20. Of 100 clock radios with digital tuners and/or CD players sold recently in a department store, 70 had digital tuners and 90 had CD players. How many radios had both digital tuners and CD players?

21. **Brand Preferences of Consumers** In a survey of 120 consumers conducted in a shopping mall, 80 consumers indicated that they buy Brand A of a certain product, 68 buy Brand B, and 42 buy both brands. How many consumers participating in the survey buy
    a. At least one of these brands?
    b. Exactly one of these brands?
    c. Only Brand A?
    d. Neither of these brands?

22. **Sports Club Survey** In a survey of 200 members of a local sports club, 100 members indicated that they plan to attend the next Summer Olympic Games, 60 indicated that they plan to attend the next Winter Olympic Games, and 40 indicated that they plan to attend both games. How many members of the club plan to attend

a. At least one of the two games?
b. Exactly one of the games?
c. The Summer Olympic Games only?
d. None of the games?

23. **Investors' Usage of Brokers** In a poll conducted among 200 active investors, it was found that 120 use discount brokers, 126 use full-service brokers, and 64 use both discount and full-service brokers. How many investors

a. Use at least one kind of broker?
b. Use exactly one kind of broker?
c. Use only discount brokers?
d. Don't use a broker?

24. **Commuter Trends** Of 50 employees of a store located in downtown Boston, 18 people take the subway to work, 12 take the bus, and 7 take both the subway and the bus. How many employees

a. Take the subway or the bus to work?
b. Take only the bus to work?
c. Take either the bus or the subway to work?
d. Get to work by some other means?

25. **Consumer Survey of Desktop and Tablet Computer Users** In a survey of 200 households regarding the ownership of desktop and tablet computers, the following information was obtained:

120 households own only desktop computers.

10 households own only tablet computers.

40 households own neither desktop nor tablet computers.

How many households own both desktop and tablet computers?

26. **Consumer Survey of HDTV and MP3 Player Owners** In a survey of 400 households regarding the ownership of HDTVs and MP3 players, the following data were obtained:

360 households own one or more HDTVs.

170 households own one or more HDTVs and one or more MP3 players.

19 households do not own a HDTV or a MP3 player.

How many households own only one or more MP3 players?

In Exercises 27 and 28, refer to the following Venn diagram.

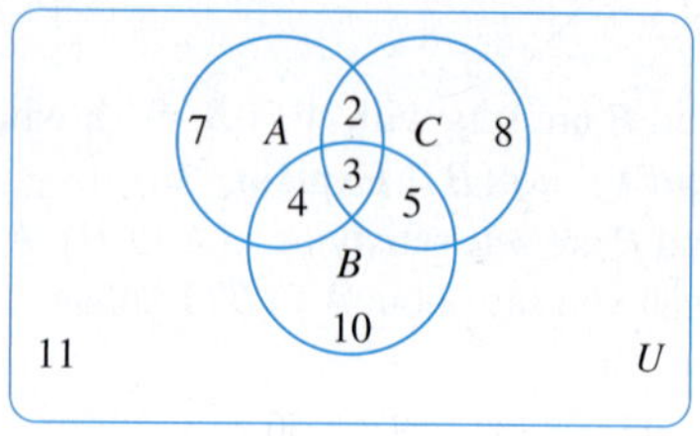

27. Find:

a. $n(A)$ **b.** $n(A \cup B)$ **c.** $n(A \cap B \cap C^c)$
d. $n[(A \cup B) \cap C^c)]$ **e.** $n[(A \cup B \cup C)^c]$

28. Find:

a. $n(A \cup B^c)$ **b.** $n[(A \cap (B \cup C)^c]$ **c.** $n(A^c)$
d. $n[(A \cap B \cap C)^c]$ **e.** $n(A^c \cup B^c \cup C^c)$

In Exercises 29–32, use the given information to draw a Venn diagram.

29. $n(A \cap B \cap C) = 3$, $n(A \cap B^c \cap C) = 8$, $n(A^c \cap B^c \cap C) = 5$, $n(A^c \cap B \cap C^c) = 10$, $n(A^c \cap B^c \cap C^c) = 60$, $n(A) = 16$, $n(B) = 17$, and $n(U) = 92$

30. $n(A \cap B \cap C^c) = 2$, $n(A \cap B^c \cap C^c) = 10$, $n(A \cap B^c \cap C) = 5$, $n(A^c \cap B \cap C) = 3$, $n(A^c \cap B^c \cap C^c) = 9$, $n(B) = 10$, $n(C) = 12$, and $n(A \cup B \cup C) = 27$

31. $n(A \cap B^c \cap C^c) = 4$, $n(A \cap B^c \cap C) = 3$, $n(A^c \cap B^c \cap C) = 14$, $n(A^c \cap B \cap C^c) = 12$, $n(A) = 15$, $n(B) = 22$, $n(C) = 24$, and $n(A^c \cap B^c \cap C^c) = 30$

32. $n(A \cap B \cap C) = 2$, $n(A \cap B \cap C^c) = 5$, $n(A^c \cap B^c \cap C) = 3$, $n(A^c \cap B \cap C) = 4$, $n(A \cup B) = 20$, $n(B \cup C) = 19$, $n(A \cup C) = 21$, and $n[(A \cup B \cup C)^c] = 20$

In Exercises 33–36, let A, B, and C be subsets of a universal set U and suppose $n(U) = 100$, $n(A) = 28$, $n(B) = 30$, $n(C) = 34$, $n(A \cap B) = 8$, $n(A \cap C) = 10$, $n(B \cap C) = 15$, and $n(A \cap B \cap C) = 5$. Compute:

33. **a.** $n(A \cup B \cup C)$ **b.** $n(A^c \cap B \cap C)$

34. **a.** $n[A \cap (B \cup C)]$ **b.** $n[A \cap (B \cup C)^c]$

35. **a.** $n(A^c \cap B^c \cap C^c)$ **b.** $n[A^c \cap (B \cup C)]$

36. **a.** $n[A \cup (B \cap C)]$ **b.** $n[(A^c \cap B^c \cap C^c)^c]$

37. **On-the-Job Distractions** In a survey of 500 advertising and marketing executives, the following question was posed: What is the most common cause of on-the-job distraction? The replies were as follows:

1: People stopping by your office

2: Phone calls

3: Email alerts

4: Text messages

5: Social media

6: Don't know

The number of respondents that gave each reply is given in the following table.

Reply	1	2	3	4	5	6
Number	135	130	95	50	40	50

Let A denote the set of replies that totaled more than 90, let B denote the set of replies that totaled more than 50 but less than 135, and let C denote the set of replies that totaled less than 130. Find:

a. $n(A), n(B), n(C)$ **b.** $n(A \cup B)$ **c.** $n(A \cap B^c)$
d. $n(B \cap C^c)$ **e.** $n(A \cap B^c \cap C)$

Source: The Creative Group.

38. Brand Switching Among Female College Students In a study of factors that influence brand switching by 18- to 24-year-old female college students, the following factors were identified as being significant as judged by the responses of the 439 participants in the survey:

1: Better price

2: Friend's recommendation

3: Seeing others use it

4: Interesting packaging

5: Buzz—people talking about the brand

6: An advertisement

7: Press stories

8: Entertainer/sports celebrity endorsement

The results of the survey follow:

Factor	1	2	3	4	5	6	7	8
Respondents (%)	68.8	60.4	22.8	20.8	16.8	16.8	14.9	6.9

Let A denote the set of factors that had a response greater than 20%, let B denote the set of factors that had a response between 15% and 25%, and let C denote the set of factors that had a response less than 25%. Find:

a. $n(A), n(B), n(C)$ **b.** $n(A \cap B)$ **c.** $n(A^c \cap C)$
d. $n(A \cap B^c)$ **e.** $n(A^c \cap C^c)$ **f.** $n[(A \cup B) \cap C]$

Source: Burst Media Research.

39. Brand Switching Among Male College Students In a study of factors that influence brand switching by 18- to 24-year-old male college students, the following factors were identified as being significant as judged by the responses of the 439 participants in the survey:

1: Better price

2: Friend's recommendation

3: Seeing others use it

4: Buzz—people talking about the brand

5: An advertisement

6: Interesting packaging

7: Press stories

8: Entertainer/sports celebrity endorsement

The results of the survey follow:

Factor	1	2	3	4	5	6	7	8
Respondents (%)	64.2	52.5	28.4	24.5	20.1	15.7	13.7	12.2

Let A denote the set of factors that had a response greater than 20%, let B denote the set of factors that had a response between 15% and 25%, and let C denote the set of factors that had a response less than 25%. Find:

a. $n(A), n(B), n(C)$ **b.** $n(A \cap B)$ **c.** $n(A^c \cap C)$
d. $n(A \cap B^c)$ **e.** $n(A^c \cap C^c)$ **f.** $n[(A \cup B) \cap C]$

Source: Burst Media Research.

40. Federal Budget Allocation According to the Office of Management and Budget, certain percentages of the $3.7 trillion U.S. 2012 budget were spent on the following items:

1: Social Security

2: National defense

3: Safety net programs

4: Medicare

5: Medicaid, CHIP, and other

6: Interest on the debt

7: Education

8: Transportation

9: Veterans' benefits and services

10: International affairs

11: Science and technology

12: Other

The percentages are as follows:

Item	1	2	3	4	5	6	7	8	9	10	11	12
Percent	21	18	15	13	9	6	5	4	3	2	1	3

Let A denote the set of items in which the expenditure was more than 10%, let B denote the set of items in which the expenditure was strictly between 5% and 15% (exclusive), and let C denote the set of all items in which the expenditure was less than 14%. Find:

a. $n(A), n(B), n(C)$ **b.** $n(A \cup C)$ **c.** $n(B \cap C)$
d. $n(A \cap B^c)$ **e.** $n(A^c \cap B^c)$

Source: Office of Management and Budget.

41. Survey of Leading Economists A survey of the opinions of ten leading economists in a certain country showed that, because oil prices were expected to drop in that country over the next 12 months,

Seven had lowered their estimate of the consumer inflation rate.

Eight had raised their estimate of the gross national product (GNP) growth rate.

Two had lowered their estimate of the consumer inflation rate but had not raised their estimate of the GNP growth rate.

How many economists had both lowered their estimate of the consumer inflation rate and raised their estimate of the GNP growth rate for that period?

42. Student Dropout Rate Data released by the Department of Education regarding the rate (percentage) of ninth-grade students who don't graduate in a certain year showed that, out of 50 states,

12 states had an increase in the dropout rate during the past 2 years.

15 states had a dropout rate of at least 30% during the past 2 years.

21 states had an increase in the dropout rate and/or a dropout rate of at least 30% during the past 2 years.

a. How many states had both a dropout rate of at least 30% and an increase in the dropout rate over the 2-year period?
b. How many states had a dropout rate that was less than 30% but that had increased over the 2-year period?

43. Student Magazine Preferences A survey of 100 college students who frequent the reading lounge of a university revealed the following results:

40 read *Time*.

30 read *The New Yorker*.

25 read *Vanity Fair*.

15 read *Time* and *The New Yorker*.

12 read *Time* and *Vanity Fair*.

10 read *The New Yorker* and *Vanity Fair*.

4 read all three magazines.

How many of the students surveyed read:
a. At least one of these magazines?
b. Exactly one of these magazines?
c. Exactly two of these magazines?
d. None of these magazines?

44. SAT Scores Results of a Department of Education survey of SAT test scores in 22 states showed that

10 states had an average composite SAT score of at least 1000 during the past 3 years.

15 states had an increase of at least 10 points in the average composite SAT score during the past 3 years.

8 states had both an average composite SAT score of at least 1000 and an increase in the average composite SAT score of at least 10 points during the past 3 years.

a. How many of the 22 states had composite SAT scores of less than 1000 and showed an increase of at least 10 points over the 3-year period?
b. How many of the 22 states had composite SAT scores of at least 1000 and did not show an increase of at least 10 points over the 3-year period?

45. Brand Preferences of Consumers The 120 consumers of Exercise 21 were also asked about their buying preferences concerning another product that is sold in the market under three labels. The results were as follows:

12 buy only those sold under Label A.

25 buy only those sold under Label B.

26 buy only those sold under Label C.

15 buy only those sold under Labels A and B.

10 buy only those sold under Labels A and C.

12 buy only those sold under Labels B and C.

8 buy the product sold under all three labels.

How many of the consumers surveyed buy the product sold under:
a. At least one of the three labels?
b. Labels A and B but not C?
c. Label A?
d. None of these labels?

46. Cafeteria Student Usage To help plan the number of meals (breakfast, lunch, and dinner) to be prepared in a college cafeteria, a survey was conducted and the following data were obtained:

130 students ate breakfast.

180 students ate lunch.

275 students ate dinner.

68 students ate breakfast and lunch.

112 students ate breakfast and dinner.

90 students ate lunch and dinner.

58 students ate all three meals.

How many of the students ate:
a. At least one meal in the cafeteria?
b. Exactly one meal in the cafeteria?
c. Only dinner in the cafeteria?
d. Exactly two meals in the cafeteria?

47. 401(k) Investments In a survey of 200 employees of a company regarding their 401(k) investments, the following data were obtained:

141 had investments in stock funds.

91 had investments in bond funds.

60 had investments in money market funds.

47 had investments in stock funds and bond funds.

36 had investments in stock funds and money market funds.

36 had investments in bond funds and money market funds.

5 had investments only in some other vehicle.

a. How many of the employees surveyed had investments in all three types of funds?
b. How many of the employees had investments in stock funds only?

48. Newspaper Preferences of Investors In a survey of 300 individual investors regarding subscriptions to the *New York Times* (*NYT*), *Wall Street Journal* (*WSJ*), and *USA Today* (*UST*), the following data were obtained:

122 subscribe to the *NYT*.

150 subscribe to the *WSJ*.

62 subscribe to the *UST*.

38 subscribe to the *NYT* and *WSJ*.

20 subscribe to the *NYT* and *UST*.

28 subscribe to the *WSJ* and *UST*.

36 do not subscribe to any of these newspapers.

a. How many of the individual investors surveyed subscribe to all three newspapers?
b. How many subscribe to only one of these newspapers?

In Exercises 49–52, determine whether the statement is true or false. If it is true, explain why it is true. If it is false, give an example to show why it is false.

49. If $A \cap B \neq \varnothing$, then $n(A \cup B) \neq n(A) + n(B)$.

50. If $A \subseteq B$, then $n(B) = n(A) + n(A^c \cap B)$.

51. If $n(A \cup B) = n(A) + n(B)$, then $A \cap B = \varnothing$.

52. If $n(A \cup B) = 0$ and $n(A \cap B) = 0$, then $A = \varnothing$.

53. Prove Equation (5).
Hint: Equation (4) can be written as $n(D \cup E) = n(D) + n(E) - n(D \cap E)$. Now, put $D = A \cup B$ and $E = C$. Use Equation (4) again if necessary.

54. Find conditions on the sets A, B, and C so that $n(A \cup B \cup C) = n(A) + n(B) + n(C)$.

6.2 Solutions to Self-Check Exercises

1. Use the given information to construct the following Venn diagram:

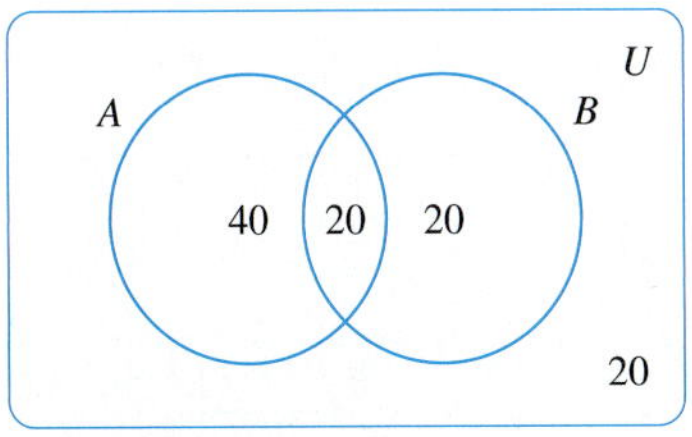

Using this diagram, we see that
a. $n(A \cup B) = 40 + 20 + 20 = 80$
b. $n(A \cap B^c) = 40$
c. $n(A^c \cap B) = 20$

2. Let U denote the set of all readers surveyed, and let

$A = \{x \in U \mid x$ owns at least one HD player in the HD-DVD format$\}$

$B = \{x \in U \mid x$ owns at least one HD player in the Blu-ray format$\}$

The fact that 22 of the readers own HD players in both formats means that $n(A \cap B) = 22$. Also, $n(A) = 166$ and $n(B) = 161$. Using this information, we obtain the following Venn diagram:

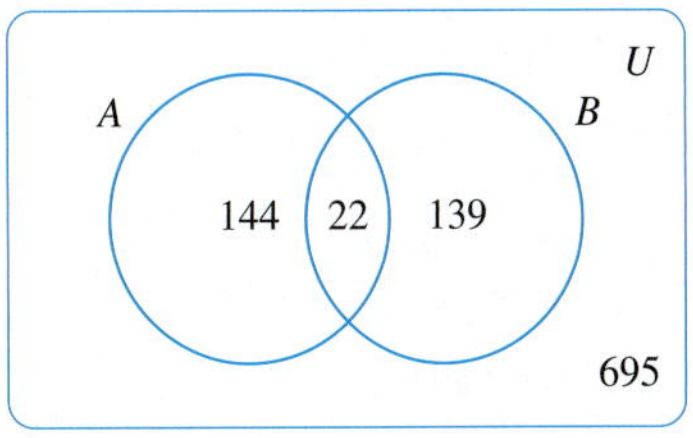

From the Venn diagram, we see that the number of readers who own HD players in only the Blu-ray format is given by

$$n(A \cap B^c) = 144$$

The number of readers who do not own an HD player in either format is given by

$$n(A^c \cap B^c) = 695$$

6.3 The Multiplication Principle

The Fundamental Principle of Counting

The solution of certain problems requires more sophisticated counting techniques than those developed in the previous section. We look at some such techniques in this and the following section. We begin by stating a fundamental principle of counting called the **multiplication principle.**

> **The Multiplication Principle**
>
> Suppose there are m ways of performing a task T_1 and n ways of performing a task T_2. Then there are mn ways of performing the task T_1 followed by the task T_2.

EXAMPLE 1 Three trunk roads connect Town A and Town B, and two trunk roads connect Town B and Town C.

a. Use the multiplication principle to find the number of ways in which a journey from Town A to Town C via Town B can be completed.
b. Verify part (a) directly by exhibiting all possible routes.

Solution

a. Since there are three ways of performing the first task (going from Town A to Town B) followed by two ways of performing the second task (going from Town B to Town C), the multiplication principle says that there are $3 \cdot 2$, or 6, ways to complete a journey from Town A to Town C via Town B.
b. Label the trunk roads connecting Town A and Town B with the Roman numerals I, II, and III, and label the trunk roads connecting Town B and Town C with the lowercase letters a and b. A schematic of this is shown in Figure 15. Then the routes from Town A to Town C via Town B can be exhibited with the aid of a **tree diagram** (Figure 16).

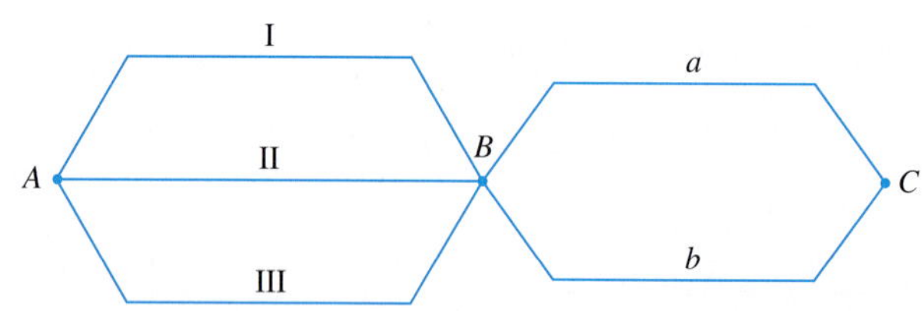

FIGURE **15**
Roads from Town A to Town C

If we follow all of the branches from the initial point A to the right-hand edge of the tree, we obtain the six routes represented by six ordered pairs:

$$(\mathrm{I}, a), (\mathrm{I}, b), (\mathrm{II}, a), (\mathrm{II}, b), (\mathrm{III}, a), (\mathrm{III}, b)$$

where (I, a) means that the journey from Town A to Town B is made on Trunk Road I with the rest of the journey, from Town B to Town C, completed on Trunk Road a, and so forth.

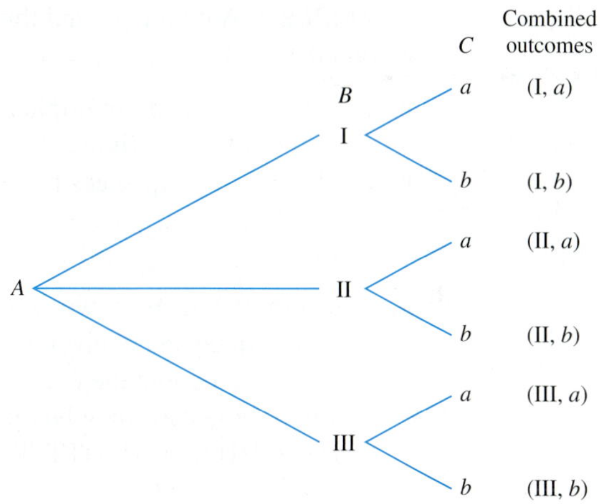

FIGURE 16
Tree diagram displaying the possible routes from Town A to Town C

Explore and Discuss

One way of gauging the performance of an airline is to track the arrival times of its flights. Suppose we denote by E, O, and L a flight that arrives early, on time, or late, respectively.

1. Use a tree diagram to exhibit the possible outcomes when you track two successive flights of the airline. How many outcomes are there?
2. How many outcomes are there if you track three successive flights? Justify your answer.

APPLIED EXAMPLE 2 Menu Choices Diners at Angelo's Spaghetti Bar can select their entree from 6 varieties of pasta and 28 choices of sauce. How many such combinations are there that consist of 1 variety of pasta and 1 kind of sauce?

Solution There are 6 ways of choosing a pasta followed by 28 ways of choosing a sauce, so by the multiplication principle, there are $6 \cdot 28$, or 168, combinations of this pasta dish.

The multiplication principle can be easily extended, which leads to the **generalized multiplication principle**.

Generalized Multiplication Principle

Suppose a task T_1 can be performed in N_1 ways, a task T_2 can be performed in N_2 ways, . . . , and, finally, a task T_m can be performed in N_m ways. Then, the number of ways of performing the tasks $T_1, T_2, \ldots, T_m$ in succession is given by the product

$$N_1 N_2 \cdots N_m$$

We now illustrate the application of the generalized multiplication principle to several diverse situations.

EXAMPLE 3 A coin is tossed three times, and the sequence of heads and tails is recorded.

a. Use the generalized multiplication principle to determine the number of possible outcomes of this activity.
b. Exhibit all the sequences by means of a tree diagram.

Solution

a. The coin may land in two ways. Therefore, in three tosses, the number of outcomes (sequences) is given by $2 \cdot 2 \cdot 2$, or 8.
b. Let H and T denote the outcomes "a head" and "a tail," respectively. Then the required sequences may be obtained as shown in Figure 17, giving the sequence as HHH, HHT, HTH, HTT, THH, THT, TTH, and TTT.

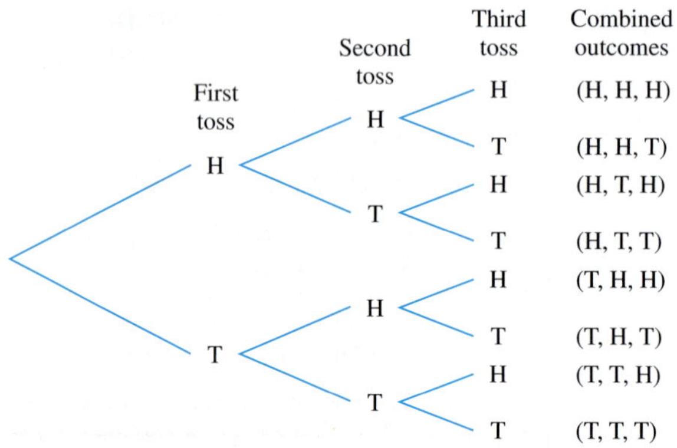

FIGURE 17
Tree diagram displaying possible outcomes of three consecutive coin tosses

APPLIED EXAMPLE 4 Combination Locks A combination lock is unlocked by dialing a sequence of numbers: first to the left, then to the right, and then to the left again.

a. If there are ten digits on the dial, determine the number of possible combinations.
b. How many combinations are possible if no digit is repeated?

Solution

a. There are ten choices for the first number, followed by ten for the second and ten for the third, so by the generalized multiplication principle there are $10 \cdot 10 \cdot 10$, or 1000, possible combinations.
b. Next, suppose that no number is repeated. Then, there are ten choices for the first number, followed by nine choices for the second, and eight for the third, so by the generalized multiplication principle, there are $10 \cdot 9 \cdot 8$, or 720, possible combinations.

APPLIED EXAMPLE 5 Investment Options An investor has decided to purchase shares in the stock of three companies: one engaged in aerospace activities, one involved in energy development, and one involved in electronics. After some research, the account executive of a brokerage firm has recommended that the investor consider stock from five aerospace companies, three energy development companies, and four electronics companies. In how many ways can the investor select the group of three companies from the executive's list?

Solution The investor has five choices for selecting an aerospace company, three choices for selecting an energy development company, and four choices for selecting an electronics company. Therefore, by the generalized multiplication principle, there are $5 \cdot 3 \cdot 4$, or 60, ways in which she can select a group of three companies, one from each industry group.

APPLIED EXAMPLE 6 Travel Options Tom is planning to leave for New York City from Washington, D.C., on Monday morning and has decided that he will either fly or take the train. There are five flights and two trains departing for New York City from Washington that morning. When he returns on Sunday afternoon, Tom plans to either fly or hitch a ride with a friend. There are two flights departing from New York City to Washington that afternoon. In how many ways can Tom complete this round trip?

Solution There are seven ways in which Tom can go from Washington, D.C., to New York City (five by plane and two by train). On the return trip, Tom can travel in three ways (two by plane and one by car). Therefore, by the multiplication principle, Tom can complete the round trip in $7 \cdot 3$, or 21, ways.

6.3 Self-Check Exercises

1. **Selecting a Travel Package** Encore Travel offers a "Theater Week in London" package originating from New York City. There is a choice of eight flights departing from New York City each week, a choice of five hotel accommodations, and a choice of one complimentary ticket to one of eight shows. How many such travel packages can a tourist choose from?

2. **Dinner Specials** The Café Napoleon offers a dinner special on Wednesdays consisting of a choice of two entrées (beef bourguignon and chicken basquaise); one dinner salad; one French roll; a choice of three vegetables; a choice of a carafe of burgundy, rosé, or chablis wine; a choice of coffee or tea; and a choice of six french pastries for dessert. How many combinations of dinner specials are there?

Solutions to Self-Check Exercises 6.3 can be found on page 368.

6.3 Concept Questions

1. Explain the multiplication principle, and illustrate it with a diagram.

2. Given the following tree diagram for an activity, what are the possible outcomes?

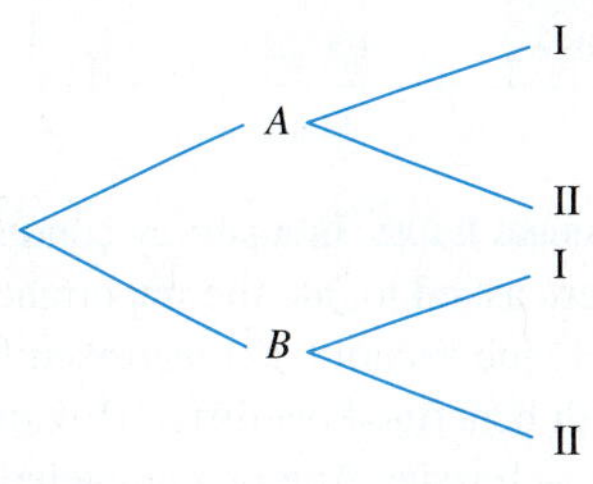

6.3 Exercises

1. **Rental Rates** Lynbrook West, an apartment complex financed by the State Housing Finance Agency, consists of one-, two-, three-, and four-bedroom units. The rental rate for each type of unit—low, moderate, or market—is determined by the income of the tenant. How many different rates are there?

2. **Commuter Passes** Five different types of monthly commuter passes are offered by a city's local transit authority for each of three different groups of passengers: youths, adults, and senior citizens. How many different kinds of passes must be printed each month?

3. **Blackjack** In the game of blackjack, a 2-card hand consisting of an ace and either a face card or a 10 is called a "blackjack." If a standard 52-card deck is used, determine how many blackjack hands can be dealt. (A "face card" is a jack, queen, or king.)

4. **Coin Tosses** A coin is tossed four times, and the sequence of heads and tails is recorded.
   a. Use the generalized multiplication principle to determine the number of outcomes of this activity.
   b. Exhibit all the sequences by means of a tree diagram.

5. **Wardrobe Selection** A female executive selecting her wardrobe purchased two blazers, four blouses, and three skirts in coordinating colors. How many ensembles consisting of a blazer, a blouse, and a skirt can she create from this collection?

6. **Commuter Options** Four commuter trains and three express buses depart from City A to City B in the morning, and three commuter trains and three express buses operate on the return trip in the evening. In how many ways can a commuter from City A to City B complete a daily round trip via bus and/or train?

7. **Psychology Experiments** A psychologist has constructed the following maze for use in an experiment. The maze is constructed so that a rat must pass through a series of one-way doors. How many different paths are there from start to finish?

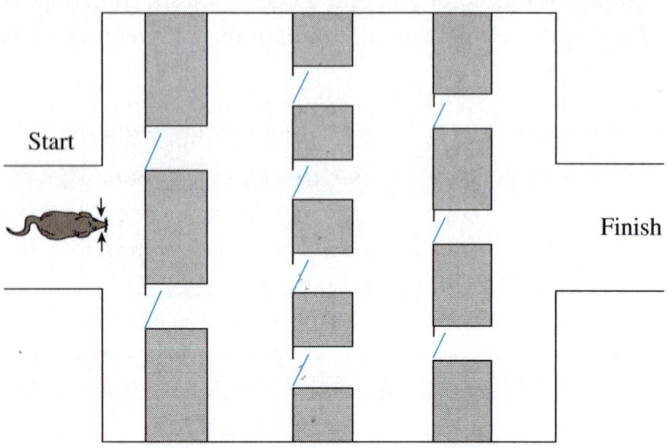

Maze

8. **Union Bargaining Issues** In a survey conducted by a union, members were asked to rate the importance of the following issues: (1) job security, (2) increased fringe benefits, and (3) health benefits. Five different responses were allowed for each issue. Among completed surveys, how many different responses to this survey were possible?

9. **Health-Care Plan Options** A new state employee is offered a choice of ten basic health plans, three dental plans, and two vision care plans. How many different health-care plans are there to choose from if one plan is selected from each category?

10. **Code Words** How many three-letter code words can be constructed from the first ten letters of the Greek alphabet if no repetitions are allowed?

11. **Social Security Numbers** A Social Security number has nine digits. How many Social Security numbers are possible?

12. **Menu Choices** The New Shanghai Restaurant offers a choice of three appetizers, two choices of soups, and ten choices of entrees for its lunch special. How many different complete lunches can be ordered from the restaurant's lunch special menu?

13. **Choosing a PIN** Janice needs to make up a personal identification number (PIN) to be used for ATM access. She has decided that the five-digit code, to be chosen from the numbers 0 through 9, should not have a 0 as the first digit and should have an odd number as the last digit. How many such PINs are possible?

14. **Menu Choices** Maria's Trattoria offers mushrooms, onions, green pepper, pepperoni, Italian sausage, and anchovies as toppings for the plain cheese base of its pizzas. How many different pizzas can be made?

15. **Binary Codes** A binary code is a method of representing text or computer processor instructions using the binary digits 0 and 1, called *bits*. How many different binary strings containing eight binary bits are possible?

16. **Building a Pickup Truck** The 2013 Sierra 1500 pickup truck by GMC comes with six body styles, two drive types, nine exterior colors, two interior colors, two styles of wheels, three choices of engines, and three choices of automatic transmissions. How many different models of the pickup truck can be built on the basis of these options?
Source: GMC.

17. **Menu Choices** The Panini Café has a special lunch menu in which the diner chooses a soup or salad, a main course, and a dessert. There are three choices of soups, three choices of salads, five choices for the main course, and four choices of desserts. How many different three-course meals can diners order for lunch?

18. **Serial Numbers** Computers manufactured by a certain company have a serial number consisting of a letter of the alphabet followed by a four-digit number. If all the serial numbers of this type have been used, how many sets have already been manufactured?

19. **Computer Dating** A computer dating service uses the results of its compatibility survey for arranging dates. The survey consists of 50 questions, each having five possible answers. How many different responses are possible if every question is answered?

20. **Automobile Color and Trim Choices** The 2014 BMW 435i Coupe is offered with a choice of 13 exterior colors (10 metallic and 3 standard), 6 interior colors, and 4 trims. How many combinations involving color and trim are available for the model?
Source: BMW.

21. **Automobile Options** The 2014 Toyota Camry comes with six grades of models, two sizes of engines, four choices of transmissions, six exterior colors, and two interior colors. How many choices of the Camry are available for a prospective buyer?
Source: Toyota.

22. **Television-Viewing Polls** An opinion poll is to be conducted among cable TV viewers. Six multiple-choice questions, each with four possible answers, will be asked. In how many different ways can a viewer complete the poll if exactly one response is given to each question?

23. **License Plate Numbers** Over the years, the state of California has used different combinations of letters of the alphabet and digits on its automobile license plates.
   a. At one time, license plates were issued that consisted of three letters followed by three digits. How many different license plates can be issued under this arrangement?
   b. Later on, license plates were issued that consisted of three digits followed by three letters. How many different license plates can be issued under this arrangement?

24. **Political Polls** An opinion poll was conducted by the Morris Polling Group. Respondents were classified according to their sex (M or F), political affiliation (D, I, R), and the region of the country in which they reside (NW, W, C, S, E, NE).
   a. Use the generalized multiplication principle to determine the number of possible classifications.
   b. Construct a tree diagram to exhibit all possible classifications of females.

25. **License Plate Numbers** In recent years, the state of California issued license plates using a combination of one letter of the alphabet followed by three digits, followed by another three letters of the alphabet. How many different license plates can be issued using this configuration?

26. **Exams** An exam consists of ten true-or-false questions. Assuming that every question is answered, in how many different ways can a student complete the exam? In how many ways can the exam be completed if a student can leave some questions unanswered because a penalty is assessed for each incorrect answer?

27. **Warranty Numbers** A warranty identification number for a certain product consists of a letter of the alphabet followed by a five-digit number. How many possible identification numbers are there if the first digit of the five-digit number must be nonzero?

28. **Choosing a Password** A password is to be made from a string of five characters chosen from the lowercase letters of the alphabet and the numbers 0 through 9.
   a. How many passwords are possible if there are no restrictions?
   b. How many passwords are possible if the characters must alternate between letters and numbers?

29. **Lotteries** In a state lottery, there are 15 finalists who are eligible for the Big Money Draw. In how many ways can the first, second, and third prizes be awarded if no ticket holder can win more than one prize?

30. **Combination Locks** A rolling combination four-digit padlock is unlocked by moving each of four rollers so as to produce the correct sequence. Each roller has ten digits.
   a. How many possible combinations are there?
   b. How many combinations are possible if no digit is repeated?

31. **Combination Locks** Lugano Leather Company makes an executive attaché case equipped with two rolling combination locks, each one with a provision for a three-digit number. The attaché case is unlocked by setting each lock to produce the correct combination. Each roller in each of the two locks has ten digits.
   a. How many possible combinations are there?
   b. How many combinations are possible if the lock on the left-hand side of the attaché case must end with an even number and the lock on the right-hand side must end with an odd number?

32. **Telephone Numbers**
   a. How many seven-digit telephone numbers are possible if the first digit must be nonzero?
   b. How many direct-dialing numbers for calls within the United States and Canada are possible if each number consists of a 1 plus a three-digit area code (the first digit of which must be nonzero) and a number of the type described in part (a)?

33. **Slot Machines** A "lucky dollar" is one of the nine symbols printed on each reel of a slot machine with three reels. A player receives one of various payouts whenever one or more "lucky dollars" appear in the window of the machine. Find the number of winning combinations for which the machine gives a payoff.
Hint: (a) Compute the number of ways in which the nine symbols on the first, second, and third reels can appear in the window slot and (b) compute the number of ways in which the eight symbols other than the "lucky dollar" can appear in the window slot. The difference $(a - b)$ is the number of ways in which the "lucky dollar" can appear in the window slot. Why?

34. **Staffing** Student Painters, which specializes in painting the exterior of residential buildings, has five people available to be organized into two-person and three-person teams.
   a. In how many ways can a two-person team be formed?
   b. In how many ways can a three-person team be formed?
   c. In how many ways can the company organize the available people into either two-person teams or three-person teams?

In Exercises 35 and 36, determine whether the statement is true or false. If it is true, explain why it is true. If it is false, give an example to show why it is false.

35. There are 32 three-digit odd numbers that can be formed from the digits 1, 2, 3, and 4.

36. If there are six toppings available, then the number of different pizzas that can be made is 2^5, or 32, pizzas.

6.3 Solutions to Self-Check Exercises

1. A tourist has a choice of eight flights, five hotel accommodations, and eight tickets. By the generalized multiplication principle, there are $8 \cdot 5 \cdot 8$, or 320, travel packages.

2. There is a choice of two entrées, one dinner salad, one French roll, three vegetables, three wines, two nonalcoholic beverages, and six pastries. Therefore, by the generalized multiplication principle, there are $2 \cdot 1 \cdot 1 \cdot 3 \cdot 3 \cdot 2 \cdot 6$, or 216, combinations of dinner specials.

6.4 Permutations and Combinations

Permutations

In this section, we apply the generalized multiplication principle to the solution of two types of counting problems. Both types involve determining the number of ways in which the elements of a set can be arranged, and both play an important role in the solution of problems in probability.

We begin by considering the permutations of a set. Specifically, given a set of distinct objects, a **permutation** of the set is an arrangement of these objects in a *definite order*. To see why the order in which objects are arranged is important in certain practical situations, suppose the winning number for the first prize in a raffle is 9237. Then the number 2973, although it contains the same digits as the winning number, cannot be the first-prize winner (Figure 18). Here, the four objects—the digits 9, 2, 3, and 7—are arranged in a different order; one arrangement is associated with the winning number for the first prize, and the other is not.

FIGURE 18
The same digits appear on each ticket, but the order of the digits is different.

EXAMPLE 1 Let $A = \{a, b, c\}$.

a. Find the number of permutations of A.

b. List all the permutations of A with the aid of a tree diagram.

Solution

a. Each permutation of A consists of a sequence of the three letters a, b, c. Therefore, we may think of such a sequence as being constructed by filling in each of the three blanks

$$\underline{\quad}\ \underline{\quad}\ \underline{\quad}$$

with one of the three letters. Now, there are three ways in which we can fill the first blank—we can choose a, b, or c. Having selected a letter for the first blank, there are two letters left for the second blank. Finally, there is but one way left to fill the third blank. Schematically, we have

$$\underline{3}\ \ \underline{2}\ \ \underline{1}$$

Invoking the generalized multiplication principle, we conclude that there are $3 \cdot 2 \cdot 1$, or 6, permutations of the set A.

b. The tree diagram associated with this problem appears in Figure 19. The six permutations of A are *abc*, *acb*, *bac*, *bca*, *cab*, and *cba*.

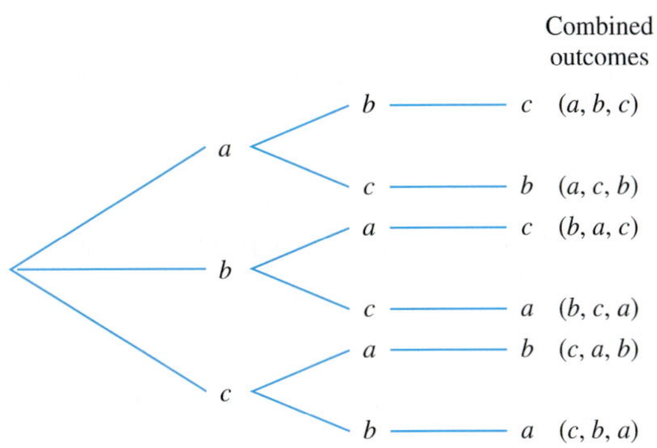

FIGURE **19**
Permutations of three objects

Note Notice that when the possible outcomes are listed in the tree diagram in Example 1, order is taken into account. Thus, (a, b, c) and (a, c, b) are two different arrangements.

APPLIED EXAMPLE 2 Taking a Group Picture Find the number of ways in which a baseball team consisting of nine people can arrange themselves in a line for a group picture.

Solution We want to determine the number of permutations of the nine members of the baseball team. Each permutation in this situation consists of an arrangement of the nine team members in a line. The nine positions can be represented by nine blanks. Thus,

$$\text{Position}\quad \underset{1}{\underline{\quad}}\ \underset{2}{\underline{\quad}}\ \underset{3}{\underline{\quad}}\ \underset{4}{\underline{\quad}}\ \underset{5}{\underline{\quad}}\ \underset{6}{\underline{\quad}}\ \underset{7}{\underline{\quad}}\ \underset{8}{\underline{\quad}}\ \underset{9}{\underline{\quad}}$$

There are nine ways to choose from among the nine players to fill the first position. When that position is filled, eight players are left, which gives us eight ways to fill the second position. Proceeding in a similar manner, we find that there are seven ways to fill the third position, and so on. Schematically, we have

$$\text{Number of ways to fill each position}\quad \underline{9}\ \ \underline{8}\ \ \underline{7}\ \ \underline{6}\ \ \underline{5}\ \ \underline{4}\ \ \underline{3}\ \ \underline{2}\ \ \underline{1}$$

Invoking the generalized multiplication principle, we conclude that there are $9 \cdot 8 \cdot 7 \cdot 6 \cdot 5 \cdot 4 \cdot 3 \cdot 2 \cdot 1$, or 362,880, ways in which the baseball team can be arranged for the picture.

Whenever we are asked to determine the number of ways in which the objects of a set can be arranged in a line, order is important. For example, if we take a picture of two baseball players, A and B, then the two players can line up for the picture in two ways, AB or BA, and the two pictures will be different.

Pursuing the same line of argument used in solving the problems in the last two examples, we can derive an expression for the number of ways of permuting a set A of n distinct objects taken n at a time. In fact, each permutation may be viewed as being obtained by filling each of n blanks with one and only one element from the set. There are n ways of filling the first blank, followed by $(n - 1)$ ways of filling the second blank, and so on. Thus, by the generalized multiplication principle, there are

$$n(n - 1)(n - 2) \cdot \cdots \cdot 3 \cdot 2 \cdot 1$$

ways of permuting the elements of the set A.

Before stating this result formally, let's introduce a notation that will enable us to write in a compact form many of the expressions that follow. We use the symbol $n!$ (read "**n-factorial**") to denote the product of the first n positive integers.

n-Factorial

For any natural number n,

$$n! = n(n - 1)(n - 2) \cdot \cdots \cdot 3 \cdot 2 \cdot 1$$
$$0! = 1$$

For example,

$$\begin{aligned} 1! &= 1 \\ 2! &= 2 \cdot 1 = 2 \\ 3! &= 3 \cdot 2 \cdot 1 = 6 \\ 4! &= 4 \cdot 3 \cdot 2 \cdot 1 = 24 \\ 5! &= 5 \cdot 4 \cdot 3 \cdot 2 \cdot 1 = 120 \\ &\vdots \\ 10! &= 10 \cdot 9 \cdot 8 \cdot 7 \cdot 6 \cdot 5 \cdot 4 \cdot 3 \cdot 2 \cdot 1 = 3{,}628{,}800 \end{aligned}$$

Using this notation, we can express *the number of permutations of n distinct objects taken n at a time, denoted by $P(n, n)$, as*

$$P(n, n) = n!$$

In many situations, we are interested in determining the number of ways of permuting n distinct objects taken r at a time, where $r \leq n$. To derive a formula for computing the number of ways of permuting a set consisting of n distinct objects taken r at a time, we observe that each such permutation may be viewed as being obtained by filling each of r blanks with precisely one element from the set. Now there are n ways of filling the first blank, followed by $(n - 1)$ ways of filling the second blank, and so on. Finally, there are $(n - r + 1)$ ways of filling the rth blank. We can represent this argument schematically:

Number of ways	n	$n - 1$	$n - 2$	$\cdots$	$n - r + 1$
Position	1st	2nd	3rd		rth

Using the generalized multiplication principle, we conclude that *the number of ways of permuting n distinct objects taken r at a time, denoted by P(n, r), is given by*

$$P(n, r) = \underbrace{n(n-1)(n-2)\cdots(n-r+1)}_{r \text{ factors}}$$

Since

$$\begin{aligned} n(n-1)(n-2)\cdots(n-r+1) &= [n(n-1)(n-2)\cdots(n-r+1)]\cdot\underbrace{\frac{(n-r)(n-r-1)\cdot\cdots\cdot 3\cdot 2\cdot 1}{(n-r)(n-r-1)\cdot\cdots\cdot 3\cdot 2\cdot 1}}_{\text{Here we are multiplying by 1.}} \\ &= \frac{[n(n-1)(n-2)\cdots(n-r+1)][(n-r)(n-r-1)\cdot\cdots\cdot 3\cdot 2\cdot 1]}{(n-r)(n-r-1)\cdot\cdots\cdot 3\cdot 2\cdot 1} \\ &= \frac{n!}{(n-r)!} \end{aligned}$$

we have the following formula.

Permutations of *n* Distinct Objects

The number of *permutations* of n distinct objects taken r at a time is

$$P(n, r) = \frac{n!}{(n-r)!} \qquad \textbf{(6)}$$

Note When $r = n$, Equation (6) reduces to

$$P(n, n) = \frac{n!}{0!} = \frac{n!}{1} = n! \qquad \text{Note that } 0! = 1.$$

In other words, the number of permutations of a set of n distinct objects, taken all together, is $n!$.

EXAMPLE 3 Compute (a) $P(4, 4)$ and (b) $P(4, 2)$, and interpret your results.

Solution

a. $P(4, 4) = \dfrac{4!}{(4-4)!} = \dfrac{4!}{0!} = \dfrac{4!}{1} = \dfrac{4\cdot 3\cdot 2\cdot 1}{1} = 24$ Note that $0! = 1$.

This gives the number of permutations of four objects taken four at a time.

b. $P(4, 2) = \dfrac{4!}{(4-2)!} = \dfrac{4!}{2!} = \dfrac{4\cdot 3\cdot 2\cdot 1}{2\cdot 1} = 4\cdot 3 = 12$

This is the number of permutations of four objects taken two at a time.

EXAMPLE 4 Let $A = \{a, b, c, d\}$.

a. Use Equation (6) to compute the number of permutations of the set A taken two at a time.

b. Display the permutations of part (a) with the aid of a tree diagram.

PORTFOLIO Lara Solanki

TITLE Los Angeles City Manager
INSTITUTION Fanfoto

Fanfoto is a photography company that focuses on a major need of sports fans and fans of the entertainment industry. At Fanfoto, we have developed a successful business model that is based on allowing customers to commemorate their trips to their favorite ballparks, stadiums, and arenas with professional photography.

As a manager, one of my primary responsibilities is to pick the right combination of employees (photographers) for various sporting events. In addition to making sure that I have enough photographers scheduled for each event, I manage labor percentage. Labor percentage is total labor divided by total sales.

The lower the labor percentage, the better for us. Most business managers would agree that 30% or lower is an acceptable percentage. I am constantly tracking and evaluating total sales, as well as the individual sales of each photographer, to ensure that the labor percentage for my staff falls in this acceptable range.

I also keep track of the number of photographs taken by each photographer at every event. I have found that there is a direct correlation between the number of photographs taken and sales. As more photos are taken, sales increase. For example, Jon shoots 500 photos every night and works for 5 hours at \$10/hour. Ben also works for 5 hours at \$10/hour but shoots only 150 photos. Jon is reaching out to more people than Ben. Jon's daily sales average is \$500 a night, whereas Ben's daily sales average is \$100 a night. Jon's labor percentage is far better than Ben's labor percentage, which increases the profit for the company.

Instead of scheduling staff at random, I create groups of staff based on their sales history and labor percentages. By scheduling a combination of higher- and lower-performing staff, I can be sure that the average labor percentage will be acceptable at any given event.

Spencer Brennan; (inset) Tom Keene

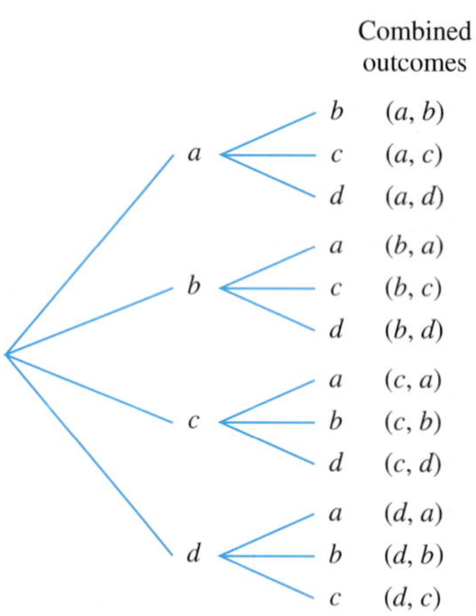

FIGURE 20
Permutations of four objects taken two at a time

Solution

a. Here, $n = 4$ and $r = 2$, so the required number of permutations is given by

$$P(4, 2) = \frac{4!}{(4-2)!} = \frac{4!}{2!} = \frac{4 \cdot 3 \cdot 2 \cdot 1}{2 \cdot 1} = 4 \cdot 3$$
$$= 12$$

b. The tree diagram associated with the problem is shown in Figure 20, and the permutations of A taken two at a time are

$$ab, ac, ad, ba, bc, bd, ca, cb, cd, da, db, dc$$

APPLIED EXAMPLE 5 Selecting a Committee Find the number of ways in which a chairman, a vice-chairman, a secretary, and a treasurer can be chosen from a committee of eight members.

Solution The problem is equivalent to finding the number of permutations of eight distinct objects taken four at a time. Therefore, there are

$$P(8, 4) = \frac{8!}{(8-4)!} = \frac{8!}{4!} = 8 \cdot 7 \cdot 6 \cdot 5 = 1680$$

ways of choosing the four officials from the committee of eight members.

The permutations considered thus far have been those involving sets of *distinct* objects. In many situations, we are interested in finding the number of permutations of a set of objects in which not all of the objects are distinct.

Permutations of n Objects, Not All Distinct

Given a set of n objects in which n_1 objects are alike and of one kind, n_2 objects are alike and of another kind, . . . , and n_m objects are alike and of yet another kind, so that

$$n_1 + n_2 + \cdots + n_m = n$$

then the number of permutations of these n objects taken n at a time is given by

$$\frac{n!}{n_1!\, n_2! \cdots n_m!} \tag{7}$$

To establish Formula (7), let's denote the number of such permutations by x. Now, if we *think* of the n_1 objects as being distinct, then they can be permuted in $n_1!$ ways. Similarly, if we *think* of the n_2 objects as being distinct, then they can be permuted in $n_2!$ ways, and so on. Therefore, if we *think* of the n objects as being distinct, then, by the generalized multiplication principle, there are $x \cdot n_1! \cdot n_2! \cdot \cdots \cdot n_m!$ permutations of these objects. But the number of permutations of a set of n distinct objects taken n at a time is just equal to $n!$. Therefore, we have

$$x(n_1! \cdot n_2! \cdot \cdots \cdot n_m!) = n!$$

from which we deduce that

$$x = \frac{n!}{n_1!\, n_2! \cdots n_m!}$$

EXAMPLE 6 Find the number of permutations that can be formed from all the letters in the word *ATLANTA*.

Solution There are seven objects (letters) involved, so $n = 7$. However, three of them are alike and of one kind (the three *A*s), while two of them are alike and of another kind (the two *T*s); hence, in this case, we have $n_1 = 3$, $n_2 = 2$, $n_3 = 1$ (the one *L*), and $n_4 = 1$ (the one *N*). Therefore, by Formula (7), there are

$$\frac{7!}{3!\,2!\,1!\,1!} = \frac{7 \cdot 6 \cdot 5 \cdot 4 \cdot 3 \cdot 2 \cdot 1}{3 \cdot 2 \cdot 1 \cdot 2 \cdot 1 \cdot 1 \cdot 1} = 420$$

permutations.

APPLIED EXAMPLE 7 Management Decisions Weaver and Kline, a stock brokerage firm, has received nine inquiries regarding new accounts. In how many ways can these inquiries be directed to any three of the firm's account executives if each account executive is to handle three inquiries?

Solution If we think of the nine inquiries as being slots arranged in a row with inquiry 1 on the left and inquiry 9 on the right, then the problem can be thought of as one of filling each slot with a business card from an account executive. Then nine business cards would be used, of which three are alike and of one kind, three are alike and of another kind, and three are alike and of yet another kind. Thus, by using Formula (7) with $n = 9$ and $n_1 = n_2 = n_3 = 3$, there are

$$\frac{9!}{3!\,3!\,3!} = \frac{9 \cdot 8 \cdot 7 \cdot 6 \cdot 5 \cdot 4 \cdot 3 \cdot 2 \cdot 1}{3 \cdot 2 \cdot 1 \cdot 3 \cdot 2 \cdot 1 \cdot 3 \cdot 2 \cdot 1} = 1680$$

ways of assigning the inquiries.

Combinations

Until now, we have dealt with permutations of a set—that is, with arrangements of the objects of the set in which the *order* of the elements is taken into consideration. In many situations, one is interested in determining the number of ways of selecting r objects from a set of n objects without any regard to the order in which the objects are selected. Such a subset is called a **combination.**

For example, if one is interested in knowing the number of 5-card poker hands that can be dealt from a standard deck of 52 cards, then the order in which the poker hand is dealt is unimportant (Figure 21). In this situation, we are interested in determining the number of combinations of 5 cards (objects) selected from a deck (set) of 52 cards (objects). (We will solve this problem in Example 10.)

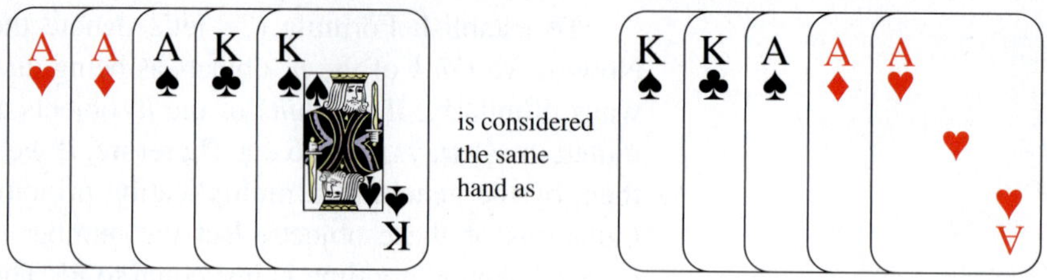

FIGURE 21

To derive a formula for determining the number of combinations of n objects taken r at a time, written

$$C(n, r) \quad \text{or} \quad \binom{n}{r}$$

we observe that each of the $C(n, r)$ combinations of r objects can be permuted in $r!$ ways (Figure 22).

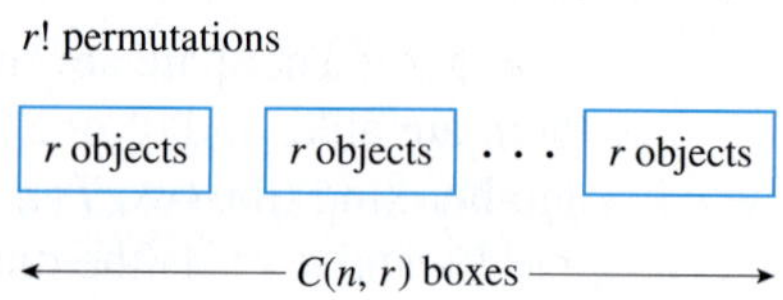

FIGURE 22

Thus, by the multiplication principle, the product $r!\,C(n, r)$ gives the number of permutations of n objects taken r at a time; that is,

$$r!\,C(n, r) = P(n, r)$$

from which we find

$$C(n, r) = \frac{P(n, r)}{r!}$$

or, using Equation (6),

$$C(n, r) = \frac{n!}{r!\,(n - r)!}$$

Combinations of n Objects

The number of combinations of n distinct objects taken r at a time is given by

$$C(n, r) = \frac{n!}{r!\,(n - r)!} \qquad (\text{where } r \le n) \qquad \textbf{(8)}$$

EXAMPLE 8 Compute and interpret the results of (a) $C(4, 4)$ and (b) $C(4, 2)$.

Solution

a. $C(4, 4) = \dfrac{4!}{4!\,(4-4)!} = \dfrac{4!}{4!\,0!} = 1$ Recall that $0! = 1$.

This gives 1 as the number of combinations of four distinct objects taken four at a time.

b. $C(4, 2) = \dfrac{4!}{2!\,(4-2)!} = \dfrac{4!}{2!\,2!} = \dfrac{4 \cdot 3 \cdot 2 \cdot 1}{2 \cdot 2} = 6$

This gives 6 as the number of combinations of four distinct objects taken two at a time.

APPLIED EXAMPLE 9 Committee Selection A Senate investigation subcommittee of four members is to be selected from a Senate committee of ten members. Determine the number of ways in which this can be done.

Solution The order in which the members of the subcommittee are selected is unimportant, so the number of ways of choosing the subcommittee is given by $C(10, 4)$, the number of combinations of ten objects taken four at a time. Hence there are

$$C(10, 4) = \frac{10!}{4!\,(10-4)!} = \frac{10!}{4!\,6!} = \frac{10 \cdot 9 \cdot 8 \cdot 7}{4 \cdot 3 \cdot 2 \cdot 1} = 210$$

ways of choosing such a subcommittee.

Note Remember, a combination is a selection of objects *without* regard to order. Thus, in Example 9, we used a combination formula rather than a permutation formula to solve the problem because the order of selection was not important; that is, it did not matter whether a member of the subcommittee was selected first, second, third, or fourth.

APPLIED EXAMPLE 10 Poker How many poker hands of 5 cards can be dealt from a standard deck of 52 cards?

Solution The order in which the 5 cards are dealt is not important. The number of ways of dealing a poker hand of 5 cards from a standard deck of 52 cards is given by $C(52, 5)$, the number of combinations of 52 objects taken five at a time. Thus, there are

$$\begin{aligned} C(52, 5) &= \frac{52!}{5!\,(52-5)!} = \frac{52!}{5!\,47!} \\ &= \frac{52 \cdot 51 \cdot 50 \cdot 49 \cdot 48}{5 \cdot 4 \cdot 3 \cdot 2 \cdot 1} \\ &= 2{,}598{,}960 \end{aligned}$$

ways of dealing such a poker hand.

The next several examples show that solving a counting problem often involves the repeated application of Equation (6) and/or (8), possibly in conjunction with the multiplication principle.

APPLIED EXAMPLE 11 Selecting Members of a Group The members of a string quartet consisting of two violinists, a violist, and a cellist are to be selected from a group of six violinists, three violists, and two cellists.

a. In how many ways can the string quartet be formed?
b. In how many ways can the string quartet be formed if one of the violinists is to be designated as the first violinist and the other is to be designated as the second violinist?

Solution

a. Since the order in which each musician is selected is not important, we use combinations. The violinists can be selected in $C(6, 2)$, or 15, ways; the violist can be selected in $C(3, 1)$, or 3, ways; and the cellist can be selected in $C(2, 1)$, or 2, ways. By the multiplication principle, there are $15 \cdot 3 \cdot 2$, or 90, ways of forming the string quartet.
b. The order in which the violinists are selected is important here. Consequently, the number of ways of selecting the violinists is given by $P(6, 2)$, or 30, ways. The number of ways of selecting the violist and the cellist remain, of course, 3 and 2, respectively. Therefore, the number of ways in which the string quartet can be formed is given by $30 \cdot 3 \cdot 2$, or 180, ways.

Note The solution of Example 11 involves both a permutation and a combination. When we select two violinists from six violinists, order is not important, and we use a combination formula to solve the problem. However, when one of the violinists is designated as a first violinist, order is important, and we use a permutation formula to solve the problem.

APPLIED EXAMPLE 12 Investment Options Refer to Example 5, page 364. Suppose the investor has decided to purchase shares in the stocks of two aerospace companies, two energy development companies, and two electronics companies. In how many ways can the investor select the group of six companies for the investment from the recommended list of five aerospace companies, three energy development companies, and four electronics companies?

Solution There are $C(5, 2)$ ways in which the investor can select the aerospace companies, $C(3, 2)$ ways in which she can select the companies involved in energy development, and $C(4, 2)$ ways in which she can select the electronics companies as investments. By the generalized multiplication principle, there are

$$C(5, 2)C(3, 2)C(4, 2) = \frac{5!}{2!\,3!} \cdot \frac{3!}{2!\,1!} \cdot \frac{4!}{2!\,2!}$$
$$= \frac{5 \cdot 4}{2} \cdot 3 \cdot \frac{4 \cdot 3}{2} = 180$$

ways of selecting the group of six companies for her investment.

APPLIED EXAMPLE 13 Scheduling Performances The Futurists, a rock group, are planning a concert tour with performances to be given in five cities: San Francisco, Los Angeles, San Diego, Denver, and Las Vegas. In how many ways can they arrange their itinerary if

a. There are no restrictions?
b. The three performances in California must be given consecutively?

Solution

a. The order is important here, and we see that there are

$$P(5, 5) = 5! = 120$$

ways of arranging their itinerary.

b. First, note that there are $P(3, 3)$ ways of choosing between performing in California and in the two cities outside that state. Next, there are $P(3, 3)$ ways of arranging their itinerary in the three cities in California. Therefore, by the multiplication principle, there are

$$P(3, 3)P(3, 3) = \frac{3!}{(3-3)!} \cdot \frac{3!}{(3-3)!} = 6 \cdot 6 = 36$$

ways of arranging their itinerary.

APPLIED EXAMPLE 14 U.N. Security Council Voting The United Nations Security Council consists of 5 permanent members and 10 nonpermanent members. Decisions made by the council require 9 votes for passage. However, any permanent member may veto a measure and thus block its passage. Assuming that there are no abstentions, in how many ways can a measure be passed if all 15 members of the Council vote?

Solution If a measure is to be passed, then all 5 permanent members must vote for passage of that measure. This can be done in $C(5, 5)$, or 1, way.

Next, observe that since 9 votes are required for passage of a measure, *at least* 4 of the 10 nonpermanent members must also vote for its passage. To determine the number of ways in which this can be done, notice that there are $C(10, 4)$ ways in which exactly 4 of the nonpermanent members can vote for passage of a measure, $C(10, 5)$ ways in which exactly 5 of them can vote for passage of a measure, and so on. Finally, there are $C(10, 10)$ ways in which all 10 nonpermanent members can vote for passage of a measure. Hence, there are

$$C(10, 4) + C(10, 5) + \cdots + C(10, 10)$$

ways in which at least 4 of the 10 nonpermanent members can vote for a measure. So by the multiplication principle, there are

$$\begin{aligned} &C(5, 5)[C(10, 4) + C(10, 5) + \cdots + C(10, 10)] \\ &= (1)\left[\frac{10!}{4!\,6!} + \frac{10!}{5!\,5!} + \cdots + \frac{10!}{10!\,0!}\right] \\ &= (1)(210 + 252 + 210 + 120 + 45 + 10 + 1) = 848 \end{aligned}$$

ways in which a measure can be passed.

6.4 Self-Check Exercises

1. Evaluate:
   a. 5! **b.** $C(7, 4)$ **c.** $P(6, 2)$

2. **Selecting a Space Shuttle Crew** A space shuttle crew consists of a shuttle commander, a pilot, three engineers, a scientist, and a civilian. The shuttle commander and pilot are to be chosen from 8 candidates, the three engineers from 12 candidates, the scientist from 5 candidates, and the civilian from 2 candidates. How many such space shuttle crews can be formed?

Solutions to Self-Check Exercises 6.4 can be found on *page 381.*

6.4 Concept Questions

1. **a.** What is a permutation of a set of distinct objects?
   b. How many permutations of a set of five distinct objects taken three at a time are there?

2. Given a set of ten objects in which three are alike and of one kind, three are alike and of another kind, and four are alike and of yet another kind, what is the formula for computing the permutation of these ten objects taken ten at a time?

3. **a.** How many combinations are there of a set of n distinct objects taken r at a time?
   b. How many combinations are there of six distinct objects taken three at a time?

6.4 Exercises

In Exercises 1–22, evaluate the given expression.

1. $3 \cdot 5!$ **2.** $2 \cdot 7!$ **3.** $\dfrac{5!}{2!\,3!}$

4. $\dfrac{6!}{4!\,2!}$ **5.** $P(5, 5)$ **6.** $P(6, 6)$

7. $P(5, 2)$ **8.** $P(5, 3)$ **9.** $P(n, 1)$

10. $P(k, 2)$ **11.** $C(6, 6)$ **12.** $C(8, 8)$

13. $C(7, 4)$ **14.** $C(9, 3)$ **15.** $C(5, 0)$

16. $C(6, 5)$ **17.** $C(9, 6)$ **18.** $C(10, 3)$

19. $C(n, 2)$ **20.** $C(7, r)$ **21.** $P(n, n - 2)$

22. $C(n, n - 2)$

In Exercises 23–30, classify each problem according to whether it involves a permutation or a combination.

23. In how many ways can the letters of the word *GLACIER* be arranged?

24. A 4-member executive committee is to be formed from a 12-member board of directors. In how many ways can it be formed?

25. As part of a quality-control program, 3 cell phones are selected at random for testing from 100 cell phones produced by the manufacturer. In how many ways can this test batch be chosen?

26. How many three-digit numbers can be formed by using the numerals in the set $\{3, 2, 7, 9\}$ if repetition is not allowed?

27. In how many ways can nine different books be arranged on a shelf?

28. A member of a book club wishes to purchase two books from a selection of eight books recommended for a certain month. In how many ways can she choose them?

29. How many five-card poker hands can be dealt consisting of three queens and a pair?

30. In how many ways can a six-letter security password be formed from letters of the alphabet if no letter is repeated?

31. How many four-letter permutations can be formed from the first four letters of the alphabet?

32. How many three-letter permutations can be formed from the first five letters of the alphabet?

33. In how many ways can four students be seated in a row of four seats?

34. In how many ways can five people line up at a checkout counter in a supermarket?

35. How many different batting orders can be formed for a nine-member baseball team?

36. In how many ways can the names of six candidates for political office be listed on a ballot?

37. In how many ways can a member of a hiring committee select 3 of 12 job applicants for further consideration?

38. In how many ways can an investor select four mutual funds for his investment portfolio from a recommended list of eight mutual funds?

39. Find the number of distinguishable permutations that can be formed from the letters of the word *ANTARCTICA*.

40. Find the number of distinguishable permutations that can be formed from the letters of the word *PHILIPPINES*.

41. In how many ways can the letters of the website *MySpace* be arranged if all of the letters are used and the vowels *a* and *e* must always stay in the order *ae*?

42. In how many ways can five people boarding a bus be seated if the bus has eight vacant seats?

43. How many distinct five-digit numbers can be made using the digits 1, 2, 2, 2, 7?

44. How many different signals can be made by hoisting two yellow flags, four green flags, and three red flags on a ship's mast at the same time?

45. **Supermarket Site Selection** In how many ways can a supermarket chain select 3 out of 12 possible sites for the construction of new supermarkets?

46. **Selecting a Reading List** A student is given a reading list of ten books from which he must select two for an outside reading requirement. In how many ways can he make his selections?

47. **Quality Control** In how many ways can a quality-control engineer select a sample of 3 microprocessors for testing from a batch of 100 microprocessors?

48. **Study Group Assignments** A group of five students studying for a bar exam has formed a study group. Each member of the group will be responsible for preparing a study outline for one of five courses. In how many different ways can the five courses be assigned to the members of the group?

49. **Television Programming** In how many ways can a television-programming director schedule six different commercials in the six time slots allocated to commercials during a 1-hr program?

50. **Waiting Lines** Seven people arrive at the ticket counter of Starlite Cinema at the same time. In how many ways can they line up to purchase their tickets?

51. **Selecting a Special Occasion Cake** Fosselman's Ice Cream Company makes two signature cakes, a white cake and a chocolate cake. The white cake comes with one of three flavors of ice cream: burgundy-cherry, coconut-pineapple, or strawberry. The chocolate cake comes with one of four flavors: chocolate chip, chocolate raspberry, cookies-and-cream, or mint chip. Jenny is considering buying one of these cakes for her daughter's birthday. How many choices of cakes does Jenny have?
Source: Fosselman's Ice Cream Company.

52. **Wedding Catering** L.A. Wedding Caterers offers a wedding reception buffet. Suppose a menu is planned around four different salads, six entrees, six side dishes, and seven desserts. There are eight different choices of salads, ten different choices of entrees, eight different choices of side dishes, and ten different choices of desserts. How many menus are possible?

53. **Management Decisions** Weaver and Kline, a stock brokerage firm, has received six inquiries regarding new accounts. In how many ways can these inquiries be directed to its 12 account executives if each executive handles no more than one inquiry?

54. **Car Pools** A company car that has a seating capacity of six is to be used by six employees who have formed a car pool. If only four of these employees can drive, how many possible seating arrangements are there for the group?

55. **Travel Wardrobe** Kaylee is planning a wardrobe for her upcoming Caribbean cruise. She has decided to bring along four blouses, four skirts, and three pairs of shorts to be selected from eight blouses, seven skirts, and six pairs of shorts. How many ensembles consisting of a blouse and either a skirt or a pair of shorts are possible?

56. **Criminology** The town of Carson employs 15 police officers. On a typical day, 6 of the officers are to be assigned to duty in patrol cars, 4 are assigned to the foot patrol, and the remaining 5 are assigned to duty at the station. How many different job configurations are there?

57. **Book Exhibitions** At a college library exhibition of faculty publications, three mathematics books, four social science books, and three biology books will be displayed on a shelf. (Assume that none of the books is alike.)
 a. In how many ways can the ten books be arranged on the shelf?
 b. In how many ways can the ten books be arranged on the shelf if books on the same subject matter are placed together?

58. **Concert Seating** In how many ways can four married couples attending a concert be seated in a row of eight seats if:
 a. There are no restrictions?
 b. Each married couple is seated together?
 c. The members of each sex are seated together?

59. **Newspaper Ads** Four items from five different departments of Metro Department Store will be featured in a one-page newspaper advertisement, as shown in the following diagram:

Advertisement

1	2	3	4
5	6	7	8
9	10	11	12
13	14	15	16
17	18	19	20

 a. In how many different ways can the 20 featured items be arranged on the page?
 b. If items from the same department must be in the same row, how many arrangements are possible?

60. **Management Decisions** C & J Realty has received 12 inquiries from prospective home buyers. In how many ways can the inquiries be directed to any four of the firm's real estate agents if each agent handles three inquiries?

61. **Selecting a Baseball Team** A Little League baseball team has 12 players available for a 9-member team (no designated team positions).
  a. How many different 9-person batting orders are possible?
  b. How many different 9-member teams are possible?
  c. How many different 9-member teams and 2 alternates are possible?

62. **Tennis Match** In the men's tennis tournament at Wimbledon, two finalists, A and B, are competing for the title, which will be awarded to the first player to win three sets. In how many different ways can the match be completed?

63. **Tennis Match** In the women's tennis tournament at Wimbledon, two finalists, A and B, are competing for the title, which will be awarded to the first player to win two sets. In how many different ways can the match be completed?

64. **Jury Selection** In how many different ways can a panel of 12 jurors and 2 alternate jurors be chosen from a group of 30 prospective jurors?

65. **U.N. Voting** Refer to Example 14. In how many ways can a measure be passed if two particular permanent and two particular nonpermanent members of the Security Council abstain from voting?

66. **Exams** A student taking an examination is required to answer exactly 10 out of 15 questions.
  a. In how many ways can the 10 questions be selected?
  b. In how many ways can the 10 questions be selected if exactly 2 of the first 3 questions must be answered?

67. **Teaching Assistantships** Twelve graduate students have applied for three available teaching assistantships. In how many ways can the assistantships be awarded among these applicants if:
  a. No preference is given to any student?
  b. One particular student must be awarded an assistantship?
  c. The group of applicants includes seven men and five women and it is stipulated that at least one woman must be awarded an assistantship?

68. **Senate Committees** In how many ways can a subcommittee of four be chosen from a Senate committee of five Democrats and four Republicans if:
  a. All members are eligible?
  b. The subcommittee must consist of two Republicans and two Democrats?

69. **Contract Bidding** UBS Television Company is considering bids submitted by seven different firms for each of three different contracts. In how many ways can the contracts be awarded among these firms if no firm is to receive more than two contracts?

70. **Personnel Selection** JCL Computers has five vacancies in its executive trainee program. In how many ways can the company select five trainees from a group of ten female and ten male applicants if the vacancies
  a. Can be filled by any combination of men and women?
  b. Must be filled by two men and three women?

71. **Course Selection** A student planning her curriculum for the upcoming year must select one of five business courses, one of three mathematics courses, two of six elective courses, and either one of four history courses or one of three social science courses. How many different curricula are available for her consideration?

72. **Drivers' Tests** A state Motor Vehicle Department requires learners to pass a written test on the motor vehicle laws of the state. The exam consists of ten true-or-false questions, of which eight must be answered correctly to qualify for a permit. In how many different ways can a learner who answers all the questions on the exam qualify for a permit?

A list of poker hands ranked in order from the highest to the lowest is shown in the following table, along with a description and example of each hand. Use the table to answer Exercises 73–78.

Hand	Description	Example
Straight flush	5 cards in sequence in the same suit	A♥ 2♥ 3♥ 4♥ 5♥
Four of a kind	4 cards of the same rank and any other card	K♥ K♦ K♠ K♣ 2♥
Full house	3 of a kind and a pair	3♥ 3♦ 3♣ 7♥ 7♦
Flush	5 cards of the same suit that are not all in sequence	5♥ 6♥ 9♥ J♥ K♥
Straight	5 cards in sequence but not all of the same suit	10♥ J♦ Q♣ K♠ A♥
Three of a kind	3 cards of the same rank and 2 unmatched cards	K♥ K♦ K♠ 2♥ 4♦
Two pair	2 cards of the same rank and 2 cards of any other rank with an unmatched card	K♥ K♦ 2♥ 2♠ 4♣
One pair	2 cards of the same rank and 3 unmatched cards	K♥ K♦ 5♥ 2♠ 4♥

If a 5-card poker hand is dealt from a well-shuffled deck of 52 cards, how many different hands consist of the following:

73. **Poker** A straight flush? (Note that an ace may be played as either a high or a low card in a straight sequence—that

is, A, 2, 3, 4, 5 or 10, J, Q, K, A. Hence there are ten possible sequences for a straight in one suit.)

74. **Poker** A straight (but not a straight flush)?

75. **Poker** A flush (but not a straight flush)?

76. **Poker** Four of a kind?

77. **Poker** A full house?

78. **Poker** Two pair?

79. **Bus Routing** The following is a schematic diagram of a city's street system between the points A and B. The City Transit Authority is in the process of selecting a route from A to B along which to provide bus service. If the company's intention is to keep the route as short as possible, how many routes must be considered?

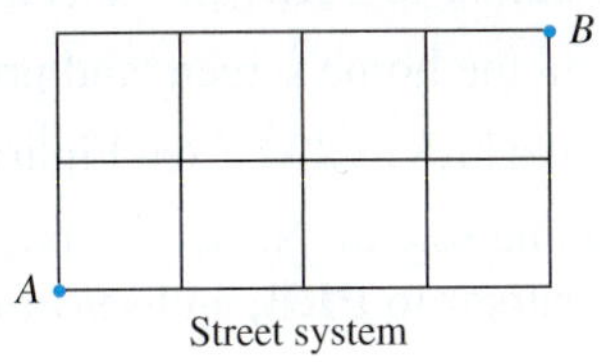

Street system

80. **World Series** In the World Series, one National League team and one American League team compete for the title, which is awarded to the first team to win four games. In how many different ways can the series be completed?

81. **Voting Quorums** A quorum (minimum) of 6 voting members is required at all meetings of the Curtis Townhomes Owners Association. If there is a total of 12 voting members in the group, find the number of ways in which this quorum can be formed.

82. **Circular Permutations** Suppose n distinct objects are arranged in a circle. Show that the number of (different) circular arrangements of the n objects is $(n - 1)!$.
Hint: Consider the arrangement of the five letters A, B, C, D, and E in the accompanying figure. The permutations $ABCDE$, $BCDEA$, $CDEAB$, $DEABC$, and $EABCD$ are not distinguishable. Generalize this observation to the case of n objects.

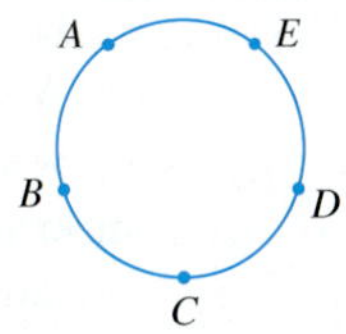

83. **Round Table** Refer to Exercise 82. In how many ways can five TV commentators be seated at a round table for a discussion?

84. **Round Table Seating** Refer to Exercise 82. In how many ways can four men and four women be seated at a round table at a dinner party if each guest is seated between members of the opposite sex?

85. At the end of Section 3.3, we mentioned that solving a linear programming problem in three variables and five constraints by the methods of corners requires that we solve 56 3×3 systems of linear equations. Verify this assertion.

86. Refer to Exercise 85. Show that to solve a linear programming problem in five variables and ten constraints, we must solve 3003 5×5 systems of linear equations. This assertion was also made at the end of Section 3.3.

In Exercises 87–90, determine whether the statement is true or false. If it is true, explain why it is true. If it is false, give an example to show why it is false.

87. The number of permutations of n distinct objects taken all together is $n!$

88. $P(n, r) = r!\, C(n, r)$

89. The number of combinations of n objects taken $n - r$ at a time is the same as the number taken r at a time.

90. If a set of n objects consists of r elements of one kind and $n - r$ elements of another kind, then the number of permutations of the n objects taken all together is $P(n, r)$.

6.4 Solutions to Self-Check Exercises

1. a. $5! = 5 \cdot 4 \cdot 3 \cdot 2 \cdot 1 = 120$

b. $C(7, 4) = \dfrac{7!}{4!\,3!} = \dfrac{7 \cdot 6 \cdot 5}{3 \cdot 2 \cdot 1} = 35$

c. $P(6, 2) = \dfrac{6!}{4!} = 6 \cdot 5 = 30$

2. There are $P(8, 2)$ ways of picking the shuttle commander and pilot (the order *is* important here), $C(12, 3)$ ways of picking the engineers (the order is not important here), $C(5, 1)$ ways of picking the scientist, and $C(2, 1)$ ways of picking the civilian. By the multiplication principle, there are

$$
\begin{aligned}
&P(8, 2) \cdot C(12, 3) \cdot C(5, 1) \cdot C(2, 1) \\
&= \frac{8!}{6!} \cdot \frac{12!}{9!\,3!} \cdot \frac{5!}{4!\,1!} \cdot \frac{2!}{1!\,1!} \\
&= \frac{8 \cdot 7 \cdot 12 \cdot 11 \cdot 10 \cdot 5 \cdot 2}{3 \cdot 2} \\
&= 123{,}200
\end{aligned}
$$

ways in which a crew can be selected.

USING TECHNOLOGY Evaluating $n!$, $P(n, r)$, and $C(n, r)$

Graphing Utility

A graphing utility can be used to calculate factorials, permutations, and combinations with relative ease. A graphing utility is therefore an indispensable tool in solving counting problems involving large numbers of objects. Here, we use the **nPr** (permutation) and **nCr** (combination) functions of a graphing utility.

EXAMPLE 1 Use a graphing utility to find (a) 12!, (b) $P(52, 5)$, and (c) $C(38, 10)$.

```
12!
          479001600
52 nPr        5
          311875200
38 nCr 10
          472733756
```

FIGURE **T1**
The TI 83/84 screen showing the entries and results for Example 1

a. To find 12!, first enter **12** on the home screen. Next, we obtain the **!** symbol on the TI-83/84 by pressing **MATH**, moving the cursor to **PRB,** and selecting **4:!** by pressing **4**. Finally, press **ENTER** to obtain 12! = 479,001,600. (See Figure T1.)

b. To find $P(52, 5)$, first enter **52** on the home screen. Next, we obtain the **nPr** symbol by pressing **MATH**, moving the cursor to **PRB,** and selecting **2: nPr** by pressing **2**. Finally, enter **5** on the home screen, and press **ENTER** to obtain $P(52, 5) = 52 \text{ nPr } 5 = 311{,}875{,}200$. (See Figure T1.)

c. To find $C(38, 10)$, first enter **38** on the home screen. Next, we obtain the **nCr** symbol by pressing **MATH**, moving the cursor to **PRB,** and selecting **3: nCr** by pressing **3**. Finally, enter **10** on the home screen, and press **ENTER** to obtain $C(38, 10) = 38 \text{ nCr } 10 = 472733756$. (See Figure T1.)

Excel

Excel has built-in functions for calculating factorials, permutations, and combinations.

EXAMPLE 2 Use Excel to calculate

a. 12! **b.** $P(52, 5)$ **c.** $C(38, 10)$

Solution

a. In cell A1, enter `=FACT(12)` and press **Shift-Enter**. The number 479001600 will appear.

b. In cell A2, enter `=PERMUT(52,5)` and press **Shift-Enter**. The number 311875200 will appear.

c. In cell A3, enter `=COMBIN(38,10)` and press **Shift-Enter**. The number 472733756 will appear.

TECHNOLOGY EXERCISES

In Exercises 1–10, evaluate the expression.

1. 15!

2. 20!

3. 4(18!)

4. $\dfrac{30!}{18!}$

5. $P(52, 7)$

6. $P(24, 8)$

7. $C(52, 7)$

8. $C(26, 8)$

9. $P(10, 4)C(12, 6)$

10. $P(20, 5)C(9, 3)C(8, 4)$

11. **EXAMS** A mathematics professor uses a computerized test bank to prepare her final exam. If 25 different problems are available for the first three exam questions, 40 differ-

---

Note: Boldfaced words/characters enclosed in a box (for example, **Enter**) indicate that an action (click, select, or press) is required. Words/characters printed blue (for example, Chart sub-type:) indicate words/characters that appear on the screen. Words/characters printed in a monospace font (for example, `=(-2/3)*A2+2)`) indicate words/characters that need to be typed and entered.

ent problems are available for the next five questions, and 30 different problems are available for the last two questions, how many different ten-question exams can she prepare? (Assume that the order of the questions within each group is not important.)

12. **Job Assignments** S & S Brokerage has received 100 inquiries from prospective clients. In how many ways can the inquiries be directed to any five of the firm's brokers if each broker handles 20 inquiries?

CHAPTER 6 Summary of Principal Formulas and Terms

FORMULAS

1. Commutative laws	$A \cup B = B \cup A$ $A \cap B = B \cap A$
2. Associative laws	$A \cup (B \cup C) = (A \cup B) \cup C$ $A \cap (B \cap C) = (A \cap B) \cap C$
3. Distributive laws	$A \cup (B \cap C) = (A \cup B) \cap (A \cup C)$ $A \cap (B \cup C) = (A \cap B) \cup (A \cap C)$
4. De Morgan's laws	$(A \cup B)^c = A^c \cap B^c$ $(A \cap B)^c = A^c \cup B^c$
5. Number of elements in the union of two finite sets	$n(A \cup B) = n(A) + n(B) - n(A \cap B)$
6. Permutation of n distinct objects, taken r at a time	$P(n, r) = \dfrac{n!}{(n-r)!}$
7. Permutation of n objects, not all distinct, taken n at a time	$\dfrac{n!}{n_1!\, n_2! \cdots n_m!}$
8. Combination of n distinct objects, taken r at a time	$C(n, r) = \dfrac{n!}{r!\,(n-r)!}$

TERMS

set (342)
element of a set (342)
roster notation (342)
set-builder notation (342)
set equality (342)
subset (343)
empty set (343)
universal set (344)
Venn diagram (344)
set union (345)
set intersection (345)
complement of a set (345)
set complementation (346)
multiplication principle (362)
generalized multiplication principle (363)
permutation (368)
n-factorial (370)
combination (374)

CHAPTER 6 Concept Review Questions

Fill in the blanks.

1. A well-defined collection of objects is called a/an _____. These objects are called _____ of the _____.

2. Two sets having exactly the same elements are said to be _____.

3. If every element of a set A is also an element of a set B, then A is a/an _____ of B.

4. **a.** The empty set $\varnothing$ is the set containing _____ elements.
   b. The universal set is the set containing _____ elements.

5. a. The set of all elements in A and/or B is called the _____ of A and B.
   b. The set of all elements in both A and B is called the _____ of A and B.

6. The set of all elements in U that are not in A is called the _____ of A.

7. Applying De Morgan's Laws, we can write $(A \cup B \cup C)^c =$ _____.

8. An arrangement of a set of distinct objects in a definite order is called a/an _____; an arrangement in which the order is not important is a/an _____.

CHAPTER 6 Review Exercises

In Exercises 1–4, list the elements of each set in roster notation.

1. $\{x \mid 3x - 2 = 7 \text{ and } x \text{ is an integer}\}$
2. $\{x \mid x \text{ is a letter of the word } TALLAHASSEE\}$
3. The set whose elements are the even numbers between 3 and 11
4. $\{x \mid (x - 3)(x + 4) = 0 \text{ and } x \text{ is a negative integer}\}$

Let $A = \{a, c, e, r\}$. In Exercises 5–8, determine whether the set is equal to A.

5. $\{r, e, c, a\}$
6. $\{x \mid x \text{ is a letter of the word } career\}$
7. $\{x \mid x \text{ is a letter of the word } racer\}$
8. $\{x \mid x \text{ is a letter of the word } cares\}$

In Exercises 9–12, shade the portion of the accompanying figure that represents the given set.

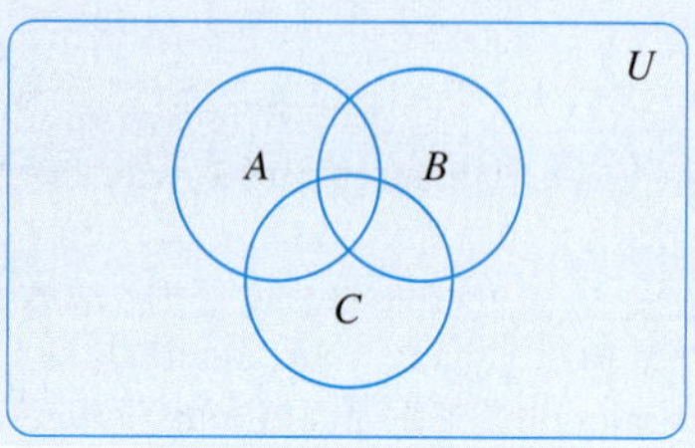

9. $A \cup (B \cap C)$
10. $(A \cap B \cap C)^c$
11. $A^c \cap B^c \cap C^c$
12. $A^c \cap (B^c \cup C^c)$

In Exercises 13–16, verify the equation by direct computation. Let $U = \{a, b, c, d, e\}$, $A = \{a, b\}$, $B = \{b, c, d\}$, and $C = \{a, d, e\}$.

13. $A \cup (B \cup C) = (A \cup B) \cup C$
14. $A \cap (B \cap C) = (A \cap B) \cap C$
15. $A \cap (B \cup C) = (A \cap B) \cup (A \cap C)$
16. $A \cup (B \cap C) = (A \cup B) \cap (A \cup C)$

For Exercises 17–20, let

U = {all participants in a consumer-behavior survey conducted by a national polling group}

A = {consumers who avoided buying a product because it is not recyclable}

B = {consumers who used cloth rather than disposable diapers}

C = {consumers who boycotted a company's products because of its record on the environment}

D = {consumers who voluntarily recycled their garbage}

Describe each set in words.

17. $A \cap C$
18. $A \cup D$
19. $B^c \cap D$
20. $C^c \cup D^c$

Let A and B be subsets of a universal set U and suppose $n(U) = 350$, $n(A) = 120$, $n(B) = 80$, and $n(A \cap B) = 50$. In Exercises 21–26, find the number of elements in each set.

21. $n(A \cup B)$
22. $n(A^c)$
23. $n(B^c)$
24. $n(A^c \cap B)$
25. $n(A \cap B^c)$
26. $n(A^c \cap B^c)$

In Exercises 27–30, evaluate each quantity.

27. $C(20, 18)$
28. $P(9, 7)$
29. $C(5, 3) \cdot P(4, 2)$
30. $4 \cdot P(5, 3) \cdot C(7, 4)$

31. **Credit Card Comparisons** A comparison of five major credit cards showed that

    3 offered cash advances.

    3 offered extended payments for all goods and services purchased.

    2 required an annual fee of less than \$35.

    2 offered both cash advances and extended payments.

    1 offered extended payments and had an annual fee less than \$35.

    No card had an annual fee less than \$35 and offered both cash advances and extended payments.

    How many cards had an annual fee less than \$35 and offered cash advances? (Assume that every card had at least one of the three mentioned features.)

32. **Student Surveys** The Department of Foreign Languages of a liberal arts college conducted a survey of its recent graduates to determine the foreign language courses they had taken while undergraduates at the college. Of the 480 graduates,

200 had at least 1 year of Spanish.

178 had at least 1 year of French.

140 had at least 1 year of German.

33 had at least 1 year of Spanish and French.

24 had at least 1 year of Spanish and German.

18 had at least 1 year of French and German.

3 had at least 1 year of all three languages.

How many of the graduates had

a. At least 1 year of at least one of the three languages?
b. At least 1 year of exactly one of the three languages?
c. Less than 1 year of any of the three languages?

33. In how many ways can six different DVDs be arranged on a shelf?

34. In how many ways can three pictures be selected from a group of six different pictures?

35. In how many ways can six different books, four of which are math books, be arranged on a shelf if the math books must be placed next to each other?

36. In how many ways can six people be arranged in a line for a group picture

a. If there are no restrictions?
b. If two people in the group insist on not standing next to each other?

37. Find the number of distinguishable permutations that can be formed from the letters of each word.

a. *CINCINNATI* **b.** *HONOLULU*

38. How many three-digit numbers can be formed from the numerals in the set $\{1, 2, 3, 4, 5\}$ if

a. Repetition of digits is not allowed?
b. Repetition of digits is allowed?

39. **Automobile Selection** An automobile manufacturer has three different subcompact cars in the line. Customers selecting one of these cars have a choice of three engine sizes, four body styles, and three color schemes. How many different selections can a customer make?

40. **Menu Selections** Two soups, five entrées, and three desserts are listed on the "Special" menu at the Neptune Restaurant. How many different selections consisting of one soup, one entrée, and one dessert can a customer choose from this menu?

41. **Investments** In a survey conducted by Helena, a financial consultant, it was revealed that of her 400 clients

300 own stocks.

180 own bonds.

160 own mutual funds.

110 own both stocks and bonds.

120 own both stocks and mutual funds.

90 own both bonds and mutual funds.

How many of Helena's clients own stocks, bonds, and mutual funds?

42. **Poker** From a standard 52-card deck, how many 5-card poker hands can be dealt consisting of

a. Five clubs? **b.** Three kings and one pair?

43. **Elections** In an election being held by the Associated Students Organization, there are six candidates for president, four for vice president, five for secretary, and six for treasurer. How many different possible outcomes are there for this election?

44. **Team Selection** There are eight seniors and six juniors in the Math Club at Jefferson High School. In how many ways can a math team consisting of four seniors and two juniors be selected from the members of the Math Club?

45. If order matters, in how many ways can two cards be drawn from a 52-card deck

a. If the first card is replaced before the second card is drawn?
b. If the second card is drawn without replacing the first card?

46. **Seating Arrangements** In how many ways can seven students be assigned seats in a row containing seven desks if:

a. There are no restrictions?
b. Two of the students must not be seated next to each other?

47. **Quality Control** From a shipment of 60 CPUs, 5 of which are defective, a sample of 4 CPUs is selected at random.

a. In how many different ways can the sample be selected?
b. How many samples contain 3 defective CPUs?
c. How many samples do not contain any defective CPUs?

48. **Random Samples** A sample of 4 balls is to be selected at random from an urn containing 15 balls numbered 1 to 15. If 6 balls are green, 5 are white, and 4 are black, then:

a. How many different samples can be selected?
b. How many samples can be selected that contain at least 1 white ball?

49. **Seating Arrangements** In how many ways can three married couples be seated in a row of six seats:

a. If there are no seating restrictions?
b. If men and women must alternate?
c. If each married couple must sit together?

50. **Team Selection** There are seven boys and five girls in the debate squad at Franklin High School. In how many ways can a four-member debating team be selected:

a. If there are no restrictions?
b. If two boys and two girls must be on the team?
c. If at least two boys must be on the team?

CHAPTER 6 Before Moving On . . .

1. Let $U = \{a, b, c, d, e, f, g\}$, $A = \{a, d, f, g\}$, $B = \{d, f, g\}$, and $C = \{b, c, e, f\}$. Find:

a. $A \cap (B \cup C)$

b. $(A \cap C) \cup (B \cup C)$

c. A^c

2. Let A, B, and C be subsets of a universal set U, and suppose that $n(U) = 120$, $n(A) = 20$, $n(A \cap B) = 10$, $n(A \cap C) = 11$, $n(B \cap C) = 9$, and $n(A \cap B \cap C) = 4$. Find $n[A \cap (B \cup C)^c]$.

3. In how many ways can four compact discs be selected from six different compact discs?

4. From a standard 52-card deck, how many 5-card poker hands can be dealt consisting of 3 deuces and 2 face cards?

5. There are six seniors and five juniors in the Chess Club at Madison High School. In how many ways can a team consisting of three seniors and two juniors be selected from the members of the Chess Club?

7 Probability

THE SYSTEMATIC STUDY of probability began in the seventeenth century when certain aristocrats wanted to discover superior strategies to use in the gaming rooms of Europe. Some of the best mathematicians of the period were engaged in this pursuit. Since then, probability has evolved in virtually every sphere of human endeavor in which an element of uncertainty is present.

We begin by introducing some of the basic terminology used in the study of the subject. Then, in Section 7.2, we give the technical meaning of the term *probability*. The rest of this chapter is devoted to the development of techniques for computing the probabilities of the occurrence of certain events.

What is the probability that a randomly chosen couple from a certain large metropolitan area is in the upper-income bracket if both spouses are working? In Example 2, page 443, we show how the income distribution for that metropolitan area can be used to determine this probability.

7.1 Experiments, Sample Spaces, and Events

Terminology

A number of specialized terms are used in the study of probability. We begin by defining the term *experiment*.

Experiment

An **experiment** is an activity with observable results.

The results of the experiment are called the **outcomes** of the experiment. Three examples of experiments are the following:

- Tossing a coin and observing whether it falls heads or tails
- Rolling a die and observing whether the number 1, 2, 3, 4, 5, or 6 shows up
- Testing a spark plug from a batch of 100 spark plugs and observing whether or not it is defective

In our discussion of experiments, we use the following terms:

Sample Point, Sample Space, and Event

Sample point: An outcome of an experiment

Sample space: The set consisting of all possible sample points of an experiment

Event: A subset of a sample space of an experiment

The sample space of an experiment is a universal set whose elements are precisely the outcomes, or the sample points, of the experiment; the events of the experiment are the subsets of the universal set. A sample space associated with an experiment that has a finite number of possible outcomes (sample points) is called a **finite sample space.**

Since the events of an experiment are subsets of a universal set (the sample space of the experiment), we may use the results for set theory given in Chapter 6 to help us study probability. The event B is said to **occur** in a trial of an experiment whenever B contains the observed outcome. We begin by explaining the roles played by the empty set and a universal set when they are viewed as events associated with an experiment. The empty set, $\varnothing$, is called the *impossible event*; it cannot occur because $\varnothing$ has no elements (outcomes). Next, the universal set S is referred to as the *certain event*; it must occur because S contains all the outcomes of the experiment.

This terminology is illustrated in the next several examples.

EXAMPLE 1 Describe the sample space associated with the experiment of tossing a coin and observing whether it falls heads or tails. What are the events of this experiment?

Solution The two outcomes are heads and tails, and the required sample space is given by $S = \{\text{H}, \text{T}\}$, where H denotes the outcome heads and T denotes the outcome tails. The events of the experiment, the subsets of S, are

$$\varnothing, \{\text{H}\}, \{\text{T}\}, S$$

Note that we have included the impossible event, $\varnothing$, and the certain event, S. ■

Since the events of an experiment are subsets of the sample space of the experiment, we may talk about the union and intersection of any two events; we can also consider the complement of an event with respect to the sample space.

> **Union of Two Events**
> The **union of two events** E and F is the event $E \cup F$.

Thus, the event $E \cup F$ contains the set of outcomes of E and/or F.

> **Intersection of Two Events**
> The **intersection of two events** E and F is the event $E \cap F$.

Thus, the event $E \cap F$ contains the set of outcomes common to E and F.

> **Complement of an Event**
> The **complement of event** E is the event E^c.

Thus, the event E^c is the set containing all the outcomes in the sample space S that are not in E.

Venn diagrams depicting the union, intersection, and complement of events are shown in Figure 1. These concepts are illustrated in the following example.

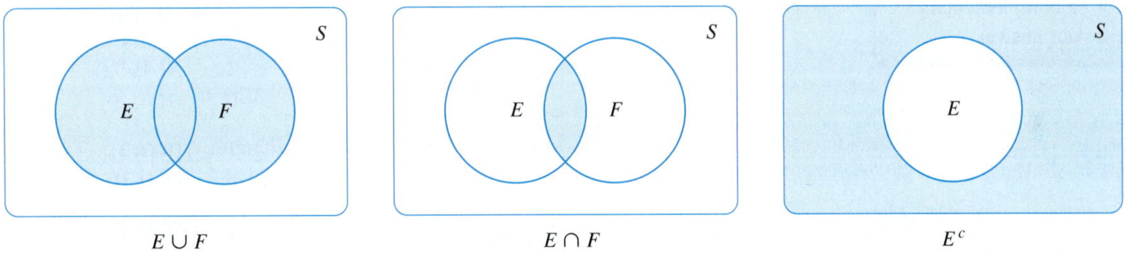

(a) The union of two events **(b)** The intersection of two events **(c)** The complement of the event E

FIGURE **1**

EXAMPLE 2 Consider the experiment of rolling a die and observing the number that falls uppermost. Let $S = \{1, 2, 3, 4, 5, 6\}$ denote the sample space of the experiment, and let $E = \{2, 4, 6\}$ and $F = \{1, 3\}$ be events of this experiment. List the set of outcomes of (a) $E \cup F$, (b) $E \cap F$, and (c) F^c. Interpret your results.

Solution

a. $E \cup F = \{1, 2, 3, 4, 6\}$ and is the event that the outcome of the experiment is a 1, a 2, a 3, a 4, or a 6.

b. $E \cap F = \varnothing$ is the impossible event; the number appearing uppermost when a die is rolled cannot be both even and odd at the same time.

c. $F^c = \{2, 4, 5, 6\}$ is precisely the event that the event F does not occur.

If two events cannot occur at the same time, they are said to be mutually exclusive. Using set notation, we have the following definition.

> **Mutually Exclusive Events**
>
> E and F are **mutually exclusive** if $E \cap F = \emptyset$.

As before, we may use Venn diagrams to illustrate these events. In this case, the two mutually exclusive events are depicted as two nonintersecting circles (Figure 2).

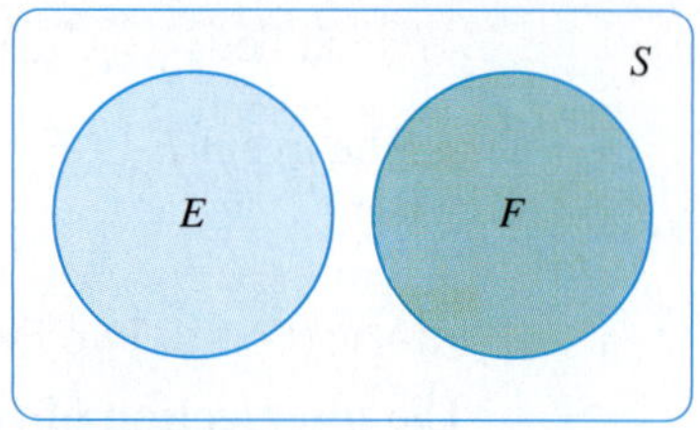

FIGURE 2
Mutually exclusive events

EXAMPLE 3 An experiment consists of tossing a coin three times and observing the resulting sequence of heads and tails.

a. Describe the sample space S of the experiment.
b. Determine the event E that exactly two heads appear.
c. Determine the event F that at least one head appears.

> **Explore and Discuss**
>
> 1. Suppose E and F are two complementary events. Must E and F be mutually exclusive? Explain your answer.
> 2. Suppose E and F are mutually exclusive events. Must E and F be complementary? Explain your answer.

Solution

a. The sample points may be obtained with the aid of a tree diagram (Figure 3).

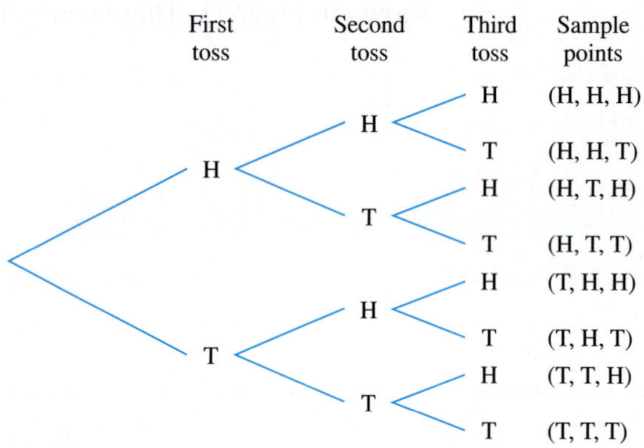

FIGURE 3

The required sample space S is given by

$$S = \{\text{HHH, HHT, HTH, HTT, THH, THT, TTH, TTT}\}$$

b. By scanning the sample space S obtained in part (a), we see that the outcomes in which exactly two heads appear are given by the event

$$E = \{\text{HHT, HTH, THH}\}$$

c. Proceeding as in part (b), we find

$$F = \{\text{HHH, HHT, HTH, HTT, THH, THT, TTH}\}$$

EXAMPLE 4 An experiment consists of rolling a pair of dice and observing the number that falls uppermost on each die.

a. Describe an appropriate sample space S for this experiment.

b. Determine the events $E_2, E_3, E_4, \ldots, E_{12}$ that the sum of the numbers falling uppermost is 2, 3, 4, . . . , 12, respectively.

Solution

a. We may represent each outcome of the experiment by an ordered pair of numbers, the first representing the number that appears uppermost on the first die and the second representing the number that appears uppermost on the second die. To distinguish between the two dice, think of the first die as being red and the second as being green. Since there are six possible outcomes for each die, the multiplication principle implies that there are 6 · 6, or 36, elements in the sample space:

$$\begin{aligned} S = \{&(1, 1), (1, 2), (1, 3), (1, 4), (1, 5), (1, 6), \\ &(2, 1), (2, 2), (2, 3), (2, 4), (2, 5), (2, 6), \\ &(3, 1), (3, 2), (3, 3), (3, 4), (3, 5), (3, 6), \\ &(4, 1), (4, 2), (4, 3), (4, 4), (4, 5), (4, 6), \\ &(5, 1), (5, 2), (5, 3), (5, 4), (5, 5), (5, 6), \\ &(6, 1), (6, 2), (6, 3), (6, 4), (6, 5), (6, 6)\} \end{aligned}$$

b. With the aid of the results of part (a), we obtain the required list of events, shown in Table 1.

TABLE 1

Sum of Uppermost Numbers	Event
2	$E_2 = \{(1, 1)\}$
3	$E_3 = \{(1, 2), (2, 1)\}$
4	$E_4 = \{(1, 3), (2, 2), (3, 1)\}$
5	$E_5 = \{(1, 4), (2, 3), (3, 2), (4, 1)\}$
6	$E_6 = \{(1, 5), (2, 4), (3, 3), (4, 2), (5, 1)\}$
7	$E_7 = \{(1, 6), (2, 5), (3, 4), (4, 3), (5, 2), (6, 1)\}$
8	$E_8 = \{(2, 6), (3, 5), (4, 4), (5, 3), (6, 2)\}$
9	$E_9 = \{(3, 6), (4, 5), (5, 4), (6, 3)\}$
10	$E_{10} = \{(4, 6), (5, 5), (6, 4)\}$
11	$E_{11} = \{(5, 6), (6, 5)\}$
12	$E_{12} = \{(6, 6)\}$

APPLIED EXAMPLE 5 Movie Attendance The manager of a local cinema records the number of patrons attending a first-run movie at the 1 P.M. screening. The theater has a seating capacity of 500.

a. What is an appropriate sample space for this experiment?
b. Describe the event E that fewer than 50 people attend the screening.
c. Describe the event F that the theater is more than half full at the screening.

Solution

a. The number of patrons at the screening (the outcome) could run from 0 to 500. Therefore, a sample space for this experiment is

$$S = \{0, 1, 2, 3, \ldots, 500\}$$

b. $E = \{0, 1, 2, 3, \ldots, 49\}$

c. $F = \{251, 252, 253, \ldots, 500\}$

APPLIED EXAMPLE 6 Family Birth Order An experiment consists of recording, in order of their births, the sex composition of a three-child family in which the children were born at different times.

a. Describe an appropriate sample space S for this experiment.
b. Describe the event E that there are two girls and a boy in the family.
c. Describe the event F that the oldest child is a girl.
d. Describe the event G that the oldest child is a girl and the youngest child is a boy.

Solution

a. The sample points of the experiment may be obtained with the aid of the tree diagram shown in Figure 4, where b denotes a boy and g denotes a girl.

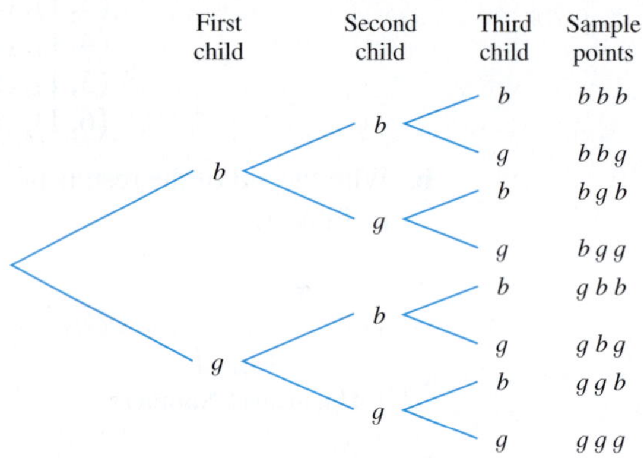

FIGURE 4
Tree diagram for three-child families

We see from the tree diagram that the required sample space is given by

$$S = \{bbb, bbg, bgb, bgg, gbb, gbg, ggb, ggg\}$$

Using the tree diagram, we find that:

b. $E = \{bgg, gbg, ggb\}$
c. $F = \{gbb, gbg, ggb, ggg\}$
d. $G = \{gbb, ggb\}$

Explore and Discuss

Think of an experiment.

1. Describe the sample point(s) and sample space of the experiment.
2. Construct two events, E and F, of the experiment.
3. Find the union and intersection of E and F and the complement of E.
4. Are E and F mutually exclusive? Explain your answer.

Sample spaces may be infinite, as illustrated in the next example.

APPLIED EXAMPLE 7 Testing New Products EverBrite is developing a high-amperage, high-capacity battery as a source for powering electric cars. The battery is tested by installing it in a prototype electric car and running the car with a fully charged battery on a test track at a constant speed of 55 mph until the car runs out of power. The distance covered by the car is then observed.

a. What is the sample space for this experiment?
b. Describe the event E that the driving range of the prototype car under test conditions is less than 150 miles.
c. Describe the event F that the driving range of the prototype car is between 200 and 250 miles, inclusive.

Solution

a. Since the distance d covered by the car in any run may be any nonnegative number, the sample space S is given by

$$S = \{d \mid d \geq 0\}$$

b. The event E is given by

$$E = \{d \mid 0 \le d < 150\}$$

c. The event F is given by

$$F = \{d \mid 200 \le d \le 250\}$$

7.1 Self-Check Exercises

1. **Sampling Fruit** A sample of three apples taken from Cavallero's Fruit Stand is examined to determine whether the apples are good or rotten.
   a. What is an appropriate sample space for this experiment?
   b. Describe the event E that exactly one of the apples picked is rotten.
   c. Describe the event F that the first apple picked is rotten.

2. **Sampling Fruit** Refer to Self-Check Exercise 1.
   a. Find $E \cup F$.
   b. Find $E \cap F$.
   c. Find F^c.
   d. Are the events E and F mutually exclusive?

Solutions to Self-Check Exercises 7.1 can be found on page 396.

7.1 Concept Questions

1. Explain what is meant by an experiment. Give an example. For the example you have chosen, describe (a) a sample point, (b) the sample space, and (c) an event of the experiment.

2. What does it mean for two events to be mutually exclusive? Give an example of two mutually exclusive events E and F. How can you prove that they are mutually exclusive?

7.1 Exercises

In Exercises 1–6, let $S = \{a, b, c, d, e, f\}$ be a sample space of an experiment and let $E = \{a, b\}$, $F = \{a, d, f\}$, and $G = \{b, c, e\}$ be events of this experiment.

1. Find the events $E \cup F$ and $E \cap F$.
2. Find the events $F \cup G$ and $F \cap G$.
3. Find the events F^c and $E \cap G^c$.
4. Find the events E^c and $F^c \cap G$.
5. Are the events E and F mutually exclusive?
6. Are the events $E \cup F$ and $E \cap F^c$ mutually exclusive?

In Exercises 7–14, let $S = \{1, 2, 3, 4, 5, 6\}$, $E = \{2, 4, 6\}$, $F = \{1, 3, 5\}$, and $G = \{5, 6\}$.

7. Find the event $E \cup F \cup G$.
8. Find the event $E \cap F \cap G$.
9. Find the event $(E \cup F \cup G)^c$.
10. Find the event $(E \cap F \cap G)^c$.
11. Are the events E and F mutually exclusive?
12. Are the events F and G mutually exclusive?
13. Are the events E and F complementary?
14. Are the events F and G complementary?

In Exercises 15–20, let S be any sample space, and let E, F, and G be any three events associated with the experiment. Describe the events using the symbols $\cup$, $\cap$, and ^{c}.

15. The event that E and/or F occurs
16. The event that both E and F occur
17. The event that G does not occur
18. The event that E but not F occurs
19. The event that none of the events E, F, and G occurs
20. The event that E occurs but neither of the events F or G occurs

21. Consider the sample space S of Example 4, page 390.
  a. Determine the event that the number that falls uppermost on the first die is greater than the number that falls uppermost on the second die.
  b. Determine the event that the number that falls uppermost on the second die is double the number that falls uppermost on the first die.

22. Consider the sample space S of Example 4, page 390.
  a. Determine the event that the sum of the numbers falling uppermost is less than or equal to 7.
  b. Determine the event that the number falling uppermost on one die is a 4 and the number falling uppermost on the other die is greater than 4.

23. Let $S = \{a, b, c\}$ be a sample space of an experiment with outcomes a, b, and c. List all the events of this experiment.

24. Let $S = \{1, 2, 3\}$ be a sample space associated with an experiment.
  a. List all the events of this experiment.
  b. How many subsets of S contain the number 3?
  c. How many subsets of S contain either the number 2 or the number 3?

25. An experiment consists of selecting a card from a standard deck of playing cards and noting whether the card is black (B) or red (R).
  a. Describe an appropriate sample space for this experiment.
  b. What are the events of this experiment?

26. An experiment consists of selecting a letter at random from the letters in the word *MASSACHUSETTS* and observing the outcomes.
  a. What is an appropriate sample space for this experiment?
  b. Describe the event "the letter selected is a vowel."

27. An experiment consists of tossing a coin, rolling a die, and observing the outcomes.
  a. Describe an appropriate sample space for this experiment.
  b. Describe the event "a head is tossed and an even number is rolled."

28. An experiment consists of spinning the hand of the numbered disc shown in the following figure and then observing the region in which the pointer stops. (If the needle stops on a line, the result is discounted, and the needle is spun again.)

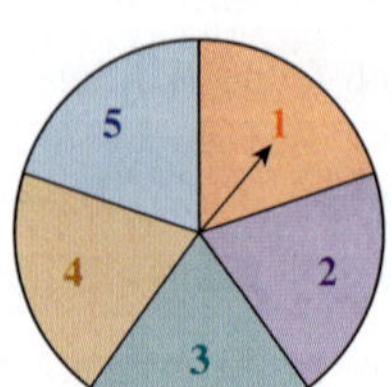

  a. What is an appropriate sample space for this experiment?
  b. Describe the event "the spinner points to the number 2."
  c. Describe the event "the spinner points to an odd number."

29. A die is rolled, and the number that falls uppermost is observed. Let E denote the event that the number shown is a 2, and let F denote the event that the number shown is an even number.
  a. Are the events E and F mutually exclusive?
  b. Are the events E and F complementary?

30. A die is rolled, and the number that falls uppermost is observed. Let E denote the event that the number shown is even, and let F denote the event that the number is an odd number.
  a. Are the events E and F mutually exclusive?
  b. Are the events E and F complementary?

31. **Quality Control** A sample of three transistors taken from a local electronics store was examined to determine whether the transistors were defective (d) or nondefective (n). What is an appropriate sample space for this experiment?

32. **Selecting Job Applicants** From a list of five applicants for a sales position, a, b, c, d, and e, two are selected for the next round of interviews.
  a. Describe an appropriate sample space S for this experiment.
  b. Describe the event E that the interviewees include applicant a.
  c. Describe the event F that the interviewees include applicants a and c.
  d. Describe the event G that the interviewees include applicants d and e.

33. **Family Birth Order** An experiment consists of recording, in order of their births, the sex composition of a four-child family in which the children were born at different times.
  a. Describe an appropriate sample space S for this experiment.
  b. Describe the event E that there are three boys and a girl in the family.
  c. Describe the event F that the youngest child is a girl.
  d. Describe the event G that the oldest and the youngest children are both girls.

34. **Blood Types** Human blood is classified by the presence or absence of three main antigens (A, B, and Rh). When a blood specimen is typed, the presence of the A and/or B antigen is indicated by listing the letter A and/or the letter B. If neither the A nor the B antigen is present, the letter O is used. The presence or absence of the Rh antigen is indicated by the symbols $+$ or $-$, respectively. Thus, if a blood specimen is classified as AB^+, it contains the A and the B antigens as well as the Rh antigen. Similarly, O^- blood contains none of the three antigens. Using this information, determine the sample space corresponding to the different blood groups.

35. **Game Shows** In a television game show, the winner is asked to select three prizes from five different prizes, A, B, C, D, and E.
   a. Describe a sample space of possible outcomes (order is not important).
   b. How many points are there in the sample space corresponding to a selection that includes A?
   c. How many points are there in the sample space corresponding to a selection that includes A and B?
   d. How many points are there in the sample space corresponding to a selection that includes either A or B?

36. **ATMs** The manager of a local bank observes how long it takes a customer to complete his transactions at the automatic bank teller machine (ATM).
   a. Describe an appropriate sample space for this experiment.
   b. Describe the event that it takes a customer between 2 and 3 min to complete his transactions at the ATM.

37. **Price Changes in Common Stocks** Robin purchased shares of a machine tool company and shares of an airline company. Let E be the event that the shares of the machine tool company increase in value over the next 6 months, and let F be the event that the shares of the airline company increase in value over the next 6 months. Using the symbols $\cup$, $\cap$, and ^{c}, describe the following events.
   a. The shares in the machine tool company do not increase in value.
   b. The shares in both the machine tool company and the airline company do not increase in value.
   c. The shares of at least one of the two companies increase in value.
   d. The shares of only one of the two companies increase in value.

38. **Quality Assurance Surveys** The customer service department of Universal Instruments, manufacturer of the Orion tablet computer, conducted a survey among customers who had returned their purchase registration cards. Purchasers of its tablet computer were asked to report the length of time t in days before service was required.
   a. Describe a sample space corresponding to this survey.
   b. Describe the event E that a tablet computer required service before a period of 90 days had elapsed.
   c. Describe the event F that a tablet computer did not require service before a period of 1 year had elapsed.

39. **Assembly-Time Studies** A time study was conducted by the production manager of Vista Vision to determine the length of time in minutes required by an assembly worker to complete a certain task during the assembly of its Pulsar HDTV sets.
   a. Describe a sample space corresponding to this time study.
   b. Describe the event E that an assembly worker took 2 min or less to complete the task.
   c. Describe the event F that an assembly worker took more than 2 min to complete the task.

40. **Political Polls** An opinion poll is conducted among a state's electorate to determine the relationship between their income levels and their stands on a proposition aimed at reducing state income taxes. Voters are classified as belonging to either the low-, middle-, or upper-income group. They are asked whether they favor, oppose, or are undecided about the proposition. Let the letters L, M, and U represent the low-, middle-, and upper-income groups, respectively, and let the letters f, o, and u represent the responses—favor, oppose, and undecided, respectively.
   a. Describe a sample space corresponding to this poll.
   b. Describe the event E_1 that a respondent favors the proposition.
   c. Describe the event E_2 that a respondent opposes the proposition and does not belong to the low-income group.
   d. Describe the event E_3 that a respondent does not favor the proposition and does not belong to the upper-income group.

41. **Quality Control** As part of a quality-control procedure, an inspector at Bristol Farms randomly selects ten eggs from each consignment of eggs he receives and records the number of broken eggs.
   a. What is an appropriate sample space for this experiment?
   b. Describe the event E that at most three eggs are broken.
   c. Describe the event F that at least five eggs are broken.

42. **Political Polls** In the opinion poll of Exercise 40, the voters were also asked to indicate their political affiliations: Democrat, Republican, or Independent. As before, let the letters L, M, and U represent the low-, middle-, and upper-income groups, respectively. Let the letters D, R, and I represent Democrat, Republican, and Independent, respectively.
   a. Describe a sample space corresponding to this poll.
   b. Describe the event E_1 that a respondent is a Democrat.
   c. Describe the event E_2 that a respondent belongs to the upper-income group and is a Republican.
   d. Describe the event E_3 that a respondent belongs to the middle-income group and is not a Democrat.

43. **Shuttle Bus Usage** A certain airport hotel operates a shuttle bus service between the hotel and the airport. The maximum capacity of a bus is 20 passengers. On alternate trips of the shuttle bus over a period of 1 week, the hotel manager kept a record of the number of passengers arriving at the hotel in each bus.
   a. What is an appropriate sample space for this experiment?
   b. Describe the event E that a shuttle bus carried fewer than ten passengers.
   c. Describe the event F that a shuttle bus arrived with a full load.

44. **Sports** Eight players, A, B, C, D, E, F, G, and H, are competing in a series of elimination matches of a tennis tournament in which the winner of each preliminary match will advance to the semifinals and the winners of the semifinals will advance to the finals. An outline of the scheduled matches follows. Describe a sample space listing the possible participants in the finals.

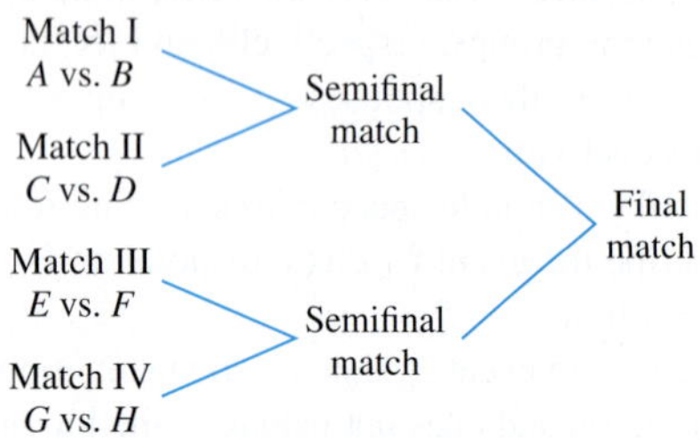

45. An experiment consists of selecting a card at random from a well-shuffled 52-card deck. Let E denote the event that an ace is drawn, and let F denote the event that a spade is drawn. Show that $n(E \cup F) = n(E) + n(F) - n(E \cap F)$.

46. Let S be a sample space for an experiment. Show that if E is any event of an experiment, then E and E^c are mutually exclusive.

47. Let S be a sample space for an experiment, and let E and F be events of this experiment. Show that the events $E \cup F$ and $E^c \cap F^c$ are mutually exclusive.
Hint: Use De Morgan's Law.

48. Let S be a sample space of an experiment with n outcomes. Determine the number of events of this experiment.

In Exercises 49 and 50, determine whether the statement is true or false. If it is true, explain why it is true. If it is false, give an example to show why it is false.

49. If E and F are mutually exclusive and E and G are mutually exclusive, then F and G are mutually exclusive.

50. The numbers 1, 2, and 3 are written separately on three pieces of paper. These slips of paper are then placed in a bowl. If you draw two slips from the bowl, one at a time and without replacement, then the sample space for this experiment consists of six elements.

7.1 Solutions to Self-Check Exercises

1. a. Let g denote a good apple, and let r denote a rotten apple. Thus, the required sample points may be obtained with the aid of a tree diagram (compare with Example 3). The required sample space is given by

$$S = \{ggg, ggr, grg, grr, rgg, rgr, rrg, rrr\}$$

b. By scanning the sample space S obtained in part (a), we identify the outcomes in which exactly one apple is rotten. We find

$$E = \{ggr, grg, rgg\}$$

c. Proceeding as in part (b), we find

$$F = \{rgg, rgr, rrg, rrr\}$$

2. Using the results of Self-Check Exercise 1, we find:
a. $E \cup F = \{ggr, grg, rgg, rgr, rrg, rrr\}$
b. $E \cap F = \{rgg\}$
c. F^c is the set of outcomes in S but not in F. Thus,

$$F^c = \{ggg, ggr, grg, grr\}$$

d. Since $E \cap F \neq \varnothing$, we conclude that E and F are not mutually exclusive.

7.2 Definition of Probability

Finding the Probability of an Event

Let's return to the coin-tossing experiment. The sample space of this experiment is given by $S = \{\text{H}, \text{T}\}$, where the sample points H and T correspond to the two possible outcomes, heads and tails. If the coin is *unbiased*, then there is *one chance out of two* of obtaining a head (or a tail), and we say that the *probability* of tossing a head (tail) is $\frac{1}{2}$, abbreviated

$$P(\text{H}) = \frac{1}{2} \quad \text{and} \quad P(\text{T}) = \frac{1}{2}$$

An alternative method of obtaining the values of $P(\text{H})$ and $P(\text{T})$ is based on continued experimentation and does not depend on the assumption that the two outcomes are equally likely. Table 2 summarizes the results of such an exercise.

TABLE 2

Tossing a Coin: As the Number of Trials Increases, the Relative Frequency Approaches .5

Number of Tosses, n	Number of Heads, m	Relative Frequency of Heads, m/n
10	4	.4000
100	58	.5800
1,000	492	.4920
10,000	5,034	.5034
20,000	10,024	.5012
40,000	20,032	.5008

Observe that the relative frequencies (column 3) differ considerably when the number of trials is small, but as the number of trials becomes very large, the relative frequency approaches the number .5. This result suggests that we assign to $P(\mathrm{H})$ the value $\frac{1}{2}$, as before.

More generally, consider an experiment that may be repeated over and over again under independent and similar conditions. Suppose that in n trials an event E occurs m times. We call the ratio m/n the **relative frequency** of the event E after n repetitions. If this relative frequency approaches some value $P(E)$ as n becomes larger and larger, then $P(E)$ is called the **empirical probability** of E. Thus, the probability $P(E)$ of an event occurring is a measure of the proportion of the time that the event E will occur in the long run. Observe that this method of computing the probability of a head occurring is effective even when a biased coin is used in the experiment. The relative frequency distribution is often referred to as an *observed* or *empirical probability distribution.*

The **probability of an event** is a number that lies between 0 and 1, inclusive. In general, the larger the probability of an event, the more likely that the event will occur. Thus, an event with a probability of .8 is more likely to occur than an event with a probability of .6. An event with a probability of $\frac{1}{2}$, or .5, has a fifty–fifty chance of occurring.

Now suppose we are given an experiment and wish to determine the probabilities associated with certain events of the experiment. This problem could be solved by computing $P(E)$ directly for each event E of interest. In practice, however, the number of events of interest is usually quite large, so this approach is not satisfactory.

The following approach is particularly suitable when the sample space of an experiment is finite.* Let S be a finite sample space with n outcomes; that is,

$$S = \{s_1, s_2, s_3, \ldots, s_n\}$$

Then the events

$$\{s_1\}, \{s_2\}, \{s_3\}, \ldots, \{s_n\}$$

that consist of exactly one point are called the **elementary**, or **simple**, **events** of the experiment. They are elementary in the sense that any (nonempty) event of the experiment may be obtained by taking a finite union of suitable elementary events. The simple events of an experiment are also *mutually exclusive;* that is, given any two simple events of the experiment, only one can occur.

By assigning probabilities to each of the simple events, we obtain the results shown in Table 3. This table is called a **probability distribution** for the experiment. The function P, which assigns a probability to each of the simple events, is called a **probability function.**

TABLE 3

A Probability Distribution

Simple Event	Probability*
$\{s_1\}$	$P(s_1)$
$\{s_2\}$	$P(s_2)$
$\{s_3\}$	$P(s_3)$
$\vdots$	$\vdots$
$\{s_n\}$	$P(s_n)$

*For simplicity, we use the notation $P(s_i)$ instead of the technically more correct $P(\{s_i\})$.

The numbers $P(s_1), P(s_2), \ldots, P(s_n)$ have the following properties:

1. $0 \le P(s_i) \le 1 \quad i = 1, 2, \ldots, n$
2. $P(s_1) + P(s_2) + \cdots + P(s_n) = 1$
3. $P(\{s_i\} \cup \{s_j\}) = P(s_i) + P(s_j)\ (i \ne j) \quad i = 1, 2, \ldots, n; j = 1, 2, \ldots, n$

*For the remainder of the chapter, we assume that all sample spaces are finite.

The first property simply states that the probability of a simple event must be between 0 and 1, inclusive. The second property states that the sum of the probabilities of all simple events of the sample space is 1. This follows from the fact that the event S is certain to occur. The third property states that the probability of the union of two simple events is given by the sum of their probabilities.

Exploring with TECHNOLOGY

We can use a graphing calculator to simulate the coin-tossing experiment described earlier. Associate the outcome "a head" with the number 1 and the outcome "a tail" with the number 0. Select the function **randInt(** on the TI-83/84. (You can find this by pressing MATH and then moving the cursor to **PRB**.) Select **randInt(** and enter `0,1)` and then press ENTER repeatedly. This generates 0s and 1s randomly, which simulates the results of tossing an unbiased coin.

As we saw earlier, there is no unique method for assigning probabilities to the simple events of an experiment. In practice, the methods that are used to determine these probabilities may range from theoretical considerations of the problem on the one extreme to the reliance on "educated guesses" on the other.

Sample spaces in which the outcomes are equally likely are called **uniform sample spaces.** Assigning probabilities to the simple events in these spaces is relatively easy.

Probability of an Event in a Uniform Sample Space

If

$$S = \{s_1, s_2, \ldots, s_n\}$$

is the sample space for an experiment in which the outcomes are equally likely, then we assign the probabilities

$$P(s_1) = P(s_2) = \cdots = P(s_n) = \frac{1}{n}$$

to each of the simple events $\{s_1\}, \{s_2\}, \ldots, \{s_n\}$.

TABLE 4

A Probability Distribution

Simple Event	Probability
$\{1\}$	$\frac{1}{6}$
$\{2\}$	$\frac{1}{6}$
$\{3\}$	$\frac{1}{6}$
$\{4\}$	$\frac{1}{6}$
$\{5\}$	$\frac{1}{6}$
$\{6\}$	$\frac{1}{6}$

EXAMPLE 1 A fair die is rolled, and the number that falls uppermost is observed. Determine the probability distribution for the experiment.

Solution The sample space for the experiment is $S = \{1, 2, 3, 4, 5, 6\}$, and the simple events are accordingly given by the sets $\{1\}$, $\{2\}$, $\{3\}$, $\{4\}$, $\{5\}$, and $\{6\}$. Since the die is assumed to be fair, the six outcomes are equally likely. We therefore assign a probability of $\frac{1}{6}$ to each of the simple events and obtain the probability distribution shown in Table 4.

Explore and Discuss

You suspect that a die is biased.

1. Describe a method you might use to show that your assertion is correct.
2. How would you assign the probability to each outcome 1 through 6 of an experiment that consists of rolling the die and observing the number that lands uppermost?

The next example shows how the *relative frequency* interpretation of probability lends itself to the computation of probabilities.

APPLIED EXAMPLE 2 Testing New Products Refer to Example 7, Section 7.1. The data shown in Table 5 were obtained in tests involving 200 test runs. Each run was made with a fully charged battery.

a. Describe an appropriate sample space for this experiment.

b. Find the empirical probability distribution for this experiment.

TABLE 5

Data Obtained During 200 Test Runs of an Electric Car

Distance Covered in Miles, x	Frequency of Occurrence
$0 < x \leq 50$	4
$50 < x \leq 100$	10
$100 < x \leq 150$	30
$150 < x \leq 200$	100
$200 < x \leq 250$	40
$250 < x$	16

Solution

a. Let s_1 denote the outcome that the distance covered by the car does not exceed 50 miles; let s_2 denote the outcome that the distance covered by the car is greater than 50 miles but does not exceed 100 miles, and so on. Finally, let s_6 denote the outcome that the distance covered by the car is greater than 250 miles. Then the required sample space is given by

$$S = \{s_1, s_2, s_3, s_4, s_5, s_6\}$$

b. To compute the empirical probability distribution for the experiment, we turn to the relative frequency interpretation of probability. Accepting the inaccuracies inherent in a relatively small number of trials (200 runs), we take the probability of s_1 occurring as

$$P(s_1) = \frac{\text{Number of trials in which } s_1 \text{ occurs}}{\text{Total number of trials}}$$

$$= \frac{4}{200} = .02$$

In a similar manner, we assign probabilities to the other simple events, obtaining the probability distribution shown in Table 6.

TABLE 6

A Probability Distribution

Simple Event	Probability
$\{s_1\}$	.02
$\{s_2\}$	.05
$\{s_3\}$	.15
$\{s_4\}$	.50
$\{s_5\}$	.20
$\{s_6\}$	.08

We are now in a position to give a procedure for computing the probability $P(E)$ of an arbitrary event E of an experiment.

Finding the Probability of an Event E

1. Determine a sample space S associated with the experiment.
2. Assign probabilities to the simple events of S.
3. If $E = \{s_1, s_2, s_3, \ldots, s_n\}$, where $\{s_1\}, \{s_2\}, \{s_3\}, \ldots, \{s_n\}$ are simple events, then

$$P(E) = P(s_1) + P(s_2) + P(s_3) + \cdots + P(s_n)$$

If E is the empty set, $\varnothing$, then $P(E) = 0$.

The principle stated in Step 3 is called the **addition principle** and is a consequence of Property 3 of the probability function (page 397). This principle allows us to find the probabilities of all other events once the probabilities of the simple events are known.

The addition rule in Step 3 applies *only* to the addition of probabilities of simple events.

APPLIED EXAMPLE 3 Rolling a Pair of Dice A pair of fair dice is rolled.

a. Calculate the probability that the two dice show the same number.
b. Calculate the probability that the sum of the numbers of the two dice is 6.

Solution From the results of Example 4, Section 7.1, page 390, we see that the sample space S of the experiment consists of 36 outcomes:

$$S = \{(1, 1), (1, 2), \ldots, (6, 5), (6, 6)\}$$

Since both dice are fair, each of the 36 outcomes is equally likely. Accordingly, we assign the probability of $\frac{1}{36}$ to each simple event.

a. The event that the two dice show the same number is given by

$$E = \{(1, 1), (2, 2), (3, 3), (4, 4), (5, 5), (6, 6)\}$$

(Figure 5). Therefore, by the addition principle, the probability that the two dice show the same number is given by

$$\begin{aligned} P(E) &= P[(1, 1)] + P[(2, 2)] + \cdots + P[(6, 6)] \\ &= \frac{1}{36} + \frac{1}{36} + \cdots + \frac{1}{36} \qquad \text{Six terms} \\ &= \frac{1}{6} \end{aligned}$$

b. The event that the sum of the numbers of the two dice is 6 is given by

$$E_6 = \{(1, 5), (2, 4), (3, 3), (4, 2), (5, 1)\}$$

(Figure 6). Therefore, the probability that the sum of the numbers on the two dice is 6 is given by

$$\begin{aligned} P(E_6) &= P[(1, 5)] + P[(2, 4)] + P[(3, 3)] + P[(4, 2)] + P[(5, 1)] \\ &= \frac{1}{36} + \frac{1}{36} + \cdots + \frac{1}{36} \qquad \text{Five terms} \\ &= \frac{5}{36} \end{aligned}$$

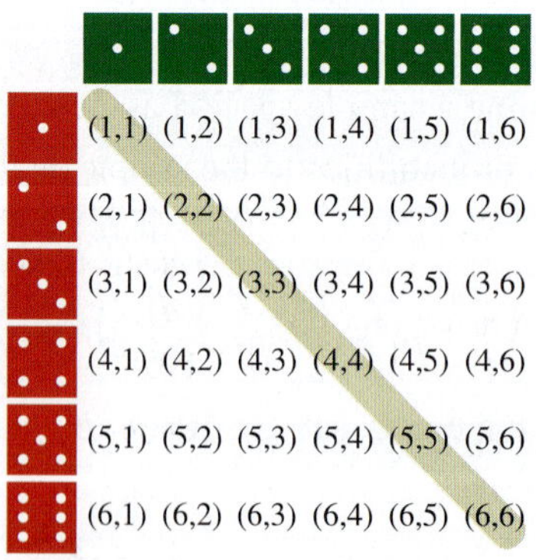

FIGURE 5
The event that the two dice show the same number

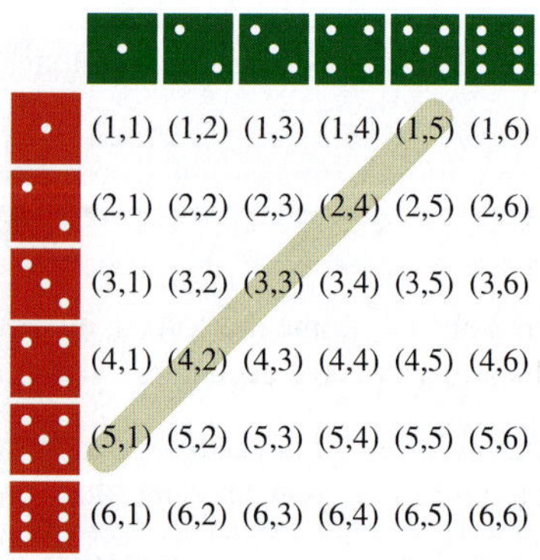

FIGURE 6
The event that the sum of the numbers on the two dice is 6

APPLIED EXAMPLE 4 Testing New Products Consider the experiment by EverBrite in Example 2. What is the probability that the prototype car will travel more than 150 miles on a fully charged battery?

Solution Using the results of Example 2, we see that the event that the car will travel more than 150 miles on a fully charged battery is given by $E = \{s_4, s_5, s_6\}$. Therefore, the probability that the car will travel more than 150 miles on one charge is given by

$$P(E) = P(s_4) + P(s_5) + P(s_6)$$

or, using the probability distribution for the experiment obtained in Example 2,

$$P(E) = .50 + .20 + .08 = .78$$

7.2 Self-Check Exercises

1. A biased die was rolled repeatedly, and the results of the experiment are summarized in the following table:

Outcome	1	2	3	4	5	6
Frequency of Occurrence	142	173	158	175	162	190

Using the relative frequency interpretation of probability, find the empirical probability distribution for this experiment.

2. **Accident Prevention** In an experiment conducted to study the effectiveness of an eye-level third brake light in the prevention of rear-end collisions, 250 of the 500 highway patrol cars of a certain state were equipped with such lights. At the end of the 1-year trial period, the records revealed that for those equipped with a third brake light there were 14 incidents of rear-end collision. There were 22 such incidents involving the cars not equipped with the accessory. On the basis of these data, what is the probability that a highway patrol car equipped with a third brake light will be rear-ended within a 1-year period? What is the probability that a car not so equipped will be rear-ended within a 1-year period?

Solutions to Self-Check Exercises 7.2 can be found on page 406.

7.2 Concept Questions

1. Define (a) a probability distribution and (b) a probability function. Give examples of each.

2. If $S = \{s_1, s_2, \ldots, s_n\}$ is the sample space for an experiment in which the outcomes are equally likely, what is the probability of each of the simple events $s_1, s_2, \ldots, s_n$? What is this type of sample space called?

3. Suppose $E = \{s_1, s_2, s_3, \ldots, s_n\}$, where E is an event of an experiment and $\{s_1\}, \{s_2\}, \{s_3\}, \ldots, \{s_n\}$ are simple events. If E is nonempty, what is $P(E)$? If E is empty, what is $P(E)$?

7.2 Exercises

In Exercises 1–8, list the simple events associated with each experiment.

1. A nickel and a dime are tossed, and the result of heads or tails is recorded for each coin.

2. A card is selected at random from a standard 52-card deck, and its suit—hearts (h), diamonds (d), spades (s), or clubs (c)—is recorded.

3. **Opinion Polls** An opinion poll is conducted among a group of registered voters. Their political affiliation—Democrat (D), Republican (R), or Independent (I)—and their sex—male (m) or female (f)—are recorded.

4. **Quality Control** As part of a quality-control procedure, eight circuit boards are checked, and the number of defective boards is recorded.

5. **Movie Attendance** In a survey conducted to determine whether movie attendance is increasing (i), decreasing (d), or holding steady (s) among various sectors of the population, participants are classified as follows:

Group 1: Those aged 10–19

Group 2: Those aged 20–29

Group 3: Those aged 30–39

Group 4: Those aged 40–49

Group 5: Those aged 50 and older

The response and age group of each participant are recorded.

6. **Durable Goods Orders** An economist obtains data concerning durable goods orders each month. A record is

kept for a 1-year period of any increase (i), decrease (d), or unchanged movement (u) in the number of durable goods orders for each month as compared with the number of such orders in the same month of the previous year.

7. **Blood Types** Blood tests are given as a part of the admission procedure at the Monterey Garden Community Hospital. The blood type of each patient (A, B, AB, or O) and the presence or absence of the Rh factor in each patient's blood (Rh^+ or Rh^-) are recorded.

8. **Meteorology** A meteorologist preparing a weather map classifies the expected average temperature in each of five neighboring states (MN, WI, IA, IL, MO) for the upcoming week as follows:
   a. More than 10° below average
   b. Normal to 10° below average
   c. Higher than normal to 10° above average
   d. More than 10° above average

   Using each state's abbreviation and the categories—(a), (b), (c), and (d)—the meteorologist records these data.

9. **Social Media Accounts** In a survey of 1000 social media account holders, the following question was asked: How many social media accounts do you have? The results of the survey are summarized below:

Answer	1–2	3–4	5 or more
Respondents	650	200	150

   a. Determine the empirical probability distribution associated with these data.
   b. What is the probability that a participant in the survey selected at random answered that he or she had three or four accounts?

   Source: AARP.

10. **Workplace** In a survey of 26,612 Parade.com visitors, the following question was asked: How do workers get ahead? The results of the survey are as follows:

Answer	Internal politics	Hard work	Initiative	Creativity
Respondents	13,572	7,185	4,790	1,065

   a. Determine the empirical probability distribution associated with these data.
   b. What is the probability that a participant in the survey selected at random answered that one gets ahead at work through hard work?

   Source: Yahoo Finance/Parade Survey.

11. **Stress Level** In a study on stress experienced by Americans, 800 adults ages 18 years and older were asked to rate their stress level as low, middle, or extreme. The results of the survey are summarized below:

Answer	Low	Middle	Extreme	No response
Respondents	272	352	160	16

   a. Determine the empirical probability distribution associated with these data.
   b. What is the probability that a participant in the survey answered that he or she had experienced an extreme stress level?

   Source: American Psychological Association.

12. **Blood Types** The percentage of the general population that has each blood type is shown in the following table. Determine the probability distribution associated with these data.

Blood Type	A	B	AB	O
Population (%)	41	12	3	44

13. **Grade Distributions** The grade distribution for a certain class is shown in the following table. Find the probability distribution associated with these data.

Grade	A	B	C	D	F
Frequency of Occurrence	4	10	18	6	2

14. **Great Recession** In a survey conducted by AlixPartners of 4980 adults 18 years old and older in June 2009, during the "Great Recession," the following question was asked: How long do you think it will take to recover your personal net worth? The results of the survey follow:

Answer (in years)	1–2	3–4	5–10	>10
Respondents	1006	1308	2113	553

   a. Determine the empirical probability distribution associated with these data.
   b. If a person who participated in the survey is selected at random, what is the probability that he or she expected that it would take 5 or more years to recover his or her personal net worth?

   Source: AlixPartners.

15. **Starting a New Job** In a survey of 420 workers, the following question was asked: What are the greatest challenges when starting a new job? The results of the survey are as follows:

Answer	New processes/ procedures	Getting to know a new boss and coworkers	New technology tools	Fitting into the corporate culture	Other
Respondents	185	84	71	50	30

   a. Determine the empirical probability distribution associated with these data.
   b. What is the probability that a participant in the survey selected at random answered that the greatest challenges when starting a new job were fitting into the corporate culture?

   Source: Accountemps.

16. **Safety of American-Made Products** The accompanying data were obtained from a survey of 1500 Americans who were asked: How safe are American-made consumer products? Determine the empirical probability distribution associated with these data.

Rating	A	B	C	D	E
Respondents	285	915	225	30	45

A: Very safe

B: Somewhat safe

C: Not too safe

D: Not safe at all

E: Don't know

17. **Political Views of College Freshmen** In a poll conducted among 2000 college freshmen to ascertain the political views of college students, the accompanying data were obtained. Determine the empirical probability distribution associated with these data.

Political Views	A	B	C	D	E
Respondents	52	398	1140	386	24

A: Far left

B: Liberal

C: Middle of the road

D: Conservative

E: Far right

18. **Red-Light Runners** In a survey of 800 likely voters, the following question was asked: Do you support using cameras to identify red-light runners? The results of the survey follow:

Answer	Strongly support	Somewhat support	Somewhat oppose	Strongly oppose	Don't know
Respondents	360	192	88	144	16

What is the probability that a person in the survey selected at random favors using cameras to identify red-light runners?

Source: Public Opinion Strategies.

19. **Cooking at Home** In an online survey of 500 adults living with children under the age of 18 years, the participants were asked how many days per week they cook at home. The results of the survey are summarized below:

Number of Days	0	1	2	3	4	5	6	7
Respondents	25	30	45	75	55	100	85	85

Determine the empirical probability distribution associated with these data.

Source: Super Target.

20. **Checking into a Hotel Room** In a survey of 3019 hotel guests, the following question was asked: What is the first thing you do after checking into a hotel room? The results of the survey follow:

Activity	Respondents
Adjust the thermostat	1027
Turn on the TV	755
Unpack	634
Check out the free toiletries	211
Plug in rechargeable electronics	60
Find the gym	30
Other	302

What is the probability that a person selected at random from the list of guests surveyed would, as a first activity:

a. Adjust the thermostat or plug in rechargeable electronics?

b. Unpack or find the gym?

Source: Tripadvisor.

21. **Traffic Surveys** The number of cars entering a tunnel leading to an airport in a major city over a period of 200 peak hours was observed, and the following data were obtained:

Number of Cars, x	Frequency of Occurrence
$0 < x \leq 200$	15
$200 < x \leq 400$	20
$400 < x \leq 600$	35
$600 < x \leq 800$	70
$800 < x \leq 1000$	45
$x > 1000$	15

a. Describe an appropriate sample space for this experiment.

b. Find the empirical probability distribution for this experiment.

22. **Arrival Times of Commuter Trains** The arrival times of the 8 A.M. Boston-based commuter train as observed in the suburban town of Sharon over 120 weekdays is summarized below:

Arrival Time, x	Frequency of Occurrence
7:56 A.M. $< x \leq$ 7:58 A.M.	4
7:58 A.M. $< x \leq$ 8:00 A.M.	18
8:00 A.M. $< x \leq$ 8:02 A.M.	50
8:02 A.M. $< x \leq$ 8:04 A.M.	32
8:04 A.M. $< x \leq$ 8:06 A.M.	9
8:06 A.M. $< x \leq$ 8:08 A.M.	4
8:08 A.M. $< x \leq$ 8:10 A.M.	3

a. Describe an appropriate sample space for this experiment.

b. Find the empirical probability distribution for this experiment.

23. **Corrective Lens Use** According to Mediamark Research, during a certain year 84 million out of 179 million adults in the United States corrected their vision by using prescription eyeglasses, bifocals, or contact lenses. (Some respondents use more than one type.) What is the probability that an adult selected at random from the adult population uses corrective lenses?
Source: Mediamark Research.

24. **Correctional Supervision** A study conducted by the Corrections Department of a certain state revealed that 163,605 people out of a total adult population of 1,778,314 were under correctional supervision (on probation, on parole, or in jail). What is the probability that a person selected at random from the adult population in that state is under correctional supervision?

25. **Lightning Deaths** According to data obtained from the National Weather Service, 376 of the 439 people killed by lightning in the United States over a 7-year period were men. (Job and recreational habits of men make them more vulnerable to lightning.) Assuming that this trend holds in the future, what is the probability that a person killed by lightning:
a. Is a male? **b.** Is a female?
Source: National Weather Service.

26. **Quality Control** One light bulb is selected at random from a lot of 120 light bulbs, of which 5% are defective. What is the probability that the light bulb selected is defective?

27. **Efforts to Stop Shoplifting** According to a survey of 176 retailers, 46% of them use electronic tags as protection against shoplifting and employee theft. If one of these retailers is selected at random, what is the probability that the retailer uses electronic tags as antitheft devices?

28. If a ball is selected at random from an urn containing three red balls, two white balls, and five blue balls, what is the probability that it will be a white ball?

29. If a card is drawn at random from a standard 52-card deck, what is the probability that the card drawn is:
a. A diamond? **b.** A black card?
c. An ace?

30. A pair of fair dice is rolled. What is the probability that:
a. The sum of the numbers shown uppermost is less than 5?
b. At least one 6 is rolled?

31. **Traffic Lights** What is the probability of arriving at a traffic light when it is red if the red signal is lit for 30 sec, the yellow signal for 5 sec, and the green signal for 45 sec?

32. **Roulette** What is the probability that a roulette ball will come to rest on an even number other than 0 or 00? (Assume that there are 38 equally likely outcomes consisting of the numbers 1–36, 0, and 00.)

In Exercises 33–35, determine whether the given experiment has a sample space with equally likely outcomes.

33. A loaded die is rolled, and the number appearing uppermost on the die is recorded.

34. Two fair dice are rolled, and the sum of the numbers appearing uppermost is recorded.

35. A ball is selected at random from an urn containing six black balls and six red balls, and the color of the ball is recorded.

36. Let $S = \{s_1, s_2, s_3, s_4, s_5, s_6\}$ be the sample space associated with an experiment having the following probability distribution:

Outcome	s_1	s_2	s_3	s_4	s_5	s_6
Probability	$\frac{1}{12}$	$\frac{1}{4}$	$\frac{1}{12}$	$\frac{1}{6}$	$\frac{1}{3}$	$\frac{1}{12}$

Find the probability of the event:
a. $A = \{s_1, s_3\}$
b. $B = \{s_2, s_4, s_5, s_6\}$
c. $C = S$

37. Let $S = \{s_1, s_2, s_3, s_4, s_5\}$ be the sample space associated with an experiment having the following probability distribution:

Outcome	s_1	s_2	s_3	s_4	s_5
Probability	$\frac{1}{14}$	$\frac{3}{14}$	$\frac{6}{14}$	$\frac{2}{14}$	$\frac{2}{14}$

Find the probability of the event:
a. $A = \{s_1, s_2, s_4\}$
b. $B = \{s_1, s_5\}$
c. $C = S$

38. A pair of fair dice is rolled, and the sum of the two numbers falling uppermost is observed. The probability of obtaining a sum of 2 is the same as that of obtaining a 7 since there is only one way of getting a 2—namely, by each die showing a 1; and there is only one way of obtaining a 7—namely, by one die showing a 3 and the other die showing a 4. What is wrong with this argument?

39. **Selecting Job Applicants** Refer to Exercises 7.1, Problem 32. From a list of five applicants for a sales position, *a*, *b*, *c*, *d*, and *e*, two are selected for the next round of interviews. If the applicants are selected at random, what is the probability that the two interviewees chosen:
a. Include applicant *a*?
b. Include applicants *a* and *c*?
c. Include applicants *d* and *e*?

40. **Family Birth Order** Refer to Exercises 7.1, Problem 33. An experiment consists of recording the sex composition, in order of their births, of a four-child family in which the children were born at different times. Assuming that a boy is equally likely as a girl to be born into a family, what is the probability that a four-child family chosen at random will have:
a. Three boys and a girl in the family?

b. A youngest child in the family who is a girl?
c. An oldest child and a youngest child in the family who are both girls?

41. **Disposition of Criminal Court Cases** Of the 98 first-degree murder cases from 2002 through the first half of 2004 in the Suffolk superior court, 9 cases were thrown out of the system, 62 cases were plea-bargained, and 27 cases went to trial. What is the probability that a case selected at random
a. Was settled through plea bargaining?
b. Went to trial?
Source: Boston Globe.

42. **Sweepstakes** In a sweepstakes sponsored by Gemini Paper Products, 100,000 entries have been received. If 1 grand prize, 5 first prizes, 25 second prizes, and 500 third prizes are to be awarded, what is the probability that a person who has submitted one entry will win:
a. The grand prize?
b. A prize?

43. **Political Polls** An opinion poll was conducted among a group of registered voters in a certain state concerning a proposition aimed at limiting state and local taxes. Results of the poll indicated that 35% of the voters favored the proposition, 32% were against it, and the remaining group were undecided. If the results of the poll are assumed to be representative of the opinions of the state's electorate, what is the probability that a registered voter selected at random from the electorate:
a. Favors the proposition?
b. Is undecided about the proposition?

44. **Security Breaches** In a survey of 106 senior information technology and data security professionals at major U.S. companies regarding their confidence that they had detected all significant security breaches in the past year, the following responses were obtained:

Answer	Very confident	Moderately confident	Not very confident	Not at all confident
Respondents	21	56	22	7

What is the probability that a respondent in the survey selected at random:
a. Had little or no confidence that he or she had detected all significant security breaches in the past year?
b. Was very confident that he or she had detected all significant security breaches in the past year?
Source: Forsythe Solutions Group.

45. **Green Companies** In a survey conducted in a certain year of 1004 adults 18 years old and older, the following question was asked: How are American companies doing on protecting the environment compared with companies in other countries? The results are summarized below:

Answer	Behind	Equal	Ahead	Don't know
Respondents	382	281	251	90

If an adult in the survey is selected at random, what is the probability that he or she said that American companies are equal or ahead on protecting the environment compared with companies in other countries?
Source: GfK Roper.

46. **Parental Influence on Children's Career Choices** In an online survey of 1962 executives from 64 countries conducted by Korn/Ferry International between August and October in a certain year, the executives were asked whether they would try to influence their children's career choices. Their replies: A (to a very great extent), B (to a great extent), C (to some extent), D (to a small extent), and E (not at all) are recorded below:

Answer	A	B	C	D	E
Respondents	135	404	1057	211	155

What is the probability that a randomly selected respondent's answer was D (to a small extent) or E (not at all)?
Source: Korn/Ferry International.

47. **Spending Methods** In a survey on consumer spending methods conducted in a certain year, the following results were obtained:

Payment Method	Checks	Cash	Credit cards	Debit/ATM cards	Other
Transactions (%)	37	14	25	15	9

If a transaction tracked in this survey is selected at random, what is the probability that the transaction was paid for:
a. With a credit card or with a debit/ATM card?
b. With cash or some method other than with a check, a credit card, or a debit/ATM card?
Source: Minute/Visa USA Research Services.

48. **Staying in Touch** In a poll conducted by the Pew Research Center in a certain year, 2000 adults ages 18 years old and older were asked how frequently they are in touch with their parents by phone. The results of the poll are as follows:

Answer	Monthly	Weekly	Daily	Don't know	Less
Respondents (%)	11	47	32	2	8

If a person who participated in the poll is selected at random, what is the probability that the person said he or she kept in touch with his or her parents:
a. Once a week?
b. At least once a week?
Source: Pew Research Center.

49. **Music Venues** In a survey designed to determine where people listen to music in their home, 1000 people were asked in which room at home they were mostly likely to listen to music. The results are tabulated below:

Room	Living room	Master bedroom	Study/home office	Kitchen	Bathroom	Other
Respondents	448	169	155	100	22	106

If a respondent is selected at random, what is the probability that he or she most likely listens to music:

a. In the living room?

b. In the study/home office or the kitchen?

Source: Phillips Electronics.

50. Retirement Benefits Versus Salary In a survey conducted in a certain year of 1402 workers 18 years old and older regarding their opinion on retirement benefits, the following data were obtained: 827 said that it was better to have excellent retirement benefits with a lower-than-expected salary, 477 said that it was better to have a higher-than-expected salary with poor retirement benefits, 42 said "neither," and 56 said "not sure." If a worker in the survey is selected at random, what is the probability that he or she answered that it was better to have:

a. Excellent retirement benefits with a lower-than-expected salary?

b. A higher-than-expected salary with poor retirement benefits?

Source: Transamerica Center for Retirement.

51. Airline Safety In an attempt to study the leading causes of airline crashes, the following data were compiled from records of airline crashes over a 35-year period (excluding sabotage and military action):

Primary Factor	Accidents
Pilot	327
Airplane	49
Maintenance	14
Weather	22
Airport/air traffic control	19
Miscellaneous/other	15

Assume that you have just learned of an airline crash and that the data give a generally good indication of the causes of airline crashes. Give an estimate of the probability that the primary cause of the crash was due to pilot error or bad weather.

Source: National Transportation Safety Board.

52. Housing Appreciation In a survey conducted in fall 2006, a year before the financial crisis of 2007–2008, 800 homeowners were asked about their expectations regarding the value of their home in the next few years; the results of the survey are as follows:

Expectations	Homeowners
Decrease	48
Stay the same	152
Increase less than 5%	232
Increase 5–10%	240
Increase more than 10%	128

If a homeowner in the survey is chosen at random, what is the probability that he or she expected his or her home to:

a. Stay the same or decrease in value in the next few years?

b. Increase 5% or more in value in the next few years?

Source: S&P, RBC Capital Markets.

In Exercises 53 and 54, determine whether the statement is true or false. If it is true, explain why it is true. If it is false, give an example to show why it is false.

53. If $S = \{s_1, s_2, \ldots, s_n\}$ is a uniform sample space with n outcomes, then $0 \le P(s_1) + P(s_2) + \cdots + P(s_n) \le 1$.

54. Let $S = \{s_1, s_2, \ldots, s_n\}$ be a uniform sample space for an experiment. If $n \ge 5$ and $E = \{s_1, s_2, s_5\}$, then $P(E) = 3/n$.

7.2 Solutions to Self-Check Exercises

1.

$$P(1) = \frac{\text{Number of trials in which a 1 appears uppermost}}{\text{Total number of trials}} = \frac{142}{1000} = .142$$

Similarly, we compute $P(2), \ldots, P(6)$, obtaining the following probability distribution:

Outcome	1	2	3	4	5	6
Probability	.142	.173	.158	.175	.162	.190

2. The probability that a highway patrol car equipped with a third brake light will be rear-ended within a 1-year period is given by

$$\frac{\text{Number of rear-end collisions involving cars equipped with a third brake light}}{\text{Total number of such cars}} = \frac{14}{250} = .056$$

The probability that a highway patrol car not equipped with a third brake light will be rear-ended within a 1-year period is given by

$$\frac{\text{Number of rear-end collisions involving cars not equipped with a third brake light}}{\text{Total number of such cars}} = \frac{22}{250} = .088$$

7.3 Rules of Probability

Properties of the Probability Function and Their Applications

In this section, we examine some of the properties of the probability function and look at the role they play in solving certain problems. We begin by looking at the generalization of the three properties of the probability function, which were stated for simple events in Section 7.2. Let S be a sample space of an experiment, and suppose E and F are events of the experiment. We have the following properties:

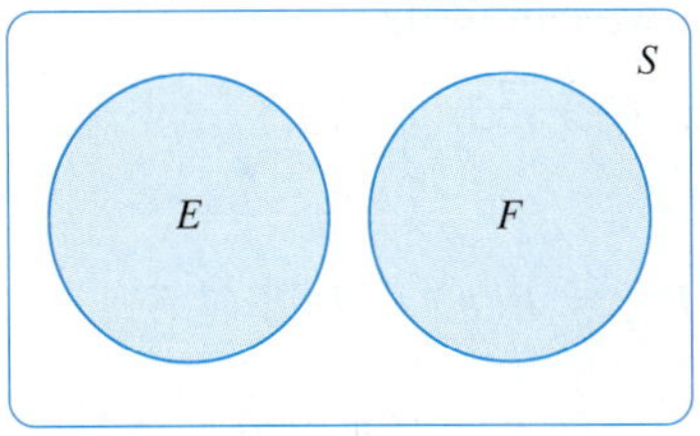

FIGURE 7
If E and F are mutually exclusive events, then $P(E \cup F) = P(E) + P(F)$.

Property 1. $P(E) \geq 0$ for any E.

Property 2. $P(S) = 1$.

Property 3. If E and F are mutually exclusive (that is, only one of them can occur or, equivalently, $E \cap F = \varnothing$), then

$$P(E \cup F) = P(E) + P(F)$$

(Figure 7).

Property 3 may be easily extended to the case involving any finite number of mutually exclusive events. Thus, if $E_1, E_2, \ldots, E_n$ are mutually exclusive events, then

$$P(E_1 \cup E_2 \cup \cdots \cup E_n) = P(E_1) + P(E_2) + \cdots + P(E_n)$$

TABLE 7

Probability Distribution

Score, x	Probability
$x > 700$	.01
$600 < x \leq 700$	.07
$500 < x \leq 600$	.19
$400 < x \leq 500$	.23
$300 < x \leq 400$	.31
$x \leq 300$	.19

APPLIED EXAMPLE 1 SAT Verbal Scores The superintendent of a metropolitan school district has estimated the probabilities associated with the SAT verbal scores of students from that district. The results are shown in Table 7. If a student is selected at random, what is the probability that his or her SAT verbal score will be:

a. More than 400?
b. Less than or equal to 500?
c. Greater than 400 but less than or equal to 600?

Solution Let A, B, C, D, E, and F denote, respectively, the event that the score is greater than 700, greater than 600 but less than or equal to 700, greater than 500 but less than or equal to 600, and so forth. Then these events are mutually exclusive. Therefore,

a. The probability that the student's score will be more than 400 is given by

$$\begin{aligned} P(D \cup C \cup B \cup A) &= P(D) + P(C) + P(B) + P(A) \\ &= .23 + .19 + .07 + .01 \\ &= .5 \end{aligned}$$

b. The probability that the student's score will be less than or equal to 500 is given by

$$\begin{aligned} P(D \cup E \cup F) &= P(D) + P(E) + P(F) \\ &= .23 + .31 + .19 = .73 \end{aligned}$$

c. The probability that the student's score will be greater than 400 but less than or equal to 600 is given by

$$\begin{aligned} P(C \cup D) &= P(C) + P(D) \\ &= .19 + .23 = .42 \end{aligned}$$

Property 3 holds if and only if E and F are mutually exclusive. In the general case, we have the following rule:

Property 4. Addition Rule

If E and F are any two events of an experiment, then

$$P(E \cup F) = P(E) + P(F) - P(E \cap F)$$

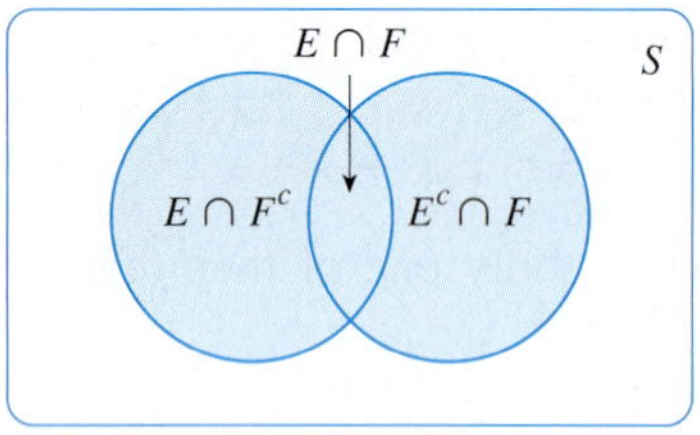

FIGURE 8
$E \cup F = (E \cap F^c) \cup (E \cap F) \cup (E^c \cap F)$

To derive this property, refer to Figure 8. Observe that we can write

$$E = (E \cap F^c) \cup (E \cap F) \quad \text{and} \quad F = (E^c \cap F) \cup (E \cap F)$$

as a union of disjoint sets. Therefore,

$$P(E) = P(E \cap F^c) + P(E \cap F) \quad \text{or} \quad P(E \cap F^c) = P(E) - P(E \cap F)$$

and

$$P(F) = P(E^c \cap F) + P(E \cap F) \quad \text{or} \quad P(E^c \cap F) = P(F) - P(E \cap F)$$

Finally, since $E \cup F = (E \cap F^c) \cup (E \cap F) \cup (E^c \cap F)$ is a union of disjoint sets, we have

$$\begin{aligned} P(E \cup F) &= P(E \cap F^c) + P(E \cap F) + P(E^c \cap F) \\ &= P(E) - P(E \cap F) + P(E \cap F) + P(F) - P(E \cap F) \quad \text{Use the earlier results.} \\ &= P(E) + P(F) - P(E \cap F) \end{aligned}$$

Note Observe that if E and F are mutually exclusive—that is, if $E \cap F = \varnothing$—then the equation of Property 4 reduces to that of Property 3. In other words, if E and F are mutually exclusive events, then $P(E \cup F) = P(E) + P(F)$. If E and F are not mutually exclusive events, then $P(E \cup F) = P(E) + P(F) - P(E \cap F)$.

EXAMPLE 2 A card is drawn from a well-shuffled deck of 52 playing cards. What is the probability that it is an ace or a spade?

Solution Let E denote the event that the card drawn is an ace and let F denote the event that the card drawn is a spade. Then

$$P(E) = \frac{4}{52} \quad \text{and} \quad P(F) = \frac{13}{52}$$

Furthermore, E and F are not mutually exclusive events. In fact, $E \cap F$ is the event that the card drawn is the ace of spades. Consequently,

$$P(E \cap F) = \frac{1}{52}$$

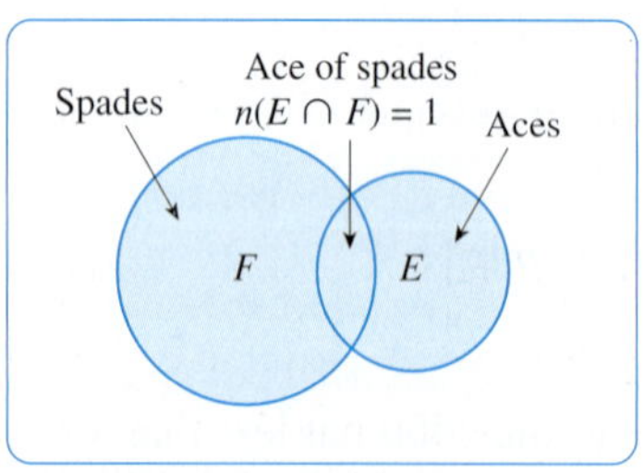

FIGURE 9
$P(E \cup F) = P(E) + P(F) - P(E \cap F)$

The event that a card drawn is an ace or a spade is $E \cup F$, with probability given by

$$\begin{aligned} P(E \cup F) &= P(E) + P(F) - P(E \cap F) \\ &= \frac{4}{52} + \frac{13}{52} - \frac{1}{52} = \frac{16}{52} = \frac{4}{13} \end{aligned}$$

(Figure 9). This result, of course, can be obtained by arguing that 16 of the 52 cards are either spades or aces of other suits.

Explore and Discuss

Let E, F, and G be any three events of an experiment. Use Equation (5) of Section 6.2 to show that

$$P(E \cup F \cup G) = P(E) + P(F) + P(G) - P(E \cap F) - P(E \cap G) - P(F \cap G) + P(E \cap F \cap G)$$

If E, F, and G are pairwise mutually exclusive, what is $P(E \cup F \cup G)$?

APPLIED EXAMPLE 3 Product Reliability The quality-control department of Vista Vision, manufacturer of the Pulsar 42-inch plasma TV, has determined from records obtained from the company's service centers that 3% of the sets sold experience video problems, 1% experience audio problems, and 0.1% experience both video and audio problems before the expiration of the 90-day warranty. Find the probability that a plasma TV purchased by a consumer will experience video or audio problems before the warranty expires.

Solution Let E denote the event that a plasma TV purchased will experience video problems within 90 days, and let F denote the event that a plasma TV purchased will experience audio problems within 90 days. Then

$$P(E) = .03 \qquad P(F) = .01 \qquad P(E \cap F) = .001$$

The event that a plasma TV purchased will experience video problems or audio problems before the warranty expires is $E \cup F$, and the probability of this event is given by

$$\begin{aligned} P(E \cup F) &= P(E) + P(F) - P(E \cap F) \\ &= .03 + .01 - .001 \\ &= .039 \end{aligned}$$

(Figure 10).

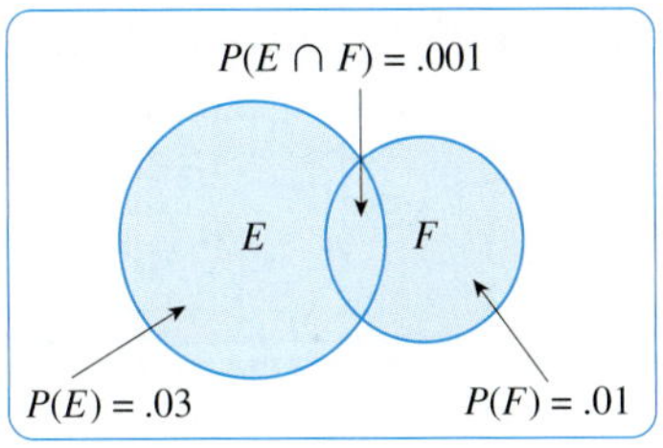

FIGURE **10**
$P(E \cup F) = P(E) + P(F) - P(E \cap F)$

Here is another property of a probability function that is of considerable aid in computing the probability of an event:

Property 5. Rule of Complements
If E is an event of an experiment and E^c denotes the complement of E, then

$$P(E^c) = 1 - P(E)$$

Property 5 is an immediate consequence of Properties 2 and 3. Indeed, we have $E \cup E^c = S$ and $E \cap E^c = \varnothing$, so

$$1 = P(S) = P(E \cup E^c) = P(E) + P(E^c)$$

and therefore,

$$P(E^c) = 1 - P(E)$$

APPLIED EXAMPLE 4 Warranties Refer to Example 3. What is the probability that a Pulsar 42-inch plasma TV bought by a consumer will *not* experience video or audio difficulties before the warranty expires?

Solution Let E denote the event that a plasma TV bought by a consumer will experience video or audio difficulties before the warranty expires. Then the event that the plasma TV will not experience either problem before the warranty expires is given by E^c, with probability

$$\begin{aligned} P(E^c) &= 1 - P(E) \\ &= 1 - .039 \\ &= .961 \end{aligned}$$

Computations Involving the Rules of Probability

We close this section by looking at two additional examples that illustrate how we use the rules of probability.

EXAMPLE 5 Let E and F be two mutually exclusive events, and suppose that $P(E) = .1$ and $P(F) = .6$. Compute:

a. $P(E \cap F)$ **b.** $P(E \cup F)$ **c.** $P(E^c)$
d. $P(E^c \cap F^c)$ **e.** $P(E^c \cup F^c)$

Solution

a. Since the events E and F are mutually exclusive—that is, $E \cap F = \varnothing$—we have $P(E \cap F) = 0$.

b. $$\begin{aligned} P(E \cup F) &= P(E) + P(F) && \text{Since } E \text{ and } F \text{ are mutually exclusive} \\ &= .1 + .6 \\ &= .7 \end{aligned}$$

c. $$\begin{aligned} P(E^c) &= 1 - P(E) && \text{Property 5} \\ &= 1 - .1 \\ &= .9 \end{aligned}$$

d. Observe that, by De Morgan's Law, $E^c \cap F^c = (E \cup F)^c$. Hence,

$$\begin{aligned} P(E^c \cap F^c) &= P[(E \cup F)^c] && \text{See Figure 11.} \\ &= 1 - P(E \cup F) && \text{Property 5} \\ &= 1 - .7 && \text{Use the result of part (b).} \\ &= .3 \end{aligned}$$

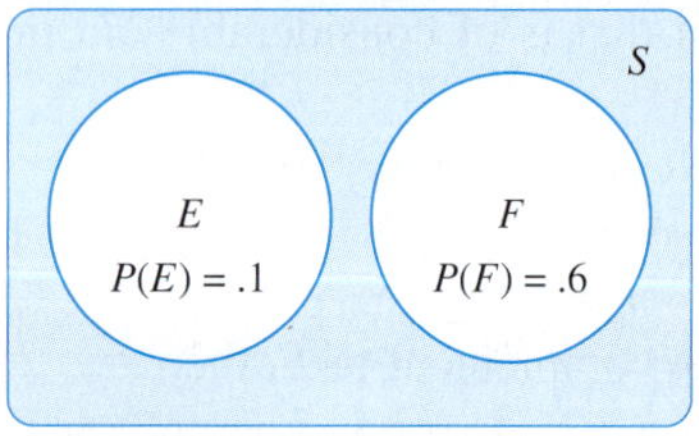

FIGURE 11
$P(E^c \cap F^c) = P[(E \cup F)^c]$

e. Again using De Morgan's Law, we find

$$\begin{aligned} P(E^c \cup F^c) &= P[(E \cap F)^c] \\ &= 1 - P(E \cap F) \\ &= 1 - 0 && \text{Use the result of part (a).} \\ &= 1 \end{aligned}$$

EXAMPLE 6 Let E and F be two events of an experiment with sample space S. Suppose $P(E) = .2$, $P(F) = .1$, and $P(E \cap F) = .05$. Compute:

a. $P(E \cup F)$

b. $P(E^c \cap F^c)$

c. $P(E^c \cap F)$ *Hint:* Draw a Venn diagram.

Solution

a. $P(E \cup F) = P(E) + P(F) - P(E \cap F)$ Property 4

$= .2 + .1 - .05$

$= .25$

b. Using De Morgan's Law, we have

$$P(E^c \cap F^c) = P[(E \cup F)^c]$$

$$= 1 - P(E \cup F) \quad \text{Property 5}$$

$$= 1 - .25 \quad \text{Use the result of part (a).}$$

$$= .75$$

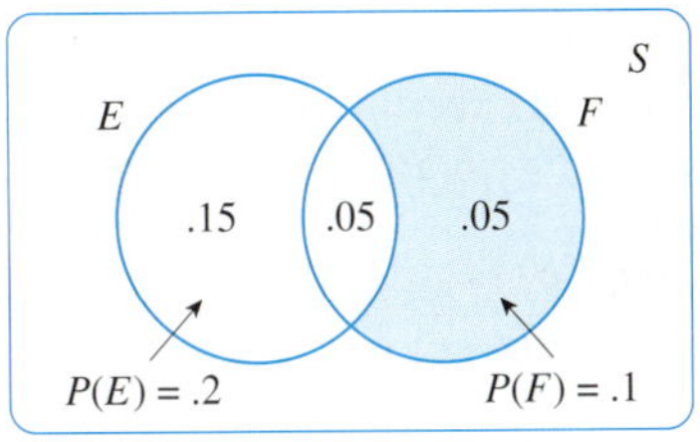

FIGURE 12
$P(E^c \cap F)$: the probability that the event F, but not the event E, will occur

c. From the Venn diagram describing the relationship among E, F, and S (Figure 12), we have

$$P(E^c \cap F) = .05 \quad \text{The shaded subset is the event } E^c \cap F.$$

This result may also be obtained by using the relationship

$$P(E^c \cap F) = P(F) - P(E \cap F)$$

$$= .1 - .05$$

$$= .05$$

7.3 Self-Check Exercises

1. Let E and F be events of an experiment with sample space S. Suppose $P(E) = .4$, $P(F) = .5$, and $P(E \cap F) = .1$. Compute:
   a. $P(E \cup F)$ **b.** $P(E \cap F^c)$

2. **Probability of a Sale or Lease** Susan Garcia wishes to sell or lease a condominium through a realty company. The realtor estimates that the probability of finding a buyer within a month of the date the property is listed for sale or lease is .3, the probability of finding a lessee is .8, and the probability of finding both a buyer and a lessee is .1. Determine the probability that the property will be sold or leased within 1 month from the date the property is listed for sale or lease.

Solutions to Self-Check Exercises 7.3 can be found on page 417.

7.3 Concept Questions

1. Suppose that S is a sample space of an experiment, E and F are events of the experiment, and P is a probability function. Give the meaning of each of the following statements:
   a. $P(E) = 0$ **b.** $P(F) = 0.5$ **c.** $P(S) = 1$
   d. $P(E \cup F) = P(E) + P(F) - P(E \cap F)$

2. Give an example, based on a real-life situation, illustrating the property $P(E^c) = 1 - P(E)$, where E is an event and E^c is the complement of E.

7.3 Exercises

A pair of dice is rolled, and the number that appears uppermost on each die is observed. In Exercises 1–6, refer to this experiment, and find the probability of the given event.

1. The sum of the numbers is an even number.
2. The sum of the numbers is either 7 or 11.
3. A pair of 1s is thrown.
4. A double is thrown.
5. One die shows a 6, and the other is a number less than 3.
6. The sum of the numbers is at least 4.

An experiment consists of selecting a card at random from a 52-card deck. In Exercises 7–12, refer to this experiment and find the probability of the event.

7. A king of diamonds is drawn.

8. A diamond or a king is drawn.

9. A face card (i.e., a jack, queen, or king) is drawn.

10. A red face card is drawn.

11. An ace is not drawn.

12. A black face card is not drawn.

13. **Raffles** Five hundred raffle tickets were sold. What is the probability that a person holding one ticket will win the first prize? What is the probability that he or she will not win the first prize?

14. **TV Households with Remote Controls** The results of a recent television survey of American TV households revealed that 87 out of every 100 TV households have at least one remote control. What is the probability that a randomly selected TV household does not have at least one remote control?

In Exercises 15–22, explain why the statement is incorrect.

15. The sample space associated with an experiment is given by $S = \{a, b, c\}$, where $P(a) = .3$, $P(b) = .4$, and $P(c) = .4$.

16. The probability that a bus will arrive late at the Civic Center is .35, and the probability that it will be on time or early is .60.

17. **Office Pools** A person participates in a weekly office pool in which he has one chance in ten of winning the purse. If he participates for 5 weeks in succession, the probability of winning at least one purse is $\frac{5}{10}$.

18. **Stock Prices** The probability that a certain stock will increase in value over a period of 1 week is .6. Therefore, the probability that the stock will decrease in value is .4.

19. A red die and a green die are tossed. The probability that a 6 will appear uppermost on the red die is $\frac{1}{6}$, and the probability that a 1 will appear uppermost on the green die is $\frac{1}{6}$. Hence, the probability that the red die will show a 6 or the green die will show a 1 is $\frac{1}{6} + \frac{1}{6}$.

20. **College Admissions** Joanne, a high school senior, has applied for admission to four colleges, A, B, C, and D. She has estimated that the probability that she will be accepted for admission by college A, B, C, and D is .5, .3, .1, and .08, respectively. Thus, the probability that she will be accepted for admission by at least one college is $P(A) + P(B) + P(C) + P(D) = .5 + .3 + .1 + .08 = .98$.

21. The sample space associated with an experiment is given by $S = \{a, b, c, d, e\}$. The events $E = \{a, b\}$ and $F = \{c, d\}$ are mutually exclusive. Hence, the events E^c and F^c are mutually exclusive.

22. **Forecasting Sales** Mark Owens, an optician, estimates that the probability that a customer coming into his store will purchase one or more pairs of glasses but not contact lenses is .40, and the probability that he will purchase one or more pairs of contact lenses but not glasses is .25. Hence Owens concludes that the probability that a customer coming into his store will purchase neither a pair of glasses nor a pair of contact lenses is .35.

23. Let E and F be two events that are mutually exclusive, and suppose $P(E) = .2$ and $P(F) = .5$. Compute:

a. $P(E \cap F)$ **b.** $P(E \cup F)$
c. $P(E^c)$ **d.** $P(E^c \cap F^c)$

24. Let E and F be two events of an experiment with sample space S. Suppose $P(E) = .6$, $P(F) = .4$, and $P(E \cap F) = .2$. Compute:

a. $P(E \cup F)$ **b.** $P(E^c)$
c. $P(F^c)$ **d.** $P(E^c \cap F)$

25. Let $S = \{s_1, s_2, s_3, s_4\}$ be the sample space associated with an experiment having the probability distribution shown in the accompanying table. If $A = \{s_1, s_2\}$ and $B = \{s_1, s_3\}$, find:

a. $P(A)$, $P(B)$ **b.** $P(A^c)$, $P(B^c)$
c. $P(A \cap B)$ **d.** $P(A \cup B)$
e. $P(A^c \cap B^c)$ **f.** $P(A^c \cup B^c)$

Outcome	Probability
s_1	$\frac{1}{8}$
s_2	$\frac{3}{8}$
s_3	$\frac{1}{4}$
s_4	$\frac{1}{4}$

26. Let $S = \{s_1, s_2, s_3, s_4, s_5, s_6\}$ be the sample space associated with an experiment having the probability distribution shown in the accompanying table. If $A = \{s_1, s_2\}$ and $B = \{s_1, s_5, s_6\}$, find:

a. $P(A)$, $P(B)$ **b.** $P(A^c)$, $P(B^c)$
c. $P(A \cap B)$ **d.** $P(A \cup B)$
e. $P(A^c \cap B^c)$ **f.** $P(A^c \cup B^c)$

Outcome	Probability
s_1	$\frac{1}{3}$
s_2	$\frac{1}{8}$
s_3	$\frac{1}{6}$
s_4	$\frac{1}{6}$
s_5	$\frac{1}{12}$
s_6	$\frac{1}{8}$

27. **Makeup of U.S. Moviegoer Audience** In a survey of 3000 Americans aged 12 through 74 years, the following makeup of the moviegoer audience was obtained:

Age	12–24	25–44	45–64	65–74
Audience (%)	30	36	28	6

If a moviegoer is selected at random from the respondents of the survey, what is the probability that he or she is between:

a. 12 and 24 years of age?
b. 25 and 64 years of age?
c. 12 and 24 years of age or 65 and 74 years of age?

Source: Nielsen.

28. Climate Change In a survey of 1089 adults conducted by Duke University between January 16, 2013 and January 22, 2013, the following question was asked: How serious a threat is climate change? The results of the survey are summarized below:

Answer	Very serious	Somewhat serious	Not that much	Not at all
Respondents (%)	38	46	15	1

On the basis of the results of the survey, what is the probability that a person chosen at random from the survey said that climate change posed:

a. A very serious threat or a somewhat serious threat?
b. No threat?
c. A somewhat serious threat or not that much of a threat?

Source: Duke University.

29. Stay When Visiting National Parks In a survey conducted by travel usatoday.com, in the week of August 20, 2012, the following question was asked: Where do you stay when you visit national parks? The results of the survey follow:

Accommodation	Respondents (%)
Nearby hotel or motel	55
In-park lodge	25
In-park campground	14
Nearby campground or RV park	6

What is the probability that a randomly chosen participant in the survey said that he or she would:

a. Stay in a nearby campground or RV park?
b. Stay in an in-park lodge or an in-park campground?
c. Not stay in a nearby hotel or motel

Source: travel.usatoday.com.

30. 401(k) Investors According to a study conducted in 2011 concerning the participation, by age, of 401(k) investors, the following data were obtained:

Age	20s	30s	40s	50s	60s
Percent	12	23	28	27	10

a. What is the probability that a 401(k) investor selected at random in 2011 was in his or her 20s or 60s?
b. What is the probability that a 401(k) investor selected at random in 2011 was under the age of 50?

Source: Investment Company Institute.

31. Recovery from the Great Recession In a survey of 1140 middle-class adults who said they were worse off now than before the recession of 2008, the following question was asked: How many years will it take you to fully recover financially? The results are as follows:

Time to Recover	Respondents (%)
Four years or less	29
More than 4 years but less than 10 years	24
Ten years or more	27
Don't know/refused	20

What is the probability that a randomly chosen participant in the survey said that he or she will:

a. Take less than 10 years to recover financially?
b. Take more than 4 years to recover financially?
c. Take 4 years or less, or 10 years or more to recover financially?

Source: Pew Research Center.

32. Electricity Generation Electricity in the United States is generated from many sources. The following table gives the sources as well as their shares in the production of electricity:

Source	Coal	Nuclear	Natural gas	Hydropower	Oil	Other
Share (%)	50.0	19.3	18.7	6.7	3.0	2.3

If a source for generating electricity is picked at random, what is the probability that it comes from:

a. Coal or natural gas?
b. Nonnuclear sources?

Source: Energy Information Administration.

Refer to Exercises 7.1, Problem 34, where you were asked to find the sample space corresponding to the different human blood groups. The following table gives the percent of the U.S. population having each of the eight possible blood types in the sample space. Note that the presence or absence of the Rh antigen is indicated by the symbols + or −, respectively.

Blood Types	A^+	A^-	B^+	B^-	AB^+	AB^-	O^+	O^-
Percent	35.7	6.3	8.5	1.5	3.4	0.6	37.4	6.6

Source: American Red Cross.

In Exercises 33 and 34, use the above table.

33. Blood Types

a. What is the probability that a person selected randomly from the U.S. population has a blood type that is Rh^-?
b. What is the probability that a person selected randomly from the U.S. population has a blood type that is type O?

34. Blood Types

a. What is the probability that a person selected randomly from the U.S. population has a blood type that is type O or Rh^+?
b. What is the probability that a person selected randomly from the U.S. population has a blood type that contains the A antigen?
c. What is the probability that a person selected randomly from the U.S. population has a blood type that is type AB or Rh^-?

35. Downloading Music The following table gives the percentage of music downloaded from the United States and other countries in a certain year by U.S. users:

Country	U.S.	Germany	Canada	Italy	U.K.	France	Japan	Other
Percent	45.1	16.5	6.9	6.1	4.2	3.8	2.5	14.9

a. Verify that the table does give a probability distribution for the experiment.
b. What is the probability that a user who downloads music, selected at random, obtained it from either the United States or Canada?
c. What is the probability that a U.S. user who downloads music, selected at random, does not obtain it from Italy, the United Kingdom (U.K.), or France?

Source: Felix Oberholtzer-Gee and Koleman Strumpf.

36. Teen Spending Behavior In a survey conducted in the spring of 2013, the following results pertaining to teen spending behavior were obtained:

Category	Share of Spending (%)
Clothing	21
Food	18
Accessories/personal care	10
Shoes	9
Car	8
Electronics	8
Music and movies	7
Video games	6
Concerts and events	6
Books	2
Furniture	2
Other	3

a. Verify that the table does give a probability distribution for the experiment.
b. If the expenditure for an item purchased by a teen is chosen at random, what is the probability that it was on clothing or shoes?

Source: Piper Jeffry & Co.

37. Tire Safety A team of automobile safety experts was asked by a television news station to conduct an experiment in which the tires of 100 randomly chosen cars of its employees were subjected to a safety inspection. It was determined that of the 100 cars inspected, 11 cars failed the tread depth test (at least one tire on the car was worn excessively), 45 cars failed the tire pressure test (at least one tire on the car was overinflated or underinflated), and 4 failed both the tread depth test and the tire pressure test. Find the probability that a car selected at random from this group of cars:
a. Failed only the tire pressure test.
b. Passed both the tread depth test and the tire pressure test.

38. Alternative Fuel Vehicles A survey was conducted by the local chapter of an environmental club regarding the ownership of alternative fuel vehicles (AFVs) among the members of the group. An AFV is a vehicle that runs on fuel other than petroleum fuels (petrol and diesel) or that does not involve solely petroleum fuels. It was found that of the 80 members of the club surveyed, 22 of them own at least one hybrid car, 12 of them own at least one electric car, and 4 of them own at least one hybrid and at least one electric car. If a member of the club selected at random is surveyed, what is the probability that he or she:
a. Owns only hybrid cars?
b. Owns no alternative fuel vehicles?

39. Forecasting Sales The probability that a shopper in a certain boutique will buy a blouse is .35, that she will buy a pair of pants is .30, and that she will buy a skirt is .27. The probability that she will buy both a blouse and a skirt is .15, the probability that she will buy both a skirt and a pair of pants is .19, and the probability that she will buy both a blouse and a pair of pants is .12. Finally, the probability that she will buy all three items is .08. What is the probability that a customer selected at random will buy:
a. Exactly one of these items?
b. None of these items?

40. Course Enrollments Among 500 freshmen pursuing a business degree at a university, 320 are enrolled in an economics course, 225 are enrolled in a mathematics course, and 140 are enrolled in both an economics and a mathematics course. What is the probability that a freshman selected at random from this group is enrolled in:
a. An economics and/or a mathematics course?
b. Exactly one of these two courses?
c. Neither an economics course nor a mathematics course?

41. Customer Surveys A leading manufacturer of kitchen appliances advertised its products in two magazines: *Good Housekeeping* and the *Ladies Home Journal*. A survey of 500 customers revealed that 140 learned of its products from *Good Housekeeping*, 130 learned of its products from the *Ladies Home Journal*, and 80 learned of its products from both magazines. What is the probability that a person selected at random from this group saw the manufacturer's advertisement in:
a. Both magazines?
b. At least one of the two magazines?
c. Exactly one magazine?

42. Assembly-Time Studies A time study was conducted by the production manager of Universal Instruments to determine how much time it took an assembly worker to complete a certain task during the assembly of its Orion tablet computers. Results of the study indicated that 20% of the workers were able to complete the task in less than 3 min, 60% of the workers were able to complete the task in 4 min or less, and 10% of the workers required more than 5 min to complete the task. If an assembly-line worker is selected at random from this group, what is the probability that:
a. He or she will be able to complete the task in 5 min or less?
b. He or she will not be able to complete the task within 4 min?
c. The time taken for the worker to complete the task will be between 3 and 4 min (inclusive)?

43. **Use of Landline Phone Versus Cell Phone** A survey of 1000 adults aged 18 years and older conducted in July 2012 found that 910 of them had a cell phone, 670 of them had a landline phone, and 580 of them had both a cell phone and a landline phone. Find the probability that a person selected at random from this group of respondents:
a. Has only a cell phone?
b. Has only a landline phone?
Source: AARP Bulletin Poll.

44. **Time on a Diet** A survey on how long dieters stay on a diet found that 26% of them stayed on the diet for a month or less, 36% of them stayed on for more than a month but less than 6 months, 11% of them stayed on for 6 or more months but less than a year, and 27% of them stayed on for a year or more. For the survey, respondents could define "diet" any way they wanted. On the basis of this survey, what is the probability that a person selected at random from the survey said that he or she stayed on a diet for:
a. A month or less?
b. More than 1 month but less than a year?
c. Six months or more?
Source: NPD Group.

45. **Teachers' Views of Educational Problems** A nonprofit organization conducted a survey of 2140 metropolitan-area teachers regarding their beliefs about educational problems. The following data were obtained:

900 said that lack of parental support is a problem.

890 said that abused or neglected children are problems.

680 said that malnutrition or students in poor health is a problem.

120 said that lack of parental support and abused or neglected children are problems.

110 said that lack of parental support and malnutrition or poor health are problems.

140 said that abused or neglected children and malnutrition or poor health are problems.

40 said that lack of parental support, abuse or neglect, and malnutrition or poor health are problems.

What is the probability that a teacher selected at random from this group said that lack of parental support is the only problem hampering a student's schooling?
Hint: Draw a Venn diagram.

46. **401(k) Investments** In a survey of 200 employees of a company regarding their 401(k) investments, the following data were obtained:

141 had investments in stock funds.

91 had investments in bond funds.

60 had investments in money market funds.

47 had investments in stock funds and bond funds.

36 had investments in stock funds and money market funds.

36 had investments in bond funds and money market funds.

22 had investments in stock funds, bond funds, and money market funds.

What is the probability that an employee of the company chosen at random:
a. Had investments in exactly two kinds of investment funds?
b. Had investments in exactly one kind of investment fund?
c. Had no investment in any of the three types of funds?

47. **Rollover Deaths** The following table gives the number of people killed in rollover crashes in various types of vehicles in 2010:

Types of Vehicles	Cars	Light Trucks	Large Trucks	Buses	Other
Deaths	2748	4814	443	10	227

Find the empirical probability distribution associated with these data. If a fatality due to a rollover crash in 2010 is picked at random, what is the probability that the victim was in:
a. A car? **b.** A light truck? **c.** A bus or a large truck?
Source: National Highway Traffic Safety Administration.

48. **Switching Jobs** Two hundred workers were asked: Would a better economy lead you to switch jobs? The results of the survey follow:

Answer	Very likely	Somewhat likely	Somewhat unlikely	Very unlikely	Don't know
Respondents	40	28	26	104	2

If a worker is chosen at random, what is the probability that he or she:
a. Is very unlikely to switch jobs?
b. Is somewhat likely or very likely to switch jobs?
Source: Accountemps.

49. **Losing Weight** In a survey, 1012 American adults were asked how many times they had tried to lose weight in their lifetime. The results of the survey follow:

Answer	Once or twice	3–10 times	More than 10 times	Never	No opinion
Respondents	253	304	81	334	40

If a person in the survey is selected at random, what is the probability that the person answered that he or she had:
a. Tried at least once to lose weight in his or her lifetime?
b. Tried between one and ten times to lose weight in his or her lifetime?
c. Never tried to lose weight in his or her lifetime?
Source: Gallup Poll.

50. **Walking on Mars** In a survey of 1000 adults conducted in 2012, the following question was asked: Will humans

walk on Mars in the next 25 years? The results of the survey are summarized in the following table:

Answer	Very likely	Somewhat likely	Not very likely	Not at all likely	Not sure
Respondents	210	290	340	80	80

If a respondent in the survey is chosen at random, what is the probability that he or she answered that:

a. It is very likely or somewhat likely that humans will walk on Mars in the next 25 years?

b. It is not very likely or not at all likely that humans will walk on Mars in the next 25 years?

Source: Rasmussen Reports.

51. **RISK OF AN AIRPLANE CRASH** According to a study of Western-built commercial jets involved in crashes over a 10-year period, the percentage of airplane crashes that occur at each stage of flight are as follows:

Phase	Percent
On ground, taxiing	4
During takeoff	10
Climbing to cruise altitude	19
En route	5
Descent and approach	31
Landing	31

If one of the doomed flights in this period is picked at random, what is the probability that it crashed:

a. While taxiing on the ground or while en route?

b. During takeoff or landing?

If the study is indicative of airplane crashes in general, when is the risk of a plane crash the highest?

Source: National Transportation Safety Board.

52. **DISTRACTED DRIVING** According to a study of 100 drivers in metropolitan Washington, D.C., whose cars were equipped with cameras with sensors, the distractions and the number of incidents (crashes, near crashes, and situations that require an evasive maneuver after the driver was distracted) caused by these distractions are as follows:

Distraction	A	B	C	D	E	F	G	H	I
Driving Incidents	668	378	194	163	133	134	111	111	89

where A = Wireless device (cell phone, PDA)

B = Passenger

C = Something inside car

D = Vehicle

E = Personal hygiene

F = Eating

G = Something outside car

H = Talking/singing

I = Other

If an incident caused by a distraction is picked at random, what is the probability that it was caused by:

a. The use of a wireless device?

b. Something other than personal hygiene or eating?

Source: Virginia Tech Transportation Institute and NHTSA.

53. **PLANS TO KEEP CARS** In a survey conducted to determine how long Americans keep their cars, 2000 automobile owners were asked how long they planned to keep their present cars. The results of the survey follow:

Years Car Is Kept, x	Respondents
$0 \leq x < 1$	60
$1 \leq x < 3$	440
$3 \leq x < 5$	360
$5 \leq x < 7$	340
$7 \leq x < 10$	240
$10 \leq x$	560

Find the probability distribution associated with these data. What is the probability that an automobile owner selected at random from those surveyed planned to keep his or her present car:

a. Less than 5 years?

b. 3 years or more?

54. **GUN-CONTROL LAWS** A poll was conducted among 250 residents of a certain city regarding tougher gun-control laws. The results of the poll are shown in the table:

	Own Only a Handgun	Own Only a Rifle	Own a Handgun and a Rifle	Own Neither	Total
Favor Tougher Laws	0	12	0	138	150
Oppose Tougher Laws	58	5	25	0	88
No Opinion	0	0	0	12	12
Total	58	17	25	150	250

If one of the participants in this poll is selected at random, what is the probability that he or she:

a. Favors tougher gun-control laws?

b. Owns a handgun?

c. Owns a handgun but not a rifle?

d. Favors tougher gun-control laws and does not own a handgun?

55. Suppose the probability that Bill can solve a problem is p_1 and the probability that Mike can solve it is p_2. Show that the probability that Bill and Mike working independently can solve the problem is $p_1 + p_2 - p_1p_2$.

56. Fifty raffle tickets are numbered 1 through 50, and one of them is drawn at random. What is the probability that the number is a multiple of 5 or 7? Consider the following "solution": Since 10 tickets bear numbers that are multiples of 5 and since 7 tickets bear numbers that are multiples of 7, we conclude that the required probability is

$$\frac{10}{50} + \frac{7}{50} = \frac{17}{50}$$

What is wrong with this argument? What is the correct answer?

In Exercises 57–62, determine whether the statement is true or false. If it is true, explain why it is true. If it is false, give an example to show why it is false.

57. If A is a subset of B and $P(B) = 0$, then $P(A) = 0$.

58. If A is a subset of B, then $P(A) \leq P(B)$.

59. If $E_1, E_2, \ldots, E_n$ are events of an experiment, then $P(E_1 \cup E_2 \cup \cdots \cup E_n) = P(E_1) + P(E_2) + \cdots + P(E_n)$.

60. If E is an event of an experiment, then $P(E) + P(E^c) = 1$.

61. If $P(E) = P(F)$ and $E \subseteq F$, then $E = F$.

62. If S is a sample space of an experiment and $E \neq S$, then $P(E) < 1$.

7.3 Solutions to Self-Check Exercises

1. a. Using Property 4, we find

$$\begin{aligned} P(E \cup F) &= P(E) + P(F) - P(E \cap F) \\ &= .4 + .5 - .1 \\ &= .8 \end{aligned}$$

b. From the accompanying Venn diagram, in which the subset $E \cap F^c$ is shaded, we see that

$$P(E \cap F^c) = .3$$

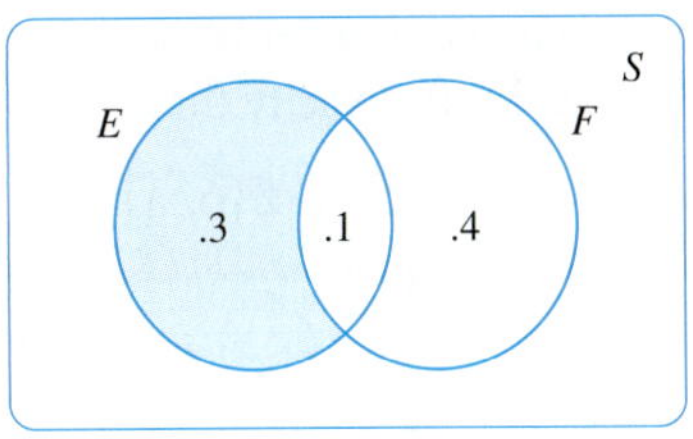

The result may also be obtained by using the relationship

$$\begin{aligned} P(E \cap F^c) &= P(E) - P(E \cap F) \\ &= .4 - .1 = .3 \end{aligned}$$

2. Let E denote the event that the realtor will find a buyer within 1 month of the date the property is listed for sale or lease, and let F denote the event that the realtor will find a lessee within the same time period. Then

$$P(E) = .3 \qquad P(F) = .8 \qquad P(E \cap F) = .1$$

The probability of the event that the realtor will find a buyer or a lessee within 1 month of the date the property is listed for sale or lease is given by

$$\begin{aligned} P(E \cup F) &= P(E) + P(F) - P(E \cap F) \\ &= .3 + .8 - .1 = 1 \end{aligned}$$

that is, a certainty.

7.4 Use of Counting Techniques in Probability

Further Applications of Counting Techniques

As we have seen many times before, a problem in which the underlying sample space has a small number of elements may be solved by first determining all such sample points. However, for problems involving sample spaces with a large number of sample points, this approach is neither practical nor desirable.

In this section, we see how the counting techniques studied in Chapter 6 may be employed to help us solve problems in which the associated sample spaces contain large numbers of sample points. In particular, we restrict our attention to the study of uniform sample spaces—that is, sample spaces in which the outcomes are equally likely. For such spaces, we have the following result:

Computing the Probability of an Event in a Uniform Sample Space

Let S be a uniform sample space, and let E be any event. Then

$$P(E) = \frac{\text{Number of outcomes in } E}{\text{Number of outcomes in } S} = \frac{n(E)}{n(S)} \qquad \textbf{(1)}$$

EXAMPLE 1 An unbiased coin is tossed six times. What is the probability that the coin will land heads:

a. Exactly three times?
b. At most three times?
c. On the first and the last toss?

Solution

a. Each outcome of the experiment may be represented as a sequence of heads and tails. Using the generalized multiplication principle, we see that the number of outcomes of this experiment is given by 2^6, or 64. Let E denote the event that the coin lands heads exactly three times. Since there are $C(6, 3)$ ways this can occur, we see that the required probability is

$$P(E) = \frac{n(E)}{n(S)} = \frac{C(6,3)}{64} = \frac{\frac{6!}{3!\,3!}}{64} \qquad S \text{ is a sample space of the experiment.}$$

$$= \frac{\frac{6 \cdot 5 \cdot 4}{3 \cdot 2}}{64} = \frac{20}{64} = \frac{5}{16} = .3125$$

b. Let F denote the event that the coin lands heads at most three times. Then $n(F)$ is given by the sum of the number of ways the coin lands heads zero times (no heads!), the number of ways it lands heads exactly once, the number of ways it lands heads exactly twice, and the number of ways it lands heads exactly three times. That is,

$$\begin{aligned} n(F) &= C(6,0) + C(6,1) + C(6,2) + C(6,3) \\ &= \frac{6!}{0!\,6!} + \frac{6!}{1!\,5!} + \frac{6!}{2!\,4!} + \frac{6!}{3!\,3!} \\ &= 1 + 6 + \frac{6 \cdot 5}{2} + \frac{6 \cdot 5 \cdot 4}{3 \cdot 2} = 42 \end{aligned}$$

Therefore, the required probability is

$$P(F) = \frac{n(F)}{n(S)} = \frac{42}{64} = \frac{21}{32} \approx .6563$$

c. Let F denote the event that the coin lands heads on the first and the last toss. Then $n(F) = 1 \cdot 2 \cdot 2 \cdot 2 \cdot 2 \cdot 1 = 2^4$, so the probability that this event occurs is

$$\begin{aligned} P(F) &= \frac{2^4}{2^6} \\ &= \frac{1}{2^2} \\ &= \frac{1}{4} \end{aligned}$$

EXAMPLE 2 Two cards are selected at random (without replacement) from a well-shuffled deck of 52 playing cards. What is the probability that:

a. They are both aces? **b.** Neither of them is an ace?

Solution

a. The experiment consists of selecting 2 cards from a pack of 52 playing cards. Since the order in which the cards are selected is immaterial, the sample points are combinations of 52 cards taken 2 at a time. Now there are $C(52, 2)$ ways of selecting 52 cards taken 2 at a time, so the number of elements in the sample space S is given by $C(52, 2)$. Next, we observe that there are $C(4, 2)$ ways of selecting 2 aces from the 4 in the deck. Therefore, if E denotes the event that the cards selected are both aces, then

$$\begin{aligned} P(E) &= \frac{n(E)}{n(S)} \\ &= \frac{C(4, 2)}{C(52, 2)} = \frac{\frac{4!}{2!\,2!}}{\frac{52!}{2!\,50!}} = \frac{4 \cdot 3}{2} \cdot \frac{2}{52 \cdot 51} \\ &= \frac{1}{221} \approx .0045 \end{aligned}$$

b. Let F denote the event that neither of the two cards selected is an ace. Since there are $C(48, 2)$ ways of selecting two cards neither of which is an ace, we find that

$$\begin{aligned} P(F) &= \frac{n(F)}{n(S)} = \frac{C(48, 2)}{C(52, 2)} = \frac{\frac{48!}{2!\,46!}}{\frac{52!}{2!\,50!}} = \frac{48 \cdot 47}{2} \cdot \frac{2}{52 \cdot 51} \\ &= \frac{188}{221} \approx .8507 \end{aligned}$$

APPLIED EXAMPLE 3 Quality Control A bin in the hi-fi department of Building 20, a bargain outlet, contains 100 blank DVDs, of which 10 are known to be defective. If a customer selects 6 of these DVDs at random, determine the probability:

a. That 2 of them are defective.
b. That at least 1 of them is defective.

Solution

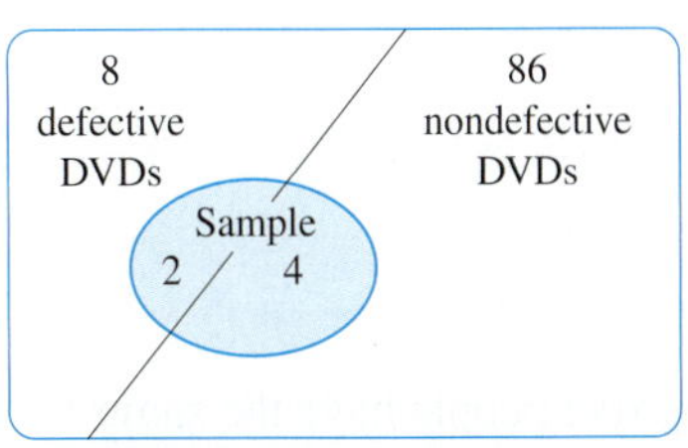

FIGURE 13
A sample of 6 DVDs selected from 90 nondefective DVDs and 10 defective DVDs

a. There are $C(100, 6)$ ways of selecting a set of 6 DVDs from the 100, and this gives $n(S)$, the number of outcomes in the sample space associated with the experiment. Next, we observe that there are $C(10, 2)$ ways of selecting a set of 2 defective DVDs from the 10 defective DVDs and $C(90, 4)$ ways of selecting a set of 4 nondefective DVDs from the 90 nondefective DVDs (Figure 13). Thus, by the multiplication principle, there are $C(10, 2) \cdot C(90, 4)$ ways of selecting 2 defective and 4 nondefective DVDs. Therefore, the probability of selecting 6 DVDs of which 2 are defective is given by

$$\begin{aligned} \frac{C(10, 2) \cdot C(90, 4)}{C(100, 6)} &= \frac{\frac{10!}{2!\,8!}\,\frac{90!}{4!\,86!}}{\frac{100!}{6!\,94!}} \\ &= \frac{10 \cdot 9}{2} \cdot \frac{90 \cdot 89 \cdot 88 \cdot 87}{4 \cdot 3 \cdot 2} \cdot \frac{6 \cdot 5 \cdot 4 \cdot 3 \cdot 2}{100 \cdot 99 \cdot 98 \cdot 97 \cdot 96 \cdot 95} \\ &\approx .096 \end{aligned}$$

b. Let E denote the event that none of the DVDs selected is defective. Then E^c gives the event that at least 1 of the DVDs is defective. By the rule of complements,

$$P(E^c) = 1 - P(E)$$

To compute $P(E)$, we observe that there are $C(90, 6)$ ways of selecting a set of 6 DVDs that are nondefective. Therefore,

$$P(E) = \frac{C(90, 6)}{C(100, 6)}$$

$$\begin{aligned} P(E^c) &= 1 - \frac{C(90, 6)}{C(100, 6)} \\ &= 1 - \frac{\dfrac{90!}{6!\,84!}}{\dfrac{100!}{6!\,94!}} \\ &= 1 - \frac{90 \cdot 89 \cdot 88 \cdot 87 \cdot 86 \cdot 85}{6 \cdot 5 \cdot 4 \cdot 3 \cdot 2} \cdot \frac{6 \cdot 5 \cdot 4 \cdot 3 \cdot 2}{100 \cdot 99 \cdot 98 \cdot 97 \cdot 96 \cdot 95} \\ &\approx .478 \end{aligned}$$

The Birthday Problem

APPLIED EXAMPLE 4 The Birthday Problem A group of five people is selected at random. What is the probability that at least two of them have the same birthday?

Solution For simplicity, we assume that none of the five people was born on February 29 of a leap year. Since the five people were selected at random, we also assume that each of them is equally likely to have any of the 365 days of a year as his or her birthday. If we let A, B, C, D, and F represent the five people, then an outcome of the experiment may be represented by (a, b, c, d, f), where the dates a, b, c, d, and f give the birthdays of A, B, C, D, and F, respectively.

We first observe that since there are 365 possibilities for each of the dates a, b, c, d, and f, the multiplication principle implies that there are

$$\underset{a}{\boxed{365}} \cdot \underset{b}{\boxed{365}} \cdot \underset{c}{\boxed{365}} \cdot \underset{d}{\boxed{365}} \cdot \underset{f}{\boxed{365}}$$

or 365^5 outcomes of the experiment. Therefore,

$$n(S) = 365^5$$

where S denotes the sample space of the experiment.

Next, let E denote the event that two or more of the five people have the same birthday. It is now necessary to compute $P(E)$. However, a direct computation of $P(E)$ is relatively difficult. It is much easier to compute $P(E^c)$, where E^c is the event that no two of the five people have the same birthday, and then use the relation

$$P(E) = 1 - P(E^c)$$

To compute $P(E^c)$, observe that there are 365 ways (corresponding to the 365 dates) on which A's birthday can occur, followed by 364 ways on which B's birthday could occur if B were not to have the same birthday as A, and so on. Therefore, by the generalized multiplication principle,

$$n(E^c) = \underset{\substack{A\text{'s}\\ \text{birthday}}}{365} \cdot \underset{\substack{B\text{'s}\\ \text{birthday}}}{364} \cdot \underset{\substack{C\text{'s}\\ \text{birthday}}}{363} \cdot \underset{\substack{D\text{'s}\\ \text{birthday}}}{362} \cdot \underset{\substack{F\text{'s}\\ \text{birthday}}}{361}$$

Thus,

$$\begin{aligned} P(E^c) &= \frac{n(E^c)}{n(S)} \\ &= \frac{365 \cdot 364 \cdot 363 \cdot 362 \cdot 361}{365^5} \\ P(E) &= 1 - P(E^c) \\ &= 1 - \frac{365 \cdot 364 \cdot 363 \cdot 362 \cdot 361}{365^5} \\ &\approx .027 \end{aligned}$$

TABLE 8

Probability That at Least Two People in a Randomly Selected Group of *r* People Have the Same Birthday

r	*P*(*E*)
5	.027
10	.117
15	.253
20	.411
22	.476
23	.507
25	.569
30	.706
40	.891
50	.970

We can extend the result obtained in Example 4 to the general case involving r people. In fact, if E denotes the event that at least two of the r people have the same birthday, an argument similar to that used in Example 4 leads to the result

$$P(E) = 1 - \frac{365 \cdot 364 \cdot 363 \cdot \cdots \cdot (365 - r + 1))}{365^r}$$

By letting r take on the values 5, 10, 15, 20, . . . , 50, in turn, we obtain the probabilities that at least 2 of 5, 10, 15, 20, . . . , 50 people, respectively, have the same birthday. These results are summarized in Table 8.

The results show that in a group of 23 randomly selected people, the chances are greater than 50% that at least 2 of them will have the same birthday. In a group of 50 people, it is an excellent bet that at least 2 people in the group will have the same birthday.

Explore and Discuss

During an episode of the *Tonight Show*, a talk show host related "The Birthday Problem" to the audience—noting that in a group of 50 or more people, probabilists have calculated that the probability of at least 2 people having the same birthday is very high. To illustrate this point, he proceeded to conduct his own experiment. A person selected at random from the audience was asked to state his birthday. The host then asked whether anyone in the audience had the same birthday. The response was negative. He repeated the experiment. Once again, the response was negative. These results, observed the host, were contrary to expectations. In a later episode of the show, the host explained why this experiment had been improperly conducted. Explain why the host failed to illustrate the point he was trying to make in the earlier episode.

7.4 Self-Check Exercises

1. Four balls are selected at random without replacement from an urn containing ten white balls and eight red balls. What is the probability that all the chosen balls are white?

2. **Quality Control** A box contains 20 microchips, of which 4 are substandard. If 2 of the chips are taken from the box, what is the probability that they are both substandard?

Solutions to Self-Check Exercises 7.4 can be found on page 424.

7.4 Concept Questions

1. What is the probability of an event E in a uniform sample space S?

2. Suppose we want to find the probability that at least two people in a group of six randomly selected people have the same birthday.
   a. If S denotes the sample space of this experiment, what is $n(S)$?
   b. If E is the event that two or more of the six people in the group have the same birthday, explain how you would use $P(E^c)$ to determine $P(E)$.

7.4 Exercises

An unbiased coin is tossed five times. In Exercises 1–4, find the probability of the given event.

1. The coin lands heads all five times.
2. The coin lands heads exactly once.
3. The coin lands heads at least once.
4. The coin lands heads more than once.

Two cards are selected at random without replacement from a well-shuffled deck of 52 playing cards. In Exercises 5–8, find the probability of the given event.

5. A pair is drawn.
6. A pair is not drawn.
7. Two black cards are drawn.
8. Two cards of the same suit are drawn.

Four balls are selected at random without replacement from an urn containing three white balls and five blue balls. In Exercises 9–12, find the probability of the given event.

9. Two of the balls are white, and two are blue.
10. All of the balls are blue.
11. Exactly three of the balls are blue.
12. Two or three of the balls are white.

Assume that the probability of a boy being born is the same as the probability of a girl being born. In Exercises 13–16, find the probability that a family with three children will have the given composition.

13. Two boys and one girl
14. At least one girl
15. No girls
16. The two oldest children are girls.
17. **Taking Exams** An exam consists of ten true-or-false questions. If a student guesses at every answer, what is the probability that he or she will answer exactly six questions correctly?
18. **Personnel Selection** Jacobs & Johnson, an accounting firm, employs 14 accountants, of whom 8 are CPAs. If a delegation of 3 accountants is randomly selected from the firm to attend a conference, what is the probability that 3 CPAs will be selected?
19. **Quality Control** Two light bulbs are selected at random from a lot of 24, of which 4 are defective. What is the probability that:
    a. Both of the light bulbs are defective?
    b. At least 1 of the light bulbs is defective?
20. A customer at Cavallaro's Fruit Stand picks a sample of 3 oranges at random from a crate containing 60 oranges, of which 4 are rotten. What is the probability that the sample contains 1 or more rotten oranges?
21. **Quality Control** A shelf in the Metro Department Store contains 80 colored ink cartridges for a popular ink-jet printer. Six of the cartridges are defective. If a customer selects 2 cartridges at random from the shelf, what is the probability that:
    a. Both are defective?
    b. At least 1 is defective?
22. **Quality Control** Electronic baseball games manufactured by Tempco Electronics are shipped in lots of 24. Before shipping, a quality-control inspector randomly selects a sample of 8 from each lot for testing. If the sample contains any defective games, the entire lot is rejected. What is the probability that a lot containing exactly 2 defective games will still be shipped?
23. **Personnel Selection** The City Transit Authority plans to hire 12 new bus drivers. From a group of 100 qualified applicants, of whom 60 are men and 40 are women, 12 names are to be selected by lot. Suppose that Mary and John Lewis are among the 100 qualified applicants.
    a. What is the probability that Mary's name will be selected? That both Mary's and John's names will be selected?
    b. If it is stipulated that an equal number of men and women are to be selected (6 men from the group of 60 men and 6 women from the group of 40 women),

what is the probability that Mary's name will be selected? That Mary's and John's names will be selected?

24. **Selection of Public Housing Applicants** The City Housing Authority has received 50 applications from qualified applicants for eight low-income apartments. Three of the apartments are on the north side of town, and five are on the south side. If the apartments are to be assigned by means of a lottery, what is the probability that:
   a. A specific qualified applicant will be selected for one of these apartments?
   b. Two specific qualified applicants will be selected for apartments on the same side of town?

25. **Exams** A student studying for a vocabulary test knows the meanings of 12 words from a list of 20 words. If the test contains 10 words from the study list, what is the probability that at least 8 of the words on the test are words that the student knows?

26. **Driving Tests** Four different written driving tests are administered by the Motor Vehicle Department. One of these four tests is selected at random for each applicant for a driver's license. If a group consisting of two women and three men apply for a license, what is the probability that:
   a. Exactly two of the five will take the same test?
   b. The two women will take the same test?

27. **Brand Selection** A druggist wishes to select three brands of aspirin to sell in his store. He has five major brands to choose from: *A*, *B*, *C*, *D*, and *E*. If he selects the three brands at random, what is the probability that he will select:
   a. Brand *B*?
   b. Brands *B* and *C*?
   c. At least one of the two brands *B* and *C*?

28. **Blackjack** In the game of blackjack, a 2-card hand consisting of an ace and a face card or a 10 is called a blackjack.
   a. If a player is dealt 2 cards from a standard deck of 52 well-shuffled cards, what is the probability that the player will receive a blackjack?
   b. If a player is dealt 2 cards from 2 well-shuffled standard decks, what is the probability that the player will receive a blackjack?

29. **Slot Machines** Refer to Exercise 33, Section 6.3, in which the "lucky dollar" slot machine was described. What is the probability that the three "lucky dollar" symbols will appear in the window of the slot machine?

30. **Roulette** In 1959, a world record was set for the longest run on an ungaffed (fair) roulette wheel at the El San Juan Hotel in Puerto Rico. The number 10 appeared six times in a row. What is the probability of the occurrence of this event? (Assume that there are 38 equally likely outcomes consisting of the numbers 1–36, 0, and 00.)

In the Numbers Game, a state lottery, four numbers are drawn with replacement from an urn containing balls numbered 0–9, inclusive. In Exercises 31–34, find the probability that a ticket holder has the indicated winning ticket.

31. **Lotteries** All four digits in exact order (the grand prize)

32. **Lotteries** Two specified, consecutive digits in exact order (the first two digits, the middle two digits, or the last two digits)

33. **Lotteries** One digit (the first, second, third, or fourth digit)

34. **Lotteries** Three digits in exact order

A list of poker hands, ranked in order from the highest to the lowest, is shown in the accompanying table along with a description and example of each hand. Use the table to answer Exercises 35–40.

Hand	Description	Example
Straight flush	5 cards in sequence in the same suit	A♥ 2♥ 3♥ 4♥ 5♥
Four of a kind	4 cards of the same rank and any other card	K♥ K♦ K♠ K♣ 2♥
Full house	3 of a kind and a pair	3♥ 3♦ 3♣ 7♥ 7♦
Flush	5 cards of the same suit that are not all in sequence	5♥ 6♥ 9♥ J♥ K♥
Straight	5 cards in sequence but not all of the same suit	10♥ J♦ Q♣ K♠ A♥
Three of a kind	3 cards of the same rank and 2 unmatched cards	K♥ K♦ K♠ 2♥ 4♦
Two pair	2 cards of the same rank and 2 cards of any other rank with an unmatched card	K♥ K♦ 2♥ 2♠ 4♣
One pair	2 cards of the same rank and 3 unmatched cards	K♥ K♦ 5♥ 2♠ 4♥

If a 5-card poker hand is dealt from a well-shuffled deck of 52 cards, what is the probability of being dealt the given hand?

35. **Poker** A straight flush (Note that an ace may be played as either a high or a low card in a straight sequence—that is, A, 2, 3, 4, 5 or 10, J, Q, K, A. Hence, there are ten possible sequences for a straight in one suit.)

36. **Poker** A straight (but not a straight flush)

37. **Poker** A flush (but not a straight flush)

38. **Poker** Four of a kind

39. **Poker** A full house

40. **Poker** Two pairs

41. **Zodiac Signs** There are 12 signs of the Zodiac: Aries, Taurus, Gemini, Cancer, Leo, Virgo, Libra, Scorpio, Sagittarius, Capricorn, Aquarius, and Pisces. Each sign corresponds to a different calendar period of approximately 1 month. Assuming that a person is just as likely to be born under one sign as another, what is the probability that in a group of five people at least two of them:
   a. Have the same sign?
   b. Were born under the sign of Aries?

42. **Birthday Problem** What is the probability that at least two of the nine justices of the U.S. Supreme Court have the same birthday?

43. **Birthday Problem** Fifty people are selected at random. What is the probability that none of the people in this group have the same birthday?

44. **Birthday Problem** A group of five people are selected at random. What is the probability that two of them were born on the same day of the week? (Assume that a person is equally likely to be born on any day of the week.)

45. **Birthday Problem** A group of twelve people are selected at random. What is the probability that at least two of them have the same birthday?

46. **Birthday Problem** There were 44 different presidents of the United States from 1789 through 2014. What is the probability that at least two of them had the same birthday? Compare your calculation with the facts by checking an almanac or some other source.

7.4 Solutions to Self-Check Exercises

1. The probability that all four balls selected are white is given by

$$\frac{\text{The number of ways of selecting 4 white balls from the 10 in the urn}}{\text{The number of ways of selecting any 4 balls from the 18 balls in the urn}}$$

$$= \frac{C(10,4)}{C(18,4)} = \frac{\frac{10!}{4!\,6!}}{\frac{18!}{4!\,14!}} = \frac{10\cdot 9\cdot 8\cdot 7}{4\cdot 3\cdot 2}\cdot\frac{4\cdot 3\cdot 2}{18\cdot 17\cdot 16\cdot 15} \approx .069$$

2. The probability that both chips are substandard is given by

$$\frac{\text{The number of ways of choosing any 2 of the 4 substandard chips}}{\text{The number of ways of choosing any 2 of the 20 chips}}$$

$$= \frac{C(4,2)}{C(20,2)} = \frac{\frac{4!}{2!\,2!}}{\frac{20!}{2!\,18!}} = \frac{4\cdot 3}{2}\cdot\frac{2}{20\cdot 19} \approx .032$$

7.5 Conditional Probability and Independent Events

Conditional Probability

Suppose that three cities, A, B, and C, are vying to play host to the Summer Olympic Games in 2020. If each city has the same chance of winning the right to host the Games, then the probability of City A hosting the Games is $\frac{1}{3}$. Now suppose City B decides to pull out of contention because of fiscal problems. Then it would seem that City A's chances of playing host will increase. In fact, if the two remaining cities have equal chances of winning, then the probability of City A playing host to the Games is $\frac{1}{2}$.

In general, the probability of an event is affected by the occurrence of other events and/or by the knowledge of information relevant to the event. Basically, the injection of conditions into a problem modifies the underlying sample space of the original problem. This in turn leads to a change in the probability of the event.

EXAMPLE 1 Two cards are drawn without replacement from a well-shuffled deck of 52 playing cards.

a. What is the probability that the first card drawn is an ace?
b. What is the probability that the second card drawn is an ace given that the first card drawn was not an ace?
c. What is the probability that the second card drawn is an ace given that the first card drawn was an ace?

Solution

a. The sample space here consists of 52 equally likely outcomes, 4 of which are aces. Therefore, the probability that the first card drawn is an ace is $\frac{4}{52}$, or $\frac{1}{13}$.
b. The first card having been drawn, there are 51 cards left in the deck. In other words, for the second phase of the experiment, we are working in a *reduced* sample space. If the first card drawn was not an ace, then this modified sample space of 51 points contains 4 "favorable" outcomes (the 4 aces), so the probability that the second card drawn is an ace is given by $\frac{4}{51}$.
c. If the first card drawn was an ace, then there are 3 aces left in the deck of 51 playing cards, so the probability that the second card drawn is an ace is given by $\frac{3}{51}$, or $\frac{1}{17}$.

Observe that in Example 1, the occurrence of the first event reduces the size of the original sample space. The information concerning the first card drawn also leads us to the consideration of modified sample spaces: In part (b), the deck contained four aces, and in part (c), the deck contained three aces.

The probability found in part (b) or part (c) of Example 1 is known as a **conditional probability,** since it is the probability of an event occurring given that another event has already occurred. For example, in part (b), we computed the probability of the event that the second card drawn is an ace *given that* the first card drawn was not an ace. In general, given two events A and B of an experiment, under certain circumstances one may compute the probability of the event B given that the event A has already occurred. This probability, denoted by $P(B|A)$, is called the **conditional probability of *B* given *A*.**

A formula for computing the conditional probability of B given A may be discovered with the aid of a Venn diagram. Consider an experiment with a uniform sample space S, and suppose that A and B are two events of the experiment (Figure 14).

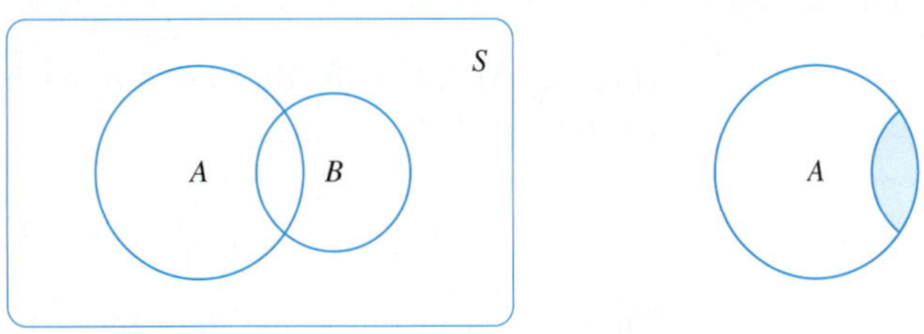

(a) Original sample space

(b) Reduced sample space *A*. The shaded area is $A \cap B$.

FIGURE **14**

The condition that the event A has occurred tells us that the possible outcomes of the experiment in the second phase are restricted to those outcomes (elements) in the set A. In other words, we may work with the reduced sample space A instead of the original sample space S in the experiment. Next we observe that, with respect to the reduced sample space A, the outcomes in the event B are precisely those elements in the set $A \cap B$. Consequently, the conditional probability of B given A is

$$P(B \mid A) = \frac{\text{Number of elements in } A \cap B}{\text{Number of elements in } A}$$

$$= \frac{n(A \cap B)}{n(A)} \qquad n(A) \neq 0$$

Dividing the numerator and the denominator by $n(S)$, the number of elements in S, we have

$$P(B \mid A) = \frac{\dfrac{n(A \cap B)}{n(S)}}{\dfrac{n(A)}{n(S)}}$$

which is equivalent to the following formula:

Conditional Probability of an Event

If A and B are events in an experiment and $P(A) \neq 0$, then the conditional probability that the event B will occur given that the event A has already occurred is

$$P(B \mid A) = \frac{P(A \cap B)}{P(A)} \tag{2}$$

EXAMPLE 2 A pair of fair dice is rolled. What is the probability that the sum of the numbers falling uppermost is 7 if it is known that one of the numbers is a 5?

Solution Let A denote the event that the sum of the numbers falling uppermost is 7, and let B denote the event that one of the numbers is a 5. From the results of Example 4, Section 7.1, we find that

$$A = \{(6, 1), (5, 2), (4, 3), (3, 4), (2, 5), (1, 6)\}$$
$$B = \{(5, 1), (5, 2), (5, 3), (5, 4), (5, 5), (5, 6), (1, 5), (2, 5), (3, 5), (4, 5), (6, 5)\}$$

so

$$A \cap B = \{(5, 2), (2, 5)\}$$

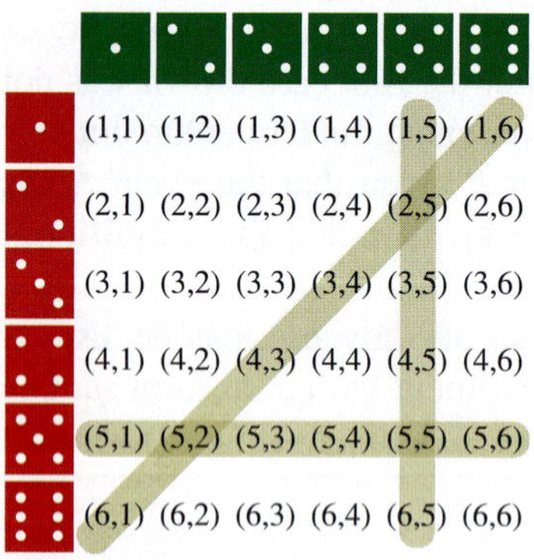

FIGURE 15
$A \cap B = \{(5, 2), (2, 5)\}$

(Figure 15). Since the dice are fair, each outcome of the experiment is equally likely; therefore,

$$P(A \cap B) = \frac{2}{36} \quad \text{and} \quad P(B) = \frac{11}{36} \qquad \text{Recall that } n(S) = 36.$$

Thus, the probability that the sum of the numbers falling uppermost is 7 given that one of the numbers is a 5 is, by virtue of Equation (2),

$$P(A \mid B) = \frac{\dfrac{2}{36}}{\dfrac{11}{36}} = \frac{2}{11}$$

APPLIED EXAMPLE 3 Color Blindness In a test conducted by the U.S. Army, it was found that of 1000 new recruits (600 men and 400 women), 50 of the men and 4 of the women were red-green color-blind. Given that a recruit selected at random from this group is red-green color-blind, what is the probability that the recruit is a male?

Solution Let C denote the event that a randomly selected subject is red-green color-blind, and let M denote the event that the subject is a male recruit. Since 54 out of the 1000 subjects are color-blind, we have

$$P(C) = \frac{54}{1000} = .054$$

Also, there are 50 male recruits who are red-green color-blind so

$$P(M \cap C) = \frac{50}{1000} = .05$$

Therefore, by Equation (2), the probability that a subject is male given that the subject is red-green color-blind is

$$\begin{aligned} P(M \mid C) &= \frac{P(M \cap C)}{P(C)} \\ &= \frac{.05}{.054} \approx .926 \end{aligned}$$

Explore and Discuss

Let A and B be events in an experiment, and suppose that $P(A) \neq 0$. In n trials, the event A occurs m times, the event B occurs k times, and the events A and B occur together l times.

1. Explain why it makes good sense to call the ratio l/m the conditional relative frequency of the event B given the event A.
2. Show that the relative frequencies l/m, m/n, and l/n satisfy the equation

$$\frac{l}{m} = \frac{\frac{l}{n}}{\frac{m}{n}}$$

3. Explain why the result of part 2 suggests that Equation (2),

$$P(B \mid A) = \frac{P(A \cap B)}{P(A)} \qquad P(A) \neq 0$$

is plausible.

In certain problems, the probability of an event B occurring given that A has occurred, written $P(B \mid A)$, is known, and we wish to find the probability of A *and* B occurring. The solution to such a problem is facilitated by the use of the following formula:

Product Rule

$$P(A \cap B) = P(A) \cdot P(B \mid A) \tag{3}$$

This formula is obtained from Equation (2) by multiplying both sides of the equation by $P(A)$. We illustrate the use of the Product Rule in the next several examples.

APPLIED EXAMPLE 4 Seniors with Driver's Licenses There are 300 seniors at Jefferson High School, of whom 140 are males. It is known that 80% of the males and 60% of the females have their driver's license. If a student is selected at random from this senior class, what is the probability that the student is:

a. A male and has a driver's license?
b. A female and does not have a driver's license?

Solution

a. Let M denote the event that the student is a male, and let D denote the event that the student has a driver's license. Then

$$P(M) = \frac{140}{300} \quad \text{and} \quad P(D \mid M) = .8$$

The event that the student selected at random is a male and has a driver's license is $M \cap D$, and by the Product Rule, the probability of this event occurring is given by

$$\begin{aligned} P(M \cap D) &= P(M) \cdot P(D \mid M) \\ &= \left(\frac{140}{300}\right)(.8) \approx .373 \end{aligned}$$

b. Let F denote the event that the student is a female. Then D^c is the event that the student does not have a driver's license. We have

$$P(F) = \frac{160}{300} \quad \text{and} \quad P(D^c \mid F) = 1 - .6 = .4$$

Note that we have used the rule of complements in the computation of $P(D^c \mid F)$. The event that the student selected at random is a female and does not have a driver's license is $F \cap D^c$, so by the Product Rule, the probability of this event occurring is given by

$$\begin{aligned} P(F \cap D^c) &= P(F) \cdot P(D^c \mid F) \\ &= \left(\frac{160}{300}\right)(.4) \approx .213 \end{aligned}$$

EXAMPLE 5 Two cards are drawn without replacement from a well-shuffled deck of 52 playing cards. What is the probability that the first card drawn is an ace and the second card drawn is a face card?

Solution Let A denote the event that the first card drawn is an ace, and let F denote the event that the second card drawn is a face card. Then $P(A) = \frac{4}{52}$. After the first card is drawn, there are 51 cards left in the deck, of which 12 are face cards. Therefore, the probability of drawing a face card given that the first card drawn was an ace is given by

$$P(F \mid A) = \frac{12}{51}$$

By the Product Rule, the probability that the first card drawn is an ace and the second card drawn is a face card is given by

$$P(A \cap F) = P(A) \cdot P(F \mid A)$$
$$= \frac{4}{52} \cdot \frac{12}{51} = \frac{4}{221} \approx .018$$

Explore and Discuss

The Product Rule can be extended to the case involving three or more events. For example, if A, B, and C are three events in an experiment, then it can be shown that

$$P(A \cap B \cap C) = P(A) \cdot P(B \mid A) \cdot P(C \mid A \cap B)$$

1. Explain the formula in words.
2. Suppose 3 cards are drawn without replacement from a well-shuffled deck of 52 playing cards. Use the given formula to find the probability that the 3 cards are aces.

The Product Rule may be generalized to the case involving any finite number of events. For example, in the case involving the three events E, F, and G, it may be shown that

$$P(E \cap F \cap G) = P(E) \cdot P(F \mid E) \cdot P(G \mid E \cap F) \quad \textbf{(4)}$$

More on Tree Diagrams

Equation (4) and its generalizations may be used to help us solve problems that involve finite stochastic processes. A **finite stochastic process** is an experiment consisting of a finite number of stages in which the outcomes and associated probabilities of each stage depend on the outcomes and associated probabilities of the preceding stages.

We can use tree diagrams to help us solve problems involving finite stochastic processes. Consider, for example, the experiment consisting of drawing 2 cards without replacement from a well-shuffled deck of 52 playing cards. What is the probability that the second card drawn is a face card?

We may think of this experiment as a stochastic process with two stages. The events associated with the first stage are F, that the card drawn is a face card, and F^c, that the card drawn is not a face card. Since there are 12 face cards, we have

$$P(F) = \frac{12}{52} \quad \text{and} \quad P(F^c) = 1 - \frac{12}{52} = \frac{40}{52}$$

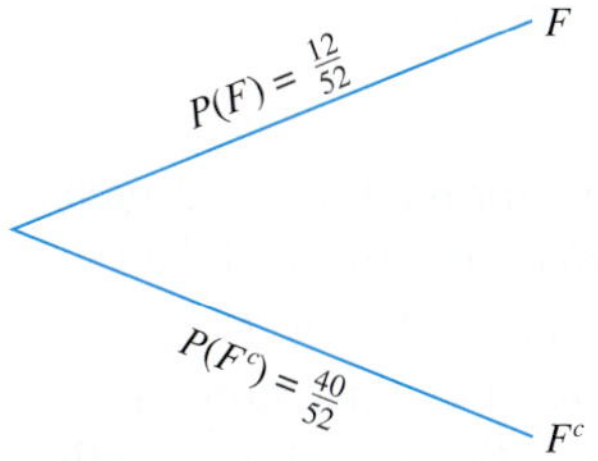

FIGURE 16
F is the event that a face card is drawn.

The outcomes of this trial, together with the associated probabilities, may be represented along two branches of a tree diagram as shown in Figure 16.

In the second trial, we again have two events: G, that the card drawn is a face card, and G^c, that the card drawn is not a face card. But the outcome of the second trial depends on the outcome of the first trial. For example, if the first card drawn was a face card, then the event G that the second card drawn is a face card has probability given by the *conditional probability* $P(G \mid F)$. Since the occurrence of a face card in the first draw leaves 11 face cards in a deck of 51 cards for the second draw, we see that

$$P(G \mid F) = \frac{11}{51}$$

The probability of drawing a face card given that a face card has already been drawn

Similarly, the occurrence of a face card in the first draw leaves 40 that are other than face cards in a deck of 51 cards for the second draw. Therefore, the probability of

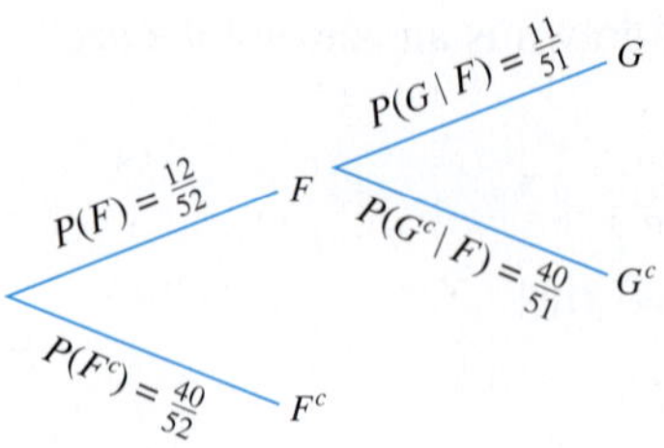

FIGURE 17
G is the event that the second card drawn is a face card.

drawing a card other than a face card in the second draw given that the first card drawn is a face card is

$$P(G^c \mid F) = \frac{40}{51}$$

Using these results, we extend the tree diagram of Figure 16 by displaying another two branches of the tree growing from its upper branch (Figure 17).

To complete the tree diagram, we compute $P(G \mid F^c)$ and $P(G^c \mid F^c)$, the conditional probabilities that the second card drawn is a face card and other than a face card, respectively, given that the first card drawn is not a face card. We find that

$$P(G \mid F^c) = \frac{12}{51} \quad \text{and} \quad P(G^c \mid F^c) = \frac{39}{51}$$

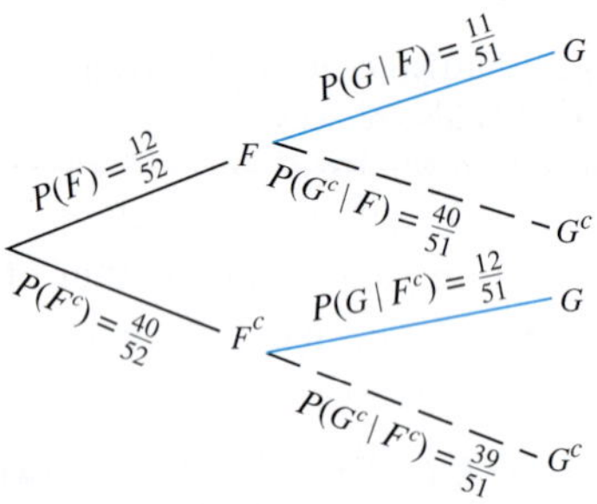

FIGURE 18
Tree diagram showing the two trials of the experiment

This leads to the completion of the tree diagram, shown in Figure 18, in which the branches of the tree that lead to the two outcomes of interest have been highlighted.

Having constructed the tree diagram associated with the problem, we are now in a position to answer the question posed earlier: What is the probability of the second card being a face card? Observe that Figure 18 shows the two ways in which a face card may result in the second draw—namely, the two Gs on the extreme right of the diagram.

Now, by the Product Rule, the probability that the second card drawn is a face card and the first card drawn is a face card (this is represented by the upper branch) is

$$P(G \cap F) = P(F) \cdot P(G \mid F)$$

Similarly, the probability that the second card drawn is a face card and the first card drawn is other than a face card (this corresponds to the other branch) is

$$P(G \cap F^c) = P(F^c) \cdot P(G \mid F^c)$$

Observe that each of these probabilities is obtained by taking the *product of the probabilities appearing on the respective branches*. Since $G \cap F$ and $G \cap F^c$ are mutually exclusive events (why?), the probability that the second card drawn is a face card is given by

$$P(G \cap F) + P(G \cap F^c) = P(F) \cdot P(G \mid F) + P(F^c) \cdot P(G \mid F^c)$$

or, upon replacing the probabilities on the right of the expression by their numerical values,

$$\begin{aligned} P(G \cap F) + P(G \cap F^c) &= \frac{12}{52} \cdot \frac{11}{51} + \frac{40}{52} \cdot \frac{12}{51} \\ &= \frac{3}{13} \end{aligned}$$

APPLIED EXAMPLE 6 Quality Control The panels for the Pulsar 32-inch widescreen LCD HDTVs are manufactured in three locations and then shipped to the main plant of Vista Vision for final assembly. Plants A, B, and C supply 50%, 30%, and 20%, respectively, of the panels used by the company. The quality-control department of the company has determined that 1% of the panels produced by Plant A are defective, whereas 2% of the panels produced by Plants B and C are defective. What is the probability that a randomly selected Pulsar 32-inch HDTV will have a defective panel?

Solution Let A, B, and C denote the events that the HDTV chosen has a panel manufactured in Plant A, Plant B, and Plant C, respectively. Also, let D denote the event

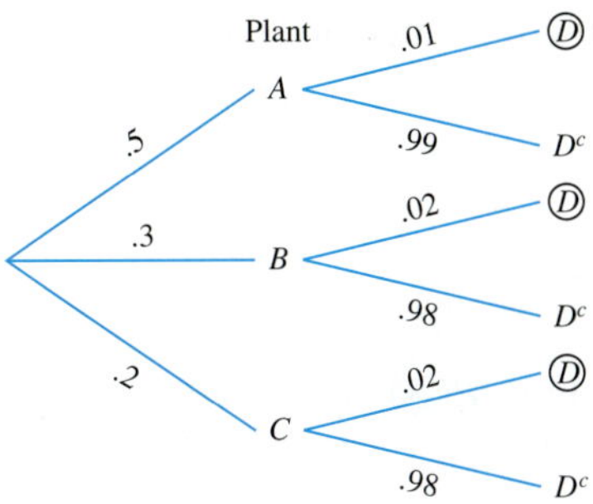

FIGURE 19
Tree diagram showing the probabilities of producing defective panels at each plant

that a HDTV has a defective panel. Using the given information, we draw the tree diagram shown in Figure 19. (The events that result in a HDTV with a defective panel being selected are circled.) Taking the product of the probabilities along each branch leading to such an event and then adding them, we obtain the probability that a HDTV chosen at random has a defective panel. Thus, the required probability is given by

$$\begin{aligned}(.5)(.01) + (.3)(.02) + (.2)(.02) &= .005 + .006 + .004 \\ &= .015\end{aligned}$$

APPLIED EXAMPLE 7 Quality Control A box contains eight 9-volt batteries, of which two are known to be defective. The batteries are selected one at a time without replacement and tested until a nondefective one is found. What is the probability that the number of batteries tested is (a) One? (b) Two? (c) Three?

Solution We may view this experiment as a multistage process with up to three stages. In the first stage, a battery is selected with a probability of $\frac{6}{8}$ of being nondefective and a probability of $\frac{2}{8}$ of being defective. If the battery selected is good, the experiment is terminated. Otherwise, a second battery is selected with probability of $\frac{6}{7}$ and $\frac{1}{7}$, respectively, of being nondefective and defective. If the second battery selected is good, the experiment is terminated. Otherwise, a third battery is selected with probability of 1 and 0, respectively, of its being nondefective and defective. The tree diagram associated with this experiment is shown in Figure 20, where N denotes the event that the battery selected is nondefective and D denotes the event that the battery selected is defective.

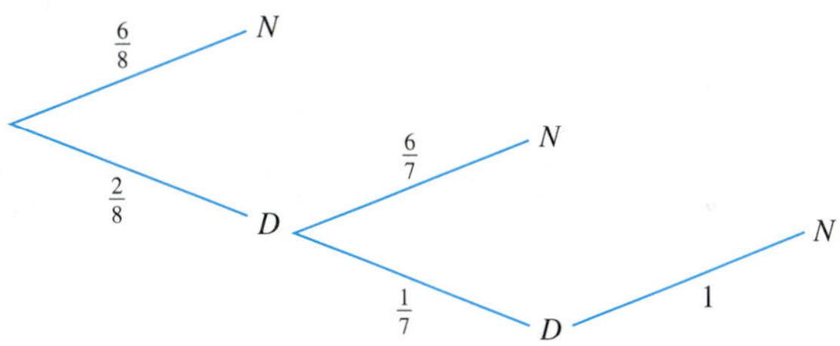

FIGURE 20
In this experiment, batteries are selected until a nondefective one is found.

With the aid of the tree diagram, we see that (a) the probability that only one battery is selected is $\frac{6}{8} = \frac{3}{4}$, (b) the probability that two batteries are selected is $\left(\frac{2}{8}\right)\left(\frac{6}{7}\right)$, or $\frac{3}{14}$, and (c) the probability that three batteries are selected is $\left(\frac{2}{8}\right)\left(\frac{1}{7}\right)(1) = \frac{1}{28}$.

Independent Events

Let's return to the experiment of drawing 2 cards in succession without replacement from a well-shuffled deck of 52 playing cards as considered in Example 5. Let E denote the event that the first card drawn is not a face card, and let F denote the event that the second card drawn is a face card. Intuitively, it is clear that the events E and F are *not* independent of each other, because whether or not the first card drawn is a face card affects the likelihood that the second card drawn is a face card.

Next, let's consider the experiment of tossing a coin twice and observing the outcomes: If H denotes the event that the first toss produces heads and T denotes the event that the second toss produces tails, then it is intuitively clear that H and T *are* independent of each other because the outcome of the first toss does not affect the outcome of the second.

In general, two events A and B are independent if the outcome of one does not affect the outcome of the other. Thus, we have

Independent Events

If A and B are **independent events,** then

$$P(A \mid B) = P(A) \quad \text{and} \quad P(B \mid A) = P(B)$$

Using the Product Rule, we can find a simple test to determine the independence of two events. Suppose that A and B are independent and that $P(A) \neq 0$and $P(B) \neq 0$. Then

$$P(B \mid A) = P(B)$$

Thus, by the Product Rule, we have

$$P(A \cap B) = P(A) \cdot P(B \mid A) = P(A) \cdot P(B)$$

Conversely, if this equation holds, then it can be seen that $P(B \mid A) = P(B)$; that is, A and B are independent. Accordingly, we have the following test for the independence of two events:

Test for the Independence of Two Events

Two events A and B are independent if and only if

$$P(A \cap B) = P(A) \cdot P(B) \tag{5}$$

Do not confuse *independent* events with *mutually exclusive* events. The former pertains to how the occurrence of one event affects the occurrence of another event, whereas the latter pertains to the question of whether the events can occur at the same time.

EXAMPLE 8 Determine whether E and F are (a) mutually exclusive and (b) independent if:

a. $P(E) = .5, P(F) = .4$, and $P(E \cap F) = .2$.
b. $P(E) = .5, P(F) = .7$, and $P(E \cap F) = .3$.
c. $P(E) = .1, P(F) = .3$, and $P(E \cap F) = 0$.

Solution

a. Here, $P(E \cap F) = .2$. Since $P(E \cap F) = .2 \neq 0$, we know that $E \cap F \neq \varnothing$, and this tells us that E and F are not mutually exclusive. Next, observe that

$$P(E)P(F) = (.5)(.4) = .2$$

Since $P(E \cap F), = P(E)P(F)$, we conclude that E and F are independent.

b. Since $P(E \cap F) = .3 \neq 0$, we know that $E \cap F \neq \varnothing$, and we see that E and F are not mutually exclusive. Next, observe that

$$P(E)P(F) = (.5)(.7) = .35$$

Since $P(E \cap F) \neq P(E)P(F)$, we conclude that E and F are not independent.

c. Here, $P(E \cap F) = 0$, and this shows that E and F have no points in common. In other words, E and F are mutually exclusive. Next, observe that

$$P(E)P(F) = (.1)(.3) = .03$$

Since

$$P(E \cap F) \neq P(E)P(F)$$

we conclude that E and F are not independent.

Note that part (c) of Example 8 shows that two events may be mutually exclusive but not independent and vice versa.

EXAMPLE 9 Consider the experiment consisting of tossing a fair coin twice and observing the outcomes. Show that the event of heads on the first toss and the event of tails on the second toss are independent events.

Solution Let A denote the event that the outcome of the first toss is a head, and let B denote the event that the outcome of the second toss is a tail. The sample space of the experiment is

$$S = \{(\text{HH}), (\text{HT}), (\text{TH}), (\text{TT})\}$$
$$A = \{(\text{HH}), (\text{HT})\}$$
$$B = \{(\text{HT}), (\text{TT})\}$$

so

$$A \cap B = \{(\text{HT})\}$$

Next, we compute

$$P(A \cap B) = \frac{1}{4} \qquad P(A) = \frac{1}{2} \qquad P(B) = \frac{1}{2}$$

and observe that Equation (5) is satisfied in this case. Hence, A and B are independent events, as we set out to show.

APPLIED EXAMPLE 10 Medical Surveys A survey conducted by an independent agency for the National Lung Society found that of 2000 women, 680 were heavy smokers and 50 had emphysema. Of those who had emphysema, 42 were also heavy smokers. Using the data in this survey, determine whether the events "being a heavy smoker" and "having emphysema" are independent events.

Solution Let A denote the event that a woman chosen at random in this survey is a heavy smoker, and let B denote the event that a woman chosen at random in this survey has emphysema. Then the probability that a woman is a heavy smoker and has emphysema is given by

$$P(A \cap B) = \frac{42}{2000} = .021$$

Next,

$$P(A) = \frac{680}{2000} = .34 \quad \text{and} \quad P(B) = \frac{50}{2000} = .025$$

so

$$P(A) \cdot P(B) = (.34)(.025) = .0085$$

Since $P(A \cap B) \neq P(A) \cdot P(B)$, we conclude that A and B are not independent events.

The solution of many practical problems involves more than two independent events. In such cases, we use the following result.

Explore and Discuss

Let E and F be independent events in a sample space S. Are E^c and F^c independent?

Independence of More Than Two Events

If $E_1, E_2, \ldots, E_n$ are independent events, then

$$P(E_1 \cap E_2 \cap \cdots \cap E_n) = P(E_1) \cdot P(E_2) \cdot \cdots \cdot P(E_n) \quad \textbf{(6)}$$

Formula (6) states that the probability of the simultaneous occurrence of n independent events is equal to the product of the probabilities of the n events.

It is important to note that the mere requirement that the n events $E_1, E_2, \ldots, E_n$ satisfy Formula (6) is not sufficient to guarantee that the n events are indeed independent. However, a criterion does exist for determining the independence of n events; it may be found in more advanced texts on probability.

EXAMPLE 11 It is known that the three events A, B, and C are independent and that $P(A) = .2$, $P(B) = .4$, and $P(C) = .5$. Compute:

a. $P(A \cap B)$ **b.** $P(A \cap B \cap C)$

Solution Using Formulas (5) and (6), we find

a. $P(A \cap B) = P(A) \cdot P(B)$
$= (.2)(.4) = .08$

b. $P(A \cap B \cap C) = P(A) \cdot P(B) \cdot P(C)$
$= (.2)(.4)(.5) = .04$

APPLIED EXAMPLE 12 Predicting Travel Weather Ron is planning to visit Paris on Thursday and Friday. A quick check of the weather report for Paris revealed that there is a probability of .2 that it will rain on Thursday and a probability of .4 that it will rain on Friday. Assuming that the likelihood of rain on Thursday is independent of the likelihood of rain on Friday, what is the probability that Ron will see rain on at least one of those two days?

Solution Let A and B denote the event that it will rain on Thursday and Friday, respectively. Then the probability that it will rain on Thursday or Friday is given by $P(A \cup B)$. Using the Addition Rule, we have

$$P(A \cup B) = P(A) + P(B) - P(A \cap B)$$

Here, $P(A) = .2$ and $P(B) = .4$, and since A and B are independent events, we see that

$$P(A \cap B) = P(A)P(B) = (.2)(.4) = .08$$

Therefore,

$$P(A \cup B) = .2 + .4 - .08 = .52$$

We conclude that there is a 52% likelihood that Ron will see rain on at least one of the two days of his stay in Paris.

APPLIED EXAMPLE 13 Quality Control The Acrosonic model F loudspeaker system has four loudspeaker components: a woofer, a mid-range, a tweeter, and an electrical crossover. The quality-control manager of Acrosonic has determined that on the average, 1% of the woofers, 0.8% of the midranges, and 0.5% of the tweeters are defective, while 1.5% of the electrical crossovers are defective. Determine the probability that a loudspeaker system selected at random as it comes off the assembly line (and before final inspection) is not defective. Assume that the defects in the manufacturing of the components are unrelated.

Solution Let A, B, C, and D denote, respectively, the events that the woofer, the midrange, the tweeter, and the electrical crossover are defective. Then

$$P(A) = .01 \qquad P(B) = .008 \qquad P(C) = .005 \qquad P(D) = .015$$

and the probabilities of the corresponding complementary events are

$$P(A^c) = .99 \qquad P(B^c) = .992 \qquad P(C^c) = .995 \qquad P(D^c) = .985$$

The event that a loudspeaker system selected at random is not defective is given by $A^c \cap B^c \cap C^c \cap D^c$. Because the events A, B, C, and D (and therefore also A^c, B^c, C^c, and D^c) are assumed to be independent, we find that the required probability is given by

$$\begin{aligned} P(A^c \cap B^c \cap C^c \cap D^c) &= P(A^c) \cdot P(B^c) \cdot P(C^c) \cdot P(D^c) \\ &= (.99)(.992)(.995)(.985) \\ &\approx .96 \end{aligned}$$

7.5 Self-Check Exercises

1. Let A and B be events in a sample space S such that $P(A) = .4$, $P(B) = .8$, and $P(A \cap B) = .3$. Find:
   a. $P(A \mid B)$ b. $P(B \mid A)$

2. **Happiness with Marriage** According to a survey cited in *Newsweek*, 29.7% of married survey respondents who married between the ages of 20 and 22 (inclusive), 26.9% of those who married between the ages of 23 and 27, and 45.1% of those who married at age 28 or older said that "their marriage was less than 'very happy.'" Suppose that a survey respondent from each of the three age groups was selected at random. What is the probability that all three respondents said that their marriage was "less than very happy"?
   Source: Marc Bain, *Newsweek*.

Solutions to Self-Check Exercises 7.5 can be found on page 440.

7.5 Concept Questions

1. What is conditional probability? Illustrate the concept with an example.

2. If A and B are events in an experiment and $P(A) \neq 0$, then what is the formula for computing $P(B \mid A)$?

3. If A and B are events in an experiment and the conditional probability $P(B \mid A)$ is known, give the formula that can be used to compute the probability of the event that both A and B will occur.

4. a. What is the test for determining the independence of two events?
   b. What is the difference between mutually exclusive events and independent events?

7.5 Exercises

1. Let A and B be two events in a sample space S such that $P(A) = .6$, $P(B) = .5$, and $P(A \cap B) = .2$. Find:
   a. $P(A \mid B)$ b. $P(B \mid A)$

2. Let A and B be two events in a sample space S such that $P(A) = .4$, $P(B) = .6$, and $P(A \cap B) = .3$. Find:
   a. $P(A \mid B)$ b. $P(B \mid A)$

3. Let A and B be two events in a sample space S such that $P(A) = .6$ and $P(B \mid A) = .5$. Find $P(A \cap B)$.

4. Let A and B be the events described in Exercise 1. Find:
   a. $P(A \mid B^c)$ b. $P(B \mid A^c)$
   Hint: $(A \cap B^c) \cup (A \cap B) = A$

In Exercises 5–8, determine whether the events A and B are independent.

5. $P(A) = .3$, $P(B) = .6$, $P(A \cap B) = .18$

6. $P(A) = .6$, $P(B) = .8$, $P(A \cap B) = .2$

7. $P(A) = .5$, $P(B) = .7$, $P(A \cup B) = .85$

8. $P(A^c) = .3$, $P(B^c) = .4$, $P(A \cap B) = .42$

9. If A and B are independent events, $P(A) = .4$, and $P(B) = .6$, find:
   a. $P(A \cap B)$ b. $P(A \cup B)$
   c. $P(A \mid B)$ d. $P(A^c \cup B^c)$

10. If A and B are independent events, $P(A) = .35$, and $P(B) = .45$, find:
    a. $P(A \cap B)$ b. $P(A \cup B)$
    c. $P(A \mid B)$ d. $P(A^c \cup B^c)$

11. The accompanying tree diagram represents an experiment consisting of two trials:

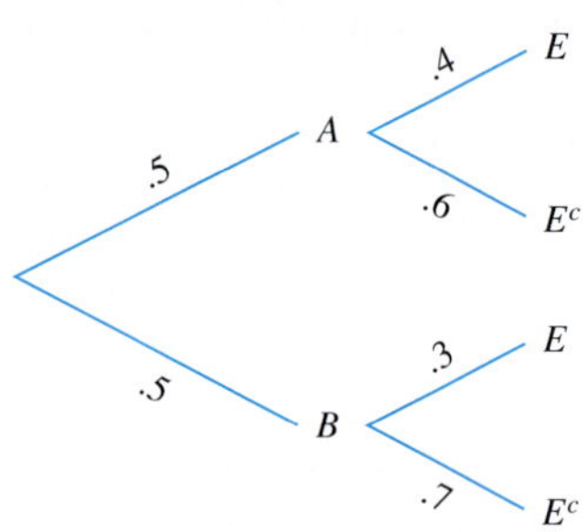

Use the diagram to find:
   a. $P(A)$ b. $P(E \mid A)$
   c. $P(A \cap E)$ d. $P(E)$
   e. Does $P(A \cap E) = P(A) \cdot P(E)$?
   f. Are A and E independent events?

12. The accompanying tree diagram represents an experiment consisting of two trials. Use the diagram to find:
    a. $P(A)$ b. $P(E \mid A)$
    c. $P(A \cap E)$ d. $P(E)$
    e. Does $P(A \cap E) = P(A) \cdot P(E)$?
    f. Are A and E independent events?

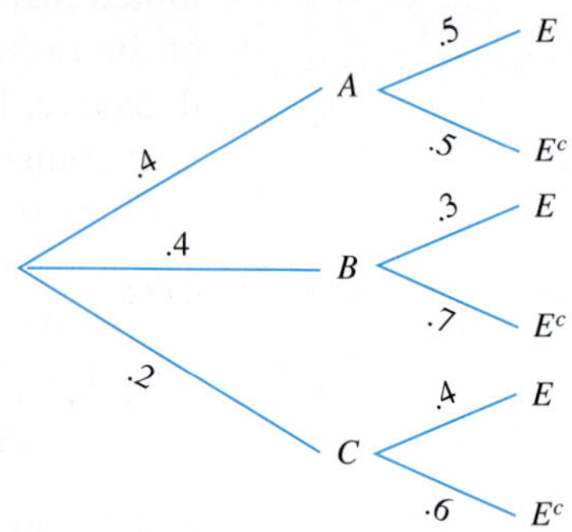

13. An experiment consists of two independent trials. The outcomes of the first trial are A and B with probabilities of occurring equal to .4 and .6. There are also two outcomes, C and D, in the second trial with probabilities of .3 and .7. Draw a tree diagram representing this experiment, and use it to find:
    a. $P(A)$ b. $P(C \mid A)$
    c. $P(A \cap C)$ d. $P(C)$
    e. Does $P(A \cap C) = P(A) \cdot P(C)$?
    f. Are A and C independent events?

14. An experiment consists of two independent trials. The outcomes of the first trial are A, B, and C, with probabilities of occurring equal to .2, .5, and .3, respectively. The outcomes of the second trial are E and F, with probabilities of occurring equal to .6 and .4. Draw a tree diagram representing this experiment. Use this diagram to find:
    a. $P(B)$ b. $P(F \mid B)$
    c. $P(B \cap F)$ d. $P(F)$
    e. Does $P(B \cap F) = P(B) \cdot P(F)$?
    f. Are B and F independent events?

15. A pair of fair dice is rolled. Let E denote the event that the number falling uppermost on the first die is 5, and let F denote the event that the sum of the numbers falling uppermost is 10.
    a. Compute $P(F)$. b. Compute $P(E \cap F)$.
    c. Compute $P(F \mid E)$. d. Compute $P(E)$.
    e. Are E and F independent events?

16. A pair of fair dice is rolled. Let E denote the event that the number falling uppermost on the first die is 4, and let F denote the event that the sum of the numbers falling uppermost is 6.
    a. Compute $P(F)$. b. Compute $P(E \cap F)$.
    c. Compute $P(F \mid E)$. d. Compute $P(E)$.
    e. Are E and F independent events?

17. A pair of fair dice is rolled. What is the probability that the sum of the numbers falling uppermost is less than 9, given that at least one of the numbers is a 6?

18. A pair of fair dice is rolled. What is the probability that the number landing uppermost on the first die is a 4 if it is known that the sum of the numbers landing uppermost is 7?

19. A pair of fair dice is rolled. Let E denote the event that the number landing uppermost on the first die is a 3, and let F denote the event that the sum of the numbers landing uppermost is 7. Determine whether E and F are independent events.

20. A pair of fair dice is rolled. Let E denote the event that the number landing uppermost on the first die is a 3, and let F denote the event that the sum of the numbers landing uppermost is 6. Determine whether E and F are independent events.

21. A card is drawn from a well-shuffled deck of 52 playing cards. Let E denote the event that the card drawn is black and let F denote the event that the card drawn is a spade. Determine whether E and F are independent events. Give an intuitive explanation for your answer.

22. A card is drawn from a well-shuffled deck of 52 playing cards. Let E denote the event that the card drawn is an ace and let F denote the event that the card drawn is a diamond. Determine whether E and F are independent events. Give an intuitive explanation for your answer.

23. **Battery Life** The probability that a battery will last 10 hr or more is .80, and the probability that it will last 15 hr or more is .15. Given that a battery has lasted 10 hr, find the probability that it will last 15 hr or more.

24. Five black balls and four white balls are placed in an urn. Two balls are then drawn in succession. What is the probability that the second ball drawn is a white ball if:

a. The second ball is drawn without replacing the first?
b. The first ball is replaced before the second is drawn?

Refer to Exercises 7.1, Problem 34, where you were asked to find the sample space corresponding to the different human blood groups. The following table gives the percent of the U.S. population having each of the eight possible blood types in the sample space. Note that the presence or absence of the Rh antigen is indicated by the symbols + or −, respectively.

Blood Types	A^+	A^-	B^+	B^-	AB^+	AB^-	O^+	O^-
Percent	35.7	6.3	8.5	1.5	3.4	0.6	37.4	6.6

Source: American Red Cross.

In Exercises 25 and 26, use the above table.

25. **Blood Types**

a. What is the probability that a person selected at random from the U.S. population has a blood type that is type A given that the person is Rh^-?
b. What is the probability that a person selected at random from the U.S. population has a blood type that is Rh^+ given that the person is type B?

26. **Blood Types**

a. What is the probability that a person selected at random from the U.S. population has a blood type that is type AB given that the person is Rh^+?
b. What is the probability that a person selected at random from the U.S. population has a blood type that is Rh^- given that the person is type O?

27. **Selling a Car** Jack has decided to advertise the sale of his car by placing flyers in the student union and the dining hall of the college. He estimates that there is a probability of .3 that a potential buyer will read the advertisement and that, if it is read, the probability that the reader will buy his car will be .2. Using these estimates, find the probability that the person who reads the ad will buy Jack's car.

28. **Auditing Tax Returns** A tax specialist has estimated that the probability that a tax return selected at random will be audited is .02. Furthermore, he estimates that the probability that an audited return will result in additional assessments being levied on the taxpayer is .60. What is the probability that a tax return selected at random will result in additional assessments being levied on the taxpayer?

29. **Student Enrollment** At a certain medical school, $\frac{1}{7}$ of the students are from a minority group. Of the students who belong to a minority group, $\frac{1}{3}$ are black.

a. What is the probability that a student selected at random from this medical school is black?
b. What is the probability that a student selected at random from this medical school is black if it is known that the student is a member of a minority group?

30. **Educational Level of Voters** In a survey of 1000 eligible voters selected at random, it was found that 80 had a college degree. Additionally, it was found that 80% of those who had a college degree voted in the last presidential election, whereas 55% of the people who did not have a college degree voted in the last presidential election. Assuming that the poll is representative of all eligible voters, find the probability that an eligible voter selected at random:

a. Had a college degree and voted in the last presidential election.
b. Did not have a college degree and did not vote in the last presidential election.
c. Voted in the last presidential election.
d. Did not vote in the last presidential election.

31. The probability that Sandy takes her daughter Olivia to the supermarket on Friday is .6. If Sandy does bring Olivia to the supermarket on Friday, the probability that she buys Olivia a popsicle is .8. What is the probability that Sandy takes Olivia to the supermarket on Friday and buys her a popsicle?

32. **Solving Problems** The probability that Art will submit the correct solution to a certain homework problem is $\frac{1}{3}$, and the probability that Candice will submit the correct solution to it is $\frac{1}{2}$. Assuming that they work independently, what is the probability that either Art or Candice or both Art and Candice will submit the correct solution?

33. **Medical Survey** A nationwide survey conducted by the National Cancer Society revealed the following information. Of 10,000 people surveyed, 3200 were "heavy coffee drinkers," and 160 had cancer of the pancreas. Of those who had cancer of the pancreas, 132 were heavy coffee drinkers. Using the data in this survey, determine whether the events "being a heavy coffee drinker" and "having cancer of the pancreas" are independent events.

34. **Employee Education and Income** The personnel department of Franklin National Life Insurance Company compiled the accompanying data regarding the income and education of its employees:

	Income $65,000 or Below	Income Above $65,000
Noncollege Graduate	2040	840
College Graduate	400	720

Let A be the event that a randomly chosen employee has a college degree, and let B be the event that the chosen employee's income is more than $65,000.

a. Find each of the following probabilities: $P(A)$, $P(B)$, $P(A \cap B)$, $P(B \mid A)$, and $P(B \mid A^c)$.
b. Are the events A and B independent events?

35. **Student Financial Aid** The accompanying data were obtained from the financial aid office of a certain university:

	Receiving Financial Aid	Not Receiving Financial Aid	Total
Undergraduates	4,222	3,898	8,120
Graduates	1,879	731	2,610
Total	6,101	4,629	10,730

Let A be the event that a student selected at random from this university is an undergraduate student, and let B be the event that a student selected at random is receiving financial aid.

a. Find each of the following probabilities: $P(A)$, $P(B)$, $P(A \cap B)$, $P(B \mid A)$, and $P(B \mid A^c)$.
b. Are the events A and B independent events?

36. Two cards are drawn without replacement from a well-shuffled deck of 52 playing cards.
a. What is the probability that the first card drawn is a heart?
b. What is the probability that the second card drawn is a heart if the first card drawn was not a heart?
c. What is the probability that the second card drawn is a heart if the first card drawn was a heart?

37. **Family Composition** In a three-child family, what is the probability that all three children are girls given that at least one of the children is a girl? (Assume that the probability of a boy being born is the same as the probability of a girl being born.)

38. A coin is tossed three times. What is the probability that the coin will land heads:
a. At least twice?
b. On the second toss, given that heads were thrown on the first toss?
c. On the third toss, given that tails were thrown on the first toss?

39. **Television Pilots** Max Productions has two pilots for the coming television season. The probability that the first pilot will be successful is estimated to be .9, and the probability that the second pilot will be successful is estimated to be .8. Assuming that the success of one pilot does not have a bearing on the success of the other, what is the probability that Max Productions will have at least one successful pilot for the coming television season.

40. **Car Theft** Figures obtained from a city's police department seem to indicate that of all motor vehicles reported as stolen, 64% were stolen by professionals, whereas 36% were stolen by amateurs (primarily for joy rides). Of the vehicles presumed stolen by professionals, 24% were recovered within 48 hr, 16% were recovered after 48 hr, and 60% were never recovered. Of the vehicles presumed stolen by amateurs, 38% were recovered within 48 hr, 58% were recovered after 48 hr, and 4% were never recovered.
a. Draw a tree diagram representing these data.
b. What is the probability that a vehicle stolen by a professional in this city will be recovered within 48 hr?
c. What is the probability that a vehicle stolen in this city will never be recovered?

41. **Switching Broadband Service** According to a survey conducted in 2010 by the Federal Communications Commission (FCC) of 3005 adults who were home broadband users, 37.5% of those surveyed had switched their service over the past 3 years. Of those who had switched service in the past 3 years, 51% were very satisfied, 39% were somewhat satisfied, and 10% were not satisfied with their service. Of the 62.5% who had not switched service in the past 3 years, 48% were very satisfied, 43% were somewhat satisfied, and 9% were not satisfied with their service.
a. What is the probability that a participant chosen at random had not switched service in the past 3 years and was very satisfied with their service?
b. What is the probability that a participant chosen at random was not satisfied with service?

Source: FCC.

42. **Probability of Transplant Rejection** The probabilities that the three patients who are scheduled to receive kidney transplants at General Hospital will suffer rejection are $\frac{1}{2}$,

$\frac{1}{3}$, and $\frac{1}{10}$. Assuming that the events (kidney rejection) are independent, find the probability that:

a. At least one patient will suffer rejection.

b. Exactly two patients will suffer rejection.

43. **Quality Control** An automobile manufacturer obtains the microprocessors used to regulate fuel consumption in its automobiles from three microelectronic firms: *A*, *B*, and *C*. The quality-control department of the company has determined that 1% of the microprocessors produced by Firm *A* are defective, 2% of those produced by Firm *B* are defective, and 1.5% of those produced by Firm *C* are defective. Firms *A*, *B*, and *C* supply 45%, 25%, and 30%, respectively, of the microprocessors used by the company. What is the probability that a randomly selected automobile manufactured by the company will have a defective microprocessor?

44. **Housing Loans** The chief loan officer of La Crosse Home Mortgage Company summarized the housing loans extended by the company in 2014 according to type and term of the loan. Her list shows that 70% of the loans were fixed-rate mortgages (F), 25% were adjustable-rate mortgages (A), and 5% belong to some other category (O) (mostly second trust-deed loans and home equity loans). Of the fixed-rate mortgages, 80% were 30-year loans and 20% were 15-year loans; of the adjustable-rate mortgages, 40% were 30-year loans and 60% were 15-year loans; finally, of the other loans extended, 30% were 20-year loans, 60% were 10-year loans, and 10% were for a term of 5 years or less.

a. Draw a tree diagram representing these data.

b. What is the probability that a home loan extended by La Crosse has an adjustable rate and is for a term of 15 years?

c. What is the probability that a home loan extended by La Crosse is for a term of 15 years?

45. Three cards are drawn without replacement from a well-shuffled deck of 52 playing cards. What is the probability that the third card drawn is a diamond?

46. **College Admissions** The admissions office of a private university released the following data for the preceding academic year: From a pool of 3900 male applicants, 40% were accepted by the university, and 40% of these subsequently enrolled. Additionally, from a pool of 3600 female applicants, 45% were accepted by the university, and 40% of these subsequently enrolled. What is the probability that:

a. A male applicant will be accepted by and subsequently will enroll in the university?

b. A student who applies for admissions will be accepted by the university?

c. A student who applies for admission will be accepted by the university and subsequently will enroll?

47. **Quality Control** A box contains two defective Christmas tree lights that have been inadvertently mixed with eight nondefective lights. If the lights are selected one at a time without replacement and tested until both defective lights are found, what is the probability that both defective lights will be found after exactly three trials?

48. **NYC Tourists** In 2011, 21% of all tourists in New York City (NYC) were international visitors, of whom an estimated 70% visited the Empire State Building. Of the international tourists who visited the Empire State Building, approximately 40% bought at least one souvenir in the gift shop. What percentage of international tourists in NYC in 2011 visited the Empire State Building and also bought at least one souvenir from the gift shop?
Source: nycgo.com.

49. **Real Estate Sales** Mark, a real estate agent, estimates that of prospective homebuyers who see an ad for a house he has listed for sale, 24% will show up for the open house. Of those who show up for the open house, 30% will return for a second showing, and of these, 75% will make an offer to buy the house. What is the probability that a prospective home buyer who has seen Mark's ad will show up for a second viewing and ultimately make an offer to buy the house?

50. **Winning Bids** Brian, a landscape architect, submitted a bid on each of three home landscaping projects. He estimates that the probabilities of winning the bid on Project *A*, Project *B*, and Project *C* are .7, .6, and .5, respectively. Assume that the probability of winning a bid on one of the three projects is independent of winning or losing the bids on the other two projects. Find the probability that Brian will:

a. Win all three of the bids.

b. Win exactly two of the bids.

c. Win exactly one bid.

51. **Family Portraits** The Jackson family of four is posing for a family portrait in a studio. The probability of Dad and Mom striking an acceptable pose is .9 each; the probability of daughter Janet striking an acceptable pose is .7; and the probability of baby Bob striking an acceptable pose is .3. Assuming that the probability of each family member striking an acceptable pose is independent of the way in which the other members of the family pose, what is the probability that an acceptable photo will result after one take?

52. Two cards are drawn without replacement from a well-shuffled deck of 52 cards. Let A be the event that the first card drawn is a heart, and let B be the event that the second card drawn is a red card. Show that the events A and B are dependent events.

53. **Reliability of Security Systems** Before being allowed to enter a maximum-security area at a military installation, a person must pass three independent identification tests: a voice-pattern test, a fingerprint test, and a handwriting test. If the reliability of the first test is 97%, that of the second test is 98.5%, and that of the third is 98.5%, what is the probability that this security system will allow an improperly identified person to enter the maximum-security area?

54. **Quality Control** Copykwik has four photocopy machines: A, B, C, and D. The probability that a given machine will break down on a particular day is

$$P(A) = \frac{1}{50} \quad P(B) = \frac{1}{60} \quad P(C) = \frac{1}{75} \quad P(D) = \frac{1}{40}$$

Assuming independence, what is the probability on a particular day that:

a. All four machines will break down?

b. None of the machines will break down?

55. **Quality Control** It is estimated that 0.80% of a large consignment of eggs in a certain supermarket is broken.

a. What is the probability that a customer who randomly selects a dozen of these eggs receives at least one broken egg?

b. What is the probability that a customer who selects these eggs at random will have to check three cartons before finding a carton without any broken eggs? (Each carton contains a dozen eggs.)

56. **Reliability of a Home Theater System** In a home theater system, the probability that the video components need repair within 1 year is .01, the probability that the electronic components need repair within 1 year is .005, and the probability that the audio components need repair within 1 year is .001. Assuming that the events are independent, find the probability that:

a. At least one of these components will need repair within 1 year.

b. Exactly one of these components will need repair within 1 year.

57. **Product Reliability** The proprietor of Cunningham's Hardware Store has decided to install floodlights on the premises as a measure against vandalism and theft. If the probability is .01 that a certain brand of floodlight will burn out within a year, find the minimum number of floodlights that must be installed to ensure that the probability that at least one of them will remain functional for the whole year is at least .99999. (Assume that the floodlights operate independently.)

58. Suppose the probability that an event will occur in one trial is p. Show that the probability that the event will occur at least once in n independent trials is $1 - (1 - p)^n$.

59. Let E be any event in a sample space S.

a. Are E and S mutually exclusive? Explain your answer.

b. Are E and $\varnothing$ mutually exclusive? Explain your answer.

60. Let E and F be events such that $F \subset E$. Find $P(E \mid F)$, and interpret your result.

61. Let E and F be mutually exclusive events, and suppose $P(F) \neq 0$. Find $P(E \mid F)$, and interpret your result.

62. Let E and F be independent events; show that E and F^c are independent.

63. Suppose that A and B are mutually exclusive events and that $P(A \cup B) \neq 0$. What is $P(A \mid A \cup B)$?

64. Prove Equation (4),
$P(A \cap B \cap C) = P(A)P(B \mid A)P(C \mid A \cap B)$.

In Exercises 65–68, determine whether the statement is true or false. If it is true, explain why it is true. If it is false, give an example to show why it is false.

65. If A and B are mutually exclusive and $P(B) \neq 0$, then $P(A \mid B) = 0$.

66. If A is an event of an experiment, then $P(A \mid A^c) \neq 0$.

67. If A and B are events of an experiment, then

$$P(A \cap B) = P(A \mid B) \cdot P(B) = P(B \mid A) \cdot P(A)$$

68. If A and B are independent events with $P(A) \neq 0$ and $P(B) \neq 0$, then $A \cap B \neq \varnothing$.

7.5 Solutions to Self-Check Exercises

1. a. $P(A \mid B) = \dfrac{P(A \cap B)}{P(B)} = \dfrac{.3}{.8} = \dfrac{3}{8}$

b. $P(B \mid A) = \dfrac{P(A \cap B)}{P(A)} = \dfrac{.3}{.4} = \dfrac{3}{4}$

2. Let A, B, and C denote the events that a respondent who married between the ages of 20 and 22, between the ages of 23 and 27, and at age 28 or older (respectively) said that his or her marriage was "less than very happy." Then the probability of each of these events occurring is $P(A) = .297$, $P(B) = .269$, and $P(C) = .451$. So the probability that all three of the respondents said that his or her marriage was "less than very happy" is

$$P(A) \cdot P(B) \cdot P(C) = (.297)(.269)(.451) \approx .036$$

7.6 Bayes' Theorem

A Posteriori Probabilities

Suppose three machines, A, B, and C, produce similar engine components. Machine A produces 45% of the total components, Machine B produces 30%, and Machine C produces 25%. For the usual production schedule, 6% of the components produced by Machine A do not meet established specifications; for Machine B and Machine C, the corresponding figures are 4% and 3%, respectively. One component is selected at random from the total output and is found to be defective. What is the probability that the component selected was produced by Machine A?

The answer to this question is found by calculating the probability *after* the outcomes of the experiment have been observed. Such probabilities are called **a posteriori probabilities** in contrast to **a priori probabilities**—probabilities that give the likelihood that an event *will* occur, the subject of the last several sections.

Returning to the example under consideration, we need to determine the a posteriori probability for the event that the component selected was produced by Machine A. Toward this end, let A, B, and C denote the events that a component is produced by Machine A, Machine B, and Machine C, respectively. We may represent this experiment with a Venn diagram (Figure 21).

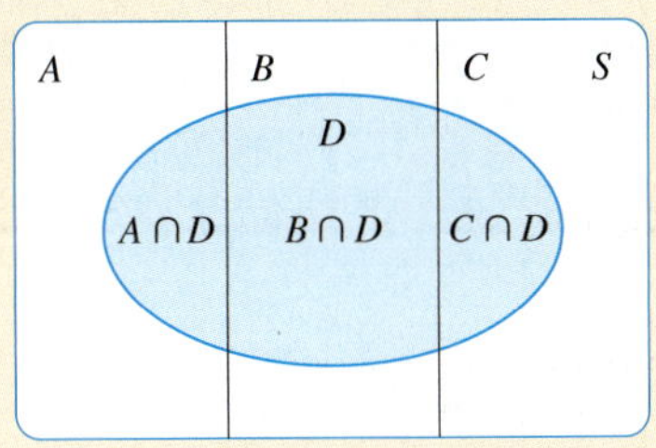

FIGURE 21
D is the event that a defective component is produced by Machine A, Machine B, or Machine C.

The three mutually exclusive events A, B, and C form a **partition** of the sample space S; that is, aside from being mutually exclusive, their union is precisely S. The event D that a component is defective is the shaded area. Again referring to Figure 21, we see that

1. The event D may be expressed as

$$D = (A \cap D) \cup (B \cap D) \cup (C \cap D)$$

2. The event that a component is defective and is produced by Machine A is given by $A \cap D$.

Thus, the a posteriori probability that a defective component selected was produced by Machine A is given by

$$P(A \mid D) = \frac{P(A \cap D)}{P(D)}$$

Upon dividing both the numerator and the denominator by $P(S)$ and observing that the events $A \cap D$, $B \cap D$, and $C \cap D$ are mutually exclusive, we obtain

$$P(A \mid D) = \frac{P(A \cap D)}{P(D)} = \frac{P(A \cap D)}{P(A \cap D) + P(B \cap D) + P(C \cap D)} \tag{7}$$

Next, using the Product Rule, we may express

$$P(A \cap D) = P(A) \cdot P(D \mid A)$$
$$P(B \cap D) = P(B) \cdot P(D \mid B)$$
$$P(C \cap D) = P(C) \cdot P(D \mid C)$$

so Equation (7) may be expressed in the form

$$P(A \mid D) = \frac{P(A) \cdot P(D \mid A)}{P(A) \cdot P(D \mid A) + P(B) \cdot P(D \mid B) + P(C) \cdot P(D \mid C)} \tag{8}$$

which is a special case of a result known as **Bayes' Theorem.**

Observe that the expression on the right of Equation (8) involves the probabilities $P(A)$, $P(B)$, and $P(C)$ as well as the conditional probabilities $P(D \mid A)$, $P(D \mid B)$, and $P(D \mid C)$. In fact, by displaying these probabilities on a tree diagram, we obtain Figure 22. We may compute the required probability by substituting the relevant quantities into Equation (8), or we may make use of the following device:

$$P(A \mid D) = \frac{\text{Product of probabilities along the branch through } A \text{ terminating at } D}{\text{Sum of products of the probabilities along each branch terminating at } D}$$

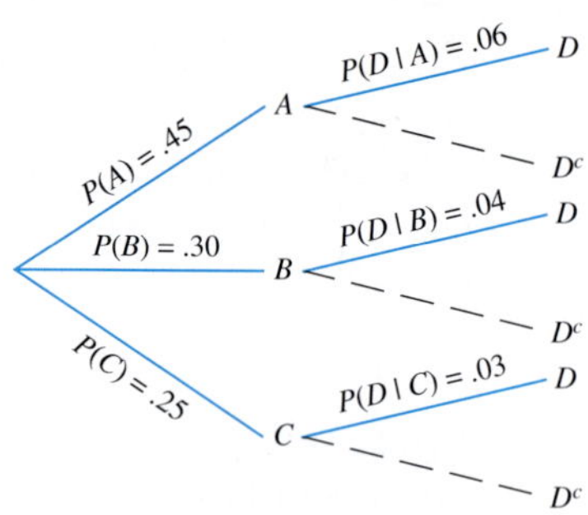

FIGURE 22
A tree diagram displaying the probabilities that a defective component is produced by Machine A, Machine B, or Machine C

In either case, we obtain

$$P(A \mid D) = \frac{(.45)(.06)}{(.45)(.06) + (.30)(.04) + (.25)(.03)} \approx .58$$

Before looking at any further examples, let's state the general form of Bayes' Theorem.

Bayes' Theorem

Let $A_1, A_2, \ldots, A_n$ be a partition of a sample space S, and let E be an event of the experiment such that $P(E) \neq 0$ and $P(A_i) \neq 0$ for $1 \leq i \leq n$. Then the a posteriori probability $P(A_i \mid E)$ $(1 \leq i \leq n)$ is given by

$$P(A_i \mid E) = \frac{P(A_i) \cdot P(E \mid A_i)}{P(A_1) \cdot P(E \mid A_1) + P(A_2) \cdot P(E \mid A_2) + \cdots + P(A_n) \cdot P(E \mid A_n)} \tag{9}$$

APPLIED EXAMPLE 1 Quality Control The panels for the Pulsar 32-inch widescreen LCD HDTVs are manufactured in three locations and then shipped to the main plant of Vista Vision for final assembly. Plants A, B, and C supply 50%, 30%, and 20%, respectively, of the panels used by Vista Vision. The quality-control department of the company has determined that 1% of the panels produced by Plant A are defective, whereas 2% of the panels produced by Plants B and C are defective. If a Pulsar 32-inch HDTV is selected at random and the panel is found to be defective, what is the probability that the panel was manufactured in Plant C? (Compare with Example 6, page 430.)

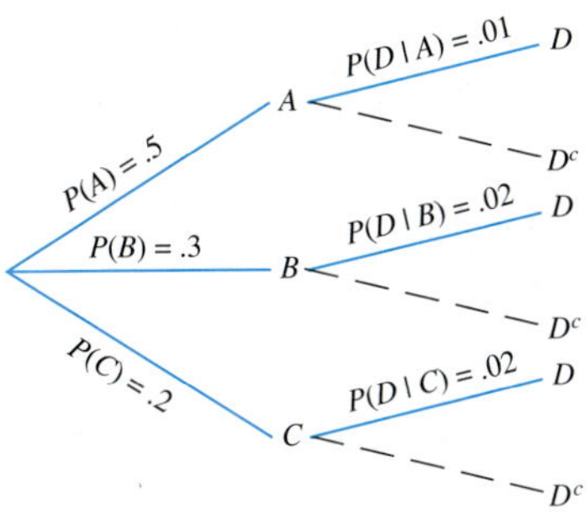

FIGURE 23

$$P(C \mid D) = \frac{\text{Product of probabilities of branches to } D \text{ through } C}{\text{Sum of product of probabilities of branches leading to } D}$$

Solution Let A, B, and C denote the events that the set chosen has a panel manufactured in Plant A, Plant B, and Plant C, respectively. Also, let D denote the event that a set has a defective panel. Using the given information, we may draw the tree diagram shown in Figure 23. Next, using Formula (9), we find that the required a posteriori probability is given by

$$\begin{aligned} P(C \mid D) &= \frac{P(C) \cdot P(D \mid C)}{P(A) \cdot P(D \mid A) + P(B) \cdot P(D \mid B) + P(C) \cdot P(D \mid C)} \\ &= \frac{(.20)(.02)}{(.50)(.01) + (.30)(.02) + (.20)(.02)} \\ &\approx .27 \end{aligned}$$

APPLIED EXAMPLE 2 Income Distributions A study was conducted in a large metropolitan area to determine the annual incomes of married couples in which the husbands were the sole providers and of those in which the husbands and wives were both employed. Table 9 gives the results of this study.

TABLE 9

Annual Family Income ($)	Married Couples (%)	Income Group with Both Spouses Employed (%)
150,000 and over	4	65
100,000–149,999	10	73
75,000–99,999	21	68
50,000–74,999	24	63
30,000–49,999	30	43
Under 30,000	11	28

a. What is the probability that a couple selected at random from this area has two incomes?

b. If a randomly chosen couple has two incomes, what is the probability that the annual income of this couple is \$150,000 or more?

c. If a randomly chosen couple has two incomes, what is the probability that the annual income of this couple is greater than \$49,999?

Solution Let A denote the event that the annual income of the couple is \$150,000 or more; let B denote the event that the annual income is between \$100,000 and \$149,999; let C denote the event that the annual income is between \$75,000 and \$99,999; and so on. Finally, let T denote the event that both spouses are employed. The probabilities of the occurrence of these events are displayed in Figure 24.

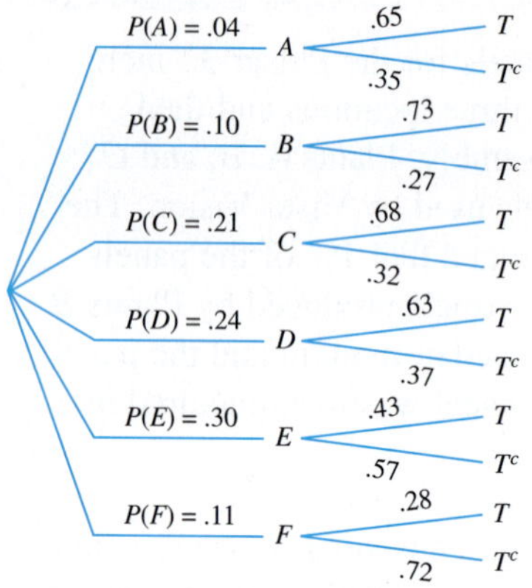

FIGURE 24

a. The probability that a couple selected at random from this group has two incomes is given by

$$\begin{aligned}P(T) &= P(A)\cdot P(T|A) + P(B)\cdot P(T|B) + P(C)\cdot P(T|C)\\&\quad + P(D)\cdot P(T|D) + P(E)\cdot P(T|E) + P(F)\cdot P(T|F)\\&= (.04)(.65) + (.10)(.73) + (.21)(.68) + (.24)(.63)\\&\quad + (.30)(.43) + (.11)(.28)\\&= .5528\end{aligned}$$

b. Using the results of part (a) and Bayes' Theorem, we find that the probability that a randomly chosen couple has an annual income of \$150,000 or more, given that both spouses are employed, is

$$P(A|T) = \frac{P(A)\cdot P(T|A)}{P(T)} = \frac{(.04)(.65)}{.5528} \approx .047$$

c. The probability that a randomly chosen couple has an annual income greater than \$49,999, given that both spouses are employed, is

$$\begin{aligned}&P(A|T) + P(B|T) + P(C|T) + P(D|T)\\&\quad = \frac{P(A)\cdot P(T|A) + P(B)\cdot P(T|B) + P(C)\cdot P(T|C) + P(D)\cdot P(T|D)}{P(T)}\\&\quad = \frac{(.04)(.65) + (.1)(.73) + (.21)(.68) + (.24)(.63)}{.5528}\\&\quad \approx .711\end{aligned}$$

7.6 Self-Check Exercises

1. The accompanying tree diagram represents a two-stage experiment. Use the diagram to find $P(B|D)$.

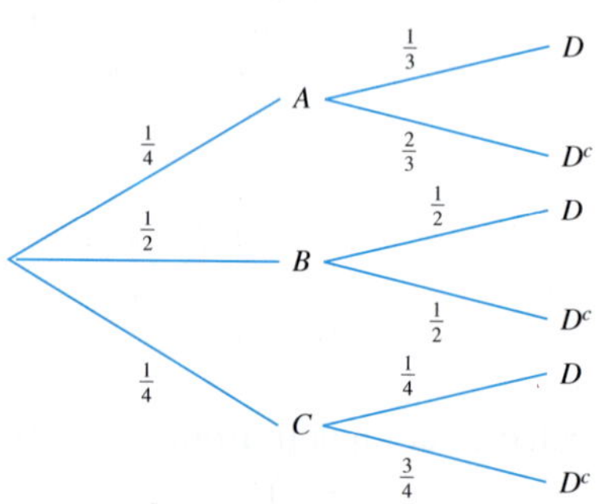

2. **Politics** In a recent presidential election, it was estimated that the probability that the Republican candidate would be elected was $\frac{3}{5}$ and therefore the probability that the Democratic candidate would be elected was $\frac{2}{5}$ (the two Independent candidates were given little chance of being elected). It was also estimated that if the Republican candidate were elected, then the probability that research for a new manned bomber would continue was $\frac{4}{5}$. But if the Democratic candidate were successful, then the probability that the research would continue was $\frac{3}{10}$. Research was terminated shortly after the successful presidential candidate took office. What is the probability that the Republican candidate won that election?

Solutions to Self-Check Exercises 7.6 can be found on page 451.

7.6 Concept Questions

1. What are a priori probabilities and a posteriori probabilities? Give an example of each.

2. Suppose the events A, B, and C form the partition of a sample space S, and suppose E is an event of an experiment such that $P(E) \neq 0$. Use Bayes' Theorem to write the formula for the a posteriori probability $P(A|E)$. (Assume that $P(A), P(B), P(C) \neq 0$.)

3. Refer to Question 2. If E is the event that a product was produced in Factory A, Factory B, or Factory C and $P(E) \neq 0$, what does $P(A|E)$ represent?

7.6 Exercises

In Exercises 1–3, refer to the accompanying Venn diagram. An experiment in which the three mutually exclusive events *A*, *B*, and *C* form a partition of the uniform sample space *S* is depicted in the diagram.

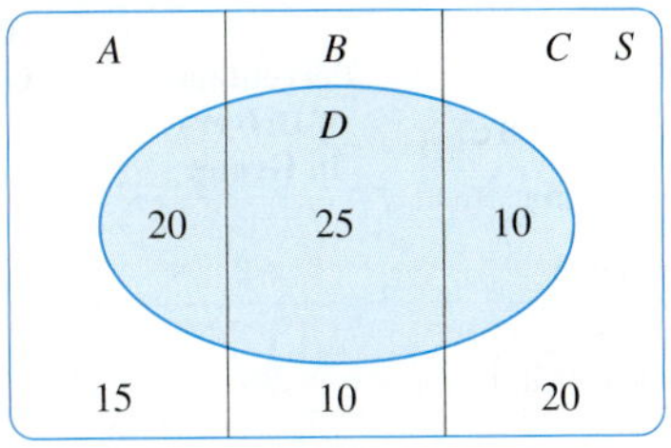

1. Using the information given in the Venn diagram, draw a tree diagram illustrating the probabilities of the events *A*, *B*, *C*, and *D*.

2. Find: **a.** $P(D)$ **b.** $P(A \mid D)$

3. Find: **a.** $P(D^c)$ **b.** $P(B \mid D^c)$

In Exercises 4–6, refer to the accompanying Venn diagram. An experiment in which the three mutually exclusive events *A*, *B*, and *C* form a partition of the uniform sample space *S* is depicted in the diagram.

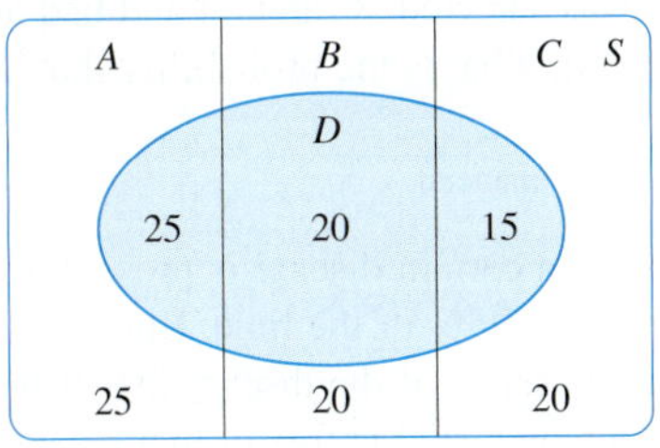

4. Using the information given in the Venn diagram, draw a tree diagram illustrating the probabilities of the events *A*, *B*, *C*, and *D*.

5. Find: **a.** $P(D)$ **b.** $P(B \mid D)$

6. Find: **a.** $P(D^c)$ **b.** $P(B \mid D^c)$

7. The accompanying tree diagram represents a two-stage experiment. Use the diagram to find:
   a. $P(A) \cdot P(D \mid A)$ **b.** $P(B) \cdot P(D \mid B)$
   c. $P(A \mid D)$

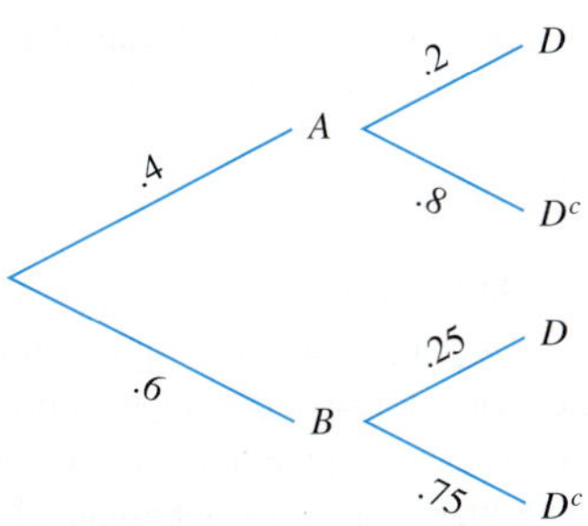

8. The accompanying tree diagram represents a two-stage experiment. Use the diagram to find:
   a. $P(A) \cdot P(D \mid A)$ **b.** $P(B) \cdot P(D \mid B)$
   c. $P(A \mid D)$

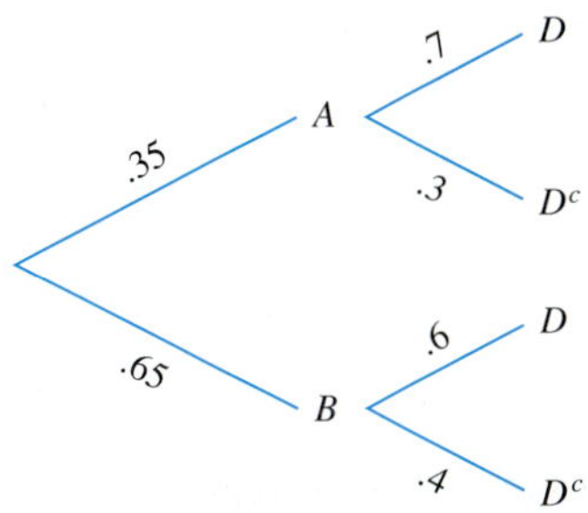

9. The accompanying tree diagram represents a two-stage experiment. Use the diagram to find:
   a. $P(A) \cdot P(D \mid A)$ **b.** $P(B) \cdot P(D \mid B)$
   c. $P(C) \cdot P(D \mid C)$ **d.** $P(A \mid D)$

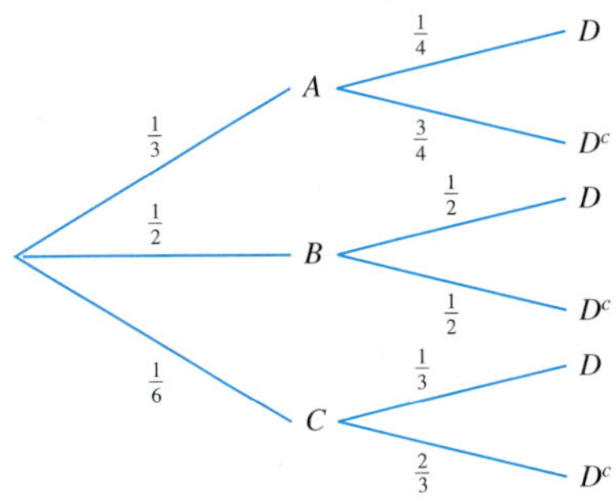

10. The accompanying tree diagram represents a two-stage experiment. Use this diagram to find:
    a. $P(A \cap D)$ **b.** $P(B \cap D)$ **c.** $P(C \cap D)$ **d.** $P(D)$
    e. Verify:

$$P(A \mid D) = \frac{P(A \cap D)}{P(D)}$$

$$= \frac{P(A) \cdot P(D \mid A)}{P(A) \cdot P(D \mid A) + P(B) \cdot P(D \mid B) + P(C) \cdot P(D \mid C)}$$

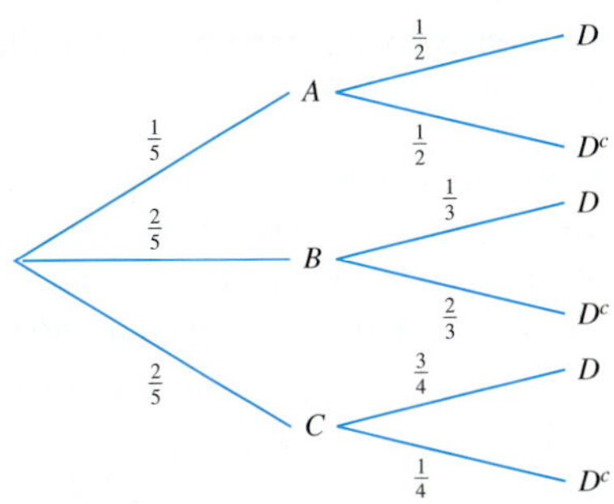

11. The accompanying diagram represents a two-stage experiment. Complete the information on the diagram, and use it to find:

a. $P(B)$ **b.** $P(A \mid B)$
c. $P(B^c)$ **d.** $P(A \mid B^c)$

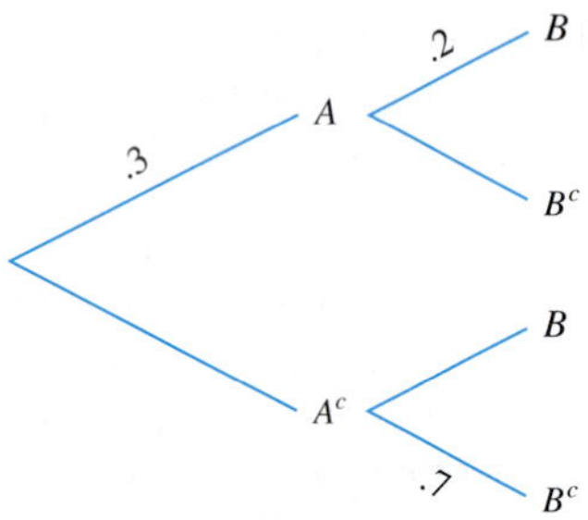

12. The accompanying diagram represents a two-stage experiment. Here, $B = (A \cup C)^c$. Complete the information on the diagram, and use it to find:

a. $P(D)$ **b.** $P(B \mid D)$
c. $P(D^c)$ **d.** $P(A \mid D^c)$

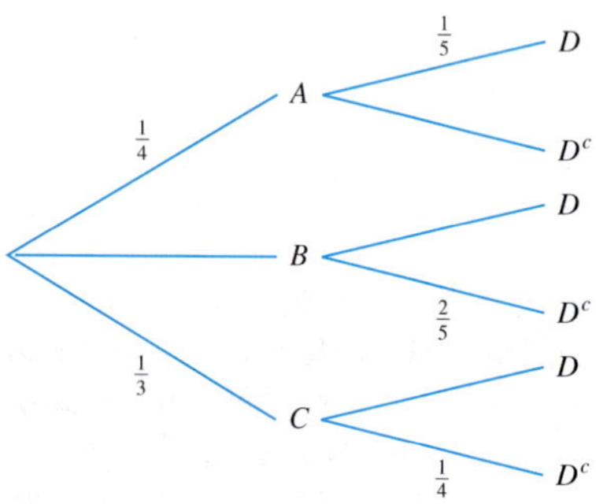

In Exercises 13–16, refer to the following experiment: Two cards are drawn in succession without replacement from a standard deck of 52 cards.

13. What is the probability that the first card is a heart given that the second card is a heart?

14. What is the probability that the first card is a heart given that the second card is a diamond?

15. What is the probability that the first card is a jack given that the second card is an ace?

16. What is the probability that the first card is a face card given that the second card is an ace?

In Exercises 17–20, refer to the following experiment: Urn *A* contains four white balls and six black balls. Urn *B* contains three white balls and five black balls. A ball is drawn from Urn *A* and then transferred to Urn *B*. A ball is then drawn from Urn *B*.

17. Represent the probabilities associated with this two-stage experiment in the form of a tree diagram.

18. What is the probability that the transferred ball was white given that the second ball drawn was white?

19. What is the probability that the transferred ball was black given that the second ball drawn was white?

20. What is the probability that the transferred ball was black given that the second ball drawn was black?

21. **Seat-Belt Compliance** Data compiled by the Highway Patrol Department regarding the use of seat-belts by drivers in a certain area after the passage of a compulsory seat-belt law are shown in the accompanying table.

Drivers	Percentage of Drivers in Group	Percentage of Group Stopped for Moving Violation
Group I (using seat-belts)	.64	.002
Group II (not using seat-belts)	.36	.005

If a driver in that area is stopped for a moving violation, what is the probability that he or she:

a. Will have a seat-belt on?
b. Will not have a seat-belt on?

22. **Retirement Needs** In a survey of 2000 adults 50 years old and older of whom 60% were retired and 40% were pre-retired, the following question was asked: Do you expect your income needs to vary from year to year in retirement? Of those who were retired, 33% answered no, and 67% answered yes. Of those who were pre-retired, 28% answered no, and 72% answered yes. If a respondent in the survey was selected at random and had answered yes to the question, what is the probability that he or she was retired?

Source: Sun Life Financial.

23. **Blood Tests** If a certain disease is present, then a blood test will reveal it 95% of the time. But the test will also indicate the presence of the disease 2% of the time when in fact the person tested is free of that disease; that is, the test gives a false positive 2% of the time. If 0.3% of the general population actually has the disease, what is the probability that a person chosen at random from the population has the disease given that he or she tested positive?

24. **Opinion Polls** In a survey to determine the opinions of Americans on health insurers, 400 baby boomers and 600 pre-boomers were asked this question: Do you believe that insurers are very responsible for high health costs? Of the baby boomers, 212 answered in the affirmative, whereas 198 of the pre-boomers answered in the affirmative. If a respondent chosen at random from those surveyed answered the question in the affirmative, what is the probability that he or she is a baby boomer? A pre-boomer?

Source: GfK Roper Consulting.

25. **Crime Rates** Data compiled by the Department of Justice on the number of people arrested in a certain year for serious crimes (murder, forcible rape, robbery, etc.) revealed that 89% were male and 11% were female. Of the males, 30% were under 18, whereas 27% of the females arrested were under 18.

a. What is the probability that a person arrested for a serious crime in that year was under 18?

b. If a person arrested for a serious crime in that year is known to be under 18, what is the probability that the person is female?

Source: Department of Justice.

26. **Gender Gap** A study of the faculty at U.S. medical schools in 2006 revealed that 32% of the faculty were women and 68% were men. Of the female faculty, 31% were full/associate professors, 47% were assistant professors, and 22% were instructors. Of the male faculty, 51% were full/associate professors, 37% were assistant professors, and 12% were instructors. If a faculty member at a U.S. medical school selected at random in 2006 held the rank of full/associate professor, what is the probability that the faculty member was female?

Source: Association of American Medical Colleges.

27. **Medical Research** On the basis of data obtained from the National Institute of Dental Research, it has been determined that 42% of 12-year-olds have never had a cavity, 34% of 13-year-olds have never had a cavity, and 28% of 14-year-olds have never had a cavity. Suppose a child is selected at random from a group of 24 junior high school students that includes six 12-year-olds, eight 13-year-olds, and ten 14-year-olds. If this child does not have a cavity, what is the probability that this child is 14 years old?

Source: National Institute of Dental Research.

28. **Voting Patterns** In a recent senatorial election, 50% of the voters in a certain district were registered as Democrats, 35% were registered as Republicans, and 15% were registered as Independents. The incumbent Democratic senator was reelected over her Republican and Independent opponents. Exit polls indicated that she gained 75% of the Democratic vote, 25% of the Republican vote, and 30% of the Independent vote. Assuming that the exit poll is accurate, what is the probability that a vote for the incumbent was cast by a registered Republican?

29. **Medical Diagnoses** A study was conducted among a certain group of union members whose health insurance policies required second opinions prior to surgery. Of those members whose doctors advised them to have surgery, 20% were informed by a second doctor that no surgery was needed. Of these, 70% took the second doctor's opinion and did not go through with the surgery. Of the members who were advised to have surgery by both doctors, 95% went through with the surgery. What is the probability that a union member who had surgery was advised to do so by a second doctor?

30. **Personnel Selection** Applicants for temporary office work at Carter Temporary Help Agency who have successfully completed an administrative assistant course are then placed in suitable positions by Nancy Dwyer and Darla Newberg. Employers who hire temporary help through the agency return a card indicating satisfaction or dissatisfaction with the work performance of those hired. From past experience it is known that 80% of the employees placed by Nancy are rated as satisfactory, and 70% of those placed by Darla are rated as satisfactory. Darla places 55% of the temporary office help at the agency, and Nancy places the remaining 45%. If a Carter office worker is rated unsatisfactory, what is the probability that he or she was placed by Darla?

31. **Impact of Gas Prices on Consumers** In a survey of 1012 adults aged 18 years and older conducted by Social Science Research Solutions, it was found that gas prices have caused financial hardship for 60% of the respondents aged 18–49 years and 55% of those who are 50 years or older. There were 742 adults aged 18–49 years and 270 adults who were 50 years or older in the survey. If a respondent in the survey selected at random reported that he or she did not experience financial hardship from gas prices, what is the probability that he or she was an adult aged 18–49 years?

Source: Social Science Research Solutions.

32. **Detecting Shoplifters** The management of Mark's Department Store estimates that approximately 2% of all shoppers enter the store with the intention of shoplifting. Of those people who engage in the act of shoplifting, 95% are apprehended by security guards using closed-circuit television (CCTV) and an electronic article surveillance system. The latter, also called a tag-and-alarm system, involves the use of a tag or label that is attached to an item. The tag is deactivated or detached at the time of purchase of the item; failure to do so will trigger an alarm when the item is carried through the gates. Of the other 98% of the shoppers, 1% are mistakenly identified as shoplifters by security guards watching them on CCTV or have triggered the alarm because the salesperson failed to deactivate or detach the tag. If a shopper is detained by security, what is the probability that the person was in fact a shoplifter?

33. **Opinion Polls** A survey involving 400 likely Democratic voters and 300 likely Republican voters asked the question: Do you support or oppose legislation that would require registration of all handguns? The following results were obtained:

Answer	Democrats (%)	Republicans (%)
Support	77	59
Oppose	14	31
Don't know/refused	9	10

If a randomly chosen respondent in the survey answered "oppose," what is the probability that he or she is a likely Democratic voter?

34. **Opinion Polls** A survey involving 400 likely Democratic voters and 300 likely Republican voters asked the question: Do you support or oppose legislation that would

require trigger locks on guns, to prevent misuse by children? The following results were obtained:

Answer	Democrats (%)	Republicans (%)
Support	88	71
Oppose	7	20
Don't know/refused	5	9

If a randomly chosen respondent in the survey answered "support," what is the probability that he or she is a likely Republican voter?

35. **Beverage Preferences** In a study of scientific research on soft drinks, juices, and milk, 50 studies were fully sponsored by the food industry, and 30 studies were conducted with no corporate ties. Of those that were fully sponsored by the food industry, 14% of the participants found the products unfavorable, 23% were neutral, and 63% found the products favorable. Of those that had no industry funding, 38% found the products unfavorable, 15% were neutral, and 47% found the products favorable.
   a. What is the probability that a participant selected at random found the products favorable?
   b. If a participant selected at random found the product favorable, what is the probability that he or she belongs to a group that participated in a corporate-sponsored study?

   Source: Children's Hospital, Boston.

36. **Selection of Supreme Court Judges** In a past presidential election, it was estimated that the probability that the Republican candidate would be elected was $\frac{3}{5}$ and therefore the probability that the Democratic candidate would be elected was $\frac{2}{5}$ (the two Independent candidates were given no chance of being elected). It was also estimated that if the Republican candidate were elected, the probability that a conservative, moderate, or liberal judge would be appointed to the Supreme Court (one retirement was expected during the presidential term) was $\frac{1}{2}$, $\frac{1}{3}$, and $\frac{1}{6}$, respectively. If the Democratic candidate were elected, the probabilities that a conservative, moderate, or liberal judge would be appointed to the Supreme Court would be $\frac{1}{8}$, $\frac{3}{8}$, and $\frac{1}{2}$, respectively. A conservative judge was appointed to the Supreme Court during the presidential term. What is the probability that the Democratic candidate was elected?

37. **Age Distribution of Renters** A study conducted by the Metro Housing Agency in a midwestern city revealed the following information concerning the age distribution of renters within the city.

Age	Adult Population (%)	Group Who Are Renters (%)
21–44	51	58
45–64	31	45
65 and over	18	60

   a. What is the probability that an adult selected at random from this population is a renter?
   b. If a renter is selected at random, what is the probability that he or she is in the 21–44 age bracket?
   c. If a renter is selected at random, what is the probability that he or she is 45 years old or older?

38. **Product Reliability** The estimated probability that a Brand *A*, a Brand *B*, and a Brand *C* plasma TV will last at least 30,000 hr is .90, .85, and .80, respectively. Of the 4500 plasma TVs that Ace TV sold in a certain year, 1000 were Brand *A*, 1500 were Brand *B*, and 2000 were Brand *C*. If a plasma TV set sold by Ace TV that year is selected at random and is still working after 30,000 hr of use:
   a. What is the probability that it was a Brand *A* TV?
   b. What is the probability that it was not a Brand *A* TV?

39. An experiment consists of randomly selecting one of three coins, tossing it, and observing the outcome—heads or tails. The first coin is a two-headed coin, the second is a biased coin such that $P(\text{H}) = .75$, and the third is a fair coin.
   a. What is the probability that the coin that is tossed will show heads?
   b. If the coin selected shows heads, what is the probability that this coin is the fair coin?

40. **Gun Owners in the Senate** As of January 3, 2013, the U.S. Senate was made up of 53 Democrats, 45 Republicans, and 2 Independents who caucus with the Democrats. In a survey of the U.S. Senate conducted at that time, every senator was asked whether he or she owned at least one gun. Of the Democrats, 16 declared themselves gun owners; of the Republicans, 26 of them declared themselves gun owners; none of the Independents owned guns. If a senator participating in that survey was picked at random and turned out to be a gun owner, what was the probability that he or she was a Democrat?

   Source: Gannett Washington Bureau.

41. **Reliability of Medical Tests** A medical test has been designed to detect the presence of a certain disease. Among people who have the disease, the probability that the disease will be detected by the test is .95. However, the probability that the test will erroneously indicate the presence of the disease in those who do not actually have it is .04. It is estimated that 4% of the population who take this test have the disease.
   a. If the test administered to an individual is positive, what is the probability that the person actually has the disease?
   b. If an individual takes the test twice and the test is positive both times, what is the probability that the person actually has the disease? (Assume that the tests are independent.)

42. **Reliability of Medical Tests** Refer to Exercise 41. Suppose 20% of the people who were referred to a clinic for the test did in fact have the disease. If the test administered to an individual from this group is positive, what is the probability that the person actually has the disease?

43. **Quality Control** Jansen Electronics has four machines that produce identical components for use in its DVD players. The proportion of the components produced by each machine and the probability of a component produced by that machine being defective are shown in the accompanying table. What is the probability that a component selected at random:

a. Is defective?

b. Was produced by Machine I, given that it is defective?

c. Was produced by Machine II, given that it is defective?

Machine	Proportion of Components Produced	Probability of Defective Component
I	.15	.04
II	.30	.02
III	.35	.02
IV	.20	.03

44. **Commuting Times** According to the U.S. Census Bureau, in 2011, 121,298,000 workers who do not work at home have travel times of less than 1 hr, and 10,979,000 workers who do not work at home have travel times of 1 hr or longer. Of those workers whose travel times are less than 1 hr, 91.4% drive alone or carpool, 1.6% take the subway or railroad, 2.1% take other public transportation, and 4.9% use other means. Of those workers whose travel times are 1 hr or longer, 74.0% drive alone or carpool, 11.8% take the subway or railroad, 11.2% take other public transportation, and 3.0% use other means. If a worker chosen at random drives or carpools, what is the probability that he or she has travel times of less than 1 hr?
Source: U.S. Census Bureau.

45. **Quality Control** A halogen desk lamp produced by Luminar was found to be defective. The company has three factories where the lamps are manufactured. The percentage of the total number of halogen desk lamps produced by each factory and the probability that a lamp manufactured by that factory is defective are shown in the accompanying table. What is the probability that the defective lamp was manufactured in Factory III?

Factory	Total Production (%)	Probability of Defective Component
I	35	.015
II	35	.01
III	30	.02

46. **Obesity in Children** Researchers weighed 1976 3-year-olds from low-income families in 20 U.S. cities. Each child was classified by race (white, black, or Hispanic) and by weight (normal weight, overweight, or obese). The results follow:

		Weight (%)		
Race	Children	Normal Weight	Overweight	Obese
White	406	68	18	14
Black	1081	68	15	17
Hispanic	489	56	20	24

If a child in the research study is selected at random and is found to be obese, what is the probability that the child is white? Hispanic?
Source: American Journal of Public Health.

47. **Auto-Accident Rates** An insurance company has compiled the accompanying data relating the age of drivers and the accident rate (the probability of being involved in an accident during a 1-year period) for drivers within that group:

Age Group	Insured Drivers (%)	Accident Rate (%)
Under 25	16	5.5
25–44	40	2.5
45–64	30	2
65 and over	14	4

What is the probability that an insured driver selected at random:

a. Will be involved in an accident during a particular 1-year period?

b. Who is involved in an accident is under 25?

48. **Personal Habits** There were 80 male guests at a party. The number of men in each of four age categories is given in the following table. The table also gives the probability that a man in the respective age category will keep his paper money in order of denomination.

Age	Men	Keep Paper Money in Order (%)
21–34	25	90
35–44	30	61
45–54	15	80
55 and over	10	80

A man's wallet was retrieved, and the paper money in it was kept in order of denomination. What is the probability that the wallet belonged to a male guest between the ages of 35 and 44?
Source: USA Today.

49. **Sleeping with Cell Phones** In a survey of 920 people aged from 18 through 84 years, of whom 84 belonged to the Millenial Generation, 224 belonged to Generation X, 200 belonged to the Baby Boom Generation, and the rest belonged to the Silent Generation, the following question was asked: Who has slept with a cell phone nearby? Of those who answered in the affirmative, 83%, 68%, 50%, and 20% were from the Millenial Generation, Generation X, the Baby Boom Generation, and the Silent Generation, respectively. If a person in the survey is selected at random and has not slept with a cell phone nearby, what is the probability that the person belongs to the Millenial Generation?
Source: Pew Research Center.

50. Voter Turnout by Income Voter turnout drops steadily as income level declines. The following table gives the percentage of eligible voters in a certain city, categorized by income, who responded with "did not vote" in the 2000 presidential election. The table also gives the number of eligible voters in the city, categorized by income.

Income Percentile	Did Not Vote (%)	Eligible Voters
0–16	52	4,000
17–33	31	11,000
34–67	30	17,500
68–95	14	12,500
96–100	12	5,000

If an eligible voter from this city who had voted in the election is selected at random, what is the probability that this person had an income in the 17–33 percentile?
Source: The National Election Studies.

51. The Social Ladder The following table summarizes the results of a poll conducted with 1154 adults.

Annual Household Income ($)	Respondents Within That Income Range (%)	Respondents Who Call Themselves Rich (%)	Middle Class (%)	Poor (%)
Less than 15,000	11.2	0	24	76
15,000–29,999	18.6	3	60	37
30,000–49,999	24.5	0	86	14
50,000–74,999	21.9	2	90	8
75,000 and higher	23.8	5	91	4

a. What is the probability that a respondent chosen at random calls himself or herself middle class?
b. If a randomly chosen respondent calls himself or herself middle class, what is the probability that the annual household income of that individual is between \$30,000 and \$49,999, inclusive?
c. If a randomly chosen respondent calls himself or herself middle class, what is the probability that the individual's income is either less than or equal to \$29,999 or greater than or equal to \$50,000?
Source: New York Times/CBS News; *Wall Street Journal Almanac.*

52. Selection of College Majors The Office of Admissions and Records of a large western university released the accompanying information concerning the contemplated majors of its freshman class:

Major	Freshmen Choosing This Major (%)	Females Choosing This Major (%)	Males Choosing This Major (%)
Business	24	38	62
Humanities	8	60	40
Education	8	66	34
Social science	7	58	42
Natural sciences	9	52	48
Other	44	48	52

What is the probability that:
a. A student selected at random from the freshman class is a female?
b. A business student selected at random from the freshman class is a male?
c. A female student selected at random from the freshman class is majoring in business?

53. Voter Turnout by Profession The following table gives the percentage of eligible voters grouped according to profession who responded with "voted" in the 2000 presidential election. The table also gives the percentage of people in a survey categorized by their profession.

Profession	Percentage Who Voted	Percentage in Each Profession
Professionals	.84	.12
White collar	.73	.24
Blue collar	.66	.32
Unskilled	.57	.10
Farmers	.68	.08
Housewives	.66	.14

If an eligible voter who participated in the survey and voted in the election is selected at random, what is the probability that this person is a housewife?
Source: The National Election Studies.

54. Smoking and Education According to the Centers for Disease Control and Prevention, the percentage of adults 25 years old and older who smoke, by educational level, is as follows:

Educational Level	No diploma	GED diploma	High school graduate	Some college	Undergraduate level	Graduate degree
Respondents (%)	27	4.3	26	24	10.7	8

In a group of 140 people, there were 8 with no diploma, 14 with GED diplomas, 40 high school graduates, 24 with some college, 42 with an undergraduate degree, and 12 with a graduate degree. (Assume that these categories are mutually exclusive.) If a person selected at random from this group was a smoker, what is the probability that he or she is a person with a graduate degree?
Source: Centers for Disease Control and Prevention.

7.6 Solutions to Self-Check Exercises

1. Using the probabilities given in the tree diagram and Bayes' Theorem, we have

$$P(B \mid D) = \frac{P(B) \cdot P(D \mid B)}{P(A) \cdot P(D \mid A) + P(B) \cdot P(D \mid B) + P(C) \cdot P(D \mid C)}$$

$$= \frac{\left(\frac{1}{2}\right)\left(\frac{1}{2}\right)}{\left(\frac{1}{4}\right)\left(\frac{1}{3}\right) + \left(\frac{1}{2}\right)\left(\frac{1}{2}\right) + \left(\frac{1}{4}\right)\left(\frac{1}{4}\right)} = \frac{12}{19}$$

2. Let R and D, respectively, denote the event that the Republican and the Democratic candidate won the presidential election. Then $P(R) = \frac{3}{5}$ and $P(D) = \frac{2}{5}$. Also, let C denote the event that research for the new manned bomber continued. These data may be exhibited as in the accompanying tree diagram:

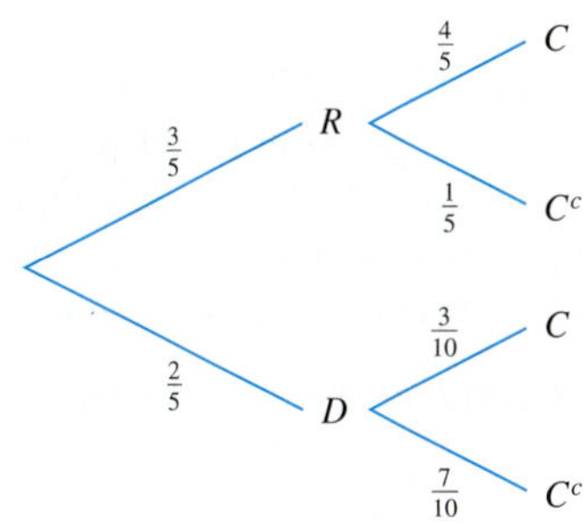

Using Bayes' Theorem, we find that the probability that the Republican candidate had won the election is given by

$$P(R \mid C^c) = \frac{P(R) \cdot P(C^c \mid R)}{P(R) \cdot P(C^c \mid R) + P(D) \cdot P(C^c \mid D)}$$

$$= \frac{\left(\frac{3}{5}\right)\left(\frac{1}{5}\right)}{\left(\frac{3}{5}\right)\left(\frac{1}{5}\right) + \left(\frac{2}{5}\right)\left(\frac{7}{10}\right)} = \frac{3}{10}$$

CHAPTER 7 Summary of Principal Formulas and Terms

FORMULAS

1. Probability of an event in a uniform sample space	$P(E) = \dfrac{n(E)}{n(S)}$
2. Probability of the union of two mutually exclusive events	$P(E \cup F) = P(E) + P(F)$
3. Addition rule	$P(E \cup F) = P(E) + P(F) - P(E \cap F)$
4. Rule of complements	$P(E^c) = 1 - P(E)$
5. Conditional probability	$P(B \mid A) = \dfrac{P(A \cap B)}{P(A)}, P(A) \neq 0$
6. Product rule	$P(A \cap B) = P(A) \cdot P(B \mid A)$
7. Test for independence	$P(A \cap B) = P(A) \cdot P(B)$

TERMS

experiment (388)
outcome (388)
sample point (388)
sample space (388)
event (388)
finite sample space (388)
union of two events (389)
intersection of two events (389)
complement of an event (389)
mutually exclusive events (390)
relative frequency (397)
empirical probability (397)
probability of an event (397)
elementary (simple) event (397)
probability distribution (397)
probability function (397)
uniform sample space (398)
addition principle (399)
conditional probability (425)
finite stochastic process (429)
independent events (432)
Bayes' Theorem (442)

CHAPTER 7 Concept Review Questions

Fill in the blanks.

1. An activity with observable results is called a/an _____; an outcome of an experiment is called a/an _____ point, and the set consisting of all possible sample points of an experiment is called a sample _____; a subset of a sample space of an experiment is called a/an _____.

2. The events E and F are mutually exclusive if $E \cap F =$ _____.

3. A sample space in which the outcomes are equally likely is called a/an _____ sample space; if such a space contains n simple events, then the probability of each simple event is _____.

4. The probability of the occurrence of event B given that the event A has already occurred is called the _____ probability of B given A.

5. If the outcome of one event does not depend on a second event, then the two events are said to be _____.

6. The probability of an event after the outcomes of an experiment have been observed is called a/an _____ _____ _____.

CHAPTER 7 Review Exercises

1. Let E and F be two mutually exclusive events, and suppose $P(E) = .4$ and $P(F) = .2$. Compute:
   a. $P(E \cap F)$ **b.** $P(E \cup F)$
   c. $P(E^c)$ **d.** $P(E^c \cap F^c)$
   e. $P(E^c \cup F^c)$

2. Let E and F be two events of an experiment with sample space S. Suppose $P(E) = .3$, $P(F) = .2$, and $P(E \cap F) = .15$. Compute:
   a. $P(E \cup F)$
   b. $P(E^c \cap F^c)$
   c. $P(E^c \cap F)$

3. Let E and F be two mutually exclusive events, and suppose $P(E) = .35$ and $P(F) = .47$. Find:
   a. $P(F^c)$ **b.** $P(E \cap F^c)$
   c. $P(E \cup F)$ **d.** $P(E^c \cap F^c)$

4. A die is loaded, and it has been determined that the probability distribution associated with the experiment of rolling the die and observing which number falls uppermost is given by the following:

Simple Event	Probability
{1}	.20
{2}	.12
{3}	.16
{4}	.18
{5}	.15
{6}	.19

   a. What is the probability of the number being even?
   b. What is the probability of the number being either a 1 or a 6?
   c. What is the probability of the number being less than 4?

5. An urn contains six red balls, five black balls, and four green balls. If two balls are selected at random without replacement from the urn, what is the probability that a red ball and a black ball will be selected?

6. **QUALITY CONTROL** The quality-control department of Starr Communications, a manufacturer of video-game DVDs, has determined from records that 1.5% of the DVDs sold have video defects, 0.8% have audio defects, and 0.4% have both audio and video defects. What is the probability that a DVD purchased by a customer:
   a. Will have a video or audio defect?
   b. Will not have a video or audio defect?

7. Let E and F be two events, and suppose that $P(E) = .35$, $P(F) = .55$, and $P(E \cup F) = .70$. Find $P(E|F)$.

8. Suppose that $P(E) = .60$, $P(F) = .32$, and $P(E \cap F) = .22$. Are E and F independent?

9. Suppose that E and F are independent events. If $P(E) = .32$ and $P(E \cap F) = .16$, what is $P(F)$?

The accompanying tree diagram represents an experiment consisting of two trials. In Exercises 10–14, use the diagram to find the given probability.

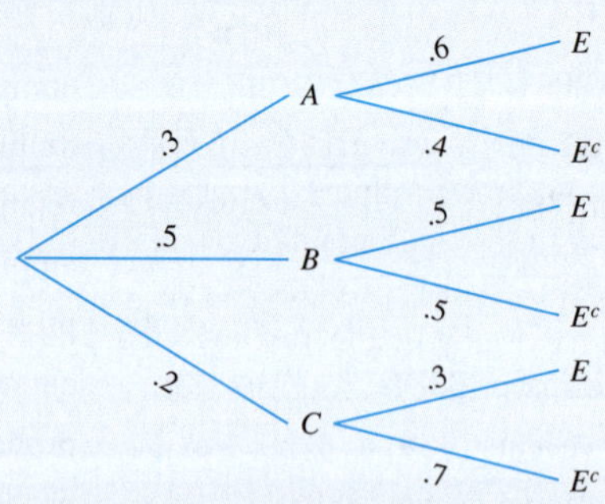

10. $P(A \cap E)$ **11.** $P(B \cap E)$

12. $P(C \cap E)$ **13.** $P(E)$

14. $P(A \mid E)$

15. An experiment consists of tossing a fair coin three times and observing the outcomes. Let A be the event that at least one head is thrown, and let B be the event that at most two tails are thrown.
a. Find $P(A)$. **b.** Find $P(B)$.
c. Are A and B independent events?

16. Quality Control In a group of 20 ballpoint pens on a shelf in the stationery department of Metro Department Store, 2 are known to be defective. If a customer selects 3 of these pens, what is the probability that:
a. At least 1 is defective?
b. No more than 1 is defective?

17. Birthday Problem Five people are selected at random. What is the probability that none of the people in this group were born on the same day of the week?

18. A pair of fair dice is rolled. What is the probability that the sum of the numbers falling uppermost is 8 if it is known that the two numbers are different?

19. A fair die is rolled three times. What is the probability that it shows an even number in the first toss, an odd number in the second toss, and a 1 on the third toss? Assume that the outcomes of the tosses are independent.

20. A fair die is rolled, a fair coin is tossed, and a card is drawn from a standard deck of 52 playing cards. Assuming these events are independent, what is the probability that the number falling uppermost on the die is a 6, the coin shows a tail, and the card drawn is a face card?

Three cards are drawn at random without replacement from a standard deck of 52 playing cards. In Exercises 21–25, find the probability of each of the given events.

21. All three cards are aces.

22. All three cards are face cards.

23. The second and third cards are red.

24. The second card is black, given that the first card was red.

25. The second card is a club, given that the first card was black.

26. Assembly-Time Studies The results of a time study conducted by the production manager of Ace Novelty are shown in the accompanying table, where the number of action figures produced each quarter hour during an 8-hour workday has been tabulated. Find the empirical probability distribution associated with this experiment.

Figures Produced (in dozens)	Frequency of Occurrence
30	4
31	0
32	6
33	8
34	6
35	4
36	4

27. Fighting Inflation In a survey of 2000 adults aged 18 years and older conducted in a certain year, the following question was asked: Is your family income keeping pace with the cost of living? The results of the survey follow:

Answer	Falling behind	Staying even	Increasing faster	Don't know
Respondents	800	880	240	80

Determine the empirical probability distribution associated with these data.
Source: Pew Research Center.

28. U.S. Income Distribution for Households According to the U.S. Census Bureau, the income distribution for households in 2011 was as follows:

Income ($)	0–24,999	25,000–49,999	50,000–74,999
Households and Families	30,337,000	30,134,000	21,294,000

Income ($)	75,000–99,999	100,000–124,999	125,000–149,999
Households and Families	13,899,000	9,130,000	5,311,000

Income ($)	150,000–199,999	200,000–249,999	250,000 or more
Households and Families	5,875,000	2,297,000	2,808,000

Find the empirical probability distribution associated with these data.
Source: U.S. Census Bureau.

29. Poker What is wrong with the following statement? A 5-card poker hand is dealt from a 52-card deck. Let A denote the event that a flush is dealt, and let B be the event that a straight is dealt. Then the events A and B are mutually exclusive.

30. What is wrong with the following statement? There are eight grades in Garfield Elementary School. If a student is selected at random from the school, then the probability that the student is in the first grade is $\frac{1}{8}$.

31. Work Habits In a survey of 7780 workers, the following question was asked: How often are you late for work? A summary of the results of the survey follow:

Answer	Never	At least once a week	Once a month	Once a year
Respondents	4746	1244	856	934

If a worker in the survey is chosen at random, what is the probability that he or she:

a. Was never late for work?

b. Was late once a year?

Source: USA Today.

32. Consumer Preferences Olivia is contemplating buying a laser printer. The probability that she will buy a printer manufactured by Epson, Brother, Canon, and Hewlett-Packard is .23, .18, .31, and .28, respectively. Find the probability that she will buy a laser printer manufactured by:

a. Epson or Canon. **b.** Epson, Brother, or Canon.

33. Transportation Fatalities The following breakdown of a total of 18,598 transportation fatalities that occurred in a certain year was obtained from records compiled by the U.S. Department of Transportation (DOT).

Mode of Transportation	Car	Train	Bicycle	Plane
Number of Fatalities	16,520	845	698	535

What is the probability that a victim randomly selected from this list of transportation fatalities for that year died in:

a. A car crash or a bicycle accident?

b. A train or a plane accident?

Source: U.S. Department of Transportation.

34. Credit Card Ownership A survey of 1020 adults aged 18–49 years found that 34% of them have no credit cards, 22% have one credit card, 28% have two or three credit cards, 11% have four or five credit cards, and 5% have more than five credit cards. What is the probability that a randomly chosen respondent in the survey has:

a. No credit card?

b. Between one and three credit cards?

c. Four or more credit cards?

Source: ORC International.

35. Retirement Expectations In a survey on retirement, participants were asked this question: Do you think that life will be better, worse, or about the same when you retire? The results of the survey follow:

Answer	Better	Worse	Same	Don't know
Respondent (%)	38	18	41	3

If a person in the survey is selected at random, what is the probability that he or she answered that life after retirement would be:

a. The same or better?

b. The same or worse?

Source: Bankrate.com

36. Family Composition Consider the composition of a three-child family in which the children were born at different times. Assume that a girl is as likely as a boy at each birth. What is the probability that:

a. There are two girls and a boy in the family?

b. The oldest child is a girl?

c. The oldest child is a girl and the youngest child is a boy?

37. Sales of Plasma TVs The records of Ace Electronics show that of the plasma TVs sold by the company, 26% were manufactured by Panasonic, 15.4% were manufactured by LG, 13.7% were manufactured by Samsung, 13.3% were manufactured by Philips, and 7.3% were manufactured by Hitachi. If a customer chosen at random purchases a plasma TV from Ace Electronics, what is the probability that the set was manufactured by:

a. Panasonic, LG, Samsung, Philips, or Hitachi?

b. A company other than those mentioned in part (a)?

38. Keeping up with the Cost of Living In a survey of 2000 adults, 18 years old and older, conducted in 2007, the following question was asked: Is your family income keeping pace with the cost of living? The results of the survey follow:

Answer	Falling behind	Staying even	Increasing faster	Don't know
Respondents	800	880	240	80

According to the survey, what percentage of the people polled said their family income is:

a. At least keeping pace with the cost of living?

b. Falling behind the cost of living?

Source: Pew Research Center.

39. Tax Preparation A survey in which people were asked how they were planning to prepare their taxes revealed the following:

Method of Preparation	Percent
Computer software	33.9
Accountant	23.6
Tax preparation service	17.4
Spouse, friend, or other relative will prepare	10.8
By hand	14.3

What is the probability that a randomly chosen participant in the survey:

a. Was planning to use an accountant or a tax preparation service to prepare his or her taxes?

b. Was not planning to use computer software to prepare his or her taxes and was not planning to do taxes by hand?

Source: National Retail Federation.

40. **Women's Apparel** In an online survey for Talbots of 1095 women ages 35 years old and older, the participants were asked what article of clothing women most want to fit perfectly. A summary of the results of the survey follows:

Article of Clothing	Respondents
Jeans	470
Black pantsuit	307
Cocktail dress	230
White shirt	22
Gown	11
Other	55

If a woman who participated in the survey is chosen at random, what is the probability that she most wants:

a. Jeans to fit perfectly?

b. A black pantsuit or a cocktail dress to fit perfectly?

Source: Market Tool's Zoom Panel.

41. **Sales of Disaster-Recovery Systems** Jay sells disaster-recovery computer systems to hedge funds. He estimates the probability of Hedge Fund A purchasing a system to be .6 and that of Hedge Fund B purchasing a system to be .5. He also estimates that the probability of both hedge funds purchasing a system is .3. What is the probability that only Hedge Fund A or only Hedge Fund B will purchase a system?

42. **Flex-Time** Of 320 male and 280 female employees at the home office of Gibraltar Insurance Company, 160 of the men and 190 of the women are on flex-time (flexible working hours). Given that an employee selected at random from this group is on flex-time, what is the probability that the employee is a man?

43. **Quality Control** In a manufacturing plant, three machines, A, B, and C, produce 40%, 35%, and 25%, respectively, of the total production. The company's quality-control department has determined that 1% of the items produced by Machine A, 1.5% of the items produced by Machine B, and 2% of the items produced by Machine C are defective. If an item is selected at random and found to be defective, what is the probability that it was produced by Machine B?

44. **College Admissions** Applicants who wish to be admitted to a certain professional school in a large university are required to take a screening test devised by an educational testing service. From past results, the testing service has estimated that 70% of all applicants are eligible for admission and that 92% of those who are eligible for admission pass the exam, whereas 12% of those who are ineligible for admission pass the exam. Using these results, what is the probability that an applicant for admission:

a. Passed the exam?

b. Passed the exam but was actually ineligible?

45. **Customer Surveys** The sales department of Thompson Drug Company released the accompanying data concerning the sales of a certain pain reliever manufactured by the company.

Pain Reliever	Drug Sold (%)	Group Sold in Extra-Strength Dosage (%)
Group I (capsule form)	57	38
Group II (tablet form)	43	31

If a customer purchased the extra-strength dosage of this drug, what is the probability that it was in capsule form?

46. **Opinion Polls** A survey involving 600 Democrats, 400 Republicans, and 200 Independents asked the question: Do you favor or oppose eliminating taxes on dividends paid to shareholders? The following results were obtained:

Answer	Democrats (%)	Republicans (%)	Independents (%)
Favor	29	66	48
Opposed	71	34	52

If a randomly chosen respondent in the survey answered "favor," what is the probability that he or she is an Independent?

Source: TechnoMetrica Market Intelligence.

47. **Opinion Polls** A poll was conducted among 500 registered voters in a certain area regarding their position on a national lottery to raise revenue for the government. The results of the poll are shown in the accompanying table.

Sex	Voters Polled (%)	Favoring Lottery (%)	Not Favoring Lottery (%)	Expressing No Opinion (%)
Male	51	62	32	6
Female	49	68	28	4

What is the probability that a registered voter who:

a. Favored a national lottery was a woman?

b. Expressed no opinion regarding the lottery was a woman?

48. **Impact of Gas Prices on Consumers** In a survey of 1012 adults aged 18 years and older conducted by Social Science Research Solutions, 24% of the respondents aged 18–49 years said that they would increase their use of public transportation, whereas only 11% of those 50 years or older said they would do so. There were 742 adults aged 18–49 years and 270 adults who were 50 years or older in the survey. If a respondent in the survey selected at random reported that he or she would not increase the use of public transportation, what is the probability that he or she was 50 years or older?

Source: Social Science Research Solutions.

49. **Gun Owners in the House of Representatives** As of January 3, 2013, the U.S. House of Representatives was made up of

200 Democrats and 232 Republicans. There were 3 seats vacant. In a survey of the House of Representatives at that time, each representative was asked whether he or she owned at least one gun. Of the Democrats, 30 declared themselves gun owners. Of the Republicans, 93 declared themselves gun owners. If a representative participating in the survey was picked at random and turned out to be a gun owner, what is the probability that he or she is a Democrat?

Source: Gannett Washington Bureau.

50. Commuting According to the 2011 U.S. Census, 121,298,000 workers who do not work at home have travel times of less than 1 hr, and 10,979,000 workers who do not work at home have travel times of 1 hr or longer. Of those workers whose travel times are less than 1 hr, 91.4% drive alone or carpool, 1.6% take the subway or railroad, 2.1% take other public transportation, and 4.9% use other means. Of those workers whose travel times are 1 hr or longer, 74.0% drive alone or carpool, 11.8% take the subway or railroad, 11.2% take other public transportation, and 3.0% use other means. If a worker chosen at random takes the subway or railroad, what is the probability that he or she has travel times of less than 1 hr?

Source: U.S. Census Bureau.

CHAPTER 7 Before Moving On . . .

1. Let $S = \{s_1, s_2, s_3, s_4, s_5, s_6\}$ be the sample space associated with an experiment having the following probability distribution:

Outcome	s_1	s_2	s_3	s_4	s_5	s_6
Probability	$\frac{1}{12}$	$\frac{2}{12}$	$\frac{3}{12}$	$\frac{2}{12}$	$\frac{3}{12}$	$\frac{1}{12}$

Find the probability of the event $A = \{s_1, s_3, s_6\}$.

2. A card is drawn from a well-shuffled 52-card deck. What is the probability that the card drawn is a deuce or a face card?

3. Let E and F be events of an experiment with sample space S. Suppose $P(E) = .5$, $P(F) = .6$, and $P(E \cap F) = .2$. Compute:

a. $P(E \cup F)$ **b.** $P(E \cap F^c)$

4. Suppose A and B are independent events with $P(A) = .3$ and $P(B) = .6$. Find $P(A \cup B)$.

5. The following tree diagram represents a two-stage experiment. Use the diagram to find $P(A \mid D)$.

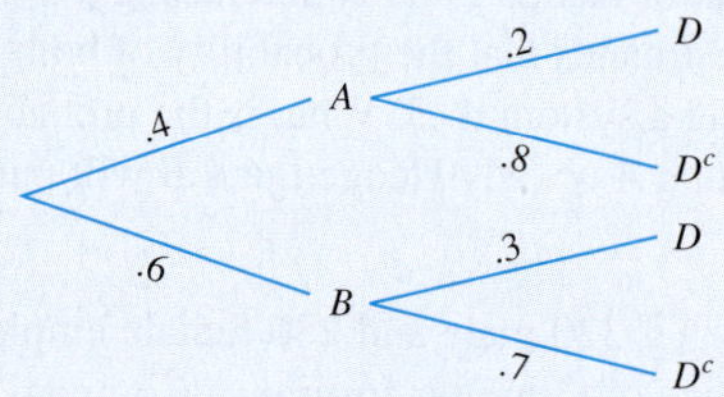

8 Probability Distributions and Statistics

STATISTICS IS that branch of mathematics concerned with the collection, analysis, and interpretation of data. In Sections 8.1–8.3, we take a look at descriptive statistics; here, our interest lies in the description and presentation of data in the form of tables and graphs. In the rest of the chapter, we briefly examine inductive statistics, and we see how mathematical tools such as those developed in Chapter 7 may be used in conjunction with these data to help us draw certain conclusions and make forecasts.

Which of two motels should a certain private equity firm purchase? In Example 4, page 475, we show how the occupancy rate and the average daily profit for each motel can be used to help us determine which motel will generate the higher daily profit.

8.1 Distributions of Random Variables

Random Variables

In many situations, it is desirable to assign numerical values to the outcomes of an experiment. For example, if an experiment consists of rolling a die and observing the face that lands uppermost, then it is natural to assign the numbers 1, 2, 3, 4, 5, and 6, respectively, to the outcomes *one, two, three, four, five*, and *six* of the experiment. If we let X denote the outcome of the experiment, then X assumes one of these numbers. Because the values assumed by X depend on the outcomes of a chance experiment, the outcome X is referred to as a random variable.

Random Variable

A **random variable** is a rule that assigns a number to each outcome of a chance experiment.

More precisely, a random variable is a function with domain given by the set of outcomes of a chance experiment and range contained in the set of real numbers.

EXAMPLE 1 A coin is tossed three times. Let the random variable X denote the number of heads that occur in the three tosses.

a. List the outcomes of the experiment; that is, find the domain of the function X.
b. Find the value assigned to each outcome of the experiment by the random variable X.
c. Find the event comprising the outcomes to which a value of 2 has been assigned by X. This event is written $(X = 2)$ and is the event consisting of the outcomes in which two heads occur.

TABLE 1
Number of Heads in Three Coin Tosses

Outcome	Value of X
HHH	3
HHT	2
HTH	2
THH	2
HTT	1
THT	1
TTH	1
TTT	0

Solution

a. From the results of Example 3, Section 7.1 (page 390), we see that the set of outcomes of the experiment is given by the sample space

$$S = \{\text{HHH, HHT, HTH, THH, HTT, THT, TTH, TTT}\}$$

b. The outcomes of the experiment are displayed in the first column of Table 1. The corresponding value assigned to each such outcome by the random variable X (the number of heads) appears in the second column.
c. With the aid of Table 1, we see that the event $(X = 2)$ is given by the set

$$\{\text{HHT, HTH, THH}\}$$

EXAMPLE 2 A coin is tossed repeatedly until a head occurs. Let the random variable Y denote the number of coin tosses in the experiment. What are the values of Y?

TABLE 2
Number of Coin Tosses Before Heads Appear

Outcome	Value of Y
H	1
TH	2
TTH	3
TTTH	4
TTTTH	5
⋮	⋮

Solution The outcomes of the experiment make up the infinite set

$$S = \{\text{H, TH, TTH, TTTH, TTTTH, . . .}\}$$

These outcomes of the experiment are displayed in the first column of Table 2. The corresponding values assumed by the random variable Y (the number of tosses) appear in the second column.

APPLIED EXAMPLE 3 Product Reliability A disposable flashlight is turned on and left on until its battery runs out. Let the random variable Z denote the length (in hours) of the life of the battery. What values may Z assume?

Solution The value of Z may be any nonnegative real number; that is, the possible values of Z make up the interval $0 \leq Z < \infty$.

One advantage of working with random variables—rather than working directly with the outcomes of an experiment—is that random variables are functions that may be added, subtracted, and multiplied. Because of this, results developed in the field of algebra and other areas of mathematics may be used freely to help us solve problems in probability and statistics.

A random variable is classified into three categories depending on the set of values it assumes. A random variable is called **finite discrete** if it assumes only finitely many values. For example, the random variable X of Example 1 is finite discrete because it may assume values only from the finite set of numbers $\{0, 1, 2, 3\}$. Next, a random variable is said to be **infinite discrete** if it takes on infinitely many values, which may be arranged in a sequence. For example, the random variable Y of Example 2 is infinite discrete because it assumes values from the set $\{1, 2, 3, 4, 5, \ldots\}$, which has been arranged in the form of an infinite sequence. Finally, a random variable is called **continuous** if the values it may assume comprise an interval of real numbers. For example, the random variable Z of Example 3 is continuous because the values it may assume make up the interval of nonnegative real numbers. For the remainder of this section, unless otherwise noted, *all random variables will be assumed to be finite discrete.*

Probability Distributions of Random Variables

In Section 7.2, we learned how to construct the probability distribution for an experiment. There, the probability distribution took the form of a table that gave the probabilities associated with the outcomes of an experiment. Since the random variable associated with an experiment is related to the outcomes of the experiment, it is clear that we should be able to construct a probability distribution associated with the *random variable* rather than one associated with the outcomes of the experiment. Such a distribution is called the **probability distribution of a random variable** and may be given in the form of a formula or displayed in a table that gives the distinct (numerical) values of the random variable X and the probabilities associated with these values. Thus, if $x_1, x_2, \ldots, x_n$ are the values assumed by the random variable X with associated probabilities $P(X = x_1), P(X = x_2), \ldots, P(X = x_n)$, respectively, then the required probability distribution of the random variable X may be expressed in the form of the table shown in Table 3, where $p_i = P(X = x_i)$, $i = 1, 2, \ldots, n$.

TABLE 3

Probability Distribution for the Random Variable X

x	$P(X = x)$
x_1	p_1
x_2	p_2
x_3	p_3
$\vdots$	$\vdots$
x_n	p_n

The probability distribution of a random variable X satisfies

1. $0 \leq p_i \leq 1 \qquad i = 1, 2, \ldots, n$
2. $p_1 + p_2 + \cdots + p_n = 1$

In the next several examples, we illustrate the construction and application of probability distributions.

EXAMPLE 4

a. Find the probability distribution of the random variable associated with the experiment of Example 1.
b. What is the probability of obtaining at least two heads in the three tosses of the coin?

Solution

a. From the results of Example 1, we see that the values assumed by the random variable X are 0, 1, 2, and 3, corresponding to the events of 0, 1, 2, and 3 heads occurring, respectively. Referring to Table 1 once again, we see that the outcome associated with the event $(X = 0)$ is given by the set {TTT}. Consequently, the probability associated with the random variable X when it assumes the value 0 is given by

$$P(X = 0) = \frac{1}{8}$$

Note that $n(S) = 8$.

TABLE 4

Probability Distribution

x	$P(X = x)$
0	$\frac{1}{8}$
1	$\frac{3}{8}$
2	$\frac{3}{8}$
3	$\frac{1}{8}$

Next, observe that the event $(X = 1)$ is given by the set {HTT, THT, TTH}, so

$$P(X = 1) = \frac{3}{8}$$

In a similar manner, we may compute $P(X = 2)$ and $P(X = 3)$, which gives the probability distribution shown in Table 4.

b. The probability of obtaining at least two heads is given by

$$\begin{aligned} P(X \geq 2) &= P(X = 2) + P(X = 3) \\ &= \frac{3}{8} + \frac{1}{8} = \frac{1}{2} \end{aligned}$$

EXAMPLE 5 Let X denote the random variable that gives the sum of the faces that fall uppermost when two fair dice are rolled.

a. Find the probability distribution of X.

b. What is the probability that the sum of the faces that fall uppermost is less than or equal to 5? Between 8 and 10, inclusive?

TABLE 5

x	$P(X = x)$
2	$\frac{1}{36}$
3	$\frac{2}{36}$
4	$\frac{3}{36}$
5	$\frac{4}{36}$
6	$\frac{5}{36}$
7	$\frac{6}{36}$
8	$\frac{5}{36}$
9	$\frac{4}{36}$
10	$\frac{3}{36}$
11	$\frac{2}{36}$
12	$\frac{1}{36}$

Solution

a. The values assumed by the random variable X are 2, 3, 4, . . . , 12, corresponding to the events $E_2, E_3, E_4, \ldots, E_{12}$ (see Example 4, Section 7.1). The probabilities associated with the random variable X when X assumes the values 2, 3, 4, . . . , 12 are precisely the probabilities $P(E_2), P(E_3), \ldots, P(E_{12})$, respectively, and may be computed in much the same way as the solution to Example 3, Section 7.2. Thus,

$$P(X = 2) = P(E_2) = \frac{1}{36}$$

$$P(X = 3) = P(E_3) = \frac{2}{36}$$

and so on. The required probability distribution of X is given in Table 5.

b. The probability that the sum of the faces that fall uppermost is less than or equal to 5 is given by

$$\begin{aligned} P(X \leq 5) &= P(X = 2) + P(X = 3) + P(X = 4) + P(X = 5) \\ &= \frac{1}{36} + \frac{2}{36} + \frac{3}{36} + \frac{4}{36} = \frac{10}{36} = \frac{5}{18} \end{aligned}$$

The probability that the sum of the faces that fall uppermost is between 8 and 10, inclusive, is given by

$$\begin{aligned} P(8 \leq X \leq 10) &= P(X = 8) + P(X = 9) + P(X = 10) \\ &= \frac{5}{36} + \frac{4}{36} + \frac{3}{36} = \frac{12}{36} = \frac{1}{3} \end{aligned}$$

APPLIED EXAMPLE 6 Waiting Lines The following data give the number of cars observed waiting in line at the beginning of 2-minute intervals between 3 P.M. and 5 P.M. on a certain Friday at the drive-in teller of Westwood Savings Bank and the corresponding frequency of occurrence:

Cars	0	1	2	3	4	5	6	7	8
Frequency of Occurrence	2	9	16	12	8	6	4	2	1

a. Find the probability distribution of the random variable X, where X denotes the number of cars observed waiting in line.

b. What is the probability that the number of cars observed waiting in line in any 2-minute interval between 3 P.M. and 5 P.M. on a Friday is less than or equal to 3? Between 2 and 4, inclusive? Greater than 6?

Solution

a. The sum of the numbers in the second row of the table given above is 60. Dividing each number in the second row of the table by 60 gives the respective probabilities associated with the random variable X when X assumes the values 0, 1, 2, . . . , 8. (Here, we use the relative frequency interpretation of probability.) For example,

$$P(X = 0) = \frac{2}{60} \approx .03$$

$$P(X = 1) = \frac{9}{60} = .15$$

and so on. The resulting probability distribution is shown in Table 6.

TABLE 6

Probability Distribution

x	$P(X = x)$
0	.03
1	.15
2	.27
3	.20
4	.13
5	.10
6	.07
7	.03
8	.02

b. The probability that the number of cars observed waiting in line is less than or equal to 3 is given by

$$\begin{aligned} P(X \le 3) &= P(X = 0) + P(X = 1) + P(X = 2) + P(X = 3) \\ &= .03 + .15 + .27 + .20 = .65 \end{aligned}$$

The probability that the number of cars observed waiting in line is between 2 and 4, inclusive, is given by

$$\begin{aligned} P(2 \le X \le 4) &= P(2) + P(3) + P(4) \\ &= .27 + .20 + .13 = .60 \end{aligned}$$

The probability that the number of cars observed waiting in line is greater than 6 is given by

$$\begin{aligned} P(X > 6) &= P(7) + P(8) \\ &= .03 + .02 = .05 \end{aligned}$$

Histograms

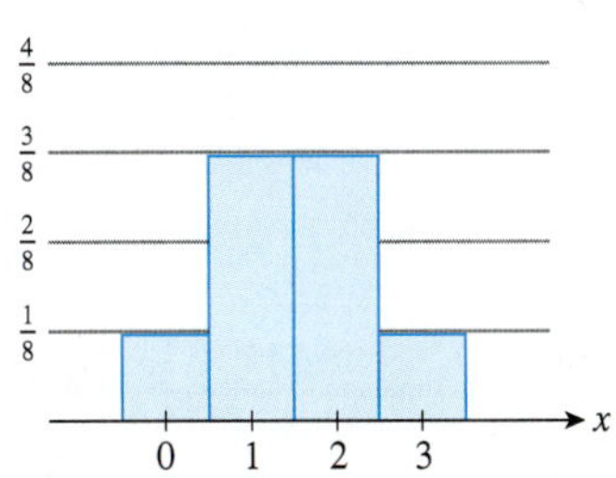

FIGURE 1
Histogram showing the probability distribution for the number of heads occurring in three coin tosses

A probability distribution of a random variable may be exhibited graphically by means of a **histogram.** To construct a histogram of a particular probability distribution, first locate the values of the random variable on a number line. Then, above each such number, draw a rectangle with width 1 and height equal to the probability associated with that value of the random variable. For example, the histogram of the probability distribution appearing in Table 4 is shown in Figure 1. The histograms of the probability distributions of Examples 5 and 6 are constructed in a similar manner and are displayed in Figures 2 and 3, respectively.

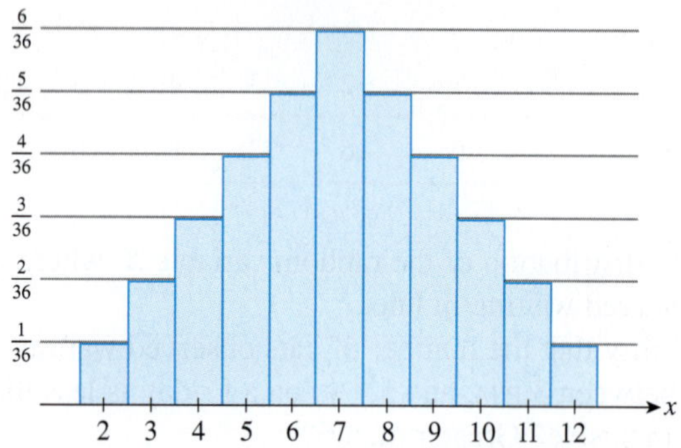

FIGURE 2
Histogram showing the probability distribution for the sum of the uppermost faces of two dice

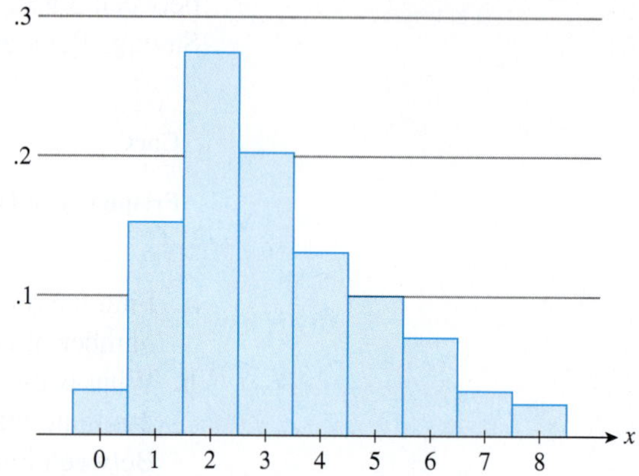

FIGURE 3
Histogram showing the probability distribution for the number of cars waiting in line

Observe that in each histogram, the area of a rectangle associated with a value of a random variable X gives precisely the probability associated with the value of X. This follows because each such rectangle, by construction, has width 1 and height corresponding to the probability associated with the value of the random variable. Another consequence arising from the method of construction of a histogram is that *the probability associated with more than one value of the random variable X is given by the sum of the areas of the rectangles associated with those values of X*. For example, in the coin-tossing experiment of Example 1, the event of obtaining at least two heads, which corresponds to the event $(X = 2)$ or $(X = 3)$, is given by

$$P(X = 2) + P(X = 3)$$

and may be obtained from the histogram depicted in Figure 1 by adding the areas associated with the values 2 and 3 of the random variable X. We obtain

$$P(X = 2) + P(X = 3) = (1)\left(\frac{3}{8}\right) + (1)\left(\frac{1}{8}\right) = \frac{1}{2}$$

This result provides us with a method of computing the probabilities of events directly from the knowledge of a histogram of the probability distribution of the random variable associated with the experiment.

EXAMPLE 7 Suppose the probability distribution of a random variable X is represented by the histogram shown in Figure 4. Identify the part of the histogram whose area gives the probability $P(10 \le X \le 20)$.

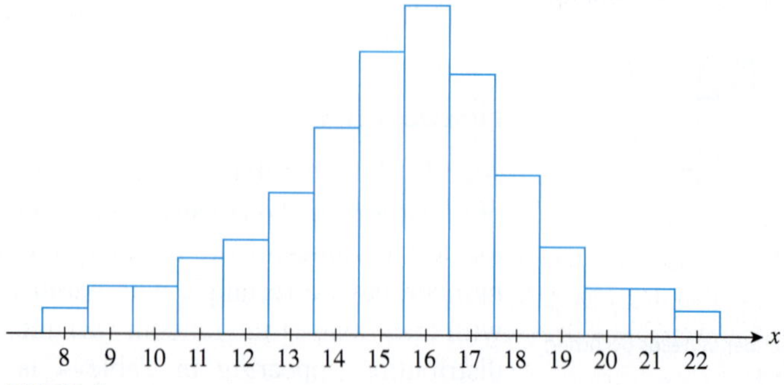

FIGURE 4

Solution The event $(10 \le X \le 20)$ is the event consisting of outcomes related to the values 10, 11, 12, . . . , 20 of the random variable X. The probability of this event $P(10 \le X \le 20)$ is therefore given by the shaded area of the histogram in Figure 5.

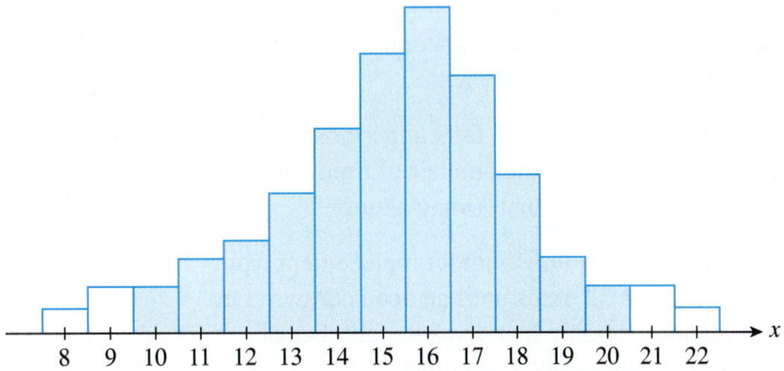

FIGURE 5
$P(10 \le X \le 20)$

Pie charts are also widely used in business and media presentations to describe numerical data. Pie charts are often used to show the relationship of the parts to the whole. We illustrate how Excel can be used to create pie charts in Using Technology, page 470.

8.1 Self-Check Exercises

1. Three balls are selected at random without replacement from an urn containing four black balls and five white balls. Let the random variable X denote the number of black balls drawn.
   a. List the outcomes of the experiment.
   b. Find the value assigned to each outcome of the experiment by the random variable X.
   c. Find the event consisting of the outcomes to which a value of 2 has been assigned by X.

2. Library Usage The following data, extracted from the records of Dover Public Library, give the number of books borrowed by the library's members over a 1-month period:

Books	0	1	2	3	4	5	6	7	8
Frequency of Occurrence	780	300	412	205	98	54	57	30	6

   a. Find the probability distribution of the random variable X, where X denotes the number of books checked out over a 1-month period by a randomly chosen member.
   b. Draw the histogram representing this probability distribution.

Solutions to Self-Check Exercises 8.1 can be found on page 467.

8.1 Concept Questions

1. What is a random variable? Give an example.
2. Give an example of (a) a finite discrete random variable, (b) an infinite discrete random variable, and (c) a continuous random variable.
3. Suppose you are given the probability distribution for a random variable X. Explain how you would construct a histogram for this probability distribution. What does the area of each rectangle in the histogram represent?

8.1 Exercises

1. Three balls are selected at random without replacement from an urn containing four green balls and six red balls. Let the random variable X denote the number of green balls drawn.
   a. List the outcomes of the experiment.
   b. Find the value assigned to each outcome of the experiment by the random variable X.
   c. Find the event consisting of the outcomes to which a value of 3 has been assigned by X.

2. A coin is tossed four times. Let the random variable X denote the number of tails that occur.
a. List the outcomes of the experiment.
b. Find the value assigned to each outcome of the experiment by the random variable X.
c. Find the event consisting of the outcomes to which a value of 2 has been assigned by X.

3. A die is rolled repeatedly until a 6 falls uppermost. Let the random variable X denote the number of times the die is rolled. What are the values that X may assume?

4. Cards are selected one at a time without replacement from a well-shuffled deck of 52 cards until an ace is drawn. Let X denote the random variable that gives the number of cards drawn. What values may X assume?

5. Let X denote the random variable that gives the sum of the faces that fall uppermost when two fair dice are rolled. Find $P(X = 7)$.

6. Two cards are drawn from a well-shuffled deck of 52 playing cards. Let X denote the number of aces drawn. Find $P(X = 2)$.

In Exercises 7–12, give the range of values that the random variable X may assume and classify the random variable as finite discrete, infinite discrete, or continuous.

7. $X =$ The number of times a die is thrown until a 2 appears

8. $X =$ The number of defective iPads in a sample of eight iPads

9. $X =$ The distance in miles a commuter travels to work

10. $X =$ The number of hours a child watches television on a given day

11. $X =$ The number of times an accountant takes the CPA examination before passing

12. $X =$ The number of boys in a four-child family

In Exercises 13–16, determine whether the table gives the probability distribution of the random variable X. Explain your answer.

13.

x	-3	-2	-1	0	1	2
$P(X = x)$	0.2	0.4	0.3	-0.2	0.1	0.1

14.

x	-2	-1	0	1	2
$P(X = x)$	0.2	0.1	0.3	0.2	0.1

15.

x	1	2	3	4	5	6
$P(X = x)$	0.3	0.1	0.2	0.2	0.1	0.2

16.

x	-1	0	1	2	3
$P(X = x)$	0.3	0.1	0.2	0.2	0.2

In Exercises 17 and 18, find conditions on the numbers a and/or b such that the table gives the probability distribution of the random variable X.

17.

x	0	2	4	6	8
$P(X = x)$	0.1	0.4	a	0.1	0.2

18.

x	-1	0	1	2	4	5
$P(X = x)$	0.3	a	0.2	0.2	b	0.1

19. The probability distribution of the random variable X is shown in the accompanying table:

x	-10	-5	0	5	10	15	20
$P(X = x)$	.20	.15	.05	.1	.25	.1	.15

Find:
a. $P(X = -10)$
b. $P(X \geq 5)$
c. $P(-5 \leq X \leq 5)$
d. $P(X \leq 20)$
e. $P(X < 5)$
f. $P(X = 3)$

20. The probability distribution of the random variable X is shown in the accompanying table:

x	-5	-3	-2	0	2	3
$P(X = x)$	.17	.13	.33	.16	.11	.10

Find:
a. $P(X \leq 0)$
b. $P(X \leq -3)$
c. $P(-2 \leq X \leq 2)$
d. $P(X = -2)$
e. $P(X > 0)$
f. $P(X = 1)$

21. Suppose that the probability distribution of a random variable X is represented by the accompanying histogram. Shade the part of the histogram whose area gives the probability $P(17 \leq X \leq 20)$.

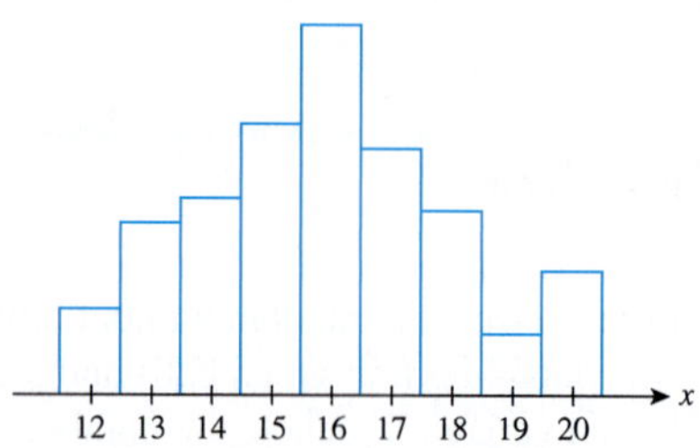

22. **Exams** An examination consisting of ten true-or-false questions was taken by a class of 100 students. The probability distribution of the random variable X, where X denotes the number of questions answered correctly by a randomly chosen student, is represented by the accompanying histogram. The rectangle with base centered on the number 8 is missing. What should be the height of this rectangle?

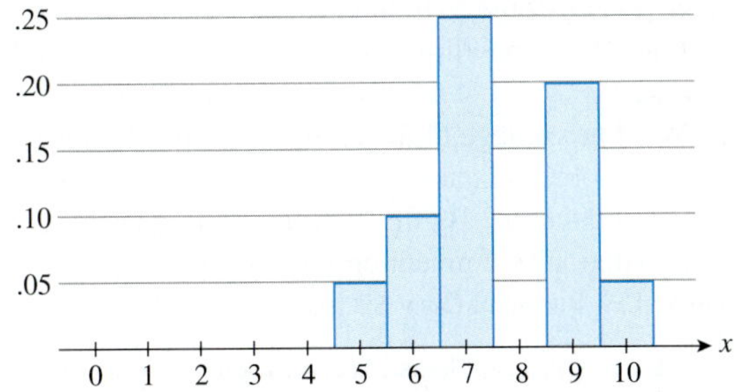

23. Two dice are rolled. Let the random variable X denote the number that falls uppermost on the first die, and let Y denote the number that falls uppermost on the second die.
   a. Find the probability distributions of X and Y.
   b. Find the probability distribution of $X + Y$.

24. **Emergency Fund Savings** In a survey of 1000 people, the following question was asked: How much emergency living expense savings do you have? Answers were expressed in the number of months the person could live on savings. The following results were obtained:

Group	1	2	3	4
Time (in months)	Less than 3	3–6	More than 6	No answer
Number	480	200	290	30

Let X denote the random variable that takes on the values 1, 2, 3, and 4, corresponding to the groups given in the table.
   a. Find the probability distribution associated with these data.
   b. If a respondent in the survey is chosen at random, what is the probability that he or she said that they had 6 or fewer months' worth of emergency living expense savings?

Source: Princeton Survey Research Associates.

25. **Money Market Rates** The interest rates paid by 30 financial institutions on a certain day for money market deposit accounts are shown in the accompanying table:

Rate (%)	2	2.25	2.55	2.56
Institutions	1	7	7	1

Rate (%)	2.58	2.60	2.65	2.85
Institutions	1	8	3	2

Let the random variable X denote the interest rate per year paid by a randomly chosen financial institution on its money market deposit accounts.
   a. Find the probability distribution associated with these data.
   b. Find the probability that the interest rate paid by a financial institution chosen at random is less than 2.56% per year.

26. **Distribution of Families by Size** The Public Housing Authority in a certain community conducted a survey of 1000 families to determine the distribution of families by size. The results follow:

Family Size	2	3	4	5	6	7	8
Frequency of Occurrence	350	200	245	125	66	10	4

   a. Find the probability distribution of the random variable X, where X denotes the number of people in a randomly chosen family.
   b. Draw the histogram corresponding to the probability distribution found in part (a).
   c. Find the probability that a family chosen at random from those surveyed has more than five members.

27. **Waiting Lines** The accompanying data were obtained in a study conducted by the manager of SavMore Supermarket. In this study, the number of customers waiting in line at the express checkout at the beginning of each 3-min interval between 9 A.M. and 12 noon on Saturday was observed.

Customers	0	1	2	3	4
Frequency of Occurrence	1	4	2	7	14

Customers	5	6	7	8	9	10
Frequency of Occurrence	8	10	6	3	4	1

   a. Find the probability distribution of the random variable X, where X denotes the number of customers observed waiting in line.
   b. Draw the histogram representing the probability distribution.
   c. Find the probability that the number of customers waiting in line in any 3-min interval between 9 A.M. and 12 noon is between 1 and 3, inclusive.

28. **Television Pilots** After the private screening of a new television pilot, audience members were asked to rate the new show on a scale of 1 to 10 (10 being the highest rating). From a group of 140 people, the following responses were obtained:

Rating	1	2	3	4	5	6	7	8	9	10
Frequency of Occurrence	1	4	3	11	23	21	28	29	16	4

Let the random variable X denote the rating given to the show by a randomly chosen audience member.
   a. Find the probability distribution associated with these data.
   b. What is the probability that the new television pilot got a rating that is higher than 5?

29. **Smartphone Ownership by Age** The following table gives the number of adults in the United States within each of four age groups who own smartphones.

Age (in years)	18–29	30–49	50–64	65+
Number of Owners (in thousands)	25,115	54,157	16,310	14,509

Let X denote the random variable that takes on the values 1, 2, 3, and 4, corresponding to the age groups 18–29, 30–49, 50–64, and 65+, respectively.

a. Find the probability distribution associated with these data.

b. If a smartphone owner in one of these age groups is chosen at random, what is the probability that he or she is between 18 and 49 years of age?

Source: Pew Research.

30. **Smartphone Ownership by Income** The following table gives the number of U.S. households within each of four income groups in which someone owns a smartphone.

Income ($)	<30,000	30,000–49,999	50,000–74,999	75,000+
Number of Owners (in thousands)	11,817	8,475	7,609	32,057

Let X denote the random variable that takes on the values 1, 2, 3, and 4, corresponding to the income groups <$30,000, $30,000–$49,999, $50,000–$74,999, and $75,000+, respectively.

a. Find the probability distribution associated with these data.

b. If a smartphone owner in one of these households is chosen at random, what is the probability that his or her household income is between $30,000 and $74,999?

Source: Pew Research.

31. **Academy Members** The following table gives the number of members in different age groups of the Academy of Motion Picture Arts and Sciences (AMPAS). All academy members are allowed to vote for a winner in most categories, including Best Picture.

Group	1	2	3	4	5
Age (in years)	Under 40	40–49	50–59	60 and over	Unknown
Number	115	634	1441	3113	462

Let X be the random variable that takes on the values 1, 2, 3, 4, and 5, corresponding to the age groups in the AMPAS membership.

a. Find the probability distribution associated with these data.

b. What is the percentage of AMPAS members who are younger than 50 years old?

Source: LATIMES.com.

32. **Food Expenditure Away from Home** According to a report of the U.S. Bureau of Labor Statistics, the average annual food expenditures away from home by quintiles of household income before taxes is as follows:

Income Quintile	Lowest 20%	Second 20%	Third 20%	Fourth 20%	Highest 20%
Expenditures ($)	1038	1569	2127	3206	5151

Let the random variable X denote a randomly chosen quintile (in income) within the set of all households.

a. Find the probability distribution associated with these data.

b. What percentage of the total food expenditures away from home is incurred by the top 40% by income of the population? By the lowest 40% by income of the population? Comment on your results.

Source: U.S. Bureau of Labor Statistics.

33. **Food Expenditure at Home** According to a report of the U.S. Bureau of Labor Statistics, the average annual food expenditures at home by quintiles of household income before taxes is as follows:

Income Quintile	Lowest 20%	Second 20%	Third 20%	Fourth 20%	Highest 20%
Expenditures ($)	2463	2999	3355	4316	5629

Let the random variable X denote a randomly chosen quintile (in income) within the set of all households.

a. Find the probability distribution associated with these data.

a. What percentage of the total food expenditures is incurred by the top 40% by income of the population? By the lowest 40% by income of the population?

Source: U.S. Bureau of Labor Statistics.

34. **U.S. Population by Age** The following table gives the 2011 age distribution of the U.S. population:

Group	1	2	3	4	5	6
Age (in years)	Under 5	5–19	20–24	25–44	45–64	65 and over
Number (in thousands)	21,265	61,776	21,525	81,426	80,938	39,178

Let x denote a random variable that takes on the values 1, 2, 3, 4, 5, and 6 corresponding to the age groups given in the table.

a. Find the probability distribution associated with these data.

b. What percentage of the U.S. population is between 5 and 24 years old, inclusive?

Source: U.S. Census Bureau.

In Exercises 35 and 36, determine whether the statement is true or false. If it is true, explain why it is true. If it is false, give an example to show why it is false.

35. Suppose X is a finite discrete random variable assuming the values $x_1, x_2, \ldots, x_n$ and associated probabilities $p_1, p_2, \ldots, p_n$. Then $p_1 + p_2 + \cdots + p_n = 1$.

36. The area of a histogram associated with a probability distribution is a number between 0 and 1.

8.1 Solutions to Self-Check Exercises

1. a. Using the accompanying tree diagram, we see that the outcomes of the experiment are

$$S = \{BBB, BBW, BWB, BWW, WBB, WBW, WWB, WWW\}$$

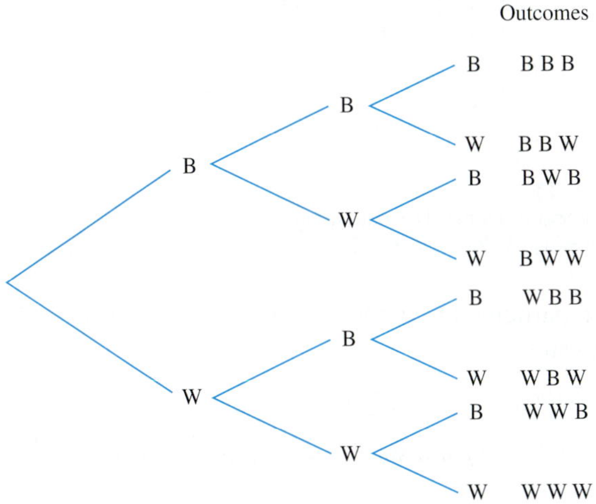

b. Using the results of part (a), we obtain the values assigned to the outcomes of the experiment as follows:

Outcome	BBB	BBW	BWB	BWW
Value	3	2	2	1

Outcome	WBB	WBW	WWB	WWW
Value	2	1	1	0

c. The required event is {BBW, BWB, WBB}.

2. a. We divide each number in the bottom row of the given table by 1942 (the sum of these numbers) to obtain the probabilities associated with the random variable X when X takes on the values 0, 1, 2, 3, 4, 5, 6, 7, and 8. For example,

$$P(X = 0) = \frac{780}{1942} \approx .402$$

$$P(X = 1) = \frac{300}{1942} \approx .154$$

The required probability distribution and histogram follow:

x	0	1	2	3	4
$P(X = x)$	.402	.154	.212	.106	.050

x	5	6	7	8
$P(X = x)$	.028	.029	.015	.003

b.

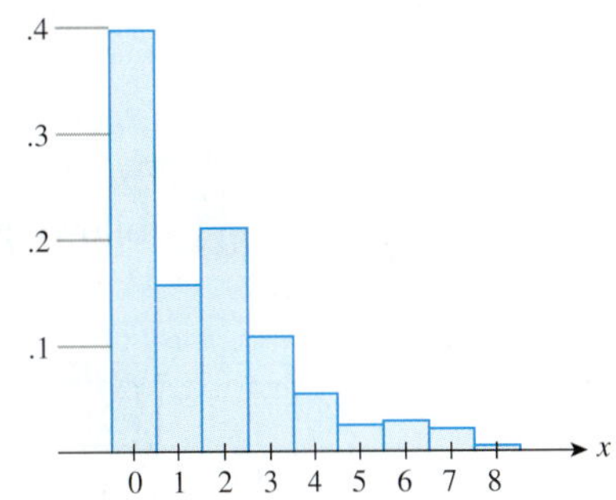

USING TECHNOLOGY Graphing a Histogram

Graphing Utility

A graphing utility can be used to plot the histogram for a given set of data, as illustrated in the following example.

APPLIED EXAMPLE 1 A survey of 90,000 households conducted in a certain year revealed the following percentage of women who wear a shoe size within the given ranges.

Shoe Size	<5	$5\text{–}5\frac{1}{2}$	$6\text{–}6\frac{1}{2}$	$7\text{–}7\frac{1}{2}$	$8\text{–}8\frac{1}{2}$	$9\text{–}9\frac{1}{2}$	$10\text{–}10\frac{1}{2}$	$>10\frac{1}{2}$
Women (%)	1	5	15	27	29	14	7	2

Let X denote the random variable taking on the values 1 through 8, where 1 corresponds to a shoe size less than 5, 2 corresponds to a shoe size of $5\text{–}5\frac{1}{2}$, and so on.

a. Plot a histogram for the given data.

b. What percentage of women in the survey wear a shoe size within the ranges $7\text{–}7\frac{1}{2}$ or $8\text{–}8\frac{1}{2}$?

Source: Footwear Market Insights survey.

Solution

a. Enter the values of X as $x_1 = 1, x_2 = 2, \ldots, x_8 = 8$ and the corresponding values of Y as $y_1 = 1, y_2 = 5, \ldots, y_8 = 2$. Then using the **DRAW** function from the Statistics menu, we draw the histogram shown in Figure T1.

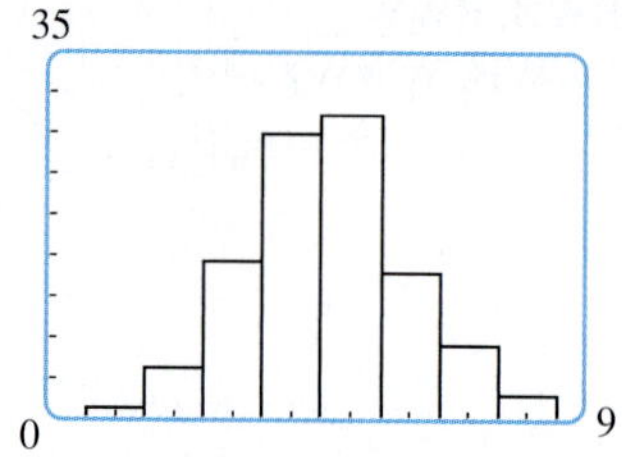

FIGURE **T1**
The histogram for the given data, using the viewing window [0, 9] × [0, 35]

b. The probability that a woman participating in the survey wears a shoe size within the ranges $7\text{–}7\frac{1}{2}$ or $8\text{–}8\frac{1}{2}$ is given by

$$P(X = 4) + P(X = 5) = .27 + .29 = .56$$

This tells us that 56% of the women wear a shoe size within the ranges $7\text{–}7\frac{1}{2}$ or $8\text{–}8\frac{1}{2}$.

Excel

Excel can be used to plot the histogram for a given set of data, as illustrated in the following example.

APPLIED EXAMPLE 2 A survey of 90,000 households conducted in a certain year revealed the following percentage of women who wear a shoe size within the given ranges.

Shoe Size	<5	$5\text{–}5\frac{1}{2}$	$6\text{–}6\frac{1}{2}$	$7\text{–}7\frac{1}{2}$	$8\text{–}8\frac{1}{2}$	$9\text{–}9\frac{1}{2}$	$10\text{–}10\frac{1}{2}$	$>10\frac{1}{2}$
Women (%)	1	5	15	27	29	14	7	2

Let X denote the random variable taking on the values 1 through 8, where 1 corresponds to a shoe size less than 5, 2 corresponds to a shoe size of $5\text{–}5\frac{1}{2}$, and so on.

a. Plot a histogram for the given data.
b. What percentage of women in the survey wear a shoe size within the ranges $7\text{–}7\frac{1}{2}$ or $8\text{–}8\frac{1}{2}$?

Source: Footwear Market Insights survey.

Solution

a. Enter the given data in columns A and B onto a spreadsheet, as shown in Figure T2. Highlight the data in column B, and select **[Σ]** from the Editing group under the Home tab. The sum of the numbers in this column (100) will appear in cell B10. In cell C2, type `=B2/100`, and then press **[Enter]**. To extend the formula

Note: Boldfaced words/characters enclosed in a box (for example, **[Enter]**) indicate an action (click, select, or press) is required. Words/characters printed blue (for example, Chart sub-type:) indicate words/characters that appear on the screen. Words/characters printed in a monospace font (for example, `=(-2/3)*A2+2`) indicates words/characters that need to be typed and entered).

to cell C9, select C2 and move the pointer to the small black box at the lower right corner of that cell. Drag the black + that appears (at the lower right corner of cell C2) through cell C9, and then release it. The probability distribution shown in cells C2 to C9 will then appear on your spreadsheet. Then highlight the data in the Probability column, and select **Column** from the Charts group under the Insert tab. Click on the first chart type (Clustered Column). Under the Layout tab, click on **Chart Title** in the Labels group, and select **Above Chart**. Enter `Histogram` as the title. Select **Primary Horizontal Axis Title** under Axis Titles from the Labels group, select **Title Below Axis**, and then enter `X` as the horizontal axis title. Next, select **Primary Vertical Axis Title** under Axis Titles followed by **Rotated Title** and enter `Probability`. Right-click a bar on the chart and select **Format Data Series**. Adjust the slider under Gap Width to 0% in the dialog box that appears, and then click **Close**. Finally, delete the Series 1 legend entry.

	A	B	C
1	X	Frequency	Probability
2	1	1	0.01
3	2	5	0.05
4	3	15	0.15
5	4	27	0.27
6	5	29	0.29
7	6	14	0.14
8	7	7	0.07
9	8	2	0.02
10		100	

FIGURE **T2**
Completed spreadsheet for Example 2

The histogram shown in Figure T3 will appear.

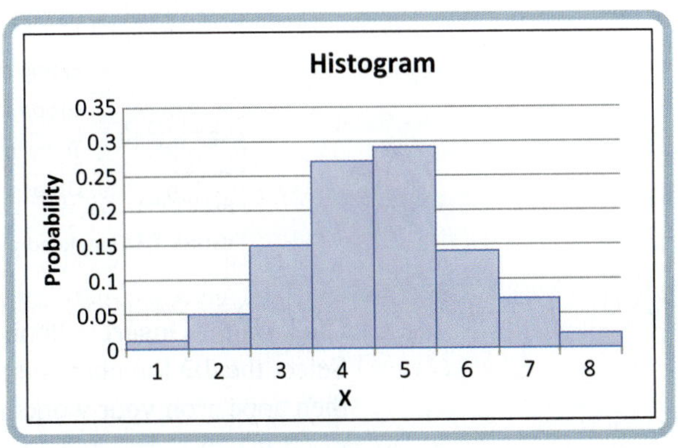

FIGURE **T3**
The histogram for the random variable *X*

b. The probability that a woman participating in the survey wears a shoe size within the ranges $7\text{–}7\frac{1}{2}$ or $8\text{–}8\frac{1}{2}$ is given by

$$P(X = 4) + P(X = 5) = .27 + .29 = .56$$

This tells us that 56% of the women wear a shoe size within the ranges $7\text{–}7\frac{1}{2}$ or $8\text{–}8\frac{1}{2}$.

Excel can also be used to create pie charts, as illustrated in the following example.

APPLIED EXAMPLE 3 Time Use of College Students Use the data given in Table 1 to construct a pie chart.

TABLE T1

Time Used on an Average Weekday for Full-Time University and College Students

Time Use	Time (in hours)
Sleeping	8.5
Leisure and sports	3.7
Working and related activities	2.9
Educational activities	3.3
Eating and drinking	1.0
Grooming	0.7
Traveling	1.5
Other	2.4

Source: Bureau of Labor Statistics.

Solution

We begin by entering the information from Table T1 in Columns A and B on a spreadsheet. Then follow these steps:

Step 1 First, highlight the data in cells A2:A9 and B2:B9 as shown in Figure T4.

	A	B
1	Time Use	Time (in hours)
2	Sleeping	8.5
3	Leisure and sports	3.7
4	Working and related activities	2.9
5	Educational activities	3.3
6	Eating and drinking	1
7	Grooming	0.7
8	Traveling	1.5
9	Other	2.4

FIGURE **T4** Completed spreadsheet

Step 2 Click on the **Insert** ribbon tab, and then select **Pie** from the Charts group. Select the 2D Pie chart subtype in the first row and first column. A chart will then appear on your worksheet.

Step 3 From the Chart Tools group that now appears on the ribbon, click the **Design** ribbon tab, and then select the chart appearing in the first row and first column of the Charts Layouts group. Note that this chart displays the percentage of a day (24 hours) spent on each activity. Next, select the chart appearing in row 2 and column 2 of the Chart Styles group.

Step 4 Click the **Layout** tab, and select **Chart Title** from the Labels group followed by **Above Chart**. Type the title of the chart and click **Enter**.

The pie chart shown in Figure T5 will appear.

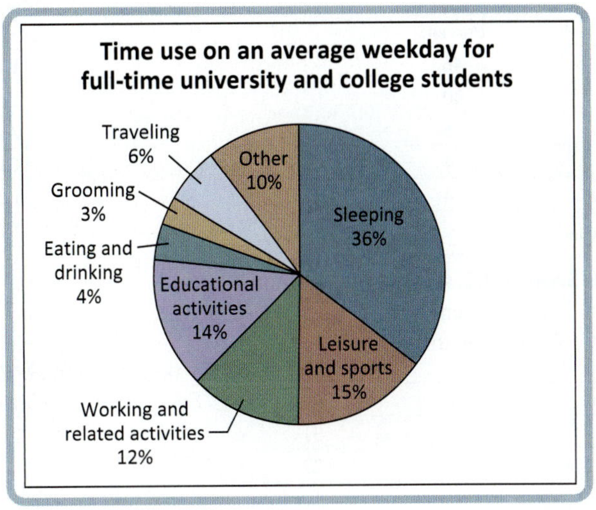

FIGURE **T5**
The pie chart describing the data in Table T1

TECHNOLOGY EXERCISES

1. Graph the histogram associated with the data given in Table 1, page 458. Compare your graph with that given in Figure 1, page 461.

2. Graph the histogram associated with the data given in Exercise 26, page 465.

3. Graph the histogram associated with the data given in Exercise 27, page 465.

4. Graph the histogram associated with the data given in Exercise 28, page 465.

In Exercises 5–10, use the data given in the table to construct a pie chart. (These exercises are for EXCEL only.)

5. **Who Pays Taxes?** Payroll taxes are used for Social Security and Medicare, two of the biggest items in the federal budget. The following facts regarding the people in the United States who pay taxes were reported in *Money* magazine in March 2012:

Taxpaying and Non-Taxpaying Groups	Percent
Pay income tax	53.6
Pay no income tax but do pay payroll taxes	28.3
Pay no income tax but are elderly	10.3
Pay no income tax but earn less than $20,000	6.9
Others	0.9

Source: Urban Brookings Tax Policy Center.

6. **Time Spent per Week on the Internet by College Students** The following table gives the time spent on the Internet by 18- to 24-year-old college students:

Time Spent	College Students (%)
20 or more hours	19.6
10 or more but fewer than 20 hours	13.4
5 or more but fewer than 10 hours	20.3
3 or more but fewer than 5 hours	22.6
Fewer than 3 hours	24.1

Source: Burst Research.

7. **Main Reasons Why Young Adults Shop Online** The results of an online survey by Bing among 1077 adults aged 18–34 years in November 2012 regarding the main reasons why they shopped online are summarized in the following table:

Main Reason	Percent
Better prices	37
Avoiding holiday crowds, hassles	29
Convenience	18
Better selection	13
Ships directly	3

Source: Impulse Research.

8. **Bringing Something to a Party** In a survey of 2008 adults conducted by American Express, the following question was asked: When invited to a party, do you contribute

something even if not asked? The responses given are shown in the following table:

Response	Percent
Always	46
Sometimes	40
Rarely	6
Never	4
Not Sure	4

Source: American Express.

9. **How Money Is Spent in U.S. for Health Care** In the United States, \$2.5 trillion was spent on health care in 2009. The following table shows how this money was spent:

Where Spent	Percent
Hospitals	31
Doctors, other professionals	27
Prescription drugs (retail)	10
Nursing home care	5
Other private revenues	27

Source: "Covering Health Issues: A Sourcebook for Journalists," 6E.

10. **Percentage of Mobile Ad Revenues by Device Type** The following table gives the percentage of mobile ad revenues in the first quarter of 2013 by device type:

Device Type	Percent
iPhone	34.2
Android phone	26.2
Other	15.4
iPad	12.6
RIM	6.2
Symbian	2.5
iPod Touch	2.4
Android tablet	0.5

Source: Opera Mediaworks.

8.2 Expected Value

Mean

The average value of a set of numbers is a familiar notion to most people. For example, to compute the average of the four numbers

$$12, 16, 23, 37$$

we simply add these numbers and divide the resulting sum by 4, giving the required average as

$$\frac{12 + 16 + 23 + 37}{4} = \frac{88}{4} = 22$$

In general, we have the following definition:

Average, or Mean

The **average**, or **mean**, of the n numbers

$$x_1, x_2, \ldots, x_n$$

is $\bar{x}$ (read "x bar"), where

$$\bar{x} = \frac{x_1 + x_2 + \cdots + x_n}{n}$$

APPLIED EXAMPLE 1 Waiting Lines Refer to Example 6, Section 8.1. Find the average number of cars waiting in line at the bank's drive-in teller at the beginning of each 2-minute interval during the period in question.

TABLE 7

Cars	Frequency of Occurrence
0	2
1	9
2	16
3	12
4	8
5	6
6	4
7	2
8	1

Solution The number of cars and the corresponding frequency of occurrence are reproduced in Table 7. Observe that the number 0 (of cars) occurs twice, the number 1 occurs 9 times, and so on. There are altogether

$$2 + 9 + 16 + 12 + 8 + 6 + 4 + 2 + 1 = 60$$

numbers to be averaged. Therefore, the required average is given by

$$\frac{(0 \cdot 2) + (1 \cdot 9) + (2 \cdot 16) + (3 \cdot 12) + (4 \cdot 8) + (5 \cdot 6) + (6 \cdot 4) + (7 \cdot 2) + (8 \cdot 1)}{60} \approx 3.1 \qquad (1)$$

or approximately 3.1 cars.

Expected Value

Let's reconsider the expression on the left-hand side of Equation (1), which gives the average of the frequency distribution shown in Table 7. Dividing each term by the denominator, we may rewrite the expression in the form

$$0 \cdot \left(\frac{2}{60}\right) + 1 \cdot \left(\frac{9}{60}\right) + 2 \cdot \left(\frac{16}{60}\right) + 3 \cdot \left(\frac{12}{60}\right) + 4 \cdot \left(\frac{8}{60}\right) + 5 \cdot \left(\frac{6}{60}\right)$$
$$+ 6 \cdot \left(\frac{4}{60}\right) + 7 \cdot \left(\frac{2}{60}\right) + 8 \cdot \left(\frac{1}{60}\right)$$

Observe that each term in the sum is a product of two factors; the first factor is the value assumed by the random variable X, where X denotes the number of cars waiting in line, and the second factor is just the probability associated with that value of the random variable. This observation suggests the following general method for calculating the expected value (that is, the average or mean) of a random variable X that assumes a finite number of values from the knowledge of its probability distribution.

Expected Value of a Random Variable X

Let X denote a random variable that assumes the values $x_1, x_2, \ldots, x_n$ with associated probabilities $p_1, p_2, \ldots, p_n$, respectively. Then the **expected value** of X, denoted by $E(X)$, is given by

$$E(X) = x_1p_1 + x_2p_2 + \cdots + x_np_n \qquad (2)$$

Note The numbers $x_1, x_2, \ldots, x_n$ may be positive, zero, or negative. For example, such a number might be positive if it represents a profit and negative if it represents a loss.

TABLE 8

Probability Distribution

x	$P(X = x)$
0	.03
1	.15
2	.27
3	.20
4	.13
5	.10
6	.07
7	.03
8	.02

APPLIED EXAMPLE 2 Waiting Lines Re-solve Example 1 by using the probability distribution associated with the experiment, which is reproduced in Table 8.

Solution Let X denote the number of cars waiting in line. Then the average number of cars waiting in line is given by the expected value of X—that is, by

$$\begin{aligned} E(X) &= (0)(.03) + (1)(.15) + (2)(.27) + (3)(.20) + (4)(.13) \\ &\quad + (5)(.10) + (6)(.07) + (7)(.03) + (8)(.02) \\ &= 3.1 \text{ cars} \end{aligned}$$

which agrees with the earlier result.

The expected value of a random variable X is a measure of the central tendency of the probability distribution associated with X. In repeated trials of an experiment with random variable X, the average of the observed values of X gets closer and closer to the expected value of X as the number of trials gets larger and larger. Geometrically, the expected value of a random variable X has the following simple interpretation: If a laminate is made of the histogram of a probability distribution associated with a random variable X, then the expected value of X corresponds to the point on the base of the laminate at which the laminate will balance perfectly when the point is directly over a fulcrum (Figure 6).

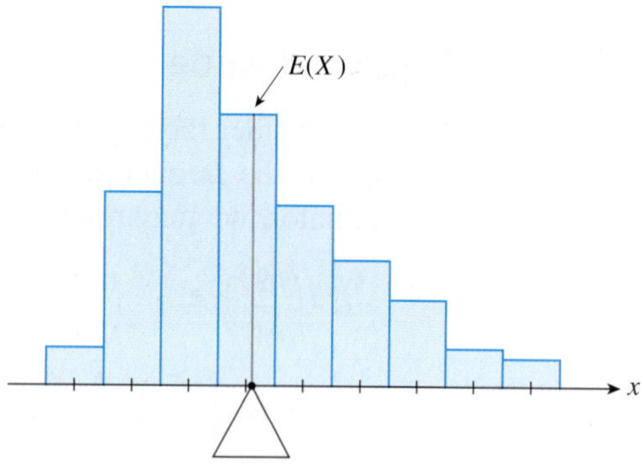

FIGURE 6
Expected value of a random variable X

EXAMPLE 3 Let X denote the random variable that gives the sum of the faces that fall uppermost when two fair dice are rolled. Find the expected value, $E(X)$, of X.

TABLE 9
Probability Distribution

x	$P(X = x)$
2	$\frac{1}{36}$
3	$\frac{2}{36}$
4	$\frac{3}{36}$
5	$\frac{4}{36}$
6	$\frac{5}{36}$
7	$\frac{6}{36}$
8	$\frac{5}{36}$
9	$\frac{4}{36}$
10	$\frac{3}{36}$
11	$\frac{2}{36}$
12	$\frac{1}{36}$

Solution The probability distribution of X, reproduced in Table 9, was found in Example 5, Section 8.1. Using this result, we find

$$\begin{aligned}E(X) &= 2\left(\frac{1}{36}\right) + 3\left(\frac{2}{36}\right) + 4\left(\frac{3}{36}\right) + 5\left(\frac{4}{36}\right) + 6\left(\frac{5}{36}\right) + 7\left(\frac{6}{36}\right)\\ &\quad + 8\left(\frac{5}{36}\right) + 9\left(\frac{4}{36}\right) + 10\left(\frac{3}{36}\right) + 11\left(\frac{2}{36}\right) + 12\left(\frac{1}{36}\right)\\ &= 7\end{aligned}$$

Note that, because of the symmetry of the histogram of the probability distribution with respect to the vertical line $x = 7$, the result could have been obtained by merely inspecting Figure 7.

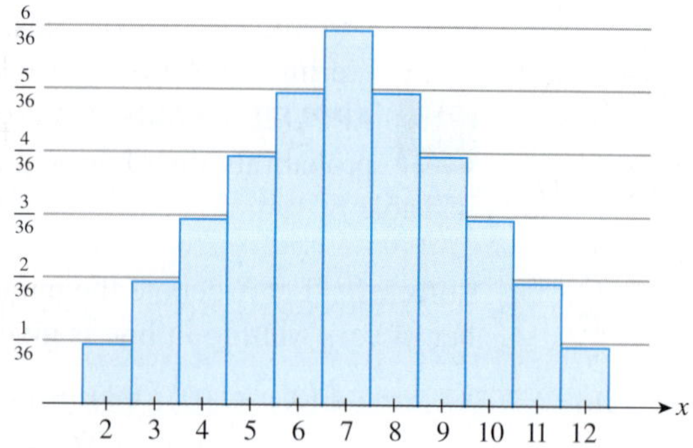

FIGURE 7
Histogram showing the probability distribution for the sum of the uppermost faces of two dice

The next example shows how we can use the concept of expected value to help us make the best investment decision.

APPLIED EXAMPLE 4 Expected Profit A private equity group intends to purchase one of two motels currently being offered for sale in a certain city. The terms of sale of the two motels are similar, although the Regina Inn has 52 rooms and is in a slightly better location than the Merlin Motor Lodge, which has 60 rooms. Records obtained for each motel reveal that the occupancy rates, with corresponding probabilities, during the May–September tourist season are as shown in the following tables:

Regina Inn					
Occupancy Rate	.80	.85	.90	.95	1.00
Probability	.19	.22	.31	.23	.05

Merlin Motor Lodge						
Occupancy Rate	.75	.80	.85	.90	.95	1.00
Probability	.35	.21	.18	.15	.09	.02

The average profit per day for each occupied room at the Regina Inn is \$40, whereas the average profit per day for each occupied room at the Merlin Motor Lodge is \$36.

a. Find the average number of rooms occupied per day at each motel.

b. If the investors' objective is to purchase the motel that generates the higher daily profit, which motel should they purchase? (Compare the expected daily profit of the two motels.)

Solution

a. Let X denote the occupancy rate at the Regina Inn. Then the average daily occupancy rate at the Regina Inn is given by the expected value of X—that is, by

$$\begin{aligned} E(X) &= (.80)(.19) + (.85)(.22) + (.90)(.31) \\ &\quad + (.95)(.23) + (1.00)(.05) \\ &= .8865 \end{aligned}$$

The average number of rooms occupied per day at the Regina Inn is

$$(.8865)(52) \approx 46.1$$

or approximately 46.1 rooms. Similarly, letting Y denote the occupancy rate at the Merlin Motor Lodge, we have

$$\begin{aligned} E(Y) &= (.75)(.35) + (.80)(.21) + (.85)(.18) + (.90)(.15) \\ &\quad + (.95)(.09) + (1.00)(.02) \\ &= .8240 \end{aligned}$$

The average number of rooms occupied per day at the Merlin Motor Lodge is

$$(.8240)(60) \approx 49.4$$

or approximately 49.4 rooms.

b. The expected daily profit at the Regina Inn is given by

$$(46.1)(40) = 1844$$

or \$1844. The expected daily profit at the Merlin Motor Lodge is given by

$$(49.4)(36) \approx 1778$$

or approximately \$1778. From these results, we conclude that the private equity group should purchase the Regina Inn, which is expected to yield a higher daily profit.

APPLIED EXAMPLE 5 Raffles The Island Club is holding a fund-raising raffle. Ten thousand tickets have been sold for \$2 each. There will be a first prize of \$3000, 3 second prizes of \$1000 each, 5 third prizes of \$500 each, and 20 consolation prizes of \$100 each. Letting X denote the net winnings (that is, winnings less the cost of the ticket) associated with a ticket, find $E(X)$. Interpret your results.

TABLE 10

Probability Distribution for a Raffle

x	$P(X = x)$
−2	.9971
98	.0020
498	.0005
998	.0003
2998	.0001

Solution The values assumed by X are $(0 - 2)$, $(100 - 2)$, $(500 - 2)$, $(1000 - 2)$, and $(3000 - 2)$—that is, −2, 98, 498, 998, and 2998—which correspond, respectively, to the value of a losing ticket, a consolation prize, a third prize, and so on. The probability distribution of X may be calculated in the usual manner and appears in Table 10. Using the table, we find

$$\begin{aligned} E(X) &= (-2)(.9971) + 98(.0020) + 498(.0005) \\ &\quad + 998(.0003) + 2998(.0001) \\ &= -0.95 \end{aligned}$$

This expected value gives the long-run average loss (negative gain) of a holder of one ticket; that is, if one participated in such a raffle by purchasing one ticket each time, in the long run, one may expect to lose, on the average, 95 cents per raffle.

APPLIED EXAMPLE 6 Roulette In the game of roulette as played in Las Vegas casinos, the wheel is divided into 38 compartments numbered 1 through 36, 0, and 00. One-half of the numbers 1 through 36 are red, the other half are black, and 0 and 00 are green (Figure 8). Of the many types of bets that may be placed, one type involves betting on the outcome of the color of the winning number. For example, one may place a certain sum of money on *red*. If the winning number is red, one wins an amount equal to the bet placed and the amount of the bet is returned; otherwise, one loses the amount of the bet. Find the expected value of the winnings on a \$1 bet placed on *red*.

FIGURE 8
Roulette wheel

Solution Let X be a random variable whose values are 1 and -1, which correspond to a win and a loss, respectively. The probabilities associated with the values 1 and -1 are $\frac{18}{38}$ and $\frac{20}{38}$, respectively. Therefore, the expected value is given by

$$E(X) = 1\left(\frac{18}{38}\right) + (-1)\left(\frac{20}{38}\right) = -\frac{2}{38}$$
$$\approx -0.053$$

Thus, if one places a \$1 bet on *red* over and over again, one may expect to lose, on the average, approximately 5 cents per bet in the long run.

Examples 5 and 6 illustrate games that are not "fair." Of course, most participants in such games are aware of this fact and participate in them for other reasons. In a fair game, neither party has an advantage, a condition that translates into the condition that $E(X) = 0$, where X takes on the values of a player's winnings.

APPLIED EXAMPLE 7 Fair Games Mike and Bill play a card game with a standard deck of 52 cards. Mike selects a card from a well-shuffled deck and receives A dollars from Bill if the card selected is a diamond; otherwise, Mike pays Bill a dollar. Determine the value of A if the game is to be fair.

Solution Let X denote a random variable whose values are associated with Mike's winnings. Then X takes on the value A with probability $P(X = A) = \frac{1}{4}$ (since there are 13 diamonds in the deck) if Mike wins and takes on the value -1 with probability $P(X = -1) = \frac{3}{4}$ if Mike loses. Since the game is to be a fair one, the expected value $E(X)$ of Mike's winnings must be equal to zero; that is,

$$E(X) = A\left(\frac{1}{4}\right) + (-1)\left(\frac{3}{4}\right) = 0$$

Solving this equation for A gives $A = 3$. Thus, the card game will be fair if Bill makes a \$3 payoff when a diamond is drawn.

Expected Value for Grouped Data

The following example illustrates the technique for finding the mean for grouped data.

APPLIED EXAMPLE 8 Commuting Times The following table gives the travel time to work of workers aged 16 years and over who do not work at home:

Travel Time, x (in minutes)	$x < 10$	$10 \le x < 15$	$15 \le x < 20$	$20 \le x < 25$	$25 \le x < 30$	$30 \le x < 35$	$35 \le x < 45$	$45 \le x < 60$	$x \ge 60$
Number (in thousands)	17,725	18,915	20,503	19,577	8,069	18,122	8,466	9,921	10,979

Source: U.S. Census Bureau.

Estimate the average time it takes for a worker who does not work at home to travel to work. (Assume that "$x < 10$" means "$0 < x < 10$" and "$x \ge 60$" means "$60 \le x \le 90$.")

Solution Let X denote the random variable that gives the time a worker takes to travel to work. Next, we find the *midpoint* of each of the nine class intervals. Thus,

$$x_1 = \frac{0 + 10}{2} = 5 \qquad x_2 = \frac{10 + 15}{2} = \frac{25}{2} \qquad x_3 = \frac{15 + 20}{2} = \frac{35}{2}$$

$$x_4 = \frac{20 + 25}{2} = \frac{45}{2} \qquad x_5 = \frac{25 + 30}{2} = \frac{55}{2} \qquad x_6 = \frac{30 + 35}{2} = \frac{65}{2}$$

$$x_7 = \frac{35 + 45}{2} = \frac{80}{2} \qquad x_8 = \frac{45 + 60}{2} = \frac{105}{2} \qquad x_9 = \frac{60 + 90}{2} = \frac{150}{2}$$

Then we use these values of $x_1, x_2, \ldots, x_8$ in Equation (2), obtaining

$$\begin{aligned}
E(X) &= x_1p_1 + x_2p_2 + \cdots + x_8p_8 \\
&= \left(\frac{10}{2}\right)\left(\frac{17{,}725}{132{,}277}\right) + \left(\frac{25}{2}\right)\left(\frac{18{,}915}{132{,}277}\right) + \left(\frac{35}{2}\right)\left(\frac{20{,}503}{132{,}277}\right) \\
&\quad + \left(\frac{45}{2}\right)\left(\frac{19{,}577}{132{,}277}\right) + \left(\frac{55}{2}\right)\left(\frac{8{,}069}{132{,}277}\right) + \left(\frac{65}{2}\right)\left(\frac{18{,}122}{132{,}277}\right) \\
&\quad + \left(\frac{80}{2}\right)\left(\frac{8{,}466}{132{,}277}\right) + \left(\frac{105}{2}\right)\left(\frac{9{,}921}{132{,}277}\right) + \left(\frac{150}{2}\right)\left(\frac{10{,}979}{132{,}277}\right) \\
&\approx 27.35
\end{aligned}$$

Thus, the average time it takes for a worker who does not work at home to travel to work is 27.35 minutes.

Odds

In everyday parlance, the probability of the occurrence of an event is often stated in terms of the *odds in favor of* (or *odds against*) the occurrence of the event. For example, one often hears statements such as "The odds that the Dodgers will win the World Series this season are 7 to 5" and "The odds that it will not rain tomorrow are 3 to 2." We will return to these examples later. But first, let us look at a definition that ties together these two concepts.

Odds In Favor Of and Odds Against

If $P(E)$ is the probability of an event E occurring, then

1. The odds in favor of E occurring are

$$\frac{P(E)}{1 - P(E)} = \frac{P(E)}{P(E^c)} \qquad P(E) \neq 1 \qquad \textbf{(3a)}$$

2. The odds against E occurring are

$$\frac{1 - P(E)}{P(E)} = \frac{P(E^c)}{P(E)} \qquad P(E) \neq 0 \qquad \textbf{(3b)}$$

Notes

1. The odds in favor of the occurrence of an event are given by the ratio of the probability of the event occurring to the probability of the event not occurring. The odds against the occurrence of an event are given by the reciprocal of the odds in favor of the occurrence of the event.
2. Whenever possible, odds are expressed as ratios of whole numbers. If the odds in favor of E are a/b, we say that the odds in favor of E are a to b. If the odds against E occurring are b/a, we say that the odds against E are b to a.

APPLIED EXAMPLE 9 Roulette Find the odds in favor of winning a bet on red in American roulette. What are the odds against winning a bet on *red*?

Solution The probability of winning a bet here—the probability that the ball lands in a red compartment—is given by $P = \frac{18}{38}$. Therefore, using Equation (3a), we see that the odds in favor of winning a bet on *red* are

$$\begin{aligned}\frac{P(E)}{1-P(E)} &= \frac{\frac{18}{38}}{1-\frac{18}{38}} \qquad E\text{, event of winning a bet on } red\\ &= \frac{\frac{18}{38}}{\frac{38-18}{38}}\\ &= \frac{18}{38}\cdot\frac{38}{20}\\ &= \frac{18}{20} = \frac{9}{10}\end{aligned}$$

or 9 to 10. Next, using Equation (3b), we see that the odds against winning a bet on *red* are $\frac{10}{9}$, or 10 to 9.

Now suppose that the odds in favor of the occurrence of an event are a to b. Then Equation (3a) gives

$$\begin{aligned}\frac{a}{b} &= \frac{P(E)}{1-P(E)}\\ a[1-P(E)] &= bP(E) \qquad \text{Cross-multiply.}\\ a - aP(E) &= bP(E)\\ a &= (a+b)P(E)\\ P(E) &= \frac{a}{a+b}\end{aligned}$$

which leads us to the following result:

Probability of an Event (Given the Odds)

If the odds in favor of an event E occurring are a to b, then the probability of E occurring is

$$P(E) = \frac{a}{a+b} \tag{4}$$

Equation (4) is often used to determine subjective probabilities, as the next example shows.

EXAMPLE 10 Consider each of the following statements.

a. "The odds that the Dodgers will win the World Series this season are 7 to 5."
b. "The odds that it will not rain tomorrow are 3 to 2."

Express each of these odds as a probability of the event occurring.

PORTFOLIO Robert H. Mason

TITLE Vice President, Wealth Management Advisor
INSTITUTION The Mason Group

The Mason Group—a team of financial advisors at a major wire house firm—acts as an interface between clients and investment markets. To meet the needs of our private clients, it is important for us to maintain constant contact with them, adjusting their investments when the markets and allocations change and when the clients' goals change.

We often help our clients determine whether their various expected sources of income in retirement will provide them with their desired retirement lifestyle. To begin, we determine the after-tax funds needed in retirement in today's dollars in consultation with the client. Using simple arithmetic, we then look at the duration and amount of their various income flows and their current portfolio allocations (stocks, bonds, cash, etc.). Once we have made this assessment, we take into consideration future possible allocations to arrive at a range of probabilities for portfolio valuation for all years up to and including retirement.

For example, we might tell a client that, on the basis of their given investment plan, there is an 80–95% probability that the expected value of his or her portfolio will increase from \$1 million to about \$1.5 million by the time the client turns 90; a 50–80% chance that it will increase from \$1 million to around \$2 million; and a 30–50% chance that it will increase to approximately \$2.8 million.

To arrive at these probabilities, we use a deterministic model that assumes that a constant annual rate of return is applied to the portfolio every year of the analysis. We also use probabilistic modeling, taking into account various factors such as economic conditions, the allocation of assets, and market volatility. By using these techniques we are able to arrive at a confidence level, without making any guarantees, that our clients will be able to attain the lifestyle in retirement they desire.

Solution

a. Using Equation (4) with $a = 7$ and $b = 5$ gives the required probability as

$$\frac{7}{7+5} = \frac{7}{12} \approx .5833$$

b. Here, the event is that it will not rain tomorrow. Using Equation (4) with $a = 3$ and $b = 2$, we conclude that the probability that it will not rain tomorrow is

$$\frac{3}{3+2} = \frac{3}{5} = .6$$

Explore and Discuss

In the movie *Casino*, the executive of the Tangiers Casino, Sam Rothstein (Robert DeNiro), fired the manager of the slot machines in the casino after three gamblers hit three "million dollar" jackpots in a span of 20 minutes. Rothstein claimed that it was a scam and that somebody had gotten into those machines to set the wheels. He was especially annoyed at the slot machine manager's assertion that there was no way to determine this. According to Rothstein, the odds of hitting a jackpot in a four-wheel machine is 1 in $1\frac{1}{2}$ million, and the probability of hitting three jackpots in a row is "in the billions." "It cannot happen! It will not happen!" To see why Rothstein was so indignant, find the odds of hitting the jackpots in three of the machines in quick succession, and comment on the likelihood of this happening.

Median and Mode

In addition to the mean, there are two other measures of central tendency of a group of numerical data: the median and the mode of a group of numbers.

Median

The **median** of a group of numbers arranged in increasing or decreasing order is (a) the middle number if there is an odd number of entries or (b) the mean of the two middle numbers if there is an even number of entries.

APPLIED EXAMPLE 11 Commuting Times

a. The times, in minutes, Susan took to go to work on nine consecutive working days were

46 42 49 40 52 48 45 43 50

What is the median of her morning commute times?

b. The times, in minutes, Susan took to return home from work on eight consecutive working days were

37 36 39 37 34 38 41 40

What is the median of her evening commute times?

Solution

a. Arranging the numbers in increasing order, we have

40 42 43 45 46 48 49 50 52

Here, we have an odd number of entries with the middle number equal to 46, and this gives the required median.

b. Arranging the numbers in increasing order, we have

34 36 37 37 38 39 40 41

Here, the number of entries is even, and the required median is

$$\frac{37 + 38}{2} = 37.5$$

Mode

The **mode** of a group of numbers is the number in the group that occurs most frequently.

Note

A group of numerical data may have no mode, a unique mode, or more than one mode.

EXAMPLE 12 Find the mode, if there is one, of the given group of numbers.

a. 1, 2, 3, 4, 6
b. 2, 3, 3, 4, 6, 8
c. 2, 3, 3, 3, 4, 4, 4, 8

Solution

a. The set has no mode because there isn't a number that occurs more frequently than the others.

b. The mode is 3 because it occurs more frequently than the others.
c. The modes are 3 and 4 because each number occurs three times.

Of the three measures of central tendency of a group of numerical data, the mean is by far the most suitable in work that requires mathematical computations.

8.2 Self-Check Exercises

1. Find the expected value of a random variable X having the following probability distribution:

x	-4	-3	-1	0	1	2
$P(X = x)$	.10	.20	.25	.10	.25	.10

2. **Forecasted Townhouse Sales** The developer of Shoreline Condominiums has provided the following estimate of the probability that 20, 25, 30, 35, 40, 45, or 50 of the townhouses will be sold within the first month they are offered for sale.

Units	20	25	30	35	40	45	50
Probability	.05	.10	.30	.25	.15	.10	.05

How many townhouses can the developer expect to sell within the first month they are put on the market?

Solutions to Self-Check Exercises 8.2 can be found on page 487.

8.2 Concept Questions

1. What is the expected value of a random variable? Give an example.
2. What is a fair game? Is the game of roulette as played in American casinos a fair game? Why or why not?
3. **a.** If the probability of an event E occurring is $P(E)$, what are the odds in favor of E occurring?
   b. If the odds in favor of an event occurring are a to b, what is the probability of E occurring?

8.2 Exercises

1. Find the expected value of a random variable X having the following probability distribution:

x	-5	-1	0	1	5	8
$P(X = x)$	.12	.16	.28	.22	.12	.10

2. Find the expected value of a random variable X having the following probability distribution:

x	0	1	2	3	4	5
$P(X = x)$	$\frac{1}{8}$	$\frac{1}{4}$	$\frac{3}{16}$	$\frac{1}{4}$	$\frac{1}{16}$	$\frac{1}{8}$

3. **Calculating GPA** During the first year at a university that uses a four-point grading system, a freshman took ten three-credit courses and received two As, three Bs, four Cs, and one D.
   a. Compute this student's grade-point average.
   b. Let the random variable X denote the number of points corresponding to a given letter grade. Find the probability distribution of the random variable X and compute $E(X)$, the expected value of X.
4. **Family Composition** In a four-child family, what is the expected number of boys? (Assume that the probability of a boy being born is the same as the probability of a girl being born.)
5. **Expected Sales** On the basis of past experience, the manager of the VideoRama Store has compiled the following table, which gives the probabilities that a customer who enters the VideoRama Store will buy 0, 1, 2, 3, or 4 DVDs. How many DVDs can a customer entering this store be expected to buy?

DVDs	0	1	2	3	4
Probability	.42	.36	.14	.05	.03

6. **Cafeteria Milk Consumption** Records kept by the chief dietitian at the university cafeteria over a 30-week period

show the following weekly consumption of milk (in gallons):

Milk	200	205	210	215	220
Weeks	3	4	6	5	4

Milk	225	230	235	240
Weeks	3	2	2	1

a. Find the average number of gallons of milk consumed per week in the cafeteria.

b. Let the random variable X denote the number of gallons of milk consumed in a week at the cafeteria. Find the probability distribution of the random variable X and compute $E(X)$, the expected value of X.

7. **Expected Earnings** The daily earnings X of an employee who works on a commission basis are given by the following probability distribution. Find the employee's expected earnings.

x ($)	0	25	50	75
$P(X = x)$	.07	.12	.17	.14

x ($)	100	125	150
$P(X = x)$	.28	.18	.04

8. **Expected Number of Defective Products** If a sample of three batteries is selected from a lot of ten, of which two are defective, what is the expected number of defective batteries?

9. **Expected Number of Auto Accidents** The numbers of accidents that occur at a certain intersection known as Five Corners on a Friday afternoon between the hours of 3 P.M. and 6 P.M., along with the corresponding probabilities, are shown in the following table. Find the expected number of accidents during the period in question.

Accidents	0	1	2	3	4
Probability	.935	.030	.020	.010	.005

10. **Expected Demand for Magazines** The owner of a newsstand in a college community estimates the weekly demand for a certain magazine as follows:

Quantity Demanded	10	11	12	13	14	15
Probability	.05	.15	.25	.30	.20	.05

Find the number of issues of the magazine that the newsstand owner can expect to sell per week.

11. **Expected ATM Reliability** A bank has two automatic teller machines at its main office and two at each of its three branches. The numbers of machines that break down on a given day, along with the corresponding probabilities, are shown in the following table:

Machines That Break Down	0	1	2	3	4
Probability	.43	.19	.12	.09	.04

Machines That Break Down	5	6	7	8
Probability	.03	.03	.02	.05

Find the expected number of machines that will break down on a given day.

12. **Expected Sales** The management of the Cambridge Company has projected the sales of its products (in millions of dollars) for the upcoming year, with the associated probabilities shown in the following table:

Sales	20	22	24	26	28	30
Probability	.05	.10	.35	.30	.15	.05

What are the expected sales for next year?

13. **Prime Interest-Rate Prediction** A panel of 50 economists was asked to predict the average prime interest rate for the upcoming year. The results of the survey follow:

Interest Rate (%)	2.9	3.0	3.1	3.2	3.3	3.4
Economists	3	8	12	14	8	5

On the basis of this survey, what does the panel expect the average prime interest rate to be next year?

14. **Forecasted Unemployment Rates** A panel of 64 economists was asked to predict the average unemployment rate for the upcoming year. The results of the survey follow:

Unemployment Rate (%)	6.5	6.6	6.7	6.8	6.9	7.0	7.1
Economists	2	4	8	20	14	12	4

On the basis of this survey, what does the panel expect the average unemployment rate to be next year?

15. **Expected Value of a Lottery Ticket** In a lottery, 5000 tickets are sold for $1 each. One first prize of $2000, 1 second prize of $500, 3 third prizes of $100, and 10 consolation prizes of $25 are to be awarded. What are the expected net earnings of a person who buys one ticket?

16. **Life Insurance Premiums** A man wishes to purchase a 5-year term-life insurance policy that will pay his beneficiary $20,000 in the event that his death occurs during the next 5 years. Using life insurance tables, he determines that the probability that he will live another 5 years is .96. What is the minimum amount that he can expect to pay for his premium?
Hint: The minimum premium occurs when the insurance company's expected profit is zero.

17. **Life Insurance Premiums** A woman purchased a $20,000, 1-year term-life insurance policy for $260. Assuming that the probability that she will live another year is .992, find the company's expected gain.

18. **Life Insurance Premiums** As a fringe benefit, Dennis Taylor receives a $50,000 life insurance policy from his employer. The probability that Dennis will live another year is .9935. If he purchases the same coverage for himself, what is the minimum amount that he can expect to pay for the policy? (See the Hint in Exercise 16.)

19. **Expected Profit of a Builder** Max built a spec house at a cost of $450,000. He estimates that he can sell the house for $580,000, $570,000, or $560,000, with probabilities .24, .40, and .36, respectively. What is Max's expected profit?

20. **Investment Analysis** The proprietor of Midland Construction Company needs to choose one of two projects. He estimates that the first project will yield a profit of $180,000 with a probability of .7 or a profit of $150,000 with a probability of .3; the second project will yield a profit of $220,000 with a probability of .6 or a profit of $80,000 with a probability of .4. Which project should the proprietor choose if he wants to maximize his expected profit?

21. **Cable Television Rights for a City** The management of MultiVision, a cable TV company, intends to submit a bid for the cable television rights in one of two cities, *A* or *B*. If the company obtains the rights to City *A*, the probability of which is .2, the estimated profit over the next 10 years is $10 million; if the company obtains the rights to City *B*, the probability of which is .3, the estimated profit over the next 10 years is $7 million. The cost of submitting a bid for rights in City *A* is $250,000, and that in City *B* is $200,000. By comparing the expected profits for each venture, determine whether the company should bid for the rights in City *A* or City *B*.

22. **Expected Auto Sales of a Dealership** Roger Hunt intends to purchase one of two car dealerships currently for sale in a certain city. Records obtained from each of the two dealers reveal that their weekly volume of sales, with corresponding probabilities, are as follows:

Dahl Motors

Cars Sold/Week	5	6	7	8
Probability	.05	.09	.14	.24

Cars Sold/Week	9	10	11	12
Probability	.18	.14	.11	.05

Farthington Auto Sales

Cars Sold/Week	5	6	7	8	9	10
Probability	.08	.21	.31	.24	.10	.06

The average profit per car at Dahl Motors is $543, and the average profit per car at Farthington Auto Sales is $654.

a. Find the average number of cars sold each week at each dealership.
b. If Roger's objective is to purchase the dealership that generates the higher weekly profit, which dealership should he purchase? (Compare the expected weekly profit for each dealership.)

23. **Expected Home Sales of a Realtor** Sally Leonard, a real estate broker, is relocating in a large metropolitan area where she has received job offers from Realty Company *A* and Realty Company *B*. The number of houses she expects to sell in a year at each firm and the associated probabilities are shown in the following tables:

Company *A*

Houses Sold	12	13	14	15	16
Probability	.02	.03	.05	.07	.07

Houses Sold	17	18	19	20
Probability	.16	.17	.13	.11

Houses Sold	21	22	23	24
Probability	.09	.06	.03	.01

Company *B*

Houses Sold	6	7	8	9	10
Probability	.01	.04	.07	.06	.11

Houses Sold	11	12	13	14
Probability	.12	.19	.17	.13

Houses Sold	15	16	17	18
Probability	.04	.03	.02	.01

The average price of a house in the locale of Company *A* is $308,000, whereas the average price of a house in the locale of Company *B* is $474,000. If Sally will receive a 3% commission on sales at either company, which job offer should she accept to maximize her expected yearly commission?

24. **Investment Analysis** Bob, the proprietor of Midway Lumber, bases his projections for the annual revenues of the company on the performance of the housing market. He rates the performance of the market as very strong, strong, normal, weak, or very weak. For the next year, Bob estimates that the probabilities for these outcomes are .18, .27, .42, .10, and .03, respectively. He also thinks that the revenues corresponding to these outcomes are $20, $18.8, $16.2, $14, and $12 million, respectively. What is Bob's expected revenue for next year?

25. **Expected Growth for a Business** Maria sees the growth of her business for the upcoming year as being tied to the gross domestic product (GDP). She believes that her business will grow (or contract) at the rate of 5%, 4.5%, 3%, 0%, or −0.5% per year if the GDP grows (or contracts) at the rate of between 2% and 2.5%, between 1.5% and 2%, between 1% and 1.5%, between 0% and 1%, and between −1% and 0%, respectively. Maria has decided to assign a

probability of .12, .24, .40, .20, and .04, respectively, to these outcomes. At what rate does Maria expect her business to grow next year?

26. **Weather Predictions** Suppose the probability that it will rain tomorrow is .3.
 a. What are the odds that it will rain tomorrow?
 b. What are the odds that it will not rain tomorrow?

27. **Expected Value of a Roulette Bet** In American roulette, as described in Example 6, a player may bet on a split (two adjacent numbers). In this case, if the player bets $1 and either number comes up, the player wins $17 and gets his $1 back. If neither comes up, he loses his $1 bet. Find the expected value of the winnings on a $1 bet placed on a split.

28. **Expected Value of a Roulette Bet** If a player placed a $1 bet on *red* and a $1 bet on *black* in a single play in American roulette, what would be the expected value of his winnings?

29. **Expected Value of a Roulette Bet** In European roulette, the wheel is divided into 37 compartments numbered 1 through 36 and 0. (In American roulette there are 38 compartments numbered 1 through 36, 0, and 00.) Find the expected value of the winnings on a $1 bet placed on *red* in European roulette.

30. **Male Commuting Times** The following table gives the travel time to work of male workers in the United States aged 16 years and over who do not work at home:

Travel Time, x (in minutes)	$x < 10$	$10 \le x < 15$	$15 \le x < 20$	$20 \le x < 25$	$25 \le x < 30$
Number (in thousands)	8,734	9,362	10,341	10,201	4,192

Travel Time, x (in minutes)	$30 \le x < 35$	$35 \le x < 45$	$45 \le x < 60$	$x \ge 60$
Number (in thousands)	9,991	4,681	5,729	6,638

Estimate the average time to work for male workers aged 16 years and over who do not work at home. (Assume that "$x < 10$" means "$0 < x < 10$" and "$x \ge 60$" means "$60 \le x \le 75$.")
Source: U.S. Census Bureau.

31. **Female Commuting Times** The following table gives the travel time to work of female workers in the United States aged 16 years and over who do not work at home:

Travel Time, x (in minutes)	$x < 10$	$10 \le x < 15$	$15 \le x < 20$	$20 \le x < 25$	$25 \le x < 30$
Number (in thousands)	9,049	9,611	10,172	9,423	3,807

Travel Time, x (in minutes)	$30 \le x < 35$	$35 \le x < 45$	$45 \le x < 60$	$x \ge 60$
Number (in thousands)	8,175	3,807	4,244	4,119

Estimate the average time to work for female workers aged 16 years and over who do not work at home. (Assume that "$x < 10$" means "$0 < x < 10$" and "$x \ge 60$" means "$60 \le x \le 75$.")
Source: U.S. Census Bureau.

32. **Population by Age in the United States** The resident population (in thousands) by age in the United States as of April 1, 2010, is summarized in the following table:

Age (in years)	Under 5	5–14	15–24	25–34	35–44	45–54
Population (in thousands)	20,201	41,026	43,626	41,064	41,071	45,007

Age (in years)	55–64	65–74	75–84	85–94	95 and over
Population (in thousands)	36,483	21,713	13,061	5,069	425

Estimate the average age of the resident population in the United States as of April 1, 2010. (Assume that "95 and over" means "95–104.")
Source: U.S. Census Bureau.

33. **Population by Age in California** The resident population (in thousands) by age in California as of April 1, 2010, is summarized in the following table:

Age (in years)	Under 5	5–14	15–24	25–34	35–44	45–54
Population (in thousands)	2531	5097	5590	5318	5183	5252

Age (in years)	55–64	65–74	75–84	85–94	95 and over
Population (in thousands)	4036	2275	1370	555	46

Estimate the average age of the resident population in California as of April 1, 2010. (Assume that "95 and over" means "95–104.")
Source: U.S. Census Bureau.

34. The probability of an event E occurring is .8. What are the odds in favor of E occurring? What are the odds against E occurring?

35. The probability of an event E not occurring is .6. What are the odds in favor of E occurring? What are the odds against E occurring?

36. The odds in favor of an event E occurring are 9 to 7. What is the probability of E occurring?

37. The odds against an event E occurring are 2 to 3. What is the probability of E not occurring?

38. **Odds of Making a Sale** Carmen, a computer sales representative, believes that the odds are 8 to 5 that she will clinch the sale of a minicomputer to a certain company. What is the (subjective) probability that Carmen will make the sale?

39. **Odds of Winning a Tennis Match** Steffi believes that the odds in favor of her winning her tennis match tomorrow are 7 to 5. What is the (subjective) probability that she will win her match tomorrow?

40. **Odds of Winning a Boxing Match** If a sports forecaster states that the odds of a certain boxer winning a match are 4 to 3, what is the (subjective) probability that the boxer will win the match?

41. **Odds of Closing a Business Deal** Bob, the proprietor of Midland Lumber, believes that the odds in favor of a business deal going through are 9 to 5. What is the (subjective) probability that this deal will *not* materialize?

42. **Expected Loss for Roulette Bet**
  a. Show that, for any number c,
$$E(cX) = cE(X)$$
  b. Use this result to find the expected loss if a gambler bets $300 on *red* in a single play in American roulette.
  Hint: Use the results of Example 6.

43. If X and Y are random variables and c is any constant, show that
  a. $E(c) = c$
  b. $E(cX) = cE(X)$
  c. $E(X + Y) = E(X) + E(Y)$
  d. $E(X - Y) = E(X) - E(Y)$

44. **Wage Rates** The frequency distribution of the hourly wage rates among blue-collar workers in a certain factory is given in the following table. Find the mean (or average) wage rate, the mode, and the median wage rate of these workers.

Wage Rate ($)	10.70	10.80	10.90	11.00	11.10	11.20
Frequency	60	90	75	120	60	45

45. **Exam Scores** In an examination given to a class of 20 students, the following test scores were obtained:

40 45 50 50 55 60 60 75 75 80
80 85 85 85 85 90 90 95 95 100

  a. Find the mean (or average) score, the mode, and the median score.
  b. Which of these three measures of central tendency do you think is the least representative of the set of scores?

46. **San Francisco Weather** The normal daily minimum temperatures in degrees Fahrenheit for the months of January through December in San Francisco follow:

46.2 48.4 48.6 49.2 50.7 52.5
53.1 54.2 55.8 54.8 51.5 47.2

Find the average and the median daily minimum temperatures in San Francisco for these months.
Source: San Francisco Convention and Visitors Bureau.

47. **Waiting Lines** Refer to Example 6, Section 8.1. Find the median of the number of cars waiting in line at the bank's drive-in teller at the beginning of each 2-min interval during the period in question. Compare your answer to the mean obtained in Example 1, Section 8.2.

48. **Boston Weather** The relative humidity, in percent, in the morning for the months of January through December in Boston follows:

68 67 69 69 71 73
74 76 79 77 74 70

Find the average and the median of these humidity readings.
Source: National Weather Service Forecast Office.

49. **Weight of Potato Chips** The weights, in ounces, of ten packages of potato chips are as follows:

16.1 16 15.8 16 15.9 16.1 15.9 16 16 16.2

Find the average and the median of these weights.

50. **Blood Types** The following table gives the top ten countries in the world whose populations have the highest concentration of type O^+ blood:

Country	Saudi Arabia	Iceland	Ireland	Taiwan	Australia
Population (%)	48.0	47.6	47.0	43.9	40.0

Country	Hong Kong	Italy	Netherlands	Canada	South Africa
Population (%)	40.0	39.5	39.0	39.0	38.0

Find the mean (average), median, and mode of these concentrations of type O^+ blood in these ten countries.
Source: Bloodbook.com.

51. **Blood Types** The following table gives the top 12 countries in the world whose populations have the highest concentration of type A^- blood.

Country	Brazil	Spain	Norway	Australia	Netherlands	Belgium
Population (%)	8.0	8.0	7.2	7.0	7.0	7.0

Country	United Kingdom	France	Denmark	Sweden	Austria	Portugal
Population (%)	7.0	7.0	7.0	7.0	7.0	6.6

Find the mean (average), median, and mode of these concentrations of type A^- blood in these 12 countries.
Source: Bloodbook.com.

In Exercises 52 and 53, determine whether the statement is true or false. If it is true, explain why it is true. If it is false, give an example to show why it is false.

52. If the odds in favor of an event E occurring are a to b, then the probability of E^c occurring is $b/(a + b)$.

53. A game between two people is fair if the expected value to both people is zero.

8.2 Solutions to Self-Check Exercises

1. $E(X) = (-4)(.10) + (-3)(.20) + (-1)(.25) + (0)(.10) + (1)(.25) + (2)(.10) = -0.8$

2. Let X denote the number of townhouses that will be sold within 1 month of being put on the market. Then the number of townhouses the developer expects to sell within 1 month is given by the expected value of X—that is, by

$$E(X) = 20(.05) + 25(.10) + 30(.30) + 35(.25) + 40(.15) + 45(.10) + 50(.05) = 34.25$$

or 34 townhouses.

8.3 Variance and Standard Deviation

Variance

The mean, or expected value, of a random variable enables us to express an important property of the probability distribution associated with the random variable in terms of a single number. But the knowledge of the location, or central tendency, of a probability distribution alone is usually not enough to give a reasonably accurate picture of the probability distribution. Consider, for example, the two probability distributions whose histograms appear in Figure 9. Both distributions have the same expected value, or mean, of $\mu = 4$ (the Greek letter μ is read "mu"). Note that the probability distribution with the histogram shown in Figure 9a is closely concentrated about its mean μ, whereas the one with the histogram shown in Figure 9b is widely dispersed or spread about its mean.

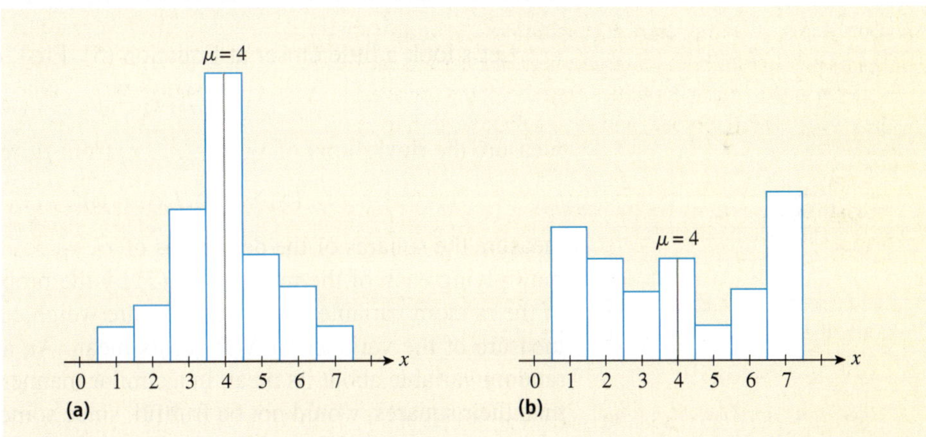

FIGURE **9**
The histograms of two probability distributions

As another example, suppose that Olivia has ten packages of Brand A potato chips and ten packages of Brand B potato chips. After carefully measuring the weights of each package, she obtains the following results:

Weight in Ounces										
Brand *A*	16.1	16	15.8	16	15.9	16.1	15.9	16	16	16.2
Brand *B*	16.3	15.7	15.8	16.2	15.9	16.1	15.7	16.2	16	16.1

In Example 3, we verify that the mean weights for each of the two brands is 16 ounces. However, a cursory examination of the data now shows that the weights of the Brand B packages exhibit much greater dispersion about the mean than do those of Brand A.

One measure of the degree of dispersion, or spread, of a probability distribution about its mean is given by the variance of the random variable associated with the probability distribution. A probability distribution with a small spread about its mean will have a small variance, whereas one with a larger spread will have a larger variance. Thus, the variance of the random variable associated with the probability distribution whose histogram appears in Figure 9a is smaller than the variance of the random variable associated with the probability distribution whose histogram is shown in Figure 9b (see Example 1). Also, as we will see in Example 3, the variance of the random variable associated with the weights of the Brand *A* potato chips is smaller than that of the random variable associated with the weights of the Brand *B* potato chips.

We now define the variance of a random variable.

Variance of a Random Variable X

Suppose a random variable has the probability distribution

x	x_1	x_2	x_3	$\cdots$	x_n
$P(X = x)$	p_1	p_2	p_3	$\cdots$	p_n

and expected value

$$E(X) = \mu$$

Then the **variance** of the random variable X is

$$\text{Var}(X) = p_1(x_1 - \mu)^2 + p_2(x_2 - \mu)^2 + \cdots + p_n(x_n - \mu)^2 \qquad \textbf{(5)}$$

Let's look a little closer at Equation (5). First, note that the numbers

$$x_1 - \mu, x_2 - \mu, \ldots, x_n - \mu \qquad \textbf{(6)}$$

measure the **deviations** of $x_1, x_2, \ldots, x_n$ from μ, respectively. Thus, the numbers

$$(x_1 - \mu)^2, (x_2 - \mu)^2, \ldots, (x_n - \mu)^2 \qquad \textbf{(7)}$$

measure the squares of the deviations of $x_1, x_2, \ldots, x_n$ from μ, respectively. Next, by multiplying each of the numbers in (7) by the probability associated with each value of the random variable X, the numbers are weighted accordingly so that their sum is a measure of the variance of X about its mean. An attempt to define the variance of a random variable about its mean in a similar manner using the deviations in (6), rather than their squares, would not be fruitful, since some of the deviations may be positive whereas others may be negative and hence (because of cancellations) the sum will not give a satisfactory measure of the variance of the random variable.

EXAMPLE 1 Find the variance of the random variable X and of the random variable Y whose probability distributions are shown in the following table. These are the probability distributions associated with the histograms shown in Figure 9a–b.

x	$P(X = x)$	y	$P(Y = y)$
1	.05	1	.2
2	.075	2	.15
3	.2	3	.1
4	.375	4	.15
5	.15	5	.05
6	.1	6	.1
7	.05	7	.25

Solution The mean of the random variable X is given by

$$\begin{aligned}\mu_X &= (1)(.05) + (2)(.075) + (3)(.2) + (4)(.375) + (5)(.15)\\ &\quad + (6)(.1) + (7)(.05)\\ &= 4\end{aligned}$$

Therefore, using Equation (5) and the data from the probability distribution of X, we find that the variance of X is given by

$$\begin{aligned}\text{Var}(X) &= (.05)(1-4)^2 + (.075)(2-4)^2 + (.2)(3-4)^2\\ &\quad + (.375)(4-4)^2 + (.15)(5-4)^2\\ &\quad + (.1)(6-4)^2 + (.05)(7-4)^2\\ &= 1.95\end{aligned}$$

Next, we find that the mean of the random variable Y is given by

$$\begin{aligned}\mu_Y &= (1)(.2) + (2)(.15) + (3)(.1) + (4)(.15) + (5)(.05)\\ &\quad + (6)(.1) + (7)(.25)\\ &= 4\end{aligned}$$

so the variance of Y is given by

$$\begin{aligned}\text{Var}(Y) &= (.2)(1-4)^2 + (.15)(2-4)^2 + (.1)(3-4)^2\\ &\quad + (.15)(4-4)^2 + (.05)(5-4)^2\\ &\quad + (.1)(6-4)^2 + (.25)(7-4)^2\\ &= 5.2\end{aligned}$$

Note that Var(X) is smaller than Var(Y), which confirms our earlier observations about the spread (or dispersion) of the probability distribution of X and Y, respectively.

Standard Deviation

Because Equation (5), which gives the variance of the random variable X, involves the squares of the deviations, the unit of measurement of Var(X) is the square of the unit of measurement of the values of X. For example, if the values assumed by the random variable X are measured in units of a gram, then Var(X) will be measured in units involving the *square* of a gram. To remedy this situation, one normally works with the square root of Var(X) rather than Var(X) itself. The former is called the standard deviation of X.

Standard Deviation of a Random Variable X

The **standard deviation** of a random variable X denoted σ (pronounced "sigma"), is defined by

$$\begin{aligned}\sigma &= \sqrt{\text{Var}(X)}\\ &= \sqrt{p_1(x_1-\mu)^2 + p_2(x_2-\mu)^2 + \cdots + p_n(x_n-\mu)^2} \qquad \textbf{(8)}\end{aligned}$$

where $x_1, x_2, \ldots, x_n$ denote the values assumed by the random variable X and $p_1 = P(X = x_1), p_2 = P(X = x_2), \ldots, p_n = P(X = x_n)$.

EXAMPLE 2 Find the standard deviations of the random variables X and Y of Example 1.

Solution From the results of Example 1, we have $\text{Var}(X) = 1.95$ and $\text{Var}(Y) = 5.2$. Taking their respective square roots, we have

$$\begin{aligned}\sigma_X &= \sqrt{1.95} \\ &\approx 1.40 \\ \sigma_Y &= \sqrt{5.2} \\ &\approx 2.28\end{aligned}$$

APPLIED EXAMPLE 3 Packaging Let X and Y denote the random variables whose values are the weights of the Brand A and Brand B potato chips, respectively (see page 487). Compute the means and standard deviations of X and Y and interpret your results.

Solution The probability distributions of X and Y may be computed from the given data as follows:

Brand A			Brand B		
x	Relative Frequency of Occurrence	$P(X = x)$	y	Relative Frequency of Occurrence	$P(Y = y)$
15.8	1	.1	15.7	2	.2
15.9	2	.2	15.8	1	.1
16.0	4	.4	15.9	1	.1
16.1	2	.2	16.0	1	.1
16.2	1	.1	16.1	2	.2
			16.2	2	.2
			16.3	1	.1

The means of X and Y are given by

$$\begin{aligned}\mu_X &= (.1)(15.8) + (.2)(15.9) + (.4)(16.0) + (.2)(16.1) \\ &\quad + (.1)(16.2) \\ &= 16 \\ \mu_Y &= (.2)(15.7) + (.1)(15.8) + (.1)(15.9) + (.1)(16.0) \\ &\quad + (.2)(16.1) + (.2)(16.2) + (.1)(16.3) \\ &= 16\end{aligned}$$

Therefore,

$$\begin{aligned}\text{Var}(X) &= (.1)(15.8 - 16)^2 + (.2)(15.9 - 16)^2 + (.4)(16 - 16)^2 \\ &\quad + (.2)(16.1 - 16)^2 + (.1)(16.2 - 16)^2 \\ &= 0.012 \\ \text{Var}(Y) &= (.2)(15.7 - 16)^2 + (.1)(15.8 - 16)^2 + (.1)(15.9 - 16)^2 \\ &\quad + (.1)(16 - 16)^2 + (.2)(16.1 - 16)^2 + (.2)(16.2 - 16)^2 \\ &\quad + (.1)(16.3 - 16)^2 \\ &= 0.042\end{aligned}$$

Explore and Discuss

A useful alternative formula for the variance is

$$\sigma^2 = E(X^2) - \mu^2$$

where $E(X^2)$ is the expected value of X^2.

1. Establish the validity of the formula.
2. Use the formula to verify the calculations in Example 3.

so the standard deviations are

$$\begin{aligned}\sigma_X &= \sqrt{\text{Var}(X)}\\ &= \sqrt{0.012}\\ &\approx 0.11\\ \sigma_Y &= \sqrt{\text{Var}(Y)}\\ &= \sqrt{0.042}\\ &\approx 0.20\end{aligned}$$

The mean of X and that of Y are both equal to 16. Therefore, the average weight of a package of potato chips of either brand is 16 ounces. However, the standard deviation of Y is greater than that of X. This tells us that the weights of the packages of Brand B potato chips are more widely dispersed about the common mean of 16 than are those of Brand A.

Explore and Discuss

Suppose the mean weight of m packages of Brand A potato chips is μ_1 and the standard deviation from the mean of their weight distribution is σ_1. Also suppose the mean weight of n packages of Brand B potato chips is μ_2 and the standard deviation from the mean of their weight distribution is σ_2.

1. Show that the mean of the combined weights of packages of Brand A and Brand B is

$$\mu = \frac{m\mu_1 + n\mu_2}{m + n}$$

2. If $\mu_1 = \mu_2$, show that the standard deviation from the mean of the combined-weight distribution is

$$\sigma = \left(\frac{m\sigma_1^2 + n\sigma_2^2}{m + n}\right)^{1/2}$$

3. Refer to Example 3, page 490. Using the results of parts 1 and 2, find the mean and the standard deviation of the combined-weight distribution.

The following example illustrates the technique for finding the standard deviation for grouped data.

APPLIED EXAMPLE 4 Married Males The following table gives the number of married males in the United States aged 15 years and over but less than 65 years in 2011:

Age (in years)	15–19	20–34	35–44	45–54	55–64
Number (in thousands)	11,220	32,206	20,308	21,990	18,346

Source: U.S. Census Bureau.

Find the mean and the standard deviation for these data.

Solution

Let X denote the random variable that measures the number of married males. Taking X to be the midpoint of a group interval, we obtain the following probability distribution:

x	17	27	39.5	49.5	59.5
$P(X = x)$	$\left(\frac{11,220}{104,070}\right)$	$\left(\frac{32,206}{104,070}\right)$	$\left(\frac{20,308}{104,070}\right)$	$\left(\frac{21,990}{104,070}\right)$	$\left(\frac{18,346}{104,070}\right)$

The mean of X is

$$\begin{aligned}\mu &= \left(\frac{11{,}220}{104{,}070}\right)(17) + \left(\frac{32{,}206}{104{,}070}\right)(27) + \left(\frac{20{,}308}{104{,}070}\right)(39.5)\\ &\quad + \left(\frac{21{,}990}{104{,}070}\right)(49.5) + \left(\frac{18{,}346}{104{,}070}\right)(59.5)\\ &\approx 38.8446\end{aligned}$$

Next, we see that

$$\begin{aligned}\text{Var}(X) &= \left(\frac{11{,}220}{104{,}070}\right)(17 - 38.8446)^2 + \left(\frac{32{,}206}{104{,}070}\right)(27 - 38.8446)^2\\ &\quad + \left(\frac{20{,}308}{104{,}070}\right)(39.5 - 38.8446)^2 + \left(\frac{21{,}990}{104{,}070}\right)(49.5 - 38.8446)^2\\ &\quad + \left(\frac{18{,}346}{104{,}070}\right)(59.5 - 38.8446)^2\\ &\approx 194.1483\end{aligned}$$

So the standard variation for these data is

$$\sigma = \sqrt{194.1483} \approx 13.93$$

Chebychev's Inequality

The standard deviation of a random variable X may be used in statistical estimations. For example, the following result, derived by the Russian mathematician P. L. Chebychev (1821–1894), gives a bound on the proportion of the values of X lying within k standard deviations of the expected value of X.

Chebychev's Inequality

Let X be a random variable with expected value μ and standard deviation σ. Then the probability that a randomly chosen outcome of the experiment lies between $\mu - k\sigma$ and $\mu + k\sigma$ is at least $1 - (1/k^2)$, where k is the number of standard deviations from the mean; that is,

$$P(\mu - k\sigma \le X \le \mu + k\sigma) \ge 1 - \frac{1}{k^2} \qquad \textbf{(9)}$$

To shed some light on this result, let's take $k = 2$ in Inequality (9) and compute

$$P(\mu - 2\sigma \le X \le \mu + 2\sigma) \ge 1 - \frac{1}{2^2} = 1 - \frac{1}{4} = .75$$

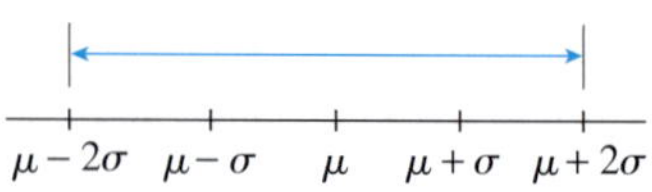

FIGURE 10
At least 75% of the outcomes fall within this interval.

This tells us that at least 75% of the outcomes of the experiment lie within 2 standard deviations of the mean (Figure 10). Taking $k = 3$ in Inequality (9), we have

$$P(\mu - 3\sigma \le X \le \mu + 3\sigma) \ge 1 - \frac{1}{3^2} = 1 - \frac{1}{9} = \frac{8}{9} \approx .89$$

This tells us that at least 89% of the outcomes of the experiment lie within 3 standard deviations of the mean (Figure 11).

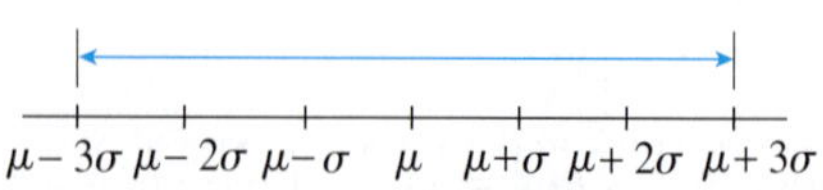

FIGURE 11
At least 89% of the outcomes fall within this interval.

EXAMPLE 5 A probability distribution has a mean of 10 and a standard deviation of 1.5. Use Chebychev's inequality to find a bound on the probability that an outcome of the experiment lies between 7 and 13.

Solution Here, $\mu = 10$ and $\sigma = 1.5$. To determine the value of k, note that $\mu - k\sigma = 7$ and $\mu + k\sigma = 13$. Substituting the appropriate values for μ and σ, we find $k = 2$. Using Chebychev's Inequality (9), we see that a bound on the probability that an outcome of the experiment lies between 7 and 13 is given by

$$P(7 \le X \le 13) \ge 1 - \left(\frac{1}{2^2}\right) = \frac{3}{4}$$

that is, at least 75%.

Note The results of Example 4 tell us that at least 75% of the outcomes of the experiment lie between $10 - 2\sigma$ and $10 + 2\sigma$—that is, between 7 and 13.

APPLIED EXAMPLE 6 Industrial Accidents Great Northwest Lumber Company employs 400 workers in its mills. It has been estimated that X, the random variable measuring the number of mill workers who have industrial accidents during a 1-year period, is distributed with a mean of 40 and a standard deviation of 6. Using Chebychev's Inequality (9), find a bound on the probability that the number of workers who will have an industrial accident over a 1-year period is between 30 and 50, inclusive.

Solution Here, $\mu = 40$ and $\sigma = 6$. We wish to estimate $P(30 \le X \le 50)$. To use Chebychev's Inequality (9), we first determine the value of k from the equation

$$\mu - k\sigma = 30 \quad \text{or} \quad \mu + k\sigma = 50$$

Since $\mu = 40$ and $\sigma = 6$ in this case, we see that k satisfies

$$40 - 6k = 30 \quad \text{and} \quad 40 + 6k = 50$$

from which we deduce that $k = \frac{5}{3}$. Thus, a bound on the probability that the number of mill workers who will have an industrial accident during a 1-year period is between 30 and 50 is given by

$$P(30 \le X \le 50) \ge 1 - \frac{1}{\left(\frac{5}{3}\right)^2} = \frac{16}{25}$$

that is, at least 64%.

8.3 Self-Check Exercises

1. Compute the mean, variance, and standard deviation of the random variable X with probability distribution as follows:

x	−4	−3	−1	0	2	5
$P(X = x)$	.1	.1	.2	.3	.1	.2

2. **Commute Times** James recorded the following commute times (the length of time in minutes it took him to drive to work) on ten consecutive days:

55 50 52 48 50 52 46 48 50 51

Calculate the mean and standard deviation of the random variable X associated with these data.

Solutions to Self-Check Exercises 8.3 can be found on page 499.

8.3 Concept Questions

1. a. What is the variance of a random variable X?
   b. What is the standard deviation of a random variable X?

2. What does Chebychev's inequality measure?

8.3 Exercises

In Exercises 1–6, the probability distribution of a random variable X is given. Compute the mean, variance, and standard deviation of X.

1.

x	1	2	3	4
$P(X = x)$	.4	.3	.2	.1

2.

x	−4	−2	0	2	4
$P(X = x)$	.1	.2	.3	.1	.3

3.

x	−2	−1	0	1	2
$P(X = x)$	1/16	4/16	6/16	4/16	1/16

4.

x	10	11	12	13	14	15
$P(X = x)$	1/8	2/8	1/8	2/8	1/8	1/8

5.

x	430	480	520	565	580
$P(X = x)$	.1	.2	.4	.2	.1

6.

x	−198	−195	−193	−188	−185
$P(X = x)$	.15	.30	.10	.25	.20

7. The following histograms represent the probability distributions of the random variables X and Y. Determine by inspection which probability distribution has the larger variance.

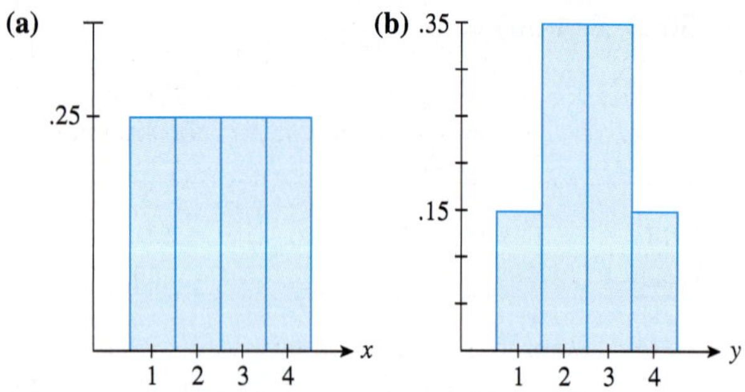

8. The following histograms represent the probability distributions of the random variables X and Y. Determine by inspection which probability distribution has the larger variance.

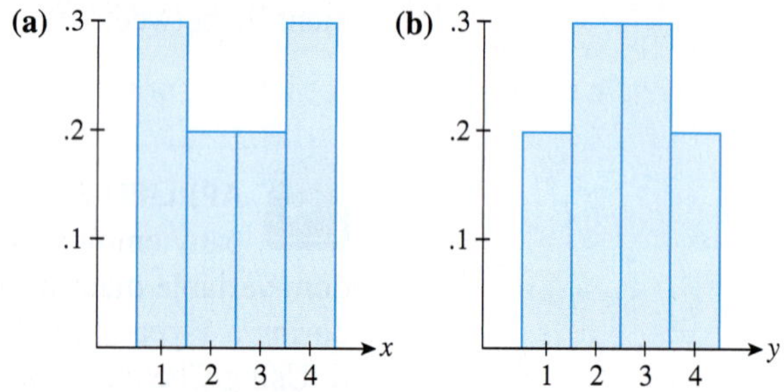

In Exercises 9 and 10, find the variance of the probability distribution for the histogram shown.

9.

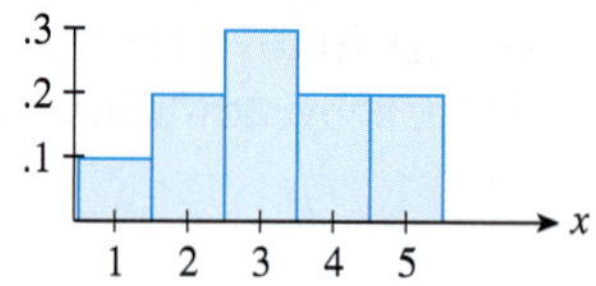

10.

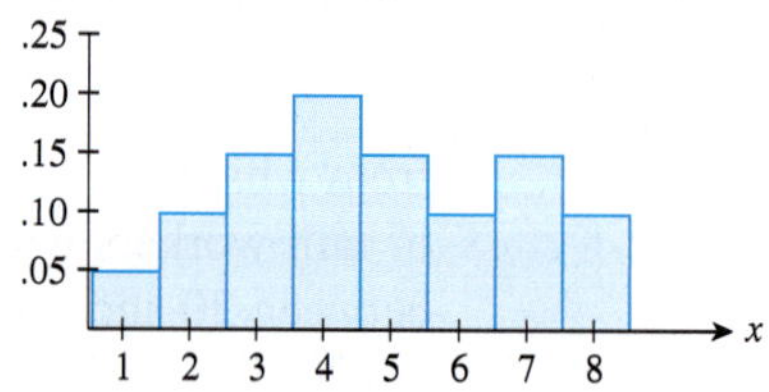

11. An experiment consists of rolling an eight-sided die (numbered 1 through 8) and observing the number that appears uppermost. Find the mean and variance of this experiment.

12. **Happiness Score** The happiness score, by generation, conducted in April 2013 is given in the following table (percent indicates the top-two box scores on a five-point happiness scale):

Generation	Average 18+	Millennials	Generation X	Boomer	Silent Generation
Score (%)	58.6	54.1	55.7	61.0	66.7

Find the average happiness score for the five generations. What is the standard deviation for these data?

Source: Prosper Insights & Analytics.

13. **Birthrates** The birthrates in the United States for the years 2003–2012 are given in the following table. (The birthrate is the number of live births/1000 population.)

Year	2003	2004	2005	2006	2007
Birthrate	14.7	14.0	14.0	14.2	14.2

Year	2008	2009	2010	2011	2012
Birthrate	14.0	13.8	13.8	13.8	13.7

a. Describe a random variable X that is associated with these data.
b. Find the probability distribution for the random variable X.
c. Compute the mean, variance, and standard deviation of X.

Source: National Center for Health Statistics.

14. **Investment Analysis** Paul Hunt is considering two business ventures. The anticipated returns (in thousands of dollars) of each venture are described by the following probability distributions:

Venture *A*

Earnings	Probability
−20	.3
40	.4
50	.3

Venture *B*

Earnings	Probability
−15	.2
30	.5
40	.3

a. Compute the mean and variance for each venture.
b. Which investment would provide Paul with the higher expected return (the greater mean)?
c. In which investment would the element of risk be less (that is, which probability distribution has the smaller variance)?

15. **Investment Analysis** Rosa Walters is considering investing \$10,000 in two mutual funds. The anticipated returns from price appreciation and dividends (in hundreds of dollars) are described by the following probability distributions:

Mutual Fund *A*

Returns	Probability
−4	.2
8	.5
10	.3

Mutual Fund *B*

Earnings	Probability
−2	.2
6	.4
8	.4

a. Compute the mean and variance associated with the returns for each mutual fund.
b. Which investment would provide Rosa with the higher expected return (the greater mean)?
c. In which investment would the element of risk be less (that is, which probability distribution has the smaller variance)?

16. The distribution of the number of chocolate chips (x) in a cookie is shown in the following table. Find the mean and the variance of the number of chocolate chips in a cookie.

x	0	1	2	3	4
$P(X = x)$	.01	.03	.05	.11	.13

x	5	6	7	8
$P(X = x)$	.24	.22	.16	.05

17. Equation (5) can also be expressed in the form

$$\text{Var}(X) = (p_1x_1^2 + p_2x_2^2 + \cdots + p_nx_n^2) - \mu^2$$

Find the variance of the distribution of Exercise 1 using this equation.

18. Find the variance of the distribution of Exercise 16 using the equation

$$\text{Var}(X) = (p_1x_1^2 + p_2x_2^2 + \cdots + p_nx_n^2) - \mu^2$$

19. **Stuck in Traffic** The following table gives the extra travel time in hours for peak-period travelers in urban areas with more than 3 million people in a certain year:

Urban Area	Annual Hours of Delay per Traveler
Los Angeles–Long Beach–Santa Ana, CA	70
San Francisco–Oakland, CA	55
Atlanta, GA	57
Washington (D.C.–VA–MD)	62
Dallas–Fort Worth–Arlington, TX	53
Houston, TX	56

Find the mean of the extra travel time in that year, in hours, for peak-period travelers in urban areas with more than 3 million people. What is the standard deviation for these data?

Source: Texas Transportation Institute.

20. **Cost of Taking Time Off** A survey was conducted of graduates of Harvard College 15 years after graduation. In the survey, the pay of graduates in different fields who had previously taken off 18 months, often to care for children, was compared with pay for graduates who had not taken time off. The average financial penalty for those who had taken time off is summarized in the following table:

Field	M.B.A.	J.D.	Ph.D	B.A. only	M.D.	Other, Masters only
Penalty (%)	−41	−29	−29	−25	−16	−13

Find the mean of the financial penalty for the graduates who had taken time off. What is the standard deviation for these data?

Source: Claudia Golden and Lawrence Katz, Harvard College.

21. **Conviction Rates** The following table gives the percentage of homicide cases in Suffolk County, Massachusetts, ending in pleas or verdicts of guilty from 2004 through 2009:

Year	2004	2005	2006	2007	2008	2009
Conviction Rate (%)	81	91	82	75	82	95

Find the mean of the percentage of homicide cases in Suffolk County ending in pleas or verdicts of guilty from 2004 through 2009. What is the standard deviation for these data?

Source: Suffolk County, Massachusetts, District Attorney's office.

22. **New York State Courts' Total Caseload** The following table gives the total caseload in the New York State courts from 2004 through 2009.

Year	2004	2005	2006	2007	2008	2009
Cases (in millions)	4.2	4.3	4.6	4.5	4.7	4.7

Find the mean of the total caseload in the New York State courts from 2004 through 2009. What is the standard deviation for these data?

Source: New York State Office of Court Administration.

23. **Hours Worked in Some Countries** The number of average hours worked per year per worker in the United States and five European countries in a certain year is given in the following table:

Country	U.S.	Spain	Great Britain	France	West Germany	Norway
Average Hours Worked	1815	1807	1707	1545	1428	1342

Find the average of the average hours worked per worker in that year for workers in the six countries. What is the standard deviation for these data?

Source: Office of Economic Cooperation and Development.

24. **Identity Fraud** The identity fraud rates in the United States for the years 2005–2011 are given in the following table:

Year	2005	2006	2007	2008	2009	2010	2011
Incidence Rate (%)	5.04	4.71	4.51	5.44	6.00	4.35	4.90

Find the average incidence rate of identity fraud for the years 2005–2011. What is the standard deviation?

Source: FBI.

25. **Health Issues in Massachusetts Cities** A random survey of health issues, conducted by the Department of Public Health of the Commonwealth of Massachusetts, examined the results from the state's seven largest cities. These cities were selected on the basis of their diverse racial and ethnic populations. The percentage of adults reporting fair or poor health for each city in the survey is given in the following table:

City	Boston	Worcester	Springfield	Lowell
Adults Reporting Fair or Poor Health (%)	16.3	15.4	22.2	17.2

City	Fall River	Lawrence	New Bedford
Adults Reporting Fair or Poor Health (%)	23.2	30.4	26.4

Find the average percentage of adults reporting fair or poor health for the seven cities. What is the standard deviation for these data?

Source: Massachusetts Department of Public Health.

26. **Diabetes in Massachusetts Cities** A random survey of health issues, conducted by the Department of Public Health of the Commonwealth of Massachusetts, examined the results from the state's seven largest cities. These cities were selected on the basis of their diverse racial and ethnic populations. The percentage of adults with diabetes in each city in the survey is given in the following table:

City	Boston	Worcester	Springfield	Lowell
Adults with Diabetes (%)	7.2	8.2	12.1	8.7

City	Fall River	Lawrence	New Bedford
Adults with Diabetes (%)	11.1	10.9	9.3

Find the average percentage of adults with diabetes in these seven cities. What is the standard deviation for these data?

Source: Massachusetts Department of Public Health.

27. **Hybrid Vehicle Mileage** The following table gives the mileage (in miles per gallon) of the eight 2013 model hybrid vehicles with the highest combined mileages:

Model	Toyota Prius	Ford C-Max	Ford Fusion	Lincoln MKZ	Volkswagen Jetta
Mileage (in mpg)	50	47	47	45	45

Model	Honda Insight	Lexus CT 200h	Lexus ES 300h
Mileage (in mpg)	42	42	40

Find the average mileage of the eight 2013 hybrid vehicles. What is the standard deviation for these data?

Source: U.S. Department of Energy.

28. **Housing Prices** The market research department of the National Real Estate Company conducted a survey among 500 prospective buyers in a suburb of a large metropolitan area to determine the maximum price a prospective buyer would be willing to pay for a house. From the data collected, the distribution that follows was obtained:

Maximum Price Considered, x	$P(X = x)$
480	$\frac{10}{500}$
490	$\frac{20}{500}$
500	$\frac{75}{500}$
510	$\frac{85}{500}$
520	$\frac{70}{500}$
550	$\frac{90}{500}$
580	$\frac{90}{500}$
600	$\frac{55}{500}$
650	$\frac{5}{500}$

Compute the mean, variance, and standard deviation of the maximum price x (in thousands of dollars) that these buyers were willing to pay for a house.

29. **Government Debt** The following table gives the projected debt as a percentage of the gross domestic product (GDP) of nine selected countries for 2011. The study was conducted by the Organization for Economic Co-operation and Development (OECD) in early 2010.

Country	Spain	U.S.	Germany	Portugal
GDP (%)	67	72	83	88

Country	U.K.	France	Japan	Italy	Greece
GDP (%)	89	91	113	121	127

Find the mean of the projected debt as a percentage of GDP of the nine countries under consideration. What is the standard deviation for these data?

Source: OECD.

30. **On-Time Arrivals** The following table gives the percentage of on-time arrivals in U.S. airports from 2004 through 2013:

Year	2004	2005	2006	2007	2008	2009	2010	2011	2012	2013
On-time Arrivals (%)	76.13	74.35	77.11	70.33	70.55	79.69	76.75	75.46	84.93	80.33

Find the average percentage of on-time arrivals over the 10 years from 2004 through 2013. What is the standard deviation for these data?

Source: U.S. Department of Transportation.

31. **Flight Cancellations** The following table gives the percentage of flights canceled by U.S. carriers from 2004 through 2013:

Year	2004	2005	2006	2007	2008	2009	2010	2011	2012	2013
Flights Canceled (%)	2.38	3.03	1.89	3.47	3.23	1.81	3.89	4.35	1.24	1.93

Find the average percentage of flights canceled over the 10 years from 2004 through 2013. What is the standard deviation for these data?

Source: U.S. Department of Transportation.

32. **Access to Capital** One of the key determinants of economic growth is access to capital. Using 54 variables to create an index of 1–7, with 7 being best possible access to capital, Milken Institute ranked the following as the top ten nations (although technically Hong Kong is not a nation) by the ability of their entrepreneurs to gain access to capital:

Country	Hong Kong	Netherlands	U.K.	Singapore	Switzerland
Index	5.70	5.59	5.57	5.56	5.55

Country	U.S.	Australia	Finland	Germany	Denmark
Index	5.55	5.31	5.24	5.23	5.22

Find the mean of the indices of the top ten nations. What is the standard deviation for these data?

Source: Milken Institute.

33. **Access to Capital** Refer to Exercise 32. Milken Institute also ranked the following as the ten worst-performing nations by the ability of their entrepreneurs to gain access to capital:

Country	Peru	Mexico	Bulgaria	Brazil	Indonesia
Index	3.76	3.70	3.66	3.50	3.46

Country	Colombia	Turkey	Argentina	Venezuela	Russia
Index	3.46	3.43	3.20	2.88	2.19

Find the mean of the indices of the ten worst-performing nations. What is the standard deviation for these data?

Source: Milken Institute.

34. **Lightning Injuries** The number of injuries due to lightning in the United States from 1999 through 2008 is given in the following table:

Year	1999	2000	2001	2002	2003	2004	2005	2006	2007	2008
Number	243	371	372	256	238	279	309	245	139	207

What is the average number of injuries per year due to lightning in the United States from 1999 through 2008? What is the standard deviation for these data?

Source: National Oceanic and Atmosphere Administration.

35. **Federal Libraries** The federal government planned to spend more than $68.7 million in 2013 on programming, operations, and maintenance of 13 presidential libraries, including $3.5 million for the central management office. The total cost to be spent on each library is shown in the following table:

Library	Hoover	Roosevelt	Truman	Eisenhower	Kennedy	Johnson
Expenditure ($) (in millions)	2.4	4.8	4.4	3.8	7.0	5.7

Library	Nixon	Ford	Carter	Reagan	G.H.W. Bush	Clinton	G.W. Bush
Expenditure ($) (in millions)	4.9	5.2	4.7	5.7	5.4	5.3	5.9

Find the mean of the federal government expenditure for the 13 presidential libraries in 2013. What is the standard deviation for these data?

Source: National Archives and Records Administration.

36. **Election Turnout** The percentage of the voting-age population who cast ballots in presidential elections from 1932 through 2008 are given in the following table:

Election Year	1932	1936	1940	1944	1948	1952	1956	1960	1964	1968
Turnout (%)	53	57	59	56	51	62	59	59	62	61

Election Year	1972	1976	1980	1984	1988	1992	1996	2000	2004	2008
Turnout (%)	55	54	53	53	50	55	49	51	55	57

Find the mean and the standard deviation of the given data.
Source: Federal Election Commission.

37. **Existing-Home Sales**

a. The monthly supply (in millions) of single-family homes for sale in 2011 is summarized in the following table:

Month	Jan.	Feb.	Mar.	Apr.	May	June	July	Aug.	Sept.	Oct.	Nov.	Dec.
Supply	7.7	8.3	8.3	8.9	8.9	9.0	8.9	8.1	7.9	7.4	7.1	6.2

Find the mean and the standard deviation for the given data.

b. The monthly supply (in millions) of single-family homes for sale in 2012 is summarized in the following table:

Month	Jan.	Feb.	Mar.	Apr.	May	June	July	Aug.	Sept.	Oct.	Nov.	Dec.
Supply	6.1	6.1	6.1	6.4	6.4	6.5	6.3	6.0	5.5	5.3	4.8	4.4

Find the mean and the standard deviation of the given data.

c. What does a comparison of the means of the housing supplies for the two years tell you about the recovery of the Great Recession of 2009?

Source: Los Angeles Times.

38. **Exam Scores** The following table gives the scores of 30 students in a mathematics examination:

Scores	90–99	80–89	70–79	60–69	50–59
Students	4	8	12	4	2

Find the mean and the standard deviation of the distribution of the given data.
Hint: Assume that all scores lying within a group interval take the middle value of that group.

39. **Marital Status of Men** The number of married men (in thousands) between the ages of 20 and 44 in the United States in 2010 is given in the following table:

Age	20–24	25–29	30–34	35–39	40–44
Men	1182	3810	6104	7124	8195

Find the mean and the standard deviation of the given data.
Hint: See the hint for Exercise 38.
Source: U.S. Census Bureau.

40. **Traffic Survey** In a survey of the distances between vehicles traveling along a stretch of Interstate Highway 5, the following data were obtained:

Distance, x (in feet)	$20 \le x < 50$	$50 \le x < 80$	$80 \le x < 110$
Frequency	232	410	143

Distance, x (in feet)	$110 \le x < 140$	$140 \le x < 170$	$170 \le x < 200$
Frequency	84	26	10

Find the mean and standard deviation for these data.

41. A probability distribution has a mean of 42 and a standard deviation of 2. Use Chebychev's inequality to find a bound on the probability that an outcome of the experiment lies between:
a. 38 and 46. b. 32 and 52.

42. A probability distribution has a mean of 20 and a standard deviation of 3. Use Chebychev's inequality to find a bound on the probability that an outcome of the experiment lies between:
a. 15 and 25. b. 10 and 30.

43. A probability distribution has a mean of 50 and a standard deviation of 1.4. Use Chebychev's inequality to find the value of c that guarantees the probability is at least 96% that an outcome of the experiment lies between $50 - c$ and $50 + c$.

44. Suppose X is a random variable with mean μ and standard deviation σ. If a large number of trials is observed, at least what percentage of these values is expected to lie between $\mu - 2\sigma$ and $\mu + 2\sigma$?

45. **Product Reliability** The deluxe ionic hair dryer produced by Roland Electric has a mean expected lifetime of 24 months with a standard deviation of 3 months. Find a bound on the probability that one of these hair dryers will last between 20 and 28 months.

46. **Product Reliability** A Christmas tree light has an expected life of 200 hr with a standard deviation of 2 hr.

a. Find a bound on the probability that one of these Christmas tree lights will last between 190 hr and 210 hr.

b. Suppose a large city uses 150,000 of these Christmas tree lights as part of its Christmas decorations. Estimate the number of lights that are likely to require replacement between 180 hr and 220 hr of use.

47. **Starting Salaries** The mean annual starting salary of a new graduate in a certain profession is \$58,000 with a standard deviation of \$500. Find a bound on the probability that the starting salary of a new graduate in this profession will be between \$56,000 and \$60,000.

48. **Quality Control** Sugar packaged by a certain machine has a mean weight of 5 lb and a standard deviation of 0.02 lb. For what values of c can the manufacturer of the machinery claim that the sugar packaged by this machine has a weight between $5 - c$ and $5 + c$ lb with probability at least 96%?

In Exercises 49 and 50, determine whether the statement is true or false. If it is true, explain why it is true. If it is false, give an example to show why it is false.

49. Both the variance and the standard deviation of a random variable measure the spread of a probability distribution.

50. Chebychev's inequality is useless when $k \leq 1$.

8.3 Solutions to Self-Check Exercises

1. The mean of the random variable X is

$$\begin{aligned}\mu &= (-4)(.1) + (-3)(.1) + (-1)(.2)\\ &\quad + (0)(.3) + (2)(.1) + (5)(.2)\\ &= 0.3\end{aligned}$$

The variance of X is

$$\begin{aligned}\text{Var}(X) &= (.1)(-4 - 0.3)^2 + (.1)(-3 - 0.3)^2\\ &\quad + (.2)(-1 - 0.3)^2 + (.3)(0 - 0.3)^2\\ &\quad + (.1)(2 - 0.3)^2 + (.2)(5 - 0.3)^2\\ &= 8.01\end{aligned}$$

The standard deviation of X is

$$\sigma = \sqrt{\text{Var}(X)} = \sqrt{8.01} \approx 2.83$$

2. We first compute the probability distribution of X from the given data as follows:

x	Relative Frequency of Occurrence	$P(X = x)$
46	1	.1
48	2	.2
50	3	.3
51	1	.1
52	2	.2
55	1	.1

The mean of X is

$$\begin{aligned}\mu &= (.1)(46) + (.2)(48) + (.3)(50)\\ &\quad + (.1)(51) + (.2)(52) + (.1)(55)\\ &= 50.2\end{aligned}$$

The variance of X is

$$\begin{aligned}\text{Var}(X) &= (.1)(46 - 50.2)^2 + (.2)(48 - 50.2)^2\\ &\quad + (.3)(50 - 50.2)^2 + (.1)(51 - 50.2)^2\\ &\quad + (.2)(52 - 50.2)^2 + (.1)(55 - 50.2)^2\\ &= 5.76\end{aligned}$$

from which we deduce the standard deviation

$$\begin{aligned}\sigma &= \sqrt{5.76}\\ &= 2.4\end{aligned}$$

USING TECHNOLOGY Finding the Mean and Standard Deviation

The calculation of the mean and standard deviation of a random variable is facilitated by the use of a graphing utility.

APPLIED EXAMPLE 1 Age Distribution of Company Directors A survey conducted in a certain year of the Fortune 1000 companies revealed the following age distribution of the company directors:

Age (in years)	20–24	25–29	30–34	35–39	40–44	45–49	50–54
Directors	1	6	28	104	277	607	1142

Age (in years)	55–59	60–64	65–69	70–74	75–79	80–84	85–89
Directors	1413	1424	494	159	62	31	5

Let X denote the random variable taking on the values 1 through 14, where 1 corresponds to the age bracket 20–24, 2 corresponds to the age bracket 25–29, and so on.

a. Plot a histogram for the given data.
b. Find the mean and the standard deviation of these data. Interpret your results.
Source: Directorship.

Solution

a. Enter the values of X as $x_1 = 1, x_2 = 2, \ldots, x_{14} = 14$ and the corresponding values of Y as $y_1 = 1, y_2 = 6, \ldots, y_{14} = 5$. Then using the **DRAW** function from the Statistics menu of a graphing utility, we obtain the histogram shown in Figure T1.

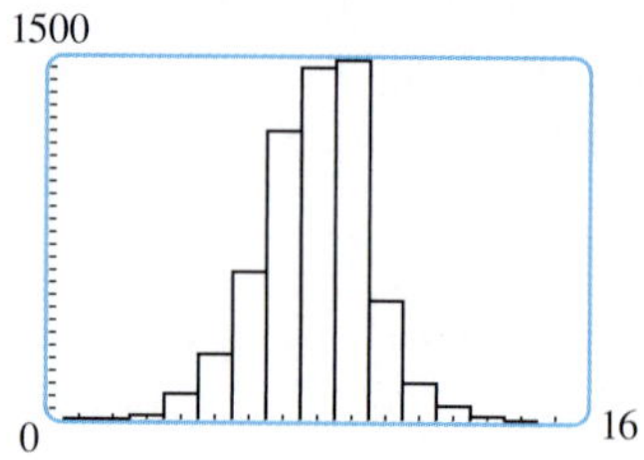

FIGURE **T1**
The histogram for the given data, using the viewing window [0, 16] × [0, 1500]

b. Using the appropriate function from the Statistics menu, we find that $\bar{x} \approx 7.9193$ and $\sigma x \approx 1.6378$; that is, the mean of X is $\mu \approx 7.9$, and the standard deviation is $\sigma \approx 1.6$. Interpreting our results, we see that the average age of the directors is in the 55- to 60-year-old bracket.

TECHNOLOGY EXERCISES

1. **a.** Graph the histogram associated with the random variable X in Example 1, page 488.
   b. Find the mean and the standard deviation for these data.

2. **a.** Graph the histogram associated with the random variable Y in Example 1, page 488.
   b. Find the mean and the standard deviation for these data.

3. **Driving Age Requirements** The minimum age requirement for a regular driver's license differs from state to state. The frequency distribution for this age requirement in the 50 states is given in the following table:

Minimum Age (in years)	15	16	17	18	19	21
Frequency of Occurrence	1	15	4	28	1	1

   a. Graph the histogram associated with the random variable X associated with these data.
   b. Find the mean and the standard deviation for these data.

4. **a.** Graph the histogram associated with the data given in Exercise 16, page 495.
   b. Find the mean and the standard deviation for these data.

5. A sugar refiner uses a machine to pack sugar in 5-lb cartons. To check the machine's accuracy, cartons are selected at random and weighed. The results follow:

4.98	5.02	4.96	4.97	5.03
4.96	4.98	5.01	5.02	5.06
4.97	5.04	5.04	5.01	4.99
4.98	5.04	5.01	5.03	5.05
4.96	4.97	5.02	5.04	4.97
5.03	5.01	5.00	5.01	4.98

   a. Describe a random variable X that is associated with these data.
   b. Find the probability distribution for the random variable X.
   c. Compute the mean and standard deviation of X.

6. The scores of 25 students in a mathematics examination are as follows:

90	85	74	92	68	94	66
87	85	70	72	68	73	72
69	66	58	70	74	88	90
98	71	75	68			

a. Describe a random variable X that is associated with these data.
b. Find the probability distribution for the random variable X.
c. Compute the mean and standard deviation of X.

7. **Heights of Women** The following data, obtained from the records of the Westwood Health Club, give the heights (to the nearest inch) of 200 female members of the club:

Height (in inches)	62	$62\frac{1}{2}$	63	$63\frac{1}{2}$	64	$64\frac{1}{2}$	65	$65\frac{1}{2}$	66	
Frequency	2	3	4	8	11	20	32	30	18	
Height (in inches)	$66\frac{1}{2}$	67	$67\frac{1}{2}$	68	$68\frac{1}{2}$	69	$69\frac{1}{2}$	70	$70\frac{1}{2}$	71
Frequency	18	16	8	10	5	5	4	3	2	1

a. Plot a histogram for the given data.
b. Find the mean and the standard deviation of these data.

8. **Age Distribution in a Town** The following table gives the distribution of the ages of the residents of the town of Monroe who are under the age of 40 years:

Age (in years)	0–3	4–7	8–11	12–15	16–19
Residents (in hundreds)	30	42	50	60	50
Age (in years)	20–23	24–27	28–31	32–35	36–39
Residents (in hundreds)	41	50	45	42	34

Let X denote the random variable taking on the values 1 through 10, where 1 corresponds to the range 0–3, . . . , and 10 corresponds to the range 36–39.
a. Plot a histogram for the given data.
b. Find the mean and the standard deviation of X.

8.4 The Binomial Distribution

Bernoulli Trials

An important class of experiments have (or may be viewed as having) two outcomes. For example, in a coin-tossing experiment, the two outcomes are *heads* and *tails*. In the card game played by Mike and Bill (Example 7, Section 8.2), one may view the selection of a diamond as a *win* (for Mike) and the selection of a card of another suit as a *loss* for Mike. For a third example, consider an experiment in which a person is inoculated with a flu vaccine. Here, the vaccine may be classified as being "effective" or "ineffective" with respect to that particular person.

In general, experiments with two outcomes are called **Bernoulli trials**, or **binomial trials.** It is standard practice to label one of the outcomes of a binomial trial a *success* and the other a *failure*. For example, in a coin-tossing experiment, the outcome a *head* may be called a success, in which case the outcome a *tail* is called a failure. Note that by using the terms *success* and *failure* in this way, we depart from their usual connotations.

A sequence of Bernoulli (binomial) trials is called a binomial experiment. More precisely, we have the following definition:

> **Binomial Experiment**
>
> A **binomial experiment** has the following properties:
>
> 1. The number of trials in the experiment is fixed.
> 2. There are two outcomes of each trial: "success" and "failure."
> 3. The probability of success in each trial is the same.
> 4. The trials are independent of each other.

In a binomial experiment, it is customary to denote the number of trials by n, the probability of a success by p, and the probability of a failure by q. Because the event of a success and the event of a failure are complementary events, we have the relationship

$$p + q = 1$$

or, equivalently,

$$q = 1 - p$$

The following example involves a binomial experiment.

EXAMPLE 1 A fair die is rolled four times. Compute the probability of obtaining exactly one 6 in the four throws.

Solution There are four trials in this experiment. Each trial consists of rolling the die once and observing the face that lands uppermost. We may view each trial as an experiment with two outcomes: a success (S) if the face that lands uppermost is a 6 and a failure (F) if it is any of the other five numbers. Letting p and q denote the probability of success and failure, respectively, of a single trial of the experiment, we find that

$$p = \frac{1}{6} \quad \text{and} \quad q = 1 - \frac{1}{6} = \frac{5}{6}$$

Furthermore, we may assume that the trials of this experiment are independent. Thus, we have a binomial experiment.

With the aid of the multiplication principle, we see that the experiment has 2^4, or 16, outcomes. We can obtain these outcomes by constructing the tree diagram associated with the experiment (see Table 11, where the outcomes are listed according to the number of successes). From the table, we see that the event of obtaining exactly one success in four trials is given by

$$E = \{\text{SFFF, FSFF, FFSF, FFFS}\}$$

with probability given by

$$P(E) = P(\text{SFFF}) + P(\text{FSFF}) + P(\text{FFSF}) + P(\text{FFFS}) \tag{10}$$

TABLE 11

0 Success	1 Success	2 Successes	3 Successes	4 Successes
FFFF	SFFF	SSFF	SSSF	SSSS
	FSFF	SFSF	SSFS	
	FFSF	SFFS	SFSS	
	FFFS	FSSF	FSSS	
		FSFS		
		FFSS		

Since the trials (throws) are independent, the terms on the right-hand side of Equation (10) may be computed as follows:

$$\begin{aligned}
P(\text{SFFF}) &= P(\text{S})P(\text{F})P(\text{F})P(\text{F}) = p \cdot q \cdot q \cdot q = pq^3 \\
P(\text{FSFF}) &= P(\text{F})P(\text{S})P(\text{F})P(\text{F}) = q \cdot p \cdot q \cdot q = pq^3 \\
P(\text{FFSF}) &= P(\text{F})P(\text{F})P(\text{S})P(\text{F}) = q \cdot q \cdot p \cdot q = pq^3 \\
P(\text{FFFS}) &= P(\text{F})P(\text{F})P(\text{F})P(\text{S}) = q \cdot q \cdot q \cdot p = pq^3
\end{aligned}$$

Therefore, upon substituting these values in Equation (10), we obtain

$$\begin{aligned} P(E) &= pq^3 + pq^3 + pq^3 + pq^3 = 4pq^3 \\ &= 4\left(\frac{1}{6}\right)\left(\frac{5}{6}\right)^3 \approx .386 \end{aligned}$$

Probabilities in Bernoulli Trials

Let's reexamine the computations we performed in the last example. There, it was found that the probability of obtaining exactly one success in a binomial experiment with four independent trials with probability of success in a single trial p is given by

$$P(E) = 4pq^3 \qquad (\text{where } q = 1 - p) \tag{11}$$

Observe that the coefficient 4 of pq^3 appearing in Equation (11) is precisely the number of outcomes of the experiment with exactly one success and three failures, the outcomes being

SFFF FSFF FFSF FFFS

Another way of obtaining this coefficient is to think of the outcomes as arrangements of the letters S and F. Then the number of ways of selecting one position for S from four possibilities is given by

$$\begin{aligned} C(4, 1) &= \frac{4!}{1!\,(4-1)!} \\ &= 4 \end{aligned}$$

Next, observe that because the trials are independent, each of the four outcomes of the experiment has the same probability, given by

$$pq^3$$

where the exponents 1 and 3 of p and q, respectively, correspond to exactly one success and three failures in the trials that make up each outcome.

As a result of the foregoing discussion, we may write Equation (11) as

$$P(E) = C(4, 1)pq^3 \tag{12}$$

We are also in a position to generalize this result. Suppose that in a binomial experiment the probability of success in any trial is p. What is the probability of obtaining exactly x successes in n independent trials? We start by counting the number of outcomes of the experiment, each of which has exactly x successes. Now, one such outcome involves x successive successes followed by $(n - x)$ failures—that is,

$$\underbrace{\text{SS}\cdots\text{S}}_{x}\,\underbrace{\text{FF}\cdots\text{F}}_{n-x} \tag{13}$$

The other outcomes, each of which has exactly x successes, are obtained by rearranging the S's (x of them) and F's ($n - x$ of them). There are $C(n, x)$ ways of arranging these letters. Next, arguing as in Example 1, we see that each such outcome has probability given by

$$p^x q^{n-x}$$

For example, for the outcome (13), we find

$$\begin{aligned} P(\underbrace{\text{SS}\cdots\text{S}}_{x}\,\underbrace{\text{FF}\cdots\text{F}}_{(n-x)}) &= \underbrace{P(\text{S})P(\text{S})\cdots P(\text{S})}_{x}\,\underbrace{P(\text{F})P(\text{F})\cdots P(\text{F})}_{(n-x)} \\ &= \underbrace{pp\cdots p}_{x}\,\underbrace{qq\cdots q}_{n-x} \\ &= p^x q^{n-x} \end{aligned}$$

Let's summarize this important result.

Computation of Probabilities in Bernoulli Trials

In a binomial experiment in which the probability of success in any trial is p, the probability of exactly x successes in n independent trials is given by

$$C(n, x)p^x q^{n-x}$$

If we let X be the random variable that gives the number of successes in a binomial experiment, then the probability of exactly x successes in n independent trials may be written

$$P(X = x) = C(n, x)p^x q^{n-x} \qquad (x = 0, 1, 2, \ldots, n) \tag{14}$$

The random variable X is called a **binomial random variable,** and the probability distribution of X is called a **binomial distribution.**

EXAMPLE 2 A fair die is rolled five times. If a 1 or a 6 lands uppermost in a trial, then the throw is considered a success. Otherwise, the throw is considered a failure.

a. Find the probabilities of obtaining exactly 0, 1, 2, 3, 4, and 5 successes in this experiment.

b. Using the results obtained in the solution to part (a), construct the binomial distribution for this experiment, and draw the histogram associated with it.

Solution

a. This is a binomial experiment with X, the binomial random variable, taking on each of the values 0, 1, 2, 3, 4, and 5 corresponding to exactly 0, 1, 2, 3, 4, and 5 successes, respectively, in five trials. Since the die is fair, the probability of a 1 or a 6 landing uppermost in any trial is given by $p = \frac{2}{6} = \frac{1}{3}$, from which it also follows that $q = 1 - p = \frac{2}{3}$. Finally, $n = 5$, since there are five trials (throws of the die) in this experiment. Using Equation (14), we find that the required probabilities are

$$P(X = 0) = C(5, 0)\left(\frac{1}{3}\right)^0\left(\frac{2}{3}\right)^5 = \frac{5!}{0!\,5!} \cdot \frac{32}{243} \approx .132$$

$$P(X = 1) = C(5, 1)\left(\frac{1}{3}\right)^1\left(\frac{2}{3}\right)^4 = \frac{5!}{1!\,4!} \cdot \frac{16}{243} \approx .329$$

$$P(X = 2) = C(5, 2)\left(\frac{1}{3}\right)^2\left(\frac{2}{3}\right)^3 = \frac{5!}{2!\,3!} \cdot \frac{8}{243} \approx .329$$

$$P(X = 3) = C(5, 3)\left(\frac{1}{3}\right)^3\left(\frac{2}{3}\right)^2 = \frac{5!}{3!\,2!} \cdot \frac{4}{243} \approx .165$$

$$P(X = 4) = C(5, 4)\left(\frac{1}{3}\right)^4\left(\frac{2}{3}\right)^1 = \frac{5!}{4!\,1!} \cdot \frac{2}{243} \approx .041$$

$$P(X = 5) = C(5, 5)\left(\frac{1}{3}\right)^5\left(\frac{2}{3}\right)^0 = \frac{5!}{5!\,0!} \cdot \frac{1}{243} \approx .004$$

b. Using these results, we find the required binomial distribution associated with this experiment given in Table 12. Next, we use this table to construct the histogram associated with the probability distribution (Figure 12).

TABLE 12

Probability Distribution

x	$P(X = x)$
0	.132
1	.329
2	.329
3	.165
4	.041
5	.004

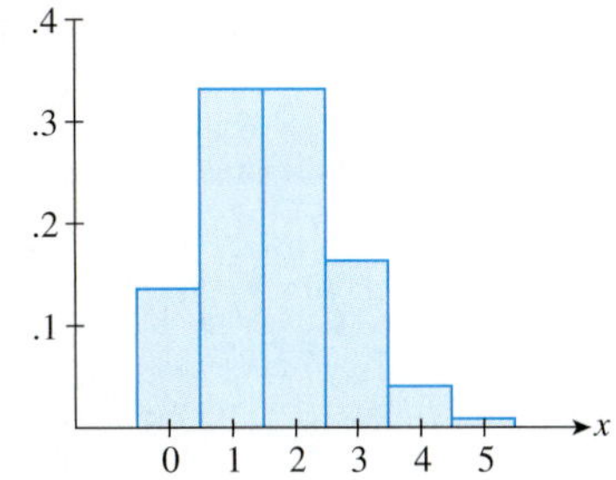

FIGURE **12**
The probability of the number of successes in five throws

EXAMPLE 3 A fair die is rolled five times. If a 1 or a 6 lands uppermost in a trial, then the throw is considered a success. Use the results from Example 2 to answer the following questions:

a. What is the probability of obtaining 0 or 1 success in the experiment?
b. What is the probability of obtaining at least 1 success in the experiment?

Solution Using Table 12, we find that

a. The probability of obtaining 0 or 1 success in the experiment is given by

$$P(X = 0) + P(X = 1) = .132 + .329 = .461$$

b. The probability of obtaining at least 1 success in the experiment is given by

$$\begin{aligned} P(X = 1) + P(X = 2) + P(X = 3) + P(X = 4) + P(X = 5) \\ = .329 + .329 + .165 + .041 + .004 \\ = .868 \end{aligned}$$

This result may also be obtained by observing that $1 - P(X = 0) = 1 - .132 = .868$.

Explore and Discuss

Consider the equation

$$P(X = x) = C(n, x)p^x q^{n-x}$$

for the binomial distribution.

1. Construct the histogram with $n = 5$ and $p = .2$, the histogram with $n = 5$ and $p = .5$, and the histogram with $n = 5$ and $p = .8$.
2. Comment on the shape of the histograms, and give an interpretation.

The following formulas (which we state without proof) will be useful in solving problems that involve binomial experiments.

Mean, Variance, and Standard Deviation of a Random Variable X

If X is a binomial random variable associated with a binomial experiment consisting of n trials with probability of success p and probability of failure q, then the **mean** (expected value), **variance,** and **standard deviation** of X are

$$\mu = E(X) = np \qquad \textbf{(15a)}$$

$$\text{Var}(X) = npq \qquad \textbf{(15b)}$$

$$\sigma_X = \sqrt{npq} \qquad \textbf{(15c)}$$

EXAMPLE 4 For the experiment in Examples 2 and 3, compute the mean, the variance, and the standard deviation of X by (a) using Equations (15a), (15b), and (15c) and (b) using the definition of each term (Sections 8.2 and 8.3).

Solution

a. We use Equations (15a), (15b), and (15c), with $p = \frac{1}{3}$, $q = \frac{2}{3}$, and $n = 5$, obtaining

$$\mu = E(X) = (5)\left(\frac{1}{3}\right) = \frac{5}{3} \approx 1.67$$

$$\text{Var}(X) = (5)\left(\frac{1}{3}\right)\left(\frac{2}{3}\right) = \frac{10}{9} \approx 1.11$$

$$\sigma_X = \sqrt{\text{Var}(X)} \approx \sqrt{1.11} \approx 1.05$$

We leave it to you to interpret the results.

b. Using the definition of expected value and the values of the probability distribution shown in Table 12, we find that

$$\begin{aligned} \mu = E(X) &\approx (0)(.132) + (1)(.329) + (2)(.329) \\ &\quad + (3)(.165) + (4)(.041) + (5)(.004) \\ &\approx 1.67 \end{aligned}$$

which agrees with the result obtained in part (a). Next, using the definition of variance and $\mu = 1.67$, we find that

$$\begin{aligned} \text{Var}(X) &= (.132)(-1.67)^2 + (.329)(-0.67)^2 + (.329)(0.33)^2 \\ &\quad + (.165)(1.33)^2 + (.041)(2.33)^2 + (.004)(3.33)^2 \\ &\approx 1.11 \\ \sigma_X &= \sqrt{\text{Var}(X)} \\ &\approx \sqrt{1.11} \approx 1.05 \end{aligned}$$

which again agrees with the preceding results.

We close this section by looking at several examples involving binomial experiments. In working through these examples, you may use a calculator, or you may consult Table 1, Binomial Probabilities, in Appendix D.

APPLIED EXAMPLE 5 Quality Control A division of Solaron manufactures photovoltaic cells to use in the company's solar energy converters. It is estimated that 5% of the cells manufactured are defective. If a random sample of 20 is selected from a large lot of cells manufactured by the company, what is the probability that it will contain at most 2 defective cells?

Solution We may view this as a binomial experiment. To see this, first note that a fixed number of trials ($n = 20$) corresponds to the selection of exactly 20 photovoltaic cells. Second, observe that there are exactly two outcomes in the experiment: defective ("success") and nondefective ("failure"). Third, the probability of success in each trial is .05 ($p = .05$), and the probability of failure in each trial is .95 ($q = .95$). This assumption is justified by virtue of the fact that the lot from which the cells are selected is "large," so the removal of a few cells will not appreciably affect the percentage of defective cells in the lot in each successive trial. Finally, the trials are independent of each other—once again because of the lot size.

Letting X denote the number of defective cells, we find that the probability of finding at most 2 defective cells in the sample of 20 is given by

$$\begin{aligned} &P(X = 0) + P(X = 1) + P(X = 2) \\ &\quad = C(20, 0)(.05)^0(.95)^{20} + C(20, 1)(.05)^1(.95)^{19} \\ &\quad\quad + C(20, 2)(.05)^2(.95)^{18} \\ &\quad \approx .3585 + .3774 + .1887 \\ &\quad = .9246 \end{aligned}$$

Thus, for lots of photovoltaic cells manufactured by Solaron, approximately 92% of the samples will have at most 2 defective cells; equivalently, approximately 8% of the samples will contain more than 2 defective cells.

APPLIED EXAMPLE 6 Success of Heart Transplants The probability that a heart transplant performed at the Medical Center is successful (that is, the patient survives 1 year or more after undergoing such an operation) is .7. Of six patients who have recently undergone such an operation, what is the probability that, 1 year from now,

a. None of the heart recipients will be alive?
b. Exactly three will be alive?
c. At least three will be alive?
d. All will be alive?

Solution Here, $n = 6, p = .7$, and $q = .3$. Let X denote the number of successful operations. Then:

a. The probability that no heart recipients will be alive after 1 year is given by

$$\begin{aligned} P(X = 0) &= C(6, 0)(.7)^0(.3)^6 \\ &= \frac{6!}{0!\,6!} \cdot 1 \cdot (.3)^6 \\ &\approx .0007 \end{aligned}$$

b. The probability that exactly three heart recipients will be alive after 1 year is given by

$$\begin{aligned} P(X = 3) &= C(6, 3)(.7)^3(.3)^3 \\ &= \frac{6!}{3!\,3!}(.7)^3(.3)^3 \\ &\approx .1852 \end{aligned}$$

c. The probability that at least three heart recipients will be alive after 1 year is given by

$$\begin{aligned} &P(X = 3) + P(X = 4) + P(X = 5) + P(X = 6) \\ &\quad = C(6, 3)(.7)^3(.3)^3 + C(6, 4)(.7)^4(.3)^2 \\ &\qquad + C(6, 5)(.7)^5(.3)^1 + C(6, 6)(.7)^6(.3)^0 \\ &\quad = \frac{6!}{3!\,3!}(.7)^3(.3)^3 + \frac{6!}{4!\,2!}(.7)^4(.3)^2 + \frac{6!}{5!\,1!}(.7)^5(.3)^1 \\ &\qquad + \frac{6!}{6!\,0!}(.7)^6 \cdot 1 \\ &\quad \approx .1852 + .3241 + .3025 + .1176 \\ &\quad = .9294 \end{aligned}$$

d. The probability that all six heart recipients will be alive after 1 year is given by

$$\begin{aligned} P(X = 6) &= C(6, 6)(.7)^6(.3)^0 = \frac{6!}{6!\,0!}(.7)^6 \cdot 1 \\ &\approx .1176 \end{aligned}$$

APPLIED EXAMPLE 7 Quality Control PAR Bearings manufactures ball bearings packaged in lots of 100 each. The company's quality-control department has determined that 2% of the ball bearings manufactured do not meet the specifications imposed by a buyer. Find the average number of ball bearings per package that fail to meet the buyer's specification.

Solution The experiment under consideration is binomial. The average number of ball bearings per package that fail to meet with the specifications is therefore given by the expected value of the associated binomial random variable. Using Equation (15a), we find that

$$\mu = E(X) = np = (100)(.02) = 2$$

substandard ball bearings in a package of 100.

Equation (15a) also provides us with a useful method for estimating the proportion of defective items in a batch by examining a few random samples, as the following example shows.

APPLIED EXAMPLE 8 Quality Control The ball bearings produced by a certain machine were checked by examining samples of 20. The following table shows the number of substandard ball bearings (they did not meet specified tolerance requirements) contained in a distribution of 100 samples:

Number of Substandard Ball Bearings in a Sample of 20	0	1	2	3	4 or more
Number of Samples	85	10	4	1	0

Find the mean number of substandard ball bearings per sample, and assuming that the distribution is binomial, estimate the percentage of substandard ball bearings in the whole batch.

Solution The mean number of substandard ball bearings per sample is

$$\mu = \frac{1}{100}[(85)(0) + (10)(1) + (4)(2) + (1)(3)] = 0.21$$

Here, $n = 20$. So Equation (15a) gives

$$0.21 = 20p$$

or $p = .0105$. Thus, approximately 1% of the ball bearings in the group are substandard.

8.4 Self-Check Exercises

1. A binomial experiment consists of four independent trials. The probability of success in each trial is .2.
   a. Find the probabilities of obtaining exactly 0, 1, 2, 3, and 4 successes in this experiment.
   b. Construct the binomial distribution, and draw the histogram associated with this experiment.
   c. Compute the mean and the standard deviation of the random variable associated with this experiment.

2. Household Microwave Ownership A survey shows that 60% of the households in a large metropolitan area have microwave ovens. If ten households are selected at random, what is the probability that five or fewer of these households have microwave ovens?

Solutions to Self-Check Exercises 8.4 can be found on page 512.

8.4 Concept Questions

1. Suppose that you are given a Bernoulli experiment.
   a. How many outcomes are there in each trial?
   b. Can the number of trials in the experiment vary, or is it fixed?
   c. Are the trials in the experiment dependent?
   d. If the probability of success in any trial is p, what is the probability of exactly x successes in n independent trials?

2. Give the formula for the mean, variance, and standard deviation of X, where X is a binomial random variable associated with a binomial experiment consisting of n trials with probability of success p and probability of failure q.

8.4 Exercises

In Exercises 1–6, determine whether the experiment is a binomial experiment. Justify your answer.

1. Rolling a fair die three times and observing the number of times a 6 is thrown

2. Rolling a fair die and observing the number of times the die is thrown until a 6 appears uppermost

3. Rolling a fair die three times and observing the number that appears uppermost

4. A card is selected from a deck of 52 cards, and its color is observed. A second card is then drawn (without replacement), and its color is observed.

5. Recording the number of accidents that occur at a given intersection on four clear days and one rainy day

6. Recording the number of hits a baseball player, whose batting average is .325, gets after being up to bat five times

In Exercises 7–10, find $C(n, x)p^x q^{n-x}$ for the given values of n, x, and p.

7. $n = 4, x = 2, p = \dfrac{1}{3}$

8. $n = 6, x = 4, p = \dfrac{1}{4}$

9. $n = 5, x = 3, p = .2$

10. $n = 6, x = 5, p = .4$

In Exercises 11–16, use the formula $C(n, x)p^x q^{n-x}$ to determine the probability of the given event.

11. The probability of exactly no successes in five trials of a binomial experiment in which $p = \frac{1}{3}$

12. The probability of exactly three successes in six trials of a binomial experiment in which $p = \frac{1}{2}$

13. The probability of at least three successes in six trials of a binomial experiment in which $p = \frac{1}{2}$

14. The probability of no successful outcomes in six trials of a binomial experiment in which $p = \frac{1}{3}$

15. The probability of no failures in five trials of a binomial experiment in which $p = \frac{1}{3}$

16. The probability of at least one failure in five trials of a binomial experiment in which $p = \frac{1}{3}$

17. A fair die is rolled four times. Calculate the probability of obtaining exactly two 6s.

18. Let X be the number of successes in five independent trials in a binomial experiment in which the probability of success is $p = \frac{2}{5}$. Find:
    a. $P(X = 4)$
    b. $P(2 \le X \le 4)$

19. A binomial experiment consists of five independent trials. The probability of success in each trial is .4.
    a. Find the probabilities of obtaining exactly 0, 1, 2, 3, 4, and 5 successes in this experiment.
    b. Construct the binomial distribution, and draw the histogram associated with this experiment.
    c. Compute the mean and the standard deviation of the random variable associated with this experiment.

20. **Family Composition** Let the random variable X denote the number of girls in a five-child family. If the probability of a female birth is .5:
    a. Find the probabilities of 0, 1, 2, 3, 4, and 5 girls in a five-child family.
    b. Construct the binomial distribution, and draw the histogram associated with this experiment.
    c. Compute the mean and the standard deviation of the random variable X.

21. **Product Reliability** The probability that a fuse produced by a certain manufacturing process will be defective is $\frac{1}{50}$. Is it correct to infer from this statement that there is at most 1 defective fuse in each lot of 50 produced by this process? Justify your answer.

22. **Sports** If the probability that a certain tennis player will serve an ace is $\frac{1}{4}$, what is the probability that he will serve exactly two aces out of five serves? (Assume that the five serves are independent.)

23. **Sports** If the probability that a certain tennis player will serve an ace is .15, what is the probability that she will

serve at least two aces out of five serves? (Assume that the five serves are independent.)

24. **Voters** In a certain congressional district, it is known that 40% of the registered voters classify themselves as conservatives. If ten registered voters are selected at random from this district, what is the probability that four of them will be conservatives?

25. **Availability of Customer Service** Mayco, a mail-order department store, has six telephone lines available for customers who wish to place their orders by phone. If the probability is $\frac{1}{4}$ that any one of the six telephone lines is engaged during business hours, find the probability that all six lines will be in use when a customer calls to place an order.

26. **Quality Control** As part of its quality-control program, the video-game DVDs produced by Starr Communications are subjected to a final inspection before shipment. A sample of six DVDs is selected at random from each lot of DVDs produced, and the lot is rejected if the sample contains one or more defective DVDs. If 1.5% of the DVDs produced by Starr is defective, find the probability that a shipment will be accepted.

27. **Restaurant Health Code Violations** Suppose 30% of the restaurants in a certain part of a town are in violation of the health code. If a health inspector randomly selects five of the restaurants for inspection, what is the probability that:
    a. None of the restaurants are in violation of the health code?
    b. One of the restaurants is in violation of the health code?
    c. At least two of the restaurants are in violation of the health code?

28. **Violations of the Building Code** Suppose that one third of the new buildings in a town are in violation of the building code. If a building inspector inspects five of the buildings chosen at random, find the probability that:
    a. The first three buildings will pass the inspection and the remaining two will fail the inspection.
    b. Exactly three of the buildings will pass inspection.

29. **Exams** A psychology quiz consists of ten true-or-false questions. If a student knows the correct answer to six of the questions but determines the answers to the remaining questions by flipping a coin, what is the probability that she will obtain a score of at least 90%?

30. **Sales Predictions** From experience, the manager of Kramer's Book Mart knows that 40% of the people who are browsing in the store will make a purchase. What is the probability that, among ten people who are browsing in the store, at least three will make a purchase?

31. **Teen Spending Behavior** According to a study on teenage shopping behavior, it was found that 78% of female teens regularly shop in stores rather than shopping online. If a group of 6 female teenagers are selected at random, what is the probability that at least 3 of them regularly do their shopping in stores?
    Source: Marketing Research.

32. **Making Friends** In a survey of 2541 adults aged 18 years and older, 64% of respondents claimed that it was easier to make friends in person, whereas 36% thought that it was easier to do so through social-networking sites. Assuming that these results apply to the entire adult population, what is the probability that at most 3 out of 6 adults chosen at random hold the opinion that it is easier to make friends through social-networking sites?
    Source: Skout.

33. **Exams** A biology quiz consists of eight multiple-choice questions. Five must be answered correctly to receive a passing grade. If each question has five possible answers, of which only one is correct, what is the probability that a student who guesses at random on each question will pass the examination?

34. **Pre-Retirees' Spending** In a survey of 1000 pre-retirees, the following question was posed: If you could afford to do only one of the following a year—set aside for retirement or save for a holiday—what would you do? Sixty-two percent of the respondents indicated that they would set aside for retirement, whereas 38% indicated that they would save for a holiday. If 8 of the participants in the survey are chosen at random, what is the probability that at least 3 of them said that they would save for a holiday?
    Source: HSBC.

35. **Blood Types** It is estimated that one third of the general population has blood type A^+. If a sample of nine people is selected at random, what is the probability that:
    a. Exactly three of them have blood type A^+?
    b. At most three of them have blood type A^+?

36. **Consumer Preferences** An advertisement for Brand *A* chicken noodle soup claims that 60% of all consumers prefer Brand *A* over Brand *B*, the chief competitor's product. To test this claim, David Horowitz, host of *The Consumer Advocate*, selected ten people at random. After tasting both soups, each person was asked to state his or her preference. Assuming that the company's claim is correct, find the probability that:
    a. Six or more people stated a preference for Brand *A*.
    b. Fewer than six people stated a preference for Brand *A*.

37. **Quality Control** The probability that a DVD player produced by VCA Television is defective is estimated to be .02. If a sample of ten players is selected at random, what is the probability that the sample contains:
    a. No defectives?
    b. At most two defectives?

38. **Lunch Expenditure** A survey of 1020 workers in February 2013 found that 63% of the respondents spend a total of \$40 or less on lunch each week. If 10 of the workers who participated in the survey were chosen at random, what is the probability that at most 3 of them spend a total of \$40 or less on lunch each week?
    Source: Braun Research for Accounting Principals.

39. **Blood Pressure** A study conducted in 2012 determined that 58% of the population has normal blood pressure

(without medication). What is the probability that at least 9 out of 10 people chosen at random have normal blood pressure (without medication)?
Source: International Food Information Council Foundation.

40. **Blood Pressure** A study conducted in 2012 determined that 23% of the population has normal blood pressure (with medication). What is the probability that at most 3 out of 10 people chosen at random have normal blood pressure (with medication)?
Source: International Food Information Council Foundation.

41. **Academy Membership** Of the 5765 members of the Academy of Motion Picture Arts and Sciences (AMPAS), 77% are male, and 23% are female. What is the probability that at least 5 out of 10 members of AMPAS chosen at random will be female?
Source: Los Angeles Times.

42. **Working with a Spouse** In a survey of 1147 small-business owners, the following question was posed: Would you recommend working with a significant other? Fifty-seven percent of those polled answered in the affirmative, whereas 43% answered in the negative. Assuming that this result reflects the view of all small-business owners, what is the probability that at least 8 out of 10 small-business owners selected at random will opine that it would be unwise to work with the significant other?
Source: Manta.

43. **Getting Ahead at Work** In a survey of 26,612 Parade.com visitors conducted by Yahoo Finance/Parade, 51% of workers said they believed that the way to get ahead at work is through internal politics. If 10 of the workers who participated in the survey were picked at random, what is the probability that all 10 of them believed that engaging in internal politics is the way to get ahead at work?
Source: Yahoo Finance/Parade.

44. **Social-Media Accounts** In a survey of 1000 social-media account holders aged 18 years and older conducted by the AARP, it was found that 65% of them have one or two such accounts. If 10 of the participants in the survey are picked at random, what is the probability that at most 7 of them have one or two social-media accounts?
Source: AARP.

45. **Robot Reliability** An automobile-manufacturing company uses ten industrial robots as welders on its assembly line. On a given working day, the probability that a robot will be inoperative is .05. What is the probability that on a given working day:
a. Exactly two robots are inoperative?
b. More than two robots are inoperative?

46. **Probability of Engine Failure** The probability that an airplane engine will fail in a transcontinental flight is .001. Assuming that engine failures are independent of each other, what is the probability that on a certain transcontinental flight, a four-engine plane will experience:
a. Exactly one engine failure?
b. Exactly two engine failures?
c. More than two engine failures? (*Note:* In this event, the airplane will crash.)

47. If X is a binomial random variable associated with a binomial experiment with $\mu = E(X) = 45$ and $\sigma_x = 6$, what are n and p?

48. **Supermarket Lines** A study of the checkout lines at Dwyer's Farms reveals that the probability that any one of ten checkout lanes is busy at an instant during peak hours is .8.
a. What is the most probable number of checkout lanes that are busy at any instant?
Hint: $P(X = x)$ will continue to increase as long as $P(X = x + 1) \geq P(X = x)$.
b. What is the probability that the number of checkout lines found in part (a) occurs?

49. **Quality Control** The McCormack Company manufactures solar panels. As a part of its quality control, the company checks the day's production by examining samples of 10. The following table shows the number of defective panels contained in a distribution of 40 samples.

Number of Defective Panels in a Sample of 20	0	1	2	3	4	5 or more
Number of Samples	33	3	2	1	1	0

Find the mean number of defective panels per sample, and assuming that the distribution is binomial, estimate the percentage of defective solar panels in the day's production.

50. **Quality Control** The panels for the Pulsar 32-inch wide-screen LCD HDTVs in a production run were checked by examining samples of 6. The following table shows the number of defective panels contained in a distribution of 30 samples.

Number of Defective Panels in a Sample of 20	0	1	2	3	4 or more
Number of Samples	26	2	1	1	0

Find the mean number of defective panels per sample, and assuming that the distribution is binomial, estimate the percentage of defective panels in the production run.

51. **Quality Control** The manager of Toy World has decided to accept a shipment of electronic games if none of a random sample of 20 is found to be defective. What is the probability that he will accept the shipment:
a. If 10% of the electronic games is defective?
b. If 5% of the electronic games is defective?

52. **Quality Control** Refer to Exercise 51. If the manager's criterion for accepting shipment is that there be no more than 1 defective electronic game in a random sample of 20, what is the probability that he will accept the shipment if 10% of the electronic games is defective?

53. **Quality Control** Refer to Exercise 51. If the manager of the store changes his sample size to 10 and decides to accept shipment only if none of the sampled games is defective, what is the probability that he will accept the shipment if 10% of the games is defective?

54. **Mail-Order Sales** Ruth, the owner of a mail-order business, estimates that the probability that a household receiving one of her catalogs will place an order with her is .10. How many catalogs must Ruth send out to ensure that the chances of obtaining at least one order is 50% or better?

55. **Flipping a Coin** How many times must a person toss a fair coin to ensure that the chances of obtaining at least one head are 99% or better?

56. **Flipping a Coin** An unbiased coin is tossed 1 million times. Show that the probability is at least .99 that the proportion of times the coin will land heads is between .495 and .505, inclusive. (*Note:* These results show that if an unbiased coin is tossed a very large number of times, the proportion of times that the coin will land heads is very close to $\frac{1}{2}$.)
Hint: Use Chebychev's inequality.

57. **Drug Testing** A new drug has been found to be effective in treating 75% of the people afflicted by a certain disease. If the drug is administered to 500 people who have this disease, what are the mean and the standard deviation of the number of people for whom the drug can be expected to be effective?

58. **Probability of Graduating** At a certain university, the probability that an entering freshman will graduate within 4 years is .6. From an incoming class of 2000 freshmen, find:
   a. The expected number of students who will graduate within 4 years.
   b. The standard deviation of the number of students who will graduate within 4 years.

In Exercises 59–62, determine whether the statement is true or false. If it is true, explain why it is true. If it is false, give an example to show why it is false.

59. In a binomial trial, the number of outcomes of the experiment may be any finite number.

60. In a binomial experiment with $n = 3$, $P(X = 1 \text{ or } 2) = 3pq$.

61. If the probability that a batter gets a hit is $\frac{1}{4}$ at each time at bat, then the batter is sure to get a hit if she bats four times.

62. The histogram associated with a binomial distribution is symmetric with respect to $x = \frac{n}{2}$ if $p = \frac{1}{2}$.

8.4 Solutions to Self-Check Exercises

1. **a.** We use Equation (14) with $n = 4$, $p = .2$, and $q = 1 - .2 = .8$, obtaining

$$P(X = 0) = C(4, 0)(.2)^0(.8)^4 = \frac{4!}{0!\,4!} \cdot 1 \cdot (.8)^4 \approx .410$$

$$P(X = 1) = C(4, 1)(.2)^1(.8)^3 = \frac{4!}{1!\,3!}(.2)(.8)^3 \approx .410$$

$$P(X = 2) = C(4, 2)(.2)^2(.8)^2 = \frac{4!}{2!\,2!}(.2)^2(.8)^2 \approx .154$$

$$P(X = 3) = C(4, 3)(.2)^3(.8)^1 = \frac{4!}{3!\,1!}(.2)^3(.8) \approx .026$$

$$P(X = 4) = C(4, 4)(.2)^4(.8)^0 = \frac{4!}{4!\,0!}(.2)^4 \cdot 1 \approx .002$$

b. The required binomial distribution and histogram are as follows:

x	$P(X = x)$
0	.410
1	.410
2	.154
3	.026
4	.002

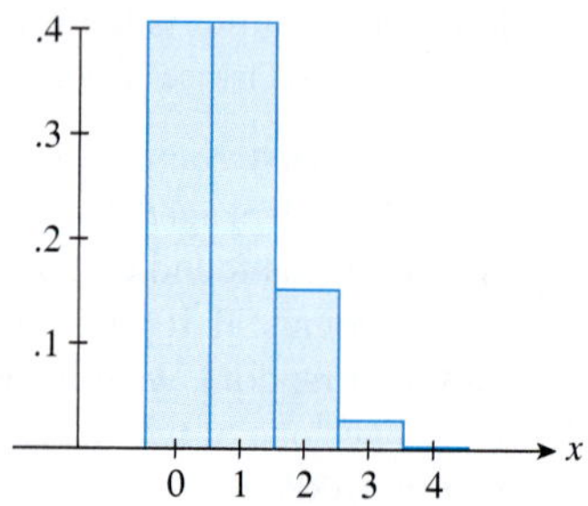

c. The mean is

$$\mu = E(X) = np = (4)(.2) = 0.8$$

and the standard deviation is

$$\sigma = \sqrt{npq} = \sqrt{(4)(.2)(.8)} = 0.8$$

2. This is a binomial experiment with $n = 10$, $p = .6$, and $q = .4$. Let X denote the number of households that have microwave ovens. Then the probability that five or fewer households have microwave ovens is given by

$$\begin{aligned} P(X = 0) &+ P(X = 1) + P(X = 2) + P(X = 3) \\ &+ P(X = 4) + P(X = 5) \\ &= C(10, 0)(.6)^0(.4)^{10} + C(10, 1)(.6)^1(.4)^9 \\ &\quad + C(10, 2)(.6)^2(.4)^8 + C(10, 3)(.6)^3(.4)^7 \\ &\quad + C(10, 4)(.6)^4(.4)^6 + C(10, 5)(.6)^5(.4)^5 \\ &\approx 0 + .002 + .011 + .042 + .111 + .201 \\ &\approx .367 \end{aligned}$$

8.5 The Normal Distribution

Probability Density Functions

The probability distributions discussed in the preceding sections were all associated with finite random variables—that is, random variables that take on finitely many values. Such probability distributions are referred to as *finite probability distributions*. In this section, we consider probability distributions associated with a continuous random variable—that is, a random variable that may take on any value lying in an interval of real numbers. Such probability distributions are called **continuous probability distributions.**

Unlike a finite probability distribution, which may be exhibited in the form of a table, a continuous probability distribution is defined by a function f whose domain coincides with the interval of values taken on by the random variable associated with the experiment. Such a function f is called the **probability density function** associated with the probability distribution, and it has the following properties:

1. $f(x)$ is nonnegative for all values of x in its domain.
2. The area of the region between the graph of f and the x-axis is equal to 1 (Figure 13).

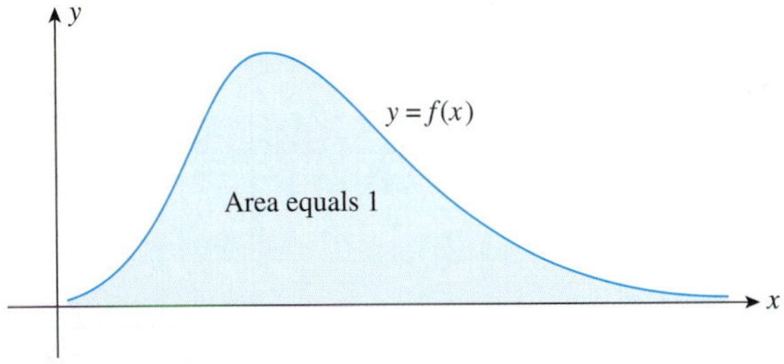

FIGURE 13
A probability density function

Now suppose we are given a continuous probability distribution defined by a probability density function f. Then the probability that the random variable X assumes a value in an interval $a < x < b$ is given by the area of the region between the graph of f and the x-axis from $x = a$ to $x = b$ (Figure 14). We denote the value of this

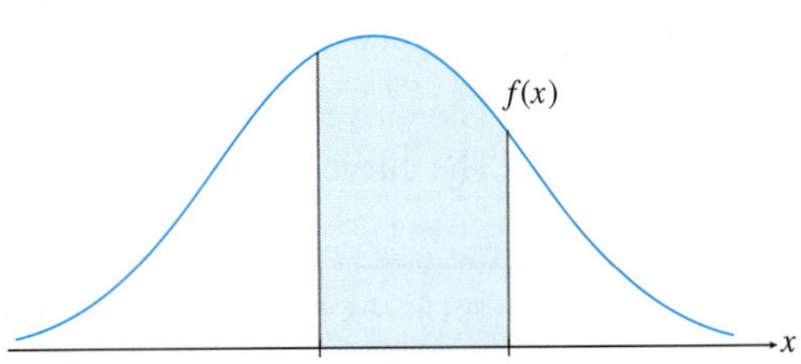

FIGURE 14
$P(a < X < b)$ is given by the area of the shaded region.

probability by $P(a < X < b)$.* Observe that property 2 of the probability density function states that the probability that a continuous random variable takes on a value lying in its range is 1, a certainty, which is expected. Note the analogy between the areas under the probability density curves and the areas of the histograms associated with finite probability distributions (see Section 8.1).

Normal Distributions

The mean μ and the standard deviation σ of a continuous probability distribution have roughly the same meanings as the mean and standard deviation of a finite probability distribution. Thus, the mean of a continuous probability distribution is a measure of the central tendency of the probability distribution, and the standard deviation of the probability distribution measures its spread about its mean. Both of these numbers will play an important role in the following discussion.

For the remainder of this section, we will discuss a special class of continuous probability distributions known as **normal distributions.** Normal distributions are without a doubt the most important of all the probability distributions. Many phenomena—such as the heights of people in a given population, the weights of newborn infants, the IQs of college students, the actual weights of 16-ounce packages of cereals, and so on—have probability distributions that are normal. The normal distribution also provides us with an accurate approximation to the distributions of many random variables associated with random-sampling problems. In fact, in the next section we will see how a normal distribution may be used to approximate a binomial distribution under certain conditions.

The graph of a normal distribution, which is bell shaped, is called a **normal curve** (Figure 15).

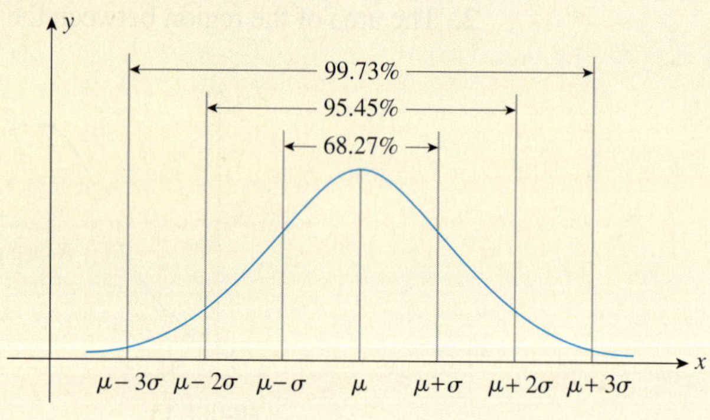

FIGURE 15
A normal curve

The normal curve (and therefore the corresponding normal distribution) is completely determined by its mean μ and standard deviation σ. In fact, the normal curve has the following characteristics, which are described in terms of these two parameters.†

1. The curve has a peak at $x = \mu$.
2. The curve is symmetric with respect to the vertical line $x = \mu$.

*Because the area under one point of the graph of f is equal to zero, we see that $P(a < X < b) = P(a < X \le b) = P(a \le X < b) = P(a \le X \le b)$.

†The probability density function associated with this normal curve is given by

$$y = \frac{1}{\sigma\sqrt{2\pi}} e^{-(1/2)[(x-\mu)/\sigma]^2}$$

but the direct use of this formula will not be required in our discussion of the normal distribution.

3. The curve always lies above the x-axis but approaches the x-axis as x extends indefinitely in either direction.
4. The area under the curve is 1.
5. For any normal curve, 68.27% of the area under the curve lies within 1 standard deviation of the mean (that is, between $\mu - \sigma$ and $\mu + \sigma$), 95.45% of the area lies within 2 standard deviations of the mean, and 99.73% of the area lies within 3 standard deviations of the mean.

Figure 16 shows two normal curves with different means μ_1 and μ_2 but the same standard deviation. Figure 17 shows two normal curves with the same mean but different standard deviations σ_1 and σ_2. (Which number is smaller?)

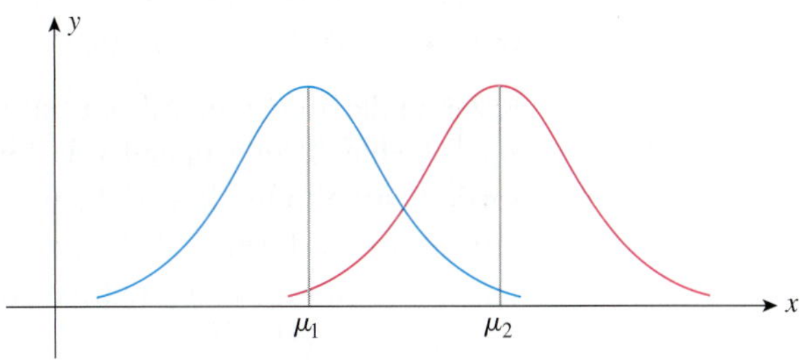

FIGURE 16
Two normal curves that have the same standard deviation but different means

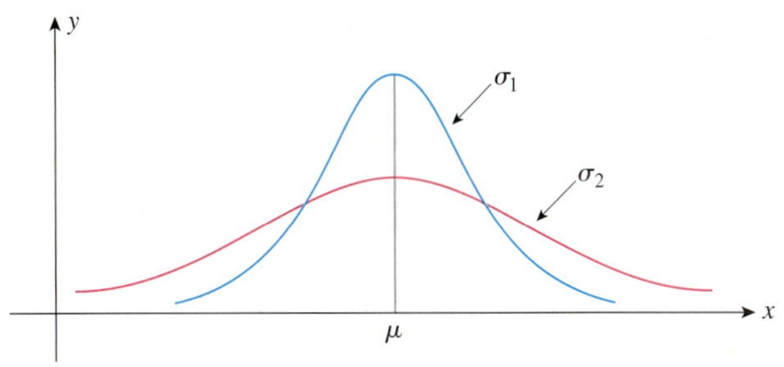

FIGURE 17
Two normal curves that have the same mean but different standard deviations

The mean μ of a normal distribution determines where the center of the curve is located, whereas the standard deviation σ of a normal distribution determines the peakedness (or flatness) of the curve.

As this discussion reveals, there are infinitely many normal curves corresponding to different choices of the parameters μ and σ that characterize such curves. Fortunately, any normal curve may be transformed into any other normal curve (as we will see later), so in the study of normal curves, it suffices to single out one such particular curve for special attention. The normal curve with mean $\mu = 0$ and standard deviation $\sigma = 1$ is called the **standard normal curve.** The corresponding distribution is called the **standard normal distribution.** The random variable itself is called the **standard normal random variable** and is commonly denoted by Z.

Exploring with TECHNOLOGY

Consider the probability density function

$$f(x) = \frac{1}{\sqrt{2\pi}} e^{-x^2/2}$$

(continued)

which is the formula given in the footnote on page 514 with $\mu = 0$ and $\sigma = 1$.

1. Use a graphing utility to plot the graph of f, using the viewing window $[-4, 4] \times [0, 0.5]$.
2. Use the numerical integration function of a graphing utility to find the area of the region under the graph of f on the intervals $[-1, 1]$, $[-2, 2]$, and $[-3, 3]$, thereby verifying property 5 of normal distributions for the special case in which $\mu = 0$ and $\sigma = 1$.

Computations of Probabilities Associated with Normal Distributions

Areas under the standard normal curve have been extensively computed and tabulated. Table 2, in Appendix D, gives the areas of the regions under the standard normal curve to the left of the number z; these areas correspond, of course, to probabilities of the form $P(Z < z)$ or $P(Z \leq z)$. The next several examples illustrate the use of this table in computations involving the probabilities associated with the standard normal variable.

EXAMPLE 1 Let Z be the standard normal variable. Make a sketch of the appropriate region under the standard normal curve, and then find the values of

a. $P(Z < 1.24)$ **b.** $P(Z > 0.5)$
c. $P(0.24 < Z < 1.48)$ **d.** $P(-1.65 < Z < 2.02)$

Solution

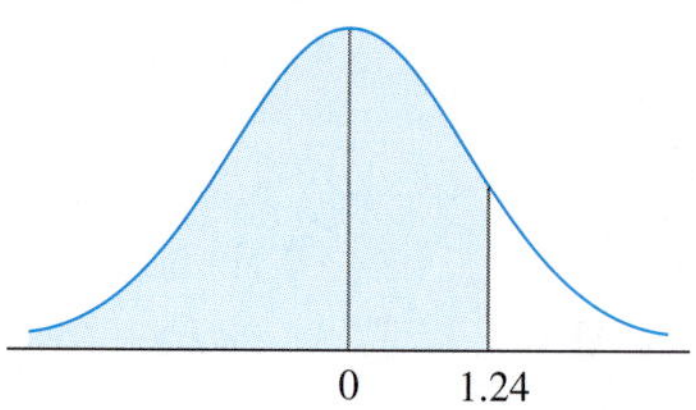

FIGURE 18
$P(Z < 1.24)$

a. The region under the standard normal curve associated with the probability $P(Z < 1.24)$ is shown in Figure 18. To find the area of the required region using Table 2 in Appendix D, we first locate the number 1.2 in the column and the number 0.04 in the row, both headed by z, and read off the number 0.8925 appearing in the body of the table. Thus,

$$P(Z < 1.24) = .8925$$

b. The region under the standard normal curve associated with the probability $P(Z > 0.5)$ is shown in Figure 19a. Observe, however, that the required area is, by virtue of the symmetry of the standard normal curve, equal to the shaded area shown in Figure 19b. Thus,

$$P(Z > 0.5) = P(Z < -0.5)$$
$$= .3085$$

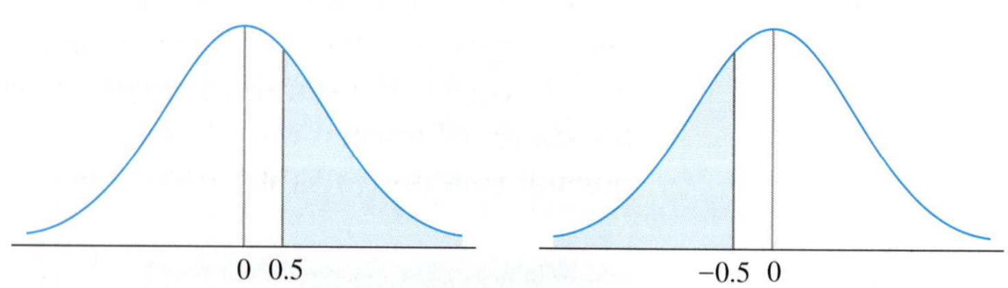

(a) $P(Z > 0.5)$ **(b)** $P(Z < -0.5)$
FIGURE 19

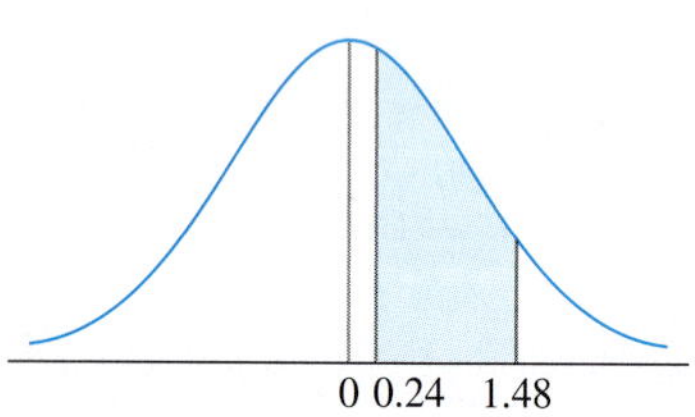

FIGURE 20
$P(0.24 < Z < 1.48)$

c. The probability $P(0.24 < Z < 1.48)$ is equal to the shaded area shown in Figure 20. This area is obtained by subtracting the area under the curve to the left of $z = 0.24$ from the area under the curve to the left of $z = 1.48$; that is,

$$\begin{aligned} P(0.24 < Z < 1.48) &= P(Z < 1.48) - P(Z < 0.24) \\ &= .9306 - .5948 \\ &= .3358 \end{aligned}$$

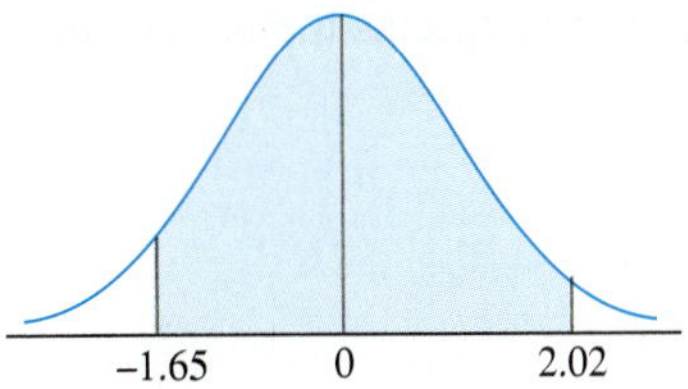

FIGURE 21
$P(-1.65 < Z < 2.02)$

d. The probability $P(-1.65 < Z < 2.02)$ is given by the shaded area shown in Figure 21. We have

$$\begin{aligned} P(-1.65 < Z < 2.02) &= P(Z < 2.02) - P(Z < -1.65) \\ &= .9783 - .0495 \\ &= .9288 \end{aligned}$$

Exploring with TECHNOLOGY

We can calculate the areas under the standard normal curve using the function normalcdf(. This will give a more accurate value than one obtained from the table. To call the function, press **2nd** **VARS** on the T1-83/84, then select `2:normal cdf(`. For example, to compute $P(0.24 < Z < 1.48)$, enter

```
normalcdf(.24,1.48)
```

The TI-83/84 screen is shown in Figure 22. The answer (to three decimal places) agrees with the result obtained in Example 1c.

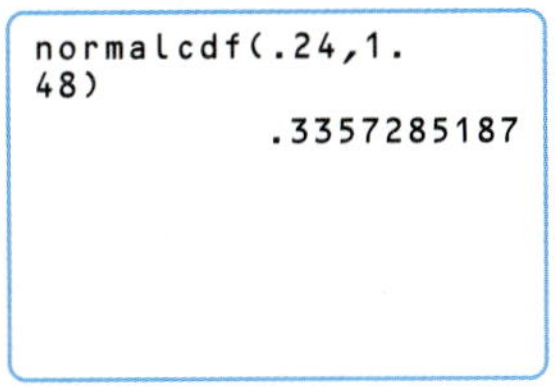

FIGURE 22

To find $P(Z < 1.24)$, we write $P(Z < 1.24) = .5 + P(0 < Z < 1.24)$ and enter

```
.5+normalcdf(0,1.24)
```

The TI-83/84 screen is shown in Figure 23. The answer agrees with the result obtained in Example 1a.

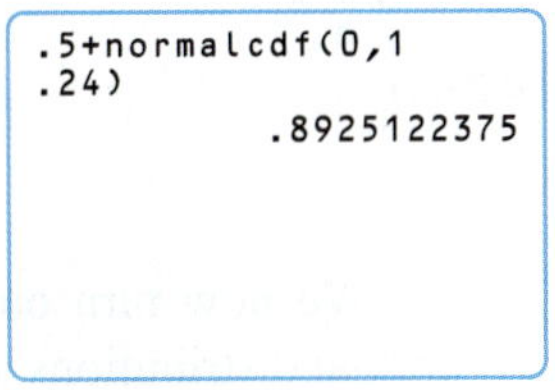

FIGURE 23

EXAMPLE 2 Let Z be the standard normal random variable. Find the value of z if z satisfies

a. $P(Z < z) = .9474$ **b.** $P(Z > z) = .9115$ **c.** $P(-z < Z < z) = .7888$

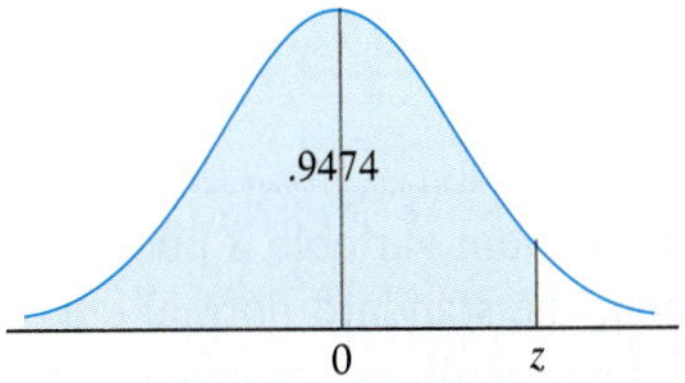

FIGURE 24
$P(Z < z) = .9474$

Solution

a. Refer to Figure 24. We want the value of Z such that the area of the region under the standard normal curve and to the left of $Z = z$ is .9474. Locating the number .9474 in Table 2, Appendix D, and reading back, we find that $z = 1.62$.

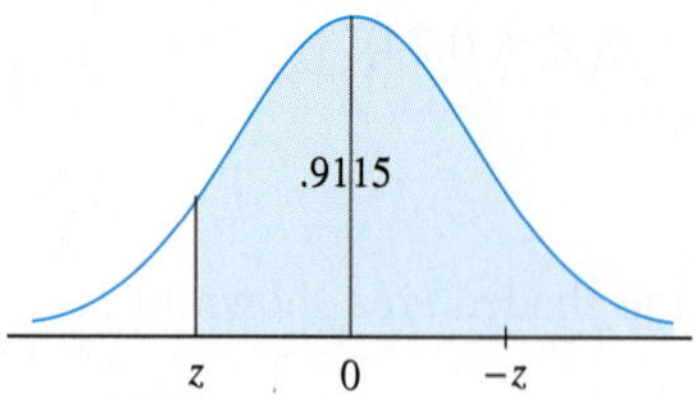

FIGURE 25
$P(Z > z) = .9115$

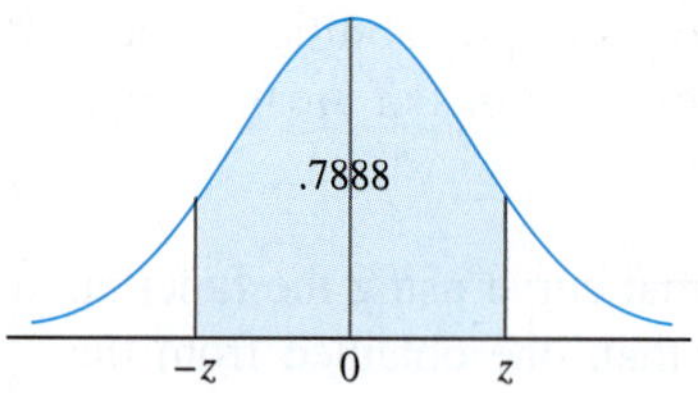

FIGURE 26
$P(-z < Z < z) = .7888$

b. Since $P(Z > z)$, or equivalently, the area of the region to the right of z is greater than 0.5, it follows that z must be negative (Figure 25); hence, $-z$ is positive. Furthermore, the area of the region to the right of z is the same as the area of the region to the left of $-z$. Therefore,

$$\begin{aligned} P(Z > z) &= P(Z < -z) \\ &= .9115 \end{aligned}$$

Looking up the table, we find $-z = 1.35$, so $z = -1.35$.

c. The region associated with $P(-z < Z < z)$ is shown in Figure 26. Observe that by symmetry, the area of this region is just double that of the area of the region between $Z = 0$ and $Z = z$; that is,

$$P(-z < Z < z) = 2P(0 < Z < z)$$

Furthermore,

$$P(0 < Z < z) = P(Z < z) - \frac{1}{2}$$

(Figure 27). Therefore,

$$\frac{1}{2}P(-z < Z < z) = P(Z < z) - \frac{1}{2}$$

or, solving for $P(Z < z)$, we have

$$\begin{aligned} P(Z < z) &= \frac{1}{2} + \frac{1}{2}P(-z < Z < z) \\ &= \frac{1}{2}(1 + .7888) \\ &= .8944 \end{aligned}$$

Consulting the table, we find $z = 1.25$.

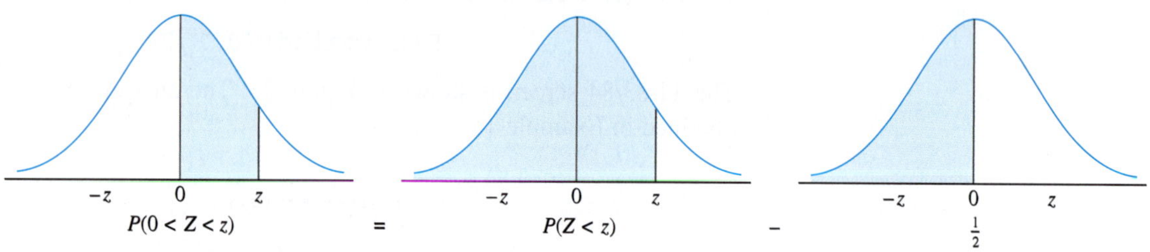

FIGURE 27

We now turn our attention to the computation of probabilities associated with normal distributions whose means and standard deviations are not necessarily equal to 0 and 1, respectively. As was mentioned earlier, any normal curve may be transformed into the standard normal curve. In particular, it may be shown that if X is a normal random variable with mean μ and standard deviation σ, then it can be transformed into the standard normal random variable Z by means of the substitution

$$Z = \frac{X - \mu}{\sigma}$$

The area of the region under the normal curve (with random variable X) between $x = a$ and $x = b$ is *equal* to the area of the region under the standard normal curve between $z = (a - \mu)/\sigma$ and $z = (b - \mu)/\sigma$. In terms of probabilities associated with these distributions, we have

$$P(a < X < b) = P\left(\frac{a-\mu}{\sigma} < Z < \frac{b-\mu}{\sigma}\right) \tag{16}$$

(Figure 28). Similarly, we have

$$P(X < b) = P\left(Z < \frac{b-\mu}{\sigma}\right) \tag{17}$$

$$P(X > a) = P\left(Z > \frac{a-\mu}{\sigma}\right) \tag{18}$$

Thus, with the help of Equations (16)–(18), computations of probabilities associated with any normal distribution may be reduced to the computations of areas of regions under the standard normal curve.

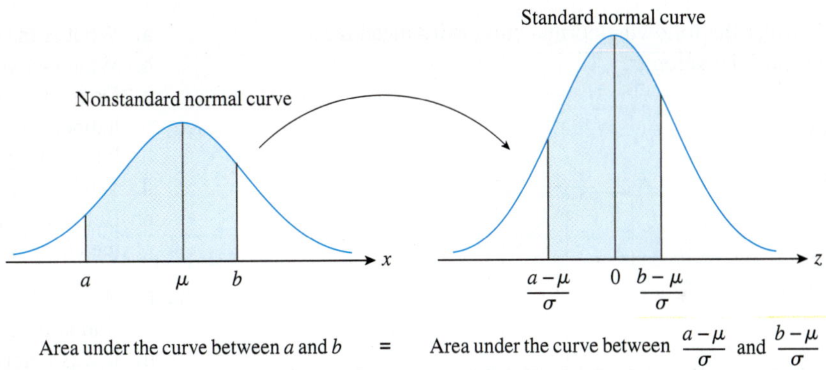

FIGURE **28**

EXAMPLE 3 Suppose X is a normal random variable with $\mu = 100$ and $\sigma = 20$. Find the values of:

a. $P(X < 120)$ **b.** $P(X > 70)$ **c.** $P(75 < X < 110)$

Solution

a. Using Equation (17) with $\mu = 100$, $\sigma = 20$, and $b = 120$, we have

$$P(X < 120) = P\left(Z < \frac{120 - 100}{20}\right)$$
$$= P(Z < 1) = .8413$$

Use the table of values of Z in Appendix D.

b. Using Equation (18) with $\mu = 100$, $\sigma = 20$, and $a = 70$, we have

$$P(X > 70)$$
$$= P\left(Z > \frac{70 - 100}{20}\right)$$
$$= P(Z > -1.5) = P(Z < 1.5) = .9332$$

c. Using Equation (16) with $\mu = 100$, $\sigma = 20$, $a = 75$, and $b = 110$, we have

$$P(75 < X < 110)$$
$$= P\left(\frac{75 - 100}{20} < Z < \frac{110 - 100}{20}\right)$$
$$= P(-1.25 < Z < 0.5)$$
$$= P(Z < 0.5) - P(Z < -1.25)$$
$$= .6915 - .1056 = .5859$$

See Figure 29.

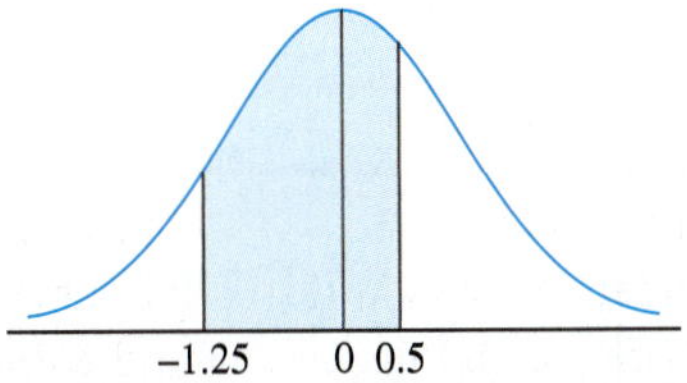

FIGURE **29**

8.5 Self-Check Exercises

1. Let Z be a standard normal variable.
   a. Find the value of $P(-1.2 < Z < 2.1)$ by first making a sketch of the appropriate region under the standard normal curve.
   b. Find the value of z if z satisfies $P(-z < Z < z) = .8764$.

2. Let X be a normal random variable with $\mu = 80$ and $\sigma = 10$. Find the values of:
   a. $P(X < 100)$ b. $P(X > 60)$ c. $P(70 < X < 90)$

Solutions to Self-Check Exercises 8.5 can be found on page 521.

8.5 Concept Questions

1. Consider the following normal curve with mean μ and standard deviation σ:

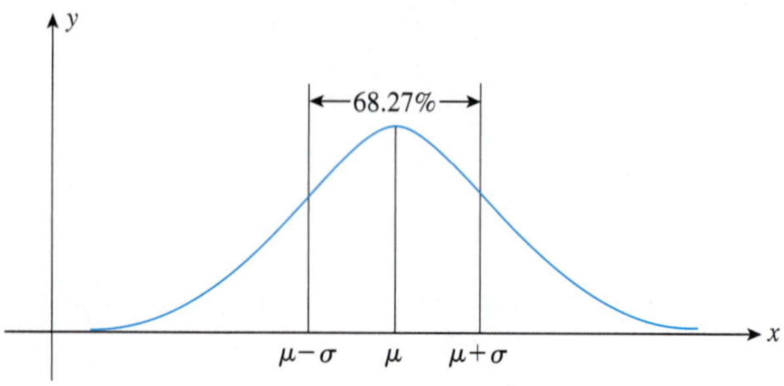

   a. What is the x-coordinate of the peak of the curve?
   b. What can you say about the symmetry of the curve?
   c. Does the curve always lie above the x-axis? What happens to the curve as x extends indefinitely to the left or right?
   d. What is the value of the area under the curve?
   e. Between what values does 68.27% of the area under the curve lie?

2. a. What is the difference between a normal curve and a standard normal curve?
   b. If X is a normal random variable with mean μ and standard deviation σ, write $P(a < X < b)$ in terms of the probabilities associated with the standard normal random variable Z.

8.5 Exercises

In Exercises 1–6, find the value of the probability of the standard normal variable Z corresponding to the shaded area under the standard normal curve.

1. $P(Z < 1.45)$

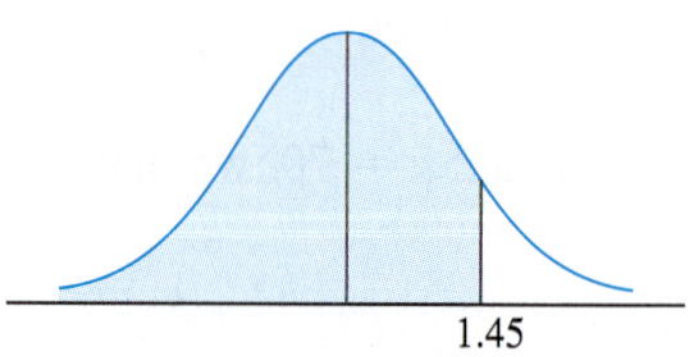

2. $P(Z > 1.11)$

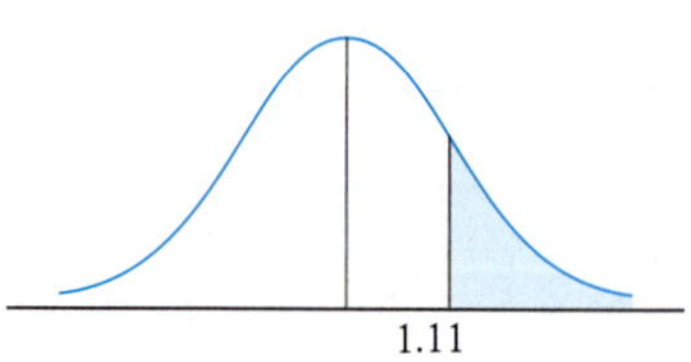

3. $P(Z < -1.75)$

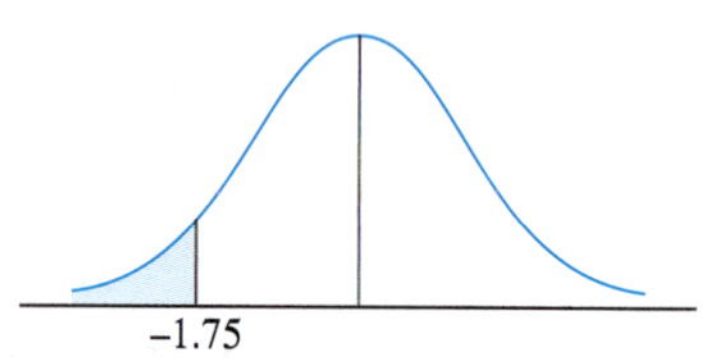

4. $P(0.3 < Z < 1.83)$

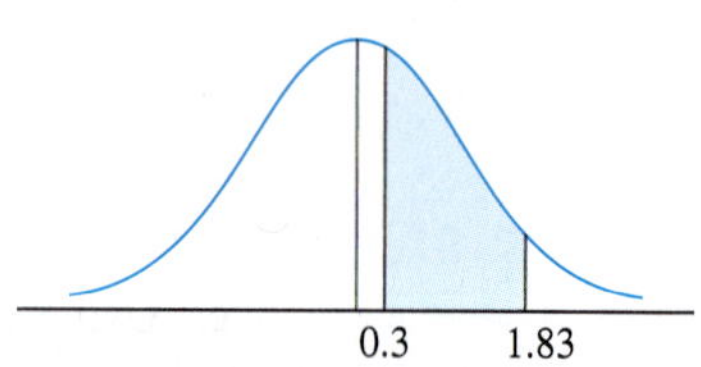

5. $P(-1.32 < Z < 1.74)$

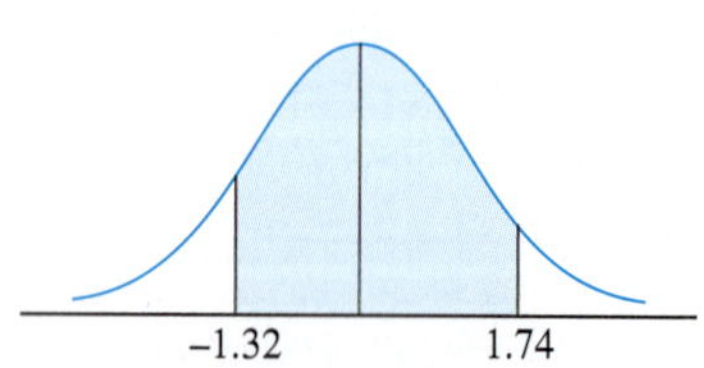

6. $P(-2.35 < Z < -0.51)$

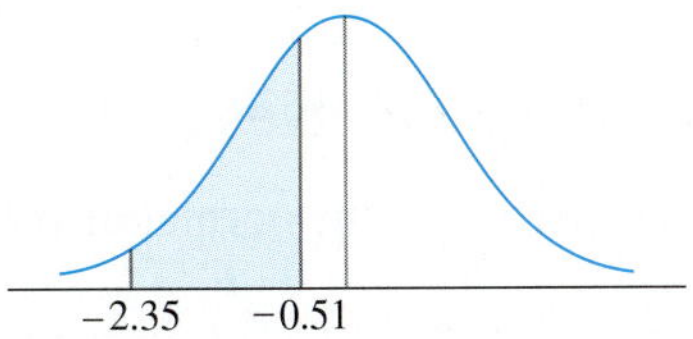

In Exercises 7–14, (a) make a sketch of the area under the standard normal curve corresponding to the probability and (b) find the value of the probability of the standard normal random variable Z corresponding to this area.

7. $P(Z < 1.38)$ **8.** $P(Z > 2.27)$

9. $P(Z < -0.64)$ **10.** $P(0.45 < Z < 1.75)$

11. $P(Z > -1.26)$ **12.** $P(-1.48 < Z < 1.54)$

13. $P(0.68 < Z < 2.02)$ **14.** $P(-1.41 < Z < -0.24)$

15. Let Z be the standard normal variable. Find the values of z if z satisfies:
a. $P(Z < z) = .8907$ **b.** $P(Z < z) = .2090$

16. Let Z be the standard normal variable. Find the values of z if z satisfies:
a. $P(Z > z) = .9678$ **b.** $P(-z < Z < z) = .8354$

17. Let Z be the standard normal variable. Find the values of z if z satisfies:
a. $P(Z > -z) = .9713$ **b.** $P(Z < -z) = .9713$

18. Suppose X is a normal random variable with $\mu = 380$ and $\sigma = 20$. Find the value of:
a. $P(X < 405)$ **b.** $P(400 < X < 430)$ **c.** $P(X > 400)$

19. Suppose X is a normal random variable with $\mu = 50$ and $\sigma = 5$. Find the value of:
a. $P(X < 60)$ **b.** $P(X > 43)$ **c.** $P(46 < X < 58)$

20. Suppose X is a normal random variable with $\mu = 500$ and $\sigma = 75$. Find the value of:
a. $P(X < 750)$ **b.** $P(X > 350)$ **c.** $P(400 < X < 600)$

8.5 Solutions to Self-Check Exercises

1. a. The probability $P(-1.2 < Z < 2.1)$ is given by the shaded area in the accompanying figure:

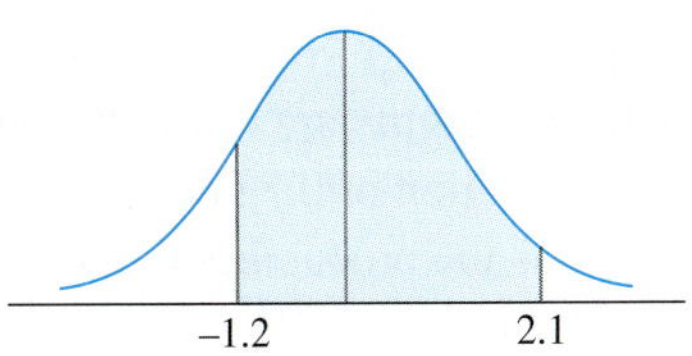

We have

$$\begin{aligned} P(-1.2 < Z < 2.1) &= P(Z < 2.1) - P(Z < -1.2) \\ &= .9821 - .1151 \\ &= .867 \end{aligned}$$

b. The region associated with $P(-z < Z < z)$ is shown in the accompanying figure:

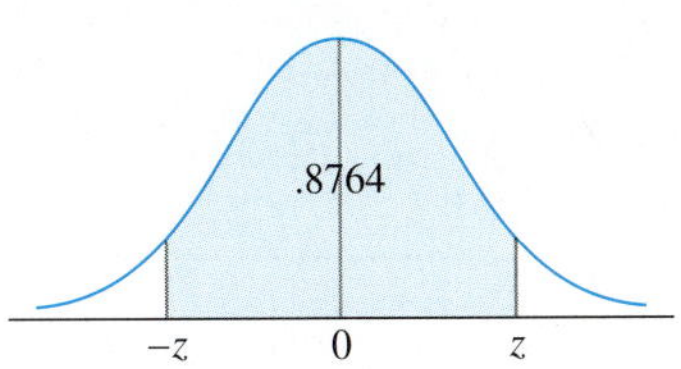

Observe that we have the following relationship:

$$P(Z < z) = \frac{1}{2}[1 + P(-z < Z < z)]$$

(see Example 2c). With $P(-z < Z < z) = .8764$, we find that

$$\begin{aligned} P(Z < z) &= \frac{1}{2}(1 + .8764) \\ &= .9382 \end{aligned}$$

Consulting the table in Appendix D, we find $z = 1.54$.

2. Using the transformation (16) and the table of values of Z, we have

a. $$\begin{aligned} P(X < 100) &= P\left(Z < \frac{100 - 80}{10}\right) \\ &= P(Z < 2) \\ &= .9772 \end{aligned}$$

b. $$\begin{aligned} P(X > 60) &= P\left(Z > \frac{60 - 80}{10}\right) \\ &= P(Z > -2) \\ &= P(Z < 2) \\ &= .9772 \end{aligned}$$

c. $$\begin{aligned} P(70 < X < 90) &= P\left(\frac{70 - 80}{10} < Z < \frac{90 - 80}{10}\right) \\ &= P(-1 < Z < 1) \\ &= P(Z < 1) - P(Z < -1) \\ &= .8413 - .1587 \\ &= .6826 \end{aligned}$$

8.6 Applications of the Normal Distribution

Applications Involving Normal Random Variables

In this section, we look at some applications involving the normal distribution.

APPLIED EXAMPLE 1 Birth Weights of Infants The medical records of infants delivered at the Kaiser Memorial Hospital show that the infants' birth weights in pounds are normally distributed with a mean of 7.4 and a standard deviation of 1.2. Find the probability that an infant selected at random from among those delivered at the hospital weighed more than 9.2 pounds at birth.

Solution Let X be the normal random variable denoting the birth weights of infants delivered at the hospital. Then the probability that an infant selected at random has a birth weight of more than 9.2 pounds is given by $P(X > 9.2)$. To compute $P(X > 9.2)$, we use Equation (18), Section 8.5, with $\mu = 7.4$, $\sigma = 1.2$, and $a = 9.2$. We find that

$$\begin{aligned} P(X > 9.2) &= P\left(Z > \frac{9.2 - 7.4}{1.2}\right) \qquad P(X > a) = P\left(Z > \frac{a - \mu}{\sigma}\right) \\ &= P(Z > 1.5) \\ &= P(Z < -1.5) \\ &= .0668 \end{aligned}$$

Thus, the probability that an infant delivered at the hospital weighs more than 9.2 pounds is .0668.

APPLIED EXAMPLE 2 Packaging Idaho Natural Produce Corporation ships potatoes to its distributors in bags whose weights are normally distributed with a mean weight of 50 pounds and standard deviation of 0.5 pound. If a bag of potatoes is selected at random from a shipment, what is the probability that it weighs:

a. More than 51 pounds?
b. Less than 49 pounds?
c. Between 49 and 51 pounds?

Solution Let X denote the weight of a bag of potatoes packed by the company. Then the mean and standard deviation of X are $\mu = 50$ and $\sigma = 0.5$, respectively.

a. The probability that a bag selected at random weighs more than 51 pounds is given by

$$\begin{aligned} P(X > 51) &= P\left(Z > \frac{51 - 50}{0.5}\right) \qquad P(X > a) = P\left(Z > \frac{a - \mu}{\sigma}\right) \\ &= P(Z > 2) \\ &= P(Z < -2) \\ &= .0228 \end{aligned}$$

b. The probability that a bag selected at random weighs less than 49 pounds is given by

$$\begin{aligned} P(X < 49) &= P\left(Z < \frac{49 - 50}{0.5}\right) \qquad P(X < b) = P\left(Z < \frac{b - \mu}{\sigma}\right) \\ &= P(Z < -2) \\ &= .0228 \end{aligned}$$

c. The probability that a bag selected at random weighs between 49 and 51 pounds is given by

$$
\begin{aligned}
&P(49 < X < 51)\\
&\quad = P\left(\frac{49 - 50}{0.5} < Z < \frac{51 - 50}{0.5}\right)\\
&\quad = P(-2 < Z < 2)\\
&\quad = P(Z < 2) - P(Z < -2)\\
&\quad = .9772 - .0228\\
&\quad = .9544
\end{aligned}
$$

$P(a < X < b)$
$= P\left(\frac{a - \mu}{\sigma} < Z < \frac{b - \mu}{\sigma}\right)$

APPLIED EXAMPLE 3 College Admissions Eligibility The grade point average (GPA) of the senior class of Jefferson High School is normally distributed with a mean of 2.7 and a standard deviation of 0.4. If a senior in the top 10% of his or her class is eligible for admission to any of the nine campuses of the state university system, what is the minimum GPA that a senior should have to ensure eligibility for university admission?

Solution Let X denote the GPA of a randomly selected senior at Jefferson High School, and let x denote the minimum GPA that will ensure his or her eligibility for admission to the university. Since only the top 10% are eligible for admission, x must satisfy the equation

$$P(X \geq x) = .1$$

Using Equation (18), Section 8.5, with $\mu = 2.7$ and $\sigma = 0.4$, we find that

$$P(X \geq x) = P\left(Z \geq \frac{x - 2.7}{0.4}\right) = .1$$

$P(X > a) = P\left(Z > \frac{a - \mu}{\sigma}\right)$

This is equivalent to the equation

$$P\left(Z < \frac{x - 2.7}{0.4}\right) = .9$$

Why?

Consulting Table 2 in Appendix D, we find that

$$\frac{x - 2.7}{0.4} = 1.28$$

Upon solving for x, we obtain

$$
\begin{aligned}
x &= (1.28)(0.4) + 2.7\\
&\approx 3.2
\end{aligned}
$$

Thus, to ensure eligibility for admission to one of the nine campuses of the state university system, a senior at Jefferson High School should have a minimum of 3.2 GPA.

Approximating Binomial Distributions

As was mentioned in Section 8.5, one important application of the normal distribution is that it provides us with an accurate approximation of other continuous probability distributions. Here, we show how a binomial distribution may be approximated by a suitable normal distribution. This technique leads to a convenient and simple solution to certain problems involving binomial probabilities.

Recall that a binomial distribution is a probability distribution of the form

$$P(X = x) = C(n, x)p^x q^{n-x} \qquad x = 0, 1, 2, \ldots, n \tag{19}$$

(See Section 8.4.) For small values of n, the arithmetic computations of the binomial probabilities may be done with relative ease. However, if n is large, then the work involved becomes prodigious, even when tables of $P(X = x)$ are available. For example, if $n = 50$, $p = .3$, and $q = .7$, then the probability of ten or more successes is given by

$$P(X \geq 10) = P(X = 10) + P(X = 11) + \cdots + P(X = 50)$$
$$= \frac{50!}{10!\,40!}(.3)^{10}(.7)^{40} + \frac{50!}{11!\,39!}(.3)^{11}(.7)^{39} + \cdots + \frac{50!}{50!\,0!}(.3)^{50}(.7)^{0}$$

TABLE 13

Probability Distribution

x	$P(X = x)$
0	.0000
1	.0000
2	.0002
3	.0011
4	.0046
5	.0148
6	.0370
7	.0739
8	.1201
9	.1602
10	.1762
11	.1602
12	.1201
.	.
.	.
.	.
20	.0000

To see how the normal distribution helps us in such situations, let's consider a coin-tossing experiment. Suppose a fair coin is tossed 20 times and we wish to compute the probability of obtaining 10 or more heads. The solution to this problem may be obtained, of course, by computing

$$P(X \geq 10) = P(X = 10) + P(X = 11) + \cdots + P(X = 20)$$

The inconvenience of this approach for solving the problem at hand has already been pointed out. As an alternative solution, let's begin by interpreting the solution in terms of finding the area of suitable rectangles of the histogram for the distribution associated with the problem. We may use Equation (19) to compute the probability of obtaining exactly x heads in 20 coin tosses. The results lead to the binomial distribution displayed in Table 13.

Using the data from the table, we next construct the histogram for the distribution (Figure 30). The probability of obtaining 10 or more heads in 20 coin tosses is equal to the sum of the areas of the shaded rectangles of the histogram of the binomial distribution shown in Figure 31.

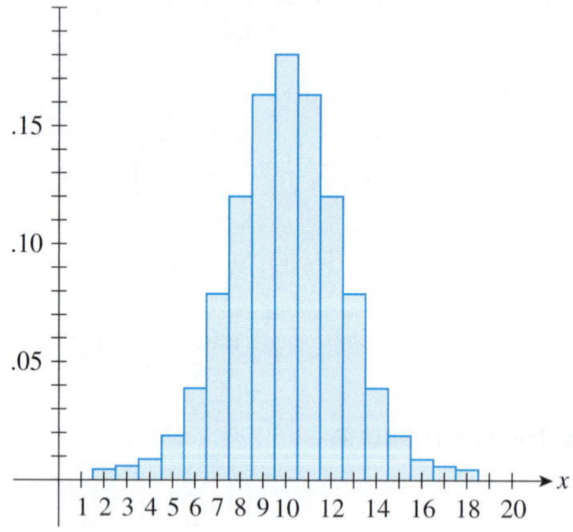

FIGURE 30
Histogram showing the probability of obtaining x heads in 20 coin tosses

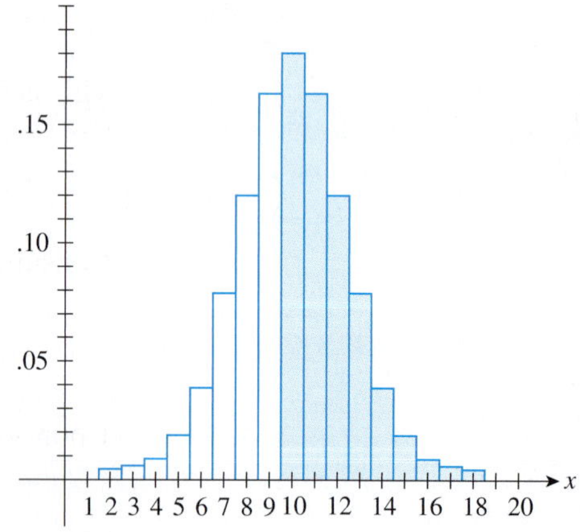

FIGURE 31
The shaded area gives the probability of obtaining 10 or more heads in 20 coin tosses.

Next, observe that the shape of the histogram suggests that the binomial distribution under consideration may be approximated by a suitable normal distribution. Since the mean and standard deviation of the binomial distribution are given by

$$\begin{aligned}\mu &= np \\ &= (20)(.5) = 10 \\ \sigma &= \sqrt{npq} \\ &= \sqrt{(20)(.5)(.5)} \\ &= 2.24\end{aligned}$$

respectively (see Section 8.4), the natural choice of a normal curve for this purpose is one with a mean of 10 and standard deviation of 2.24. Figure 32 shows such a normal curve superimposed on the histogram of the binomial distribution.

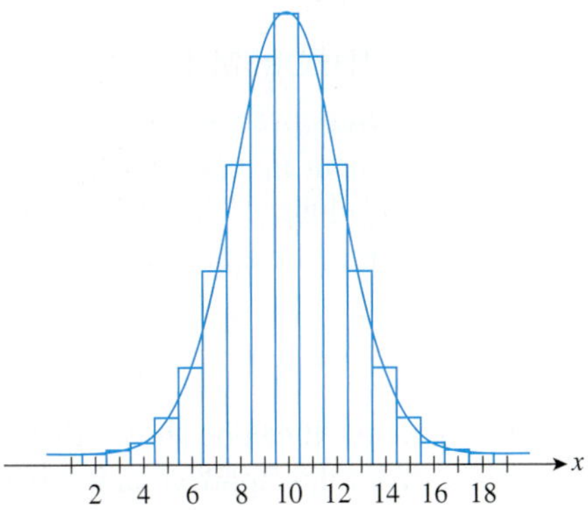

FIGURE **32**
Normal curve superimposed on the histogram for a binomial distribution

The good fit suggests that the sum of the areas of the rectangles representing $P(X \geq 10)$, the probability of obtaining 10 or more heads in 20 coin tosses, may be approximated by the area of an appropriate region under the normal curve. To determine this region, let's note that the base of the portion of the histogram representing the required probability extends from $x = 9.5$ on, since the base of the leftmost rectangle in the shaded region is centered at $x = 10$ and the base of each rectangle has length 1 (Figure 33). Therefore, the required region under the normal curve should also have $x \geq 9.5$. Letting Y denote the continuous normal variable, we obtain

$$\begin{aligned} P(X \geq 10) &\approx P(Y \geq 9.5) \\ &= P(Y > 9.5) \\ &\approx P\left(Z > \frac{9.5 - 10}{2.24}\right) \qquad P(X > a) = P\left(Z > \frac{a - \mu}{\sigma}\right) \\ &\approx P(Z > -0.22) \\ &= P(Z < 0.22) \\ &= .5871 \qquad \text{Use the table of values of } Z \text{ in Appendix D.} \end{aligned}$$

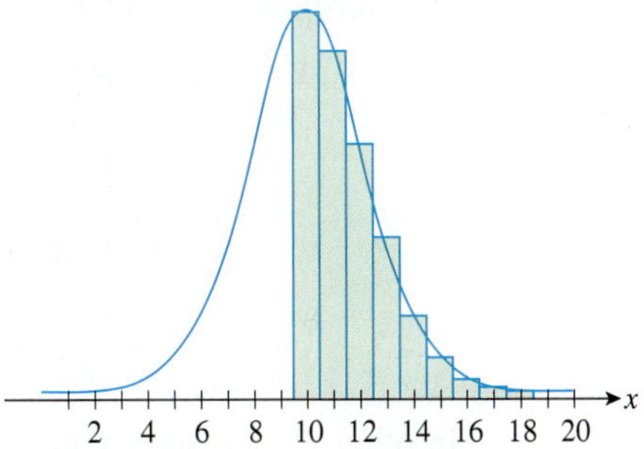

FIGURE **33**
$P(X \geq 10)$ is approximated by the area under the normal curve.

The exact value of $P(X \geq 10)$ may be found by computing

$$P(X = 10) + P(X = 11) + \cdots + P(X = 20)$$

in the usual fashion and is equal to .5881. Thus, the normal distribution with suitably chosen mean and standard deviation does provide us with a good approximation of the binomial distribution.

In the general case, the following result, which is a special case of the *central limit theorem*, guarantees the accuracy of the approximation of a binomial distribution by a normal distribution under certain conditions.

THEOREM 1

Suppose we are given a binomial distribution associated with a binomial experiment involving n trials, each with a probability of success p and probability of failure q. Then, if n is large and p is not close to 0 or 1, the binomial distribution may be approximated by a normal distribution with

$$\mu = np \quad \text{and} \quad \sigma = \sqrt{npq}$$

Note It can be shown that if both np and nq are greater than 5, then the error resulting from this approximation is negligible.

Applications Involving Binomial Random Variables

Next, we look at some applications involving binomial random variables.

APPLIED EXAMPLE 4 Quality Control An automobile manufacturer receives the microprocessors that are used to regulate fuel consumption in its automobiles in shipments of 1000 each from a certain supplier. It has been estimated that, on the average, 1% of the microprocessors manufactured by the supplier are defective. Determine the probability that more than 20 of the microprocessors in a single shipment are defective.

Solution Let X denote the number of defective microprocessors in a single shipment. Then X has a binomial distribution with $n = 1000$, $p = .01$, and $q = .99$, so

$$\begin{aligned} \mu &= (1000)(.01) = 10 \\ \sigma &= \sqrt{(1000)(.01)(.99)} \\ &\approx 3.15 \end{aligned}$$

Approximating the binomial distribution by a normal distribution with a mean of 10 and a standard deviation of 3.15, we find that the probability that more than 20 microprocessors in a shipment are defective is given by

$$\begin{aligned} P(X > 20) &\approx P(Y > 20.5) && \text{Where } Y \text{ denotes the normal random variable} \\ &= P\left(Z > \frac{20.5 - 10}{3.15}\right) && P(X > a) = P\left(Z > \frac{a - \mu}{\sigma}\right) \\ &\approx P(Z > 3.33) \\ &\approx P(Z < -3.33) \\ &= .0004 \end{aligned}$$

In other words, approximately 0.04% of the shipments containing 1000 microprocessors each will contain more than 20 defective units.

APPLIED EXAMPLE 5 Heart Transplant Survival Rate The probability that a heart transplant performed at the Medical Center is successful (that is, the patient survives 1 year or more after undergoing the surgery) is .7. Of 100 patients who have undergone such an operation, what is the probability that:

a. Fewer than 75 will survive 1 year or more after the operation?
b. Between 80 and 90, inclusive, will survive 1 year or more after the operation?

Solution Let X denote the number of patients who survive 1 year or more after undergoing a heart transplant at the Medical Center; then X is a binomial random variable. Also, $n = 100$, $p = .7$, and $q = .3$, so

$$\mu = (100)(.7) = 70$$
$$\sigma = \sqrt{(100)(.7)(.3)} \approx 4.58$$

Approximating the binomial distribution by a normal distribution with a mean of 70 and a standard deviation of 4.58, we find, upon letting Y denote the associated normal random variable:

a. The probability that fewer than 75 patients will survive 1 year or more is given by

$$\begin{aligned} P(X < 75) &\approx P(Y < 74.5) && \text{Why?} \\ &\approx P\left(Z < \frac{74.5 - 70}{4.58}\right) && P(X < b) = P\left(Z < \frac{b - \mu}{\sigma}\right) \\ &\approx P(Z < 0.98) \\ &= .8365 \end{aligned}$$

b. The probability that between 80 and 90 patients, inclusive, will survive 1 year or more is given by

$$\begin{aligned} &P(80 \le X \le 90) \\ &\quad \approx P(79.5 < Y < 90.5) && P(a < X < b) \\ &\quad \approx P\left(\frac{79.5 - 70}{4.58} < Z < \frac{90.5 - 70}{4.58}\right) && = P\left(\frac{a - \mu}{\sigma} < Z < \frac{b - \mu}{\sigma}\right) \\ &\quad \approx P(2.07 < Z < 4.48) \\ &\quad = P(Z < 4.48) - P(Z < 2.07) \\ &\quad = 1 - .9808 && \text{Note: } P(Z < 4.48) \approx 1 \\ &\quad = .0192 \end{aligned}$$

8.6 Self-Check Exercises

1. **Cholesterol Levels** The serum cholesterol levels in milligrams per decaliter (mg/dL) in a current Mediterranean population are found to be normally distributed with a mean of 160 and a standard deviation of 50. Scientists at the National Heart, Lung, and Blood Institute consider this pattern ideal for a minimal risk of heart attacks. Find the percentage of the population having blood cholesterol levels between 160 and 180 mg/dL.

2. **Classifying Seconds** It has been estimated that 4% of the luggage manufactured by The Luggage Company fails to meet the standards established by the company and is sold as "seconds" to discount and outlet stores. If 500 bags are produced, what is the probability that more than 30 will be classified as "seconds"?

Solutions to Self-Check Exercises 8.6 can be found on page 529.

8.6 Concept Questions

1. What does Theorem 1, page 526, allow us to do?

2. Suppose a binomial distribution is associated with a binomial experiment involving n trials, each with a probability of success p and probability of failure q, and suppose n and p satisfy the other conditions given in Theorem 1. What formulas for μ and σ can be used to approximate this binomial distribution by a normal distribution?

8.6 Exercises

1. **Medical Records** The medical records of infants delivered at Kaiser Memorial Hospital show that the infants' lengths at birth (in inches) are normally distributed with a mean of 20 and a standard deviation of 2.6. Find the probability that an infant selected at random from among those delivered at the hospital measures:
   a. More than 22 in. **b.** Less than 18 in.
   c. Between 19 and 21 in.

2. **Factory Workers' Wages** According to the data released by the Chamber of Commerce of a certain city, the weekly wages of factory workers are normally distributed with a mean of \$720 and a standard deviation of \$60. What is the probability that a factory worker selected at random from the city makes a weekly wage:
   a. Of less than \$720? **b.** Of more than \$912?
   c. Between \$660 and \$780?

3. **Product Reliability** TKK Products manufactures 50-, 60-, 75-, and 100-watt electric light bulbs. Laboratory tests show that the lives of these light bulbs are normally distributed with a mean of 750 hr and a standard deviation of 75 hr. What is the probability that a TKK light bulb selected at random will burn:
   a. For more than 900 hr?
   b. For less than 600 hr?
   c. Between 750 and 900 hr?
   d. Between 600 and 800 hr?

4. **Education** On average, a student takes dictation at a speed of 100 words/minute midway through an advanced court reporting course at the American Institute of Court Reporting. Assuming that the dictation speeds of the students are normally distributed and that the standard deviation is 20 words/minute, what is the probability that a student randomly selected from the course can take dictation at a speed:
   a. Of more than 120 words/minute?
   b. Between 80 and 120 words/minute?
   c. Of less than 80 words/minute?

5. **IQs** The IQs of students at Wilson Elementary School were measured recently and found to be normally distributed with a mean of 100 and a standard deviation of 15. What is the probability that a student selected at random will have an IQ:
   a. Of 140 or higher? **b.** Of 120 or higher?
   c. Between 100 and 120? **d.** Of 90 or less?

6. **Product Reliability** The tread lives of the Super Titan radial tires under normal driving conditions are normally distributed with a mean of 40,000 mi and a standard deviation of 2000 mi. What is the probability that a tire selected at random will have a tread life of more than 35,000 mi? Determine the probability that four tires selected at random still have useful tread lives after 35,000 mi of driving. (Assume that the tread lives of the tires are independent of each other.)

7. **Female Factory Workers' Wages** According to data released by the Chamber of Commerce of a certain city, the weekly wages (in dollars) of female factory workers are normally distributed with a mean of 675 and a standard deviation of 50. Find the probability that a female factory worker selected at random from the city makes a weekly wage of \$650 to \$750.

8. **Civil Service Exams** To be eligible for further consideration, applicants for certain civil service positions must first pass a written qualifying examination on which a score of 70 or more must be obtained. In a recent examination, it was found that the scores were normally distributed with a mean of 60 points and a standard deviation of 10 points. Determine the percentage of applicants who passed the written qualifying examination.

9. **Warranties** The general manager of the service department of MCA Television has estimated that the time that elapses between the dates of purchase and the dates on which the 50-in. plasma TVs manufactured by the company first require service is normally distributed with a mean of 22 months and a standard deviation of 4 months. If the company gives a 1-year warranty on parts and labor for these TVs, determine the percentage of these TVs manufactured and sold by the company that will require service before the warranty period runs out.

10. **Grade Distributions** The scores on an economics examination are normally distributed with a mean of 72 and a standard deviation of 16. If the instructor assigns a grade of A to 10% of the class, what is the lowest score a student may have and still obtain an A?

11. **Grade Distributions** The scores on a sociology examination are normally distributed with a mean of 70 and a standard deviation of 10. If the instructor assigns As to 15%, Bs to 25%, Cs to 40%, Ds to 15%, and Fs to 5% of the class, find the cutoff points for grades A–D.

12. **Highway Speeds** The speeds (in miles per hour) of motor vehicles on a certain stretch of Route 3A as clocked at a certain place along the highway are normally distributed with a mean of 64.2 mph and a standard deviation of 8.44 mph. What is the probability that a motor vehicle selected at random is traveling at:
    a. More than 65 mph?
    b. Less than 60 mph?
    c. Between 65 and 70 mph?

In Exercises 13–24, use the appropriate normal distributions to approximate the resulting binomial distributions.

13. A coin is weighted so that the probability of obtaining a head in a single toss is .4. If the coin is tossed 25 times, what is the probability of obtaining:
    a. Fewer than 10 heads?
    b. Between 10 and 12 heads, inclusive?
    c. More than 15 heads?

14. A fair coin is tossed 20 times. What is the probability of obtaining:
    a. Fewer than 8 heads?
    b. More than 6 heads?
    c. Between 6 and 10 heads inclusive?

15. **Chances of Hitting a Target** A marksman's chance of hitting a target with each of his shots is 60%. (Assume that the shots are independent of each other.) If he fires 30 shots, what is the probability of his hitting the target:
    a. At least 20 times?
    b. Fewer than 10 times?
    c. Between 15 and 20 times, inclusive?

16. **Chances of Making a Free Throw** A basketball player has a 75% chance of making a free throw. (Assume that the throws are independent of each other.) What is the probability of her making 100 or more free throws in 120 trials?

17. **Classifying Seconds** The manager of C & R Clothiers, a manufacturer of men's dress shirts, has determined that 3% of C & R's shirts do not meet company standards and are sold as "seconds" to discount and outlet stores. What is the probability that in a production run of 200 shirts, fewer than 10 will be classified as "seconds"?

18. **Telemarketing** Jorge sells magazine subscriptions over the phone. He estimates that the probability of his making a sale with each attempt is .12. What is the probability of Jorge making more than 10 sales if he makes 80 calls?

19. **Industrial Accidents** Colorado Mining and Mineral has 800 employees engaged in its mining operations. It has been estimated that the probability of a worker meeting with an accident during a 1-year period is .1. What is the probability that more than 70 workers will meet with an accident during the 1-year period?

20. **Quality Control** PAR Bearings is the principal supplier of ball bearings for the Sperry Gyroscope Company. It has been determined that 6% of the ball bearings shipped are rejected because they fail to meet tolerance requirements. What is the probability that a shipment of 200 ball bearings contains more than 10 rejects?

21. **Drug Testing** An experiment was conducted to test the effectiveness of a new drug in treating a certain disease. The drug was administered to 50 mice that had been previously exposed to the disease. It was found that 35 mice subsequently recovered from the disease. It has been determined that the natural recovery rate from the disease is 0.5.
    a. Determine the probability that 35 or more of the mice not treated with the drug would recover from the disease.
    b. Using the results obtained in part (a), comment on the effectiveness of the drug in the treatment of the disease.

22. **Loan Delinquencies** The manager of Madison Finance Company has estimated that, because of a recession, 5% of its 400 loan accounts will be delinquent. If the manager's estimate is correct, what is the probability that 25 or more of the accounts will be delinquent?

23. **Cruise Ship Bookings** Because of late cancellations, Neptune Lines, an operator of cruise ships, has a policy of accepting more reservations than there are accommodations available. From experience, 8% of the bookings for the 90-day around-the-world cruise on the S.S. *Drion*, which has accommodations for 2000 passengers, are subsequently canceled. If the management of Neptune Lines has decided, for public relations reasons, that 99% of all booked passengers will obtain accommodation on the ship, determine the largest number of reservations that should be taken for this cruise on the S.S. *Drion*.

24. **Theater Overbookings** Preview Showcase, a research firm, screens pilots of new TV shows before a randomly selected audience and then solicits the audience members' opinions of the shows. Based on past experience, 20% of those who receive complimentary tickets are "no-shows." The theater has a seating capacity of 500. Management has decided, for public relations reasons, that 99% of all ticket holders will be seated. How many tickets should the company send out to prospective viewers for each screening?

8.6 Solutions to Self-Check Exercises

1. Let X be the normal random variable denoting the serum cholesterol levels in milligrams per deciliter in the current Mediterranean population under consideration. Then the percentage of the population having blood cholesterol levels between 160 and 180 mg/dL is given by $P(160 < X < 180)$. To compute $P(160 < X < 180)$, we use Equation (16), Section 8.5, with $\mu = 160$, $\sigma = 50$, $a = 160$, and $b = 180$. We find

$$\begin{aligned} P(160 < X < 180) &= P\left(\frac{160 - 160}{50} < Z < \frac{180 - 160}{50}\right) \\ &= P(0 < Z < 0.4) \\ &= P(Z < 0.4) - P(Z < 0) \\ &= .6554 - .5000 \\ &= .1554 \end{aligned}$$

Thus, approximately 15.5% of the population has blood cholesterol levels between 160 and 180 mg/dL.

2. Let X denote the number of substandard bags in the production. Then X has a binomial distribution with $n = 500$, $p = .04$, and $q = .96$, so

$$\mu = (500)(.04) = 20$$
$$\sigma = \sqrt{(500)(.04)(.96)} \approx 4.38$$

Approximating the binomial distribution by a normal distribution with a mean of 20 and standard deviation of 4.38, we find that the probability that more than 30 bags in the production of 500 will be classified as "seconds" is given by

$$\begin{aligned} P(X > 30) &\approx P(Y > 30.5) \quad \text{Where } Y \text{ denotes the normal random variable} \\ &\approx P\left(Z > \frac{30.5 - 20}{4.38}\right) \\ &\approx P(Z > 2.40) \\ &= P(Z < -2.40) \\ &= .0082 \end{aligned}$$

or approximately 0.8%.

CHAPTER 8 Summary of Principal Formulas and Terms

FORMULAS

1. Mean of n numbers	$\bar{x} = \dfrac{x_1 + x_2 + \cdots + x_n}{n}$
2. Expected value	$E(X) = x_1p_1 + x_2p_2 + \cdots + x_np_n$
3. Odds in favor of E occurring	$\dfrac{P(E)}{P(E^c)}$
4. Odds against E occurring	$\dfrac{P(E^c)}{P(E)}$
5. Probability of an event occurring given the odds	$\dfrac{a}{a + b}$
6. Variance of a random variable	$\text{Var}(X) = p_1(x_1 - \mu)^2 + p_2(x_2 - \mu)^2 + \cdots + p_n(x_n - \mu)^2$
7. Standard deviation of a random variable	$\sigma = \sqrt{\text{Var}(X)}$
8. Chebychev's inequality	$P(\mu - k\sigma \le X \le \mu + k\sigma) \ge 1 - \dfrac{1}{k^2}$
9. Probability of x successes in n Bernoulli trials	$C(n, x)p^xq^{n-x}$
10. Binomial random variable: Mean Variance Standard deviation	$\mu = E(X) = np$ $\text{Var}(X) = npq$ $\sigma_X = \sqrt{npq}$

TERMS

random variable (458)
finite discrete random variable (459)
infinite discrete random variable (459)
continuous random variable (459)
probability distribution of a random variable (459)
histogram (461)
average (mean) (472)
expected value (473)
median (481)
mode (481)
variance (488)
standard deviation (489)
Bernoulli (binomial) trial (501)
binomial experiment (501)
binomial random variable (504)
binomial distribution (504)
probability density function (513)
normal distribution (514)

CHAPTER 8 Concept Review Questions

Fill in the blanks.

1. A rule that assigns a number to each outcome of a chance experiment is called a/an ______ variable.

2. If a random variable assumes only finitely many values, then it is called ______ discrete; if it takes on infinitely many values that can be arranged in a sequence, then it is called ______ discrete; if it takes on all real numbers in an interval, then it is said to be ______.

3. The expected value of a random variable X is given by the ______ of the products of the values assumed by the random variable and their associated probabilities. For example, if X assumes the values -2, 3, and 4 with associated probabilities $\frac{1}{2}$, $\frac{1}{4}$, and $\frac{1}{4}$, then its expected value is ______.

4. **a.** If the probability of an event E occurring is $P(E)$, then the odds in favor of E occurring are ______.
   b. If the odds in favor of an event E occurring are a to b, then the probability of E occurring is ______.

5. Suppose a random variable X takes on the values $x_1, x_2, \ldots, x_n$ with probabilities $p_1, p_2, \ldots, p_n$ and has a mean of μ. Then the variance of X is ______, and the standard deviation of X is ______.

6. In a binomial experiment, the number of trials is ______, there are exactly ______ outcomes in each trial, the probability of "success" in each trial is the ______, and the trials are ______ of each other.

7. A probability distribution that is associated with a continuous random variable is called a/an ______ probability distribution. Such a probability distribution is defined by a/an ______ ______ ______ whose domain is the ______ of values taken on by the random variable associated with the experiment.

8. A binomial distribution may be approximated by a/an ______ distribution with $\mu = np$ and $\sigma = \sqrt{npq}$ if n is ______ and p is not close to ______ or ______.

CHAPTER 8 Review Exercises

1. Three balls are selected at random without replacement from an urn containing three white balls and four blue balls. Let the random variable X denote the number of blue balls drawn.
   a. List the outcomes of this experiment.
   b. Find the value assigned to each outcome of this experiment by the random variable X.
   c. Find the probability distribution of the random variable associated with this experiment.
   d. Draw the histogram representing this distribution.

2. **Life Insurance Policies** A man purchased a \$25,000, 1-year term-life insurance policy for \$375. Assuming that the probability that he will live for another year is .989, find the company's expected gain.

3. The probability distribution of a random variable X is shown in the following table:

x	$P(X = x)$
0	.1
1	.1
2	.2
3	.3
4	.2
5	.1

   a. Compute $P(1 \leq X \leq 4)$.
   b. Compute the mean and standard deviation of X.

4. A binomial experiment consists of four trials in which the probability of success in any one trial is $\frac{2}{5}$.
   a. Construct the probability distribution for the experiment.
   b. Compute the mean and standard deviation of the probability distribution.

In Exercises 5–8, let Z be the standard normal variable. Make a rough sketch of the appropriate region under the standard normal curve, and find the probability.

5. $P(Z < 0.5)$
6. $P(Z < -0.75)$
7. $P(-0.75 < Z < 0.5)$
8. $P(-0.42 < Z < 0.66)$

In Exercises 9–12, let Z be the standard normal variable. Find z if z satisfies the given value.

9. $P(Z < z) = .9922$
10. $P(Z < z) = .1469$
11. $P(Z > z) = .9788$
12. $P(-z < Z < z) = .8444$

In Exercises 13–16, let X be a normal random variable with $\mu = 10$ and $\sigma = 2$. Find the value of the given probability.

13. $P(X < 11)$
14. $P(X > 8)$
15. $P(7 < X < 9)$
16. $P(6.5 < X < 11.5)$

17. **Driving Age Requirements** The minimum age requirement for a regular driver's license differs from state to state.

The frequency distribution for this age requirement in the 50 states is given in the following table:

Minimum Age (in years)	15	16	17	18	19	21
Frequency of Occurrence	1	15	4	28	1	1

a. Describe a random variable X that is associated with these data.
b. Find the probability distribution for the random variable X.
c. Compute the mean, variance, and standard deviation of X.

18. Annual Food Expenditure According to a report of the U.S. Bureau of Labor Statistics, the average annual food expenditures by quintiles of household income before taxes are as follows:

Income Quintile	Lowest 20%	Second 20%	Third 20%	Fourth 20%	Highest 20%
Expenditures ($)	3501	4568	5482	7522	10,780

Let the random variable X denote a randomly chosen quintile (in income) within the set of all households.
a. Find the probability distribution associated with these data.
b. What percentage of the total food expenditures is incurred by the top 40% by income of the population? By the lowest 40% by income of the population?
Source: U.S. Bureau of Labor Statistics.

19. Traffic A traffic survey of the speeds of vehicles traveling along a stretch of Hampton Road between 4 P.M. and 6 P.M. yielded the following results:

Speed (in mph)	30–34	35–39	40–44	45–49	50–54
Probability	.07	.28	.42	.18	.05

Find the average speed of the vehicles.

20. Expected Profit A buyer for Discount Fashions, an outlet for women's apparel, is considering buying a batch of clothing for $64,000. She estimates that the company will be able to sell it for $80,000, $75,000, or $70,000 with probabilities of .30, .60, and .10, respectively. On the basis of these estimates, what will be the company's expected gross profit?

21. Bowling a Strike If the probability that a bowler will bowl a strike is .7, what is the probability that he will get exactly two strikes in four attempts? At least two strikes in four attempts? (Assume that the attempts to bowl a strike are independent of each other.)

22. Heights of Women The heights of 4000 women who participated in a recent survey were found to be normally distributed with a mean of 64.5 in. and a standard deviation of 2.5 in. What percentage of these women have heights of 67 in. or greater?

23. Heights of Women Refer to Exercise 22. Use Chebychev's inequality to estimate the probability that the height of a woman who participated in the survey will fall within 2 standard deviations of the mean—that is, that her height will be between 59.5 and 69.5 in.

24. Netflix Revenue from Streaming Subscribers The revenue of Netflix from its streaming subscribers (in millions of dollars) for the five quarters beginning with the first quarter of 2012 are summarized in the following table:

	2012				2013
Quarter	Q1	Q2	Q3	Q4	Q1
Revenue	26.2	27.5	28.4	35.8	37.1

Find the average quarterly revenue of Netflix from its streaming subscribers for the five quarters in question. What is the standard deviation?
Source: Company reports.

25. Marital Status of Women The number of single women between the ages of 20 and 44 in the United States in 2010 is given in the following table:

Age (in years)	20–24	25–29	30–34	35–39	40–44
Women (in thousands)	8296	5026	2678	1768	1430

Find the mean and the standard deviation of the given data.
Hint: Assume that all values lying within a group interval take the middle value of that group.
Source: U.S. Census Bureau.

26. Quality Control The proprietor of a hardware store will accept a shipment of ceramic wall tiles if no more than 2 tiles of a random sample of 20 are found to be defective. What is the probability that he will accept shipment if exactly 10% of the tiles in a certain shipment is defective?

27. Drug Effectiveness An experimental drug has been found to be effective in treating 15% of the people afflicted by a certain disease. If the drug is administered to 800 people who have this disease, what are the mean and standard deviation of the number of people for whom the drug can be expected to be effective?

28. Quality Control Dayton Iron Works manufactures steel rods to a specification of 1-in. diameter. These rods are accepted by the buyer if they fall within the tolerance limits of 0.995 and 1.005. Assuming that the diameter of the rods is normally distributed about a mean of 1 in. and has a standard deviation of 0.002 in., estimate the percentage of rods that will be rejected by the buyer.

29. Expected Sales The division manager of a company claims that 80% of her sales representatives will make or exceed their sales quota in the following month. Assuming that the manager's assessment is correct, what is the probability that the sales quota will be made or exceeded by:

a. Four of the six sales representatives?

b. At least four of the sales representatives?

30. **On-Time Arrivals** Diane, who commutes regularly between Los Angeles and San Francisco for business, estimates that the probability that her flight will arrive on time or earlier is .92. Assuming that her assessment is accurate, what is the probability that her flight will arrive on time or earlier:

a. In three of her next five flights?

b. In at least three of her next five flights?

31. **Coin Tosses** A coin is biased so that the probability of it landing heads is .6. If the coin is tossed 100 times, what is the probability that heads will appear more than 50 times in the 100 tosses?

32. **Quality Control** A division of Solaron Corporation manufactures photovoltaic cells for use in the company's solar energy converters. It is estimated that 5% of the cells manufactured are defective. In a batch of 200 cells manufactured by the company, what is the probability that it will contain at most 20 defective units?

CHAPTER 8 Before Moving On . . .

1. The values taken on by a random variable X and the frequency of their occurrence are shown in the following table. Find the probability distribution of X.

x	-3	-2	0	1	2	3
Frequency of Occurrence	4	8	20	24	16	8

2. The probability distribution of the random variable X is shown in the following table:

x	-4	-3	-1	0	1	3
$P(X = x)$	.06	.14	.32	.28	.12	.08

Find:

a. $P(X \leq 0)$ **b.** $P(-4 \leq X \leq 1)$

3. Find the mean, variance, and standard deviation of a random variable X having the following probability distribution:

x	-3	-1	0	1	3	5
$P(X = x)$	.08	.24	.32	.16	.12	.08

4. A binomial experiment consists of four independent trials, and the probability of success in each trial is 0.3.

a. Find the probability of obtaining 0, 1, 2, 3, and 4 successes, respectively.

b. Compute the mean and standard deviation of the random variable associated with this experiment.

5. Let X be a normal random variable with $\mu = 60$ and $\sigma = 5$. Find the values of:

a. $P(X < 70)$

b. $P(X > 50)$

c. $P(50 < X < 70)$

6. A fair coin is tossed 30 times. Using the appropriate normal distribution to approximate a binomial distribution, find the probability of obtaining:

a. Fewer than 10 heads.

b. Between 12 and 16 heads, inclusive.

c. More than 20 heads.

9 Markov Chains and the Theory of Games

IN THIS CHAPTER, we look at two important applications of mathematics that are based primarily on matrix theory and the theory of probability. Both of these applications, *Markov chains* and the *theory of games,* though relatively recent developments in the field of mathematics, have wide applications in many practical areas.

After the successful implementation of an urban renewal program, what percentage of the population of a metropolitan area will live in the city, and what percentage of the population will live in the suburbs? In Examples 4 and 5, pages 539–540, we show how we can find the population distribution over the next few years for a population that can be described by a Markov process.

9.1 Markov Chains

Transitional Probabilities

A finite stochastic process, you may recall, is an experiment consisting of a finite number of stages in which the outcomes and associated probabilities at each stage depend on the outcomes and associated probabilities of the *preceding stages*. In this chapter, we are concerned with a special class of stochastic processes—namely, those in which the probabilities associated with the outcomes at any stage of the experiment depend only on the outcomes of the *preceding stage*. Such a process is called a **Markov process,** or a **Markov chain,** named after the Russian mathematician A. A. Markov (1856–1922).

The outcome at any stage of the experiment in a Markov process is called the **state** of the experiment. In particular, the outcome at the current stage of the experiment is called the **current state** of the process. Here is a typical problem involving a Markov chain:

Starting from one state of a process (the current state), determine the probability that the process will be at a particular state at some future time.

APPLIED EXAMPLE 1 Common Stocks An analyst at Weaver and Kline, a stock brokerage firm, observes that the closing price of the preferred stock of an airline company over a short span of time depends only on the stock's previous closing price. At the end of each trading day, the analyst makes a note of the stock's performance for that day, recording the closing price as "higher," "unchanged," or "lower" according to whether the stock closes higher, unchanged, or lower than the previous day's closing price. This sequence of observations may be viewed as a Markov chain. ■

The transition from one state to another in a Markov chain may be studied with the aid of tree diagrams, as in the next example.

APPLIED EXAMPLE 2 Common Stocks Refer to Example 1. If on a certain day, the stock's closing price is higher than that of the previous day, then the probability that the stock closes higher, unchanged, or lower on the next trading day is .2, .3, and .5, respectively. Next, if the stock's closing price is unchanged from the previous day, then the probability that the stock closes higher, unchanged, or lower on the next trading day is .5, .2, and .3, respectively. Finally, if the stock's closing price is lower than that of the previous day, then the probability that the stock closes higher, unchanged, or lower on the next trading day is .4, .4, and .2, respectively. With the aid of tree diagrams, describe the transition between states and the probabilities associated with these transitions.

Solution The Markov chain being described has three states: higher, unchanged, and lower. If the current state is higher, then the transition to the other states from this state may be displayed by constructing a tree diagram in which the associated

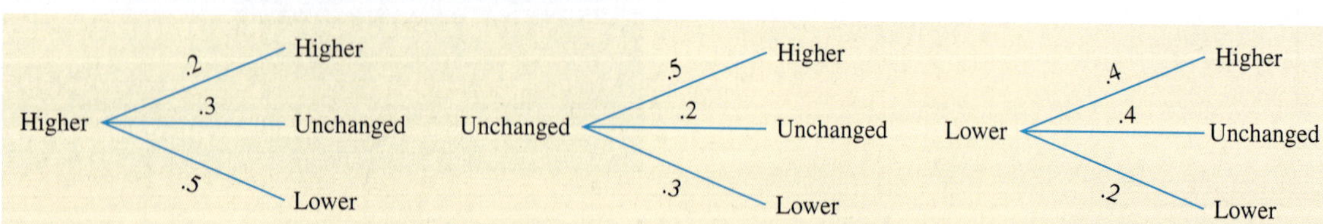

FIGURE 1
Tree diagrams showing transition probabilities between states

probabilities are shown on the appropriate limbs (Figure 1). Tree diagrams describing the transition from each of the other two possible current states, unchanged and lower, to the other states are constructed in a similar manner. ■

The probabilities encountered in this example are called **transition probabilities** because they are associated with the transition from one state to the next in the Markov process. These transition probabilities may be conveniently represented in the form of a matrix. Suppose for simplicity that we have a Markov chain with three possible outcomes at each stage of the experiment. Let's refer to these outcomes as state 1, state 2, and state 3. Then the transition probabilities associated with the transition from state 1 to each of the states 1, 2, and 3 in the next phase of the experiment are precisely the respective conditional probabilities that the outcome is state 1, state 2, and state 3 *given* that the outcome state 1 has occurred. In short, the desired transition probabilities are $P(\text{state } 1 \mid \text{state } 1)$, $P(\text{state } 2 \mid \text{state } 1)$, and $P(\text{state } 3 \mid \text{state } 1)$, respectively. Thus, we write

Next state

$$\begin{aligned} a_{11} &= P(\text{state } 1 \mid \text{state } 1) \\ a_{21} &= P(\text{state } 2 \mid \text{state } 1) \\ a_{31} &= P(\text{state } 3 \mid \text{state } 1) \end{aligned}$$

Current state

Note that the first subscript in this notation refers to the state in the next stage of the experiment, and the second subscript refers to the current state. Using a tree diagram, we have the following representation:

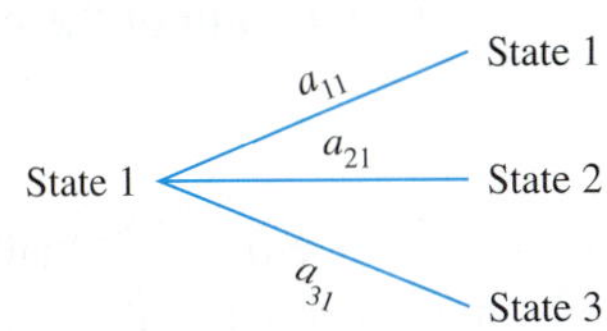

Similarly, the transition probabilities associated with the transition from state 2 and state 3 to each of the states 1, 2, and 3 are

$$\begin{aligned} a_{12} &= P(\text{state } 1 \mid \text{state } 2) \quad \text{and} \quad & a_{13} &= P(\text{state } 1 \mid \text{state } 3) \\ a_{22} &= P(\text{state } 2 \mid \text{state } 2) & a_{23} &= P(\text{state } 2 \mid \text{state } 3) \\ a_{32} &= P(\text{state } 3 \mid \text{state } 2) & a_{33} &= P(\text{state } 3 \mid \text{state } 3) \end{aligned}$$

These observations lead to the following matrix representation of the transition probabilities:

		Current state		
		State 1	State 2	State 3
Next state	State 1	a_{11}	a_{12}	a_{13}
	State 2	a_{21}	a_{22}	a_{23}
	State 3	a_{31}	a_{32}	a_{33}

EXAMPLE 3 Use a matrix to represent the transition probabilities obtained in Example 2.

Solution There are three states at each stage of the Markov chain under consideration. Letting state 1, state 2, and state 3 denote the states "higher," "unchanged," and "lower," respectively, we find that

$$a_{11} = .2 \qquad a_{21} = .3 \qquad a_{31} = .5$$

and so on, so the required matrix representation is given by

$$T = \begin{bmatrix} .2 & .5 & .4 \\ .3 & .2 & .4 \\ .5 & .3 & .2 \end{bmatrix}$$

The matrix obtained in Example 3 is a transition matrix. In the general case, we have the following definition:

Transition Matrix

A **transition matrix** associated with a Markov chain with n states is an $n \times n$ matrix T with entries a_{ij} $(1 \le i \le n, 1 \le j \le n)$

$$\begin{array}{c} \text{Current state} \\ \begin{array}{ccccc} \text{State 1} & \text{State 2} & \cdots\ \text{State } j & \cdots & \text{State } n \end{array} \end{array}$$

$$T = \text{Next state} \begin{array}{c} \text{State 1} \\ \text{State 2} \\ \vdots \\ \text{State } i \\ \vdots \\ \text{State } n \end{array} \begin{bmatrix} a_{11} & a_{12} & \cdots & a_{1j} & \cdots & a_{1n} \\ a_{21} & a_{22} & \cdots & a_{2j} & \cdots & a_{2n} \\ \vdots & \vdots & & \vdots & & \vdots \\ a_{i1} & a_{i2} & \cdots & a_{ij} & \cdots & a_{in} \\ \vdots & \vdots & & \vdots & & \vdots \\ a_{n1} & a_{n2} & \cdots & a_{nj} & \cdots & a_{nn} \end{bmatrix}$$

having the following properties:

1. $a_{ij} \ge 0$ for all i and j.

2. The sum of the entries in each column of T is 1.

Since $a_{ij} = P(\text{state } i \mid \text{state } j)$ is the probability of the occurrence of an event, it must be nonnegative, and this is precisely what Property 1 implies. Property 2 follows from the fact that the transition from any one of the current states must terminate in one of the n states in the next stage of the experiment. Any square matrix that satisfies properties 1 and 2 is referred to as a **stochastic matrix.**

One advantage in representing the transition probabilities in the form of a matrix is that we may use the results from matrix theory to help us solve problems involving Markov processes, as we will see in the next several sections.

Explore and Discuss

Let

$$A = \begin{bmatrix} p & q \\ 1-p & 1-q \end{bmatrix} \quad \text{and} \quad B = \begin{bmatrix} r & s \\ 1-r & 1-s \end{bmatrix}$$

be two 2×2 stochastic matrices, where $0 \le p \le 1$, $0 \le q \le 1$, $0 \le r \le 1$, and $0 \le s \le 1$.

1. Show that AB is a 2×2 stochastic matrix.

2. Use the result of part (a) to explain why $A^2, A^3, \ldots, A^n$, where n is a positive integer, are also 2×2 stochastic matrices.

Next, for simplicity, let's consider the following Markov process, in which each stage of the experiment has precisely two possible states.

APPLIED EXAMPLE 4 Urban–Suburban Population Flow Because of the continued successful implementation of an urban renewal program, it is expected that each year, 3% of the population currently residing in the city will move to the suburbs, and 6% of the population currently residing in the suburbs will move into the city. At present, 65% of the total population of the metropolitan area lives in the city itself, while the remaining 35% lives in the suburbs. Assuming that the total population of the metropolitan area remains constant, what will be the distribution of the population 1 year from now?

Solution This problem may be solved with the aid of a tree diagram and the techniques of Chapter 7. The required tree diagram describing this process is shown in Figure 2. Using the method of Section 7.5, we find that the probability that a person selected at random will be a city dweller 1 year from now is given by

$$(.65)(.97) + (.35)(.06) = .6515$$

In a similar manner, we find that the probability that a person selected at random will reside in the suburbs 1 year from now is given by

$$(.65)(.03) + (.35)(.94) = .3485$$

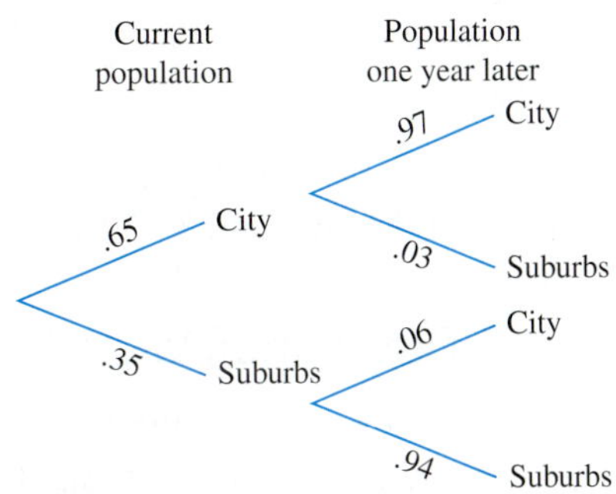

FIGURE 2
Tree diagram showing a Markov process with two states: living in the city and living in the suburbs

Thus, the population of the area 1 year from now may be expected to be distributed as follows: 65.15% living in the city and 34.85% residing in the suburbs.

Let's reexamine the solution to this problem. As we noted earlier, the process under consideration may be viewed as a Markov chain with two possible states at each stage of the experiment: "living in the city" (state 1) and "living in the suburbs" (state 2). The transition matrix associated with this Markov chain is

$$T = \begin{array}{c} \\ \text{State 1} \\ \text{State 2} \end{array} \begin{array}{c} \begin{array}{cc} \text{State 1} & \text{State 2} \end{array} \\ \begin{bmatrix} .97 & .06 \\ .03 & .94 \end{bmatrix} \end{array} \qquad \text{Transition matrix}$$

Next, observe that the initial (current) probability distribution of the population may be summarized in the form of a column vector of dimension 2 (that is, a 2×1 matrix). Thus,

$$X_0 = \begin{array}{c} \text{State 1} \\ \text{State 2} \end{array} \begin{bmatrix} .65 \\ .35 \end{bmatrix} \qquad \text{Initial-state matrix}$$

Using the results of Example 4, we may write the population distribution 1 year later as

$$X_1 = \begin{array}{c} \text{State 1} \\ \text{State 2} \end{array} \begin{bmatrix} .6515 \\ .3485 \end{bmatrix} \qquad \text{Distribution after 1 year}$$

You may now verify that

$$TX_0 = \begin{bmatrix} .97 & .06 \\ .03 & .94 \end{bmatrix} \begin{bmatrix} .65 \\ .35 \end{bmatrix} = \begin{bmatrix} .6515 \\ .3485 \end{bmatrix} = X_1$$

so this problem may be solved by using matrix multiplication.

APPLIED EXAMPLE 5 Urban–Suburban Population Flow Refer to Example 4. What is the population distribution of the city after 2 years? After 3 years?

Solution Let X_2 be the column vector representing the population distribution of the metropolitan area after 2 years. We may view X_1, the vector representing the population distribution of the metropolitan area after 1 year, as representing the "initial" probability distribution in this part of our calculation. Thus,

$$X_2 = TX_1 = \begin{bmatrix} .97 & .06 \\ .03 & .94 \end{bmatrix} \begin{bmatrix} .6515 \\ .3485 \end{bmatrix} \approx \begin{bmatrix} .6529 \\ .3471 \end{bmatrix}$$

The vector representing the probability distribution of the metropolitan area after 3 years is given by

$$X_3 = TX_2 \approx \begin{bmatrix} .97 & .06 \\ .03 & .94 \end{bmatrix} \begin{bmatrix} .6529 \\ .3471 \end{bmatrix} \approx \begin{bmatrix} .6541 \\ .3459 \end{bmatrix}$$

That is, after 3 years, the population will be distributed as follows: 65.41% will live in the city, and 34.59% will live in the suburbs.

Distribution Vectors

Observe that in the foregoing computations, we have $X_1 = TX_0$, $X_2 = TX_1 = T^2X_0$, and $X_3 = TX_2 = T^3X_0$. These results are easily generalized. To see this, suppose we have a Markov process in which there are n possible states at each stage of the experiment. Suppose further that the probability of the system initially being in state 1, state 2, . . . , state n is given by $p_1, p_2, \ldots, p_n$, respectively. This distribution may be represented as an n-dimensional vector

$$X_0 = \begin{bmatrix} p_1 \\ p_2 \\ \vdots \\ p_n \end{bmatrix}$$

called a **distribution vector.** If T represents the $n \times n$ transition matrix associated with the Markov process, then the probability distribution of the system after m observations is given by

$$X_m = T^m X_0 \tag{1}$$

APPLIED EXAMPLE 6 Taxi Movement Between Zones To keep track of the location of its cabs, Zephyr Cab has divided a town into three zones: Zone I, Zone II, and Zone III. Zephyr's management has determined from company records that of the passengers picked up in Zone I, 60% are discharged in the same zone,

30% are discharged in Zone II, and 10% are discharged in Zone III. Of those picked up in Zone II, 40% are discharged in Zone I, 30% are discharged in Zone II, and 30% are discharged in Zone III. Of those picked up in Zone III, 30% are discharged in Zone I, 30% are discharged in Zone II, and 40% are discharged in Zone III. At the beginning of the day, 80% of the cabs are in Zone I, 15% are in Zone II, and 5% are in Zone III. Furthermore, a taxi without a passenger will cruise within the zone it is currently in until a pickup is made.

a. Find the transition matrix for the Markov chain that describes the successive locations of a cab.

b. What is the distribution of the cabs after all of them have made one pickup and discharge?

c. What is the distribution of the cabs after all of them have made two pickups and discharges?

Solution Let Zone I, Zone II, and Zone III correspond to state 1, state 2, and state 3, respectively, of the Markov chain.

a. The required transition matrix is given by

$$T = \begin{bmatrix} .6 & .4 & .3 \\ .3 & .3 & .3 \\ .1 & .3 & .4 \end{bmatrix}$$

b. The initial distribution vector associated with the problem is

$$X_0 = \begin{bmatrix} .80 \\ .15 \\ .05 \end{bmatrix}$$

If X_1 denotes the distribution vector after one observation—that is, after all the cabs have made one pickup and discharge—then

$$\begin{aligned} X_1 &= TX_0 \\ &= \begin{bmatrix} .6 & .4 & .3 \\ .3 & .3 & .3 \\ .1 & .3 & .4 \end{bmatrix} \begin{bmatrix} .80 \\ .15 \\ .05 \end{bmatrix} = \begin{bmatrix} .555 \\ .300 \\ .145 \end{bmatrix} \end{aligned}$$

That is, 55.5% of the cabs are in Zone I, 30% are in Zone II, and 14.5% are in Zone III.

c. Let X_2 denote the distribution vector after all the cabs have made two pickups and discharges. Then

$$\begin{aligned} X_2 &= TX_1 \\ &= \begin{bmatrix} .6 & .4 & .3 \\ .3 & .3 & .3 \\ .1 & .3 & .4 \end{bmatrix} \begin{bmatrix} .555 \\ .300 \\ .145 \end{bmatrix} = \begin{bmatrix} .4965 \\ .3000 \\ .2035 \end{bmatrix} \end{aligned}$$

That is, 49.65% of the cabs are in Zone I, 30% are in Zone II, and 20.35% are in Zone III. You should verify that the same result may be obtained by computing T^2X_0.

Note In this simplified model, we do not take into consideration variable demand and variable delivery time.

9.1 Self-Check Exercises

1. **Market Share of Supermarkets** Three supermarkets serve a certain section of a city. During the upcoming year, Supermarket *A* is expected to retain 80% of its customers, lose 5% of its customers to Supermarket *B*, and lose 15% to Supermarket *C*. Supermarket *B* is expected to retain 90% of its customers and lose 5% of its customers to each of Supermarkets *A* and *C*. Supermarket *C* is expected to retain 75% of its customers, lose 10% to Supermarket *A*, and lose 15% to Supermarket *B*. Construct the transition matrix for the Markov chain that describes the expected change in the market share of the three supermarkets.

2. **Market Share of Supermarkets** Refer to Self-Check Exercise 1. Currently the market shares of Supermarket *A*, Supermarket *B*, and Supermarket *C* are 0.4, 0.3, and 0.3, respectively.
   a. Find the initial distribution vector for this Markov chain.
   b. What share of the market will be held by each supermarket after 1 year? Assuming that the trend continues, what will be the market share after 2 years?

Solutions to Self-Check Exercises 9.1 can be found on page 545.

9.1 Concept Questions

1. What is a finite stochastic process? What can you say about the finite stochastic processes in a Markov chain?

2. Define the following terms for a Markov chain:
   a. State
   b. Current state
   c. Transition probabilities

3. Consider a transition matrix T for a Markov chain with entries a_{ij}, where $1 \le i \le n$ and $1 \le j \le n$.
   a. If there are n states associated with the Markov chain, what is the size of the matrix T?
   b. Describe the probability that each entry represents. Can an entry be negative?
   c. What is the sum of the entries in each column of T?

9.1 Exercises

In Exercises 1–10, determine which of the matrices are stochastic.

1. $\begin{bmatrix} .4 & .7 \\ .6 & .3 \end{bmatrix}$

2. $\begin{bmatrix} .8 & .2 \\ .3 & .7 \end{bmatrix}$

3. $\begin{bmatrix} \frac{1}{4} & \frac{1}{8} \\ \frac{3}{4} & \frac{7}{8} \end{bmatrix}$

4. $\begin{bmatrix} \frac{1}{3} & 0 & \frac{1}{2} \\ \frac{1}{2} & 1 & 0 \\ \frac{1}{4} & 0 & \frac{1}{2} \end{bmatrix}$

5. $\begin{bmatrix} .3 & .2 & .4 \\ .4 & .7 & .3 \\ .3 & .1 & .2 \end{bmatrix}$

6. $\begin{bmatrix} \frac{1}{3} & \frac{1}{4} & \frac{1}{2} \\ \frac{1}{3} & 0 & -\frac{1}{2} \\ \frac{1}{4} & \frac{3}{4} & \frac{1}{2} \end{bmatrix}$

7. $\begin{bmatrix} .1 & .4 & .3 \\ .7 & .2 & .1 \\ .2 & .4 & .6 \end{bmatrix}$

8. $\begin{bmatrix} 1 & 0 & 0 \\ 0 & 0 & 1 \\ 0 & 1 & 0 \end{bmatrix}$

9. $\begin{bmatrix} .2 & .3 \\ .3 & .1 \\ .5 & .6 \end{bmatrix}$

10. $\begin{bmatrix} .5 & .2 & .3 \\ .2 & .3 & .2 \\ .3 & .4 & .1 \\ 0 & .1 & .4 \end{bmatrix}$

11. The transition matrix for a Markov process is given by

$$T = \begin{array}{c} \\ \text{State 1} \\ \text{State 2} \end{array} \begin{array}{c} \text{State} \\ \begin{array}{cc} 1 & 2 \end{array} \\ \begin{bmatrix} .3 & .6 \\ .7 & .4 \end{bmatrix} \end{array}$$

   a. What does the entry $a_{11} = .3$ represent?
   b. Given that the outcome state 1 has occurred, what is the probability that the next outcome of the experiment will be state 2?
   c. If the initial-state distribution vector is given by

$$X_0 = \begin{array}{c} \text{State 1} \\ \text{State 2} \end{array} \begin{bmatrix} .4 \\ .6 \end{bmatrix}$$

   find TX_0, the probability distribution of the system after one observation.

12. The transition matrix for a Markov process is given by

$$T = \begin{array}{c} \\ \text{State 1} \\ \text{State 2} \end{array} \begin{array}{c} \text{State} \\ \begin{array}{cc} 1 & 2 \end{array} \\ \begin{bmatrix} \frac{1}{6} & \frac{2}{3} \\ \frac{5}{6} & \frac{1}{3} \end{bmatrix} \end{array}$$

   a. What does the entry $a_{22} = \frac{1}{3}$ represent?
   b. Given that the outcome state 1 has occurred, what is the probability that the next outcome of the experiment will be state 2?

c. If the initial-state distribution vector is given by

$$X_0 = \begin{matrix} \text{State 1} \\ \text{State 2} \end{matrix} \begin{bmatrix} \frac{1}{4} \\ \frac{3}{4} \end{bmatrix}$$

find TX_0, the probability distribution of the system after one observation.

13. The transition matrix for a Markov process is given by

$$T = \begin{matrix} \\ \text{State 1} \\ \text{State 2} \end{matrix} \begin{matrix} \text{State} \\ \begin{matrix} 1 & 2 \end{matrix} \\ \begin{bmatrix} .6 & .2 \\ .4 & .8 \end{bmatrix} \end{matrix}$$

and the initial-state distribution vector is given by

$$X_0 = \begin{matrix} \text{State 1} \\ \text{State 2} \end{matrix} \begin{bmatrix} .5 \\ .5 \end{bmatrix}$$

Find TX_0, and interpret your result with the aid of a tree diagram.

14. The transition matrix for a Markov process is given by

$$T = \begin{matrix} \\ \text{State 1} \\ \text{State 2} \end{matrix} \begin{matrix} \text{State} \\ \begin{matrix} 1 & 2 \end{matrix} \\ \begin{bmatrix} \frac{1}{2} & \frac{3}{4} \\ \frac{1}{2} & \frac{1}{4} \end{bmatrix} \end{matrix}$$

and the initial-state distribution vector is given by

$$X_0 = \begin{matrix} \text{State 1} \\ \text{State 2} \end{matrix} \begin{bmatrix} \frac{1}{3} \\ \frac{2}{3} \end{bmatrix}$$

Find TX_0, and interpret your result with the aid of a tree diagram.

In Exercises 15–18, find X_2 (the probability distribution of the system after two observations) for the distribution vector X_0 and the transition matrix T.

15. $X_0 = \begin{bmatrix} .6 \\ .4 \end{bmatrix}, T = \begin{bmatrix} .4 & .8 \\ .6 & .2 \end{bmatrix}$

16. $X_0 = \begin{bmatrix} \frac{1}{2} \\ \frac{1}{2} \\ 0 \end{bmatrix}, T = \begin{bmatrix} \frac{1}{2} & \frac{1}{3} & \frac{1}{2} \\ 0 & \frac{1}{3} & \frac{1}{4} \\ \frac{1}{2} & \frac{1}{3} & \frac{1}{4} \end{bmatrix}$

17. $X_0 = \begin{bmatrix} \frac{1}{4} \\ \frac{1}{2} \\ \frac{1}{4} \end{bmatrix}, T = \begin{bmatrix} \frac{1}{4} & \frac{1}{4} & \frac{1}{2} \\ \frac{1}{4} & \frac{1}{2} & \frac{1}{2} \\ \frac{1}{2} & \frac{1}{4} & 0 \end{bmatrix}$

18. $X_0 = \begin{bmatrix} .25 \\ .40 \\ .35 \end{bmatrix}, T = \begin{bmatrix} .1 & .1 & .3 \\ .8 & .7 & .2 \\ .1 & .2 & .5 \end{bmatrix}$

19. **Psychology Experiments** A psychologist conducts an experiment in which a mouse is placed in a T-maze, where it has a choice at the T-junction of turning left and receiving a reward (cheese) or turning right and receiving a mild electric shock (see the accompanying figure). At the end of each trial, a record is kept of the mouse's response. The psychologist observes that the mouse is as likely to turn left (state 1) as right (state 2) during the first trial. In subsequent trials, however, the psychologist observes that if the mouse turned left in the previous trial, then on the next trial, the probability that it will turn left is .8, whereas the probability that it will turn right is .2. If the mouse turned right in the previous trial, then the probability that it will turn right on the next trial is .1, whereas the probability that it will turn left is .9.

a. Using a tree diagram, describe the transitions between states and the probabilities associated with these transitions.

b. Represent the transition probabilities obtained in part (a) in terms of a matrix.

c. What is the initial-state probability vector?

d. Use the results of parts (b) and (c) to find the probability that a mouse will turn left on the second trial.

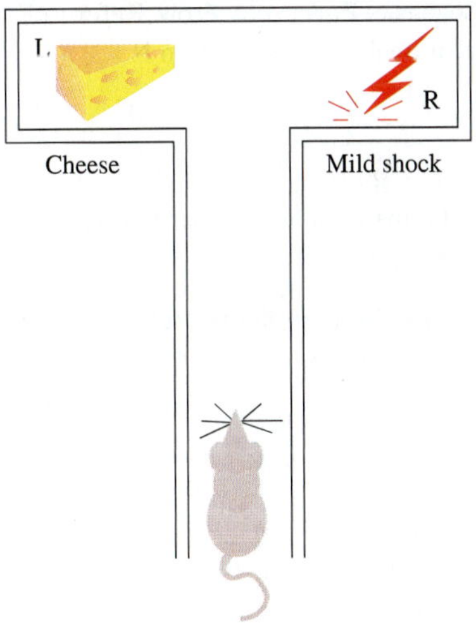

20. **Small-Town Revival** At the beginning of 2004, the population of a certain state was 55.4% rural and 44.6% urban. On the basis of past trends, it is expected that 10% of the population currently residing in the rural areas will move into the urban areas, while 17% of the population currently residing in the urban areas will move into the rural areas in the next decade. What was the population distribution in that state at the beginning of 2014?

21. **Political Polls** Morris Polling conducted a poll 6 months before an election in a state in which a Democrat and a Republican were running for governor and found that 60% of the voters intended to vote for the Republican and 40% intended to vote for the Democrat. A poll conducted 3 months later found that 70% of those who had earlier stated a preference for the Republican candidate still maintained that preference, whereas 30% of these voters now preferred the Democratic candidate. Of those who had earlier stated a preference for the Democrat, 80% still maintained that preference, whereas 20% now preferred the Republican candidate.

a. If the election were held at this time, who would win?

b. Assuming that this trend continues, which candidate is expected to win the election?

22. **Commuter Trends** In a large metropolitan area, 20% of the commuters currently use the public transportation system, whereas the remaining 80% commute via automobile. The city has recently revitalized and expanded its public transportation system. It is expected that 6 months from now, 30% of those who are now commuting to work via automobile will switch to public transportation, and 70% will continue to commute via automobile. At the same time, it is expected that 20% of those now using public transportation will commute via automobile and 80% will continue to use public transportation.
    a. Construct the transition matrix for the Markov chain that describes the change in the mode of transportation used by these commuters.
    b. Find the initial distribution vector for this Markov chain.
    c. What percentage of the commuters are expected to use public transportation 6 months from now?

23. **Urban–Suburban Population Flow** Refer to Example 4. If the initial probability distribution is

$$X_0 = \begin{matrix} \text{City} \\ \text{Suburb} \end{matrix} \begin{bmatrix} .80 \\ .20 \end{bmatrix}$$

what will be the population distribution of the city after 1 year? After 2 years?

24. **Taxi Movement Between Zones** Refer to Example 6. If the initial distribution vector for the location of the taxis is

$$X_0 = \begin{matrix} \text{Zone I} \\ \text{Zone II} \\ \text{Zone III} \end{matrix} \begin{bmatrix} .6 \\ .2 \\ .2 \end{bmatrix}$$

what will be the distribution after all of them have made one pickup and discharge?

25. **Market Share of Bookstores** At a certain university, three bookstores—the University Bookstore, the Campus Bookstore, and the Book Mart—currently serve the university community. A survey conducted at the beginning of the fall quarter found that the University Bookstore and the Campus Bookstore each had 40% of the market, whereas the Book Mart had 20% of the market. Each quarter, the University Bookstore retains 80% of its customers but loses 10% to the Campus Bookstore and 10% to the Book Mart. The Campus Bookstore retains 75% of its customers but loses 10% to the University Bookstore and 15% to the Book Mart. The Book Mart retains 90% of its customers but loses 5% to the University Bookstore and 5% to the Campus Bookstore. If these trends continue, what percentage of the market will each store have at the beginning of the second quarter? The third quarter?

26. **Market Share of Auto Manufacturers** In a study of the domestic market share of the three major automobile manufacturers A, B, and C in a certain country, it was found that their current market shares were 60%, 30%, and 10%, respectively. Furthermore, it was found that of the customers who bought a car manufactured by A, 75% would again buy a car manufactured by A, 15% would buy a car manufactured by B, and 10% would buy a car manufactured by C. Of the customers who bought a car manufactured by B, 90% would again buy a car manufactured by B, whereas 5% each would buy cars manufactured by A and C. Finally, of the customers who bought a car manufactured by C, 85% would again buy a car manufactured by C, 5% would buy a car manufactured by A, and 10% would buy a car manufactured by B. Assuming that these sentiments reflect the buying habits of customers in the future, determine the market share that will be held by each manufacturer after the next two model years.

27. **Change of Majors by Students** Records compiled by the admissions office at a state university indicating the percentage of students who change their major each year are shown in the following transition matrix. Of the freshmen now at the university, 30% have chosen their major field in business, 30% in the humanities, 20% in education, and 20% in the natural sciences and other fields. Assuming that this trend continues, find the percentage of these students that will be majoring in each of the given areas in their senior year.
Hint: Find T^3X_0.

	Bus.	Hum.	Educ.	Nat. sci. and others
Business	.80	.10	.20	.10
Humanities	.10	.70	.10	.05
Education	.05	.10	.60	.05
Nat. sci. and others	.05	.10	.10	.80

28. **Homeowners' Choice of Energy** A study conducted by the Urban Energy Commission in a large metropolitan area indicates the probabilities that homeowners within the area will use certain heating fuels or solar energy during the next 10 years as the major source of heat for their homes. The transition matrix representing the transition probabilities from one state to another is

	Elec.	Gas	Oil	Solar
Electricity	.70	0	0	0
Natural gas	.15	.90	.20	.05
Fuel oil	.05	.02	.75	0
Solar energy	.10	.08	.05	.95

Among homeowners within the area, 20% currently use electricity, 35% use natural gas, 40% use oil, and 5% use solar energy as the major source of heat for their homes. What is the expected distribution of the homeowners who will be using each type of heating fuel or solar energy within the next decade?

In Exercises 29 and 30, determine whether the statement is true or false. If it is true, explain why it is true. If it is false, give an example to show why it is false.

29. A Markov chain is a process in which the outcomes at any stage of the experiment depend on the outcomes of the preceding stages.

30. The sum of the entries in each column of a transition matrix must not exceed 1.

9.1 Solutions to Self-Check Exercises

1. The required transition matrix is

$$T = \begin{bmatrix} .80 & .05 & .10 \\ .05 & .90 & .15 \\ .15 & .05 & .75 \end{bmatrix}$$

2. a. The initial distribution vector is

$$X_0 = \begin{bmatrix} .4 \\ .3 \\ .3 \end{bmatrix}$$

b. The vector representing the market share of each supermarket after 1 year is

$$X_1 = TX_0 = \begin{bmatrix} .80 & .05 & .10 \\ .05 & .90 & .15 \\ .15 & .05 & .75 \end{bmatrix} \begin{bmatrix} .4 \\ .3 \\ .3 \end{bmatrix} = \begin{bmatrix} .365 \\ .335 \\ .300 \end{bmatrix}$$

That is, after 1 year, Supermarket A will command a 36.5% market share, Supermarket B will have a 33.5% share, and Supermarket C will have a 30% market share.

The vector representing the market share of the supermarkets after 2 years is

$$X_2 = TX_1 = \begin{bmatrix} .80 & .05 & .10 \\ .05 & .90 & .15 \\ .15 & .05 & .75 \end{bmatrix} \begin{bmatrix} .365 \\ .335 \\ .300 \end{bmatrix} \approx \begin{bmatrix} .33875 \\ .36475 \\ .29650 \end{bmatrix}$$

That is, 2 years later, the market shares of Supermarkets A, B, and C will be 33.88%, 36.48%, and 29.65%, respectively.

USING TECHNOLOGY Finding Distribution Vectors

Since the computation of the probability distribution of a system after a certain number of observations involves matrix multiplication, a graphing utility may be used to facilitate the work.

APPLIED EXAMPLE 1 Taxi Movement Between Zones Consider the problem posed in Example 6, page 540, where

$$T = \begin{bmatrix} .6 & .4 & .3 \\ .3 & .3 & .3 \\ .1 & .3 & .4 \end{bmatrix} \quad \text{and} \quad X_0 = \begin{bmatrix} .80 \\ .15 \\ .05 \end{bmatrix}$$

Verify that

$$X_2 = \begin{bmatrix} .4965 \\ .3000 \\ .2035 \end{bmatrix}$$

as obtained in that example.

Solution First, we enter the matrix X_0 as the matrix A and the matrix T as the matrix B. Then, performing the indicated multiplication, we find that

$$\text{B^2*A} = \begin{bmatrix} .4965 \\ .3000 \\ .2035 \end{bmatrix}$$

That is,

$$X_2 = T^2X_0 = \begin{bmatrix} .4965 \\ .3000 \\ .2035 \end{bmatrix}$$

as was to be shown.

TECHNOLOGY EXERCISES

In Exercises 1–2, find X_5 (the probability distribution of the system after five observations) for the distribution vector X_0 and the transition matrix T.

1. $X_0 = \begin{bmatrix} .2 \\ .3 \\ .2 \\ .1 \\ .2 \end{bmatrix}, T = \begin{bmatrix} .2 & .2 & .3 & .2 & .1 \\ .1 & .2 & .1 & .2 & .1 \\ .3 & .4 & .1 & .3 & .3 \\ .2 & .1 & .2 & .2 & .2 \\ .2 & .1 & .3 & .1 & .3 \end{bmatrix}$

2. $X_0 = \begin{bmatrix} .1 \\ .2 \\ .2 \\ .3 \\ .2 \end{bmatrix}, T = \begin{bmatrix} .3 & .2 & .1 & .3 & .1 \\ .2 & .1 & .2 & .1 & .2 \\ .1 & .2 & .3 & .2 & .2 \\ .1 & .3 & .2 & .3 & .2 \\ .3 & .2 & .2 & .1 & .3 \end{bmatrix}$

3. **Market Share of Auto Manufacturers** Refer to Exercise 26 on page 544. Using the same data, determine the market share that will be held by each manufacturer five model years after the study began.

4. **Market Share of Bookstores** Refer to Exercise 25 on page 544. Using the same data, determine the expected market share that each store will have at the beginning of the fourth quarter.

9.2 Regular Markov Chains

Steady-State Distribution Vectors

In Section 9.1, we derived a formula for computing the likelihood that a physical system will be in any one of the possible states associated with each stage of a Markov process describing the system. In this section, we use this formula to help us investigate the long-term trends of certain Markov processes.

APPLIED EXAMPLE 1 Educational Status of Women A survey conducted by the National Commission on the Educational Status of Women reveals that 70% of the daughters of women who have completed 2 or more years of college have also completed 2 or more years of college, whereas 20% of the daughters of women who have had less than 2 years of college have completed 2 or more years of college. If this trend continues, determine, in the long run, the percentage of women in the population who will have completed at least 2 years of college given that currently only 20% of the women have completed at least 2 years of college.

Solution This problem may be viewed as a Markov process with two possible states: "completed 2 or more years of college" (state 1) and "completed less than 2 years of college" (state 2). The transition matrix associated with this Markov chain is given by

$$T = \begin{bmatrix} .7 & .2 \\ .3 & .8 \end{bmatrix}$$

The initial distribution vector is given by

$$X_0 = \begin{bmatrix} .2 \\ .8 \end{bmatrix}$$

To study the long-term trend pertaining to this aspect of the educational status of women, let's compute $X_1, X_2, \ldots,$ the distribution vectors associated with the

Markov process under consideration. These vectors give the percentage of women with 2 or more years of college and that of women with less than 2 years of college after one generation, after two generations, and so on. With the aid of Equation (1), Section 9.1, we find (to four decimal places)

After one generation $$X_1 = TX_0 = \begin{bmatrix} .7 & .2 \\ .3 & .8 \end{bmatrix} \begin{bmatrix} .2 \\ .8 \end{bmatrix} = \begin{bmatrix} .3 \\ .7 \end{bmatrix}$$

After two generations $$X_2 = TX_1 = \begin{bmatrix} .7 & .2 \\ .3 & .8 \end{bmatrix} \begin{bmatrix} .3 \\ .7 \end{bmatrix} = \begin{bmatrix} .35 \\ .65 \end{bmatrix}$$

After three generations $$X_3 = TX_2 = \begin{bmatrix} .7 & .2 \\ .3 & .8 \end{bmatrix} \begin{bmatrix} .35 \\ .65 \end{bmatrix} = \begin{bmatrix} .375 \\ .625 \end{bmatrix}$$

Proceeding further, we obtain the following sequence of vectors:

$$X_4 = \begin{bmatrix} .3875 \\ .6125 \end{bmatrix} \qquad X_5 \approx \begin{bmatrix} .3938 \\ .6062 \end{bmatrix}$$

$$X_6 \approx \begin{bmatrix} .3969 \\ .6031 \end{bmatrix} \qquad X_7 \approx \begin{bmatrix} .3984 \\ .6016 \end{bmatrix}$$

$$X_8 \approx \begin{bmatrix} .3992 \\ .6008 \end{bmatrix} \qquad X_9 \approx \begin{bmatrix} .3996 \\ .6004 \end{bmatrix}$$

After ten generations $$X_{10} \approx \begin{bmatrix} .3998 \\ .6002 \end{bmatrix}$$

From the results of these computations, we see that as m increases, the probability distribution vector X_m approaches the probability distribution vector

$$\begin{bmatrix} .4 \\ .6 \end{bmatrix} \quad \text{or} \quad \begin{bmatrix} \frac{2}{5} \\ \frac{3}{5} \end{bmatrix}$$

Such a vector is called the **limiting,** or **steady-state, distribution vector** for the system. We interpret these results in the following way: Initially, 20% of the women in the population have completed 2 or more years of college, whereas 80% have completed less than 2 years of college. After one generation, the former has increased to 30% of the population, and the latter has dropped to 70% of the population. The trend continues, and eventually, 40% of all women in future generations will have completed 2 or more years of college, whereas 60% will have completed less than 2 years of college.

To explain the foregoing result, let's analyze Equation (1), Section 9.1, more closely. Now, the initial distribution vector X_0 is a constant; that is, it remains fixed throughout our computation of $X_1, X_2, \ldots$. It appears reasonable, therefore, to conjecture that this phenomenon is a result of the behavior of the powers, T^m, of the transition matrix T. Pursuing this line of investigation, we compute

$$T^2 = \begin{bmatrix} .7 & .2 \\ .3 & .8 \end{bmatrix} \begin{bmatrix} .7 & .2 \\ .3 & .8 \end{bmatrix} = \begin{bmatrix} .55 & .3 \\ .45 & .7 \end{bmatrix}$$

$$T^3 = \begin{bmatrix} .7 & .2 \\ .3 & .8 \end{bmatrix} \begin{bmatrix} .55 & .3 \\ .45 & .7 \end{bmatrix} = \begin{bmatrix} .475 & .35 \\ .525 & .65 \end{bmatrix}$$

Proceeding further, we obtain the following sequence of matrices:

$$T^4 = \begin{bmatrix} .4375 & .375 \\ .5625 & .625 \end{bmatrix} \qquad T^5 \approx \begin{bmatrix} .4188 & .3875 \\ .5813 & .6125 \end{bmatrix}$$

$$T^6 \approx \begin{bmatrix} .4094 & .3938 \\ .5906 & .6062 \end{bmatrix} \qquad T^7 \approx \begin{bmatrix} .4047 & .3969 \\ .5953 & .6031 \end{bmatrix}$$

$$T^8 \approx \begin{bmatrix} .4023 & .3984 \\ .5977 & .6016 \end{bmatrix} \qquad T^9 \approx \begin{bmatrix} .4012 & .3992 \\ .5988 & .6008 \end{bmatrix}$$

$$T^{10} \approx \begin{bmatrix} .4006 & .3996 \\ .5994 & .6004 \end{bmatrix} \qquad T^{11} \approx \begin{bmatrix} .4003 & .3998 \\ .5997 & .6002 \end{bmatrix}$$

These results show that the powers T^m of the transition matrix T tend toward a fixed matrix as m gets larger and larger. In this case, the "limiting matrix" is the matrix

$$L = \begin{bmatrix} .40 & .40 \\ .60 & .60 \end{bmatrix} \quad \text{or} \quad \begin{bmatrix} \frac{2}{5} & \frac{2}{5} \\ \frac{3}{5} & \frac{3}{5} \end{bmatrix}$$

Such a matrix is called the **steady-state matrix** for the system. Thus, as was suspected, the long-term behavior of a Markov process such as the one in this example depends on the behavior of the limiting matrix of the powers of the transition matrix—the steady-state matrix for the system. In view of this, the long-term (steady-state) distribution vector for this problem may be found by computing the product

$$LX_0 = \begin{bmatrix} .40 & .40 \\ .60 & .60 \end{bmatrix} \begin{bmatrix} .2 \\ .8 \end{bmatrix} = \begin{bmatrix} .40 \\ .60 \end{bmatrix}$$

which agrees with the result obtained earlier.

Next, since the transition matrix T in this situation seems to have a stabilizing effect over the long term, we are led to wonder whether the steady state would be reached regardless of the initial state of the system. To answer this question, suppose the initial distribution vector is

$$X_0 = \begin{bmatrix} p \\ 1 - p \end{bmatrix}$$

Then, as before, the steady-state distribution vector is given by

$$LX_0 = \begin{bmatrix} .40 & .40 \\ .60 & .60 \end{bmatrix} \begin{bmatrix} p \\ 1 - p \end{bmatrix} = \begin{bmatrix} .40 \\ .60 \end{bmatrix}$$

Thus, the steady state is reached regardless of the initial state of the system!

Regular Markov Chains

The transition matrix T of Example 1 has several important properties, which we emphasized in the foregoing discussion. First, the sequence $T, T^2, T^3, \ldots$ approaches a steady-state matrix in which the columns of the limiting matrix are all equal and all entries are positive. A matrix T having this property is called a *regular* Markov chain.

Regular Markov Chain

A stochastic matrix T is a **regular Markov chain** if the sequence

$$T, T^2, T^3, \ldots$$

approaches a steady-state matrix in which the columns of the limiting matrix are all equal and all the entries are positive.

It can be shown that *a stochastic matrix T is regular if and only if some power of T has entries that are all positive.* Second, as in the case of Example 1, a Markov chain with a regular transition matrix has a steady-state distribution vector whose column is identical to a column in the steady-state matrix; thus, this steady-state distribution vector is always reached regardless of the initial distribution vector.

We will return to computations involving regular Markov chains, but for the moment, let's see how one may determine whether a given matrix is indeed regular.

EXAMPLE 2 Determine which of the following matrices are regular.

a. $\begin{bmatrix} .7 & .2 \\ .3 & .8 \end{bmatrix}$ **b.** $\begin{bmatrix} .4 & 1 \\ .6 & 0 \end{bmatrix}$ **c.** $\begin{bmatrix} 0 & 1 \\ 1 & 0 \end{bmatrix}$

Solution

a. Since all the entries of the matrix are positive, the given matrix is regular. Note that this is the transition matrix of Example 1.

b. In this case, one of the entries of the given matrix is equal to zero. Let's compute

$$\begin{bmatrix} .4 & 1 \\ .6 & 0 \end{bmatrix}^2 = \begin{bmatrix} .4 & 1 \\ .6 & 0 \end{bmatrix}\begin{bmatrix} .4 & 1 \\ .6 & 0 \end{bmatrix} = \begin{bmatrix} .76 & .4 \\ .24 & .6 \end{bmatrix}$$

All entries are positive.

Since the second power of the matrix has entries that are all positive, we conclude that the given matrix is in fact regular.

c. Denote the given matrix by A. Then

$$A = \begin{bmatrix} 0 & 1 \\ 1 & 0 \end{bmatrix}$$

$$A^2 = \begin{bmatrix} 0 & 1 \\ 1 & 0 \end{bmatrix}\begin{bmatrix} 0 & 1 \\ 1 & 0 \end{bmatrix} = \begin{bmatrix} 1 & 0 \\ 0 & 1 \end{bmatrix}$$

$$A^3 = \begin{bmatrix} 0 & 1 \\ 1 & 0 \end{bmatrix}\begin{bmatrix} 1 & 0 \\ 0 & 1 \end{bmatrix} = \begin{bmatrix} 0 & 1 \\ 1 & 0 \end{bmatrix}$$

Not all entries are positive.

Observe that $A^3 = A$. It therefore follows that $A^4 = A^2$, $A^5 = A$, and so on. In other words, any power of A must coincide with either A or A^2. Since not all entries of A and A^2 are positive, the same is true of any power of A. We conclude accordingly that the given matrix is not regular.

> *Explore and Discuss*
>
> Find the set of all 2×2 stochastic matrices with elements that are either 0 or 1.

We now return to the study of regular Markov chains. In Example 1, we found the steady-state distribution vector associated with a regular Markov chain by studying the limiting behavior of a sequence of distribution vectors. Alternatively, as was pointed out in the subsequent discussion, the steady-state distribution vector may also be obtained by first determining the steady-state matrix associated with the regular Markov chain.

Fortunately, there is a relatively simple procedure for finding the steady-state distribution vector associated with a regular Markov process. It does not involve the rather tedious computations required to obtain the sequences in Example 1. The procedure follows.

Finding the Steady-State Distribution Vector

Let T be a regular stochastic matrix. Then the steady-state distribution vector X may be found by solving the vector equation

$$TX = X$$

together with the condition that the sum of the elements of the vector X be equal to 1.

A justification of the foregoing procedure is given in Exercise 29.

EXAMPLE 3 Find the steady-state distribution vector for the regular Markov chain whose transition matrix is

$$T = \begin{bmatrix} .7 & .2 \\ .3 & .8 \end{bmatrix}$$

See Example 1.

Solution Let

$$X = \begin{bmatrix} x \\ y \end{bmatrix}$$

be the steady-state distribution vector associated with the Markov process, where the numbers x and y are to be determined. The condition $TX = X$ translates into the matrix equation

$$\begin{bmatrix} .7 & .2 \\ .3 & .8 \end{bmatrix} \begin{bmatrix} x \\ y \end{bmatrix} = \begin{bmatrix} x \\ y \end{bmatrix}$$

or, equivalently, the system of linear equations

$$\begin{aligned} 0.7x + 0.2y &= x \\ 0.3x + 0.8y &= y \end{aligned}$$

But each of the equations that make up this system of equations is equivalent to the single equation

$$0.3x - 0.2y = 0$$

$0.7x - x + 0.2y = 0$
$0.3x + 0.8y - y = 0$

Next, the condition that the sum of the elements of X is 1 gives

$$x + y = 1$$

Thus, the simultaneous fulfillment of the two conditions implies that x and y are the solutions of the system

$$\begin{aligned} 0.3x - 0.2y &= 0 \\ x + \quad y &= 1 \end{aligned}$$

Solving the first equation for x, we obtain

$$x = \frac{2}{3}y$$

which, upon substitution into the second, yields

$$\frac{2}{3}y + y = 1$$

$$y = \frac{3}{5}$$

Thus, $x = \frac{2}{5}$, and the required steady-state distribution vector is given by

$$X = \begin{bmatrix} \frac{2}{5} \\ \frac{3}{5} \end{bmatrix}$$

which agrees with the result obtained earlier.

APPLIED EXAMPLE 4 Taxi Movement Between Zones In Example 6, Section 9.1, we showed that the transition matrix that described the movement of taxis from zone to zone was given by the regular stochastic matrix

$$T = \begin{bmatrix} .6 & .4 & .3 \\ .3 & .3 & .3 \\ .1 & .3 & .4 \end{bmatrix}$$

Use this information to determine the long-term distribution of the taxis in the three zones.

Solution Let

$$X = \begin{bmatrix} x \\ y \\ z \end{bmatrix}$$

be the steady state distribution vector associated with the Markov process under consideration, where x, y, and z are to be determined. The condition $TX = X$ translates into the matrix equation

$$\begin{bmatrix} .6 & .4 & .3 \\ .3 & .3 & .3 \\ .1 & .3 & .4 \end{bmatrix} \begin{bmatrix} x \\ y \\ z \end{bmatrix} = \begin{bmatrix} x \\ y \\ z \end{bmatrix}$$

or, equivalently, the system of linear equations

$$\begin{aligned} 0.6x + 0.4y + 0.3z &= x \\ 0.3x + 0.3y + 0.3z &= y \\ 0.1x + 0.3y + 0.4z &= z \end{aligned}$$

The equations in this system are equivalent to the system

$$\begin{aligned} 4x - 4y - 3z &= 0 \\ 3x - 7y + 3z &= 0 \\ x + 3y - 6z &= 0 \end{aligned}$$

Since $x + y + z = 1$ as well, we are required to solve the system

$$\begin{aligned} x + y + z &= 1 \\ 4x - 4y - 3z &= 0 \\ 3x - 7y + 3z &= 0 \\ x + 3y - 6z &= 0 \end{aligned}$$

Using the Gauss–Jordan elimination method of Chapter 2, we find that

$$x = \frac{33}{70} \qquad y = \frac{3}{10} \qquad z = \frac{8}{35}$$

or $x \approx 0.47$, $y = 0.30$, and $z \approx 0.23$. Thus, in the long run, approximately 47% of the taxis will be in Zone I, 30% in Zone II, and 23% in Zone III.

9.2 Self-Check Exercises

1. Find the steady-state distribution vector for the regular Markov chain whose transition matrix is

$$T = \begin{bmatrix} .5 & .8 \\ .5 & .2 \end{bmatrix}$$

2. **Market Share of Supermarkets** Three supermarkets serve a certain section of a city. During the year, Supermarket A is expected to retain 80% of its customers, lose 5% of its customers to Supermarket B, and lose 15% to Supermarket C. Supermarket B is expected to retain 90% of its customers and lose 5% to each of Supermarket A and Supermarket C. Supermarket C is expected to retain 75% of its customers, lose 10% to Supermarket A, and lose 15% to Supermarket B. If these trends continue, what will be the market share of each supermarket in the long run?

Solutions to Self-Check Exercises 9.2 can be found on page 554.

9.2 Concept Questions

1. Explain (a) a steady-state distribution vector, (b) a steady-state matrix, and (c) a regular Markov chain.

2. How do you find the steady-state distribution vector given a regular stochastic matrix T associated with a Markov process?

9.2 Exercises

In Exercises 1–8, determine which of the matrices are regular.

1. $\begin{bmatrix} \frac{2}{5} & \frac{3}{4} \\ \frac{3}{5} & \frac{1}{4} \end{bmatrix}$

2. $\begin{bmatrix} 0 & .3 \\ 1 & .7 \end{bmatrix}$

3. $\begin{bmatrix} 1 & .8 \\ 0 & .2 \end{bmatrix}$

4. $\begin{bmatrix} \frac{1}{3} & 0 \\ \frac{2}{3} & 1 \end{bmatrix}$

5. $\begin{bmatrix} \frac{1}{2} & \frac{3}{4} & 0 \\ \frac{1}{2} & 0 & \frac{1}{2} \\ 0 & \frac{1}{4} & \frac{1}{2} \end{bmatrix}$

6. $\begin{bmatrix} 1 & .3 & .1 \\ 0 & .4 & .8 \\ 0 & .3 & .1 \end{bmatrix}$

7. $\begin{bmatrix} .7 & .2 & .3 \\ .3 & .8 & .3 \\ 0 & 0 & .4 \end{bmatrix}$

8. $\begin{bmatrix} 0 & 0 & \frac{1}{4} \\ 1 & 0 & 0 \\ 0 & 1 & \frac{3}{4} \end{bmatrix}$

In Exercises 9–16, find the steady-state vector for the transition matrix.

9. $\begin{bmatrix} \frac{1}{3} & \frac{1}{4} \\ \frac{2}{3} & \frac{3}{4} \end{bmatrix}$

10. $\begin{bmatrix} \frac{4}{5} & \frac{3}{5} \\ \frac{1}{5} & \frac{2}{5} \end{bmatrix}$

11. $\begin{bmatrix} .5 & .2 \\ .5 & .8 \end{bmatrix}$

12. $\begin{bmatrix} .9 & 1 \\ .1 & 0 \end{bmatrix}$

13. $\begin{bmatrix} 0 & \frac{1}{8} & 1 \\ 1 & \frac{5}{8} & 0 \\ 0 & \frac{1}{4} & 0 \end{bmatrix}$

14. $\begin{bmatrix} .6 & .3 & 0 \\ .4 & .4 & .6 \\ 0 & .3 & .4 \end{bmatrix}$

15. $\begin{bmatrix} .2 & 0 & .3 \\ 0 & .6 & .4 \\ .8 & .4 & .3 \end{bmatrix}$

16. $\begin{bmatrix} .1 & .2 & .3 \\ .1 & .2 & .3 \\ .8 & .6 & .4 \end{bmatrix}$

17. **Psychology Experiments** A psychologist conducts an experiment in which a mouse is placed in a T-maze, where it has a choice at the T-junction of turning left and receiving a reward (cheese) or turning right and receiving a mild shock. At the end of each trial, a record is kept of the mouse's response. The psychologist observes that the mouse is as likely to turn left (state 1) as right (state 2) during the first trial. In subsequent trials, however, the psychologist observes that if the mouse turned left in the previous trial, then the probability that it will turn left in the next trial is .8, whereas the probability that it will turn right is .2. If the mouse turned right in the previous trial, then the probability that it will turn right in the next trial is .1, whereas the probability that it will turn left is .9. In the long run, what percentage of the time will the mouse turn left at the T-junction?

18. **Commuter Trends** Within a large metropolitan area, 20% of the commuters currently use the public transportation system, whereas the remaining 80% commute via automobile. The city has recently revitalized and expanded its public transportation system. It is expected that 6 months from now, 30% of those who are now commuting to work via automobile will switch to public transportation, and 70% will continue to commute via automobile. At the same time, it is expected that 20% of those now using public transportation will commute via automobile, and 80% will continue to use public transportation. In the long run, what percentage of the commuters will be using public transportation?

19. **One- and Two-Income Families** From data compiled over a 10-year period by Manpower, Inc., in a statewide study of married couples in which at least one spouse was working, the following transition matrix was constructed. It gives the transitional probabilities for one and two wage earners among married couples.

		Current State	
		1 Wage Earner	2 Wage Earners
Next State	1 Wage Earner	.72	.12
	2 Wage Earners	.28	.88

At the present time, 48% of the married couples (in which at least one spouse is working) have one wage earner, and 52% have two wage earners. Assuming that this trend continues, what will be the distribution of one- and two-wage-earner families among married couples in this area 10 years from now? Over the long run?

20. **Professional Women** From data compiled over a 5-year period by *Women's Daily* in a study of the number of women in the professions, the following transition matrix was constructed. It gives the transitional probabilities for the number of men and women in the professions.

		Current State	
		Men	Women
Next State	Men	.95	.04
	Women	.05	.96

As of the beginning of 1986, 52.9% of professional jobs were held by men. If this trend continues, what percentage of professional jobs will be held by women in the long run?

21. **Buying Trends of Home Buyers** From data collected over the past decade by the Association of Realtors of a certain city, the following transition matrix was obtained. The matrix describes the buying pattern of home buyers who buy single-family homes (S) or condominiums (C).

		Current State	
		S	C
Next State	S	.85	.35
	C	.15	.65

Currently, 80% of the homeowners live in single-family homes, whereas 20% live in condominiums. If this trend continues, what percentage of homeowners in this city will own single-family homes and condominiums two decades from now? In the long run?

22. **Homeowners' Choice of Energy** A study conducted by the Urban Energy Commission in a large metropolitan area indicates the probabilities that homeowners within the area will use certain heating fuels or solar energy during the next 10 years as the major source of heat for their homes. The following transition matrix represents the transition probabilities from one state to another:

$$\begin{array}{l} \\ \text{Electricity} \\ \text{Natural gas} \\ \text{Fuel oil} \\ \text{Solar energy} \end{array} \begin{array}{c} \begin{array}{cccc} \text{Elec.} & \text{Gas} & \text{Oil} & \text{Solar} \end{array} \\ \begin{bmatrix} .70 & 0 & .10 & 0 \\ .15 & .90 & .10 & .05 \\ .05 & .02 & .75 & .05 \\ .10 & .08 & .05 & .90 \end{bmatrix} \end{array}$$

Among the homeowners within the area, 20% currently use electricity, 35% use natural gas, 40% use oil, and 5% use solar energy as the major source of heat for their homes. In the long run, what percentage of homeowners within the area will be using solar energy as their major source of heating fuel?

23. **Network News Viewership** A television poll was conducted among regular viewers of the national news in a certain region where the three national networks share the same time slot for the evening news. Results of the poll indicate that 30% of the viewers watch the ABC evening news, 40% watch the CBS evening news, and 30% watch the NBC evening news. Furthermore, it was found that of those viewers who watched the ABC evening news during one week, 80% would again watch the ABC evening news during the next week, 10% would watch the CBS news, and 10% would watch the NBC news. Of those viewers who watched the CBS evening news during one week, 85% would again watch the CBS evening news during the next week, 10% would watch the ABC news, and 5% would watch the NBC news. Of those viewers who watched the NBC evening news during one week, 85% would again watch the NBC news during the next week, 10% would watch ABC, and 5% would watch CBS.

a. What share of the audience consisting of regular viewers of the national news will each network command after 2 weeks?

b. In the long run, what share of the audience will each network command?

24. **Network News Viewership** Refer to Exercise 23. If the initial distribution vector is

$$X_0 = \begin{array}{l} \text{ABC} \\ \text{CBS} \\ \text{NBC} \end{array} \begin{bmatrix} .40 \\ .40 \\ .20 \end{bmatrix}$$

what share of the audience will each network command in the long run?

25. **Genetics** In a certain species of roses, a plant with genotype (genetic makeup) AA has red flowers, a plant with genotype Aa has pink flowers, and a plant with genotype aa has white flowers, where A is the dominant gene and a is the recessive gene for color. If a plant with one genotype is crossed with another plant, then the color of the offspring's flowers is determined by the genotype of the parent plants. If a plant of each genotype is crossed with a

pink-flowered plant, then the transition matrix used to determine the color of the offspring's flowers is given by

		Parent		
		Red	Pink	White
	Red (AA)	$\frac{1}{2}$	$\frac{1}{4}$	0
Offspring	Pink (Aa) or (aA)	$\frac{1}{2}$	$\frac{1}{2}$	$\frac{1}{2}$
	White (aa)	0	$\frac{1}{4}$	$\frac{1}{2}$

If the offspring of each generation are crossed only with pink-flowered plants, what percentage of the plants will have red flowers in the long run? Pink flowers? White flowers?

26. **Market Share of Auto Manufacturers** In a study of the domestic market share of the three major automobile manufacturers A, B, and C in a certain country, it was found that of the customers who bought a car manufactured by A, 75% would again buy a car manufactured by A, 15% would buy a car manufactured by B, and 10% would buy a car manufactured by C. Of the customers who bought a car manufactured by B, 90% would again buy a car manufactured by B, whereas 5% each would buy cars manufactured by A and C. Finally, of the customers who bought a car manufactured by C, 85% would again buy a car manufactured by C, 5% would buy a car manufactured by A, and 10% would buy a car manufactured by B. Assuming that these sentiments reflect the buying habits of customers in the future model years, determine the market share that will be held by each manufacturer in the long run.

In Exercises 27 and 28, determine whether the statement is true or false. If it is true, explain why it is true. If it is false, give an example to show why it is false.

27. A stochastic matrix T is a regular Markov chain if the powers of T approach a fixed matrix whose columns are all equal.

28. To find the steady-state distribution vector X, we solve the system

$$TX = X$$
$$x_1 + x_2 + \cdots + x_n = 1$$

where T is the regular stochastic matrix associated with the Markov process and

$$X = \begin{bmatrix} x_1 \\ x_2 \\ \vdots \\ x_n \end{bmatrix}$$

29. Let T be a regular stochastic matrix. Show that the steady-state distribution vector X may be found by solving the vector equation $TX = X$ together with the condition that the sum of the elements of X is 1.
Hint: Take the initial distribution to be X, the steady-state distribution vector. Then, when n is large, $X \approx T^nX$. (Why?) Multiply both sides of the last equation by T (on the left), and consider the resulting equation when n is large.

9.2 Solutions to Self-Check Exercises

1. Let

$$X = \begin{bmatrix} x \\ y \end{bmatrix}$$

be the steady-state distribution vector associated with the Markov process, where the numbers x and y are to be determined. The condition $TX = X$ translates into the matrix equation

$$\begin{bmatrix} .5 & .8 \\ .5 & .2 \end{bmatrix}\begin{bmatrix} x \\ y \end{bmatrix} = \begin{bmatrix} x \\ y \end{bmatrix}$$

which is equivalent to the system of linear equations

$$0.5x + 0.8y = x$$
$$0.5x + 0.2y = y$$

Each equation in the system is equivalent to the equation

$$0.5x - 0.8y = 0$$

Next, the condition that the sum of the elements of X is 1 gives

$$x + y = 1$$

Thus, the simultaneous fulfillment of the two conditions implies that x and y are the solutions of the system

$$0.5x - 0.8y = 0$$
$$x + \quad y = 1$$

Solving the first equation for x, we obtain

$$x = \frac{8}{5}y$$

which, upon substitution into the second, yields

$$\frac{8}{5}y + y = 1$$

$$y = \frac{5}{13}$$

Therefore, $x = \frac{8}{13}$, and the required steady-state distribution vector is

$$\begin{bmatrix} \frac{8}{13} \\ \frac{5}{13} \end{bmatrix}$$

2. The transition matrix for the Markov process under consideration is

$$T = \begin{bmatrix} .80 & .05 & .10 \\ .05 & .90 & .15 \\ .15 & .05 & .75 \end{bmatrix}$$

Now, let

$$X = \begin{bmatrix} x \\ y \\ z \end{bmatrix}$$

be the steady-state distribution vector associated with the Markov process under consideration, where x, y, and z are to be determined. The condition $TX = X$ is

$$\begin{bmatrix} .80 & .05 & .10 \\ .05 & .90 & .15 \\ .15 & .05 & .75 \end{bmatrix} \begin{bmatrix} x \\ y \\ z \end{bmatrix} = \begin{bmatrix} x \\ y \\ z \end{bmatrix}$$

or, equivalently, the following system of linear equations.

$$\begin{aligned} 0.80x + 0.05y + 0.10z &= x \\ 0.05x + 0.90y + 0.15z &= y \\ 0.15x + 0.05y + 0.75z &= z \end{aligned}$$

This system simplifies to

$$\begin{aligned} 4x - y - 2z &= 0 \\ x - 2y + 3z &= 0 \\ 3x + y - 5z &= 0 \end{aligned}$$

Since $x + y + z = 1$ as well, we are required to solve the system

$$\begin{aligned} 4x - y - 2z &= 0 \\ x - 2y + 3z &= 0 \\ 3x + y - 5z &= 0 \\ x + y + z &= 1 \end{aligned}$$

Using the Gauss–Jordan elimination method, we find

$$x = \frac{1}{4} \qquad y = \frac{1}{2} \qquad z = \frac{1}{4}$$

Therefore, in the long run, Supermarkets A and C will each have 25% of the customers, and Supermarket B will have 50% of the customers.

USING TECHNOLOGY Finding the Long-Term Distribution Vector

The problem of finding the long-term distribution vector for a regular Markov chain ultimately rests on the problem of solving a system of linear equations. As such, the **rref** or equivalent function of a graphing utility proves indispensable, as the following example shows.

EXAMPLE 1 Find the steady-state distribution vector for the regular Markov chain whose transition matrix is

$$T = \begin{bmatrix} .4 & .2 & .1 \\ .3 & .4 & .5 \\ .3 & .4 & .4 \end{bmatrix}$$

Solution Let

$$\begin{bmatrix} x \\ y \\ z \end{bmatrix}$$

be the steady-state distribution vector, where x, y, and z are to be determined. The condition $TX = X$ translates into the matrix equation

$$\begin{bmatrix} .4 & .2 & .1 \\ .3 & .4 & .5 \\ .3 & .4 & .4 \end{bmatrix} \begin{bmatrix} x \\ y \\ z \end{bmatrix} = \begin{bmatrix} x \\ y \\ z \end{bmatrix}$$

or, equivalently, the system of linear equations

$$\begin{aligned} 0.4x + 0.2y + 0.1z &= x \\ 0.3x + 0.4y + 0.5z &= y \\ 0.3x + 0.4y + 0.4z &= z \end{aligned}$$

Because $x + y + z = 1$, we are required to solve the system

$$\begin{aligned} -0.6x + 0.2y + 0.1z &= 0 \\ 0.3x - 0.6y + 0.5z &= 0 \\ 0.3x + 0.4y - 0.6z &= 0 \\ x + \quad y + \quad z &= 1 \end{aligned}$$

Entering this system into the graphing calculator as the augmented matrix

$$A = \left[\begin{array}{ccc|c} -.6 & .2 & .1 & 0 \\ .3 & -.6 & .5 & 0 \\ .3 & .4 & -.6 & 0 \\ 1 & 1 & 1 & 1 \end{array}\right]$$

and then using the **rref** function, we obtain the equivalent system (to two decimal places)

$$\left[\begin{array}{ccc|c} 1 & 0 & 0 & .20 \\ 0 & 1 & 0 & .42 \\ 0 & 0 & 1 & .38 \\ 0 & 0 & 0 & 0 \end{array}\right]$$

Therefore, $x \approx 0.20$, $y \approx 0.42$, and $z \approx 0.38$, so the required steady-state distribution vector is approximately

$$\begin{bmatrix} .20 \\ .42 \\ .38 \end{bmatrix}$$

TECHNOLOGY EXERCISES

In Exercises 1 and 2, find the steady-state vector for the matrix T.

1. $$T = \begin{bmatrix} .2 & .2 & .3 & .2 & .1 \\ .1 & .2 & .1 & .2 & .1 \\ .3 & .4 & .1 & .3 & .3 \\ .2 & .1 & .2 & .2 & .2 \\ .2 & .1 & .3 & .1 & .3 \end{bmatrix}$$

2. $$T = \begin{bmatrix} .3 & .2 & .1 & .3 & .1 \\ .2 & .1 & .2 & .1 & .2 \\ .1 & .2 & .3 & .2 & .2 \\ .1 & .3 & .2 & .3 & .2 \\ .3 & .2 & .2 & .1 & .3 \end{bmatrix}$$

3. Verify that the steady-state vector for Example 4, page 551, is (to two decimal places)

$$X = \begin{bmatrix} .47 \\ .30 \\ .23 \end{bmatrix}$$

9.3 Absorbing Markov Chains

Absorbing Markov Chains

In this section, we investigate the long-term trends of a certain class of Markov chains that involve transition matrices that are not regular. In particular, we study Markov chains in which the transition matrices, known as absorbing stochastic matrices, have the special properties that we now describe.

Consider the stochastic matrix

$$\begin{bmatrix} 1 & 0 & .2 & 0 \\ 0 & 1 & .3 & 1 \\ 0 & 0 & .5 & 0 \\ 0 & 0 & 0 & 0 \end{bmatrix}$$

associated with a Markov process. Interpreting it in the usual fashion, we see that after one observation, the probability is 1 (a certainty) that an object previously in state 1 will remain in state 1. Similarly, we see that an object that was previously in state 2 must remain in state 2. Next, we find that an object that was previously in state 3 has a probability of .2 of going to state 1, a probability of .3 of going to state 2, a probability of .5 of remaining in state 3, and no chance of going to state 4. Finally, an object that was previously in state 4 must, after one observation, end up in state 2.

This stochastic matrix exhibits certain special characteristics. First, as was observed earlier, an object in state 1 or state 2 must stay in state 1 or state 2, respectively. Such states are called absorbing states. In general, an **absorbing state** is one from which it is impossible for an object to leave. To identify the absorbing states of a stochastic matrix, we simply examine each column of the matrix. If column i has a 1 in the a_{ii} position (that is, on the main diagonal of the matrix) and zeros elsewhere in that column, then and only then is state i an absorbing state.

Second, observe that states 3 and 4, although not absorbing states, have the property that an object in each of these states has a possibility of going to an absorbing state. For example, an object that is currently in state 3 has a probability of .2 of ending up in state 1, an absorbing state, and an object in state 4 must end up in state 2, also an absorbing state, after one transition.

Absorbing Stochastic Matrix

An **absorbing stochastic matrix** has the following properties:

1. There is at least one absorbing state.
2. It is possible to go from each nonabsorbing state to an absorbing state in one or more steps.

A Markov chain is said to be an **absorbing Markov chain** if the transition matrix associated with the process is an absorbing stochastic matrix.

EXAMPLE 1 Determine whether the following matrices are absorbing stochastic matrices.

a. $\begin{bmatrix} .7 & 0 & .1 & 0 \\ 0 & 1 & .5 & 0 \\ .3 & 0 & .2 & 0 \\ 0 & 0 & .2 & 1 \end{bmatrix}$ **b.** $\begin{bmatrix} 1 & 0 & 0 & 0 \\ 0 & 1 & 0 & 0 \\ 0 & 0 & .5 & .4 \\ 0 & 0 & .5 & .6 \end{bmatrix}$

Solution

a. States 2 and 4 are both absorbing states. Furthermore, even though state 1 is not an absorbing state, there is a possibility (with probability .3) that an object may go from this state to state 3. State 3 itself is nonabsorbing, but an object in that state has a probability of .5 of going to the absorbing state 2 and a probability of .2 of going to the absorbing state 4. Thus, the given matrix is an absorbing stochastic matrix.

b. States 1 and 2 are absorbing states. However, it is impossible for an object to go from the nonabsorbing states 3 and 4 to either or both of the absorbing states. Thus, the given matrix is not an absorbing stochastic matrix.

Given an absorbing stochastic matrix, it is always possible, by suitably reordering the states if necessary, to rewrite it so that the absorbing states appear first. Then the resulting matrix can be partitioned into four submatrices,

$$\begin{array}{cc} \text{Absorbing} & \text{Nonabsorbing} \end{array}$$
$$\left[\begin{array}{c|c} I & S \\ \hline O & R \end{array}\right]$$

where I is an identity matrix whose order is determined by the number of absorbing states and O is a zero matrix. The submatrices R and S correspond to the nonabsorbing states. As an example, the absorbing stochastic matrix of Example 1(a),

$$\begin{array}{c} \\ 1 \\ 2 \\ 3 \\ 4 \end{array}\begin{array}{c} \begin{array}{cccc} 1 & 2 & 3 & 4 \end{array} \\ \begin{bmatrix} .7 & 0 & .1 & 0 \\ 0 & 1 & .5 & 0 \\ .3 & 0 & .2 & 0 \\ 0 & 0 & .2 & 1 \end{bmatrix} \end{array} \quad \text{may be written as} \quad \begin{array}{c} \\ 4 \\ 2 \\ 1 \\ 3 \end{array}\begin{array}{c} \begin{array}{cccc} 4 & 2 & 1 & 3 \end{array} \\ \left[\begin{array}{cc|cc} 1 & 0 & 0 & .2 \\ 0 & 1 & 0 & .5 \\ \hline 0 & 0 & .7 & .1 \\ 0 & 0 & .3 & .2 \end{array}\right] \end{array}$$

upon reordering the states as indicated.

APPLIED EXAMPLE 2 Gambler's Ruin John has decided to risk \$2 in the following game of chance. He places a \$1 bet on each repeated play of the game in which the probability of his winning \$1 is .4, and he continues to play until he has accumulated a total of \$3 or he has lost all of his money. Write the transition matrix for the related absorbing Markov chain.

Solution There are four states in this Markov chain, which correspond to John accumulating a total of \$0, \$1, \$2, and \$3. Since the first and last states listed are absorbing states, we will list these states first, resulting in the transition matrix

$$\begin{array}{c} \\ \\ \$0 \\ \$3 \\ \$1 \\ \$2 \end{array}\begin{array}{c} \begin{array}{cc} \overbrace{\$0 \quad \$3}^{\text{Absorbing}} & \overbrace{\$1 \quad \$2}^{\text{Nonabsorbing}} \end{array} \\ \begin{bmatrix} 1 & 0 & .6 & 0 \\ 0 & 1 & 0 & .4 \\ 0 & 0 & 0 & .6 \\ 0 & 0 & .4 & 0 \end{bmatrix} \end{array}$$

which is constructed as follows: Since the state "\$0" is an absorbing state, we see that $a_{11} = 1$, $a_{21} = a_{31} = a_{41} = 0$. Similarly, the state "\$3" is an absorbing state, so $a_{22} = 1$, and $a_{12} = a_{32} = a_{42} = 0$. To construct the column corresponding to the nonabsorbing state "\$1," we note that there is a probability of .6 (John loses) in going from an accumulated amount of \$1 to \$0, so $a_{13} = .6$; $a_{23} = a_{33} = 0$ because it is not feasible to go from an accumulated amount of \$1 to either an accumulated amount of \$3 or \$1 in one transition (play). Finally, there is a probability of .4 (John wins) in going from an accumulated amount of \$1 to an accumulated amount of \$2, so $a_{43} = .4$. The last column of the transition matrix is constructed by reasoning in a similar manner.

The following question arises in connection with the last example: If John continues to play the game as originally planned, what is the probability that he will depart from the game victorious—that is, leave with an accumulated amount of \$3?

To answer this question, we have to look at the long-term trend of the relevant Markov chain. Taking a cue from our work in the last section, we may compute the powers of the transition matrix associated with the Markov chain. Just as in the case of regular stochastic matrices, it turns out that the powers of an absorbing stochastic matrix approach a steady-state matrix. However, instead of demonstrating this, we use the following result, which we state without proof, for computing the steady-state matrix:

Finding the Steady-State Matrix for an Absorbing Stochastic Matrix

Suppose an absorbing stochastic matrix A has been partitioned into submatrices

$$A = \left[\begin{array}{c|c} I & S \\ \hline O & R \end{array}\right]$$

Then the *steady-state matrix* of A is given by

$$\left[\begin{array}{c|c} I & S(I-R)^{-1} \\ \hline O & O \end{array}\right]$$

where the order of the identity matrix appearing in the expression $(I - R)^{-1}$ is chosen to have the same order as R.

APPLIED EXAMPLE 3 Gambler's Ruin (continued) Refer to Example 2. If John continues to play the game until either he has accumulated a sum of \$3 or he has lost all of his money, what is the probability that he will accumulate \$3?

Solution The transition matrix associated with the Markov process is (see Example 2)

$$A = \left[\begin{array}{cc|cc} 1 & 0 & .6 & 0 \\ 0 & 1 & 0 & .4 \\ \hline 0 & 0 & 0 & .6 \\ 0 & 0 & .4 & 0 \end{array}\right]$$

We need to find the steady-state matrix of A. In this case,

$$R = \begin{bmatrix} 0 & .6 \\ .4 & 0 \end{bmatrix} \quad \text{and} \quad S = \begin{bmatrix} .6 & 0 \\ 0 & .4 \end{bmatrix}$$

so

$$I - R = \begin{bmatrix} 1 & 0 \\ 0 & 1 \end{bmatrix} - \begin{bmatrix} 0 & .6 \\ .4 & 0 \end{bmatrix} = \begin{bmatrix} 1 & -.6 \\ -.4 & 1 \end{bmatrix}$$

Explore and Discuss

Consider the stochastic matrix

$$A = \begin{bmatrix} 1 & 0 & a \\ 0 & 1 & b \\ 0 & 0 & 1-a-b \end{bmatrix}$$

where a and b satisfy $0 < a < 1$, $0 < b < 1$, and $0 < a + b < 1$.

1. Find the steady-state matrix.
2. What is the probability that state 3 will be absorbed in state 2?

Using the formula in Section 2.6 for finding the inverse of a 2×2 matrix, we find that (to two decimal places)

$$(I - R)^{-1} = \begin{bmatrix} 1.32 & .79 \\ .53 & 1.32 \end{bmatrix}$$

so

$$S(I - R)^{-1} = \begin{bmatrix} .6 & 0 \\ 0 & .4 \end{bmatrix} \begin{bmatrix} 1.32 & .79 \\ .53 & 1.32 \end{bmatrix} = \begin{bmatrix} .79 & .47 \\ .21 & .53 \end{bmatrix}$$

Therefore, the required steady-state matrix of A is given by

$$\left[\begin{array}{c:c} I & S(I-R)^{-1} \\ \hdashline O & O \end{array}\right] = \begin{array}{c} \\ \$0 \\ \$3 \\ \$1 \\ \$2 \end{array} \begin{array}{c} \begin{array}{cccc} \$0 & \$3 & \$1 & \$2 \end{array} \\ \left[\begin{array}{cc:cc} 1 & 0 & .79 & .47 \\ 0 & 1 & .21 & .53 \\ \hdashline 0 & 0 & 0 & 0 \\ 0 & 0 & 0 & 0 \end{array}\right] \end{array}$$

From this result, we see that if John starts with \$2, the probability is .53 that John will leave the game with an accumulated amount of \$3—that is, that he wins \$1.

Our last example shows an application of Markov chains in the field of genetics.

APPLIED EXAMPLE 4 Genetics In a certain species of flowers, a plant of genotype (genetic makeup) AA has red flowers, a plant of genotype Aa has pink flowers, and a plant of genotype aa has white flowers, where A is the dominant gene and a is the recessive gene for color. If a plant of one genotype is crossed with another plant, then the color of the offspring's flowers is determined by the genotype of the parent plants. If the offspring are crossed successively with plants of genotype AA only, show that in the long run, all the flowers produced by the plants will be red.

Solution First, let's construct the transition matrix associated with the resulting Markov chain. In crossing a plant of genotype AA with another of the same genotype AA, the offspring will inherit one dominant gene from each parent and thus will have genotype AA. Therefore, the probabilities of the offspring being genotype AA, Aa, and aa are 1, 0, and 0, respectively.

Next, in crossing a plant of genotype AA with one of genotype Aa, the probability of the offspring having genotype AA (inheriting an A gene from the first parent and an A from the second) is $\frac{1}{2}$; the probability of the offspring having genotype Aa (inheriting an A gene from the first parent and an a gene from the second parent) is $\frac{1}{2}$; finally, the probability of the offspring being of genotype aa is 0 since this is clearly impossible.

A similar argument shows that when a plant of genotype AA is crossed with one of genotype aa, the probabilities of the offspring having genotype AA, Aa, and aa are 0, 1, and 0, respectively.

The required transition matrix is thus given by

$$T = \begin{array}{c} \\ AA \\ Aa \\ aa \end{array} \begin{array}{c} \overset{\text{Absorbing state}}{\downarrow} \\ \begin{array}{ccc} AA & Aa & aa \end{array} \\ \begin{bmatrix} 1 & \frac{1}{2} & 0 \\ 0 & \frac{1}{2} & 1 \\ 0 & 0 & 0 \end{bmatrix} \end{array}$$

Observe that the state AA is an absorbing state. Furthermore, it is possible to go from each of the other two nonabsorbing states to the absorbing state AA. Thus, the Markov chain is an absorbing Markov chain. To determine the long-term effects of this experiment, let's compute the steady-state matrix of T. Partitioning T in the usual manner, we find

$$T = \left[\begin{array}{c|cc} 1 & \frac{1}{2} & 0 \\ \hline 0 & \frac{1}{2} & 1 \\ 0 & 0 & 0 \end{array}\right]$$

so

$$R = \begin{bmatrix} \frac{1}{2} & 1 \\ 0 & 0 \end{bmatrix} \quad \text{and} \quad S = \begin{bmatrix} \frac{1}{2} & 0 \end{bmatrix}$$

Next, we compute

$$I - R = \begin{bmatrix} 1 & 0 \\ 0 & 1 \end{bmatrix} - \begin{bmatrix} \frac{1}{2} & 1 \\ 0 & 0 \end{bmatrix} = \begin{bmatrix} \frac{1}{2} & -1 \\ 0 & 1 \end{bmatrix}$$

and, using the formula for finding the inverse of a 2×2 matrix in Section 2.6,

$$(I - R)^{-1} = \begin{bmatrix} 2 & 2 \\ 0 & 1 \end{bmatrix}$$

Thus,

$$S(I - R)^{-1} = \begin{bmatrix} \frac{1}{2} & 0 \end{bmatrix} \begin{bmatrix} 2 & 2 \\ 0 & 1 \end{bmatrix} = \begin{bmatrix} 1 & 1 \end{bmatrix}$$

Therefore, the steady-state matrix of T is given by

$$\left[\begin{array}{c|c} I & S(I - R)^{-1} \\ \hline O & O \end{array}\right] = \begin{array}{c} \\ AA \\ Aa \\ aa \end{array} \begin{array}{c} \begin{array}{ccc} AA & Aa & aa \end{array} \\ \left[\begin{array}{c|cc} 1 & 1 & 1 \\ \hline 0 & 0 & 0 \\ 0 & 0 & 0 \end{array}\right] \end{array}$$

Interpreting the steady-state matrix of T, we see that the long-term result of crossing the offspring with plants of genotype AA leads only to the absorbing state AA. In other words, such a procedure will result in the production of plants that will bear only red flowers, as we set out to demonstrate.

9.3 Self-Check Exercises

1. Let

$$T = \begin{bmatrix} .2 & 0 & 0 \\ .3 & 1 & .6 \\ .5 & 0 & .4 \end{bmatrix}$$

   a. Show that T is an absorbing stochastic matrix.
   b. Rewrite T so that the absorbing states appear first, partition the resulting matrix, and identify the submatrices R and S.
   c. Compute the steady-state matrix of T.

2. **Computer-Aided Transcription** There is a trend toward increased use of computer-aided transcription (CAT) and electronic recording (ER) as alternatives to manual transcription (MT) of court proceedings by court reporters in a certain state. Suppose the following stochastic matrix gives the transition matrix associated with the Markov process over the past decade:

$$T = \begin{array}{c} \\ \text{CAT} \\ \text{ER} \\ \text{MT} \end{array} \begin{array}{c} \begin{array}{ccc} \text{CAT} & \text{ER} & \text{MT} \end{array} \\ \begin{bmatrix} 1 & .3 & .2 \\ 0 & .6 & .3 \\ 0 & .1 & .5 \end{bmatrix} \end{array}$$

Determine the probability that a court now using electronic recording or manual transcribing of its proceedings will eventually change to CAT.

Solutions to Self-Check Exercises 9.3 can be found on page 563.

9.3 Concept Questions

1. What is an absorbing stochastic matrix?

2. Suppose that the absorbing stochastic matrix A has been partitioned into submatrices

$$\left[\begin{array}{c|c} I & S \\ \hline O & R \end{array}\right]$$

Write the expression representing the steady-state matrix of A.

9.3 Exercises

In Exercises 1–8, determine whether the matrix is an absorbing stochastic matrix.

1. $\begin{bmatrix} \frac{2}{5} & 0 \\ \frac{3}{5} & 1 \end{bmatrix}$

2. $\begin{bmatrix} 1 & 0 \\ 0 & 1 \end{bmatrix}$

3. $\begin{bmatrix} 1 & .5 & 0 \\ 0 & 0 & 1 \\ 0 & .5 & 0 \end{bmatrix}$

4. $\begin{bmatrix} 1 & 0 & 0 \\ 0 & .7 & .2 \\ 0 & .3 & .8 \end{bmatrix}$

5. $\begin{bmatrix} \frac{1}{8} & 0 & 0 \\ \frac{1}{4} & 1 & 0 \\ \frac{5}{8} & 0 & 1 \end{bmatrix}$

6. $\begin{bmatrix} 1 & 0 & 0 & 0 \\ 0 & \frac{5}{8} & 0 & \frac{1}{6} \\ 0 & \frac{1}{8} & 1 & 0 \\ 0 & \frac{1}{4} & 0 & \frac{5}{6} \end{bmatrix}$

7. $\begin{bmatrix} 1 & 0 & .3 & 0 \\ 0 & 1 & .2 & 0 \\ 0 & 0 & .1 & .5 \\ 0 & 0 & .4 & .5 \end{bmatrix}$

8. $\begin{bmatrix} 1 & 0 & 0 & 0 \\ 0 & 1 & 0 & 0 \\ 0 & 0 & .2 & .6 \\ 0 & 0 & .8 & .4 \end{bmatrix}$

In Exercises 9–14, rewrite each absorbing stochastic matrix so that the absorbing states appear first, partition the resulting matrix, and identify the submatrices R and S.

9. $\begin{bmatrix} .6 & 0 \\ .4 & 1 \end{bmatrix}$

10. $\begin{bmatrix} \frac{1}{4} & 0 & 0 \\ \frac{1}{4} & 1 & 0 \\ \frac{1}{2} & 0 & 1 \end{bmatrix}$

11. $\begin{bmatrix} 0 & .2 & 0 \\ .5 & .4 & 0 \\ .5 & .4 & 1 \end{bmatrix}$

12. $\begin{bmatrix} .5 & 0 & .3 \\ 0 & 1 & .1 \\ .5 & 0 & .6 \end{bmatrix}$

13. $\begin{bmatrix} .4 & .2 & 0 & 0 \\ .2 & .3 & 0 & 0 \\ 0 & .3 & 1 & 0 \\ .4 & .2 & 0 & 1 \end{bmatrix}$

14. $\begin{bmatrix} .1 & 0 & 0 & 0 \\ .2 & 1 & 0 & .2 \\ .3 & 0 & 1 & 0 \\ .4 & 0 & 0 & .8 \end{bmatrix}$

In Exercises 15–24, compute the steady-state matrix of each stochastic matrix.

15. $\begin{bmatrix} .55 & 0 \\ .45 & 1 \end{bmatrix}$

16. $\begin{bmatrix} \frac{3}{5} & 0 \\ \frac{2}{5} & 1 \end{bmatrix}$

17. $\begin{bmatrix} 1 & .2 & .3 \\ 0 & .4 & .2 \\ 0 & .4 & .5 \end{bmatrix}$

18. $\begin{bmatrix} \frac{1}{5} & 0 & 0 \\ 0 & 1 & \frac{3}{8} \\ \frac{4}{5} & 0 & \frac{5}{8} \end{bmatrix}$

19. $\begin{bmatrix} \frac{1}{2} & 0 & \frac{1}{3} & 0 \\ \frac{1}{2} & 1 & 0 & 0 \\ 0 & 0 & \frac{2}{3} & 0 \\ 0 & 0 & 0 & 1 \end{bmatrix}$

20. $\begin{bmatrix} 1 & \frac{1}{8} & \frac{1}{3} & 0 \\ 0 & \frac{1}{8} & 0 & 0 \\ 0 & \frac{1}{4} & \frac{2}{3} & 0 \\ 0 & \frac{1}{2} & 0 & 1 \end{bmatrix}$

21. $\begin{bmatrix} 1 & 0 & \frac{1}{4} & \frac{1}{3} \\ 0 & 1 & \frac{1}{4} & \frac{1}{3} \\ 0 & 0 & \frac{1}{2} & 0 \\ 0 & 0 & 0 & \frac{1}{3} \end{bmatrix}$

22. $\begin{bmatrix} 1 & 0 & .2 & .1 \\ 0 & 1 & .4 & .2 \\ 0 & 0 & 0 & .4 \\ 0 & 0 & .4 & .3 \end{bmatrix}$

23. $\begin{bmatrix} 1 & 0 & 0 & .2 & .1 \\ 0 & 1 & 0 & .1 & .2 \\ 0 & 0 & 1 & .3 & .1 \\ 0 & 0 & 0 & .2 & .2 \\ 0 & 0 & 0 & .2 & .4 \end{bmatrix}$

24. $\begin{bmatrix} 1 & 0 & \frac{1}{4} & \frac{1}{3} & 0 \\ 0 & 1 & 0 & \frac{1}{3} & \frac{1}{2} \\ 0 & 0 & \frac{1}{4} & \frac{1}{3} & 0 \\ 0 & 0 & \frac{1}{2} & 0 & \frac{1}{2} \\ 0 & 0 & 0 & 0 & 0 \end{bmatrix}$

25. Broadband Internet Service As more and more people switched to broadband Internet service, the demand for dial-up Internet service continued to drop. Suppose the transition matrix

$$A = \begin{array}{c} \\ \text{D} \\ \text{B} \end{array}\begin{array}{c} \begin{array}{cc} \text{D} & \text{B} \end{array} \\ \begin{bmatrix} .80 & 0 \\ .20 & 1 \end{bmatrix} \end{array}$$

describes this Markov process, where B denotes broadband Internet service and D denotes dial-up Internet service.

a. Show that A is an absorbing stochastic matrix, and rewrite it so that the absorbing state appears first. Partition the resulting matrix, and identify the submatrices R and S.

b. Compute the steady-state matrix of A, and interpret your results.

26. Game of Chance Diane has decided to play the following game of chance. She places a \$1 bet on each repeated play of the game in which the probability of her winning \$1 is .5. She has further decided to continue playing the game until she either has accumulated a total of \$3 or has lost all her money. What is the probability that Diane will eventually leave the game a winner if she started with a capital of \$1? Of \$2?

27. **GAME OF CHANCE** Refer to Exercise 26. Suppose Diane has decided to stop playing after she has accumulated a sum of \$4 or has lost all her money. All other conditions being the same, what is the probability that Diane will leave the game a winner if she started with a capital of \$1? Of \$2? Of \$3?

28. **VIDEO RECORDERS** Over the years, consumers turned more and more to newer and much improved video-recording devices. The following transition matrix describes the Markov chain associated with this process. Here V stands for VHS recorders, D stands for DVD recorders, and H stands for high-definition video recorders.

$$A = \begin{array}{c} \\ V \\ D \\ H \end{array} \begin{array}{c} \begin{array}{ccc} V & D & H \end{array} \\ \begin{bmatrix} .10 & 0 & 0 \\ .70 & .60 & 0 \\ .20 & .40 & 1 \end{bmatrix} \end{array}$$

   a. Show that A is an absorbing stochastic matrix, and rewrite it so that the absorbing state appears first. Partition the resulting matrix, and identify the submatrices R and S.

   b. Compute the steady-state matrix of A and interpret your results.

29. **COLLEGE GRADUATION RATE** The registrar of Computronics Institute has compiled the following statistics on the progress of the school's students in the 2-year computer programming course leading to an associate degree: Of beginning students in a particular year, 75% successfully complete their first year of study and move on to the second year, whereas 25% drop out of the program; of second-year students in a particular year, 90% go on to graduate at the end of the year, whereas 10% drop out of the program.

   a. Construct the transition matrix associated with this Markov process.

   b. Compute the steady-state matrix.

   c. Determine the probability that a beginning student enrolled in the program will complete the course successfully.

30. **COLLEGE GRADUATION RATE** The registrar of a law school has compiled the following statistics on the progress of the school's students working toward the LLB degree: Of the first-year students in a particular year, 85% successfully complete their course of studies and move on to the second year, whereas 15% drop out of the program; of the second-year students in a particular year, 92% go on to the third year, whereas 8% drop out of the program; of the third-year students in a particular year, 98% go on to graduate at the end of the year, whereas 2% drop out of the program.

   a. Construct the transition matrix associated with the Markov process.

   b. Find the steady-state matrix.

   c. Determine the probability that a beginning law student enrolled in the program will go on to graduate.

31. **GENETICS** Refer to Example 4. If the offspring are crossed successively with plants of genotype *aa* only, show that in the long run, all the flowers produced by the plants will be white.

In Exercises 32 and 33, determine whether the statement is true or false. If it is true, explain why it is true. If it is false, give an example to show why it is false.

32. An absorbing stochastic matrix need not contain an absorbing state.

33. In partitioning an absorbing matrix into subdivisions,

$$A = \left[\begin{array}{c|c} I & S \\ \hline O & R \end{array}\right]$$

the identity matrix I is chosen to have the same order as R.

9.3 Solutions to Self-Check Exercises

1. **a.** State 2 is an absorbing state. States 1 and 3 are not absorbing, but each has a possibility (with probability .3 and .6) that an object may go from these states to state 2. Therefore, the matrix T is an absorbing stochastic matrix.

   b. Denoting the states as indicated, we rewrite

$$\begin{array}{c} \\ 1 \\ 2 \\ 3 \end{array} \begin{array}{c} \begin{array}{ccc} 1 & 2 & 3 \end{array} \\ \begin{bmatrix} .2 & 0 & 0 \\ .3 & 1 & .6 \\ .5 & 0 & .4 \end{bmatrix} \end{array}$$

in the form

$$\begin{array}{c} \\ 2 \\ 3 \\ 1 \end{array} \begin{array}{c} \begin{array}{ccc} 2 & 3 & 1 \end{array} \\ \left[\begin{array}{c|cc} 1 & .6 & .3 \\ \hline 0 & .4 & .5 \\ 0 & 0 & .2 \end{array}\right] \end{array}$$

We see that

$$S = [.6 \quad .3] \quad \text{and} \quad R = \begin{bmatrix} .4 & .5 \\ 0 & .2 \end{bmatrix}$$

c. We compute

$$I - R = \begin{bmatrix} 1 & 0 \\ 0 & 1 \end{bmatrix} - \begin{bmatrix} .4 & .5 \\ 0 & .2 \end{bmatrix} = \begin{bmatrix} .6 & -.5 \\ 0 & .8 \end{bmatrix}$$

and, using the formula for finding the inverse of a 2×2 matrix in Section 2.6,

$$(I - R)^{-1} = \begin{bmatrix} 1.67 & 1.04 \\ 0 & 1.25 \end{bmatrix}$$

so

$$S(I - R)^{-1} = [.6 \quad .3]\begin{bmatrix} 1.67 & 1.04 \\ 0 & 1.25 \end{bmatrix} = [1 \quad 1]$$

Therefore, the steady-state matrix of T is

$$\left[\begin{array}{c|cc} 1 & 1 & 1 \\ \hline 0 & 0 & 0 \\ 0 & 0 & 0 \end{array}\right]$$

2. We want to compute the steady-state matrix of T. Note that T is in the form

$$\left[\begin{array}{c|c} I & S \\ \hline O & R \end{array}\right]$$

where

$$S = [.3 \quad .2] \quad \text{and} \quad R = \begin{bmatrix} .6 & .3 \\ .1 & .5 \end{bmatrix}$$

We compute

$$I - R = \begin{bmatrix} 1 & 0 \\ 0 & 1 \end{bmatrix} - \begin{bmatrix} .6 & .3 \\ .1 & .5 \end{bmatrix} = \begin{bmatrix} .4 & -.3 \\ -.1 & .5 \end{bmatrix}$$

and, using the inverse formula in Section 2.6,

$$(I - R)^{-1} = \begin{bmatrix} 2.94 & 1.76 \\ 0.59 & 2.36 \end{bmatrix}$$

so

$$S(I - R)^{-1} = [.3 \quad .2]\begin{bmatrix} 2.94 & 1.76 \\ 0.59 & 2.36 \end{bmatrix} = [1 \quad 1]$$

Therefore, the steady-state matrix of T is

$$\begin{array}{c} \\ \text{CAT} \\ \text{ER} \\ \text{MT} \end{array} \begin{array}{c} \begin{array}{ccc} \text{CAT} & \text{ER} & \text{MT} \end{array} \\ \left[\begin{array}{c|cc} 1 & 1 & 1 \\ \hline 0 & 0 & 0 \\ 0 & 0 & 0 \end{array}\right] \end{array}$$

Interpreting the steady-state matrix of T, we see that in the long run all courts in this state will use computer-aided transcription.

9.4 Game Theory and Strictly Determined Games

The theory of games is a relatively new branch of mathematics and owes much of its development to John von Neumann (1903–1957), one of the mathematical giants of the 20th century. John Harsanyi, John Nash, and Reinhard Selten won the Nobel Prize in Economics in 1994 for their work in this field. Basically, the theory of games combines matrix methods with the theory of probability to determine the optimal strategies to be employed by two or more opponents involved in a competitive situation, with each opponent seeking to maximize his or her "gains," or, equivalently, to minimize his or her "losses." The players may be poker players, managers of rival corporations seeking to extend their share of the market, campaign managers, or generals of opposing armies, to name a few.

For simplicity, we limit our discussion to games with two players. Such games are, naturally enough, called two-person games.

Two-Person Games

APPLIED EXAMPLE 1 Coin-Matching Game Richie and Chuck play a coin-matching game in which each player selects a side of a penny without prior knowledge of the other's choice. Then, upon a predetermined signal, the players disclose their choices simultaneously. Chuck agrees to pay Richie a sum of \$3 if both choose heads; if Richie chooses heads and Chuck chooses tails, then

Richie pays Chuck \$6; if Richie chooses tails and Chuck chooses heads, then Chuck pays Richie \$2; finally, if both Richie and Chuck choose tails, then Chuck pays Richie \$1. In this game, the objective of each player is to discover a strategy that will ensure that his winnings are maximized (equivalently, that his losses are minimized).

The coin-matching game is an example of a **zero-sum game**—that is, a game in which the payoff to one party results in an equal loss to the other. For such games, the sum of the payments made by both players at the end of each play adds up to zero.

To help us analyze the coin-matching game in Example 1, we represent the given data in the form of a matrix called a **payoff matrix:**

$$\begin{array}{cc} & \text{C's moves} \\ & \begin{array}{cc}\text{Heads} & \text{Tails}\end{array} \\ \text{R's moves} \begin{array}{c}\text{Heads}\\ \text{Tails}\end{array} & \begin{bmatrix} 3 & -6 \\ 2 & 1 \end{bmatrix} \end{array}$$

Each row of the matrix corresponds to one of the two possible moves by Richie (referred to as the row player, R), whereas each column corresponds to one of the two possible moves by Chuck (referred to as the column player, C). Each entry in the matrix represents the payoff from C to R. For example, the entry $a_{11} = 3$ represents a \$3 payoff from Chuck to Richie (C to R) when Richie chooses to play row 1 (heads) and Chuck chooses to play column 1 (heads). On the other hand, the entry $a_{12} = -6$ represents (because it's negative) a \$6 payoff to C (from R) when R chooses to play row 1 (heads) and C chooses to play column 2 (tails). (Interpret the meaning of $a_{21} = 2$ and $a_{22} = 1$ for yourself.)

More generally, suppose we are given a two-person game with two players R and C. Furthermore, suppose that R has m possible moves $R_1, R_2, \ldots, R_m$ and that C has n possible moves $C_1, C_2, \ldots, C_n$. Then we can represent the game in terms of an $m \times n$ matrix in which each row of the matrix represents one of the m possible moves of R and each column of the matrix represents one of the n possible moves of C:

$$\begin{array}{cc} & \text{C's moves} \\ & \begin{array}{ccccccc} C_1 & C_2 & \cdots & C_j & \cdots & C_n \end{array} \\ \text{R's moves} \begin{array}{c} R_1 \\ R_2 \\ \vdots \\ R_i \\ \vdots \\ R_m \end{array} & \begin{bmatrix} a_{11} & a_{12} & \cdots & a_{1j} & \cdots & a_{1n} \\ a_{21} & a_{22} & \cdots & a_{2j} & \cdots & a_{2n} \\ \vdots & \vdots & & \vdots & & \vdots \\ a_{i1} & a_{i2} & \cdots & a_{ij} & \cdots & a_{in} \\ \vdots & \vdots & & \vdots & & \vdots \\ a_{m1} & a_{m2} & \cdots & a_{mj} & \cdots & a_{mn} \end{bmatrix} \end{array}$$

The entry a_{ij} in the ith row and jth column of the (payoff) matrix represents the payoff to R when R chooses move R_i and C chooses move C_j. In this context, note that a payoff to R means, in actuality, a payoff to C in the event that the value of a_{ij} is negative.

EXAMPLE 2 The payoff matrix associated with a game is given by

$$\begin{array}{cc} & \text{C's moves} \\ & \begin{array}{ccc} C_1 & C_2 & C_3 \end{array} \\ \text{R's moves} \begin{array}{c} R_1 \\ R_2 \end{array} & \begin{bmatrix} 1 & -2 & 3 \\ 4 & -5 & -1 \end{bmatrix} \end{array}$$

Give an interpretation of this payoff matrix.

PORTFOLIO Christian Derrick

TITLE Technical Director, Europe
INSTITUTION SpiderCloud Wireless

SpiderCloud Wireless is the start-up company that introduced the first enterprise radio access network (E-RAN) systems into mobile networks. E-RAN technology allows for superior indoor coverage and capacity as well as high-value services for a mobile operator's most important customer, the business user. As the Technical Director, Europe, I am responsible for working with European mobile operators to develop and deliver new wireless technologies.

At SpiderCloud, we draw on our knowledge of mathematics to understand the strategic decisions of our mobile partners. Applying our knowledge of game theory, for example, we can begin to answer a variety of questions such as: How will mobile operators price their voice, data, and SMS plans? And how will suppliers price their cell phone equipment?

Let's consider the first question. If each mobile operator offered identical pricing plans, one would predict that a price reduction by an operator would lead to an increase in that operator's market share. But price, of course, is not the only factor a consumer takes into consideration when selecting a plan. Consumers also consider other variables such as the operator's brand name and reputation, the device that is bundled with the plan, the geographical coverage of the operator's network, and a range of add-on options.

Game theory helps mobile operators to factor the perceived value of these factors into their pricing plans. Operators use game theory to evaluate the likely outcomes of changes in pricing in response to both their competitors' actions and their competitors' responses to their own actions. This helps operators to maximize their profits. Working at SpiderCloud, I have learned that an in-depth knowledge of game theory, including Nash equilibrium and the prisoner's dilemma, is extremely useful for predicting the behavior of players in a complex business environment.

Michelle Preen; (inset) © dani3315/ShutterStock.com

Solution In this two-person game, player R has two possible moves, whereas player C has three possible moves. The payoffs are determined as follows: If R chooses R_1, then

R wins 1 unit if C chooses C_1.
R loses 2 units if C chooses C_2.
R wins 3 units if C chooses C_3.

If R chooses R_2, then

R wins 4 units if C chooses C_1.
R loses 5 units if C chooses C_2.
R loses 1 unit if C chooses C_3.

Optimal Strategies

Let's return to the payoff matrix of Example 1 and see how it may be used to help us determine the "best" strategy for each of the two players R and C. For convenience, this matrix is reproduced here:

$$\text{R's moves} \begin{array}{c} R_1 \\ R_2 \end{array} \overset{\begin{array}{c}\text{C's moves}\\ C_1 \quad\quad C_2\end{array}}{\begin{bmatrix} 3 & -6 \\ 2 & 1 \end{bmatrix}}$$

Let's first consider the game from R's point of view. Since the entries in the payoff matrix represent payoffs to R, his initial reaction might be to seek out the largest entry in the matrix and consider the row containing such an entry as a possible move. Thus, he is led to the consideration of R_1 as a possible move.

Let's examine this choice a little more closely. To be sure, R would realize the largest possible payoff to himself ($3) if C chose C_1; but if C chose C_2, then R would lose $6! Since R does not know beforehand what C's move will be, a more prudent approach on R's part would be to assume that no matter what row he chooses, C will counter with a move (column) that will result in the smallest payoff to R. To maximize the payoff to himself under these circumstances, R would then select from among the moves (rows) the one in which the smallest payoff is as large as possible. This strategy for R, which is called, for obvious reasons, the **maximin strategy,** may be summarized as follows:

Maximin Strategy

1. For each row of the payoff matrix, find the smallest entry in that row.
2. Choose the row for which the entry found in step 1 is as large as possible. This row constitutes R's "best" move.

For the problem under consideration, we can organize our work as follows:

$$\begin{bmatrix} 3 & -6 \\ 2 & 1 \end{bmatrix} \quad \begin{matrix} \text{Row minima} \\ -6 \\ \textcircled{1} \leftarrow \text{Larger of the row minima} \end{matrix}$$

From these results, we can see that R's "best" move is row 2. By choosing this move, R stands to win at least $1.

Next, let's consider the game from C's point of view. His objective is to minimize the payoff to R. This is accomplished by choosing the column whose largest payoff is as small as possible. This strategy for C, which is called the **minimax strategy,** may be summarized as follows:

Minimax Strategy

1. For each column of the payoff matrix, find the largest entry in that column.
2. Choose the column for which the entry found in step 1 is as small as possible. This column constitutes C's "best" move.

We can organize the work involved in determining C's "best" move as follows:

$$\begin{bmatrix} 3 & -6 \\ 2 & 1 \end{bmatrix}$$

$$\text{Column maxima} \quad 3 \quad \textcircled{1}$$

$$\uparrow \text{ Smaller of the column maxima}$$

From these results, we see that C's "best" move is column 2. By choosing this move, C stands to lose at most $1.

EXAMPLE 3 For the game with the following payoff matrix, determine the maximin and minimax strategies for each player.

$$\begin{bmatrix} -3 & -2 & 4 \\ -2 & 0 & 3 \\ 6 & -1 & 1 \end{bmatrix}$$

Solution We determine the minimum of each row and the maximum of each column of the payoff matrix, and we then display these numbers by circling the largest of the row minima and the smallest of the column maxima:

$$\begin{array}{r} \\ \\ \\ \text{Column maxima} \end{array}\begin{array}{c} \begin{bmatrix} -3 & -2 & 4 \\ -2 & 0 & 3 \\ 6 & -1 & 1 \end{bmatrix} \\ \begin{matrix} 6 & \enclose{circle}{0} & 4 \end{matrix} \\ \uparrow \\ \text{Smallest of the column maxima} \end{array}\begin{array}{c} \text{Row minima} \\ -3 \\ -2 \\ \enclose{circle}{-1} \leftarrow \text{Largest of the row minima} \end{array}$$

From these results, we conclude that the maximin strategy (for the row player) is to play row 3, whereas the minimax strategy (for the column player) is to play column 2.

EXAMPLE 4 Determine the maximin and minimax strategies for each player in a game whose payoff matrix is given by

$$\begin{bmatrix} 3 & 4 & -4 \\ 2 & -1 & -3 \end{bmatrix}$$

Solution Proceeding as in Example 3, we obtain the following:

$$\begin{array}{r} \\ \\ \text{Column maxima} \end{array}\begin{array}{c} \begin{bmatrix} 3 & 4 & -4 \\ 2 & -1 & -3 \end{bmatrix} \\ \begin{matrix} 3 & 4 & \enclose{circle}{-3} \end{matrix} \\ \uparrow \\ \text{Smallest of the column maxima} \end{array}\begin{array}{c} \text{Row minima} \\ -4 \\ \enclose{circle}{-3} \leftarrow \text{Largest of the row minima} \end{array}$$

from which we conclude that the maximin strategy for the row player is to play row 2, whereas the minimax strategy for the column player is to play column 3.

In arriving at the maximin and minimax strategies for the respective players, we have assumed that both players always act rationally, with the knowledge that their opponents will always act rationally. This means that each player adopts a strategy of always making the same move and assumes that his opponent is always going to counter that move with a move that will maximize the payoff to the opponent. Thus, each player adopts the pure strategy of always making the move that will minimize the payoff his opponent can receive and thereby maximize the payoff to himself.

This raises the following question: Suppose a game is played repeatedly and one of the players realizes that the opponent is employing his maximin (or minimax) strategy. Can this knowledge be used to the player's advantage? To obtain a partial answer to this question, let's consider the game posed in Example 3. There, the minimax strategy for the column player is to play column 2. Suppose, in repeated plays of the game, a player consistently plays that column and this strategy becomes known to the row player. The row player may then change the strategy from playing row 3 (the maximin strategy) to playing row 2, thereby reducing losses from 1 unit to zero. Thus, at least for this game, the knowledge that a player is employing the maximin or minimax strategy can be used to the opponent's advantage.

There is, however, a class of games in which the knowledge that a player is using the maximin (or minimax) strategy proves of no help to the opponent. Consider the game of Example 4. There, the row player's maximin strategy is to play row 2, and the column player's minimax strategy is to play column 3. Suppose, in repeated plays of the game, R (the row player) has discovered that C (the column player) consistently chooses to play column 3 (the minimax strategy). Can this knowledge be used to R's advantage? Now, other than playing row 2 (the maximin strategy), R may choose to play row 1. But if R makes this choice, then he would lose 4 units instead of 3 units! Clearly, in this case, the knowledge that C is using the minimax strategy cannot be used to advantage. R's optimal (best) strategy is the maximin strategy.

Optimal Strategy

The **optimal strategy** in a game is the strategy that is most profitable to a particular player.

Next, suppose that in repeated plays of the game, C has discovered that R consistently plays row 2 (his optimal strategy). Can this knowledge be used to C's advantage? Another glance at the payoff matrix reveals that by playing column 1, C stands to lose 2 units, and by playing column 2, he stands to win 1 unit, as compared with winning 3 units by playing column 3, as called for by the minimax strategy. Thus, as in the case of R, C does not benefit from knowing his opponent's move. Furthermore, his optimal strategy coincides with the minimax strategy.

This game, in which the row player cannot benefit from knowing that his or her opponent is using the minimax strategy and the column player cannot benefit from knowing that his or her opponent is using the maximin strategy, is said to be strictly determined.

Strictly Determined Game

A **strictly determined game** is characterized by the following properties:

1. There is an entry in the payoff matrix that is *simultaneously* the smallest entry in its row and the largest entry in its column. This entry is called the **saddle point** for the game.
2. The optimal strategy for the row player is the maximin strategy, obtained by choosing the row containing the saddle point. The optimal strategy for the column player is the minimax strategy, obtained by choosing the column containing the saddle point.

The saddle point of a strictly determined game is also referred to as the **value of the game.** If the value of a strictly determined game is positive, then the game favors the row player. If the value is negative, the game favors the column player. If the value of the game is zero, the game is called a **fair game.**

Returning to the coin-matching game discussed earlier, we conclude that Richie's optimal strategy consists of playing row 2 repeatedly, whereas Chuck's optimal strategy consists of playing column 2 repeatedly. Furthermore, the value of the game is 1, implying that the game favors the row player, Richie.

EXAMPLE 5 A two-person, zero-sum game is defined by the payoff matrix

$$A = \begin{bmatrix} 1 & 2 & -3 \\ -1 & 2 & -2 \\ 2 & 3 & -4 \end{bmatrix}$$

a. Show that the game is strictly determined, and find the saddle point(s) for the game.
b. What is the optimal strategy for each player?
c. What is the value of the game? Does the game favor one player over the other?

Solution

a. First, we determine the minimum of each row and the maximum of each column of the payoff matrix A and display these minima and maxima as follows:

$$\begin{array}{r} \\ \\ \\ \text{Column maxima} \end{array}\begin{array}{c} \left[\begin{array}{rrr} 1 & 2 & -3 \\ -1 & 2 & \boxed{-2} \\ 2 & 3 & -4 \end{array}\right] \\ \begin{array}{rrr} 2 & 3 & -2 \end{array} \end{array} \begin{array}{c} \text{Row minima} \\ -3 \\ -2 \leftarrow \text{Largest of the row minima} \\ -4 \\ \end{array}$$

↑ Smallest of the column maxima

From these results, we see that the circled entry, -2, is simultaneously the smallest entry in its row and the largest entry in its column. Therefore, the game is strictly determined, with the entry $a_{23} = -2$ as its saddle point.

b. From these results, we see that the optimal strategy for the row player is to make the move represented by the second row of the matrix, and the optimal strategy for the column player is to make the move represented by the third column.

c. The value of the game is -2, which implies that if both players adopt their best strategy, the column player will win 2 units in a play. Consequently, the game favors the column player.

A game may have more than one saddle point, as the next example shows.

EXAMPLE 6 A two-person, zero-sum game is defined by the payoff matrix

$$A = \begin{bmatrix} 4 & 5 & 4 \\ -2 & -5 & -3 \\ 4 & 6 & 8 \end{bmatrix}$$

a. Show that the game is strictly determined, and find the saddle points for the game.

b. Discuss the optimal strategies for the players.

c. Does the game favor one player over the other?

Solution

a. Proceeding as in the previous example, we obtain the following information:

$$\begin{array}{r} \\ \\ \\ \text{Column maxima} \end{array}\begin{array}{c} \left[\begin{array}{rrr} \boxed{4} & 5 & 4 \\ -2 & -5 & -3 \\ \boxed{4} & 6 & 8 \end{array}\right] \\ \begin{array}{rrr} 4 & 6 & 8 \end{array} \end{array} \begin{array}{c} \text{Row minima} \\ 4 \\ -5 \quad \text{Largest of the row minima} \\ 4 \\ \end{array}$$

↑ Smallest of the column maxima

We see that each of the circled entries, 4, is simultaneously the smallest entry in the row and the largest entry in the column containing it. Therefore, the game is strictly determined. In this case, it has two saddle points: the entry $a_{11} = 4$ and the entry $a_{31} = 4$. In general, it can be shown that every saddle point of a payoff matrix must have the same value.

b. Since the game has two saddle points, both lying in the first column and in the first and third rows of the payoff matrix, we see that the row player's optimal strategy consists of playing either row 1 or row 3 consistently, whereas the column player's optimal strategy consists of playing column 1 repeatedly.

c. The value of the game is 4, which implies that it favors the row player.

APPLIED EXAMPLE 7 Bidding for Rights Two television subscription companies, UBS and Telerama, are planning to extend their operations to a certain city. Each has the option of making its services available to prospective subscribers with a special introductory subscription rate. It is estimated that if both UBS and Telerama offer the special subscription rate, each will get 50% of the market, whereas if UBS offers the special subscription rate and Telerama does not, UBS will get 70% of the market. If Telerama offers the special subscription rate and UBS does not, it is estimated that UBS will get 40% of the market. It is estimated that if both companies elect not to offer the special subscription rate, UBS will get 60% of the market.

a. Construct the payoff matrix for the game.
b. Show that the game is strictly determined.
c. Determine the optimal strategy for each company, and find the value of the game.

Solution

a. The required payoff matrix is given by

		Telerama: Intro. rate	Telerama: Usual rate
UBS	Intro. rate	.50	.70
	Usual rate	.40	.60

b. The entry $a_{11} = .50$ is the smaller entry in its row and the larger entry in its column. Therefore, the entry $a_{11} = .50$ is a saddle point of the game, and the game is strictly determined.

c. The entry $a_{11} = .50$ is the only saddle point of the game, so UBS's optimal strategy is to choose row 1, and Telerama's optimal strategy is to choose column 1. In other words, both companies should offer their potential customers their respective introductory subscription rates.

Explore and Discuss

A two-person, zero-sum game is defined by the payoff matrix

$$A = \begin{bmatrix} a & a \\ c & d \end{bmatrix}$$

1. Show that the game is strictly determined.
2. What can you say about the game with the following payoff matrix?

$$B = \begin{bmatrix} a & b \\ c & c \end{bmatrix}$$

9.4 Self-Check Exercises

1. A two-person, zero-sum game is defined by the payoff matrix

$$A = \begin{bmatrix} -2 & 1 & 3 \\ 3 & 2 & 2 \\ 2 & -1 & 4 \end{bmatrix}$$

   a. Show that the game is strictly determined, and find the saddle point(s) for the game.
   b. What is the optimal strategy for each player?
   c. What is the value of the game? Does the game favor one player over the other?

2. **Financial Analysis** The management of Delta Corporation, a construction and development company, is deciding whether to go ahead with the construction of a large condominium complex. A financial analysis of the project indicates that if Delta goes ahead with the development and the home mortgage rate drops 1 point or more by next year, when the complex is expected to be completed, it will stand to make a profit of \$750,000. If Delta goes ahead with the development and the mortgage rate stays within 1 point of the current rate by next year, Delta will stand to make a profit of \$600,000. If Delta goes ahead with the development and the mortgage rate increases 1 point or more by next year, Delta will stand to make a profit of \$350,000. If Delta does not go ahead with the development and the mortgage rate drops 1 point or more by next year, Delta will stand to make a profit of \$400,000. If Delta does not go ahead with the development and the mortgage rate stays within 1 point of the current rate by next year, Delta will stand to make a profit of \$350,000. Finally, if Delta does not go ahead with the development and the mortgage rate increases 1 point or more by next year, Delta stands to make \$250,000.
   a. Represent this information in the form of a payoff matrix.
   b. Assuming that the home mortgage rate trend is volatile over the next year, determine whether or not Delta should go ahead with the project.

Solutions to Self-Check Exercises 9.4 can be found on page 574.

9.4 Concept Questions

1. a. What is the maximin strategy for the row player in a two-person game that is represented by a payoff matrix?
   b. What is the minimax strategy for the column player in a two-person game that is represented by a payoff matrix?

2. a. How do you find the saddle point in the payoff matrix for a strictly determined game?
   b. What is the optimal strategy for the row player in a strictly determined game? The column player?

9.4 Exercises

In Exercises 1–8, determine the maximin and minimax strategies for each two-person, zero-sum matrix game.

1. $\begin{bmatrix} 2 & 3 \\ 4 & 1 \end{bmatrix}$

2. $\begin{bmatrix} -1 & 3 \\ 2 & 5 \end{bmatrix}$

3. $\begin{bmatrix} 1 & 3 & 2 \\ 0 & -1 & 4 \end{bmatrix}$

4. $\begin{bmatrix} 1 & 4 & -2 \\ 4 & 6 & -3 \end{bmatrix}$

5. $\begin{bmatrix} 3 & 2 & 1 \\ 1 & -2 & 3 \\ 6 & 4 & 1 \end{bmatrix}$

6. $\begin{bmatrix} 1 & 4 \\ 2 & -2 \\ 3 & 0 \end{bmatrix}$

7. $\begin{bmatrix} 4 & 2 & 1 \\ 1 & 0 & -1 \\ 2 & 1 & 3 \end{bmatrix}$

8. $\begin{bmatrix} -1 & 1 & 2 \\ 3 & 1 & 1 \\ -1 & 1 & 2 \\ 3 & 2 & -1 \end{bmatrix}$

In Exercises 9–18, determine whether the two-person, zero-sum matrix game is strictly determined. If a game is strictly determined,
a. Find the saddle point(s) of the game.
b. Find the optimal strategy for each player.
c. Find the value of the game.
d. Determine whether the game favors one player over the other.

9. $\begin{bmatrix} 2 & 3 \\ 1 & -4 \end{bmatrix}$

10. $\begin{bmatrix} 1 & 0 \\ 0 & -1 \end{bmatrix}$

11. $\begin{bmatrix} 1 & 3 & 2 \\ -1 & 4 & -6 \end{bmatrix}$

12. $\begin{bmatrix} 3 & 2 \\ -1 & -2 \\ 4 & 1 \end{bmatrix}$

13. $\begin{bmatrix} 1 & 3 & 4 & 2 \\ 0 & 2 & 6 & -4 \\ -1 & -3 & -2 & 1 \end{bmatrix}$ **14.** $\begin{bmatrix} 2 & 4 & 2 \\ 0 & 3 & 0 \\ -1 & -2 & 1 \end{bmatrix}$

15. $\begin{bmatrix} 1 & 2 \\ 0 & 3 \\ -1 & 2 \\ 2 & -2 \end{bmatrix}$ **16.** $\begin{bmatrix} -1 & 2 & 4 \\ 2 & 3 & 5 \\ 0 & 1 & -3 \\ -2 & 4 & -2 \end{bmatrix}$

17. $\begin{bmatrix} 1 & -1 & 3 & 2 \\ 1 & 0 & 2 & 2 \\ -2 & 2 & 3 & -1 \end{bmatrix}$ **18.** $\begin{bmatrix} 3 & -1 & 0 & -4 \\ 2 & 1 & 0 & 2 \\ -3 & 1 & -2 & 1 \\ -1 & -1 & -2 & 1 \end{bmatrix}$

19. Game of Matching Fingers Robin and Cathy play a game of matching fingers. On a predetermined signal, both players simultaneously extend one, two, or three fingers from a closed fist. If the sum of the number of fingers extended is even, then Robin receives an amount in dollars equal to that sum from Cathy. If the sum of the number of fingers extended is odd, then Cathy receives an amount in dollars equal to that sum from Robin.

a. Construct the payoff matrix for the game.
b. Find the maximin and the minimax strategies for Robin and Cathy, respectively.
c. Is the game strictly determined?
d. If the answer to part (c) is yes, what is the value of the game?

20. Management Decisions Brady's, a conventional department store, and ValueMart, a discount department store, are both considering opening new stores at one of two possible sites: the Civic Center and North Shore Plaza. The strategies available to the management of each store are given in the following payoff matrix, where each entry represents the amounts (in hundreds of thousands of dollars) either gained or lost by one business from or to the other as a result of the sites selected.

		ValueMart Center	ValueMart Plaza
Brady's	Civic Center	2	−2
	North Shore Plaza	3	−4

a. Show that the game is strictly determined.
b. What is the value of the game?
c. Determine the best strategy for the management of each store (that is, determine the ideal locations for each store).

21. Financial Analysis The management of Acrosonic is faced with the problem of deciding whether to expand the production of its line of electrostatic loudspeaker systems. It has been estimated that an expansion will result in an annual profit of \$200,000 for Acrosonic if the general economic climate is good. On the other hand, an expansion during a period of economic recession will cut Acrosonic's annual profit to \$120,000. As an alternative, Acrosonic may hold the production of its electrostatic loudspeaker systems at the current level and expand its line of conventional loudspeaker systems. In this event, the company will make a profit of \$50,000 in an expanding economy (because many potential customers will be expected to buy electrostatic loudspeaker systems from other competitors) and a profit of \$150,000 in a recessionary economy.

a. Construct the payoff matrix for this game.
Hint: The row player is the management of the company, and the column player is the economy.
b. Should management recommend expanding the company's line of electrostatic loudspeaker systems?

22. Financial Analysis The proprietor of Belvedere's is faced with the problem of deciding whether to expand her restaurant facilities now or to wait until some future date to do so. If she expands the facilities now and the economy experiences a period of growth during the coming year, she will make a net profit of \$442,000; if she expands now and a period of zero growth follows, then she will make a net profit of \$40,000; and if she expands now and an economic recession follows, she will suffer a net loss of \$108,000. If she does not expand the restaurant now and the economy experiences a period of growth during the coming year, she will make a net profit of \$280,000; if she does not expand now and a period of zero growth follows, she will make a net profit of \$190,000. Finally, if she does not expand now and an economic recession follows, she will make a net profit of \$100,000.

a. Represent this information in the form of a payoff matrix.
b. Determine whether the owner of the restaurant should expand her facilities at this time.

23. Market Share Roland's Barber Shop and Charley's Barber Shop are both located in the business district of a certain town. Roland estimates that if he raises the price of a haircut by \$1, he will increase his market share by 3% if Charley raises his price by the same amount; he will decrease his market share by 1% if Charley holds his price at the same level; and he will decrease his market share by 3% if Charley lowers his price by \$1. If Roland keeps his price the same, he will increase his market share by 2% if Charley raises his price by \$1; he will keep the same market share if Charley holds the price at the same level; and he will decrease his market share by 2% if Charley lowers his price by \$1. Finally, if Roland lowers the price he charges by \$1, his market share will increase by 5% if Charley raises his prices by the same amount; he will increase his market share by 2% if Charley holds his price at the same level; and he will increase his market share by 1% if Charley lowers his price by \$1.

a. Construct the payoff matrix for this game.
b. Show that the game is strictly determined.
c. If neither party is willing to lower the price he charges for a haircut, show that both should keep their present price structures.

In Exercises 24–26, determine whether the statement is true or false. If it is true, explain why it is true. If it is false, give an example to show why it is false.

24. In a zero-sum game, the payments made by the players at the end of each play add up to zero.

25. In a strictly determined game, the value of the game is given by the saddle point of the game.

26. If the value of a strictly determined game is not negative, it favors the row player.

9.4 Solutions to Self-Check Exercises

1. a. Displaying the minimum of each row and the maximum of each column of the payoff matrix A, we obtain

				Row minima	
	−2	1	3	−2	
	3	(2)	2	2	← Largest of the row minima
	2	−1	4	−1	
Column maxima	3	2	4		
		↑ Smallest of the column maxima			

From these results, we see that the circled entry, 2, is simultaneously the smallest entry in its row and the largest entry in its column. Therefore, the game is strictly determined, with the entry $a_{22} = 2$ as its saddle point.

b. From these results, we see that the optimal strategy for the row player is to make the move represented by the second row of the matrix, and the optimal strategy for the column player is to make the move represented by the second column.

c. The value of the game is 2, which implies that if both players adopt their best strategy, the row player will win 2 units in a play. Consequently, the game favors the row player.

2. a. We may review this situation as a game in which the row player is Delta and the column player is the home mortgage rate. The required payoff matrix is

		Mortgage rate		
		Decrease	Steady	Increase
Delta Corp.	Project	750	600	350
	No project	400	350	250

(All figures are in thousands of dollars.)

b. From part (a), the payoff matrix under consideration is

$$\begin{bmatrix} 750 & 600 & 350 \\ 400 & 350 & 250 \end{bmatrix}$$

Proceeding in the usual manner, we find

				Row minima	
	750	600	(350)	350	← Largest of the row minima
	400	350	250	250	
Column maxima	750	600	350		
			↑ Smallest of the column maxima		

From these results, we see that the entry $a_{13} = 350$ is a saddle point, and the game is strictly determined. We can also conclude that the company should go ahead with the project.

9.5 Games with Mixed Strategies

In Section 9.4, we discussed strictly determined games and found that the optimal strategy for the row player is to select the row containing a saddle point for the game, and the optimal strategy for the column player is to select the column containing a saddle point. Furthermore, in repeated plays of the game, each player's optimal strategy consists of making the same move over and over again, since the discovery of the opponent's optimal strategy cannot be used to advantage. Such strategies are called **pure strategies.** In this section, we look at games that are not strictly determined and the strategies associated with such games.

Mixed Strategies

As a simple example of a game that is not strictly determined, let's consider the following slightly modified version of the coin-matching game played by Richie and

Chuck (see Example 1, Section 9.4). Suppose Richie wins \$3 if both parties choose heads and \$1 if both choose tails and loses \$2 if one chooses heads and the other tails. Then the payoff matrix for this game is given by

$$\begin{array}{c c} & \begin{array}{cc} & C\text{'s moves} \\ & \begin{array}{cc} C_1 & C_2 \\ \text{(heads)} & \text{(tails)} \end{array} \end{array} \\ R\text{'s moves} \begin{array}{c} R_1 \text{ (heads)} \\ R_2 \text{ (tails)} \end{array} & \begin{bmatrix} 3 & -2 \\ -2 & 1 \end{bmatrix} \end{array}$$

A quick examination of this matrix reveals that it does not contain an entry that is simultaneously the smallest entry in its row and the largest entry in its column; that is, the game has no saddle point and is therefore not strictly determined. What strategy might Richie adopt for the game? Offhand, it would seem that he should consistently select row 1, since he stands to win \$3 by playing this row and only \$1 by playing row 2 at a risk, in either case, of losing \$2. However, if Chuck discovers that Richie is playing row 1 consistently, he would counter this strategy by playing column 2, causing Richie to lose \$2 on each play! In view of this, Richie is led to consider a strategy whereby he chooses row 1 some of the time and row 2 at other times. A similar analysis of the game from Chuck's point of view suggests that he might consider choosing column 1 some of the time and column 2 at other times. Such strategies are called **mixed strategies.**

From a practical point of view, there are many ways in which a player may choose moves in a game with mixed strategies. For example, in the game just mentioned, if Richie decides to play heads half the time and tails the other half of the time, he could toss an unbiased coin before each move and let the outcome of the toss determine which move he should make. Here is another more general but less practical way of deciding on the choice of a move: Having determined beforehand the proportion of the time row 1 is to be chosen (and therefore the proportion of the time row 2 is to be chosen), Richie might construct a spinner (Figure 3) in which the areas of the two sectors reflect these proportions and let the move be decided by the outcome of a spin. These two methods for determining a player's move in a game with mixed strategies guarantee that the strategy will not fall into a pattern that can be discovered by the opponent.

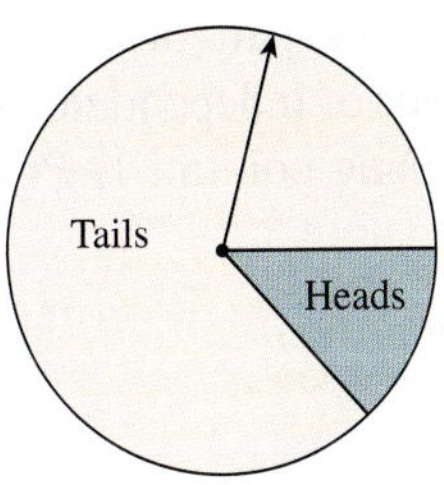

FIGURE 3

From a mathematical point of view, we may describe the mixed strategy of a row player in terms of a row vector whose dimension coincides with the number of possible moves the player has. For example, if Richie had decided on a strategy in which he chose to play row 1 half the time and row 2 the other half of the time, then this strategy is represented by the row vector

$$[.5 \quad .5]$$

Similarly, the mixed strategy for a column player may be represented by a column vector of appropriate dimension. For example, returning to our illustration, suppose Chuck has decided that 20% of the time he will choose column 1 and 80% of the time he will choose column 2. This strategy is represented by the column vector

$$\begin{bmatrix} .2 \\ .8 \end{bmatrix}$$

Expected Value of a Game

For the purpose of comparing the merits of a player's different mixed strategies in a game, it is convenient to introduce a number called the expected value of a game. The **expected value of a game** measures the average payoff to the row player when both players adopt a particular set of mixed strategies. We now explain this notion using a 2×2 matrix game whose payoff matrix has the general form

$$A = \begin{bmatrix} a_{11} & a_{12} \\ a_{21} & a_{22} \end{bmatrix}$$

Suppose that in repeated plays of the game, the row player R adopts the mixed strategy

$$P = [p_1 \quad p_2]$$

(that is, the player selects row 1 with probability p_1 and row 2 with probability p_2), and the column player C adopts the mixed strategy

$$Q = \begin{bmatrix} q_1 \\ q_2 \end{bmatrix}$$

(that is, the column player selects column 1 with probability q_1 and column 2 with probability q_2). Now, in each play of the game, there are four possible outcomes, which may be represented by the ordered pairs

(row 1, column 1)
(row 1, column 2)
(row 2, column 1)
(row 2, column 2)

where the first number of each ordered pair represents R's selection and the second number of each ordered pair represents C's selection. Since the choice of moves is made by one player without knowing the other's choice, each pair of events (for example, the events "row 1" and "column 1") constitutes a pair of independent events. Therefore, the probability of R choosing row 1 and C choosing column 1, P(row 1, column 1), is given by

$$\begin{aligned} P(\text{row 1, column 1}) &= P(\text{row 1}) \cdot P(\text{column 1}) \\ &= p_1 q_1 \end{aligned}$$

In a similar manner, we compute the probability of each of the other three outcomes. These calculations, together with the payoffs associated with each of the four possible outcomes, may be summarized as follows:

Outcome	Probability	Payoff
(row 1, column 1)	p_1q_1	a_{11}
(row 1, column 2)	p_1q_2	a_{12}
(row 2, column 1)	p_2q_1	a_{21}
(row 2, column 2)	p_2q_2	a_{22}

Then the *expected payoff E* of the game is the sum of the products of the payoffs and the corresponding probabilities (see Section 8.2). Thus,

$$E = p_1q_1a_{11} + p_1q_2a_{12} + p_2q_1a_{21} + p_2q_2a_{22}$$

In terms of the matrices P, A, and Q, we have the following relatively simple expression for E:

$$E = PAQ$$

which you may verify (Exercise 22). This result may be generalized as follows:

Expected Value of a Game

Let

$$P = [p_1 \ \ p_2 \ \ \cdots \ \ p_m] \quad \text{and} \quad Q = \begin{bmatrix} q_1 \\ q_2 \\ \vdots \\ q_n \end{bmatrix}$$

be the vectors representing the mixed strategies for the row player R and the column player C, respectively, in a game with the $m \times n$ payoff matrix

$$A = \begin{bmatrix} a_{11} & a_{12} & \cdots & a_{1n} \\ a_{21} & a_{22} & \cdots & a_{2n} \\ \vdots & \vdots & & \vdots \\ a_{m1} & a_{m2} & \cdots & a_{mn} \end{bmatrix}$$

Then the expected value of the game is given by

$$E = PAQ = [p_1 \ \ p_2 \ \ \cdots \ \ p_m] \begin{bmatrix} a_{11} & a_{12} & \cdots & a_{1n} \\ a_{21} & a_{22} & \cdots & a_{2n} \\ \vdots & \vdots & & \vdots \\ a_{m1} & a_{m2} & \cdots & a_{mn} \end{bmatrix} \begin{bmatrix} q_1 \\ q_2 \\ \vdots \\ q_n \end{bmatrix}$$

We now look at several examples involving the computation of the expected value of a game.

APPLIED EXAMPLE 1 Coin-Matching Game Consider a coin-matching game played by Richie and Chuck with the payoff matrix (in dollars) given by

$$A = \begin{bmatrix} 3 & -2 \\ -2 & 1 \end{bmatrix}$$

Compute the expected payoff of the game if Richie adopts the mixed strategy P and Chuck adopts the mixed strategy Q, where

a. $P = [.5 \quad .5]$ and $Q = \begin{bmatrix} .5 \\ .5 \end{bmatrix}$

b. $P = [.8 \quad .2]$ and $Q = \begin{bmatrix} .1 \\ .9 \end{bmatrix}$

Solution

a. We compute

$$\begin{aligned} E = PAQ &= [.5 \quad .5] \begin{bmatrix} 3 & -2 \\ -2 & 1 \end{bmatrix} \begin{bmatrix} .5 \\ .5 \end{bmatrix} \\ &= [.5 \quad -.5] \begin{bmatrix} .5 \\ .5 \end{bmatrix} \\ &= 0 \end{aligned}$$

Thus, in repeated plays of the game, it may be expected that in the long term, the payoff to each player is 0.

b. We compute

$$\begin{aligned} E = PAQ &= [.8 \quad .2]\begin{bmatrix} 3 & -2 \\ -2 & 1 \end{bmatrix}\begin{bmatrix} .1 \\ .9 \end{bmatrix} \\ &= [2 \quad -1.4]\begin{bmatrix} .1 \\ .9 \end{bmatrix} \\ &= -1.06 \end{aligned}$$

That is, in the long run, Richie may be expected to lose \$1.06 on the average in each play.

EXAMPLE 2 The payoff matrix for a certain game is given by

$$A = \begin{bmatrix} 1 & -2 \\ -1 & 2 \\ 3 & -3 \end{bmatrix}$$

a. Find the expected payoff to the row player if the row player R uses her maximin pure strategy and the column player C uses her minimax pure strategy.

b. Find the expected payoff to the row player if R uses her maximin strategy 50% of the time and chooses each of the other two rows 25% of the time, while C chooses each column 50% of the time.

Solution

a. The maximin and minimax strategies for the row and column players, respectively, may be found by using the method of Section 9.4. Thus,

			Row minima	
	1	−2	−2	
	−1	2	(−1)	← Largest of the row minima
	3	−3	−3	
Column maxima	3	(2)		
		↑ Smallest of the column maxima		

From these results, we see that R's optimal pure strategy is to choose row 2, whereas C's optimal pure strategy is to choose column 2. Furthermore, if both players use these strategies, then the expected payoff to R is 2 units.

b. In this case, R's mixed strategy may be represented by the row vector

$$P = [.25 \quad .50 \quad .25]$$

and C's mixed strategy may be represented by the column vector

$$Q = \begin{bmatrix} .5 \\ .5 \end{bmatrix}$$

The expected payoff to the row player will then be given by

$$\begin{aligned} E = PAQ &= [.25 \quad .50 \quad .25]\begin{bmatrix} 1 & -2 \\ -1 & 2 \\ 3 & -3 \end{bmatrix}\begin{bmatrix} .5 \\ .5 \end{bmatrix} \\ &= [.5 \quad -.25]\begin{bmatrix} .5 \\ .5 \end{bmatrix} \\ &= .125 \end{aligned}$$

In Section 9.4, we studied optimal strategies associated with strictly determined games and found them to be precisely the maximin and minimax pure strategies adopted by the row and column players. We now look at optimal mixed strategies associated with matrix games that are not strictly determined. In particular, we consider the optimal mixed strategies to be adopted by the players in a 2×2 matrix game.

As we saw earlier, a player in a nonstrictly determined game should adopt a mixed strategy, since a pure strategy can soon be detected by the opponent, who may then use this knowledge to his advantage in devising a counterstrategy. Since there are infinitely many mixed strategies for each player in such a game, the question arises as to how an optimal mixed strategy may be discovered for each player. An optimal mixed strategy for a player is one in which the row player maximizes his expected payoff and the column player simultaneously minimizes the row player's expected payoff.

More precisely, the optimal mixed strategy for the row player is arrived at by using the following argument: The row player anticipates that any mixed strategy he adopts will be met by a counterstrategy by the column player that will minimize the row player's payoff. Consequently, the row player adopts the mixed strategy for which the expected payoff to the row player (when the column player uses his best counterstrategy) is maximized.

Similarly, the optimal mixed strategy for the column player is arrived at by using the following argument: The column player anticipates that the row player will choose a counterstrategy that will maximize the row player's payoff regardless of the mixed strategy he (the column player) chooses. Consequently, the column player adopts the mixed strategy for which the expected payoff to the row player (who will use his best counterstrategy) is minimized.

The problem of finding the optimal mixed strategies for the players in a nonstrictly determined game is equivalent to the problem of solving a related linear programming problem. However, for a 2×2 nonstrictly determined game, the optimal mixed strategies for the players may be found by employing the formulas contained in the following result, which we state without proof.

Optimal Strategies for Nonstrictly Determined Games

Let

$$\begin{bmatrix} a & b \\ c & d \end{bmatrix}$$

be the payoff matrix for a nonstrictly determined game. Then the **optimal mixed strategy for the row player** is given by

$$P = [p_1 \quad p_2] \tag{2a}$$

where

$$p_1 = \frac{d - c}{a + d - b - c} \quad \text{and} \quad p_2 = 1 - p_1$$

and the **optimal mixed strategy for the column player** is given by

$$Q = \begin{bmatrix} q_1 \\ q_2 \end{bmatrix} \tag{2b}$$

where

$$q_1 = \frac{d - b}{a + d - b - c} \quad \text{and} \quad q_2 = 1 - q_1$$

(continued)

Furthermore, the **value of the game** is given by the expected value of the game, $E = PAQ$, where P and Q are the optimal mixed strategies for the row and column players, respectively. Thus,

$$\begin{aligned} E &= PAQ \\ &= \frac{ad - bc}{a + d - b - c} \end{aligned} \qquad \textbf{(2c)}$$

The next example illustrates the use of these formulas in finding the optimal mixed strategies and in finding the value of a 2×2 (nonstrictly determined) game.

APPLIED EXAMPLE 3 Coin-Matching Game (continued) Consider the coin-matching game played by Richie and Chuck with the payoff matrix

$$A = \begin{bmatrix} 3 & -2 \\ -2 & 1 \end{bmatrix}$$ See Example 1.

a. Find the optimal mixed strategies for both Richie and Chuck.
b. Find the value of the game. Does it favor one player over the other?

Solution

a. The game under consideration has no saddle point and is accordingly nonstrictly determined. Using Equation (2a) with $a = 3$, $b = -2$, $c = -2$, and $d = 1$, we find that

$$\begin{aligned} p_1 &= \frac{d - c}{a + d - b - c} = \frac{1 - (-2)}{3 + 1 - (-2) - (-2)} = \frac{3}{8} \\ p_2 &= 1 - p_1 \\ &= 1 - \frac{3}{8} \\ &= \frac{5}{8} \end{aligned}$$

so Richie's optimal mixed strategy is given by

$$\begin{aligned} P &= [p_1 \quad p_2] \\ &= [\tfrac{3}{8} \quad \tfrac{5}{8}] \end{aligned}$$

Using Equation (2b), we find that

$$\begin{aligned} q_1 &= \frac{d - b}{a + d - b - c} = \frac{1 - (-2)}{3 + 1 - (-2) - (-2)} = \frac{3}{8} \\ q_2 &= 1 - q_1 \\ &= 1 - \frac{3}{8} \\ &= \frac{5}{8} \end{aligned}$$

giving Chuck's optimal mixed strategy as

$$Q = \begin{bmatrix} \frac{3}{8} \\ \frac{5}{8} \end{bmatrix}$$

b. The value of the game may be found by computing the matrix product PAQ, where P and Q are the vectors found in part (a). Equivalently, using Equation (2c), we find that

$$\begin{aligned} E &= \frac{ad - bc}{a + d - b - c} \\ &= \frac{(3)(1) - (-2)(-2)}{3 + 1 - (-2) - (-2)} \\ &= -\frac{1}{8} \end{aligned}$$

Since the value of the game is negative, we conclude that the coin-matching game with the given payoff matrix favors Chuck (the column player) over Richie. Over the long run, in repeated plays of the game, where each player uses his optimal strategy, Chuck is expected to win $\frac{1}{8}$, or 12.5¢, on the average per play.

APPLIED EXAMPLE 4 Investment Strategies As part of their investment strategy, the Carringtons have earmarked $40,000 for short-term investments in the stock market and the money market. The performance of the investments depends on the prime rate (that is, the interest rate that banks charge their best customers). An increase in the prime rate generally favors their investment in the money market, whereas a decrease in the prime rate generally favors their investment in the stock market. Suppose the following payoff matrix gives the percentage increase or decrease in the value of each investment for each state of the prime rate:

	Prime rate up	Prime rate down
Money market investment	15	10
Stock market investment	−5	25

a. Determine the optimal investment strategy for the Carringtons' short-term investment of $40,000.

b. What short-term profit can the Carringtons expect to make on their investments?

Solution

a. We treat the problem as a matrix game in which the Carringtons are the row player. Letting $p = [p_1 \quad p_2]$ denote their optimal strategy, we find that

$$\begin{aligned} p_1 &= \frac{d - c}{a + d - b - c} = \frac{25 - (-5)}{15 + 25 - 10 - (-5)} \qquad a = 15, b = 10, c = -5, \text{ and } d = 25 \\ &= \frac{30}{35} = \frac{6}{7} \\ p_2 &= 1 - p_1 = 1 - \frac{6}{7} = \frac{1}{7} \end{aligned}$$

Thus, the Carringtons should put $(\frac{6}{7})$($40,000), or approximately $34,300, into the money market and $(\frac{1}{7})$($40,000), or approximately $5700, into the stock market.

b. The expected value of the game is given by

$$\begin{aligned} E &= \frac{ad - bc}{a + d - b - c} \\ &= \frac{(15)(25) - (10)(-5)}{15 + 25 - 10 - (-5)} = \frac{425}{35} \\ &\approx 12.14 \end{aligned}$$

Thus, the Carringtons can expect to make a short-term profit of 12.14% on their total investment of $40,000—that is, a profit of (0.1214)(40,000), or $4856.

Explore and Discuss

A two-person, zero-sum game is defined by the payoff matrix

$$A = \begin{bmatrix} x & 1 - x \\ 1 - x & x \end{bmatrix}$$

1. For what value(s) of x is the game strictly determined? For what value(s) of x is the game not strictly determined?
2. What is the value of the game?

9.5 Self-Check Exercises

1. The payoff matrix for a game is given by

$$A = \begin{bmatrix} 2 & 3 & -1 \\ -3 & 2 & -2 \\ 3 & -2 & 2 \end{bmatrix}$$

   a. Find the expected payoff to the row player if the row player R uses the maximin pure strategy and the column player C uses the minimax pure strategy.
   b. Find the expected payoff to the row player if R uses the maximin strategy 40% of the time and chooses each of the other two rows 30% of the time while C uses the minimax strategy 50% of the time and chooses each of the other two columns 25% of the time.
   c. Which pair of strategies favors the row player?

2. **Crop Planning** A farmer has allocated 2000 acres of her farm for planting two crops. Crop A is more susceptible to frost than Crop B is. If there is no frost in the growing season, she can expect to make $40/acre from Crop A and $25/acre from Crop B. If there is mild frost, the expected profits are $20/acre from Crop A and $30/acre from Crop B. How many acres of each crop should the farmer cultivate to maximize her profits? What profit could she expect to make using this optimal strategy?

Solutions to Self-Check Exercises 9.5 can be found on page 585.

9.5 Concept Questions

1. What does the expected value of a game measure?
2. Suppose

$$\begin{bmatrix} a & b \\ c & d \end{bmatrix}$$

is the payoff matrix for a nonstrictly determined game.
   a. What is the optimal mixed strategy for the column player?
   b. What is the optimal mixed strategy for the row player?
   c. What is the value of the game?

9.5 Exercises

In Exercises 1–6, the payoff matrix and strategies P and Q (for the row and column players, respectively) are given. Find the expected payoff E of each game.

1. $\begin{bmatrix} 3 & 1 \\ -4 & 2 \end{bmatrix}$, $P = [\frac{1}{2} \ \ \frac{1}{2}]$, $Q = \begin{bmatrix} \frac{3}{5} \\ \frac{2}{5} \end{bmatrix}$

2. $\begin{bmatrix} -1 & 4 \\ 3 & -2 \end{bmatrix}$, $P = [.8 \ \ .2]$, $Q = \begin{bmatrix} .6 \\ .4 \end{bmatrix}$

3. $\begin{bmatrix} -4 & 3 \\ 2 & 1 \end{bmatrix}$, $P = [\frac{1}{3} \ \ \frac{2}{3}]$, $Q = \begin{bmatrix} \frac{3}{4} \\ \frac{1}{4} \end{bmatrix}$

4. $\begin{bmatrix} 1 & 2 \\ -3 & 1 \end{bmatrix}$, $P = [\frac{3}{5} \ \ \frac{2}{5}]$, $Q = \begin{bmatrix} \frac{1}{3} \\ \frac{2}{3} \end{bmatrix}$

5. $\begin{bmatrix} 2 & 0 & -2 \\ 1 & -1 & 3 \\ 2 & 1 & -4 \end{bmatrix}$, $P = [.2 \ \ .6 \ \ .2]$, $Q = \begin{bmatrix} .2 \\ .6 \\ .2 \end{bmatrix}$

6. $\begin{bmatrix} 1 & -4 & 2 \\ 2 & 1 & -1 \\ 2 & -2 & 0 \end{bmatrix}$, $P = [.2 \ \ .3 \ \ .5]$, $Q = \begin{bmatrix} .6 \\ .2 \\ .2 \end{bmatrix}$

7. The payoff matrix for a game is given by

$$\begin{bmatrix} 1 & -2 \\ -2 & 3 \end{bmatrix}$$

Compute the expected payoffs of the game for the pairs of strategies in parts (a–d). Which of these pairs of strategies is most advantageous to R?

a. $P = [1 \ \ 0]$, $Q = \begin{bmatrix} 1 \\ 0 \end{bmatrix}$

b. $P = [0 \ \ 1]$, $Q = \begin{bmatrix} 1 \\ 0 \end{bmatrix}$

c. $P = [\frac{1}{2} \ \ \frac{1}{2}]$, $Q = \begin{bmatrix} \frac{1}{2} \\ \frac{1}{2} \end{bmatrix}$

d. $P = [.5 \ \ .5]$, $Q = \begin{bmatrix} .8 \\ .2 \end{bmatrix}$

8. The payoff matrix for a game is

$$\begin{bmatrix} 3 & 1 & 1 \\ 0 & 2 & 0 \\ -1 & 0 & 2 \end{bmatrix}$$

Compute the expected payoffs of the game for the pairs of strategies in parts (a–d). Which of these pairs of strategies is most advantageous to R?

a. $P = [\frac{1}{3} \ \ \frac{1}{3} \ \ \frac{1}{3}]$, $Q = \begin{bmatrix} \frac{1}{3} \\ \frac{1}{3} \\ \frac{1}{3} \end{bmatrix}$

b. $P = [\frac{1}{4} \ \ \frac{1}{2} \ \ \frac{1}{4}]$, $Q = \begin{bmatrix} \frac{1}{8} \\ \frac{3}{8} \\ \frac{1}{2} \end{bmatrix}$

c. $P = [.4 \ \ .3 \ \ .3]$, $Q = \begin{bmatrix} .6 \\ .2 \\ .2 \end{bmatrix}$

d. $P = [.1 \ \ .5 \ \ .4]$, $Q = \begin{bmatrix} .3 \\ .3 \\ .4 \end{bmatrix}$

9. The payoff matrix for a game is

$$\begin{bmatrix} -3 & 3 & 2 \\ -3 & 1 & 1 \\ 1 & -2 & 1 \end{bmatrix}$$

a. Find the expected payoff to the row player if the row player R uses the maximin pure strategy and the column player C uses the minimax pure strategy.
b. Find the expected payoff to the row player if R uses the maximin strategy 50% of the time and chooses each of the other two rows 25% of the time, while C uses the minimax strategy 60% of the time and chooses each of the other columns 20% of the time.
c. Which of these pairs of strategies is more advantageous to the row player?

10. The payoff matrix for a game is

$$\begin{bmatrix} 4 & -3 & 3 \\ -4 & 2 & 1 \\ 3 & -5 & 2 \end{bmatrix}$$

a. Find the expected payoff to the row player if the row player R uses the maximin pure strategy and the column player C uses the minimax pure strategy.
b. Find the expected payoff to the row player if R uses the maximin strategy 40% of the time and chooses each of the other two rows 30% of the time, while C uses the minimax strategy 50% of the time and chooses each of the other columns 25% of the time.
c. Which of these pairs of strategies is more advantageous to the row player?

In Exercises 11–16, find the optimal strategies, P and Q, for the row and column players, respectively. Also compute the expected payoff E of each matrix game and determine which player it favors, if any, if the row and column players use their optimal strategies.

11. $\begin{bmatrix} 4 & 1 \\ 2 & 3 \end{bmatrix}$

12. $\begin{bmatrix} 2 & 5 \\ 3 & -6 \end{bmatrix}$

13. $\begin{bmatrix} -1 & 2 \\ 1 & -3 \end{bmatrix}$

14. $\begin{bmatrix} -1 & 3 \\ 2 & 0 \end{bmatrix}$

15. $\begin{bmatrix} -2 & -6 \\ -8 & -4 \end{bmatrix}$

16. $\begin{bmatrix} 2 & 5 \\ -2 & 4 \end{bmatrix}$

17. Coin-Matching Game Consider the coin-matching game played by Richie and Chuck (see Examples 1 and 3) with the payoff matrix

$$A = \begin{bmatrix} 4 & -2 \\ -2 & 1 \end{bmatrix}$$

a. Find the optimal strategies for Richie and Chuck.
b. Find the value of the game. Does it favor one player over the other?

18. Investment Strategies As part of their investment strategy, the Carringtons have decided to put \$100,000 into stock market investments and also into purchasing precious metals. The performance of the investments depends on the state of the economy in the next year. In an expanding economy, it is expected that their stock market investment will outperform their investment in precious metals, whereas an economic recession will have precisely the opposite effect. Suppose the following payoff matrix gives the expected percentage increase or decrease in the value of each investment for each state of the economy:

	Expanding economy	Economic recession
Stock market investment	20	−5
Commodity investment	10	15

a. Determine the optimal investment strategy for the Carringtons' investment of \$100,000.
b. What profit can the Carringtons expect to make on their investments over the year if they use their optimal investment strategy?

19. Investment Strategies The Maxwells have decided to invest \$40,000 in the common stocks of two companies listed on the New York Stock Exchange. One of the companies derives its revenue mainly from its worldwide operation of a chain of hotels, whereas the other company is a domestic major brewery. It is expected that if the economy is in a state of growth, then the hotel stock should outperform the brewery stock; however, the brewery stock is expected to hold its own better than the hotel stock in a recessionary period. Suppose the following payoff matrix gives the expected percentage increase or decrease in the value of each investment for each state of the economy:

	Expanding economy	Economic recession
Investment in hotel stock	25	−5
Investment in brewery stock	10	15

a. Determine the optimal investment strategy for the Maxwells' investment of \$40,000.
b. What profit can the Maxwells expect to make on their investments if they use their optimal investment strategy?

20. Campaign Strategies Bella Robinson and Steve Carson are running for a seat in the U.S. Senate. If both candidates campaign only in the major cities of the state, then Robinson will get 60% of the votes; if both candidates campaign only in the rural areas, then Robinson will get 55% of the votes; if Robinson campaigns exclusively in the city and Carson campaigns exclusively in the rural areas, then Robinson will get 40% of the votes; finally, if Robinson campaigns exclusively in the rural areas and Carson campaigns exclusively in the city, then Robinson will get 45% of the votes.
a. Construct the payoff matrix for the game, and show that it is not strictly determined.
b. Find the optimal strategy for both Robinson and Carson.

21. Marketing Strategies Two dentists, Lydia Russell and Jerry Carlton, are planning to establish practices in a newly developed community. Both have allocated approximately the same total budget for advertising in the local newspaper and for the distribution of fliers announcing their practices. Because of the location of their offices, Russell will get 48% of the business if both dentists advertise only in the local newspaper; if both dentists advertise through fliers, then Russell will get 45% of the business; if Russell advertises exclusively in the local newspaper and Carlton advertises exclusively through fliers, then Russell will get 65% of the business. Finally, if Russell advertises through fliers exclusively and Carlton advertises exclusively in the local newspaper, then Russell will get 50% of the business.
a. Construct the payoff matrix for the game, and show that it is not strictly determined.
b. Find the optimal strategy for both Russell and Carlton.

22. Let

$$\begin{bmatrix} a_{11} & a_{12} \\ a_{21} & a_{22} \end{bmatrix}$$

be the payoff matrix with a 2×2 matrix game. Assume that either the row player uses the optimal mixed strategy $P = [p_1 \quad p_2]$, where

$$p_1 = \frac{d - c}{a + d - b - c} \quad \text{and} \quad p_2 = 1 - p_1$$

or the column player uses the optimal mixed strategy

$$Q = \begin{bmatrix} q_1 \\ q_2 \end{bmatrix}$$

where

$$q_1 = \frac{d - b}{a + d - b - c} \quad \text{and} \quad q_2 = 1 - q_1$$

Show by direct computation that the expected value of the game is given by $E = PAQ$.

23. Let

$$\begin{bmatrix} a & b \\ c & d \end{bmatrix}$$

be the payoff matrix associated with a nonstrictly determined 2×2 matrix game. Prove that the expected payoff of the game is given by

$$E = \frac{ad - bc}{a + d - b - c}$$

Hint: Compute $E = PAQ$, where P and Q are the optimal strategies for the row and column players, respectively.

9.5 Solutions to Self-Check Exercises

1. a. From the following calculations,

$$\begin{bmatrix} 2 & 3 & -1 \\ -3 & 2 & -2 \\ 3 & -2 & 2 \end{bmatrix} \quad \begin{matrix} \text{Row minima} \\ \textcircled{-1} \leftarrow \text{Largest of the row minima} \\ -3 \\ -2 \end{matrix}$$

Column maxima 3 3 (2)
↑ Smallest of the column maxima

we see that R's optimal pure strategy is to choose row 1, whereas C's optimal pure strategy is to choose column 3. Furthermore, if both players use their respective optimal strategies, then the expected payoff to R is -1 unit.

b. R's mixed strategy may be represented by the row vector

$$P = [.4 \quad .3 \quad .3]$$

and C's mixed strategy may be represented by the column vector

$$Q = \begin{bmatrix} .25 \\ .25 \\ .50 \end{bmatrix}$$

The expected payoff to the row player will then be given by

$$E = PAQ = [.4 \quad .3 \quad .3] \begin{bmatrix} 2 & 3 & -1 \\ -3 & 2 & -2 \\ 3 & -2 & 2 \end{bmatrix} \begin{bmatrix} .25 \\ .25 \\ .50 \end{bmatrix}$$

$$= [.4 \quad .3 \quad .3] \begin{bmatrix} .75 \\ -1.25 \\ 1.25 \end{bmatrix}$$

$$= .3$$

c. From the results of parts (a) and (b), we see that the mixed strategies of part (b) will be better for R.

2. We may view this problem as a matrix game with the farmer as the row player and the weather as the column player. The payoff matrix for the game is

	No frost	Mild frost
Crop *A*	40	20
Crop *B*	25	30

The game under consideration has no saddle point and is accordingly nonstrictly determined. Letting $p = [p_1 \quad p_2]$ denote the farmer's optimal strategy and using the formula for determining the optimal mixed strategies for a 2×2 game with $a = 40$, $b = 20$, $c = 25$, and $d = 30$, we find

$$p_1 = \frac{d - c}{a + d - b - c} = \frac{30 - 25}{40 + 30 - 20 - 25} = \frac{5}{25} = \frac{1}{5}$$

$$p_2 = 1 - p_1 = 1 - \frac{1}{5} = \frac{4}{5}$$

Therefore, the farmer should cultivate $\left(\frac{1}{5}\right)(2000)$, or 400, acres of Crop A and 1600 acres of Crop B. By using her optimal strategy, the farmer can expect to realize a profit of

$$E = \frac{ad - bc}{a + d - b - c}$$

$$= \frac{(40)(30) - (20)(25)}{40 + 30 - 20 - 25}$$

$$= 28$$

or \$28/acre—that is, a total profit of (28)(2000), or \$56,000.

CHAPTER 9 Summary of Principal Formulas and Terms

FORMULAS

1. Steady-state matrix for an absorbing stochastic matrix	If $A = \left[\begin{array}{c:c} I & S \\ \hdashline O & R \end{array}\right]$ then the steady-state matrix of A is $\left[\begin{array}{c:c} I & S(I - R)^{-1} \\ \hdashline O & O \end{array}\right]$

2. Expected value of a game	$E = PAQ = [p_1 \ p_2 \ \cdots \ p_m] \begin{bmatrix} a_{11} & a_{12} & \cdots & a_{1n} \\ a_{21} & a_{22} & \cdots & a_{2n} \\ \vdots & \vdots & & \vdots \\ a_{m1} & a_{m2} & \cdots & a_{mn} \end{bmatrix} \begin{bmatrix} q_1 \\ q_2 \\ \vdots \\ q_n \end{bmatrix}$
3. Optimal strategy for a nonstrictly determined game	$P = [p_1 \ \ p_2]$, where $p_1 = \dfrac{d - c}{a + d - b - c}$ and $p_2 = 1 - p_1$ and $Q = \begin{bmatrix} q_1 \\ q_2 \end{bmatrix}$ where $q_1 = \dfrac{d - b}{a + d - b - c}$ and $q_2 = 1 - q_1$ The expected value of the game is $E = PAQ = \dfrac{ad - bc}{a + d - b - c}$

TERMS

Markov chain (process) (536)
transition matrix (538)
stochastic matrix (538)
steady-state (limiting) distribution vector (547)
steady-state matrix (548)
regular Markov chain (548)
absorbing state (557)
absorbing stochastic matrix (557)
absorbing Markov chain (557)
zero-sum game (565)
maximin strategy (567)
minimax strategy (567)
optimal strategy (569)
strictly determined game (569)
saddle point (569)
value of the game (569)
fair game (569)
pure strategy (574)
mixed strategy (575)
expected value of a game (575)

CHAPTER 9 Concept Review Questions

Fill in the blanks.

1. A Markov chain is a stochastic process in which the _____ associated with the outcomes at any stage of the experiment depend only on the outcomes of the _____ stage.

2. The outcome at any stage of the experiment in a Markov process is called the _____ of the experiment; the outcome at the current stage of the experiment is called the current _____.

3. The probabilities in a Markov chain are called _____ probabilities because they are associated with the transition from one state to the next in the Markov process.

4. A transition matrix associated with a Markov chain with n states is a/an _____ matrix T with entries satisfying the following conditions: (a) All entries are _____, and (b) the sum of the entries in each column of T is _____.

5. If the probability distribution vector X_N associated with a Markov process approaches a fixed vector as N gets larger and larger, then the fixed vector is called the steady-state _____ vector for the system. To find this vector, we are led to finding the limit of T^m, which (if it exists) is called the _____ matrix.

6. A stochastic matrix T is a/an _____ Markov chain if T^m approaches a steady-state matrix in which the _____ of the limiting matrix are all _____ and all the entries are _____. To find the steady-state distribution vector X, we solve the vector equation _____ together with the condition that the sum of the _____ of the vector X is equal to _____.

7. In an absorbing stochastic matrix, (a) there is at least one _____ state, a state in which it is impossible for an object to _____, and (b) it is possible to go from each nonabsorbing state to an absorbing state in one or more _____.

8. **a.** A game in which the payoff to one party results in an equal loss to the other is called a/an _____ game.

b. The strategy employed by the row player in which he or she selects from among the rows one in which the smallest payoff is as large as possible is called the _____ strategy. The strategy in which C chooses from among the columns one in which the largest payoff is as small as possible, is called the _____ strategy.

9. A strategy that is most profitable to a particular player is called a/an _____ strategy.

10. In a strictly determined game, an entry in the payoff matrix that is simultaneously the smallest entry in the row and the largest entry in the column is called a/an _____ _____; the optimal strategy for the row player in a strictly determined game is the _____ strategy, obtained by choosing the _____ containing the _____ point; the optimal strategy for the column player is the _____ strategy, obtained by choosing the _____ containing the _____ point.

CHAPTER 9 Review Exercises

In Exercises 1–4, determine which of the following are regular stochastic matrices.

1. $\begin{bmatrix} 1 & -2 \\ 0 & -8 \end{bmatrix}$ **2.** $\begin{bmatrix} .3 & 1 \\ .7 & 0 \end{bmatrix}$

3. $\begin{bmatrix} \frac{1}{2} & 0 & \frac{1}{3} \\ 0 & 0 & \frac{1}{3} \\ \frac{1}{2} & 1 & \frac{1}{3} \end{bmatrix}$ **4.** $\begin{bmatrix} .3 & 0 & .5 \\ .2 & 1 & 0 \\ .1 & 0 & .5 \end{bmatrix}$

In Exercises 5 and 6, find X_2 (the probability distribution of the system after two observations) for the distribution vector X_0 and the transition matrix T.

5. $X_0 = \begin{bmatrix} \frac{1}{2} \\ \frac{1}{2} \\ 0 \end{bmatrix}, T = \begin{bmatrix} 0 & \frac{1}{4} & \frac{3}{5} \\ \frac{2}{5} & \frac{1}{2} & \frac{1}{5} \\ \frac{3}{5} & \frac{1}{4} & \frac{1}{5} \end{bmatrix}$

6. $X_0 = \begin{bmatrix} .35 \\ .25 \\ .40 \end{bmatrix}, T = \begin{bmatrix} .2 & .1 & .3 \\ .5 & .4 & .4 \\ .3 & .5 & .3 \end{bmatrix}$

In Exercises 7–10, determine whether the matrix is an absorbing stochastic matrix.

7. $\begin{bmatrix} 1 & .6 & .1 \\ 0 & .2 & .6 \\ 0 & .2 & .3 \end{bmatrix}$ **8.** $\begin{bmatrix} .3 & .2 & .1 \\ .7 & .5 & .3 \\ 0 & .3 & .6 \end{bmatrix}$

9. $\begin{bmatrix} .32 & .22 & .44 \\ .68 & .78 & .56 \\ 0 & 0 & 0 \end{bmatrix}$ **10.** $\begin{bmatrix} .31 & .35 & 0 \\ .32 & .40 & 0 \\ .37 & .25 & 1 \end{bmatrix}$

In Exercises 11–14, find the steady-state matrix for the transition matrix.

11. $\begin{bmatrix} .6 & .3 \\ .4 & .7 \end{bmatrix}$ **12.** $\begin{bmatrix} .5 & .4 \\ .5 & .6 \end{bmatrix}$

13. $\begin{bmatrix} .6 & .4 & .3 \\ .2 & .2 & .2 \\ .2 & .4 & .5 \end{bmatrix}$ **14.** $\begin{bmatrix} .1 & .2 & .6 \\ .3 & .4 & .2 \\ .6 & .4 & .2 \end{bmatrix}$

15. **Urbanization of Farmland** A study conducted by the State Department of Agriculture in a Sunbelt state reveals an increasing trend toward urbanization of the farmland within the state. Ten years ago, 50% of the land within the state was used for agricultural purposes (A), 15% had been urbanized (U), and the remaining 35% was neither agricultural nor urban (N). Since that time, 10% of the agricultural land has been converted to urban land, 5% has been used for other purposes, and the remaining 85% is still agricultural. Of the urban land, 95% has remained urban, whereas 5% of it has been used for other nonagricultural purposes. Of the land that was neither agricultural nor urban, 10% has been converted to agricultural land, 5% has been urbanized, and the remaining 85% remains unchanged.

a. Construct the transition matrix for the Markov chain that describes the shift in land use within the state.

b. Find the probability vector describing the distribution of land within the state 10 years ago.

c. Assuming that this trend continues, find the probability vector describing the distribution of land within the state 10 years from now.

16. **Automobile Preferences** *Auto Trend* magazine conducted a survey among automobile owners in a certain area of the country to determine what type of car they now own and what type of car they expect to own 4 years from now. For purposes of classification, automobiles mentioned in the survey were placed into three categories: large, intermediate, and small. Results of the survey follow:

		Present car		
		Large	Intermediate	Small
Future car	Large	.3	.1	.1
	Intermediate	.3	.5	.2
	Small	.4	.4	.7

Assuming that these results indicate the long-term buying trend of car owners in the area, what will be the distribution of cars (relative to size) in this area over the long run?

In Exercises 17–20, determine whether each game within the given payoff matrix is strictly determined. If so, give the optimal pure strategies for the row player and the column player and also give the value of the game.

17. $\begin{bmatrix} 1 & 2 \\ 3 & 5 \\ 4 & 6 \end{bmatrix}$

18. $\begin{bmatrix} 1 & 0 & 3 \\ 2 & -1 & -2 \end{bmatrix}$

19. $\begin{bmatrix} 1 & 3 & 6 \\ -2 & 4 & 3 \\ -5 & -4 & -2 \end{bmatrix}$

20. $\begin{bmatrix} 4 & 3 & 2 \\ -6 & 3 & -1 \\ 2 & 3 & 4 \end{bmatrix}$

In Exercises 21–24, find the expected payoff E of each game whose payoff matrix and strategies P and Q (for the row and column players, respectively) are given.

21. $\begin{bmatrix} 4 & 8 \\ 6 & -12 \end{bmatrix}$, $P = [\frac{1}{2} \quad \frac{1}{2}]$, $Q = \begin{bmatrix} \frac{1}{4} \\ \frac{3}{4} \end{bmatrix}$

22. $\begin{bmatrix} 3 & 0 & -3 \\ 2 & 1 & 2 \end{bmatrix}$, $P = [\frac{1}{3} \quad \frac{2}{3}]$, $Q = \begin{bmatrix} \frac{1}{3} \\ \frac{1}{3} \\ \frac{1}{3} \end{bmatrix}$

23. $\begin{bmatrix} 3 & -1 & 2 \\ 1 & 2 & 4 \\ -2 & 3 & 6 \end{bmatrix}$, $P = [.2 \quad .4 \quad .4]$, $Q = \begin{bmatrix} .2 \\ .6 \\ .2 \end{bmatrix}$

24. $\begin{bmatrix} 2 & -2 & 3 \\ 1 & 2 & -1 \\ -1 & 2 & 3 \end{bmatrix}$, $P = [.2 \quad .4 \quad .4]$, $Q = \begin{bmatrix} .3 \\ .3 \\ .4 \end{bmatrix}$

In Exercises 25–28, find the optimal strategies, P and Q, for the row player and the column player, respectively. Also compute the expected payoff E of each matrix game if the row and column players adopt their optimal strategies and determine which player it favors, if any.

25. $\begin{bmatrix} 1 & -2 \\ 0 & 3 \end{bmatrix}$

26. $\begin{bmatrix} 4 & -7 \\ -5 & 6 \end{bmatrix}$

27. $\begin{bmatrix} 3 & -6 \\ 1 & 2 \end{bmatrix}$

28. $\begin{bmatrix} 12 & 10 \\ 6 & 14 \end{bmatrix}$

29. **Competitive Pricing** Two competing music stores, DiscoMart and Stereo World, each have the option of selling a certain popular compact disc (CD) label at a price of either \$7/CD or \$8/CD. If both sell the label at the same price, they are each expected to get 50% of the business. If DiscoMart sells the label at \$7/CD and Stereo World sells the label at \$8/CD, DiscoMart is expected to get 70% of the business; if DiscoMart sells the label at \$8/CD and Stereo World sells the label at \$7/CD, DiscoMart is expected to get 40% of the business.
   a. Represent this information in the form of a payoff matrix.
   b. Determine the optimal price that each company should sell the CD label for to ensure that it captures the largest possible expected market share.

30. **Optimizing Demand** The management of a division of National Motor Corporation that produces compact and subcompact cars has estimated that the quantity demanded of their compact models is 1500 units/week if the price of oil increases at a higher than normal rate, whereas the quantity demanded of their subcompact models is 2500 units/week under similar conditions. However, the quantity demanded of their compact models and subcompact models is 3000 units and 2000 units/week, respectively, if the price of oil increases at a normal rate. Determine the percentages of compact and subcompact cars the division should plan to manufacture to maximize the expected number of cars demanded each week.

CHAPTER 9 Before Moving On . . .

1. The transition matrix for a Markov process is

$$T = \begin{array}{c} \\ \text{State 1} \\ \text{State 2} \end{array} \begin{array}{c} \text{State} \\ \begin{array}{cc} 1 & 2 \end{array} \\ \begin{bmatrix} .3 & .4 \\ .7 & .6 \end{bmatrix} \end{array}$$

and the initial-state distribution vector is

$$X_0 = \begin{array}{c} \text{State 1} \\ \text{State 2} \end{array} \begin{bmatrix} .6 \\ .4 \end{bmatrix}$$

Find X_2.

2. Find the steady-state vector for the transition matrix

$$T = \begin{bmatrix} \frac{1}{3} & \frac{1}{4} \\ \frac{2}{3} & \frac{3}{4} \end{bmatrix}$$

3. Compute the steady-state matrix of the absorbing stochastic matrix

$$\begin{bmatrix} \frac{1}{3} & 0 & 0 \\ 0 & 1 & \frac{1}{4} \\ \frac{2}{3} & 0 & \frac{3}{4} \end{bmatrix}$$

4. A two-person, zero-sum game is defined by the matrix

$$A = \begin{bmatrix} 2 & 3 & -1 \\ -1 & 2 & -3 \\ 3 & 4 & -2 \end{bmatrix}$$

a. Show that the game is strictly determined, and find the saddle point for the game.
b. What is the optimal strategy for each player?
c. What is the value of the game? Does the game favor one player over the other?

5. The payoff matrix for a certain game is

$$A = \begin{bmatrix} 2 & -1 \\ 3 & 2 \\ -3 & 4 \end{bmatrix}$$

a. Find the expected payoff to the row player if the row player R uses her maximin pure strategy and the column player C uses his minimax pure strategy.
b. Find the expected payoff to the row player if R uses her maximin strategy 40% of the time and chooses each of the other two rows 30% of the time, while C chooses the minimax strategy 60% of the time.

6. The payoff matrix for a certain game is

$$\begin{bmatrix} 3 & 1 \\ -2 & 2 \end{bmatrix}$$

a. Find the optimal strategies, P and Q, for the row and column players, respectively.
b. Find the expected payoff E of the game, and determine which player it favors, if any, if the row and column players use their optimal strategies.

APPENDIX A

Introduction to Logic

ONE OF THE earliest records of humanity's efforts to understand the reasoning process can be found in Aristotle's work *Organon* (third century B.C.E.), in which he presented his ideas on logical arguments. Up until the present day, Aristotelian logic has been the basis for traditional logic, the systematic study of valid inferences.

The field of mathematics known as symbolic logic, in which symbols are used to replace ordinary language, had its beginnings in the eighteenth and nineteenth centuries with the works of the German mathematician Gottfried Wilhelm Leibniz (1646–1716) and the English mathematician George Boole (1815–1864). Boole introduced algebraic-type operations to the field of logic, thereby providing us with a systematic method of combining statements. Boolean algebra is the basis of modern-day computer technology as well as being central to the study of pure mathematics.

In this appendix, we introduce the fundamental concepts of symbolic logic. Beginning with the definitions of statements and their truth values, we proceed to a discussion of the combination of statements and valid arguments. In Section A.6, we present an application of symbolic logic that is widely used in computer technology: switching networks.

A.1 Propositions and Connectives

We use deductive reasoning in many of the things we do. Whether in a formal debate or in an expository article, we use language to express our thoughts. In English, sentences are used to express assertions, questions, commands, and wishes. In our study of logic, we will be concerned with only one type of sentence, the declarative sentence.

> **Proposition**
>
> A **proposition,** or *statement,* is a declarative sentence that can be classified as either true or false but not both.

Commands, requests, questions, and exclamations are examples of sentences that are not propositions.

EXAMPLE 1 Which of the following are propositions?

a. Toronto is the capital of Ontario.
b. Close the door!
c. There are 100 trillion connections among the neurons in the human brain.
d. Who is the chief justice of the Supreme Court?
e. The new television sit-com is a successful show.
f. The 2014 Summer Olympic Games were held in Montreal.
g. $x + 3 = 8$
h. How wonderful!
i. Either seven is an odd number or it is even.

Solution Statements (a), (c), (e), (f), and (i) are propositions. Statement (c) would be difficult to verify, but the validity of the statement can at least theoretically be determined. Statement (e) is a proposition if we assume that the meaning of the term *successful* has been defined. For example, one might use the Nielsen ratings to determine the success of a new show. Statement (f) is a proposition that is false.

Statements (b), (d), (g), and (h) are *not* propositions. Statement (b) is a command, (d) is a question, and (h) is an exclamation. Statement (g) is an open sentence that cannot be classified as true or false. For example, if $x = 5$, then $5 + 3 = 8$ and the sentence is true. On the other hand, if $x = 4$, then $4 + 3 \neq 8$ and the sentence is false. ■

Having considered what is meant by a proposition, we now discuss the ways in which propositions may be combined. For example, the two propositions

a. Toronto is the capital of Ontario

and

b. Toronto is the largest city in Canada

may be joined to form the proposition

c. Toronto is the capital of Ontario and is the largest city in Canada.

Propositions (a) and (b) are called **prime,** or *simple,* **propositions** because they are simple statements expressing a single complete thought. Propositions that are combinations of two or more propositions, such as proposition (c), are called **compound propositions.** In the ensuing discussion, we use the lowercase letters p, q, r, and so on to denote prime propositions, and the uppercase letters P, Q, R, and so on to denote compound propositions.

The words that are used to combine propositions are called **logical connectives.** The basic connectives that we will consider are given in Table 1 along with their symbols. We discuss the first three in this section and the remaining two in Section A.3.

TABLE 1

Name	Logical Connective	Symbol
Conjunction	and	$\wedge$
Disjunction	or	$\vee$
Negation	not	$\sim$
Conditional	if . . . then	$\rightarrow$
Biconditional	if and only if	$\leftrightarrow$

Conjunction

A **conjunction** is a statement of the form "p and q" and is represented symbolically by

$$p \wedge q$$

The conjunction $p \wedge q$ is true if *both* p and q are true; it is false otherwise.

We have already encountered a conjunction in an earlier example. The two propositions

p: Toronto is the capital of Ontario

q: Toronto is the largest city in Canada

were combined to form the conjunction

$p \wedge q$: Toronto is the capital of Ontario and is the largest city in Canada.

Disjunction

A **disjunction** is a proposition of the form "p or q" and is represented symbolically by

$$p \vee q$$

The disjunction $p \vee q$ is false if *both* statements p and q are false; it is true in all other cases.

The use of the word *or* in this definition is meant to convey the meaning "one or the other, or both." Since this disjunction is true when *both* p and q are true, as well as when only one of p and q is true, it is often referred to as an **inclusive disjunction.**

EXAMPLE 2 Consider the propositions

p: Dorm residents can purchase meal plans.

q: Dorm residents can purchase à la carte cards.

The disjunction is

$p \vee q$: Dorm residents can purchase meal plans or à la carte cards. ■

The disjunction in Example 2 is true when dorm residents can purchase either meal plans or à la carte cards or both meal plans and à la carte cards.

Exclusive Disjunction

An **exclusive disjunction** is a proposition of the form either "p or q" and is denoted by

$$p \veebar q$$

The disjunction $p \veebar q$ is true if *either* p or q is true but not both.

In contrast to an inclusive disjunction, an exclusive disjunction is *false* when both p and q are true. In Example 2, the exclusive disjunction would not be true if dorm residents could buy both meal plans and à la carte cards. The difference between exclusive and inclusive disjunctions is further illustrated in the next example.

EXAMPLE 3 Consider the propositions

p: The base price of each condominium unit includes a private deck.

q: The base price of each condominium unit includes a private patio.

Find the exclusive disjunction $p \veebar q$.

Solution The exclusive disjunction is

$p \veebar q$: The base price of each condominium unit includes either a private deck or a patio, but not both.

Observe that this statement is not true when both p and q are true. In other words, the base price of each condominium unit does not include both a private deck and a patio.

In our everyday use of language, the meaning of the word *or* is not always clear. In legal documents, the two cases are distinguished by the words *and/or* (inclusive) and *either/or* (exclusive). In mathematics, we use the word *or* in the inclusive sense unless otherwise specified.

Negation

A **negation** is a proposition of the form "not p" and is represented symbolically by

$$\sim p$$

The proposition $\sim p$ is true if p is false and vice versa.

EXAMPLE 4 Form the negation of the proposition

p: The New York Stock Exchange Composite Index rose today.

Solution The negation is

$\sim p$: The New York Stock Exchange Composite Index did not rise today.

EXAMPLE 5 Consider the two propositions

p: The birthrate declined in the United States last year.

q: The population of the United States increased last year.

Write the following statements in symbolic form:

a. Last year, the birthrate declined, and the population increased in the United States.

b. Either the birthrate declined or the population increased in the United States last year, but not both.

c. It is not true that the birthrate declined and the population increased in the United States last year.

Solution The symbolic form of each statement is given by

a. $p \wedge q$ **b.** $p \veebar q$ **c.** $\sim(p \wedge q)$

A.1 Exercises

In Exercises 1–14, determine whether the statement is a proposition.

1. The defendant was convicted of grand larceny.
2. Who won the 2012 presidential election?
3. The first month of the year is February.
4. The number 2 is odd.
5. $x - 1 \geq 0$
6. Keep off the grass!
7. Coughing may be caused by a lack of water in the air.
8. The first McDonald's fast-food restaurant was opened in California.
9. Exercise protects men and women from sudden heart attacks.
10. If voters do not pass the proposition, then taxes will be increased.
11. Don't swim immediately after eating!
12. What ever happened to Baby Jane?
13. $x + 1 = 1$
14. A major corporation will enter or expand operations into the home-security products market this year.

In Exercises 15–20, identify the logical connective that is used in the statement.

15. Housing starts in the United States did not increase last month.
16. Mel Bieber's will is valid, and he was of sound mind and memory when he made the will.
17. Americans are spending less and saving more this year.
18. You traveled on your job, and you were not reimbursed.
19. Both loss of appetite and irritability are symptoms of mental stress.
20. Prices for many imported goods have either stayed flat or dropped slightly this year.

In Exercises 21–26, state the negation of the proposition.

21. New orders for manufactured goods fell last month.
22. The space shuttle was not launched on January 24.
23. Drinking alcohol during pregnancy affects the size and weight of babies.
24. Not all patients suffering from influenza lose weight.
25. The commuter airline industry is now undergoing a shakeup.
26. The Dow Jones Industrial Average registered its fourth consecutive decline today.
27. Let p and q denote the propositions

    p: Domestic car sales increased over the past year.

    q: Foreign car sales decreased over the past year.

    Express the following compound propositions in words:

    a. $p \vee q$ **b.** $p \wedge q$ **c.** $p \veebar q$
    d. $\sim p$ **e.** $\sim p \vee q$ **f.** $\sim p \vee \sim q$

28. Let p and q denote the propositions

    p: Every employee is required to be fingerprinted.

    q: Every employee is required to take an oath of allegiance.

    Express the following compound propositions in words:

    a. $p \vee q$ **b.** $p \wedge q$ **c.** $\sim p \vee \sim q$
    d. $\sim p \wedge \sim q$ **e.** $p \vee \sim q$

29. Let p and q denote the propositions

    p: The doctor recommended surgery to treat Sam's hyperthyroidism.

    q: The doctor recommended radioactive iodine to treat Sam's hyperthyroidism.

    a. State the exclusive disjunction for these propositions in words.

    b. State the inclusive disjunction for these propositions in words.

30. Let p and q denote the propositions

p: The investment newsletter recommended buying bond mutual funds.

q: The investment newsletter recommended buying stock mutual funds.

a. State the exclusive disjunction for these propositions in words.
b. State the inclusive disjunction for these propositions in words.

31. Let p and q denote the propositions

p: The SAT verbal scores improved in this school district last year.

q: The SAT math scores improved in this school district last year.

Express each of the following statements symbolically.
a. The SAT verbal scores and the SAT math scores improved in this school district last year.
b. Either the SAT verbal scores or the SAT math scores improved in this school district last year.
c. Neither the SAT verbal scores nor the SAT math scores improved in this school district last year.
d. It is not true that the SAT math scores did not improve in this school district last year.

32. Let p and q denote the propositions

p: Laura purchased an LED television.

q: Laura did not purchase a Blu-ray player.

Express each of the following statements symbolically.
a. Laura purchased either an LED television or a Blu-ray player or both an LED television and a Blu-ray player.
b. Laura purchased an LED television and a Blu-ray player.
c. It is not true that Laura purchased a Blu-ray player.
d. Laura purchased neither an LED television nor a Blu-ray player.

33. Let p, q, and r denote the propositions

p: The popularity of prime-time soaps increased this year.

q: The popularity of prime-time situation comedies increased this year.

r: The popularity of prime-time detective shows decreased this year.

Express each of the following propositions in words.
a. $\sim p \wedge \sim q$ **b.** $\sim p \vee r$
c. $\sim(\sim r) \vee \sim q$ **d.** $\sim p \veebar \sim q$

A.2 Truth Tables

One of the primary objectives in the study of logic is to determine whether a given proposition is true or false. In other words, we wish to determine the **truth value** of a given statement.

Consider, for example, the conjunction $p \wedge q$ formed from the prime propositions p and q. There are four possible cases:

1. p is true and q is true.
2. p is true and q is false.
3. p is false and q is true.
4. p is false and q is false.

By definition, the conjunction is true when both p and q are true, and it is false otherwise. We can summarize this information in the form of a truth table, as shown in Table 2.

TABLE 2

Conjunction

p	q	$p \wedge q$
T	T	T
T	F	F
F	T	F
F	F	F

In a similar manner, we can construct the truth tables for inclusive disjunction, exclusive disjunction, and negation (Tables 3a–c).

TABLE 3

p	q	$p \vee q$
T	T	T
T	F	T
F	T	T
F	F	F

(a) Inclusive disjunction

p	q	$p \veebar q$
T	T	F
T	F	T
F	T	T
F	F	F

(b) Exclusive disjunction

p	$\sim p$
T	F
F	T

(c) Negation

In general, if we are given a compound proposition, we construct a truth table to exhibit all possible truth values for the given proposition. The next several examples illustrate this method.

EXAMPLE 1 Construct the truth table for the proposition $\sim p \vee q$.

Solution The truth table is constructed in the following manner.

1. The two prime propositions p and q are placed at the head of the first two columns (Table 4).
2. The two propositions containing the connectives, $\sim p$ and $\sim p \vee q$, are placed at the head of the next two columns.
3. The possible truth values for p are entered in the column headed by p, and then the possible truth values for q are entered in the column headed by q. Notice that the *possible* T *values are always exhausted first.*
4. The possible truth values for the negation $\sim p$ are entered in the column headed by $\sim p$, and then the possible truth values for the disjunction $\sim p \vee q$ are entered in the column headed by $\sim p \vee q$.

TABLE 4

p	q	$\sim p$	$\sim p \vee q$
T	T	F	T
T	F	F	F
F	T	T	T
F	F	T	T

EXAMPLE 2 Construct the truth table for the proposition $\sim p \vee (p \wedge q)$.

Solution Proceeding as in the previous example, we construct the truth table as shown in Table 5. Since the given proposition is the disjunction of the propositions $\sim p$ and $(p \wedge q)$, columns are introduced for $\sim p$, $p \wedge q$, and $\sim p \vee (p \wedge q)$.

TABLE 5

p	q	$\sim p$	$p \wedge q$	$\sim p \vee (p \wedge q)$
T	T	F	T	T
T	F	F	F	F
F	T	T	F	T
F	F	T	F	T

The following is an example involving three prime propositions.

EXAMPLE 3 Construct the truth table for the proposition $(p \vee q) \vee (r \wedge \sim p)$.

Solution Following the method of the previous example, we construct the truth table as shown in Table 6.

TABLE 6

p	q	r	$p \vee q$	$\sim p$	$r \wedge \sim p$	$(p \vee q) \vee (r \wedge p)$
T	T	T	T	F	F	T
T	T	F	T	F	F	T
T	F	T	T	F	F	T
T	F	F	T	F	F	T
F	T	T	T	T	T	T
F	T	F	T	T	F	T
F	F	T	F	T	T	T
F	F	F	F	T	F	F

Observe that the truth table in Example 3 contains eight rows, whereas the truth tables in Examples 1 and 2 contain four rows. In general, *if a compound proposition* $P(p, q, \ldots)$ *contains the n prime propositions* $p, q, \ldots,$ *then the corresponding truth table contains* 2^n *rows.* For example, since the compound proposition in Example 3 contains three prime propositions, its corresponding truth table contains 2^3, or 8, rows.

A.2 Exercises

In Exercises 1–18, construct a truth table for each compound proposition.

1. $p \vee \sim q$
2. $\sim p \wedge \sim q$
3. $\sim(\sim p)$
4. $\sim(p \wedge q)$
5. $p \vee \sim p$
6. $\sim(\sim p \vee \sim q)$
7. $\sim p \wedge (p \vee q)$
8. $(p \vee \sim q) \wedge q$
9. $(p \vee q) \wedge (p \wedge \sim q)$
10. $(p \vee q) \wedge \sim p$
11. $(p \vee q) \wedge \sim(p \vee q)$
12. $(p \vee q) \vee (\sim p \wedge q)$
13. $(p \vee q) \wedge (p \vee r)$
14. $p \wedge (q \vee r)$
15. $(p \wedge q) \vee \sim r$
16. $(\sim p \vee q) \wedge \sim r$
17. $(p \wedge \sim q) \vee (p \wedge r)$
18. $\sim(p \wedge q) \vee (q \wedge r)$
19. If a compound proposition consists of the prime propositions p, q, r, and s, how many rows does its corresponding truth table contain?

A.3 The Conditional and Biconditional Connectives

In this section, we introduce two other connectives: the conditional and the biconditional. We also discuss three variations of conditional statements: the inverse, the contrapositive, and the converse.

We often use expressions of the form

If it rains, *then* the baseball game will be postponed

to specify the conditions under which a statement will be true. The "if . . . then" statement is the building block on which deductive reasoning is based, and it is important to understand its use in forming logical proofs.

Conditional Statement

A **conditional statement** is a proposition of the form "if p, then q" and is represented symbolically by

$$p \rightarrow q$$

The connective "if . . . then" is called the **conditional connective**; the proposition p is called the **hypothesis**; and the proposition q is called the **conclusion**. A conditional statement is false if the hypothesis is true and the conclusion is false; it is true in all other cases.

The truth table determined by the conditional $p \rightarrow q$ is shown in Table 7.

TABLE 7

p	q	$p \rightarrow q$
T	T	T
T	F	F
F	T	T
F	F	T

One question that is often asked is: Why is a conditional statement true when its hypothesis is false? We can answer this question by considering the following conditional statement made by a mother to her son:

If you do your homework, then you may watch TV.

Think of the statement consisting of the two prime propositions

p: You do your homework

q: You may watch TV

as a *promise* made by the mother to her son. Four cases arise:

1. The son does his homework, and his mother lets him watch TV.
2. The son does his homework, and his mother does not let him watch TV.
3. The son does not do his homework, and his mother lets him watch TV.
4. The son does not do his homework, and his mother does not let him watch TV.

In case 1, p and q are both true, the promise has been kept, and consequently, $p \to q$ is true. In case 2, p is true and q is false, and the mother has broken her promise. Therefore, $p \to q$ is false. In cases 3 and 4, p is not true, and the promise is *not* broken. Thus, $p \to q$ is regarded as a true statement. In other words, the conditional statement is regarded as false only if the "promise" is broken.

There are several equivalent expressions for the conditional connective "if p then q." Among them are the following:

1. p implies q
2. p only if q
3. q if p
4. q whenever p
5. Suppose p, then q

Care should be taken not to confuse the conditional "q if p" with the conditional $q \to p$, because the two statements have quite different meanings. For example, the conditional $p \to q$ formed from the two propositions

p: There is a fire

q: You call the fire department

is

If there is a fire, then you call the fire department.

The conditional $q \to p$ is

If you call the fire department, then there is a fire.

Obviously, the two statements have quite different meanings.

We refer to statements that are variations of the conditional $p \to q$ as **logical variants.** We define three logical variants of the conditional $p \to q$ as follows:

1. The **converse** is the compound statement "if q, then p" and is represented symbolically by

$$q \to p$$

2. The **contrapositive** is the compound statement "if not q, then not p" and is represented symbolically by

$$\sim q \to \sim p$$

3. The **inverse** is the compound statement "if not p, then not q" and is represented symbolically by

$$\sim p \to \sim q$$

EXAMPLE 1 Given the two propositions

p: You vote in the presidential election.

q: You are a registered voter.

a. State the conditional $p \to q$.
b. State the converse, the contrapositive, and the inverse of $p \to q$.

Solution

a. The conditional is "If you vote in the presidential election, then you are a registered voter."

b. The converse is "If you are a registered voter, then you vote in the presidential election." The contrapositive is "If you are not a registered voter, then you do not vote in the presidential election." The inverse is "If you do not vote in the presidential election, then you are not a registered voter."

The truth table for the conditional $p \to q$ and its three logical variants is shown in Table 8. Notice that the conditional $p \to q$ and its contrapositive $\sim q \to \sim p$ have identical truth tables. In other words, the conditional statement and its contrapositive have the same meaning.

TABLE 8

p	q	Conditional $p \to q$	Converse $q \to p$	$\sim p$	$\sim q$	Contrapositive $\sim q \to \sim p$	Inverse $\sim p \to \sim q$
T	T	T	T	F	F	T	T
T	F	F	T	F	T	F	T
F	T	T	F	T	F	T	F
F	F	T	T	T	T	T	T

Logical Equivalence

Two propositions P and Q are **logically equivalent,** denoted by

$$P \Leftrightarrow Q$$

if they have identical truth tables.

Referring once again to the truth table in Table 8, we see that $p \to q$ is logically equivalent to its contrapositive. Similarly, the converse of a conditional statement is logically equivalent to the inverse.

It is sometimes easier to prove the contrapositive of a conditional statement than it is to prove the conditional itself. We may use this to our advantage in establishing a proof, as shown in the next example.

EXAMPLE 2 Prove that if n^2 is an odd number, then n is an odd number.

Solution Let p and q be the propositions

p: n^2 is odd.

q: n is an odd number.

Since $p \to q$ is logically equivalent to $\sim q \to \sim p$, it suffices to prove $\sim q \to \sim p$. Thus, we wish to prove that if n is not an odd number, then n^2 is an even number and consequently is not odd. If n is even, then $n = 2k$, where k is an integer. Therefore,

$$n^2 = (2k)(2k) = 2(2k^2)$$

Since $2(2k^2)$ is a multiple of 2, it is an even number. Thus, we have shown that the contrapositive $\sim q \to \sim p$ is true, and it follows that the conditional $p \to q$ is also true.

We now turn our attention to the last of the five basic connectives.

Biconditional Propositions

Statements of the form "p if and only if q" are called **biconditional propositions** and are represented symbolically by

$$p \longleftrightarrow q$$

The connective "if and only if" is called the *biconditional connective.* The biconditional $p \longleftrightarrow q$ is true whenever p and q are *both true* or *both false.*

TABLE 9

p	q	$p \longleftrightarrow q$
T	T	T
T	F	F
F	T	F
F	F	T

The truth table for $p \longleftrightarrow q$ is shown in Table 9.

EXAMPLE 3 Let p and q be the propositions

p: Mark is going to the senior prom.

q: Linda is going to the senior prom.

State the biconditional $p \longleftrightarrow q$.

Solution The required statement is

Mark is going to the senior prom
if and only if
Linda is going to the senior prom.

As suggested by its name, the biconditional statement is actually composed of two conditional statements. For instance, an *equivalent expression* for the biconditional statement in Example 3 is given by the conjunction of the two conditional statements

Mark is going to the senior prom
if Linda is going to the senior prom.

Linda is going to the senior prom
if Mark is going to the senior prom.

Thus, "p if q, and q if p" is equivalent to "p if and only if q."

The equivalent expressions that are most commonly used in forming conditional and biconditional statements are summarized in Table 10. The words *necessary* and *sufficient* occur frequently in mathematical proofs. When we say that p is **sufficient** for q we mean that *when p is true, q is also true*—that is, $p \to q$. In like manner, when we say that p is **necessary** for q, we mean that *if p is not true, then q is not true*—that is, $\sim p \to \sim q$. But this last expression is logically equivalent to $q \to p$. Thus, when we say that p is necessary and sufficient for q, we mean $p \to q$ and $q \to p$. Example 4 illustrates the use of these expressions.

TABLE 10

		Equivalent Form
Conditional Statement	If p, then q	p is sufficient for q p only if q q is necessary for p q, if p
Biconditional Statement	p if and only if q	If p then q; if q then p p is necessary and sufficient for q

EXAMPLE 4 Let p and q be the propositions

p: The stock market goes up.

q: Interest rates decrease.

Represent the following statements symbolically:

a. If interest rates decrease, then the stock market goes up.
b. If interest rates do not decrease, then the stock market does not go up.
c. The stock market will go up if and only if interest rates decrease.
d. Decreasing interest rates is a sufficient condition for the stock market to go up.
e. Decreasing interest rates is a necessary and sufficient condition for a rising stock market.

Solution

a. $q \to p$ **b.** $\sim q \to \sim p$, or $p \to q$ **c.** $p \leftrightarrow q$
d. $q \to p$ **e.** $q \leftrightarrow p$

Just as there is an order of precedence when we use the arithmetical operations $\times$, $\div$, $+$, and $-$, we must also observe an order of precedence when we use the logical connectives. The following list dictates the order of precedence for logical connectives.

Order of Precedence of Logical Connectives

$$\sim,\ \wedge,\ \vee,\ \to,\ \leftrightarrow$$

Thus, the connective $\sim$ should be applied first, followed by $\wedge$, and so on. For example, using the order of precedence, we see that $p \wedge r \vee s$ is $(p \wedge r) \vee s$.

Also, as in the case of arithmetical operations, parentheses can be used to override the order of precedence.

A.3 Exercises

In Exercises 1–4, write the converse, the contrapositive, and the inverse of the conditional statement.

1. $p \to \sim q$
2. $\sim p \to \sim q$
3. $q \to p$
4. $\sim p \to q$

In Exercises 5 and 6, refer to the following propositions p and q:

p: It is snowing

q: The temperature is below freezing.

5. Express the conditional and the biconditional of p and q in words.

6. Express the converse, contrapositive, and inverse of the conditional $p \to q$ in words.

In Exercises 7 and 8, refer to the following propositions p and q:

p: The company's union and management reach a settlement.

q: The workers do not strike.

7. Express the conditional and biconditional of p and q in words.

8. Express the converse, contrapositive, and inverse of the conditional $p \to q$ in words.

In Exercises 9–12, determine whether the statement is true or false.

9. A conditional proposition and its converse are logically equivalent.

10. The converse and the inverse of a conditional proposition are logically equivalent.

11. A conditional proposition and its inverse are logically equivalent.

12. The converse and the contrapositive of a conditional proposition are logically equivalent.

13. Consider the conditional statement "If the owner lowers the selling price of the house, then I will buy it." Under what conditions is the conditional statement false?

14. Consider the biconditional statement "I will buy the house if and only if the owner lowers the selling price." Under what conditions is the biconditional statement false?

In Exercises 15–28, construct a truth table for the compound proposition.

15. $\sim(p \to q)$

16. $\sim(q \to \sim p)$

17. $\sim(p \to q) \wedge p$

18. $(p \to q) \vee (q \to p)$

19. $(p \to \sim q) \veebar \sim p$

20. $(p \to q) \wedge (\sim p \vee q)$

21. $(p \to q) \leftrightarrow (\sim q \to \sim p)$

22. $(p \to q) \leftrightarrow (\sim p \vee q)$

23. $(p \wedge q) \to (p \vee q)$

24. $\sim q \to (\sim p \wedge \sim q)$

25. $(p \vee q) \to \sim r$

26. $(p \leftrightarrow q) \vee r$

27. $p \to (q \vee r)$

28. $[(p \to q) \vee (q \to r)] \to (p \to r)$

In Exercises 29–36, determine whether the compound propositions are logically equivalent.

29. $\sim p \vee q$; $p \to q$

30. $\sim(p \vee q)$; $\sim p \wedge \sim q$

31. $q \to p$; $\sim p \to \sim q$

32. $\sim p \to q$; $\sim p \vee q$

33. $p \wedge q$; $p \to \sim q$

34. $\sim(p \wedge \sim q)$; $\sim p \vee q$

35. $(p \to q) \to r$; $(p \vee q) \vee r$

36. $p \vee (q \wedge r)$; $(p \vee q) \wedge (p \vee r)$

37. Let p and q denote the following propositions:

p: Taxes are increased.

q: The federal deficit increases.

Represent the following statements symbolically.

a. If taxes are increased, then the federal deficit will not increase.

b. If taxes are not increased, then the federal deficit will increase.

c. The federal deficit will not increase if and only if taxes are increased.

d. Increased taxation is a sufficient condition for halting the growth of the federal deficit.

e. Increased taxation is a necessary and sufficient condition for halting the growth of the federal deficit.

38. Let p and q denote the following propositions:

p: The unemployment rate decreases.

q: Consumer confidence improves.

Represent the following statements symbolically.

a. If the unemployment rate does not decrease, consumer confidence will not improve.

b. Consumer confidence will improve if and only if the unemployment rate does not decrease.

c. A decreasing unemployment rate is a sufficient condition for consumer confidence to improve.

d. A decreasing unemployment rate is a necessary and sufficient condition for consumer confidence to improve.

A.4 Laws of Logic

Just as the laws of algebra help us to perform operations with real numbers, the **laws of logic** provide us with a systematic method of simplifying statements.

Laws of Logic

Let p, q, and r be any three propositions. Then

1.	$p \wedge p \Leftrightarrow p$	*Idempotent law for conjunction*
2.	$p \vee p \Leftrightarrow p$	*Idempotent law for disjunction*
3.	$(p \wedge q) \wedge r \Leftrightarrow p \wedge (q \wedge r)$	*Associative law for conjunction*
4.	$(p \vee q) \vee r \Leftrightarrow p \vee (q \vee r)$	*Associative law for disjunction*
5.	$p \wedge q \Leftrightarrow q \wedge p$	*Commutative law for conjunction*
6.	$p \vee q \Leftrightarrow q \vee p$	*Commutative law for disjunction*
7.	$p \wedge (q \vee r) \Leftrightarrow (p \wedge q) \vee (p \wedge r)$	*Distributive law for conjunction*
8.	$p \vee (q \wedge r) \Leftrightarrow (p \vee q) \wedge (p \vee r)$	*Distributive law for disjunction*
9.	$\sim(p \vee q) \Leftrightarrow \sim p \wedge \sim q$	*De Morgan's Law*
10.	$\sim(p \wedge q) \Leftrightarrow \sim p \vee \sim q$	*De Morgan's Law*

To verify any of these laws, we need only construct a truth table to show that the given statements are logically equivalent. We illustrate this procedure in the next example.

EXAMPLE 1 Prove the distributive law for conjunction.

Solution We wish to prove that $p \wedge (q \vee r)$ is logically equivalent to $(p \wedge q) \vee (p \wedge r)$. This is easily done by constructing the associated truth table, as shown in Table 11. Since the entries in the last two columns are the same, we conclude that $p \wedge (q \vee r) \Leftrightarrow (p \wedge q) \vee (p \wedge r)$.

TABLE 11

p	q	r	$q \vee r$	$p \wedge q$	$p \wedge r$	$p \wedge (q \vee r)$	$(p \wedge q) \vee (p \wedge r)$
T	T	T	T	T	T	T	T
T	T	F	T	T	F	T	T
T	F	T	T	F	T	T	T
T	F	F	F	F	F	F	F
F	T	T	T	F	F	F	F
F	T	F	T	F	F	F	F
F	F	T	T	F	F	F	F
F	F	F	F	F	F	F	F

The proofs of the other laws are left to you as an exercise.

De Morgan's Laws, Laws 9 and 10, are useful in forming the negation of a statement. For example, if we wish to state the negation of the proposition

Steve plans to major in business administration or economics

we can represent the statement symbolically by $p \vee q$, where the prime propositions are

p: Steve plans to major in business administration.

q: Steve plans to major in economics.

Then the negation of $p \vee q$ is $\sim(p \vee q)$. Using De Morgan's Laws, we have

$$\sim(p \vee q) \Leftrightarrow \sim p \wedge \sim q$$

Thus, the required statement is

Steve does not plan to major in business administration, *and* he does not plan to major in economics.

Up to this point, we have considered propositions that have both true and false entries in their truth tables. Some statements have the property that they are always true; other propositions have the property that they are always false.

> **Tautologies and Contradictions**
>
> A **tautology** is a statement that is always true.
>
> A **contradiction** is a statement that is always false.

EXAMPLE 2 Show that

a. $p \vee \sim p$ is a tautology. **b.** $p \wedge \sim p$ is a contradiction.

Solution

a. The truth table associated with $p \vee \sim p$ is shown in Table 12a. Since all the entries in the last column are Ts, the proposition is a tautology.

b. The truth table for $p \wedge \sim p$ is shown in Table 12b. Since all the entries in the last column are Fs, the proposition is a contradiction.

TABLE 12

p	$\sim p$	$p \vee \sim p$
T	F	T
F	T	T

(a)

p	$\sim p$	$p \wedge \sim p$
T	F	F
F	T	F

(b)

In Section A.3, we defined logically equivalent statements as statements that have identical truth tables. The definition of equivalence may also be restated in terms of a tautology.

Logical Equivalence

Two propositions P and Q are **logically equivalent** if the biconditional $P \longleftrightarrow Q$ is a tautology.

Consider, for example, the prime propositions p and q. Let's look more closely at the truth table for $p \longleftrightarrow q$ (Table 13a).

TABLE 13

	p	q	$p \longleftrightarrow q$
Case 1	T	T	T
Case 2	T	F	F
Case 3	F	T	F
Case 4	F	F	T

(a)

p	q	$p \longleftrightarrow q$
T	T	T
F	F	T

(b)

If the biconditional is always true, then the second and third cases shown in the truth table are excluded, and we are left with the truth table in Table 13b. Notice that the entries in each row of the p and q columns are identical. In other words, p and q have identical truth values and hence are logically equivalent.

In addition to the ten laws stated earlier, we have the following six laws involving tautologies and contradictions.

Laws of Logic

Let t be a tautology, and let c be a contradiction. Then

11. $p \vee \sim p \Leftrightarrow t$ **12.** $p \wedge \sim p \Leftrightarrow c$ **13.** $p \vee t \Leftrightarrow t$

14. $p \wedge t \Leftrightarrow p$ **15.** $p \vee c \Leftrightarrow p$ **16.** $p \wedge c \Leftrightarrow c$

It now remains only to show how these laws are used to simplify proofs.

EXAMPLE 3 Using the laws of logic, show that

$$p \vee (\sim p \wedge q) \Leftrightarrow (p \vee q)$$

Solution

$$\begin{aligned} p \vee (\sim p \wedge q) &\Leftrightarrow (p \vee \sim p) \wedge (p \vee q) && \text{By Law 8} \\ &\Leftrightarrow t \wedge (p \vee q) && \text{By Law 11} \\ &\Leftrightarrow p \vee q && \text{By Law 14} \end{aligned}$$

EXAMPLE 4 Using the laws of logic, show that

$$\sim(p \vee q) \vee (\sim p \wedge q) \Leftrightarrow \sim p$$

Solution

$$\begin{aligned} &\sim(p \vee q) \vee (\sim p \wedge q) \\ &\Leftrightarrow (\sim p \wedge \sim q) \vee (\sim p \wedge q) && \text{By Law 9} \\ &\Leftrightarrow \sim p \wedge (\sim q \vee q) && \text{By Law 7} \\ &\Leftrightarrow \sim p \wedge t && \text{By Law 11} \\ &\Leftrightarrow \sim p && \text{By Law 14} \end{aligned}$$

A.4 Exercises

1. Prove the idempotent law for conjunction, $p \wedge p \Leftrightarrow p$.
2. Prove the idempotent law for disjunction, $p \vee p \Leftrightarrow p$.
3. Prove the associative law for conjunction, $(p \wedge q) \wedge r \Leftrightarrow p \wedge (q \wedge r)$.
4. Prove the associative law for disjunction, $(p \vee q) \vee r \Leftrightarrow p \vee (q \vee r)$.
5. Prove the commutative law for conjunction, $p \wedge q \Leftrightarrow q \wedge p$.
6. Prove the commutative law for disjunction, $p \vee q \Leftrightarrow q \vee p$.
7. Prove the distributive law for disjunction, $p \vee (q \wedge r) \Leftrightarrow (p \vee q) \wedge (p \vee r)$.
8. Prove De Morgan's Laws
   a. $\sim(p \vee q) \Leftrightarrow \sim p \wedge \sim q$ **b.** $\sim(p \wedge q) \Leftrightarrow \sim p \vee \sim q$

In Exercises 9–18, determine whether the statement is a tautology, a contradiction, or neither.

9. $(p \to q) \leftrightarrow (\sim p \vee q)$
10. $(p \veebar q) \wedge (p \leftrightarrow q)$
11. $p \to (p \vee q)$
12. $(p \to q) \vee (q \to p)$
13. $(p \to q) \leftrightarrow (\sim q \to \sim p)$
14. $p \wedge (p \to q) \to q$
15. $(p \to q) \wedge (\sim q) \to (\sim p)$
16. $[(p \to q) \wedge (q \to r)] \to (p \to r)$
17. $[(p \to q) \vee (q \to r)] \to (p \to r)$
18. $[p \wedge (q \vee r)] \leftrightarrow [(p \wedge q) \vee (p \wedge r)]$
19. Let p and q denote the statements

    p: The candidate opposes changes in the Social Security system.

    q: The candidate supports immigration reform.

    Use De Morgan's Laws to state the negation of $p \wedge q$ and the negation of $p \vee q$.
20. Let p and q denote the statements

    p: The recycling bill was passed by the voters.

    q: The tax on oil and hazardous materials was not approved by the voters.

    Use De Morgan's Laws to state the negation of $p \wedge q$ and the negation of $p \vee q$.

In Exercises 21–26, use the laws of logic to prove the propositions.

21. $[p \wedge (q \vee \sim q) \vee (p \wedge q)] \Leftrightarrow p \vee (p \wedge q)$
22. $p \vee (\sim p \wedge \sim q) \Leftrightarrow p \vee \sim q$
23. $(p \wedge \sim q) \vee (p \wedge \sim r) \Leftrightarrow p \wedge (\sim q \vee \sim r)$
24. $(p \vee q) \vee \sim q \Leftrightarrow t$
25. $p \wedge \sim(q \wedge r) \Leftrightarrow (p \wedge \sim q) \vee (p \wedge \sim r)$
26. $p \vee (q \vee r) \Leftrightarrow r \vee (q \vee p)$

A.5 Arguments

In this section, we discuss arguments and the methods that are used to determine the validity of arguments.

Argument

An **argument** or **proof** consists of a set of propositions $p_1, p_2, \ldots, p_n$, called the *premises*, and a proposition q, called the *conclusion*. An argument is *valid* if and only if the conclusion is true whenever the premises are all true. An argument that is *not* valid is called a *fallacy*, or an *invalid* argument.

The next example illustrates the form in which an argument is presented. Notice that the premises are written separately above the horizontal line and the conclusion is written below the line.

EXAMPLE 1 An argument is presented in the following way:

Premises: If Pam studies diligently, she passes her exams.
Pam studies diligently.
____
Conclusion: Pam passes her exams.

To determine the validity of an argument, we first write the argument in symbolic form and then construct the associated truth table containing the prime propositions, the premises, and the conclusion. We then *check the rows in which the premises are all true. If the conclusion in each of these rows is also true, then the argument is valid. Otherwise, it is a fallacy.*

EXAMPLE 2 Determine the validity of the argument in Example 1.

Solution The symbolic form of the argument is

$$\begin{array}{l} p \rightarrow q \\ \underline{p \qquad} \\ \therefore q \end{array}$$

Observe that the conclusion is preceded by the symbol $\therefore$, which is used to represent the word *therefore*. The truth table associated with this argument is shown in Table 14. Observe that only row 1 contains true values for both premises. Since the conclusion is also true in this row, we conclude that the argument is valid.

TABLE 14

Propositions		**Premises**		**Conclusion**
p	q	$p \rightarrow q$	p	q
T	T	T	T	T
T	F	F	T	F
F	T	T	F	T
F	F	T	F	F

EXAMPLE 3 Determine the validity of the argument

If Michael is overtired, then he is grumpy.
Michael is grumpy.
Therefore, Michael is overtired.

Solution The argument is written symbolically as follows:

$$\begin{array}{l} p \to q \\ q \\ \hline \therefore p \end{array}$$

The associated truth table is shown in Table 15. Observe that the entries for the premises in the third row are both true, but the corresponding entry for the conclusion is false. We conclude that the argument is a fallacy.

TABLE 15

p	q	$p \to q$	q	p
T	T	T	T	T
T	F	F	F	T
F	T	T	T	F
F	F	T	F	F

When we consider the validity of an argument, we are concerned only with the *form* of the argument and not the truth or falsity of the premises. In other words, the conclusion of an argument may follow validly from the premises, but the premises themselves may be false. The next example demonstrates this point.

EXAMPLE 4 Determine the validity of the argument

If I drive a car, then I eat lunch.
If I eat lunch, then I play Wii.
Therefore, if I drive a car, then I play Wii.

Solution The symbolic form of the argument is

$$\begin{array}{l} p \to q \\ q \to r \\ \hline \therefore p \to r \end{array}$$

The associated truth table is shown in Table 16. Since the conclusion is true in each of the rows that contain true values for both premises, we conclude that the argument is valid. Note that the validity of the argument is not affected by the truth or falsity of the premise "If I drive a car, then I eat lunch."

TABLE 16

p	q	r	$p \to q$	$q \to r$	$p \to r$	
T	T	T	T	T	T	(✓)
T	T	F	T	F	F	
T	F	T	F	T	T	
T	F	F	F	T	F	
F	T	T	T	T	T	(✓)
F	T	F	T	F	T	
F	F	T	T	T	T	(✓)
F	F	F	T	T	T	(✓)

A question that may already have arisen in your mind is: How does one determine the validity of an argument if the associated truth table does not contain true values for all the premises? The next example provides the answer to this question.

EXAMPLE 5 Determine the validity of the argument

The door is locked.
The door is unlocked.
―――
Therefore, the door is locked.

Solution The symbolic form of the argument is

$$\begin{array}{c} p \\ \sim p \\ \hline \therefore p \end{array}$$

TABLE 17

p	$\sim p$	p
T	F	T
F	T	F

The associated truth table is shown in Table 17. Observe that no rows contain true values for both premises. Nevertheless, the argument is considered to be valid, since the condition that the conclusion is true whenever the premises are all true is not violated. Again, we remind you that the validity of an argument is not determined by the truth or falsity of its premises; rather, it is determined only by the form of the argument. ■

The next proposition provides us with an alternative method for determining the validity of an argument.

> **Proposition**
>
> Suppose an argument consists of the premises $p_1, p_2, \ldots, p_n$ and conclusion q. Then, the argument is valid if and only if the proposition
>
> $$(p_1 \wedge p_2 \wedge \cdots \wedge p_n) \rightarrow q$$
>
> is a tautology.

To prove this proposition, we must show that (a) if an argument is valid, then the given proposition is a tautology, and (b) if the given proposition is a tautology, then the argument is valid.

Proof of (a) Since the argument is valid, it follows that q is true whenever all of the premises $p_1, p_2, \ldots, p_n$ are true. But the conjunction $(p_1 \wedge p_2 \wedge \cdots \wedge p_n)$ is true when all the premises are true (by the definition of conjunction). Hence, q is true whenever the conjunction is true, and we conclude that the conditional $(p_1 \wedge p_2 \wedge \cdots \wedge p_n) \rightarrow q$ is true. (Recall that a conditional statement is always true when its hypothesis is false. Hence to show that a conditional is a tautology, we need only prove that its conclusion is always true when its hypothesis is true.)

Proof of (b) Since the given proposition is a tautology, it is always true. Therefore, q is true when the conjunction

$$p_1 \wedge p_2 \wedge \cdots \wedge p_n$$

is true. But the conjunction is true when all of the premises $p_1, p_2, \ldots, p_n$ are true. Hence q is true whenever the premises are all true, and the argument is valid. ■

Having proved this proposition, we can now use it to determine the validity of an argument. If we are given an argument, we construct a truth table for $(p_1 \wedge p_2 \wedge \cdots \wedge p_n) \rightarrow q$. If the truth table contains all Ts in its last column, then the argument is valid; otherwise, it is invalid. This method of proof is illustrated in the next example.

EXAMPLE 6 Determine the validity of the argument

You return the book on time, or you will have to pay a fine.
You do not return the book on time.
———
Therefore, you will have to pay a fine.

Solution The symbolic form of the argument is

$$\begin{array}{l} p \vee q \\ \underline{\sim p} \\ \therefore q \end{array}$$

Following the method described, we construct the truth table for

$$[(p \vee q) \wedge \sim p] \rightarrow q$$

as shown in Table 18. Since the entries in the last column are all Ts, the proposition is a tautology, and we conclude that the argument is valid.

TABLE 18

p	q	$p \vee q$	$\sim p$	$(p \vee q) \wedge \sim p$	$[(p \vee q) \wedge \sim p] \rightarrow q$
T	T	T	F	F	T
T	F	T	F	F	T
F	T	T	T	T	T
F	F	F	T	F	T

By familiarizing ourselves with a few of the most commonly used argument forms, we can simplify the problem of determining the validity of an argument. Some of the argument forms most commonly used are the following:

1. **Modus ponens** (a manner of affirming), or rule of detachment

$$\begin{array}{l} p \rightarrow q \\ \underline{p} \\ \therefore q \end{array}$$

2. **Modus tollens** (a manner of denying)

$$\begin{array}{l} p \rightarrow q \\ \underline{\sim q} \\ \therefore \sim p \end{array}$$

3. **Law of syllogisms**

$$\begin{array}{l} p \rightarrow q \\ \underline{q \rightarrow r} \\ \therefore p \rightarrow r \end{array}$$

The truth tables verifying modus ponens and the law of syllogisms have already been constructed (see Tables 14 and 16, respectively). The verification of modus tollens is left as an exercise. The next two examples illustrate the use of these argument forms.

EXAMPLE 7 Determine the validity of the argument

If the battery is dead, the car will not start.
The car starts.
———
Therefore, the battery is not dead.

Solution As before, we express the argument in symbolic form. Thus,

$$\begin{array}{l} p \rightarrow q \\ \underline{\sim q} \\ \therefore \sim p \end{array}$$

Identifying the form of this argument as modus tollens, we conclude that the given argument is valid.

EXAMPLE 8 Determine the validity of the argument

If mortgage rates are lowered, housing sales increase.
If housing sales increase, then the prices of houses increase.
Therefore, if mortgage rates are lowered, the prices of houses increase.

Solution The symbolic form of the argument is

$$\begin{array}{l} p \rightarrow q \\ \underline{q \rightarrow r} \\ \therefore p \rightarrow r \end{array}$$

Using the law of syllogisms, we conclude that the argument is valid.

A.5 Exercises

In Exercises 1–16, determine whether the argument is valid.

1. $\begin{array}{l} p \rightarrow q \\ \underline{q \rightarrow r} \\ \therefore p \rightarrow r \end{array}$

2. $\begin{array}{l} p \vee q \\ \underline{\sim p} \\ \therefore q \end{array}$

3. $\begin{array}{l} p \wedge q \\ \underline{\sim p} \\ \therefore q \end{array}$

4. $\begin{array}{l} p \rightarrow q \\ \underline{\sim q} \\ \therefore \sim p \end{array}$

5. $\begin{array}{l} p \rightarrow q \\ \underline{\sim p} \\ \therefore \sim q \end{array}$

6. $\begin{array}{l} p \rightarrow q \\ \underline{q \wedge r} \\ \therefore p \vee r \end{array}$

7. $\begin{array}{l} p \leftrightarrow q \\ \underline{q} \\ \therefore p \end{array}$

8. $\begin{array}{l} p \wedge q \\ \underline{\sim p \rightarrow \sim q} \\ \therefore p \wedge \sim q \end{array}$

9. $\begin{array}{l} p \rightarrow q \\ \underline{q \rightarrow p} \\ \therefore p \leftrightarrow q \end{array}$

10. $\begin{array}{l} p \rightarrow \sim q \\ \underline{p \veebar q} \\ \therefore \sim q \end{array}$

11. $\begin{array}{l} p \leftrightarrow q \\ \underline{q \leftrightarrow r} \\ \therefore p \leftrightarrow r \end{array}$

12. $\begin{array}{l} p \rightarrow q \\ \underline{q \leftrightarrow r} \\ \therefore p \wedge r \end{array}$

13. $\begin{array}{l} p \veebar r \\ \underline{q \wedge r} \\ \therefore p \rightarrow r \end{array}$

14. $\begin{array}{l} p \leftrightarrow q \\ q \vee r \\ \underline{\sim r} \\ \therefore p \rightarrow \sim r \end{array}$

15. $\begin{array}{l} p \leftrightarrow q \\ q \vee r \\ \underline{\sim p} \\ \therefore \sim p \rightarrow \sim r \end{array}$

16. $\begin{array}{l} p \rightarrow q \\ r \rightarrow q \\ \underline{p \wedge q} \\ \therefore p \vee r \end{array}$

In Exercises 17–22, represent the argument symbolically, and determine whether it is a valid argument.

17. If Carla studies, then she passes her exams.
    Carla did not study.
    Therefore, Carla did not pass her exams.

18. If Tony is wealthy, he is either intelligent or a good businessman.
    Tony is intelligent and he is not a good businessman.
    Therefore, Tony is not wealthy.

19. Steve will attend the matinee and/or the evening show.
    If Steve doesn't go to the matinee show, then he will not go to the evening show.
    Therefore, Steve will attend the matinee show.

20. If Mary wins the race, then Stacy loses the race.
    Neither Mary nor Linda won the race.
    Therefore, Stacy won the race.

21. If mortgage rates go up, then housing prices will go up.
    If housing prices go up, more people will rent houses.
    More people are renting houses.
    Therefore, mortgage rates went up.

22. If taxes are cut, then retail sales increase.

    If retail sales increase, then the unemployment rate will decrease.

    If the unemployment rate decreases, then the incumbent will win the election.

    Therefore, if taxes are not cut, the incumbent will not win the election.

23. Show that the following argument is valid.

    If George prepares gourmet food, then he is a good cook.

    If Brenda does not prepare gourmet food, then Sarah buys fast food.

    Sarah does not buy fast food.

    Therefore, George does not prepare gourmet food, and Brenda does not prepare gourmet food.

24. What conclusion can be drawn from the following statements?

    His date is pretty, or she is tall and skinny.

    If his date is tall, then she is a brunette.

    His date is not a brunette.

25. Show that modus tollens is a valid form of argument.

A.6 Applications of Logic to Switching Networks

In this section, we see how the principles of logic can be used in the design and analysis of switching networks. A **switching network** is an arrangement of wires and switches connecting two terminals. These networks are used extensively in digital computers.

A switch may be open or closed. If a switch is closed, current will flow through the wire. If it is open, no current will flow through the wire. Because a switch has exactly two states, it can be represented by a proposition p that is true if the switch is closed and false if the switch is open.

Now let's consider a circuit with two switches p and q. If the circuit is constructed as shown in Figure 1, the switches p and q are said to be **in series.**

FIGURE 1
Two switches connected in series

For such a network, current will flow from A to B if both p and q are closed, but no current will flow if one or more of the switches is open. Thinking of p and q as propositions, we have the truth table shown in Table 19. (Recall that T corresponds to the situation in which the switch is closed.) From the truth table, we see that two switches p and q connected in series are analogous to the conjunction $p \wedge q$ of the two propositions p and q.

TABLE 19

p	q	
T	T	T
T	F	F
F	T	F
F	F	F

If a circuit is constructed as shown in Figure 2, the switches p and q are said to be connected **in parallel.**

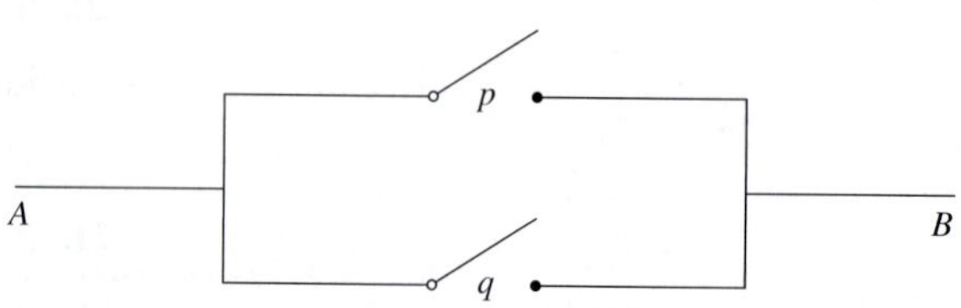

FIGURE 2
Two switches connected in parallel

For such a network, current will flow from A to B if and only if one or more of the switches p or q is closed. Once again, thinking of p and q as propositions, we have the truth table shown in Table 20. From the truth table, we conclude that two switches p and q connected in parallel are analogous to the inclusive disjunction $p \vee q$ of the two propositions p and q.

TABLE 20

p	q	
T	T	T
T	F	T
F	T	T
F	F	F

EXAMPLE 1 Find a logic statement that represents the network shown in Figure 3. By constructing the truth table for this logic statement, determine the conditions under which current will flow from A to B in the network.

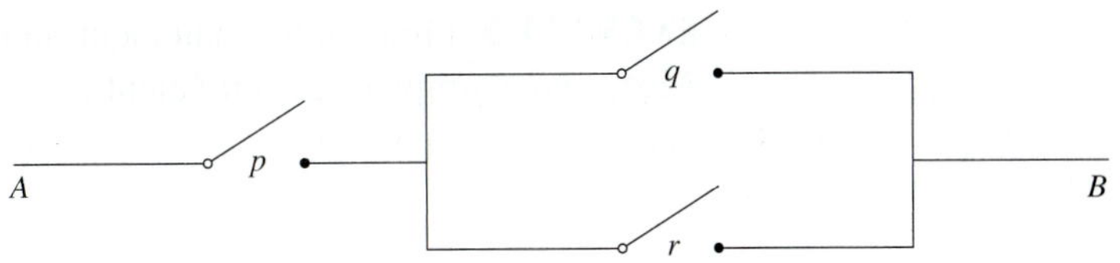

FIGURE 3
We want to determine when the current will flow from A to B.

Solution The required logic statement is $p \wedge (q \vee r)$. Next, we construct the truth table for the logic statement $p \wedge (q \vee r)$ (Table 21).

TABLE 21

p	q	r	$q \vee r$	$p \wedge (q \vee r)$
T	T	T	T	T
T	T	F	T	T
T	F	T	T	T
T	F	F	F	F
F	T	T	T	F
F	T	F	T	F
F	F	T	T	F
F	F	F	F	F

From the truth table, we conclude that current will flow from A to B if and only if one of the following conditions is satisfied:

1. p, q, and r are all closed.
2. p and q are closed, but r is open.
3. p and r are closed, but q is open.

In other words, current will flow from A to B if and only if p is closed and either q or r or both q and r are closed.

Before looking at some additional examples, let's remark that $\sim p$ represents a switch that is open when p is closed and vice versa. Furthermore, a circuit that is always closed is represented by a tautology, $p \vee \sim p$, whereas a circuit that is always open is represented by a contradiction, $p \wedge \sim p$. (Why?)

EXAMPLE 2 Given the logic statement $(p \vee q) \wedge (r \vee s \vee \sim t)$, draw the corresponding network.

Solution Recalling that the disjunction $p \vee q$ of the propositions p and q represents two switches p and q connected in parallel and the conjunction $p \wedge q$ represents two switches connected in series, we obtain the following network (Figure 4).

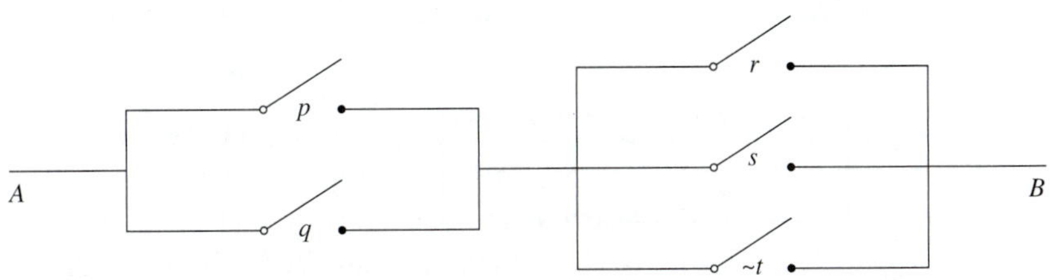

FIGURE 4
The network that corresponds to the logic statement $(p \vee q) \wedge (r \vee s \vee \sim t)$

When used in conjunction with the laws of logic, the theory of networks developed so far is a useful tool in network analysis. In particular, network analysis enables us to find equivalent, and often simpler, networks, as the following example shows.

EXAMPLE 3 Find a logic statement representing the network shown in Figure 5. Also, find a simpler but equivalent network.

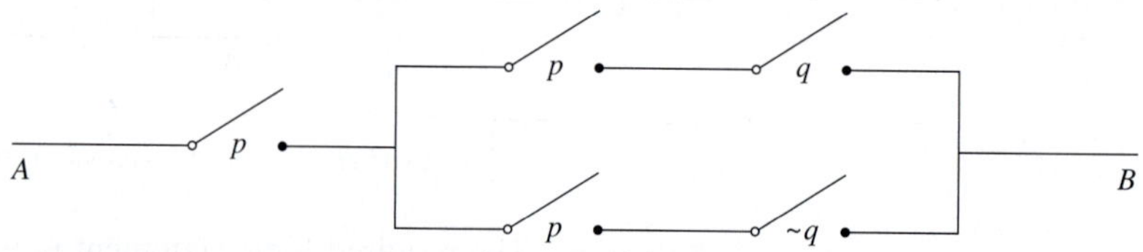

FIGURE 5
We want to find the logic statement that corresponds to the network shown in the figure.

Solution The logic statement corresponding to the given network is $p \wedge [(p \wedge q) \vee (p \wedge {\sim}q)]$. Next, using the rules of logic to simplify this statement, we obtain

$$\begin{aligned} p \wedge [(p \wedge q) \vee (p \wedge {\sim}q)] &\Leftrightarrow p \wedge [p \wedge (q \vee {\sim}q)] && \text{Distributive law} \\ &\Leftrightarrow p \wedge p && \text{Tautology} \\ &\Leftrightarrow p \end{aligned}$$

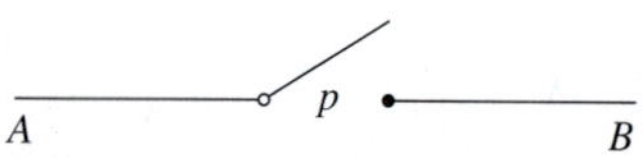

FIGURE 6
The network is equivalent to the one shown in Figure 5.

Thus, the network shown in Figure 6 is equivalent to the one shown in Figure 5.

A.6 Exercises

In Exercises 1–5, find a logic statement corresponding to the network. Determine the conditions under which current will flow from *A* to *B*.

1.

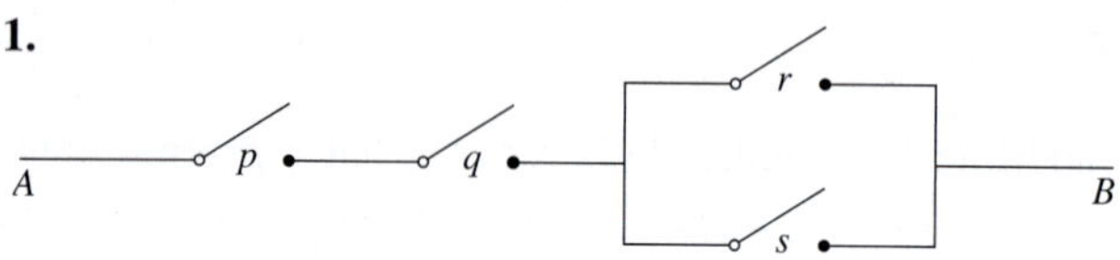

2.

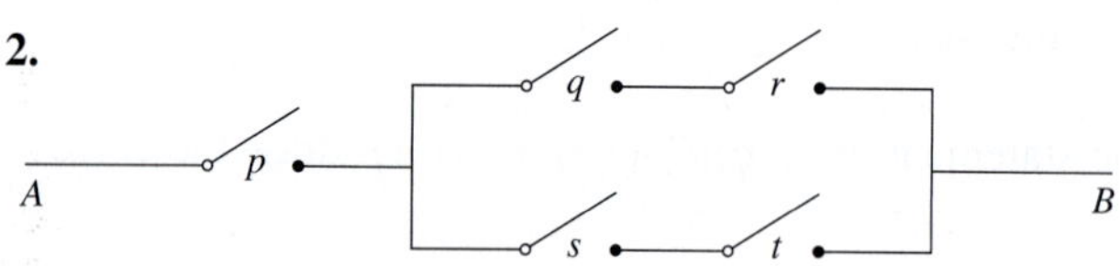

3.

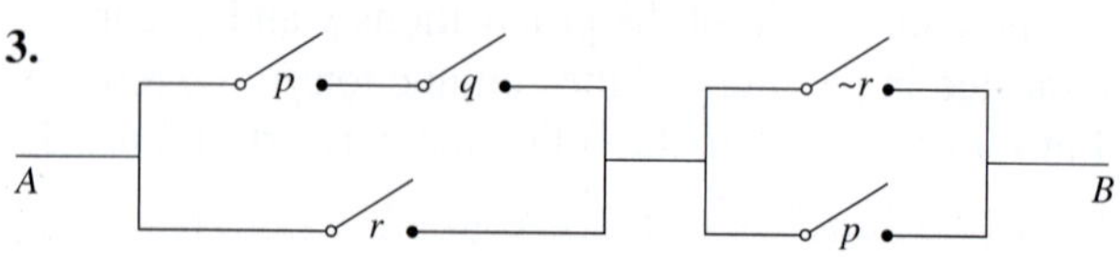

4.

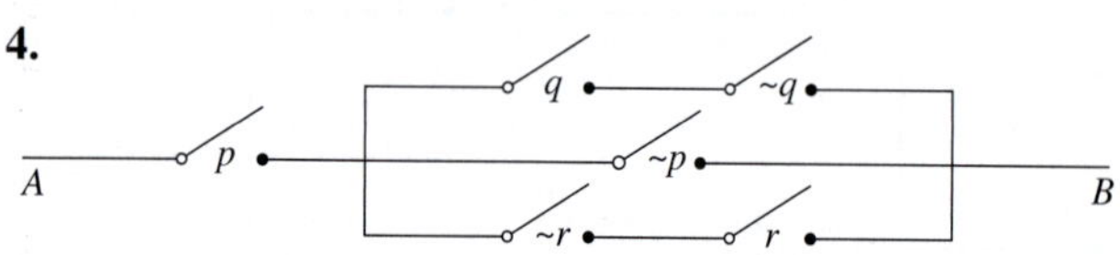

5.

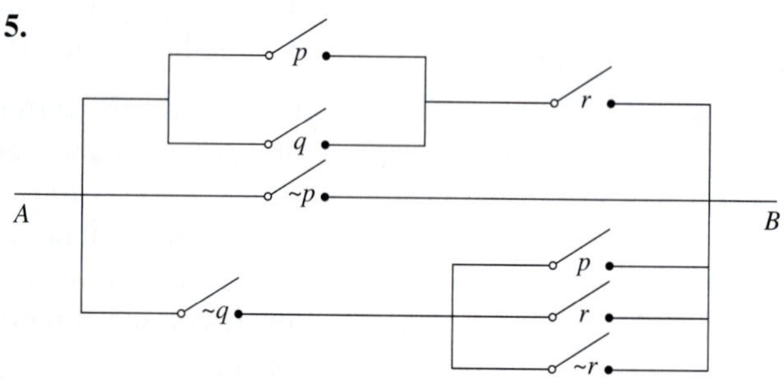

In Exercises 6–11, draw the network corresponding to the logic statement.

6. $p \vee (q \wedge r)$

7. $(p \wedge q) \wedge r$

8. $[p \vee (q \wedge r)] \wedge {\sim}q$

9. $(p \vee q) \vee [r \wedge ({\sim}r \vee {\sim}p)]$

10. $({\sim}p \vee q) \wedge (p \vee {\sim}q)$

11. $(p \wedge q) \vee [(r \vee {\sim}q) \wedge (s \vee {\sim}p)]$

In Exercises 12–15, find a logic statement corresponding to the network. Then find a simpler but equivalent logic statement.

12.

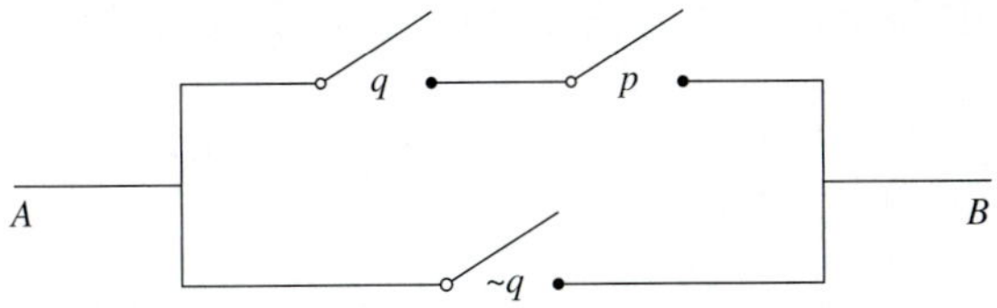

13.

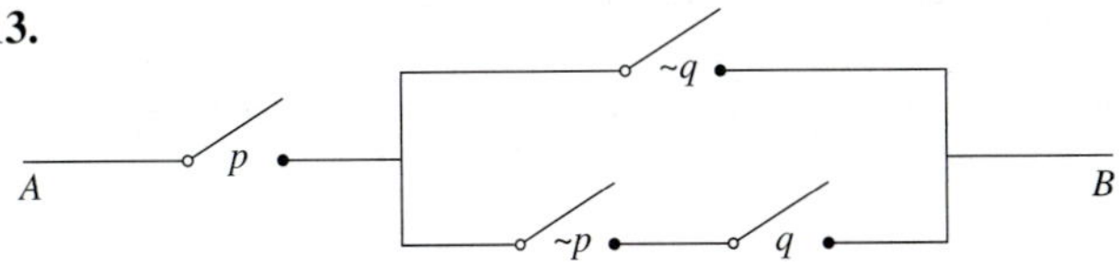

14.

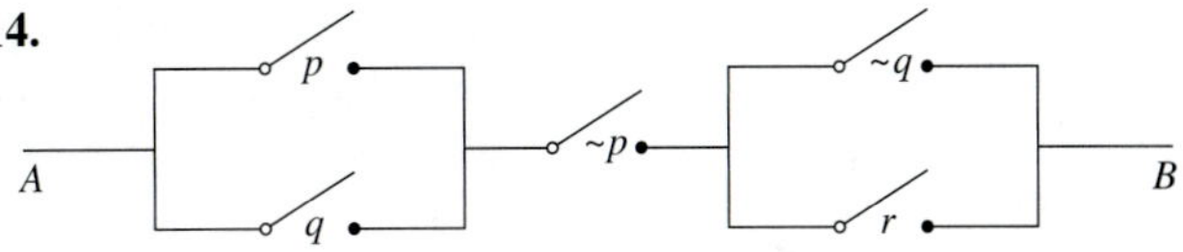

15.

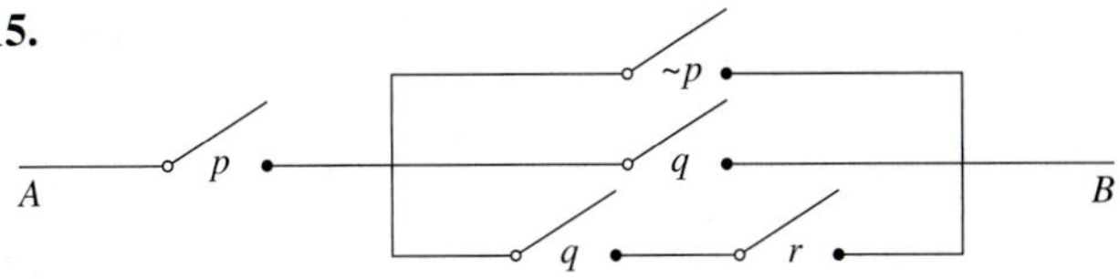

16. Designing a Logic Circuit A hallway light is to be operated by two switches, one located at the bottom of the staircase and the other located at the top of the staircase. Design a suitable network.

Hint: Let p and q be the switches. Construct a truth table for the associated propositions.

APPENDIX

B The System of Real Numbers

In this appendix, we briefly review the **system of real numbers.** This system consists of a set of objects called real numbers together with two operations, addition and multiplication, that enable us to combine two or more real numbers to obtain other real numbers. These operations are subject to certain rules that we will state after first recalling the set of real numbers.

The set of real numbers may be constructed from the set of **natural** (also called **counting**) **numbers**

$$N = \{1, 2, 3, \ldots\}$$

by adjoining other objects (numbers) to it. Thus, the set

$$W = \{0, 1, 2, 3, \ldots\}$$

obtained by adjoining the single number 0 to N is called the set of **whole numbers.** By adjoining *negatives* of the numbers 1, 2, 3, . . . to the set W of whole numbers, we obtain the set of **integers**

$$I = \{\ldots, -3, -2, -1, 0, 1, 2, 3, \ldots\}$$

Next, consider the set

$$Q = \left\{\frac{a}{b} \middle| a \text{ and } b \text{ are integers with } b \neq 0\right\}$$

Now, the set I of integers is contained in the set Q of **rational numbers.** To see this, observe that each integer may be written in the form a/b with $b = 1$, thus qualifying as a member of the set Q. The converse, however, is false, for the rational numbers (fractions) such as

$$\frac{1}{2}, \quad \frac{23}{25}, \quad \text{and so on}$$

are clearly not integers.

The sets N, W, I, and Q constructed thus far have the relationship

$$N \subset W \subset I \subset Q$$

That is, N is a proper subset of W, W is a proper subset of I, and I is a proper subset of Q.

Finally, consider the set Ir of all real numbers that cannot be expressed in the form a/b, where a, b are integers $(b \neq 0)$. The members of this set, called the set of **irrational numbers,** include $\sqrt{2}$, $\sqrt{3}$, π, and so on. The set

$$R = Q \cup Ir$$

which is the set of all rational and irrational numbers, is called the set of all **real numbers** (Figure 1).

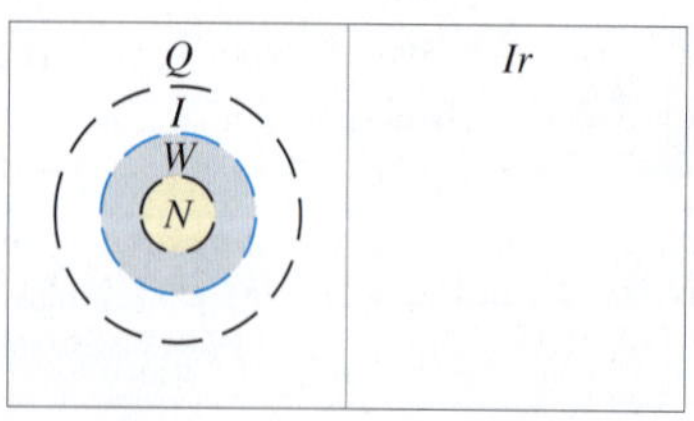

Q = Rationals
I = Integers
W = Whole numbers
N = Natural numbers
Ir = Irrationals

FIGURE 1
The set of all real numbers consists of the set of rational numbers and the set of irrational numbers.

Note the following important representation of real numbers: Every real number has a decimal representation; a rational number has a representation as a terminating or repeating decimal. For example, $\frac{1}{40} = 0.025$, and

$$\frac{1}{7} = 0.142857142857142857...$$

Note that the block of integers 142857 repeats.

On the other hand, the irrational number $\sqrt{2}$ has a representation as a nonrepeating decimal. Thus,

$$\sqrt{2} = 1.41421...$$

As was mentioned earlier, any two real numbers may be combined to obtain another real number. The operation of *addition*, written $+$, enables us to combine any two numbers a and b to obtain their sum, denoted by $a + b$. Another operation, called *multiplication* and written $\cdot$, enables us to combine any two real numbers a and b to form their product, denoted by $a \cdot b$ or, more simply, ab. These two operations are subjected to the following rules of operation: Given any three real numbers a, b, and c, we have

I. Under addition

1. $a + b = b + a$ — Commutative law of addition
2. $a + (b + c) = (a + b) + c$ — Associative law of addition
3. $a + 0 = a$ — Identity law of addition
4. $a + (-a) = 0$ — Inverse law of addition

II. Under multiplication

1. $ab = ba$ — Commutative law of multiplication
2. $a(bc) = (ab)c$ — Associative law of multiplication
3. $a \cdot 1 = a$ — Identity law of multiplication
4. $a(1/a) = 1 \quad (a \neq 0)$ — Inverse law of multiplication

III. Under addition and multiplication

1. $a(b + c) = ab + ac$ — Distributive law for multiplication with respect to addition

APPENDIX C

Review of Logarithms

Logarithms

You are already familiar with exponential equations of the form

$$b^y = x \qquad (b > 0, b \neq 1)$$

where the variable x is expressed in terms of a real number b and a variable y. But what about solving this same equation for y? You may recall from your study of algebra that the number y is called the **logarithm of x to the base b** and is denoted by **$\log_b x$.** It is the power to which the base b must be raised to obtain the number x.

> **Logarithm of x to the Base b, $b > 0$, $b \neq 1$**
>
> $$y = \log_b x \quad \text{if and only if} \quad x = b^y \qquad (x > 0)$$

⚠ Observe that the logarithm $\log_b x$ is defined only for positive values of x.

EXAMPLE 1

a. $\log_{10} 100 = 2$ since $100 = 10^2$
b. $\log_5 125 = 3$ since $125 = 5^3$
c. $\log_3 \frac{1}{27} = -3$ since $\frac{1}{27} = \frac{1}{3^3} = 3^{-3}$
d. $\log_{20} 20 = 1$ since $20 = 20^1$

EXAMPLE 2 Solve each of the following equations for x.

a. $\log_3 x = 4$ **b.** $\log_{16} 4 = x$ **c.** $\log_x 8 = 3$

Solution

a. By definition, $\log_3 x = 4$ implies $x = 3^4 = 81$.
b. $\log_{16} 4 = x$ is equivalent to $4 = 16^x = (4^2)^x = 4^{2x}$, or $4^1 = 4^{2x}$, from which we deduce that

$$2x = 1 \qquad b^m = b^n \Rightarrow m = n$$
$$x = \frac{1}{2}$$

c. Referring once again to the definition, we see that the equation $\log_x 8 = 3$ is equivalent to

$$8 = 2^3 = x^3$$
$$x = 2 \qquad a^m = b^m \Rightarrow a = b$$

The two widely used systems of logarithms are the system of **common logarithms**, which uses the number 10 as the base, and the system of **natural logarithms**, which uses the irrational number $e = 2.71828\ldots$ as the base. Also, it is standard practice to write **log** for $\log_{10}$ and **ln** for $\log_e$.

> **Logarithmic Notation**
>
> $\log x = \log_{10} x$ Common logarithm
>
> $\ln x = \log_e x$ Natural logarithm

The system of natural logarithms is widely used in theoretical work. Using natural logarithms rather than logarithms to other bases often leads to simpler expressions.

Laws of Logarithms

Computations involving logarithms are facilitated by the following **laws of logarithms.**

Laws of Logarithms

If m and n are positive numbers and $b > 0$, $b \neq 1$, then

1. $\log_b mn = \log_b m + \log_b n$
2. $\log_b \dfrac{m}{n} = \log_b m - \log_b n$
3. $\log_b m^r = r \log_b m$ for any real number r
4. $\log_b 1 = 0$
5. $\log_b b = 1$

Do not confuse the expression $\log m/n$ (Law 2) with the expression $\log m/\log n$. For example,

$$\log \frac{100}{10} = \log 100 - \log 10 = 2 - 1 = 1 \neq \frac{\log 100}{\log 10} = \frac{2}{1} = 2$$

The following examples illustrate the properties of logarithms.

EXAMPLE 3

a. $\log(2 \cdot 3) = \log 2 + \log 3$ **b.** $\ln \dfrac{5}{3} = \ln 5 - \ln 3$

c. $\log \sqrt{7} = \log 7^{1/2} = \dfrac{1}{2} \log 7$ **d.** $\log_5 1 = 0$

e. $\log_{45} 45 = 1$

EXAMPLE 4 Given that $\log 2 \approx 0.3010$, $\log 3 \approx 0.4771$, and $\log 5 \approx 0.6990$, use the laws of logarithms to find

a. $\log 15$ **b.** $\log 7.5$ **c.** $\log 81$

Solution

a. Note that $15 = 3 \cdot 5$, so by Law 1 for logarithms,

$$\begin{aligned}\log 15 &= \log 3 \cdot 5\\ &= \log 3 + \log 5\\ &\approx 0.4771 + 0.6990\\ &= 1.1761\end{aligned}$$

b. Observing that $7.5 = 15/2 = (3 \cdot 5)/2$, we apply Laws 1 and 2, obtaining

$$\begin{aligned}\log 7.5 &= \log \frac{(3)(5)}{2}\\ &= \log 3 + \log 5 - \log 2\\ &\approx 0.4771 + 0.6990 - 0.3010\\ &= 0.8751\end{aligned}$$

c. Since $81 = 3^4$, we apply Law 3 to obtain

$$\begin{aligned} \log 81 &= \log 3^4 \\ &= 4 \log 3 \\ &\approx 4(0.4771) \\ &= 1.9084 \end{aligned}$$

Examples 5 and 6 illustrate how the properties of logarithms are used to solve equations.

EXAMPLE 5 Solve $\log_3(x + 1) - \log_3(x - 1) = 1$ for x.

Solution Using the properties of logarithms, we obtain

$$\begin{aligned} \log_3(x + 1) - \log_3(x - 1) &= 1 \\ \log_3 \frac{x + 1}{x - 1} &= 1 && \text{Law 2} \\ \frac{x + 1}{x - 1} &= 3^1 = 3 && \text{Definition of logarithms} \end{aligned}$$

So

$$\begin{aligned} x + 1 &= 3(x - 1) \\ x + 1 &= 3x - 3 \\ 4 &= 2x \\ x &= 2 \end{aligned}$$

Remember to check to see whether the original equation holds true for $x = 2$.

EXAMPLE 6 Solve $\log x + \log(2x - 1) = \log 6$.

Solution We have

$$\begin{aligned} \log x + \log(2x - 1) &= \log 6 \\ \log x + \log(2x - 1) - \log 6 &= 0 \\ \log\left[\frac{x(2x - 1)}{6}\right] &= 0 && \text{Laws 1 and 2} \\ \frac{x(2x - 1)}{6} &= 10^0 = 1 && \text{Definition of logarithms} \end{aligned}$$

So

$$\begin{aligned} x(2x - 1) &= 6 \\ 2x^2 - x - 6 &= 0 \\ (2x + 3)(x - 2) &= 0 \\ x &= -\frac{3}{2} \quad \text{or} \quad 2 \end{aligned}$$

Since $(2x - 1)$ must be positive, the domain of $\log(2x - 1)$ is the interval $(\frac{1}{2}, \infty)$. So we reject the root $-\frac{3}{2}$ of the quadratic equation and conclude that the solution of the given equation is $x = 2$.

Note Using the fact that $\log a = \log b$ if and only if $a = b$, we can also solve the equation of Example 6 in the following manner:

$$\begin{aligned} \log x + \log(2x - 1) &= \log 6 \\ \log x(2x - 1) &= \log 6 \\ x(2x - 1) &= 6 \end{aligned}$$

The rest of the solution is the same as that in Example 6.

Exercises

In Exercises 1–6, express each equation in logarithmic form.

1. $2^6 = 64$

2. $3^5 = 243$

3. $3^{-2} = \frac{1}{9}$

4. $5^{-3} = \frac{1}{125}$

5. $32^{3/5} - 8$

6. $81^{3/4} = 27$

In Exercises 7–10, use the facts that $\log 3 \approx 0.4771$ and $\log 4 \approx 0.6021$ to find the value of each logarithm.

7. $\log 12$

8. $\log \frac{3}{4}$

9. $\log 16$

10. $\log \sqrt{3}$

In Exercises 11–14, write the expression as the logarithm of a single quantity.

11. $2 \ln a + 3 \ln b$

12. $\frac{1}{2} \ln x + 2 \ln y - 3 \ln z$

13. $\ln 3 + \frac{1}{2} \ln x + \ln y - \frac{1}{3} \ln z$

14. $\ln 2 + \frac{1}{2} \ln(x + 1) - 2 \ln(1 + \sqrt{x})$

In Exercises 15–22, use the laws of logarithms to solve the equation.

15. $\log_2 x = 3$

16. $\log_2 8 = x$

17. $\log_x 10^3 = 3$

18. $\log_x \frac{1}{16} = -2$

19. $\log_2(2x + 5) = 3$

20. $\log_4(5x - 4) = 2$

21. $\log_5(2x + 1) - \log_5(x - 2) = 1$

22. $\log(x + 7) - \log(x - 2) = 1$

In Exercises 23–30, use logarithms to solve the equation for t.

23. $e^{0.4t} = 8$

24. $\frac{1}{3} e^{-3t} = 0.9$

25. $5e^{-2t} = 6$

26. $4e^{t-1} = 4$

27. $2e^{-0.2t} - 4 = 6$

28. $12 - e^{0.4t} = 3$

29. $\frac{50}{1 + 4e^{0.2t}} = 20$

30. $\frac{200}{1 + 3e^{-0.3t}} = 100$

APPENDIX D

Tables

TABLE 1

Binomial Probabilities

		p										
n	*x*	**0.05**	**0.1**	**0.2**	**0.3**	**0.4**	**0.5**	**0.6**	**0.7**	**0.8**	**0.9**	**0.95**
2	0	0.902	0.810	0.640	0.490	0.360	0.250	0.160	0.090	0.040	0.010	0.002
	1	0.095	0.180	0.320	0.420	0.480	0.500	0.480	0.420	0.320	0.180	0.095
	2	0.002	0.010	0.040	0.090	0.160	0.250	0.360	0.490	0.640	0.810	0.902
3	0	0.857	0.729	0.512	0.343	0.216	0.125	0.064	0.027	0.008	0.001	
	1	0.135	0.243	0.384	0.441	0.432	0.375	0.288	0.189	0.096	0.027	0.007
	2	0.007	0.027	0.096	0.189	0.288	0.375	0.432	0.441	0.384	0.243	0.135
	3		0.001	0.008	0.027	0.064	0.125	0.216	0.343	0.512	0.729	0.857
4	0	0.815	0.656	0.410	0.240	0.130	0.062	0.026	0.008	0.002		
	1	0.171	0.292	0.410	0.412	0.346	0.250	0.154	0.076	0.026	0.004	
	2	0.014	0.049	0.154	0.265	0.346	0.375	0.346	0.265	0.154	0.049	0.014
	3		0.004	0.026	0.076	0.154	0.250	0.346	0.412	0.410	0.292	0.171
	4			0.002	0.008	0.026	0.062	0.130	0.240	0.410	0.656	0.815
5	0	0.774	0.590	0.328	0.168	0.078	0.031	0.010	0.002			
	1	0.204	0.328	0.410	0.360	0.259	0.156	0.077	0.028	0.006		
	2	0.021	0.073	0.205	0.309	0.346	0.312	0.230	0.132	0.051	0.008	0.001
	3	0.001	0.008	0.051	0.132	0.230	0.312	0.346	0.309	0.205	0.073	0.021
	4			0.006	0.028	0.077	0.156	0.259	0.360	0.410	0.328	0.204
	5				0.002	0.010	0.031	0.078	0.168	0.328	0.590	0.774
6	0	0.735	0.531	0.262	0.118	0.047	0.016	0.004	0.001			
	1	0.232	0.354	0.393	0.303	0.187	0.094	0.037	0.010	0.002		
	2	0.031	0.098	0.246	0.324	0.311	0.234	0.138	0.060	0.015	0.001	
	3	0.002	0.015	0.082	0.185	0.276	0.312	0.276	0.185	0.082	0.015	0.002
	4		0.001	0.015	0.060	0.138	0.234	0.311	0.324	0.246	0.098	0.031
	5			0.002	0.010	0.037	0.094	0.187	0.303	0.393	0.354	0.232
	6				0.001	0.004	0.016	0.047	0.118	0.262	0.531	0.735
7	0	0.698	0.478	0.210	0.082	0.028	0.008	0.002				
	1	0.257	0.372	0.367	0.247	0.131	0.055	0.017	0.004			
	2	0.041	0.124	0.275	0.318	0.261	0.164	0.077	0.025	0.004		
	3	0.004	0.023	0.115	0.227	0.290	0.273	0.194	0.097	0.029	0.003	
	4		0.003	0.029	0.097	0.194	0.273	0.290	0.227	0.115	0.023	0.004
	5			0.004	0.025	0.077	0.164	0.261	0.318	0.275	0.124	0.041
	6				0.004	0.017	0.055	0.131	0.247	0.367	0.372	0.257
	7					0.002	0.008	0.028	0.082	0.210	0.478	0.698
8	0	0.663	0.430	0.168	0.058	0.017	0.004	0.001				
	1	0.279	0.383	0.336	0.198	0.090	0.031	0.008	0.001			
	2	0.051	0.149	0.294	0.296	0.209	0.109	0.041	0.010	0.001		
	3	0.005	0.033	0.147	0.254	0.279	0.219	0.124	0.047	0.009		
	4		0.005	0.046	0.136	0.232	0.273	0.232	0.136	0.046	0.005	
	5			0.009	0.047	0.124	0.219	0.279	0.254	0.147	0.033	0.005
	6			0.001	0.010	0.041	0.109	0.209	0.296	0.294	0.149	0.051
	7				0.001	0.008	0.031	0.090	0.198	0.336	0.383	0.279
	8					0.001	0.004	0.017	0.058	0.168	0.430	0.663

TABLE 1 (*continued*)

Binomial Probabilities

		p										
n	*x*	**0.05**	**0.1**	**0.2**	**0.3**	**0.4**	**0.5**	**0.6**	**0.7**	**0.8**	**0.9**	**0.95**
9	0	0.630	0.387	0.134	0.040	0.010	0.002					
	1	0.299	0.387	0.302	0.156	0.060	0.018	0.004				
	2	0.063	0.172	0.302	0.267	0.161	0.070	0.021	0.004			
	3	0.008	0.045	0.176	0.267	0.251	0.164	0.074	0.021	0.003		
	4	0.001	0.007	0.066	0.172	0.251	0.246	0.167	0.074	0.017	0.001	
	5		0.001	0.017	0.074	0.167	0.246	0.251	0.172	0.066	0.007	0.001
	6			0.003	0.021	0.074	0.164	0.251	0.267	0.176	0.045	0.008
	7				0.004	0.021	0.070	0.161	0.267	0.302	0.172	0.063
	8					0.004	0.018	0.060	0.156	0.302	0.387	0.299
	9						0.002	0.010	0.040	0.134	0.387	0.630
10	0	0.599	0.349	0.107	0.028	0.006	0.001					
	1	0.315	0.387	0.268	0.121	0.040	0.010	0.002				
	2	0.075	0.194	0.302	0.233	0.121	0.044	0.011	0.001			
	3	0.010	0.057	0.201	0.267	0.215	0.117	0.042	0.009	0.001		
	4	0.001	0.011	0.088	0.200	0.251	0.205	0.111	0.037	0.006		
	5		0.001	0.026	0.103	0.201	0.246	0.201	0.103	0.026	0.001	
	6			0.006	0.037	0.111	0.205	0.251	0.200	0.088	0.011	0.001
	7			0.001	0.009	0.042	0.117	0.215	0.267	0.201	0.057	0.010
	8				0.001	0.011	0.044	0.121	0.233	0.302	0.194	0.075
	9					0.002	0.010	0.040	0.121	0.268	0.387	0.315
	10						0.001	0.006	0.028	0.107	0.349	0.599
11	0	0.569	0.314	0.086	0.020	0.004						
	1	0.329	0.384	0.236	0.093	0.027	0.005	0.001				
	2	0.087	0.213	0.295	0.200	0.089	0.027	0.005	0.001			
	3	0.014	0.071	0.221	0.257	0.177	0.081	0.023	0.004			
	4	0.001	0.016	0.111	0.220	0.236	0.161	0.070	0.017	0.002		
	5		0.002	0.039	0.132	0.221	0.226	0.147	0.057	0.010		
	6			0.010	0.057	0.147	0.226	0.221	0.132	0.039	0.002	
	7			0.002	0.017	0.070	0.161	0.236	0.220	0.111	0.016	0.001
	8				0.004	0.023	0.081	0.177	0.257	0.221	0.071	0.014
	9				0.001	0.005	0.027	0.089	0.200	0.295	0.213	0.087
	10					0.001	0.005	0.027	0.093	0.236	0.384	0.329
	11							0.004	0.020	0.086	0.314	0.569
12	0	0.540	0.282	0.069	0.014	0.002						
	1	0.341	0.377	0.206	0.071	0.017	0.003					
	2	0.099	0.230	0.283	0.168	0.064	0.016	0.002				
	3	0.017	0.085	0.236	0.240	0.142	0.054	0.012	0.001			
	4	0.002	0.021	0.133	0.231	0.213	0.121	0.042	0.008	0.001		
	5		0.004	0.053	0.158	0.227	0.193	0.101	0.029	0.003		
	6			0.016	0.079	0.177	0.226	0.177	0.079	0.016		
	7			0.003	0.029	0.101	0.193	0.227	0.158	0.053	0.004	
	8			0.001	0.008	0.042	0.121	0.213	0.231	0.133	0.021	0.002
	9				0.001	0.012	0.054	0.142	0.240	0.236	0.085	0.017
	10					0.002	0.016	0.064	0.168	0.283	0.230	0.099
	11						0.003	0.017	0.071	0.206	0.377	0.341
	12							0.002	0.014	0.069	0.282	0.540

TABLE 1 (*continued*)

Binomial Probabilities

		p										
n	*x*	**0.05**	**0.1**	**0.2**	**0.3**	**0.4**	**0.5**	**0.6**	**0.7**	**0.8**	**0.9**	**0.95**
13	0	0.513	0.254	0.055	0.010	0.001						
	1	0.351	0.367	0.179	0.054	0.011	0.002					
	2	0.111	0.245	0.268	0.139	0.045	0.010	0.001				
	3	0.021	0.100	0.246	0.218	0.111	0.035	0.006	0.001			
	4	0.003	0.028	0.154	0.234	0.184	0.087	0.024	0.003			
	5		0.006	0.069	0.180	0.221	0.157	0.066	0.014	0.001		
	6		0.001	0.023	0.103	0.197	0.209	0.131	0.044	0.006		
	7			0.006	0.044	0.131	0.209	0.197	0.103	0.023	0.001	
	8			0.001	0.014	0.066	0.157	0.221	0.180	0.069	0.006	
	9				0.003	0.024	0.087	0.184	0.234	0.154	0.028	0.003
	10				0.001	0.006	0.035	0.111	0.218	0.246	0.100	0.021
	11					0.001	0.010	0.045	0.139	0.268	0.245	0.111
	12						0.002	0.011	0.054	0.179	0.367	0.351
	13							0.001	0.010	0.055	0.254	0.513
14	0	0.488	0.229	0.044	0.007	0.001						
	1	0.359	0.356	0.154	0.041	0.007	0.001					
	2	0.123	0.257	0.250	0.113	0.032	0.006	0.001				
	3	0.026	0.114	0.250	0.194	0.085	0.022	0.003				
	4	0.004	0.035	0.172	0.229	0.155	0.061	0.014	0.001			
	5		0.008	0.086	0.196	0.207	0.122	0.041	0.007			
	6		0.001	0.032	0.126	0.207	0.183	0.092	0.023	0.002		
	7			0.009	0.062	0.157	0.209	0.157	0.062	0.009		
	8			0.002	0.023	0.092	0.183	0.207	0.126	0.032	0.001	
	9				0.007	0.041	0.122	0.207	0.196	0.086	0.008	
	10				0.001	0.014	0.061	0.155	0.229	0.172	0.035	0.004
	11					0.003	0.022	0.085	0.194	0.250	0.114	0.026
	12					0.001	0.006	0.032	0.113	0.250	0.257	0.123
	13						0.001	0.007	0.041	0.154	0.356	0.359
	14							0.001	0.007	0.044	0.229	0.488
15	0	0.463	0.206	0.035	0.005							
	1	0.366	0.343	0.132	0.031	0.005						
	2	0.135	0.267	0.231	0.092	0.022	0.003					
	3	0.031	0.129	0.250	0.170	0.063	0.014	0.002				
	4	0.005	0.043	0.188	0.219	0.127	0.042	0.007	0.001			
	5	0.001	0.010	0.103	0.206	0.186	0.092	0.024	0.003			
	6		0.002	0.043	0.147	0.207	0.153	0.061	0.012	0.001		
	7			0.014	0.081	0.177	0.196	0.118	0.035	0.003		
	8			0.003	0.035	0.118	0.196	0.177	0.081	0.014		
	9			0.001	0.012	0.061	0.153	0.207	0.147	0.043	0.002	
	10				0.003	0.024	0.092	0.186	0.206	0.103	0.010	0.001
	11				0.001	0.007	0.042	0.127	0.219	0.188	0.043	0.005
	12					0.002	0.014	0.063	0.170	0.250	0.129	0.031
	13						0.003	0.022	0.092	0.231	0.267	0.135
	14							0.005	0.031	0.132	0.343	0.366
	15								0.005	0.035	0.206	0.463

TABLE 2

The Standard Normal Distribution

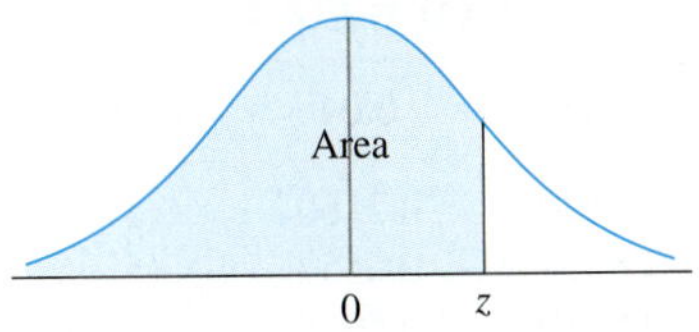

$$F_z(z) = P(Z \leq z)$$

z	**0.00**	**0.01**	**0.02**	**0.03**	**0.04**	**0.05**	**0.06**	**0.07**	**0.08**	**0.09**
−3.4	0.0003	0.0003	0.0003	0.0003	0.0003	0.0003	0.0003	0.0003	0.0003	0.0002
−3.3	0.0005	0.0005	0.0005	0.0004	0.0004	0.0004	0.0004	0.0004	0.0004	0.0003
−3.2	0.0007	0.0007	0.0006	0.0006	0.0006	0.0006	0.0006	0.0005	0.0005	0.0005
−3.1	0.0010	0.0009	0.0009	0.0009	0.0008	0.0008	0.0008	0.0008	0.0007	0.0007
−3.0	0.0013	0.0013	0.0013	0.0012	0.0012	0.0011	0.0011	0.0011	0.0010	0.0010
−2.9	0.0019	0.0018	0.0017	0.0017	0.0016	0.0016	0.0015	0.0015	0.0014	0.0014
−2.8	0.0026	0.0025	0.0024	0.0023	0.0023	0.0022	0.0021	0.0021	0.0020	0.0019
−2.7	0.0035	0.0034	0.0033	0.0032	0.0031	0.0030	0.0029	0.0028	0.0027	0.0026
−2.6	0.0047	0.0045	0.0044	0.0043	0.0041	0.0040	0.0039	0.0038	0.0037	0.0036
−2.5	0.0062	0.0060	0.0059	0.0057	0.0055	0.0054	0.0052	0.0051	0.0049	0.0048
−2.4	0.0082	0.0080	0.0078	0.0075	0.0073	0.0071	0.0069	0.0068	0.0066	0.0064
−2.3	0.0107	0.0104	0.0102	0.0099	0.0096	0.0094	0.0091	0.0089	0.0087	0.0084
−2.2	0.0139	0.0136	0.0132	0.0129	0.0125	0.0122	0.0119	0.0116	0.0113	0.0110
−2.1	0.0179	0.0174	0.0170	0.0166	0.0162	0.0158	0.0154	0.0150	0.0146	0.0143
−2.0	0.0228	0.0222	0.0217	0.0212	0.0207	0.0202	0.0197	0.0192	0.0188	0.0183
−1.9	0.0287	0.0281	0.0274	0.0268	0.0262	0.0256	0.0250	0.0244	0.0239	0.0233
−1.8	0.0359	0.0352	0.0344	0.0336	0.0329	0.0322	0.0314	0.0307	0.0301	0.0294
−1.7	0.0446	0.0436	0.0427	0.0418	0.0409	0.0401	0.0392	0.0384	0.0375	0.0367
−1.6	0.0548	0.0537	0.0526	0.0516	0.0505	0.0495	0.0485	0.0475	0.0465	0.0455
−1.5	0.0668	0.0655	0.0643	0.0630	0.0618	0.0606	0.0594	0.0582	0.0571	0.0559
−1.4	0.0808	0.0793	0.0778	0.0764	0.0749	0.0735	0.0722	0.0708	0.0694	0.0681
−1.3	0.0968	0.0951	0.0934	0.0918	0.0901	0.0885	0.0869	0.0853	0.0838	0.0823
−1.2	0.1151	0.1131	0.1112	0.1093	0.1075	0.1056	0.1038	0.1020	0.1003	0.0985
−1.1	0.1357	0.1335	0.1314	0.1292	0.1271	0.1251	0.1230	0.1210	0.1190	0.1170
−1.0	0.1587	0.1562	0.1539	0.1515	0.1492	0.1469	0.1446	0.1423	0.1401	0.1379
−0.9	0.1841	0.1814	0.1788	0.1762	0.1736	0.1711	0.1685	0.1660	0.1635	0.1611
−0.8	0.2119	0.2090	0.2061	0.2033	0.2005	0.1977	0.1949	0.1922	0.1894	0.1867
−0.7	0.2420	0.2389	0.2358	0.2327	0.2296	0.2266	0.2236	0.2206	0.2177	0.2148
−0.6	0.2743	0.2709	0.2676	0.2643	0.2611	0.2578	0.2546	0.2514	0.2483	0.2451
−0.5	0.3085	0.3050	0.3015	0.2981	0.2946	0.2912	0.2877	0.2843	0.2810	0.2776
−0.4	0.3446	0.3409	0.3372	0.3336	0.3300	0.3264	0.3228	0.3192	0.3156	0.3121
−0.3	0.3821	0.3783	0.3745	0.3707	0.3669	0.3632	0.3594	0.3557	0.3520	0.3483
−0.2	0.4207	0.4168	0.4129	0.4090	0.4052	0.4013	0.3974	0.3936	0.3897	0.3859
−0.1	0.4602	0.4562	0.4522	0.4483	0.4443	0.4404	0.4364	0.4325	0.4286	0.4247
−0.0	0.5000	0.4960	0.4920	0.4880	0.4840	0.4801	0.4761	0.4721	0.4681	0.4641

TABLE 2 (*continued*)

The Standard Normal Distribution

$$F_z(z) = P(Z \leq z)$$

z	**0.00**	**0.01**	**0.02**	**0.03**	**0.04**	**0.05**	**0.06**	**0.07**	**0.08**	**0.09**
0.0	0.5000	0.5040	0.5080	0.5120	0.5160	0.5199	0.5239	0.5279	0.5319	0.5359
0.1	0.5398	0.5438	0.5478	0.5517	0.5557	0.5596	0.5636	0.5675	0.5714	0.5753
0.2	0.5793	0.5832	0.5871	0.5910	0.5948	0.5987	0.6026	0.6064	0.6103	0.6141
0.3	0.6179	0.6217	0.6255	0.6293	0.6331	0.6368	0.6406	0.6443	0.6480	0.6517
0.4	0.6554	0.6591	0.6628	0.6664	0.6700	0.6736	0.6772	0.6808	0.6844	0.6879
0.5	0.6915	0.6950	0.6985	0.7019	0.7054	0.7088	0.7123	0.7157	0.7190	0.7224
0.6	0.7257	0.7291	0.7324	0.7357	0.7389	0.7422	0.7454	0.7486	0.7517	0.7549
0.7	0.7580	0.7611	0.7642	0.7673	0.7704	0.7734	0.7764	0.7794	0.7823	0.7852
0.8	0.7881	0.7910	0.7939	0.7967	0.7995	0.8023	0.8051	0.8078	0.8106	0.8133
0.9	0.8159	0.8186	0.8212	0.8238	0.8264	0.8289	0.8315	0.8340	0.8365	0.8389
1.0	0.8413	0.8438	0.8461	0.8485	0.8508	0.8531	0.8554	0.8577	0.8599	0.8621
1.1	0.8643	0.8665	0.8686	0.8708	0.8729	0.8749	0.8770	0.8790	0.8810	0.8830
1.2	0.8849	0.8869	0.8888	0.8907	0.8925	0.8944	0.8962	0.8980	0.8997	0.9015
1.3	0.9032	0.9049	0.9066	0.9082	0.9099	0.9115	0.9131	0.9147	0.9162	0.9177
1.4	0.9192	0.9207	0.9222	0.9236	0.9251	0.9265	0.9278	0.9292	0.9306	0.9319
1.5	0.9332	0.9345	0.9357	0.9370	0.9382	0.9394	0.9406	0.9418	0.9429	0.9441
1.6	0.9452	0.9463	0.9474	0.9484	0.9495	0.9505	0.9515	0.9525	0.9535	0.9545
1.7	0.9554	0.9564	0.9573	0.9582	0.9591	0.9599	0.9608	0.9616	0.9625	0.9633
1.8	0.9641	0.9649	0.9656	0.9664	0.9671	0.9678	0.9686	0.9693	0.9699	0.9706
1.9	0.9713	0.9719	0.9726	0.9732	0.9738	0.9744	0.9750	0.9756	0.9761	0.9767
2.0	0.9772	0.9778	0.9783	0.9788	0.9793	0.9798	0.9803	0.9808	0.9812	0.9817
2.1	0.9821	0.9826	0.9830	0.9834	0.9838	0.9842	0.9846	0.9850	0.9854	0.9857
2.2	0.9861	0.9864	0.9868	0.9871	0.9875	0.9878	0.9881	0.9884	0.9887	0.9890
2.3	0.9893	0.9896	0.9898	0.9901	0.9904	0.9906	0.9909	0.9911	0.9913	0.9916
2.4	0.9918	0.9920	0.9922	0.9925	0.9927	0.9929	0.9931	0.9932	0.9934	0.9936
2.5	0.9938	0.9940	0.9951	0.9943	0.9945	0.9946	0.9948	0.9949	0.9951	0.9952
2.6	0.9953	0.9955	0.9956	0.9957	0.9959	0.9960	0.9961	0.9962	0.9963	0.9964
2.7	0.9965	0.9966	0.9967	0.9968	0.9969	0.9970	0.9971	0.9972	0.9973	0.9974
2.8	0.9974	0.9975	0.9976	0.9977	0.9977	0.9978	0.9979	0.9979	0.9980	0.9981
2.9	0.9981	0.9982	0.9982	0.9983	0.9984	0.9984	0.9985	0.9985	0.9986	0.9986
3.0	0.9987	0.9987	0.9987	0.9988	0.9988	0.9989	0.9989	0.9989	0.9990	0.9990
3.1	0.9990	0.9991	0.9991	0.9991	0.9992	0.9992	0.9992	0.9992	0.9993	0.9993
3.2	0.9993	0.9993	0.9994	0.9994	0.9994	0.9994	0.9994	0.9995	0.9995	0.9995
3.3	0.9995	0.9995	0.9995	0.9996	0.9996	0.9996	0.9996	0.9996	0.9996	0.9997
3.4	0.9997	0.9997	0.9997	0.9997	0.9997	0.9997	0.9997	0.9997	0.9997	0.9998

Answers

CHAPTER 1

Exercises 1.1, page 7

1. (3, 3); Quadrant I **3.** (2, −2); Quadrant IV

5. (−4, −6); Quadrant III **7.** *A*

9. *E*, *F*, and *G* **11.** *F*

13–19. See the following figure.

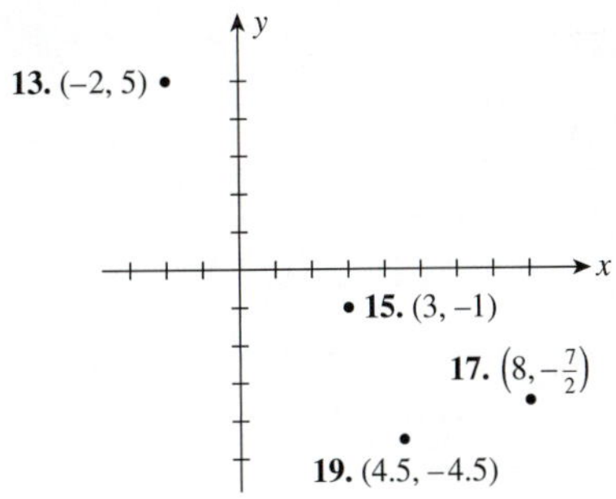

21. 5 **23.** $\sqrt{61}$ **25.** (−8, −6) and (8, −6)

29. $(x - 2)^2 + (y + 3)^2 = 25$

31. $x^2 + y^2 = 25$

33. $(x - 2)^2 + (y + 3)^2 = 34$

35. **a.** (4, 4) **b.** 10 mi **c.** 5.66 mi

37. No **39.** freight train; \$4,400 **41.** Model C

43. **a.** $\sqrt{400t^2 + 625(t + \frac{1}{2})^2}$ mi **b.** 58.31 mi

45. **b.** $\left(\frac{1}{2}, -\frac{3}{2}\right)$ **47.** False

Exercises 1.2, page 19

1. $\frac{1}{2}$ **3.** Not defined **5.** 5 **7.** $\frac{5}{6}$

9. $\dfrac{d - b}{c - a}$ $(a \neq c)$ **11.** **a.** 4 **b.** −8

13. (e) **15.** (a) **17.** (f) **19.** Parallel **21.** $a = -5$

23. $y = 2x - 10$ **25.** $y = 2$ **27.** $y = 3x - 2$ **29.** $y = x + 1$

31. Perpendicular **33.** $y = 3x + 4$ **35.** $y = 5$

37. $y = \frac{1}{2}x$; $m = \frac{1}{2}$; $b = 0$

39. $y = \frac{2}{3}x - 3$; $m = \frac{2}{3}$; $b = -3$

41. $y = -\frac{1}{2}x + \frac{7}{2}$; $m = -\frac{1}{2}$; $b = \frac{7}{2}$ **43.** $y = -3$

45. $y = \frac{1}{2}x + 3$ **47.** $y = \frac{4}{3}x + \frac{4}{3}$ **49.** $y = -2x + 2$

51. $y = -6$ **53.** $y = b$ **55.** $y = \frac{2}{3}x - \frac{2}{3}$ **57.** $k = 8$

59.

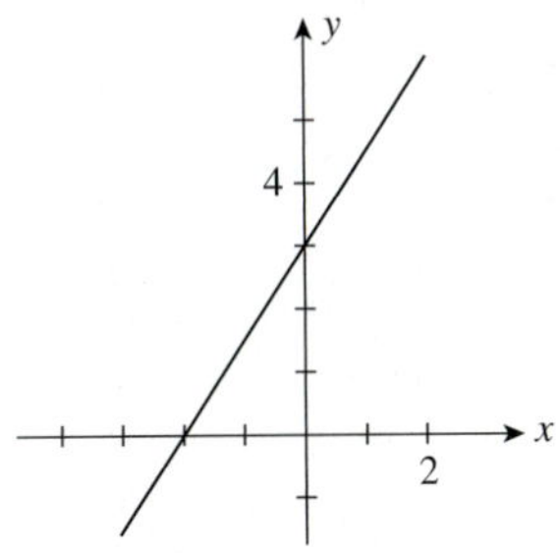

61.

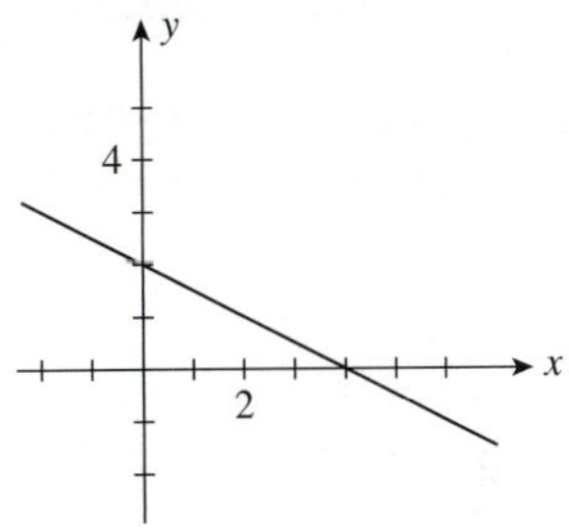

63.

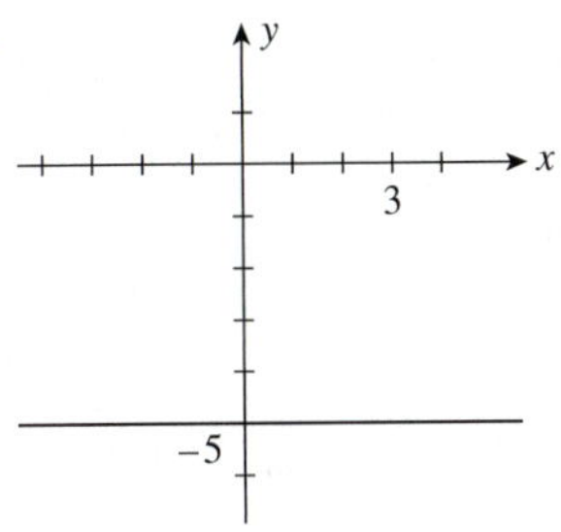

67. $y = -2x - 4$ **69.** $y = \frac{1}{8}x - \frac{1}{2}$ **71.** Yes

73. The points do not lie on a straight line.

75. **a.**

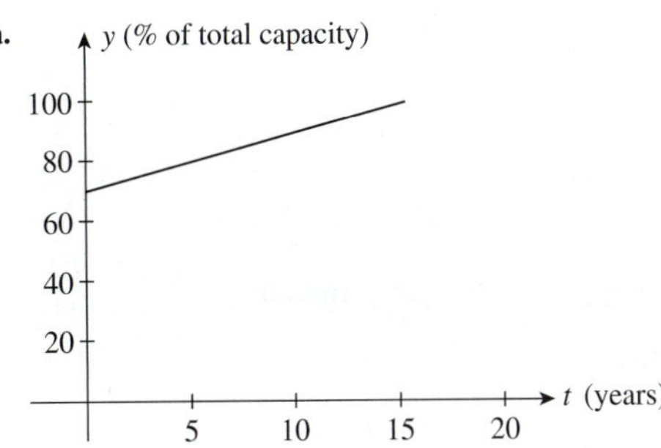

b. 1.9467; 70.082

c. The capacity utilization has been increasing by 1.9467% each year since 1990 when it stood at 70.082%.

d. In the first half of 2005

77. **a.** $y = 0.55x$ **b.** 2000 **79.** 84.8%

81. **a.** and **b.**

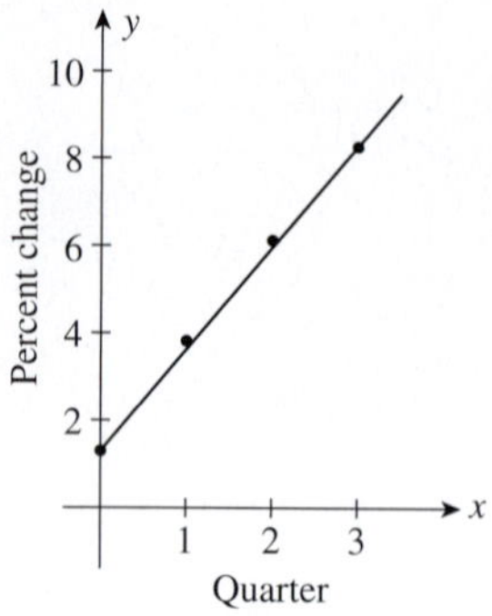

c. $y = 2.3x + 1.3$
d. 10.5%

83. **a.** and **b.**

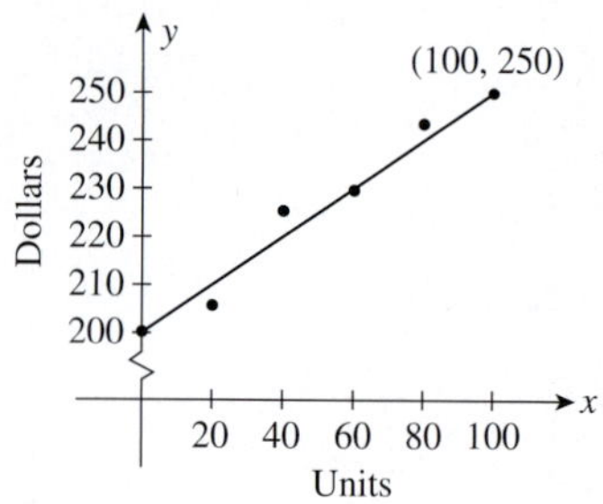

c. $y = \frac{1}{2}x + 200$
d. \$227

85. **a.** and **b.**

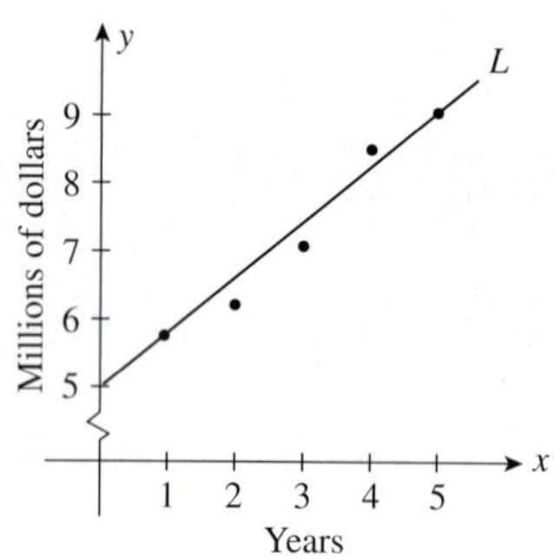

c. $y = 0.8x + 5$
d. \$12.2 million

87. Yes **89.** True **91.** True **93.** True

Using Technology Exercises 1.2, page 28

Graphing Utility

1.

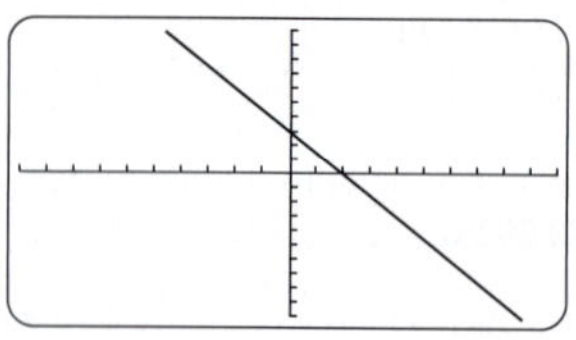

3.

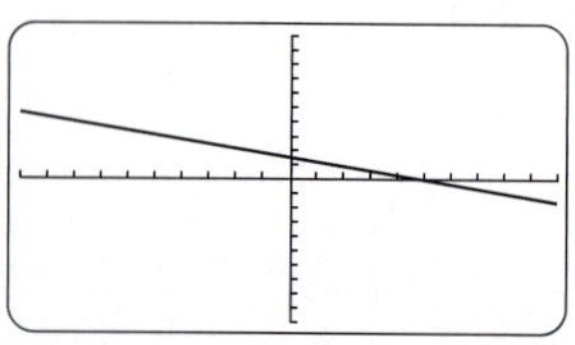

5. **a.**

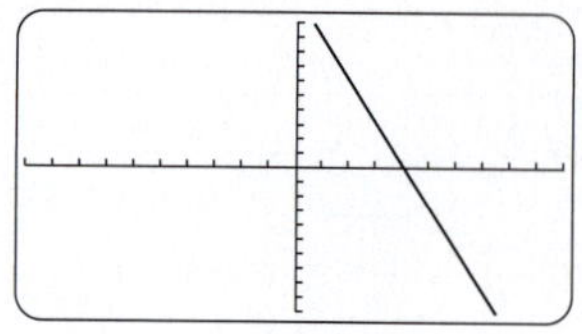

b.

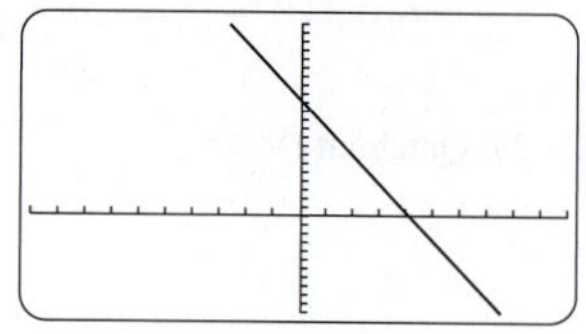

7. **a.**

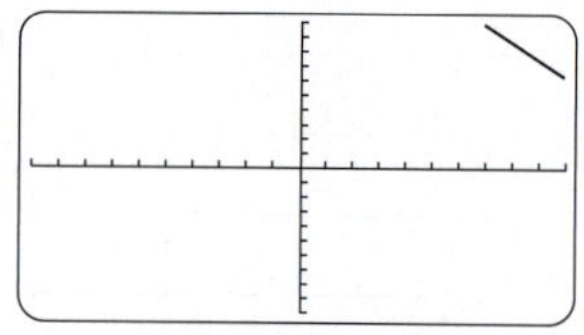

b.

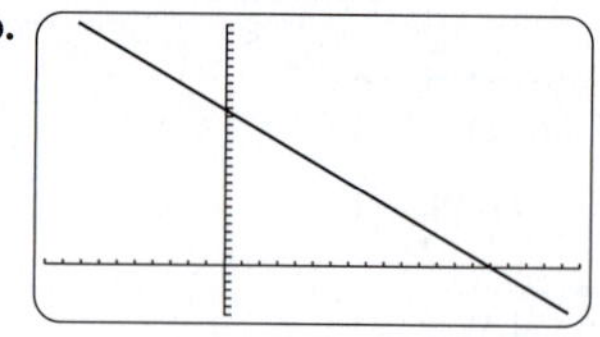

9.

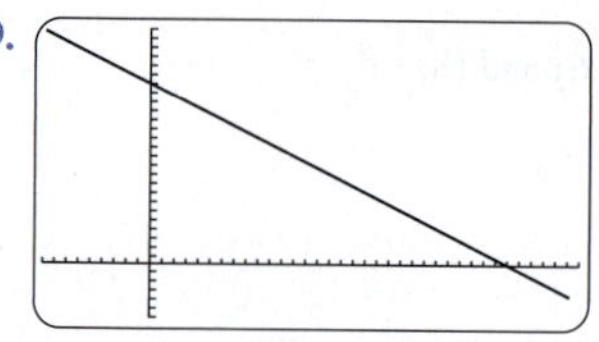

11.

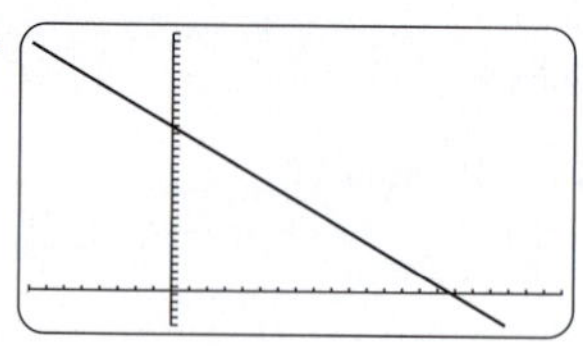

Excel

1.

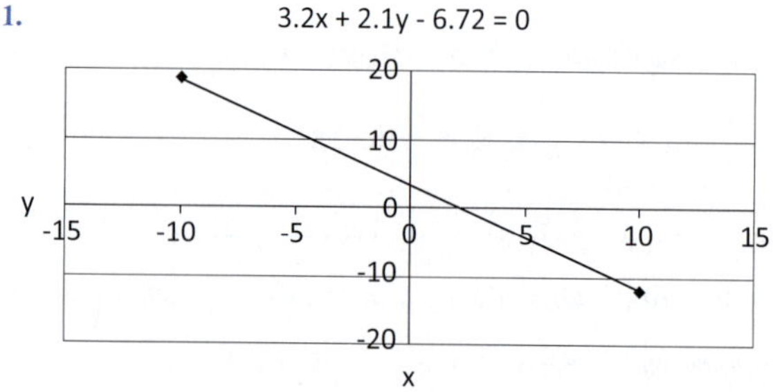

3.

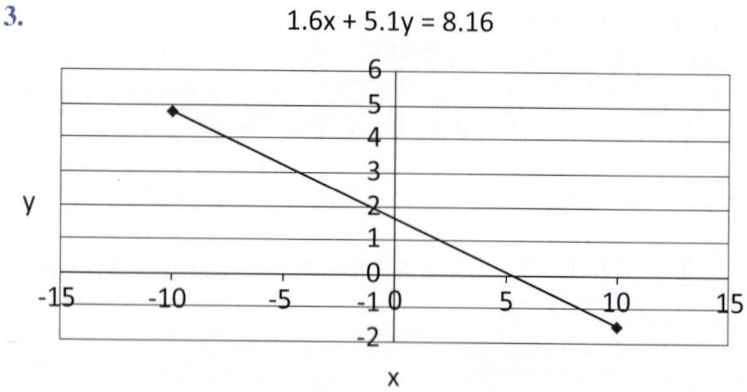

5.

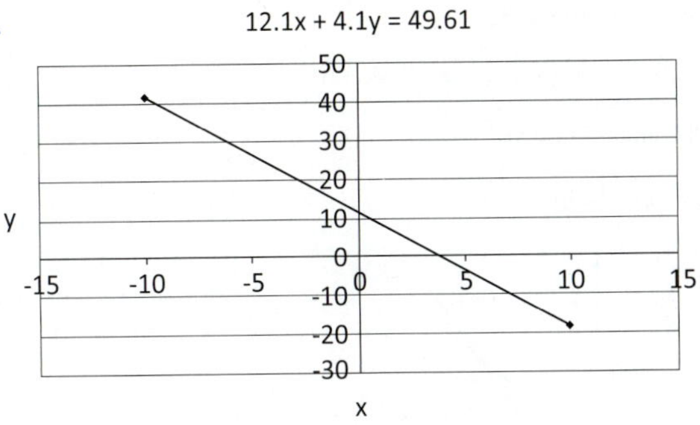

7.

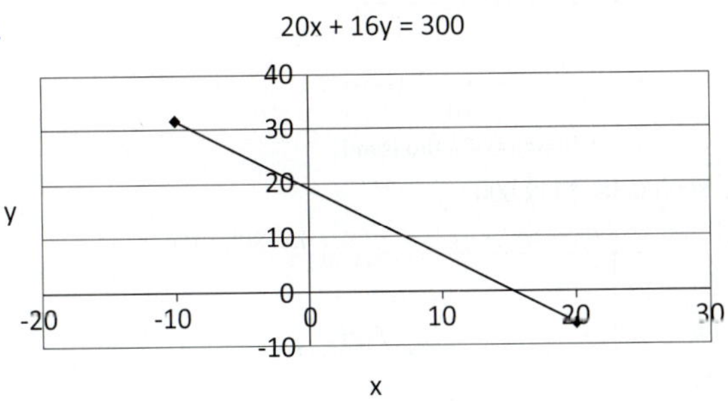

9.

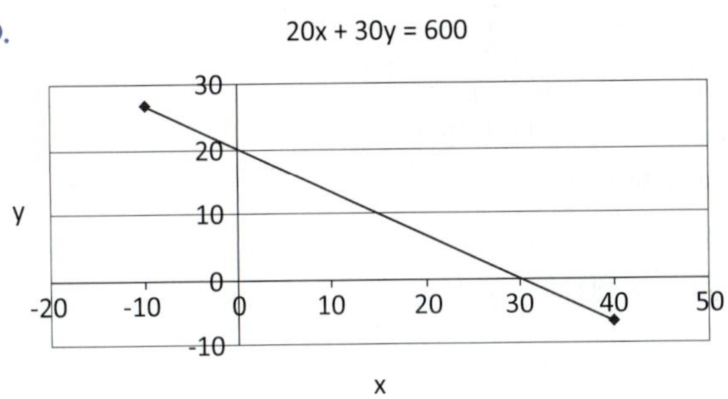

11.

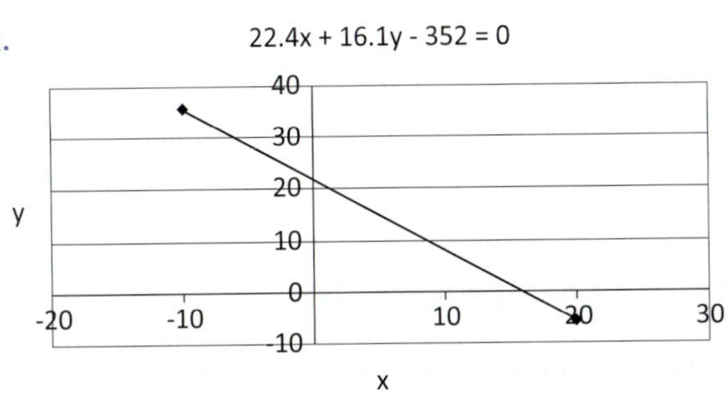

Exercises 1.3, page 36

1. Yes; $y = -\frac{2}{3}x + 2$ **3.** Yes; $y = \frac{1}{2}x + 2$

5. Yes; $y = \frac{1}{2}x + \frac{9}{4}$ **7.** No **9.** No

11. **a.** $C(x) = 8x + 40{,}000$
b. $R(x) = 12x$
c. $P(x) = 4x - 40{,}000$
d. Loss of \$8000; profit of \$8000

13. $m = -1$; $b = 2$

15. \$900,000; \$800,000

17. \$6 billion; \$43.5 billion; \$81 billion

19. **a.** $y = 1.033x$ **b.** \$1260.26

21. $C(x) = 0.6x + 12{,}100$; $R(x) = 1.15x$; $P(x) = 0.55x - 12{,}100$

23. **a.** \$12,000/year **b.** $V = 60{,}000 - 12{,}000t$

c.

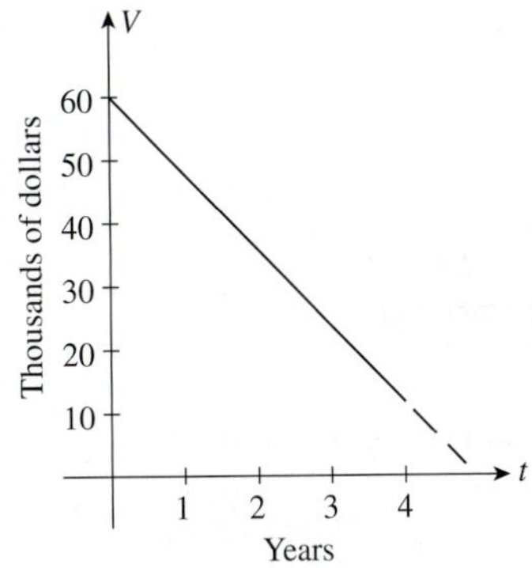

d. \$24,000

25. \$900,000; \$800,000

27. **a.** $m = a/1.7$; $b = 0$ **b.** 117.65 mg

29. **a.** $f(t) = -0.72t + 17.5$ **b.** 8.14%

31. **a.** $f(t) = -2.5t + 61$ **b.** 48.5%

33. **a.** $F = \frac{9}{5}C + 32$ **b.** 68°F **c.** 21.1°C

35. **a.**

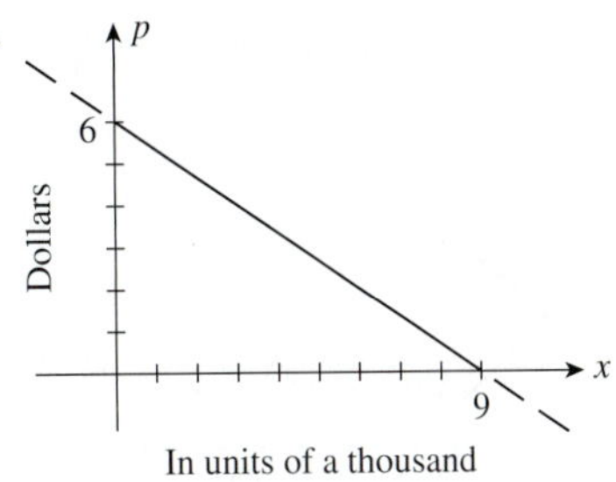

b. 3000

37. **a.**

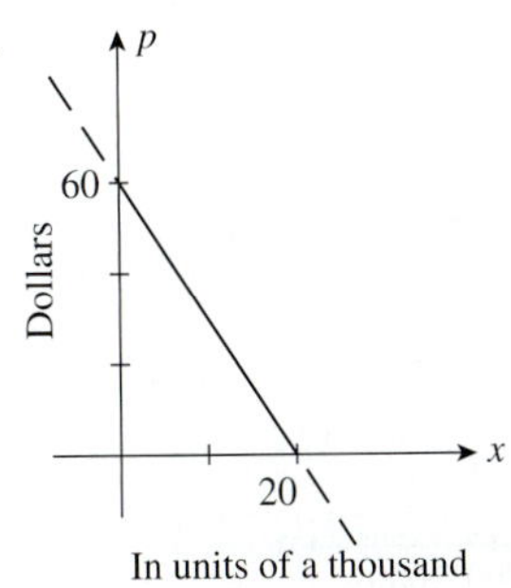

b. 10,000

39. $p = -\frac{3}{40}x + 130$; \$130; 1733 **41.** 2500 units

43. **a.**

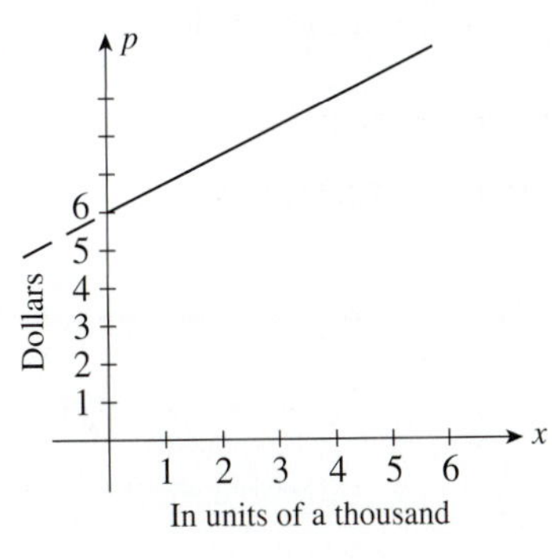

b. 2667 units

45. a.

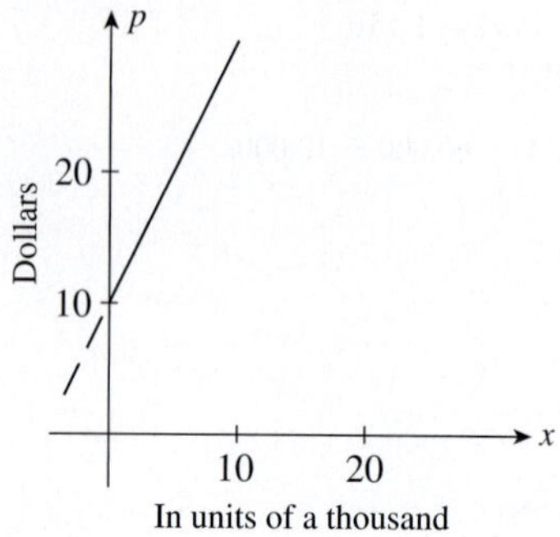

b. 2000 units

47. $p = \frac{1}{2}x + 40$ (x is measured in units of a thousand)

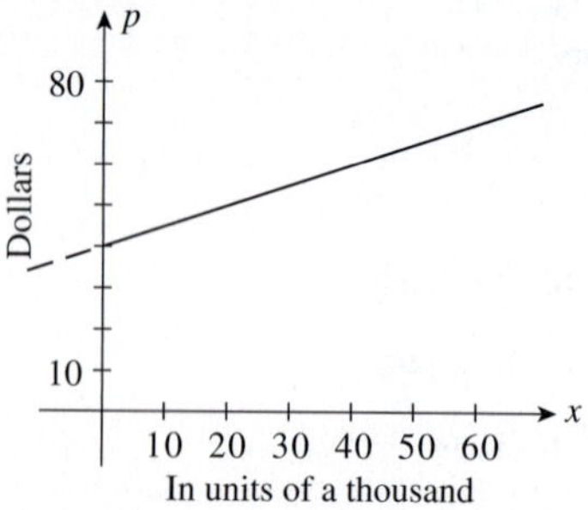

60,000 units

49. a.

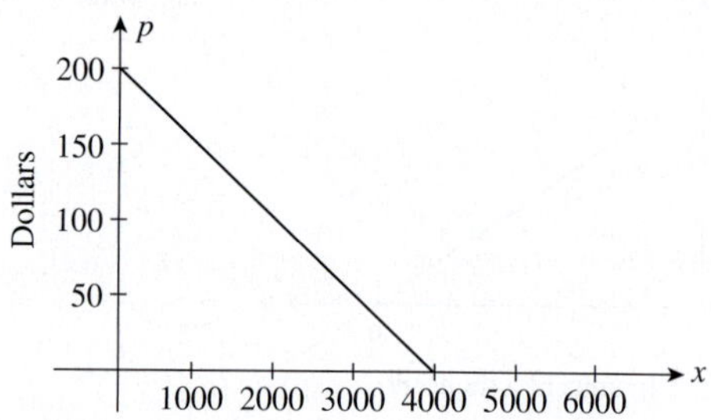

b. \$200 **c.** 2000

51. a.

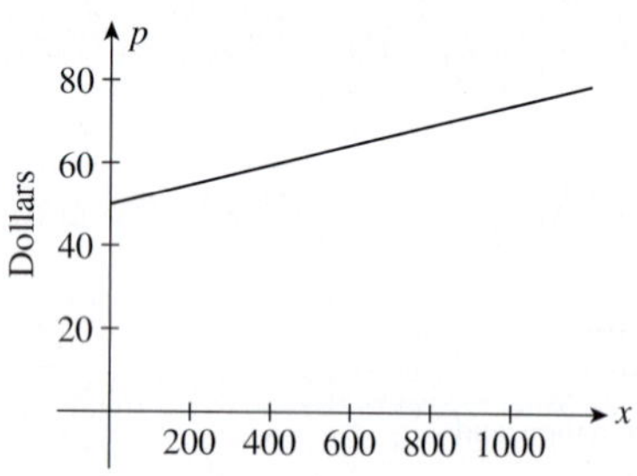

b. \$50 **c.** 2000/month

53. False

Using Technology Exercises 1.3, page 43

1. 2.2875 **3.** 2.880952381 **5.** 7.2851648352

7. 2.4680851064

Exercises 1.4, page 50

1. (2, 10) **3.** $(4, \frac{2}{3})$ **5.** $(-4, -6)$

7. 1000 units; \$15,000

9. 600 units; \$240

11. a.

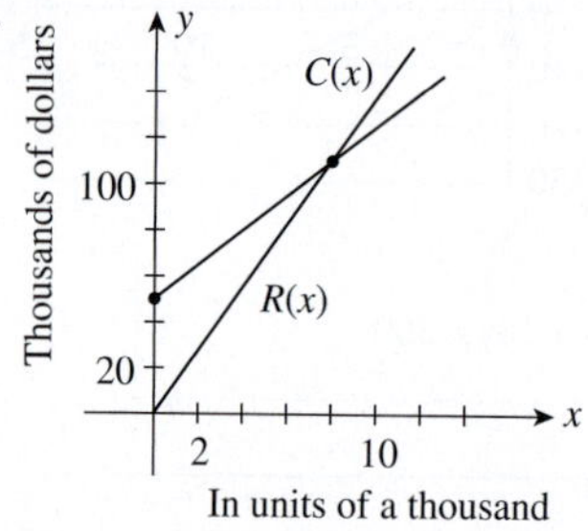

b. 8000 units; \$112,000

c.

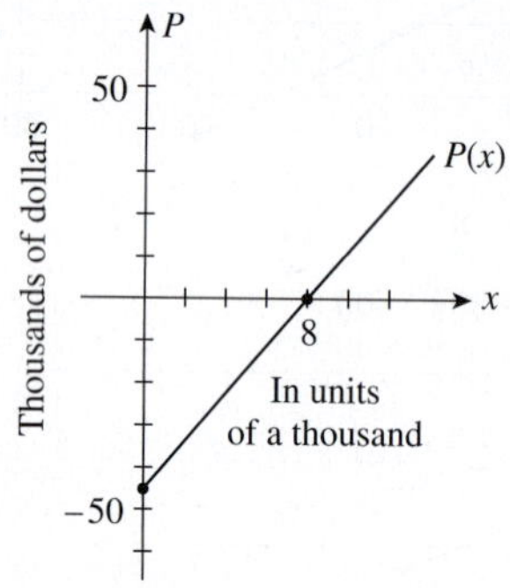

d. (8000, 0)

13. 9259 units; \$83,331

15. a. $C_1(x) = 18{,}000 + 15x$
$C_2(x) = 15{,}000 + 20x$

b.

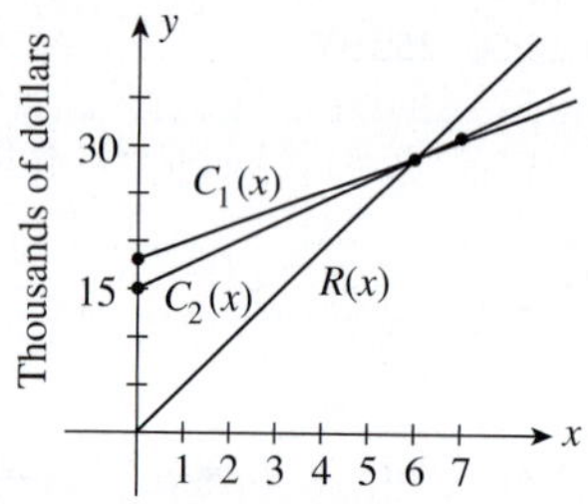

c. Machine II; machine II; machine I
d. (\$1500); \$1500; \$4750

17. Middle of 2003

19. a.

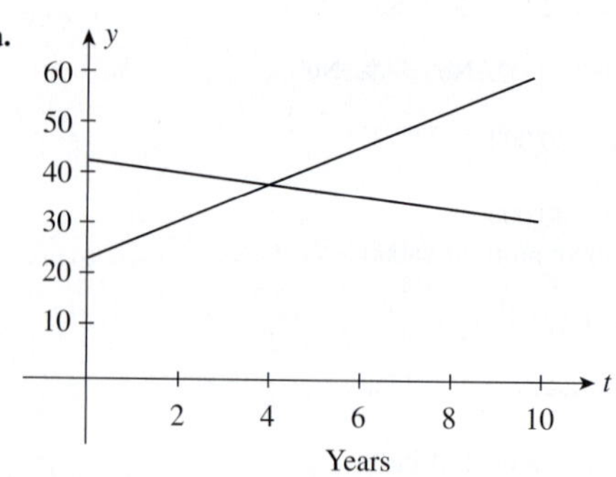

b. Feb. 2005

21. 8000 units; \$9 23. 2000 units; \$18

25. a. $p = -0.08x + 725$
b. $p = 0.09x + 300$
c. 2500 DVD players; \$525

27. 300 tablet computers; \$600

29. 2000/month; \$100/speaker

31. a. $\dfrac{b-d}{c-a}; \dfrac{bc-ad}{c-a}$
b. If c is increased, x gets smaller and p gets larger.
c. If b is decreased, x decreases and p decreases.

33. True

35. a. $m_1 = m_2$ and $b_2 \neq b_1$
b. $m_1 \neq m_2$
c. $m_1 = m_2$ and $b_1 = b_2$

Using Technology Exercises 1.4, page 54

1. (0.6, 6.2) 3. (3.8261, 0.1304)

5. (386.9091, 145.3939)

7. a.

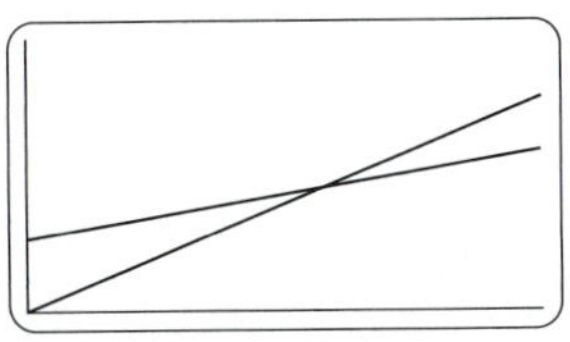

b. (3548, 27,997)

c.

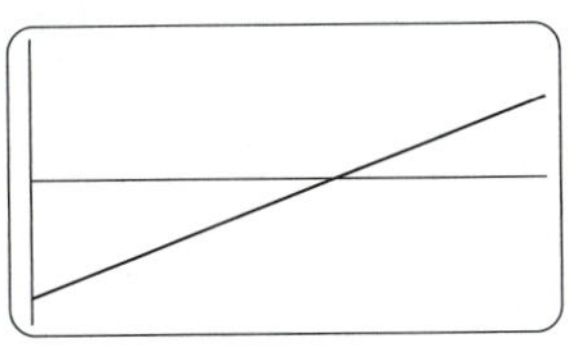

x-intercept: 3548

9. a. $C_1(x) = 34 + 0.18x$; $C_2(x) = 28 + 0.22x$

b.

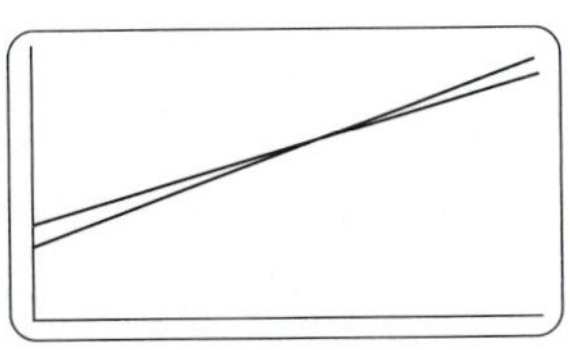

c. (150, 61)

d. If the distance driven is less than or equal to 150 mi, rent from Acme Truck Leasing; if the distance driven is more than 150 mi, rent from Ace Truck Leasing.

11. a. $p = -\frac{1}{10}x + 284$; $p = \frac{1}{60}x + 60$

b.

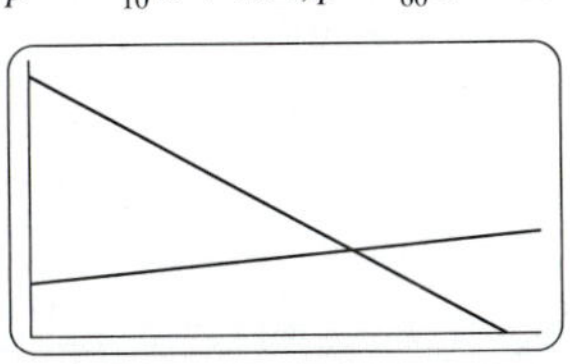

(1920, 92)

c. 1920/week; \$92/radio

Exercises 1.5, page 60

1. a. $y = 2.3x + 1.5$

b.

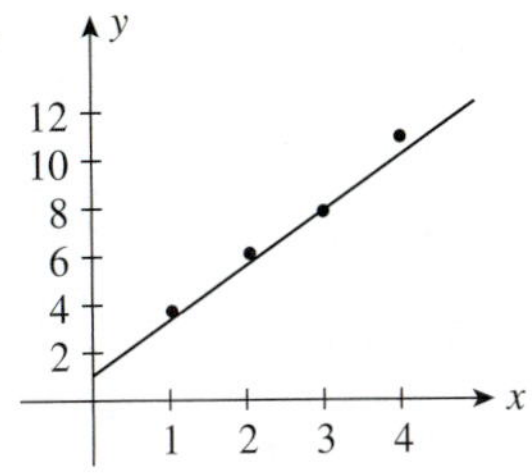

3. a. $y = -0.77x + 5.74$

b.

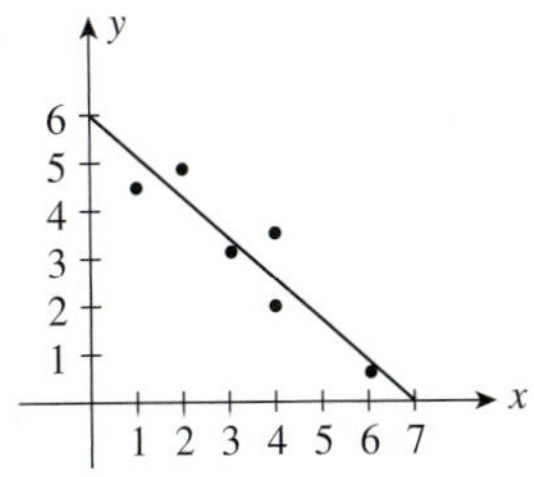

5. a. $y = 1.2x + 2$

b.

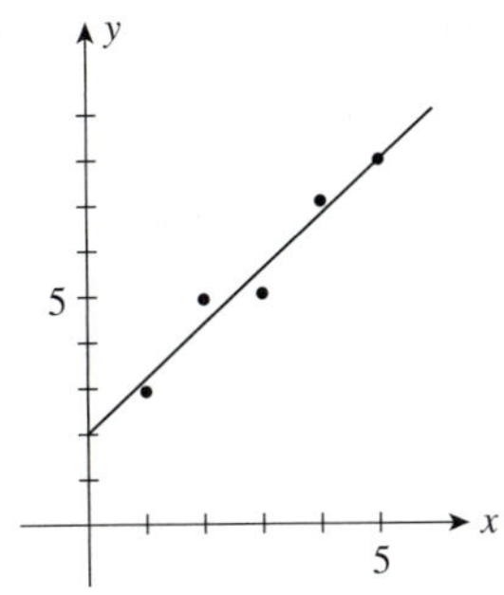

7. a. $y = 0.34x - 0.9$

b.

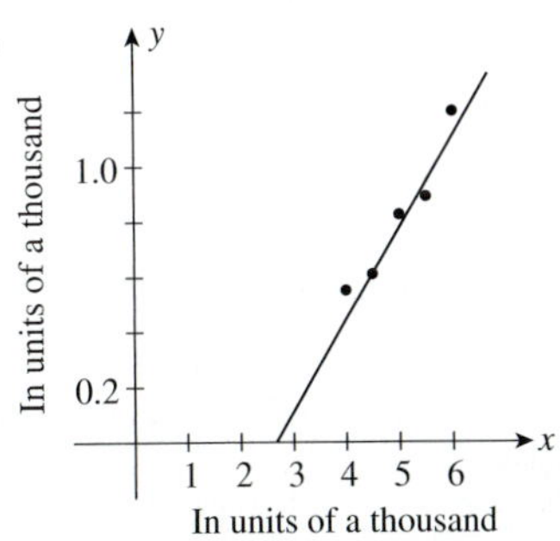

c. 1276 applications

9. a. $y = -2.8x + 440$

b.

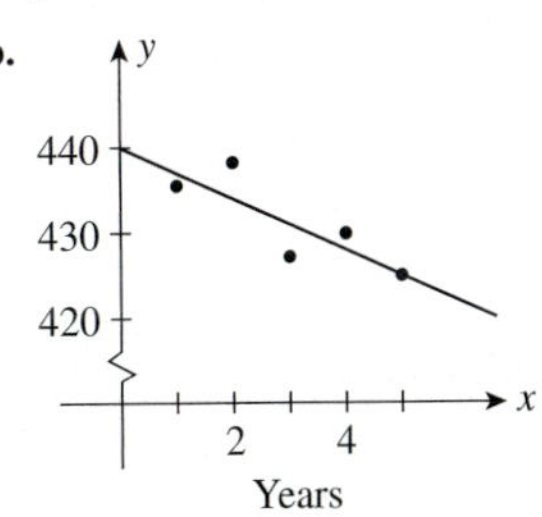

c. 420

11. **a.** $y = 234.4x + 157.3$ **b.** 1798.1 million

13. **a.** $y = 2.8x + 17.6$ **b.** \$40 million

15. **a.** $y = 1.69x + 24.11$ **b.** \$37.6 billion

17. **a.** $y = 2.46x + 82.1$ **b.** 94.4 million

19. **a.** $y = 2.75x + 29.4$ **b.** 2.75 million/year

21. **a.** $y = 0.3x + 6.46$ **b.** \$0.3 billion/year

23. **a.** $y = 2.19x + 18.38$ **b.** \$2.19 billion/year **c.** \$42.5 billion

25. **a.** $y = 7.25x + 60.21$ **b.** \$139.96 billion **c.** \$7.25 billion/year

27. False 29. True

Using Technology Exercises 1.5, page 67

1. $y = 2.3596x + 3.8639$

3. $y = -1.1948x + 3.5525$

5. **a.** $y = 22.3x + 143.5$ **b.** \$22.3 billion/year **c.** \$366.5 billion

7. **a.** $y = 1.5857t + 6.6857$ **b.** \$19.4 billion

9. **a.** $y = 1.7571x + 7.9143$ **b.** \$22 billion

Chapter 1 Concept Review Questions, page 68

1. ordered; abscissa (x-coordinate); ordinate (y-coordinate)

2. **a.** x-; y- **b.** third

3. $\sqrt{(c-a)^2 + (d-b)^2}$

4. $(x-a)^2 + (y-b)^2 = r^2$

5. **a.** $\dfrac{y_2 - y_1}{x_2 - x_1}$ **b.** undefined **c.** 0 **d.** positive

6. $m_1 = m_2$; $m_1 = -\dfrac{1}{m_2}$

7. **a.** $y - y_1 = m(x - x_1)$; point-slope
b. $y = mx + b$; slope-intercept

8. **a.** $Ax + By + C = 0$ (A, B, not both zero) **b.** $-\dfrac{a}{b}$

9. $mx + b$

10. **a.** price; demanded; demand
b. price; supplied; supply

11. break-even 12. demand; supply

Chapter 1 Review Exercises, page 69

1. 5 2. 5 3. 5 4. 2 5. No

6. $x = -2$ 7. $y = 4$

8. $x + 10y - 38 = 0$ 9. $y = -\frac{4}{5}x + \frac{12}{5}$

10. $5x - 2y + 18 = 0$ 11. $y = \frac{3}{4}x + \frac{11}{2}$

12. $y = -\frac{1}{2}x - 3$ 13. $\frac{3}{5}$; $-\frac{6}{5}$

14. $3x + 4y - 18 = 0$

15. $y = -\frac{3}{5}x + \frac{12}{5}$ 16. $3x + 2y + 14 = 0$

17. $k = 3$ 18. $m = -\frac{5}{2}$; $b = \frac{11}{2}$

19.
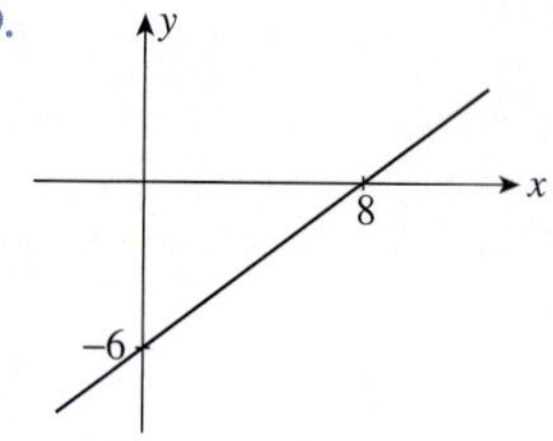

20.
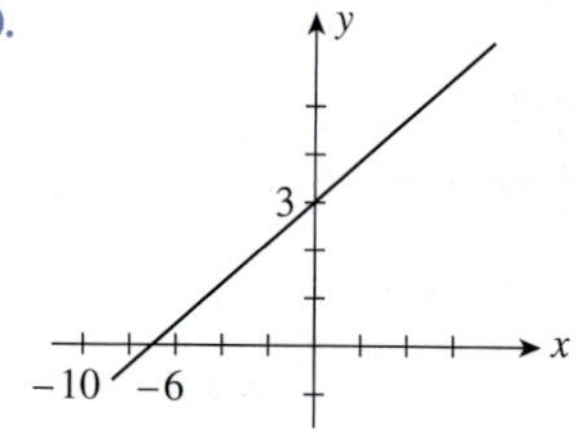

21. 60,000 22. **a.** $f(x) = x + 2.4$ **b.** \$5.4 million

24. **b.** 117 mg 25. **a.** \$200,000/year **b.** \$4,000,000

26. **a.** \$22,500/year **b.** $V = -22{,}500t + 300{,}000$

27. **a.** $6x + 30{,}000$ **b.** $10x$ **c.** $4x - 30{,}000$
d. (\$6,000); \$2000; \$18,000

28. **a.** $f(t) = 0.45t + 23.4$ **b.** 26.1%

29. **a.** and **b.**
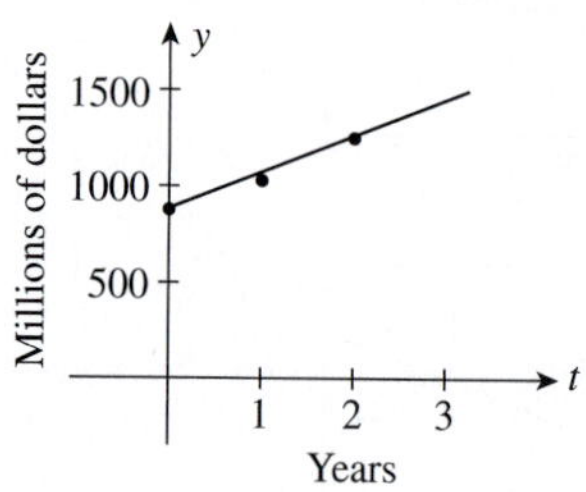

c. $y = 182t + 887$ **d.** \$1797 million

30. $p = -0.05x + 200$
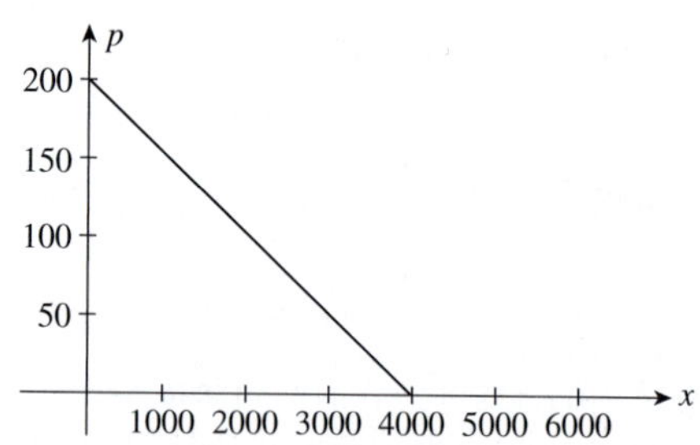

31. $p = \frac{1}{36}x + \frac{400}{9}$

32. **a.**
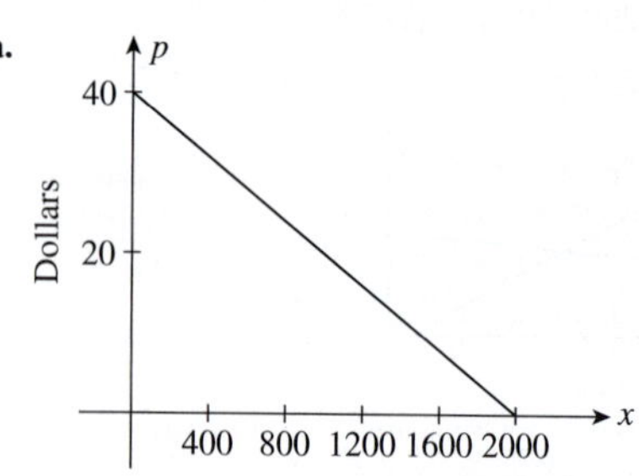

b. \$40 **c.** 1000

33. a.

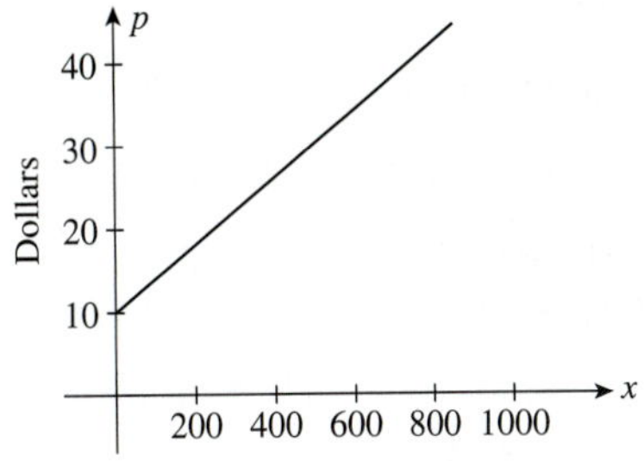

b. \$10 c. 250 headphones/week

34. $(2, -3)$ 35. $\left(6, \frac{21}{2}\right)$ 36. $(2500, 50{,}000)$ 37. 6000; \$22

38. a. $y = 0.25x$ b. 1600 39. 600; \$80

40. a. $y = 3.87x + 82.94$ b. \$117,770 41. 500; \$30

42. a. $y = 0.059x + 19.50$ b. 21.9 years; almost the same
c. 21.3 years

Chapter 1 Before Moving On, page 71

1.

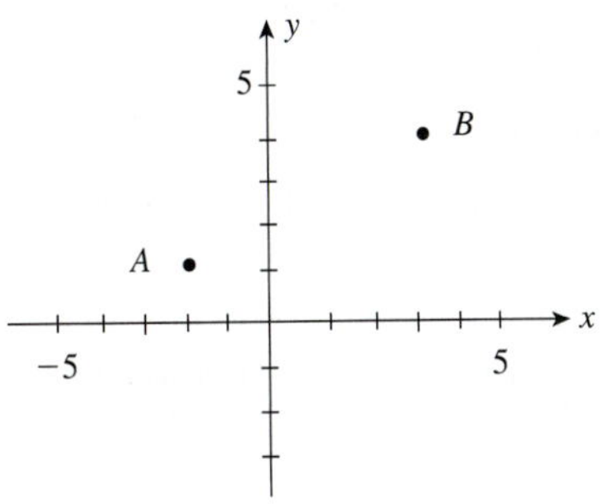

$\sqrt{34}$

2. $y = 3x - 8$ 3. Yes

4. a. \$15 b. \$22,000 c. \$18

5. $\left(1, \frac{4}{3}\right)$ 6. After 5 years

CHAPTER 2

Exercises 2.1, page 79

1. Unique solution; $(2, 1)$ 3. No solution

5. Unique solution; $(3, 2)$

7. Infinitely many solutions; $\left(t, \frac{2}{5}t - 2\right)$; t, a parameter

9. Unique solution; $(1, -2)$

11. No solution 13. Unique solution; $\left(\frac{1}{2}, \frac{1}{2}\right)$

15. Infinitely many solutions; $\left(2t + \frac{2}{3}, t\right)$; t, a parameter

17. $(-1, 2)$ 19. $k = -2$

21. $a \neq 0$ and $b \neq 0$; $\left(\dfrac{c+d}{2a}, \dfrac{d-c}{2b}\right)$

23. $\begin{aligned} x + y &= 500 \\ 42x + 30y &= 18{,}600 \end{aligned}$ 25. $\begin{aligned} x + y &= 100 \\ 8x + 9y &= 860 \end{aligned}$

27. $\begin{aligned} x + y &= 1000 \\ 0.5x + 1.5y &= 1300 \end{aligned}$ 29. $\begin{aligned} x + y &= 110 \\ y - x &= 100 \end{aligned}$

31. $\begin{aligned} 0.06x + 0.08y + 0.12z &= 21{,}600 \\ z &= 2x \\ 0.12z &= 0.08y \end{aligned}$

33. $\begin{aligned} x + y + z &= 100 \\ x + y \quad &= 67 \\ x \quad - z &= 17 \end{aligned}$

35. $\begin{aligned} 18x + 20y + 24z &= 26{,}400 \\ 4x + 4y + 3z &= 4{,}900 \\ 5x + 4y + 6z &= 6{,}200 \end{aligned}$

37. $\begin{aligned} 18{,}000x + 27{,}000y + 36{,}000z &= 2{,}250{,}000 \\ x &= 2y \\ x + y + z &= 100 \end{aligned}$

39. $\begin{aligned} 10x + 6y + 8z &= 100 \\ 10x + 12y + 6z &= 100 \\ 5x + 4y + 12z &= 100 \end{aligned}$

41. $\begin{aligned} x + y + z &= 1{,}000 \\ 80x + 60y + 50z &= 62{,}800 \\ x + y - 2z &= 400 \end{aligned}$ 43. $\begin{aligned} 280x + 330y + 260z &= 4060 \\ 130x + 140y + 110z &= 1800 \\ x - y - z &= 0 \end{aligned}$

45. True 47. False

Exercises 2.2, page 93

1. $\left[\begin{array}{rr|r} 2 & -3 & 7 \\ 3 & 1 & 4 \end{array}\right]$

3. $\left[\begin{array}{rrr|r} 0 & -1 & 2 & 5 \\ 2 & 2 & -8 & 4 \\ 0 & 3 & 4 & 0 \end{array}\right]$

5. $\begin{aligned} 3x + 2y &= -4 \\ x - y &= 5 \end{aligned}$ 7. $\begin{aligned} x + 3y + 2z &= 4 \\ 2x \quad &= 5 \\ 3x - 3y + 2z &= 6 \end{aligned}$

9. Yes 11. No 13. Yes 15. No 17. No

19. $\left[\begin{array}{rr|r} 1 & 3 & 4 \\ 0 & -2 & -2 \end{array}\right]$ 21. $\left[\begin{array}{rr|r} 1 & -2 & -3 \\ 0 & 20 & 20 \end{array}\right]$

23. $\left[\begin{array}{rrr|r} 1 & 2 & 3 & 6 \\ 0 & -1 & -5 & -7 \\ 0 & -7 & -7 & -14 \end{array}\right]$

25. $\left[\begin{array}{rrr|r} -6 & -11 & 0 & -5 \\ 2 & 4 & 1 & 3 \\ 1 & -2 & 0 & -10 \end{array}\right]$

27. $\left[\begin{array}{rr|r} 3 & 9 & 6 \\ 2 & 1 & 4 \end{array}\right] \xrightarrow{\frac{1}{3}R_1} \left[\begin{array}{rr|r} 1 & 3 & 2 \\ 2 & 1 & 4 \end{array}\right]$

$\xrightarrow{R_2 - 2R_1} \left[\begin{array}{rr|r} 1 & 3 & 2 \\ 0 & -5 & 0 \end{array}\right] \xrightarrow{-\frac{1}{5}R_2}$

$\left[\begin{array}{rr|r} 1 & 3 & 2 \\ 0 & 1 & 0 \end{array}\right] \xrightarrow{R_1 - 3R_2} \left[\begin{array}{rr|r} 1 & 0 & 2 \\ 0 & 1 & 0 \end{array}\right]$

29. $\left[\begin{array}{ccc|c} 1 & 3 & 1 & 3 \\ 3 & 8 & 3 & 7 \\ 2 & -3 & 1 & -10 \end{array}\right] \xrightarrow[R_3 - 2R_1]{R_2 - 3R_1}$

$\left[\begin{array}{ccc|c} 1 & 3 & 1 & 3 \\ 0 & -1 & 0 & -2 \\ 0 & -9 & -1 & -16 \end{array}\right] \xrightarrow{-R_2}$

$\left[\begin{array}{ccc|c} 1 & 3 & 1 & 3 \\ 0 & 1 & 0 & 2 \\ 0 & -9 & -1 & -16 \end{array}\right] \xrightarrow[R_3 + 9R_2]{R_1 - 3R_2}$

$\left[\begin{array}{ccc|c} 1 & 0 & 1 & -3 \\ 0 & 1 & 0 & 2 \\ 0 & 0 & -1 & 2 \end{array}\right] \xrightarrow[-R_3]{R_1 + R_3}$

$\left[\begin{array}{ccc|c} 1 & 0 & 0 & -1 \\ 0 & 1 & 0 & 2 \\ 0 & 0 & 1 & -2 \end{array}\right]$

31. $(2, 0)$ **33.** $(-1, 2, -2)$ **35.** $(2, 1)$

37. $(4, -2)$ **39.** $(-1, 2)$ **41.** $(\frac{1}{2}, \frac{3}{2})$

43. $(\frac{1}{2}, \frac{1}{4})$ **45.** $(6, -2, 3)$ **47.** $(19, -7, -15)$

49. $(3, 0, 2)$ **51.** $(1, -2, 1)$

53. $(-20, -28, 13)$ **55.** $(4, -1, 3)$

57. $k \neq \frac{15}{4}; x = \dfrac{3k - 50}{4k - 15}, y = \dfrac{31}{4k - 15}$

59. 300 acres of corn, 200 acres of wheat

61. In 100 lb of blended coffee, use 40 lb of the \$8/lb coffee and 60 lb of the \$9/lb coffee.

63. 200 children and 800 adults

65. The bat costs \$105, and the ball costs \$5.

67. \$40,000 in a savings account, \$120,000 in mutual funds, \$80,000 in bonds

69. $x = 50, y = 17$, and $z = 33$

71. 400 bags of grade A fertilizer; 600 bags of grade B fertilizer; 300 bags of grade C fertilizer

73. 60 compact, 30 intermediate-size, and 10 full-size cars

75. 4 oz of Food I, 2 oz of Food II, 6 oz of Food III

77. 240 front orchestra seats, 560 rear orchestra seats, 200 front balcony seats

79. 7 days in London, 4 days in Paris, and 3 days in Rome

81. False

Using Technology Exercises 2.2, page 99

1. $x_1 = 3; x_2 = 1; x_3 = -1; x_4 = 2$

3. $x_1 = 5; x_2 = 4; x_3 = -3; x_4 = -4$

5. $x_1 = 1; x_2 = -1; x_3 = 2; x_4 = 0; x_5 = 3$

Exercises 2.3, page 106

1. a. One solution **b.** $(3, -1, 2)$

3. a. One solution **b.** $(2, 5)$ **5.** No solution

7. a. Infinitely many solutions
b. $(4 - t, -2, t)$; t, a parameter

9. a. No solution

11. a. Infinitely many solutions
b. $(4, -1, 3 - t, t)$; t, a parameter

13. a. Infinitely many solutions
b. $(2 - 3s, 1 + s, s, t)$; s, t, parameters

15. $(2, 1)$ **17.** No solution **19.** $(1, -1)$

21. $(2 + 2t, t)$; t, a parameter **23.** $(4 + t, -3 - t, t)$

25. $(\frac{4}{3} - \frac{2}{3}t, t)$; t, a parameter **27.** No solution

29. $(-1, \frac{17}{7}, \frac{23}{7})$

31. $(1 - \frac{1}{4}s + \frac{1}{4}t, s, t)$; s, t, parameters

33. No solution **35.** $(2, -1, 4)$

37. $x = 20 + z, y = 40 - 2z$; 25 compact cars, 30 mid-sized cars, and 5 full-sized cars; 30 compact cars, 20 mid-sized cars, and 10 full-sized cars

41. \$60,000 in stocks, \$30,000 in bonds, and \$10,000 in money-market account; \$70,000 in stocks, \$10,000 in bonds, and \$20,000 in money-market account

43. a.
$$\begin{aligned} x_1 \qquad\qquad\qquad\qquad + x_6 \qquad &= 1700 \\ x_1 - x_2 \qquad\qquad\qquad\qquad + x_7 &= 700 \\ x_2 - x_3 \qquad\qquad\qquad &= 300 \\ - x_3 + x_4 \qquad\qquad &= 400 \\ - x_4 + x_5 \qquad + x_7 &= 700 \\ x_5 + x_6 \qquad &= 1800 \end{aligned}$$
b. $(1700 - s, 1000 - s + t, 700 - s + t, 1100 - s + t, 1800 - s, s, t)$; $(900, 1000, 700, 1100, 1000, 800, 800)$; $(1000, 1100, 800, 1200, 1100, 700, 800)$
c. x_6 must have at least 300 cars/hr.

45. $k = 6$ **47.** $k = 1$; $(2 + a, 1 - a, a)$; a, a parameter

49. $k = 12$ **51.** $x = -1$ and $y = \frac{1}{3}$ **53.** False

Using Technology Exercises 2.3, page 110

1. $(1 + t, 2 + t, t)$; t, a parameter

3. $(-\frac{17}{7} + \frac{6}{7}t, 3 - t, -\frac{18}{7} + \frac{1}{7}t, t)$; t, a parameter

5. No solution

Exercises 2.4, page 118

1. 4×4; 4×3; 1×5; 4×1 **3.** 2; 3; 8

5. D; $D^T = [1 \quad 3 \quad -2 \quad 0]$ **7.** 3×2; 3×2; 3×3; 3×3

9. $\begin{bmatrix} 1 & 6 \\ 6 & -1 \\ 2 & 2 \end{bmatrix}$ **11.** $\begin{bmatrix} 1 & 1 & -4 \\ -1 & -8 & 1 \\ 6 & 3 & 1 \end{bmatrix}$ **13.** $\begin{bmatrix} 1 & 1 & 2 \\ 3 & 6 & 3 \end{bmatrix}$

15. $\begin{bmatrix} 6 & 0 & 2 & -5 \\ 9 & 3 & 0 & -3 \end{bmatrix}$ **17.** $\begin{bmatrix} -1.9 & 3.0 & -0.6 \\ 6.0 & 9.6 & 1.2 \end{bmatrix}$

19. $\begin{bmatrix} -5 & -1 & 23 \\ 25 & 14 & 17 \\ 33 & 21 & 30 \end{bmatrix}$ **21.** $\begin{bmatrix} \frac{7}{2} & 3 & -1 & \frac{10}{3} \\ -\frac{19}{6} & \frac{2}{3} & -\frac{17}{2} & \frac{23}{3} \\ \frac{29}{3} & \frac{17}{6} & -1 & -2 \end{bmatrix}$

23. $u = 3$, $x = \frac{5}{2}$, $y = 7$, and $z = 2$

25. $x = 2$, $y = 2$, $z = -\frac{7}{3}$, and $u = 15$ **27.** $\begin{bmatrix} 10 & -11 \\ 3 & -12 \end{bmatrix}$

35. $\begin{bmatrix} 3 \\ 2 \\ -1 \\ 5 \end{bmatrix}$ **37.** $\begin{bmatrix} 1 & 3 & 0 \\ -1 & 4 & 1 \\ 2 & 2 & 0 \end{bmatrix}$

39. $\begin{bmatrix} 220 & 215 & 210 & 205 \\ 220 & 210 & 200 & 195 \\ 215 & 205 & 195 & 190 \end{bmatrix}$

41.

	White	Black	Hispanic
Women	82.6	80.5	91.2
Men	78.0	73.9	84.8

	Women	Men
White	82.6	78.0
Black	80.5	73.9
Hispanic	91.2	84.8

43. a.

	6-month	1-year	$2\frac{1}{2}$-year	5-year
Current week	0.17	0.27	0.41	0.87
Previous week	0.17	0.27	0.42	0.88
1 year ago	0.22	0.34	0.52	1.15

b. $a_{12} = 0.27, a_{22} = 0.27$ **c.** $a_{13} = 0.41, a_{23} = 0.42$
d. $a_{33} = 0.52, a_{34} = 1.15$

45. a. $A =$

	Textbooks	Fiction	Nonfiction	Reference
Hardcover	5280	1680	2320	1890
Paperback	1940	2810	1490	2070

b. $B =$

	Textbooks	Fiction	Nonfiction	Reference
Hardcover	6340	2220	1790	1980
Paperback	2050	3100	1720	2710

c. $C =$

	Textbooks	Fiction	Nonfiction	Reference
Hardcover	11,620	3900	4110	3870
Paperback	3,990	5910	3210	4780

47. a. $A =$

	Large-cap	Small-cap	International	Bonds	Cash
Conservative	15	0	5	50	30
Moderately conservative	25	5	10	50	10
Moderate	35	10	15	35	5
Moderately aggressive	45	15	20	15	5
Aggressive	50	20	25	0	5

b. $a_{12} = 0$ **c.** $a_{13} = 5, a_{23} = 10, a_{33} = 15, a_{43} = 20$, and $a_{53} = 25$
d. 100

49. a. $A =$

	NY	NY Co-ops	NJ	CT
30-year fixed	3.83	3.67	3.78	3.79
15-year fixed	3.16	2.98	3.03	3.03
Adjustable	2.99	2.96	2.97	2.45

$B =$

	NY	NY Co-ops	NJ	CT
30-year fixed	3.84	3.75	3.81	3.80
15-year fixed	3.15	2.97	3.04	3.02
Adjustable	2.99	2.95	2.96	2.44

b. $a_{12} = 3.67, b_{12} = 3.75$ **c.** $a_{33} = 2.97, b_{33} = 2.96$
d. 3.7675 **e.** 2.445

51. $B = \begin{bmatrix} 350.2 & 370.8 & 391.4 \\ 422.3 & 442.9 & 453.2 \\ 638.6 & 679.8 & 721 \end{bmatrix}$ **53.** True **55.** False

Using Technology Exercises 2.4, page 125

1. $\begin{bmatrix} 15 & 38.75 & -67.5 & 33.75 \\ 51.25 & 40 & 52.5 & -38.75 \\ 21.25 & 35 & -65 & 105 \end{bmatrix}$

3. $\begin{bmatrix} -5 & 6.3 & -6.8 & 3.9 \\ 1 & 0.5 & 5.4 & -4.8 \\ 0.5 & 4.2 & -3.5 & 5.6 \end{bmatrix}$

5. $\begin{bmatrix} 16.44 & -3.65 & -3.66 & 0.63 \\ 12.77 & 10.64 & 2.58 & 0.05 \\ 5.09 & 0.28 & -10.84 & 17.64 \end{bmatrix}$

7. $\begin{bmatrix} 22.2 & -0.3 & -12 & 4.5 \\ 21.6 & 17.7 & 9 & -4.2 \\ 8.7 & 4.2 & -20.7 & 33.6 \end{bmatrix}$

Exercises 2.5, page 132

1. 2×5; not defined **3.** 1×1; 7×7

5. $n = s$; $m = t$ **7.** $\begin{bmatrix} -1 \\ 3 \end{bmatrix}$

9. $\begin{bmatrix} 13 \\ -10 \end{bmatrix}$ **11.** $\begin{bmatrix} 4 & -2 \\ 9 & 13 \end{bmatrix}$

13. $\begin{bmatrix} 2 & 9 \\ 5 & 16 \end{bmatrix}$ **15.** $\begin{bmatrix} 0.57 & 1.93 \\ 0.64 & 1.76 \end{bmatrix}$

17. $\begin{bmatrix} 6 & -3 & 0 \\ -2 & 1 & -8 \\ 4 & -4 & 9 \end{bmatrix}$ **19.** $\begin{bmatrix} 5 & 1 & -6 \\ 1 & 7 & -4 \end{bmatrix}$

21. $\begin{bmatrix} -4 & -20 & 4 \\ 4 & 12 & 0 \\ 12 & 32 & 20 \end{bmatrix}$ **23.** $\begin{bmatrix} 4 & -3 & 2 \\ 7 & 1 & -5 \end{bmatrix}$

27. $AB = \begin{bmatrix} 10 & 7 \\ 22 & 15 \end{bmatrix}$; $BA = \begin{bmatrix} 5 & 8 \\ 13 & 20 \end{bmatrix}$ **31.** $A = \begin{bmatrix} -2 & -1 \\ 5 & 2 \end{bmatrix}$

33. $B = \begin{bmatrix} \frac{1}{3} & -\frac{1}{6} \\ \frac{1}{3} & \frac{1}{3} \end{bmatrix}$ **35.** b. No **37.** **a.** $A^T = \begin{bmatrix} 2 & 5 \\ 4 & -6 \end{bmatrix}$

39. $AX = B$, where $A = \begin{bmatrix} 2 & -3 \\ 3 & -4 \end{bmatrix}$, $X = \begin{bmatrix} x \\ y \end{bmatrix}$, and $B = \begin{bmatrix} 7 \\ 8 \end{bmatrix}$

41. $AX = B$, where $A = \begin{bmatrix} 2 & -3 & 4 \\ 0 & 2 & -3 \\ 1 & -1 & 2 \end{bmatrix}$, $X = \begin{bmatrix} x \\ y \\ z \end{bmatrix}$,

and $B = \begin{bmatrix} 6 \\ 7 \\ 4 \end{bmatrix}$

43. $AX = B$, where $A = \begin{bmatrix} -1 & 1 & 1 \\ 2 & -1 & -1 \\ -3 & 2 & 4 \end{bmatrix}$, $X = \begin{bmatrix} x_1 \\ x_2 \\ x_3 \end{bmatrix}$,

and $B = \begin{bmatrix} 0 \\ 2 \\ 4 \end{bmatrix}$

45. a. $AB = \begin{bmatrix} 30{,}100 \\ 32{,}700 \end{bmatrix}$

b. The first entry shows that the value of Olivia's total stock holdings are \$30,100; the second shows that the value of Isabella's stock holdings are \$32,700.

47. a. $AB = \begin{bmatrix} 1070 & 1640 & 2550 \end{bmatrix}$ **b.** $ABC = \begin{bmatrix} 42{,}810{,}000 \end{bmatrix}$

49. $C = [374{,}051.60 \quad 378{,}524.00 \quad 376{,}775.80]$

51. a.

$$A = \begin{array}{c} \\ \\ \text{Ava} \\ \text{Ella} \end{array} \begin{array}{c} \text{N} \\ \text{kroner} \\ 82 \\ 64 \end{array} \begin{array}{c} \text{S} \\ \text{kronor} \\ 68 \\ 74 \end{array} \begin{array}{c} \text{D} \\ \text{kroner} \\ 62 \\ 44 \end{array} \begin{array}{c} \text{R} \\ \text{rubles} \\ 1200 \\ 1600 \end{array}$$

b. $B = \begin{array}{c} \text{N} \\ \text{S} \\ \text{D} \\ \text{R} \end{array} \begin{bmatrix} 0.1751 \\ 0.1560 \\ 0.1747 \\ 0.0325 \end{bmatrix}$ **c.** Ava: \$74.80; Ella: \$82.44

53. a. [90 125 210 55]; the entries give the respective total number of Model I, II, III, and IV houses built in the three states.

b. $\begin{bmatrix} 300 \\ 120 \\ 60 \end{bmatrix}$; the entries give the respective total number of Model I, II, III, and IV houses built in all three states.

55. $B = \begin{bmatrix} 4 \\ 6 \\ 8 \end{bmatrix}$; $AB = \begin{bmatrix} 1960 \\ 3180 \\ 2510 \\ 3300 \end{bmatrix}$; \$10,950

57. $BA = \begin{array}{ccc} \text{Dem.} & \text{Rep.} & \text{Ind.} \\ [41{,}000 & 35{,}000 & 14{,}000] \end{array}$

59. $AB = \begin{bmatrix} 1575 & 1590 & 1560 & 975 \\ 410 & 405 & 415 & 270 \\ 215 & 205 & 225 & 155 \end{bmatrix}$

61. [136.80]; it represents Cindy's long-distance bill for phone calls to London, Tokyo, and Hong Kong.

63. a. $\begin{bmatrix} 8800 \\ 3380 \\ 1020 \end{bmatrix}$ **b.** $\begin{bmatrix} 8800 \\ 3380 \\ 1020 \end{bmatrix}$ **c.** $\begin{bmatrix} 17{,}600 \\ 6{,}760 \\ 2{,}040 \end{bmatrix}$

65. False **67.** True

Using Technology Exercises 2.5, page 140

1. $\begin{bmatrix} 18.66 & 15.2 & -12 \\ 24.48 & 41.88 & 89.82 \\ 15.39 & 7.16 & -1.25 \end{bmatrix}$

3. $\begin{bmatrix} 20.09 & 20.61 & -1.3 \\ 44.42 & 71.6 & 64.89 \\ 20.97 & 7.17 & -60.65 \end{bmatrix}$

5. $\begin{bmatrix} 32.89 & 13.63 & -57.17 \\ -12.85 & -8.37 & 256.92 \\ 13.48 & 14.29 & 181.64 \end{bmatrix}$

7. $\begin{bmatrix} 128.59 & 123.08 & -32.50 \\ 246.73 & 403.12 & 481.52 \\ 125.06 & 47.01 & -264.81 \end{bmatrix}$

9. $\begin{bmatrix} 87 & 68 & 110 & 82 \\ 119 & 176 & 221 & 143 \\ 51 & 128 & 142 & 94 \\ 28 & 174 & 174 & 112 \end{bmatrix}$

$\begin{bmatrix} 113 & 117 & 72 & 101 & 90 \\ 72 & 85 & 36 & 72 & 76 \\ 81 & 69 & 76 & 87 & 30 \\ 133 & 157 & 56 & 121 & 146 \\ 154 & 157 & 94 & 127 & 122 \end{bmatrix}$

11. $\begin{bmatrix} 170 & 18.1 & 133.1 & -106.3 & 341.3 \\ 349 & 226.5 & 324.1 & 164 & 506.4 \\ 245.2 & 157.7 & 231.5 & 125.5 & 312.9 \\ 310 & 245.2 & 291 & 274.3 & 354.2 \end{bmatrix}$

Exercises 2.6, page 148

5. $\begin{bmatrix} 3 & -5 \\ -1 & 2 \end{bmatrix}$ **7.** Does not exist

9. $\begin{bmatrix} 2 & -11 & -3 \\ 1 & -6 & -2 \\ 0 & -1 & 0 \end{bmatrix}$ **11.** Does not exist

13. $\begin{bmatrix} -\frac{13}{10} & \frac{7}{5} & \frac{1}{2} \\ \frac{2}{5} & -\frac{1}{5} & 0 \\ -\frac{7}{10} & \frac{3}{5} & \frac{1}{2} \end{bmatrix}$

15. $\begin{bmatrix} 3 & 4 & -6 & 1 \\ -2 & -3 & 5 & -1 \\ -4 & -4 & 7 & -1 \\ -4 & -5 & 8 & -1 \end{bmatrix}$

17. a. $AX = B$, where $A = \begin{bmatrix} 2 & 5 \\ 1 & 3 \end{bmatrix}$; $X = \begin{bmatrix} x \\ y \end{bmatrix}$; $B = \begin{bmatrix} 3 \\ 2 \end{bmatrix}$

b. $x = -1$; $y = 1$

19. a. $AX = B$, where $A = \begin{bmatrix} 2 & -3 & -4 \\ 0 & 0 & -1 \\ 1 & -2 & 1 \end{bmatrix}$; $X = \begin{bmatrix} x \\ y \\ z \end{bmatrix}$; $B = \begin{bmatrix} 4 \\ 3 \\ -8 \end{bmatrix}$

b. $x = -1$; $y = 2$; $z = -3$

21. a. $AX = B$, where $A = \begin{bmatrix} 1 & 4 & -1 \\ 2 & 3 & -2 \\ -1 & 2 & 3 \end{bmatrix}$; $X = \begin{bmatrix} x \\ y \\ z \end{bmatrix}$; $B = \begin{bmatrix} 3 \\ 1 \\ 7 \end{bmatrix}$

b. $x = 1$; $y = 1$; $z = 2$

23. a. $AX = B$, where $A = \begin{bmatrix} 1 & 1 & -1 & 1 \\ 2 & 1 & 1 & 0 \\ 2 & 1 & 0 & 1 \\ 2 & -1 & -1 & 3 \end{bmatrix}$; $X = \begin{bmatrix} x_1 \\ x_2 \\ x_3 \\ x_4 \end{bmatrix}$; $B = \begin{bmatrix} 6 \\ 4 \\ 7 \\ 9 \end{bmatrix}$

b. $x_1 = 1$; $x_2 = 2$; $x_3 = 0$; $x_4 = 3$

25. b. (i) $x = 4.8$ and $y = 4.6$
(ii) $x = 0.4$ and $y = 1.8$

27. b. (i) $x = -1$; $y = 3$; $z = 2$
(ii) $x = 1$; $y = 8$; $z = -12$

29. b. (i) $x = -\frac{2}{17}$; $y = -\frac{10}{17}$; $z = -\frac{60}{17}$
(ii) $x = 1$; $y = 0$; $z = -5$

31. b. (i) $x_1 = 1$; $x_2 = -4$; $x_3 = 5$; $x_4 = -1$
(ii) $x_1 = 12$; $x_2 = -24$; $x_3 = 21$; $x_4 = -7$

33. a. $A^{-1} = \begin{bmatrix} -\frac{5}{2} & -\frac{3}{2} \\ 2 & 1 \end{bmatrix}$

35. a. $ABC = \begin{bmatrix} 4 & 10 \\ 2 & 3 \end{bmatrix}$; $A^{-1} = \begin{bmatrix} 3 & -5 \\ 1 & -2 \end{bmatrix}$;

$B^{-1} = \begin{bmatrix} 1 & -3 \\ -1 & 4 \end{bmatrix}$; $C^{-1} = \begin{bmatrix} \frac{1}{8} & -\frac{3}{8} \\ \frac{1}{4} & \frac{1}{4} \end{bmatrix}$

37. $\begin{bmatrix} \frac{5}{7} & \frac{3}{7} \\ -\frac{3}{7} & \frac{8}{7} \end{bmatrix}$

39. a. 3214; 3929 b. 4286; 3571 c. 3929; 5357

41. a. 400 acres of soybeans; 300 acres of corn; 300 acres of wheat
b. 500 acres of soybeans; 400 acres of corn; 300 acres of wheat

43. a. \$80,000 in high-risk stocks; \$20,000 in medium-risk stocks; \$100,000 in low-risk stocks
b. \$88,000 in high-risk stocks; \$22,000 in medium-risk stocks; \$110,000 in low-risk stocks
c. \$56,000 in high-risk stocks; \$64,000 in medium-risk stocks; \$120,000 in low-risk stocks

45. All values of k except $k = \frac{3}{2}$; $\frac{1}{3 - 2k}\begin{bmatrix} 3 & -2 \\ -k & 1 \end{bmatrix}$

47. A^{-1} exists provided that $ad \neq 0$; every entry along the main diagonal is not equal to zero.

49. True 51. True

Using Technology Exercises 2.6, page 155

1. $\begin{bmatrix} 0.36 & 0.04 & -0.36 \\ 0.06 & 0.05 & 0.20 \\ -0.19 & 0.10 & 0.09 \end{bmatrix}$

3. $\begin{bmatrix} 0.01 & -0.09 & 0.31 & -0.11 \\ -0.25 & 0.58 & -0.15 & -0.02 \\ 0.86 & -0.42 & 0.07 & -0.37 \\ -0.27 & 0.01 & -0.05 & 0.31 \end{bmatrix}$

5. $\begin{bmatrix} 0.30 & 0.85 & -0.10 & -0.77 & -0.11 \\ -0.21 & 0.10 & 0.01 & -0.26 & 0.21 \\ 0.03 & -0.16 & 0.12 & -0.01 & 0.03 \\ -0.14 & -0.46 & 0.13 & 0.71 & -0.05 \\ 0.10 & -0.05 & -0.10 & -0.03 & 0.11 \end{bmatrix}$

7. $x = 1.2$; $y = 3.6$; $z = 2.7$

9. $x_1 = 2.50$; $x_2 = -0.88$; $x_3 = 0.70$; $x_4 = 0.51$

Exercises 2.7, page 160

1. a. \$10 million b. \$160 million
c. Agricultural; manufacturing and transportation

3. $x = 23.75$ and $y = 21.25$

5. $x = 42.85$ and $y = 57.14$

9. a. \$318.2 million worth of agricultural products and \$336.4 million worth of manufactured goods
b. \$198.2 million worth of agricultural products and \$196.4 million worth of manufactured goods

11. a. \$443.72 million, \$381.28 million, and \$281.28 million worth of agricultural products, manufactured goods, and transportation, respectively
b. \$243.72 million, \$281.28 million, and \$221.28 million worth of agricultural products, manufactured goods, and transportation, respectively

13. \$45 million and \$75 million

15. \$34.4 million, \$33 million, and \$21.6 million

Using Technology Exercises 2.7, page 164

1. The final outputs of the first, second, third, and fourth industries are 602.62, 502.30, 572.57, and 523.46 million dollars, respectively.

3. The final outputs of the first, second, third, and fourth industries are 143.06, 132.98, 188.59, and 125.53 million dollars, respectively.

Chapter 2 Concept Review Questions, page 165

1. a. one; many; no b. one; many; no 2. equations

3. $R_i \leftrightarrow R_j$; cR_i; $R_i + aR_j$; solution

4. a. unique b. no; infinitely many; unique

5. size; entries 6. size; corresponding

7. $m \times n$; $n \times m$; a_{ji} 8. cA; c

9. a. columns; rows b. $m \times p$

10. a. $A(BC)$; $AB + AC$ b. $n \times r$

11. $A^{-1}A$; AA^{-1}; singular 12. $A^{-1}B$

Chapter 2 Review Exercises, page 166

1. $\begin{bmatrix} 2 & 2 \\ -1 & 4 \\ 3 & 3 \end{bmatrix}$ 2. $\begin{bmatrix} -2 & 0 \\ -2 & 6 \end{bmatrix}$ 3. $[-6 \quad -2]$ 4. $\begin{bmatrix} 17 \\ 13 \end{bmatrix}$

5. $x = 2$; $y = 3$; $z = 1$; $w = 3$ 6. $x = 2$; $y = -2$

7. $a = 3$; $b = 4$; $c = -2$; $d = 2$; $e = -3$

8. $x = -1$; $y = -2$; $z = 1$

9. $\begin{bmatrix} 8 & 9 & 11 \\ -10 & -1 & 3 \\ 11 & 12 & 10 \end{bmatrix}$ 10. $\begin{bmatrix} -1 & 7 & -3 \\ -2 & 5 & 11 \\ 10 & -8 & 2 \end{bmatrix}$

11. $\begin{bmatrix} 6 & 18 & 6 \\ -12 & 6 & 18 \\ 24 & 0 & 12 \end{bmatrix}$ 12. $\begin{bmatrix} -10 & 10 & -18 \\ 4 & 14 & 26 \\ 16 & -32 & -4 \end{bmatrix}$

13. $\begin{bmatrix} -11 & -16 & -15 \\ -4 & -2 & -10 \\ -6 & 14 & 2 \end{bmatrix}$ 14. $\begin{bmatrix} 5 & 20 & 19 \\ -2 & 20 & 8 \\ 26 & 10 & 30 \end{bmatrix}$

15. $\begin{bmatrix} -3 & 17 & 8 \\ -2 & 56 & 27 \\ 74 & 78 & 116 \end{bmatrix}$ 16. $\begin{bmatrix} \frac{3}{2} & -2 & -5 \\ \frac{11}{2} & -1 & 11 \\ \frac{7}{2} & -3 & 0 \end{bmatrix}$

17. $x = 1$; $y = -1$ 18. $x = -1$; $y = 3$

19. $x = 1$; $y = 2$; $z = 3$

20. $(2, 2t - 5, t)$; t, a parameter 21. No solution

22. $x = 1$; $y = -1$; $z = 2$; $w = 2$

23. $x = 1$; $y = 0$; $z = 1$ 24. $x = 2$; $y = -1$; $z = 3$

25. $\begin{bmatrix} \frac{2}{5} & -\frac{1}{5} \\ -\frac{1}{5} & \frac{3}{5} \end{bmatrix}$ **26.** $\begin{bmatrix} \frac{3}{4} & -\frac{1}{2} \\ -\frac{1}{8} & \frac{1}{4} \end{bmatrix}$

27. $\begin{bmatrix} -1 & 2 \\ 1 & -\frac{3}{2} \end{bmatrix}$ **28.** $\begin{bmatrix} \frac{1}{4} & \frac{1}{2} \\ \frac{1}{8} & -\frac{1}{4} \end{bmatrix}$

29. $\begin{bmatrix} \frac{5}{4} & \frac{1}{4} & -\frac{7}{4} \\ -\frac{1}{4} & -\frac{1}{4} & \frac{3}{4} \\ -\frac{3}{4} & \frac{1}{4} & \frac{5}{4} \end{bmatrix}$ **30.** $\begin{bmatrix} -\frac{1}{4} & \frac{1}{2} & -\frac{1}{4} \\ \frac{7}{8} & -\frac{3}{4} & -\frac{5}{8} \\ -\frac{1}{8} & \frac{1}{4} & \frac{3}{8} \end{bmatrix}$

31. $\begin{bmatrix} -\frac{1}{5} & \frac{2}{5} & 0 \\ \frac{2}{3} & -\frac{1}{3} & \frac{1}{3} \\ -\frac{1}{30} & \frac{1}{15} & -\frac{1}{6} \end{bmatrix}$ **32.** $\begin{bmatrix} 0 & -\frac{1}{5} & \frac{2}{5} \\ -2 & 1 & 1 \\ -1 & \frac{1}{5} & \frac{3}{5} \end{bmatrix}$

33. $\begin{bmatrix} \frac{3}{2} & 1 \\ -\frac{7}{2} & -1 \end{bmatrix}$ **34.** $\begin{bmatrix} \frac{11}{24} & -\frac{7}{8} \\ -\frac{1}{12} & \frac{1}{4} \end{bmatrix}$

35. $\begin{bmatrix} \frac{2}{5} & -\frac{3}{5} \\ \frac{1}{5} & \frac{1}{5} \end{bmatrix}$ **36.** $\begin{bmatrix} \frac{4}{7} & -\frac{3}{7} \\ -\frac{3}{7} & \frac{4}{7} \end{bmatrix}$

37. $A^{-1} = \begin{bmatrix} \frac{2}{7} & \frac{3}{7} \\ \frac{1}{7} & -\frac{2}{7} \end{bmatrix}$; $x = -1$; $y = -2$

38. $A^{-1} = \begin{bmatrix} \frac{2}{5} & \frac{3}{10} \\ -\frac{1}{5} & \frac{1}{10} \end{bmatrix}$; $x = 2$; $y = 1$

39. $A^{-1} = \begin{bmatrix} 1 & -\frac{2}{5} & \frac{4}{5} \\ -1 & 1 & -1 \\ -\frac{1}{2} & \frac{3}{5} & -\frac{7}{10} \end{bmatrix}$; $x = 1$; $y = 2$; $z = 4$

40. $A^{-1} = \begin{bmatrix} 0 & \frac{1}{7} & \frac{2}{7} \\ -1 & -\frac{4}{7} & \frac{6}{7} \\ -\frac{1}{2} & -\frac{1}{2} & \frac{1}{2} \end{bmatrix}$; $x = 3$; $y = -1$; $z = 2$

41. \$11,150, \$10,380, and \$13,660

42. \$2,300,000; \$2,450,000; an increase of \$150,000

43. a. $A = \begin{bmatrix} 800 & 1200 & 250 & 1500 \\ 600 & 1400 & 300 & 1200 \end{bmatrix}$ **b.** $B = \begin{bmatrix} 12.57 \\ 28.21 \\ 214.92 \\ 36.34 \end{bmatrix}$

b. Josh: \$152,148; Hannah: \$155,120

44. a. A =

	IBM	Google	Boeing	GM
Jennifer	800	500	1200	1500
Max	500	600	2000	800

;

B =

	IBM	Google	Boeing	GM
Jennifer	900	600	1000	1200
Max	700	500	2100	900

b. C =

	IBM	Google	Boeing	GM
Jennifer	100	100	−200	−300
Max	200	−100	100	100

45. 30 of each type

46. Houston: 100,000 gal; Tulsa: 600,000 gal

47. a. \$30 million **b.** \$75 million **c.** Agricultural; manufacturing

48. \$31.18 million, \$31.79 million, and \$36.31 million worth of the first, second, and third industry, respectively

49. a. \$145.86 million worth of agricultural products; \$111.28 million worth of manufactured goods
b. \$45.86 million worth of agricultural products and \$31.28 million worth of manufactured goods

Chapter 2 Before Moving On, page 168

1. $(\frac{2}{3}, -\frac{2}{3}, \frac{5}{3})$

2. a. $(2, -3, 1)$ **b.** No solution **c.** $(2, 1 - 3t, t)$, t, a parameter
d. $(0, 0, 0, 0)$ **e.** $(2 + t, 3 - 2t, t)$, t, a parameter

3. a. $(-1, 2)$ **b.** $(\frac{4}{7}, -\frac{5}{7} + 2t, t)$, t, a parameter

4. a. $\begin{bmatrix} 3 & 1 & 4 \\ 5 & -2 & 6 \end{bmatrix}$ **b.** $\begin{bmatrix} 14 & 3 & 7 \\ 14 & 5 & 1 \end{bmatrix}$ **c.** $\begin{bmatrix} 0 & 5 & 3 \\ 4 & -1 & -11 \end{bmatrix}$

5. $\begin{bmatrix} 3 & -2 & -5 \\ -3 & 2 & 6 \\ -1 & 1 & 2 \end{bmatrix}$ **6.** $(1, -1, 2)$

CHAPTER 3

Exercises 3.1, page 178

1.

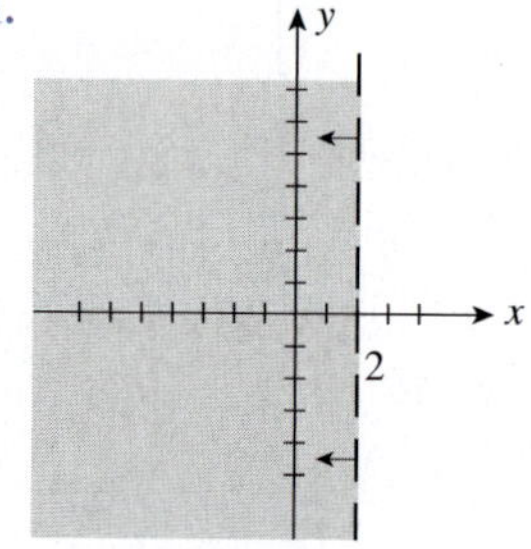

3.

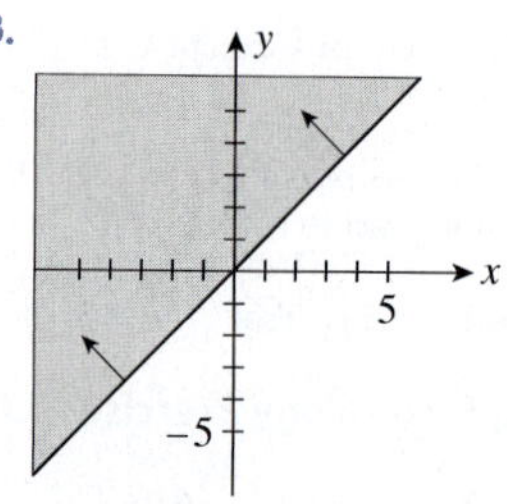

5.

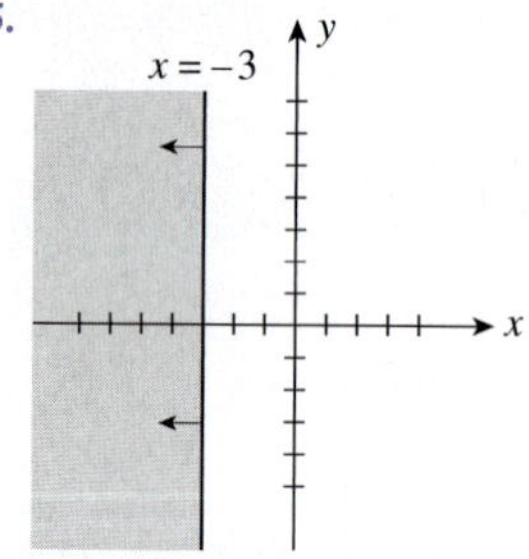

7.

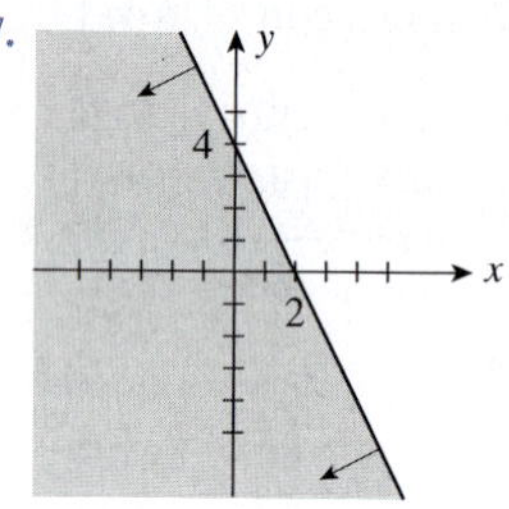

9.

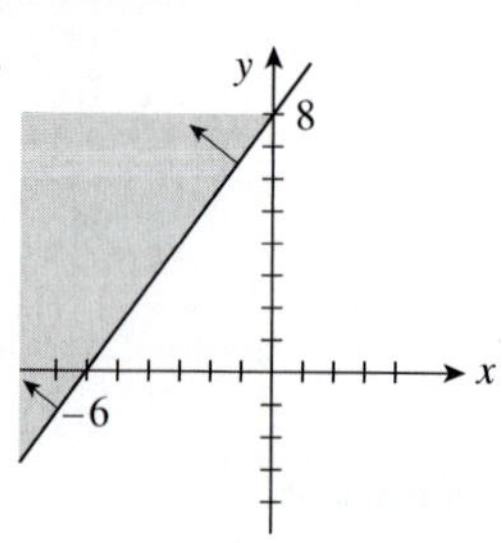

11. $x \geq 1$; $x \leq 5$; $y \geq 2$; $y \leq 4$

13. $2x - y \geq 2$; $5x + 7y \geq 35$; $x \leq 4$

15. $x - y \geq -10$; $7x + 4y \leq 140$; $x + 3y \geq 30$

17. $x + y \geq 7$; $x \geq 2$; $y \geq 3$; $y \leq 7$

19. $(3, 3)$ lies in S. **21.** $(10, 10)$ lies in S.

23. 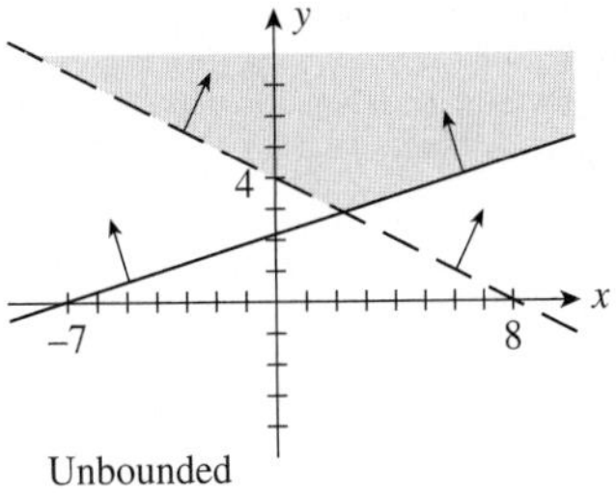

Unbounded

25. 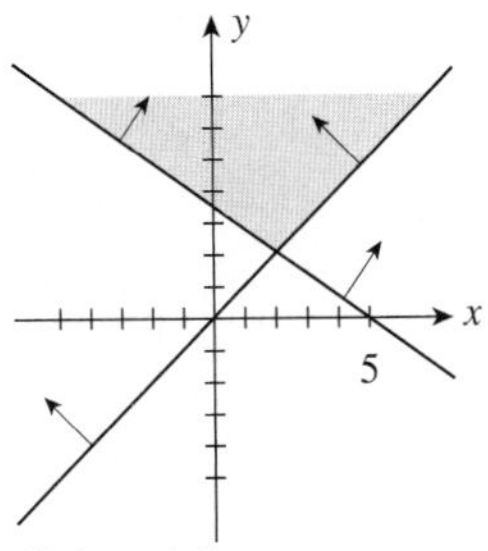

Unbounded

27. 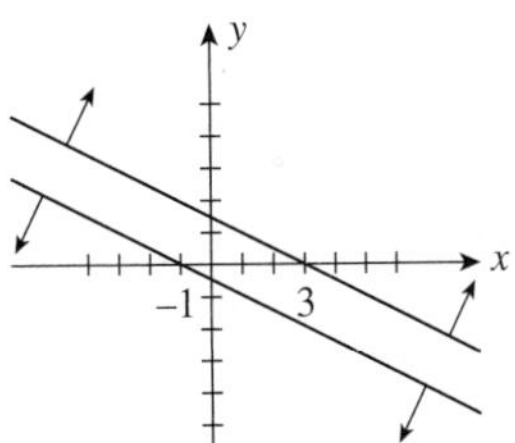

Bounded No solution

29. 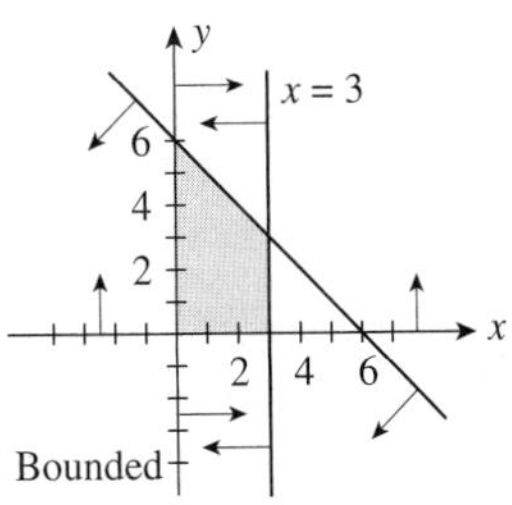

Bounded

31. 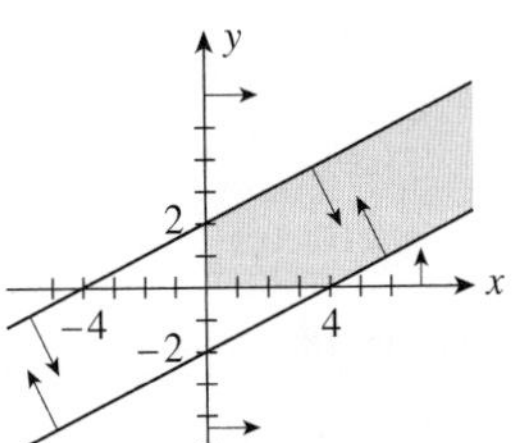

Unbounded

33. 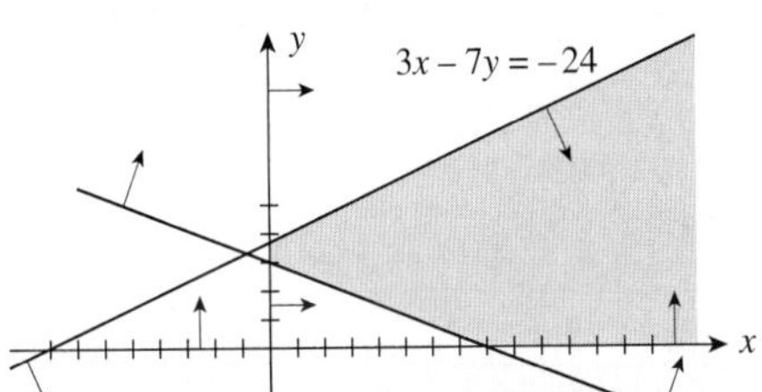

Unbounded

35. 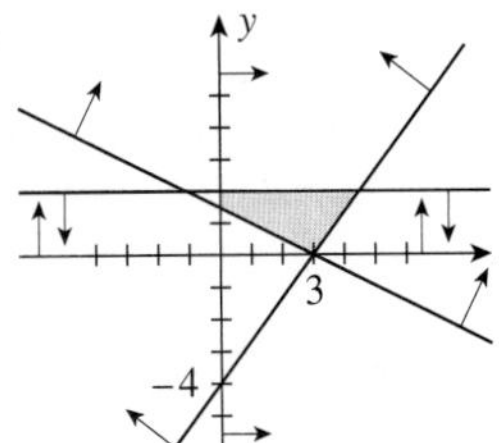

Bounded

37. 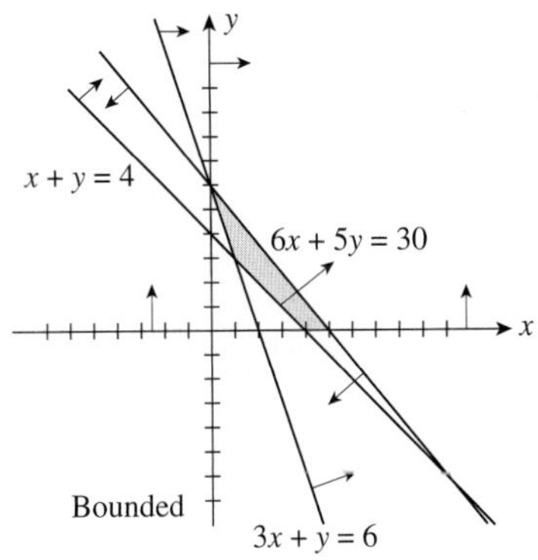

Bounded

39. 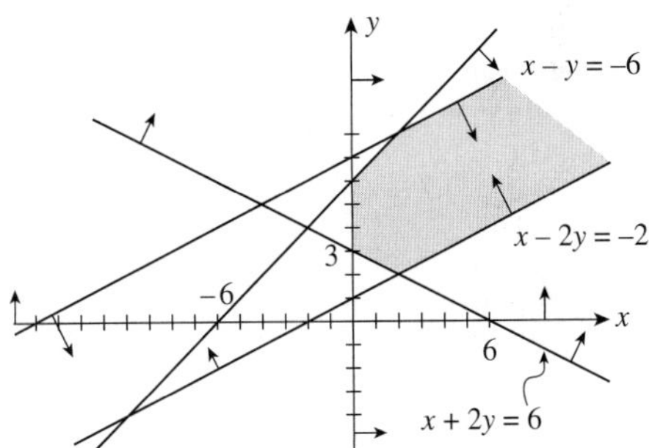

Unbounded

41. **a.** $x + y \leq 500$
$x \geq 200$
$y \geq 100$

b. 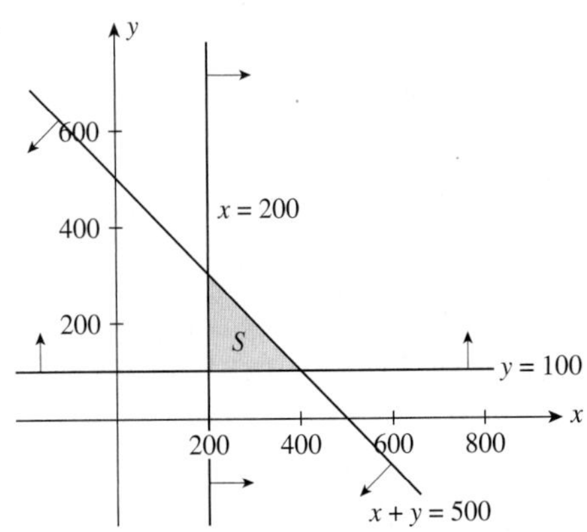

c. Yes

43. a. $$\begin{aligned} x + y &\le 250{,}000 \\ x &\ge 50{,}000 \\ y &\ge 50{,}000 \\ 5x - 6y &\le 0 \end{aligned}$$

b.

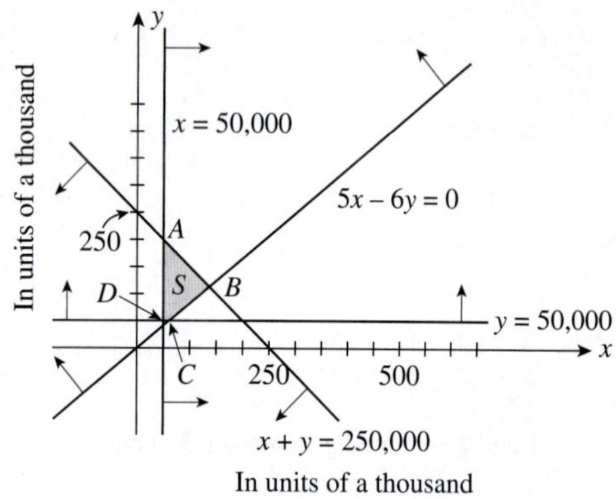

c. No

45. False 47. True

Exercises 3.2, page 185

1. Maximize $P = 3x + 4y$
subject to $$\begin{aligned} 6x + 9y &\le 300 \\ 5x + 4y &\le 180 \\ x \ge 0, y &\ge 0 \end{aligned}$$

3. Maximize $P = 2x + 1.5y$
subject to $$\begin{aligned} 3x + 4y &\le 1000 \\ 6x + 3y &\le 1200 \\ x \ge 0, y &\ge 0 \end{aligned}$$

5. Maximize $P = 45x + 20y$
subject to $$\begin{aligned} 40x + 16y &\le 3200 \\ 3x + 4y &\le 520 \\ x \ge 0, y &\ge 0 \end{aligned}$$

7. Maximize $P = 0.1x + 0.12y$
subject to $$\begin{aligned} x + y &\le 20 \\ x - 4y &\ge 0 \\ x \ge 0, y &\ge 0 \end{aligned}$$

9. Maximize $P = 0.12x + 0.20y$
subject to $$\begin{aligned} x + y &\le 60{,}000 \\ 3x - 2y &\ge 0 \\ x - 4y &\le 0 \\ x \ge 0, y &\ge 0 \end{aligned}$$

11. Minimize $C = 14{,}000x + 16{,}000y$
subject to $$\begin{aligned} 50x + 75y &\ge 650 \\ 3000x + 1000y &\ge 18{,}000 \\ x \ge 0, y &\ge 0 \end{aligned}$$

13. Maximize $P = 50x + 40y$
subject to $$\begin{aligned} \tfrac{1}{200}x + \tfrac{1}{200}y &\le 1 \\ \tfrac{1}{100}x + \tfrac{1}{300}y &\le 1 \\ x \ge 0, y &\ge 0 \end{aligned}$$

15. Minimize $C = 300x + 500y$
subject to $$\begin{aligned} x + y &\ge 10 \\ x &\le 5 \\ y &\le 10 \\ y &\ge 6 \\ x &\ge 0 \end{aligned}$$

17. Minimize $C = 2x + 5y$
subject to $$\begin{aligned} 30x + 25y &\ge 400 \\ x + 0.5y &\ge 10 \\ 2x + 5y &\ge 40 \\ x \ge 0, y &\ge 0 \end{aligned}$$

19. Minimize $C = 64{,}000 - 2x - 6y$
subject to $$\begin{aligned} x + y &\le 6000 \\ x + y &\ge 2000 \\ x &\le 3000 \\ y &\le 4000 \\ x \ge 0, y &\ge 0 \end{aligned}$$

21. Maximize $P = 18x + 12y + 15z$
subject to $$\begin{aligned} 2x + y + 2z &\le 900 \\ 3x + y + 2z &\le 1080 \\ 2x + 2y + z &\le 840 \\ x \ge 0, y \ge 0, z &\ge 0 \end{aligned}$$

23. Maximize $P = 26x + 28y + 24z$
subject to $$\begin{aligned} \tfrac{5}{4}x + \tfrac{3}{2}y + \tfrac{3}{2}z &\le 310 \\ x + y + \tfrac{3}{4}z &\le 205 \\ x + y + \tfrac{1}{2}z &\le 190 \\ x \ge 0, y \ge 0, z &\ge 0 \end{aligned}$$

25. Maximize $P = 0.06x + 0.1y + 0.15z$
subject to $$\begin{aligned} x + y + z &\le 250{,}000 \\ -x - y + 3z &\le 0 \\ -x + y - z &\le 0 \\ x \ge 0, y \ge 0, z &\ge 0 \end{aligned}$$

27. Minimize $C = 16x_1 + 20x_2 + 22x_3 + 18x_4 + 16x_5 + 14x_6$
subject to $$\begin{aligned} x_1 + x_2 + x_3 &\le 800 \\ x_4 + x_5 + x_6 &\le 600 \\ x_1 + x_4 &\ge 500 \\ x_2 + x_5 &\ge 400 \\ x_3 + x_6 &\ge 400 \\ x_1, x_2, \ldots, x_6 &\ge 0 \end{aligned}$$

29. Maximize $P = x + 0.8y + 0.9z$
subject to $$\begin{aligned} 8x + 4z &\le 16{,}000 \\ 8x + 12y + 8z &\le 24{,}000 \\ 4y + 4z &\le 5000 \\ z &\le 800 \\ x \ge 0, y \ge 0, z &\ge 0 \end{aligned}$$

31. False

Exercises 3.3, page 196

1. Max: 35; min: 5 3. No max. value; min: 18

5. Max: 44; min: 15 7. $x = 3$; $y = 3$; $P = 15$

9. Any point (x, y) lying on the line segment joining $\left(\frac{5}{2}, 0\right)$ and $(1, 3)$; $P = 5$

11. $x = 0$; $y = 8$; $P = 64$

13. $x = 0$; $y = 4$; $P = 12$

15. $x = 4$; $y = 0$; $C = 8$

17. Any point (x, y) lying on the line segment joining $(20, 10)$ and $(40, 0)$; $C = 120$

19. $x = 14$; $y = 3$; $C = 58$

21. $x = 3$; $y = 3$; $C = 75$

23. $x = 15$; $y = 17.5$; $P = 115$

25. $x = 10$; $y = 38$; $P = 134$

27. Min: $x = 9$; $y = 9$; $P = 117$
Max: $x = 15$; $y = 3$; $P = 135$

29. 20 Product A, 20 Product B; \$140

31. 120 model A, 160 model B; \$480

33. 40 tables; 100 chairs; \$3800

35. \$16 million in homeowner loans, \$4 million in auto loans; \$2.08 million

37. \$24,000 in medium-risk stocks; \$36,000 in high-risk stocks; \$10,080

39. Saddle Mine: 4 days; Horseshoe Mine: 6 days; \$152,000

41. 50 fully assembled units, 150 kits; \$8500

43. Reservoir: 4 million gallons; pipeline: 6 million gallons; \$4200

45. Infinitely many solutions; 10 oz of Food A and 4 oz of Food B or 20 oz of Food A and 0 oz of Food B, etc., with a minimum value of 40 mg of cholesterol

47. 2000 televisions from Location I to City A and 4000 televisions from Location I to City B; 1000 televisions from Location II to City A and 0 televisions from Location II to City B; \$36,000

49. 80 from I to A, 20 from I to B, 0 from II to A, 50 from II to B; \$12,700

51. \$22,500 in growth stocks and \$7500 in speculative stocks; maximum return; \$5250

53. 750 urban, 750 suburban; \$10,950

55. False 57. **a.** True **b.** True

61. **a.**

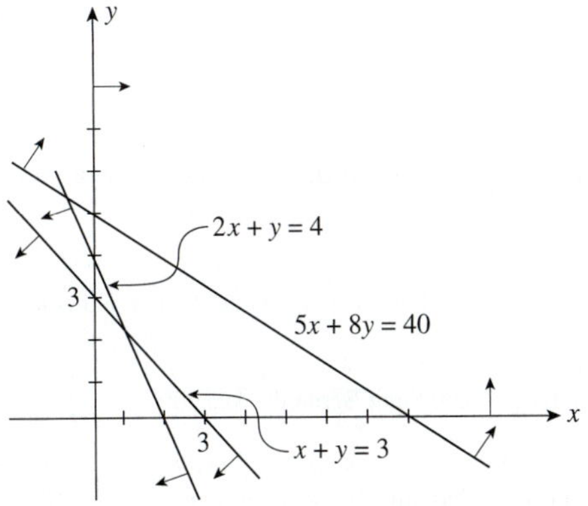

b. No solution

Exercises 3.4, page 212

1. **c.** \$216

3. **a.** Between 750 and 1425
b. It cannot be decreased by more than 60 units.

5. **a.** $x = 3$; $y = 2$; $P = 17$ **b.** $\frac{8}{3} \le c \le 8$
c. $8 \le b \le 24$ **d.** $\frac{5}{4}$
e. Both constraints are binding.

7. **a.** $x = 4$; $y = 0$; $C = 8$ **b.** $0 \le c \le \frac{5}{2}$
c. $b \ge 3$ **d.** 2
e. Constraint 1 is binding; constraint 2 is nonbinding.

9. **a.** $x = 3$; $y = 5$; $P = 27$ **b.** $2 \le c \le 5$
c. $21 \le b \le 33$ **d.** $\frac{2}{3}$
e. The first two constraints are binding; the third is nonbinding.

11. **a.** 20 units of each product **b.** $\frac{8}{3} \le c \le 5$
c. $216 \le b \le 405$ **d.** $\frac{8}{21}$

13. **a.** Operate Saddle Mine for 4 days, Horseshoe Mine for 6 days
b. $10{,}666\frac{2}{3} \le c \le 48{,}000$ **c.** $300 \le b \le 1350$
d. \$194.29

15. **a.** Produce 60 of each; maximum profit of \$1320
b. $8\frac{1}{3} \le c \le 15$ **c.** $1100 \le b \le 1633\frac{1}{3}$
d. \$0.55
e. Constraints 1 and 2 are binding; constraint 3 is not.

17. **a.** 120 model A and 160 model B grates; maximum profit of \$480
b. $1.125 \le c \le 3$ **c.** $600 \le b \le 1100$
d. \$0.20
e. Constraints 1 and 2 are binding; constraint 3 is nonbinding.

Chapter 3 Concept Review Questions, page 215

1. **a.** half-plane; line **b.** $ax + by \le c$; $ax + by = c$

2. **a.** points; each **b.** bounded; enclosed

3. objective function; maximized; minimized; linear; inequalities

4. **a.** corner point **b.** line

5. parameters; optimal

6. resource; amount; value; improved; increased

Chapter 3 Review Exercises, page 216

1. Max: 18—any point (x, y) lying on the line segment joining $(0, 6)$ and $(3, 4)$; min: 0

2. Max: 27—any point (x, y) lying on the line segment joining $(3, 5)$ and $(6, 1)$; min: 7

3. $x = 0, y = 4$; $P = 20$

4. $x = 0, y = 12$; $P = 36$

5. $x = 3, y = 4$; $C = 26$

6. $x = 1.25, y = 1.5$; $C = 9.75$

7. $x = 3, y = 10$; $P = 29$

8. $x = 8, y = 0$; $P = 48$

9. $x = 20, y = 0$; $C = 40$

10. $x = 2, y = 4$; $C = 14$

11. $x = 2, y = 6$; $C = 14$

12. Max: $x = \frac{100}{11}, y = \frac{210}{11}$; $Q = \frac{1140}{11}$; min: $x = 0, y = 10$; $Q = 40$

13. Max: $x = 22, y = 0$; $Q = 22$; min: $x = 3, y = \frac{5}{2}$; $Q = \frac{11}{2}$

14. Max: $x = 12, y = 6$; $Q = 54$; min: $x = 4, y = 0$; $Q = 8$

15. \$40,000 in each company; \$13,600

16. 60 model A satellite radios; 60 model B satellite radios; \$1320

17. 93 model A, 180 model B; \$456

18. 600 to Warehouse I and 400 to Warehouse II; \$11,200

Chapter 3 Before Moving On, page 217

1. a.

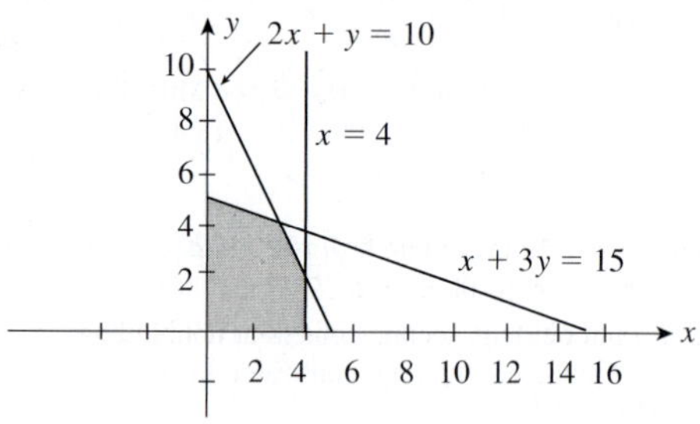

b.

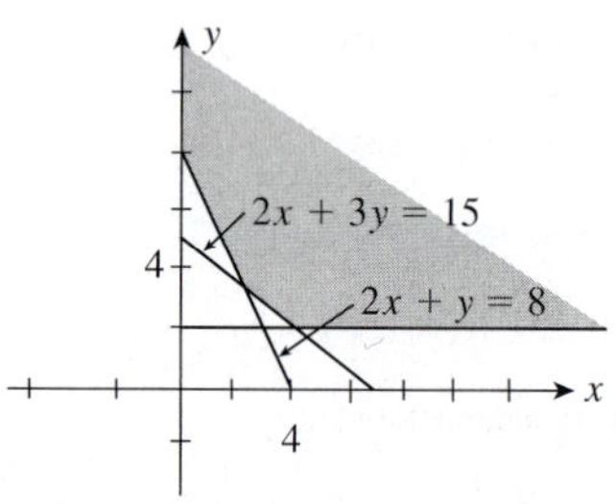

2. Min: $C = -7$; max: $P = 76$

3. Max: $x = 0, y = \frac{24}{7}; P = \frac{72}{7}$

4. Min: $x = 0, y = 10; C = 10$

5. a. Max: $x = 4, y = 6; P = 26$ **b.** Between 1.5 and 4.5
c. Between 8 and 24 d. 1.25
e. Constraints 1 and 2 are binding.

CHAPTER 4

Exercises 4.1, page 236

1. **a.** It is already in standard form.
b.

x	y	u	v	P	Constant
1	4	1	0	0	12
1	3	0	1	0	10
−2	−4	0	0	1	0

3. **a.** Maximize $P = 2x + 3y$
subject to
$x + y \le 10$
$x + 2y \le 12$
$2x + y \le 12$
$x \ge 0, y \ge 0$

b.

x	y	u	v	w	P	Constant
1	1	1	0	0	0	10
1	2	0	1	0	0	12
2	1	0	0	1	0	12
−2	−3	0	0	0	1	0

5. **a.** Maximize $P = x + 3y + 4z$
subject to
$x + 2y + z \le 40$
$x + y + z \le 30$
$x \ge 0, y \ge 0, z \ge 0$

b.

x	y	z	u	v	P	Constant
1	2	1	1	0	0	40
1	1	1	0	1	0	30
−1	−3	−4	0	0	1	0

7. In final form; $x = \frac{30}{7}, y = \frac{20}{7}, u = 0, v = 0; P = \frac{220}{7}$

9. Not in final form; pivot element is $\frac{1}{2}$, lying in the first row, second column.

11. In final form; $x = \frac{1}{3}, y = 0, z = \frac{13}{3}, u = 0, v = 6, w = 0;$ $P = 17$

13. Not in final form; pivot element is 1, lying in the third row, second column.

15. In final form; $x = 30, y = 10, z = 0, u = 0, v = 0; P = 60;$ $x = 30, y = 0, z = 0, u = 10, v = 0; P = 60$; among others

17. $x = 0, y = 4, u = 0, v = 1; P = 16$

19. $x = 6, y = 3, u = 0, v = 0; P = 96$

21. $x = 6, y = 6, u = 0, v = 0, w = 0; P = 60$

23. $x = 0, y = 3, z = 0, u = 90, v = 0, w = 75; P = 12$

25. $x = 0, y = 4, z = 4, u = 0, v = 0; P = 36$

27. $x = 15, y = 3, z = 0, u = 2, v = 0, w = 0; P = 78$

29. $x = \frac{5}{4}, y = \frac{15}{2}, z = 0, u = 0, v = \frac{15}{4}, w = 0; P = 90$

31. $x = 2, y = 1, z = 1, u = 0, v = 0, w = 0; P = 87$

33. $x = 30, y = 0, z = 0, P = 60$, and $x = 0, y = 30, z = 0, P = 60$, among others

35. No model A, 2500 model B; \$100,000

37. Medium-risk stocks: \$24,000; high-risk stocks: \$36,000; \$10,080

39. 40 tables; 100 chairs; \$3800

41. 180 units of Product A, 140 units of Product B, and 200 units of Product C; \$7920; no

43. 800 Giant Pandas; 1200 Saint Bernards; \$26,000

45. 80 units of model A, 80 units of model B, and 60 units of model C; maximum profit: \$5760; no

47. Growth funds: \$50,000; balanced funds: \$50,000; income funds: \$100,000; \$17,000

49. 9000 bottles of Formula I, 7833 bottles of Formula II, 6000 bottles of Formula III; maximum profit: \$4986.60; yes, ingredients for 4167 bottles of Formula II

51. a.

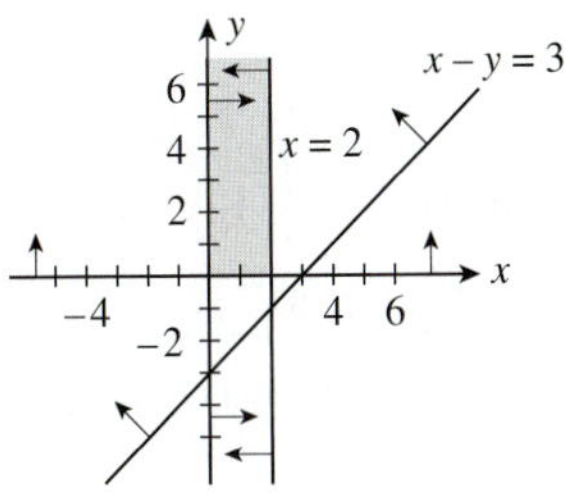

c. The ratios cannot be computed.

53. True 55. True

Using Technology Exercises 4.1, page 246

1. $x = 1.2, y = 0, z = 1.6, w = 0; P = 8.8$

3. $x = 1.6, y = 0, z = 0, w = 3.6; P = 12.4$

Exercises 4.2, page 257

1. $x = 4, y = 0; C = -8$

3. $x = 4, y = 3; C = -18$

5. $x = 0, y = 13, z = 18, w = 14; C = -111$

7. $x = \frac{5}{4}, y = \frac{1}{4}, u = 2, v = 3;\ C = P = 13$

9. $x = 5, y = 10, z = 0, u = 1, v = 2; C = P = 80$

11. Maximize $P = 90u + 120v$
subject to $2u + 3v \le 3$
$3u + 2v \le 2$; $x = 0, y = 60; C = 120$
$u \ge 0, v \ge 0$

13. Maximize $P = 60u + 40v + 30w$
subject to $6u + 2v + w \le 6$
$u + v + w \le 4$; $x = 10, y = 20; C = 140$
$u \ge 0, v \ge 0, w \ge 0$

15. Maximize $P = 10u + 20v$
subject to $20u + v \le 200$
$10u + v \le 150$; $x = 0, y = 0, z = 10; C = 1200$
$u + 2v \le 120$
$u \ge 0, v \ge 0$

17. Maximize $P = 10u + 24v + 16w$
subject to $u + 2v + w \le 6$
$2u + v + w \le 8$; $x = 8, y = 0, z = 8; C = 80$
$2u + v + w \le 4$
$u \ge 0, v \ge 0, w \ge 0$

19. Maximize $P = 6u + 2v + 4w$
subject to $2u + 6v \le 30$
$4u + 6w \le 12$; $x = \frac{1}{3}, y = \frac{4}{3}, z = 0; C = 26$
$3u + v + 2w \le 20$
$u \ge 0, v \ge 0, w \ge 0$

21. 2 type A vessels; 3 type B vessels; \$250,000

23. 8 oz of orange juice; 6 oz of pink grapefruit juice; 178 calories

25. Plant I: 500 to Warehouse *A* and 200 to Warehouse *B*; Plant II: 200 to Warehouse *B* and 400 to Warehouse *C*; \$20,800

27. Operate Refinery I for 2 days, and operate Refinery II for 6 days; \$2200

29. False

Using Technology Exercises 4.2, page 263

1. $x = 1.333333, y = 3.333333, z = 0$; and $C = 4.66667$

3. $x = 0.9524, y = 4.2857, z = 0; C = 6.0952$

5. a. $x = 3, y = 2, P = 17$ b. $\frac{8}{3} \le c_1 \le 8; \frac{3}{2} \le c_2 \le \frac{9}{2}$
c. $8 \le b_1 \le 24; 4 \le b_2 \le 12$ d. $\frac{5}{4}; \frac{1}{4}$
e. Both constraints are binding.

7. a. $x = 4, y = 0, C = 8$ b. $0 \le c_1 \le \frac{5}{2}; 4 \le c_2 < \infty$
c. $3 \le b_1 < \infty; -\infty < b_2 \le 4$ d. 2; 0
e. First constraint binding; second constraint nonbinding

Exercises 4.3, page 273

1. Maximize $P = -C = -2x + 3y$
subject to $-3x - 5y \le -20$
$3x + y \le 16$
$-2x + y \le 1$
$x \ge 0, y \ge 0$

3. Maximize $P = -C = -5x - 10y - z$
subject to $-2x - y - z \le -4$
$-x - 2y - 2z \le -2$
$2x + 4y + 3z \le 12$
$x \ge 0, y \ge 0, z \ge 0$

5. $x = 5, y = 2; P = 9$

7. $x = 4, y = 0; C = -8$

9. $x = 4, y = \frac{2}{3}; P = \frac{20}{3}$

11. $x = 3, y = 2; P = 7$

13. $x = 6, y = 0, z = 0; P = 6$

15. $x = 0, y = 3, z = 0; C = -9$

17. $x = \frac{46}{7}, y = 0, z = \frac{50}{7}; P = \frac{142}{7}$

19. $x = 0, y = 0, z = 10; P = 30$

21. \$35,000 in Company *A* stock, \$15,000 in Company *B* stock; \$6500

23. \$50 million worth of home loans, \$10 million worth of commercial-development loans; \$4.6 million

25. 0 units of Product *A*, 280 units of Product *B*, 280 units of Product *C*; \$7560

27. 10 oz of Food *A*, 4 oz of Food *B*, 40 mg of cholesterol; infinitely many other solutions

Chapter 4 Concept Review Questions, page 276

1. maximized; nonnegative; less than; equal to

2. equations; slack variables; $-c_1x_1 - c_2x_2 - \cdots - c_nx_n + P = 0$; below; augmented

3. minimized; nonnegative; greater than; equal to

4. dual; objective; optimal value

Chapter 4 Review Exercises, page 277

1. $x = 3, y = 4, u = 0, v = 0; P = 25$

2. $x = 3, y = 6, u = 4, v = 0, w = 0; P = 36$

3. $x = \frac{11}{3}, y = 4, u = \frac{7}{3}, v = 0, w = 0; P = 19$

4. $x = 8, y = 4, u = 2, v = 0, w = 0; P = 52$

5. $x = \frac{56}{5}, y = \frac{2}{5}, z = 0, u = 0, v = 0; P = \frac{118}{5}$

6. $x = 0, y = \frac{11}{3}, z = \frac{25}{6}, u = \frac{37}{6}, v = 0, w = 0; P = \frac{119}{6}$

7. $x = 2, y = 2, u = 0, v = 0; C = -22$

8. $x = 0, y = 4, z = 0, u = 20, v = 0, w = 20; C = -12$

9. Primal problem: $x = 2, y = 1; C = 9$; dual problem: $u = \frac{3}{10}, v = \frac{11}{10}; P = 9$

10. Maximize $P = 14u + 21v + 12w$ subject to

$$\begin{aligned} u + 4v - 3w &\le 2 \\ 2u + v - 2w &\le 4 \\ u + 2v + 5w &\le 3 \\ u \ge 0, v \ge 0, w &\ge 0 \end{aligned}$$

11. $x = \frac{3}{2}, y = 1; C = \frac{13}{2}$

12. $x = \frac{32}{11}, y = \frac{36}{11}; C = \frac{104}{11}$

13. $x = \frac{3}{4}, y = 0, z = \frac{7}{4}; C = 60$

14. $x = 0, y = 2, z = 0; C = 4$

15. $x = 45, y = 0, u = 0, v = 35; P = 135$

16. $x = 4, y = 4, u = 2, v = 0, w = 0; C = 20$

17. $x = 5, y = 2, u = 0, v = 0; P = 16$

18. $x = 20, y = 25, z = 0, u = 0, v = 30, w = 10; C = -160$

19. Saddle Mine: 4 days; Horseshoe Mine: 6 days; \$152,000

20. Ship *A*: 10 trips; Ship *B*: 20 trips; \$9,400,000

21. Reservoir: 4 million gallons; pipeline: 6 million gallons; \$4200

22. \$70,000 in blue-chip stocks; \$0 in growth stocks; \$30,000 in speculative stocks; maximum return: \$13,000

23. 0 unit of Product *A*, 30 units of Product *B*, 0 unit of Product *C*; $P = \$180$

24. \$50,000 in stocks, \$100,000 in bonds, \$50,000 in money market funds; $P = \$21,500$

Chapter 4 Before Moving On, page 278

1.

x	y	z	u	v	w	P	Constant
2	(1)	−1	1	0	0	0	3
1	−2	3	0	1	0	0	1
3	2	4	0	0	1	0	17
−1	−2	3	0	0	0	1	0

2. $x = 2, y = 0, z = 11, u = 2, v = 0, w = 0; P = 28$

3. $x = 6, y = 2; u = 0, v = 0; P = 34$ 4. $x = 3, y = 0; C = 3$

5. $x = 10, y = 0; P = 20$

CHAPTER 5

Exercises 5.1, page 294

1. \$80; \$580 3. \$836 5. \$1000 7. 292 days

9. 10%/year 11. \$1368.57 13. \$3714.87

15. \$20,219.60 17. \$175, 979.80 19. \$196,488.13

21. 6.09%/year 23. 4.07%/year 25. \$34,139.61

27. \$35,482.13 29. \$6356.25 31. 24%/year

33. \$123,000 35. 5%/year 37. 2.39%/year

39. No. It is less than its original value. 41. 11.8%/year

43. \$999.14/day 45. \$316,032 47. \$1.92 million

49. \$20,471.64 51. \$33,603.85

53. **a.** \$56,427.16 **b.** \$56,073 **c.** \$55,892.84

55. Acme Mutual Fund 57. \$23,227.22 59. \$5994.86

61. \$339.79 billion 63. Investment *A*

65. \$75,888.25; \$75,602.68 67. $80,000e^{(\sqrt{t}/2 - 0.05t)}$; \$178,043.27

71. 3.4% 73. 5.83% 75. True 77. True 79. 4.4 years

81. 13.9 years 83. 6.08%/year 85. 2.8 years

Using Technology Exercises 5.1, page 300

1. \$5872.78 3. \$475.49 5. 8.95%/year

7. 10.20%/year 9. \$29,743.30 11. \$53,303.25

Exercises 5.2, page 308

1. \$12,577.89 3. \$17,213.24 5. \$27,069.30 7. \$100,289.96

9. \$31,048.97 11. \$12,309.32 13. \$18,181.37

15. \$62,468.86 17. \$445.54 19. Karen 21. \$753,031.24

23. \$5643.08 25. \$21,805.83

27. Between \$462,692 and \$568,365

29. Between \$343,493 and \$419,366 31. \$16,883.30

33. \$61,464.19 35. False

Using Technology Exercises 5.2, page 312

1. \$59,622.15 3. \$8453.59 5. \$35,607.23

7. \$13,828.60

Exercises 5.3, page 320

1. \$13,586.80 3. \$444.24 5. \$622.13

7. \$454.23 9. \$1491.19 11. \$516.76

13. \$333.85 15. \$2216.02 17. \$3033.55

19. \$12,950.46 21. **a.** \$243.70 **b.** \$805.23; \$268.41

23. **a.** \$456.33; \$352.28 **b.** \$1427.88; \$1909.44

25. \$1438.92; \$46,669.74; \$69,154.44; \$140,391.51 **27.** \$83,954.38

29. \$3135.48 **31.** \$452.08 **33.** \$447.98

35. \$1014.94; \$1947.04 **37.** \$1,111.63 **39.** \$25,645.51

41. \$152,670.69 **43.** \$194.65 **45.** \$1418.72; \$243,673.79

47. **a.** \$1264.14 **b.** \$190,119.14 **c.** \$991.75 **d.** \$272.39

49. \$10,413.60 **51.** \$725.43 **53.** \$29,658 **55.** Yes

57. **a.** \$599,088.30 **b.** \$779,220.78

Using Technology Exercises 5.3, page 327

1. \$628.02 **3.** \$1379.28 **5.** \$1988.41

7. \$894.12 **9.** \$15,165.46

Exercises 5.4, page 333

1. 30 **3.** $-\frac{9}{2}$ **5.** $-3, 8, 19, 30, 41$ **7.** $x + 6y$

9. 795 **11.** 792 **13.** 671

15. **a.** 275 **b.** -280 **17.** At the beginning of the 37th week

19. \$15.80 **21.** **b.** \$800 **23.** GP; 256; 508 **25.** Not a GP

27. GP; $\frac{1}{3}$; $364\frac{1}{3}$ **29.** 3; 0 **31.** 293,866

33. \$56,284 **35.** Annual raise of 8%/year

37. **a.** \$20,113.57 **b.** \$87,537.38 **39.** \$25,165.82

41. \$39,321.60; \$110,678.40 **43.** True

Chapter 5 Concept Review Questions, page 336

1. **a.** original; $P(1 + rt)$ **b.** interest; $P(1 + i)^n$; $A(1 + i)^{-n}$

2. simple; one; nominal; m; $\left(1 + \frac{r}{m}\right)^m - 1$

3. annuity; ordinary annuity; simple annuity.

4. **a.** $R\left[\frac{(1 + i)^n - 1}{i}\right]$ **b.** $R\left[\frac{1 - (1 + i)^{-n}}{i}\right]$

5. $\frac{Pi}{1 - (1 + i)^{-n}}$ **6.** future; $\frac{iS}{(1 + i)^n - 1}$

7. constant d; $a + (n - 1)d$; $\frac{n}{2}[2a + (n - 1)d]$

8. constant r; ar^{n-1}; $\frac{a(1 - r^n)}{1 - r}$

Chapter 5 Review Exercises, page 337

1. **a.** \$6077.53 **b.** \$6092.01 **c.** \$6099.45 **d.** \$6104.48

2. **a.** \$15,801.71 **b.** \$15,839.15 **c.** \$15,858.23 **d.** \$15,871.09

3. **a.** 6% **b.** 6.09% **c.** 6.1363% **d.** 6.168%

4. **a.** 5.5% **b.** 5.576% **c.** 5.6145% **d.** 5.6408%

5. \$33,110.52 **6.** \$50,789.23 **7.** \$4991.91

8. \$18,143.77 **9.** \$8404.23 **10.** \$232,624.14 **11.** \$644.65

12. \$297.92 **13.** \$353.42 **14.** \$228.55 **15.** 3.660%

16. 4.907% **17.** \$97,712.24 **18.** \$2,143,825.25; \$7,769,900.25

19. \$5,491,922 **20.** \$2346.91 **21.** \$16,904.04 **22.** \$5000

23. 7.6% **24.** \$278.88 **25.** \$62,112.91 **26.** \$13,026.89

27. \$2099.44 **28.** **a.** \$608.02 **b.** \$98,887.20 **c.** \$53,893.05

29. **a.** \$917.99 **b.** \$45,238.20 **c.** \$100,760 **30.** \$22,160.19

31. \$4727.67 **32.** \$191.15; 9.707%/year **33.** \$2203.83

Chapter 5 Before Moving On, page 339

1. \$2540.47 **2.** 6.18%/year **3.** \$569,565.47 **4.** \$1213.28

5. \$35.13 **6.** **a.** 210 **b.** 127.5

CHAPTER 6

Exercises 6.1, page 348

1. $\{x \mid x$ is a gold medalist in the 2014 Winter Olympic Games$\}$

3. $\{x \mid x$ is an integer greater than 2 and less than 8$\}$

5. $\{2, 3, 4, 5, 6\}$ **7.** $\{-2\}$

9. **a.** True **b.** False **11.** **a.** False **b.** False

13. True **15.** **a.** True **b.** False **17.** **a.** and **b.**

19. **a.** $\varnothing, \{1\}, \{2\}, \{1, 2\}$
b. $\varnothing, \{1\}, \{2\}, \{3\}, \{1, 2\}, \{1, 3\}, \{2, 3\}, \{1, 2, 3\}$
c. $\varnothing, \{1\}, \{2\}, \{3\}, \{4\}, \{1, 2\}, \{1, 3\}, \{1, 4\}, \{2, 3\}, \{2, 4\}, \{3, 4\}, \{1, 2, 3\}, \{1, 2, 4\}, \{1, 3, 4\}, \{2, 3, 4\}, \{1, 2, 3, 4\}$

21. $\{1, 2, 3, 4, 6, 8, 10\}$

23. {Jill, John, Jack, Susan, Sharon}

25. **a.**

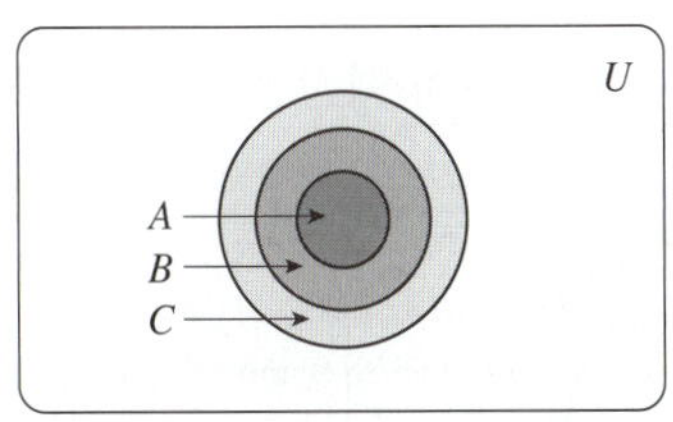

b.

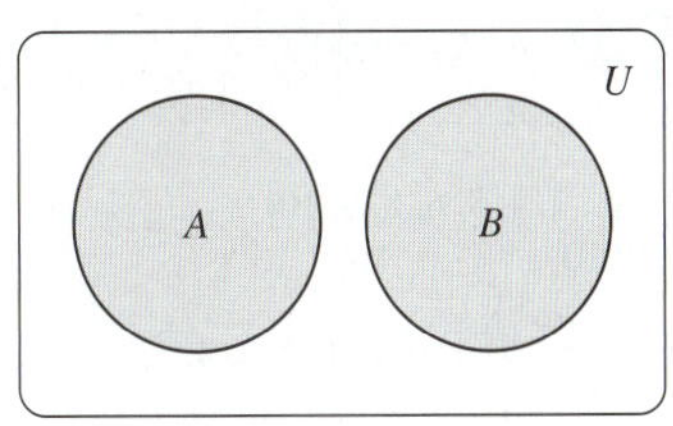

c.

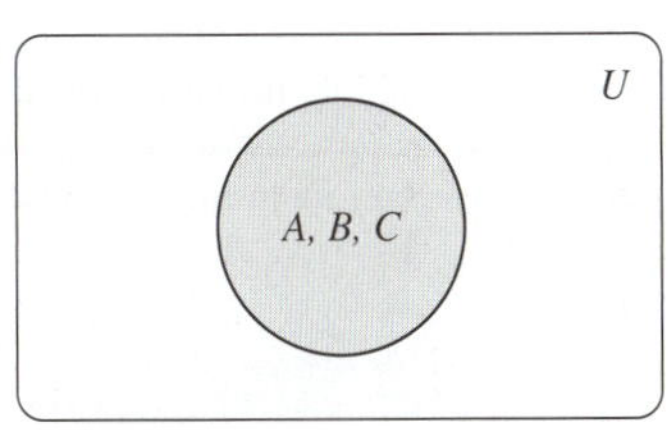

27. **a.** $(A \cap B^c) \cup (A^c \cap B)$ **b.** $(A \cap B)^c$ or $A^c \cup B^c$

29. **a.**

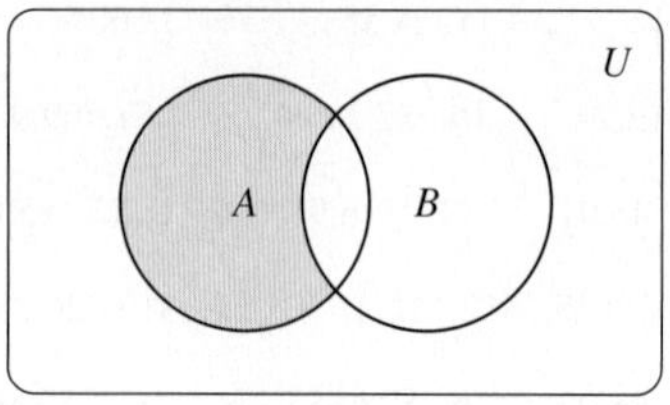

b.

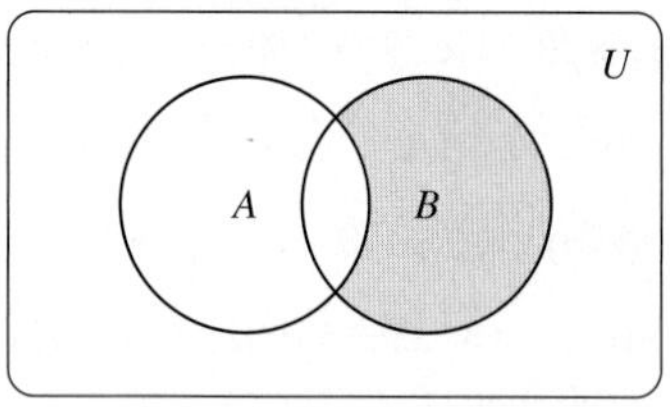

31. **a.**

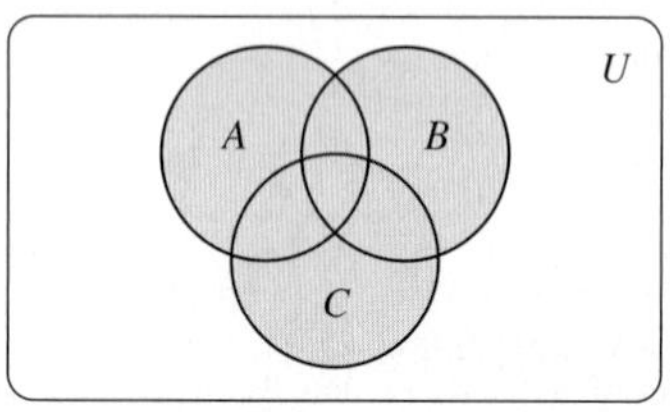

b.

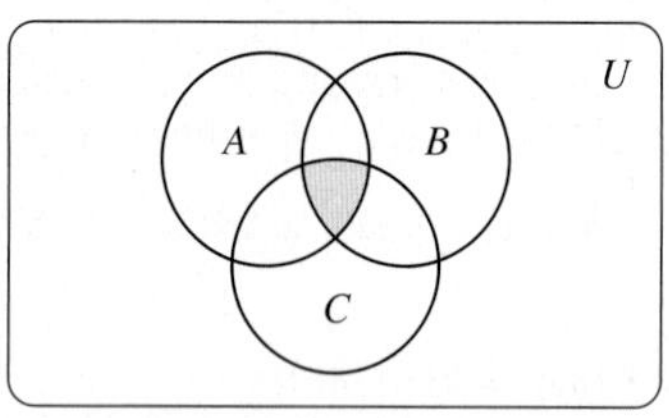

33. **a.**

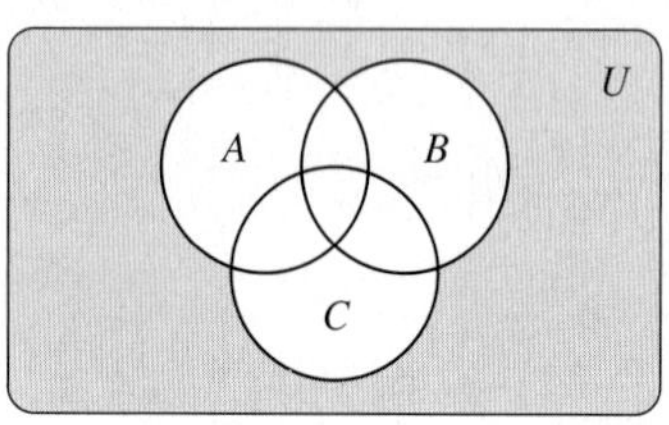

b.

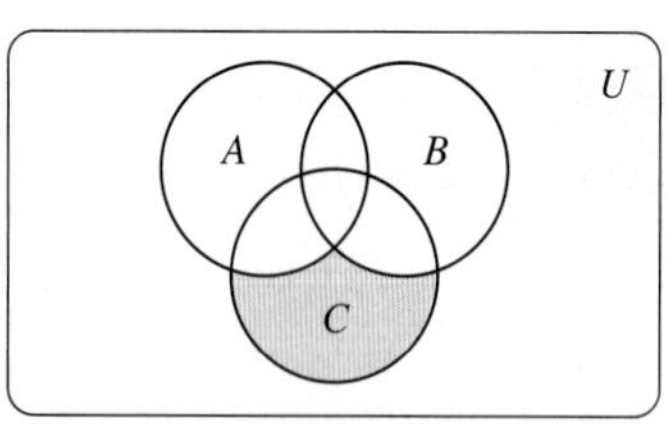

35. **a.** {2, 4, 6, 8, 10} **b.** {1, 2, 4, 5, 6, 8, 9, 10}
c. {1, 2, 3, 4, 5, 6, 7, 8, 9, 10}

37. **a.** $C = \{1, 2, 4, 5, 8, 9\}$ **b.** $\varnothing$ **c.** {1, 2, 3, 4, 5, 6, 7, 8, 9, 10}

39. **a.** Not disjoint **b.** Disjoint

41. **a.** The set of all employees at Universal Life Insurance who do not drink tea
b. The set of all employees at Universal Life Insurance who do not drink coffee

43. **a.** The set of all employees at Universal Life Insurance who drink tea but not coffee
b. The set of all employees at Universal Life Insurance who drink coffee but not tea

45. **a.** The set of all employees in a hospital who are not doctors
b. The set of all employees in a hospital who are not nurses

47. **a.** The set of all employees in a hospital who are female doctors
b. The set of all employees in a hospital who are both doctors and administrators

49. **a.** $D \cap F$ **b.** $R \cap F^c \cap L^c$

51. **a.** B^c **b.** $A \cap B$ **c.** $A \cap B \cap C^c$

53. **a.** A = {New York, Chicago, Boston}; B = {Chicago, Boston}; C = {Las Vegas, San Francisco}
b. {New York, Chicago, Boston}
c. {Chicago, Boston} **d.** $\varnothing$ **e.** {New York}
f. {Las Vegas, San Francisco}

55. **a.** $A \cap B \cap C$; the set of tourists who have taken the underground, a cab, and a bus over a 1-week period in London
b. $A \cap C$; the set of tourists who have taken the underground and a bus over a 1-week period in London
c. B^c; the set of tourists who have not taken a cab over a 1-week period in London

57.

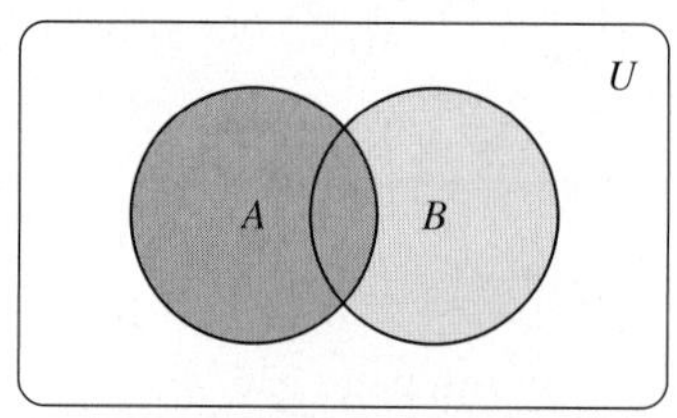

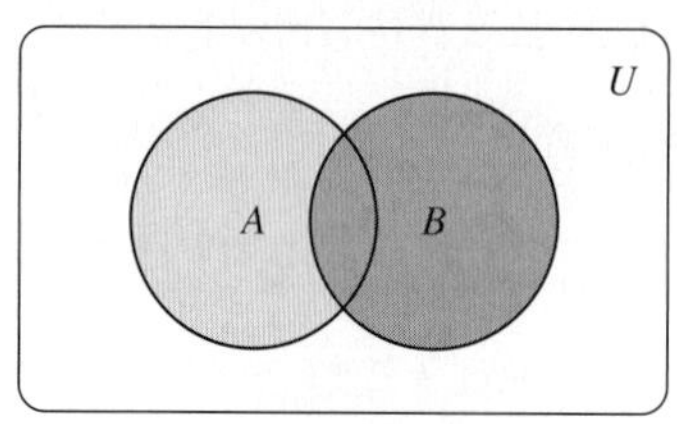

59.

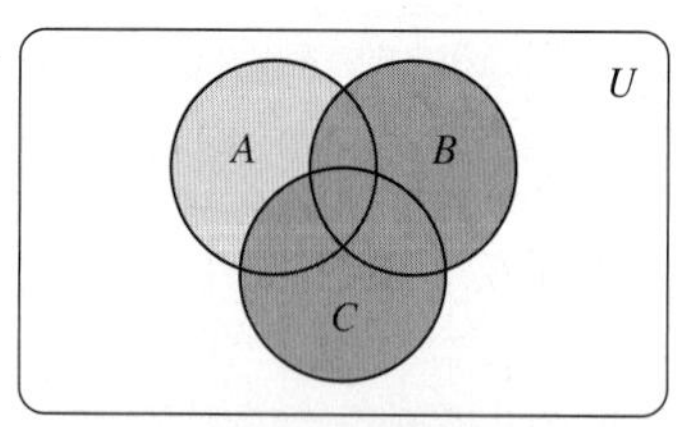

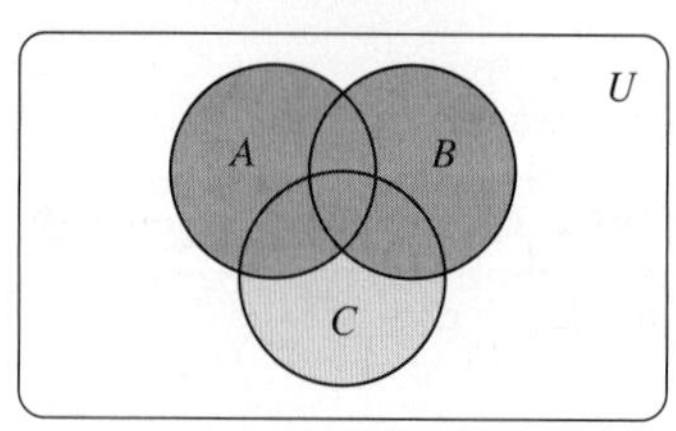

61.

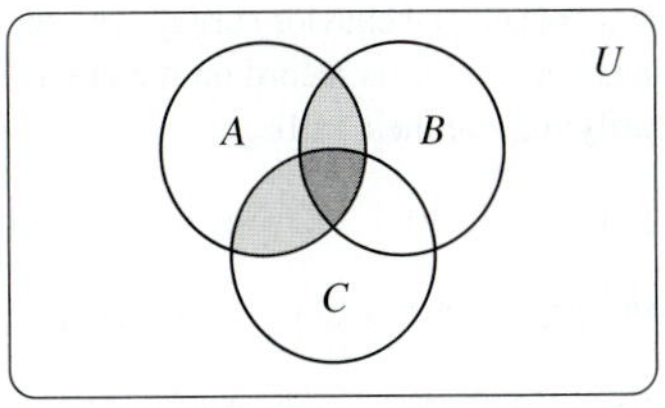

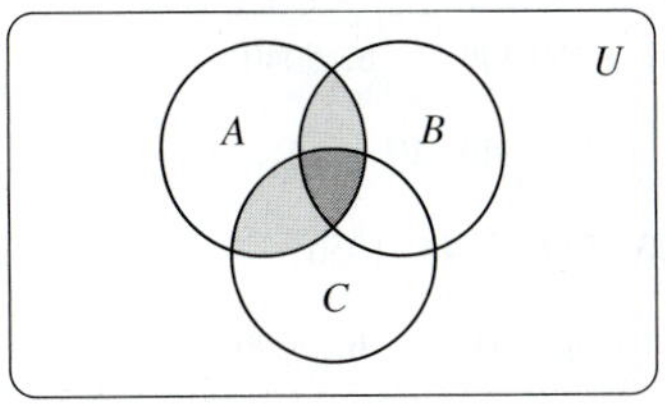

65. **a.** x, y, v, r, w, u **b.** v, r

67. **a.** s, t, y **b.** t, z, w, x, s 69. $A \subset C$

71. False 73. True 75. True 77. True 79. True

Exercises 6.2, page 357

3. **a.** 4 **b.** 5 **c.** 7 **d.** 2 7. 20

9. **a.** 15 **b.** 30 **c.** 15 **d.** 12 **e.** 50 **f.** 20

11. **a.** 140 **b.** 100 **c.** 60

13. 13 15. 0 17. 13 19. 61

21. **a.** 106 **b.** 64 **c.** 38 **d.** 14

23. **a.** 182 **b.** 118 **c.** 56 **d.** 18 25. 30

27. **a.** 16 **b.** 31 **c.** 4 **d.** 21 **e.** 11

29.

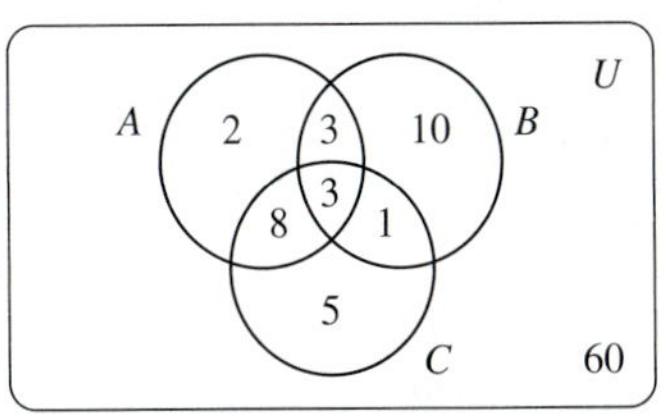

31.

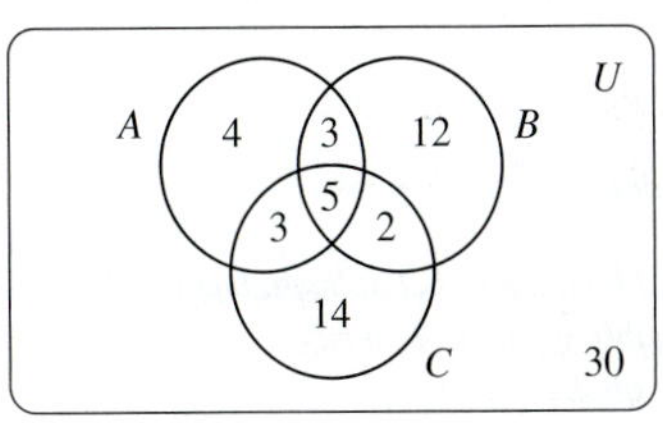

33. **a.** 64 **b.** 10 35. **a.** 36 **b.** 36

37. **a.** 3, 2, 4 **b.** 3 **c.** 1 **d.** 1 **e.** 0

39. **a.** 5, 3, 5 **b.** 2 **c.** 3 **d.** 3 **e.** 0 **f.** 3

41. 5 43. **a.** 62 **b.** 33 **c.** 25 **d.** 38

45. **a.** 108 **b.** 15 **c.** 45 **d.** 12

47. **a.** 22 **b.** 80

49. True 51. True

Exercises 6.3, page 365

1. 12 3. 64 5. 24 7. 24 9. 60

11. 1 billion 13. 45,000 15. $2^8 = 256$ 17. 120

19. 5^{50} 21. 576 23. **a.** 17,576,000 **b.** 17,576,000

25. 456,976,000 27. 2,340,000 29. 2730

31. **a.** $10^6 = 1{,}000{,}000$ **b.** 250,000 33. 217 35. True

Exercises 6.4, page 378

1. 360 3. 10 5. 120 7. 20 9. n 11. 1

13. 35 15. 1 17. 84 19. $\dfrac{n(n-1)}{2}$ 21. $\dfrac{n!}{2}$

23. Permutation 25. Combination

27. Permutation 29. Combination

31. $P(4, 4) = 24$ 33. $P(4, 4) = 24$ 35. $P(9, 9) = 362{,}880$

37. $C(12, 3) = 220$ 39. 151,200 41. 2520 43. 20

45. $C(12, 3) = 220$ 47. $C(100, 3) = 161{,}700$

49. $P(6, 6) = 720$ 51. $C(3, 1) + C(4, 1) = 7$

53. $P(12, 6) = 665{,}280$ 55. $C(8, 4)[C(7, 4) + C(6, 3)] = 3850$

57. **a.** $P(10, 10) = 3{,}628{,}800$
b. $P(3, 3)P(4, 4)P(3, 3)P(3, 3) = 5184$

59. **a.** $P(20, 20) = 20!$
b. $P(5, 5)P[(4, 4)]^5 = 5!(4!)^5 = 955{,}514{,}880$

61. **a.** $P(12, 9) = 79{,}833{,}600$
b. $C(12, 9) = 220$ **c.** $C(12, 9) \cdot C(3, 2) = 660$

63. $2\{C(2, 2) + [C(3, 2) - C(2, 2)]\} = 6$

65. $C(3, 3)[C(8, 6) + C(8, 7) + C(8, 8)] = 37$

67. **a.** $C(12, 3) = 220$ **b.** $C(11, 2) = 55$
c. $C(5, 1)C(7, 2) + C(5, 2)C(7, 1) + C(5, 3) = 185$

69. $P(7, 3) + C(7, 2)P(3, 2) = 336$

71. $C(5, 1)C(3, 1)C(6, 2)[C(4, 1) + C(3, 1)] = 1575$

73. $10C(4, 1) = 40$ 75. $C(4, 1)C(13, 5) - 40 = 5108$

77. $13C(4, 3) \cdot 12C(4, 2) = 3744$ 79. $C(6, 2) = 15$

81. $C(12, 6) + C(12, 7) + C(12, 8) + C(12, 9) + C(12, 10) + C(12, 11) + C(12, 12) = 2510$

83. $4! = 24$ 87. True 89. True

Using Technology Exercises 6.4, page 382

1. $1.307674368 \times 10^{12}$ 3. $2.56094948229 \times 10^{16}$

5. 674,274,182,400 7. 133,784,560

9. 4,656,960 11. 658,337,004,000

Chapter 6 Concept Review Questions, page 383

1. set; elements; set 2. equal 3. subset

4. a. no b. all 5. a. union b. intersection

6. complement 7. $A^c \cap B^c \cap C^c$ 8. permutation; combination

Chapter 6 Review Exercises, page 384

1. $\{3\}$ 2. $\{A, E, H, L, S, T\}$

3. $\{4, 6, 8, 10\}$ 4. $\{-4\}$ 5. Yes

6. Yes 7. Yes 8. No

9.

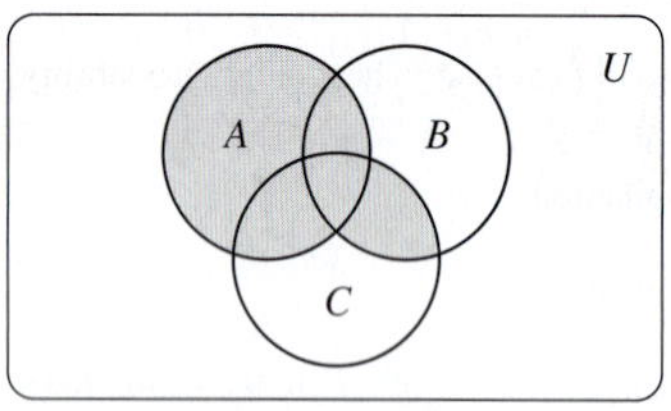

$A \cup (B \cap C)$

10.

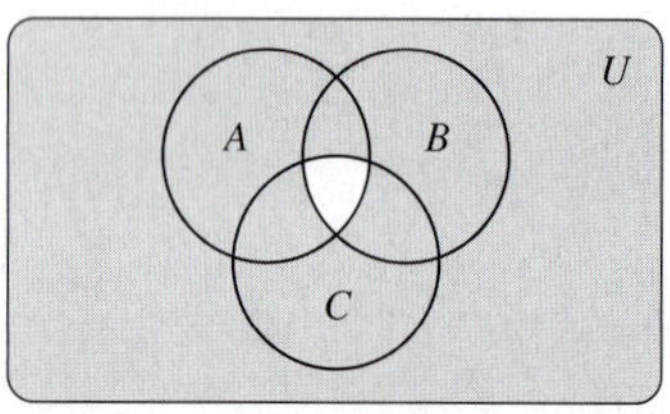

$(A \cap B \cap C)^c$

11.

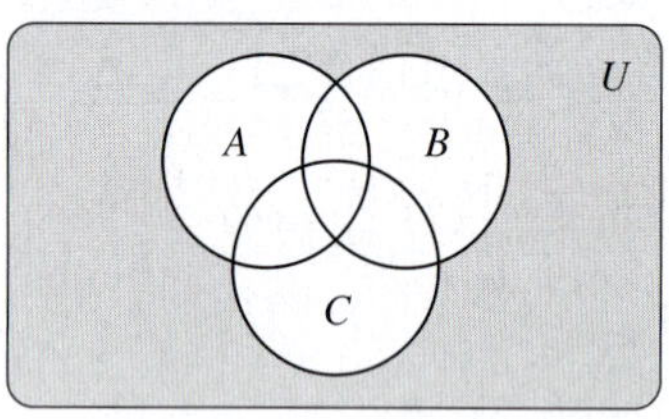

$A^c \cap B^c \cap C^c$

12.

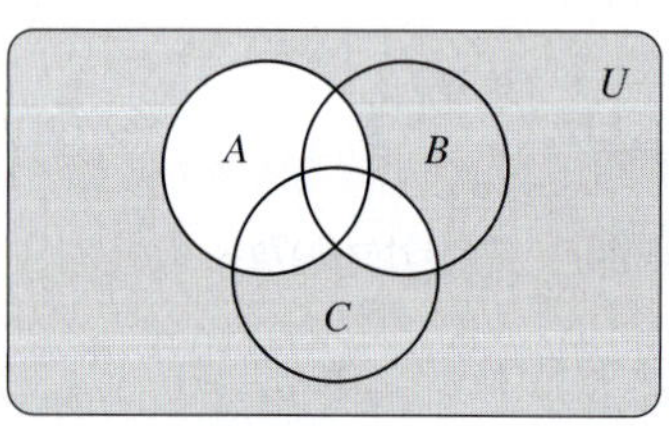

$A^c \cap (B^c \cup C^c)$

17. The set of all participants in a consumer-behavior survey who both avoided buying a product because it is not recyclable and boycotted a company's products because of its record on the environment.

18. The set of all participants in a consumer-behavior survey who avoided buying a product because it is not recyclable and/or voluntarily recycled their garbage.

19. The set of all participants in a consumer-behavior survey who both did not use cloth diapers rather than disposable diapers and voluntarily recycled their garbage.

20. The set of all participants in a consumer-behavior survey who did not boycott a company's products because of its record on the environment and/or did not voluntarily recycle their garbage.

21. 150 22. 230 23. 270 24. 30 25. 70 26. 200

27. 190 28. 181,440 29. 120 30. 8400 31. None

32. a. 446 b. 377 c. 34 33. 720 34. 20 35. 144

36. a. 720 b. 480 37. a. 50,400 b. 5040

38. a. 60 b. 125 39. 108 40. 30 41. 80

42. a. 1287 b. 288 43. 720 44. 1050

45. a. 2704 b. 2652 46. a. 5040 b. 3600

47. a. 487,635 b. 550 c. 341,055 48. a. 1365 b. 1155

49. a. 720 b. 72 c. 48 50. a. 495 b. 210 c. 420

Chapter 6 Before Moving On, page 386

1. a. $\{d, f, g\}$ b. $\{b, c, d, e, f, g\}$ c. $\{b, c, e\}$

2. 3 3. 15 4. 264 5. 200

CHAPTER 7

Exercises 7.1, page 393

1. $\{a, b, d, f\}$; $\{a\}$ 3. $\{b, c, e\}$; $\{a\}$ 5. No 7. S

9. $\varnothing$ 11. Yes 13. Yes 15. $E \cup F$ 17. G^c

19. $(E \cup F \cup G)^c$

21. a. $\{(2, 1), (3, 1), (4, 1), (5, 1), (6, 1), (3, 2), (4, 2), (5, 2), (6, 2), (4, 3), (5, 3), (6, 3), (5, 4), (6, 4), (6, 5)\}$
b. $\{(1, 2), (2, 4), (3, 6)\}$

23. $\varnothing$, $\{a\}$, $\{b\}$, $\{c\}$, $\{a, b\}$, $\{a, c\}$, $\{b, c\}$, S

25. a. $S = \{\text{B}, \text{R}\}$ b. $\varnothing$, $\{\text{B}\}$, $\{\text{R}\}$, S

27. a. $S = \{(\text{H}, 1), (\text{H}, 2), (\text{H}, 3), (\text{H}, 4), (\text{H}, 5), (\text{H}, 6), (\text{T}, 1), (\text{T}, 2), (\text{T}, 3), (\text{T}, 4), (\text{T}, 5), (\text{T}, 6)\}$
b. $\{(\text{H}, 2), (\text{H}, 4), (\text{H}, 6)\}$

29. a. No b. No

31. $S = \{ddd, ddn, dnd, ndd, dnn, ndn, nnd, nnn\}$

33. a. $S = \{bbbb, bbbg, bbgb, bbgg, bgbb, bgbg, bggb, bggg, gbbb, gbbg, gbgb, gbgg, ggbb, ggbg, gggb, gggg\}$
b. $E = \{bbbg, bbgb, bgbb, gbbb\}$
c. $F = \{bbbg, bbgg, bgbg, bggg, gbbg, gbgg, ggbg, gggg\}$
d. $G = \{gbbg, gbgg, ggbg, gggg\}$

35. a. {ABC, ABD, ABE, ACD, ACE, ADE, BCD, BCE, BDE, CDE}
b. 6 c. 3 d. 6

37. a. E^c b. $E^c \cap F^c$ c. $E \cup F$
d. $(E \cap F^c) \cup (E^c \cap F)$

39. a. $\{t \mid t > 0\}$ b. $\{t \mid 0 < t \le 2\}$ c. $\{t \mid t > 2\}$

41. a. $S = \{0, 1, 2, 3, \ldots, 10\}$ b. $E = \{0, 1, 2, 3\}$
c. $F = \{5, 6, 7, 8, 9, 10\}$

43. **a.** $S = \{0, 1, 2, \ldots, 20\}$
b. $E = \{0, 1, 2, \ldots, 9\}$ **c.** $F = \{20\}$

49. False

Exercises 7.2, page 401

1. $\{(H, H)\}, \{(H, T)\}, \{(T, H)\}, \{(T, T)\}$

3. $\{(D, m)\}, \{(D, f)\}, \{(R, m)\}, \{(R, f)\}, \{(I, m)\}, \{(I, f)\}$

5. $\{(1, i)\}, \{(1, d)\}, \{(1, s)\}, \{(2, i)\}, \{(2, d)\}, \{(2, s)\}, \ldots, \{(5, i)\}, \{(5, d)\}, \{(5, s)\}$

7. $\{(A, Rh^+)\}, \{(A, Rh^-)\}, \{(B, Rh^+)\}, \{(B, Rh^-)\}, \{(AB, Rh^+)\}, (AB, Rh^-)\}, \{(O, Rh^+)\}, \{(O, Rh^-)\}$

9. **a.**

Answer	1–2	3–4	5 or more
Probability	.65	.20	.15

b. .20

11. **a.**

Answer	Low	Middle	Extreme	No Response
Probability	.34	.44	.20	.02

b. .20

13.

Grade	A	B	C	D	F
Probability	.10	.25	.45	.15	.05

15. **a.**

Answer	New processes/ procedures	Getting to know a new boss and coworkers	New technology tools	Fitting into the corporate culture	Other
Probability	.44	.20	.17	.12	.07

b. .12

17.

Event	A	B	C	D	E
Probability	.026	.199	.570	.193	.012

19.

Number of Days	0	1	2	3	4	5	6	7
Probability	.05	.06	.09	.15	.11	.20	.17	.17

21. **a.** $S = \{(0 < x \leq 200), (200 < x \leq 400), (400 < x \leq 600), (600 < x \leq 800), (800 < x \leq 1000), (x > 1000)\}$

b.

Cars, x	Probability
$0 < x \leq 200$	.075
$200 < x \leq 400$	.1
$400 < x \leq 600$	.175
$600 < x \leq 800$	.35
$800 < x \leq 1000$	.225
$x > 1000$	.075

23. .469 25. **a.** .856 **b.** .144 27. .46

29. **a.** $\frac{1}{4}$ **b.** $\frac{1}{2}$ **c.** $\frac{1}{13}$ 31. $\frac{3}{8}$ 33. No 35. Yes

37. **a.** $\frac{3}{7}$ **b.** $\frac{3}{14}$ **c.** 1 39. **a.** .4 **b.** .1 **c.** .1

41. **a.** .633 **b.** .276 43. **a.** .35 **b.** .33

45. .530 47. **a.** .4 **b.** .23

49. **a.** .448 **b.** .255 51. .783 53. True

Exercises 7.3, page 411

1. $\frac{1}{2}$ 3. $\frac{1}{36}$ 5. $\frac{1}{9}$ 7. $\frac{1}{52}$

9. $\frac{3}{13}$ 11. $\frac{12}{13}$ 13. .002; .998

15. $P(a) + P(b) + P(c) \neq 1$

17. Since the five events are not mutually exclusive, Property 3 cannot be used; that is, he could win more than one purse.

19. The two events are not mutually exclusive; hence, the probability of the given event is $\frac{1}{6} + \frac{1}{6} - \frac{1}{36} = \frac{11}{36}$.

21. $E^c \cap F^c = \{e\} \neq \varnothing$

23. **a.** 0 **b.** .7 **c.** .8 **d.** .3

25. **a.** $\frac{1}{2}, \frac{3}{8}$ **b.** $\frac{1}{2}, \frac{5}{8}$ **c.** $\frac{1}{8}$ **d.** $\frac{3}{4}$ **e.** $\frac{1}{4}$ **f.** $\frac{7}{8}$

27. **a.** .30 **b.** .64 **c.** .36 29. **a.** .06 **b.** .39 **c.** .45

31. **a.** .53 **b.** .51 **c.** .56 33. **a.** .15 **b.** .44

35. **b.** .52 **c.** .859 37. **a.** .41 **b.** .48

39. **a.** .24 **b.** .46 41. **a.** .16 **b.** .38 **c.** .22

43. **a.** .33 **b.** .09 45. .332

47. **a.** .333 **b.** .584 **c.** .055 49. **a.** .63 **b.** .55 **c.** .33

51. **a.** .09 **b.** .41; during descent and approach and landing

53. **a.** .43 **b.** .75 57. True 59. False 61. True

Exercises 7.4, page 422

1. $\frac{1}{32}$ 3. $\frac{31}{32}$

5. $P(E) = 13C(4, 2)/C(52, 2) \approx .059$

7. $C(26, 2)/C(52, 2) \approx .245$

9. $[C(3, 2)C(5, 2)]/C(8, 4) = 3/7$

11. $[C(5, 3)C(3, 1)]/C(8, 4) = 3/7$ 13. $C(3, 2)/8 = 3/8$

15. $1/8$ 17. $C(10, 6)/2^{10} \approx .205$

19. **a.** $C(4, 2)/C(24, 2) \approx .022$
b. $1 - C(20, 2)/C(24, 2) \approx .312$

21. **a.** $C(6, 2)/C(80, 2) \approx .005$
b. $1 - C(74, 2)/C(80, 2) \approx .145$

23. **a.** .12; $C(98, 10)/C(100, 12) \approx .013$
b. .15; .015

25. $[C(12, 8)C(8, 2) + C(12, 9)C(8, 1) + C(12, 10)]/C(20, 10) \approx .085$

27. **a.** $\frac{3}{5}$ **b.** $C(3, 1)/C(5, 3) = .3$ **c.** $1 - C(3, 3)/C(5, 3) = .9$

29. $\frac{1}{729}$ 31. .0001 33. .1 35. $40/C(52, 5) \approx .0000154$

37. $[4C(13, 5) - 40]/C(52, 5) \approx .00197$

39. $[13C(4, 3) \cdot 12C(4, 2)]/C(52, 5) \approx .00144$

41. **a.** .618 **b.** .059 **43.** .030 **45.** .167

Exercises 7.5, page 436

1. **a.** .4 **b.** .33 **3.** .3 **5.** Independent

7. Independent **9.** **a.** .24 **b.** .76 **c.** .4 **d.** .76

11. **a.** .5 **b.** .4 **c.** .2 **d.** .35 **e.** No **f.** No

13. **a.** .4 **b.** .3 **c.** .12 **d.** .30 **e.** Yes **f.** Yes

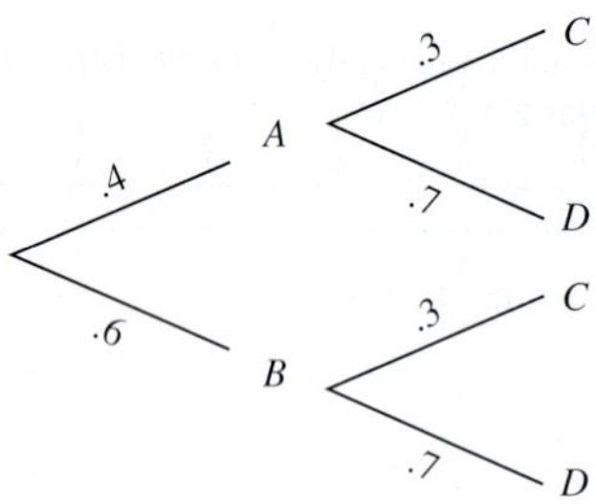

15. **a.** $\frac{1}{12}$ **b.** $\frac{1}{36}$ **c.** $\frac{1}{6}$ **d.** $\frac{1}{6}$ **e.** No

17. $\frac{4}{11}$ **19.** Independent **21.** Not independent **23.** .1875

25. **a.** .42 **b.** .85 **27.** .06 **29.** **a.** $\frac{1}{21}$ **b.** $\frac{1}{3}$

31. .48 **33.** Not independent

35. **a.** .757; .569; .393; .520; .720 **b.** Not independent

37. $\frac{1}{7}$ **39.** .98 **41.** **a.** .3 **b.** .09375 **43.** .014

45. .25 **47.** $\frac{1}{15}$ **49.** .054 **51.** .1701 **53.** .0000068

55. **a.** .092 **b.** .008 **57.** 3 **59.** **a.** No **b.** Yes

61. 0 **63.** $\dfrac{P(A)}{P(A) + P(B)}$ **65.** True **67.** True

Exercises 7.6, page 445

1.

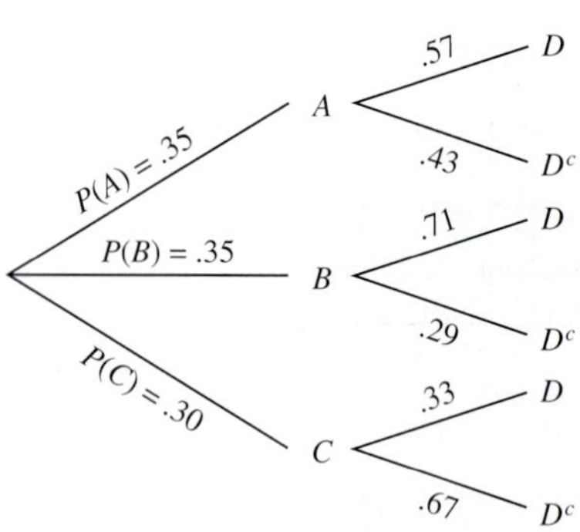

3. **a.** .45 **b.** .22 **5.** **a.** .48 **b.** .33

7. **a.** .08 **b.** .15 **c.** .348

9. **a.** $\frac{1}{12}$ **b.** $\frac{1}{4}$ **c.** $\frac{1}{18}$ **d.** $\frac{3}{14}$

11.

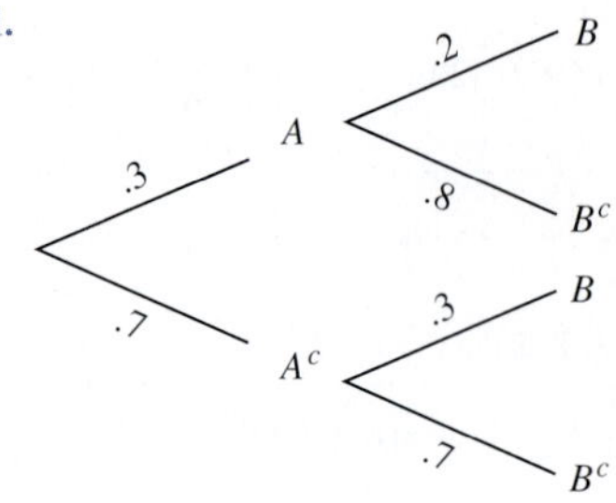

a. .27 **b.** .22 **c.** .73 **d.** .33

13. $\frac{4}{17}$ **15.** $\frac{4}{51}$

17.

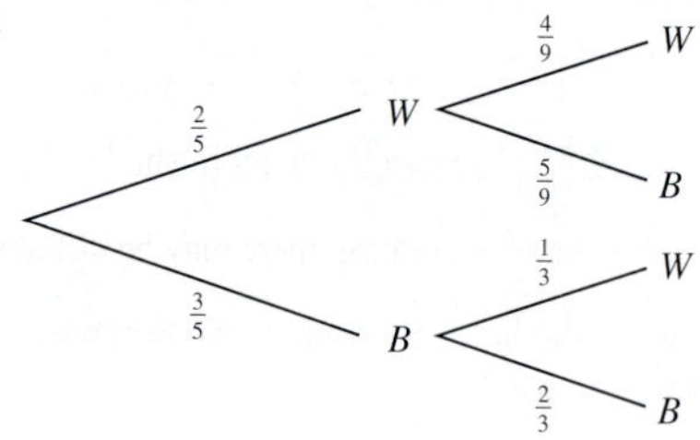

19. $\frac{9}{17}$ **21.** **a.** .416 **b.** .584 **23.** .125

25. **a.** .297 **b.** .10 **27.** .348 **29.** .927 **31.** .710

33. .3758 **35.** **a.** .57 **b.** .691

37. **a.** .543 **b.** .545 **c.** .455 **39.** **a.** $\frac{3}{4}$ b. $\frac{2}{9}$

41. **a.** .497 **b.** .959 **43.** **a.** .025 **b.** .24 **c.** .24

45. .407 **47.** **a.** .03 **b.** .29 **49.** .028

51. **a.** .763 **b.** .276 **c.** .724 **53.** .1337

Chapter 7 Concept Review Questions, page 452

1. experiment; sample; space; event **2.** $\varnothing$ **3.** uniform; $\frac{1}{n}$

4. conditional **5.** independent **6.** a posteriori probability

Chapter 7 Review Exercises, page 452

1. **a.** 0 **b.** .6 **c.** .6 **d.** .4 **e.** 1

2. **a.** .35 **b.** .65 **c.** .05

3. **a.** .53 **b.** .35 **c.** .82 **d.** .18

4. **a.** .49 **b.** .39 **c.** .48

5. $\frac{2}{7}$ **6.** **a.** .019 **b.** .981 **7.** .364 **8.** No **9.** .5

10. .18 **11.** .25 **12.** .06 **13.** .49 **14.** .37

15. **a.** $\frac{7}{8}$ **b.** $\frac{7}{8}$ **c.** No **16.** **a.** .284 **b.** .984

17. .150 **18.** $\frac{2}{15}$ **19.** $\frac{1}{24}$ **20.** $\frac{1}{52}$ **21.** .00018

22. .00995 **23.** .245 **24.** .510 **25.** .245

26.

Figures Produced (in dozens)	30	31	32	33	34	35	36
Probability	.125	0	.1875	.25	.1875	.125	.125

27.

Answer	Falling behind	Staying even	Increasing faster	Don't know
Probability	.40	.44	.12	.04

28.

Income ($)	0–24,999	25,000–49,999	50,000–74,999
Probability	.251	.249	.176

Income ($)	75,000–99,999	100,000–124,999	125,000–149,999
Probability	.115	.075	.044

Income ($)	150,000–199,999	200,000–249,999	250,000 or more
Probability	.049	.019	.023

29. $A \cap B \neq \varnothing$, since it is possible to obtain a straight flush.

30. The sample space is not necessarily uniform; there may be different numbers of students in each grade.

31. a. .61 b. .12 32. a. .54 b. .72

33. a. .926 b. .074 34. a. .34 b. .50 c. .16

35. a. .79 b. .59 36. a. $\frac{3}{8}$ b. $\frac{1}{2}$ c. $\frac{1}{4}$

37. a. .757 b. .243 38. a. .56 b. .4

39. a. .41 b. .518 40. a. .429 b. .490

41. .5 42. .457 43. .368

44. a. .68 b. .053 45. .619

46. .180 47. a. .513 b. .390

48. .30 49. .244 50. .60

Chapter 7 Before Moving On, page 456

1. $\frac{5}{12}$ 2. $\frac{4}{13}$ 3. a. .9 b. .3 4. .72 5. .308

CHAPTER 8

Exercises 8.1, page 463

1. a. See part (b).

b.

Outcome	GGG	GGR	GRG	RGG
Value	3	2	2	2

Outcome	GRR	RGR	RRG	RRR
Value	1	1	1	0

c. {GGG}

3. Any positive integer 5. $\frac{1}{6}$

7. Any positive integer; infinite discrete

9. $x \geq 0$; continuous

11. Any positive integer; infinite discrete

13. No. The probability assigned to a value of the random variable X cannot be negative.

15. No. The sum of the probabilities exceeds 1.

17. $a = .2$

19. a. .20 b. .60 c. .30 d. 1 e. .40 f. 0

21.

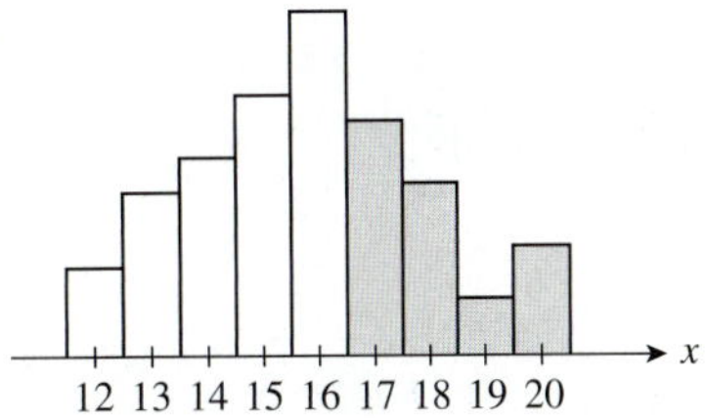

23. a.

x	1	2	3	4	5	6
$P(X = x)$	$\frac{1}{6}$	$\frac{1}{6}$	$\frac{1}{6}$	$\frac{1}{6}$	$\frac{1}{6}$	$\frac{1}{6}$

y	1	2	3	4	5	6
$P(Y = y)$	$\frac{1}{6}$	$\frac{1}{6}$	$\frac{1}{6}$	$\frac{1}{6}$	$\frac{1}{6}$	$\frac{1}{6}$

b.

$x + y$	2	3	4	5	6	7
$P(X + Y = x + y)$	$\frac{1}{36}$	$\frac{2}{36}$	$\frac{3}{36}$	$\frac{4}{36}$	$\frac{5}{36}$	$\frac{6}{36}$

$x + y$	8	9	10	11	12
$P(X + Y = x + y)$	$\frac{5}{36}$	$\frac{4}{36}$	$\frac{3}{36}$	$\frac{2}{36}$	$\frac{1}{36}$

25. a.

x	2	2.25	2.55	2.56	2.58	2.6	2.65	2.85
$P(X = x)$	$\frac{1}{30}$	$\frac{7}{30}$	$\frac{7}{30}$	$\frac{1}{30}$	$\frac{1}{30}$	$\frac{8}{30}$	$\frac{3}{30}$	$\frac{2}{30}$

b. $\frac{1}{2}$

27. a.

x	0	1	2	3	4
$P(X = x)$	.017	.067	.033	.117	.233

x	5	6	7	8	9	10
$P(X = x)$	.133	.167	.100	.050	.067	.017

b.

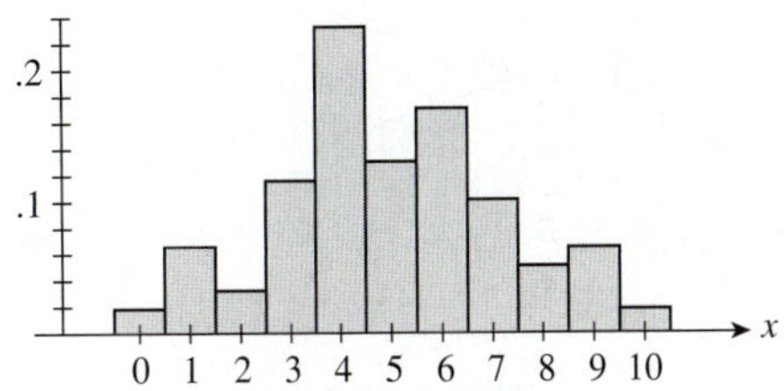

c. .217

29. a.

x	1	2	3	4
$P(X = x)$	.228	.492	.148	.132

b. .72

31. a.

x	1	2	3	4	5
$P(X = x)$	.020	.110	.250	.540	.080

b. .13

33. a.

x	1	2	3	4	5
$P(X = x)$	.131	.160	.179	.230	.300

b. .530; .291

35. True

Using Technology Exercises 8.1, page 471

Graphing Utility

1.

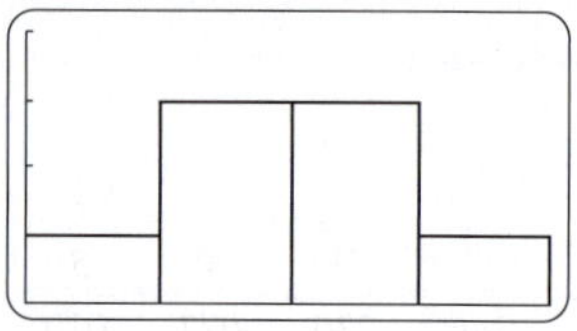

3.

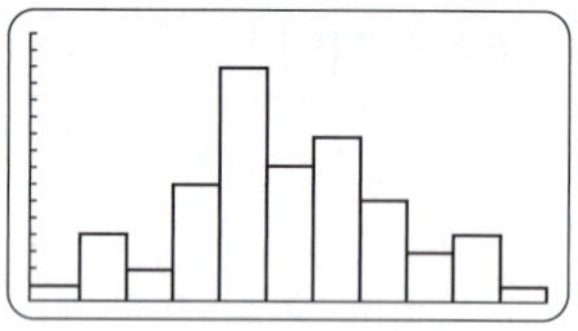

Excel

1.

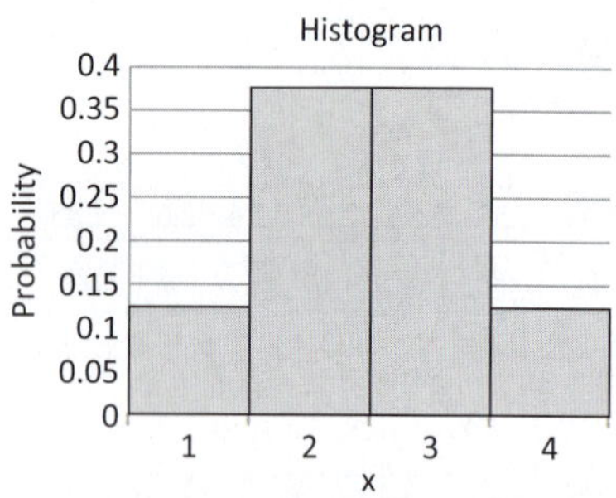

3.

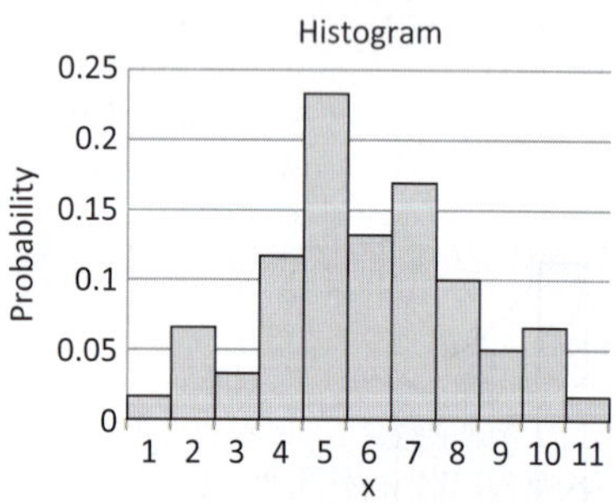

5.

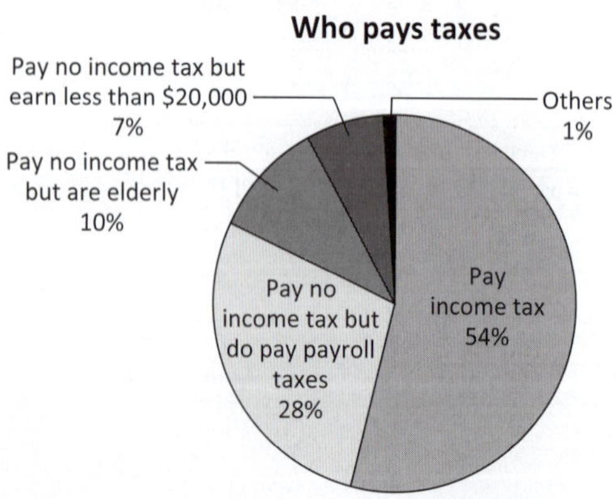

7.

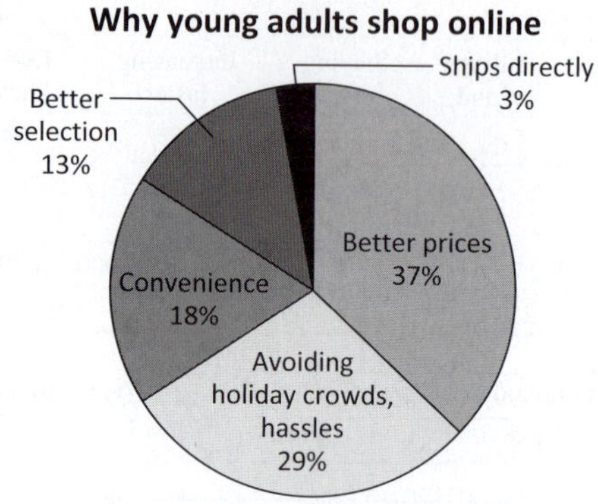

9.

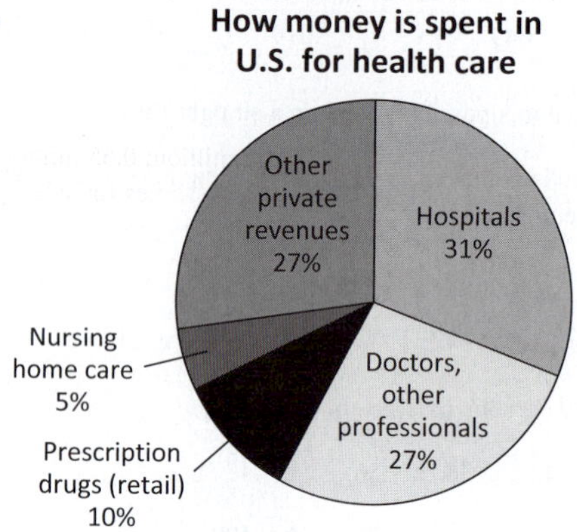

Exercises 8.2, page 482

1. 0.86

3. a. 2.6

b.

x	0	1	2	3	4
$P(X = x)$	0	.1	.4	.3	.2

; 2.6

5. 0.91 7. \$78.50 9. 0.12 11. 1.73 13. 3.16%

15. −39¢ 17. \$100 19. \$118,800

21. City *B* 23. Company *B* 25. 2.86%

27. −5.3¢ 29. −2.7¢ 31. 25.3 min 33. 36.2 years

35. 2 to 3; 3 to 2 37. .4 39. $\frac{7}{12}$ 41. $\frac{5}{14}$

45. a. Mean: 74; mode: 85; median: 80 b. Mode

47. 3; close 49. 16; 16; 16

51. Mean: 7.15; median: 7; mode: 7 53. True

Exercises 8.3, page 494

1. $\mu = 2$, $\text{Var}(X) = 1$, $\sigma = 1$

3. $\mu = 0$, $\text{Var}(X) = 1$, $\sigma = 1$

5. $\mu = 518$, $\text{Var}(X) = 1891$, $\sigma \approx 43.5$

7. Figure (a) 9. 1.56

11. $\mu = 4.5$, $\text{Var}(X) = 5.25$

13. a. Let X = the annual birthrate during the years 2003–2012.
b.

x	13.7	13.8	14.0	14.2	14.7
$P(X = x)$	.1	.3	.3	.2	.1

c. $\mu = 14.02$, $\text{Var}(X) = 0.0776$, $\sigma \approx 0.2786$

15. a. Mutual Fund A: $\mu = \$620$, $\text{Var}(X) = 267{,}600$;
Mutual Fund B: $\mu = \$520$, $\text{Var}(X) = 137{,}600$
b. Mutual Fund A
c. Mutual Fund B

17. 1 19. 58.833 hr; 5.70 hr 21. 84.33%; 6.67%

23. 1607 hr; 182 hr 25. 21.59%; 5.20% 27. 44.75 mi; 3.07 mi

29. 94.56%; 19.94% 31. 2.722%; 0.969% 33. 3.324; 0.4497

35. \$5.0154 million; \$1.07 million

37. a. 8.0583 million; 0.82 million b. 5.825 million; 0.65 million
c. The average monthly supply of single-family homes for sale dropped from 2011 to 2012 as the economy recovered from the Great Recession.

39. 35.28 thousand; 5.92 thousand

41. a. At least .75
b. At least .96

43. 7 45. At least $\frac{7}{16}$

47. At least $\frac{15}{16}$ 49. True

Using Technology Exercises 8.3, page 500

1. a.
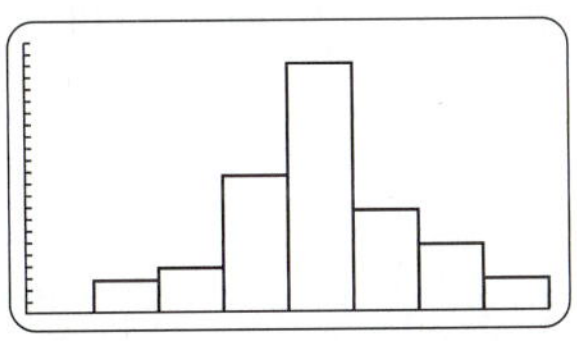
b. $\mu = 4$, $\sigma \approx 1.40$

3. a.
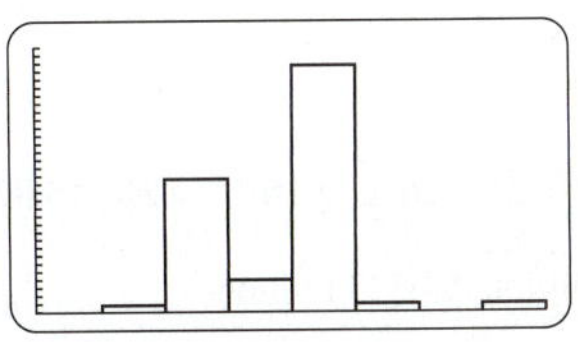
b. $\mu = 17.34$, $\sigma \approx 1.11$

5. a. Let X denote the random variable that gives the weight of a carton of sugar.
b.

x	4.96	4.97	4.98	4.99	5.00	5.01
$P(X = x)$	$\frac{3}{30}$	$\frac{4}{30}$	$\frac{4}{30}$	$\frac{1}{30}$	$\frac{1}{30}$	$\frac{5}{30}$

x	5.02	5.03	5.04	5.05	5.06
$P(X = x)$	$\frac{3}{30}$	$\frac{3}{30}$	$\frac{4}{30}$	$\frac{1}{30}$	$\frac{1}{30}$

c. $\mu \approx 5.00$; $\sigma \approx 0.03$

7. a.
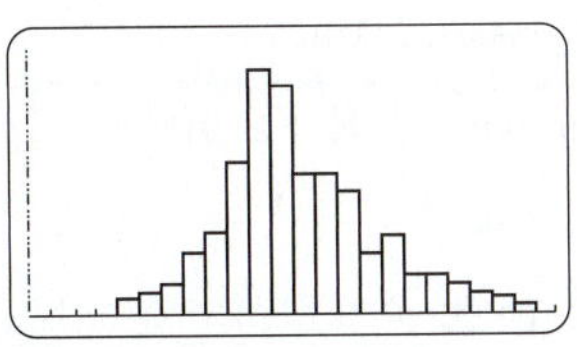
b. 65.875; 1.73

Exercises 8.4, page 509

1. Yes

3. No. There are more than two outcomes to each trial.

5. No. The probability of an accident on a clear day is not the same as the probability of an accident on a rainy day.

7. .296 9. .051 11. .132

13. .656 15. .004 17. .116

19. a. $P(X = 0) \approx .078$; $P(X = 1) \approx .259$;
$P(X = 2) \approx .346$; $P(X = 3) \approx .230$;
$P(X = 4) \approx .077$; $P(X = 5) \approx .010$
b.

x	0	1	2	3	4	5
$P(X = x)$	.078	.259	.346	.230	.077	.010

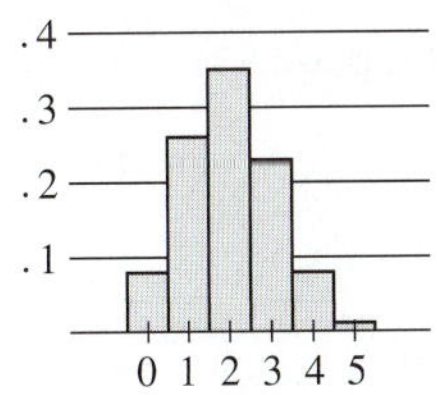

c. $\mu = 2$; $\sigma \approx 1.1$

21. No. The probability that at most 1 is defective is $P(X = 0) + P(X = 1) = .74$.

23. .165 25. .0002 27. a. .168 b. .360 c. .472

29. .313 31. .976 33. .01 35. a. .273 b. .650

37. a. .817 b. .999 39. .036 41. .0569 43. .0012

45. a. .075 b. .012 47. $n = 225$; $p = \frac{1}{5}$ 49. 0.35; 3.5%

51. a. .122 b. .358 53. .349 55. At least 7 times

57. $\mu = 375$; $\sigma \approx 9.68$ 59. False 61. False

Exercises 8.5, page 520

1. .9265 3. .0401 5. .8657

7. a.
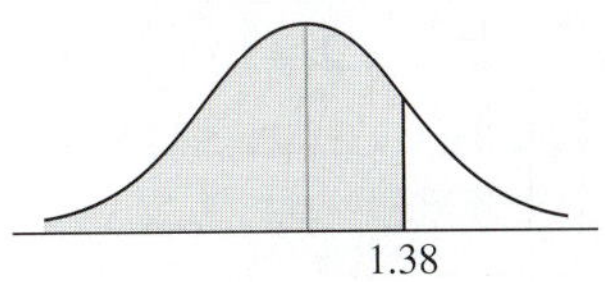

b. .9162

9. a.
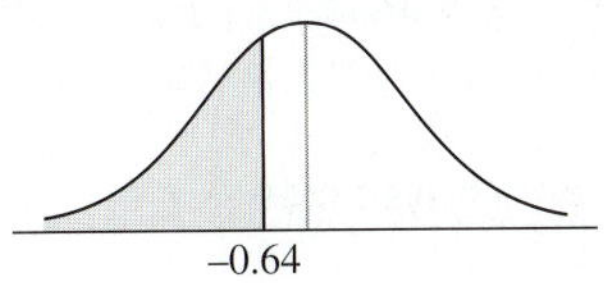

b. .2611

11. a.
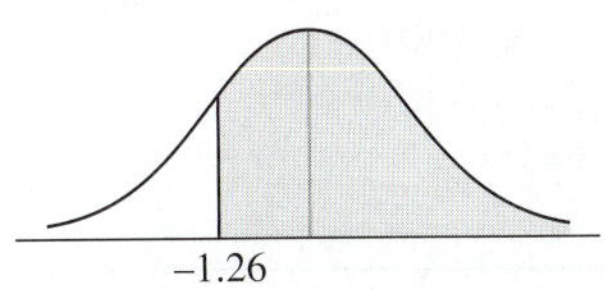

b. .8962

13. a.
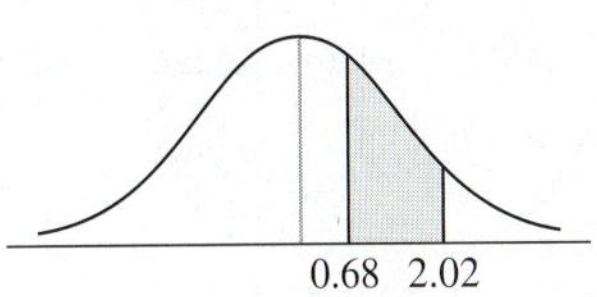

b. .2266

15. a. 1.23 **b.** −0.81 **17. a.** 1.9 **b.** −1.9

19. a. .9772 **b.** .9192 **c.** .7333

Exercises 8.6, page 528

1. a. .2206 **b.** .2206 **c.** .2960

3. a. .0228 **b.** .0228 **c.** .4772 **d.** .7258

5. a. .0038 **b.** .0918 **c.** .4082 **d.** .2514

7. .6247 **9.** 0.62% **11.** A: 80; B: 73; C: 62; D: 54

13. a. .4207 **b.** .4254 **c.** .0125

15. a. .2877 **b.** .0008 **c.** .7287

17. .9265 **19.** .8686

21. a. .0037 **b.** The drug is effective. **23.** 2142

Chapter 8 Concept Review Questions, page 531

1. random **2.** finite; infinite; continuous **3.** sum; .75

4. a. $\dfrac{P(E)}{P(E^c)}$ **b.** $\dfrac{a}{a+b}$

5. $p_1(x_1-\mu)^2 + p_2(x_2-\mu)^2 + \cdots + p_n(x_n-\mu)^2$; $\sqrt{\text{Var}(X)}$

6. fixed; two; same; independent

7. continuous; probability density function; set

8. normal; large; 0; 1

Chapter 8 Review Exercises, page 531

1. a. {WWW, BWW, WBW, WWB, BBW, BWB, WBB, BBB}

b.

Outcome	WWW	BWW	WBW	WWB
Value of X	0	1	1	1

Outcome	BBW	BWB	WBB	BBB
Value of X	2	2	2	3

c.

x	0	1	2	3
$P(X = x)$	$\frac{1}{35}$	$\frac{12}{35}$	$\frac{18}{35}$	$\frac{4}{35}$

d.

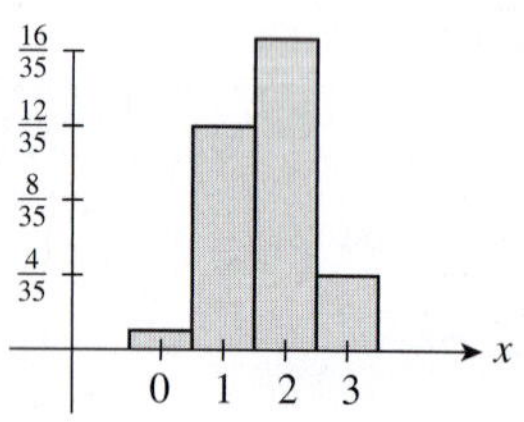

2. $100

3. a. .8 **b.** $\mu = 2.7$; $\sigma \approx 1.42$

4. a.

x	0	1	2	3	4
$P(X = x)$	.1296	.3456	.3456	.1536	.0256

b. $\mu = 1.6$; $\text{Var}(X) = 0.96$; $\sigma \approx 0.9798$

5. .6915

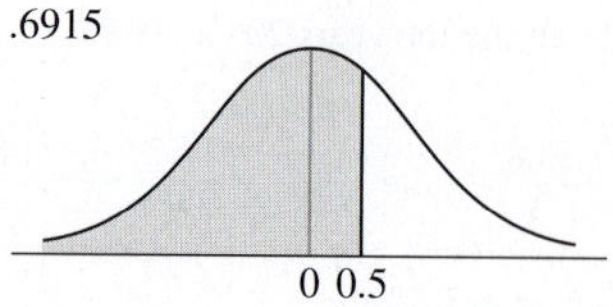

6. .2266

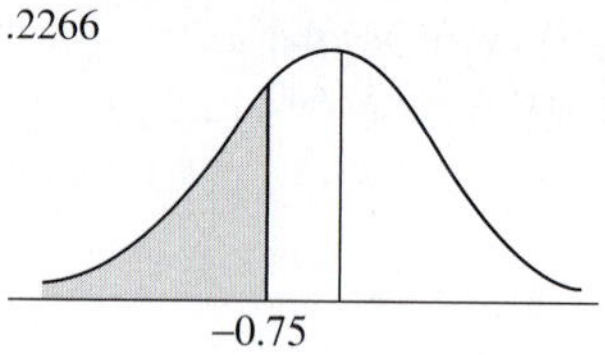

7. .4649

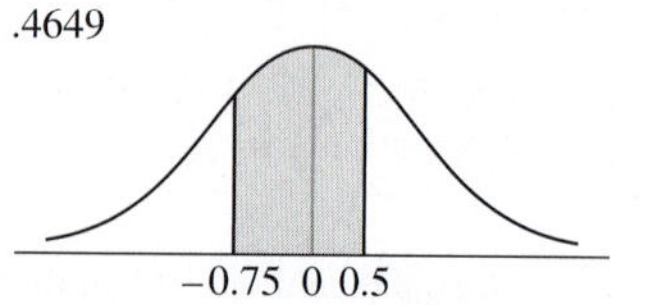

8. .4082

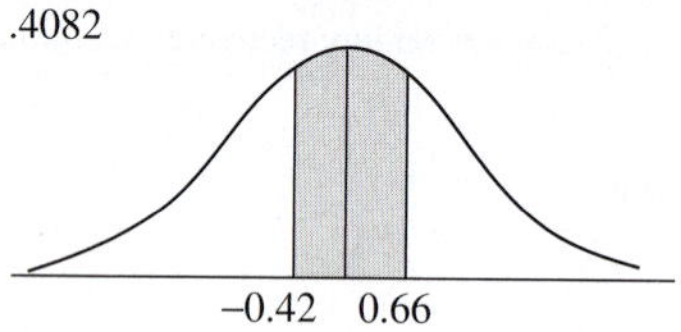

9. 2.42 **10.** −1.05 **11.** −2.03 **12.** 1.42 **13.** .6915

14. .8413 **15.** .2417 **16.** .7333

17. a. X gives the minimum age requirement for a regular driver's license.

b.

x	15	16	17	18	19	21
$P(X = x)$	.02	.30	.08	.56	.02	.02

c. 17.34; 1.2244; 1.11

18. a.

x	1	2	3	4	5
$P(X = x)$	.110	.143	.172	.236	.338

b. 57.4%; 25.3%

19. 41.3 mph **20.** $12,000 **21.** .2646; .9163 **22.** 15.87%

23. At least .75 **24.** $31.0 million; $4.523 million

25. $\mu = 27.58$; $\sigma = 6.32$ **26.** .677 **27.** $\mu = 120$; $\sigma \approx 10.1$

28. 0.6% **29. a.** .246 **b.** .901 **30. a.** .050 **b.** .995

31. .9738 **32.** .9997

Chapter 8 Before Moving On, page 533

1.

x	−3	−2	0	1	2	3
$P(X = x)$	.05	.1	.25	.3	.2	.1

2. a. .8 **b.** .92 **3.** 0.44; 4.0064; 2.0016

4. a. .2401; .4116; .2646; .0756; .0081 **b.** 1.2; .917

5. a. .9772 **b.** .9772 **c.** .9544

6. a. .0222 **b.** .6085 **c.** .0222

CHAPTER 9

Exercises 9.1, page 542

1. Yes **3.** Yes **5.** No **7.** Yes **9.** No

11. a. Given that the outcome state 1 has occurred, the conditional probability that the outcome state 1 will occur is .3.

b. .7 **c.** $\begin{bmatrix} .48 \\ .52 \end{bmatrix}$

13. $TX_0 = \begin{bmatrix} .4 \\ .6 \end{bmatrix}$;

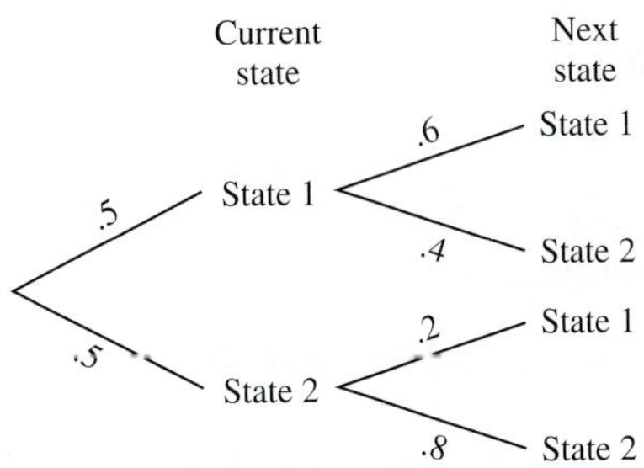

15. $X_2 = \begin{bmatrix} .576 \\ .424 \end{bmatrix}$ **17.** $X_2 = \begin{bmatrix} \frac{5}{16} \\ \frac{27}{64} \\ \frac{17}{64} \end{bmatrix}$

19. a.

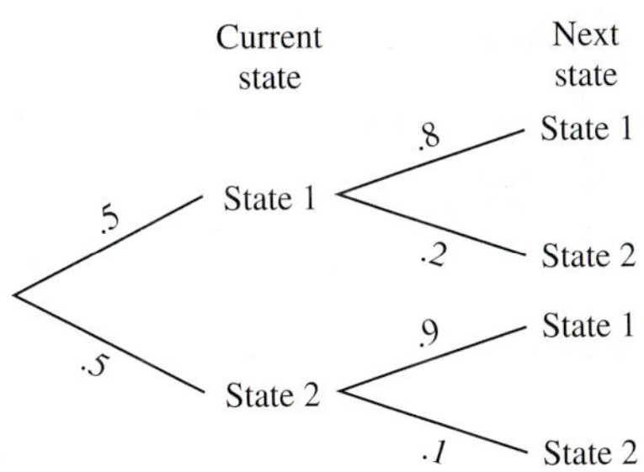

b. $T = \begin{matrix} & \text{L} & \text{R} \\ \text{L} & .8 & .9 \\ \text{R} & .2 & .1 \end{matrix}$ **c.** $X_0 = \begin{matrix} \text{L} \\ \text{R} \end{matrix} \begin{bmatrix} .5 \\ .5 \end{bmatrix}$ **d.** .85

21. a. Vote is evenly split. **b.** Democrat

23. After 1 year: 78.8% in the city and 21.2% in the suburbs. After 2 years: 77.7% in the city and 22.3% in the suburbs.

25. University: 37%, Campus: 35%, Book Mart: 28%; University: 34.5%, Campus: 31.35%, Book Mart: 34.15%

27. Business: 36.0%, humanities: 23.8%, education: 15.0%, natural sciences and others: 25.1%

29. False

Using Technology Exercises 9.1, page 546

1. $X_5 = \begin{bmatrix} .204489 \\ .131869 \\ .261028 \\ .186814 \\ .215800 \end{bmatrix}$

3. Manufacturer A will have 23.95% of the market share, Manufacturer B will have 49.71% of the market share, and Manufacturer C will have 26.34% of the market share.

Exercises 9.2, page 552

1. Regular **3.** Not regular **5.** Regular

7. Not regular **9.** $\begin{bmatrix} \frac{3}{11} \\ \frac{8}{11} \end{bmatrix}$ **11.** $\begin{bmatrix} \frac{2}{7} \\ \frac{5}{7} \end{bmatrix}$

13. $\begin{bmatrix} \frac{3}{13} \\ \frac{8}{13} \\ \frac{2}{13} \end{bmatrix}$ **15.** $\begin{bmatrix} \frac{3}{19} \\ \frac{8}{19} \\ \frac{8}{19} \end{bmatrix}$ **17.** 81.8%

19. 40.8% one wage earner and 59.2% two wage earners; 30% one wage earner and 70% two wage earners

21. 72.5% in single-family homes and 27.5% in condominiums; 70% in single-family homes and 30% in condominiums

23. a. 31.7% ABC, 37.35% CBS, 30.95% NBC
b. $33\frac{1}{3}$% ABC, $33\frac{1}{3}$% CBS, $33\frac{1}{3}$% NBC

25. 25% red, 50% pink, 25% white **27.** False

Using Technology Exercises 9.2, page 556

1. $\begin{bmatrix} .2045 \\ .1319 \\ .2610 \\ .1868 \\ .2158 \end{bmatrix}$

Exercises 9.3, page 562

1. Yes **3.** Yes **5.** Yes **7.** Yes

9. $\left[\begin{array}{c:c} 1 & .4 \\ \hdashline 0 & .6 \end{array}\right]$, $R = [.6]$, and $S = [.4]$

11. $\left[\begin{array}{c:cc} 1 & .4 & .5 \\ \hdashline 0 & .4 & .5 \\ 0 & .2 & 0 \end{array}\right]$, $R = \begin{bmatrix} .4 & .5 \\ .2 & 0 \end{bmatrix}$, and $S = [.4 \quad .5]$, or

$\left[\begin{array}{c:cc} 1 & .5 & .4 \\ \hdashline 0 & 0 & .2 \\ 0 & .5 & .4 \end{array}\right]$, $R = \begin{bmatrix} 0 & .2 \\ .5 & .4 \end{bmatrix}$, and $S = [.5 \quad .4]$

13. $\left[\begin{array}{cc:cc} 1 & 0 & .2 & .4 \\ 0 & 1 & .3 & 0 \\ \hdashline 0 & 0 & .3 & .2 \\ 0 & 0 & .2 & .4 \end{array}\right]$,

$R = \begin{bmatrix} .3 & .2 \\ .2 & .4 \end{bmatrix}$ and $S = \begin{bmatrix} .2 & .4 \\ .3 & 0 \end{bmatrix}$, or

$\left[\begin{array}{cc:cc} 1 & 0 & .4 & .2 \\ 0 & 1 & 0 & .3 \\ \hdashline 0 & 0 & .4 & .2 \\ 0 & 0 & .2 & .3 \end{array}\right]$,

$R = \begin{bmatrix} .4 & .2 \\ .2 & .3 \end{bmatrix}$, $S = \begin{bmatrix} .4 & .2 \\ 0 & .3 \end{bmatrix}$, and so forth

15. $\left[\begin{array}{c:c} 1 & 1 \\ \hdashline 0 & 0 \end{array}\right]$ **17.** $\left[\begin{array}{c:cc} 1 & 1 & 1 \\ \hdashline 0 & 0 & 0 \\ 0 & 0 & 0 \end{array}\right]$

19. $\left[\begin{array}{cc|cc} 1 & 0 & 1 & 1 \\ 0 & 1 & 0 & 0 \\ \hline 0 & 0 & 0 & 0 \\ 0 & 0 & 0 & 0 \end{array}\right]$ **21.** $\left[\begin{array}{cc|cc} 1 & 0 & \frac{1}{2} & \frac{1}{2} \\ 0 & 1 & \frac{1}{2} & \frac{1}{2} \\ \hline 0 & 0 & 0 & 0 \\ 0 & 0 & 0 & 0 \end{array}\right]$

23. $\left[\begin{array}{ccc|cc} 1 & 0 & 0 & \frac{7}{22} & \frac{3}{11} \\ 0 & 1 & 0 & \frac{5}{22} & \frac{9}{22} \\ 0 & 0 & 1 & \frac{5}{11} & \frac{7}{22} \\ \hline 0 & 0 & 0 & 0 & 0 \\ 0 & 0 & 0 & 0 & 0 \end{array}\right]$

25. a.
$$\begin{array}{c} \\ B \\ D \end{array}\begin{array}{c} \begin{array}{cc} B & D \end{array} \\ \left[\begin{array}{c|c} 1 & .2 \\ \hline 0 & .8 \end{array}\right] \end{array}, R = [.8], \text{ and } S = [.2]$$

b. $\left[\begin{array}{c|c} 1 & 1 \\ \hline 0 & 0 \end{array}\right]$; eventually, only broadband Internet service will be used.

27. .25; .50; .75

29. a.
$$\begin{array}{c} \\ D \\ G \\ 1 \\ 2 \end{array}\begin{array}{c} \begin{array}{cccc} D & G & 1 & 2 \end{array} \\ \left[\begin{array}{cc|cc} 1 & 0 & .25 & .1 \\ 0 & 1 & 0 & .9 \\ \hline 0 & 0 & 0 & 0 \\ 0 & 0 & .75 & 0 \end{array}\right] \end{array}$$

b. $\left[\begin{array}{cc|cc} 1 & 0 & .325 & .1 \\ 0 & 1 & .675 & .9 \\ \hline 0 & 0 & 0 & 0 \\ 0 & 0 & 0 & 0 \end{array}\right]$

c. .675

33. True

Exercises 9.4, page 572

1. R: row 1; C: column 2 **3.** R: row 1; C: column 1

5. R: row 1 or row 3; C: column 3

7. R: row 1 or row 3; C: column 2

9. Strictly determined;
a. 2 **b.** R: row 1; C: column 1
c. 2 **d.** Favors row player

11. Strictly determined;
a. 1 **b.** R: row 1; C: column 1
c. 1 **d.** Favors row player

13. Strictly determined;
a. 1 **b.** R: row 1; C: column 1
c. 1 **d.** Favors row player

15. Not strictly determined **17.** Not strictly determined

19. a. $\begin{bmatrix} 2 & -3 & 4 \\ -3 & 4 & -5 \\ 4 & -5 & 6 \end{bmatrix}$

b. Robin: row 1; Cathy: column 1 or column 2
c. Not strictly determined
d. Not strictly determined

21. a.

		Economy	
		Good	Recess.
Mgmt.	Expand	200,000	120,000
	Not exp.	50,000	150,000

b. Yes

23. a.

		Charley		
		Raises	Holds	Lowers
Roland	Raises	3	−1	−3
	Holds	2	0	−2
	Lowers	5	2	1

25. True

Exercises 9.5, page 583

1. $\frac{3}{10}$ **3.** $\frac{5}{12}$ **5.** 0.16

7. a. 1 **b.** −2 **c.** 0
d. −.3; (a) is most advantageous

9. a. 1 **b.** −.35 **c.** The first pair of strategies

11. $P = [\frac{1}{4} \quad \frac{3}{4}]$, $Q = \begin{bmatrix} \frac{1}{2} \\ \frac{1}{2} \end{bmatrix}$, and $E = 2.5$; favors row player

13. $P = [\frac{4}{7} \quad \frac{3}{7}]$, $Q = \begin{bmatrix} \frac{5}{7} \\ \frac{2}{7} \end{bmatrix}$, and $E = -\frac{1}{7}$; favors column player

15. $P = [\frac{1}{2} \quad \frac{1}{2}]$, $Q = \begin{bmatrix} \frac{1}{4} \\ \frac{3}{4} \end{bmatrix}$, and $E = -5$; favors column player

17. a. $P = [\frac{1}{3} \quad \frac{2}{3}]$ and $Q = \begin{bmatrix} \frac{1}{3} \\ \frac{2}{3} \end{bmatrix}$

b. $E = 0$; no

19. a. \$5714 in hotel stock; \$34,286 in brewery stock
b. \$4857

21. a.

		C	
		N	F
R	N	.48	.65
	F	.50	.45

C = Carlton; R = Russell
N = local newspaper; F = flier

b. Russell's strategy: $P \approx [.23 \quad .77]$

Carlton's strategy: $Q \approx \begin{bmatrix} .91 \\ .09 \end{bmatrix}$

Chapter 9 Concept Review Questions, page 586

1. probabilities; preceding **2.** state; state

3. transition **4.** $n \times n$; nonnegative; 1

5. distribution; steady-state

6. regular; columns; equal; positive; $TX = X$; elements; 1

7. absorbing; leave; steps

8. a. zero-sum **b.** maximin; minimax **9.** optimal

10. saddle point; maximin; row; saddle; minimax; column; saddle

Chapter 9 Review Exercises, page 587

1. Not regular 2. Regular

3. Regular 4. Not regular

5. $\begin{bmatrix} .3675 \\ .36 \\ .2725 \end{bmatrix}$ 6. $\begin{bmatrix} .1915 \\ .4215 \\ .387 \end{bmatrix}$

7. Yes 8. No 9. No 10. Yes

11. $\begin{bmatrix} \frac{3}{7} & \frac{3}{7} \\ \frac{4}{7} & \frac{4}{7} \end{bmatrix}$ 12. $\begin{bmatrix} \frac{4}{9} & \frac{4}{9} \\ \frac{5}{9} & \frac{5}{9} \end{bmatrix}$

13. $\begin{bmatrix} .457 & .457 & .457 \\ .200 & .200 & .200 \\ .343 & .343 & .343 \end{bmatrix}$ 14. $\begin{bmatrix} .323 & .323 & .323 \\ .290 & .290 & .290 \\ .387 & .387 & .387 \end{bmatrix}$

15. **a.** $\begin{array}{c} \\ A \\ U \\ N \end{array} \begin{array}{c} \begin{array}{ccc} A & U & N \end{array} \\ \begin{bmatrix} .85 & 0 & .10 \\ .10 & .95 & .05 \\ .05 & .05 & .85 \end{bmatrix} \end{array}$ A = Agriculture, U = Urban, N = Nonagricultural

    b. $\begin{array}{c} A \\ U \\ N \end{array} \begin{bmatrix} .50 \\ .15 \\ .35 \end{bmatrix}$ **c.** $\begin{array}{c} A \\ U \\ N \end{array} \begin{bmatrix} .424 \\ .262 \\ .314 \end{bmatrix}$

16. 12.5% large cars, 30.36% intermediate cars, 57.14% small cars

17. Strictly determined; R: row 3; C: column 1; value is 4.

18. Strictly determined; R: row 1; C: column 2; value is 0.

19. Strictly determined; R: row 1; C: column 1; value is 1.

20. Not strictly determined

21. $-\frac{1}{4}$ 22. $\frac{10}{9}$ 23. 2 24. 1.04

25. $P = [\frac{1}{2} \quad \frac{1}{2}]$, $Q = \begin{bmatrix} \frac{5}{6} \\ \frac{1}{6} \end{bmatrix}$, and $E = \frac{1}{2}$; favors row player

26. $P = [\frac{1}{2} \quad \frac{1}{2}]$, $Q = \begin{bmatrix} \frac{13}{22} \\ \frac{9}{22} \end{bmatrix}$, and $E = -\frac{1}{2}$, favors column player

27. $P = [\frac{1}{10} \quad \frac{9}{10}]$, $Q = \begin{bmatrix} \frac{4}{5} \\ \frac{1}{5} \end{bmatrix}$, and $E = 1.2$; favors row player

28. $P = [\frac{4}{5} \quad \frac{1}{5}]$, $Q = \begin{bmatrix} \frac{2}{5} \\ \frac{3}{5} \end{bmatrix}$, and $E = 10.8$; favors row player

29. **a.** $\begin{bmatrix} .5 & .7 \\ .4 & .5 \end{bmatrix}$ **b.** \$7

30. 25% compact models; 75% subcompact models

Chapter 9 Before Moving On, page 589

1. $\begin{bmatrix} .366 \\ .634 \end{bmatrix}$ 2. $\begin{bmatrix} \frac{3}{11} \\ \frac{8}{11} \end{bmatrix}$ 3. $\left[\begin{array}{c:cc} 1 & 1 & 1 \\ \hdashline 0 & 0 & 0 \\ 0 & 0 & 0 \end{array}\right]$

4. **a.** -1 **b.** R: row 1; C; column 3 **c.** -1; column player

5. **a.** 3 units **b.** 1.22 units

6. **a.** $P = [\frac{2}{3} \quad \frac{1}{3}]$, $Q = \begin{bmatrix} \frac{1}{6} \\ \frac{5}{6} \end{bmatrix}$ **b.** $\frac{4}{3}$; row player

APPENDIX A

Exercises A.1, page 595

1. Yes 3. Yes 5. No

7. Yes 9. Yes 11. No

13. No 15. Negation 17. Conjunction

19. Conjunction

21. New orders for manufactured goods did not fall last month.

23. Drinking alcohol during pregnancy does not affect both the size and weight of babies.

25. The commuter airline industry is not now undergoing a shakeup.

27. **a.** Domestic car sales increased over the past year, or foreign car sales decreased over the past year, or both.
    b. Domestic car sales increased over the past year, and foreign car sales decreased over the past year.
    c. Either domestic car sales increased over the past year or foreign car sales decreased over the past year.
    d. Domestic car sales did not increase over the past year.
    e. Domestic car sales did not increase over the past year, or foreign car sales decreased over the past year, or both.
    f. Domestic car sales did not increase over the past year, or foreign car sales did not decrease over the past year, or both.

29. **a.** Either the doctor recommended surgery to treat Sam's hyperthyroidism or the doctor recommended radioactive iodine to treat Sam's hyperthyroidism.
    b. The doctor recommended surgery to treat Sam's hyperthyroidism, or the doctor recommended radioactive iodine to treat Sam's hyperthyroidism, or both.

31. **a.** $p \wedge q$ **b.** $p \veebar q$ **c.** $\sim p \wedge \sim q$ **d.** $\sim(\sim q)$

33. **a.** Both the popularity of prime-time soaps and the popularity of prime-time situation comedies did not increase this year.
    b. The popularity of prime-time soaps did not increase this year, or the popularity of prime-time detective shows decreased this year, or both.
    c. The popularity of prime-time detective shows decreased this year, or the popularity of prime-time situation comedies did not increase this year, or both.
    d. Either the popularity of prime-time soaps did not increase this year or the popularity of prime-time situation comedies did not increase this year.

Exercises A.2, page 598

1.

p	q	$\sim q$	$p \vee \sim q$
T	T	F	T
T	F	T	T
F	T	F	F
F	F	T	T

3.

p	$\sim p$	$\sim(\sim p)$
T	F	T
F	T	F

5.

p	$\sim p$	$p \vee \sim p$
T	F	T
F	T	T

7.

p	q	$\sim p$	$p \vee q$	$\sim p \wedge (p \vee q)$
T	T	F	T	F
T	F	F	T	F
F	T	T	T	T
F	F	T	F	F

9.

p	q	$\sim q$	$p \vee q$	$p \wedge \sim q$	$(p \vee q) \wedge (p \wedge \sim q)$
T	T	F	T	F	F
T	F	T	T	T	T
F	T	F	T	F	F
F	F	T	F	F	F

11.

p	q	$p \vee q$	$\sim(p \vee q)$	$(p \vee q) \wedge \sim(p \vee q)$
T	T	T	F	F
T	F	T	F	F
F	T	T	F	F
F	F	F	T	F

13.

p	q	r	$p \vee q$	$p \vee r$	$(p \vee q) \wedge (p \vee r)$
T	T	T	T	T	T
T	T	F	T	T	T
T	F	T	T	T	T
T	F	F	T	T	T
F	T	T	T	T	T
F	T	F	T	F	F
F	F	T	F	T	F
F	F	F	F	F	F

15.

p	q	r	$p \wedge q$	$\sim r$	$(p \wedge q) \vee \sim r$
T	T	T	T	F	T
T	T	F	T	T	T
T	F	T	F	F	F
T	F	F	F	T	T
F	T	T	F	F	F
F	T	F	F	T	T
F	F	T	F	F	F
F	F	F	F	T	T

17.

p	q	r	$\sim q$	$p \wedge \sim q$	$p \wedge r$	$(p \wedge \sim q) \vee (p \wedge r)$
T	T	T	F	F	T	T
T	T	F	F	F	F	F
T	F	T	T	T	T	T
T	F	F	T	T	F	T
F	T	T	F	F	F	F
F	T	F	F	F	F	F
F	F	T	T	F	F	F
F	F	F	T	F	F	F

19. 16 rows

Exercises A.3, page 602

1. $\sim q \to p$; $q \to \sim p$; $\sim p \to q$

3. $p \to q$; $\sim p \to \sim q$; $\sim q \to \sim p$

5. Conditional: If it is snowing, then the temperature is below freezing.
Biconditional: It is snowing if and only if the temperature is below freezing.

7. Conditional: If the company's union and management reach a settlement, then the workers do not strike.
Biconditional: The company's union and management will reach a settlement if and only if the workers do not strike.

9. False 11. False

13. It is false when I do not buy the house after the owner lowers the selling price.

15.

p	q	$p \to q$	$\sim(p \to q)$
T	T	T	F
T	F	F	T
F	T	T	F
F	F	T	F

17.

p	q	$p \to q$	$\sim(p \to q)$	$\sim(p \to q) \wedge p$
T	T	T	F	F
T	F	F	T	T
F	T	T	F	F
F	F	T	F	F

19.

p	q	$\sim p$	$\sim q$	$p \to \sim q$	$(p \to \sim q) \veebar \sim p$
T	T	F	F	F	F
T	F	F	T	T	T
F	T	T	F	T	F
F	F	T	T	T	F

21.

p	q	$\sim p$	$\sim q$	$p \to q$	$\sim q \to \sim p$	$(p \to q) \leftrightarrow (\sim q \to \sim p)$
T	T	F	F	T	T	T
T	F	F	T	F	F	T
F	T	T	F	T	T	T
F	F	T	T	T	T	T

23.

p	q	$p \wedge q$	$p \vee q$	$(p \wedge q) \to (p \vee q)$
T	T	T	T	T
T	F	F	T	T
F	T	F	T	T
F	F	F	F	T

25.

p	q	r	$p \vee q$	$\sim r$	$(p \vee q) \to \sim r$
T	T	T	T	F	F
T	T	F	T	T	T
T	F	T	T	F	F
T	F	F	T	T	T
F	T	T	T	F	F
F	T	F	T	T	T
F	F	T	F	F	T
F	F	F	F	T	T

27.

p	q	r	$q \vee r$	$p \to (q \vee r)$
T	T	T	T	T
T	T	F	T	T
T	F	T	T	T
T	F	F	F	F
F	T	T	T	T
F	T	F	T	T
F	F	T	T	T
F	F	F	F	T

29. Logically equivalent

31. Logically equivalent

33. Not logically equivalent

35. Not logically equivalent

37. a. $p \to \sim q$ b. $\sim p \to q$ c. $\sim q \leftrightarrow p$
d. $p \to \sim q$ e. $p \leftrightarrow \sim q$

Exercises A.4, page 606

1.

p	p	$p \wedge p$
T	T	T
F	F	F

3.

p	q	r	$p \wedge q$
T	T	T	T
T	T	F	T
T	F	T	F
T	F	F	F
F	T	T	F
F	T	F	F
F	F	T	F
F	F	F	F

$(p \wedge q) \wedge r$	$q \wedge r$	$p \wedge (q \wedge r)$
T	T	T
F	F	F
F	F	F
F	F	F
F	T	F
F	F	F
F	F	F
F	F	F

5.

p	q	$p \wedge q$	$q \wedge p$
T	T	T	T
T	F	F	F
F	T	F	F
F	F	F	F

7.

p	q	r	$q \wedge r$	$p \vee (q \wedge r)$
T	T	T	T	T
T	T	F	F	T
T	F	T	F	T
T	F	F	F	T
F	T	T	T	T
F	T	F	F	F
F	F	T	F	F
F	F	F	F	F

$p \vee q$	$p \vee r$	$(p \vee q) \wedge (p \vee r)$
T	T	T
T	T	T
T	T	T
T	T	T
T	T	T
T	F	F
F	T	F
F	F	F

9. Tautology 11. Tautology 13. Tautology

15. Tautology 17. Neither

19. $\sim(p \wedge q)$: The candidate does not oppose changes in the Social Security system, or the candidate does not support immigration reform.

$\sim(p \vee q)$: The candidate does not oppose changes in the Social Security system, and the candidate does not support immigration reform.

21. $[p \wedge (q \vee \sim q) \vee (p \wedge q)]$
$\Leftrightarrow p \wedge t \vee (p \wedge q)]$ By Law 11
$\Leftrightarrow p \vee (p \wedge q)$ By Law 14

23. $(p \wedge \sim q) \vee (p \wedge \sim r)$
$\Leftrightarrow p \wedge (\sim q \vee \sim r)$ By Law 7

25. $(p \wedge \sim(q \wedge r)$
$\Leftrightarrow p \wedge (\sim q \vee \sim r)$ By Law 10
$\Leftrightarrow (p \wedge \sim q) \vee (p \wedge \sim r)$ By Law 7

Exercises A.5, page 611

1. Valid 3. Valid 5. Invalid 7. Valid

9. Valid 11. Valid 13. Valid 15. Invalid

17. $p \to q$; invalid
$\underline{\sim p}$
$\therefore \sim q$

19. $p \vee q$; valid
$\underline{\sim p \to \sim q}$
$\therefore p$

21. $p \to q$; invalid
$q \to r$
$\underline{r}$
$\therefore p$

Exercises A.6, page 614

1. $p \wedge q \wedge (r \vee s)$

3. $[(p \wedge q) \vee r] \wedge (\sim r \vee p)$

5. $[(p \vee q) \wedge r] \vee (\sim p) \vee [\sim q \wedge (p \vee r \vee \sim r)]$

7.

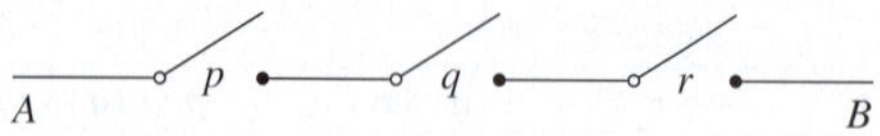

9.

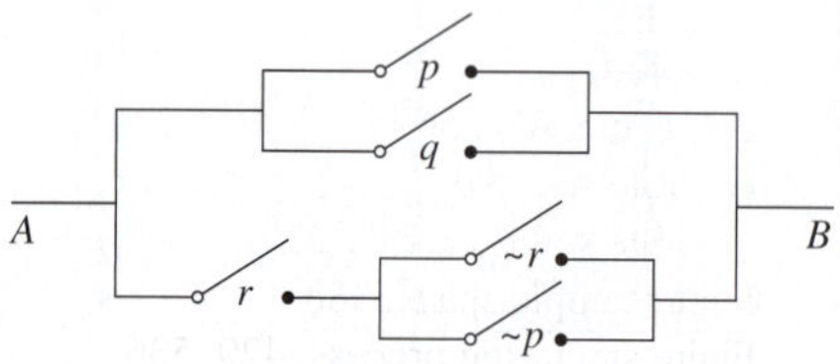

11.

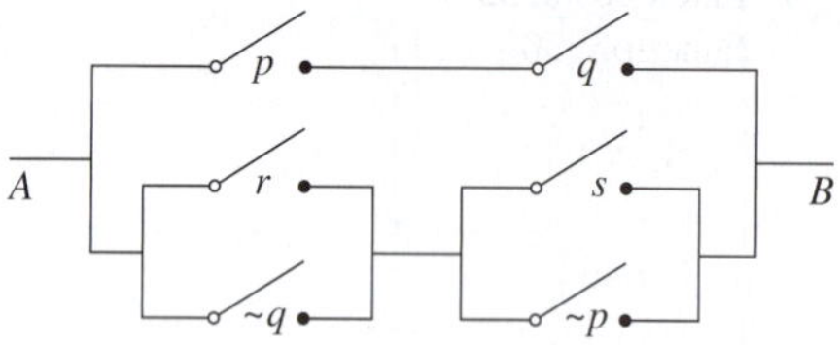

13. $p \wedge [\sim q \vee (\sim p \wedge q)]$; $p \wedge \sim q$

15. $p \wedge [\sim p \vee q \vee (q \wedge r)]$; $p \wedge q$

APPENDIX C

Exercises, page 625

1. $\log_2 64 = 6$ 3. $\log_3 \frac{1}{9} = -2$ 5. $\log_{32} 8 = \frac{3}{5}$

7. 1.0792 9. 1.2042 11. $\ln a^2b^3$

13. $\ln \dfrac{3\sqrt{x}\,y}{\sqrt[3]{z}}$ 15. $x = 8$ 17. $x = 10$

19. $x = \frac{3}{2}$ 21. $x = \frac{11}{3}$ 23. 5.1986

25. -0.0912 27. -8.0472 29. -4.9041

INDEX

How-To Technology Index

Formulas

Equation of a Straight Line

a. point-slope form: $y - y_1 = m(x - x_1)$
b. slope-intercept form: $y = mx + b$
c. general form: $Ax + By + C = 0$

Equation of the Least-Squares Line

$$y = mx + b$$

where m and b satisfy the **normal equations**

$$nb + (x_1 + x_2 + \cdots + x_n)m = y_1 + y_2 + \cdots + y_n$$
$$(x_1 + x_2 + \cdots + x_n)b + (x_1^2 + x_2^2 + \cdots + x_n^2)m = x_1y_1 + x_2y_2 + \cdots + x_ny_n$$

Compound Interest

$$A = P(1 + i)^n \qquad (i = r/m,\ n = mt)$$

where A is the accumulated amount at the end of n conversion periods, P is the principal, r is the interest rate per year, m is the number of conversion periods per year, and t is the number of years.

Effective Rate of Interest

$$r_{\text{eff}} = \left(1 + \frac{r}{m}\right)^m - 1$$

where r_{eff} is the effective rate of interest, r is the nominal interest rate per year, and m is the number of conversion periods per year.

Future Value of an Annuity

$$S = R\left[\frac{(1 + i)^n - 1}{i}\right]$$

Present Value of an Annuity

$$P = R\left[\frac{1 - (1 + i)^{-n}}{i}\right]$$

Amortization Formula

$$R = \frac{Pi}{1 - (1 + i)^{-n}}$$

Sinking Fund Payment

$$R = \frac{iS}{(1+i)^n - 1}$$

The Number of Permutations of *n* Distinct Objects Taken *r* at a Time

$$P(n, r) = \frac{n!}{(n-r)!}$$

The Number of Permutations of *n* Objects, Not All Distinct

$$\frac{n!}{n_1!\, n_2! \cdots n_m!}, \quad \text{where } n_1 + n_2 + \cdots + n_m = n$$

The Number of Combinations of *n* Distinct Objects Taken *r* at a Time

$$C(n, r) = \frac{n!}{r!\,(n-r)!}$$

The Product Rule for Probability

$$P(A \cap B) = P(A) \cdot P(B \mid A)$$

Bayes' Formula

$$P(A_i \mid E) = \frac{P(A_i) \cdot P(E \mid A_i)}{P(A_1) \cdot P(E \mid A_1) + P(A_2) \cdot P(E \mid A_2) + \cdots + P(A_n) \cdot P(E \mid A_n)}$$

Expected Value of a Random Variable

$$E(X) = x_1p_1 + x_2p_2 + \cdots + x_np_n$$

Bernoulli Trials

$$P(X = x) = C(n, x)p^x q^{n-x} \qquad (x = 0, 1, 2, \ldots, n)$$

$$E(X) = np$$

$$\text{Var}(X) = npq$$

$$\sigma_X = \sqrt{npq}$$

List of Applications

BUSINESS AND ECONOMICS

(*continued*)

SOCIAL SCIENCES

A Sampling of New Applications

Drawn from diverse fields of interest and situations that occur in the real world

BUSINESS

- Online Video Advertising *Use a trend line to project spending on Web video advertising.* p. 62
- Switching Broadband Service p. 438
- Satellite TV Subscribers p. 62
- Market Share of Motorcycles p. 120

COLLEGE LIFE

- Time Spent per Week on the Internet p. 471
- Time Use of College Students p. 470
- Brand Switching Among Female College Students p. 359

CRIMINAL JUSTICE

- Corporate Fraud *Use a linear equation to estimate the number of pending corporate fraud cases.* p. 23
- Detecting Shoplifters p. 447
- Identity Fraud p. 496

DEMOGRAPHICS

- U.S. Population by Age p. 485
- Percent of Population Enrolled in School p. 61
- Households with Someone Under 18 p. 61

ECONOMICS

- Recovery from the Great Recession p. 413
- Existing Home Sales p. 498
- Impact of Gas Prices on Consumers p. 447
- Pre-Retirees' Spending p. 510

ENTERTAINMENT

- Concert Attendance *Use a system of linear inequalities to determine the types of ticket holders at a concert.* p. 179
- Makeup of U.S. Moviegoer Audience p. 412
- Academy Membership p. 511

ENVIRONMENT

- California Emissions Caps p. 37
- Climate Change p. 413

FINANCE

- Model Investment Portfolios p. 120
- Mortgage Rates p. 120
- Asset Allocation p. 238
- Ability-to-Repay Rule *Apply the rule to determine if a potential homebuyer will qualify for a mortgage.* p. 323

GOVERNMENT

- Gun Owners in the Senate p. 448
- Federal Budget Allocation p. 359
- Federal Libraries p. 497

PERSONAL FINANCE

- Comparative Shopping p. 134
- Credit Card Ownership *Calculate probabilities using survey data about credit card ownership.* p. 454
- Emergency Fund Savings p. 465

HEALTH

- Smokers in the United States p. 19
- Stress Level p. 402
- Diet Planning *Use a system of linear equations and linear inequalities to help plan a meal that satisfies certain nutritional requirements.* p. 179

SOCIAL MEDIA

- Facebook Users p. 61
- Cyber Privacy *Use operations on sets to rate companies on how they keep personal information secure.* p. 347
- Social Media Accounts p. 402
- Smartphone Ownership p. 466

IN THE NEWS

- Erosion of the Middle Class *Use a linear function to model the percentage of U.S. middle-income adults.* p. 38
- Terrorism Poll p. 119
- Who Pays Taxes? p. 471

TRANSPORTATION

- Hybrid Vehicle Mileage p. 496
- Flight Cancellations p. 497
- Traffic Survey p. 498

ON THE JOB

- Starting a New Job p. 402
- On-the-Job Distractions p. 358
- Work Habits p. 454
- Getting Ahead at Work p. 511

TRAVEL

- N.Y.C. Tourists p. 439
- Best U.S. City for Italian Restaurants p. 351
- Stay When Visiting National Parks p. 413

See the complete Index of Applications at the back of the text to find out more ...